DRILL HALL LIBRARY
MEDWAY

MicroRNAs in Medicine

MicroRNAs in Medicine

EDITED BY

Charles H. Lawrie

WILEY Blackwell

Copyright © 2014 by John Wiley & Sons, Inc. All rights reserved.

Published by John Wiley & Sons, Inc., Hoboken, New Jersey.
Published simultaneously in Canada.

No part of this publication may be reproduced, stored in a retrieval system, or transmitted in any form or by any means, electronic, mechanical, photocopying, recording, scanning, or otherwise, except as permitted under Section 107 or 108 of the 1976 United States Copyright Act, without either the prior written permission of the Publisher, or authorization through payment of the appropriate per-copy fee to the Copyright Clearance Center, Inc., 222 Rosewood Drive, Danvers, MA 01923, (978) 750-8400, fax (978) 750-4470, or on the web at www.copyright.com. Requests to the Publisher for permission should be addressed to the Permissions Department, John Wiley & Sons, Inc., 111 River Street, Hoboken, NJ 07030, (201) 748-6011, fax (201) 748-6008, or online at http://www.wiley.com/go/permissions.

Limit of Liability/Disclaimer of Warranty: While the publisher and author have used their best efforts in preparing this book, they make no representations or warranties with respect to the accuracy or completeness of the contents of this book and specifically disclaim any implied warranties of merchantability or fitness for a particular purpose. No warranty may be created or extended by sales representatives or written sales materials. The advice and strategies contained herein may not be suitable for your situation. You should consult with a professional where appropriate. Neither the publisher nor author shall be liable for any loss of profit or any other commercial damages, including but not limited to special, incidental, consequential, or other damages.

For general information on our other products and services or for technical support, please contact our Customer Care Department within the United States at (800) 762-2974, outside the United States at (317) 572-3993 or fax (317) 572-4002.

Wiley also publishes its books in a variety of electronic formats. Some content that appears in print may not be available in electronic formats. For more information about Wiley products, visit our web site at www.wiley.com.

Library of Congress Cataloging-in-Publication Data:

MicroRNAs in medicine / edited by Charles H. Lawrie.
     p. ; cm.
  Includes bibliographical references and index.
    ISBN 978-1-118-30039-8 (cloth : alk. paper)
  I. Lawrie, Charles H., editor of compilation.
  [DNLM: 1. MicroRNAs. QU 58.7]
  QP623.5.S63
  572.8'8–dc23
                            2013038052

Printed in Singapore.

10 9 8 7 6 5 4 3 2 1

CONTENTS

Foreword   ix
Sir David Baulcombe

Preface   xi

Contributors   xiii

1 MicroRNAs: A Brief Introduction   1
Charles H. Lawrie

PART I: MicroRNAs as Physiological Regulators   25

2 MicroRNA Regulation of Stem Cell Fate and Reprogramming   27
Erika Lorenzo Vivas, Gustavo Tiscornia, and Juan Carlos Izpisua Belmonte

3 MicroRNAs as Regulators of Immunity   41
Donald T. Gracias and Peter D. Katsikis

4 Regulation of Senescence by MicroRNAs   59
Ioannis Grammatikakis and Myriam Gorospe

5 The Emergence of GeroMIRs: A Group of MicroRNAs Implicated in Aging   77
Alejandro P. Ugalde, Agnieszka Kwarciak, Xurde M. Caravia, Carlos López-Otín, and Andrew J. Ramsay

6 MicroRNAs and Hematopoiesis   91
Sukhinder K. Sandhu and Ramiro Garzon

7 MicroRNAs in Platelet Production and Activation   101
Leonard C. Edelstein, Srikanth Nagalla, and Paul F. Bray

PART II: MicroRNAs in Infectious Disease: Host–Pathogen Interactions   117

8 MicroRNAs as Key Players in Host-Virus Interactions   119
Aurélie Fender and Sébastien Pfeffer

9 MicroRNA Expression in Avian Herpesviruses   137
Yongxiu Yao and Venugopal Nair

10 FUNCTION OF HUMAN CYTOMEGALOVIRUS MicroRNAs AND POTENTIAL ROLES IN LATENCY   153
Natalie L. Reynolds, Jon A. Pavelin, and Finn E. Grey

11 INVOLVEMENT OF SMALL NON-CODING RNA IN HIV-1 INFECTION   165
Guihua Sun, John J. Rossi, and Daniela Castanotto

12 MicroRNA IN MALARIA   183
Panote Prapansilp and Gareth D.H. Turner

PART III: CANCER   199

13 THE MicroRNA DECALOGUE OF CANCER INVOLVEMENT   201
Tanja Kunej, Irena Godnic, Minja Zorc, Simon Horvat, and George A. Calin

14 MicroRNAs AS ONCOGENES AND TUMOR SUPPRESSORS   223
Eva E. Rufino-Palomares, Fernando J. Reyes-Zurita, Jose Antonio Lupiáñez, and Pedro P. Medina

15 LONG NON-CODING RNAs AND THEIR ROLES IN CANCER   245
Yolanda Sánchez and Maite Huarte

16 REGULATION OF HYPOXIA RESPONSES BY MicroRNA EXPRESSION   267
Carme Camps, Adrian L. Harris, and Jiannis Ragoussis

17 CONTROL OF RECEPTOR FUNCTION BY MicroRNAs IN BREAST CANCER   287
Claudia Piovan and Marilena V. Iorio

18 MicroRNAs IN HUMAN PROSTATE CANCER: FROM PATHOGENESIS TO THERAPEUTIC IMPLICATIONS   311
Mustafa Ozen and Omer Faruk Karatas

19 MicroRNA SIGNATURES AS BIOMARKERS OF COLORECTAL CANCER   329
Katrin Pfütze, Xiaoya Luo, and Barbara Burwinkel

20 GENETIC VARIATIONS IN MicroRNA-ENCODING SEQUENCES AND MicroRNA TARGET SITES ALTER LUNG CANCER SUSCEPTIBILITY AND SURVIVAL   343
Ming Yang and Dongxin Lin

21 MicroRNA IN MYELOPOIESIS AND MYELOID DISORDERS   353
Sara E. Meyer and H. Leighton Grimes

22 MicroRNA DEREGULATION BY ABERRANT DNA METHYLATION IN ACUTE LYMPHOBLASTIC LEUKEMIA   371
Xabier Agirre and Felipe Prósper

23 ROLE OF miRNAs IN THE PATHOGENESIS OF CHRONIC LYMPHOCYTIC LEUKEMIA 383
Veronica Balatti, Yuri Pekarsky, Lara Rizzotto, and Carlo M. Croce

24 MicroRNA IN B-CELL NON-HODGKIN'S LYMPHOMA: DIAGNOSTIC MARKERS AND THERAPEUTIC TARGETS 403
Nerea Martínez, Lorena Di Lisio, and Miguel Angel Piris

25 MicroRNAs IN DIFFUSE LARGE B-CELL LYMPHOMA 419
Izidore S. Lossos and Alvaro J. Alencar

26 THE ROLE OF MicroRNAs IN HODGKIN'S LYMPHOMA 435
Wouter Plattel, Joost Kluiver, Arjan Diepstra, Lydia Visser, and Anke van den Berg

27 MicroRNA EXPRESSION IN CUTANEOUS T-CELL LYMPHOMAS 449
Cornelis P. Tensen

PART IV: HEREDITARY AND OTHER NON-INFECTIOUS DISEASES 463

28 MicroRNAs AND HEREDITARY DISORDERS 465
Matías Morín and Miguel A. Moreno-Pelayo

29 MicroRNAs AND CARDIOVASCULAR DISEASES 477
Koh Ono

30 MicroRNAs AND DIABETES 495
Romano Regazzi

31 MicroRNAs IN LIVER DISEASES 509
Patricia Munoz-Garrido, Marco Marzioni, Elizabeth Hijona, Luis Bujanda, and Jesus M. Banales

32 MicroRNA REGULATION IN MULTIPLE SCLEROSIS 523
Andreas Junker

33 THE ROLE OF MicroRNAs IN ALZHEIMER'S DISEASE 539
Shahar Barbash and Hermona Soreq

34 CURRENT VIEWS ON THE ROLE OF MicroRNAs IN PSYCHOSIS 553
Aoife Kearney, Javier A. Bravo, and Timothy G. Dinan

PART V: CIRCULATING MicroRNAs AS CELLULAR MESSENGERS AND NOVEL BIOMARKERS 567

35 CIRCULATING MicroRNAs AS NON-INVASIVE BIOMARKERS 569
Heidi Schwarzenbach and Klaus Pantel

36 CIRCULATING MicroRNAs AS CELLULAR MESSENGERS 589
Kasey C. Vickers

37 RELEASE OF MicroRNA-CONTAINING VESICLES CAN STIMULATE ANGIOGENESIS AND METASTASIS IN RENAL CARCINOMA 607
Federica Collino, Cristina Grange, and Giovanni Camussi

PART VI: THERAPEUTIC USES OF MicroRNAs: CURRENT PERSPECTIVES AND FUTURE DIRECTIONS 623

38 MicroRNA REGULATION OF CANCER STEM CELLS AND MicroRNAs AS POTENTIAL CANCER STEM CELL THERAPEUTICS 625
Can Liu and Dean G. Tang

39 THERAPEUTIC MODULATION OF MicroRNAs 639
Achim Aigner and Hannelore Dassow

40 LOCKED NUCLEIC ACIDS AS MicroRNA THERAPEUTICS 663
Henrik Ørum

Index 673

FOREWORD

The history of microRNA (miRNA) starts with an elegant genetic analysis by Ambros and Ruvkun that led to the discovery of a small non-coding RNA regulator of developmental timing. Eventually these two collaborators realised, during a late night phone conversation, that their RNA regulator binds, by Watson-Crick base pairing, to its target mRNA. Later work in the 1990s identified a second similar regulatory RNA, but I do not think that anyone would have predicted at that time that these RNA regulators would be the pioneers of a large class of RNA—the miRNAs—that affects the expression of a very large number of mRNAs.

In my laboratory, we work on plants and, in 1997, we tried to make a connection with the work of Ambros and Ruvkun. We had discovered small RNA that has a role in post-transcriptional gene silencing of transgenic and virus-infected plants. Like most biologists, we are always keen to make connections between different branches of the tree of life, and we hoped that our plant RNA would be similar to the regulatory RNA of worms. However, our silencing phenomena clearly operated at the level of RNA turnover, whereas the Ambros and Ruvkun RNAs mediated translational suppression. Our initial reluctant conclusion was that the worm RNAs and transgene silencing are separate phenomena.

Two later developments caused us, and others, to change our minds. First, there was use of sequencing to characterize the small RNA populations in several animals. This analysis revealed that the original Ambros and Ruvkun miRNAs are highly conserved from worms to man and that there are many similar RNAs that also bind to the 3′-UTR of their target RNA. Second, from genetic analyses, it was clear that the enzymes involved in the biogenesis and the effector activity of these regulatory RNAs—Dicer and Slicer—are implicated in many regulatory processes throughout development, as well as with gene silencing in transgenic and virus-infected plants. It was clear that the RNAs of Ambros and Ruvkun do not represent a specialized regulatory mechanism of early development in worms: they are part of a large family of silencing RNAs that includes the short RNAs that we had seen in plants. RNA silencing is common to both animal and plant kingdoms, and it can have many different biological effects.

The diversity of RNA silencing is indicated by the multiplicity of effector mechanisms involving RNA turnover and chromatin modification in addition to translational effects. This diversification is manifested even among miRNAs. They can act on target RNA stability, as well as on translation and they can both block and activate translation. Adding to the complexity of miRNA regulation there are "sponge" RNAs that are decoys of the natural miRNA target and miRNAs feature in regulatory systems with negative feedback loops. Some miRNAs are found in circulating blood, and they may act both outside and inside the cell. Clearly, there is the potential for great diversity and complexity in miRNA-mediated regulation.

Given this diversity and complexity, it is not surprising that there is great interest in clinical application of miRNAs. There is a good prospect that, even with the present level of understanding, miRNAs will feature in novel diagnostic tools, and that they will help identify targets for pharmaceutical and other inventions. Key areas for research include the targeting specificity of miRNAs and their place in networks of genetic regulation. New analytic methods based on next-generation sequencing will accelerate this research, and computational approaches for data analysis and systems modeling will be important drivers of progress.

The other, as yet relatively underexplored potential of miRNA, is as a therapeutic agent. A set of artificial miRNAs could be designed that would target one or more motifs in disease genes, and these RNAs could then be delivered so that they are taken up and have an effect in cells. In plants, the use of artificial miRNAs is a routine tool, although the targeting mechanism is simpler than in animals and delivery can be via transgenes rather than through uptake of RNA molecules into cells. Delivery is the major challenge for this therapeutic application of miRNAs, but there are early indications that it can be overcome for liver and possibly superficial tissues.

The translation pathway from basic research to the clinic and patient care is always complicated. Practical requirements often thwart the good intentions or clever ideas of the researchers. However, in the case of miRNAs, as with other applications related to RNA silencing, we can be more than usually optimistic for two reasons. The first reason is because a single set of miRNA mechanisms are involved in many aspects of growth, development, and responses to external stimuli. There is, therefore, a good prospect that miRNA research findings will have general relevance to many clinical applications. The second reason follows from the finding that miRNAs interact with their target through Watson-Crick base pairing. Such interactions are more predictable and computable than processes involving, for example, proteins or lipids or small molecules. Over the next decade, I anticipate that miRNAs will feature in many different clinical applications.

PROF. SIR DAVID BAULCOMBE
University of Cambridge
Corecipient (along with Victor Ambros and Gary Ruvkun)
of the 2008 Lasker Award for work on siRNA and miRNA

PREFACE

Since their formal recognition just over 10 years ago, microRNAs (miRNAs) have become one of the hottest topics in biology, not least of all because during this short time they have been found to act as crucial regulators of many, if not all, physiological and pathological processes. Nowhere has this increasing interest in miRNAs been more pronounced than within the medical field. Yet surprisingly, until now, there has been no book that attempts to cover this subject in any significant depth. Therefore, the primary aim of this project was to fill this gap by putting together a comprehensive collection of reviews from some of the leading lights in the miRNA world; for the first time, combining areas of medicine as diverse as stem cells, immunology, aging, infectious disease, cancer, psychiatric disease, and hereditary disorders are united by the central theme of miRNA involvement.

A criticism often leveled at a project like this is that it covers such a fast-moving subject that the book is out of date before it even hits the shelf. Had the aim of the book been solely to provide a collection of up-to-date reviews, then this criticism would indeed have been well founded; instead, we have tried to highlight general concepts of miRNA involvement as applied to well-established areas within medicine. While the specific roles for miRNAs described within these chapters will surely change and expand in the future, it is believed that the field is now sufficiently mature that these central concepts will stand the test of time and consequently this book will provide an invaluable resource for many years to come. Moreover, although specialist scientific journals can provide the reader with the very latest developments in the miRNA arena, in general, these texts are presented within a very narrow context and are not readily accessible to non-experts. A central goal of this endeavor was to provide each chapter with sufficient background context in order to open it up to readers outside of their specialist field, and in doing so, allow the reader to draw comparisons of the role of miRNAs between differing disciplines. For example, the hematologist may recognize the central role of *miR-181* in lymphoid differentiation and malignancy, but may not yet realize its importance to other pathologies, such as breast cancer, colorectal carcinoma, or even schizophrenia. This book attempts to offer a *"one-stop shop"* for information related to miRNA involvement in differing areas of medicine, and it is hoped that this cross-fertilization of ideas will stimulate novel research directions as a consequence. Another important role for this book was to serve as a preparatory text to the world of miRNAs for the uninitiated. With this in mind, an introductory chapter that aims to cover the FAQs of miRNAs has been included in order to provide a general framework for appreciating the subsequent chapters.

In summary, *MicroRNAs in Medicine* aspires to provide experts and non-experts alike with an understanding of the excitement, importance, breadth, and potential of miRNAs to modern medicine, and is aimed to appeal to clinicians, researchers, students, and journalists, as well as the interested public. It is hoped that this book marks the beginning (or continuation) of the readers journey into the miRNA world, and although comprehensive, the book makes no claim to be an exhaustive authority on the subject; rather, it is intended to serve as a foundation for further investigation.

I am indebted to the many contributing authors who have given so much of their valuable time to make this project a success. The involvement of such a high caliber of contributors, including some of the true pioneers of the field, have made the editorial role a pleasure, and it has been a great honor to work alongside many of the people that inspired my original foray into the miRNA world.

Special thanks should be given to Dr. Chris Hatton (Director of Clinical Medicine at the John Radcliffe Hospital, Oxford) for his inspiration and continual support over the years. This book is dedicated to my two beautiful children, Julia and Carlos, and my wonderful and understanding wife, María.

CHARLES H. LAWRIE
Biodonostia Research Institute, San Sebastián, Spain

CONTRIBUTORS

Xabier Agirre, Oncology Division, Foundation for Applied Medical Research, University of Navarra, Pamplona, Spain

Achim Aigner, Rudolf-Boehm-Institute for Pharmacology and Toxicology Clinical Pharmacology, University of Leipzig, Leipzig, Germany

Alvaro J. Alencar, Department of Medicine, Division of Hematology-Oncology and Molecular and Cellular Pharmacology, Sylvester Comprehensive Cancer Center, University of Miami, Miami, FL, USA

Veronica Balatti, Department of Molecular Virology, Immunology and Medical Genetics, Comprehensive Cancer Center and the Wexner Medical Center, The Ohio State University, Columbus, OH, USA

Jesus M. Banales, Division of Hepatology and Gastroenterology, Biodonostia Research Institute, San Sebastián, Spain; and IKERBASQUE, Basque Foundation of Science, Bilbao, Spain

Shahar Barbash, Department of Biological Chemistry and The Edmond and Lily Safra Center for Brain Sciences, The Hebrew University of Jerusalem, Jerusalem, Israel

Javier A. Bravo, Department of Psychiatry, University College Cork, Cork, Ireland

Paul F. Bray, The Cardeza Foundation for Hematologic Research and the Department of Medicine, Jefferson Medical College, Thomas Jefferson University, Philadelphia, PA, USA

Luis Bujanda, Division of Hepatology and Gastroenterology, Biodonostia Research Institute, San Sebastián, Spain

Barbara Burwinkel, Molecular Epidemiology (C080), German Cancer Research Center, Heidelberg, Germany; and Molecular Biology of Breast Cancer, Department of Obstetrics and Gynecology, University of Heidelberg, Heidelberg, Germany

George A. Calin, Experimental Therapeutics Department and Center for RNA Interference and Non-Coding RNA, The University of Texas MD Anderson Cancer Center, Houston, TX, USA

Carme Camps, Genomics Research Group, The Wellcome Trust Centre for Human Genetics, University of Oxford, Oxford, UK

Giovanni Camussi, Department of Internal Medicine, Molecular Biotechnology Center (MBC) and Centre for Research in Experimental Medicine (CeRMS), Torino, Italy

Xurde M. Caravia, Departamento de Bioquímica y Biología Molecular, Instituto Universitario de Oncología-IUOPA, Universidad de Oviedo, Oviedo, Spain

Daniela Castanotto, Department of Molecular and Cellular Biology, Beckman Research Institute of the City of Hope, Duarte, CA, USA

Federica Collino, Department of Internal Medicine, Molecular Biotechnology Center (MBC) and Centre for Research in Experimental Medicine (CeRMS), Torino, Italy

Carlo M. Croce, Department of Molecular Virology, Immunology and Medical Genetics, Comprehensive Cancer Center and the Wexner Medical Center, The Ohio State University, Columbus, OH, USA

Hannelore Dassow, Rudolf-Boehm-Institute for Pharmacology and Toxicology Clinical Pharmacology, University of Leipzig, Leipzig, Germany

Arjan Diepstra, Department of Pathology and Medical Biology, University of Groningen, University Medical Center Groningen, Groningen, The Netherlands

Lorena Di Lisio, Cancer Genomics Laboratory, IFIMAV, Santander, Spain

Timothy G. Dinan, Department of Psychiatry, University College Cork, Cork, Ireland

Leonard C. Edelstein, The Cardeza Foundation for Hematologic Research and the Department of Medicine, Jefferson Medical College, Thomas Jefferson University, Philadelphia, PA, USA

Aurélie Fender, Architecture et Réactivité de l'ARN, Institut de biologie moléculaire et cellulaire du CNRS, Université de Strasbourg, Strasbourg, France

Ramiro Garzon, Division of Hematology, Department of Internal Medicine, The Ohio State University Wexner Medical Center, Columbus, OH, USA

Irena Godnic, Department of Animal Science, Biotechnical Faculty, University of Ljubljana, Domzale, Slovenia

Myriam Gorospe, Laboratory of Genetics, NIA-IRP, NIH, Baltimore, MD, USA

Donald T. Gracias, Department of Microbiology and Immunology, Drexel University College of Medicine, Philadelphia, PA, USA

Ioannis Grammatikakis, Laboratory of Genetics, NIA-IRP, NIH, Baltimore, MD, USA

Cristina Grange, Department of Internal Medicine, Molecular Biotechnology Center (MBC) and Centre for Research in Experimental Medicine (CeRMS), Torino, Italy

Finn E. Grey, Division of Infection and Immunity, The Roslin Institute, University of Edinburgh, Easter Bush, Midlothian, UK

H. Leighton Grimes, Division of Cellular and Molecular Immunology, Cincinnati Children's Hospital Medical Center, Cincinnati, OH, USA; and Division of Experimental Hematology and Cancer Biology, Cincinnati Children's Hospital Medical Center, Cincinnati, OH, USA

Adrian L. Harris, Growth Factor Group, Cancer Research UK, Molecular Oncology Laboratories, Weatherall Institute of Molecular Medicine, John Radcliffe Hospital, University of Oxford, Oxford, UK

Elizabeth Hijona, Division of Hepatology and Gastroenterology, Biodonostia Research Institute, San Sebastián, Spain

Simon Horvat, Department of Animal Science, Biotechnical Faculty, University of Ljubljana, Domzale, Slovenia; and Department of Biotechnology, National Institute of Chemistry, Ljubljana, Slovenia

Maite Huarte, Center for Applied Medical Research (CIMA), Division of Oncology, University of Navarra, Pamplona, Spain

Marilena V. Iorio, Start Up Unit, Department of Experimental Oncology, Fondazione IRCCS, Istituto Nazionale Tumori, Milano, Italy

Juan Carlos Izpisua Belmonte, Centre for Regenerative Medicine in Barcelona, Barcelona, Spain; and Salk Institute for Biological Studies, La Jolla, CA, USA

Andreas Junker, Department of Neuropathology, University Medical Center Goettingen, Georg-August University, Gottingen, Germany

Omer Faruk Karatas, Molecular Biology and Genetics Department, Erzurum Technical University, Erzurum, Turkey

Peter D. Katsikis, Department of Microbiology and Immunology, Drexel University College of Medicine, Philadelphia, PA, USA

Aoife Kearney, Department of Psychiatry, University College Cork, Cork, Ireland

Joost Kluiver, Department of Pathology and Medical Biology, University Medical Center Groningen, Groningen, The Netherlands

Tanja Kunej, Department of Animal Science, Biotechnical Faculty, University of Ljubljana, Domzale, Slovenia

Agnieszka Kwarciak, Departamento de Bioquímica y Biología Molecular, Instituto Universitario de Oncología-IUOPA, Universidad de Oviedo, Oviedo, Spain

Charles H. Lawrie, Biodonostia Research Institute, San Sebastián, Spain; and Nuffield Department of Clinical Laboratory Sciences, University of Oxford, Oxford, UK

Dongxin Lin, State Key Laboratory of Molecular Oncology and Beijing Key Laboratory of Carcinogenesis and Cancer Prevention, Cancer Institute and Hospital, Chinese Academy of Medical Sciences and Peking Union Medical College, Beijing, China

Can Liu, Department of Molecular Carcinogenesis, The University of Texas MD Anderson Cancer Center, Smithville, TX, USA

Carlos López-Otín, Departamento de Bioquímica y Biología Molecular, Instituto Universitario de Oncología-IUOPA, Universidad de Oviedo, Oviedo, Spain

Erika Lorenzo Vivas, Centre for Regenerative Medicine in Barcelona, Barcelona, Spain

Izidore S. Lossos, Department of Medicine, Division of Hematology-Oncology and Molecular and Cellular Pharmacology, Sylvester Comprehensive Cancer Center, University of Miami, Miami, FL, USA

Xiaoya Luo, Division of Clinical Epidemiology and Aging Research (C070), German Cancer Research Center, Heidelberg, Germany

Jose Antonio Lupiáñez, Department of Biochemistry and Molecular Biology, Faculty of Sciences, University of Granada, Granada, Spain

Nerea Martínez, Cancer Genomics Laboratory, IFIMAV, Santander, Spain

Marco Marzioni, Department of Gastroenterology, "Università Politecnica delle Marche," Ancona, Italy

Pedro P. Medina, Department of Biochemistry and Molecular Biology, Faculty of Sciences, University of Granada, Granada, Spain; and Centre for Genomics and Oncological Research (GENYO), Granada, Spain

Sara E. Meyer, Division of Cellular and Molecular Immunology, Cincinnati Children's Hospital Medical Center, Cincinnati, OH, USA

Miguel A. Moreno-Pelayo, Unidad de Genética Molecular, Ramón y Cajal Institute of Health Research (IRYCIS) and Biomedical Network Research Centre on Rare Diseases (CIBERER), Madrid, Spain

Matías Morín, Unidad de Genética Molecular, Ramón y Cajal Institute of Health Research (IRYCIS) and Biomedical Network Research Centre on Rare Diseases (CIBERER), Madrid, Spain

Patricia Munoz-Garrido, Division of Hepatology and Gastroenterology, Biodonostia Research Institute, San Sebastián, Spain

Srikanth Nagalla, The Cardeza Foundation for Hematologic Research and the Department of Medicine, Jefferson Medical College, Thomas Jefferson University, Philadelphia, PA, USA

Venugopal Nair, Avian Viral Diseases Programme, The Pirbright Institute, Compton Laboratory, Compton, Berkshire, UK

Koh Ono, Department of Cardiovascular Medicine, Graduate School of Medicine, Kyoto University, Kyoto, Japan

Henrik Ørum, Santaris Pharma, Hørsholm, Denmark

Mustafa Ozen, Department of Medical Genetics, Istanbul University Cerrahpasa Medical School, Istanbul, Turkey; Bezmialem Vakif University, Istanbul, Turkey; and Department of Pathology & Immunology, Baylor College of Medicine, Houston, TX, USA

Klaus Pantel, Department of Tumor Biology, University Medical Center Hamburg-Eppendorf, Hamburg, Germany

Jon A. Pavelin, Division of Infection and Immunity, The Roslin Institute, University of Edinburgh, Easter Bush, Midlothian, UK

Yuri Pekarsky, Department of Molecular Virology, Immunology and Medical Genetics, Comprehensive Cancer Center and the Wexner Medical Center, The Ohio State University, Columbus, OH, USA

Sébastien Pfeffer, Architecture et Réactivité de l'ARN, Institut de biologie moléculaire et cellulaire du CNRS, Université de Strasbourg, Strasbourg, France

Katrin Pfütze, Molecular Epidemiology (C080), German Cancer Research Center, Heidelberg, Germany; and Molecular Biology of Breast Cancer, Department of Obstetrics and Gynecology, University of Heidelberg, Heidelberg, Germany

Claudia Piovan, Department of Molecular Virology, Immunology and Medical Genetics and Comprehensive Cancer Center, Ohio State University, Columbus, OH, USA; and Start Up Unit, Department of Experimental Oncology, Fondazione IRCCS, Istituto Nazionale Tumori, Milano, Italy

Miguel Angel Piris, Cancer Genomics Laboratory, IFIMAV, Santander, Spain; and Department of Pathology, Hospital U. Marqués de Valdecilla, Santander, Spain

Wouter Plattel, Department of Pathology and Medical Biology, Department of Hematology, University of Groningen, University Medical Center Groningen, Groningen, The Netherlands

Panote Prapansilp, Department of Laboratory Medicine and WHO Collaborating Center for Research and Training on Viral Zoonoses, Faculty of Medicine, Chulalongkorn University, Bangkok, Thailand

CONTRIBUTORS

Felipe Prósper, Oncology Division, Foundation for Applied Medical Research, University of Navarra, Pamplona, Spain; and Hematology Service and Area of Cell Therapy, Clínica Universidad de Navarra, University of Navarra, Pamplona, Spain

Jiannis Ragoussis, Genomics Research Group, The Wellcome Trust Centre for Human Genetics, University of Oxford, Oxford, UK

Andrew J. Ramsay, Departamento de Bioquímica y Biología Molecular, Instituto Universitario de Oncología-IUOPA, Universidad de Oviedo, Oviedo, Spain

Romano Regazzi, Department of Fundamental Neurosciences, University of Lausanne, Lausanne, Switzerland

Fernando J. Reyes-Zurita, Department of Biochemistry and Molecular Biology, Faculty of Sciences, University of Granada, Granada, Spain

Natalie L. Reynolds, Division of Infection and Immunity, The Roslin Institute, University of Edinburgh, Easter Bush, Midlothian, UK

Lara Rizzotto, Department of Molecular Virology, Immunology and Medical Genetics, Comprehensive Cancer Center and the Wexner Medical Center, The Ohio State University, Columbus, OH, USA

John J. Rossi, Department of Molecular and Cellular Biology, Beckman Research Institute of the City of Hope, Duarte, CA, USA

Eva E. Rufino-Palomares, Department of Biochemistry and Molecular Biology, Faculty of Sciences, University of Granada, Granada, Spain

Yolanda Sánchez, Center for Applied Medical Research (CIMA), Division of Oncology, University of Navarra, Pamplona, Spain

Sukhinder K. Sandhu, Molecular Virology, Immunology and Medical Genetics, Comprehensive Cancer Center, The Ohio State University Wexner Medical Center, Columbus, OH, USA

Heidi Schwarzenbach, Department of Tumor Biology, University Medical Center Hamburg-Eppendorf, Hamburg, Germany

Hermona Soreq, Department of Biological Chemistry and The Edmond and Lily Safra Center for Brain Sciences, The Hebrew University of Jerusalem, Jerusalem, Israel

Guihua Sun, Department of Molecular and Cellular Biology, Beckman Research Institute of the City of Hope, Duarte, CA, USA

Dean G. Tang, Department of Molecular Carcinogenesis, The University of Texas MD Anderson Cancer Center, Smithville, TX, USA

Cornelis P. Tensen, Department of Dermatology, Leiden University Medical Center, Leiden, The Netherlands

Gustavo Tiscornia, Department of Biomedical Sciences and Medicine, University of Algarve, Faro, Portugal

Gareth D.H. Turner, Mahidol-Oxford Tropical Medicine Research Unit and Department of Tropical Pathology, Faculty of Tropical Medicine, Mahidol University, Bangkok, Thailand; and Centre for Tropical Medicine, Nuffield Department of Clinical Medicine, Oxford University, Oxford, UK

Alejandro P. Ugalde, Departamento de Bioquímica y Biología Molecular, Instituto Universitario de Oncología-IUOPA, Universidad de Oviedo, Oviedo, Spain

Anke van den Berg, Department of Pathology and Medical Biology, University of Groningen, University Medical Center Groningen, Groningen, The Netherlands

Kasey C. Vickers, Division of Cardiovascular Medicine, Department of Medicine, Vanderbilt University School of Medicine, Nashville, TN, USA

Lydia Visser, Department of Pathology and Medical Biology, University of Groningen, University Medical Center Groningen, Groningen, The Netherlands

Ming Yang, College of Life Science and Technology, Beijing University of Chemical Technology, Beijing, China

Yongxiu Yao, Avian Viral Diseases Programme, The Pirbright Institute, Compton Laboratory, Compton, Berkshire, UK

Minja Zorc, Department of Animal Science, Biotechnical Faculty, University of Ljubljana, Domzale, Slovenia

MicroRNAs: A Brief Introduction

Charles H. Lawrie

Biodonostia Research Institute, San Sebastián, Spain
Nuffield Department of Clinical Laboratory Sciences, University of Oxford, Oxford, UK

I. A Short History of Small RNAs	2
II. Biogenesis of miRNAs	3
A. miRNA Nomenclature: What's in a Name?	5
III. miRNA Function: Controlling mRNA Stability, Degradation, and/or Translation	6
IV. Regulating the Regulators: miRNA Control and Dysfunction in Disease	7
A. Genetic Dysregulation of miRNA Expression	7
B. Epigenetic Regulation	8
C. Transcription Factors and miRNA Regulatory Networks	9
D. Regulating miRNA Synthesis and Processing	10
E. Control of miRNA Function	11
V. Present and Future Perspectives for miRNAs in Medicine	12
A. Deciphering the miRNA Targetome: Understanding the Functional Consequences of miRNA Dysregulation in Disease	12
B. Tip of the Non-Coding RNA Iceberg	13
C. Are miRNAs Clinically Useful Molecules?	14
References	15

MicroRNAs in Medicine, First Edition. Edited by Charles H. Lawrie.
© 2014 John Wiley & Sons, Inc. Published 2014 by John Wiley & Sons, Inc.

ABBREVIATIONS

ADAR	adenosine deaminases that act on RNA
Ago	Argonaute
ALL	acute lymphoblastic leukemia
CLL	chronic lymphocytic leukemia
DGCR8	DiGeorge critical region 8
dsRNA	double-stranded RNA
Exp-5	exportin 5
HITS-CLIP	high-throughput sequencing of RNA isolated by cross-linking immunoprecipitation
IP	immunoprecipitation
lncRNA	long non-coding RNA
miRISC	miRNA RNA interference silencing complex
miRNA	microRNA
mRNA	messenger RNA
ncRNA	non-coding RNA
nt	nucleotide
PACT	protein activator of the interferon-induced protein kinase
PAR-CLIP	photoactivatable-ribonucleoside-enhanced cross-linking and immunoprecipitation
PASR	promoter-associated small RNAs
piRNA	piwi-interacting RNA
PROMPT	promoter upstream transcripts
RBP	RNA-binding protein
RNAi	RNA interference
snoRNA	small nucleolar RNA
ssRNA	single-stranded RNA
TF	transcription factor
tiRNA	tiny RNA
TRBP	HIV-1 TAR RNA binding protein
tRNA	transfer RNA
TSSa-RNA	TSS-associated RNA
T-UCR	transcribed ultraconserved regions
UTR	untranslated region

I. A SHORT HISTORY OF SMALL RNAs

The central dogma of molecular biology, first postulated by Francis Crick in 1958 and later refined in 1970, states that biological information flows unidirectionally from DNA to RNA to protein (Crick 1970). This view implies that non-coding RNA (ncRNA) has little or no intrinsic value, despite accounting for more than 90% of eukaryotic transcriptional output (Mattick 2001). Consequently, it is perhaps not surprising that microRNAs (miRNAs) were unknown to the scientific community until very recently. Indeed, it was only in 1993 when the first, what we now know to be a miRNA, was announced by the Ambros and Ruvkun laboratories simultaneously in the December edition of the journal *Cell* (Lee et al. 1993; Wightman et al. 1993). The Ambros group had identified and cloned a *Caenorhabditis elegans* developmental regulatory locus, *lin-4*, that did not contain conventional start and stop codons. Furthermore, introducing mutations that disrupted the putative open reading frame in this 700-nt fragment did not affect function, suggesting that *lin-4* did not encode for a protein at all (Lee et al. 1993). At the same time, the Ruvkun lab were working on another temporal regulator of *C. elegans*, *lin-14*. They had found that *lin-14* was regulated posttranscriptionally via a repeat sequence in the 3'-UTR (untranslated region) of the gene (Wightman et al. 1993). The two labs shared their unpublished findings and realized that the small transcripts of *lin-4* (22 nt and 61 nt in length) contained complementary sequences to the 3'-UTR sequence of *lin-14*, and could regulate this gene via an entirely new regulatory mechanism involving non-coding RNA. However, as *lin-4* has no clear homologue outside of worms, the biological significance of this discovery was not realized until many years later.

Although RNA silencing had been known in plants since the beginning of the 1990s (Napoli et al. 1990), the connection with small RNAs was not made until 1999, when the Baulcombe laboratory identified small (25-nt) non-coding RNA species complementary to the target gene that were responsible for gene silencing (Hamilton and Baulcombe 1999). A few months later, it was demonstrated that dsRNA, the trigger for RNA interference (RNAi) (Fire et al. 1998), was sequentially processed into 21–23 nt ssRNA fragments (Zamore et al. 2000). Soon after, another publication from the Ruvkun laboratory described a heterochronic gene of *C. elegans*, *let-7*, that controls juvenile to adult transition in larval development (Reinhart et al. 2000). *let-7* shared many of the characteristics of *lin-4* as it encoded for a small (21 nt) ncRNA transcript that negatively regulated the mRNA of *lin* family members through complementary RNA-RNA interactions at the 3'-UTR of these genes. Unlike *lin-4*, however, the sequence of *let-7* was found to be conserved in most eukaryotic organisms (Pasquinelli et al. 2000; Lagos-Quintana et al. 2001). Together, these discoveries instigated the start of the miRNA revolution, a term first coined by Lee and Ambros in 2001 (Lee and Ambros 2001). Since this time, over 25,000 miRNAs (including more than 2000 human miRNAs) have been identified from a diverse range of more than 190 different species, including algae, plants, mycetozoa, arthropods, nematodes, protozoa, vertebrates, plants, and viruses (Griffiths-Jones et al. 2006). For a current list of annotated miRNAs, see the miRBase database (http://www.mirbase.org/).

MiRNAs primarily function as posttranscriptional (negative) regulators of gene expression via binding to complementary sequences located mainly within the UTRs of target genes. Because a single miRNA can target several hundred genes, it is believed ~60% of all human genes are a potential target for miRNA regulation (Friedman et al. 2009). In addition, a single target gene often contains binding sites for multiple miRNAs that can bind cooperatively (Lewis et al. 2003), allowing miRNAs to form complex regulatory control networks. Perhaps, unsurprisingly, miRNAs have been shown to play key regulatory roles in virtually every aspect of biology, including the many physiological and pathological processes described in the chapters of this book.

II. BIOGENESIS OF MIRNAS

The majority of human miRNAs are encoded within introns of coding mRNAs, while others are located exgenically, in non-coding mRNAs or within the 3'-UTR sequence of coding mRNA (Rodriguez et al. 2004). MiRNAs are transcribed as 5'-capped large polyadenylated transcripts (pri-microRNA) primarily in a Pol II-dependent manner, although the involvement of Pol-III transcription has also been postulated for miRNAs encoded within Alu repeat sequences (Borchert et al. 2006).

Adenosine deaminases that act on RNA (ADARs) can alter the specificity and binding capacity of miRNA transcripts by changing adenosine bases to inosine post-transcriptionally. For example, ADAR-mediated changes to the *pre-miR-151* sequence cause accumulation of the pre-miRNA by blocking Dicer processing (Kawahara et al. 2007a). A selective change to the seed sequence of *miR-376* by ADARs causes it to additionally target *PRPS1* (Kawahara et al. 2007b). Deep sequencing of mouse brain tissue has identified a number of "edited" miRNAs, increasing the potential repertoire of miRNA targets available for regulation (Chiang et al. 2010).

Approximately 40% of human miRNAs are cotranscribed as clusters encoding multiple miRNA sequences in a single pri-microRNA transcript (Altuvia et al. 2005; Hertel et al. 2006). Pri-miRNAs are cleaved within the nucleus by Drosha, an RNaseIII-type nuclease, to form 60–110 nucleotide hairpin structures (pre-microRNA) (Figure 1.1). Drosha by itself possesses little enzymatic activity and requires the cofactor DiGeorge syndrome critical region 8 gene (DGCR8) in humans (Pasha in *Drosophila*) to form the microprocessor complex (Yeom et al. 2006).

Once produced, pre-miRNAs are exported from the nucleus to the cytoplasm by Exportin-5 (Exp-5) in a Ran-GTP dependent manner (Zeng 2006). The cytoplasmic pre-miRNA is further cleaved by Dicer, another RNaseIII-type enzyme, to form an asymmetric duplex intermediate (miRNA:miRNA*), consisting of the mature miRNA sequence and the antisense miRNA passenger strand (miRNA*). Similar to Drosha, cofactors, such as

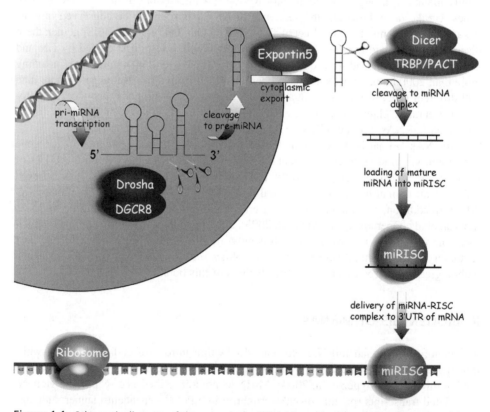

Figure 1.1. Schematic diagram of the canonical miRNA biosynthetic pathway. Reproduced from Lawrie, C.H. (2007) *Br J Haematol* **136** (6):503–512. See color insert.

TRBP and PACT (in humans), are necessary for Dicer activity (Lee et al. 2006). The miRNA:miRNA* duplex is, in turn, loaded into the miRISC complex in which Argonaut (Ago) proteins are the key effector molecules. The strand that becomes the active mature miRNA appears to be dependent upon which has the lowest free energy 5′ end and is retained by the miRISC complex, while the passenger strand is generally believed to be degraded by an unknown nuclease (Khvorova et al. 2003; Schwarz et al. 2003). It should be noted, however, that many miRNA passenger strands are also capable of silencing target transcripts and probably play a more important biological role than was previously realized (Okamura et al. 2008; Ghildiyal et al. 2010).

The loaded miRISC is guided by the mature miRNA sequence (19–24 nucleotide) to complementary sequences located primarily within the 3′-UTR of the target gene mRNA, although binding sites have additionally been identified in both 5′-UTR (Lytle et al. 2007) and coding regions of genes (Tay et al. 2008). In contrast to plant miRNAs that contain extensive regions of complementarity with their target genes, animal miRNAs are only partially complementary and have a propensity to recognize targets via 6–8 nt "seed" sequences, usually located at nt position 2–8 of the 5′-end of the miRNA (Bartel 2009), although sometimes also in the center of the miRNA sequence (Shin et al. 2010). There are rare examples of animal miRNAs (e.g., *miR-196* and *HOXB8*) that do share near-perfect complementarity, resulting in direct cleavage of the mRNA (Yekta et al. 2004).

While the vast majority of animal miRNAs are generated by the canonical miRNA biosynthetic pathway described above (Ghildiyal and Zamore 2009) (Figure 1.1), alternative Drosha-independent and Dicer-independent pathways do exist (for detailed review, see Yang and Lai 2011). For example, mirtrons are short RNA duplexes derived from the splice acceptor and donor sites of introns (Okamura et al. 2007). Splicing occurs independently of Drosha cleavage, and mirtrons can enter the canonical pathway via Exp-5 export and can regulate typical seed-matching targets. While mirtrons appear to be prevalent in both *D. melanogaster* and *C. elegans* genomes (Chung et al. 2011), they are little studied in vertebrates, although *drosha* and *dgcr8* murine knockouts maintain expression of mirtrons, suggesting that a similar mechanism does exist in mammals (Berezikov et al. 2007; Chan and Slack 2007; Babiarz et al. 2011; Ladewig et al. 2012). In addition, functional miRNAs can also be derived from larger ncRNA molecules, such as snoRNAs and tRNAs (Babiarz et al. 2008), and tRNaseZ (Bogerd et al. 2010), in a Drosha-independent Dicer-dependent manner. In contrast, *miR-451* is processed independently of Dicer but requires Drosha activity (Cheloufi et al. 2010; Cifuentes et al. 2010).

A. miRNA Nomenclature: What's in a Name?

The need for a rationalized system of nomenclature for miRNAs was realized soon after their discovery (Ambros et al. 2003). While the first miRNAs were named on the basis of position or function (i.e., *let-7* is "lethal phenotype-7" and *lin-4* represents a previously defined *C. elegans* genetic locus), it became apparent that such an approach would quickly become unmanageable. Novel miRNA sequences are therefore assigned a number that reflects the order of their discovery by the central miRNA database, miRBase (http://www.mirbase.org/) (Griffiths-Jones et al. 2006). Closely related miRNA sequences are followed by a letter (e.g., *miR-34a*, *miR-34b*, and *miR-34c* [*miR-34* was the 34th discovered miRNA]). The same sequence may be encoded at multiple genomic locations, in which case a further suffix is added (e.g., *hsa-mir-16-1* is encoded at chromosome 13 and *hsa-mir-16-2* at chromosome 3). The first prefix in the earlier example indicates species (e.g., *hsa-* is *Homo sapiens*) and is followed by either "*mir*" or "*miR*," the former indicating a pri-miRNA, pre-miRNA, or genomic locus, and the latter (with a capitalized R), the mature miRNA sequence. Each pre-miRNA encodes for two possible mature miRNA forms derived from either the 3′ or 5′ arm of the hairpin sequence (Figure 1.1). Previously, these were designated the "miR" sequence, for the major mature product and the "miR*" sequence, for the minor mature product (e.g., *miR-155* and *miR-155**). This has recently been replaced in miRBase by the use of the -5p or -3p suffix to reflect their origin (e.g., *miR-17-3p* and *miR-17-5p*). This change also acknowledges increasing evidence that both strands can be functionally important (Czech and Hannon 2011), and that the frequency of a particular form may differ between different cell types (Griffiths-Jones et al. 2011).

III. MIRNA FUNCTION: CONTROLLING MRNA STABILITY, DEGRADATION, AND/OR TRANSLATION

Although repression of translation without mRNA degradation was originally believed to be the *modus operandi* of animal miRNAs, the situation appears to be more complex than previously thought, and there is now compelling evidence that miRNAs can also affect transcriptional levels through deadenylation, degradation, and/or destabilization of target mRNAs (Giraldez et al. 2006). Indeed, it has been suggested that translational inhibition has only a modest role to play in miRNA function, and that mRNA destabilization is the predominant mechanism used to inhibit target genes in mammals (Guo et al. 2010). MiRNAs appear capable of utilizing a wide range of methods to regulate gene expression; however, the relative contribution of each of these in mammalian cells remains unclear, not least of all because most studies have been carried out in invertebrates.

In perhaps the simplest mechanism, near-perfect pairing of miRNA and target gene sequences, a common occurrence in plants (Llave et al. 2002) but exceedingly rare in animals (Yekta et al. 2004), allows Ago-mediated endonucleotic cleavage of the target mRNA to take place (Figure 1.2A). More commonly in animals, miRNA binding results in destabilization of the target mRNA via recruitment of deadenylation factors, making the mRNA more susceptible to degradation (Figure 1.2B). Deadenylation is mediated by GW182 proteins that form part of the miRISC complex. The carboxy-terminus of GW182 interacts with poly(A)-binding protein (PABP) and recruits CCR4 and CAF1 deadenylases (Wu et al. 2006; Huntzinger and Izaurralde 2011).

How translational repression in the absence of mRNA degradation operates remains controversial, and several mechanisms have been proposed. There is evidence for inhibition of translation occurring at the stage of initiation via repression of eukaryotic initiation factors (eIF) (Humphreys et al. 2005) (Figure 1.2C). Alternatively, other studies demonstrate that miRNAs can inhibit translation after initiation during elongation of the nascent peptide (Olsen and Ambros 1999) (Figure 1.2D). It has also been suggested that miRNA-bound mRNA can be sequestered away from the translational machinery in P-bodies that additionally act in concert with enzymes to remove the 5′ cap, hence preventing translation (Liu et al. 2005; Sen and Blau 2005) (Figure 1.2E), or may prevent recognition of the 5′ cap by translation factors (Pillai et al. 2005).

In addition to negative regulation, miRNAs are also able to function as translational activators of some genes (Vasudevan et al. 2007) (Figure 1.2F). To complicate matters further, miRNAs can also possess decoy activity that interferes with the function of regulatory proteins. For example, *miR-328* can bind directly to hnERP E2, preventing its interaction with *CEBP* mRNA and therefore function (Eiring et al. 2010).

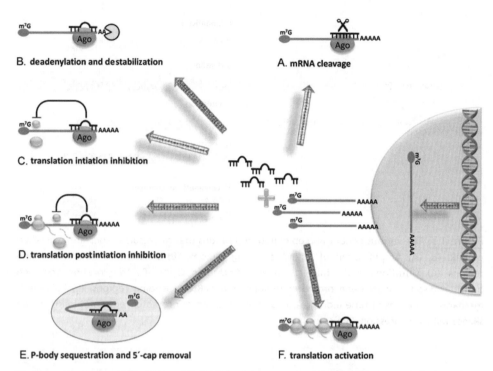

Figure 1.2. Schematic diagram of proposed mechanisms for miRNA function. (A) Ago-mediated cleavage of mRNA can occur when the miRNA sequence is complementary to the target gene-binding site. (B) Removal of poly(A) tail by deadenylases causes destabilization and degradation of mRNA. (C) Translation initiation inhibited by miRISC interactions with eukaryotic translation initiation factors (eIFs). (D) Inhibition of translation postinitiation. (E) Sequestration of mRNA in P-bodies. (F) miRNA-mediated translational activation. See color insert.

IV. REGULATING THE REGULATORS: miRNA CONTROL AND DYSFUNCTION IN DISEASE

Under physiological conditions the range of mechanisms available to the cell to control miRNA expression and function are every bit as varied as those involved in regulating protein-encoding genes, and can occur at the level of miRNA expression or posttranscriptionally via modulation of miRNA processing or miRNA function. As with protein-encoding genes, dysfunction of these regulatory mechanisms is often associated with disease, as are genetic alterations that result in dysregulated expression of miRNA-associated regions.

A. Genetic Dysregulation of miRNA Expression

Changes to the genetic structure of miRNA-associated regions have been linked with many different pathologies, most notably cancer. Such chromosomal aberrations include amplifications (e.g., *miR-26a* in glioma [Huse et al. 2009] [Figure 1.3A]), deletions (e.g., *miR-15a/16-1* in chronic lymphocytic leukemia [CLL] [Calin et al. 2002] [Figure 1.3B]), mutations (e.g., *miR-125a* in breast cancer [Li et al. 2009] [Figure 1.3C]), translocations

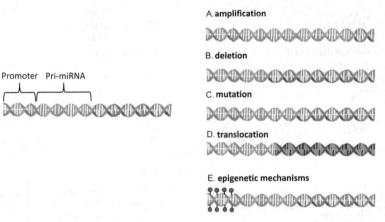

Figure 1.3. The various genetic and epigenetic alterations that can result in aberrant expression of miRNAs. (A) Amplification of miRNA-encoding regions. (B) Deletion of miRNA-encoding regions. (C) Mutations in the miRNA sequence (including SNPs). (D) Translocation occurring between distal, usually gene promoter regions, and miRNA-encoding regions. (E) Epigenetic mechanisms, such as histone modification and methylation of promoter regions of miRNAs, can silence miRNA expression. See color insert.

(e.g., *miR-125b* in leukemia [Bousquet et al. 2008] [Figure 1.3D]), single nucleotide polymorphisms (e.g., *miR-608* in lung cancer [Lin et al. 2012]), and loss of heterozygosity (e.g., 14q32 cluster in acute lymphoblastic leukemia [ALL] [Agueli et al. 2010]). These alterations can occur not only in the sequence of the mature miRNA itself, but also in the promoter region/pri-miRNA sequence (Calin et al. 2002, 2005) or at the miRNA-binding sites of target genes (Abelson et al. 2005), and can occur somatically or hereditarily.

B. Epigenetic Regulation

In addition to genetic alterations, aberrant miRNA expression can also result from epigenetic mechanisms, such as DNA methylation and histone modification (Figure 1.3E). While the importance of these mechanisms in disease is now apparent, the role of epigenetic regulation of miRNA expression under physiological conditions is at present much less clear.

DNA methylation occurs primarily at cytosine residues (changing it to 5-methylcytosine) that form part of the CpG dinucleotide motif, which is most commonly found at the proximal end of the promoter region of genes. Under physiological conditions, most genes are unmethylated; however, in tumor cells, the majority of genes are hypermethylated leading to aberrant silencing of multiple tumor suppressor genes (Esteller 2008). As with protein-encoding genes, approximately half of miRNA genes are associated with CpG islands (Weber et al. 2007). It has been shown that malignancy-associated changes in the methylation status of miRNAs can specifically regulate the expression of genes directly implicated in tumorigenesis (Bueno et al. 2008; Agirre et al. 2009). To complicate matters further, miRNAs can themselves regulate the expression of important components of the epigenetic machinery, including DNA methyltransferases (Benetti et al. 2008) and histone deacetylases (Roccaro et al. 2010).

Epigenetic gene silencing can also occur via histone modification. The amino acids of histone proteins (particularly their N-terminal "tails") can be posttranslationally modified in a number of ways that can repress transcriptional activity, including an increase in closed chromatin marks, such as trimethylation of 3mK9H3 and 3mK27H3, and a decrease in marks of open chromatin, for example, acetylation of AcH3 and AcH4 or trimethylation of 3mK4H3. The importance of histone regulation for miRNA expression is implied by the fact that histone modulators, such as HDAC-inhibitors, fundamentally alter the miRNA profile of treated cells (Scott et al. 2006; Barski et al. 2009). Indeed, the use of epigenetic drugs, such as DNA-demethylating agents (e.g., 5-aza-2'-deoxycytidine and zebularine) and histone deacetylase inhibitors (e.g., trichostatin A), may be of potential therapeutic use in reexpressing epigenetically silenced miRNAs, such as *miR-124a*, which is associated with poor prognosis in acute lymphoblastic leukemia (ALL) (Agirre et al. 2009). For a much more detailed description of the role of epigenetic regulation of miRNAs in cancer, the reader is directed to Chapter 22.

C. Transcription Factors and miRNA Regulatory Networks

Like protein-encoding genes, a myriad of transcription factors (TFs) can influence miRNA expression levels, and this form of regulation appears to be particularly important in controlling tissue specificity and developmental stage. The promoter regions of autonomously expressed miRNA genes are very similar to that of protein-encoding genes and can contain initiation and response elements, TATA boxes, and CpG islands, and so on (Corcoran et al. 2009). TFs can regulate miRNA expression by direct binding to promoter regions either positively (e.g., p53 stimulates *miR-34* expression [Christoffersen et al. 2010]) or negatively (e.g., ERα inhibits *miR-221* and *miR-222* expression [Di Leva et al. 2010]), or even both positively and negatively. MYC, for example, an important oncogene, up-regulates the *miR-17~92* cluster (O'Donnell et al. 2005), but can also down-regulate *miR-15a/16-1* (Chang et al. 2008) (Figure 1.4A).

In addition to direct control of miRNA expression, TFs are commonly involved in regulatory feedforward and feedback loops, whereby miRNAs autoregulate their expression in a negative (type I) or positive (type II) manner (Tsang et al. 2007). For example, *miR-27a* inhibits expression of *RUNX1*, which, in turn, stimulates expression of *miR-27a* in a simple unilateral negative feedback loop (Ben-Ami et al. 2009) (Figure 1.4B). *miR-15a* and *MYB* operate in a similar manner in hematopoietic cells (Zhao et al. 2009). Alternatively, a reciprocal-negative (positive) feedback loop can exist when the TF inhibits the miRNA, which, in turn, inhibits the TF. This, for example, is the relationship between *HBL1* and *let-7* (Roush and Slack 2009), and *ZEB1* and *miR-200* (Bracken et al. 2008) (Figure 1.4C).

Further levels of control can be achieved by adding extra components to the loops. In its simplest form, this results in a double-negative (or positive) loop, such as what occurs in the establishment of left-right asymmetry for the ASE chemosensory neurons of *C. elegans*. In this system, COG-1 inhibits the left ASE while stimulating expression of *miR-273* in the right ASE, which, in turn, inhibits *DIE-1* that up-regulates lys-6 expression promoting the left ASE-specific pathway. In the left, ASE *COG-1* is blocked by lys-6 (Johnston et al. 2005).

In reality, such loops represent the basic network motifs that make up much more complicated TF/miRNA regulatory networks that are a common feature of mammalian cell physiology (Shalgi et al. 2007; Tsang et al. 2007). These networks allow for

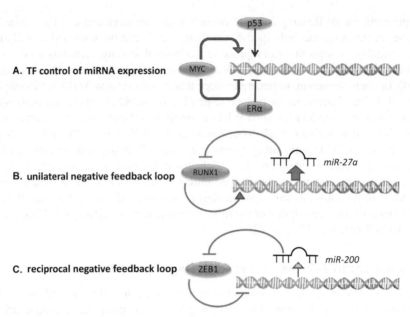

Figure 1.4. Examples of simple regulatory loop motifs involving miRNAs and transcription factors (TFs). (A) TFs can bind directly to the promoter region of miRNAs, either up-regulating or down-regulating expression, or a single TF (e.g., MYC) can up-regulate expression of one miRNA but down-regulate another. (B) When the TF that regulates the miRNA is itself regulated by that miRNA, a simple loop motif results. Up-regulation of the miRNA by the TF decreases its own expression in a unilateral negative feedback loop. (C) When the TF down-regulates the miRNA, this can increase TF expression in a reciprocal (positive) feedback loop. See color insert.

acceleration of transcriptional response times and dampening of fluctuating protein/miRNA levels within the cellular environment (Becskei and Serrano 2000; Rosenfeld et al. 2002).

In addition to employing TFs to regulate miRNA expression, it has recently been shown that miRNAs can autoregulate their own expression. *Let-7* bound to miRISC was found to directly bind and regulate its own primary transcript in a feedforward loop (Zisoulis et al. 2012).

D. Regulating miRNA Synthesis and Processing

The importance of correct functioning of the miRNA biosynthetic canonical pathway to mammalian biology is implied by the severity of phenotypes obtained when components of this pathway are deleted. Indeed, when the first mouse models were created using constitutive deletion of *Dicer* (Bernstein et al. 2003), *DGCR8* (Wang et al. 2007), *Drosha* (Fukuda et al. 2007), or *Ago2* (Morita et al. 2007), all resulting progeny were nonviable and died during early gestation with severe developmental defects. Needless to say subsequent studies utilized Cre-inducible conditional knockout mice in a more targeted approach. For example, targeted deletion of *Dicer* activity in astroglial cells (Tao et al. 2011) or oligodendrocytes (Shin et al. 2009) resulted in severe neuronal dysfunction. When *Dicer* was deleted specifically in the myogenic compartment this caused severe disruption to early skeletal muscle development (O'Rourke et al. 2007), while when deleted in the female reproductive tract resulted in a loss of fertility and adenomyosis (Gonzalez and Behringer 2009). Further studies have demonstrated the essential nature of Dicer to the development of the heart (da Costa Martins et al. 2008) and the lung (Harris et al. 2006). In a similar manner, conditional knockouts of *DGCR8* in vascular smooth muscle cells resulted in severe liver hemorrhage (Chen et al. 2012), while deletion in cardiac neural crest cells lead to severe cardiac malformations (Chapnik et al. 2012). Interestingly, when *Dicer* was conditionally mutated in the lung tissue of K-ras mice, this resulted in an enhancement of lung tumor development (Kumar et al. 2007).

The levels and activities of the components of the miRNA biosynthetic pathway are subject to careful regulation. For example, Drosha is stabilized by its co-factor DGCR8, and conversely *DGCR8* mRNA levels are controlled by Drosha-mediated degradation (Han et al. 2009). Similarly, a decrease in levels of co-factor TRBP can destabilize Dicer (Chendrimada et al. 2005). In addition, a multitude of factors interacting with Drosha, Dicer or miRNA precursors can control the kinetics of miRNA biogenesis. For example, *LIN-28* can bind to the terminal loop of *pri-let-7*, interfering with Drosha cleavage and also Dicer processing (Viswanathan and Daley 2010). In contrast, SF2/ASF promotes cleavage of *pri-let-7* by Drosha (Winter et al. 2009).

Defects in the functioning of the miRNA biosynthetic pathway have frequently been associated with disease and cancer in particular. For example, low levels of Dicer and/or Drosha have been linked with poor clinical outcome for patients with ovarian cancer (Merritt et al. 2008), nasopharyngeal cancer (Guo et al. 2012), neuroblastoma (Lin et al. 2010), breast cancer (Khoshnaw et al. 2012), and lung cancer (Karube et al. 2005). In contrast, high levels of Dicer and/or Drosha have been associated with poor prognosis for esophageal cancer (Sugito et al. 2006), colorectal carcinoma (Faber et al. 2011), and prostate cancer (Chiosea et al. 2006). Whether these apparently contradictory findings represent true biological differences between the cancer types or are artifactual remains to be determined. Nevertheless, such studies underline the fundamental importance of the global miRNA machinery to both physiological and pathological processes, and suggest intriguing therapeutic possibilities based upon restoration or inhibition of components of this pathway.

E. Control of miRNA Function

The miRNA pathway downstream of synthesis of the mature miRNA is also subject to extensive regulation. For example, members of the TRIM-NHL family of proteins can regulate the ability of miRISC to repress target gene expression by directly associating with Ago proteins. TRIM71 can ubiquitinate Ago proteins, targeting them for proteasome-mediated degradation, which, in turn, regulates miRNA stability and decreases miRNA-mediated repression (Chatterjee and Grosshans 2009; Rybak et al. 2009). Mei-26 has also been demonstrated to repress miRISC function, although the exact mechanism involved is currently unclear (Neumuller et al. 2008). In contrast, NHL2 and TRIM32 can positively regulate miRISC function (Hammell et al. 2009; Schwamborn et al. 2009).

RBPs (RNA-binding proteins), such as HuR and Deadend1 (DND1), can bind target mRNA and therefore interfere with miRNA functioning. For example, HuR binding to the mRNA of *CAT1* can inhibit repression of this gene by *miR-122* (Bhattacharyya et al. 2006). In contrast, HuR binding to the 3'-UTR of *MYC* is necessary for *let-7*-mediated repression (Kim et al. 2009). HuR mediated control of miRNA repression may be a widespread phenomena, as 75% of 3'-UTRs with miRNA-binding sites also contain binding sites for HuR (Mukherjee et al. 2011). DND1 can also bind in the vicinity of miRNA-binding sites apparently functioning by reducing the accessibility of the mRNA to miRISC (Kedde et al. 2007). *DND1*, in turn, can be regulated by *miR-24*-mediated inhibition (Liu et al. 2010).

Besides the regulatory mechanisms described earlier, other levels of control of miRNA function exist. For example, control of intracellular compartmentalization occurs when miRISC-bound mRNA is sequestered in P-bodies and stress granules (Liu et al. 2005). Regulation of the kinetics of miRNA decay have also been observed (Hwang et al. 2007).

V. PRESENT AND FUTURE PERSPECTIVES FOR miRNAs IN MEDICINE

The aim of the following section is to cover some of the current controversies in the miRNA field and how they are being addressed, and additionally speculate about future developments for miRNAs, in particular their potential usefulness to medicine.

A. Deciphering the miRNA Targetome: Understanding the Functional Consequences of miRNA Dysregulation in Disease

The first indication that miRNA dysfunction was directly associated with disease originated from the laboratory of Carlo Croce in 2002, whose seminal publication by Calin et al., made the connection between 13q14, a frequently deleted locus in CLL, and downregulation of the *mir-15a/16-1* cluster that is encoded within this region (Calin et al. 2002) (see Chapter 23 for more details). The same group later highlighted the potential importance of miRNAs in cancer, with the somewhat surprising finding that the majority of human miRNAs are in fact located at cancer-associated genomic regions (Calin et al. 2004). There is now overwhelming evidence that dysfunctional expression of miRNAs is a major contributor to the pathogenesis of most, if not all, human malignancies (Croce 2009; Iorio and Croce 2012). Besides cancer, evidence is rapidly accumulating that the dysregulation of miRNAs is fundamental to the pathogenesis of many non-neoplastic diseases as well, including infectious diseases, autoimmune conditions, cardiovascular, neuropathologies, hereditary, and inflammatory diseases, many of which are discussed in detail in subsequent chapters of this book.

However, while great effort has been put into identifying and cataloging aberrantly expressed miRNAs in disease, very little is known about the functional consequences of this dysregulation, and understanding the biological function of identified miRNAs is perhaps the biggest challenge facing the miRNA field at the moment. The primary reason for this is a paucity of knowledge about which genes are actually targeted by individual miRNAs and which of these genes are functionally important in specific cellular settings.

With very few functionally annotated exceptions, current approaches to this problem primarily rely upon the use of the many predictive computational algorithms available (Krek et al. 2005; Lewis et al. 2005; Miranda et al. 2006; Grimson et al. 2007; Kertesz et al. 2007). However, these algorithms typically predict hundreds or even thousands of target genes for each miRNA, and in reality perform very poorly. When the most widely used algorithms were tested against experimentally validated miRNA–target gene interactions, sensitivity ranged from just 1.3% to 48.8% (Sethupathy et al. 2006). Additionally, the degree of overlap between predictions (three algorithms) was found to range from 3.6% to 28.6%, and surprisingly, no commonly predicted genes were identified at all when the five most commonly used algorithms were compared. Importantly, this study showed that even when all five algorithms were used in union, only 72% of experimentally validated miRNA–target gene interactions were predicted. For example, *KRAS* and *HRAS* targeting by let-7 (Johnson et al. 2005), or *E2F2* and *MYC* targeting by miR-24 (Lal et al. 2009) are not predicted targets of these algorithms. To compound matters further, the function of a particular miRNA is dependent upon cellular context. Indeed, the same miRNA can act as both tumor suppressor and oncogene depending upon the cell type. For example, *miR-222* is overexpressed in hepatocarcinoma, where it targets tumor suppressor *PTEN* (Garofalo et al. 2009), but is down-regulated in erythroblastic leukemias, where it targets the *KIT* oncogene (Felli et al. 2005).

Consequently, much effort has been expended to resolve this issue. Particularly hopeful is the development of techniques to directly measure the miRNA:target gene interface, the so-called *targetome*, in cells under physiologically relevant conditions. A strategy that has frequently been employed to this end, is the use of gene expression arrays to elucidate which genes change in response to permutations of individual microRNAs (Lim et al. 2005; Johnson et al. 2007). A similar approach has been used to measure differences in protein levels using state-of-the-art proteomic techniques, such as stable isotope labeling by amino acids in culture (SILAC) (Yang et al. 2010; Kaller et al. 2011). A major drawback of these techniques, however, is their inability to distinguish between direct and indirect targets of miRNAs.

A more promising tactic is the use of immunoprecipitation (IP)-based techniques that allow for recovery of RNA that is directly bound to miRISC using antibodies against Ago proteins (Karginov et al. 2007). However, in the first incarnations of this approach, only the mRNA fraction could be recovered, making subsequent target gene identification difficult. Development of the HITS-CLIP technique overcame this problem by adding a UV cross-linking step to the IP procedure so that the miRNA fraction is also retained (Chi et al. 2009). Additionally, RNAse degradation of non-miRISC bound RNA, as well as the use of next generation sequencing to elucidate the recovered miRNA:target mRNA population, helped to increase the specificity of this approach. Further refinement has been obtained with the PAR-CLIP technique that utilizes incorporation of photoreactive ribonucleoside analogues, such as 4-thiouridine or 6-thioguanosine, into living cells, and allows direct identification of miRISC-bound miRNA:mRNA by mapping characteristic T-to-C mutations (Hafner et al. 2010). There are still limitations to this approach, however, as it does not address whether or not changes to protein levels occur, and because a target gene is bound to miRISC does not necessarily mean it is regulated. In addition, the choice of a particular antibody can greatly influence the results obtained. When IP experiments were compared using either Ago1 or Ago2 antibodies, there was only a partial overlap in the resultant targetome profiles (Karginov et al. 2007).

Nevertheless, techniques such as PAR-CLIP can provide a potential means to decipher the targetome of cells under physiologically relevant conditions and hence determine the true sphere of influence of dysregulated miRNAs in disease. However, these experiments are currently difficult and expensive to perform and consequently remain little used outside of specialist laboratories. In order to progress the miRNA field beyond the "stamp-collecting" phase and truly understand the function of these molecules in disease, we must apply such techniques in a much more concerted and systematic manner similar to that currently employed by the genomic, transcriptomic, and epigenomic fields.

B. Tip of the Non-Coding RNA Iceberg

One of the major shocks that arose from completion of the human genome project was that there are far fewer protein-encoding genes (~20,000) than was first envisaged (Schmutz et al. 2004). Indeed, although ~75% of the human genome is transcribed (Djebali et al. 2012), the protein-encoding portion of the genome only accounts for 1.5% (Alexander et al. 2010). Despite the level of excitement generated in the scientific/medical world, it is worth remembering that miRNAs only represent a small proportion of the so-called dark matter of the genome, the non-coding transcriptome (Yamada et al. 2003). The most recent data from the Encyclopedia of DNA Elements (ENCODE) project annotated 1756 miRNAs in the human genome, representing just 1.8% of the transcriptional output (Djebali et al. 2012). So what about the other >95% of the transcriptome? While we cannot expect all of the remaining ncRNA to be functional, particularly because transcripts are present at very low levels and permissive transcription appears to be a common feature in eukaryotes (Ebisuya et al. 2008), there is emerging evidence that ncRNA other than miRNA is essential for both physiological function and development, as well as playing a fundamental role in disease (Mercer et al. 2009; Esteller 2011). Although relative to miRNAs, the study of other ncRNA molecules is very much in its infancy, many classes of ncRNAs are now recognized, including short ncRNAs, such as miRNAs, piRNAs, and tiRNAs; mid-size ncRNAs, such as snoRNAs, PASRs, TSSa-RNAs, and PROMPTs; and long ncRNAs (lncRNAs) (Esteller 2011; Harries 2012). For a much more detailed description of ncRNAs, and in particular lncRNAs and their role in cancer, please see Chapter 15.

The current emphasis of this field is on understanding the contribution and function of ncRNA to disease. In terms of this first goal, the development of next-generation sequencing technology and large concerted efforts, such as the ENCODE project, have made great strides over the last couple of years that will surely only accelerate in the future. In parallel, many of the approaches and techniques developed by the miRNA field are applicable to other classes of ncRNA and in all probability will uncover new insights concerning disease in the next couple of years.

C. Are miRNAs Clinically Useful Molecules?

While there is now overwhelming evidence that miRNAs play a fundamental role in the pathogenesis of many, if not all, diseases, the obvious question remains: What practical use are miRNAs likely to have for future clinical practice? MiRNAs show perhaps their greatest, and certainly most immediate potential, as novel biomarkers of diagnosis and prognosis, and as predictors of treatment response. MiRNA expression profiling can distinguish cancers according to diagnosis and developmental stage of the tumor to a greater degree of accuracy than traditional gene expression analysis (Lu et al. 2005). The use of miRNAs as novel classifiers of disease is extensively discussed within the chapters of this book. A particularly attractive characteristic of miRNAs in this regard is their stability to chemical and enzymatic degradation. This means they can be purified and robustly measured from routinely prepared formalin-fixed paraffin embedded (FFPE) biopsy material (Lawrie et al. 2007). Consequently, miRNA expression studies can be carried out (albeit retrospectively) on the vast resources of the world's pathology departments in a way that is simply not feasible for traditional gene expression studies, and miRNA expression studies have now been carried out for all except the rarest of pathologies.

A further manifestation of the stability of miRNAs, as well as their potential clinical usefulness, is their presence in extracellular biological fluids including blood. Tumor-associated miRNAs have been found at higher levels in sera and plasma of cancer patients than healthy controls (Lawrie et al. 2008). Subsequently, there has been a great deal of interest in the use of miRNAs as non-invasive biomarkers of disease, and miRNAs have now been detected in many biological fluids, including plasma, serum, tears, urine, cerebral spinal fluid, breast milk, and saliva (Weber et al. 2010). This has particular clinical potential for cancer, where typically the diagnostic gold standards are invasive biopsy procedures that are expensive, uncomfortable, and sometimes risky for patients. Furthermore, a reliable blood test for cancer could pave the way for screening programs, leading to better detection rates and helping to increase cancer prevention. Further discussion on the use of miRNAs as noninvasive biomarkers can be found in Chapter 35.

Remarkably, it has even been suggested that exogenous miRNAs can be acquired in the blood as a result of the food that we eat (Zhang et al. 2012). Furthermore, these food-derived plant-specific miRNAs were demonstrated to be functional in human cells. This leads to the intriguing possibility that therapeutic miRNAs could be administered by incorporating them in food directly, or even that GM (genetically modified) crops could be engineered to express miRNAs (or antimiRs) with, for example, anti-cancer properties.

Of course, perhaps the most promising clinical aspect of miRNAs is their potential as novel therapeutic molecules, either as a tool to modulate target genes associated with disease or by correcting dysfunctional expression of the miRNAs themselves. The former approach is particularly attractive in that a single agent (i.e., a miRNA) can be used against multiple targets in a disease pathway or even against the whole pathway (Bui and Mendell 2010). There are two major strategies to therapeutically modulate dysregulated miRNAs in disease: using miRNA mimics to restore physiological levels of miRNAs that are downregulated (e.g., tumor suppressor miRNAs, such as *let-7* or *miR-34*), or the use of miRNA inhibitors targeted against overexpressed miRNAs (e.g., oncomirs, such as *miR-21* or *miR-155*). There is now a wealth of *in vivo* animal experiments that have established the proof-of-principle for the therapeutic efficacy of miRNAs in disease; however, at present, all bar a couple of these studies are still at the preclinical stage. The major hurdles to be overcome in order to translate these results into the clinic include the effective targeting of therapy (e.g., tissue-specific delivery, dosage, and pharmacodynamics) and safety concerns (e.g., off-target effects, RNA-mediated immunostimulation, and the use of viral vectors). That said, this is an area very much still in its infancy that is almost certain to flourish in the near future as the field matures, and promises to add to the current arsenal of therapies available to the clinician in the continual fight against disease. The use of miRNAs as therapeutics are covered in greater detail in Chapters 38–40 of this book.

In summary, although it is clear that the functional importance of microRNAs in medicine is gaining momentum rapidly, it is equally obvious that we still have much to learn from these tiny molecules. To answer the question, "Are miRNAs *really* clinically useful molecules?," readers are encouraged to explore further chapters of this book, which includes contributions from many of the most respected pioneers and experts in the miRNA field, and make up their own mind.

REFERENCES

Abelson, J.F., K.Y. Kwan, B.J. O'Roak, D.Y. Baek, A.A. Stillman, T.M. Morgan, C.A. Mathews, D.L. Pauls, M.R. Rasin, M. Gunel, N.R. Davis, A.G. Ercan-Sencicek, et al. 2005. Sequence variants in SLITRK1 are associated with Tourette's syndrome. *Science* **310**:317–320.

Agirre, X., A. Vilas-Zornoza, A. Jimenez-Velasco, J.I. Martin-Subero, L. Cordeu, L. Garate, E. San Jose-Eneriz, G. Abizanda, P. Rodriguez-Otero, P. Fortes, J. Rifon, E. Bandres, et al. 2009. Epigenetic silencing of the tumor suppressor microRNA Hsa-miR-124a regulates CDK6 expression and confers a poor prognosis in acute lymphoblastic leukemia. *Cancer Res* **69**:4443–4453.

Agueli, C., G. Cammarata, D. Salemi, L. Dagnino, R. Nicoletti, M. La Rosa, F. Messana, A. Marfia, M.G. Bica, M.L. Coniglio, M. Pagano, F. Fabbiano, et al. 2010. 14q32/miRNA clusters loss of heterozygosity in acute lymphoblastic leukemia is associated with up-regulation of BCL11a. *Am J Hematol* **85**:575–578.

Alexander, R.P., G. Fang, J. Rozowsky, M. Snyder, and M.B. Gerstein. 2010. Annotating non-coding regions of the genome. *Nat Rev Genet* **11**:559–571.

Altuvia, Y., P. Landgraf, G. Lithwick, N. Elefant, S. Pfeffer, A. Aravin, M.J. Brownstein, T. Tuschl, and H. Margalit. 2005. Clustering and conservation patterns of human microRNAs. *Nucleic Acids Res* **33**:2697–2706.

Ambros, V., B. Bartel, D.P. Bartel, C.B. Burge, J.C. Carrington, X. Chen, G. Dreyfuss, S.R. Eddy, S. Griffiths-Jones, M. Marshall, M. Matzke, G. Ruvkun, et al. 2003. A uniform system for microRNA annotation. *RNA* **9**:277–279.

Babiarz, J.E., R. Hsu, C. Melton, M. Thomas, E.M. Ullian, and R. Blelloch. 2011. A role for non-canonical microRNAs in the mammalian brain revealed by phenotypic differences in Dgcr8 versus Dicer1 knockouts and small RNA sequencing. *RNA* **17**:1489–1501.

Babiarz, J.E., J.G. Ruby, Y. Wang, D.P. Bartel, and R. Blelloch. 2008. Mouse ES cells express endogenous shRNAs, siRNAs, and other Microprocessor-independent, Dicer-dependent small RNAs. *Genes Dev* **22**:2773–2785.

Barski, A., R. Jothi, S. Cuddapah, K. Cui, T.Y. Roh, D.E. Schones, and K. Zhao. 2009. Chromatin poises miRNA- and protein-coding genes for expression. *Genome Res* **19**:1742–1751.

Bartel, D.P. 2009. MicroRNAs: target recognition and regulatory functions. *Cell* **136**:215–233.

Becskei, A. and L. Serrano. 2000. Engineering stability in gene networks by autoregulation. *Nature* **405**:590–593.

Ben-Ami, O., N. Pencovich, J. Lotem, D. Levanon, and Y. Groner. 2009. A regulatory interplay between miR-27a and Runx1 during megakaryopoiesis. *Proc Natl Acad Sci U S A* **106**:238–243.

Benetti, R., S. Gonzalo, I. Jaco, P. Munoz, S. Gonzalez, S. Schoeftner, E. Murchison, T. Andl, T. Chen, P. Klatt, E. Li, M. Serrano, et al. 2008. A mammalian microRNA cluster controls DNA methylation and telomere recombination via Rbl2-dependent regulation of DNA methyltransferases. *Nat Struct Mol Biol* **15**:268–279.

Berezikov, E., W.J. Chung, J. Willis, E. Cuppen, and E.C. Lai. 2007. Mammalian mirtron genes. *Mol Cell* **28**:328–336.

Bernstein, E., S.Y. Kim, M.A. Carmell, E.P. Murchison, H. Alcorn, M.Z. Li, A.A. Mills, S.J. Elledge, K.V. Anderson, and G.J. Hannon. 2003. Dicer is essential for mouse development. *Nat Genet* **35**:215–217.

Bhattacharyya, S.N., R. Habermacher, U. Martine, E.I. Closs, and W. Filipowicz. 2006. Relief of microRNA-mediated translational repression in human cells subjected to stress. *Cell* **125**:1111–1124.

Bogerd, H.P., H.W. Karnowski, X. Cai, J. Shin, M. Pohlers, and B.R. Cullen. 2010. A mammalian herpesvirus uses noncanonical expression and processing mechanisms to generate viral MicroRNAs. *Mol Cell* **37**:135–142.

Borchert, G.M., W. Lanier, and B.L. Davidson. 2006. RNA polymerase III transcribes human microRNAs. *Nat Struct Mol Biol* **13**:1097–1101.

Bousquet, M., C. Quelen, R. Rosati, V. Mansat-De Mas, R. La Starza, C. Bastard, E. Lippert, P. Talmant, M. Lafage-Pochitaloff, C. Leroux, C. Gervais, F. Viguie, et al. 2008. Myeloid cell differentiation arrest by miR-125b-1 in myelodysplastic syndrome and acute myeloid leukemia with the t(2;11)(p21;q23) translocation. *J Exp Med* **205**:2499–2506.

Bracken, C.P., P.A. Gregory, N. Kolesnikoff, A.G. Bert, J. Wang, M.F. Shannon, and G.J. Goodall. 2008. A double-negative feedback loop between ZEB1-SIP1 and the microRNA-200 family regulates epithelial-mesenchymal transition. *Cancer Res* **68**:7846–7854.

REFERENCES

Bueno, M.J., I. Perez de Castro, M. Gomez de Cedron, J. Santos, G.A. Calin, J.C. Cigudosa, C.M. Croce, J. Fernandez-Piqueras, and M. Malumbres. 2008. Genetic and epigenetic silencing of microRNA-203 enhances ABL1 and BCR-ABL1 oncogene expression. *Cancer Cell* **13**: 496–506.

Bui, T.V. and J.T. Mendell. 2010. Myc: maestro of MicroRNAs. *Genes Cancer* **1**:568–575.

Calin, G.A., C.D. Dumitru, M. Shimizu, R. Bichi, S. Zupo, E. Noch, H. Aldler, S. Rattan, M. Keating, K. Rai, L. Rassenti, T. Kipps, et al. 2002. Frequent deletions and down-regulation of micro- RNA genes miR15 and miR16 at 13q14 in chronic lymphocytic leukemia. *Proc Natl Acad Sci U S A* **99**:15524–15529.

Calin, G.A., M. Ferracin, A. Cimmino, G. Di Leva, M. Shimizu, S.E. Wojcik, M.V. Iorio, R. Visone, N.I. Sever, M. Fabbri, R. Iuliano, T. Palumbo, et al. 2005. A MicroRNA signature associated with prognosis and progression in chronic lymphocytic leukemia. *N Engl J Med* **353**:1793–1801.

Calin, G.A., C. Sevignani, C.D. Dumitru, T. Hyslop, E. Noch, S. Yendamuri, M. Shimizu, S. Rattan, F. Bullrich, M. Negrini, and C.M. Croce. 2004. Human microRNA genes are frequently located at fragile sites and genomic regions involved in cancers. *Proc Natl Acad Sci U S A* **101**: 2999–3004.

Chan, S.P. and F.J. Slack. 2007. And now introducing mammalian mirtrons. *Dev Cell* **13**:605–607.

Chang, T.C., D. Yu, Y.S. Lee, E.A. Wentzel, D.E. Arking, K.M. West, C.V. Dang, A. Thomas-Tikhonenko, and J.T. Mendell. 2008. Widespread microRNA repression by Myc contributes to tumorigenesis. *Nat Genet* **40**:43–50.

Chapnik, E., V. Sasson, R. Blelloch, and E. Hornstein. 2012. Dgcr8 controls neural crest cells survival in cardiovascular development. *Dev Biol* **362**:50–56.

Chatterjee, S. and H. Grosshans. 2009. Active turnover modulates mature microRNA activity in *Caenorhabditis elegans*. *Nature* **461**:546–549.

Cheloufi, S., C.O. Dos Santos, M.M. Chong, and G.J. Hannon. 2010. A dicer-independent miRNA biogenesis pathway that requires Ago catalysis. *Nature* **465**:584–589.

Chen, Z., J. Wu, C. Yang, P. Fan, L. Balazs, Y. Jiao, M. Lu, W. Gu, C. Li, L.M. Pfeffer, G. Tigyi, and J. Yue. 2012. DiGeorge syndrome critical region 8 (DGCR8) protein-mediated microRNA biogenesis is essential for vascular smooth muscle cell development in mice. *J Biol Chem* **287**:19018–19028.

Chendrimada, T.P., R.I. Gregory, E. Kumaraswamy, J. Norman, N. Cooch, K. Nishikura, and R. Shiekhattar. 2005. TRBP recruits the Dicer complex to Ago2 for microRNA processing and gene silencing. *Nature* **436**:740–744.

Chi, S.W., J.B. Zang, A. Mele, and R.B. Darnell. 2009. Argonaute HITS-CLIP decodes microRNA-mRNA interaction maps. *Nature* **460**:479–486.

Chiang, H.R., L.W. Schoenfeld, J.G. Ruby, V.C. Auyeung, N. Spies, D. Baek, W.K. Johnston, C. Russ, S. Luo, J.E. Babiarz, R. Blelloch, G.P. Schroth, et al. 2010. Mammalian microRNAs: experimental evaluation of novel and previously annotated genes. *Genes Dev* **24**:992–1009.

Chiosea, S., E. Jelezcova, U. Chandran, M. Acquafondata, T. McHale, R.W. Sobol, and R. Dhir. 2006. Up-regulation of dicer, a component of the MicroRNA machinery, in prostate adenocarcinoma. *Am J Pathol* **169**:1812–1820.

Christoffersen, N.R., R. Shalgi, L.B. Frankel, E. Leucci, M. Lees, M. Klausen, Y. Pilpel, F.C. Nielsen, M. Oren, and A.H. Lund. 2010. p53-independent upregulation of miR-34a during oncogene-induced senescence represses MYC. *Cell Death Differ* **17**:236–245.

Chung, W.J., P. Agius, J.O. Westholm, M. Chen, K. Okamura, N. Robine, C.S. Leslie, and E.C. Lai. 2011. Computational and experimental identification of mirtrons in Drosophila melanogaster and *Caenorhabditis elegans*. *Genome Res* **21**:286–300.

Cifuentes, D., H. Xue, D.W. Taylor, H. Patnode, Y. Mishima, S. Cheloufi, E. Ma, S. Mane, G.J. Hannon, N.D. Lawson, S.A. Wolfe, and A.J. Giraldez. 2010. A novel miRNA processing pathway independent of Dicer requires Argonaute2 catalytic activity. *Science* **328**:1694–1698.

Corcoran, D.L., K.V. Pandit, B. Gordon, A. Bhattacharjee, N. Kaminski, and P.V. Benos. 2009. Features of mammalian microRNA promoters emerge from polymerase II chromatin immunoprecipitation data. *PLoS ONE* **4**:e5279.

Crick, F. 1970. Central dogma of molecular biology. *Nature* **227**:561–563.

Croce, C.M. 2009. Causes and consequences of microRNA dysregulation in cancer. *Nat Rev Genet* **10**:704–714.

Czech, B. and G.J. Hannon. 2011. Small RNA sorting: matchmaking for Argonautes. *Nat Rev Genet* **12**:19–31.

da Costa Martins, P.A., M. Bourajjaj, M. Gladka, M. Kortland, R.J. van Oort, Y.M. Pinto, J.D. Molkentin, and L.J. De Windt. 2008. Conditional dicer gene deletion in the postnatal myocardium provokes spontaneous cardiac remodeling. *Circulation* **118**:1567–1576.

Di Leva, G., P. Gasparini, C. Piovan, A. Ngankeu, M. Garofalo, C. Taccioli, M.V. Iorio, M. Li, S. Volinia, H. Alder, T. Nakamura, G. Nuovo, et al. 2010. MicroRNA cluster 221–222 and estrogen receptor alpha interactions in breast cancer. *J Natl Cancer Inst* **102**:706–721.

Djebali, S., C.A. Davis, A. Merkel, A. Dobin, T. Lassmann, A. Mortazavi, A. Tanzer, J. Lagarde, W. Lin, F. Schlesinger, C. Xue, G.K. Marinov, et al. 2012. Landscape of transcription in human cells. *Nature* **489**:101–108.

Ebisuya, M., T. Yamamoto, M. Nakajima, and E. Nishida. 2008. Ripples from neighbouring transcription. *Nat Cell Biol* **10**:1106–1113.

Eiring, A.M., J.G. Harb, P. Neviani, C. Garton, J.J. Oaks, R. Spizzo, S. Liu, S. Schwind, R. Santhanam, C.J. Hickey, H. Becker, J.C. Chandler, et al. 2010. miR-328 functions as an RNA decoy to modulate hnRNP E2 regulation of mRNA translation in leukemic blasts. *Cell* **140**:652–665.

Esteller, M. 2008. Epigenetics in cancer. *N Engl J Med* **358**:1148–1159.

Esteller, M. 2011. Non-coding RNAs in human disease. *Nat Rev Genet* **12**:861–874.

Faber, C., D. Horst, F. Hlubek, and T. Kirchner. 2011. Overexpression of Dicer predicts poor survival in colorectal cancer. *Eur J Cancer* **47**:1414–1419.

Felli, N., L. Fontana, E. Pelosi, R. Botta, D. Bonci, F. Facchiano, F. Liuzzi, V. Lulli, O. Morsilli, S. Santoro, M. Valtieri, G.A. Calin, et al. 2005. MicroRNAs 221 and 222 inhibit normal erythropoiesis and erythroleukemic cell growth via kit receptor down-modulation. *Proc Natl Acad Sci U S A* **102**:18081–18086.

Fire, A., S. Xu, M.K. Montgomery, S.A. Kostas, S.E. Driver, and C.C. Mello. 1998. Potent and specific genetic interference by double-stranded RNA in *Caenorhabditis elegans*. *Nature* **391**:806–811.

Friedman, R.C., K.K. Farh, C.B. Burge, and D.P. Bartel. 2009. Most mammalian mRNAs are conserved targets of microRNAs. *Genome Res* **19**:92–105.

Fukuda, T., K. Yamagata, S. Fujiyama, T. Matsumoto, I. Koshida, K. Yoshimura, M. Mihara, M. Naitou, H. Endoh, T. Nakamura, C. Akimoto, Y. Yamamoto, et al. 2007. DEAD-box RNA helicase subunits of the Drosha complex are required for processing of rRNA and a subset of microRNAs. *Nat Cell Biol* **9**:604–611.

Garofalo, M., G. Di Leva, G. Romano, G. Nuovo, S.S. Suh, A. Ngankeu, C. Taccioli, F. Pichiorri, H. Alder, P. Secchiero, P. Gasparini, A. Gonelli, et al. 2009. miR-221&222 regulate TRAIL resistance and enhance tumorigenicity through PTEN and TIMP3 downregulation. *Cancer Cell* **16**:498–509.

Ghildiyal, M., J. Xu, H. Seitz, Z. Weng, and P.D. Zamore. 2010. Sorting of Drosophila small silencing RNAs partitions microRNA* strands into the RNA interference pathway. *RNA* **16**:43–56.

Ghildiyal, M. and P.D. Zamore. 2009. Small silencing RNAs: an expanding universe. *Nat Rev Genet* **10**:94–108.

REFERENCES

Giraldez, A.J., Y. Mishima, J. Rihel, R.J. Grocock, S. Van Dongen, K. Inoue, A.J. Enright, and A.F. Schier. 2006. Zebrafish miR-430 promotes deadenylation and clearance of maternal mRNAs. *Science* **312**:75–79.

Gonzalez, G. and R.R. Behringer. 2009. Dicer is required for female reproductive tract development and fertility in the mouse. *Mol Reprod Dev* **76**:678–688.

Griffiths-Jones, S., R.J. Grocock, S. van Dongen, A. Bateman, and A.J. Enright. 2006. miRBase: microRNA sequences, targets and gene nomenclature. *Nucleic Acids Res* **34**:D140–D144.

Griffiths-Jones, S., J.H. Hui, A. Marco, and M. Ronshaugen. 2011. MicroRNA evolution by arm switching. *EMBO Rep* **12**:172–177.

Grimson, A., K.K. Farh, W.K. Johnston, P. Garrett-Engele, L.P. Lim, and D.P. Bartel. 2007. MicroRNA targeting specificity in mammals: determinants beyond seed pairing. *Mol Cell* **27**:91–105.

Guo, H., N.T. Ingolia, J.S. Weissman, and D.P. Bartel. 2010. Mammalian microRNAs predominantly act to decrease target mRNA levels. *Nature* **466**:835–840.

Guo, X., Q. Liao, P. Chen, X. Li, W. Xiong, J. Ma, Z. Luo, H. Tang, M. Deng, Y. Zheng, R. Wang, W. Zhang, et al. 2012. The microRNA-processing enzymes: Drosha and Dicer can predict prognosis of nasopharyngeal carcinoma. *J Cancer Res Clin Oncol* **138**:49–56.

Hafner, M., M. Landthaler, L. Burger, M. Khorshid, J. Hausser, P. Berninger, A. Rothballer, M. Ascano, Jr., A.C. Jungkamp, M. Munschauer, A. Ulrich, G.S. Wardle, S. Dewell, M. Zavolan, T. Tuschl. 2010. Transcriptome-wide identification of RNA-binding protein and microRNA target sites by PAR-CLIP. *Cell* **141**(1):129–141.

Hamilton, A.J. and D.C. Baulcombe. 1999. A species of small antisense RNA in posttranscriptional gene silencing in plants. *Science* **286**:950–952.

Hammell, C.M., I. Lubin, P.R. Boag, T.K. Blackwell, and V. Ambros. 2009. nhl-2 Modulates microRNA activity in *Caenorhabditis elegans*. *Cell* **136**:926–938.

Han, J., J.S. Pedersen, S.C. Kwon, C.D. Belair, Y.K. Kim, K.H. Yeom, W.Y. Yang, D. Haussler, R. Blelloch, and V.N. Kim. 2009. Posttranscriptional crossregulation between Drosha and DGCR8. *Cell* **136**:75–84.

Harries, L.W. 2012. Long non-coding RNAs and human disease. *Biochem Soc Trans* **40**:902–906.

Harris, K.S., Z. Zhang, M.T. McManus, B.D. Harfe, and X. Sun. 2006. Dicer function is essential for lung epithelium morphogenesis. *Proc Natl Acad Sci U S A* **103**:2208–2213.

Hertel, J., M. Lindemeyer, K. Missal, C. Fried, A. Tanzer, C. Flamm, I.L. Hofacker, and P.F. Stadler. 2006. The expansion of the metazoan microRNA repertoire. *BMC Genomics* **7**:25.

Humphreys, D.T., B.J. Westman, D.I. Martin, and T. Preiss. 2005. MicroRNAs control translation initiation by inhibiting eukaryotic initiation factor 4E/cap and poly(A) tail function. *Proc Natl Acad Sci U S A* **102**:16961–16966.

Huntzinger, E. and E. Izaurralde. 2011. Gene silencing by microRNAs: contributions of translational repression and mRNA decay. *Nat Rev Genet* **12**:99–110.

Huse, J.T., C. Brennan, D. Hambardzumyan, B. Wee, J. Pena, S.H. Rouhanifard, C. Sohn-Lee, C. le Sage, R. Agami, T. Tuschl, and E.C. Holland. 2009. The PTEN-regulating microRNA miR-26a is amplified in high-grade glioma and facilitates gliomagenesis in vivo. *Genes Dev* **23**:1327–1337.

Hwang, H.W., E.A. Wentzel, and J.T. Mendell. 2007. A hexanucleotide element directs microRNA nuclear import. *Science* **315**:97–100.

Iorio, M.V. and C.M. Croce. 2012. Causes and consequences of microRNA dysregulation. *Cancer J* **18**:215–222.

Johnson, C.D., A. Esquela-Kerscher, G. Stefani, M. Byrom, K. Kelnar, D. Ovcharenko, M. Wilson, X. Wang, J. Shelton, J. Shingara, L. Chin, D. Brown, et al. 2007. The let-7 microRNA represses cell proliferation pathways in human cells. *Cancer Res* **67**:7713–7722.

Johnson, S.M., H. Grosshans, J. Shingara, M. Byrom, R. Jarvis, A. Cheng, E. Labourier, K.L. Reinert, D. Brown, and F.J. Slack. 2005. RAS is regulated by the let-7 microRNA family. *Cell* **120**:635–647.

Johnston, R.J., Jr., S. Chang, J.F. Etchberger, C.O. Ortiz, and O. Hobert. 2005. MicroRNAs acting in a double-negative feedback loop to control a neuronal cell fate decision. *Proc Natl Acad Sci U S A* **102**:12449-12454.

Kaller, M., S.T. Liffers, S. Oeljeklaus, K. Kuhlmann, S. Roh, R. Hoffmann, B. Warscheid, and H. Hermeking. 2011. Genome-wide characterization of miR-34a induced changes in protein and mRNA expression by a combined pulsed SILAC and microarray analysis. *Mol Cell Proteomics* **10**:M111 010462.

Karginov, F.V., C. Conaco, Z. Xuan, B.H. Schmidt, J.S. Parker, G. Mandel, and G.J. Hannon. 2007. A biochemical approach to identifying microRNA targets. *Proc Natl Acad Sci U S A* **104**: 19291-19296.

Karube, Y., H. Tanaka, H. Osada, S. Tomida, Y. Tatematsu, K. Yanagisawa, Y. Yatabe, J. Takamizawa, S. Miyoshi, T. Mitsudomi, and T. Takahashi. 2005. Reduced expression of Dicer associated with poor prognosis in lung cancer patients. *Cancer Sci* **96**:111-115.

Kawahara, Y., B. Zinshteyn, T.P. Chendrimada, R. Shiekhattar, and K. Nishikura. 2007a. RNA editing of the microRNA-151 precursor blocks cleavage by the Dicer-TRBP complex. *EMBO Rep* **8**:763-769.

Kawahara, Y., B. Zinshteyn, P. Sethupathy, H. Iizasa, A.G. Hatzigeorgiou, and K. Nishikura. 2007b. Redirection of silencing targets by adenosine-to-inosine editing of miRNAs. *Science* **315**: 1137-1140.

Kedde, M., M.J. Strasser, B. Boldajipour, J.A. Oude Vrielink, K. Slanchev, C. le Sage, R. Nagel, P.M. Voorhoeve, J. van Duijse, U.A. Orom, A.H. Lund, A. Perrakis, et al. 2007. RNA-binding protein Dnd1 inhibits microRNA access to target mRNA. *Cell* **131**:1273-1286.

Kertesz, M., N. Iovino, U. Unnerstall, U. Gaul, and E. Segal. 2007. The role of site accessibility in microRNA target recognition. *Nat Genet* **39**:1278-1284.

Khoshnaw, S.M., E.A. Rakha, T.M. Abdel-Fatah, C.C. Nolan, Z. Hodi, D.R. Macmillan, I.O. Ellis, and A.R. Green. 2012. Loss of Dicer expression is associated with breast cancer progression and recurrence. *Breast Cancer Res Treat* **135**:403-413.

Khvorova, A., A. Reynolds, and S.D. Jayasena. 2003. Functional siRNAs and miRNAs exhibit strand bias. *Cell* **115**:209-216.

Kim, H.H., Y. Kuwano, S. Srikantan, E.K. Lee, J.L. Martindale, and M. Gorospe. 2009. HuR recruits let-7/RISC to repress c-Myc expression. *Genes Dev* **23**:1743-1748.

Krek, A., D. Grun, M.N. Poy, R. Wolf, L. Rosenberg, E.J. Epstein, P. MacMenamin, I. da Piedade, K.C. Gunsalus, M. Stoffel, and N. Rajewsky. 2005. Combinatorial microRNA target predictions. *Nat Genet* **37**:495-500.

Kumar, M.S., J. Lu, K.L. Mercer, T.R. Golub, and T. Jacks. 2007. Impaired microRNA processing enhances cellular transformation and tumorigenesis. *Nat Genet* **39**:673-677.

Ladewig, E., K. Okamura, A.S. Flynt, J.O. Westholm, and E.C. Lai. 2012. Discovery of hundreds of mirtrons in mouse and human small RNA data. *Genome Res* **22**:1634-1645.

Lagos-Quintana, M., R. Rauhut, W. Lendeckel, and T. Tuschl. 2001. Identification of novel genes coding for small expressed RNAs. *Science* **294**:853-858.

Lal, A., F. Navarro, C.A. Maher, L.E. Maliszewski, N. Yan, E. O'Day, D. Chowdhury, D.M. Dykxhoorn, P. Tsai, O. Hofmann, K.G. Becker, M. Gorospe, et al. 2009. miR-24 inhibits cell proliferation by targeting E2F2, MYC, and other cell-cycle genes via binding to "seedless" 3'UTR microRNA recognition elements. *Mol Cell* **35**:610-625.

Lawrie, C.H., S. Gal, H.M. Dunlop, B. Pushkaran, A.P. Liggins, K. Pulford, A.H. Banham, F. Pezzella, J. Boultwood, J.S. Wainscoat, C.S. Hatton, and A.L. Harris. 2008. Detection of elevated levels of tumour-associated microRNAs in serum of patients with diffuse large B-cell lymphoma. *Br J Haematol* **141**:672-675.

Lawrie, C.H., S. Soneji, T. Marafioti, C.D. Cooper, S. Palazzo, J.C. Paterson, H. Cattan, T. Enver, R. Mager, J. Boultwood, J.S. Wainscoat, and C.S. Hatton. 2007. MicroRNA expression distin-

REFERENCES

guishes between germinal center B cell-like and activated B cell-like subtypes of diffuse large B cell lymphoma. *Int J Cancer* **121**:1156–1161.

Lee, R.C. and V. Ambros. 2001. An extensive class of small RNAs in *Caenorhabditis elegans*. *Science* **294**:862–864.

Lee, R.C., R.L. Feinbaum, and V. Ambros. 1993. The C. elegans heterochronic gene lin-4 encodes small RNAs with antisense complementarity to lin-14. *Cell* **75**:843–854.

Lee, Y., I. Hur, S.Y. Park, Y.K. Kim, M.R. Suh, and V.N. Kim. 2006. The role of PACT in the RNA silencing pathway. *Embo J* **25**:522–532.

Lewis, B.P., C.B. Burge, and D.P. Bartel. 2005. Conserved seed pairing, often flanked by adenosines, indicates that thousands of human genes are microRNA targets. *Cell* **120**:15–20.

Lewis, B.P., I.H. Shih, M.W. Jones-Rhoades, D.P. Bartel, and C.B. Burge. 2003. Prediction of mammalian microRNA targets. *Cell* **115**:787–798.

Li, W., R. Duan, F. Kooy, S.L. Sherman, W. Zhou, and P. Jin. 2009. Germline mutation of microRNA-125a is associated with breast cancer. *J Med Genet* **46**:358–360.

Lim, L.P., N.C. Lau, P. Garrett-Engele, A. Grimson, J.M. Schelter, J. Castle, D.P. Bartel, P.S. Linsley, and J.M. Johnson. 2005. Microarray analysis shows that some microRNAs downregulate large numbers of target mRNAs. *Nature* **433**:769–773.

Lin, M., J. Gu, C. Eng, L.M. Ellis, M.A. Hildebrandt, J. Lin, M. Huang, G.A. Calin, D. Wang, R.N. Dubois, E.T. Hawk, and X. Wu. 2012. Genetic polymorphisms in MicroRNA-related genes as predictors of clinical outcomes in colorectal adenocarcinoma patients. *Clin Cancer Res* **18**:3982–3991.

Lin, R.J., Y.C. Lin, J. Chen, H.H. Kuo, Y.Y. Chen, M.B. Diccianni, W.B. London, C.H. Chang, and A.L. Yu. 2010. microRNA signature and expression of Dicer and Drosha can predict prognosis and delineate risk groups in neuroblastoma. *Cancer Res* **70**:7841–7850.

Liu, J., M.A. Valencia-Sanchez, G.J. Hannon, and R. Parker. 2005. MicroRNA-dependent localization of targeted mRNAs to mammalian P-bodies. *Nat Cell Biol* **7**:719–723.

Liu, X., A. Wang, C.E. Heidbreder, L. Jiang, J. Yu, A. Kolokythas, L. Huang, Y. Dai, and X. Zhou. 2010. MicroRNA-24 targeting RNA-binding protein DND1 in tongue squamous cell carcinoma. *FEBS Lett* **584**:4115–4120.

Llave, C., Z. Xie, K.D. Kasschau, and J.C. Carrington. 2002. Cleavage of Scarecrow-like mRNA targets directed by a class of Arabidopsis miRNA. *Science* **297**:2053–2056.

Lu, J., G. Getz, E.A. Miska, E. Alvarez-Saavedra, J. Lamb, D. Peck, A. Sweet-Cordero, B.L. Ebert, R.H. Mak, A.A. Ferrando, J.R. Downing, T. Jacks, et al. 2005. MicroRNA expression profiles classify human cancers. *Nature* **435**:834–838.

Lytle, J.R., T.A. Yario, and J.A. Steitz. 2007. Target mRNAs are repressed as efficiently by microRNA-binding sites in the 5′ UTR as in the 3′ UTR. *Proc Natl Acad Sci U S A* **104**:9667–9672.

Mattick, J.S. 2001. Non-coding RNAs: the architects of eukaryotic complexity. *EMBO Rep* **2**:986–991.

Mercer, T.R., M.E. Dinger, and J.S. Mattick. 2009. Long non-coding RNAs: insights into functions. *Nat Rev Genet* **10**:155–159.

Merritt, W.M., Y.G. Lin, L.Y. Han, A.A. Kamat, W.A. Spannuth, R. Schmandt, D. Urbauer, L.A. Pennacchio, J.F. Cheng, A.M. Nick, M.T. Deavers, A. Mourad-Zeidan, et al. 2008. Dicer, Drosha, and outcomes in patients with ovarian cancer. *N Engl J Med* **359**:2641–2650.

Miranda, K.C., T. Huynh, Y. Tay, Y.S. Ang, W.L. Tam, A.M. Thomson, B. Lim, and I. Rigoutsos. 2006. A pattern-based method for the identification of MicroRNA binding sites and their corresponding heteroduplexes. *Cell* **126**:1203–1217.

Morita, S., T. Horii, M. Kimura, Y. Goto, T. Ochiya, and I. Hatada. 2007. One Argonaute family member, Eif2c2 (Ago2), is essential for development and appears not to be involved in DNA methylation. *Genomics* **89**:687–696.

Mukherjee, N., D.L. Corcoran, J.D. Nusbaum, D.W. Reid, S. Georgiev, M. Hafner, M. Ascano, Jr., T. Tuschl, U. Ohler, and J.D. Keene. 2011. Integrative regulatory mapping indicates that the RNA-binding protein HuR couples pre-mRNA processing and mRNA stability. *Mol Cell* **43**:327–339.

Napoli, C., C. Lemieux, and R. Jorgensen. 1990. Introduction of a Chimeric Chalcone Synthase Gene into Petunia Results in Reversible Co-Suppression of Homologous Genes in trans. *Plant Cell* **2**:279–289.

Neumuller, R.A., J. Betschinger, A. Fischer, N. Bushati, I. Poernbacher, K. Mechtler, S.M. Cohen, and J.A. Knoblich. 2008. Mei-P26 regulates microRNAs and cell growth in the Drosophila ovarian stem cell lineage. *Nature* **454**:241–245.

O'Donnell, K.A., E.A. Wentzel, K.I. Zeller, C.V. Dang, and J.T. Mendell. 2005. c-Myc-regulated microRNAs modulate E2F1 expression. *Nature* **435**:839–843.

Okamura, K., J.W. Hagen, H. Duan, D.M. Tyler, and E.C. Lai. 2007. The mirtron pathway generates microRNA-class regulatory RNAs in Drosophila. *Cell* **130**:89–100.

Okamura, K., M.D. Phillips, D.M. Tyler, H. Duan, Y.T. Chou, and E.C. Lai. 2008. The regulatory activity of microRNA* species has substantial influence on microRNA and 3′ UTR evolution. *Nat Struct Mol Biol* **15**:354–363.

Olsen, P.H. and V. Ambros. 1999. The lin-4 regulatory RNA controls developmental timing in *Caenorhabditis elegans* by blocking LIN-14 protein synthesis after the initiation of translation. *Dev Biol* **216**:671–680.

O'Rourke, J.R., S.A. Georges, H.R. Seay, S.J. Tapscott, M.T. McManus, D.J. Goldhamer, M.S. Swanson, and B.D. Harfe. 2007. Essential role for Dicer during skeletal muscle development. *Dev Biol* **311**:359–368.

Pasquinelli, A.E., B.J. Reinhart, F. Slack, M.Q. Martindale, M.I. Kuroda, B. Maller, D.C. Hayward, E.E. Ball, B. Degnan, P. Muller, J. Spring, A. Srinivasan, et al. 2000. Conservation of the sequence and temporal expression of let-7 heterochronic regulatory RNA. *Nature* **408**:86–89.

Pillai, R.S., S.N. Bhattacharyya, C.G. Artus, T. Zoller, N. Cougot, E. Basyuk, E. Bertrand, and W. Filipowicz. 2005. Inhibition of translational initiation by Let-7 MicroRNA in human cells. *Science* **309**:1573–1576.

Reinhart, B.J., F.J. Slack, M. Basson, A.E. Pasquinelli, J.C. Bettinger, A.E. Rougvie, H.R. Horvitz, and G. Ruvkun. 2000. The 21-nucleotide let-7 RNA regulates developmental timing in *Caenorhabditis elegans*. *Nature* **403**:901–906.

Roccaro, A.M., A. Sacco, X. Jia, A.K. Azab, P. Maiso, H.T. Ngo, F. Azab, J. Runnels, P. Quang, and I.M. Ghobrial. 2010. microRNA-dependent modulation of histone acetylation in Waldenstrom macroglobulinemia. *Blood* **116**:1506–1514.

Rodriguez, A., S. Griffiths-Jones, J.L. Ashurst, and A. Bradley. 2004. Identification of mammalian microRNA host genes and transcription units. *Genome Res* **14**:1902–1910.

Rosenfeld, N., M.B. Elowitz, and U. Alon. 2002. Negative autoregulation speeds the response times of transcription networks. *J Mol Biol* **323**:785–793.

Roush, S.F. and F.J. Slack. 2009. Transcription of the C. elegans let-7 microRNA is temporally regulated by one of its targets, hbl-1. *Dev Biol* **334**:523–534.

Rybak, A., H. Fuchs, K. Hadian, L. Smirnova, E.A. Wulczyn, G. Michel, R. Nitsch, D. Krappmann, and F.G. Wulczyn. 2009. The let-7 target gene mouse lin-41 is a stem cell specific E3 ubiquitin ligase for the miRNA pathway protein Ago2. *Nat Cell Biol* **11**:1411–1420.

Schmutz, J., J. Wheeler, J. Grimwood, M. Dickson, J. Yang, C. Caoile, E. Bajorek, S. Black, Y.M. Chan, M. Denys, J. Escobar, D. Flowers, et al. 2004. Quality assessment of the human genome sequence. *Nature* **429**:365–368.

Schwamborn, J.C., E. Berezikov, and J.A. Knoblich. 2009. The TRIM-NHL protein TRIM32 activates microRNAs and prevents self-renewal in mouse neural progenitors. *Cell* **136**:913–925.

Schwarz, D.S., G. Hutvagner, T. Du, Z. Xu, N. Aronin, and P.D. Zamore. 2003. Asymmetry in the assembly of the RNAi enzyme complex. *Cell* **115**:199–208.

Scott, G.K., M.D. Mattie, C.E. Berger, S.C. Benz, and C.C. Benz. 2006. Rapid alteration of microRNA levels by histone deacetylase inhibition. *Cancer Res* **66**:1277–1281.

Sen, G.L. and H.M. Blau. 2005. Argonaute 2/RISC resides in sites of mammalian mRNA decay known as cytoplasmic bodies. *Nat Cell Biol* **7**:633–636.

Sethupathy, P., M. Megraw, and A.G. Hatzigeorgiou. 2006. A guide through present computational approaches for the identification of mammalian microRNA targets. *Nat Methods* **3**:881–886.

Shalgi, R., D. Lieber, M. Oren, and Y. Pilpel. 2007. Global and local architecture of the mammalian microRNA-transcription factor regulatory network. *PLoS Comput Biol* **3**:e131.

Shin, C., J.W. Nam, K.K. Farh, H.R. Chiang, A. Shkumatava, and D.P. Bartel. 2010. Expanding the microRNA targeting code: functional sites with centered pairing. *Mol Cell* **38**:789–802.

Shin, D., J.Y. Shin, M.T. McManus, L.J. Ptacek, and Y.H. Fu. 2009. Dicer ablation in oligodendrocytes provokes neuronal impairment in mice. *Ann Neurol* **66**:843–857.

Sugito, N., H. Ishiguro, Y. Kuwabara, M. Kimura, A. Mitsui, H. Kurehara, T. Ando, R. Mori, N. Takashima, R. Ogawa, and Y. Fujii. 2006. RNASEN regulates cell proliferation and affects survival in esophageal cancer patients. *Clin Cancer Res* **12**:7322–7328.

Tao, J., H. Wu, Q. Lin, W. Wei, X.H. Lu, J.P. Cantle, Y. Ao, R.W. Olsen, X.W. Yang, I. Mody, M.V. Sofroniew, and Y.E. Sun. 2011. Deletion of astroglial Dicer causes non-cell-autonomous neuronal dysfunction and degeneration. *J Neurosci* **31**:8306–8319.

Tay, Y., J. Zhang, A.M. Thomson, B. Lim, and I. Rigoutsos. 2008. MicroRNAs to Nanog, Oct4 and Sox2 coding regions modulate embryonic stem cell differentiation. *Nature* **455**:1124–1128.

Tsang, J., J. Zhu, and A. van Oudenaarden. 2007. MicroRNA-mediated feedback and feedforward loops are recurrent network motifs in mammals. *Mol Cell* **26**:753–767.

Vasudevan, S., Y. Tong, and J.A. Steitz. 2007. Switching from repression to activation: microRNAs can up-regulate translation. *Science* **318**:1931–1934.

Viswanathan, S.R. and G.Q. Daley. 2010. Lin28: a microRNA regulator with a macro role. *Cell* **140**:445–449.

Wang, Y., R. Medvid, C. Melton, R. Jaenisch, and R. Blelloch. 2007. DGCR8 is essential for microRNA biogenesis and silencing of embryonic stem cell self-renewal. *Nat Genet* **39**:380–385.

Weber, B., C. Stresemann, B. Brueckner, and F. Lyko. 2007. Methylation of human microRNA genes in normal and neoplastic cells. *Cell Cycle* **6**:1001–1005.

Weber, J.A., D.H. Baxter, S. Zhang, D.Y. Huang, K.H. Huang, M.J. Lee, D.J. Galas, and K. Wang. 2010. The microRNA spectrum in 12 body fluids. *Clin Chem* **56**:1733–1741.

Wightman, B., I. Ha, and G. Ruvkun. 1993. Posttranscriptional regulation of the heterochronic gene lin-14 by lin-4 mediates temporal pattern formation in C. elegans. *Cell* **75**:855–862.

Winter, J., S. Jung, S. Keller, R.I. Gregory, and S. Diederichs. 2009. Many roads to maturity: microRNA biogenesis pathways and their regulation. *Nat Cell Biol* **11**:228–234.

Wu, L., J. Fan, and J.G. Belasco. 2006. MicroRNAs direct rapid deadenylation of mRNA. *Proc Natl Acad Sci U S A* **103**:4034–4039.

Yamada, K., J. Lim, J.M. Dale, H. Chen, P. Shinn, C.J. Palm, A.M. Southwick, H.C. Wu, C. Kim, M. Nguyen, P. Pham, R. Cheuk, et al. 2003. Empirical analysis of transcriptional activity in the Arabidopsis genome. *Science* **302**:842–846.

Yang, J.S. and E.C. Lai. 2011. Alternative miRNA biogenesis pathways and the interpretation of core miRNA pathway mutants. *Mol Cell* **43**:892–903.

Yang, Y., R. Chaerkady, K. Kandasamy, T.C. Huang, L.D. Selvan, S.B. Dwivedi, O.A. Kent, J.T. Mendell, and A. Pandey. 2010. Identifying targets of miR-143 using a SILAC-based proteomic approach. *Mol Biosyst* **6**:1873–1882.

Yekta, S., I.H. Shih, and D.P. Bartel. 2004. MicroRNA-directed cleavage of HOXB8 mRNA. *Science* **304**:594–596.

Yeom, K.H., Y. Lee, J. Han, M.R. Suh, and V.N. Kim. 2006. Characterization of DGCR8/Pasha, the essential cofactor for Drosha in primary miRNA processing. *Nucleic Acids Res* **34**:4622–4629.

Zamore, P.D., T. Tuschl, P.A. Sharp, and D.P. Bartel. 2000. RNAi: double-stranded RNA directs the ATP-dependent cleavage of mRNA at 21 to 23 nucleotide intervals. *Cell* **101**:25–33.

Zeng, Y. 2006. Principles of micro-RNA production and maturation. *Oncogene* **25**:6156–6162.

Zhang, L., D. Hou, X. Chen, D. Li, L. Zhu, Y. Zhang, J. Li, Z. Bian, X. Liang, X. Cai, Y. Yin, C. Wang, et al. 2012. Exogenous plant MIR168a specifically targets mammalian LDLRAP1: evidence of cross-kingdom regulation by microRNA. *Cell Res* **22**:107–126.

Zhao, H., A. Kalota, S. Jin, and A.M. Gewirtz. 2009. The c-myb proto-oncogene and microRNA-15a comprise an active autoregulatory feedback loop in human hematopoietic cells. *Blood* **113**: 505–516.

Zisoulis, D.G., Z.S. Kai, R.K. Chang, and A.E. Pasquinelli. 2012. Autoregulation of microRNA biogenesis by let-7 and Argonaute. *Nature* **486**:541–544.

PART I

MicroRNAs as Physiological Regulators

MicroRNA REGULATION OF STEM CELL FATE AND REPROGRAMMING

Erika Lorenzo Vivas,[1] Gustavo Tiscornia,[2] and Juan Carlos Izpisua Belmonte[1,3]

[1]*Centre for Regenerative Medicine in Barcelona, Barcelona, Spain*
[2]*Department of Biomedical Sciences and Medicine, University of Algarve, Faro, Portugal*
[3]*Salk Institute for Biological Studies, La Jolla, CA, USA*

I.	Introduction	28
	A. About Stem Cells	28
II.	MicroRNAs in Pluripotency and Establishment of Different Lineages	30
	A. MicroRNAs in Embryonic Stem Cells (ESc)	30
	B. Maintenance of the Pluripotent State through Controlling the Cell Cycle	30
	C. MicroRNAs in Pluripotency versus Differentiation Fate Choice	31
III.	MicroRNAs, Induced Pluripotency, and Transdifferentiation	33
	A. Induced Pluripotency and MicroRNAs	33
	B. Underlying Mechanisms in microRNA iPSc Generation	34
	C. Transdifferentiation and miRNAs	35
IV.	Conclusion	37
	Acknowledgments	37
	References	37

ABBREVATIONS

DNA	deoxyribonucleic acid
DNMT	DNA methyltransferase
EMT	epithelial to mesenchymal transition
ESc	embryonic stem cell

MicroRNAs in Medicine, First Edition. Edited by Charles H. Lawrie.
© 2014 John Wiley & Sons, Inc. Published 2014 by John Wiley & Sons, Inc.

ESCC	embryonic stem cell-specific cell cycle regulating microRNAs
HSC	hematopoietic stem cell
iPSc	induced pluripotent stem cells
KO	knockout
MEFs	mouse embryonic fibroblasts
mESc	mouse embryonic stem cell
MET	mesenchymal to epithelial transition
miRNA	microRNA
mRNA	messenger RNA
NPCs	neural progenitor cells
OSK	Oct4, Sox2, Klf4 = reprogramming factors
OSKM	Oct4, Sox2, Klf4, c-Myc = reprogramming factors
RNA	ribonucleic acid

I. INTRODUCTION

A. About Stem Cells

Historically, the regenerative capacities of the human body have been well recognized, even since ancient times. In Greek mythology, Zeus punished Prometheus's gift of fire to man by having an eagle devour his liver every day, only to have it regrow every night. In the last century, we have appreciated the cellular replenishment ongoing in the skin, the gut, and the hematopoietic system. But is has only been during the last 30 years that we have developed a conceptual biological framework to study these phenomena: the stem cell and its associated concepts (Table 2.1).

Today, stem cells are classified as either being adult or embryonic. Adult stem cells are a cell type typically found in a number of tissues, whose function involves high cell turnover. For example, the bone marrow produces 2.5 billion red cells, 2.5 billion platelets, and 1 billion granulocytes per kilogram of body weight daily (Molineux and Testa 1993). This remarkable regenerative capacity derives from a relatively small population of hematopoietic stem cells (HSC) residing in the bone marrow. Adult stem cells have a number of unique characteristics. They are capable of remaining quiescent for long periods, avoiding cell division and its potential to generate mutations. They can divide asymmetrically, giving rise to a daughter cell that will remain in the stem cell state and a sister cell that will undergo differentiation. In mice, a single HSC can, when transplanted into a recipient mouse, regenerate the entire hematopoietic system (Matsuzaki et al. 2004). They are multipotent, giving rise to all the different cell types of the blood. It does so by giving rise to a hierarchical tree of progenitors, such as the myeloid lineage progenitor or the lymphoid lineage progenitor. Each progenitor type can, by cell division and differentiation, give rise to all the cell types found in its lineage, but not those of other blood lineages. Its potential, compared with that of the HSC, has been restricted. Thus, the hematopoietic system is formed by a collection of cellular compartments related to one another in a hierarchical manner. The overall processes of cell division and differentiation (and their control) are required to keep these compartments in balance, and any alteration in this equilibrium can result in disease. Insufficient differentiation or exhaustion of progenitor compartments can result in cytopenias of different kinds; excess cell division can cause cancer, as in the case of early lymphoblast progenitors in acute lymphoblastic leukemia.

INTRODUCTION

TABLE 2.1. Common Terms Associated with Stem Cells and Their Meanings

Stem cell	A cell with the capability to differentiate into different cell types. Stem cells are characterized by *self-renewal* and *potency*.
Self-renewal	Capacity of stem cells to maintain their potency as undifferentiated cells through an unlimited number of cell divisions.
Potency	Capability of a stem cell to differentiate into different cell types. The zygote is a *totipotent* cell because it can give rise to all cell types of both the embryo and the extraembryonic tissues. A *pluripotent* cell can differentiate in all cell types of the embryo, but cannot produce extraembryonic tissues (ESc). *Multipotent* cells are committed to a specific differentiation program that can produce a limited repertoire of cell types.
Asymmetric cell division	A mitotic event in which a cell divides into two daughter cells with different characteristics; in stem cells, one of the daughter cells will retain the original cell's potency, while the other will be committed to a differentiation pathway in which it will eventually lose its potency.
Embryonic stem cell (ESc)	Cell type derived from the inner cell mass of the blastocyst and can be maintained in cell culture indefinitely. They are pluripotent, and if reintroduced in a blastocyst, can produce a whole individual.
Adult/somatic stem cells	Undifferentiated cells found in living beings after birth. Adult stem cells are multipotent.
Plasticity	Capability of a cell to adapt its phenotype to its environment crossing lineage barriers. In general, once a cell has entered one of the three germ layer lineages (ectoderm, mesoderm, and endoderm), it does not switch to another lineage *in vivo*. However, under specific conditions, some cells may undergo lineage switching. *In vitro*, lineage switching is called *transdifferentiation*.

Embryonic stem cells are derived from the inner cell mass of the blastocyst. The inner cell mass is a transient compartment in development; it quickly continues on to gastrulation and organogenesis, giving rise to the entire organism. ESc are considered an *in vitro* phenomenon that "captures" and stabilizes the ESc state. They are capable of self-renewal and are pluripotent, capable of giving rise to all cell types of the adult organism. They can be maintained in culture indefinitely, can be differentiated *in vitro*, and are capable of participating in development normally if reintroduced into a blastocyst. Stem cells have generated great interest due to their potential in regenerative medicine.

The remarkable characteristics and abilities of stem cells have made them the object of intense study for over five decades, much of which has been dedicated to understand the underlying molecular organization that enables their unique capabilities. This ESc regulatory state is achieved when a number of cellular parameters are in the right range. It is a complex interaction of particular signaling pathways, a specific transcription factor profile, chromatin remodeling proteins, and non-coding RNAs, all coming together to create and maintain this peculiar cell state.

In this chapter, we focus on the role of microRNAs (miRNAs) in regulating the ESc regulatory state and in differentiation.

II. MicroRNAs IN PLURIPOTENCY AND ESTABLISHMENT OF DIFFERENT LINEAGES

A. MicroRNAs in Embryonic Stem Cells (ESc)

Oct4, Sox2, and Nanog are the core transcription factors that control the establishment and maintenance of pluripotency in ESc. Each of these three transcription factors bind to their own promoter, as well as the promoter of the other two, forming a positive feedback loop enhancing their transcription and stabilizing their protein levels (Boyer et al. 2005). The three of them have been detected cooccupying regulatory regions of hundreds of genes, enhancing their expression, recruiting chromatin remodeling factors, or silencing them when polycomb proteins are recruited. MiRNAs are not an exception of this regulation. Human and mouse embryonic stem cells (mESc) express their own miRnome (Houbaviy et al. 2003; Suh et al. 2004). Oct4, Sox2, Nanog, and Tcf3 cooccupy the promoter region of miRNA clusters in mESc, enhancing their expression or silencing it by recruiting polycomb proteins (Marson et al. 2008). Such precise regulation in miRNAs underlines the importance of miRNAs in pluripotency and differentiation establishment.

The role of miRNAs in ESc function has been investigated by knocking out the microprocessor complex. When the miRNA biogenesis pathway is truncated by knocking out Dicer or DGRC8, mature miRNAs expression is eliminated. Mouse ESc lacking Dicer or DGCR8 are viable but present lower proliferation rates than wt ESc and are unable to fully down-regulate the pluripotency program when exposed to differentiation cues (Kanellopoulou et al. 2005; Murchison et al. 2005; Wang et al. 2007). Dicer knockout mESc undergo a complete proliferation block that after continuous cell culture, is partially reversed, giving rise to ESc colonies with a lower proliferation rate than wild-type mESc. In contrast, DGCR8 deficient cells present a milder phenotype having a lower cycling rate due to an arrest in G1/S phase, without the first proliferation block suffered by Dicer KO mESc. This difference between the two phenotypes might be due to the fact that while DGCR8 is exclusively involved in pri-miRNA processing, Dicer also has roles in telomere maintenance, DNA methylation (Benetti et al. 2008), and in the processing of other microprocessor-independent Dicer-dependent small RNAs, such as mirtrons (Yang and Lai 2011).

In short, lack of mature miRNA in the mESc causes the loss of the two main hallmark characteristics of the ESc: self-renewal and pluripotency, highlighting the importance of miRNAs in the maintenance of the ESc phenotype.

B. Maintenance of the Pluripotent State through Controlling the Cell Cycle

In mitosis, G1 phase length relies upon the G1/S transition governed by the restriction checkpoint (R checkpoint), which is dependent on external promitotic and antimitotic signals. So, one hypothesis that has been proposed is that differentiation starts during G1 mitotic phase (Burdon et al. 2002), in which the cell receives external inputs from mitogens and differentiation signals that produce an interplay between prodifferentiation and propluripotency networks, resulting in a choice of either starting the differentiation program or staying undifferentiated. In contrast to somatic cells, ESc present a shortened G1 phase without R checkpoint, making cell division independent from external mitogens. This is achieved through a high expression and activation of cycle-promoting complexes Cdk2-Cyclin E, with no presence of Cdk4/6-Cyclin D and an absence of Cdk-Cyclin inhibitors, such as INK and CIP families, that inhibit overcoming the G1/S transition through the R checkpoint. During differentiation, the R checkpoint is established, elongating G1 phase and becoming mitogen dependent. In somatic cells, during G1 phase, D-type cyclins are expressed and form a complex, with Cdk4/6 activating it and triggering the phosphorylation and inactivation of the inhibitor of the G1/S transition pRb. In ESc, pRb is normally hyperphosphorylated and inactivated. Inhibition of pRb enables the release of the transcription factor E2F, which activates the transcription of cyclin E, forming a complex with Cdk2 and promoting the progression to the S phase (Wang and Blelloch 2009). It has been observed that a lack of pRb in mesenchymal stem cells biased the choice between adipocyte and osteogenic toward adipocyte (Calo et al. 2010), making it a good example of how the cell cycle can be involved in the regulation of cell fate.

DGCR8–/– mESc present a low rate of proliferation due to an elongation of the G1 phase, suggesting a role of the ESc-specific miRNAs in the G1/S transition. Several members of different miRNA families, with similar seed sequence, were demonstrated to partially rescue the cycling phenotype in DGCR8–/– mESc (miRNAs 291a-3p, 291b-3p, 294, 295 from *miR-290* family; miRNAs 302b, 302c, 302d from *miR-302* family and miRNAs 20a, 20b, 93, and 106a from *miR-17* family) (Wang et al. 2008). This group of miRNAs was named as ESCC (ES cell-specific cell cycle-regulating miRNAs). ESCC are up-regulated in ESc and decrease their expression upon differentiation. The members of these families are organized in clusters; the whole cluster is transcribed at the same time, having a combined effect on the cell.

The *miR-290* cluster is the most abundantly expressed of all the miRNAs present in mESc. Further characterization of the members of this cluster in DGCR8-/- mESc shows a complete rescue of the G1 arrest phenotype, pointing to their role in promoting G1/S transition. In mESc, members of the *miR-290* cluster ensure a rapid progression through the R checkpoint and a rapid G1/S transition targeting the mRNA of Lats2, p21, and pRb, inhibitors of the complex Cdk2-cyclin E and, therefore, G1/S transition inhibitors (Wang et al. 2008). This is consistent with the observation that p21 protein levels increase upon differentiation, maintaining a constant mRNA concentration (Sabapathy et al. 1997). *miR-372*, a homologue of the *miR-290* family in human ESc, also targets p21 (Qi et al. 2009). On the other hand, *miR-302* has been demonstrated to target Cyclin D1 (Card et al. 2008), suggesting that lack of expression of Cdk 4/6-Cyclin D complex in ESc is mediated, at least in part, by this mechanism.

c-Myc, a transcription factor involved in the G1/S transition, directly enhances ESCC miRNA transcription; in turn, ESCC miRNAs indirectly enhance *c-Myc* expression, forming a positive feedback loop that stabilizes the ESc cell cycle profile. *let-7* is a miRNA family widely expressed in differentiated tissues and well conserved across species. The mature form of *let-7* miRNA is expressed upon differentiation and directly targets c-Myc (Kumar et al. 2007), switching off the self-renewal program and promoting differentiation. *let-7* and ESCC miRNAs form part of the pluripotency-differentiation network, but with opposing effects, supplying a molecular switch between self-renewal and differentiation (Melton et al. 2010).

C. MicroRNAs in Pluripotency versus Differentiation Fate Choice

For differentiation to take place, the self-renewal program must be turned off, while the differentiation program must be turned on. Under differentiation conditions, DGCR8-/- mESc can up-regulate a number of differentiation markers, but are unable to fully suppress the pluripotency markers (Wang et al. 2007).

MiRNAs can regulate differentiation through direct or indirect effects. The *miR-290* family, although highly expressed in ESc and down-regulated upon differentiation, is indirectly involved in the establishment of differentiation program. The *miR-290* family represses pRb, a transcriptional repressor of DNA methyltransferases (DNMT) involved in the *de novo* DNA methylation required for full silencing of the pluripotency markers (Benetti et al. 2008; Sinkkonen et al. 2008). On the other hand, *miR-290* family by itself cannot induce differentiation in DGCR8 -/- mESc. The *miR-290* family is required for DNMT activity, but it is not enough to overcome the effect of miRNA depletion in differentiation. This suggests miRNAs must be involved in other pathways required for the establishment of a differentiation program.

One mechanism by which miRs mediate the transition from self-renewal to differentiation is through direct targeting of core self-renewal transcription factors. miRNAs 134, 296, and 470 are up-regulated via retinoic acid differentiation of mESc to ectoderm and directly target Oct4, Nanog, and Sox2 (Tay et al. 2008). Simlarly, miRNAs 200c, 203, and 183 repress Sox2 and Klf4 (Wellner et al. 2009). In hESc, *miR-145* is up-regulated in the vascular differentiation program and directly targets and represses Oct4, Sox2, and Klf4 (Xu et al. 2009). Therefore, miR expression is directly responsible for shutting down the pluripotency regulatory network upon differentiation.

Lin-28 is a protein highly expressed in ESc, where it binds to pre-miRNAs and targets them for degradation (Heo et al. 2009). Lin-28 binds to pre-*let-7* (Heo et al. 2008), inhibiting its processing, and therefore, the *let-7* mediated repression on the self-renewal promoting activities of n-Myc and c-Myc. In turn, mature *let-7* targets Lin-28 mRNA down-regulating it (Rybak et al. 2008). As soon as differentiation starts, levels of Oct4, Sox2, and Nanog begin to fall, releasing promoter regions, such as *Lin-28* (Marson et al. 2008), down-regulating its expression and giving way to differentiation miRNAs to accumulate and further down-regulate both, Lin-28 by *let-7* up-regulation, and the core pluripotency transcription factors by the miRNAs up-regulated in the different differentiation programmes. Hence, the equilibrium between self-renewal and differentiation is governed by the mutually opposing effects of ESCC and differentiation promoting miRs, such as *let-7*.

Another player in the outcome of this transition is *miR-302*, which is expressed in ESc (Houbaviy et al. 2003). Oct4 and *miR-302* are involved in the inhibition of the transcriptional repressor NR2F2 at the transcriptional and post-transcriptional level, respectively. NR2F2 is up-regulated in mesoderm differentiation and directly inhibits Oct4 expression. As explained earlier, Oct4 promotes *miR-302* cluster, establishing a regulatory circuitry in which, in ESc, NR2F2 is down-regulated by *miR-302*, and in differentiation, NR2F2 inhibits Oct4, down-regulating also the expression of *miR-302*, leading to a higher expression of NR2F2 (Rosa and Brivanlou 2011). In order to maintain pluripotency, *miR-302* is also involved in the targeting and repression of Lefty1, an early marker of differentiation (Barroso-Del Jesús et al. 2011).

MiRNAs quick response upon differentiation can be explained by the regulation of their promoters. Some miRNAs which are up-regulated during differentiation have bivalent domains in their promoter regions that can silence the miRNAs in ESc and are activated only in their specific lineage (Marson et al. 2008). This is the case of *miRNA-9*, which is a marker of neural progenitor cells (NPCs). This miRNA maintains in its promoter both signals of activation (histone H3K4me3) and inactivation (Histone H3K27me3) in ESc. In differentiation towards NPCs, the promoter loses the repressing mark, leaving only the H3K4me3 activation mark leading to *miRNA-9* up-regulation; conversely, in differentiation toward mouse embryonic fibroblasts (MEFs), the active mark disappears while

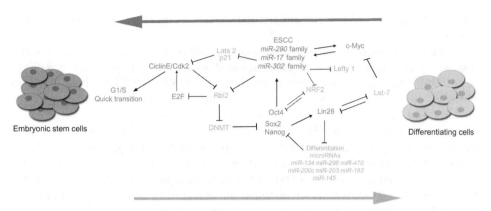

Figure 2.1. The pluripotency/differentiation regulatory network. Propluripotency elements are in blue and prodifferentiation in orange. There is a balance between ESc state and differentiation. ESCC miRNAs target cell cycle inhibitors and early differentiation markers, promoting self-renewal and the maintenance of the pluripotent state. On the other hand, differentiation miRNAs directly target the core pluripotency transcription factors, destabilizing the network and promoting differentiation. The final choice will depend on external signals, which will up-regulate the expression of ones, causing the repression of the others. See color insert.

the H3K27me3 inactivation mark remains, locking down expression of *miRNA-9* in MEFs (Mikkelsen et al. 2007).

Another strategy by which miRs are used to guide differentiation programs is to express miRs that preferentially target mRNAs not present in their lineage, thereby inhibiting alternative differentiation pathways. For example, *miRNA-1* and *miRNA-133* are expressed in the developing mesoderm and act to down-regulate ectodermal and endodermal genes (Ivey et al. 2008).

MiRNAs are an integral part of the pluripotency-differentiation network providing an additional layer of regulation of transcription factors and epigenetic remodeling proteins involved in starting new differentiation programs or stabilizing existing ones (Figure 2.1).

III. MicroRNAs, INDUCED PLURIPOTENCY, AND TRANSDIFFERENTIATION

A. Induced Pluripotency and MicroRNAs

In 2006, Takahashi et al. showed for the first time that exogenous expression of four transcription factors (Oct4, Sox2, Klf4, and c-Myc, termed as OSKM) could produce cells similar to ESc (Takahashi and Yamanaka 2006). This remarkable result initiated the new field of direct reprogramming that has exploded in recent years, creating intense interest due to its conceptual importance and potential applications for regenerative medicine. Induced pluripotent stem cells (iPSc) offer a virtually unlimited source of cells for research and therapeutic applications, completely circumventing the ethical and immunological caveats of working with human ESc.

Given the importance of miRNAs in the establishment and maintenance of the pluripotent state and their close relationship with the transcription factors involved in reprogramming, the question of if and how miRNAs were involved in the reprogramming process arose.

Initially, cotransfection of embryonic specific miRNAs with reprogramming transcription factors was shown to raise the efficiency of iPSc derivation. In particular, codelivery of *miR-294* with a retrovirus expressing Oct4, Sox2, and Klf4 (OSK) produced iPSc with higher efficiency than OSK alone (Judson et al. 2009). *miR-294* is related to the *miR-290* cluster whose expression is directly enhanced by c-Myc, the fourth Yamanaka reprogramming factor. The addition of c-Myc to the pluripotency cocktail did not improve efficiency further, suggesting that an important role of c-Myc in reprogramming is the up-regulation of the *miR-290* cluster. In contrast, transfection of embryonic-specific miRNAs *miR-302b* and *miR-372* together with OSK or OSKM enhanced the reprogramming efficiency of both transcription factor combinations (Subramanyam et al. 2011), highlighting that miRNAs are involved in different pathways to produce iPSc. Indeed, not only overexpression of embryonic specific miRNAs enhances reprogramming, but also the down-regulation of tissue-specific miRNAs improves iPSc derivation efficiency. Inhibition of *let-7* in mouse embryonic fibroblasts (MEFs) together with the retrovirus expressing OSK, raised the efficiency of reprogramming 4.3-fold times but, again, addition of c-Myc to the cocktail only improved to 1.75-fold (Melton et al. 2010), which is consistent with the observation that c-Myc is a direct target of *let-7*.

Further highlighting the importance of miRs in ESc biology, it has been shown that iPSc can be derived by ectopic expression of miRs in the absence of reprogramming transcription factors. In 2008, overexpression of *miR-302* family members in human skin cancer cells resulted in derivation of iPSc (Lin et al. 2008), a result that was later extended to human hair follicle cells (Lin et al. 2011). Lentiviral transfection of the *miR-302/367* cluster reprograms human fibroblasts. Interestingly, the same cluster can only reprogram MEFs to iPSc if Hdac2 (histone deacetylase2) is inhibited with valproic acid, suggesting that Hdac2-mediated chromatin remodeling represents a barrier for reprogramming in the mouse system (Anokye-Danso et al. 2011). Furthermore, overexpression of mature miRNAs *miR-200c* plus family members of *miR-302* and *miR-369* in mouse adiposal stromal cells, human adiposal stromal cells, and human dermal fibroblasts also resulted in iPSc derivation, avoiding DNA-based factors that can be integrated in the genome (Miyoshi et al. 2011).

B. Underlying Mechanisms in microRNA iPSc Generation

The low efficiency in reprogramming suggests a resistance by the somatic cell to change its identity to become pluripotent. During reprogramming, the cells pass through two phases. The first one involves the establishment of a prepluripotent state through the increase of cell cycle rate and completion of a mesenchymal to epithelial transition (MET). The second step in reprogramming is the consolidation of the pluripotent state, in which the endogenous pluripotency network becomes independent from the ectopic transcription factor expression. The examples discussed earlier, together with the finding that lack of miRNA activity in somatic cells impedes their dedifferentiation (Li et al. 2011; Kim et al. 2012), suggest an active role for miRNAs in overcoming the reprogramming barriers.

Somatic cell cycle is highly regulated, which presents an obstacle for reprogramming. As explained above, *miR-290* family and *let-7* have opposing roles in self-renewal in ESc through their different regulation of c-Myc, a transcription factor involved in G1/S transition progression. *miR-290* cluster up-regulate c-Myc transcription, and is in turn enhanced by c-Myc. Conversely, c-Myc is a direct target of *let-7*. Therefore, ectopic expression of *miR-290* cluster and inhibition of *let-7* increase reprogramming efficiency by promoting cell cycle progression (Melton et al. 2010). Introduction of miRNAs *miR-130b, miR-301b,* and *miR-721* in MEFs also enhanced iPSc production promoting cell cycle by targeting Meox2, which up-regulates p21 and p16, both inhibitors of the G1/S transition (Pfaff et al. 2011).

Mesenchymal cells, such as fibroblasts or keratinocytes, must undergo a MET as part of the reprogramming process to become iPSc, which show epithelial morphology. *miR-302* and *miR-372,* orthologs of mouse *miR-302* and *miR-290,* respectively, have been shown to improve reprogramming efficiency in part by accelerating MET (Liao et al. 2011; Subramanyam et al. 2011) through direct inhibition of TGFBR2 and RHOC, which are involved in EMT (epithelial to mesenchymal transition), reverting it to MET. *miR-302* also promotes MET in an indirect manner by targeting BMP inhibitors (Lipchina et al. 2011). The ensuing up-regulation of BMP combined with the expression of *miR-205,* and the *miR-200* family, which target Zeb1 and Zeb2, inhibitors of E-cadherin expression, lead to increased levels of E-cadherin, one of the hallmarks of MET (Gregory et al. 2008; Samavarchi-Tehrani et al. 2010).

There is a set of miRNA clusters specifically induced during reprogramming of MEFs, among them *miR-106b* and *miR-93* encoded by the *miR-106-25* cluster which target TGFBR2 and p21 (Li et al. 2011) enhancing reprogramming through both cell cycle progression and MET, indicating that not only ectopically expressed miRNAs can have a role in the reprogramming process lowering reprogramming barriers.

The consolidation of the pluripotent state, the second phase of the reprogramming, includes the epigenetic remodeling and the establishment of the pluripotency network, two processes in which *miR-302* is an active player. DNA demethylation is needed for establishing pluripotency and naturally occurs after fertilization. *miR-302* indirectly destabilizes DNA methyltransferase 1 (DNMT1) by targeting epigenetic factors, such as lysine-specific histone demethylases 1 and 2 (AOF1 and AOF2) and methyl CpG-binding proteins 1 and 2 (MECP1-p66 and MCEP2) (Lin et al. 2011). DNMT1 methylates newly replicated DNA, so, with its destabilization, daughter cells suffer a passive demethylation that erases the marks on the promoters and facilitates the reprogramming. In addition, as explained earlier, *miR-302* is involved in a positive feedback loop with the main pluripotency factors and can trigger the transcription of endogenous Oct4 by the inhibition of the transcriptional repressor NR2F2; in turn, then, Oct4 stimulates transcription of *miR-302,* closing the loop and establishing the endogenous pluripotency network (Rosa and Brivanlou 2011) (Figure 2.2).

C. Transdifferentiation and miRNAs

Reprogramming in its wide sense is the conversion of one cell type into another. In addition to reverting somatic cells to the pluripotent state, cells can also be reprogrammed directly to alternative differentiated cell fates, a process called transdifferentiation. Recent examples of transdifferentiation include the conversion of mouse B-lymphocytes into macrophages (Xie et al. 2004), pancreatic acinar cells to beta islet cells (Zhou et al. 2008), and fibroblast to neurons (Vierbuchen et al. 2010).

As in iPSc research, the majority of experiments in transdifferentiation have been achieved using different combinations of transcription factors, a mixture of transcription factors with miRNAs and only miRNAs. *miR-124,* highly expressed in neuronal tissue, has been used in combination with neuronal transcription factors in the direct conversion of fibroblasts to neurons (Ambasudhan et al. 2011); *miR-124* and *miR-9* can produce

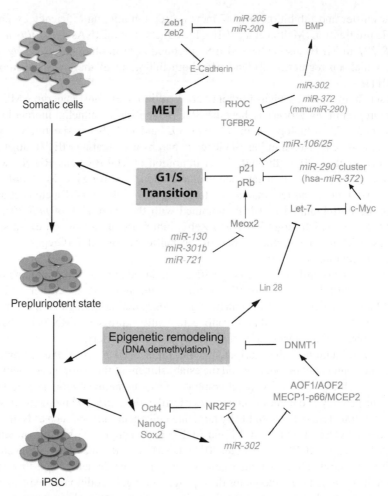

Figure 2.2. MiRNAs involved in reprogramming: In the figure are represented the two phases in the reprogramming process. Embryonic-specific miRNAs (*miR-290* cluster and *miR-302* family) target inhibitors of MET (RHOC and TGFBR2) and G1/S transition (p21 and pRb). Indirectly can up-regulate MET promoting factors, such as BMP. In the reprogramming process, some endogenous miRNAs are up-regulated helping in overcoming MET (*miR-205* and *miR-200* target MET inhibitors Zeb1 and 2) or promoting G1/S transition (*miR-130*, *miR-301b*, and *miR-721* inhibit Meox2, a G1/S inhibitor). *miR-302* is also involved in DNA demethylation needed for the epigenetic remodeling in the cell, as well as in the consolidation of the pluripotency network, together with the endogenous pluripotency markers (Oct4, Nanog, Sox2, and Lin-28). Proreprogramming players are colored in green; antireprogramming ones are colored in red. See color insert.

neuronal fate by their own, but with low efficiency and functional characterization of the resulting neurons has not been reported (Yoo et al. 2011). Expression of NeuroD1 with *miR-124* and *miR-9* raised the efficiency and produced functional neurons. The latest demonstration of miRNA-directed phenotype conversions is the use of *miR-1*, *miR-133*, and *miR-208* to convert cardiac fibroblast into cardiomyocytes (Jayawardena et al. 2012).

Much remains unknown regarding the mechanisms underlying miRNA-mediated transdifferentiation, and in each different case of transdifferentiation, the targets of the used miRNAs will differ, but in general, the rationale is that the ectopic expression of tissue-specific miRNAs will down-regulate non-related transcripts (down-regulating non-neuronal transcripts, for example) or, alternatively, target inhibitors of the related cell type leading to the activation of the proposed cell fate.

IV. CONCLUSION

In this chapter, we have summarized the role of miRNAs as important regulators in pluripotency maintenance, as well as in directing cell fates. Due to their effects on transcription factors and chromatin remodeling proteins, miRNAs are revealed as key components of regulatory networks in cell identity. MiRNAs participate in regulatory feedback loops conferring robustness to the network and reinforcing the cell fate decisions either by suppressing parallel differentiation pathways or by inhibiting the immediately previous state. MiRNAs can also function as switches, as is the case of *miR-290* and *let-7*, in which the activation of one implies the repression of the other. As a result, miRNAs can fine-tune the transcriptome defining the molecular scenery in the cell. Further studies in the miRNA-pluripotency field will reveal new insights in the regulation of pluripotency and differentiation network, as well as contributing to the development of new strategies for regenerative medicine.

ACKNOWLEDGMENTS

Work in the laboratory of J.C.I.B. was supported by MINECO, Fundacion Cellex, G. Harold and Leila Y. Mathers Charitable Foundation, The Leona M. and Harry B. Helmsley Charitable Trust, and The Ellison Medical Foundation.

REFERENCES

Ambasudhan, R., M. Talantova, R. Coleman, X. Yuan, S. Zhu, S.A. Lipton, and S. Ding. 2011. Direct reprogramming of adult human fibroblasts to functional neurons under defined conditions. *Cell Stem Cell* **9**:113–118.

Anokye-Danso, F., C.M. Trivedi, D. Juhr, M. Gupta, Z. Cui, Y. Tian, Y. Zhang, W. Yang, P.J. Gruber, J.A. Epstein, and E.E. Morrisey. 2011. Highly efficient miRNA-mediated reprogramming of mouse and human somatic cells to pluripotency. *Cell Stem Cell* **8**:376–388.

Barroso-Del Jesús, A., G. Lucena-Aguilar, L. Sanchez, G. Ligero, I. Gutierrez-Aranda, and P. Menendez. 2011. The Nodal inhibitor Lefty is negatively modulated by the microRNA *miR-302* in human embryonic stem cells. *FASEB J* **25**:1497–1508.

Benetti, R., S. Gonzalo, I. Jaco, P. Munoz, S. Gonzalez, S. Schoeftner, E. Murchison, T. Andl, T. Chen, P. Klatt, E. Li, M. Serrano, et al. 2008. A mammalian microRNA cluster controls DNA methylation and telomere recombination via Rbl2-dependent regulation of DNA methyltransferases. *Nat Struct Mol Biol* **15**:268–279.

Boyer, L.A., T.I. Lee, M.F. Cole, S.E. Johnstone, S.S. Levine, J.P. Zucker, M.G. Guenther, R.M. Kumar, H.L. Murray, R.G. Jenner, D.K. Gifford, D.A. Melton, et al. 2005. Core transcriptional regulatory circuitry in human embryonic stem cells. *Cell* **122**:947–956.

Burdon, T., A. Smith, and P. Savatier. 2002. Signalling, cell cycle and pluripotency in embryonic stem cells. *Trends Cell Biol* **12**:432–438.

Calo, E., J.A. Quintero-Estades, P.S. Danielian, S. Nedelcu, S.D. Berman, and J.A. Lees. 2010. Rb regulates fate choice and lineage commitment *in vivo*. *Nature* **466**:1110–1114.

Card, D.A., P.B. Hebbar, L. Li, K.W. Trotter, Y. Komatsu, Y. Mishina, and T.K. Archer. 2008. Oct4/Sox2-regulated *miR-302* targets cyclin D1 in human embryonic stem cells. *Mol Cell Biol* **28**:6426–6438.

Gregory, P.A., A.G. Bert, E.L. Paterson, S.C. Barry, A. Tsykin, G. Farshid, M.A. Vadas, Y. Khew-Goodall, and G.J. Goodall. 2008. The *miR-200* family and *miR-205* regulate epithelial to mesenchymal transition by targeting ZEB1 and SIP1. *Nat Cell Biol* **10**:593–601.

Heo, I., C. Joo, J. Cho, M. Ha, J. Han, and V.N. Kim. 2008. Lin28 mediates the terminal uridylation of *let-7* precursor microRNA. *Mol Cell* **32**:276–284.

Heo, I., C. Joo, Y.K. Kim, M. Ha, M.J. Yoon, J. Cho, K.H. Yeom, J. Han, and V.N. Kim. 2009. TUT4 in concert with Lin28 suppresses microRNA biogenesis through pre-microRNA uridylation. *Cell* **138**:696–708.

Houbaviy, H.B., M.F. Murray, and P.A. Sharp. 2003. Embryonic stem cell-specific microRNAs. *Dev Cell* **5**:351–358.

Ivey, K.N., A. Muth, J. Arnold, F.W. King, R.F. Yeh, J.E. Fish, E.C. Hsiao, R.J. Schwartz, B.R. Conklin, H.S. Bernstein, and D. Srivastava. 2008. MicroRNA regulation of cell lineages in mouse and human embryonic stem cells. *Cell Stem Cell* **2**:219–229.

Jayawardena, T.M., B. Egemnazarov, E.A. Finch, L. Zhang, J.A. Payne, K. Pandya, Z. Zhang, P. Rosenberg, M. Mirotsou, and V.J. Dzau. 2012. MicroRNA-mediated *in vitro* and *in vivo* direct reprogramming of cardiac fibroblasts to cardiomyocytes. *Circ Res* **110**:1465–1473.

Judson, R.L., J.E. Babiarz, M. Venere, and R. Blelloch. 2009. Embryonic stem cell-specific microRNAs promote induced pluripotency. *Nat Biotechnol* **27**:459–461.

Kanellopoulou, C., S.A. Muljo, A.L. Kung, S. Ganesan, R. Drapkin, T. Jenuwein, D.M. Livingston, and K. Rajewsky. 2005. Dicer-deficient mouse embryonic stem cells are defective in differentiation and centromeric silencing. *Genes Dev* **19**:489–501.

Kim, B.M., M.C. Thier, S. Oh, R. Sherwood, C. Kanellopoulou, F. Edenhofer, and M.Y. Choi. 2012. MicroRNAs are indispensable for reprogramming mouse embryonic fibroblasts into induced stem cell-like cells. *PLoS ONE* **7**:e39239.

Kumar, M.S., J. Lu, K.L. Mercer, T.R. Golub, and T. Jacks. 2007. Impaired microRNA processing enhances cellular transformation and tumorigenesis. *Nat Genet* **39**:673–677.

Li, Z., C.S. Yang, K. Nakashima, and T.M. Rana. 2011. Small RNA-mediated regulation of iPS cell generation. *EMBO J* **30**:823–834.

Liao, B., X. Bao, L. Liu, S. Feng, A. Zovoilis, W. Liu, Y. Xue, J. Cai, X. Guo, B. Qin, R. Zhang, J. Wu, et al. 2011. MicroRNA cluster 302-367 enhances somatic cell reprogramming by accelerating a mesenchymal-to-epithelial transition. *J Biol Chem* **286**:17359–17364.

Lin, S.L., D.C. Chang, S. Chang-Lin, C.H. Lin, D.T. Wu, D.T. Chen, and S.Y. Ying. 2008. Mir-302 reprograms human skin cancer cells into a pluripotent ES-cell-like state. *RNA* **14**:2115–2124.

Lin, S.L., D.C. Chang, C.H. Lin, S.Y. Ying, D. Leu, and D.T. Wu. 2011. Regulation of somatic cell reprogramming through inducible mir-302 expression. *Nucleic Acids Res* **39**:1054–1065.

Lipchina, I., Y. Elkabetz, M. Hafner, R. Sheridan, A. Mihailovic, T. Tuschl, C. Sander, L. Studer, and D. Betel. 2011. Genome-wide identification of microRNA targets in human ES cells reveals a role for *miR-302* in modulating BMP response. *Genes Dev* **25**:2173–2186.

Marson, A., S.S. Levine, M.F. Cole, G.M. Frampton, T. Brambrink, S. Johnstone, M.G. Guenther, W.K. Johnston, M. Wernig, J. Newman, J.M. Calabrese, L.M. Dennis, et al. 2008. Connecting microRNA genes to the core transcriptional regulatory circuitry of embryonic stem cells. *Cell* **134**:521–533.

Matsuzaki, Y., K. Kinjo, R.C. Mulligan, and H. Okano. 2004. Unexpectedly efficient homing capacity of purified murine hematopoietic stem cells. *Immunity* **20**:87–93.

REFERENCES

Melton, C., R.L. Judson, and R. Blelloch. 2010. Opposing microRNA families regulate self-renewal in mouse embryonic stem cells. *Nature* **463**:621–626.

Mikkelsen, T.S., M. Ku, D.B. Jaffe, B. Issac, E. Lieberman, G. Giannoukos, P. Alvarez, W. Brockman, T.K. Kim, R.P. Koche, W. Lee, E. Mendenhall, et al. 2007. Genome-wide maps of chromatin state in pluripotent and lineage-committed cells. *Nature* **448**:553–560.

Miyoshi, N., H. Ishii, H. Nagano, N. Haraguchi, D.L. Dewi, Y. Kano, S. Nishikawa, M. Tanemura, K. Mimori, F. Tanaka, T. Saito, J. Nishimura, et al. 2011. Reprogramming of mouse and human cells to pluripotency using mature MicroRNAs. *Cell Stem Cell* **8**:633–638.

Molineux, G. and N.G. Testa, Eds. 1993. *Haemopiesis: a practical approach*. Practical approach series. Oxford University Press, Oxford; New York.

Murchison, E.P., J.F. Partridge, O.H. Tam, S. Cheloufi, and G.J. Hannon. 2005. Characterization of Dicer-deficient murine embryonic stem cells. *Proc Natl Acad Sci U S A* **102**:12135–12140.

Pfaff, N., J. Fiedler, A. Holzmann, A. Schambach, T. Moritz, T. Cantz, and T. Thum. 2011. miRNA screening reveals a new miRNA family stimulating iPS cell generation via regulation of Meox2. *EMBO Rep* **12**:1153–1159.

Qi, J., J.Y. Yu, H.R. Shcherbata, J. Mathieu, A.J. Wang, S. Seal, W. Zhou, B.M. Stadler, D. Bourgin, L. Wang, A. Nelson, C. Ware, et al. 2009. microRNAs regulate human embryonic stem cell division. *Cell Cycle* **8**:3729–3741.

Rosa, A. and A.H. Brivanlou. 2011. A regulatory circuitry comprised of *miR-302* and the transcription factors OCT4 and NR2F2 regulates human embryonic stem cell differentiation. *EMBO J* **30**:237–248.

Rybak, A., H. Fuchs, L. Smirnova, C. Brandt, E.E. Pohl, R. Nitsch, and F.G. Wulczyn. 2008. A feedback loop comprising lin-28 and *let-7* controls pre-*let-7* maturation during neural stem-cell commitment. *Nat Cell Biol* **10**:987–993.

Sabapathy, K., M. Klemm, R. Jaenisch, and E.F. Wagner. 1997. Regulation of ES cell differentiation by functional and conformational modulation of p53. *EMBO J* **16**:6217–6229.

Samavarchi-Tehrani, P., A. Golipour, L. David, H.K. Sung, T.A. Beyer, A. Datti, K. Woltjen, A. Nagy, and J.L. Wrana. 2010. Functional genomics reveals a BMP-driven mesenchymal-to-epithelial transition in the initiation of somatic cell reprogramming. *Cell Stem Cell* **7**:64–77.

Sinkkonen, L., T. Hugenschmidt, P. Berninger, D. Gaidatzis, F. Mohn, C.G. Artus-Revel, M. Zavolan, P. Svoboda, and W. Filipowicz. 2008. MicroRNAs control de novo DNA methylation through regulation of transcriptional repressors in mouse embryonic stem cells. *Nat Struct Mol Biol* **15**:259–267.

Subramanyam, D., S. Lamouille, R.L. Judson, J.Y. Liu, N. Bucay, R. Derynck, and R. Blelloch. 2011. Multiple targets of *miR-302* and *miR-372* promote reprogramming of human fibroblasts to induced pluripotent stem cells. *Nat Biotechnol* **29**:443–448.

Suh, M.R., Y. Lee, J.Y. Kim, S.K. Kim, S.H. Moon, J.Y. Lee, K.Y. Cha, H.M. Chung, H.S. Yoon, S.Y. Moon, V.N. Kim, and K.S. Kim. 2004. Human embryonic stem cells express a unique set of microRNAs. *Dev Biol* **270**:488–498.

Takahashi, K. and S. Yamanaka. 2006. Induction of pluripotent stem cells from mouse embryonic and adult fibroblast cultures by defined factors. *Cell* **126**:663–676.

Tay, Y., J. Zhang, A.M. Thomson, B. Lim, and I. Rigoutsos. 2008. MicroRNAs to Nanog, Oct4 and Sox2 coding regions modulate embryonic stem cell differentiation. *Nature* **455**:1124–1128.

Vierbuchen, T., A. Ostermeier, Z.P. Pang, Y. Kokubu, T.C. Sudhof, and M. Wernig. 2010. Direct conversion of fibroblasts to functional neurons by defined factors. *Nature* **463**:1035–1041.

Wang, Y., S. Baskerville, A. Shenoy, J.E. Babiarz, L. Baehner, and R. Blelloch. 2008. Embryonic stem cell-specific microRNAs regulate the G1-S transition and promote rapid proliferation. *Nat Genet* **40**:1478–1483.

Wang, Y. and R. Blelloch. 2009. Cell cycle regulation by microRNAs in stem cells. *Results Probl Cell Differ* **53**:459–472.

Wang, Y., R. Medvid, C. Melton, R. Jaenisch, and R. Blelloch. 2007. DGCR8 is essential for microRNA biogenesis and silencing of embryonic stem cell self-renewal. *Nat Genet* **39**: 380–385.

Wellner, U., J. Schubert, U.C. Burk, O. Schmalhofer, F. Zhu, A. Sonntag, B. Waldvogel, C. Vannier, D. Darling, A. zur Hausen, V.G. Brunton, J. Morton, et al. 2009. The EMT-activator ZEB1 promotes tumorigenicity by repressing stemness-inhibiting microRNAs. *Nat Cell Biol* **11**:1487–1495.

Xie, H., M. Ye, R. Feng, and T. Graf. 2004. Stepwise reprogramming of B cells into macrophages. *Cell* **117**:663–676.

Xu, N., T. Papagiannakopoulos, G. Pan, J.A. Thomson, and K.S. Kosik. 2009. MicroRNA-145 regulates OCT4, SOX2, and KLF4 and represses pluripotency in human embryonic stem cells. *Cell* **137**:647–658.

Yang, J.S. and E.C. Lai. 2011. Alternative miRNA biogenesis pathways and the interpretation of core miRNA pathway mutants. *Mol Cell* **43**:892–903.

Yoo, A.S., A.X. Sun, L. Li, A. Shcheglovitov, T. Portmann, Y. Li, C. Lee-Messer, R.E. Dolmetsch, R.W. Tsien, and G.R. Crabtree. 2011. MicroRNA-mediated conversion of human fibroblasts to neurons. *Nature* **476**:228–231.

Zhou, Q., J. Brown, A. Kanarek, J. Rajagopal, and D.A. Melton. 2008. In vivo reprogramming of adult pancreatic exocrine cells to beta-cells. *Nature* **455**:627–632.

3

MicroRNAs AS REGULATORS OF IMMUNITY

Donald T. Gracias and Peter D. Katsikis

Department of Microbiology and Immunology, Drexel University College of Medicine, Philadelphia, PA, USA

I. Introduction — 42
II. Innate Immunity and miRNAs — 43
    A. Granulocytes — 43
    B. Monocytes and Macrophages — 44
    C. Dendritic Cells — 45
    D. NK and NKT Cells — 47
    E. Mast Cells — 48
III. Adaptive Immunity and miRNAs — 48
    A. B Lymphocytes — 48
    B. T Lymphocytes — 49
IV. Conclusions and Future Directions — 52
References — 53

ABBREVIATIONS

3′-UTR	3′-untranslated region
AID	activation-induced cytidinedeaminase
AP-1	activator protein 1
BTK	bruton tyrosine kinase
DC	dendritic cell

MicroRNAs in Medicine, First Edition. Edited by Charles H. Lawrie.
© 2014 John Wiley & Sons, Inc. Published 2014 by John Wiley & Sons, Inc.

EAE	experimental autoimmune encephalomyelitis
FOXP3	forkhead box P3
GC	germinal center
IFN	interferon
IKKε	IκB kinase ε
IRAK1	IL-1 receptor-associated kinase 1
LMO2	LIM domain only 2
LPS	lipopolysaccharide
MDDC	monocyte-derived dendritic cells
MiRNA	microRNA
NK	natural killer
PRR	pathogen recognition receptor
SHIP1	SH2 domain-containing inositol 5'-phosphatase 1
SOCS1	suppressor of cytokine signaling 1
TCR	T-cell receptor
TLR	toll-like receptor
TRAF6	TNF receptor associated factor 6
T_{REG}	regulatory T cells

I. INTRODUCTION

MicroRNAs (miRNAs) are an important class of small non-coding RNA molecules that can posttranscriptionally regulate gene expression by pairing to target mRNAs. While a single miRNA may only achieve a moderate degree of gene silencing, it may instead function to fine-tune gene expression (Lodish et al. 2008). Some studies have suggested that individual miRNA may regulate the expression of large numbers of genes, albeit modestly, but this modest regulation of individual genes results in significant effects on entire signaling pathways and a strong biological outcome (Li et al. 2007; Linsley et al. 2007; Tsang et al. 2007). Thus, miRNAs could have a significant impact on distinct molecular signaling pathways affecting various cellular processes. While studies have shown miRNAs to be crucial in oncogenesis (Croce 2009), metabolism (Poy et al. 2007), and cell development (Stefani and Slack 2008), their role in the immune systems has only recently started to be characterized.

Recent literature has shown various miRNAs to be key players in regulating the development, differentiation, functionality, and survival of immune cells, such as DCs, macrophages, and B and T lymphocytes, thus influencing both innate and adaptive immunity. The importance of miRNAs in immune cells was first gauged by studies of selective deletion of Dicer, a critical enzyme required for miRNA production. Such studies demonstrated that Dicer deletion in different immune cells results in impaired survival and function. This has been observed in natural killer (NK) cells (Bezman et al. 2010), B cells (Koralov et al. 2008), different T-cell subsets (Cobb et al. 2005; Muljo et al. 2005; Zhou et al. 2008; Zhang and Bevan 2010), as well as Langerhans cells (Kuipers et al. 2010). Although such Dicer experiments highlight the global importance of miRNAs in immunity, these profound effects are not unexpected, as presumably deletion of Dicer results in impaired production of all miRNAs, and as a consequence, the expression of very large numbers of genes is perturbed. These studies do however point out the importance of miRNA in immunity and suggest that individual miRNAs may modulate inflammation and other immune responses during infections, cancer, and autoimmune disorders.

This review will focus on the impact of miRNAs on innate and adaptive immunity, their role in different immune cell types, and their implications for various immune-mediated diseases.

II. INNATE IMMUNITY AND miRNAs

The innate immune system plays a crucial role in providing the first line of defense against different pathogens and involves a number of innate immune cell types. Cells such as granulocytes, NK cells, dendritic cells, and macrophages constitute the major players of the innate immune response. Many of these cell types are involved in detecting pathogen-associated molecular patterns on different microbial pathogens by use of several conserved pathogen recognition receptors (PRRs). Of the known PRRs, Toll-like receptors (TLRs) have been well studied in mediating inflammatory responses during microbial infections. Recent studies have suggested that miRNAs such as *miR-9, miR-125b, miR-146a*, and *miR-155* may play important roles in regulating TLR signaling and subsequent inflammatory responses downstream of TLR activation (Tili et al. 2007; Bazzoni et al. 2009; Jurkin et al. 2010). Other studies have also indicated an absolute requirement for miRNAs in the different innate immune cells, especially for their development, differentiation and function.

A. Granulocytes

Neutrophils constitute the largest subset of granulocytes and are often the first to arrive at sites invaded by pathogens. Interestingly, miRNAs may regulate several cellular processes in granulocytes, such as their differentiation, proliferation, and function (Johnnidis et al. 2008). In *miR-223*-deficient mice, increased proliferation of myeloid progenitor cells and enhanced neutrophil responsiveness to pathogenic stimuli resulting in exaggerated inflammation has been observed. The transcription factor MEF2C was shown to be directly targeted by *miR-223,* and was deemed critical for the increase in granulocyte proliferation, but not the effect of *miR-223* on granulocyte function. This suggests that *miR-223* regulates additional targets that modulate neutrophil functions, such as respiratory burst and *in vitro* pathogen killing activity. Therefore, *miR-223* expression in granulocytes may act to fine-tune both the generation and function of granulocytes during activation, preventing excessive expansion and activation and aberrant inflammatory responses that could result in immune pathology.

A role for *miR-155* in the expansion of granulocytes has also been identified during inflammation and in certain cases of acute myeloid leukemia (O'Connell et al. 2009). Overexpression of *miR-155* in mice results in enhanced myeloproliferation, potentially as a major consequence of directly targeting and downregulating SH2 domain-containing inositol 5′-phosphatase 1 (SHIP1), a negative regulator of the phosphoinositide 3-kinase pathway. Additionally, lipopolysaccharide (LPS) can induce *miR-9* in human neutrophils in a MYD88-dependent manner, to potentially prevent the inhibition of NF-κB signaling through a negative feedback loop (Bazzoni et al. 2009). *miR-9* modulates the levels of NFKB1 gene to prevent the accumulation of its active form p50, which at higher concentrations may form homodimers, that have an inhibitory effect on NF-κB signaling (Ghosh and Hayden 2008). These observations suggest a tightly regulated expression of miRNAs, allowing for initiation of granulocytic/neutrophil responses while simultaneously preventing excessive inflammation.

B. Monocytes and Macrophages

Monocytes and macrophages are myeloid cell types essential for innate immune response as they phagocytose microbial pathogens and produce inflammatory cytokines. Analysis of the miRNA expression profile of human monocytes revealed increased expression of *miR-132*, *miR-146a/b*, and *miR-155* in response to LPS stimulation (Taganov et al. 2006). Some miRNAs have been observed to have a regulatory effect, acting like a rheostat to modulate inflammatory responses. One such miRNA, *miR-146a/b*, may directly down-regulate the expression of TNF receptor-associated factor 6 (TRAF6) and IL-1 receptor-associated kinase 1 (IRAK1), which are adapter molecules acting downstream of the TLR signaling pathway (Taganov et al. 2006), suggesting *miR-146a/b* may act as a negative feedback regulator of TLR and cytokine receptor signaling. *miR-146a* may also inhibit type I interferons (IFNs) produced in response to VSV infection by regulating the RIG-I pathway by targeting TRAF6, IRAK1, and IRAK2 in macrophages (Hou et al. 2009). Additionally, *miR-146a*-deficient murine macrophages were found to be hyperresponsive to LPS, indicating that it may function to down-regulate TLR signaling and cytokine production (Boldin et al. 2011). Interestingly, in this study, *miR-146a*-deficient mice have massive myeloproliferation and develop autoimmune disorders and tumors with age. This is most likely is due to excessive TLR signaling and the subsequent chronic production of inflammatory cytokines. This raises the possibility that dysregulation of *miR-146a* may contribute to autoimmune pathology in humans. It is important to note that the distal region of the chromosome 5q containing the *miR-146a* gene (5q33) has been suggested to harbor the susceptibility loci for Crohn's disease (Rioux et al. 2001), rheumatoid arthritis (Tokuhiro et al. 2003), and psoriasis (Friberg et al. 2006).

Other miRNAs, such as *miR-132,and miR-147*, may also down-regulate TLR and cytokine signaling in monocytes. *miR-132* was highly up-regulated in monocytes after viral infection, resulting in suppression of the p300 transcriptional co-activator (Lagos et al. 2010). This suppression has a negative effect on the expression of interferon-inducible genes and IFN-β, leading to increased viral replication. Similarly, a TLR4-induced miRNA, *miR-147*, dampened inflammatory responses in murine macrophages (Liu et al. 2009).

On the other hand, *miR-155* has been shown to play a critical role in mediating inflammatory responses, with exposure to poly (I:C) and IFN-β significantly up-regulating its expression in murine macrophages (O'Connell et al. 2007). In response to LPS stimulation, *miR-155* was found to increase TNF-α production and directly regulate expression of IκB kinase ε (IKKε), Fas-associated death domain (FADD), and receptor (TNFR superfamily)-interacting serine threonine kinase I (Tili et al. 2007). SHIP1 and SMAD2 were also directly regulated by *miR-155*, driving the inflammatory response in macrophages (O'Connell et al. 2009; Louafi et al. 2010). Interestingly, *miR-155* was found to be up-regulated in synovial membrane and synovial fluid macrophages from patients with rheumatoid arthritis and was associated with lower expression of SHIP1 (Kurowska-Stolarska et al. 2011). *miR-155* deficiency in mice reduces inflammatory cytokines and protects from collagen-induced arthritis (Kurowska-Stolarska et al. 2011). Similar to its role in human neutrophils, regulation of NFKB1 by induced *miR-9* in monocytes also prevents negative feedback control of NF-κB-dependent responses after activation by either TLRs or by cytokines TNF-α and IL-1β, by preventing the accumulation and formation of p50 homodimers (Bazzoni et al. 2009). Additionally, expression of *miR-125b* in macrophages promotes greater activation, augments responsiveness to IFN-γ, and enhances function by targeting IRF4 (Chaudhuri et al. 2011). It also promotes the ability of the macrophage to present antigen and induce T-cell activation, thus amplifying both innate and adaptive immune responses. From the previous discussion, it is apparent that some miRNAs, such as *miR-146*, *miR-132*, and *miR-147*, play an anti-inflammatory role in macrophages, while others, such as *miR-155*, *miR-9*, and *miR-125b*, play a proinflammatory role and may promote inflammation mediated by macrophages (Table 3.1).

C. Dendritic Cells

DCs are derived from common myeloid progenitors in the bone marrow, as well as from blood monocytes. These cells are quite specialized in their ability to uptake and present antigen, and are thus crucial for the induction of adaptive T- and B-cell immune responses. Knocking out Dicer in specific immune subsets resulted in impaired generation of NK, B, and T cells (Cobb et al. 2005; Muljo et al. 2005; Koralov et al. 2008; Zhou et al. 2008; Bezman et al. 2010; Zhang and Bevan 2010), hinting toward miRNAs playing key roles in development. However, a DC-specific deletion of Dicer (CD11c-specific) only resulted in slightly reduced miRNA levels in lymph node and splenic DCs, with no apparent phenotype (Kuipers et al. 2010). Since most DCs are short lived and the miRNA half-life is ~5 days, it is likely that miRNAs inherited by DC from their progenitor cells may be sufficient to allow for normal differentiation and DC function. In contrast, in CD11c cell-specific Dicer knockout mice, skin-resident Langerhans cells, which have a half-life of several weeks in mice, were reduced and were inefficient in priming CD4+ T cells.

One study involving microarray profiling of human monocyte-derived dendritic cells (MDDC) has revealed differential expression of about 20 miRNAs during a course of 5 days, when differentiated *in vitro* (Hashimi et al. 2009). Importantly, *miR-21* and *miR-34a* were found to be critical during MDDC differentiation by down-regulating *Wnt1* and *Jagged-1* genes. Inhibition of these miRNAs or addition of exogenous WNT1 and JAG-1 (ligands for WNT-1 and JAG-1, respectively) stalled MDDC differentiation. Another study extended this analysis further by activating DCs with LPS to induce maturation and then comparing their miRNA profile to that of immature DCs and undifferentiated monocytes (C. Lu et al. 2011), where expression of *miR-221* and *miR-155* was found to be associated with DC maturation. In this study, immature DCs were found to highly up-regulate expression of *miR-221* upon differentiation, which correlated with decreased P27KIP1 protein levels and DC apoptosis. Additionally, they also deemed *miR-155* to be important, with its up-regulation being characteristic of DC maturation. This was confirmed by another group using *miR-155*-deficient mice, where it was found to be required for optimal functioning of mature DCs (Rodriguez et al. 2007). In this study, *miR-155*-deficient DCs could not efficiently activate T cells, suggesting impaired costimulatory signaling or antigen presentation. Pathogen uptake by DCs may also be controlled by *miR-155* by directly repressing PU.1 expression (Martinez-Nunez et al. 2009), which in turn inhibits DC-SIGN expression. Targeting of Suppresor of cytokine signaling 1 (SOCS1) and c-Fos by *miR-155* may allow these cells to mediate inflammatory responses (Dunand-Sauthier et al. 2011; C. Lu et al. 2011). Interestingly, *miR-155* may also form part of a negative feedback loop at later stages of DC activation. It does this by directly targeting TAB2 (Ceppi et al. 2009), decreasing inflammatory cytokine production, as well by inhibiting KPC1 (C. Lu et al. 2011), causing increased apoptosis. Thus, at earlier stages of DC activation, *miR-155* may promote inflammation, while at later stages, it serves to dampen the inflammatory response by DC. Therefore, *miR-155* may play a dual role in both initiating and dampening DC responses, depending on its stage of activation.

In some cases, miRNAs like *miR-21, miR-142-3p,* and *miR-146a* have been shown to have regulatory roles in DCs preventing excessive inflammation. DCs from mice lacking

TABLE 3.1. Key miRNAs with Multifunctional Roles in Innate and Adaptive Immunity

miRNA	Cell Type(s)	Biological Effect	Target(s)
miR-9	Neutrophils Monocytes	Prevents feedback control of NF-κB dependent responses.	NFKB1
miR-21	CD4+ T cells	Increased IL-2 and IFN-γ production.	BLIMP1/PRDM1, BCL-6
	Monocytes	Promotes differentiation of MDDCs.	WNT1
	Dendritic cells	Inhibits IL-12 production and promotes a Th2 responses.	IL-12P35
	CD4+ T cells	Down-regulates IFN-γ production.	Not determined
	T_REG cells	Positive regulator of FOXP3 expression.	
miR-125b	Macrophages	Regulates macrophage activation and IFN-γ responsiveness.	IRF4
	B cells (GC)	Inhibits premature plasma cell differentiation.	BLIMP1/PRDM1
miR-146a	Monocytes Macrophages Langerhans cells	Negative feedback regulator of TLR and cytokine receptor signaling modulating inflammation.	TRAF6, IRAK1, IRAK2
	CD4+ and CD8+ T cells	Modulates IL-2 production and reduces AICD	FADD
	T_REG cells	Maintains suppressor functions of T_REG cells by regulating IFN-γ signaling.	STAT1
miR-150	NK cells	Required for maturation of NK cells.	C-MYB
	NKT cells	Negatively regulates iNKT cell generation.	
	B cells	Inhibits pro- to pre-B-cell transition and B1 cell expansion.	
miR-155	Granulocytes	Increases proliferation.	SHIP1
	Macrophages	Required for increased inflammatory responses in response to TLR and cytokine stimulation.	SHIP1, SMAD2, IKKε, FADD, RIPK1
	Dendritic cells	Regulates DC pathogen uptake and maturation. Mediates negative feedback loop at later stages of DC activation.	PU.1, SOCS1, C-FOS, TAB2, KPC1
	NK cells	IFN-γ production	SHIP1
	B cells	Plasma B cell formation and production of high affinity antibodies.	SHIP1, AID, PU.1
	CD4+ T cells	Th1 and Th17 differentiation.	C-MAF, IFNγRα
	T_REG cells	Promotes T_REG development.	SOCS1
miR-181a/b	NK cells	Promotes NK cell development and IFN-γ production	NLK
	B cells	Regulates class switch recombination which restricts B cell lymphomagenesis, increased expansion of B lineage cells	AID
	CD4+ and CD8+ T cells	Regulates strength of TCR signaling and sensitivity. May negatively regulate function of CD8+ T cells.	SHP-2, PTPN22, DUSP5, DUSP6
miR-223	Granulocytes	Regulates expansion and activation to prevent aberrant inflammatory responses.	MEF2C
	Memory B cells	B-cell differentiation.	LMO2

miR-21 produced more IL-12 after LPS stimulation, serving as a regulator of Th1 versus Th2 responses (T.X. Lu et al. 2011). Similarly, a recent study has shown *miR-142-3p* to directly down-regulate IL-6 gene expression in DCs, where endogenous IL-6 was found to protect mice from endotoxin-induced mortality (Sun et al. 2011). Therefore, blocking *miR-142-3p* may serve as a therapeutic option during endotoxin-induced septic shock. Additionally, myeloid-derived DC subsets have been shown to differentially express *miR-146a*, with constitutively higher expression in Langerhans cells (Jurkin et al. 2010). The transcription factor PU.1 can induce *miR-146a* in response to TGF-β1 to dampen TLR signaling. Consequentially, *miR-146a*-deficient BMDMs proliferate rapidly in response to M-CSF, with its dysregulation potentially playing a contributory role in certain autoimmune and tumorogenic disorders (Boldin et al. 2011). Thus, miRNAs play an essential role in regulating DC differentiation and function, allowing DCs to be key players in regulating innate and adaptive immunity.

D. NK and NKT Cells

NK cells constitute an important component of the innate immune system, playing an essential role in early host defense by either producing cytokines or by mediating cytotoxicity. Deletion of Dicer resulted in defects in NK cell activation, function, and survival during mouse cytomegalovirus infection (Bezman et al. 2010). Genome-wide miRNA analysis of resting and type I IFN-activated human NK cells found *miR-378* and *miR-30e* levels to be markedly decreased with activation (Wang et al. 2012). This study also revealed that these two miRNAs were negative regulators of human NK cell cytotoxicity by directly targeting granzyme B and perforin, respectively. Another study has implicated *miR-155* as being critical for IFN-γ production in NK cells (Trotta et al. 2012). Some miRNAs, such as *miR-150* and *miR-181*, may also promote NK cell development (Bezman et al. 2011; Cichocki et al. 2011). Mice with a targeted deletion of *miR-150* have reduced NK cell numbers with impaired maturation, potentially as a consequence of increased c-Myb and decreased Ly49A expression during NK cell development (Bezman et al. 2011). Interestingly, in this study, overexpression of *miR-150* lead to an increased accumulation of mature NK cells. It was postulated that c-Myb might cause a reduction of the Ly49A subset of NK cells by potentially binding and inhibiting the Klra1promoter (required for Ly49A expression), which contains a c-Myb-binding sequence in its upstream region (Tanamachi et al. 2004). It may also target and inhibit GATA3 (Maurice et al. 2007), with its deficiency resulting in a reduction of the Ly49A subset of NK cells (Samson et al. 2003). In the case of *miR-181*, it functions by targeting nemo-like kinase (NLK), a negative regulator of Notch signaling (Cichocki et al. 2011). Notch signaling is thought to play an important role during the early stages of NK cell development; inducing the generation of CD7+ early lymphoid precursors and NK cell commitment (Bachanova et al. 2009).

Natural killer T cells (NKT cells) are a distinct class of innate cells that express receptors of the NK cell lineage, such as NK1.1 and Ly49 family markers, along with a T-cell receptor (TCR). Most NKT cells express an invariant TCRα chain that pairs with a limited repertoire of TCRβ chains and are often referred to as iNKT cells. Interestingly, iNKT cells have a distinct miRNA expression profile relative to other T cells, with 13 miRNAs being down-regulated and with only *miR-21* levels being up-regulated (Fedeli et al. 2009). Deletion of Dicer also resulted in a significant reduction in thymic and peripheral iNKT cells, due to impaired homeostasis (Zhou et al. 2009). While *miR-150* was found to be critical for NK cell development, its overexpression resulted in a significant reduction of iNKT cells (Bezman et al. 2011).

E. Mast Cells

Mast cells are tissue resident cells that are prevalent near exposed surfaces, such as the lung airways, gastrointestinal tract, and skin. Mast cells usually mediate multiple functions during parasitic infections and allergic responses, which may be regulated by miRNAs. Two such miRNAs, *miR-221* and *miR-222*, were shown to be highly up-regulated in mast cells upon activation (Mayoral et al. 2009), with this up-regulation reducing proliferation and regulating cell cycle checkpoints through partial inhibition of *p27kip1* gene. On the other hand, *miR-126* enhances the proliferation and function of mast cells, through suppression of SPRED1 (Ishizaki et al. 2011). SPRED1 has been shown to negatively regulate mast cell proliferation by suppressing stem cell factor and IL-3-induced ERK activation (Nonami et al. 2004). MiRNA regulation of mast cells therefore may control responses to parasites and allergic reactions, suggesting novel ways to enhance or dampen such responses.

The inhibitory and enhancing effect of miRNAs on innate immune cells, such as granulocytes, macrophages, DC, NK cells, NKT cells, and mast cells, suggests that they may be crucial in regulating innate immune responses by striking a fine balance that allows for optimal inflammation for adaptive immunity and pathogen clearance while simultaneously preventing exacerbated immune pathology.

III. ADAPTIVE IMMUNITY AND miRNAs

The adaptive immune response is characterized by the increased expansion of B and T lymphocytes, followed by effector functions, such as antibody production, cytokine production, and cytotoxicity. These effector responses mediate protective immunity to microbial infections and malignant cells. Importantly, after clearance of antigen and pathogen, these immune cells contract to leave only a small population referred to as "memory cells." These memory cells are responsible for rapidly mounting immune responses to secondary infections, thereby preventing reinfection with the same pathogen. In cases of dysregulation, however, adaptive immunity may also result in either immune responses against self-antigens causing autoimmune disorders, in exaggerated responses causing inflammatory diseases or impaired responses that lead to chronic infections. A number of studies have shown miRNAs to be critically required for distinct processes, such as B- and T-cell development, expansion, and effector functions.

A. B Lymphocytes

B lymphocytes form a major component of the adaptive immune response, resulting in the production of antibodies and long-lasting immunity. The overall importance of miRNAs was established by B cell-specific deletion of Dicer in early B cells where the development of B cells was blocked past the pro-B cell stage (Koralov et al. 2008). This was thought to be a consequence of increased BIM and PTEN expression, which were the top predicted *miR-17-92* targets. Mice with a later deletion of Dicer in B cells resulted in an overrepresentation of transitional and marginal zone B cells and a decrease in follicular B cells (Belver et al. 2010). This study also hinted that bruton tyrosine kinase (BTK) may be a major target in this phenotype, as BTK was increased. BTK may be targeted and suppressed by *miR-185*. In the absence of miRNAs, these mice showed increased production of autoantibodies that lead to the development of autoimmune disease, especially in aged females (Belver et al. 2010). Interestingly, BTK overexpression in B cells in mice also resulted in an increased production of autoantibodies, leading to the development of a systemic autoimmune disease (Kil et al. 2012). The genome-wide miRNA expression profile of different B-cell subsets revealed temporal changes in distinct miRNAs during B-cell differentiation (Malumbres et al. 2009). The germinal center (GC) lymphocytes (centroblasts) were found to have increased expression of *miR-125b*, targeting IRF4 and BLIMP1/PRDM1 expression. These factors may repress BCL6, which is required for B-cell differentiation into plasma and memory B cells. Memory B cells instead express *miR-223* that could down-regulate LIM domain only 2 (LMO2) expression. Centroblasts were found to preferentially express LMO2, though its function remains elusive. However, in diffuse large B-cell lymphoma patients, LMO2 expression has been associated with increased survival of patients (Natkunam et al. 2008).

During B-cell development, *miR-150* is constitutively expressed to inhibit expression of c-Myb, which when deleted leads to a severe block in the pro- to pre-B transition and the disappearance of B1 cells (Xiao et al. 2007). Therefore, with *miR-150* deletion, there is enhanced B1-cell expansion and humoral responses, including an increased antibody response to T cell-dependent antigens (which is usually mediated by B2 cells). Other miRNAs may also function during different periods of B-cell differentiation. *miR-181a* was found to be preferentially expressed by bone marrow cells of the B lymphoid lineage in mice, and its ectopic expression in hematopoietic progenitors resulted in a selective doubling of B lymphoid cells (Chen et al. 2004). Another related miRNA, *miR-181b*, could directly down-regulate expression of activation-induced cytidinedeaminase (AID), which typically causes mutations leading to increased antibody diversity, but sometimes can cause B-cell lymphomas (de Yebenes et al. 2008). *miR-181b* may therefore restrict the development of B-cell malignancies. Additionally, *miR-17-92* overexpression was found to cause lymphoproliferative disease and autoimmunity. This was due to suppression of BIM and PTEN, leading to increased proliferation and decreased activation-induced cell death (Xiao et al. 2008). On the flip side, the DLE2/*miR-15a/16-1* cluster may regulate B-cell proliferation by targeting G0/G1-S phase-related genes (Klein et al. 2010). In a mouse model of lupus, *miR-15a* was up-regulated in different B-cell subsets by IFN-α treatment and correlated with elevated autoantibody levels, hinting that this miRNA may play a role in autoimmune disease (Yuan et al. 2012).

One of the best-characterized miRNAs in B-cell responses is *miR-155*, the deletion of which results in reduced numbers of GC and extrafollicular B cells (Rodriguez et al. 2007; Thai et al. 2007). This is accompanied by impaired production of high affinity IgG1 antibodies. *miR-155* may regulate the expression of many target genes in B cells, which may include SHIP1 (Pedersen et al. 2009), AID (Dorsett et al. 2008), and PU.1 (Vigorito et al. 2007). Regulation of PU.1 by *miR-155* was found to be critical for the formation of plasma cells (Vigorito et al. 2007). In addition, its suppression of AID-mediated Myc-Igh translocation may allow it to act as a tumor suppressor, potentially reducing tumorogenic transformations (Dorsett et al. 2008). Importantly, *miR-155* was shown to be increased in human B-cell lymphomas (Eis et al. 2005) and was critical for driving the development of B-cell malignancy (Costinean et al. 2006; Pedersen et al. 2009). These studies have therefore revealed an important role for miRNAs in B cells, with its dysregulation being associated with increased autoimmune phenotypes but also B-cell malignancies.

B. T Lymphocytes

T lymphocytes tend to play important roles in initiating potent responses against invading pathogens. They can perform diverse functions and can cause both, the activation as well as the suppression of various facets of the immune system. Different signaling cascades are required for proper development of T cells, and these cascades are in turn regulated by various miRNAs. Different stages of T-cell development show distinct patterns of miRNA expression, with the changes in miRNA levels regulating processes, such as differentiation and activation (Neilson et al. 2007). Additionally, dynamic changes in the levels of miRNA were observed between naïve, effector, and memory CD8+ T cells, as revealed by expression profiling (Wu et al. 2007). When compared with naïve CD8+ T cells, effector CD8+ T cells were observed to have a global down-regulation of miRNAs, with their level again increasing in memory CD8+ T cells. This would hint toward most miRNAs having regulatory roles in T cells to prevent aberrant responses. In support of this is the finding that as CD4+ T cells proliferate in response to activation, they were found to express mRNAs with shorter 3′-UTRs, thus making them less prone to regulation by miRNAs (Sandberg et al. 2008). These observations would therefore suggest that miRNAs might modulate gene expression during T-cell differentiation and activation.

1. Helper CD4+ T Cells. A requirement for miRNAs was revealed by a study where T cell-specific deletion of Dicer resulted in impaired T-cell development, T helper cell differentiation, and function (Cobb et al. 2005). Mature T cells were also reduced in numbers as a consequence of reduced survival and proliferation (Muljo et al. 2005). One of the first studies investigating the role of miRNAs in T cells was done with *miR-181a*, which was found to be important for T-cell development (Li et al. 2007; Ebert et al. 2009). *miR-181a* could regulate the strength of transduced TCR signaling by suppressing multiple phosphatases (Li et al. 2007), with its absence allowing for T-cell reactivity against self-antigens (Ebert et al. 2009). Thymic *miR-181a* expression increases TCR sensitivity, which is important for deletion of moderate-affinity interacting self-reactive thymocytes, thus allowing for maintenance of central tolerance (Ebert et al. 2009). Another miRNA that has also been implicated in T-cell development is the *miR-17-92* cluster (Xiao et al. 2008). Its expression during the early stages of thymopoiesis may regulate T-cell survival during development by impairing expression of BIM and PTEN.

In addition to miRNAs role in development, several studies have also indicated a critical need for them in T-cell differentiation, activation, and function, especially in CD4+ T cells. This was quite evident with absence of *miR-155*, where CD4+ T cells increasingly produced cytokines like IL-4, IL-5, and IL-10, as they were more prone toward Th2 differentiation (Rodriguez et al. 2007). In the absence of *miR-155*, the expression of C-MAF, a repressor of IL-4 production, is increased and this promotes the development of Th2 type cells (Rodriguez et al. 2007). Additionally, elevated *miR-155* levels may make CD4+ T cells less sensitive to inhibition by regulatory T cells (T_{REG}) (Stahl et al. 2009). It also suppresses IFNγRα expression on CD4+ T cells (Banerjee et al. 2010), potentially making them more resistant to the antiproliferative effects of IFN-γ. *miR-155* expression was specifically found to drive the formation of Th1 and Th17 cells, and was found to be critical for inducing experimental autoimmune encephalomyelitis (EAE) and colitis (O'Connell et al. 2010; Oertli et al. 2011). Similarly, *miR-301a* and *miR-326* were found to promote Th17 differentiation by targeting PIAS3 and ETS-1, respectively, exacerbating the severity of EAE (Du et al. 2009; Mycko et al. 2012) a mouse model of multiple sclerosis. These miRNAs may therefore play a contributory role in the pathogenesis of multiple sclerosis. On the other hand, *miR-146a* may have a modulatory role, impairing activator protein 1 (AP-1) and IL-2 production. Its expression also protected T cells from activation-induced cell death through direct targeting of FADD expression (Curtale et al. 2010).

miR-17-92 was found to regulate T follicular helper cell differentiation, which are essential for maintenance of germinal center B cells and resultant antibody responses (Yu et al. 2009). Overexpression of this miRNA in T cells also enhances their ability to proliferate and survive. Moreover, mice overexpressing *miR-17-92* developed lymphoproliferative disease implicating its dysregulation as a potential factor in inducing autoimmune diseases (Xiao et al. 2008). Signaling through costimulatory molecules are required for T-cell activation, and may involve regulating expression of miRNAs, such as *miR-214* (Jindra et al. 2010). CD28 costimulation can increase *miR-214* levels, allowing for enhanced proliferation through suppression of PTEN. Cytokine production can also be regulated by miRNA in CD4+ T cells, such as *miR-29*, which can directly target IFN-γ, thereby suppressing immune responses to intracellular pathogens (Ma et al. 2011). On the other hand, *miR-9* has been shown to enhance IL-2 production by inhibiting BLIMP1 in activated CD4+ T cells (Thiele et al. 2012). Interestingly, IL-2 can in turn induce *miR-182* expression, promoting clonal expansion of activated CD4+ T cells (Stittrich et al. 2010). Therefore miRNAs could be said to regulate every facet of T-cell activation, expansion, and effector function.

2. Regulatory T Cells. miRNAs have also been implicated in the development and function of regulatory T cells (T_{REG}). Selective ablation of Dicer in Forkhead box P3 (FOXP3)+ T_{REG} cells leads to down-regulation of transcription factor FOXP3 as well as loss of its suppressor functions *in vivo* (Zhou et al. 2008). T_{REG} cells have elevated levels of *miR-146a*, which were critical for their suppressor function (Lu et al. 2010). Deficiency of *miR-146a* in T_{REG}s resulted in the breakdown of immunological tolerance as a consequence of increased IFN-γ signaling through augmented expression and activation of its target gene, STAT1. FOXP3 has been shown to upregulate expression of *miR-155* in T_{REG} cells and is critical for T_{REG} development (Kohlhaas et al. 2009), as well as for maintaining its proliferative activity by regulating SOCS1 (Lu et al. 2009). Although the development of T_{REG} is inhibited with *miR-155* deficiency (Kohlhaas et al. 2009; Lu et al. 2009), T_{REG} suppressor function does not seem to depend on *miR-155* (Kohlhaas et al. 2009). FOXP3 can also down-regulate *miR-142-3p* in CD4+CD25+ T_{REG} cells, which targets adenylyl cyclase 9 mRNA (Huang et al. 2009). This allows for increased levels of cAMP to build up in T_{REG} cells, with which it can exert its suppressive function. Other miRNAs, such as *miR-21* and *miR-31*, may also function to modulate FOXP3 expression in T cells (Rouas et al. 2009). The earlier observations indicate that miRNAs have essential roles in mediating T_{REG}-cell responses.

3. CD8+ T Cells. CD8+ T cells play an essential role in controlling various intracellular infections and malignancies by producing inflammatory cytokines and mediating cytotoxicity. As mentioned earlier, microarray analysis has revealed that miRNAs levels to be differentially expressed within naïve, effector, and memory CD8+ T cells (Wu et al. 2007). Interestingly, fate determination to central memory CD8+ T cells is also characterized by a balancing effect of a number of miRNAs, including *miR-150*, *miR-155*, and the *let-7* family (Almanza et al. 2010). Deletion of Dicer in CD8+ T cells impairs its migration and survival, resulting in reduced CD8+ effector T cell expansion (Zhang and Bevan 2010). miRNA cluster *miR-17-92* regulates both, effector and memory CD8+ T-cell differentiation during acute viral infections, as shown by a recent study (Wu et al. 2012). While *miR-17-92* deficiency was found to impair effector CD8+ T cell proliferation, its overexpression lead to a gradual loss of memory cells due to skewing of differentiation toward short-lived terminal effector cells. In addition, earlier-mentioned miRNAs, such as *miR-146a*, *miR-29*, and *miR-181a*, may also play important roles in regulating different aspects of CD8+ T-cell responses (Curtale et al. 2010; Ma et al. 2011; Schietinger et al. 2012). Expression levels of *miR-146a* is up-regulated on human effector and central memory CD8+ T cells when compared with naïve CD8+ T cells (Curtale et al. 2010). IFN-γ production in response to microbial infections is also regulated by *miR-29* in CD8+ T cells (Ma et al. 2011). Interestingly, while *miR-181a* expression in CD4+ T cells has been associated with TCR signaling (Li et al. 2007; Ebert et al. 2009), a recent study may suggest it to play a role in negatively regulating cell function in CD8+ T cells (Schietinger et al. 2012). In this study, which uses a murine model of CD8+ T-cell tolerance to self antigens, *miR-181a* was highly expressed in tolerant and retolerized CD8+ T cells, but low in memory and rescued CD8+ T cells, with high expression correlating with a tolerance-specific gene signature. Our own studies have shown that *miR-155* is essential for *in vivo* primary and memory CD8+ T-cell responses regulating immunity against viruses and bacteria (Gracias et al. 2013). miRNAs may therefore be crucial in controlling the development, differentiation and functionality of different T-cell subsets modulating immune responses to infections and tumors and regulating autoimmunity.

IV. CONCLUSIONS AND FUTURE DIRECTIONS

In recent years, a great deal of progress has been made in the understanding the role of miRNAs in the immune system. Several miRNAs have been identified in different immune cells regulating processes, such as development, differentiation, and function. It is quite interesting to note that a single miRNA, such as *miR-146a* and *miR-155*, may have a global effect on immune responses through its effect on dendritic cells, macrophages, and B and T cells (Table 3.1). Additionally, such miRNAs may have different functions in different lineages based on the regulation and expression level of distinct target genes (Table 3.1). By regulating multiple target genes in a given cell type, a miRNA may function to fine-tune responses, thus functioning more like a rheostat (Li et al. 2007; Linsley et al. 2007; Tsang et al. 2007). Based on the observations made from several studies done so far, it is now apparent that dysregulation of miRNA may play an important role in various immunological disorders. This could be true, especially for inflammatory diseases, such as sepsis, as well as autoimmune disorders, such as multiple sclerosis, lupus, and colitis. Further studies examining the expression profile of miRNA in autoimmune and inflammatory diseases may identify miRNA that affect multiple signaling pathways in these diseases. Targeting such miRNA would have a higher likelihood of inhibiting the inflammatory process than targeting individual signaling pathways. Research into exploring such therapeutic applications of miRNAs is already progressing. The development of antagomirs (Krutzfeldt et al. 2005) and locked nucleic acids oligonucleotides (Orom et al. 2006) have made it possible to use these therapeutics as a means of inhibiting inappropriate immune responses. Additionally, modulating the expression or inhibition of specific miRNAs in cell types like B and T cells could boost immunity, specifically during chronic infections (like HIV or HCV) or against cancers.

In conclusion, elucidating the role of miRNAs in immunity will further our understanding on the regulation of immune responses to pathogens and tumors, the maintenance of host immunity and the balance between an optimal response and exacerbated responses that lead to immunopathological disorders. Future studies are needed to help delineate the precise role of specific miRNA in immunity and disease and explore their potential therapeutic application in the treatment of immune-mediated diseases.

REFERENCES

Almanza, G., A. Fernandez, S. Volinia, X. Cortez-Gonzalez, C.M. Croce, and M. Zanetti. 2010. Selected microRNAs define cell fate determination of murine central memory CD8 T cells. *PLoS ONE* **5**:e11243.

Bachanova, V., V. McCullar, T. Lenvik, R. Wangen, K.A. Peterson, D.E. Ankarlo, A. Panoskaltsis-Mortari, J.E. Wagner, and J.S. Miller. 2009. Activated notch supports development of cytokine producing NK cells which are hyporesponsive and fail to acquire NK cell effector functions. *Biol Blood Marrow Transplant* **15**:183–194.

Banerjee, A., F. Schambach, C.S. DeJong, S.M. Hammond, and S.L. Reiner. 2010. Micro-RNA-155 inhibits IFN-gamma signaling in CD4+ T cells. *Eur J Immunol* **40**:225–231.

Bazzoni, F., M. Rossato, M. Fabbri, D. Gaudiosi, M. Mirolo, L. Mori, N. Tamassia, A. Mantovani, M.A. Cassatella, and M. Locati. 2009. Induction and regulatory function of miR-9 in human monocytes and neutrophils exposed to proinflammatory signals. *Proc Natl Acad Sci U S A* **106**: 5282–5287.

Belver, L., V.G. de Yebenes, and A.R. Ramiro. 2010. MicroRNAs prevent the generation of autoreactive antibodies. *Immunity* **33**:713–722.

Bezman, N.A., E. Cedars, D.F. Steiner, R. Blelloch, D.G. Hesslein, and L.L. Lanier. 2010. Distinct requirements of microRNAs in NK cell activation, survival, and function. *J Immunol* **185**:3835–3846.

Bezman, N.A., T. Chakraborty, T. Bender, and L.L. Lanier. 2011. miR-150 regulates the development of NK and iNKT cells. *J Exp Med* **208**:2717–2731.

Boldin, M.P., K.D. Taganov, D.S. Rao, L. Yang, J.L. Zhao, M. Kalwani, Y. Garcia-Flores, M. Luong, A. Devrekanli, J. Xu, G. Sun, J. Tay, et al. 2011. miR-146a is a significant brake on autoimmunity, myeloproliferation, and cancer in mice. *J Exp Med* **208**:1189–1201.

Ceppi, M., P.M. Pereira, I. Dunand-Sauthier, E. Barras, W. Reith, M.A. Santos, and P. Pierre. 2009. MicroRNA-155 modulates the interleukin-1 signaling pathway in activated human monocyte-derived dendritic cells. *Proc Natl Acad Sci U S A* **106**:2735–2740.

Chaudhuri, A.A., A.Y. So, N. Sinha, W.S. Gibson, K.D. Taganov, R.M. O'Connell, and D. Baltimore. 2011. MicroRNA-125b potentiates macrophage activation. *J Immunol* **187**:5062–5068.

Chen, C.Z., L. Li, H.F. Lodish, and D.P. Bartel. 2004. MicroRNAs modulate hematopoietic lineage differentiation. *Science* **303**:83–86.

Cichocki, F., M. Felices, V. McCullar, S.R. Presnell, A. Al-Attar, C.T. Lutz, and J.S. Miller. 2011. Cutting edge: microRNA-181 promotes human NK cell development by regulating Notch signaling. *J Immunol* **187**:6171–6175.

Cobb, B.S., T.B. Nesterova, E. Thompson, A. Hertweck, E. O'Connor, J. Godwin, C.B. Wilson, N. Brockdorff, A.G. Fisher, S.T. Smale, and M. Merkenschlager. 2005. T cell lineage choice and differentiation in the absence of the RNase III enzyme Dicer. *J Exp Med* **201**:1367–1373.

Costinean, S., N. Zanesi, Y. Pekarsky, E. Tili, S. Volinia, N. Heerema, and C.M. Croce. 2006. Pre-B cell proliferation and lymphoblastic leukemia/high-grade lymphoma in E(mu)-miR155 transgenic mice. *Proc Natl Acad Sci U S A* **103**:7024–7029.

Croce, C.M. 2009. Causes and consequences of microRNA dysregulation in cancer. *Nat Rev Genet* **10**:704–714.

Curtale, G., F. Citarella, C. Carissimi, M. Goldoni, N. Carucci, V. Fulci, D. Franceschini, F. Meloni, V. Barnaba, and G. Macino. 2010. An emerging player in the adaptive immune response: microRNA-146a is a modulator of IL-2 expression and activation-induced cell death in T lymphocytes. *Blood* **115**:265–273.

Dorsett, Y., K.M. McBride, M. Jankovic, A. Gazumyan, T.H. Thai, D.F. Robbiani, M. Di Virgilio, B.R. San-Martin, G. Heidkamp, T.A. Schwickert, T. Eisenreich, K. Rajewsky, et al. 2008.

MicroRNA-155 suppresses activation-induced cytidine deaminase-mediated Myc-Igh translocation. *Immunity* **28**:630–638.

Du, C., C. Liu, J. Kang, G. Zhao, Z. Ye, S. Huang, Z. Li, Z. Wu, and G. Pei. 2009. MicroRNA *miR-326* regulates TH-17 differentiation and is associated with the pathogenesis of multiple sclerosis. *Nat Immunol* **10**:1252–1259.

Dunand-Sauthier, I., M.L. Santiago-Raber, L. Capponi, C.E. Vejnar, O. Schaad, M. Irla, Q. Seguin-Estevez, P. Descombes, E.M. Zdobnov, H. Acha-Orbea, and W. Reith. 2011. Silencing of c-Fos expression by microRNA-155 is critical for dendritic cell maturation and function. *Blood* **117**:4490–4500.

Ebert, P.J., S. Jiang, J. Xie, Q.J. Li, and M.M. Davis. 2009. An endogenous positively selecting peptide enhances mature T cell responses and becomes an autoantigen in the absence of microRNA miR-181a. *Nat Immunol* **10**:1162–1169.

Eis, P.S., W. Tam, L. Sun, A. Chadburn, Z. Li, M.F. Gomez, E. Lund, and J.E. Dahlberg. 2005. Accumulation of *miR-155* and BIC RNA in human B cell lymphomas. *Proc Natl Acad Sci U S A* **102**:3627–3632.

Fedeli, M., A. Napolitano, M.P. Wong, A. Marcais, C. de Lalla, F. Colucci, M. Merkenschlager, P. Dellabona, and G. Casorati. 2009. Dicer-dependent microRNA pathway controls invariant NKT cell development. *J Immunol* **183**:2506–2512.

Friberg, C., K. Bjorck, S. Nilsson, A. Inerot, J. Wahlstrom, and L. Samuelsson. 2006. Analysis of chromosome 5q31-32 and psoriasis: confirmation of a susceptibility locus but no association with SNPs within SLC22A4 and SLC22A5. *J Invest Dermatol* **126**:998–1002.

Ghosh, S. and M.S. Hayden. 2008. New regulators of NF-kappaB in inflammation. *Nat Rev Immunol* **8**:837–848.

Gracias, D.T., E. Stelekati, J.L. Hope, A.C. Boesteanu, J.A. Fraietta, T. Doering, J. Norton, Y.M. Mueller, E.J. Wherry, M. Turner, and P.D. Katsikis. 2013. MicroRNA-155 controls CD8+ T cell responses by regulating interferon signaling. *Nat Immunol* **14**:593–602.

Hashimi, S.T., J.A. Fulcher, M.H. Chang, L. Gov, S. Wang, and B. Lee. 2009. MicroRNA profiling identifies *miR-34a* and *miR-21* and their target genes JAG1 and WNT1 in the coordinate regulation of dendritic cell differentiation. *Blood* **114**:404–414.

Hou, J., P. Wang, L. Lin, X. Liu, F. Ma, H. An, Z. Wang, and X. Cao. 2009. MicroRNA-146a feedback inhibits RIG-I-dependent Type I IFN production in macrophages by targeting TRAF6, IRAK1, and IRAK2. *J Immunol* **183**:2150–2158.

Huang, B., J. Zhao, Z. Lei, S. Shen, D. Li, G.X. Shen, G.M. Zhang, and Z.H. Feng. 2009. *miR-142-3p* restricts cAMP production in CD4+CD25- T cells and CD4+CD25+ TREG cells by targeting AC9 mRNA. *EMBO Rep* **10**:180–185.

Ishizaki, T., T. Tamiya, K. Taniguchi, R. Morita, R. Kato, F. Okamoto, K. Saeki, M. Nomura, Y. Nojima, and A. Yoshimura. 2011. *miR126* positively regulates mast cell proliferation and cytokine production through suppressing Spred1. *Genes Cells* **16**:803–814.

Jindra, P.T., J. Bagley, J.G. Godwin, and J. Iacomini. 2010. Costimulation-dependent expression of microRNA-214 increases the ability of T cells to proliferate by targeting Pten. *J Immunol* **185**:990–997.

Johnnidis, J.B., M.H. Harris, R.T. Wheeler, S. Stehling-Sun, M.H. Lam, O. Kirak, T.R. Brummelkamp, M.D. Fleming, and F.D. Camargo. 2008. Regulation of progenitor cell proliferation and granulocyte function by microRNA-223. *Nature* **451**:1125–1129.

Jurkin, J., Y.M. Schichl, R. Koeffel, T. Bauer, S. Richter, S. Konradi, B. Gesslbauer, and H. Strobl. 2010. *miR-146a* is differentially expressed by myeloid dendritic cell subsets and desensitizes cells to TLR2-dependent activation. *J Immunol* **184**:4955–4965.

Kil, L.P., M.J. de Bruijn, M. van Nimwegen, O.B. Corneth, J.P. van Hamburg, G.M. Dingjan, F. Thaiss, G.F. Rimmelzwaan, D. Elewaut, D. Delsing, P.F. van Loo, and R.W. Hendriks. 2012.

REFERENCES

Btk levels set the threshold for B-cell activation and negative selection of autoreactive B cells in mice. *Blood* **119**:3744–3756.

Klein, U., M. Lia, M. Crespo, R. Siegel, Q. Shen, T. Mo, A. Ambesi-Impiombato, A. Califano, A. Migliazza, G. Bhagat, and R. Dalla-Favera. 2010. The *DLEU2/miR-15a/16-1* cluster controls B cell proliferation and its deletion leads to chronic lymphocytic leukemia. *Cancer Cell* **17**:28–40.

Kohlhaas, S., O.A. Garden, C. Scudamore, M. Turner, K. Okkenhaug, and E. Vigorito. 2009. Cutting edge: the Foxp3 target *miR-155* contributes to the development of regulatory T cells. *J Immunol* **182**:2578–2582.

Koralov, S.B., S.A. Muljo, G.R. Galler, A. Krek, T. Chakraborty, C. Kanellopoulou, K. Jensen, B.S. Cobb, M. Merkenschlager, N. Rajewsky, and K. Rajewsky. 2008. Dicer ablation affects antibody diversity and cell survival in the B lymphocyte lineage. *Cell* **132**:860–874.

Krutzfeldt, J., N. Rajewsky, R. Braich, K.G. Rajeev, T. Tuschl, M. Manoharan, and M. Stoffel. 2005. Silencing of microRNAs in vivo with "antagomirs." *Nature* **438**:685–689.

Kuipers, H., F.M. Schnorfeil, H.J. Fehling, H. Bartels, and T. Brocker. 2010. Dicer-dependent microRNAs control maturation, function, and maintenance of Langerhans cells in vivo. *J Immunol* **185**:400–409.

Kurowska-Stolarska, M., S. Alivernini, L.E. Ballantine, D.L. Asquith, N.L. Millar, D.S. Gilchrist, J. Reilly, M. Ierna, A.R. Fraser, B. Stolarski, C. McSharry, A.J. Hueber, et al. 2011. MicroRNA-155 as a proinflammatory regulator in clinical and experimental arthritis. *Proc Natl Acad Sci U S A* **108**:11193–11198.

Lagos, D., G. Pollara, S. Henderson, F. Gratrix, M. Fabani, R.S. Milne, F. Gotch, and C. Boshoff. 2010. *miR-132* regulates antiviral innate immunity through suppression of the p300 transcriptional co-activator. *Nat Cell Biol* **12**:513–519.

Li, Q.J., J. Chau, P.J. Ebert, G. Sylvester, H. Min, G. Liu, R. Braich, M. Manoharan, J. Soutschek, P. Skare, L.O. Klein, M.M. Davis, et al. 2007. *miR-181a* is an intrinsic modulator of T cell sensitivity and selection. *Cell* **129**:147–161.

Linsley, P.S., J. Schelter, J. Burchard, M. Kibukawa, M.M. Martin, S.R. Bartz, J.M. Johnson, J.M. Cummins, C.K. Raymond, H. Dai, N. Chau, M. Cleary, et al. 2007. Transcripts targeted by the microRNA-16 family cooperatively regulate cell cycle progression. *Mol Cell Biol* **27**:2240–2252.

Liu, G., A. Friggeri, Y. Yang, Y.J. Park, Y. Tsuruta, and E. Abraham. 2009. *miR-147*, a microRNA that is induced upon Toll-like receptor stimulation, regulates murine macrophage inflammatory responses. *Proc Natl Acad Sci U S A* **106**:15819–15824.

Lodish, H.F., B. Zhou, G. Liu, and C.Z. Chen. 2008. Micromanagement of the immune system by microRNAs. *Nat Rev Immunol* **8**:120–130.

Louafi, F., R.T. Martinez-Nunez, and T. Sanchez-Elsner. 2010. MicroRNA-155 targets SMAD2 and modulates the response of macrophages to transforming growth factor-{beta}. *J Biol Chem* **285**:41328–41336.

Lu, C., X. Huang, X. Zhang, K. Roensch, Q. Cao, K.I. Nakayama, B.R. Blazar, Y. Zeng, and X. Zhou. 2011. *miR-221* and *miR-155* regulate human dendritic cell development, apoptosis, and IL-12 production through targeting of p27kip1, KPC1, and SOCS-1. *Blood* **117**:4293–4303.

Lu, L.F., T.H. Thai, D.P. Calado, A. Chaudhry, M. Kubo, K. Tanaka, G.B. Loeb, H. Lee, A. Yoshimura, K. Rajewsky, and A.Y. Rudensky. 2009. Foxp3-dependent microRNA155 confers competitive fitness to regulatory T cells by targeting SOCS1 protein. *Immunity* **30**:80–91.

Lu, L.F., M.P. Boldin, A. Chaudhry, L.L. Lin, K.D. Taganov, T. Hanada, A. Yoshimura, D. Baltimore, and A.Y. Rudensky. 2010. Function of miR-146a in controlling Treg cell-mediated regulation of Th1 responses. *Cell* **142**:914–929.

Lu, T.X., J. Hartner, E.J. Lim, V. Fabry, M.K. Mingler, E.T. Cole, S.H. Orkin, B.J. Aronow, and M.E. Rothenberg. 2011. MicroRNA-21 limits in vivo immune response-mediated activation of the IL-12/IFN-gamma pathway, Th1 polarization, and the severity of delayed-type hypersensitivity. *J Immunol* **187**:3362–3373.

Ma, F., S. Xu, X. Liu, Q. Zhang, X. Xu, M. Liu, M. Hua, N. Li, H. Yao, and X. Cao. 2011. The microRNA miR-29 controls innate and adaptive immune responses to intracellular bacterial infection by targeting interferon-gamma. *Nat Immunol* **12**:861–869.

Malumbres, R., K.A. Sarosiek, E. Cubedo, J.W. Ruiz, X. Jiang, R.D. Gascoyne, R. Tibshirani, and I.S. Lossos. 2009. Differentiation stage-specific expression of microRNAs in B lymphocytes and diffuse large B-cell lymphomas. *Blood* **113**:3754–3764.

Martinez-Nunez, R.T., F. Louafi, P.S. Friedmann, and T. Sanchez-Elsner. 2009. MicroRNA-155 modulates the pathogen binding ability of dendritic cells (DCs) by down-regulation of DC-specific intercellular adhesion molecule-3 grabbing non-integrin (DC-SIGN). *J Biol Chem* **284**:16334–16342.

Maurice, D., J. Hooper, G. Lang, and K. Weston. 2007. c-Myb regulates lineage choice in developing thymocytes via its target gene Gata3. *EMBO J* **26**:3629–3640.

Mayoral, R.J., M.E. Pipkin, M. Pachkov, E. van Nimwegen, A. Rao, and S. Monticelli. 2009. MicroRNA-221-222 regulate the cell cycle in mast cells. *J Immunol* **182**:433–445.

Muljo, S.A., K.M. Ansel, C. Kanellopoulou, D.M. Livingston, A. Rao, and K. Rajewsky. 2005. Aberrant T cell differentiation in the absence of Dicer. *J Exp Med* **202**:261–269.

Mycko, M.P., M. Cichalewska, A. Machlanska, H. Cwiklinska, M. Mariasiewicz, and K.W. Selmaj. 2012. MicroRNA-301a regulation of a T-helper 17 immune response controls autoimmune demyelination. *Proc Natl Acad Sci U S A* **109**:E1248–E1257.

Natkunam, Y., P. Farinha, E.D. Hsi, C.P. Hans, R. Tibshirani, L.H. Sehn, J.M. Connors, D. Gratzinger, M. Rosado, S. Zhao, B. Pohlman, N. Wongchaowart, et al. 2008. LMO2 protein expression predicts survival in patients with diffuse large B-cell lymphoma treated with anthracycline-based chemotherapy with and without rituximab. *J Clin Oncol* **26**:447–454.

Neilson, J.R., G.X. Zheng, C.B. Burge, and P.A. Sharp. 2007. Dynamic regulation of miRNA expression in ordered stages of cellular development. *Genes Dev* **21**:578–589.

Nonami, A., R. Kato, K. Taniguchi, D. Yoshiga, T. Taketomi, S. Fukuyama, M. Harada, A. Sasaki, and A. Yoshimura. 2004. Spred-1 negatively regulates interleukin-3-mediated ERK/mitogen-activated protein (MAP) kinase activation in hematopoietic cells. *J Biol Chem* **279**: 52543–52551.

O'Connell, R.M., K.D. Taganov, M.P. Boldin, G. Cheng, and D. Baltimore. 2007. MicroRNA-155 is induced during the macrophage inflammatory response. *Proc Natl Acad Sci U S A* **104**: 1604–1609.

O'Connell, R.M., A.A. Chaudhuri, D.S. Rao, and D. Baltimore. 2009. Inositol phosphatase SHIP1 is a primary target of miR-155. *Proc Natl Acad Sci U S A* **106**:7113–7118.

O'Connell, R.M., D. Kahn, W.S. Gibson, J.L. Round, R.L. Scholz, A.A. Chaudhuri, M.E. Kahn, D.S. Rao, and D. Baltimore. 2010. MicroRNA-155 promotes autoimmune inflammation by enhancing inflammatory T cell development. *Immunity* **33**:607–619.

Oertli, M., D.B. Engler, E. Kohler, M. Koch, T.F. Meyer, and A. Muller. 2011. MicroRNA-155 is essential for the T cell-mediated control of Helicobacter pylori infection and for the induction of chronic gastritis and colitis. *J Immunol* **187**:3578–3586.

Orom, U.A., S. Kauppinen, and A.H. Lund. 2006. LNA-modified oligonucleotides mediate specific inhibition of microRNA function. *Gene* **372**:137–141.

Pedersen, I.M., D. Otero, E. Kao, A.V. Miletic, C. Hother, E. Ralfkiaer, R.C. Rickert, K. Gronbaek, and M. David. 2009. Onco-miR-155 targets SHIP1 to promote TNFalpha-dependent growth of B cell lymphomas. *EMBO Mol Med* **1**:288–295.

Poy, M.N., M. Spranger, and M. Stoffel. 2007. microRNAs and the regulation of glucose and lipid metabolism. *Diabetes Obes Metab* **9 Suppl 2**:67–73.

REFERENCES

Rioux, J.D., M.J. Daly, M.S. Silverberg, K. Lindblad, H. Steinhart, Z. Cohen, T. Delmonte, K. Kocher, K. Miller, S. Guschwan, E.J. Kulbokas, S. O'Leary, et al. 2001. Genetic variation in the 5q31 cytokine gene cluster confers susceptibility to Crohn disease. *Nat Genet* **29**:223–228.

Rodriguez, A., E. Vigorito, S. Clare, M.V. Warren, P. Couttet, D.R. Soond, S. van Dongen, R.J. Grocock, P.P. Das, E.A. Miska, D. Vetrie, K. Okkenhaug, et al. 2007. Requirement of bic/microRNA-155 for normal immune function. *Science* **316**:608–611.

Rouas, R., H. Fayyad-Kazan, N. El Zein, P. Lewalle, F. Rothe, A. Simion, H. Akl, M. Mourtada, M. El Rifai, A. Burny, P. Romero, P. Martiat, et al. 2009. Human natural Treg microRNA signature: role of microRNA-31 and microRNA-21 in FOXP3 expression. *Eur J Immunol* **39**:1608–1618.

Samson, S.I., O. Richard, M. Tavian, T. Ranson, C.A. Vosshenrich, F. Colucci, J. Buer, F. Grosveld, I. Godin, and J.P. Di Santo. 2003. GATA-3 promotes maturation, IFN-gamma production, and liver-specific homing of NK cells. *Immunity* **19**:701–711.

Sandberg, R., J.R. Neilson, A. Sarma, P.A. Sharp, and C.B. Burge. 2008. Proliferating cells express mRNAs with shortened 3' untranslated regions and fewer microRNA target sites. *Science* **320**:1643–1647.

Schietinger, A., J.J. Delrow, R.S. Basom, J.N. Blattman, and P.D. Greenberg. 2012. Rescued tolerant CD8 T cells are preprogrammed to reestablish the tolerant state. *Science* **335**:723–727.

Stahl, H.F., T. Fauti, N. Ullrich, T. Bopp, J. Kubach, W. Rust, P. Labhart, V. Alexiadis, C. Becker, M. Hafner, A. Weith, M.C. Lenter, et al. 2009. *miR-155* inhibition sensitizes CD4+ Th cells for TREG mediated suppression. *PLoS ONE* **4**:e7158.

Stefani, G. and F.J. Slack. 2008. Small non-coding RNAs in animal development. *Nat Rev Mol Cell Biol* **9**:219–230.

Stittrich, A.B., C. Haftmann, E. Sgouroudis, A.A. Kuhl, A.N. Hegazy, I. Panse, R. Riedel, M. Flossdorf, J. Dong, F. Fuhrmann, G.A. Heinz, Z. Fang, et al. 2010. The microRNA *miR-182* is induced by IL-2 and promotes clonal expansion of activated helper T lymphocytes. *Nat Immunol* **11**:1057–1062.

Sun, Y., S. Varambally, C.A. Maher, Q. Cao, P. Chockley, T. Toubai, C. Malter, E. Nieves, I. Tawara, Y. Wang, P.A. Ward, A. Chinnaiyan, et al. 2011. Targeting of microRNA-142-3p in dendritic cells regulates endotoxin-induced mortality. *Blood* **117**:6172–6183.

Taganov, K.D., M.P. Boldin, K.J. Chang, and D. Baltimore. 2006. NF-kappaB-dependent induction of microRNA miR-146, an inhibitor targeted to signaling proteins of innate immune responses. *Proc Natl Acad Sci U S A* **103**:12481–12486.

Tanamachi, D.M., D.C. Moniot, D. Cado, S.D. Liu, J.K. Hsia, and D.H. Raulet. 2004. Genomic Ly49A transgenes: basis of variegated Ly49A gene expression and identification of a critical regulatory element. *J Immunol* **172**:1074–1082.

Thai, T.H., D.P. Calado, S. Casola, K.M. Ansel, C. Xiao, Y. Xue, A. Murphy, D. Frendewey, D. Valenzuela, J.L. Kutok, M. Schmidt-Supprian, N. Rajewsky, et al. 2007. Regulation of the germinal center response by *microRNA-155*. *Science* **316**:604–608.

Thiele, S., J. Wittmann, H.M. Jack, and A. Pahl. 2012. miR-9 enhances IL-2 production in activated human CD4(+) T cells by repressing Blimp-1. *Eur J Immunol* **42**:2100–2108.

Tili, E., J.J. Michaille, A. Cimino, S. Costinean, C.D. Dumitru, B. Adair, M. Fabbri, H. Alder, C.G. Liu, G.A. Calin, and C.M. Croce. 2007. Modulation of *miR-155* and *miR-125b* levels following lipopolysaccharide/TNF-alpha stimulation and their possible roles in regulating the response to endotoxin shock. *J Immunol* **179**:5082–5089.

Tokuhiro, S., R. Yamada, X. Chang, A. Suzuki, Y. Kochi, T. Sawada, M. Suzuki, M. Nagasaki, M. Ohtsuki, M. Ono, H. Furukawa, M. Nagashima, et al. 2003. An intronic SNP in a RUNX1 binding site of SLC22A4, encoding an organic cation transporter, is associated with rheumatoid arthritis. *Nat Genet* **35**:341–348.

Trotta, R., L. Chen, D. Ciarlariello, S. Josyula, C. Mao, S. Costinean, L. Yu, J.P. Butchar, S. Tridandapani, C.M. Croce, and M.A. Caligiuri. 2012. *miR-155* regulates IFN-gamma production in natural killer cells. *Blood* **119**:3478–3485.

Tsang, J., J. Zhu, and A. van Oudenaarden. 2007. MicroRNA-mediated feedback and feedforward loops are recurrent network motifs in mammals. *Mol Cell* **26**:753–767.

Vigorito, E., K.L. Perks, C. Abreu-Goodger, S. Bunting, Z. Xiang, S. Kohlhaas, P.P. Das, E.A. Miska, A. Rodriguez, A. Bradley, K.G. Smith, C. Rada, et al. 2007. microRNA-155 regulates the generation of immunoglobulin class-switched plasma cells. *Immunity* **27**:847–859.

Wang, P., Y. Gu, Q. Zhang, Y. Han, J. Hou, L. Lin, C. Wu, Y. Bao, X. Su, M. Jiang, Q. Wang, N. Li, et al. 2012. Identification of resting and type I IFN-activated human NK cell miRNomes reveals microRNA-378 and microRNA-30e as negative regulators of NK cell cytotoxicity. *J Immunol* **189**:211–221.

Wu, H., J.R. Neilson, P. Kumar, M. Manocha, P. Shankar, P.A. Sharp, and N. Manjunath. 2007. miRNA profiling of naive, effector and memory CD8 T cells. *PLoS ONE* **2**:e1020.

Wu, T., A. Wieland, K. Araki, C.W. Davis, L. Ye, J.S. Hale, and R. Ahmed. 2012. Temporal expression of microRNA cluster *miR-17-92* regulates effector and memory CD8+ T-cell differentiation. *Proc Natl Acad Sci U S A* **109**:9965–9970.

Xiao, C., D.P. Calado, G. Galler, T.H. Thai, H.C. Patterson, J. Wang, N. Rajewsky, T.P. Bender, and K. Rajewsky. 2007. MiR-150 controls B cell differentiation by targeting the transcription factor c-Myb. *Cell* **131**:146–159.

Xiao, C., L. Srinivasan, D.P. Calado, H.C. Patterson, B. Zhang, J. Wang, J.M. Henderson, J.L. Kutok, and K. Rajewsky. 2008. Lymphoproliferative disease and autoimmunity in mice with increased *miR-17-92* expression in lymphocytes. *Nat Immunol* **9**:405–414.

de Yebenes, V.G., L. Belver, D.G. Pisano, S. Gonzalez, A. Villasante, C. Croce, L. He, and A.R. Ramiro. 2008. *miR-181b* negatively regulates activation-induced cytidine deaminase in B cells. *J Exp Med* **205**:2199–2206.

Yu, D., S. Rao, L.M. Tsai, S.K. Lee, Y. He, E.L. Sutcliffe, M. Srivastava, M. Linterman, L. Zheng, N. Simpson, J.I. Ellyard, I.A. Parish, et al. 2009. The transcriptional repressor Bcl-6 directs T follicular helper cell lineage commitment. *Immunity* **31**:457–468.

Yuan, Y., S. Kasar, C. Underbayev, D. Vollenweider, E. Salerno, S.V. Kotenko, and E. Raveche. 2012. Role of microRNA-15a in autoantibody production in interferon-augmented murine model of lupus. *Mol Immunol* **52**:61–70.

Zhang, N. and M.J. Bevan. 2010. Dicer controls CD8+ T-cell activation, migration, and survival. *Proc Natl Acad Sci U S A* **107**:21629–21634.

Zhou, L., K.H. Seo, H.Z. He, R. Pacholczyk, D.M. Meng, C.G. Li, J. Xu, J.X. She, Z. Dong, and Q.S. Mi. 2009. Tie2cre-induced inactivation of the miRNA-processing enzyme Dicer disrupts invariant NKT cell development. *Proc Natl Acad Sci U S A* **106**:10266–10271.

Zhou, X., L.T. Jeker, B.T. Fife, S. Zhu, M.S. Anderson, M.T. McManus, and J.A. Bluestone. 2008. Selective miRNA disruption in T reg cells leads to uncontrolled autoimmunity. *J Exp Med* **205**:1983–1991.

4

REGULATION OF SENESCENCE BY MicroRNAs

Ioannis Grammatikakis and Myriam Gorospe

Laboratory of Genetics, NIA-IRP, NIH, Baltimore, MD, USA

I.	Introduction: Senescence	60
	A. Senescence of Cultured Cells	60
	B. Senescence *In Vivo*	61
	C. Detection of Senescent Cells	63
II.	Posttranscriptional Regulation of Senescence by miRNAs	64
	A. miRNA Regulation of the p53/p21/Arf Pathway	64
	B. miRNA Regulation of the p16/Rb Pathway	65
	C. miRNA Regulation of SASP	65
	D. miRNAs Modulating Transcriptional and Posttranscriptional Regulators of Senescence	65
	E. Senescence Regulation of miRNA Biogenesis	66
III.	Impact of SA-miRNAs on *In Vivo* Senescence	67
IV.	Concluding Remarks and Future Perspectives	68
	Acknowledgments	69
	References	69

ABBREVIATIONS

Cdk	cyclin-dependent kinase
DDR	DNA damage response
ECM	extracellular matrix
HDFs	human diploid fibroblasts

MicroRNAs in Medicine, First Edition. Edited by Charles H. Lawrie.
© 2014 John Wiley & Sons, Inc. Published 2014 by John Wiley & Sons, Inc.

miRNAs	microRNAs
RB	retinoblastoma
RBP	RNA-binding protein
RISC	RNA-induced silencing complex
SASP	senescence-associated secretory phenotype
UTR	untranslated region

I. INTRODUCTION: SENESCENCE

A. Senescence of Cultured Cells

As first described by Hayflick 50 years ago, somatic cells explanted from human tissue divide for a finite number of times, whereupon they cease proliferation but remain viable and metabolically active in a state known as senescence (Hayflick and Moorhead 1961; Hayflick 1965). Cells can become senescent via two related but distinct mechanisms: replicative senescence, which develops through genetic programming, and premature senescence, which arises in response to damaging conditions (Kuilman et al. 2010).

1. Replicative Senescence. This form of senescence is achieved when the protective ends of chromosomes (the telomeres) of proliferating cells become critically shortened, as DNA polymerase cannot completely replicate the lagging strands. Accordingly, this form of senescence can be prevented by expression of the enzyme telomerase (Bodnar et al. 1998; Vaziri and Benchimol 1998). After reaching a critically short size, the structure of the telomere is lost, triggering a DNA damage response (DDR) that is associated with the appearance of nuclear foci enriched in DDR proteins (e.g., histone H2AX phosphorylated on serine 139 [γ-H2AX], the Nijmegen breakage syndrome 1 protein [NBS1], mediator of DNA-damage checkpoint 1 [MDC1], and the p53-binding protein 1 [53BP1]), and with the sequential activation of upstream kinases ataxia telangiectasia mutated (ATM) and ATM-related (ATR), downstream effectors (the checkpoint kinases CHK1 and CHK2), and cell cycle inhibitors that include cell division cycle (CDC)25, and the transcription factor and tumor suppressor p53. Importantly, p53, its transcriptional target p21$^{Cip1/Waf1}$ (p21), a broad inhibitor of cyclin-dependent kinases (cdks), and the p53 regulator Arf, are key components of a major senescence regulatory mechanism, the *p53/p21/Arf* senescence pathway (Figure 4.1). Replicative senescence can also be triggered via activation of the retinoblastoma (RB) tumor suppressor protein through its upstream inducer, the cdk inhibitor p16^{INK4a} (p16), and related proteins p15, p18, and p19 (Ben-Porath and Weinberg 2005; Campisi 2005); these proteins comprise the *p16/RB* senescence pathway, a second major mechanism to elicit senescence (Figure 4.1). In addition, replicative senescence can also be triggered through gene expression programs central to the senescent phenotype, some related to the *p53/p21/Arf* and *p16/RB* pathways, some independent of these pathways (Figure 4.1).

2. Premature Senescence. Also known as "stress-induced senescence," this form of senescence is triggered by harmful stimuli without apparent loss of telomere function. For example, cells explanted from a tissue can become senescent rapidly due to challenging culture conditions, such as their maintenance in supraphysiological oxygen that causes oxidative damage or their culture in the absence of a proper extracellular matrix (ECM), neighboring cells, or a suitable culture medium (Wright and Shay 2002). Premature

INTRODUCTION: SENESCENCE

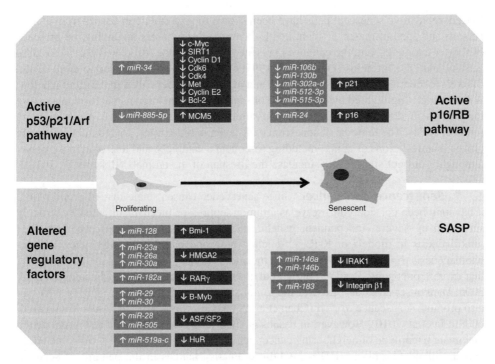

Figure 4.1. miRNAs that influence senescence-relevant pathways. Proliferating cells progress to senescence by acquiring several prominent phenotypes: an active p53/p21/Arf pathway (green), an active pRB/p16 pathway (blue), a senescence-associated secretory phenotype (SASP, yellow), and changes in senescence-associated gene regulatory factors (pink). Dark green, the main senescence-associated (SA)miRNAs identified to date as mediating each senescence phenotype. Dark blue, the principal target proteins influenced by SA-miRNAs. The major downstream consequences of activation of p53/p21/Arf and p16/RB and gene regulatory factors are the inhibition of the cell division machinery and the implementation of senescence-associated gene expression patterns. The main consequences of SASP is the secretion of factors that cause inflammation and compromise the integrity of the ECM. See color insert.

senescence can also be induced in untransformed cells by the loss of tumor suppressors, such as phosphatase and tensin homologue (PTEN), neurofibromatosis (NF)1, or von Hippel-Lindau (pVHL) proteins (Chen et al. 2005; Courtois-Cox et al. 2006; Young et al. 2008), as well as by oncoproteins, such as Ras^{V12} and the v-Raf murine sarcoma viral oncogene homologue B1 $(Braf)^{E600}$ (Serrano et al. 1997; Ben-Porath and Weinberg 2005). These senescence-triggering mechanisms typically cannot be rescued by ectopic restoration of telomerase function (Wei and Sedivy 1999).

B. Senescence *In Vivo*

Initially, there was concern that senescence of cultured cells could simply be an artifact of non-physiological maintenance in culture. However, this early skepticism has largely disappeared; evidence gathered over the past decade that senescence occurs *in vivo* and underlies several physiological and pathological processes, as discussed later in the text.

Indeed, replicative senescent cells have been shown to accumulate in tissues from elderly persons and aged primates, as identified by a number of markers, including the presence of shorter telomeres, and the enhanced presence of senescence-associated β-galactosidase activity and DDR proteins (Dimri et al. 1995; Cristofalo et al. 2004; Herbig et al. 2006). Markers to identify senescent cells *in vivo* and in culture are described in the section below. A recent study highlighted the importance of senescence at the organismal level in a mouse model, where drug-induced caspase activation specifically eliminated p16-expressing senescent cells. The removal of senescent cells, even in older mice, was found to attenuate classic manifestations of aging, including sarcopenia, cataracts, and loss of adipose tissue, although it did not significantly increase the lifespan of the animals (Baker et al. 2011).

1. Senescence and Cancer. Since senescence causes stable loss of proliferation, it has long been considered to serve as a tumor suppressor mechanism, a theory strongly supported by several independent genetic studies. For example, senescence repressed tumorigenesis in models of K-Ras^{V12}-elicited lung and pancreas malignancies, in lymphomagenesis triggered by N-Ras, and in melanoma arising from benign melanocytic nevi that express oncogenic Braf^{E600} (Braig et al. 2005; Collado et al. 2005; Michaloglou et al. 2005; Dankort et al. 2009). Loss of the tumor suppressors PTEN, VHL, RB, or NF1 can also promote senescence *in vivo* (Chen et al. 2005; Courtois-Cox et al. 2006; Young et al. 2008; Lin et al. 2010). However, in tissues *in vivo*, senescent cells could also promote the malignant phenotype of neighboring cancer cells. In this regard, senescent cells, including human diploid fibroblasts (HDFs), exhibit dramatic changes in the patterns of secreted proteins, a phenomenon named senescence-associated secretory phenotype (SASP) (Coppé et al. 2008; Rodier et al. 2009). This phenotype includes secretion of numerous chemokines and cytokines, notably interleukin (IL)-6, IL-8, IL-1α, granulocyte-macrophage colony stimulating factor (GM-CSF), the growth-regulated oncogene α (GROa), monocyte chemotactic protein (MCP)-2, MCP-3, matrix metalloprotease (MMP)-1, MMP-3, and many insulin-like growth factor (IGF)-binding proteins (Freund et al. 2010). Co-culture of untransformed or precancer cells or with senescent fibroblasts or with their conditioned medium was shown to enhance oncogenesis (Krtolica et al. 2001; Parrinello et al. 2003; Yang et al. 2006). Nonetheless, the influence of SASP upon cancer cells is complex, since SASP components promoted cancer cell migration by degrading the ECM, were pro- or antioncogenic, depending on the tumor stage, and were capable of promoting the clearance of tumor cells by immune cells (Xue et al. 2007; Acosta et al. 2008; Kuilman et al. 2008; Massagué 2008).

2. Senescence and Other Pathologies. Besides cancer, senescence has been implicated in several other diseases. They include pathologies in tissues in which age-related diseases develop (e.g., atherosclerosis), in renewable tissues in which senescent cells accumulate in an age-related manner (e.g., the stroma, the epithelium of different organs, and the hematopoietic system), and in hyperproliferative lesions, such as nevi (Dimri et al. 1995; Choi et al. 2000; Krishnamurthy et al. 2004; Jeyapalan et al. 2007; Krizhanovsky et al. 2008). A protective role was identified for hepatic stellate cells, as their senescence following liver damage increased ECM deposition on fibrotic scars and preserved liver function (Krizhanovsky et al. 2008). However, in many other instances, senescent cells appeared to have detrimental consequences. Since SASP elevates chemokine and cytokine production (Freund et al. 2010), the accumulation of senescent cells during aging could contribute to the chronic proinflammatory phenotype seen in the elderly. This chronic proinflammatory state is believed to exacerbate age-related pathologies, such as diabetes, cancer, neurodegeneration, and cardiovascular disease (Freund et al. 2010), and is also believed to promote immunosenescence, an age-related decline of the adaptive immune system (McElhaney and Effros 2009). Senescent brain immune cells (microglia) also accumulate with aging, and this process is linked to neurodegenerative pathologies, such as Parkinson's and Alzheimer's diseases (Luo et al. 2010). In addition, senescent endothelial cells in atherosclerotic lesions were proposed to contribute to impaired vascular function in elderly patients (Foreman and Tang 2003). In chronic obstructive pulmonary disease, the improvement of symptoms seen in p21-deficient mice was suggested as evidence that senescence contributed to this pathology (Yao et al. 2008), although some controversy remains in this regard. In diabetes, both p53 and p21 were proposed to enhance endothelial progenitor cell senescence, impairing vascularization in diabetes, while senescence of skeletal muscle cells and muscle cell precursors (satellite cells) were linked to age-related frailty and sarcopenia (Navarro et al. 2001; Rosso et al. 2006).

In summary, the past few years have firmly established that *in vivo* senescence underlies a growing number of physiological processes and pathologies. Thus, there is escalating interest in understanding the molecular mechanisms that govern the senescent phenotype.

C. Detection of Senescent Cells

Senescent cells can be identified by a number of markers. In reality, since there is not a single definitive marker of senescence in culture or *in vivo*, several markers must be assessed together in order to identify senescent cells.

1. Changes in Cellular Morphology, Organization, and Enzymatic Activity.
Depending on the cell type and senescence trigger, senescent cells often become flat and enlarged. In the cytoplasm, senescent cells can accumulate vacuoles and autophagosomes, while in the nucleus, they show altered chromatin structure and senescence-associated heterochromatic foci (SAHF). The lysosomal senescence-associated β-galactosidase (SA-βgal) activity increases in senescent cells and is a popular marker of senescence (Dimri et al. 1995); however, other cellular states can also produce high SA-βgal-positive cells, and there is no evidence that SA-βgal activity influences senescence. Thus, it is important to assess SA-βgal activity in conjunction with other markers.

2. Markers of Cell Damage and Tumor Suppressor Networks.
Given the key roles of p53/p21/Arf and p16/RB pathways in triggering senescence, several mediators of these pathways are used as senescence markers. Hypophosphorylated (active) RB, high levels of cdk inhibitors (p16, p15, and p21, which block RB phosphorylation), Arf (an inducer of p53), and p53 itself are all informative senescence markers. These markers also help to identify cells in long-term cell cycle arrest, a hallmark of senescence. However, it is important to note that senescence growth arrest is not irreversible; contrary to earlier views, senescent cells may reenter proliferation via inactivation of the p16 and p53 pathways (Beauséjour et al. 2003; Dirac and Bernards 2003). Senescent cells are also characterized by the presence of damage from reactive oxygen species (ROS), although little is known about the factors that elevate ROS or the cellular targets affected by ROS, and by DNA damage caused by telomere attrition and other internal and external agents. Thus, DDR proteins, such as p53, γ-H2AX, NBS1, MDC1, and 53BP1, are important markers of senescence, although exposure to DNA damaging agents may not always trigger senescence.

3. SASP. Senescent cells exhibit a senescence-associated secretory phenotype (SASP), consisting in the secretion of numerous proteins, including many cytokines and chemokines (Coppé et al. 2008), with broad implications in oncogenesis and other pathologies, as discussed earlier. SASP factors IL-6, IL-8, and CXCR2 are valid indicators of senescence in some instances, but it is unknown if they are broadly useful markers.

II. POSTTRANSCRIPTIONAL REGULATION OF SENESCENCE BY miRNAs

Gene expression in cellular senescence is regulated at multiple levels. At the level of transcription, factors such as p53, activation protein (AP)-1, E2F, Id, and Ets control gene activation or repression of proteins playing a role in cell cycle, growth arrest, and senescence (Fridman and Tainsky 2008). However, at the posttranscriptional level, lesser-known senescence-regulatory factors have also emerged in recent years. RNA-binding proteins (RBPs), such as HuR, AUF1, and TTP, have the ability to bind to the 3′-untranslated region (UTR) of mRNAs encoding proteins that modulate senescence, and influence their stability and translation rate (Wang et al. 2001, 2005; Sanduja et al. 2009; Verduci et al. 2010; Pont et al. 2012). Noncoding RNA, particularly microRNAs (miRNAs, ~22-nt in length), are increasingly recognized as modulators of cellular senescence by inhibiting the translation and/or promoting the decay of target mRNAs (Fabian et al. 2010), although under specific circumstances (e.g., cellular quiescence), they can also promote translation (Vasudevan et al. 2007). miRNAs associate with different degrees of complementarity with specific mRNAs, typically at the mRNA 3′- UTR, forming a partial hybrid through the miRNA "seed" region (nucleotides 2–7). It is believed that multiple miRNAs can work in concert to repress expression of a shared target mRNA. The role of miRNAs as regulators of cellular senescence is described in the next section.

A. miRNA Regulation of the p53/p21/Arf Pathway

As indicated earlier, the p53/p21/Arf pathway is one of the major pathways affecting cellular senescence (Rodier et al. 2007). Numerous miRNAs can modulate the expression and activity of p53 through direct and indirect mechanisms (reviewed in Jones and Lal 2012). Furthermore, p53 has the ability to regulate transcription of several senescence-associated (SA)-miRNAs, prominent among them *miR-34,* a major effector of the p53 response (Bommer et al. 2007; Chang et al. 2007; Corney et al. 2007; He et al. 2007; Raver-Shapira et al. 2007; Tarasov et al. 2007; Tazawa et al. 2007; Sun et al. 2008). Over the past 5 years, *miR-34* has emerged as a central regulator of senescence and a tumor suppressor miRNA. The senescence-promoting function of *miR-34* was first shown in colon cancer cells and in HDFs (He et al. 2007; Tarasov et al. 2007). Its tumor-suppressor activity was found to be mediated through *miR-34-repressed* translation of transcripts encoding cell cycle regulators and transcription factors, including E2F, c-Myc, sirtuin 1 (SIRT1), Cdk4, Cdk6, B-cell leukemia (Bcl)-2, hepatocyte growth factor receptor (Met), and cyclins D1 and E2 (Fujita et al. 2009; Cannell and Bushell 2010; Hermeking 2010; Fabbri et al. 2011; Wang et al. 2011). Transcription of *miR-34* transcription can also be regulated by hypermethylation and silencing of the *miR-34* locus, a modification associated with oncogenic alteration in various cancer cell types (Lodygin et al. 2008). Other transcription factors that regulate *miR-34* expression include the Ets-like gene 1 (Elk1) in BRAF-triggered HDF senescence (18), and the transcription factor CCAAT enhancer-binding protein (C/EBP)α, which induces *miR-34* expression during granulocyte senescence (Pulikkan et al. 2010). By inhibiting SIRT1 translation, *miR-34* was also linked to the inhibition of endothelial cell proliferation and angiogenesis in senescent endothelial cell progenitors (Yamakuchi et al. 2008; Ito et al. 2010); in endothelial cells, *miR-217* also inhibits SIRT1 expression and promotes endothelial senescence (Menghinim et al. 2009). ΔNp63, a member of the p53 family, represses transcription of *miR-138, miR-181a*, and *miR-181b*, SA-miRNAs that target SIRT1 to induce senescence in keratinocytes and show elevated abundance with aging (Rivetti di Val Cervo et al. 2012). Finally, *miR-885-p* triggered neuroblastoma cell senescence by inhibiting expression of cdk2 and minichromosome maintenance complex component 5 (MCM5), thereby increasing p53 expression and p53-regulated genes (Afanasyeva et al. 2011) (Figure 4.1).

B. miRNA Regulation of the p16/Rb Pathway

The p16/Rb system targets cdks and hence the cell division cycle. *miR-24* was shown to inhibit translation of p16 in HDFs and human cervical cancer cells (31), but in order to enhance senescence, *miR-24* needed to function in combination with other SA-miRNAs (*miR-15b, miR-25,* and *miR-141*) (Marasa et al. 2009), perhaps because *miR-24* can downregulate factors that either promote (MYC, E2F2) or inhibit (p27 and VHL) cell cycle progression (Lal et al. 2009). *miR-34* also participates in this pathway by inhibiting cyclin D1 and Cdk6 expression to induce cell cycle arrest (Sun et al. 2008).

The p53/p21/Arf pathway intersects extensively with the p16/Rb pathway. Oncogenic RasG12V triggered senescence and growth arrest of primary cells in a p21-dependent manner; this phenotype was rescued by overexpression of *miR-106b* family miRNAs, as well as others that target p21 with similar seed sequences (*miR-130b, miR-302a-d, miR-512-3p,* and *miR-515-3p*) (Borgdorff et al. 2010). In another model of stress-induced senescence, the increased levels of p21 were attributed, at least in part, to the reduced levels of *miR-106b* (Li et al. 2009), while the lowered expression of *miR-15* led to the accumulation of Bcl-2 and increased resistance of senescent cells to apoptosis (Cimmino et al. 2005) (Figure 4.1).

C. miRNA Regulation of SASP

SASP defines a trait of senescent cells whereby they show enhanced secretion of growth factors, ECM-degrading enzymes and cytokines (Coppé et al. 2008). Given the relatively recent discovery of SASP, few regulators of SASP gene expression are known to date. Among them, *miR-146a/b* shows enhanced expression during senescence, causing reduced expression of the IL-1 receptor-associated kinase 1 (IRAK1), a key component of the IL-1 signal transduction pathway. As this leads to reduced IL-6 and IL-8 expression, two major SASP components that mediate a proinflammatory response, *miR-146a/b* is seen as a component of a negative feedback loop to avoid excessive SASP (Bhaumik et al. 2009). *miR-146a* was elevated during extended cell culture in the absence of senescence, suggesting that SASP can also be regulated in a senescence-independent manner (Bonifacio and Jarstfer 2010). Finally, in a model of H_2O_2-triggered senescence, *miR-183* was discovered as a negative regulator of integrin β1, a cell surface protein that interacts dynamically with the ECM and affects HDF senescence (Li et al. 2010) (Figure 4.1).

D. miRNAs Modulating Transcriptional and Posttranscriptional Regulators of Senescence

On another level, SA-miRNAs control senescence by influencing the expression of transcription factors and posttranscriptional regulators (Figure 4.1). One representative example is *miR-128a*, which downregulates expression of Bmi-1, a transcriptional repressor of p16, in turn inhibiting growth and triggering senescence of neuroblastoma cells (Venkataraman et al. 2010). In human umbilical cord blood-derived multipotent stem cells (hUCB-MSCs), senescence was triggered by inhibition of histone deacetylase (HDAC), an intervention that elevated expression of *miR-23a*, *miR-26a*, and *miR-30a* (Lee et al. 2011). Consequently, these SA-miRNAs decreased the levels of high-mobility group A2 (HMGA2) leading to elevated p16, p21, and p27 levels (Lee et al. 2011). During senescence of human fibroblasts (HDF) and trabecular meshwork cells (HTM), high *miR-182* levels reduced the levels of retinoic acid receptor γ (RARγ) (Li et al. 2009). Another study showed that RB activation during induced and replicative senescence leads to the upregulation of *miR-29* and *miR-30*, two SA-miRNAs that target the 3′-UTR of the transcription factor B-Myb, required for cell cycle progression and for overcoming Ras-induced senescence (Masselink et al. 2001; Martinez et al. 2011). Doxorubicin-induced senescence in p53- and p16-lacking K562 cells through upregulation of *miR-375*, which targets 14-3-3 and SP1 (Yang et al. 2012). Finally, *miR-22* can induce senescence in normal and cancer human cells and can reduce tumor growth and metastasis. The effect takes place through downregulation of SIRT1, CDK1, and SP1 (Xu et al. 2011).

miRNAs are also implicated in controlling the expression of posttranscriptional regulators. *miR-28* and *miR-505* inhibited expression of the splicing factor ASF/SF2, which coordinates endothelial cell senescence (Karni et al. 2007; Blanco and Bernabéu 2012). In mouse embryo fibroblasts (MEFs), *miR-28* and *miR-505* expression was inhibited by the pro-oncogenic transcriptional repressor LRF leukemia/lymphoma-related factor) leading to prevention of senescence (Verduci et al. 2010). In another prominent example, the RBP HuR, which targets and stabilizes mRNAs encoding cell cycle regulatory factors (e.g., cyclin B1, cyclin A, and c-fos), reduces cellular senescence (Abdelmohsen and Gorospe 2010). HuR was found to be a target of *miR-519*, whose abundance increased during HDF replicative senescence (Marasa et al. 2010) and caused inhibition of HuR translation, decreasing the proliferation of HeLa and colon carcinoma cells, and lowering tumor growth (Abdelmohsen et al. 2008, 2010; Abdelmohsen and Gorospe 2010).

E. Senescence Regulation of miRNA Biogenesis

A number of transcriptional and posttranscriptional factors control SA-miRNA biosynthesis (Newman and Hammond 2010). The mechanisms that control the expression of SA-miRNAs were recently reviewed (Abdelmohsen et al. 2012). These regulators include transcription factors that govern the synthesis of the primary transcripts from which miRNAs are synthesized (pri-miRNAs), such as p53, c-Myc, E2F1, C/EBPα, and the estrogen receptor (ER)α. Following transcription, several proteins control the nuclear processing of pri-miRNAs into precursor pre-miRNAs, for example, RNA editing enzymes such as ADAR (adenosine deaminases acting on RNA) and components of the microprocessor complex, which include the nuclear RBP DiGeorge critical region 8 (DGCR8, also known as Pasha) and the RNase Drosha (Beezhold et al. 2010), as well as the SMAD protein system, which may prevent senescence by inducing *miR-300* and *miR-21* (Bruna et al. 2007; Terao et al. 2011) to generate miRNA precursors (pre-miRNA), which are then exported to the cytoplasm by exportin 5. The exported pre-miRNAs are cleaved by the RNase Dicer, yielding duplex RNAs ~22-nucleotides long; one strand of each duplex is loaded into the miRNA-containing ribonucleoprotein complex (RISC), which contains Argonaute (Ago) proteins (Kim et al. 2009). Also in the nucleus, the DEAD-box (DDX) helicases associate with the microprocessor complex and regulate the expression numerous miRNAs; DDX5 and DDX17 in particular are likely involved in suppressing senescence (Fukuda et al. 2007; Fuller-Pace and Moore 2011). Gemin 3 (DDX20) and gemin 4 (DDX42), DEAD-box putative RNA helicases that interact with the RISC, were also linked to senescence via their functional association with telomerase (Mourelatos et al. 2002). The RBPs KSRP, hnRNP A1, SRF1, Lin28, nuclear factor (NF)90 and NF45, and nucleolin have also been linked to the processing of pri-miRNAs implicated in senescence (reviewed by Abdelmohsen et al. 2012).

Cytoplasmic regulators of miRNA processing have also been linked to senescence. Dicer binds to exported pre-miRNAs and aids in the formation of the mature miRNAs that target mRNAs for silencing. Dicer ablation triggered senescence in primary cells, accompanied by high levels of p53 and p19ARF (Mudhasani et al. 2008; Srikantan et al. 2011). TAp63 (which contains the N-terminal transactivation [TA] domain of p63) transcriptionally regulated Dicer expression and influenced miRNA levels, while TAp63 deficiency induced senescence and inhibited the metastatic potential of osteosarcomas (Kimura et al. 2010). Ago proteins bind the mature miRNA and become part of the RISC, which contains additional proteins, such as the dsRNA-binding protein TRBP (human immunodeficiency virus [HIV] transactivating response RNA [TAR]-binding protein), and PACT (protein activator of the interferon induced protein kinase). The complex Ago/miRNA/RISC is then directed to specific subcellular locations to cleave or suppress the translation of target RNAs (reviewed by Cenik and Zamore 2011).

Finally, miRNA action on senescence can be influenced by RBPs in cooperative or competitive interactions. For example, the senescence-inhibitory protein HuR can compete with *miR-122* for binding to the *CAT1* mRNA, encoding the stress-response protein cationic amino acid transporter 1, and with *miR-494* for binding to *NCL* mRNA, which encodes nucleolin, also a senescence-associated protein. At the same time, HuR can cooperate with *let-7* for binding to *MYC* mRNA and with *miR-19* to regulate *RHOB* mRNA. Gemin 3 (DDX20) and gemin 4 (DDX42), DEAD-box putative RNA helicases that interact with the RISC, were also linked to senescence via their functional association with telomerase (Mourelatos et al. 2002).

III. IMPACT OF SA-miRNAs ON *IN VIVO* SENESCENCE

A few examples are beginning to emerge of SA-miRs affecting senescence and aging *in vivo*. Primary melanoma cells showed CpG hypermethylation on the *miR-34a* promoter, which correlated with lower levels of expression of *miR-34a* (Lodygin et al. 2008). These findings supported the notions that p53 was capable of upregulating *miR-34a* production and that *miR-34a* elicited senescence *in vivo*. *miR-34* is highly expressed in adult fly brain; loss of the *miR-34* gene led to accelerated brain aging, brain degeneration, and reduced survival (Liu et al. 2012).

Dicer was studied recently in the context of senescence and aging. In HDFs, Dicer downregulation induces senescence (Srikantan et al. 2011); in mice, however, the loss of both Dicer alleles caused embryonic lethality, highlighting the vital role of miRNAs in vertebrate development (Bernstein et al. 2003). Dicer ablation in a conditional knockout mouse revealed that MEFs underwent premature senescence and growth arrest with elevated levels of p53, p21 and Arf, and higher SA-βgal activity; the phenomenon was p53- and Arf/p16-dependent, since crossing to mice lacking these proteins was capable of rescuing senescence (Mudhasani et al. 2008). In developing mouse limbs, where Dicer regulates morphogenesis, conditional ablation of Dicer also led to senescence (Harfe et al. 2005; Mudhasani et al. 2008). In adult mice, where hair follicle development is Dicer dependent, Dicer ablation caused hair loss and rough skin associated with morphological and biochemical signs of senescence (Andl et al. 2006; Mudhasani et al. 2008). The specific miRNAs that prevent premature senescence in these two developmental scenarios remain to be identified.

In human tissues, expression of the mitogen-activated protein kinase (MAPK) MKK4, an upstream activator of stress response and senescence-associated response MAPKs p38 and JNK (c-Jun N-terminal kinase), was higher in tissues from older individual donors than from younger donors (Marasa et al. 2009). This expression pattern recapitulated the elevated MKK4 expression in senescent HDFs relative to proliferating, early-passage HDFs. In culture, *miR-15b, miR-24, miR-25,* and *miR-141* were elevated in early-passage HDFs and coordinately downregulated MKK4 expression levels, while in senescent HDFs, the reduced levels of these miRNAs contributed to upregulating MKK4 (Marasa et al. 2009).

IV. CONCLUDING REMARKS AND FUTURE PERSPECTIVES

The past 10 years have uncovered much information about the triggers, markers, and consequences of cellular senescence. Accordingly, it is now solidly accepted that senescence occurs *in vivo* and that it underlies many physiological declines that characterize normal aging, and also many pathological processes, including cancer, atherosclerosis, diabetes, sarcopenia, neurodegeneration, and cardiovascular and pulmonary diseases. The past 5 years have also revealed that miRNAs are pivotal regulators of senescence. As described in this chapter, key senescence-associated processes are modulated by miRNAs. Activation of p53 increases the transcription of *miR-34a*, a repressor of numerous senescence-regulatory proteins, and p21. Several SA-miRNAs that diminish p21 and p16 levels show reduced abundance in senescent cells, allowing p21 and p16 to accumulate, inhibit cdks, and activate RB. Senescent cells also display altered levels of transcriptional and posttranscriptional regulatory proteins (e.g., transcription factors and RBPs), which influence key proteins in the p53/p21/Arf and p16/RB senescence pathways. Acting together, these groups of factors halt cell cycle progression and critically modulate senescence-associated gene expression programs. Finally, miRNAs can also promote SASP, facilitating the secretion of IL-6 and IL-8 and in turn increasing local and systemic inflammation and reducing ECM integrity.

We are rapidly learning how miRNAs influence senescence, but we still know little about the mechanisms that control the levels of SA-miRNAs. The transcription of *miR-34* by p53 and *miR-29/miR-30* transcription by RB are among the few documented examples. Future studies are needed to uncover the transcriptional and posttranscriptional mechanisms that underlie SA-miRNA expression levels, including transcription factors that control SA-pri-miRNA transcription, and the RBPs that modulate SA-pre-miRNA and SA-miRNA processing, localization, and function (Abdelmohsen et al. 2012).

It is now established that senescence is not an artifact of *in vitro* culture, occuring *in vivo* as well and having a vital role in aging. Specific removal of senescent cells in a mouse model had a great impact toward ameliorating the aging phenotype (Baker et al. 2011). The studies showing the role of *miR-34a in vivo*, as well as the phenotypes of the genetic mouse models with conditionally ablated Dicer, provide solid evidence that miRNAs regulate senescence in living organisms. The cummulative data from *in vitro* and *in vivo* studies are providing an increasingly clearer image of senescence and how it is implicated in various states of disease and in aging.

As aberrant senescence underlies numerous physiological and pathological states, many questions still need to be addressed. The identification of universal markers of senescence is a key area for future work. The development of genetic models with over-expressing or deleted miRNAs will provide vital information of SA-miRNAs within the framework of the entire animal. The elucidation of transcriptional and posttranscriptional regulators of SA-miRNAs during senescence and aging also promises to be a fruitful area of pursuit. As our understanding of these areas advances, we can increasingly envision the development of therapies to modulate senescence in order to intervene in age-related declines, as well as in cancer and other diseases.

ACKNOWLEDGMENTS

We thank Jennifer L. Martindale for critical reading of this chapter. M.G. and I.G. are supported by the National Institute on Aging—Intramural Research Program of the National Institutes of Health.

REFERENCES

Abdelmohsen, K. and M. Gorospe. 2010. Post-transcriptional regulation of cancer traits by HuR. *Wiley Interdiscip Rev RNA* **1**:214–229.

Abdelmohsen, K., M.M. Kim, S. Srikantan, E.M. Mercken, S.E. Brennan, G.M. Wilson, R. Cabo, and M. Gorospe. 2010. *miR-519* suppresses tumor growth by reducing HuR levels. *Cell Cycle* **9**:1354–1359.

Abdelmohsen, K., S. Srikantan, M.J. Kang, and M. Gorospe. 2012. Regulation of senescence by microRNA biogenesis factors. *Ageing Res Rev* **11**:491–500.

Abdelmohsen, K., S. Srikantan, Y. Kuwano, and M. Gorospe. 2008. *miR-519* reduces cell proliferation by lowering RNA-binding protein HuR levels. *Proc Natl Acad Sci U S A* **105**:20297–20302.

Acosta, J.C., A. O'Loghlen, A. Banito, M.V. Guijarro, A. Augert, S. Raguz, M. Fumagalli, M. Da Costa, C. Brown, N. Popov, Y. Takatsu, J. Melamed, F. d'Adda di Fagagna, D. Bernard, E. Hernando, and J. Gil. 2008. Chemokine signaling via the CXCR2 receptor reinforces senescence. *Cell* **133**:1006–1018.

Afanasyeva, E.A., P. Mestdagh, C. Kumps, J. Vandesompele, V. Ehemann, J. Theissen, M. Fischer, M. Zapatka, B. Brors, L. Savelyeva, V. Sagulenko, F. Speleman, M. Schwab, and F. Westermann. 2011. MicroRNA *miR-885-5p* targets CDK2 and MCM5, activates p53 and inhibits proliferation and survival. *Cell Death Differ* **18**:974–984.

Andl, T., E.P. Murchison, F. Liu, Y. Zhang, M. Yunta-Gonzalez, J.W. Tobias, C.D. Andl, J.T. Seykora, G.J. Hannon, and S.E. Millar. 2006. The miRNA-processing enzyme dicer is essential for the morphogenesis and maintenance of hair follicles. *Curr Biol* **16**:1041–1049.

Baker, D.J., T. Wijshake, T. Tchkonia, N.K. LeBrasseur, B.G. Childs, B. van de Sluis, J.L. Kirkland, and J.M. van Deursen. 2011. Clearance of p16Ink4a-positive senescent cells delays ageing-associated disorders. *Nature* **479**:232–236.

Beauséjour, C.M., A. Krtolica, F. Galimi, M. Narita, S.W. Lowe, P. Yaswen, and J. Campisi. 2003. Reversal of human cellular senescence: roles of the p53 and p16 pathways. *EMBO J* **22**:4212–4222.

Beezhold, K.J., V. Castranova, and F. Chen. 2010. Microprocessor of microRNAs: regulation and potential for therapeutic intervention. *Mol Cancer* **9**:134.

Ben-Porath, I. and R.A. Weinberg. 2005. The signals and pathways activating cellular senescence. *Int J Biochem Cell Biol* **37**:961–976.

Bernstein, E., S.Y. Kim, M.A. Carmell, E.P. Murchison, H. Alcorn, M.Z. Li, A.A. Mills, S.J. Elledge, K.V. Anderson, and G.J. Hannon. 2003. Dicer is essential for mouse development. *Nat Genet* **35**:215–217.

Bhaumik, D., G.K. Scott, S. Schokrpur, C.K. Patil, A.V. Orjalo, F. Rodier, G.J. Lithgow, and J. Campisi. 2009. MicroRNAs *miR-146a/b* negatively modulate the senescence-associated inflammatory mediators IL-6 and IL-8. *Aging* **1**:402–411.

Blanco, F.J. and C. Bernabéu. 2012. The splicing factor SRSF1 as a marker for endothelial senescence. *Front Physiol* **3**:54.

Bodnar, A.G., M. Ouellette, M. Frolkis, S.E. Holt, C.P. Chiu, G.B. Morin, C.B. Harley, J.W. Shay, S. Lichtsteiner, and W.E. Wright. 1998. Extension of life-span by introduction of telomerase into normal human cells. *Science* **279**:349–352.

Bommer, G.T., I. Gerin, Y. Feng, A.J. Kaczorowski, R. Kuick, R.E. Love, Y. Zhai, T.J. Giordano, Z.S. Qin, B.B. Moore, O.A. MacDougald, K.R. Cho, and E.R. Fearon. 2007. p53-mediated activation of miRNA34 candidate tumor-suppressor genes. *Curr Biol* **17**:1298–3107.

Bonifacio, L.N. and M.B. Jarstfer. 2010. MiRNA profile associated with replicative senescence, extended cell culture, and ectopic telomerase expression in human foreskin fibroblasts. *PLoS ONE* **5**:e12519.

Borgdorff, V., M.E. Lleonart, C.L. Bishop, D. Fessart, A.H. Bergin, M.G. Overhoff, and D.H. Beach. 2010. Multiple microRNAs rescue from Ras induced senescence by inhibiting p21(Waf1/Cip1). *Oncogene* **29**:2262–2271.

Braig, M., S. Lee, C. Loddenkemper, C. Rudolph, A.H. Peters, B. Schlegelberger, H. Stein, B. Dörken, T. Jenuwein, and C.A. Schmitt. 2005. Oncogene-induced senescence as an initial barrier in lymphoma development. *Nature* **436**:660–665.

Bruna, A., R.S. Darken, F. Rojo, A. Ocana, S. Penuelas, A. Arias, R. Paris, A. Tortosa, J. Mora, J. Baselga, and J. Seoane. 2007. High TGFβ-Smad activity confers poor prognosis in glioma patients and promotes cell proliferation depending on the methylation of the PDGF-B gene. *Cancer Cell* **11**:147–160.

Campisi, J. 2005. Senescent cells, tumor suppression, and organismal aging: good citizens, bad neighbors. *Cell* **120**:513–522.

Cannell, I.G. and M. Bushell. 2010. Regulation of Myc by *miR-34c:* a mechanism to prevent genomic instability? *Cell Cycle* **9**:2726–2730.

Cenik, E.S. and P.D. Zamore. 2011. Argonaute proteins. *Curr Biol* **2112**:R446–R449.

Chang, T.C., E.A. Wentzel, O.A. Kent, K. Ramachandran, M. Mullendore, K.H. Lee, G. Feldmann, M. Yamakuchi, M. Ferlito, C.J. Lowenstein, D.E. Arking, M.A. Beer, A. Maitra, and J.T. Mendell. 2007. Transactivation of *miR-34a* by p53 broadly influences gene expression and promotes apoptosis. *Mol Cell* **26**:745–752.

Chen, Z., L.C. Trotman, D. Shaffer, H.K. Lin, Z.A. Dotan, M. Niki, J.A. Koutcher, H.I. Scher, T. Ludwig, W. Gerald, C. Cordon-Cardo, and P.P. Pandolfi. 2005. Crucial role of p53-dependent cellular senescence in suppression of Pten-deficient tumorigenesis. *Nature* **436**:725–730.

Choi, J., I. Shendrik, M. Peacocke, D. Peehl, R. Buttyan, E.F. Ikeguchi, A.E. Katz, and M.C. Benson. 2000. Expression of senescence-associated β-galactosidase in enlarged prostates from men with benign prostatic hyperplasia. *Urology* **56**:160–166.

Cimmino, A., G.A. Calin, M. Fabbri, M.V. Iorio, M. Ferracin, M. Shimizu, S.E. Wojcik, R.I. Aqeilan, S. Zupo, M. Dono, L. Rassenti, H. Alder, S. Volinia, C.G. Liu, T.J. Kipps, M. Negrini, and C.M. Croce. 2005. *miR-15* and *miR-16* induce apoptosis by targeting BCL2. *Proc Natl Acad Sci U S A* **102**:13944–13949.

REFERENCES

Collado, M., J. Gil, A. Efeyan, C. Guerra, A.J. Schuhmacher, M. Barradas, A. Benguría, A. Zaballos, J.M. Flores, M. Barbacid, D. Beach, and M. Serrano. 2005. Tumour biology: senescence in premalignant tumours. *Nature* **436**:642.

Coppé, J.P., C.K. Patil, F. Rodier, Y. Sun, D.P. Muñoz, J. Goldstein, P.S. Nelson, P.Y. Desprez, and J. Campisi. 2008. Senescence-associated secretory phenotypes reveal cell-nonautonomous functions of oncogenic RAS and the p53 tumor suppressor. *PLoS Biol* **6**:2853–2868.

Corney, D.C., A. Flesken-Nikitin, A.K. Godwin, W. Wang, and A.Y. Nikitin. 2007. MicroRNA-34b and microRNA-34c are targets of p53 and cooperate in control of cell proliferation and adhesion-independent growth. *Cancer Res* **67**:8433–8438.

Courtois-Cox, S., S.M. Genther Williams, E.E. Reczek, B.W. Johnson, L.T. McGillicuddy, C.M. Johannessen, P.E. Hollstein, M. MacCollin, and K. Cichowski. 2006. A negative feedback signaling network underlies oncogene-induced senescence. *Cancer Cell* **10**:459–472.

Cristofalo, V.J., A. Lorenzini, R.G. Allen, C. Torres, and M. Tresini. 2004. Replicative senescence: a critical review. *Mech Ageing Dev* **125**:827–848.

Dankort, D., D.P. Curley, R.A. Cartlidge, B. Nelson, A.N. Karnezis, W.E. Damsky Jr., M.J. You, R.A. DePinho, M. McMahon, and M. Bosenberg. 2009. Braf(V600E) cooperates with Pten loss to induce metastatic melanoma. *Nat Genet* **41**:544–552.

Dimri, G.P., X. Lee, G. Basile, M. Acosta, G. Scott, C. Roskelley, E.E. Medrano, M. Linskens, I. Rubelj, O. Pereira-Smith, M. Peacocke, and J. Campisi. 1995. A biomarker that identifies senescent human cells in culture and in aging skin in vivo. *Proc Natl Acad Sci U S A* **92**:9363–9367.

Dirac, A.M.G. and R. Bernards. 2003. Reversal of senescence in mouse fibroblasts through lentiviral suppression of p53. *J Biol Chem* **278**:11731–11734.

Fabbri, M., A. Bottoni, M. Shimizu, R. Spizzo, M.S. Nicoloso, S. Rossi, E. Barbarotto, A. Cimmino, B. Adair, S.E. Wojcik, N. Valeri, F. Calore, D. Sampath, F. Fanini, I. Vannini, G. Musuraca, M. Dell'Aquila, H. Alder, R.V. Davuluri, L.Z. Rassenti, M. Negrini, T. Nakamura, D. Amadori, N.E. Kay, K.R. Rai, M.J. Keating, T.J. Kipps, G.A. Calin, and C.M. Croce. 2011. Association of a microRNA/TP53 feedback circuitry with pathogenesis and outcome of B-cell chronic lymphocytic leukemia. *JAMA* **305**:59–67.

Fabian, M.R., N. Sonenbert, and W. Filipowicz. 2010. Regulation of mRNA translation and stability by microRNAs. Regulation of mRNA translation and stability by microRNAs. *Annu Rev Biochem* **79**:351–379.

Foreman, K.E. and J. Tang. 2003. Molecular mechanisms of replicative senescence in endothelial cells. *Exp Gerontol* **38**:1251–1257.

Freund, A., A.V. Orjalo, P.Y. Desprez, and J. Campisi. 2010. Inflammatory networks during cellular senescence: causes and consequences. *Trends Mol Med* **16**:238–246.

Fridman, A.L. and M.A. Tainsky. 2008. Critical pathways in cellular senescence and immortalization revealed by gene expression profiling. *Oncogene* **27**:5975–5987.

Fujita, K., A.M. Mondal, I. Horikawa, G.H. Nguyen, K. Kumamoto, J.J. Sohn, E.D. Bowman, E.A. Mathe, A.J. Schetter, S.R. Pine, H. Ji, B. Vojtesek, J.C. Bourdon, D.P. Lane, and C.C. Harris. 2009. p53 isoforms Delta133p53 and p53beta are endogenous regulators of replicative cellular senescence. *Nat Cell Biol* **119**:1135–1142.

Fukuda, T., K. Yamagata, S. Fujiyama, T. Matsumoto, I. Koshida, K. Yoshimura, M. Mihara, M. Naitou, H. Endoh, T. Nakamura, C. Akimoto, Y. Yamamoto, T. Katagiri, C. Foulds, S. Takezawa, H. Kitagawa, K. Takeyama, B.W. O'Malley, and S. Kato. 2007. DEADbox RNA helicase subunits of the Drosha complex are required for processing of rRNA and a subset of microRNAs. *Nat Cell Biol* **9**:604–611.

Fuller-Pace, F.V. and H.C. Moore. 2011. RNA helicases p68 and p72: multifunctional proteins with important implications for cancer development. *Future Oncol* **7**:239–251.

Harfe, B.D., M.T. McManus, J.H. Mansfield, E. Hornstein, and C.J. Tabin. 2005. The RNaseIII enzyme Dicer is required for morphogenesis but not patterning of the vertebrate limb. *Proc Natl Acad Sci U S A* **102**:10898–10903.

Hayflick, L. 1965. The limited in vitro lifetime of human diploid cell strains. *Exp Cell Res* **37**:614–636.

Hayflick, L. and P.S. Moorhead. 1961. The serial cultivation of human diploid cell strains. *Exp Cell Res* **25**:585–621.

He, L., X. He, L.P. Lim, E. de Stanchina, Z. Xuan, Y. Liang, W. Xue, L. Zender, J. Magnus, D. Ridzon, A.L. Jackson, P.S. Linsley, C. Chen, S.W. Lowe, M.A. Cleary, and G.J. Hannon. 2007. A microRNA component of the p53 tumour suppressor network. *Nature* **447**:1130–1134.

Herbig, U., M. Ferreira, L. Condel, D. Carey, and J.M. Sedivy. 2006. Cellular senescence in aging primates. *Science* **311**:1257.

Hermeking, H. 2010. The *miR-34* family in cancer and apoptosis. *Cell Death Differ* **17**:193–199.

Ito, T., S. Yagi, and M. Yamakuchi. 2010. MicroRNA-34a regulation of endothelial senescence. *Biochem Biophys Res Commun* **398**:735–740.

Jeyapalan, J.C., M. Ferreira, J.M. Sedivy, and U. Herbig. 2007. Accumulation of senescent cells in mitotic tissue of aging primates. *Mech Ageing Dev* **128**:36–44.

Jones, M.F. and A. Lal. 2012. MicroRNAs, wild-type and mutant p53: more questions than answers. *RNA Biol* **9**:781–791.

Karni, R., E. de Stanchina, S.W. Lowe, R. Sinha, D. Mu, and A.R. Krainer. 2007. The gene encoding the splicing factor SF2/ASF is a proto-oncogene. *Nat Struct Mol Biol* **14**:185–193.

Kim, V.N., J. Han, and M.C. Siomi. 2009. Biogenesis of small RNAs in animals. *Nat Rev Mol Cell Biol* **10**:126–139.

Kimura, S., S. Naganuma, D. Susuki, Y. Hirono, A. Yamaguchi, S. Fujieda, K. Sano, and H. Itoh. 2010. Expression of microRNAs in squamous cell carcinoma of human head and neck and the esophagus: *miR-205* and *miR-21* are specific markers for HNSCC and ESCC. *Oncol Rep* **23**:1625–1633.

Krishnamurthy, J., C. Torrice, M.R. Ramsey, G.I. Kovalev, K. Al-Regaiey, L. Su, and N.E. Sharpless. 2004. Ink4a/Arf expression is a biomarker of aging. *J Clin Invest* **114**:1299–1307.

Krizhanovsky, V., M. Yon, R.A. Dickins, S. Hearn, J. Simon, C. Miething, H. Yee, L. Zender, and S.W. Lowe. 2008. Senescence of activated stellate cells limits liver fibrosis. *Cell* **134**:657–667.

Krtolica, A., S. Parrinello, S. Lockett, P.Y. Desprez, and J. Campisi. 2001. Senescent fibroblasts promote epithelial cell growth and tumorigenesis: a link between cancer and aging. *Proc Natl Acad Sci U S A* **98**:12072–12077.

Kuilman, T., C. Michaloglou, W.J. Mooi, and D.S. Peeper. 2010. The essence of senescence. *Genes Dev* **24**:2463–2479.

Kuilman, T., C. Michaloglou, L.C. Vredeveld, S. Douma, R. van Doorn, C.J. Desmet, L.A. Aarden, W.J. Mooi, and D.S. Peeper. 2008. Oncogene-induced senescence relayed by an interleukin-dependent inflammatory network. *Cell* **133**:1019–1031.

Lal, A., F. Navarro, C.A. Maher, L.E. Maliszewski, N. Yan, E. O'Day, D. Chowdhury, D.M. Dykxhoorn, P. Tsai, O. Hofmann, K.G. Becker, M. Gorospe, W. Hide, and J. Lieberman. 2009. *miR-24* Inhibits cell proliferation by targeting E2F2, MYC, and other cell-cycle genes via binding to "seedless" 3'UTR microRNA recognition elements. *Mol Cell* **35**:610–625.

Lee, S., J.W. Jung, S.B. Park, K. Roh, S.Y. Lee, J.H. Kim, S.K. Kang, and K.S. Kang. 2011. Histone deacetylase regulates high mobility group A2-targeting microRNAs in human cord blood-derived multipotent stem cell aging. *Cell Mol Life Sci* **68**:325–336.

Li, G., C. Luna, J. Qiu, D.L. Epstein, and P. Gonzalez. 2009. Alterations in microRNA expression in stress induced cellular senescence. *Mech Ageing Dev* **130**:731–741.

Li, G., C. Luna, J. Qiu, D.L. Epstein, and P. Gonzalez. 2010. Targeting of integrin beta1 and kinesin 2alpha by microRNA 183. *J Biol Chem* **285**:5461–5471.

Lin, H.-K., Z. Chen, G. Wang, C. Nardella, S.W. Lee, C.H. Chan, W.L. Yang, J. Wang, A. Egia, K.I. Nakayama, C. Cordon-Cardo, J. Teruya-Feldstein, and P.P. Pandolfi. 2010. Skp2 targeting suppresses tumorigenesis by Arf-p53-independent cellular senescence. *Nature* **464**:374–379.

Liu, N., M. Landreh, K. Cao, M. Abe, G.J. Hendriks, J.R. Kennerdell, Y. Zhu, L.S. Wang, and N.M. Bonini. 2012. The microRNA *miR-34* modulates ageing and neurodegeneration in Drosophila. *Nature* **482**:519–523.

Lodygin, D., V. Tarasov, A. Epanchintsev, C. Berking, T. Knyazeva, H. Körner, P. Knyazev, J. Diebold, and H. Hermeking. 2008. Inactivation of *miR-34a* by aberrant CpG methylation in multiple types of cancer. *Cell Cycle* **7**:2591–2600.

Luo, X.G., J.Q. Ding, and S.D. Chen. 2010. Microglia in the aging brain: relevance to neurodegeneration. *Mol Neurodegener* **5**:12.

Marasa, B.S., S. Srikantan, J.L. Martindale, M.M. Kim, E.K. Lee, M. Gorospe, and K. Abdelmohsen. 2010. MicroRNA profiling in human diploid fibroblasts uncovers *miR-519* role in replicative senescence. *Aging* **2**:333–343.

Marasa, B.S., S. Srikantan, K. Masuda, K. Abdelmohsen, Y. Kuwano, X. Yang, J.L. Martindale, C.W. Rinker-Schaeffer, and M. Gorospe. 2009. Increased MKK4 abundance with replicative senescence is linked to the joint reduction of multiple microRNAs. *Sci Signal* **2**:ra69.

Martinez, I., D. Cazalla, L.L. Almstead, J.A. Steitz, and D. DiMaio. 2011. *miR-29* and *miR-30* regulate B-Myb expression during cellular senescence. *Proc Natl Acad Sci U S A* **108**:522–527.

Massagué, J. 2008. TGFβ in cancer. *Cell* **134**:215–230.

Masselink, H., N. Vastenhouw, and R. Bernards. 2001. B-myb rescues ras-induced premature senescence, which requires its transactivation domain. *Cancer Lett* **171**:87–101.

McElhaney, J.E. and R.B. Effros. 2009. Immunosenescence: what does it mean to health outcomes in older adults? *Curr Opin Immunol* **21**:418–424.

Menghinim, R., V. Casagrande, M. Cardellini, E. Martelli, A. Terrinoni, F. Amati, M. Vasa-Nicotera, A. Ippoliti, G. Novelli, G. Melino, R. Lauro, and M. Federici. 2009. MicroRNA-217 modulates endothelial cell senescence via silent information regulator 1. *Circulation* **120**:1524–1532.

Michaloglou, C., L.C. Vredeveld, M.S. Soengas, C. Denoyelle, T. Kuilman, C.M. van der Horst, D.M. Majoor, J.W. Shay, W.J. Mooi, and D.S. Peeper. 2005. BRAFE600-associated senescence-like cell cycle arrest of human naevi. *Nature* **436**:720–724.

Mourelatos, Z., J. Dostie, S. Paushkin, A. Sharma, B. Charroux, L. Abel, J. Rappsilber, M. Mann, and G. Dreyfuss. 2002. miRNPs: a novel class of ribonucleoproteins containing numerous microRNAs. *Genes Dev* **16**:720–728.

Mudhasani, R., Z. Zhu, G. Hutvagner, C.M. Eischen, S. Lyle, L.L. Hall, J.B. Lawrence, A.N. Imbalzano, and S.N. Jones. 2008. Loss of miRNA biogenesis induces p19Arfp53 signaling and senescence in primary cells. *J Cell Biol* **181**:1055–1063.

Navarro, A., J.M. López-Cepero, and M.J. Sánchez del Pino. 2001. Skeletal muscle and aging. *Front Biosci* **6**:D26–D44.

Newman, M.A. and S.M. Hammond. 2010. Emerging paradigms of regulated microRNA processing. *Genes Dev* **24**:1086–1092.

Parrinello, S., E. Samper, A. Krtolica, J. Goldstein, S. Melov, and J. Campisi. 2003. Oxygen sensitivity severely limits the replicative lifespan of murine fibroblasts. *Nat Cell Biol* **5**:741–747.

Pont, A.R., N. Sadri, S.J. Hsiao, S. Smith, and R.J. Schneider. 2012. mRNA decay factor AUF1 maintains normal aging, telomere maintenance, and suppression of senescence by activation of telomerase transcription. *Mol Cell* **47**:5–15.

Pulikkan, J.A., P.S. Peramangalam, V. Dengler, P.A. Ho, C. Preudhomme, S. Meshinchi, M. Christopeit, O. Nibourel, C. Müller-Tidow, S.K. Bohlander, D.G. Tenen, and G. Behre. 2010.

C/EBPa regulated microRNA-34a targets E2F3 during granulopoiesis and is down-regulated in AML with CEBPA mutations. *Blood* **116**:5638–5649.

Raver-Shapira, N., E. Marciano, E. Meiri, Y. Spector, N. Rosenfeld, N. Moskovits, Z. Bentwich, and M. Oren. 2007. Transcriptional activation of *miR-34a* contributes to p53-mediated apoptosis. *Mol Cell* **26**:731–743.

Rivetti di Val Cervo, P., A.M. Lena, M. Nicoloso, S. Rossi, M. Mancini, H. Zhou, G. Saintigny, E. Dellambra, T. Odorisio, C. Mahé, G.A. Calin, E. Candi, and G. Melino. 2012. p63–microRNA feedback in keratinocyte senescence. *Proc Natl Acad Sci U S A* **109**:1133–1138.

Rodier, F., J. Campisi, and D. Bhaumik. 2007. Two faces of p53: aging and tumor suppression. *Nucleic Acids Res* **35**:7475–7484.

Rodier, F., J.P. Coppé, C.K. Patil, W.A. Hoeijmakers, D.P. Muñoz, S.R. Raza, A. Freund, E. Campeau, A.R. Davalos, and J. Campisi. 2009. Persistent DNA damage signalling triggers senescence-associated inflammatory cytokine secretion. *Nat Cell Biol* **11**:973–979.

Rosso, A., A. Balsamo, R. Gambino, P. Dentelli, R. Falcioni, M. Cassader, L. Pegoraro, G. Pagano, and M.F. Brizzi. 2006. p53 Mediates the accelerated onset of senescence of endothelial progenitor cells in diabetes. *J Biol Chem* **281**:4339–4347.

Sanduja, S., V. Kaza, and D.A. Dixon. 2009. The mRNA decay factor tristetraprolin (TTP) induces senescence in human papillomavirus-transformed cervical cancer cells by targeting E6-AP ubiquitin ligase. *Aging* **1**:803–817.

Serrano, M., A.W. Lin, M.E. McCurrach, D. Beach, and S.W. Lowe. 1997. Oncogenic ras provokes premature cell senescence associated with accumulation of p53 and p16INK4a. *Cell* **88**: 593–602.

Srikantan, S., B.S. Marasa, K.G. Becker, M. Gorospe, and K. Abdelmohsen. 2011. Paradoxical microRNAs: individual gene repressors, global translation enhancers. *Cell Cycle* **10**:751–759.

Sun, F., H. Fu, Q. Liu, Y. Tie, J. Zhu, R. Xing, Z. Sun, and X. Zheng. 2008. Downregulation of CCND1 and CDK6 by *miR-34a* induces cell cycle arrest. *FEBS Lett* **582**:1564–1568.

Tarasov, V., P. Jung, B. Verdoodt, D. Lodygin, A. Epanchintsev, A. Menssen, G. Meister, and H. Hermeking. 2007. Differential regulation of microRNAs by p53 revealed by massively parallel sequencing: *miR-34a* is a p53 target that induces apoptosis and G1-arrest. *Cell Cycle* **6**: 1586–1593.

Tazawa, H., N. Tsuchiya, M. Izumiya, and H. Nakagama. 2007. Tumor-suppressive *miR-34a* induces senescence-like growth arrest through modulation of the E2F pathway in human colon cancer cells. *Proc Natl Acad Sci U S A* **104**:15472–15477.

Terao, M., M. Fratelli, M. Kurosaki, A. Zanetti, V. Guarnaccia, G. Paroni, A. Tsykin, M. Lupi, M. Gianni, G.J. Goodall, and E. Garattini. 2011. Induction of *miR-21* by retinoic acid in estrogen receptor-positive breast carcinoma cells: biological correlates and molecular targets. *J Biol Chem* **286**:4027–4042.

Vasudevan, S., Y. Tong, and J.A. Steitz. 2007. Switching from repression to activation: microRNAs can up-regulate translation. *Science* **318**:1931–1934.

Vaziri, H. and S. Benchimol. 1998. Reconstitution of telomerase activity in normal human cells leads to elongation of telomeres and extended replicative life span. *Curr Biol* **8**:279–282.

Venkataraman, S., I. Alimova, R. Fan, P. Harris, N. Foreman, and R. Vibhakar. 2010. MicroRNA 128a increases intracellular ROS level by targeting Bmi-1 and inhibits medulloblastoma cancer cell growth by promoting senescence. *PLoS ONE* **5**:e10748.

Verduci, L., M. Simili, M. Rizzo, A. Mercatanti, M. Evangelista, L. Mariani, G. Rainaldi, and L. Pitto. 2010. MicroRNA (miRNA)-mediated interaction between leukemia/lymphoma-related factor (LRF) and alternative splicing factor/splicing factor 2 (ASF/SF2) affects mouse embryonic fibroblast senescence and apoptosis. *J Biol Chem* **285**:39551–39563.

Wang, W., J.L. Martindale, X. Yang, F.J. Chrest, and M. Gorospe. 2005. Increased stability of the p16 mRNA with replicative senescence. *EMBO Rep* **6**:158–164.

Wang, W., X. Yang, V.J. Cristofalo, N.J. Holbrook, and M. Gorospe. 2001. Loss of HuR is linked to reduced expression of proliferative genes during replicative senescence. *Mol Cell Biol* **21**:5889–5898.

Wang, Y., M.N. Scheiber, C. Neumann, G.A. Calin, and D. Zhou. 2011. MicroRNA regulation of ionizing radiation induced premature senescence. *Int J Radiat Oncol Biol Phys* **81**:839–848.

Wei, S. and J.M. Sedivy. 1999. Expression of catalytically active telomerase does not prevent premature senescence caused by overexpression of oncogenic Ha-Ras in normal human fibroblasts. *Cancer Res* **59**:1539–1543.

Wright, W.E. and J.W. Shay. 2002. Historical claims and current interpretations of replicative aging. *Nat Biotechnol* **20**:682–688.

Xu, D., F. Takeshita, Y. Hino, S. Fukunaga, Y. Kudo, A. Tamaki, J. Matsunaga, R.U. Takahashi, T. Takata, A. Shimamoto, T. Ochiya, and H. Tahara. 2011. *miR-22* represses cancer progression by inducing cellular senescence. *J Cell Biol* **193**:409–424.

Xue, W., L. Zender, C. Miething, R.A. Dickins, E. Hernando, V. Krizhanovsky, C. Cordon-Cardo, and S.W. Lowe. 2007. Senescence and tumour clearance is triggered by p53 restoration in murine liver carcinomas. *Nature* **445**:656–660.

Yamakuchi, M., M. Ferlito, and C.J. Lowenstein. 2008. *miR-34a* repression of SIRT1 regulates apoptosis. *Proc Natl Acad Sci U S A* **105**:13421–13426.

Yang, G., D.G. Rosen, Z. Zhang, R.C. Bast Jr., G.B. Mills, J.A. Colacino, I. Mercado-Uribe, and J. Liu. 2006. The chemokine growth-regulated oncogene 1 (Gro-1) links RAS signaling to the senescence of stromal fibroblasts and ovarian tumorigenesis. *Proc Natl Acad Sci U S A* **103**:16472–16477.

Yang, M.Y., P.M. Lin, Y.C. Liu, H.H. Hsiao, W.C. Yang, J.F. Hsu, C.M. Hsu, and S.F. Lin. 2012. Induction of cellular senescence by doxorubicin is associated with upregulated *miR-375* and induction of autophagy in K562 cells. *PLOS ONE* **7**:e37205.

Yao, H., S.R. Yang, I. Edirisinghe, S. Rajendrasozhan, S. Caito, D. Adenuga, M.A. O'Reilly, and I. Rahman. 2008. Disruption of p21 attenuates lung inflammation induced by cigarette smoke, LPS, and fMLP in mice. *Am J Respir Cell Mol Biol* **39**:7–18.

Young, A.P., S. Schlisio, Y.A. Minamishima, Q. Zhang, L. Li, C. Grisanzio, S. Signoretti, and W.G. Kaelin Jr. 2008. VHL loss actuates a HIF-independent senescence programme mediated by Rb and p400. *Nat Cell Biol* **10**:361–369.

THE EMERGENCE OF GeroMIRs: A GROUP OF MicroRNAs IMPLICATED IN AGING

Alejandro P. Ugalde, Agnieszka Kwarciak, Xurde M. Caravia, Carlos López-Otín, and Andrew J. Ramsay

Departamento de Bioquímica y Biología Molecular, Instituto Universitario de Oncología-IUOPA, Universidad de Oviedo, Oviedo, Spain

I. Introduction 77
II. miRNAs and Aging: Lessons from Invertebrates 78
III. Changes in miRNA Expression during Mammalian Aging 80
IV. miRNA Modulation of Mammalian DNA Damage 81
V. Micromanaging of Other Aging-Associated Pathways in Mammals 83
VI. Conclusions and Future Perspectives 85
References 85

ABBREVIATIONS

DDR   DNA damage response
IGF-1   insulin-like growth factor 1
TOR   target of rapamycin

I. INTRODUCTION

Inevitably, all of us will experience a progressive deterioration of our body's fitness due to the process known as aging. This dramatic phenomenon has long preoccupied scientists, and with the extraordinary lifespan increase of human populations over the last century,

MicroRNAs in Medicine, First Edition. Edited by Charles H. Lawrie.
© 2014 John Wiley & Sons, Inc. Published 2014 by John Wiley & Sons, Inc.

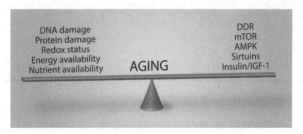

Figure 5.1. Aging results from a balance between stressors and stress-response pathways.

there is rekindled interest for understanding this complex biological occurrence. From a physiological standpoint, there is a general agreement that aging results from the accumulation of macromolecular and cellular damage, which progressively compromises tissue function and increases the risk of disease and death (Vijg and Campisi 2008). One of the great advances in aging research was the discovery of aging plasticity, which means that this process can be modulated by genetic and environmental factors. The first evidence of aging plasticity came from the discovery that caloric restriction—underfeeding without malnutrition—extended the lifespan in multiple organism models (Kenyon 2010). Likewise, other external perturbations, such as temperature or oxygen levels, were found to influence lifespan in several organisms. On the other hand, advances in genetics and molecular biology over the last few decades have identified a subset of genes whose mutation affect longevity in several organisms. Interestingly, most of these genes fall into a few evolutionarily conserved molecular circuits, including the DNA damage response (DDR), insulin/insulin-like growth factor 1 (IGF-1), the AMP-activated protein kinase (AMPK), the sirtuins, and the target of rapamycin (TOR) pathways (Haigis and Yankner 2010). These signaling circuits constitute adaptive response mechanisms that sense changes in factors such as temperature, nutrient and oxygen availability, or DNA and protein damage, and elaborate appropriate molecular responses aimed to preserve cellular and organismal function. Collectively, all these observations indicate that aging is a dynamic process that depends on both genetic and environmental factors, so that the rate of aging and the appearance of age-related symptoms result from a balance between stressors and stress response pathways (Figure 5.1).

II. miRNAs AND AGING: LESSONS FROM INVERTEBRATES

The discovery of miRNAs more than a decade ago has considerably changed the classical view of gene expression regulation, revealing a new group of molecules that can contribute to the complex process of aging. As addressed in other chapters of this book, miRNAs participate in almost every process within the cell and in numerous physiological and pathological processes. Accordingly, from a theoretical point of view, it is logical to speculate that these molecules could contribute to aging through the repression of target genes belonging to the earlier-mentioned aging signaling circuits. In practice, several works made in invertebrates have confirmed that at least a small proportion of these molecules—that we have named geromiRs (Ugalde et al. 2011a)- are able to modulate aging. The first landmark study that identified a geromiR was carried out by Boehm et al., who found that *lin-4* miRNA loss-of-function *Caenorhabditis elegans* mutants had a shortened longevity compared with wild-type animals, while overexpression of this miRNA extended organism lifespan (Boehm and Slack 2005). Subsequent expression profiles also evidenced that a high proportion of the *C. elegans* miRNAs display changes in their expression levels during adulthood, including *lin-4* miRNA (Ibanez-Ventoso et al. 2006; Kato et al. 2011). Similarly, de Lencastre et al. analyzed the expression levels of miRNAs during *C. elegans* lifespan and demonstrated that loss-of-function mutations in four of the most up-regulated miRNAs, *miR-71, miR-238, miR-239,* and *mir-246*, considerably extended or shortened the organisms lifespan (de Lencastre et al. 2010). A recent study identified *miR-34,* which is one of the most up-regulated miRNAs during *C. elegans* aging, as another geromiR whose mutation extends lifespan, although another work suggested that this miRNA might be a neutral determinant for lifespan (de Lencastre et al. 2010; Yang et al. 2011).

Different approaches have confirmed that these miRNAs influence lifespan by targeting well-known aging signaling pathways. For example, genetic approaches have demonstrated that *lin-4* controls postnatal development by regulating the insulin/IGF pathway through repression of *lin-14* mRNA. Thus, *lin-14* loss-of-function worms display an extended lifespan, while *lin-14* gain-of-function phenocopies the *lin-4* mutation. Moreover, the aging phenotype conferred by different alterations in these genes depend on *hsf-1* and *daf-16*, two downstream targets of the insulin/IGF pathway, and subsequent studies have shown that *lin-14* mRNA encodes a transcription factor that regulates the activity of the insulin/IGF gene *ins-33* (Hristova et al. 2005). Similarly, genetic studies have also shown that *miR-239* and *miR-71* function through this key pathway, as *daf-16* knockdown abolishes, or diminishes, the longer lifespan of worms deficient in these miRNAs, respectively (de Lencastre et al. 2010). These studies have also revealed that *miR-71* might participate in lifespan regulation through the modulation of the DDR pathway. Likewise, *miR-34* is essential for the DDR in *C. elegans* and is an important regulator of the autophagy flux, another process strongly associated with aging (Kato et al. 2009; Marino et al. 2010a; Yang et al. 2011). Short-lived *miR-71* and *miR-238* mutants also exhibit decreased survival in response to oxidative stress and heat shock, whereas long-lived *miR-239* and *miR-34* mutant animals are more resistance to these stressors, which supports the contribution of the stress-response pathways to the modulation of aging (Boehm and Slack 2005; de Lencastre et al. 2010). It is worth noting that the expression of various age-associated miRNAs is altered in short- and long-lived animals, reinforcing that these molecules are tightly regulated during aging. For example, long-lived animals induced by low temperature conditions or by harboring mutations in the insulin/IGF-1 receptor *daf-2*, display a delay in the up-regulation of some age-associated miRNAs, such as *miR-239* or *miR-34,* while their expression is exacerbated in short-lived animals (de Lencastre et al. 2010; Kato et al. 2011).

Similar to works on *C. elegans*, studies on *Drosophila melanogaster* have also provided evidence for the contribution of miRNAs to the aging process. For example, flies harboring a hypomorphic mutation in *loquacious (loqs),* a key component of the miRNA machinery, develop normally but show late-onset brain degeneration and a reduced lifespan, implicating miRNA involvement in age-associated pathologies (Liu et al. 2012). A detailed analysis of brain miRNAs in this organism also revealed that *miR-34* miRNA is strongly up-regulated in aged flies. Notably, flies deficient in this miRNA phenocopy the *loqs* mutation, developing an accelerated aging phenotype that is characterized by progressive neurodegeneration, increased stress sensitivity, and locomotion alterations. Excitingly, this work identified translational repression of *E74A* as the target responsible for *miR-34* effects on aging. Curiously, E74A is a component of steroid hormone signaling pathways, a molecular network that is already associated with aging modulation (Simon et al. 2003). In an elegant example of antagonistic pleiotropy (Liu et al. 2012), the authors have proposed that although E74A is required during juvenile development, silencing of E74A by *miR-34* in adulthood is critical to avoid the harmful effects of this protein with sharply opposing functions on animal fitness at different life stages.

III. CHANGES IN miRNA EXPRESSION DURING MAMMALIAN AGING

In contrast to invertebrates, current knowledge about the role of miRNAs in mammalian aging is still very limited due to the increased complexity, technical difficulties, and longer lifespan of these organisms. To date, no study has been able to demonstrate the ability of a single miRNA to modulate the rate of aging in mammals by loss- or gain-of function modifications. However, cumulative observations during the last decade of miRNA research strongly indicate that these molecules contribute to mammalian aging. First, similar to invertebrates, numerous works have reported significant age-related changes in miRNA expression during rodent and primate lifespan. Second, a growing number of miRNAs have been demonstrated to be able to influence most of the well-known aging pathways. In this section, we will discuss the current evidence that supports a role in aging modulation of the most promising "geromiRs."

Profiling studies in young and aged rodents have revealed age-related changes in miRNA expression in the liver and brain, whereas unclear results were obtained in the lung (Williams et al. 2007; Izzotti et al. 2009). In liver, *miR-93, miR-214, miR-669c,* and *miR-709* were found up-regulated in aged mice, and a similar study conducted in rats also identified the up-regulation of *miR-93* and *miR-34*, while *miR-16* and *miR-27a* were down-regulated (Maes et al. 2008; Li et al. 2011b). In murine brain, there is also an age-dependent deregulation of 70 miRNAs, and comparative analysis with previous liver profiles revealed the common up-regulation of *miR-30d, -34a, -468, -669b,* and *-709*, whereas a subset of age-related miRNAs were found to be brain specific, such as *miR-22, -101a, -720,* and *-721* (Li et al. 2011a). Data about *miR-34* was also confirmed by an independent work that reported the progressive increment with age of this miRNA in brain samples, as well as in peripheral blood mononuclear cells and in plasma (X. Li et al. 2011). In humans and primates, several works also support miRNA contribution to the process of aging. Thus, Somel et al. analyzed the mRNA, miRNA, and protein expression in human and macaque brain, finding evidence of regulatory relationships between miRNAs and mRNAs during aging of these species (Somel et al. 2010). Another work also reported the differential expression of miRNAs in cortical and cerebellar samples from aged rhesus macaques, chimpanzees, and humans, discovering that miRNA changes showed little interspecies and tissue conservation, with the exception of *miR-144*, which was found commonly up-regulated in all aged samples. Human peripheral blood mononuclear cells also present an altered miRNA expression profile during aging, characterized by the down-regulation of *miR-103, miR-107, miR-128, miR-130a, miR-155, miR-24, miR-221, miR-496,* and *miR-1538*. Similarly, analysis of miRNA expression in bone marrow mesenchymal stem cells from rhesus monkeys revealed a reduction in the expression levels of several miRNAs, including *let-7f, miR-125b, mir-221, mir-222, miR-199-3p,* and *miR-23a* (Noren Hooten et al. 2010; Yu et al. 2011). Also of particular relevance, the work of Hackl et al. found the common down-regulation of *miR-17, miR-19a, miR-20a,* and *miR-106a* expression in multiple models of replicative and organismal human aging (Hackl et al. 2010).

Perhaps the strongest evidence supporting the putative role of miRNAs in mammalian aging is the activity of specific miRNAs in both short- and long-life animal models. Thus, a study in Ames dwarf mice, which lives 70% longer than wild type mice due to deficiencies in pituitary hormones, have suggested a critical role for *miR-27a* in the lifespan extension of these mice (Bates et al. 2010). This miRNA is significantly increased in the liver of Ames mice, and is inversely correlated to its target gene, ornithine decarboxylase, an important enzyme of the aging-associated polyamine biosynthetic pathway. Conversely, the up-regulation of *miR-1* and the *miR-29* family have also been linked to the progeroid phenotype of a mouse model of the human Hutchinson–Gilford progeria syndrome by controlling multiple overlapping aging pathways (Marino et al. 2010b; Ugalde et al. 2011b). Collectively, these data strongly indicate that miRNAs might be important modulators of mammalian aging, although there is a high variability in age-related expression that suggests that their contribution to aging may be largely tissue and species specific. Despite this, some miRNAs stand out as *bona fide* geromiRs. One of clearest geromiRs is *miR-34*, whose sequence is highly conserved throughout evolution. This miRNA has been demonstrated to modulate lifespan in worms and flies, in addition to being found deregulated in multiple tissues during mammalian aging, and has been associated with multiple aging pathways, such as sirtuins, the DDR, and autophagy (Kato et al. 2009; Yamakuchi and Lowenstein 2009; Minones-Moyano et al. 2011; Yang et al. 2011).

IV. MIRNA MODULATION OF MAMMALIAN DNA DAMAGE

Even under the most controlled environmental conditions, an organism ages, illustrating that intrinsic sources of damage exist and impart a strong influence on the aging process. Although cells display a broad repertoire of macromolecule turnover and repair systems, the age-related decline, or the intrinsic limited efficiency of these systems, leads to the progressive accumulation of damaged molecules in the cell. In principle, proteins and lipids are continually renewed, but defects in DNA are permanent and can produce dramatic consequences in cell behavior. To preserve the integrity and counteract the undesirable consequences of DNA alterations, cells have acquired a variety of DNA maintenance and checkpoint mechanisms. These mechanisms include molecular circuits that detect damage and transduce signals that activate pathways aimed to repair the damage and/or prevent abnormal cellular behaviors, in a process termed the DDR (Harper and Elledge 2007). Inborn defects in components of the DDR underlie a subset of rare human disorders that result in accelerated aging, also known as progeria, cancer predisposition, or a combination of both phenotypes (Garinis et al. 2008). In this section, we discuss the functional relevance of miRNAs in the DDR for understanding the molecular basis of aging in humans (Figure 5.2).

The essential regulatory roles of miRNA in virtually all biological processes (Kim et al. 2009), includes modulating the expression of genes detecting, transducing, or affecting the DDR. This regulatory interaction is bidirectional, as antagonism of the DDR via UV radiation exposure invokes global transcriptional alterations in the miRNome (Pothof et al. 2009). DDR transcription factors, such as the p53 family, have established roles in modulating miRNA transcription. The *miR-34* family was the first group of miRNAs described to be transcriptionally regulated by p53 in response to DNA damage (Chang et al. 2007; Corney et al. 2007; He et al. 2007). Intriguingly, an apparent feedback loop exists between p53, *miR-34a*, and SIRT1, whereby, following p53 mediated transcription, *miR-34a* ablates SIRT1, which in turn prevents p53 deacetylation by SIRT1 (Yamakuchi

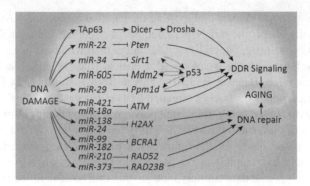

Figure 5.2. miRNAs regulate a wide variety of DNA damage response components. Dotted arrows indicate feedback regulation loops (see text for details).

et al. 2008). In *Zmpste24*-deficient mice, a model of the human Hutchinson–Gilford progeria syndrome (Varela et al. 2005), altered chromatin architecture mediates the transcription of the *miR-29* family in a p53-dependent manner (Ugalde et al. 2011b). In an analogous situation to the p53-dependent *miR-34* feedback loop, *miR-29* targets and represses *Ppm1d*, a phosphatase that fine-tunes the DDR through inhibition of the activity and stability of p53. A more recent miRNA DDR feedback mechanism has been described, in which p53 transcriptional activation of *miR-605* decreases the expression of *MDM2*, a E3 ubiquitin ligase that negatively regulates p53 (Xiao et al. 2011). p53 also positively regulates *miR-192* and *miR-215*, which in turn attenuates cell growth (Braun et al. 2008; Georges et al. 2008). Conversely, p53 reduces the transcription of the *miR-17-92* cluster by excluding the TATA-binding protein from the promoter (Yan et al. 2009). Accumulating evidence also suggests posttranscriptional control of miRNA biogenesis as an important regulatory element of the DDR pathway. In response to DNA damage, p53 interacts with the Drosha processing complex through an association with p68, and coordinates the processing of several miRNAs with growth-suppressive function, including *miR-16-1*, *miR-143*, and *miR-145* (Suzuki et al. 2009). The fine-tuning of Dicer levels is also targeted by DNA damage. TAp63, a member of the p53 family, directly binds to the Dicer promoter and activates its transcription (Su et al. 2010).

Bidirectional interplay between miRNAs and the DDR pathway intertwines miRNA biogenesis with modulation by miRNAs in the levels of proteins that sense, transmit, and repair genome damage. Accordingly, suppression of the DNA damage kinase, ataxia telangiectasia mutated (*ATM*), by *miR-421* and *miR-18a* increases cellular sensitivity to ionizing radiation and cell cycle arrest (Hu et al. 2010; Song et al. 2011). During postmitotic differentiation in hematopoietic cell lines, *miR-24* is upregulated and reduces H2AX levels—a substrate of ATM that is phosphorylated and recruited to DNA breaks—rendering the cells vulnerable to DNA damage (Lal et al. 2009). Similarly, *miR-138* ablates *H2AX* expression and enhances cellular sensitivity to DNA-damaging agents (Wang et al. 2011). miRNAs also control the expression of genes that mediate the DNA repair process. The miRNAs *miR-210* and *miR-373* suppress the levels of *RAD52* and *RAD23B*, key components in homology-dependent repair (HDR) and nucleotide excision repair (NER) pathways, respectively (Crosby et al. 2009). Interestingly, a decrease in *miR-31* levels has been demonstrated to underlie radioresistance in an esophageal adenocarcinoma cell model due to unopposed expression of genes involved in DNA repair (Lynam-Lennon et al. 2012).

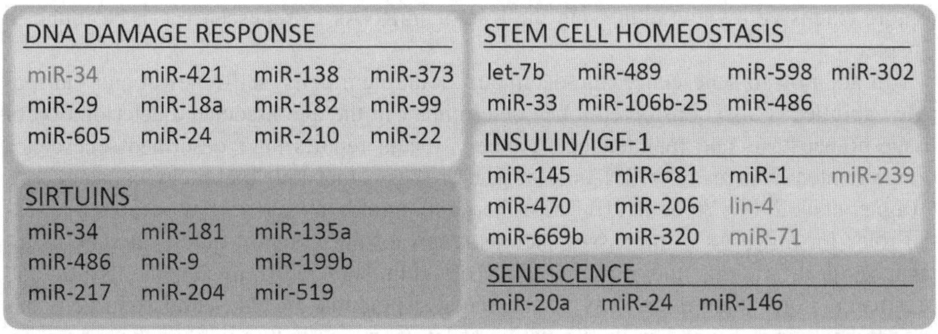

DNA DAMAGE RESPONSE				STEM CELL HOMEOSTASIS			
miR-34	miR-421	miR-138	miR-373	let-7b	miR-489	miR-598	miR-302
miR-29	miR-18a	miR-182	miR-99	miR-33	miR-106b-25	miR-486	
miR-605	miR-24	miR-210	miR-22	**INSULIN/IGF-1**			
				miR-145	miR-681	miR-1	miR-239
SIRTUINS				miR-470	miR-206	lin-4	
miR-34	miR-181	miR-135a		miR-669b	miR-320	miR-71	
miR-486	miR-9	miR-199b		**SENESCENCE**			
miR-217	miR-204	mir-519		miR-20a	miR-24	miR-146	

Figure 5.3. Comprehensive list of the most promising geromiRs that have been linked to aging-associated pathways.

Further, in response to UV radiation, the *miR-99* family is transiently induced and targets the SWI/SNF chromatin remodeling factor *SNF2H/SMARCA5*, reducing the recruitment of the tumor suppressor BRCA1—a nuclear phosphoprotein that is essential for genome stability—to the sites of DNA damage (Mueller et al. 2012). *BRCA1* expression is also ablated by *miR-182* overexpression in breast tumor cells, rendering the cells hypersensitive to irradiation (Moskwa et al. 2011). Another archetypal tumor suppressor, phosphatase and tensin homologue (PTEN), regulates genome integrity and is reduced by UV-dependent expression of *miR-22* (Tan et al. 2012).

V. MICROMANAGING OF OTHER AGING-ASSOCIATED PATHWAYS IN MAMMALS

In addition to the widespread role of miRNAs in DDR regulation, a growing number of miRNAs are emerging as important modulators of conserved pathways or processes that have been extensively linked to aging (Figure 5.3). One of these processes widely associated with aging is cellular senescence. This irreversible state of cellular growth arrest is an important effector of the cellular response against DNA damage that prevents the malignant proliferation of cells harboring oncogenic DNA mutations. However, as in the case of DDR, some aspects of cellular senescence have led to the consideration that it has a dual role through lifespan, protecting from cancer development but stimulating inflammation and tissue exhaustion, which promotes age-related alterations (Rodier and Campisi 2011). Numerous works have revealed functional roles for miRNAs in senescence through a variety of mechanisms. Thus, senescence is controlled by several DDR-associated signaling circuits that have been reported to be extensively influenced by miRNAs, such as the p53/p21/Arf and the p16/Rb pathways and their associated regulators. For example, *miR-20a* repression of *LRF* (leukemia/lymphoma related factor) in murine fibroblasts activates p19ARF and p16INKa and triggers cellular senescence (Poliseno et al. 2008), whereas the *miR-24* decrease in senescent human fibroblasts is in part responsible for p16INKa up-regulation (Lal et al. 2008). Other aspects of cellular senescence, such as the senescence-associated secretory phenotype (SASP), are also finely regulated by miRNAs. A clear example of SASP regulation is *miR-146a/b*, which takes part of a negative feedback loop that controls *IL-6* and *IL-8* expression (Bhaumik et al. 2009). Although the miRNAs exemplified here constitute some of the most prominent senescence-associated miRNAs, a more extensive review addressing the role of miRNAs in senescence can be found in Chapter 4.

In addition to senescence, the decline of adult stem cells' self-renewal and pluripotency abilities is also considered a key determinant in the age-associated deterioration of tissue homeostasis and maintenance. Notably, recent reports have described senescence or age-related changes in miRNAs of human or rhesus macaque mesenchymal stem cells (Wagner et al. 2008; Yu et al. 2011). Furthermore, numerous studies have reported essential roles for miRNAs in processes such as renewal, pluripotency, quiescent state maintenance, proliferation, and differentiation of adult stem cells in several tissues and organisms. For example, the loss of self-renewal potential in old neural stem cells has been linked to age-dependent up-regulation of *let-7b,* which in turn inhibits the expression of *HMGA2*, a repressor of the INK4a/ARF locus (Nishino et al. 2008). Hematopoietic stem cell self-renewal in mice is also regulated by *miR-33* repression of *TP53* (Herrera-Merchan et al. 2010) and the maintenance of quiescent state in human muscle adult stem cells has been shown to be highly dependent on miRNA activity, being *miR-489* one of the most prominent effectors (Cheung et al. 2012). Stem cell proliferation and neuronal differentiation in mice is also finely regulated by miRNAs of the *miR-106b-25* cluster, which are in turn controlled by the aging-associated FoxO transcription factors (Brett et al. 2011). Additionally, the expression of *miR-486-5p* in human adipose tissue-derived mesenchymal stem cells (hADSCs) progressively increases with aging and regulates the expression of SIRT1, inducing a premature senescence-like phenotype and inhibiting adipogenic and osteogenic differentiation (Kim et al. 2012). Indeed, studies of human embryonic stem cells (hESCs) have described a nonfunctional p53-p21 axis of the G1/S checkpoint pathway, which has recently been reported to be regulated by the *miR-302* family (Dolezalova et al. 2012).

The sirtuin pathway, which extends longevity in yeast, worms, and flies (Kenyon 2010), is also influenced by miRNAs. Among the sirtuins, the NAD$^+$-dependent deacetylase SIRT1 is widely recognized as a crucial regulator of metabolism, stress responses, replicative senescence, and inflammation (Haigis and Sinclair 2010). In fact, SIRT1 is an important mediator of the beneficial metabolic effects of CR, and is also the target effector of the antiaging molecule resveratrol. Among miRNAs that could account for the regulation of SIRT1 during aging, the earlier-mentioned age-related *miR-34* is one of the best examples, as assessed by its ability to directly targeting *SIRT1* mRNA in several *in vitro* and *in vivo* experiments (Lee and Kemper 2010). Likewise, *miR-486-5p* directly targets *SIRT1* in hADSCs (see previous discussion), and *miR-217* up-regulation in human endothelial cells during aging reduces SIRT1 activity and promotes senescence (Menghini et al. 2009). There is also evidence that SIRT1 is highly expressed in mESCs, but its levels decrease during their differentiation to different tissues through direct targeting by several miRNAs, including *miR-181a/b, miR-9, miR-204, miR-135a,* and *miR-199b* (Saunders et al. 2010). Alternatively, other miRNAs modulate SIRT1 activity indirectly, such as *miR-519,* which contributes to human fibroblast senescence by decreasing the protein levels of SIRT1 through direct targeting of the RNA-binding protein *HuR* (Marasa et al. 2010).

Similar to the sirtuin pathway, the insulin/IGF-1 signaling is also susceptible to miRNA regulation during aging. For example, the anomalous up-regulation of *miR-1,* which targets *Igf-1* mRNA, is associated with the systemic deregulation of the somatotroph axis in premature aging mice (Marino et al. 2010b). Conversely, *miR-470, miR-669b,* and *miR-681* are significantly up-regulated in brain of long-lived Ames dwarf mice, and their expression inversely correlates with several genes of the insulin/IGF-1 pathway.

Functional studies have demonstrated that these miRNAs target IGF-1 receptor and contribute to reduced levels of phosphorylated AKT and FOXO3a, two downstream targets of this signaling pathway, in the brain of mutant mice (Liang et al. 2011). In addition, human *miR-145* also represses the expression of IGF-1 receptor and its substrate, *IRS-1* (La Rocca et al. 2009), while *miR-206* and *miR-320* target this somatotroph axis in rats (Shan et al. 2009; Wang et al. 2009).

VI. CONCLUSIONS AND FUTURE PERSPECTIVES

The discovery of miRNAs has opened a new chapter in aging research that could help to achieve a deeper knowledge of the molecular network underlying this complex process. Although we are far from understanding the precise involvement of these molecules in age-related alterations, solid evidence from the literature supports an important role for the growing group of *geromiRs* in aging modulation (Figure 5.3). With the likely emergence of mammalian *in vivo* models for ablation and gain of specific miRNAs in the near future, we will undoubtedly see many important questions pertaining to how individual geromiRs regulate tissue aging and organism lifespan answered. Moreover, the recent advances in strategies to effectively block specific miRNA activities *in vivo* may also facilitate new therapeutic opportunities to delay or ameliorate age-related alterations, as well as premature aging syndromes. Alternatively, a promising new area for miRNAs is in diagnostics, where miRNAs have great potential as molecular biomarkers of aging.

REFERENCES

Bates, D.J., N. Li, R. Liang, H. Sarojini, J. An, M.M. Masternak, A. Bartke, and E. Wang. 2010. MicroRNA regulation in Ames dwarf mouse liver may contribute to delayed aging. *Aging Cell* **9**:1–18.

Bhaumik, D., G.K. Scott, S. Schokrpur, C.K. Patil, A.V. Orjalo, F. Rodier, G.J. Lithgow, and J. Campisi. 2009. MicroRNAs *miR-146a/b* negatively modulate the senescence-associated inflammatory mediators IL-6 and IL-8. *Aging* **1**:402–411.

Boehm, M. and F. Slack. 2005. A developmental timing microRNA and its target regulate life span in *C. elegans. Science* **310**:1954–1957.

Braun, C.J., X. Zhang, I. Savelyeva, S. Wolff, U.M. Moll, T. Schepeler, T.F. Orntoft, C.L. Andersen, and M. Dobbelstein. 2008. p53-Responsive micrornas 192 and 215 are capable of inducing cell cycle arrest. *Cancer Res* **68**:10094–10104.

Brett, J.O., V.M. Renault, V.A. Rafalski, A.E. Webb, and A. Brunet. 2011. The microRNA cluster miR-106b~25 regulates adult neural stem/progenitor cell proliferation and neuronal differentiation. *Aging* **3**:108–124.

Chang, T.C., E.A. Wentzel, O.A. Kent, K. Ramachandran, M. Mullendore, K.H. Lee, G. Feldmann, M. Yamakuchi, M. Ferlito, C.J. Lowenstein, D.E. Arking, M.A. Beer, et al. 2007. Transactivation of *miR-34a* by p53 broadly influences gene expression and promotes apoptosis. *Mol Cell* **26**:745–752.

Cheung, T.H., N.L. Quach, G.W. Charville, L. Liu, L. Park, A. Edalati, B. Yoo, P. Hoang, and T.A. Rando. 2012. Maintenance of muscle stem-cell quiescence by microRNA-489. *Nature* **482**: 524–528.

Corney, D.C., A. Flesken-Nikitin, A.K. Godwin, W. Wang, and A.Y. Nikitin. 2007. MicroRNA-34b and MicroRNA-34c are targets of p53 and cooperate in control of cell proliferation and adhesion-independent growth. *Cancer Res* **67**:8433–8438.

Crosby, M.E., R. Kulshreshtha, M. Ivan, and P.M. Glazer. 2009. MicroRNA regulation of DNA repair gene expression in hypoxic stress. *Cancer Res* **69**:1221–1229.

de Lencastre, A., Z. Pincus, K. Zhou, M. Kato, S.S. Lee, and F.J. Slack. 2010. MicroRNAs both promote and antagonize longevity in *C. elegans*. *Curr Biol* **20**:2159–2168.

Dolezalova, D., M. Mraz, T. Barta, K. Plevova, V. Vinarsky, Z. Holubcova, J. Jaros, P. Dvorak, S. Pospisilova, and A. Hampl. 2012. MicroRNAs regulate p21(Waf1/Cip1) protein expression and the DNA damage response in human embryonic stem cells. *Stem Cells* **30**:1362–1372.

Garinis, G.A., G.T. van der Horst, J. Vijg, and J.H. Hoeijmakers. 2008. DNA damage and ageing: new-age ideas for an age-old problem. *Nat Cell Biol* **10**:1241–1247.

Georges, S.A., M.C. Biery, S.Y. Kim, J.M. Schelter, J. Guo, A.N. Chang, A.L. Jackson, M.O. Carleton, P.S. Linsley, M.A. Cleary, and B.N. Chau. 2008. Coordinated regulation of cell cycle transcripts by p53-Inducible microRNAs, *miR-192* and *miR-215*. *Cancer Res* **68**:10105–10112.

Hackl, M., S. Brunner, K. Fortschegger, C. Schreiner, L. Micutkova, C. Muck, G.T. Laschober, G. Lepperdinger, N. Sampson, P. Berger, D. Herndler-Brandstetter, M. Wieser, et al. 2010. *miR-17*, *miR-19b*, *miR-20a*, and *miR-106a* are down-regulated in human aging. *Aging Cell* **9**:291–296.

Haigis, M.C. and D.A. Sinclair. 2010. Mammalian sirtuins: biological insights and disease relevance. *Annu Rev Pathol* **5**:253–295.

Haigis, M.C. and B.A. Yankner. 2010. The aging stress response. *Mol Cell* **40**:333–344.

Harper, J.W. and S.J. Elledge. 2007. The DNA damage response: ten years after. *Mol Cell* **28**: 739–745.

He, L., X. He, L.P. Lim, E. de Stanchina, Z. Xuan, Y. Liang, W. Xue, L. Zender, J. Magnus, D. Ridzon, A.L. Jackson, P.S. Linsley, et al. 2007. A microRNA component of the p53 tumour suppressor network. *Nature* **447**:1130–1134.

Herrera-Merchan, A., C. Cerrato, G. Luengo, O. Dominguez, M.A. Piris, M. Serrano, and S. Gonzalez. 2010. *miR-33*-mediated downregulation of p53 controls hematopoietic stem cell self-renewal. *Cell Cycle* **9**:3277–3285.

Hristova, M., D. Birse, Y. Hong, and V. Ambros. 2005. The *Caenorhabditis elegans* heterochronic regulator LIN-14 is a novel transcription factor that controls the developmental timing of transcription from the insulin/insulin-like growth factor gene ins-33 by direct DNA binding. *Mol Cell Biol* **25**:11059–11072.

Hu, H., L. Du, G. Nagabayashi, R.C. Seeger, and R.A. Gatti. 2010. ATM is down-regulated by N-Myc-regulated microRNA-421. *Proc Natl Acad Sci U S A* **107**:1506–1511.

Ibanez-Ventoso, C., M. Yang, S. Guo, H. Robins, R.W. Padgett, and M. Driscoll. 2006. Modulated microRNA expression during adult lifespan in *Caenorhabditis elegans*. *Aging Cell* **5**:235–246.

Izzotti, A., G.A. Calin, V.E. Steele, C.M. Croce, and S. De Flora. 2009. Relationships of microRNA expression in mouse lung with age and exposure to cigarette smoke and light. *FASEB J* **23**: 3243–3250.

Kato, M., X. Chen, S. Inukai, H. Zhao, and F.J. Slack. 2011. Age-associated changes in expression of small, noncoding RNAs, including microRNAs, in *C. elegans*. *RNA* **17**:1804–1820.

Kato, M., T. Paranjape, R.U. Muller, S. Nallur, E. Gillespie, K. Keane, A. Esquela-Kerscher, J.B. Weidhaas, and F.J. Slack. 2009. The mir-34 microRNA is required for the DNA damage response *in vivo* in *C. elegans* and *in vitro* in human breast cancer cells. *Oncogene* **28**:2419–2424.

Kenyon, C.J. 2010. The genetics of ageing. *Nature* **464**:504–512.

Kim, V.N., J. Han, and M.C. Siomi. 2009. Biogenesis of small RNAs in animals. *Nat Rev Mol Cell Biol* **10**:126–139.

Kim, Y.J., S.H. Hwang, S.Y. Lee, K.K. Shin, H.H. Cho, Y.C. Bae, and J.S. Jung. 2012. *miR-486-5p* induces replicative senescence of human adipose tissue-derived mesenchymal stem cells and its expression is controlled by high glucose. *Stem Cells Dev* **21**:1749–1760.

La Rocca, G., M. Badin, B. Shi, S.Q. Xu, T. Deangelis, L. Sepp-Lorenzinoi, and R. Baserga. 2009. Mechanism of growth inhibition by MicroRNA 145: the role of the IGF-I receptor signaling pathway. *J Cell Physiol* **220**:485–491.

Lal, A., H.H. Kim, K. Abdelmohsen, Y. Kuwano, R. Pullmann, Jr., S. Srikantan, R. Subrahmanyam, J.L. Martindale, X. Yang, F. Ahmed, F. Navarro, D. Dykxhoorn, et al. 2008. p16(INK4a) translation suppressed by miR-24. *PLoS ONE* **3**:e1864.

Lal, A., Y. Pan, F. Navarro, D.M. Dykxhoorn, L. Moreau, E. Meire, Z. Bentwich, J. Lieberman, and D. Chowdhury. 2009. *miR-24*-mediated downregulation of H2AX suppresses DNA repair in terminally differentiated blood cells. *Nat Struct Mol Biol* **16**:492–498.

Lee, J. and J.K. Kemper. 2010. Controlling SIRT1 expression by microRNAs in health and metabolic disease. *Aging* **2**:527–534.

Li, N., D.J. Bates, J. An, D.A. Terry, and E. Wang. 2011a. Up-regulation of key microRNAs, and inverse down-regulation of their predicted oxidative phosphorylation target genes, during aging in mouse brain. *Neurobiol Aging* **32**:944–955.

Li, N., S. Muthusamy, R. Liang, H. Sarojini, and E. Wang. 2011b. Increased expression of *miR-34a* and *miR-93* in rat liver during aging, and their impact on the expression of Mgst1 and Sirt1. *Mech Ageing Dev* **132**:75–85.

Li, X., A. Khanna, N. Li, and E. Wang. 2011. Circulatory *miR34a* as an RNAbased, noninvasive biomarker for brain aging. *Aging* **3**:985–1002.

Liang, R., A. Khanna, S. Muthusamy, N. Li, H. Sarojini, J.J. Kopchick, M.M. Masternak, A. Bartke, and E. Wang. 2011. Post-transcriptional regulation of IGF1R by key microRNAs in long-lived mutant mice. *Aging Cell* **10**:1080–1088.

Liu, N., M. Landreh, K. Cao, M. Abe, G.J. Hendriks, J.R. Kennerdell, Y. Zhu, L.S. Wang, and N.M. Bonini. 2012. The microRNA *miR-34* modulates ageing and neurodegeneration in Drosophila. *Nature* **482**:519–523.

Lynam-Lennon, N., J.V. Reynolds, L. Marignol, O.M. Sheils, G.P. Pidgeon, and S.G. Maher. 2012. MicroRNA-31 modulates tumour sensitivity to radiation in oesophageal adenocarcinoma. *J Mol Med (Berl)* **90**:1449–1458.

Maes, O.C., J. An, H. Sarojini, and E. Wang. 2008. Murine microRNAs implicated in liver functions and aging process. *Mech Ageing Dev* **129**:534–541.

Marasa, B.S., S. Srikantan, J.L. Martindale, M.M. Kim, E.K. Lee, M. Gorospe, and K. Abdelmohsen. 2010. MicroRNA profiling in human diploid fibroblasts uncovers miR-519 role in replicative senescence. *Aging* **2**:333–343.

Marino, G., A.F. Fernandez, and C. Lopez-Otin. 2010a. Autophagy and aging: lessons from progeria models. *Adv Exp Med Biol* **694**:61–68.

Marino, G., A.P. Ugalde, A.F. Fernandez, F.G. Osorio, A. Fueyo, J.M. Freije, and C. Lopez-Otin. 2010b. Insulin-like growth factor 1 treatment extends longevity in a mouse model of human premature aging by restoring somatotroph axis function. *Proc Natl Acad Sci U S A* **107**:16268–16273.

Menghini, R., V. Casagrande, M. Cardellini, E. Martelli, A. Terrinoni, F. Amati, M. Vasa-Nicotera, A. Ippoliti, G. Novelli, G. Melino, R. Lauro, and M. Federici. 2009. MicroRNA 217 modulates endothelial cell senescence via silent information regulator 1. *Circulation* **120**:1524–1532.

Minones-Moyano, E., S. Porta, G. Escaramis, R. Rabionet, S. Iraola, B. Kagerbauer, Y. Espinosa-Parrilla, I. Ferrer, X. Estivill, and E. Marti. 2011. MicroRNA profiling of Parkinson's disease brains identifies early downregulation of *miR-34b/c* which modulate mitochondrial function. *Hum Mol Genet* **20**:3067–3078.

Moskwa, P., F.M. Buffa, Y. Pan, R. Panchakshari, P. Gottipati, R.J. Muschel, J. Beech, R. Kulshrestha, K. Abdelmohsen, D.M. Weinstock, M. Gorospe, A.L. Harris, et al. 2011. *miR-182*-mediated downregulation of BRCA1 impacts DNA repair and sensitivity to PARP inhibitors. *Mol Cell* **41**:210–220.

Mueller, A.C., D. Sun, and A. Dutta. 2012. The *miR-99* family regulates the DNA damage response through its target SNF2H. *Oncogene* **32**(9):1164–1172.

Nishino, J., I. Kim, K. Chada, and S.J. Morrison. 2008. Hmga2 promotes neural stem cell self-renewal in young but not old mice by reducing p16Ink4a and p19Arf Expression. *Cell* **135**:227–239.

Noren Hooten, N., K. Abdelmohsen, M. Gorospe, N. Ejiogu, A.B. Zonderman, and M.K. Evans. 2010. microRNA expression patterns reveal differential expression of target genes with age. *PLoS ONE* **5**:e10724.

Poliseno, L., L. Pitto, M. Simili, L. Mariani, L. Riccardi, A. Ciucci, M. Rizzo, M. Evangelista, A. Mercatanti, P.P. Pandolfi, and G. Rainaldi. 2008. The proto-oncogene LRF is under post-transcriptional control of MiR-20a: implications for senescence. *PLoS ONE* **3**:e2542.

Pothof, J., N.S. Verkaik, W. van IJcken, E.A. Wiemer, V.T. Ta, G.T. van der Horst, N.G. Jaspers, D.C. van Gent, J.H. Hoeijmakers, and S.P. Persengiev. 2009. MicroRNA-mediated gene silencing modulates the UV-induced DNA-damage response. *EMBO J* **28**:2090–2099.

Rodier, F. and J. Campisi. 2011. Four faces of cellular senescence. *J Cell Biol* **192**:547–556.

Saunders, L.R., A.D. Sharma, J. Tawney, M. Nakagawa, K. Okita, S. Yamanaka, H. Willenbring, and E. Verdin. 2010. miRNAs regulate SIRT1 expression during mouse embryonic stem cell differentiation and in adult mouse tissues. *Aging* **2**:415–431.

Shan, Z.X., Q.X. Lin, Y.H. Fu, C.Y. Deng, Z.L. Zhou, J.N. Zhu, X.Y. Liu, Y.Y. Zhang, Y. Li, S.G. Lin, and X.Y. Yu. 2009. Upregulated expression of *miR-1/miR-206* in a rat model of myocardial infarction. *Biochem Biophys Res Commun* **381**:597–601.

Simon, A.F., C. Shih, A. Mack, and S. Benzer. 2003. Steroid control of longevity in *Drosophila melanogaster*. *Science* **299**:1407–1410.

Somel, M., S. Guo, N. Fu, Z. Yan, H.Y. Hu, Y. Xu, Y. Yuan, Z. Ning, Y. Hu, C. Menzel, H. Hu, M. Lachmann, et al. 2010. MicroRNA, mRNA, and protein expression link development and aging in human and macaque brain. *Genome Res* **20**:1207–1218.

Song, L., C. Lin, Z. Wu, H. Gong, Y. Zeng, J. Wu, M. Li, and J. Li. 2011. *miR-18a* impairs DNA damage response through downregulation of ataxia telangiectasia mutated (ATM) kinase. *PLoS ONE* **6**:e25454.

Su, X., D. Chakravarti, M.S. Cho, L. Liu, Y.J. Gi, Y.L. Lin, M.L. Leung, A. El-Naggar, C.J. Creighton, M.B. Suraokar, I. Wistuba, and E.R. Flores. 2010. TAp63 suppresses metastasis through coordinate regulation of Dicer and miRNAs. *Nature* **467**:986–990.

Suzuki, H.I., K. Yamagata, K. Sugimoto, T. Iwamoto, S. Kato, and K. Miyazono. 2009. Modulation of microRNA processing by p53. *Nature* **460**:529–533.

Tan, G., Y. Shi, and Z.H. Wu. 2012. MicroRNA-22 promotes cell survival upon UV radiation by repressing PTEN. *Biochem Biophys Res Commun* **417**:546–551.

Ugalde, A.P., Y. Espanol, and C. Lopez-Otin. 2011a. Micromanaging aging with miRNAs: new messages from the nuclear envelope. *Nucleus* **2**:549–555.

Ugalde, A.P., A.J. Ramsay, J. de la Rosa, I. Varela, G. Marino, J. Cadinanos, J. Lu, J.M. Freije, and C. Lopez-Otin. 2011b. Aging and chronic DNA damage response activate a regulatory pathway involving *miR-29* and p53. *EMBO J* **30**:2219–2232.

Varela, I., J. Cadinanos, A.M. Pendas, A. Gutierrez-Fernandez, A.R. Folgueras, L.M. Sanchez, Z. Zhou, F.J. Rodriguez, C.L. Stewart, J.A. Vega, K. Tryggvason, J.M. Freije, et al. 2005. Accelerated ageing in mice deficient in Zmpste24 protease is linked to p53 signalling activation. *Nature* **437**:564–568.

Vijg, J. and J. Campisi. 2008. Puzzles, promises and a cure for ageing. *Nature* **454**:1065–1071.

Wagner, W., P. Horn, M. Castoldi, A. Diehlmann, S. Bork, R. Saffrich, V. Benes, J. Blake, S. Pfister, V. Eckstein, and A.D. Ho. 2008. Replicative senescence of mesenchymal stem cells: a continuous and organized process. *PLoS ONE* **3**:e2213.

REFERENCES

Wang, X.H., R.Z. Qian, W. Zhang, S.F. Chen, H.M. Jin, and R.M. Hu. 2009. MicroRNA-320 expression in myocardial microvascular endothelial cells and its relationship with insulin-like growth factor-1 in type 2 diabetic rats. *Clin Exp Pharmacol Physiol* **36**:181–188.

Wang, Y., J.W. Huang, M. Li, W.K. Cavenee, P.S. Mitchell, X. Zhou, M. Tewari, F.B. Furnari, and T. Taniguchi. 2011. MicroRNA-138 modulates DNA damage response by repressing histone H2AX expression. *Mol Cancer Res* **9**:1100–1111.

Williams, A.E., M.M. Perry, S.A. Moschos, and M.A. Lindsay. 2007. microRNA expression in the aging mouse lung. *BMC Genomics* **8**:172.

Xiao, J., H. Lin, X. Luo, and Z. Wang. 2011. miR-605 joins p53 network to form a p53:*miR-605*:Mdm2 positive feedback loop in response to stress. *EMBO J* **30**:524–532.

Yamakuchi, M., M. Ferlito, and C.J. Lowenstein. 2008. miR-34a repression of SIRT1 regulates apoptosis. *Proc Natl Acad Sci U S A* **105**:13421–13426.

Yamakuchi, M. and C.J. Lowenstein. 2009. MiR-34, SIRT1 and p53: the feedback loop. *Cell Cycle* **8**:712–715.

Yan, H.L., G. Xue, Q. Mei, Y.Z. Wang, F.X. Ding, M.F. Liu, M.H. Lu, Y. Tang, H.Y. Yu, and S.H. Sun. 2009. Repression of the *miR-17-92* cluster by p53 has an important function in hypoxia-induced apoptosis. *EMBO J* **28**:2719–2732.

Yang, J., D. Chen, Y. He, A. Melendez, Z. Feng, Q. Hong, X. Bai, Q. Li, G. Cai, J. Wang, and X. Chen. 2011. MiR-34 modulates *Caenorhabditis elegans* lifespan via repressing the autophagy gene atg9. *Age* **35**:11–22.

Yu, J.M., X. Wu, J.M. Gimble, X. Guan, M.A. Freitas, and B.A. Bunnell. 2011. Age-related changes in mesenchymal stem cells derived from rhesus macaque bone marrow. *Aging Cell* **10**:66–79.

6

MicroRNAs and Hematopoiesis

Sukhinder K. Sandhu[1] and Ramiro Garzon[2]

[1]*Molecular Virology, Immunology and Medical Genetics, Comprehensive Cancer Center, The Ohio State University Wexner Medical Center, Columbus, OH, USA*
[2]*Division of Hematology, Department of Internal Medicine, The Ohio State University Wexner Medical Center, Columbus, OH, USA*

I. Introduction	92
A. MicroRNAs and Hematopoeisis	92
B. Hematopoietic Lineages	92
II. Lymphocyte Development	94
III. Monocyte and Granulocyte Development	96
IV. Erythrocyte and Megakaryocyte Development	97
V. Conclusions	97
References	98

ABBREVIATIONS

Ago	*Argonaute*
AICDA or AID	activation-induced cytidine deaminase
B-CLL	B-cell chronic lymphocytic leukemia
C/EBP	CCAAT/enhancer binding protein
cMYB	myeloblastosis oncogene
DLBCL	diffuse large B-cell lymphoma
EMP	erythrocyte–megakaryocyte precursor

MicroRNAs in Medicine, First Edition. Edited by Charles H. Lawrie.
© 2014 John Wiley & Sons, Inc. Published 2014 by John Wiley & Sons, Inc.

FL	follicular lymphoma
GMP	granulocyte–monocyte precursor
HPCs	hematopoeitic progenitor cells
HSCs	hematopoeitic stem cells
IL7R-α	interleukin-7 receptor-α
MEF2c	myocyte enhancer factor 2c
miRNAs	microRNAs
RBCs	red blood cells
RISC	RNA-induced silencing complex
TNF	tumor necrosis factor

I. INTRODUCTION

A. MicroRNAs and Hematopoiesis

MicroRNAs (MiRNAs) play critical roles in a wide array of cellular processes, including apoptosis, cell cycle, cell differentiation, and metabolism (Bartel 2004). One of the first studies to identify miRNAs relevant to hematopoiesis was reported by Chen and colleagues. In this study, 100 miRNAs from mouse bone marrow were cloned, and they found that three miRNAs: *miR-223, miR-142*, and *miR-181,* were preferentially expressed in hematopoietic tissues (Chen et al. 2004). Subsequently, researchers have identified many miRNAs with prominent expression in hematopoietic tissues. Importance of miRNAs in hematopoiesis was further reinforced by the impaired hematopoietic reconstitution and immune cell differentiation in mice, where the miRNA biogenesis machinery was knocked out by deletion of *Dicer* or *Argonaute (Ago)* (Cobb et al. 2005; O'Carroll et al. 2007). *Dicer* is a RNase III enzyme that is involved in processing of pre-miRNA into the mature miRNA, which is loaded into the *Ago* containing RNA-induced silencing complex (RISC), which then regulates target mRNA degradation. The process by which the hematopoietic stem cells (HSCs) can maintain its pluripotency and at the same time respond to lineage determining signals to differentiate into the several hematopoietic lineages is carefully orchestrated by master transcription factors and miRNAs. The multipotent HSC responds to various immune signals regulated by multiple miRNAs and miRNA-regulated transcription factors to generate the various blood cell lineages.

Functional studies, including the development of knock-in and knockout mouse models, provided useful insight into the functions of miRNAs during hematopoiesis. Here, we will discuss the expression and function of relevant miRNAs in the development of major hematopoietic lineages (Table 6.1).

B. Hematopoietic Lineages

The different hematopoietic lineages stem from a single HSC in the bone marrow that give rise to a common lymphoid progenitor (CLP) and a common myeloid progenitor (CMP) in response to different cytokines and transcription factors. The CLP gives rise to B, T, and natural killer (NK) cells after activation of specific transcription factors. The CMP undergoes another diversification into erythrocyte–megakaryocyte precursor (EMP) and granulocyte–monocyte precursor (GMP), which then differentiates into respective lineages (Figure 6.1). All the hematopoietic lineages are characterized by the presence of specific surface markers and have a specific miRNA expression signature.

INTRODUCTION

TABLE 6.1. Key Hematopoietic miRNAs and Lineage Regulation

microRNA	Target	Lineage	Effects	Reference
miR-181	BCL2, CD69, TCRα DUSP5, SHP2, PTPN22	T lymphopoiesis	$CD4^+$ $CD8^+$ development TCR sensitivity	Li et al. (2007); Neilson et al. (2007)
		B lymphopoiesis	B-cell development	Chen et al. (2004)
miR-223	MEF2C	Granulopoiesis	Inhibits granulocytic production and activation	Johnnidis et al. (2008)
miR-142s	Unknown	T lymphopoiesis	Increase T cell numbers	Chen et al. (2004)
miR-150	cMyb	B lymphopoiesis	Blocks progenitors B-cell development	Xiao et al. (2007)
		Megakaryopoiesis	Drives MEP differentiation toward megakaryocytes	Lu et al. (2008)
miR-155	SOCS1,c-MAF, CTLA4	Lymphopoiesis	T- and B-cell function	Rodriguez et al. (2007)
miR-125b	LIN28A	Myelopoiesis	Induces myeloid proliferation	Chaudhuri et al. (2012)
miR-17~92	Bim, PTEN, PP2A	B lymphopoiesis	Transition of pre- to pro-B	Ventura et al. (2008); Xiao et al. (2008)
miR-221-222	KIT	Erythropoiesis	Inhibits erythrocyte growth	Felli et al. (2005)
miR-144/451	GATA-1	Erythropoiesis		Dore et al. (2008)
miR-424	NF1A	Monopoiesis	Induces monocytic diff.	Fontana et al. (2007)

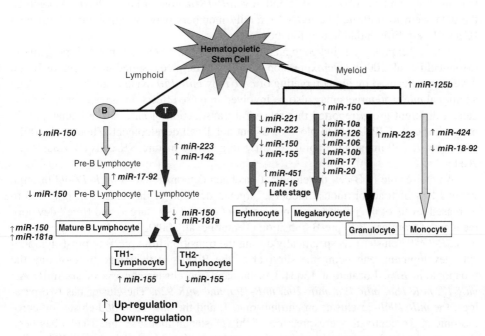

Figure 6.1. Expression levels of miRNAs during hematopoiesis. See color insert.

II. LYMPHOCYTE DEVELOPMENT

Lymphocytes are mainly characterized as B, T, and Natural killer/NK cells and develop from the CLP, which is characterized by the surface expression of interleukin-7 receptor alpha chain (*IL7R-α*). Among the miRNAs that have role in B- and T-cell development are *miR-181, miR-150, miR-17~92, miR-155*, and *miR-142. miR-181, miR-150,* and *miR-155* also plays a key role in NK cell development and function (Bezman et al. 2011; Cichocki et al. 2011; Trotta et al. 2012). NK cell transcriptome analysis by Liu et al. provided additional miRNAs, such as *miR-15a, miR-1246,* and *miR-331-3p*, which also regulate NK cell development (Liu et al. 2012).

miR-181 is detectable in bone marrow and spleen, but is most highly expressed in the thymus, which is the site of T-cell development and proliferation (Chen et al. 2004). It is also reported to have a strong expression in the brain and lungs (Chen et al. 2004). The members of the *miR-181* family are comprised of four mature miRNAs: *miR-181a, miR-181b, miR-181c,* and *miR-181d,* that are encoded from three polycistronic genes, *mir-181a-1/b-1, mir-181a-2/b-2,* and *mir-181c/d,* respectively. Reconstitution of lethally irradiated mice with *miR-181*-infected lineage negative (lin⁻) bone marrow hematopoietic progenitor cells led to significant increase in B cells and cytotoxic CD8$^+$ T cells (Chen et al. 2004). Neilson et al. showed that *miR-181* is up-regulated in the double-positive CD4$^+$ CD8$^+$ stage of thymocyte development and its expression correlated inversely with *Bcl-2, CD69,* and *TCRα* expression. Further experiments confirmed that *miR-181* family members target *Bcl-2, CD69,*and *TCRα* and regulate their levels at this stage of thymocyte development (Neilson et al. 2007). *miR-181* was also shown to have a critical role in the regulation of T cell receptor sensitivity and signaling strength at the posttranscriptional level by targeting multiple phosphatases (Table 6.1). More importantly, the authors showed that this task can be carried out efficiently by *miR-181a* alone (Li et al. 2007). Altogether, the data seem to indicate that *miR-181* family members play an important role in both B- and T-cell differentiation and function.

miR-150 expression is linked to mature, resting B and T cells, but not their progenitors (Monticelli et al. 2005). Overexpression of *miR-150* in mice resulted in a block of B-cell development caused by direct targeting of *c-Myb* by *miR-150* (Xiao et al. 2007), although additional *miR-150* targets may also be involved in these effects. In agreement with these data, a different group reported that retroviral transduction of *miR-150* in hematopoietic progenitor cells (HPCs) led to block of B- but not T-cell development (Zhou et al. 2007). Lack of *miR-150* impairs the ability of mice to generate mature NK cells (Bezman et al. 2011).

A precise role in B-cell development has been demonstrated for *miR-17~92* through various loss-of-function mouse models. Targeted deletion of this cluster in mice led to increased levels of the proapoptotic protein *Bim* (a *miR-17~92* target) and B-cell development arrest from pro-B to pre-B transition (Ventura et al. 2008). This clearly suggests that the *miR-17~92* cluster acts specifically during the transition from pre-B to pro-B lymphocyte development, enhancing the survival of the B-cells at this stage by targeting the proapoptotic *Bim*. Located at 13q31.3 in humans, the cluster consists of six miRNAs: *miR-17, miR-18a, miR-19a, miR-20a, miR-19b,* and *miR-92a*. The cluster has two paralogs: the *miR-106b-25* cluster on chromosome 7, and the *miR-106a-363* cluster on chromosome X. In contrast, overexpression of this cluster in mouse B and T-cells has been shown to cause a lymphoproliferative disease and autoimmunity (Xiao et al. 2008). The cluster is frequently over-expressed in various B-cell lymphomas including follicular lymphoma (FL) and diffuse large cell lymphoma (DLBCL). Genetic dissection of the roles of individual miRNAs in this cluster in B-cell lymphomas led to the identification of *miR-19* as the key player (Mu et al. 2009). The *miR-17~92* cluster also cooperates with the *c-Myc* oncogene to induce B-cell lymphoma in mice (He et al. 2005). Remarkably, *c-Myc* has also been shown to induce transcription of this cluster and hence contribute to its oncogenic activities (O'Donnell et al. 2005). In addition to *Bim*, some of the key targets of this cluster include the tumor suppressor phosphatases *PTEN*, *PP2A*, and *AMP-activated kinase (PRKAA1)*, which play important roles in immune cell development (Ventura et al. 2008; Xiao et al. 2008; Mu et al. 2009; Mavrakis et al. 2010).

miR-155 is encoded from the non-coding RNA called b-cell integration cluster (bic), which is the site of integration for the avian leukosis virus in chickens, and in humans is located at chromosome 21. *MiR-155* is moderately expressed in HSCs or mature hematopoietic cells, but its expression is dramatically induced on antigen receptor activation of mature B and T lymphocytes (reviewed in Baltimore et al. 2008). *MiR-155* is frequently overexpressed in several solid and lymphoid malignancies (Eis et al. 2005; Volinia et al. 2006; Fulci et al. 2007; Kluiver et al. 2007). It may seem like the most widely studied miRNA owing to the availability of multiple genetic mouse models (knockout, knock-in, and B-cell transgenic), but the mechanisms of *miR-155*-induced disease are just beginning to be unveiled. Overexpression of *miR-155* in mouse B-cells under the immunoglobulin promoter and eμ enhancer has been shown to induce pre-B cell leukemia/lymphoma (Costinean et al. 2006). At the same time, retroviral mediated expression in the HSC caused myeloproliferation (O'Connell et al. 2008). We recently showed that part of *miR-155*-induced malignant transformation is attributable to its targeting of *HDAC4* (*histone deacetylase 4*) and indirect regulation of a key transcriptional repressor *BCL6* (*B-cell lymphoma 6*), which leads to derepression of its oncogenic targets (Sandhu et al. 2012). At the same time, another study showed that *BCL6* directly represses *miR-155* and induces expression of its target, AICDA or AID (activation-induced cytidine deaminase) to positively regulate germinal center gene expression (Basso et al. 2012). Hence, this provides interesting autoregulatory mechanisms by which *miR-155* and *BCL6* regulate each other to maintain physiological immune balance.

Although loss of *miR-155* has not been linked to any gross defects in lymphoid or myeloid development, the mice have impaired immunity partly due to the defective germinal center response (Rodriguez et al. 2007; Thai et al. 2007). *MiR-155/bic*[-/-]-deleted B-cells are impaired in production of *tumor necrosis factor (TNF)* and lymphotoxin-a, cytokines that are required for the germinal center response during antigen exposure (Thai et al. 2007). T-cells from these mice showed a bias toward Th2 (T-helper 2) differentiation and higher number of *interleukin-10 (IL-10)* producing T-cells (Thai et al. 2007). Thus, it seems that the physiological role of *miR-155* is mainly to regulate T and B cell responses.

The *miR-15a/16-1* cluster is encoded from the 13q14 locus which is commonly deleted in chronic lymphocytic leukemia and has been specifically designated as a minimal deleted region (Liu et al. 1997; Migliazza et al. 2001). The *miR-15a/16-1* cluster is normally expressed in the CD5$^+$ B-cells. However, its expression is lost in a subset of CD5+ B-cell chronic lymphocytic leukemia (CLL) (Calin et al. 2002). Deletion of *miR-15a/16-1* in mice induces a mature B-cell expansion resembling human CLL (Klein et al. 2010). *miR-15a/16-1* has been mainly proposed to regulate the G_0/G_1-S phase transition in the cell cycle through targeting the key cell cycle checkpoint genes, such as the cyclins: *cyclin D1* (*CCND1*), *cyclin D2* (*CCND2*), *cyclin D3* (*CCND3*), *Cyclin E1* (*CCNE1*), and cyclin-dependent kinases (CDKs): *CDK4, CDK6, CHK1, MCM5* and *CDC25* (Liu et al. 2008; Klein et al. 2010). In addition, the cluster targets the antiapoptotic protein *Bcl2* (Cimmino et al. 2005).

In mice, B220⁺ B-lymphoid and Gr1⁺/Mac1⁺ myeloid cells express high amounts of *miR-142*, establishing a role for it in both lymphoid and myeloid development. Among the hematopoietic tissues, *miR-142* is highly expressed in bone marrow, spleen, and thymus (Chen et al. 2004), though minimal expression is found in erythroid (Ter119⁺) and CD3e⁺ T cells. *MiR-142,* encoded from 17q22 in humans and 11qC in mice, is a site of translocation associated with aggressive B-cell leukemia (Gauwerky et al. 1989; Chen et al. 2004). It lies in the antisense RNA of the BZRAP1 (*Homo sapiens* benzodiazepine receptor-associated protein 1) gene, which has multiple isoforms due to different transcription start sites. Ectopic overexpression of *miR-142* in HSC resulted in 30-40% higher T-cells without much reduction in B-cells (Chen et al. 2004).

III. MONOCYTE AND GRANULOCYTE DEVELOPMENT

The monocyte and granulocyte lineages develop from the GMP, which has differentiated from CMP along with EMP under the influence of specific factors, such as *PU.1, C/EBP*, and *GATA1*. The myeloid transcription factors, *PU.1* and *C/EBP*, have been shown to activate *miR-223* through direct binding to its promoter sequence, while the erythroid transcription factor, GATA1, suppressed its expression (Fukao et al. 2007). *miR-223,* transcribed from LOC389865 (XM_374329) at chromosome Xq12 in humans and XqC3 in mice, is expressed at high levels in the bone marrow (Chen et al. 2004). Specifically, *miR-223* is expressed at low levels in the CD34⁺ progenitors and common myeloid progenitors, increasing steadily in the granulocyte compartment, while it is repressed in the monocyte lineage. Johnnidis et al. reported that *miR-223*-deficient mice have a significant increase in the number of circulating and bone marrow neutrophils/granulocytes (Johnnidis et al. 2008). The neutrophils displayed unusual morphology, aberrant pattern of lineage-specific markers expression and increased reactivity to activating stimuli, including evidence of spontaneous inflammatory lung pathology. Furthermore, the authors showed that *MEF2c*, a transcription factor that promotes myeloid progenitor differentiation, is a *bona fide* target of *miR-223*, and the phenotype of the *miR-223*-deficient mice (neutrophilia and progenitor expansion) was corrected when the *MEF2c* gene was genetically ablated (Johnnidis et al. 2008). Altogether, the data indicate that *miR-223* decreases granulocyte production and dampens its activation.

In monocyte development, *PU.1* also activates *miR-424*, which in turn induces monocytic/macrophage differentiation in acute myeloid leukemia (AML) cell lines and CD34⁺ HPCs at least in part by repressing the transcription factor *NFIA* (Fontana et al. 2007). Another example of miRNA involvement in monocytic differentiation includes *miR-17-5p, -20a*, and *-106a*. By targeting the transcription factor acute myeloid leukemia (AML) 1, *miR-17-5p,-20a,* and *-106a* induce down-regulation of the macrophage colony-stimulating factor receptor, enhance blast proliferation, and inhibit monocytic differentiation and maturation (Fontana et al. 2007). Interestingly, AML1 can silence the expression of these miRNAs in a negative feedback loop by binding to the promoter region of the paralogous clusters *miR-17-92* and *miR-106a-92*.

Finally, there is also evidence that *miR-125b* plays a role in myelopoiesis, since *miR-125b-1* overexpression was shown to arrest myeloid cell differentiation in myelodysplastic syndrome and AML (Bousquet et al. 2008). *miR-125a* and *miR-125b* were originally cloned from brain tissue (Lagos-Quintana et al. 2002). *miR-125b* is encoded from a non-coding RNA, which is also the host for the *miR-100* and *let-7a-2* cluster of miRNAs and is located at 11q24.1 in humans and 9qA5.1 in mouse. *miR-125b* is up-regulated in acute leukemia patients and its ectopic overexpression in mice bone marrow is sufficient to cause leukemia (Bousquet et al. 2010). To investigate the role of *miR-125b* in normal hematopoiesis, Chaudhuri and colleagues used a loss of function system, where a sponge decoy was generated to competitively inhibit *miR-125b* binding with its targets. Remarkably, mice reconstituted with sponge-transduced bone marrow cells had significantly lower myeloid cells compared with controls. The authors further showed that these effects are partially caused by targeting of *LIN28A*, since *LIN28A* overexpression and knockdown mimic important aspects of *miR-125b* loss of function and gain of function (Chaudhuri et al. 2012).

IV. ERYTHROCYTE AND MEGAKARYOCYTE DEVELOPMENT

A systematic analysis of miRNA expression of erythrocyte precursors, obtained from peripheral blood mononuclear cells cultured in a three-phase liquid system, revealed a progressive down-regulation of *miR-150, miR-155, miR-221*, and *miR-222,* up-regulation of *miR-16* and *miR-451* at late stages of differentiation, and a biphasic regulation of *miR-339* and *miR-378* (Bruchova et al. 2007).

Felli et al. identified the down-regulation of *miR-221* and *miR-222* during erythrocyte differentiation of CD34$^+$ HPCs (Felli et al. 2005). The ectopic expression of these two miRNAs in CD34$^+$ HPCs inhibits erythrocyte growth and *cKIT* protein expression. Further experiments indicated that the decline in *miR-221* and *miR-222* unblocks *cKIT* expression, facilitating the expansion of early erythroblasts (Felli et al. 2005). The *miR-221/222* cluster is located at Xp11.3 in humans and XqA1.3 in mouse.

Erythropoiesis-specific *miR-144/451* cluster was identified to be under the transcriptional control of the master erythrocyte regulator GATA-1 using a conditional GATA-1 cell line (Dore et al. 2008). To further investigate the role of this cluster in erythrocytic differentiation, the authors used antisense morpholino oligomers to interfere with the expression of this cluster in zebrafish embryos. Embryos injected with *miR-451* antisense exhibited normal erythroid precursors, but their development into mature circulating red cells was strongly impaired. In contrast, no alterations were observed in zebrafish embryos injected with *miR-144* antisense (Dore et al. 2008). *MiR-144/451* is located at 17q11.2 in humans and 11qB5 in mice.

The commitment of CD34+ HPCs to the megakaryocyte lineage occurs in parallel with the down-regulation of a panel of 20 miRNAs. Among them, *miR-10a* and *miR-130a* were found to down-regulate *MafB* and *HOXA*-1, both genes up-regulated during this process, suggesting that perhaps miRNAs unblock their expression (Garzon et al. 2006).

Megakaryopoiesis has also been linked to *miR-150*. Through gain- and loss-of-function experiments, Lu et al. demonstrated that *miR-150* drives megakaryocyte–erythrocyte progenitors differentiation toward megakaryocytes at the expense of erythroid cells (Lu et al. 2008). Further experiments identified the transcription factor c*MYB* as a critical target of *miR-150* in this regulation (Lu et al. 2008).

V. CONCLUSIONS

The specific set of miRNAs expressed in the hematopoietic niches of the bone marrow, spleen, and thymus regulate the mRNA expression in these tissues and together with transcription factors regulate the complex process of hematopoiesis. The intricate balance between the miRNAs and transcription factor expression drives the lineage determination decisions and subsequent maturation to functional immune cells. We are starting to understand the crosstalk between these players. The combination of novel profiling techniques, such as deep sequencing and sophisticated animal knock-in and knockout models, will provide the tools needed to unravel these interactions.

REFERENCES

Baltimore, D., M.P. Boldin, R.M. O'Connell, D.S. Rao, and K.D. Taganov. 2008. MicroRNAs: new regulators of immune cell development and function. *Nat Immunol* **9**:839–845.

Bartel, D.P. 2004. MicroRNAs: genomics, biogenesis, mechanism, and function. *Cell* **116**:281–297.

Basso, K., C. Schneider, Q. Shen, A.B. Holmes, M. Setty, C. Leslie, and R. Dalla-Favera. 2012. BCL6 positively regulates AID and germinal center gene expression via repression of *miR-155*. *J Exp Med* **209**:2455–2465.

Bezman, N.A., T. Chakraborty, T. Bender, and L.L. Lanier. 2011. *miR-150* regulates the development of NK and iNKT cells. *J Exp Med* **208**:2717–2731.

Bousquet, M., M.H. Harris, B. Zhou, and H.F. Lodish. 2010. MicroRNA *miR-125b* causes leukemia. *Proc Natl Acad Sci U S A* **107**:21558–21563.

Bousquet, M., C. Quelen, R. Rosati, V. Mansat-De Mas, R. La Starza, C. Bastard, E. Lippert, P. Talmant, M. Lafage-Pochitaloff, D. Leroux, C. Gervais, F. Viguie, et al. 2008. Myeloid cell differentiation arrest by *miR-125b-1* in myelodysplastic syndrome and acute myeloid leukemia with the t(2;11)(p21;q23) translocation. *J Exp Med* **205**:2499–2506.

Bruchova, H., D. Yoon, A.M. Agarwal, J. Mendell, and J.T. Prchal. 2007. Regulated expression of microRNAs in normal and polycythemia vera erythropoiesis. *Exp Hematol* **35**:1657–1667.

Calin, G.A., C.D. Dumitru, M. Shimizu, R. Bichi, S. Zupo, E. Noch, H. Aldler, S. Rattan, M. Keating, K. Rai, L. Rassenti, T. Kipps, et al. 2002. Frequent deletions and down-regulation of micro- RNA genes miR15 and miR16 at 13q14 in chronic lymphocytic leukemia. *Proc Natl Acad Sci U S A* **99**:15524–15529.

Chaudhuri, A.A., A.Y. So, A. Mehta, A. Minisandram, N. Sinha, V.D. Jonsson, D.S. Rao, R.M. O'Connell, D. Baltimore. 2012. Oncomir miR-125b regulates hematopoiesis by targeting the gene Lin28A. *Proc Natl Acad Sci U S A* **109**:4233–4238.

Chen, C.Z., L. Li, H.F. Lodish, and D.P. Bartel. 2004. MicroRNAs modulate hematopoietic lineage differentiation. *Science (New York, N.Y.)* **303**:83–86.

Cichocki, F., M. Felices, V. McCullar, S.R. Presnell, A. Al-Attar, C.T. Lutz, and J.S. Miller. 2011. Cutting edge: microRNA-181 promotes human NK cell development by regulating Notch signaling. *J Immunol* **187**:6171–6175.

Cimmino, A., G.A. Calin, M. Fabbri, M.V. Iorio, M. Ferracin, M. Shimizu, S.E. Wojcik, R.I. Aqeilan, S. Zupo, M. Dono, L. Rassenti, H. Alder, et al. 2005. miR-15 and miR-16 induce apoptosis by targeting BCL2. *Proc Natl Acad Sci U S A* **102**:13944–13949.

Cobb, B.S., T.B. Nesterova, E. Thompson, A. Hertweck, E. O'Connor, J. Godwin, C.B. Wilson, N. Brockdorff, A.G. Fisher, S.T. Smale, and M. Merkenschlager. 2005. T cell lineage choice and differentiation in the absence of the RNase III enzyme Dicer. *J Exp Med* **201**:1367–1373.

Costinean, S., N. Zanesi, Y. Pekarsky, E. Tili, S. Volinia, N. Heerema, and C.M. Croce. 2006. Pre-B cell proliferation and lymphoblastic leukemia/high-grade lymphoma in E(mu)-miR155 transgenic mice. *Proc Natl Acad Sci U S A* **103**:7024–7029.

Dore, L.C., J.D. Amigo, C.O. Dos Santos, Z. Zhang, X. Gai, J.W. Tobias, D. Yu, A.M. Klein, C. Dorman, W. Wu, R.C. Hardison, B.H. Paw, et al. 2008. A GATA-1-regulated microRNA locus essential for erythropoiesis. *Proc Natl Acad Sci U S A* **105**:3333–3338.

Eis, P.S., W. Tam, L. Sun, A. Chadburn, Z. Li, M.F. Gomez, E. Lund, and J.E. Dahlberg. 2005. Accumulation of *miR-155* and BIC RNA in human B cell lymphomas. *Proc Natl Acad Sci U S A* **102**:3627–3632.

REFERENCES

Felli, N., L. Fontana, E. Pelosi, R. Botta, D. Bonci, F. Facchiano, F. Liuzzi, V. Lulli, O. Morsilli, S. Santoro, M. Valtieri, G.A. Calin, et al. 2005. MicroRNAs 221 and 222 inhibit normal erythropoiesis and erythroleukemic cell growth via kit receptor down-modulation. *Proc Natl Acad Sci U S A* **102**:18081–18086.

Fontana, L., E. Pelosi, P. Greco, S. Racanicchi, U. Testa, F. Liuzzi, C.M. Croce, E. Brunetti, F. Grignani, and C. Peschle. 2007. MicroRNAs 17-5p-20a-106a control monocytopoiesis through AML1 targeting and M-CSF receptor upregulation. *Nat Cell Biol* **9**:775–787.

Fukao, T., Y. Fukuda, K. Kiga, J. Sharif, K. Hino, Y. Enomoto, A. Kawamura, K. Nakamura, T. Takeuchi, and M. Tanabe. 2007. An evolutionarily conserved mechanism for microRNA-223 expression revealed by microRNA gene profiling. *Cell* **129**:617–631.

Fulci, V., S. Chiaretti, M. Goldoni, G. Azzalin, N. Carucci, S. Tavolaro, L. Castellano, A. Magrelli, F. Citarella, M. Messina, R. Maggio, N. Peragine, et al. 2007. Quantitative technologies establish a novel microRNA profile of chronic lymphocytic leukemia. *Blood* **109**:4944–4951.

Garzon, R., F. Pichiorri, T. Palumbo, R. Iuliano, A. Cimmino, R. Aqeilan, S. Volinia, D. Bhatt, H. Alder, G. Marcucci, G.A. Calin, C.G. Liu, et al. 2006. MicroRNA fingerprints during human megakaryocytopoiesis. *Proc Natl Acad Sci U S A* **103**:5078–5083.

Gauwerky, C.E., K. Huebner, M. Isobe, P.C. Nowell, and C.M. Croce. 1989. Activation of MYC in a masked t(8;17) translocation results in an aggressive B-cell leukemia. *Proc Natl Acad Sci U S A* **86**:8867–8871.

He, L., J.M. Thomson, M.T. Hemann, E. Hernando-Monge, D. Mu, S. Goodson, S. Powers, C. Cordon-Cardo, S.W. Lowe, G.J. Hannon, and S.M. Hammond. 2005. A microRNA polycistron as a potential human oncogene. *Nature* **435**:828–833.

Johnnidis, J.B., M.H. Harris, R.T. Wheeler, S. Stehling-Sun, M.H. Lam, O. Kirak, T.R. Brummelkamp, M.D. Fleming, and F.D. Camargo. 2008. Regulation of progenitor cell proliferation and granulocyte function by microRNA-223. *Nature* **451**:1125–1129.

Klein, U., M. Lia, M. Crespo, R. Siegel, Q. Shen, T. Mo, A. Ambesi-Impiombato, A. Califano, A. Migliazza, G. Bhagat, and R. Dalla-Favera. 2010. The DLEU2/*miR-15a/16-1* cluster controls B cell proliferation and its deletion leads to chronic lymphocytic leukemia. *Cancer Cell* **17**: 28–40.

Kluiver, J., A. van den Berg, D. de Jong, T. Blokzijl, G. Harms, E. Bouwman, S. Jacobs, S. Poppema, and B.J. Kroesen. 2007. Regulation of pri-microRNA BIC transcription and processing in Burkitt lymphoma. *Oncogene* **26**:3769–3776.

Lagos-Quintana, M., R. Rauhut, A. Yalcin, J. Meyer, W. Lendeckel, and T. Tuschl. 2002. Identification of tissue-specific microRNAs from mouse. *Curr Biol* **12**:735–739.

Li, Q.J., J. Chau, P.J. Ebert, G. Sylvester, H. Min, G. Liu, R. Braich, M. Manoharan, J. Soutschek, P. Skare, L.O. Klein, M.M. Davis, et al. 2007. miR-181a is an intrinsic modulator of T cell sensitivity and selection. *Cell* **129**:147–161.

Liu, Q., H. Fu, F. Sun, H. Zhang, Y. Tie, J. Zhu, R. Xing, Z. Sun, and X. Zheng. 2008. *miR-16* family induces cell cycle arrest by regulating multiple cell cycle genes. *Nucleic Acids Res* **36**:5391–5404.

Liu, X., Y. Wang, Q. Sun, J. Yan, J. Huang, S. Zhu, and J. Yu. 2012. Identification of microRNA transcriptome involved in human natural killer cell activation. *Immunol Lett* **143**:208–217.

Liu, Y., M. Corcoran, O. Rasool, G. Ivanova, R. Ibbotson, D. Grander, A. Iyengar, A. Baranova, V. Kashuba, M. Merup, X. Wu, A. Gardiner, et al. 1997. Cloning of two candidate tumor suppressor genes within a 10 kb region on chromosome 13q14, frequently deleted in chronic lymphocytic leukemia. *Oncogene* **15**:2463–2473.

Lu, J., S. Guo, B.L. Ebert, H. Zhang, X. Peng, J. Bosco, J. Pretz, R. Schlanger, J.Y. Wang, R.H. Mak, D.M. Dombkowski, F.I. Preffer, et al. 2008. MicroRNA-mediated control of cell fate in megakaryocyte-erythrocyte progenitors. *Dev Cell* **14**:843–853.

Mavrakis, K.J., A.L. Wolfe, E. Oricchio, T. Palomero, K. de Keersmaecker, K. McJunkin, J. Zuber, T. James, A.A. Khan, C.S. Leslie, J.S. Parker, P.J. Paddison, et al. 2010. Genome-wide RNA-mediated interference screen identifies *miR-19* targets in Notch-induced T-cell acute lymphoblastic leukaemia. *Nat Cell Biol* **12**:372–379.

Migliazza, A., F. Bosch, H. Komatsu, E. Cayanis, S. Martinotti, E. Toniato, E. Guccione, X. Qu, M. Chien, V.V. Murty, G. Gaidano, G. Inghirami, et al. 2001. Nucleotide sequence, transcription map, and mutation analysis of the 13q14 chromosomal region deleted in B-cell chronic lymphocytic leukemia. *Blood* **97**:2098–2104.

Monticelli, S., K.M. Ansel, C. Xiao, N.D. Socci, A.M. Krichevsky, T.H. Thai, N. Rajewsky, D.S. Marks, C. Sander, K. Rajewsky, A. Rao, and K.S. Kosik. 2005. MicroRNA profiling of the murine hematopoietic system. *Genome Biol* **6**:R71.

Mu, P., Y.C. Han, D. Betel, E. Yao, M. Squatrito, P. Ogrodowski, E. de Stanchina, A. D'Andrea, C. Sander, and A. Ventura. 2009. Genetic dissection of the *miR-17~92* cluster of microRNAs in Myc-induced B-cell lymphomas. *Genes Dev* **23**:2806–2811.

Neilson, R., G.X.Y. Zheng, C.B. Burge, and P.A. Sharpj. 2007. Dynamic regulation of miRNA expression in ordered stages of cellular development. *Genes Dev* **21**:578–589.

O'Carroll, D., I. Mecklenbrauker, P.P. Das, A. Santana, U. Koenig, A.J. Enright, E.A. Miska, and A. Tarakhovsky. 2007. A Slicer-independent role for Argonaute 2 in hematopoiesis and the microRNA pathway. *Genes Dev* **21**:1999–2004.

O'Connell, R.M., D.S. Rao, A.A. Chaudhuri, M.P. Boldin, K.D. Taganov, J. Nicoll, R.L. Paquette, and D. Baltimore. 2008. Sustained expression of microRNA-155 in hematopoietic stem cells causes a myeloproliferative disorder. *J Exp Med* **205**:585–594.

O'Donnell, K.A., E.A. Wentzel, K.I. Zeller, C.V. Dang, and J.T. Mendell. 2005. c-Myc-regulated microRNAs modulate E2F1 expression. *Nature* **435**:839–843.

Rodriguez, A., E. Vigorito, S. Clare, M.V. Warren, P. Couttet, D.R. Soond, S. van Dongen, R.J. Grocock, P.P. Das, E.A. Miska, D. Vetrie, K. Okkenhaug, et al. 2007. Requirement of bic/microRNA-155 for normal immune function. *Science (New York, N.Y.)* **316**:608–611.

Sandhu, S.K., S. Volinia, S. Costinean, M. Galasso, R. Neinast, R. Santhanam, M.R. Parthun, D. Perrotti, G. Marcucci, R. Garzon, and C.M. Croce. 2012. *miR-155* targets histone deacetylase 4 (HDAC4) and impairs transcriptional activity of B-cell lymphoma 6 (BCL6) in the Emu-*miR-155* transgenic mouse model. *Proc Natl Acad Sci U S A* **109**(49):20047–20052.

Thai, T.H., D.P. Calado, S. Casola, K.M. Ansel, C. Xiao, Y. Xue, A. Murphy, D. Frendewey, D. Valenzuela, J.L. Kutok, M. Schmidt-Supprian, N. Rajewsky, et al. 2007. Regulation of the germinal center response by microRNA-155. *Science (New York, N.Y.)* **316**:604–608.

Trotta, R., L. Chen, D. Ciarlariello, S. Josyula, C. Mao, S. Costinean, L. Yu, J.P. Butchar, S. Tridandapani, C.M. Croce, and M.A. Caligiuri. 2012. *miR-155* regulates IFN-gamma production in natural killer cells. *Blood* **119**:3478–3485.

Ventura, A., A.G. Young, M.M. Winslow, L. Lintault, A. Meissner, S.J. Erkeland, J. Newman, R.T. Bronson, D. Crowley, J.R. Stone, R. Jaenisch, P.A. Sharp, et al. 2008. Targeted deletion reveals essential and overlapping functions of the *miR-17* through 92 family of miRNA clusters. *Cell* **132**:875–886.

Volinia, S., G.A. Calin, C.G. Liu, S. Ambs, A. Cimmino, F. Petrocca, R. Visone, M. Iorio, C. Roldo, M. Ferracin, R.L. Prueitt, N. Yanaihara, et al. 2006. A microRNA expression signature of human solid tumors defines cancer gene targets. *Proc Natl Acad Sci U S A* **103**:2257–2261.

Xiao, C., D.P. Calado, G. Galler, T.H. Thai, H.C. Patterson, J. Wang, N. Rajewsky, T.P. Bender, and K. Rajewsky. 2007. *MiR-150* controls B cell differentiation by targeting the transcription factor c-Myb. *Cell* **131**:146–159.

Xiao, C., L. Srinivasan, D.P. Calado, H.C. Patterson, B. Zhang, J. Wang, J.M. Henderson, J.L. Kutok, and K. Rajewsky. 2008. Lymphoproliferative disease and autoimmunity in mice with increased *miR-17-92* expression in lymphocytes. *Nat Immunol* **9**:405–414.

Zhou, B., S. Wang, C. Mayr, D.P. Bartel, and H.F. Lodish. 2007. *miR-150,* a microRNA expressed in mature B and T cells, blocks early B cell development when expressed prematurely. *Proc Natl Acad Sci U S A* **104**:7080–7085.

7

MicroRNAs in Platelet Production and Activation

Leonard C. Edelstein, Srikanth Nagalla, and Paul F. Bray

The Cardeza Foundation for Hematologic Research and the Department of Medicine, Jefferson Medical College, Thomas Jefferson University, Philadelphia, PA, USA

I.	Introduction	102
II.	miRNA Biogenesis and Function	102
III.	miRNAs and Megakaryocytopoiesis	103
	A. *miR-155*	104
	B. *miR-150*	104
	C. *miR-146a*	106
	D. *miR-34a*	107
	E. Other miRNAs	107
IV.	Platelet miRNAs	108
V.	Platelet miRNAs as Biomarkers	109
VI.	Platelet Microparticles and miRNAs	110
VII.	Summary and Future Directions	110
	References	111

ABBREVIATIONS

C. elegans	*Caenorhabditis elegans*
CFU-Mk	colony-forming unit—megakaryocytic
CLP	common lymphoid progenitor
CMP	common myeloid progenitor

MicroRNAs in Medicine, First Edition. Edited by Charles H. Lawrie.
© 2014 John Wiley & Sons, Inc. Published 2014 by John Wiley & Sons, Inc.

DNA	deoxyribonucleic acid
ET	essential thrombocythemia
GWAS	genome-wide association study
H. sapiens	*Homo sapiens*
HSC	hematopoietic stem cell
IL-6	interleukin-6
LNA	locked nucleaic acid
MEP	megakaryocyte-erythrocyte progenitor
miRNA	micro RNA
mRNA	messenger RNA
NK	natural killer cells
PMP	platelet-derived microparticle
PV	polycythemia vera
RNA	ribonucleic acid
SNP	single-nucleotide polymorphism
TPA	12-O-tetradecanoylphorbol-13-acetate
UTR	untranslated region

I. INTRODUCTION

The completion of the Human Genome Project in 2003 produced an unexpectedly low estimate of the number of human protein coding genes. It was not immediately clear how the high degree of human complexity could derive from "only" ~22,000 genes. Since this number was similar to that in much simpler organisms, there had to be other mechanisms to account for the diversity in human anatomy and physiology. In the past 10 years, many additional sources of variation have been documented, including posttranslational modification of proteins, alternative splicing of protein coding mRNAs, and epigenetic regulation of the timing and quantity of those proteins. Among the most interesting discoveries of variation is the realization that at least a portion of non-protein coding "junk" DNA, ~70% of the genome, contains information for the formation of non-coding RNAs. This non-protein coding portion of the genome may be a great source of diversity, as evidenced by the fact that while there is only a 21% increase in the number of protein coding genes in *Homo sapiens* as compare with *Caenorhabditis elegans,* there is almost 17-fold higher ratio of non-protein coding to protein coding sequence (Shabalina and Spiridonov 2004). While there are several classes of non-coding RNAs (reviewed in Esteller 2011), the most intensively studied are microRNAs (miRNAs).

MiRNAs were discovered in *C. elegans* in 1993 (Lee et al. 1993; Wightman et al. 1993), and have been shown to play important roles in developmental biology, cellular stress, circadian rhythm, and immunology, as well as numerous disease states, including Alzheimer's disease, cancer, and heart failure. Abundant evidence demonstrates a critical role for miRNAs in normal human hematopoiesis, and dysregulated miRNA biology has been associated with and shown to cause numerous hematologic diseases (Table 7.1).

II. miRNA BIOGENESIS AND FUNCTION

miRNAs are 21–23 nucleotide regulatory RNAs expressed in multicellular organisms, from viruses to plants to humans (Bartel 2004). The latest version of miRBase (as of

TABLE 7.1. MicroRNA Dysregulation and Hematological Disease

Hematologic Disorder	Implicated miRNA	References
Acute leukemia	miR-125b-2, miR-29b, miR-146a, miR-181a, miR-204	Debernardi et al. (2007), Garzon et al. (2008, 2009a, 2009b), Klusmann et al. (2010), and Starczynowski et al. (2011b)
Chronic leukemia	miR-15a, miR-16-1, miR-155, miR-29b, miR-181a, miR-328	Calin et al. (2004), Bottoni et al. (2005), Fulci et al. (2007), Xu and Li (2007), Marton et al. (2008), Visone et al. (2009), Eiring et al. (2010), and Ichimura et al. (2010)
Myeloproliferative disorders	miR-28, miR-150, let-7a, miR-182, miR-26b, miR-143, miR-145, miR-223,miR-26b, miR-30b, miR-30c	Bruchova et al. (2007, 2008) and Girardot et al. (2010)
Myelodysplasia	miR-145, miR-146a	Starczynowski et al. (2010)
Lymphoma	miR-21, miR-155, miR-17-92 cluster	Eis et al. (2005), He et al. (2005), and Lawrie et al. (2007)

August 14, 2013, version 20, http://www.mirbase.org/) lists 2578 mature human miRNAs. MiRNAs regulate most (>60%) mammalian protein coding genes primarily by repressing gene expression (Friedman et al. 2009). Some miRNAs are expressed ubiquitously, but many are tissue and/or developmental stage specific (Wienholds et al. 2005). Cell miRNA content is highly variable and ranges from 1 to 10,000 copies (Chen et al. 2005).

A fundamental aspect of miRNA function relates to mRNA targeting: most miRNAs are predicted to target multiple mRNAs and most mRNAs have predicted binding sites for multiple miRNAs. Different web-based prediction tools are publicly available that predict miRNA binding sites, including TargetScan, Miranda, PicTar, RNA22, and Microcosm. However, these tools differ fundamentally in their algorithms for prediction, and there is a lack of consensus on the optimal method (Rigoutsos and Tsirigos 2011). Work by Guo et al. indicates that miRNA knockdown of mammalian protein expression is primarily via mRNA degradation, although evidence in *Drosophila* and zebrafish indicate translational inhibition precedes mRNA degradation (Guo et al. 2010; Bazzini et al. 2012; Djuranovic et al. 2012). MiRNAs have been aptly referred to as "rheostats" because their regulatory impact is generally to fine-tune protein expression (Baek et al. 2008; Selbach et al. 2008). Importantly, small differences (as little as a 20% change) in miRNA levels have been shown to cause autoimmune disease and predispose to malignancy (Alimonti et al. 2010). The significance of proper miRNA synthesis and function is underscored by diseases caused by genetic defects at virtually all steps in miRNA biogenesis and targeting, including miRNA gene deletions/duplications, SNPs in the target mRNA 3'-UTR, miRNA "decoys," and genetic variants in the proteins mediating miRNA biogenesis (Sethupathy and Collins 2008; Bandiera et al. 2010; Poliseno et al. 2010).

III. miRNAs AND MEGAKARYOCYTOPOIESIS

Hematopoiesis is the process by which the various blood cell types are derived from a progenitor hematopoietic stem cells (HSCs). HSCs differentiate into common myeloid progenitors (CMPs) and common lymphoid progenitors (CLPs). Megakaryocytes, erythrocytes, granulocytes, and monocytes are derived from CMPs, while lymphocytes and natural killer cells (NKs) are derived from CLPs. Platelets are formed from the cytokinesis of long megakaryocytic processes called proplatelets. The process by which megakaryocytes are formed is called megakaryopoiesis, and defects in it can result in a deficit of platelets (thrombocytopenia) or an excess of platelets (thrombocytosis), both of which can have clinical consequences.

Numerous laboratories have established the essential role of *Dicer1* (Raaijmakers et al. 2010) and miRNAs in hematopoiesis (Chen et al. 2004; Monticelli et al. 2005; Georgantas et al. 2007), including embryonic stem cell differentiation (Tay et al. 2008), erythropoiesis (Felli et al. 2005; Bruchova et al. 2007; Choong et al. 2007; Dore et al. 2008; Pase et al. 2009; Zhao et al. 2009), granulocytopoiesis/monocytopoiesis (Fazi et al. 2005; Fontana et al. 2007; Johnnidis et al. 2008; O'Connell et al. 2008), and lymphopoiesis (Chen et al. 2004; Neilson et al. 2007; Xiao et al. 2007; Zhou et al. 2007; Ventura et al. 2008). Since the first report by Garzon et al. in 2006, (Garzon et al. 2006), many additional studies have addressed various aspects of miRNAs in megakaryocytopoiesis using megakaryocytes generated from *in vitro* cultured CD34$^+$ HSCs or transformed cell lines with megakaryocytic properties (Table 7.2). Both unbiased miRNA profiling and candidate miRNA studies have established associations between specific miRNAs and the developmental stage of megakaryocyte progenitors. Functionality has been tested by assessing the effects of candidate miRNAs on *in vitro* and *in vivo* proliferation and differentiation after overexpression or knockdown. Mechanistic assessment of these miRNAs has included target protein knockdown and reporter gene assays using constructs with the putative target 3′-UTR.

A. *miR-155*

Georgantas and colleagues performed both mRNA and miRNA expression profiling on CD34$^+$ HSCs from healthy subjects (Georgantas et al. 2007), and used a bioinformatic approach to predict candidate miRNAs that target mRNAs encoding transcription factors associated with hematopoietic differentiation. *miR-155* was predicted to repress expression of eight different CD34$^+$-associated transcription factor mRNAs that regulate myelopoiesis, C/EBPβ, CREBP, JUN, MEIS1, PU.1, AGTR1, AGTR2, and FOS. *miR-155* expression was dramatically reduced when CD34$^+$ HSCs were differentiated along the megakaryocyte lineage (Georgantas et al. 2007; Romania et al. 2008). Forced overexpression of *miR-155* inhibited K562, a chronic myelogenous leukemia cell line, differentiation and reduced CD34$^+$ HSC-derived myeloid and erythroid colony formation *in vitro* (Georgantas et al. 2007). Transplantation of HSCs overexpressing *miR-155* into irradiated mice caused reduced numbers of megakaryocytes in recipient bone marrow (O'Connell et al. 2008). These studies suggest that the targets of these transcription factors drive the cell toward differentiation. Although the effect of *miR-155* on platelet count has not been studied, these studies provide strong support that *miR-155* inhibits megakaryocytopoiesis.

B. *miR-150*

By profiling primary cells from human umbilical cord blood, Lu and colleagues discovered *miR-150* levels increased as megakaryocyte–erythrocyte progenitors (MEPs) differentiated toward the megakaryocyte lineage, but not the erythroid lineage (Lu et al. 2008).

TABLE 7.2. MiRNAs Involved in Megakaryocytopoiesis

miRNA	Reference
Overexpression of *miR-155* in K562 cells caused a block in megakaryocytic differentiation.	Georgantas et al. (2007)
MiRNA profile of *in vitro* differentiated CD34$^+$ bone marrow (BM) cells. Identified *miR-130a* targets MAFB and *mir-10a* targets HOXA1.	Garzon et al. (2006)
Overexpression of *miR-155* in transplanted BM lead to a decrease in erythrocytes, MKs, and lymphocytes.	O'Connell et al. (2008)
Calcium release suppresses pre-*miR-181a* in Meg-01 cells. Addition on a *miR-181a* analog blocked Ca$^+$-induced differentiation and induced apoptosis.	Guimaraes-Sternberg et al. (2006)
The PLZF transcription factor represses levels of *mir-146a*, which in turn targets CXCR4 expression. Forced expression or repression alters megakaryocyte (MK) development (See text for details).	Labbaye et al. (2008)
Expression of *miR-150* drives MK-erythroid precursors toward the megakaryocytic fate in murine BM transplant experiments.	Lu et al. (2008)
Thrombopoietin induces expression of *miR-150* in UT-7/TPO cells, which in turn targets expression of the transcription factor c-Myb.	Barroga et al. (2008)
miR-155 levels decrease during megakaryopoiesis in cultured human cord blood. Enforced expression of *miR-155* impairs MK proliferation and development. *miR-155* targets the expression of Ets-1 and Meis transcription factors.	Romania et al. (2008)
MiRNA profile of laser-dissected MKs from primary myelofibrosis essential thrombocythemia patients.	Hussein et al. (2009)
miR-34a is rapidly increased during TPA-induced differentiation of K562 cells into MKs. Over expression of *miR-34a* enhances MK differentiation in HSCs and regulates c-Myb expression.	Navarro et al. (2009)
TPA induces Runx1 binding to the *mir-27a* regulatory region and causes an increase in *mir-27a* in K562 cells. *mir-27a*, in turn, targets and suppresses Runx1 levels.	Ben-Ami et al. (2009)
miR-28 targets the thrombopoietin receptor, MPL. Enforced expression on *miR-28* prevents MK differentiation in CD34$^+$ hematopoietic precursors and is overexpressed in a fraction of platelets in patients with myeloproliferative neoplasms.	Girardot et al. (2010)
miR-146a increases during MK development. See text for details.	Opalinska et al. (2010)
Enforced expression of *mir-146a* in transduced BM cells has no effect on platelet number. See text for details.	Starczynowski et al. (2011a)
miR-125b-2 is overexpressed in Down syndrome–acute megakaryoblastic leukemia patients. *MiR-125b-2* overexpression induces proliferation and differentiation of MK and MK/erythroid precursors.	Klusmann et al. (2010)
PMA treatment in K562 cells increases *miR-34a* levels. *miR-34a*, in turn, targets mitogen-activated protein kinase kinase 1(MEK1) and represses proliferation.	Ichimura et al. (2010)
miR-145 and *miR-146a* are mediators of the 5q-syndrome phenotype. See text for details.	Starczynowski et al. (2010)
Loss of *miR-145* in 5q syndrome leads to an increase MK production via an increase in the Fli-1 transcription factor	Kumar et al. (2011)
Overexpression *miR-181* targets Lin28, which is a negative regulator *Let-7*, which leads to an increase in megakaryocytic differentiation in K562 cells.	Li et al. (2012)

Overexpression of *miR-150* enhanced both *in vitro* and *in vivo* megakaryocyte differentiation at the expense of erythroid differentiation, suggesting a critical switching function at the level of the MEP. *miR-150* also knocked down expression of MYB via its 3'-UTR, consistent with data showing low c-Myb levels promote megakaryocytopoiesis (Emambokus et al. 2003). These findings, coupled with work from the Kaushansky laboratory showing thrombopoietin up-regulates *miR-150* (Barroga et al. 2008), underscore a critical role for *miR-150* in promoting megakaryocytopoiesis.

C. *miR-146a*

miR-146a levels have been reported to dramatically change upon megakaryocytic differentiation of HSCs, but there is conflicting evidence as to the direction of change (Labbaye et al. 2008; Opalinska et al. 2010; Starczynowski et al. 2010, 2011a). Opalinska et al. reported that in murine and human hematopoietic stem cells induced to differentiate into megakaryocytes, the expression level of *miR-146a* increased. This is consistent with previous studies by Landry et al. (Landry et al. 2009) and more recent work by our group in which we observe that *miR-146a* is in the top quartile of expressed miRNAs (Figure 7.1). However, forced expression of *miR-146a* was found to have no effect on megakaryocyte colony number (CFU-Mk), marker expression, or platelet activation (Opalinska et al. 2010). In contrast, Labbaye et al. found that *miR-146a* decreased in human cord blood stem cells induced to differentiate into megakaryocytes, and that forced expression of this miRNA caused a reduction in the number of polyploid cells, and inhibition of *miR-146a* by an antagomir resulted in an increase in the number of polyploid cells (Labbaye et al. 2008). Labbaye et al. also reported that *miR-146a* targeted mRNA coding for CXCR4, the receptor for SDF-1, a hematopoietic mobilization/homing factor. Two studies by Starcynowski et al. did not clarify the matter either. They reported that the level of *miR-146a* is lower in megakaryocyte/erythroid precursors in mice relative to hematopoietic

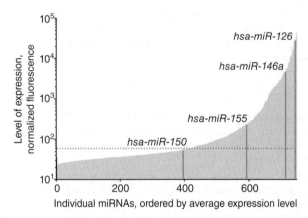

Figure 7.1. Variation in levels of platelet miRNAs. Levels of individual human platelet miRNAs are displayed, arbitrarily ordered from lowest to highest expression (unpublished data). As examples, the expression levels of *miR-150*, *miR-155*, *miR-126*, and *miR-146a* are indicated. 750 human miRNAs were profiled from leukocyte depleted platelet RNAs from 19 human volunteers using the miRCURY™ LNA Array Version 11.0 (Exiqon, Vedbaek, Denmark). Expression levels ranged over four orders on magnitude. The background level is indicated by the horizontal dashed line.

stem cells, similar to what Labbaye et al. found in human cells, but that forced expression in mice had no effect on platelet number, similar to Opalinska et al. (Opalinska et al. 2010; Starczynowski et al. 2011a). However, when down-regulated in mice by decoy targets or anti-miRNA locked nucleic acids (LNA), the CFU-Mk and platelet number increases, and the ploidy of the megakaryocytes decreases, again similar to Labbaye et al. (Labbaye et al. 2008; Starczynowski et al. 2010). Starczynowski reported that *miR-146a* targeted TRAF6, a component of the IL-6 signal pathway, which leads to megakaryocytic survival, differentiation, and platelet formation (Kishimoto 2005). These apparently conflicting findings may stem from species differences or differing experimental culture conditions (since *miR-146a* levels vary according to lineage). It is also important to note that the overexpression experiments performed by Opalinska et al. and Starcynowski et al., in which no effect on megakaryopoiesis was observed, the readout was reconstitution of hematopoiesis *in vivo* in a mouse after bone marrow transplant, as opposed to *in vitro* differentiation in culture used by Labbaye et al. when an effect *was* observed. It is also possible that different experimental conditions result in different levels of *miR-146a* with differing functional consequences. These differences could arise because Ago proteins preferentially associate with transcripts that contain targets for the highly expressed miRNAs, and mRNAs with a greater number of target sites in their 3′-UTR are subject to stricter control (Schmitter et al. 2006; Landthaler et al. 2008). Therefore, differing levels of miRNA expression, as can be generated by overexpression experiments, can result in different mRNA levels and experimental results. Alternatively, since SNPs in the 3′-UTR binding site, as well as the surrounding sequence, can significantly alter miRNA effects (Sethupathy and Collins 2008), it may be necessary to consider the precise mRNA target sequences in these different studies.

D. *miR-34a*

Increased levels of *miR-34a* have consistently been observed in K562 cells stimulated to differentiate with PMA (Navarro et al. 2009; Ichimura et al. 2010). When overexpressed in K562 cells, *miR-34a*-inhibited K562 cell proliferation and promoted differentiation. Importantly, *miR-34a* increased megakaryocyte colony formation from CD34$^+$ HSCs (Navarro et al. 2009). Targets of *miR-34a* include MYB, a negative regulator of megakaryocytopoiesis, and CDK4 and CDK6, regulators of the cell cycle (Navarro et al. 2009). Thus, *miR-34a* appears to enhance megakaryocytopoiesis, at least *in vitro*.

E. Other miRNAs

A number of other miRNAs may regulate megakaryocytopoiesis, although these miRNAs have been less well studied. Several investigators have observed that more miRNAs were down-regulated than up-regulated during megakaryocytopoiesis, but biphasic patterns of expression have also been observed (Garzon et al. 2006; Bruchova et al. 2007; Opalinska et al. 2010). Several studies have identified negative correlations between miRNA levels and target mRNAs that encode transcription factors known to be involved in hematopoiesis. Examples include *miR-130a* and MAFB, *miR-10a* and HOXA1, and *miR-27a* and Runx1 (Garzon et al. 2006; Ben-Ami et al. 2009). Forced expression of the oncomiR *miR-125b-2* increased proliferation and self-renewal of MEP and megakaryocyte progenitors (Klusmann et al. 2010). *miR-28* appears to exert a negative effect on megakaryocyte differentiation via targeting the thrombopoietin receptor, c-Mpl (Girardot et al. 2010). Emmrich and colleagues found that overexpressing *miR-125b* or *miR-660* enhanced polyploidization of CD34⁺ cell induced to differentiate in a cell culture system, while the *miR-23a/27a/24-2* cluster blocked maturation and platelet formation (Emmrich et al. 2012).

In megakaryocytic cell lines, calcium release in Meg-01 cells has been associated with a decrease in *miR-181a* levels and an increase in induced differentiation. (Guimaraes-Sternberg et al. 2006). In K562 cells, Lin28 prevents biogenesis of *Let-7,* which promotes cell differentiation. During TPA-induced differentiation, increased *miR-181* expression results in decreased Lin28, leading to greater levels of *Let-7* and differentiation (Li et al. 2012). Phorbol esther (TPA) treatment of K562 cells, which induces their differentiation into megakaryocytes, also resulted in the transcription factor Runx1 increasing expression of *miR-27a. miR-27a,* in turn, targets *Runx1* mRNAs, resulting in a regulatory feedback loop (Ben-Ami et al. 2009). In the case of megakaryocytic cell lines, such associations should be viewed cautiously since the correlation between miRNA expression in primary megakaryocytes and cell lines may not always be strong (Ramkissoon et al. 2006).

IV. PLATELET miRNAs

Platelets are anucleate blood cells that are the primary mediators of hemostasis. In addition to their role in hemostasis, platelets also function in the immune response and angiogenesis. Emerging evidence over the past several years suggests platelet miRNAs are biologically and clinically relevant as (1) potential regulators of platelet protein translation and expression, (2) markers of mature megakaryocyte miRNA levels, (3) biomarkers for hematologic disease and platelet reactivity, and (4) as a tool for understanding basic mechanisms of megakaryocyte/platelet gene expression (Edelstein and Bray 2011).

Despite the absence of a nucleus, it is well known that platelets have mRNA, mRNA splicing machinery, and translate mRNA into proteins relevant to hemostasis and inflammation (Warshaw et al. 1966; Denis et al. 2005; Weyrich et al. 2009). Notably, platelets stored in blood banks have been shown to increase synthesis of integrin β (Thon and Devine 2007). Work from the Provost laboratory has demonstrated that human platelets also contain miRNA processing machinery, including Dicer, TRBP2, and Ago2, and that platelets are able to process pre-miRNA into mature miRNA (Landry et al. 2009). Our *in silico* analysis indicates that each platelet miRNA targets an average of 174 distinct mRNAs, consistent with prior predictions (Miranda et al. 2006). Platelets contain miRNAs (Bruchova et al. 2007, 2008; Merkerova et al. 2008; Kannan et al. 2009; Landry et al. 2009; Kondkar et al. 2010), and we and others have used genome-wide profiling to demonstrate that normal human platelets express high levels of miRNA (Figure 7.1) (Landry et al. 2009). Thus, platelet miRNAs have ample opportunity to regulate platelet function, although direct evidence has yet to be reported.

As described earlier, functional miRNA levels change during megakaryocyte differentiation of cultured CD34⁺ HSCs. It is not difficult to imagine similar changes occurring *in vivo*, but this has not been formally studied. *In vitro* culture conditions lack numerous *in vivo* factors that could alter miRNA levels, including other marrow niche cells, innervation, plasma components, spatial constraints, environmental factors, and so on. Hussein et al. attempted to directly assess *in vivo* megakaryocyte miRNA levels by using laser microdissection to isolate mature megakaryocytes from patient bone marrow biopsies (Hussein et al. 2009). The extent to which platelet and mature megakaryocyte miRNA profiles correlate is unknown. As an initial attempt to address this issue, we compared platelet miRNA profiles from 19 healthy subjects with the megakaryocyte miRNA profiles

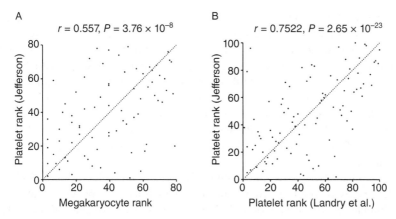

Figure 7.2. Correlation between human platelet and megakaryocyte miRNAs. MiRNA expression levels were determined in platelets from 19 healthy donors by microarray (our unpublished data) and in patient megakaryocytes by QRT-PCR (Hussein et al. 2009). The miRNAs that were queried in both studies were rank ordered and plotted. Statistical significance was calculated by Pearson Rank Correlation. Dotted line represents 1:1 ratio of rank. Panel A, platelet miRNAs correlated with megakaryocyte miRNAs. Panel B, platelet miRNAs from the author's laboratory correlated with platelet miRNAs from the Provost laboratory (Landry et al. 2009).

from the two patients reported by Hussein. As shown in Figure 7.2A, there was a significant correlation between the platelet and megakaryocyte miRNAs. We suspect that the true correlation is even stronger because different profiling platforms and RNA preparations were utilized, and only two megakaryocyte samples were assessed. Furthermore, the megakaryocyte samples were from patients, whereas the platelets were from healthy donors. As expected, when we compared our data set with the one reported by Landry et al. (Figure 7.2B), we found a greater degree of correlation, indicating a changing miRNA repertoire from megakaryocyte to platelet as well as a general concordance between the data sets.

V. PLATELET miRNAs AS BIOMARKERS

Platelet microRNAs have the potential to serve as biomarkers for platelet-mediated diseases, and may also have therapeutic application in the future. Biomarkers can be diagnostic, prognostic, or predictive in nature. Platelet miRNAs demonstrate little intraindividual variability (Stratz et al. 2012). In addition, miRNAs are very stable, and compared with mRNAs, have superior performance characteristics as biomarkers for disease activity (Lu et al. 2008; Kai and Pasquinelli 2010; Scholer et al. 2010).

Polycythemia vera (PV) and essential thrombocythemia (ET) are two of three BCR-ABL negative myeloproliferative neoplasms as classified by the World Health Organization in 2008 (Swerdlow et al. 2008). The JAK2 V617F mutation is positive in about 97% of PV patients and 55% of ET patients (Tefferi 2012). PV and ET are clonal disorders deriving from hematopoietic stem cells (Bruchova et al. 2008). It is difficult to distinguish PV and ET from non-clonal conditions, such as secondary erythrocytosis and reactive thrombocytosis, due to the variability of the causative mutations in ET and PV (Bruchova et al. 2008). Better biomarkers will help with the diagnosis of patients who have JAK2V617F negative ET. Platelet mRNA profiling has demonstrated that as few as 4 mRNA transcripts can be used to predict JAK2V617F negative ET in more than 85% of the samples (Gnatenko et al. 2010). Given their unique characteristics mentioned earlier, it is tempting to believe that platelet miRNAs may be even better than mRNA transcripts as diagnostic biomarkers for ET. Indeed, a study with PV patients demonstrated that *miR-26b* is up-regulated in the platelets of patients with PV, but not in the control subjects (Bruchova et al. 2008). Also platelet miRNA associations have been reported with both coronary artery disease and ulcerative colitis (Sondermeijer et al. 2011; Duttagupta et al. 2012), although the purity of the platelet preparations in these studies were not state-of-the-art, and the link between platelet biology and ulcerative colitis is currently unclear.

Our group has investigated the potential for platelet miRNAs to serve as biomarkers for platelet reactivity, both to better understand megakaryocyte/platelet gene expression and as a biomarker for thrombotic risk. We previously reported that VAMP8 mRNA is differentially expressed between platelets of differing reactivity (Kondkar et al. 2010). A prior genome-wide association study (GWAS), genetically linking single-nucleotide polymorphisms (SNPs) with phenotypes, had found a SNP in the VAMP8 3′-UTR to be associated with coronary artery disease. We therefore considered whether a miRNA-binding site in this region might affect VAMP8 expression. Indeed, we found that *miR-96* could target this gene, and that its expression down-regulated VAMP8 mRNA and protein levels (Kondkar et al. 2010). Furthermore, in a small number of subjects, *miR-96* was differentially expressed between platelets of differing reactivity in a manner that was consistent with its effect on VAMP8 expression. Recent data from our laboratory demonstrated the use of differentially expressed mRNA-miRNA pairs for identifying functional miRNAs, as assessed by the ability of the miRNA to target the mRNA was confirmed in cell culture (Nagalla et al. 2010).

VI. PLATELET MICROPARTICLES AND miRNAs

Microparticles are small (0.1–1 μm) vesicles shed by many cells types, including platelets, upon activation (Italiano et al. 2010). Among their protein contents, they also contain a significant quantity of miRNAs (Hunter et al. 2008). Diehl and colleagues reported that certain miRNAs had different expression levels in platelet-derived microparticles (PMPs) as compared with platelets (Diehl et al. 2012). *MiR-126* and *miR-133* were reported as being higher in PMPsm and *miR-19*, *-146*, *-223*, *-451*, and *-1246* were lower. This suggested that platelets may direct the packaging of specific miRNAs into PMPs. This is interesting because microparticles have been reported to be able to transfer miRNAs to other cells types, raising the possibility that miRNAs released upon platelet activation could regulate gene expression in other cells, such as endothelial or smooth muscle cells, at sites of vascular injury (Hunter et al. 2008; Yuan et al. 2009).

VII. SUMMARY AND FUTURE DIRECTIONS

MiRNAs have an established role in hematopoiesis and megakaryocytopoiesis, and platelet miRNAs have potential as tools for understanding basic mechanisms of megakaryocyte/platelet gene expression. MiRNAs significance in human genetics was demonstrated by genome-wide association studies that identified phenotype-associated SNPs not in or near protein-coding genes, but, in at least one such variant, was a causative mutation in a miRNA gene (Modamio-Hoybjor et al. 2004; Mencia et al. 2009). It is therefore likely that genetic variants in miRNA biogenesis will be identified as affecting megakaryocyte/platelet diseases.

In addition to potentially uncovering the causes of megakaryocyte/platelet disorders, miRNA research could lead to very exciting opportunities for future translational research in megakaryocytopoiesis and platelet biology. However, it should be realized that this pathway is far from straightforward.

REFERENCES

Alimonti, A., A. Carracedo, J.G. Clohessy, L.C. Trotman, C. Nardella, A. Egia, L. Salmena, K. Sampieri, W.J. Haveman, E. Brogi, A.L. Richardson, J. Zhang, et al. 2010. Subtle variations in Pten dose determine cancer susceptibility. *Nat Genet* **42**:454–458.

Baek, D., J. Villen, C. Shin, F.D. Camargo, S.P. Gygi, and D.P. Bartel. 2008. The impact of microRNAs on protein output. *Nature* **455**:64–71.

Bandiera, S., E. Hatem, S. Lyonnet, and A. Henrion-Caude. 2010. microRNAs in diseases: from candidate to modifier genes. *Clin Genet* **77**:306–313.

Barroga, C.F., H. Pham, and K. Kaushansky. 2008. Thrombopoietin regulates c-Myb expression by modulating micro RNA 150 expression. *Exp Hematol* **36**:1585–1592.

Bartel, D.P. 2004. MicroRNAs: genomics, biogenesis, mechanism, and function. *Cell* **116**:281–297.

Bazzini, A.A., M.T. Lee, and A.J. Giraldez. 2012. Ribosome profiling shows that miR-430 reduces translation before causing mRNA decay in zebrafish. *Science* **336**:233–237.

Ben-Ami, O., N. Pencovich, J. Lotem, D. Levanon, and Y. Groner. 2009. A regulatory interplay between miR-27a and Runx1 during megakaryopoiesis. *Proc Natl Acad Sci U S A* **106**:238–243.

Bottoni, A., D. Piccin, F. Tagliati, A. Luchin, M.C. Zatelli, and E.C. degli Uberti. 2005. miR-15a and miR-16-1 down-regulation in pituitary adenomas. *J Cell Physiol* **204**:280–285.

Bruchova, H., M. Merkerova, and J.T. Prchal. 2008. Aberrant expression of microRNA in polycythemia vera. *Haematologica* **93**:1009–1016.

Bruchova, H., D. Yoon, A.M. Agarwal, J. Mendell, and J.T. Prchal. 2007. Regulated expression of microRNAs in normal and polycythemia vera erythropoiesis. *Exp Hematol* **35**:1657–1667.

Calin, G.A., C. Sevignani, C.D. Dumitru, T. Hyslop, E. Noch, S. Yendamuri, M. Shimizu, S. Rattan, F. Bullrich, M. Negrini, and C.M. Croce. 2004. Human microRNA genes are frequently located at fragile sites and genomic regions involved in cancers. *Proc Natl Acad Sci U S A* **101**:2999–3004.

Chen, C., D.A. Ridzon, A.J. Broomer, Z. Zhou, D.H. Lee, J.T. Nguyen, M. Barbisin, N.L. Xu, V.R. Mahuvakar, M.R. Andersen, K.Q. Lao, K.J. Livak, et al. 2005. Real-time quantification of microRNAs by stem-loop RT-PCR. *Nucleic Acids Res* **33**:e179–e179.

Chen, C.-Z., L. Li, H.F. Lodish, and D.P. Bartel. 2004. MicroRNAs modulate hematopoietic lineage differentiation. *Science (New York, N.Y.)* **303**:83–86.

Choong, M.L., H.H. Yang, and I. McNiece. 2007. MicroRNA expression profiling during human cord blood-derived CD34 cell erythropoiesis. *Exp Hematol* **35**:551–564.

Debernardi, S., S. Skoulakis, G. Molloy, T. Chaplin, A. Dixon-McIver, and B.D. Young. 2007. MicroRNA miR-181a correlates with morphological sub-class of acute myeloid leukaemia and the expression of its target genes in global genome-wide analysis. *Leukemia* **21**:912–916.

Denis, M.M., N.D. Tolley, M. Bunting, H. Schwertz, H. Jiang, S. Lindemann, C.C. Yost, F.J. Rubner, K.H. Albertine, K.J. Swoboda, C.M. Fratto, E. Tolley, et al. 2005. Escaping the nuclear confines: signal-dependent pre-mRNA splicing in anucleate platelets. *Cell* **122**:379–391.

Diehl, P., A. Fricke, L. Sander, J. Stamm, N. Bassler, N. Htun, M. Ziemann, T. Helbing, A. El-Osta, J.B. Jowett, and K. Peter. 2012. Microparticles: major transport vehicles for distinct microRNAs in circulation. *Cardiovasc Res* **93**:633–644.

Djuranovic, S., A. Nahvi, and R. Green. 2012. miRNA-mediated gene silencing by translational repression followed by mRNA deadenylation and decay. *Science* **336**:237–240.

Dore, L.C., J.D. Amigo, O. Camila, Z. Zhang, X. Gai, J.W. Tobias, D. Yu, A.M. Klein, C. Dorman, W. Wu, R.C. Hardison, B.H. Paw, et al. 2008. A GATA-1-regulated microRNA locus essential for erythropoiesis. *Proc Natl Acad Sci U S A* **105**:3333–3338.

Duttagupta, R., S. DiRienzo, R. Jiang, J. Bowers, J. Gollub, J. Kao, K. Kearney, D. Rudolph, N.B. Dawany, M.K. Showe, T. Stamato, R.C. Getts, et al. 2012. Genome-wide maps of circulating miRNA biomarkers for ulcerative colitis. *PLoS ONE* **7**:e31241.

Edelstein, L.C. and P.F. Bray. 2011. MicroRNAs in platelet production and activation. *Blood* **117**:5289–5296.

Eiring, A.M., J.G. Harb, P. Neviani, C. Garton, J.J. Oaks, R. Spizzo, S. Liu, S. Schwind, R. Santhanam, C.J. Hickey, H. Becker, J.C. Chandler, et al. 2010. miR-328 functions as an RNA decoy to modulate hnRNP E2 regulation of mRNA translation in leukemic blasts. *Cell* **140**:652–665.

Eis, P.S., W. Tam, L. Sun, A. Chadburn, Z. Li, M.F. Gomez, E. Lund, and J.E. Dahlberg. 2005. Accumulation of miR-155 and BIC RNA in human B cell lymphomas. *Proc Natl Acad Sci U S A* **102**:3627–3632.

Emambokus, N., A. Vegiopoulos, B. Harman, E. Jenkinson, G. Anderson, and J. Frampton. 2003. Progression through key stages of haemopoiesis is dependent on distinct threshold levels of c-Myb. *EMBO J* **22**:4478–4488.

Emmrich, S., K. Henke, J. Hegermann, M. Ochs, D. Reinhardt, and J.H. Klusmann. 2012. miRNAs can increase the efficiency of ex vivo platelet generation. *Ann Hematol* **91**:1673–1684.

Esteller, M. 2011. Non-coding RNAs in human disease. *Nat Rev Genet* **12**:861–874.

Fazi, F., A. Rosa, A. Fatica, V. Gelmetti, M. Laura, C. Nervi, and I. Bozzoni. 2005. A minicircuitry comprised of microRNA-223 and transcription factors NFI-A and C/EBPalpha regulates human granulopoiesis. *Cell* **123**:819–831.

Felli, N., L. Fontana, E. Pelosi, R. Botta, D. Bonci, F. Facchiano, F. Liuzzi, V. Lulli, O. Morsilli, S. Santoro, M. Valtieri, G.A. Calin, et al. 2005. MicroRNAs 221 and 222 inhibit normal erythropoiesis and erythroleukemic cell growth via kit receptor down-modulation. *Proc Natl Acad Sci U S A* **102**:18081–18086.

Fontana, L., E. Pelosi, P. Greco, S. Racanicchi, U. Testa, F. Liuzzi, C.M. Croce, E. Brunetti, F. Grignani, and C. Peschle. 2007. MicroRNAs 17-5p-20a-106a control monocytopoiesis through AML1 targeting and M-CSF receptor upregulation. *Nat Cell Biol* **9**:775–787.

Friedman, R.C., K.K. Farh, C.B. Burge, and D.P. Bartel. 2009. Most mammalian mRNAs are conserved targets of microRNAs. *Genome Res* **19**:92–105.

Fulci, V., S. Chiaretti, M. Goldoni, G. Azzalin, N. Carucci, S. Tavolaro, L. Castellano, A. Magrelli, F. Citarella, M. Messina, R. Maggio, N. Peragine, et al. 2007. Quantitative technologies establish a novel microRNA profile of chronic lymphocytic leukemia. *Blood* **109**:4944–4951.

Garzon, R., M. Garofalo, M.P. Martelli, R. Briesewitz, L. Wang, C. Fernandez-Cymering, S. Volinia, C.G. Liu, S. Schnittger, T. Haferlach, A. Liso, D. Diverio, et al. 2008. Distinctive microRNA signature of acute myeloid leukemia bearing cytoplasmic mutated nucleophosmin. *Proc Natl Acad Sci U S A* **105**:3945–3950.

Garzon, R., C.E. Heaphy, V. Havelange, M. Fabbri, S. Volinia, T. Tsao, N. Zanesi, S.M. Kornblau, G. Marcucci, G.A. Calin, M. Andreeff, and C.M. Croce. 2009a. MicroRNA 29b functions in acute myeloid leukemia. *Blood* **114**:5331–5341.

Garzon, R., S. Liu, M. Fabbri, Z. Liu, C.E. Heaphy, E. Callegari, S. Schwind, J. Pang, J. Yu, N. Muthusamy, V. Havelange, S. Volinia, et al. 2009b. MicroRNA-29b induces global DNA

REFERENCES

hypomethylation and tumor suppressor gene reexpression in acute myeloid leukemia by targeting directly DNMT3A and 3B and indirectly DNMT1. *Blood* **113**:6411–6418.

Garzon, R., F. Pichiorri, T. Palumbo, R. Iuliano, A. Cimmino, R. Aqeilan, S. Volinia, D. Bhatt, H. Alder, G. Marcucci, G.A. Calin, C.G. Liu, et al. 2006. MicroRNA fingerprints during human megakaryocytopoiesis. *Proc Natl Acad Sci U S A* **103**:5078–5083.

Georgantas, R.W., III, R. Hildreth, S. Morisot, J. Alder, C.G. Liu, S. Heimfeld, G.A. Calin, C.M. Croce, and C.I. Civin. 2007. CD34+ hematopoietic stem-progenitor cell microRNA expression and function: a circuit diagram of differentiation control. *Proc Natl Acad Sci U S A* **104**: 2750–2755.

Girardot, M., C. Pecquet, S. Boukour, L. Knoops, A. Ferrant, W. Vainchenker, S. Giraudier, and S.N. Constantinescu. 2010. miR-28 is a thrombopoietin receptor targeting microRNA detected in a fraction of myeloproliferative neoplasm patient platelets. *Blood* **116**:437–445.

Gnatenko, D.V., W. Zhu, X. Xu, E.T. Samuel, M. Monaghan, M.H. Zarrabi, C. Kim, A. Dhundale, and W.F. Bahou. 2010. Class prediction models of thrombocytosis using genetic biomarkers. *Blood* **115**:7–14.

Guimaraes-Sternberg, C., A. Meerson, I. Shaked, and H. Soreq. 2006. MicroRNA modulation of megakaryoblast fate involves cholinergic signaling. *Leuk Res* **30**:583–595.

Guo, H., N.T. Ingolia, J.S. Weissman, and D.P. Bartel. 2010. Mammalian microRNAs predominantly act to decrease target mRNA levels. *Nature* **466**:835–840.

He, L., J.M. Thomson, M.T. Hemann, E. Hernando-Monge, D. Mu, S. Goodson, S. Powers, C. Cordon-Cardo, S.W. Lowe, G.J. Hannon, and S.M. Hammond. 2005. A microRNA polycistron as a potential human oncogene. *Nature* **435**:828–833.

Hunter, M.P., N. Ismail, X. Zhang, B.D. Aguda, E.J. Lee, L. Yu, T. Xiao, J. Schafer, M.L. Lee, T.D. Schmittgen, S.P. Nana-Sinkam, D. Jarjoura, et al. 2008. Detection of microRNA expression in human peripheral blood microvesicles. *PLoS ONE* **3**:e3694.

Hussein, K., K. Theophile, W. Dralle, B. Wiese, H. Kreipe, and O. Bock. 2009. MicroRNA expression profiling of megakaryocytes in primary myelofibrosis and essential thrombocythemia. *Platelets* **20**:391–400.

Ichimura, A., Y. Ruike, K. Terasawa, K. Shimizu, and G. Tsujimoto. 2010. MicroRNA-34a inhibits cell proliferation by repressing mitogen-activated protein kinase kinase 1 during megakaryocytic differentiation of K562 cells. *Mol Pharmacol* **77**:1016–1024.

Italiano, J.E., Jr., A.T. Mairuhu, and R. Flaumenhaft. 2010. Clinical relevance of microparticles from platelets and megakaryocytes. *Curr Opin Hematol* **17**:578–584.

Johnnidis, J.B., M.H. Harris, R.T. Wheeler, S. Stehling-Sun, M.H. Lam, O. Kirak, T.R. Brummelkamp, M.D. Fleming, and F.D. Camargo. 2008. Regulation of progenitor cell proliferation and granulocyte function by microRNA-223. *Nature* **451**:1125–1129.

Kai, Z.S. and A.E. Pasquinelli. 2010. MicroRNA assassins: factors that regulate the disappearance of miRNAs. *Nat Struct Mol Biol* **17**:5–10.

Kannan, M., K.V. Mohan, S. Kulkarni, and C. Atreya. 2009. Membrane array-based differential profiling of platelets during storage for 52 miRNAs associated with apoptosis. *Transfusion* **49**:1443–1450.

Kishimoto, T. 2005. Interleukin-6: from basic science to medicine–40 years in immunology. *Annu Rev Immunol* **23**:1–21.

Klusmann, J.H., Z. Li, K. Bohmer, A. Maroz, M.L. Koch, S. Emmrich, F.J. Godinho, S.H. Orkin, and D. Reinhardt. 2010. miR-125b-2 is a potential oncomiR on human chromosome 21 in megakaryoblastic leukemia. *Genes Dev* **24**:478–490.

Kondkar, A.A., M.S. Bray, S.M. Leal, S. Nagalla, D.J. Liu, Y. Jin, J.F. Dong, Q. Ren, S.W. Whiteheart, C. Shaw, and P.F. Bray. 2010. VAMP8/endobrevin is overexpressed in hyperreactive human platelets: suggested role for platelet microRNA. *J Thromb Haemost* **8**:369–378.

Kumar, M.S., A. Narla, A. Nonami, A. Mullally, N. Dimitrova, B. Ball, J.R. McAuley, L. Poveromo, J.L. Kutok, N. Galili, A. Raza, E. Attar, et al. 2011. Coordinate loss of a microRNA and protein-coding gene cooperate in the pathogenesis of 5q- syndrome. *Blood* **118**:4666–4673.

Labbaye, C., I. Spinello, M.T. Quaranta, E. Pelosi, L. Pasquini, E. Petrucci, M. Biffoni, E.R. Nuzzolo, M. Billi, R. Foa, E. Brunetti, F. Grignani, et al. 2008. A three-step pathway comprising PLZF/miR-146a/CXCR4 controls megakaryopoiesis. *Nat Cell Biol* **10**:788–801.

Landry, P., I. Plante, D.L. Ouellet, M.P. Perron, G. Rousseau, and P. Provost. 2009. Existence of a microRNA pathway in anucleate platelets. *Nat Struct Mol Biol* **16**:961–966.

Landthaler, M., D. Gaidatzis, A. Rothballer, P.Y. Chen, S.J. Soll, L. Dinic, T. Ojo, M. Hafner, M. Zavolan, and T. Tuschl. 2008. Molecular characterization of human Argonaute-containing ribonucleoprotein complexes and their bound target mRNAs. *RNA* **14**:2580–2596.

Lawrie, C.H., S. Soneji, T. Marafioti, C.D. Cooper, S. Palazzo, J.C. Paterson, H. Cattan, T. Enver, R. Mager, J. Boultwood, J.S. Wainscoat, and C.S. Hatton. 2007. MicroRNA expression distinguishes between germinal center B cell-like and activated B cell-like subtypes of diffuse large B cell lymphoma. *Int J Cancer* **121**:1156–1161.

Lee, R.C., R.L. Feinbaum, and V. Ambros. 1993. The *C. elegans* heterochronic gene lin-4 encodes small RNAs with antisense complementarity to lin-14. *Cell* **75**:843–854.

Li, X., J. Zhang, L. Gao, S. McClellan, M.A. Finan, T.W. Butler, L.B. Owen, G.A. Piazza, and Y. Xi. 2012. MiR-181 mediates cell differentiation by interrupting the Lin28 and let-7 feedback circuit. *Cell Death Differ* **19**:378–386.

Lu, J., S. Guo, B.L. Ebert, H. Zhang, X. Peng, J. Bosco, J. Pretz, R. Schlanger, J.Y. Wang, R.H. Mak, D.M. Dombkowski, F.I. Preffer, et al. 2008. MicroRNA-mediated control of cell fate in megakaryocyte-erythrocyte progenitors. *Dev Cell* **14**:843–853.

Marton, S., M.R. Garcia, C. Robello, H. Persson, F. Trajtenberg, O. Pritsch, C. Rovira, H. Naya, G. Dighiero, and A. Cayota. 2008. Small RNAs analysis in CLL reveals a deregulation of miRNA expression and novel miRNA candidates of putative relevance in CLL pathogenesis. *Leukemia* **22**:330–338.

Mencia, A., S. Modamio-Hoybjor, N. Redshaw, M. Morin, F. Mayo-Merino, L. Olavarrieta, L.A. Aguirre, I. del Castillo, K.P. Steel, T. Dalmay, F. Moreno, and M.A. Moreno-Pelayo. 2009. Mutations in the seed region of human miR-96 are responsible for nonsyndromic progressive hearing loss. *Nat Genet* **41**:609–613.

Merkerova, M., M. Belickova, and H. Bruchova. 2008. Differential expression of microRNAs in hematopoietic cell lineages. *Eur J Haematol* **81**:304–310.

Miranda, K.C., T. Huynh, Y. Tay, Y.-S. Ang, W.-L. Tam, A.M. Thomson, B. Lim, and I. Rigoutsos. 2006. A pattern-based method for the identification of MicroRNA binding sites and their corresponding heteroduplexes. *Cell* **126**:1203–1217.

Modamio-Hoybjor, S., M.A. Moreno-Pelayo, A. Mencia, I. del Castillo, S. Chardenoux, D. Morais, M. Lathrop, C. Petit, and F. Moreno. 2004. A novel locus for autosomal dominant nonsyndromic hearing loss, DFNA50, maps to chromosome 7q32 between the DFNB17 and DFNB13 deafness loci. *J Med Genet* **41**:e14.

Monticelli, S., K.M. Ansel, C. Xiao, N.D. Socci, A.M. Krichevsky, T.-H. Thai, N. Rajewsky, D.S. Marks, C. Sander, K. Rajewsky, A. Rao, and K.S. Kosik. 2005. MicroRNA profiling of the murine hematopoietic system. *Genome Biol* **6**:R71–R71.

Nagalla, S., C. Shaw, X. Kong, L. Ma, A.A. Kondkar, M.S. Bray, S.M. Leal, Y. Jin, J. Dong, and P.F. Bray. 2010. Platelet microrna-mRNA co-expression profiles correlate with platelet reactivity. *52nd ASH Annual Meeting and Exposition*. Orlando, FL.

Navarro, F., D. Gutman, E. Meire, M. Cáceres, I. Rigoutsos, Z. Bentwich, and J. Lieberman. 2009. miR-34a contributes to megakaryocytic differentiation of K562 cells independently of p53. *Blood* **114**:2181–2192.

REFERENCES

Neilson, J.R., G.X.Y. Zheng, C.B. Burge, and P.A. Sharp. 2007. Dynamic regulation of miRNA expression in ordered stages of cellular development. *Genes Dev* **21**:578–589.

O'Connell, R.M., D.S. Rao, A.A. Chaudhuri, M.P. Boldin, K.D. Taganov, J. Nicoll, R.L. Paquette, and D. Baltimore. 2008. Sustained expression of microRNA-155 in hematopoietic stem cells causes a myeloproliferative disorder. *J Exp Med* **205**:585–594.

Opalinska, J.B., A. Bersenev, Z. Zhang, A.A. Schmaier, J. Choi, Y. Yao, J. D'Souza, W. Tong, and M.J. Weiss. 2010. MicroRNA expression in maturing murine megakaryocytes. *Blood* **116**: e128–e138.

Pase, L., J.E. Layton, W.P. Kloosterman, D. Carradice, P.M. Waterhouse, and G.J. Lieschke. 2009. miR-451 regulates zebrafish erythroid maturation *in vivo* via its target gata2. *Blood* **113**: 1794–1804.

Poliseno, L., L. Salmena, J. Zhang, B. Carver, W.J. Haveman, and P.P. Pandolfi. 2010. A coding-independent function of gene and pseudogene mRNAs regulates tumour biology. *Nature* **465**:1033–1038.

Raaijmakers, M.H.G.P., S. Mukherjee, S. Guo, S. Zhang, T. Kobayashi, J.A. Schoonmaker, B.L. Ebert, F. Al-Shahrour, R.P. Hasserjian, E.O. Scadden, Z. Aung, M. Matza, et al. 2010. Bone progenitor dysfunction induces myelodysplasia and secondary leukaemia. *Nature* **464**: 852–857.

Ramkissoon, S.H., L.A. Mainwaring, Y. Ogasawara, K. Keyvanfar, J.P. McCoy, E.M. Sloand, S. Kajigaya, and N.S. Young. 2006. Hematopoietic-specific microRNA expression in human cells. *Leuk Res* **30**:643–647.

Rigoutsos, I. and A. Tsirigos. 2011. MicroRNA target prediction. In *MicroRNAs in development and cancer*. F.J. Slack, Ed. Imperial College Press, Hackensack, NJ. pp. 237–264.

Romania, P., V. Lulli, E. Pelosi, M. Biffoni, C. Peschle, and G. Marziali. 2008. MicroRNA 155 modulates megakaryopoiesis at progenitor and precursor level by targeting Ets-1 and Meis1 transcription factors. *Br J Haematol* **143**:570–580.

Schmitter, D., J. Filkowski, A. Sewer, R.S. Pillai, E.J. Oakeley, M. Zavolan, P. Svoboda, and W. Filipowicz. 2006. Effects of Dicer and Argonaute down-regulation on mRNA levels in human HEK293 cells. *Nucleic Acids Res* **34**:4801–4815.

Scholer, N., C. Langer, H. Dohner, C. Buske, and F. Kuchenbauer. 2010. Serum microRNAs as a novel class of biomarkers: a comprehensive review of the literature. *Exp Hematol* **38**:1126–1130.

Selbach, M., B. Schwanhausser, N. Thierfelder, Z. Fang, R. Khanin, and N. Rajewsky. 2008. Widespread changes in protein synthesis induced by microRNAs. *Nature* **455**:58–63.

Sethupathy, P. and F.S. Collins. 2008. MicroRNA target site polymorphisms and human disease. *Trends Genet* **24**:489–497.

Shabalina, S.A. and N.A. Spiridonov. 2004. The mammalian transcriptome and the function of non-coding DNA sequences. *Genome Biol* **5**:105.

Sondermeijer, B.M., A. Bakker, A. Halliani, M.W. de Ronde, A.A. Marquart, A.J. Tijsen, T.A. Mulders, M.G. Kok, S. Battjes, S. Maiwald, S. Sivapalaratnam, M.D. Trip, et al. 2011. Platelets in patients with premature coronary artery disease exhibit upregulation of miRNA340* and miRNA624*. *PLoS ONE* **6**:e25946.

Starczynowski, D.T., F. Kuchenbauer, B. Argiropoulos, S. Sung, R. Morin, A. Muranyi, M. Hirst, D. Hogge, M. Marra, R.A. Wells, R. Buckstein, W. Lam, et al. 2010. Identification of miR-145 and miR-146a as mediators of the 5q- syndrome phenotype. *Nat Med* **16**:49–58.

Starczynowski, D.T., F. Kuchenbauer, J. Wegrzyn, A. Rouhi, O. Petriv, C.L. Hansen, R.K. Humphries, and A. Karsan. 2011a. MicroRNA-146a disrupts hematopoietic differentiation and survival. *Exp Hematol* **39**:167–178 e164.

Starczynowski, D.T., R. Morin, A. McPherson, J. Lam, R. Chari, J. Wegrzyn, F. Kuchenbauer, M. Hirst, K. Tohyama, R.K. Humphries, W.L. Lam, M. Marra, et al. 2011b. Genome-wide identification of human microRNAs located in leukemia-associated genomic alterations. *Blood* **117**:595–607.

Stratz, C., T.G. Nuhrenberg, H. Binder, C.M. Valina, D. Trenk, W. Hochholzer, F.J. Neumann, and B.L. Fiebich. 2012. Micro-array profiling exhibits remarkable intra-individual stability of human platelet micro-RNA. *Thromb Haemost* **107**:634–641.

Swerdlow, S.H., E. Campo, N.L. Harris, E.S. Jaffe, S.A. Pileri, H. Stein, J. Thiele, and J.W. Vardiman. 2008. *WHO classification of tumours of haematopoietic and lymphoid tissues*, Fourth edition. World Health Organization, Geneva, Switzerland.

Tay, Y., J. Zhang, A.M. Thomson, B. Lim, and I. Rigoutsos. 2008. MicroRNAs to Nanog, Oct4 and Sox2 coding regions modulate embryonic stem cell differentiation. *Nature* **455**:1124–1128.

Tefferi, A. 2012. Polycythemia vera and essential thrombocythemia: 2012 update on diagnosis, risk stratification, and management. *Am J Hematol* **87**:285–293.

Thon, J.N. and D.V. Devine. 2007. Translation of glycoprotein IIIa in stored blood platelets. *Transfusion* **47**:2260–2270.

Ventura, A., A.G. Young, M.M. Winslow, L. Lintault, A. Meissner, S.J. Erkeland, J. Newman, R.T. Bronson, D. Crowley, J.R. Stone, R. Jaenisch, P.A. Sharp, et al. 2008. Targeted deletion reveals essential and overlapping functions of the miR-17 through 92 family of miRNA clusters. *Cell* **132**:875–886.

Visone, R., L.Z. Rassenti, A. Veronese, C. Taccioli, S. Costinean, B.D. Aguda, S. Volinia, M. Ferracin, J. Palatini, V. Balatti, H. Alder, M. Negrini, et al. 2009. Karyotype-specific microRNA signature in chronic lymphocytic leukemia. *Blood* **114**:3872–3879.

Warshaw, A.L., L. Laster, and N.R. Shulman. 1966. The stimulation by thrombin of glucose oxidation in human platelets. *J Clin Invest* **45**:1923–1934.

Weyrich, A.S., H. Schwertz, L.W. Kraiss, and G.A. Zimmerman. 2009. Protein synthesis by platelets: historical and new perspectives. *J Thromb Haemost* **7**:241–246.

Wienholds, E., W.P. Kloosterman, E. Miska, E. Alvarez-Saavedra, E. Berezikov, E. de Bruijn, H.R. Horvitz, S. Kauppinen, and R.H. Plasterk. 2005. MicroRNA expression in zebrafish embryonic development. *Science* **309**:310–311.

Wightman, B., I. Ha, and G. Ruvkun. 1993. Posttranscriptional regulation of the heterochronic gene lin-14 by lin-4 mediates temporal pattern formation in *C. elegans*. *Cell* **75**:855–862.

Xiao, C., D.P. Calado, G. Galler, T.-H. Thai, H.C. Patterson, J. Wang, N. Rajewsky, T.P. Bender, and K. Rajewsky. 2007. MiR-150 controls B cell differentiation by targeting the transcription factor c-Myb. *Cell* **131**:146–159.

Xu, W. and J.Y. Li. 2007. MicroRNA gene expression in malignant lymphoproliferative disorders. *Chin Med J (Engl)* **120**:996–999.

Yuan, A., E.L. Farber, A.L. Rapoport, D. Tejada, R. Deniskin, N.B. Akhmedov, and D.B. Farber. 2009. Transfer of microRNAs by embryonic stem cell microvesicles. *PLoS ONE* **4**:e4722–e4722.

Zhao, H., A. Kalota, S. Jin, and A.M. Gewirtz. 2009. The c-myb proto-oncogene and microRNA-15a comprise an active autoregulatory feedback loop in human hematopoietic cells. *Blood* **113**:505–516.

Zhou, B., S. Wang, C. Mayr, D.P. Bartel, and H.F. Lodish. 2007. miR-150, a microRNA expressed in mature B and T cells, blocks early B cell development when expressed prematurely. *Proc Natl Acad Sci U S A* **104**:7080–7085.

PART II

MicroRNAs in Infectious Disease: Host–Pathogen Interactions

8

MICRORNAS AS KEY PLAYERS IN HOST-VIRUS INTERACTIONS

Aurélie Fender and Sébastien Pfeffer

Architecture et Réactivité de l'ARN, Institut de biologie moléculaire et cellulaire du CNRS, Université de Strasbourg, Strasbourg, France

I.	Introduction	120
II.	Host miRNAs and Virus Infection	121
III.	Viral miRNAs	121
	A. MicroRNAs Encoded by DNA Viruses	122
	B. MicroRNAs Encoded by RNA Viruses	123
	C. Expression of Viral miRNAs	124
IV.	Biological Roles of miRNAs	126
	A. Functional Convergence of Viral miRNAs	127
	B. Toward Complete Viral miRNA Targetomes	127
	C. Role of Viral miRNAs *In Vivo*	129
V.	Conclusions and Future Prospects	129
	Acknowledgments	130
	References	130

ABBREVIATIONS

Ago	Argonaute
BLV	bovine leukemia retrovirus
EBV	Epstein–Barr herpesvirus
HCMV	human cytomegalovirus
HHV	human herpesvirus

MicroRNAs in Medicine, First Edition. Edited by Charles H. Lawrie.
© 2014 John Wiley & Sons, Inc. Published 2014 by John Wiley & Sons, Inc.

HITS-CLIP	high-throughput sequencing and crosslinking immunoprecipitation
HIV	human immunodeficiency virus
HPV	human papillomavirus
HSV	herpes simplex virus
HVS	herpesvirus saimiri
ICP0	infected cell protein 0
ICP4	infected cell protein 4
KSHV	Kaposi's sarcoma-associated herpesvirus
LAT	latency-associated transcript
MCMV	mouse cytomegalovirus
MHV68	murine gammaherpesvirus 68
MICB	MHC class I polypeptide-related sequence B
miRNA	microRNA
NK cell	natural killer cell
ORF	open reading frame
PAR-CLIP	photoactivatable-ribonucleoside-enhanced CLIP
pol	polymerase
pre-miRNA	precursor miRNA
pri-miRNA	primary miRNA
RCMV	rhesus cytomegalovirus
RIP-Chip	ribonucleoprotein immunoprecipitation-gene Chip
RISC	RNA-induced silencing complex
rLCV	rhesus lymphocryptovirus
RNAi	RNA interference
RTA	replication and transcription activator
SINV	Sindbis virus
SV40	simian virus 40
UTR	untranslated region
VA RNA	viral associated RNA
VZV	varicella zoster virus
WNV	West Nile virus

I. INTRODUCTION

In recent years, small regulatory RNAs have gained increased attention in a wide variety of biology-related research fields. In particular, miRNAs, which were originally believed to be an oddity of the *C. elegans* worm, have revolutionized many aspects of molecular biology. The field of virology and host–pathogen interactions is no exception, and more and more researchers realize that we have only partially lifted the veil off an intricate regulatory network involving host and viral proteins, as well as coding and non-coding RNAs. In humans alone, there are more than 1500 miRNA genes currently listed in the miRBase repository (Kozomara and Griffiths-Jones 2010), and it is estimated that more than 60% of the genome are regulated by miRNAs (Friedman et al. 2009). The recent discovery of virus-encoded miRNAs clearly adds to this complexity. In this chapter, we will mainly focus on the roles of miRNAs in virus infections in mammals. Although it is well known that the RNA silencing machinery (RNAi) play an antiviral role in other organisms, such as plants and insects, this cannot be translated to mammals, and these aspects have been reviewed elsewhere (see e.g., [Umbach and Cullen 2009]). What is clear however is that there are multiple levels of interactions between vertebrate viruses and the miRNA machinery, and these will be discussed here. Thus, we will present how viruses can perturb the expression of cellular miRNAs, both in a direct or indirect way. An important part will be devoted to the presentation of the repertoire of viral miRNAs and their expression during infection. We will also especially emphasize the diversity of strategies used by viruses to express their own miRNAs and to regulate specific pathways with the ultimate goal to fine-tune their environment while staying undetected. As we will see, it can be postulated that in many cases, the convergent evolution of viruses and their hosts helped different viruses to design separate strategies to achieve similar goals.

II. HOST miRNAs AND VIRUS INFECTION

As with any biotic stress, viral infections trigger multiple changes in the host gene expression. In this respect, host miRNAs are not any different, and several reports have shown that viruses do have a measurable impact on their expression. The effect of the infection on cellular miRNAs can be either direct or can alternatively result from a signaling cascade linked to the triggering of the innate immune response. Being among the most studied viruses, the human immunodeficiency virus (HIV) was the first one to be reported to perturb host miRNA expression. Thus, several studies found that this virus could globally down-regulate miRNA accumulation (Yeung et al. 2005), or more specifically, act negatively on the *miR-17-92* cluster (Triboulet et al. 2007). More recently others reported that *miR-34a* was induced in neuronal cells upon HIV infection (Mukerjee et al. 2011). Interestingly, cells infected by Epstein–Barr herpesvirus (EBV) also show an increase in this miRNA, which appears to have tumor suppressor activity (Forte et al. 2012). It is not uncommon that different viruses regulate the same miRNA. For example, both herpes simplex virus 1 (HSV-1) and Kaposi's sarcoma-associated herpesvirus (KSHV) induce expression of *miR-132*, which regulates antiviral immunity (Lagos et al. 2010; Mulik et al. 2012). Another example is *miR-27*, which is down-regulated by the mouse cytomegalovirus (MCMV) and the herpesvirus saimiri (HVS) (Cazalla et al. 2010; Marcinowski et al. 2012). In the latter case, the regulation occurs directly at the level of the mature miRNA stability via pairing to a viral transcript. It has been postulated that in the case of MCMV, the regulation of *miR-27* could prevent an antiviral effect mediated by the miRNA. This has also been observed for the human cytomegalovirus (HCMV), which down-regulates the cellular *miR-100* (Wang et al. 2008). There are numerous other examples of specific miRNAs being regulated by viruses, and often this participates in the associated pathology, such as the induction of the oncogenic *miR-155* by EBV (Gatto et al. 2008). It is less frequent to observe a global regulation of host miRNAs, but there are few viruses that can impact negatively on essential factors of the miRNA biogenesis machinery. Hence, HIV can down-regulate the key enzyme Dicer (Coley et al. 2010; Bennasser et al. 2011), the adenovirus viral-associated (VA) RNA can saturate the machinery at the miRNA precursor cytoplasmic export level (Lu and Cullen 2004) and the vaccinia virus has been shown to shut down the miRNA machinery (Grinberg et al. 2012). In the latter case, it has been reported very recently that it was the viral poly(A) polymerase that was involved in the degradation of cellular miRNAs (Backes et al. 2012).

III. VIRAL miRNAs

Viruses have a well-known propensity to hijack host genes and cellular machinery in order to function optimally. Knowing the key roles of miRNAs in eukaryotic cells, it seemed logical that viruses have also evolved to express their own miRNAs. First discovered in EBV (Pfeffer et al. 2004), viral miRNAs have now been shown to be encoded by a multitude of viruses. Recently, the development of high-throughput sequencing of small RNA-derived libraries has been very useful in establishing viral miRNomes. To date, more than 260 miRNAs of viral origin are described in the literature, some of which are listed in Table 8.1.

A. MicroRNAs Encoded by DNA Viruses

Up to the present time, most viral miRNAs have been identified in DNA viruses infecting vertebrates, especially the herpesvirus family (Table 8.1). These viruses have the peculiarity of being able to establish lifelong persistent infection, switching from viral production (lytic phase) to the latent phase. It makes perfect sense that they would express miRNAs,

TABLE 8.1. Some Examples of Viral miRNAs in Representative Viruses

Virus family	Genus	Species		Host	No. of miRNAs
		dsDNA			
α-Herpesvirus	Simplexvirus	Herpes simplex virus	HSV-1	Human	26 (17)
			HSV-2		24 (18)
	Mardivirus	Marek's disease virus	MDV-1	Avian	26 (14)
			MDV-2		36 (18)
β-Herpesvirus	Cytomegalovirus	Human cytomegalovirus	HCMV	Human	22 (12)
		Rhesus cytomegalovirus[a]	RhCMV	Simian	17 (17)
		Mouse cytomegalovirus	MCMV	Murine	29 (18)
	Roseolovirus	Human herpesvirus 6	HHV-6B	Human	5 (4)
γ-Herpesvirus	Lymphocryptovirus	Epstein–Barr virus	EBV	Human	44 (25)
		Rhesus lymphocryptovirus	rLCV	Simian	68 (36)
	Rhadinovirus	Kaposi's sarcoma herpesvirus	KSHV	Human	25 (12)
		Rhesus monkey rhadinovirus	RRV	Simian	25 (15)
		Herpesvirus saimiri	HVS	Simian	6 (3)
		Mouse gammaherpesvirus 68	MHV68	Murine	28 (15)
Polyomavirus		BK polyomavirus	BKV	Human	2 (1)
		JC polyomavirus	JCV	Human	2 (1)
		Merkel cell polyomavirus	MCPyV	Human	2 (1)
		Simian virus 40	SV40	Simian	2 (1)
		Murine polyomavirus[b]	muPyV	Murine	2 (1)
Polyomavirus-like		Bandicoot papillomatosis carcinomatosis viruses	BPCV1 BPCV2	Marsupial	1 (1)
Adenovirus		Human adenovirus[c]	AV	Human	3 (2[e])
		ssRNA			
Retrovirus	Deltaretrovirus	Bovine leukemia virus	BLV	Bovine	7 (5)
Flavivirus		West Nile virus[d]	WNZ	Insect	1 (1)

In parentheses: number of miRNA precursors.
Adapted from miRBase (http://www.mirbase.org). Other references: [a]Hancock et al. (2012), [b]Sullivan et al. (2009), [c]Andersson et al. (2005), Aparicio et al. (2006), Sano et al. (2006), Xu et al. (2007), and [d]Hussain et al. (2012). [e]Pre-miRNA-like structured VAI and VAII RNAs.

as these represent an ideal way to down-regulate the host system without triggering deleterious immune responses. Indeed, many viral miRNAs are expressed during latency, and some of them can control the entry into the immunogenic productive viral stage. Thus, a set of HSV-1 and -2 miRNAs are expressed from the latency-associated transcript (LAT) and are involved in the control of early gene expression (see Section III.C.2 and Section IV).

Virus-encoded miRNAs have been found in six of the eight human herpesvirus (HHV) species. Deep sequencing approaches failed to identify any viral sequences corresponding to potential miRNAs in HHV-3, also known as varicella zoster virus (VZV) (Umbach et al. 2010a), and HHV-7 has not been tested yet. Viral miRNAs were also identified in murine, simian, and in a large number of avian herpesviruses. In addition, the polyomavirus and adenovirus families also encode for their own miRNAs. In the case of human papillomaviruses (HPV), the existence of miRNAs is controversial. Whereas Cai et al. failed to detect miRNAs expressed by HPV-31 (Cai et al. 2006a), Gu et al. predicted a handful of miRNAs encoded in eight other HPV genotypes (Gu et al. 2011). These findings clearly require further validation. Finally, DNA viruses infecting fish (Iridovirus) and insects (ascovirus, baculovirus, and nudivirus) have also been shown to express miRNAs, expanding an exponentially growing list of virally encoded miRNAs (Hussain et al. 2008; Singh et al. 2010; Wu et al. 2011; Yan et al. 2011).

The number of viral miRNA genes varies greatly among the different viruses, ranging from one in polyomaviruses to up to 36 in the rhesus lymphocryptovirus (rLCV). Although closely related viruses, such as EBV and rLCV or HSV-1 and -2 share miRNA homologues (Cai et al. 2006b; Jurak et al. 2010; Walz et al. 2010), miRNA sequences are usually not conserved, reflecting different evolution histories and host specificity of viruses. However, particular genomic regions are revealed as hot spots for miRNA genes, and their location is generally conserved within related viruses. Figure 8.1 illustrates the genomic location of miRNA genes in some human herpesviruses examples. MiRNA genes can be either distributed all over the genome (e.g., in cytomegaloviruses) or clustered at predominantly one or two loci in the genome (like for the alpha and gammaherpesviruses). They can also be found in genomic repeats, either inside the genome, such as IR_L and IR_S in HSV-1 and -2, and Marek's disease viruses 1 and 2, or at the genome extremities, such as DR_L and DR_R in HHV-6B. Finally, miRNAs or other yet uncharacterized small RNAs are sometimes expressed from the lytic origin of replication (Jurak et al. 2010; Meyer et al. 2011; Tuddenham et al. 2012).

B. MicroRNAs Encoded by RNA Viruses

Examples of miRNAs deriving from RNA viruses are still scarce (Table 8.1), which most probably reflects the fact that miRNA biogenesis from an RNA genome would be deleterious for its integrity. One RNA virus, which might express miRNAs, is the retrovirus HIV-1 (reviewed in Houzet and Jeang 2011), although their validity remains to be formally proven (Pfeffer et al. 2005; Lin and Cullen 2007). Low- or high-throughput sequencing of RNAs extracted from cells infected with RNA viruses, such as hepatitis C virus, yellow fever virus, and influenza A virus failed to identify virally encoded miRNAs (Pfeffer et al. 2005; Umbach et al. 2010b). However, several independent groups did manage to genetically engineer RNA viruses to produce functional miRNAs without impeding their viral replication (Shapiro et al. 2010; Varble et al. 2010). Following these studies, the first example of a natural miRNA expressed in insect cells from the viral subgenomic RNA of the West Nile virus (WNV) was reported (Hussain et al. 2012).

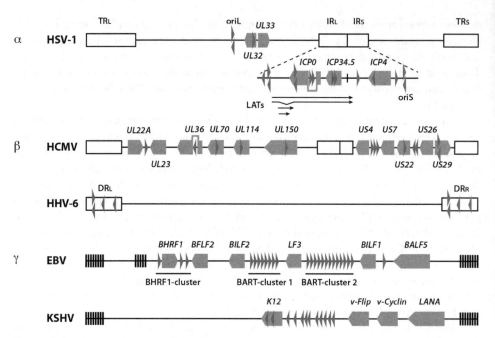

Figure 8.1. Genomic location of miRNAs encoded by human herpesviruses. Almost all human herpesviruses code for miRNA genes with the exception of VZV and HHV-7. For simplicity, only HSV-1 genome was represented, HSV-2 being very similar. The genomes were not drawn to scale, and only ORF close to or embedding miRNA genes (in green) were pictured. White boxes and successive black bars schematize repeats. TR, IR, and DR stand for terminal, internal, and direct repeats, respectively; $_L$ for long and $_S$ for short; oriL for lytic origin of replication; LAT for latency-associated transcript. See color insert.

Very recently, Kincaid et al. showed that the bovine leukemia retrovirus (BLV) evolved an ingenious way around genomic RNA destabilization by precursor miRNA (pre-miRNA) excision (Kincaid et al. 2012). Indeed, this retrovirus expresses a cluster of five miRNAs, which are only expressed from a subgenomic polymerase (pol) III transcript, but not from the genomic RNA. The authors predicted pol III-transcribed miRNAs in other retroviruses, such as spumaviruses, which will require further validation to know if this mode of expression could be a common feature of retroviruses.

C. Expression of Viral miRNAs

1. Biogenesis of Viral miRNAs.
In animals, miRNAs are canonically produced from a large primary transcript (pri-miRNA), which is sequentially processed by the nuclear Drosha and the cytoplasmic Dicer into the pre-miRNA and the miRNA/miRNA* duplex (Figure 8.2). Beside the canonical pathway, certain pri-miRNAs are processed independently of Drosha (e.g., mirtrons are produced by splicing) or of Dicer (e.g., *miR-451* is matured by argonaute (Ago) 2) (for a review, see Yang and Lai 2011). Finally, one of the two strands of the duplex, the guide strand, is preferentially incorporated into the Ago-containing effector complex RISC (RNA induced silencing complex) (Bartel 2004, 2009). To date, there is no evidence of the existence of a specific viral machinery or cofactors for the biogenesis of viral miRNAs. However, some viral miRNAs do not rely on the

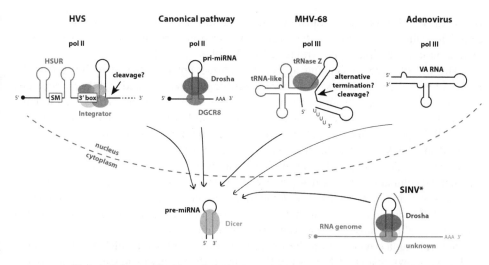

Figure 8.2. Peculiarities of viral miRNA biogenesis. See text for details.

classical biogenesis pathway. Mostly, this involves a Drosha-independent processing (Figure 8.2). For example, the murine gammaherpesvirus 68 (MHV-68) encodes miRNAs under the control of an RNA pol III promoter and downstream of a tRNA-like structure. The 5' end of the pre-miRNA is cleaved off the tRNA by tRNase Z (Bogerd et al. 2010). Similarly, HVS miRNA genes are expressed downstream of Sm class U RNA genes which, after transcription, are processed by the Integrator complex, thus generating, independently of Drosha, the 5' end of the pre-miRNAs (Cazalla et al. 2011). In both cases, the mechanism of generation of the pre-miRNA 3' end remains unknown. The retrovirus BLV also relies on pol III to express its pre-miRNAs, which do not seem to require Drosha for their processing (Kincaid et al. 2012). However, the enzyme responsible for their cleavage has not yet been identified. Finally, several independent studies showed that RNA viruses, which are restricted to the cytoplasm, could express miRNAs (either naturally, such as WNV, or artificially, such as influenza A virus, Sindbis virus [SINV], and the tick-borne encephalitis virus). This would involve either a new and unknown Drosha-independent cleavage of the pri-miRNA in the cytoplasm or the relocalization of Drosha from the nucleus to the cytoplasm. The latter hypothesis recently gained support when Shapiro et al. showed that this relocalization indeed happened in cells infected with an engineered SINV expressing the cellular pre-*miR-124* (Shapiro et al. 2012).

2. Tropism and Kinetics of Expression of Viral miRNAs. Expression of viral miRNAs has been studied using various systems, ranging from cultured cell lines (latently or lytically infected) to patient tissues and animal models. Viral miRNAs are expressed at different stages of the viral infection (productive and/or latent stages), can be detected in infected tissues or lesions associated to the pathology of the virus (e.g., Kaposi's sarcoma for KSHV miRNAs in human [Hansen et al. 2010]). Interestingly, EBV miRNAs are incorporated into viral particles and might play a role in the early phase of the infection (Jochum et al. 2012). This could be a more general feature, since it has been shown that viral particles from many other herpesviruses, such as HCMV or KSHV, contain RNA molecules (Greijer et al. 2000; Bechtel et al. 2005). However, the presence of miRNAs was not explored in these cases and must be verified. Another recent and exciting discovery is that EBV miRNAs can be transferred from an infected cell to uninfected neighboring cells. These miRNAs are transported via microvesicles and are functional in targeting gene regulation in the recipient cells (reviewed in Pegtel et al. 2011).

The use of deep sequencing approaches allows not only the identification of the miRNA repertoire of a given virus under specific conditions of infection, but also provides an insight into the expression levels of these miRNAs. For example, Dölken et al. have shown that in lytically infected fibroblasts MCMV miRNAs constituted 35–60% of the total miRNA pool (Dölken et al. 2007). The overexpression of viral miRNAs can also be detected in KSHV latently infected cells, where viral miRNAs represent from 30% to about 70% of all miRNAs (Lin et al. 2010; Gottwein et al. 2011). One could wonder whether the cellular processing machinery can cope with this additional pool of molecules. The answer is that there seems to be some room to accommodate a reasonable number of extra miRNA precursors, and with few exceptions (see Section II), there is no global decrease in cellular miRNA expression upon viral infection.

Several studies reported on the expression kinetics of viral miRNAs. Thus, Jurak et al. have shown that HSV 1 and -2 miRNAs are differentially expressed during productive and latent infection (Jurak et al. 2010). Along the same line, three independent *in vivo* studies have shown that the expression of MCMV, rhesus cytomegalovirus (RCMV), or MDV-1 miRNAs differ between tissues and timing of infection (Dölken et al. 2007, 2010a; Luo et al. 2011; Meyer et al. 2011). MiRNAs may also serve as biomarkers of a particular disease or disease state associated to the virus. Another important conclusion of these studies is that viral miRNA expression is under tight regulation in the course of infection. Elucidating how this regulation occurs and whether it involves viral factors remains a challenging research avenue.

IV. BIOLOGICAL ROLES OF VIRAL miRNAs

Whereas the identification of new viral miRNAs has been exponential, the elucidation of their targets and biological roles is still lagging behind. In a few cases, the target assignment is evident given the antisense location of miRNA genes to open reading frames in the viral genome (Figure 8.1). Indeed, one of the first identified targets was the simian virus 40 (SV40) large tumor antigen (TAg) transcript, which is cleaved by the antisense miRNA *miR-S1* (Sullivan et al. 2005). This function has been conserved in other polyomaviruses, such as the BK and JC viruses (Seo et al. 2008). Similarly, the EBV DNA polymerase BALF5 or HSV-1 and 2 immediate-early transactivator ICP0 are both targeted by antisense miRNAs (Barth et al. 2008; Umbach et al. 2008; Tang et al. 2009). In certain cases, viral miRNAs share the same seed region, and thus some common targets, with cellular miRNAs. The best examples are the oncogenic herpesviruses KSHV and MDV-1, which encode both for orthologues of the oncogenic cellular *miR-155* (Gottwein et al. 2007; Morgan et al. 2008). Finally, viral miRNAs coded in the sense strand of ORFs and/or in the UTRs of viral transcripts may be cis-regulatory elements. For example, processing of two KSHV miRNAs present in the *Kaposin* mRNA leads to down-regulation of the expression of the corresponding protein (Lin and Sullivan 2011).

For the vast majority of viral miRNAs, mRNA target assignment remains particularly challenging and is based mainly on computational predictions of miRNA-binding sites. These predictions are especially problematic for these small RNAs, which usually show poor conservation and only restricted complementarity to their targets. However, recently developed high-throughput techniques now allow for the direct and genome-wide identification of miRNA targets (see Section IV.B). This will greatly facilitate the elucidation of viral miRNAs functions.

A. Functional Convergence of Viral miRNAs

Table 8.2 lists examples of both viral and cellular mRNA targets of viral miRNAs. Strikingly, evolutionary divergent viruses appear to have deployed the same miRNA regulatory function to control pathways, such as cellular immunity, cell cycle regulation, cell survival, or control of the viral cycle. This points out the key role of virus-encoded miRNAs during the infection and in associated pathologies. For instance, the cancer-associated viruses KSHV, EBV, BLV, and MDV-1 all encode miRNAs that regulate the expression of tumor suppressor or cell cycle regulatory proteins (see Table 8.2). Many viruses have also evolved miRNAs that target components of the immune system in order to avoid clearance of latently infected cells. For example, the natural killer (NK) cell ligand MICB is targeted by non-conserved miRNAs from EBV, KSHV, and HCMV (Stern-Ginossar et al. 2007; Nachmani et al. 2009). The three viruses have therefore evolved to express functionally convergent miRNAs to modulate recognition of infected cells by the NKG2D ligand. Human BK and JC polyomaviruses miRNAs also regulate the expression of another NKG2D ligand, ULBP2, to attenuate the host innate immunity response (Bauman and Mandelboim 2011; Bauman et al. 2011). Another feature of viral miRNAs is their involvement in the control of viral early genes to regulate the entry into productive stage. Hence, several viruses encode miRNAs within the latency region, and these can down-regulate viral transactivators, such as ICP0 and ICP4 (infected cell proteins 0 and 4) for alphaherpesviruses or RTA (replication and transcription activator) for gammaherpesviruses (Table 8.2).

B. Toward Complete Viral miRNA Targetomes

Innovative biochemical methods, such as RIP-Chip (ribonucleoprotein immunoprecipitation-gene Chip), HITS-CLIP (high-throughput sequencing and cross-linking immunoprecipitation), and PAR-CLIP (photoactivatable-ribonucleoside-enhanced CLIP) (reviewed in Thomson et al. 2011), are very useful for establishing comprehensive lists of putative miRNA targets. Briefly, target mRNAs are coimmunoprecipitated within RISC complexes (via Ago2 protein), and are either analyzed by microarrays or RNA sequencing. Several groups have taken advantage of these methods to identify potential targets of EBV and KSHV miRNAs in infected B-cell lines (Dölken et al. 2010b; Gottwein et al. 2011; Riley et al. 2012; Skalsky et al. 2012). Data provided extensive lists of mRNA containing binding sites for both viral and cellular miRNAs, ranging from 44 (Dölken et al. 2010b) to >2000 targets (Gottwein et al. 2011). Collectively, these studies show that viral miRNAs regulate pathways, such as transcriptional regulation, signal transduction, regulation of cell cycle, and apoptosis. Interestingly, Ago2 PAR-CLIP analysis of B-cells infected either by KSHV alone or in combination with EBV showed that miRNAs of these viruses shared a common set of targets. Ago2 HITS-CLIP analysis of EBV-transformed cells revealed that a large number of EBV miRNA targets were also bound by cellular miRNAs. This confirms the convergent evolution of viral and cellular miRNAs.

These new approaches are very powerful and reliable to retrieve miRNA-binding sites on a large scale (Gottwein et al. tested 36 targets and validated almost 80% of them). However, the biological relevance of most of the targets identified remains to be formally proven, and among the vast amount of miRNA-binding sites identified, some may be

TABLE 8.2. Examples of Viral and Cellular mRNA Targets of Virus-Encoded miRNAs

Function	Target	Virus (miRNA)	References
Viral Cycle and Latency Maintenance			
Viral immediate-early transactivators	ICP0	HSV-1 (H2-3p), HSV-2 (III)	Umbach et al. (2008) and Tang et al. (2009)
	ICP4	HSV-1 (H6), MDV-1 (M7-5p), ILTV (I5)	Umbach et al. (2008), Waidner et al. (2011), and Strassheim et al. (2012)
Viral transactivator	RTA	KSHV (K12-5[a], -7-5p, -9[c]), EBV (BART6-5p)	Bellare and Ganem (2009), Iizasa et al. (2010), Lu et al. (2010), and Lin et al. (2011)
Viral DNA polymerases	BALF5	EBV (BART2)	Barth et al. (2008)
Viral genes involved in cleavage/packaging of viral DNA	UL28, UL32	MDV-1 (M4)	Muylkens et al. (2010)
Host transcription factor	p53	EBV (BHRF1-1)	Li et al. (2012)
Cell-Mediated Immunity			
Viral large tumor antigen	TAg	BKV (B1-5p, -3p), JCV (J1-5p, -3p), SV40 (S1-5p, 3p), BPCV1 (B1), BPCV2 (B1)	Sullivan et al. (2005), Seo et al. (2008), and Chen et al. (2011)
Host stress-induced NKG2D ligands	MICB	HCMV (UL-112-1), EBV (BART2-5p), KSHV (K12-7)	Stern-Ginossar et al. (2007) and Nachmani et al. (2009)
	ULBP3	BKV (B1-3p), JCV (J1-3p)	Bauman and Mandelboim (2011) and Bauman et al. (2011)
Host chemokines	CXCL-16	MCMV (M23-2)	Dölken et al. (2010a)
	RANTES	HCMV (UL-148D)	Kim et al. (2012)
	CXCL-11	EBV (BHRF1-3)	Xia et al. (2008)
Cell Cycle Regulator, Oncogenesis			
Viral transforming factor	LMP1[b]	EBV (BART cluster 1)	Lo et al. (2007)
Host cyclin	G1/S Cyclin E2	HCMV (US25-1)	Grey et al. (2010)
Host cell cycle inhibitor	p21	KSHV (K12-1)	Gottwein and Cullen (2010)
Host mediator in TGF-β signaling	Smad5	KSHV (K12-11)	Liu et al. (2012)
Host tumor suppressor and antiangiogenic factor	THBS1	KSHV (K12-1, -3-3p, -6-3p, -11)	Samols et al. (2007)
Host tumor suppressor	HBP1	BLV (B4)	Kincaid et al. (2012)
Cell Survival			
Host regulator of transcription	BACH1	KSHV (K12-11)	Gottwein et al. (2007) and Skalsky et al. (2007)
	BCL6	EBV (BART3, 9, 17-5p)	Martín-Pérez et al. (2012)
Host proapoptotic factors	Casp3	KSHV (K12-1, -3, -4-3p)	Suffert et al. (2011)
	PUMA	EBV (BART5)	Choy et al. (2008)

In parentheses, next to virus names, are names of miRNAs (e.g., K12-1 means *miR-K12-1*).
[a]The repression observed was partial and may be indirect since there are only low-probability seed matches for *miR-K12-5* in *Rta* mRNA sequence.
[b]Down-regulation of LMP1 expression may also play a role in immune evasion.
[c]Stands for the miRNA star/passenger sequence.

C. Role of Viral miRNAs In Vivo

Up to now, most of the investigations to decipher viral miRNA roles have been performed in cultured cells. The ultimate goal will be to assess this crucial question *in vivo* during a physiological infection, where the virus encounters a broad array of cell types and immune responses that cannot be addressed *in vitro*. Dölken et al. reported on the first functional *in vivo* phenotype of a miRNA mutant MCMV (Dölken et al. 2010a). Deletion and point mutations were introduced in the viral genome to prevent expression of two miRNAs, *miR-m21-1* and *miR-M23-2*. Interestingly, viral titers were decreased by roughly 100-fold specifically in salivary glands of C57BL/6 mice infected by the mutant virus. This phenotype could be reverted by depleting the animals of both NK and T cells. These results suggest that these miRNAs are important in controlling the immune response, and that they might be important for the establishment of a persistent infection in this particular organ.

MDV-1 herpesvirus is the causative agent of Marek's disease, a deadly T cell-based lymphoma occurring in infected chickens. The Nair laboratory investigated the role of the viral ortholog of cellular *miR-155* (*miR-M4*), in inducing lymphoma (Zhao et al. 2011). By using a series of mutant viruses, they demonstrated that deletion or mutations of this viral miRNA did not affect replication but completely abolished oncogenicity of the virus. Interestingly, this phenotype could be rescued by expressing *miR-155*, showing the conservation of oncogenic functions of the two miRNAs. This was the first time a viral miRNA was proven to induce cancer in an *in vivo* model. The authors also showed that an attenuated virus deleted from several miRNAs, including *miR-M4*, could function as effective vaccine against a virulent strain, pointing toward the therapeutic potential of viral miRNAs.

Along the same line, Boss et al. investigated the role of the KSHV orthologue of *miR-155*, *miR-K12-11*, in humanized NOD/LtSz-scid IL2Rγ^{null} mice (Boss et al. 2011). The authors showed that ectopic expression of either *miR-K12-11* or *miR-155* led to an increased expansion of B cells in spleen. This was accompanied by B-cells infiltration of the splenic red pulp, disrupting the normal architecture of the periarteriolar lymphoid sheaths. This phenotype could be partly explained by the down-regulation of the transcription factor C/EBPβ, which is involved in B-cell lymphomagenesis.

V. CONCLUSIONS AND FUTURE PROSPECTS

Although we are still far from getting a global picture of how important miRNAs are in the timecourse of the infection, it becomes increasingly clear that these tiny regulators represent yet another weapon in the arsenal deployed by viruses to successfully infect their hosts. For example, there clearly is an evolutionary pressure to maintain miRNA expression in a wide range of different viruses. Even if these non-coding RNAs are not conserved in sequence, they can sometimes target the same pathways, and even sometimes the same genes. This is evidenced by striking examples, such as the existence of orthologues of cellular miRNAs, which share the same seed sequence, but also by completely different sequences that can recognize distinct regions of a target transcript. In addition, even if some viruses do not express miRNAs, for example, papillomaviruses, they can take advantage of the host ones by subverting their function and/or modulating their expression. With the advent of large-scale approaches to obtain the full repertoires of viral miRNAs, regulated cellular miRNAs, and their targets, we should quickly progress in the elucidation of their biological roles. However, there are still some challenges lying ahead of us. The most important one will be the construction of good *in vivo* models to fully grasp the importance of miRNAs in a physiologically relevant context. Equally challenging will be the understanding of how the expression of miRNAs themselves can be modulated, and this will surely keep researchers busy for years to come. There might also be some unexpected function of viral miRNAs. Indeed, some of the miRNA effector proteins have been shown to shuttle from the cytoplasm to the nucleus, and it could be that, similar to some cellular miRNAs, virally encoded miRNAs might have a nuclear role. Finally, we are only beginning to envision the therapeutic potential of these small RNA molecules, and it will be really interesting to see whether by blocking specific miRNAs, we can modify the outcome of a viral infection.

ACKNOWLEDGMENTS

We would like to thank Béatrice Chane-Woon-Ming, Gabrielle Haas, and Erika Girardi for critical reading of the manuscript. Work in our laboratory is supported by CNRS, the European Research Council (ERC Starting Grant ncRNAVIR 260767), the Agence Nationale de la Recherche (ANR-08-MIEN-005), and by the Institut National du Cancer (INCa-PLBIO-2009-173). A.F. is supported by a postdoctoral fellowship from the Fondation pour la Recherche Médicale and a Marie Curie Reintegration Grant (FP7-PEOPLE-2010-RG-268301).

REFERENCES

Andersson, M.G., P.C.J. Haasnoot, N. Xu, S. Berenjian, B. Berkhout, and G. Akusjärvi. 2005. Suppression of RNA interference by adenovirus virus-associated RNA. *J Virol* **79**:9556–9565.

Aparicio, O., N. Razquin, M. Zaratiegui, I. Narvaiza, and P. Fortes. 2006. Adenovirus virus-associated RNA is processed to functional interfering RNAs involved in virus production. *J Virol* **80**:1376–1384.

Backes, S., J.S. Shapiro, L.R. Sabin, A.M. Pham, I. Reyes, B. Moss, S. Cherry, and B.R. Tenoever. 2012. Degradation of host microRNAs by poxvirus poly(A) polymerase reveals terminal RNA methylation as a protective antiviral mechanism. *Cell Host Microbe* **12**:200–210.

Bartel, D.P. 2004. MicroRNAs: genomics, biogenesis, mechanism, and function. *Cell* **116**:281–297.

Bartel, D.P. 2009. MicroRNAs: target recognition and regulatory functions. *Cell* **136**:215–233.

Barth, S., T. Pfuhl, A. Mamiani, C. Ehses, K. Roemer, E. Kremmer, C. Jaker, J. Hock, G. Meister, and F.A. Grasser. 2008. Epstein–Barr virus-encoded microRNA *miR-BART2* down-regulates the viral DNA polymerase BALF5. *Nucleic Acids Res* **36**:666–675.

Bauman, Y. and O. Mandelboim. 2011. MicroRNA based immunoevasion mechanism of human polyomaviruses. *RNA Biol* **8**:591–594.

Bauman, Y., D. Nachmani, A. Vitenshtein, P. Tsukerman, N. Drayman, N. Stern-Ginossar, D. Lankry, R. Gruda, and O. Mandelboim. 2011. An identical miRNA of the human JC and BK polyoma viruses targets the stress-induced ligand ULBP3 to escape immune elimination. *Cell Host Microbe* **9**:93–102.

REFERENCES

Bechtel, J., A. Grundhoff, and D. Ganem. 2005. RNAs in the virion of Kaposi's sarcoma-associated herpesvirus. *J Virol* **79**:10138–10146.

Bellare, P. and D. Ganem. 2009. Regulation of KSHV lytic switch protein expression by a virus-encoded microRNA: an evolutionary adaptation that fine-tunes lytic reactivation. *Cell Host Microbe* **6**:570–575.

Bennasser, Y., C. Chable-Bessia, R. Triboulet, D. Gibbings, C. Gwizdek, C. Dargemont, E.J. Kremer, O. Voinnet, and M. Benkirane. 2011. Competition for XPO5 binding between Dicer mRNA, pre-miRNA and viral RNA regulates human Dicer levels. *Nat Struct Mol Biol* **18**:323–327.

Bogerd, H.P., H.W. Karnowski, X. Cai, J. Shin, M. Pohlers, and B.R. Cullen. 2010. A mammalian herpesvirus uses noncanonical expression and processing mechanisms to generate viral MicroRNAs. *Mol Cell* **37**:135–142.

Boss, I.W., P.E. Nadeau, J.R. Abbott, Y. Yang, A. Mergia, and R. Renne. 2011. A Kaposi's sarcoma-associated herpesvirus-encoded ortholog of microRNA *miR-155* induces human splenic B-cell expansion in NOD/LtSz-scid IL2Rγnull mice. *J Virol* **85**:9877–9886.

Cai, X., G. Li, L.A. Laimins, and B.R. Cullen. 2006a. Human papillomavirus genotype 31 does not express detectable microRNA levels during latent or productive virus replication. *J Virol* **80**:10890–10893.

Cai, X., A. Schäfer, S. Lu, J.P. Bilello, R.C. Desrosiers, R. Edwards, N. Raab-Traub, and B.R. Cullen. 2006b. Epstein–Barr virus microRNAs are evolutionarily conserved and differentially expressed. *PLoS Pathog* **2**:e23.

Cazalla, D., M. Xie, and J.A. Steitz. 2011. A primate herpesvirus uses the integrator complex to generate viral microRNAs. *Mol Cell* **43**:982–992.

Cazalla, D., T. Yario, and J.A. Steitz. 2010. Down-regulation of a host microRNA by a Herpesvirus saimiri noncoding RNA. *Science* **328**:1563–1566.

Chen, C.J., R.P. Kincaid, G.J. Seo, M.D. Bennett, and C.S. Sullivan. 2011. Insights into Polyomaviridae microRNA function derived from study of the bandicoot papillomatosis carcinomatosis viruses. *J Virol* **85**:4487–4500.

Choy, E.Y., K.L. Siu, K.H. Kok, R.W. Lung, C.M. Tsang, K.F. To, D.L. Kwong, S.W. Tsao, and D.Y. Jin. 2008. An Epstein–Barr virus-encoded microRNA targets PUMA to promote host cell survival. *J Exp Med* **205**:2551–2560.

Coley, W., R. Van Duyne, L. Carpio, I. Guendel, K. Kehn-Hall, S. Chevalier, A. Narayanan, T. Luu, N. Lee, Z. Klase, and F. Kashanchi. 2010. Absence of DICER in monocytes and its regulation by HIV-1. *J Biol Chem* **285**:31930–31943.

Dölken, L., A. Krmpotic, S. Kothe, L. Tuddenham, M. Tanguy, L. Marcinowski, Z. Ruzsics, N. Elefant, Y. Altuvia, H. Margalit, U.H. Koszinowski, S. Jonjic, and S. Pfeffer. 2010a. Cytomegalovirus microRNAs facilitate persistent virus infection in salivary glands. *PLoS Pathog* **6**:e1001150.

Dölken, L., G. Malterer, F. Erhard, S. Kothe, C.C. Friedel, G. Suffert, L. Marcinowski, N. Motsch, S. Barth, M. Beitzinger, D. Lieber, S.M. Bailer, R. Hoffmann, Z. Ruzsics, E. Kremmer, S. Pfeffer, R. Zimmer, U.H. Koszinowski, F. Grässer, G. Meister, and J. Haas. 2010b. Systematic analysis of viral and cellular microRNA targets in cells latently infected with human gamma-herpesviruses by RISC immunoprecipitation assay. *Cell Host Microbe* **7**:324–334.

Dölken, L., J. Perot, V. Cognat, A. Alioua, M. John, J. Soutschek, Z. Ruzsics, U. Koszinowski, O. Voinnet, and S. Pfeffer. 2007. Mouse cytomegalovirus microRNAs dominate the cellular small RNA profile during lytic infection and show features of posttranscriptional regulation. *J Virol* **81**:13771–13782.

Forte, E., R.E. Salinas, C. Chang, T. Zhou, S.D. Linnstaedt, E. Gottwein, C. Jacobs, D. Jima, Q.-J. Li, S.S. Dave, and M.A. Luftig. 2012. The Epstein–Barr Virus (EBV)-induced tumor suppressor microRNA *MiR-34a* is growth promoting in EBV-infected B cells. *J Virol* **86**:6889–6898.

Friedman, R.C., K.K. Farh, C.B. Burge, and D.P. Bartel. 2009. Most mammalian mRNAs are conserved targets of microRNAs. *Genome Res* **19**:92–105.

Gatto, G., A. Rossi, D. Rossi, S. Kroening, S. Bonatti, and M. Mallardo. 2008. Epstein–Barr virus latent membrane protein 1 trans-activates *miR-155* transcription through the NF-kappaB pathway. *Nucleic Acids Res* **36**:6608–6619.

Gottwein, E., D.L. Corcoran, N. Mukherjee, R.L. Skalsky, M. Hafner, J.D. Nusbaum, P. Shamulailatpam, C.L. Love, S.S. Dave, T. Tuschl, U. Ohler, and B.R. Cullen. 2011. Viral microRNA targetome of KSHV-infected primary effusion lymphoma cell lines. *Cell Host Microbe* **10**:515–526.

Gottwein, E. and B.R. Cullen. 2010. A human herpesvirus microRNA inhibits p21 expression and attenuates p21-mediated cell cycle arrest. *J Virol* **84**:5229–5237.

Gottwein, E., N. Mukherjee, C. Sachse, C. Frenzel, W.H. Majoros, J.-T.A. Chi, R. Braich, M. Manoharan, J. Soutschek, U. Ohler, and B.R. Cullen. 2007. A viral microRNA functions as an orthologue of cellular *miR-155*. *Nature* **450**:1096–1099.

Greijer, A.E., C.A. Dekkers, and J.M. Middeldorp. 2000. Human cytomegalovirus virions differentially incorporate viral and host cell RNA during the assembly process. *J Virol* **74**:9078–9082.

Grey, F., R. Tirabassi, H. Meyers, G. Wu, S. McWeeney, L. Hook, and J.A. Nelson. 2010. A viral microRNA down-regulates multiple cell cycle genes through mRNA 5'UTRs. *PLoS Pathog* **6**:e1000967.

Grinberg, M., S. Gilad, E. Meiri, A. Levy, O. Isakov, R. Ronen, N. Shomron, Z. Bentwich, and Y. Shemer-Avni. 2012. Vaccinia virus infection suppresses the cell microRNA machinery. *Arch Virol* **157**(9):1719–1727.

Gu, W., J. An, P. Ye, K.-N. Zhao, and A. Antonsson. 2011. Prediction of conserved microRNAs from skin and mucosal human papillomaviruses. *Arch Virol* **156**:1161–1171.

Hancock, M.H., R.S. Tirabassi, and J.A. Nelson. 2012. Rhesus cytomegalovirus encodes seventeen microRNAs that are differentially expressed in vitro and in vivo. *Virology* **425**:133–142.

Hansen, A., S. Henderson, D. Lagos, L. Nikitenko, E. Coulter, S. Roberts, F. Gratrix, K. Plaisance, R. Renne, M. Bower, P. Kellam, and C. Boshoff. 2010. KSHV-encoded miRNAs target MAF to induce endothelial cell reprogramming. *Genes Dev* **24**:195–205.

Houzet, L. and K.-T. Jeang. 2011. MicroRNAs and human retroviruses. *Biochim Biophys Acta* **1809**:686–693.

Hussain, M., R.J. Taft, and S. Asgari. 2008. An insect virus-encoded microRNA regulates viral replication. *J Virol* **82**:9164–9170.

Hussain, M., S. Torres, E. Schnettler, A. Funk, A. Grundhoff, G.P. Pijlman, A.A. Khromykh, and S. Asgari. 2012. West Nile virus encodes a microRNA-like small RNA in the 3' untranslated region which up-regulates GATA4 mRNA and facilitates virus replication in mosquito cells. *Nucleic Acids Res* **40**:2210–2223.

Iizasa, H., B.-E. Wulff, N.R. Alla, M. Maragkakis, M. Megraw, A. Hatzigeorgiou, D. Iwakiri, K. Takada, A. Wiedmer, L. Showe, P. Lieberman, and K. Nishikura. 2010. Editing of Epstein–Barr virus-encoded BART6 microRNAs controls their dicer targeting and consequently affects viral latency. *J Biol Chem* **285**:33358–33370.

Jochum, S., R. Ruiss, A. Moosmann, W. Hammerschmidt, and R. Zeidler. 2012. RNAs in Epstein–Barr virions control early steps of infection. *Proc Natl Acad Sci U S A* **109**:E1396–E1404.

Jurak, I., M.F. Kramer, J.C. Mellor, A.L. van Lint, F.P. Roth, D.M. Knipe, and D.M. Coen. 2010. Numerous conserved and divergent microRNAs expressed by herpes simplex viruses 1 and 2. *J Virol* **84**:4659–4672.

Kim, Y., S. Lee, S. Kim, D. Kim, J.-H. Ahn, and K. Ahn. 2012. Human cytomegalovirus clinical strain-specific microRNA *miR-UL148D* targets the human chemokine RANTES during infection. *PLoS Pathog* **8**:e1002577.

REFERENCES

Kincaid, R.P., J.M. Burke, and C.S. Sullivan. 2012. RNA virus microRNA that mimics a B-cell oncomiR. *Proc Natl Acad Sci U S A* **109**:3077–3082.

Kozomara, A. and S. Griffiths-Jones. 2010. miRBase: integrating microRNA annotation and deep-sequencing data. *Nucleic Acids Res* **39**:D152–157.

Lagos, D., G. Pollara, S. Henderson, F. Gratrix, M. Fabani, R.S. Milne, F. Gotch, and C. Boshoff. 2010. *miR-132* regulates antiviral innate immunity through suppression of the p300 transcriptional co-activator. *Nat Cell Biol* **12**:513–519.

Li, Z., X. Chen, L. Li, S. Liu, L. Yang, X. Ma, M. Tang, A.M. Bode, Z. Dong, L. Sun, and Y. Cao. 2012. EBV encoded *miR-BHRF1-1* potentiates viral lytic replication by downregulating host p53 in nasopharyngeal carcinoma. *Int J Biochem Cell Biol* **44**:275–279.

Lin, J. and B.R. Cullen. 2007. Analysis of the interaction of primate retroviruses with the human RNA interference machinery. *J Virol* **81**:12218–12226.

Lin, X., D. Liang, Z. He, Q. Deng, E.S. Robertson, and K. Lan. 2011. *miR-K12-7-5p* encoded by Kaposi's sarcoma-associated herpesvirus stabilizes the latent state by targeting viral ORF50/RTA. *PLoS ONE* **6**:e16224.

Lin, Y.-T., R.P. Kincaid, D. Arasappan, S.E. Dowd, S.P. Hunicke-Smith, and C.S. Sullivan. 2010. Small RNA profiling reveals antisense transcription throughout the KSHV genome and novel small RNAs. *RNA* **16**:1540–1558.

Lin, Y.-T. and C.S. Sullivan. 2011. Expanding the role of Drosha to the regulation of viral gene expression. *Proc Natl Acad Sci U S A* **108**:11229–11234.

Liu, Y., R. Sun, X. Lin, D. Liang, Q. Deng, and K. Lan. 2012. Kaposi's sarcoma-associated herpesvirus-encoded microRNA *miR-K12-11* attenuates transforming growth factor beta signaling through suppression of SMAD5. *J Virol* **86**:1372–1381.

Lo, A.K., K.F. To, K.W. Lo, R.W. Lung, J.W. Hui, G. Liao, and S.D. Hayward. 2007. Modulation of LMP1 protein expression by EBV-encoded microRNAs. *Proc Natl Acad Sci U S A* **104**:16164–16169.

Lu, F., W. Stedman, M. Yousef, R. Renne, and P.M. Lieberman. 2010. Epigenetic regulation of Kaposi's sarcoma-associated herpesvirus latency by virus-encoded microRNAs that target Rta and the cellular Rbl2-DNMT pathway. *J Virol* **84**:2697–2706.

Lu, S. and B.R. Cullen. 2004. Adenovirus VA1 noncoding RNA can inhibit small interfering RNA and MicroRNA biogenesis. *J Virol* **78**:12868–12876.

Luo, J., A.-J. Sun, M. Teng, H. Zhou, Z.-Z. Cui, L.-H. Qu, and G.-P. Zhang. 2011. Expression profiles of microRNAs encoded by the oncogenic Marek's disease virus reveal two distinct expression patterns in vivo during different phases of disease. *J Gen Virol* **92**:608–620.

Marcinowski, L., M. Tanguy, A. Krmpotic, B. Rädle, V.J. Lisnić, L. Tuddenham, B. Chane-Woon-Ming, Z. Ruzsics, F. Erhard, C. Benkartek, M. Babic, R. Zimmer, J. Trgovcich, U.H. Koszinowski, S. Jonjic, S. Pfeffer, and L. Dölken. 2012. Degradation of cellular mir-27 by a novel, highly abundant viral transcript is important for efficient virus replication in vivo. *PLoS Pathog* **8**:e1002510.

Martín-Pérez, D., P. Vargiu, S. Montes-Moreno, E.A. León, S.M. Rodríguez-Pinilla, L.D. Lisio, N. Martínez, R. Rodríguez, M. Mollejo, J. Castellvi, D.G. Pisano, M. Sánchez-Beato, and M.A. Piris. 2012. Epstein–Barr virus microRNAs repress BCL6 expression in diffuse large B-cell lymphoma. *Leukemia* **26**:180–183.

Meyer, C., F. Grey, C.N. Kreklywich, T.F. Andoh, R.S. Tirabassi, S.L. Orloff, and D.N. Streblow. 2011. Cytomegalovirus microRNA expression is tissue specific and is associated with persistence. *J Virol* **85**:378–389.

Morgan, R., A. Anderson, E. Bernberg, S. Kamboj, E. Huang, G. Lagasse, G. Isaacs, M. Parcells, B.C. Meyers, P.J. Green, and J. Burnside. 2008. Sequence conservation and differential expression of Marek's disease virus microRNAs. *J Virol* **82**:12213–12220.

Mukerjee, R., J.R. Chang, L. Del Valle, A. Bagashev, M.M. Gayed, R.B. Lyde, B.J. Hawkins, E. Brailoiu, E. Cohen, C. Power, S.A. Azizi, B.B. Gelman, and B.E. Sawaya. 2011. Deregulation of microRNAs by HIV-1 Vpr protein leads to the development of neurocognitive disorders. *J Biol Chem* **286**:34976–34985.

Mulik, S., J. Xu, P.B.J. Reddy, N.K. Rajasagi, F. Gimenez, S. Sharma, P.Y. Lu, and B.T. Rouse. 2012. Role of *miR-132* in angiogenesis after ocular infection with herpes simplex virus. *Am J Pathol* **181**:525–534.

Muylkens, B., D. Coupeau, G. Dambrine, S. Trapp, and D. Rasschaert. 2010. Marek's disease virus microRNA designated Mdv1-pre-*miR-M4* targets both cellular and viral genes. *Arch Virol* **155**:1823–1837.

Nachmani, D., N. Stern-Ginossar, R. Sarid, and O. Mandelboim. 2009. Diverse herpesvirus microRNAs target the stress-induced immune ligand MICB to escape recognition by natural killer cells. *Cell Host Microbe* **5**:376–385.

Pegtel, D.M., M.D.B. van de Garde, and J.M. Middeldorp. 2011. Viral miRNAs exploiting the endosomal-exosomal pathway for intercellular cross-talk and immune evasion. *Biochim Biophys Acta* **1809**:715–721.

Pfeffer, S., A. Sewer, M. Lagos-Quintana, R. Sheridan, C. Sander, F.A. Grasser, L.F. van Dyk, C.K. Ho, S. Shuman, M. Chien, J.J. Russo, J. Ju, G. Randall, B.D. Lindenbach, C.M. Rice, V. Simon, D.D. Ho, M. Zavolan, and T. Tuschl. 2005. Identification of microRNAs of the herpesvirus family. *Nat Methods* **2**:269–276.

Pfeffer, S., M. Zavolan, F.A. Grasser, M. Chien, J.J. Russo, J. Ju, B. John, A.J. Enright, D. Marks, C. Sander, and T. Tuschl. 2004. Identification of virus-encoded microRNAs. *Science* **304**:734–736.

Riley, K.J., G.S. Rabinowitz, T.A. Yario, J.M. Luna, R.B. Darnell, and J.A. Steitz. 2012. EBV and human microRNAs co-target oncogenic and apoptotic viral and human genes during latency. *EMBO J* **31**:2207–2221.

Samols, M.A., R.L. Skalsky, A.M. Maldonado, A. Riva, M.C. Lopez, H.V. Baker, and R. Renne. 2007. Identification of cellular genes targeted by KSHV-encoded microRNAs. *PLoS Pathog* **3**:e65.

Sano, M., Y. Kato, and K. Taira. 2006. Sequence-specific interference by small RNAs derived from adenovirus VAI RNA. *FEBS Lett* **580**:1553–1564.

Seo, G.J., L.H.L. Fink, B. O'Hara, W.J. Atwood, and C.S. Sullivan. 2008. Evolutionarily conserved function of a viral microRNA. *J Virol* **82**:9823–9828.

Shapiro, J.S., R.A. Langlois, A.M. Pham, and B.R. Tenoever. 2012. Evidence for a cytoplasmic microprocessor of pri-miRNAs. *RNA* **18**:1338–1346.

Shapiro, J.S., A. Varble, A.M. Pham, and B.R. Tenoever. 2010. Noncanonical cytoplasmic processing of viral microRNAs. *RNA* **16**:2068–2074.

Singh, J., C.P. Singh, A. Bhavani, and J. Nagaraju. 2010. Discovering microRNAs from Bombyx mori nucleopolyhedrosis virus. *Virology* **407**:120–128.

Skalsky, R.L., D.L. Corcoran, E. Gottwein, C.L. Frank, D. Kang, M. Hafner, J.D. Nusbaum, R. Feederle, H.-J. Delecluse, M.A. Luftig, T. Tuschl, U. Ohler, and B.R. Cullen. 2012. The viral and cellular microRNA targetome in lymphoblastoid cell lines. *PLoS Pathog* **8**:e1002484.

Skalsky, R.L., M.A. Samols, K.B. Plaisance, I.W. Boss, A. Riva, M.C. Lopez, H.V. Baker, and R. Renne. 2007. Kaposi's sarcoma-associated herpesvirus encodes an ortholog of *miR-155*. *J Virol* **81**:12836–12845.

Stern-Ginossar, N., N. Elefant, A. Zimmermann, D.G. Wolf, N. Saleh, M. Biton, E. Horwitz, Z. Prokocimer, M. Prichard, G. Hahn, D. Goldman-Wohl, C. Greenfield, S. Yagel, H. Hengel, Y. Altuvia, H. Margalit, and O. Mandelboim. 2007. Host immune system gene targeting by a viral miRNA. *Science* **317**:376–381.

Strassheim, S., G. Stik, D. Rasschaert, and S. Laurent. 2012. mdv1-*miR-M7-5p*, located in the newly identified first intron of the latency-associated transcript of Marek's disease virus, targets the immediate-early genes ICP4 and ICP27. *J Gen Virol* **93**:1731–1742.

REFERENCES

Suffert, G., G. Malterer, J. Hausser, J. Viiliäinen, A. Fender, M. Contrant, T. Ivacevic, V. Benes, F. Gros, O. Voinnet, M. Zavolan, P.M. Ojala, J.G. Haas, and S. Pfeffer. 2011. Kaposi's sarcoma herpesvirus microRNAs target caspase 3 and regulate apoptosis. *PLoS Pathog* 7:e1002405.

Sullivan, C.S., A.T. Grundhoff, S. Tevethia, J.M. Pipas, and D. Ganem. 2005. SV40-encoded microRNAs regulate viral gene expression and reduce susceptibility to cytotoxic T cells. *Nature* 435:682–686.

Sullivan, C.S., C.K. Sung, C.D. Pack, A. Grundhoff, A.E. Lukacher, T.L. Benjamin, and D. Ganem. 2009. Murine polyomavirus encodes a microRNA that cleaves early RNA transcripts but is not essential for experimental infection. *Virology* 387:157–167.

Tang, S., A. Patel, and P.R. Krause. 2009. Novel less-abundant viral microRNAs encoded by herpes simplex virus 2 latency-associated transcript and their roles in regulating ICP34.5 and ICP0 mRNAs. *J Virol* 83:1433–1442.

Thomson, D.W., C.P. Bracken, and G.J. Goodall. 2011. Experimental strategies for microRNA target identification. *Nucleic Acids Res* 39:6845–6853.

Triboulet, R., B. Mari, Y.-L. Lin, C. Chable-Bessia, Y. Bennasser, K. Lebrigand, B. Cardinaud, T. Maurin, P. Barbry, V. Baillat, J. Reynes, P. Corbeau, K.-T. Jeang, and M. Benkirane. 2007. Suppression of microRNA-silencing pathway by HIV-1 during virus replication. *Science* 315:1579–1582.

Tuddenham, L., J.S. Jung, B. Chane-Woon-Ming, L. Dölken, and S. Pfeffer. 2012. Small RNA deep sequencing identifies microRNAs and other small noncoding RNAs from human herpesvirus 6B. *J Virol* 86:1638–1649.

Umbach, J.L. and B.R. Cullen. 2009. The role of RNAi and microRNAs in animal virus replication and antiviral immunity. *Genes Dev* 23:1151–1164.

Umbach, J.L., M.F. Kramer, I. Jurak, H.W. Karnowski, D.M. Coen, and B.R. Cullen. 2008. MicroRNAs expressed by herpes simplex virus 1 during latent infection regulate viral mRNAs. *Nature* 454:780–783.

Umbach, J.L., K. Wang, S. Tang, P.R. Krause, E.K. Mont, J.I. Cohen, and B.R. Cullen. 2010a. Identification of viral microRNAs expressed in human sacral ganglia latently infected with herpes simplex virus 2. *J Virol* 84:1189–1192.

Umbach, J.L., H.-L. Yen, L.L.M. Poon, and B.R. Cullen. 2010b. Influenza A virus expresses high levels of an unusual class of small viral leader RNAs in infected cells. *MBio* 1(4):e00204–10.

Varble, A., M.A. Chua, J.T. Perez, B. Manicassamy, A. García-Sastre, and B.R. tenOever. 2010. Engineered RNA viral synthesis of microRNAs. *Proc Natl Acad Sci U S A* 107:11519–11524.

Waidner, L.A., J. Burnside, A.S. Anderson, E.L. Bernberg, M.A. German, B.C. Meyers, P.J. Green, and R.W. Morgan. 2011. A microRNA of infectious laryngotracheitis virus can downregulate and direct cleavage of ICP4 mRNA. *Virology* 411:25–31.

Walz, N., T. Christalla, U. Tessmer, and A. Grundhoff. 2010. A global analysis of evolutionary conservation among known and predicted gammaherpesvirus microRNAs. *J Virol* 84:716–728.

Wang, F.-Z., F. Weber, C. Croce, C.-G. Liu, X. Liao, and P.E. Pellett. 2008. Human cytomegalovirus infection alters the expression of cellular microRNA species that affect its replication. *J Virol* 82:9065–9074.

Wu, Y.-L., C.P. Wu, C.Y.Y. Liu, P.W.-C. Hsu, E.C. Wu, and Y.-C. Chao. 2011. A non-coding RNA of insect HzNV-1 virus establishes latent viral infection through microRNA. *Sci Rep* 1:60.

Xia, T., A. O'Hara, I. Araujo, J. Barreto, E. Carvalho, J.B. Sapucaia, J.C. Ramos, E. Luz, C. Pedroso, M. Manrique, N.L. Toomey, C. Brites, D.P. Dittmer, and W.J. Harrington, Jr. 2008. EBV microRNAs in primary lymphomas and targeting of CXCL-11 by ebv-mir-BHRF1-3. *Cancer Res* 68:1436–1442.

Xu, N., B. Segerman, X. Zhou, and G. Akusjärvi. 2007. Adenovirus virus-associated RNAII-derived small RNAs are efficiently incorporated into the rna-induced silencing complex and associate with polyribosomes. *J Virol* 81:10540–10549.

Yan, Y., H. Cui, S. Jiang, Y. Huang, X. Huang, S. Wei, W. Xu, and Q. Qin. 2011. Identification of a novel marine fish virus, Singapore grouper iridovirus-encoded microRNAs expressed in grouper cells by Solexa sequencing. *PLoS ONE* **6**:e19148.

Yang, J.-S. and E.C. Lai. 2011. Alternative miRNA biogenesis pathways and the interpretation of core miRNA pathway mutants. *Mol Cell* **43**:892–903.

Yeung, M.L., Y. Bennasser, T. Myers, G. Jiang, M. Benkirane, and K.T. Jeang. 2005. Changes in microRNA expression profiles in HIV-1-transfected human cells. *Retrovirology* **2**:81.

Zhao, Y., H. Xu, Y. Yao, L.P. Smith, L. Kgosana, J. Green, L. Petherbridge, S.J. Baigent, and V. Nair. 2011. Critical role of the virus-encoded microRNA-155 ortholog in the induction of Marek's disease lymphomas. *PLoS Pathog* **7**:e1001305.

9

MicroRNA Expression in Avian Herpesviruses

Yongxiu Yao and Venugopal Nair

Avian Viral Diseases Programme, The Pirbright Institute, Compton Laboratory, Compton, Berkshire, UK

I.	Introduction	138
II.	Avian Herpesviruses and Associated Diseases	138
III.	Identification of miRNAs Encoded by Avian Herpesviruses	139
	A. MDV-1 miRNAs	139
	B. MDV-2 miRNAs	142
	C. HVT miRNAs	142
	D. ILTV-miRNAs	143
	E. DEV miRNAs	143
IV.	Viral Orthologues of Host miRNAs	143
V.	Target Identification of Avian Herpesvirus miRNAs	145
	A. Viral Targets of Viral miRNAs	146
	B. Cellular Targets of Viral miRNAs	147
VI.	Conclusions	148
	Acknowledgment	148
	References	148

ABBREVIATIONS

DEV   duck enteritis virus
EBV   Eptein–Barr virus

MicroRNAs in Medicine, First Edition. Edited by Charles H. Lawrie.
© 2014 John Wiley & Sons, Inc. Published 2014 by John Wiley & Sons, Inc.

HVT	herpesvirus of turkeys
ICP4	infected cell polypeptide 4
ILTV	infectious laryngotracheitis virus
KSHV	Kaposi's sarcoma herpesvirus
MDV	Marek's disease virus
miRNA	microRNA
vMDV	virulent MDV
vvMDV	very virulent MDV
vv+MDV	very virulent plus MDV

I. INTRODUCTION

With an obligatory intracellular lifestyle, viruses have to deal with a hostile cell environment and use of the cellular machinery for survival and replication. In the case of herpesviruses, where the virus-host interaction is characterized by long-term survival as latent infection, this demands sophisticated methods of survival without being detected by the innate and adaptive immune mechanisms of the host. Herpesviruses achieve this using a variety of mechanisms through restricted gene expression, epigenetic control of viral/host gene expression, and translational control (Norman and Sarnow 2010). Posttranslational regulation of gene expression by non-coding RNAs, such as short interfering RNA (siRNA) and microRNA (miRNA), is now well recognized in a number of species and biological systems. The small size of the miRNAs, combined with their ability for specific repression of the expression of multiple transcript targets, make them ideal tools for herpesviruses to reshape the gene expression in an infected cell to favor viral replication. Hence, it is not surprising that herpesviruses currently account for nearly 95% of virus-encoded miRNAs identified to date (http://www.mirbase.org). Since the first report of Epstein–Barr virus (EBV)-encoded miRNAs (Pfeffer et al. 2004), numerous miRNAs have been identified in different herpesviruses (Boss et al. 2009; Cullen 2011). The number of miRNAs encoded by different herpesviruses varies from as few as three miRNAs in herpes B virus to 68 mature miRNAs in rhesus lymphocryptovirus.

II. AVIAN HERPESVIRUSES AND ASSOCIATED DISEASES

Avian herpesviruses are a major group of pathogens associated with a number of diseases in different species of poultry. All of the pathogenic members of avian herpesviruses belong to the same subfamily *Alphaherpesvirinae* in the family *Herpesviridae*. *Alphaherpesvirinae* consists of four genera, two of which comprise of avian herpesviruses. *Mardivirus* genus consists of the pathogenic Marek's disease virus-1 (MDV-1, Gallid herpesvirus 2), and attenuated Marek's disease virus-2 (MDV-2, Gallid herpesvirus 3) and herpesvirus of turkey (HVT, Meleagrid herpesvirus 1). MDV-1 is further classified into different pathotypes referred to as virulent (vMDV), very virulent MDV (vvMDV), and very virulent plus (vv+MDV) on the basis of their pathogenicity (Witter 1997). *Iltovirus* genus consists of infectious laryngotracheitis virus (ILTV, Gallid herpesvirus 1). The family *Herpesviridae* subfamily also includes other unassigned viruses, such as the duck enteritis virus (DEV), that induces acute disease in waterfowl species. Marek's disease (MD), caused by MDV-1, is a widespread lymphoproliferative disease of chickens characterized by rapid-onset lymphomas in multiple organs, and infiltration into peripheral nerves causing paralysis. MD is widespread in the poultry population around the world and causes annual economic losses up to US$ 2 billion (Davison and Nair 2004). Although controlled by the use of live attenuated MDV-1 strains and antigenically related, but non-pathogenic MDV-2 and HVT vaccines (Bublot and Sharma 2004), there is evidence of continued evolution of the virus toward greater virulence, challenging the sustainability of the vaccination strategy (Gimeno 2008).

Infectious laryngotracheitis (ILT) is a contagious viral respiratory tract infection caused by ILTV that results in high mortality and severe losses in egg production in infected poultry flocks. Vaccination using live attenuated vaccines is used for the control of this disease. Duck enteritis caused by DEV is an acute contagious infection of waterfowl species associated with very high mortality, and is controlled by vaccination (Liu et al. 2011).

III. IDENTIFICATION OF miRNAs ENCODED BY AVIAN HERPESVIRUSES

Most viral miRNAs had initially been identified by a protocol previously developed for the identification of host-encoded miRNAs, a procedure that involves RNA size fractionation, ligation of linkers, reverse transcription, concatamerization, and Sanger sequencing (Pfeffer et al. 2005a). The computational approaches that rely on commonalities in the predicated secondary structures of pre-miRNAs to identify miRNA-encoding loci specifically in viral genome have also been developed. With the advent of massively parallel sequencing technologies it is now possible to explore libraries of the cloned small RNAs with a higher degree of reliability and unprecedented depth. We and others have reported the identification of miRNAs from a number of avian herpesvirus, including 14 miRNAs (26 mature sequences) from MDV-1 (Burnside et al. 2006; Yao et al. 2008), 18 (36 mature sequences) from MDV-2 (Yao et al. 2007; Waidner et al. 2009), 17 (28 mature sequences) from HVT (Waidner et al. 2009; Zhao et al. 2009), 7 (10 mature sequences) from ILTV (Rachamadugu et al. 2009; Waidner et al. 2009), and 24 (33 mature sequences) from DEV (Yao et al. 2012) (Figure 9.1 and Table 9.1).

A. MDV-1 miRNAs

MDV-1-encoded miRNAs were first reported by Burnside et al. describing the identification of eight pre-miRNAs encoded by MDV-1 using 454 Life Sciences sequencing technology on small RNA libraries made from chicken embryo fibroblast (CEF) infected with the RB1B strain of MDV-1 (Burnside et al. 2006). Six additional MDV-1 miRNAs were discovered subsequently by analysis of small RNA library of MSB-1, a lymphoblastoid cell line established from an MDV-induced lymphoma of the spleen and tumor sample (Morgan et al. 2008; Yao et al. 2008). In total, MDV-1 encodes for 14 precursor miRNAs, which produce 26 mature miRNAs (miRBase v19, http://www.mirbase.org). The MDV-1 miRNA coding sequences are clustered into three separate genomic loci: cluster 1 (*mdv1-mir-M9, 5, 12, 3, 2,* and *4*) and cluster 2 (*mdv1-mir-M11, 31,* and *1*) map upstream and downstream from the Meq gene respectively; cluster 3 (*mdv1-mir-M8, 13, 6, 7,* and *10*) maps near the 5′ end of the LAT antisense to the downstream region of the ICP4 gene (Burnside et al. 2006; Morgan et al. 2008; Yao et al. 2008) (Figure 9.1). All three clusters are located in the inverted repeat regions of the MDV-1 genome. The expression of these miRNAs was confirmed using Northern hybridization to RNA isolated from MDV-1-infected CEF, MDV-1-induced tumors, and MDV-1-induced MSB-1 lymphoblastoid cell

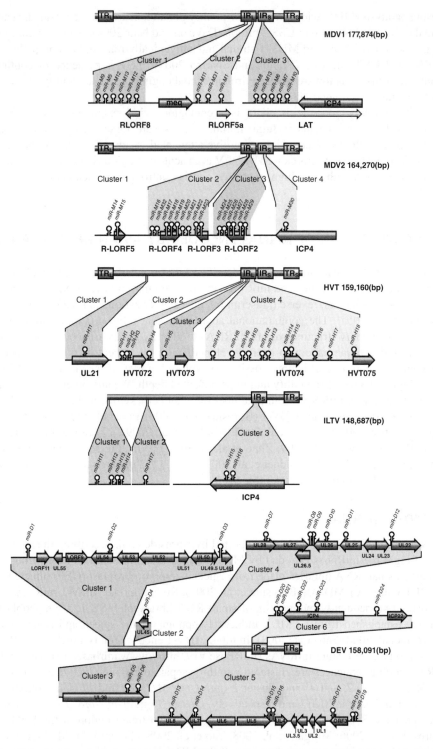

Figure 9.1. Diagrammatic representation of the viral genomes showing the positions of miRNAs. Position and orientation of selected transcripts are shown. The genome size (base pairs) of viruses (MDV1, MDV2, HVT, ILTV, and DEV) is shown on the right. See color insert.

TABLE 9.1. Avian Herpesvirus-Encoded miRNAs

Virus	Number of Pre-miRNAs	Number of Mature miRNAs	Name of miRNAs	Reference
MDV-1	14	26	mdv1-mir-M1-M13 and -M31	Burnside et al. (2006), Morgan et al. (2008), and Yao et al. (2008)
MDV-2	18	36	mdv2-mir-M14-M30 and -M32	Yao et al. (2007) and Waidner et al. (2009)
HVT	17	28	hvt-mir-H1-H5, H7-H18	Waidner et al. (2009) and Yao et al. (2009b)
ILTV	7	10	iltv-mir-I1-I7	Rachamadugu et al. (2009) and Waidner et al. (2009)
DEV	24	33	dev-mir-D1-D24	Yao et al. (2012)

line. In MDV-1, the sequence of all miRNAs was conserved among 23 different strains representing three pathotypes (Morgan et al. 2008; Burnside and Morgan 2011). The viral miRNAs were highly expressed in MD tumors and transformed cell lines (Yao et al. 2008). The cluster 1 miRNAs in tumors are expressed at higher levels with highly virulent strains compared with less virulent strains, and this differential expression has been linked to a polymorphism in the promoter region of these viral miRNAs (Morgan et al. 2008; Coupeau et al. 2012). In contrast, levels of cluster 3 miRNAs were equivalent in tumors produced by vvMDV and vv+MDV strains. These results indicate that cluster 1 miRNAs could play an important role in MDV-1 pathogenicity. Indeed, this hypothesis was proved by demonstration that the deletion of cluster 1 from the viral genome abolished the oncogenicity of the virus (Zhao et al. 2011). *mdv1-mir-M4-5p,* a member of cluster 1 miRNA and a functional orthologue of gga-*mir-155*, which plays a major role in lymphoid malignancies and the modulation of immune responses, is the most highly expressed viral miRNA in tumors, in some cases accounting for even up to 72% of all MDV miRNAs. This miRNA was shown to play a key role in MDV-1-induced oncogenesis (Zhao et al. 2011). By reverse-transcription polymerase chain reaction (RT-PCR), rapid amplification of cDNA ends (RACE)-PCR and semi-quantitative RT-PCR on latently or productively infected cells, the prmiRM9M4 promoter, corresponding to the 1300 bp immediately upstream from *mdv1-mir-M9* was identified to drive the transcription of cluster 1 and 2 miRNAs during the latent phase and the transcription of cluster 1 during the lytic phase. Indeed, this promoter has been shown to be active by both active histone marks and DNA hypomethylation during MDV-1 latency (Brown et al. 2011), confirming its transcriptional activity.

In order to identify the miRNA expression signature specific to MDV-transformed cells, Yao et al. examined the global miRNA expression profiles in seven distinct MDV-transformed cell lines by microarray analysis (Yao et al. 2009a). This study revealed that MD tumor-derived lymphoblastoid cell lines showed altered expression of several host-encoded miRNAs. Comparison of the miRNA expression profiles of these cell lines with the MDV-negative, retrovirus-transformed AVOL-1 cell line showed that *mir-150* and *mir-223* are down-regulated irrespective of the viral etiology, whereas down-regulation of *mir-155* was specific for MDV transformed tumor cells. Thus, increased expression of MDV-encoded miRNAs with specific down-regulation of *mir-155* can be considered as unique expression signatures for MD tumor cells. Analysis of the functional targets of these miRNAs would contribute to the understanding of the molecular pathways of MD oncogenicity.

B. MDV-2 miRNAs

MSB-1 is a MDV-transformed lymphoblastoid cell line co-infected with both MDV-1 strain BC-1 and MDV-2 strain HPRS24 (Hirai et al. 1990). The analysis of the small RNA library from MSB-1 allowed the identification of 17 novel MDV-2-specific miRNAs (Yao et al. 2007). Out of these, 16 were clustered in a 4.2-kb-long repeat region that encodes R-LORF2 to R-LORF5 and are expressed in the same direction, suggesting that they may be derived from a common primary transcript. However, differences in the number of sequencing reads among these miRNAs indicate that they vary with regard to stability or processing. Eight miRNAs in this cluster were located in the coding regions of these ORFs. These included *mdv2-mir-M15* (R-LORF5); *mir-M17, mir-M18,* and *mir-M19* (R-LORF4); *mir-M23* (R-LORF3) and *mir-M26, mir-M27,* and *mir-M28* (R-LORF2). The orientations of *mir-M15* and *mir-M17, -M18,* and *-M19* are the same as that for R-LORF5 and R-LORF4, respectively. On the other hand, *mir-M23* and *mir-M26, -27,* and *-28* are located in an orientation opposite to that of R-LORF3 and R-LORF2, respectively. The single miRNA outside the cluster was located in the short repeat region, within the C-terminal region of the ICP4 homologue. The expression of these miRNAs in MSB-1 cells and MDV-2-infected chicken embryo fibroblasts was further confirmed by Northern blotting analysis. The identification of miRNA clusters within the repeat regions of MDV-2 demonstrates conservation of the relative genomic positions of miRNA clusters in MDV-1 and MDV-2, despite the lack of sequence homology among the miRNAs of the two viruses. A subsequent study using high throughput sequencing of small RNAs from MSB1 cells reported an additional miRNA, *mdv2-mir-M32,* and the expression of this miRNA was confirmed by Northern hybridization (Waidner et al. 2009). The unique feature for MDV-2-encoded miRNAs is that all 18 precursors give rise to two mature forms, representing both strands of the duplex (miRBase 19; http://www.mirbase.org/), resulting in 36 mature MDV-2 miRNAs.

C. HVT miRNAs

HVT-encoded miRNAs were identified from HVT-infected CEF using both high throughput sequencing technology (Waidner et al. 2009) and traditional cloning and sequencing of a small RNA library (Yao et al. 2009b). Seventeen HVT miRNAs were reported. Sixteen of these miRNAs were clustered together within the repeat long region of the viral genome, demonstrating some degree of positional conservation with MDV-1 and MDV-2. The only HVT miRNA outside this cluster, *hvt-mir-H11,* was located in the U_L region in the same orientation to the coding region of tegument protein U_L21. Like MDV-2 miRNAs, the 16 HVT miRNAs in the repeat long region are also expressed in the same direction. Ten of these miRNAs were located in a region of the HVT genome that contains two tandem repeats, and multiple sequence alignment of the precursor of these miRNAs indicated small sequence variations, suggesting evolution by duplication. The sequence of the precursors of *mir-H9* and *mir-H12* showed high sequence homology (95.3%) with identical sequences in the loop and the mature -3p regions, while the -5p mature miRNA region showed three substitutions, including the one in the seed region. Similarly, the precursors of *mir-H16* and *mir-H17* are highly homologous, with only a single nucleotide difference in both -5p and -3p mature miRNA sequences, as well as in the loop regions. HVT-encoded miRNAs represent the first clear example of evolution of miRNAs by duplication among viruses.

D. ILTV-miRNAs

Waidner et al. first reported six miRNAs of ILTV (*iltv-mir-I1-I6*) using high-throughput sequencing from ILTV-infected chicken embryo kidney (CEK) cells (Waidner et al. 2009). An additional miRNA, *iltv-mir-I7*, was reported from infected CEK and leghorn male hepatoma (LMH) cell line by the 454 FLX sequencing method (Rachamadugu et al. 2009). Four of the miRNAs (*iltv-mir-I1-I4*) were located at the extreme terminus of the genome and are not associated with any annotated ORFs. The *iltv-mir-I7* was mapped in the replication origin (oriL) of the palindrome stem loop sequence. Two of the miRNAs, *iltv-mir-I5* and *-I6*, are located antisense to the 3′ end of the ILTV ICP4 coding region. Although the expression of all ILTV miRNAs were confirmed by the end-point PCR using small RNA libraries generated from ILTV-infected CEK, only three (*iltv-mir-I3*, *-I5*, and *-I6*) have been confirmed by Northern hybridizations.

E. DEV miRNAs

Using deep sequencing approach on RNA from infected CEF cultures, Yao et al. identified 24 DEV-encoded miRNAs (Yao et al. 2012). Unlike most *Mardivirus*-encoded miRNAs, which are located at the repeat regions, the majority of the DEV miRNAs were encoded within the unique long region as six clusters from both the coding and non-coding regions of the 15809-bp viral genome. The precursors of DEV *mir-D18* and *mir-D19* overlapped with each other, suggesting similarities to miRNA-offset RNAs, although only the dev-*mir-D18-3p* was functional in reporter assays. Using a computational approach, 12 putative DEV miRNAs have been reported (Xiang et al. 2012), although none of these miRNAs overlapped with the 24 DEV miRNAs described previously (Yao et al. 2012).

IV. VIRAL ORTHOLOGUES OF HOST miRNAs

Compared with the metazoan miRNAs, which are often highly conserved between species, virus-encoded miRNAs generally do not share sequence homologies with other virus- or host- encoded miRNAs (Cai et al. 2006; Nair and Zavolan 2006; Yao et al. 2007). However, partial sharing of sequences, particularly in the target interaction region (seed region), can result in the conservation of miRNA functions between virus- and host-encoded miRNAs.

Following the finding that kshv-*mir-K12-11*, a miRNA encoded by the human pathogen Kaposi's sarcoma herpesvirus (KSHV) is a functional orthologue of hsa-*mir-155* (Gottwein et al. 2007; Skalsky et al. 2007), *mdv1-mir-M4-5p* has been observed to share the identical seed sequences as that of the gga-*mir-155* (Figure 9.2), although the rest of the region showed little homology with each other. In subsequent experiments using reporter constructs, as well as measuring the levels of some of the putative targets, such as Pu.1, *mdv1-mir-M4-5p* was indeed shown to be a functional orthologue of *mir-155* (Zhao et al. 2009). Interestingly, in both MDV and KSHV-induced tumors, there is down-regulation of endogenous levels of *mir-155* (Gottwein et al. 2007; McClure and Sullivan 2008), although the mechanisms for such down-regulation is not fully understood. There

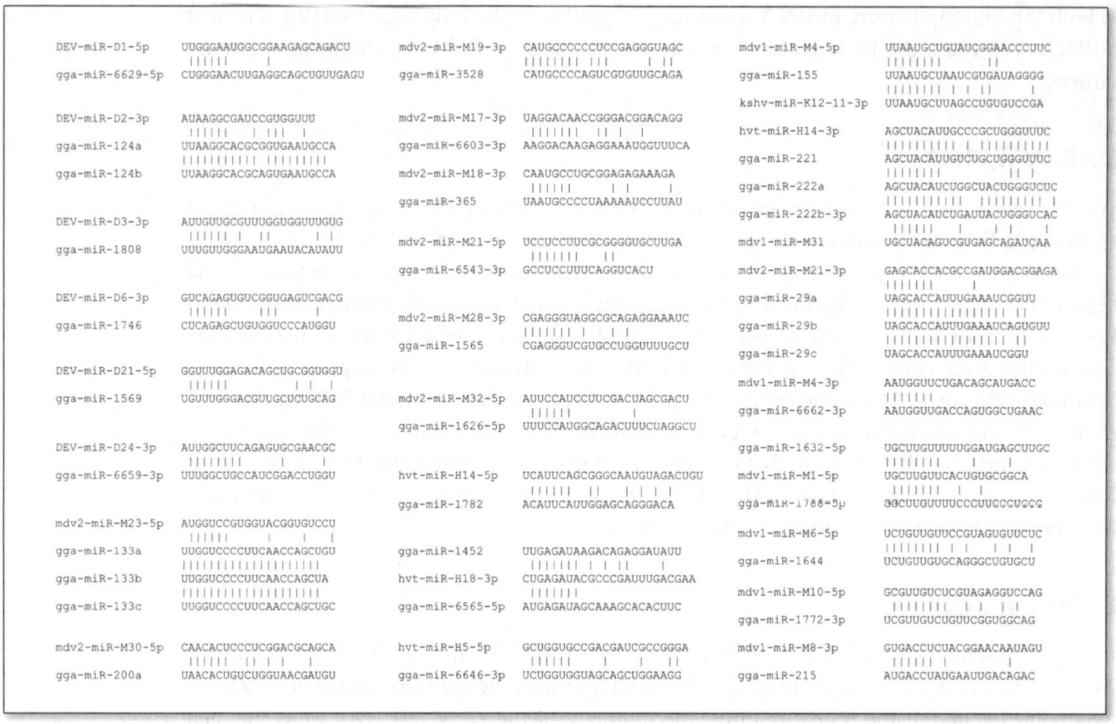

```
DEV-miR-D1-5p     UUGGGAAUGGCGGAAGAGCAGACU     mdv2-miR-M19-3p   CAUGCCCCCUCCGAGGGUAGC     mdv1-miR-M4-5p    UUAAUGCUGUAUCGGAACCCUUC
                  ||||||| |                                      |||||||| ||| | ||                           |||||||| ||
gga-miR-6629-5p   CUGGGAACUUGAGGCAGCUGUUGAGU   gga-miR-3528      CAUGCCCCAGUCGUGUUGCAGA    gga-miR-155       UUAAUGCUAAUCGUGAUAGGGG
                                                                                                              |||||||| ||| | |
                                                                                           kshv-miR-K12-11-3p UUAAUGCUUAGCCUGUGUCCGA
DEV-miR-D2-3p     AUAAGGCGAUCCGUGGUUU          mdv2-miR-M17-3p   UAGGACAACCGGGACGGACAGG
                  ||||||| | ||| |                                |||||||| ||| |           hvt-miR-H14-3p    AGCUACAUUGCCCGCUGGGUUUC
gga-miR-124a      UUAAGGCACGCGGUGAAUGCCA       gga-miR-6603-3p   AAGGACAAGAGGAAAUGGUUUCA                     |||||||||||||| |||||||
                  ||||||||||| ||||||||                                                     gga-miR-221       AGCUACAUUGCUCUGGGUUUC
gga-miR-124b      UUAAGGCACGCAGUGAAUGCCA       mdv2-miR-M18-3p   CAAUGCCUGCGGAGAGAAAGA                       |||||||| | |       |
                                                                 |||||||| || |             gga-miR-222a      AGCUACAUCUGGCUACUGGGCUC
                                               gga-miR-365       UAAUGCCCCUAAAAAAUCCUUAU                     |||||||||||| |||||||
                                                                                           gga-miR-222b-3p   AGCUACAUCUGAUUACUGGGUCAC
DEV-miR-D3-3p     AUUGUUGCGUUUGGUGGUUUGUG
                  ||||||  | ||  |               mdv2-miR-M21-5p  UCCUCCUUCGCGGGGUGCUUGA    mdv1-miR-M31      UGCUACAGUCGUGAGCAGAUCAA
gga-miR-1808      UUUGUUGGGAAUGAAUACAUAUU                         |||||||  ||
                                                gga-miR-6543-3p  GCCUCCUUUCAGGUCACU        mdv2-miR-M21-3p   GAGCACCACGCCGAUGGACGGAGA
                                                                                                             |||||||| |
DEV-miR-D6-3p     GUCAGAGUGUCGGUGAGUCGACG                                                  gga-miR-29a       UAGCACCAUUUGAAAUCGGUU
                  ||||||  |||  |                mdv2-miR-M28-3p  CGAGGGUAGGCGCAGAGGAAAUC                     |||||||||||||||||| ||
gga-miR-1746      CUCAGAGCUGUGGUCCCAUGGU                         |||||||| | |              gga-miR-29b      UAGCACCAUUUGAAAUCAGUGUU
                                                gga-miR-1565     CGAGGGUCGUGCCUGGUUUUGCU                     ||||||||||||||||||| |
                                                                                           gga-miR-29c       UAGCACCAUUUGAAAUCGGUU
DEV-miR-D21-5p    GGUUUGGAGACAGCUGCGGUGGU
                  ||||||       | | |             mdv2-miR-M32-5p  AUUCCAUCCUUCGACUAGCGACU  mdv1-miR-M4-3p    AAUGGUUCUGACAGCAUGACC
gga-miR-1569      UGUUUGGGACGUUGCUCUGCAG                          |||||| |                  gga-miR-6662-3p  AAUGGUUGACCAGUGGCUGAAC
                                                 gga-miR-1626-5p  UUUCCAUGGCAGACUUUCUAGGCU
DEV-miR-D24-3p    AUUGGCUUCAGAGUGCGAACGC                                                    gga-miR-1632-5p   UGCUUGUUUUUGGAUGAGCUUGC
                  |||||||| | | |                                                                             |||||||| | |
gga-miR-6659-3p   UUUGGCUGCCAUCGGACCUGGU        hvt-miR-H14-5p   UCAUUCAGCGGGCAAUGUAGACUGU  mdv1-miR-M1-5p    UGCUUGUUCACUGUGCGCA
                                                                 |||||| || || ||                              |||||||| |
                                                gga-miR-1782     ACAUUCAUUGGAGCAGGGACA      gga-miR-1788-5p   GGCUUGUUUUUCCCUUCCCUGCG
mdv2-miR-M23-5p   AUGGUCCGUGGUACGGUGUCCU                                                    mdv1-miR-M6-5p    UCUGUUGUUCCGUAGUGUUCUC
                  ||||||      |                                                                              |||||||| |  |
gga-miR-133a      UUGGUCCCCUUCAACCAGCUGU        gga-miR-1452     UUGAGAUAAGACAGAGGAUAUU     gga-miR-1644      UCUGUUGUGCAGGGCUGUGCU
                  ||||||||||||||||||||                            |||||||| | ||| |
gga-miR-133b      UUGGUCCCCUUCAACCAGCUA         hvt-miR-H18-3p   CUGAGAUACGCCCGAUUUGACGAA   mdv1-miR-M10-5p   GCGUUGUCUCGUAGAGGUCCAG
                  ||||||||||||||||||||                                                                         |||||||| |  |  |
gga-miR-133c      UUGGUCCCCUUCAACCAGCUGC        gga-miR-6565-5p  AUGAGAUAGCAAAGCACACUUC     gga-miR-1772-3p   UCGUUGUCUGUUCGGUGGCAG

mdv2-miR-M30-5p   CAACACUCCCUCGGACGCAGCA        hvt-miR-H5-5p    GCUGGUGCCGACGAUCGCCGGGA    mdv1-miR-M8-3p    GUGACCUCUACGGAACAAUAGU
                  ||||||  || | |                                  |||||| |  |  | ||                            |||||||| |   |
gga-miR-200a      UAACACUGUCGGUAACGAUGU         gga-miR-6646-3p  UCGGUGGUAGCAGCUGGAAGG      gga-miR-215       AUGACCUAUGAAUUGACAGAC
```

Figure 9.2. Sequence alignment of miRNA homologs. Vertical lines indicate nucleotide homology.

has been a number of studies showing that *mir-155* has direct role of oncogenesis (Faraoni et al. 2009; Tili et al. 2009) and molecular mechanisms of cancer pathogenesis (Rai et al. 2010; Valeri et al. 2010). Furthermore, overexpression of *mir-155* has been shown to be associated with lymphocyte transformation by viruses, such as EBV (Lu et al. 2008) and reticuloendotheliosis virus strain T (Bolisetty et al. 2009). Given the role of *mir-155* in many malignancies (Xiao and Rajewsky 2009), the functional role of *mdv1-mir-M4-5p* in inducing T-cell lymphomas was explored using a series of mutant viruses generated by reverse genetics techniques on the full-length infectious bacterial artificial chromosome (BAC) clone of the highly oncogenic RB-1B virus (Zhao et al. 2011). The effect of the mutations on the oncogenicity of the virus was tested in natural infection models of the disease using 1-day-old SPF chickens. The results showed that deletion/mutation of the entire set of miRNAs in cluster 1 resulted in the total loss of oncogenicity. Furthermore, deletion or seed region mutagenesis of *mdv1-mir-M4-5p* prevents lymphoma induction in infected birds, showing the importance of this single miRNA in the induction of tumours. In support of the previous finding that *mdv1-mir-M4-5p* is a functional orthologue of gga-*mir-155*, the loss of oncogenicity could be rescued using a chimeric MDV that expressed *mir-155* instead of *mdv1-mir-M4-5p*. However, insertion of *mdv1-mir-M4* into HVT genome did not induce tumors in infected birds (Burnside and Morgan 2011), suggesting that other factors are also needed for viral transformation.

One of the HVT miRNAs, hvt-*mir-H14-3p*, shares extended sequence conservation with the cellular miRNA gga-*mir-221* with perfect match of the 21/23 nucleotides, including identical sequence of the seed region (Figure 9.2). This is the first virus-encoded miRNA that shows such close and extended sequence identity with a host miRNA.

Although it has been suggested that novel miRNAs can arise *de novo* from the existing hairpin structures (Liu et al. 2008), the close sequence identity extending nearly the full length of these two miRNAs, strongly suggested that hvt-*mir-H14-3p* is most likely to have been acquired from the host genome. Further evidence for this comes from the demonstration of partial sequence conservation between the downstream flanking region of hvt-*mir-H14-3p* in the HVT genome and the gga-*mir-221* locus on chromosome 1 of the chicken genome (Waidner et al. 2009). If this is true, hvt-*mir-H14-3p* is the first example of a mature miRNA "pirated" by the virus from the host. Interestingly, *mir-221* targets the cyclin-dependent kinase inhibitor 1B (p27, Kip1), a regulator of the cell cycle G1-to-S phase transition. The repression of p27 by *mir-221* is thought to play an important role in cancer progression (Galardi et al. 2007; Fornari et al. 2008), and could play a role in MDV-1-induced tumorigenesis (Lambeth et al. 2009). The expression of a *mir-221* orthologue and down-regulation of p27 could move the cell cycle to the S phase in order to support replication of the viral genome as well as to increase growth of infected cells for additional viral production (Burnside and Morgan 2011). *mdv1-mir-M31* also shares seed sequence with *mir-221* (Morgan et al. 2008), but is limited to the nucleotide positions 2–7, corresponding to the minimal miRNA seed region. Thus, targets that depend on base pairing with the nucleotide position 8 of the miRNA may not be shared between these miRNAs.

Seed sequence homology to cellular miRNAs is also observed for *mdv2-mir-M21* with *miR-29a/b/c* (Figure 9.2), a conserved miRNA with roles in apoptosis (Pfeffer et al. 2005b; Cai et al. 2006; Waidner et al. 2009). Human *miR-29b* has been previously shown to target the "*de novo*" DNA methyltranferase, DNMT3b in humans (Fabbri et al. 2007; Garzon et al. 2009). Due to the conserved nature between chicken and human *mir-29b*, it is hypothesized that gga-*mir-29b* could also target DNMT3b. The hypothesis that *mdv2-mir-M21* acting as an orthologue to cellular *mir-29b* targeting DNMT3b has been verified by reporter assay (unpublished data). DNMT3b is responsible for active changes in DNA methylation and therefore the epigenetic control of gene expression (Kato et al. 2007). Changes in the methylation profile within cells have dramatic effect on cell phenotype, and this mechanism of methylation-induced silencing of genes could be crucial in MDV-2 infection.

With the growing list of miRNA entries in miRBase (http://www.mirbase.org), more avian herpesvirus-encoded miRNAs have been found to share their seed sequences with the host miRNAs (Figure 9.2). Future deep sequencing efforts to identify new viral and cellular miRNAs are likely to identify more viral orthologues of cellular miRNAs. Computationally predicted targets of these miRNAs may provide a useful starting point for functional analysis.

V. TARGET IDENTIFICATION OF AVIAN HERPESVIRUS MiRNAs

Although increasing data on the targets of miRNAs and associated functional pathways are beginning to emerge, it is clear that virus-encoded miRNAs can target both viral and cellular transcripts. The regulation of viral protein-coding genes by viral miRNAs probably may have a beneficiary effect on the viral biological steps in the host cell, including its replication, latency, and for evading the host immune system. The modulatory effect of viral miRNAs on cellular transcripts is generally thought to promote a cellular environment favorable to completion of the viral life cycle. A number of targets of avian herpesvirus-encoded miRNAs have been validated (Table 9.2).

TABLE 9.2. Viral and Cellular Targets of Avian Herpesvirus miRNAs

Virus	miRNAs	Target	Proposed function	Method of Confirmation	Reference
			Viral Targets		
MDV-1	*mdv1-mir-M4-5p*	UL28	Prevent lytic replication/	RA & WB	Muylkens et al. (2010)
	mdv1-mir-M4-3p	UL32	promote latency	WB	
MDV-1	*mdv1-mir-M7-5p*	ICP4	Establish and/or	RA & WB	Strassheim et al. (2012)
		ICP27	maintain latency		
ILTV	*iltv-mir-I5*	ICP4	Establish and/or maintain latency	RA, qRT-PCR, & RACE	Waidner et al. (2011)
			Cellular Targets		
MDV-1	*mdv1-mir-M3*	Smad2	Antiapoptotic	RA & WB	Xu et al. (2011)
MDV-1	*mdv1-mir-M4-5p*	Pu.1	Mimics cellular *mir-155*	RA & WB	Zhao et al. (2009) and Muylkens et al. (2010)
		CEBPβ		RA	Zhao et al. (2009) and Muylkens et al. (2010)
		HIVEP2		RA	Zhao et al. (2009)
		BCL2L13		RA	Zhao et al. (2009)
		PDCD6		RA	Zhao et al. (2009)
		GPM6B		RA	Muylkens et al. (2010)
		RREB1		RA	Muylkens et al. (2010)
		c-Myb		RA	Muylkens et al. (2010)
		MAP3K7IP2		RA	Muylkens et al. (2010)

RA, reporter assay; WB, western blot.

A. Viral Targets of Viral miRNAs

Identifying viral targets of viral miRNAs is more straightforward compared with identification of cellular targets as viral genomes encode fewer candidate mRNAs. Known examples of viral mRNA targets include transcripts that are transcribed antisense to the viral miRNA precursor and transcripts with imperfect matches. *iltv-mir-I5*, for example, lies antisense to ICP4 and directs cleavage of ICP4 mRNA (Waidner et al. 2011). *iltv-mir-I5* down-regulated luciferase activity by 60% of luciferase constructs bearing a portion of the ICP4 coding sequence containing complementary sites. The *iltv-mir-I5* mimic, when cotransfected with a plasmid expressing ICP4, reduced ICP4 transcript levels by approximately 50%, and inhibition was relieved by an *iltv-mir-I5* antagomir. In addition, it has been shown that *iltv-mir-I5*-mediated cleavage at the canonical site by modified RACE analysis (Waidner et al. 2011). ICP4 is an immediate early viral transactivator with a key role in the induction of lytic replication. The targeting of ICP4 by viral miRNAs is thought to mediate entry into latency and render the latent state more robust (Umbach et al. 2008). In addition to *iltv-mir-I5*, *iltv-mir-I6* also maps antisense to the ICP4 gene. However, repression of luciferase activity observed for *iltv-mir-I6* was not significant (Waidner et al. 2011). The blockage of accessibility to the binding region has been proposed by performing *in silico* folding of RNA containing the targets for *iltv-mir-I5* and *iltv-mir-I6*. This is consistent with the finding that target RNA folding is a key determinant of the efficacy of designed siRNAs (Brown et al. 2005). Sequences transcribed antisense to known miRNA stem-loop structures may have an increased propensity to fold into stem-loop structures themselves. Although MDV-1, MDV-2, and DEV also encode miRNAs that are antisense to certain viral transcripts (Burnside et al. 2006; Yao et al. 2007, 2008, 2009b; Waidner et al. 2009), possible regulatory relationships between miRNAs and their antisense mRNA transcripts need to be individually verified.

mdv1-mir-M4-5p is a functional orthologue of cellular *mir-155*. Apart from several cellular targets shared with *mir-155*, *mdv1-mir-M4-5p* also targets viral mRNA UL28, making it the first avian herpesviral miRNA known to target both viral and cellular mRNAs. *mdv1-mir-M4-3p*, the other strand derived from precursor *mdv1-mir-M4*, was shown to target viral mRNA UL32 (Muylkens et al. 2010). Both target sequences are located in the coding region rather than 3′-UTR. UL28 and UL32 are homologous to human herpesvirus 1 (HHV-1) proteins required for the cleavage and packaging of virion DNA. Their role in MDV-1 replication has yet to be investigated. It is possible that *mdv1-miR-M4* helps to maintain MDV-1 latency by down-regulating the production of UL28 and UL32 and impairing late MDV morphogenesis and reactivation. Interestingly, although gga-*mir-155* and *mdv1-mir-M4-5p* share the same seed sequence, only *mdv1-mir-M4-5p* targeted UL28. The precise nucleotide requirements for the functional binding of a miRNA to a target sequence are not fully understood. *mdv1-mir-M4-5p* displays 2–12 Watson–Crick base pairings with the UL28 mRNA, corresponding to four base pairings in addition to those of the seed sequence, whereas gga-*mir-155* displays only canonical 7mer-m8 base-pairing, involving exclusively the seed region. Thus, the outside sequence of the seed region also is thought to play a role in the miRNA binding to the target.

MDV immediate-early (IE) genes ICP4 and ICP27 have been identified as putative targets of *mdv1-mir-M7-5p*, a member of the MDV-1 cluster 3 (Strassheim et al. 2012). Again, the target sequences are located in the coding region rather than the 3′UTR. Endogenously expressed *mdv1-mir-M7-5p* in MSB-1 cells reduced luciferase activity significantly when miRNA-responsive elements from ICP4 or ICP27 were cloned in the 3′-UTR of the firefly luciferase gene. ICP27 protein levels were decreased by 70% when the *mdv1-mir-M7-5p* precursor was coexpressed with an ICP27 expression plasmid. Additionally, a negative correlation between the decreased expression of *mdv1-mir-M7-5p* and an increase in ICP27 expression during virus reactivation has also been observed. By targeting two IE genes, MDV miRNAs produced from LAT transcripts may contribute to establish and/ or maintain latency. These findings further support the idea that herpesvirus miRNAs play a key role in controlling the switch between viral lytic and latent infection.

B. Cellular Targets of Viral miRNAs

To complete the viral life cycle, a virus needs to keep a host cell alive long enough. It is shown that viral miRNAs can promote virus replication by targeting cellular genes to prolong cell survival and evade immune recognition.

As observed for its kshv-encoded counterpart, several cellular targets were shown to be commonly regulated by *mdv1-mir-M4-5p* and cellular *mir-155* (Zhao et al. 2009; Muylkens et al. 2010). Zhao et al. first reported that *mdv1-mir-M4-5p* shares common targets with *mir-155*, such as Pu.1, CEBPβ, HIVEP2, BCL2L13, and PDCD6 by luciferase reporter assay, as well as western blot for Pu.1. The repression of transcriptional factors, such as Pu.1, can have wide-ranging effects on the cellular milieu and global gene expression profiles in lymphocytes. Similarly, the repression of the other target genes is also likely to contribute to the induction of hematopoietic cell malignancy. Using a similar approach, Muylkens and colleagues identified four more cellular targets shared by *mdv1-mir-M4-5p* and *mir-155*, which include MAP3K7IP2, GPM6B, RREB1, and c-Myb. These observations provide additional evidence for an impact of *mir-155* and its orthologues on pathways regulating lymphocyte activation, differentiation, and immune tolerance.

Smad2 has been identified as a target for *mdv1-mir-M3*, demonstrating that this miRNA can promote cell survival from Smad2-mediated cisplatin-induced apoptosis (Xu et al. 2011). Smad2 is a member of the transforming growth factor-β signal pathway, and suppression of Smad2 by *mdv1-mir-M3* would suppress apoptosis of infected cells. These data suggest that latent/oncogenic viruses may encode miRNAs to directly target cellular factors involved in antiviral processes, including apoptosis, thus proactively creating a cellular environment beneficial to viral latency and oncogenesis. Surely, with the advances in high-throughput technologies, we will continue to identify more targets of miRNAs encoded by other herpesviruses. Once more targets of these virus-encoded miRNAs are discovered, and an integrated approach for demonstrating the functions and molecular pathways are developed, we should be able to get a more clear understanding of the role played by the small and highly effective modulators of gene expression.

VI. CONCLUSIONS

Recent advances in sequencing technology have led to the identification of a number of miRNAs encoded by avian herpesviruses. Although we have a long way to go to gain significant understanding on how these miRNAs function and the portfolio of their targets, it is clear that these small but effective regulators of gene expression play a key role in herpesvirus biology. By directly targeting key viral lytic genes or indirectly modulating cellular regulatory pathways, virus-encoded miRNAs could contribute significantly toward switching between lytic and latent infections by herpesviruses, thereby regulating viral pathogenesis *in vivo*. One key question that remains to be answered is whether these miRNAs have a small number of critical targets, or whether multiple, possibly hundreds, of miRNA-target interactions have functional significance.

ACKNOWLEDGMENT

The authors would like to acknowledge funding from the Biotechnology and Biological Sciences Research Council (BBSRC) United Kingdom.

REFERENCES

Bolisetty, M.T., G. Dy, W. Tam, and K.L. Beemon. 2009. Reticuloendotheliosis virus strain T induces miR-155, which targets JARID2 and promotes cell survival. *J Virol* **83**:12009–12017.

Boss, I.W., K.B. Plaisance, and R. Renne. 2009. Role of virus-encoded microRNAs in herpesvirus biology. *Trends Microbiol* **17**:544–553.

Brown, A.C., V. Nair, and M.J. Allday. 2011. Epigenetic regulation of the latency-associated region of Marek's disease virus in tumor-derived T-cell lines and primary lymphoma. *J Virol* **86**: 1683–1695.

Brown, K.M., C.Y. Chu, and T.M. Rana. 2005. Target accessibility dictates the potency of human RISC. *Nat Struct Mol Biol* **12**:469–470.

Bublot, M. and J. Sharma. 2004. Vaccination against Marek's disease. In *Marek's disease, an evolving problem*. F. Davison and V. Nair, Eds. Elsevier Academic Press, Amsterdam. pp. 168–185.

REFERENCES

Burnside, J., E. Bernberg, A. Anderson, C. Lu, B.C. Meyers, P.J. Green, N. Jain, G. Isaacs, and R.W. Morgan. 2006. Marek's disease virus encodes MicroRNAs that map to meq and the latency-associated transcript. *J Virol* **80**:8778–8786.

Burnside, J. and R. Morgan. 2011. Emerging roles of chicken and viral microRNAs in avian disease. *BMC Proc* **5 Suppl 4**:S2.

Cai, X., A. Schafer, S. Lu, J.P. Bilello, R.C. Desrosiers, R. Edwards, N. Raab-Traub, and B.R. Cullen. 2006. Epstein-Barr virus microRNAs are evolutionarily conserved and differentially expressed. *PLoS Pathog* **2**:e23.

Coupeau, D., G. Dambrine, and D. Rasschaert. 2012. Kinetic expression analysis of the cluster mdv1-mir-M9-M4, genes meq and vIL-8 differs between the lytic and latent phases of Marek's disease virus infection. *J Gen Virol* **93**:1519–1529.

Cullen, B.R. 2011. Herpesvirus microRNAs: phenotypes and functions. *Curr Opin Virol* **1**:211–215.

Davison, F. and V. Nair. 2004. *Marek's disease: an evolving problem*. Academic Press, London.

Fabbri, M., R. Garzon, A. Cimmino, Z. Liu, N. Zanesi, E. Callegari, S. Liu, H. Alder, S. Costinean, C. Fernandez-Cymering, S. Volinia, G. Guler, et al. 2007. MicroRNA-29 family reverts aberrant methylation in lung cancer by targeting DNA methyltransferases 3A and 3B. *Proc Natl Acad Sci U S A* **104**:15805–15810.

Faraoni, I., F.R. Antonetti, J. Cardone, and E. Bonmassar. 2009. miR-155 gene: a typical multifunctional microRNA. *Biochim Biophys Acta* **1792**:497–505.

Fornari, F., L. Gramantieri, M. Ferracin, A. Veronese, S. Sabbioni, G.A. Calin, G.L. Grazi, C. Giovannini, C.M. Croce, L. Bolondi, and M. Negrini. 2008. MiR-221 controls CDKN1C/p57 and CDKN1B/p27 expression in human hepatocellular carcinoma. *Oncogene* **27**:5651–5661.

Galardi, S., N. Mercatelli, E. Giorda, S. Massalini, G.V. Frajese, S.A. Ciafre, and M.G. Farace. 2007. miR-221 and miR-222 expression affects the proliferation potential of human prostate carcinoma cell lines by targeting p27Kip1. *J Biol Chem* **282**:23716–23724.

Garzon, R., S. Liu, M. Fabbri, Z. Liu, C.E. Heaphy, E. Callegari, S. Schwind, J. Pang, J. Yu, N. Muthusamy, V. Havelange, S. Volinia, et al. 2009. MicroRNA-29b induces global DNA hypomethylation and tumor suppressor gene reexpression in acute myeloid leukemia by targeting directly DNMT3A and 3B and indirectly DNMT1. *Blood* **113**:6411–6418.

Gimeno, I.M. 2008. Marek's disease vaccines: a solution for today but a worry for tomorrow? *Vaccine* **26 Suppl 3**:C31–C41.

Gottwein, E., N. Mukherjee, C. Sachse, C. Frenzel, W.H. Majoros, J.T. Chi, R. Braich, M. Manoharan, J. Soutschek, U. Ohler, and B.R. Cullen. 2007. A viral microRNA functions as an orthologue of cellular *miR-155*. *Nature* **450**:1096–1099.

Hirai, K., M. Yamada, Y. Arao, S. Kato, and S. Nii. 1990. Replicating Marek's disease virus (MDV) serotype 2 DNA with inserted MDV serotype 1 DNA sequences in a Marek's disease lymphoblastoid cell line MSB1-41C. *Arch Virol* **114**:153–165.

Kato, Y., M. Kaneda, K. Hata, K. Kumaki, M. Hisano, Y. Kohara, M. Okano, E. Li, M. Nozaki, and H. Sasaki. 2007. Role of the Dnmt3 family in de novo methylation of imprinted and repetitive sequences during male germ cell development in the mouse. *Hum Mol Genet* **16**:2272–2280.

Lambeth, L.S., Y. Yao, L.P. Smith, Y. Zhao, and V. Nair. 2009. MicroRNAs 221 and 222 target p27Kip1 in Marek's disease virus-transformed tumour cell line MSB-1. *J Gen Virol* **90**:1164–1171.

Liu, J., P. Chen, Y. Jiang, L. Wu, X. Zeng, G. Tian, J. Ge, Y. Kawaoka, Z. Bu, and H. Chen. 2011. A duck enteritis virus-vectored bivalent live vaccine provides fast and complete protection against H5N1 avian influenza virus infection in ducks. *J Virol* **85**:10989–10998.

Liu, N., K. Okamura, D.M. Tyler, M.D. Phillips, W.J. Chung, and E.C. Lai. 2008. The evolution and functional diversification of animal microRNA genes. *Cell Res* **18**:985–996.

Lu, F., A. Weidmer, C.G. Liu, S. Volinia, C.M. Croce, and P.M. Lieberman. 2008. Epstein-Barr virus-induced miR-155 attenuates NF-kappaB signaling and stabilizes latent virus persistence. *J Virol* **82**:10436–10443.

McClure, L.V. and C.S. Sullivan. 2008. Kaposi's sarcoma herpes virus taps into a host microRNA regulatory network. *Cell Host Microbe* **3**:1–3.

Morgan, R., A. Anderson, E. Bernberg, S. Kamboj, E. Huang, G. Lagasse, G. Isaacs, M. Parcells, B.C. Meyers, P.J. Green, and J. Burnside. 2008. Sequence conservation and differential expression of Marek's disease virus microRNAs. *J Virol* **82**(24):12213–12220.

Muylkens, B., D. Coupeau, G. Dambrine, S. Trapp, and D. Rasschaert. 2010. Marek's disease virus microRNA designated Mdv1-pre-miR-M4 targets both cellular and viral genes. *Arch Virol* **155**:1823–1837.

Nair, V. and M. Zavolan. 2006. Virus-encoded microRNAs: novel regulators of gene expression. *Trends Microbiol* **14**:169–175.

Norman, K.L. and P. Sarnow. 2010. Herpes simplex virus is Akt-ing in translational control. *Genes Dev* **24**:2583–2586.

Pfeffer, S., M. Zavolan, F.A. Grasser, M. Chien, J.J. Russo, J. Ju, B. John, A.J. Enright, D. Marks, C. Sander, and T. Tuschl. 2004. Identification of virus-encoded microRNAs. *Science* **304**:734–736.

Pfeffer, S., M. Lagos-Quintana, and T. Tuschl. 2005a. Cloning of small RNA molecules. *Curr Protoc Mol Biol* **72**:26.4.1–26.4.18.

Pfeffer, S., A. Sewer, M. Lagos-Quintana, R. Sheridan, C. Sander, F.A. Grasser, L.F. van Dyk, C.K. Ho, S. Shuman, M. Chien, J.J. Russo, J. Ju, et al. 2005b. Identification of microRNAs of the herpesvirus family. *Nat Methods* **2**:269–276.

Rachamadugu, R., J.Y. Lee, A. Wooming, and B.W. Kong. 2009. Identification and expression analysis of infectious laryngotracheitis virus encoding microRNAs. *Virus Genes* **39**:301–308.

Rai, D., S.W. Kim, M.R. McKeller, P.L. Dahia, and R.C. Aguiar. 2010. Targeting of SMAD5 links microRNA-155 to the TGF-beta pathway and lymphomagenesis. *Proc Natl Acad Sci U S A* **107**:3111–3116.

Skalsky, R.L., M.A. Samols, K.B. Plaisance, I.W. Boss, A. Riva, M.C. Lopez, H.V. Baker, and R. Renne. 2007. Kaposi's sarcoma-associated herpesvirus encodes an ortholog of miR-155. *J Virol* **81**:12836–12845.

Strassheim, S., G. Stik, D. Rasschaert, and S. Laurent. 2012. mdv1-miR-M7-5p, located in the newly identified first intron of the latency-associated transcript of Marek's disease virus, targets the immediate-early genes ICP4 and ICP27. *J Gen Virol* **93**:1731–1742.

Tili, E., C.M. Croce, and J.J. Michaille. 2009. miR-155: on the crosstalk between inflammation and cancer. *Int Rev Immunol* **28**:264–284.

Umbach, J.L., M.F. Kramer, I. Jurak, H.W. Karnowski, D.M. Coen, and B.R. Cullen. 2008. MicroRNAs expressed by herpes simplex virus 1 during latent infection regulate viral mRNAs. *Nature* **454**:780–783.

Valeri, N., P. Gasparini, M. Fabbri, C. Braconi, A. Veronese, F. Lovat, B. Adair, I. Vannini, F. Fanini, A. Bottoni, S. Costinean, S.K. Sandhu, et al. 2010. Modulation of mismatch repair and genomic stability by miR-155. *Proc Natl Acad Sci U S A* **107**:6982–6987.

Waidner, L.A., J. Burnside, A.S. Anderson, E.L. Bernberg, M.A. German, B.C. Meyers, P.J. Green, and R.W. Morgan. 2011. A microRNA of infectious laryngotracheitis virus can downregulate and direct cleavage of ICP4 mRNA. *Virology* **411**:25–31.

Waidner, L.A., R.W. Morgan, A.S. Anderson, E.L. Bernberg, S. Kamboj, M. Garcia, S.M. Riblet, M. Ouyang, G.K. Isaacs, M. Markis, B.C. Meyers, P.J. Green, et al. 2009. MicroRNAs of Gallid and Meleagrid herpesviruses show generally conserved genomic locations and are virus-specific. *Virology* **388**:128–136.

REFERENCES

Witter, R.L. 1997. Increased virulence of Marek's disease virus field isolates. *Avian Dis* **41**: 149–163.

Xiang, J., A. Cheng, M. Wang, S. Zhang, D. Zhu, R. Jia, S. Chen, Y. Zhou, X. Wang, and X. Chen. 2012. Computational identification of microRNAs in anatid herpesvirus 1 genome. *Virol J* **9**:93.

Xiao, C. and K. Rajewsky. 2009. MicroRNA control in the immune system: basic principles. *Cell* **136**:26–36.

Xu, S., C. Xue, J. Li, Y. Bi, and Y. Cao. 2011. Marek's disease virus type 1 microRNA miR-M3 suppresses cisplatin-induced apoptosis by targeting Smad2 of the transforming growth factor beta signal pathway. *J Virol* **85**:276–285.

Yao, Y., L.P. Smith, L. Petherbridge, M. Watson, and V. Nair. 2012. Novel microRNAs encoded by duck enteritis virus. *J Gen Virol* **93**:1530–1536.

Yao, Y., Y. Zhao, L.P. Smith, C.H. Lawrie, N.J. Saunders, M. Watson, and V. Nair. 2009a. Differential expression of microRNAs in Marek's disease virus-transformed T-lymphoma cell lines. *J Gen Virol* **90**:1551–1559.

Yao, Y., Y. Zhao, L.P. Smith, M. Watson, and V. Nair. 2009b. Novel microRNAs (miRNAs) encoded by herpesvirus of Turkeys: evidence of miRNA evolution by duplication. *J Virol* **83**:6969–6973.

Yao, Y., Y. Zhao, H. Xu, L.P. Smith, C.H. Lawrie, A. Sewer, M. Zavolan, and V. Nair. 2007. Marek's disease virus type 2 (MDV-2)-encoded microRNAs show no sequence conservation with those encoded by MDV-1. *J Virol* **81**:7164–7170.

Yao, Y., Y. Zhao, H. Xu, L.P. Smith, C.H. Lawrie, M. Watson, and V. Nair. 2008. MicroRNA profile of Marek's disease virus-transformed T-cell line MSB-1: predominance of virus-encoded microRNAs. *J Virol* **82**:4007–4015.

Zhao, Y., H. Xu, Y. Yao, L.P. Smith, L. Kgosana, J. Green, L. Petherbridge, S.J. Baigent, and V. Nair. 2011. Critical role of the virus-encoded microRNA-155 ortholog in the induction of Marek's disease lymphomas. *PLoS Pathog* **7**:e1001305.

Zhao, Y., Y. Yao, H. Xu, L. Lambeth, L.P. Smith, L. Kgosana, X. Wang, and V. Nair. 2009. A functional MicroRNA-155 ortholog encoded by the oncogenic Marek's disease virus. *J Virol* **83**:489–492.

FUNCTION OF HUMAN CYTOMEGALOVIRUS MicroRNAs AND POTENTIAL ROLES IN LATENCY

Natalie L. Reynolds, Jon A. Pavelin, and Finn E. Grey

Division of Infection and Immunity, The Roslin Institute, University of Edinburgh, Easter Bush, Midlothian, UK

I. Introduction to Human Cytomegalovirus	154
II. Identification of Human Cytomegalovirus miRNAs	155
III. Human Cytomegalovirus miRNAs: Creating an Environment Conducive to Latency	155
A. HCMV miRNAs and Repression of IE Gene Expression	157
B. HCMV miRNAs and Host Immune Evasion	158
C. Cellular miRNAs and HCMV	159
D. The Future of HCMV miRNA Target Identification	160
IV. Conclusions	160
References	162

ABBREVIATIONS

AIDS	acquired immunodeficiency syndrome
CCMV	chimpanzee cytomegalovirus
CDK	cyclin-dependent kinase
DNA	deoxyribonucleic acid
EBV	Epstein–Barr virus
HCMV	human cytomegalovirus
HITS-CLIP	high-throughput sequencing of RNA isolated by cross-linking immunoprecipitation

MicroRNAs in Medicine, First Edition. Edited by Charles H. Lawrie.
© 2014 John Wiley & Sons, Inc. Published 2014 by John Wiley & Sons, Inc.

HSV-1	herpes simplex virus 1
IP	immunoprecipitation
KSHV	Kaposi's sarcoma-associated herpesvirus
MHC I	major histocompatibility complex class I
MIE	major immediate early
miRNA	micro RNA
NK cells	natural killer cells
PAR-CLIP	photoactivatable ribonucleoside-enhanced cross-linking and immunoprecipitation
RISC	RNA-induced silencing complex
RNA	ribonucleic acid
RNAi	RNA interference
UTR	untranslated region
UV	ultraviolet

I. INTRODUCTION TO HUMAN CYTOMEGALOVIRUS

The discovery of RNA interference (RNAi) and microRNAs (miRNAs) is undoubtedly one of the most significant recent advances in the field of biology. Given the widespread prevalence and influential effects of miRNAs on eukaryotic gene expression, it is unsurprising that viruses exploit RNAi pathways by expressing their own small RNAs. miRNAs have been detected in a number of families of double-stranded DNA viruses, of which *Herpesviridae* has the largest number of documented species. Although this mode of viral regulation is in a relatively early stage of investigation, there is a growing body of research that suggests that miRNAs play a significant role in the control of both host and viral gene expression during infection.

Human cytomegalovirus (HCMV; human herpes virus 5; HHV-5) is a double-stranded DNA virus of the *Herpesviridae* family that infects hosts in a highly species-specific manner. Directly after transmission, HCMV replicates locally in infected mucosa before disseminating to develop a widespread systemic infection during which high levels of viral shedding and infiltration of the solid organs occurs. Replication of the virus is limited by a healthy immune system, resulting in mild or asymptomatic carriage before HCMV enters a dormant but persistent latent phase. Latency is clinically associated with low or undetectable levels of viral shedding, largely with no corresponding host symptoms.

However, in immune-compromised individuals, such as organ donor recipients and AIDS patients, HCMV infection represents a significant clinical cause of morbidity, as well as a major global financial burden. While primary infections can be devastating in such individuals, a significant proportion of pathologies occur as a result of reactivation of latent infections. Such reactivations can result in a number of pathologies, including pneumonitis, retinitis, and hepatitis (Emery 2001). By therapeutically targeting HCMV reactivation and encouraging reentry into latency, it is possible that these pathologies may be treated or prevented.

There is growing evidence to suggest that HCMV miRNAs are involved in the subtle regulation of a number of cellular and viral pathways, many of which can be related to the establishment and maintenance of latency. In this chapter, we will outline the techniques used to discover novel miRNA targets and review what is currently known about their function, focusing on how the interplay between viral and cellular miRNAs may work to create an environment conducive to latency.

II. IDENTIFICATION OF HUMAN CYTOMEGALOVIRUS miRNAs

The identification of novel HCMV miRNAs to date has been achieved through a combination of bioinformatic techniques whose predictions have then been verified experimentally, as well as via high-throughput sequencing methods.

Programs used for the detection of novel HCMV miRNAs have relied on algorithms that predict potential miRNAs based on sequence composition and the likely secondary structures that any pre-miRNA hairpins might adopt (Pfeffer et al. 2005). However, studies have also used comparative algorithms to predict conserved stem loops between HCMV and its closest known relative, chimp cytomegalovirus (CCMV), as well as those conserved between different strains of HCMV (Dunn et al. 2005; Grey et al. 2005). This combination of predictive techniques initially led to the discovery of 11 HCMV miRNAs.

Over recent years, miRNA discovery has been accelerated by the development and availability of deep sequencing technology. This technique allows for the high-throughput, unbiased sequencing of all transcripts in a sample, without any prior sequence knowledge of those transcripts. The power of this technique was demonstrated in the first application of deep sequencing to HCMV-infected cells, where the application of Illumina technology led to the detection of all previously reported HCMV miRNAs as well as two novel HCMV miRNAs, bringing the total number of identified precursors to 12, encoding 14 mature miRNAs (Stark et al. 2012). One previously predicted miRNA, *miR-UL70-1*, was not detected, and is likely to represent an artifact due to cross-reactivity with a cellular small RNA, while the exact sequence of *miR-US4-1* was found to be shifted by five bases compared with the originally predicted miRNA sequence.

A summary of the HCMV miRNAs discovered to date is shown in Table 10.1.

III. HUMAN CYTOMEGALOVIRUS miRNAs: CREATING AN ENVIRONMENT CONDUCIVE TO LATENCY

HCMV latency is defined as a state where viral DNA can still be detected, but infectious virions are no longer produced. There are two major aspects that are important for the viability of viral latency. The virus must maintain a restrictive gene expression profile while maintaining the ability to respond to subtle external signals that trigger reactivation during appropriate conditions. Second, the virus must promote survival of the infected cell, and ultimately, the long-term survival of the host, thereby enhancing the likelihood of viral persistence and dissemination. Although it is not fully understood how HCMV achieves these two main aspects, there are number of cellular and viral processes that are likely to be involved. With respect to establishing and maintaining a gene expression profile conducive to latency it has been suggested that the major immediate early genes (MIE), IE72 and IE86, may play a pivotal role. The MIE genes are expressed immediately following infection of permissive cells and play a crucial role in driving acute replication of the virus. Blocking expression of these genes severely curtails acute replication of the virus, a process necessary for establishing effective latency. Expression of the MIE genes has also been shown to be subdued in cells thought to harbor latent HCMV in a mechanism that is linked to both cell cycle regulation and cellular differentiation. Latent viruses can be detected in pluripotent $CD34^+$ hematopoietic stem cells (Taylor-Wiedeman et al. 1991; Goodrum et al. 2002), which differentiate to form a variety of hematopoietic cells. These stem cells do not support the expression of

TABLE 10.1. Reported HCMV miRNAs

miRNA hairpin	Mature miRNAs	Sequence	Annotated in miRbase?	Reported Targets	Reference
UL22A	*miR-UL22A-5p*	(C)UAACUAGCCUUCCGUGAGA	Y	—	(Dunn et al. 2005; Pfeffer et al. 2005; Stark et al. 2012)
	miR-UL22A-3p	UCACCAGAAUGCUAGUUGUAG	Y		
UL36	*miR-UL36-5p*	UCGUUGAAGACACCUGGAAAGA(A)	Y	—	(Grey et al. 2005; Pfeffer et al. 2005; Stark et al. 2012)
	miR-UL36-3p	UUUCCAGGUGUUUCAACGUGC	Y		
UL112	*miR-UL112-5p*	CCUCCGAUCACAUGGUUACUCA(G)	N	MICB (Stern-Ginossar et al. 2007)	(Pfeffer et al. 2005; Stark et al. 2012)
	miR-UL112-3p	AAGUGACGGUGAGAUCCAGGCU(U)	Y	IE72 (Grey et al. 2007; Murphy et al. 2008) UL114 (Stern-Ginossar et al. 2009)	
UL148D	*miR-UL148D*	UCGUCCUCCCCUUCUUCACCG(U)	Y	RANTES (Kim et al. 2012)	(Pfeffer et al. 2005; Stark et al. 2012)
US4	*miR-US4-5p*	CGAC<u>A</u>UGGACGUGCAGGGGAU(GUCUGU)	Y	ERAP1 (Kim et al. 2011)	(Grey et al. 2005; Stark et al. 2012)
	miR-US4-3p	UGACAGCCCGCUACACCUCUG	N		
US5-1	*miR-US5-1*	UGACAAGCCUGACGAGAGCGU	Y	US7 (Kim et al. 2011)	(Grey et al. 2005; Pfeffer et al. 2005; Stark et al. 2012)
US5-2	*miR-US5-2-5p*	(G)CUUUCGCCACACCUAUCCUGAAAG(U)	N	US7 (Tirabassi et al. 2011)	(Grey et al. 2005; Pfeffer et al. 2005; Stark et al. 2012)
	miR-US5-2-3p	UUAUGAUAGGUGUGACGAUGUC	Y		
US22	*miR-US22-5p*	UGUUUCAGCGUGUGUCCGCGG(C)	N	—	(Stark et al. 2012)
	miR-US22-3p	UCGCCGGCCGCGCGUGUAACCAGG(U)	N		
US25-1	*miR-US25-1-5p*	AACCGCUCAGUGGCUCGGACC(G)	Y	CCNE2 (Grey et al. 2010) TRIM28 (Grey et al. 2010)	(Dunn et al. 2005; Pfeffer et al. 2005; Stark et al. 2012)
	miR-US25-1-3p	(G)UCCGAACGCUAGGUCGGUUCUC	Y		
US25-2	*miR-US25-2-5p*	(U)AGCGGUCUGUUCAGGUGGAUGA	Y	—	(Pfeffer et al. 2005; Stark et al. 2012)
	miR-US25-2-3p	AUCCACUUGGAGAGCUCCCGCGG(U)	Y		
US33	*miR-US33-5p*	GAUUGUGCCCGACCGUGGGCG(C)	Y	—	(Pfeffer et al. 2005; Stark et al. 2012)
	miR-US33-3p	UCACGGUCCGAGCACAUCCA(A)	Y		
US33A	*miR-US33A-5p*	(U)GGAUGUGCUCGGACCGUGACG(GUGU)	N	—	(Stark et al. 2012)
	miR-US33A-3p	CCCACGGUCCGGGCACAAUCAA(U)	N		

The 12 reported HCMV pre-miRNA hairpins are shown, along with the corresponding predicted/experimentally validated mature miRNA sequences. Where the cataloged sequence in miRBase disagrees with the most recent HCMV miRNA data (Stark et al. 2012), sequences are shown in brackets. The five base pair discrepancy between the predicted and experimentally verified sequence for *miR-US4-5p* is depicted in *italics* (predicted sequence) and brackets (experimentally verified, Stark et al. 2012). US4, difference in 2012 Stark paper versus miRbase. Italicized extra sequence on miRbase and (BRACKETS) for sequence discrepancies within Stark paper.

immediate early (IE) viral transcripts, but differentiation to macrophages and dendritic cells has been shown to result in reactivation of MIE expression and subsequent viral replication (Taylor-Wiedeman et al. 1994). In addition, recent studies have shown that MIE gene expression is blocked during specific stages of cell cycle, implicating this process in the control of HCMV latency. Recent studies by our group have suggested that HCMV miRNAs may inhibit MIE gene expression by both directly targeting MIE genes and through indirect means by manipulation cell cycle regulation, thereby promoting latency and controlling reactivation.

A. HCMV miRNAs and Repression of IE Gene Expression

Using comparative bioinformatics analysis, our group identified a number of targets of *miR-UL112-1* that were conserved between Human cytomegalovirus and chimpanzee cytomegalovirus. One of these targets, the MIE gene IE72, is a crucial transactivating gene involved in up-regulation of early and late viral transcripts, and its deletion has previously been shown to result in attenuation of acute viral replication. Expression of *miR-UL112-1* was shown to directly reduce expression of IE72 in both plasmid-based assays, as well as during HCMV infection. Furthermore, transfection of *miR-UL112-1* mimic RNA prior to infection also significantly reduced HCMV replication in permissive fibroblast cells, suggesting *miR-UL112-1* may be involved in subduing acute replication of the virus and in so doing promoting establishment and maintenance of latency (Grey et al. 2007; Murphy et al. 2008).

In addition to directly targeting MIE gene expression, it is possible that herpesvirus miRNAs may regulate latency by manipulating cellular pathways. More recent studies by our group have shown that *miR-US25-1* targets the cell cycle regulator cyclin E2 (CCNE2). HCMV infection is cell cycle-dependent, and infection dynamics are known to be closely linked with the cell cycle stage. HCMV alters expression of both the cyclins and their binding partners (cyclin-dependent kinases; CDKs), and while infection has been shown to trigger a G1/S phase block (Kalejta and Shenk 2002; Payton and Coats 2002), IE gene expression is inhibited at S and G2 phase (Fortunato et al. 2002). This block in IE gene expression has recently been linked to Cyclin A2 (CCNA2)/CDK2 complex, which is also expressed at high levels in nonpermissive cell lines that support quiescent infection of HCMV (Zydek et al. 2010; Oduro et al. 2012). The authors postulate that expression of CCNA2 in these cell types may contribute to the establishment and latency of HCMV by suppressing MIE gene expression. Infection with HCMV is known to induce expression of CCNE2 proteins during acute infection, and it is likely that CCNE2 competes with CCNA2 as both proteins can complex with CDK2;by outcompeting CCNA2 for binding with CDK2, CCNE2 may induce MIE gene expression, thereby promoting acute replication (and potentially reactivation). Reducing CCNE2 expression through targeting by *miR-US25-1*, would therefore promote the CCNA2/CDK2 complex, resulting in reduced MIE gene expression. Furthermore, it has been reported that *miR-US25-1* target key host genes important for the replication of DNA viruses (Stern-Ginossar et al. 2009). The ectopic expression of these miRNAs results in impaired replication of a number of DNA viruses, indicating that they target a pathway essential for the replication of all DNA viruses, rather than HCMV alone.

Through the combined actions of directly targeting MIE gene expression (*miR-UL112-1*) and through indirect mechanisms of cell cycle manipulation (*miR-US25-1*), HCMV miRNAs may function to contribute to an environment conducive to viral latency.

B. HCMV miRNAs and Host Immune Evasion

With respect to cell survival, HCMV must counter the destruction of the infected cell, which can occur through the actions of intrinsic, innate, and adaptive mechanism of host defense. Several HCMV miRNAs have been shown to play a role in host immune system evasion, in particular in the alteration of antigen recognition pathways. The control of these pathways is vital to the maintenance of latency so that viral transcripts and the viral genome remain undetected within the cell, and thus persist in the host.

As well as targeting viral genes, *miR-UL112-1* has been shown to disrupt the recruitment of host natural killer (NK) cells through its targeting of MICB (Stern-Ginossar et al. 2007). MICB is a stress-induced ligand for NK cells and is known to be targeted by HCMV protein UL16 (Dunn et al. 2003). This "two-hit" down-regulation of NK cell recruitment could impair the recognition and destruction of HCMV-infected cells.

The targeting of MICB by *miR-UL112-1* is not an isolated interaction. It has been demonstrated that MICB is also targeted by a group of host miRNAs with binding sequences overlapping that of *miR UL112-1* (Stern-Ginossar et al. 2008), and it has been suggested that the *miR-UL112-1* interaction has evolved to exploit these existing host-binding sites. In addition, further studies have shown that while the ectopic coexpression of multiple host miRNAs results in the abrogation of MICB repression, coexpression of *miR-UL112-1* with host miRNAs results in effective and synergistic repression (Nachmani et al. 2010). Although the physiological relevance of these interactions have not been addressed, the demonstration that the effects of a viral miRNA can be enhanced through cooperative regulation with host miRNAs is significant. Moreover, the repression of MICB by *miR-UL112-1* is thought to be an important mechanism for multiple herpesviruses, not limited to HCMV. It has been demonstrated that both KSHV and EBV express miRNAs that also target MICB, thus showing functional conservation (Nachmani et al. 2009), which suggests that regulation of this stress ligand is an important strategy for immune evasion.

Another key process in the cellular antiviral response is antigen presentation, where viral proteins are degraded and processed to form antigenic peptides within the cell, before being bound by major histocompatibility complex class I (MHC I) peptides for transport to the cell surface. They are then recognized, and infected cells are targeted by cytotoxic CD8[+] T cells. An HCMV miRNA, *miR-US4-1*, has been reported to target a key component of this pathway, ERAP1, a processing enzyme involved in foreign peptide "trimming" prior to "presentation" at the cell surface (Kim et al. 2011). Interestingly, this interaction was shown to have a direct effect *in vitro*, where ectopic expression of *miR-US4-1* combined with HCMV infection resulted in the inhibition of lysis of infected cells by cytotoxic T cells. It should be noted, however, that since the publication of this work the predicted sequence for *miR-US4-1* upon which this study was based has recently been tested experimentally and proved to be different from the original prediction (Stark et al. 2012). It is difficult to reconcile how this study identified ERAP1 as a target of *miR-US4-1*, given the miRNA sequence used was incorrect, especially as the correct sequence for this miRNA contains a completely different seed sequence. Further studies are required to determine whether this finding stands up to additional scrutiny.

The same authors also report that *miR-UL148D* has a role in immune evasion through the blocking of immune cell recruitment. RANTES is a chemokine secreted by cells during infection, where it is chemotactic for T cells, basophils, eosinophils, and monocytes. It has been shown that although a *miR-UL148D* knockout virus displayed a similar phenotype to the wild type, levels of secreted RANTES in cells infected with mutant virus were significantly higher than those infected by the wild type (Kim et al. 2012). It was also noted that HCMV suppresses RANTES expression via the immediate early protein IE2 (IE86), so the interaction of *miR-UL148D* with RANTES might cooperate with IE2 in order to adjust expression levels. This interaction is of particular interest due to the non-canonical nature of the *miR-UL112-1-RANTES* binding. In this interaction, the "seed" sequence of *miR-UL112-1* does not show fully complementary binding to RANTES mRNA, instead showing a binding "shift" of three base pairs into the miRNA sequence, with GU "wobble" pairs also included in the region.

These studies show that the targeting of the host immune system is a common strategy for those HCMV miRNAs that have been investigated to date. In particular, the targeting of proteins already known to be blocked by HCMV glycoproteins supports the proposal that HCMV miRNAs function as subtle "fine-tuners" of gene regulation. In addition, the presence of multiple independent viral mechanisms through which gene expression may be targeted presents a "tighter" evasion strategy for HCMV, and the host is less likely to be able to adapt quickly to undermine viral subversion. This supports HCMV immune evasion during both lytic and latent infection.

C. Cellular miRNAs and HCMV

Although HCMV expresses a range of miRNAs that have varied roles in the viral life cycle, there is also growing evidence to suggest that HCMV regulates cellular miRNAs in order to facilitate lytic and latent infection by manipulating the host cell environment. Induction of host miRNA-mediated control holds the advantage over viral miRNAs in that they can begin to affect the cellular environment before viral genomic transcription begins. By augmenting or reducing host miRNAs in this way, a more favorable environment for viral replication and host immune system evasion may be established.

Microarray analysis has demonstrated that although the majority of host miRNAs are unaffected by HCMV infection, distinct subsets are both up- and down-regulated (Wang et al. 2008). In addition, host miRNA changes have been shown to have a measurable effect both on HCMV infection and on that of other herpesviruses; a screening study, quantifying viral growth following treatment with host miRNA inhibitors or mimics, revealed at least four potent antiviral and three proviral miRNAs (Santhakumar et al. 2010). Following initial studies of *miR-199a-3p*, a host miRNA identified as being broadly antiviral, it was found that this miRNA targets a number of pathways that are activated following HCMV infection, including the ERK/MAPK and PI3K/AKT signaling cascades (Santhakumar et al. 2010).

Studies have also demonstrated how the manipulation of host miRNA expression might serve to benefit HCMV in both the lytic and latent growth phases. *miR-132* has been identified as a cellular miRNA that is induced upon infection with both HCMV and KSHV (Lagos et al. 2010). Its importance to infection was demonstrated when a *miR-132* inhibitor led to a decreased viral load in infected cells, and it was determined that this corresponded to the suppression of the transcriptional coactivator EP300, which regulates the interferon response (Lagos et al. 2010), thus improving immune system evasion.

It has also been demonstrated that a distinct pool of cellular miRNAs are regulated during latent infection with HCMV (Poole et al. 2011). A striking example of the potential for regulation can be seen with the finding that *miR-92a* is reduced upon latent infection with HCMV. *miR-92a* was shown to target the transcription factor GATA2 in CD34$^+$ myeloid cells (Poole et al. 2011). This transcription factor is involved in the proliferation of haematopoietic cells and promotes cell survival, so by reducing the levels of its inhibitor *miR-92a*, HCMV may improve the cellular environment for latency by increasing the pool of cells in which it can lie latent.

D. The Future of HCMV miRNA Target Identification

While the function of the viral miRNAs is gradually being elucidated, a full understanding will not be achieved until comprehensive host and viral target identification studies have been performed. The use of prediction algorithms has proved invaluable, but this technique has a number of disadvantages: numerous studies have shown that miRNAs bind target mRNAs via imperfect sequence complementarity; however it has also been shown that this targeting is dependent on the near perfect complementarity between the first six 5′ bases of the miRNA, the "seed" region. However, it has also been shown that while this is often the case, exceptions can occur (Kim et al. 2012), and studies have also shown that viral miRNAs can target transcripts outwith the 3′-UTR (Grey et al. 2010). This variability can lead to miRNA target prediction algorithms overlooking targets that fall outwith the 3′-UTR, perfect seed match paradigm. In addition, the use of prediction algorithms encourages the study of single miRNAs acting alone, whereas it has been demonstrated in numerous studies that viral miRNAs often work synergistically with other viral and host miRNAs. A novel way to approach miRNA target identification is through the use of high-throughput systems that may uncover novel targets of the HCMV miRNAs.

The biochemical techniques PAR-CLIP (photoactivatable-ribonucleoside-enhanced cross-linking and immunoprecipitation) and HITS-CLIP (high-throughput sequencing of RNA isolated by cross-linking immunoprecipitation) are high-throughput assays that link immunoprecipitation with deep sequencing to indiscriminately identify miRNA target transcripts.

The PAR-CLIP technique requires the delivery of photoreactive ribonucleoside uracil analogues into living cells. Upon harvesting, these analogues facilitate the UV-cross-linking of mRNA to the miRNA-RISC (RNA-induced silencing complex), and immunoprecipitation of Argonaute 2 (Ago2) will isolate any mRNAs bound to RISC. Subsequent deep sequencing of these mRNAs originally revealed that the UV-cross-linking step causes thymine to cytosine mutations during the sequencing process (Hafner et al. 2010). By analyzing the distribution of these sequence changes, it is possible to deduce the location of the miRNA–mRNA-binding site. This technique may prove useful in the future in mapping the HCMV miRNA-binding sites in target mRNAs, as evidenced by its recent effective application during EBV and KSHV infections. By using this technique, Gottwein and colleagues demonstrated the direct targeting of more than 2000 cellular mRNAs by KSHV miRNAs (Gottwein 2012), and similarly, Skalsky and colleagues demonstrated more than 500 EBV miRNA cellular targets (Skalsky et al. 2012).

Techniques such as PAR-CLIP and HITS-CLIP will likely be used in future studies to elucidate further details of the miRNA–mRNA-binding sequences and target transcripts.

IV. CONCLUSIONS

Future studies will no doubt reveal the extent to which host and HCMV miRNAs impact on both lytic and latent viral replication; however, current evidence suggests that miRNAs and their regulation may have a significant effect on the induction and maintenance of latency. Our knowledge of the miRNAs implicated to date is summarized in Figure 10.1.

CONCLUSIONS

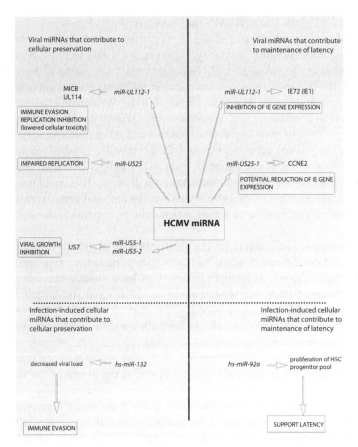

Figure 10.1. Summary of the known functions of HCMV miRNAs. The HCMV miRNAs can be grouped according to their role in the viral life cycle, with roles extending to both the protection of the host cell during lytic infection and the maintenance of the latent state. In addition, HCMV infection is known to affect the expression of key host miRNAs, which also favorably influence both lytic and latent infection.

The concept of miRNAs regulating latency is not limited to HCMV; studies have since revealed that the EBV, HSV-1 (herpes simplex virus 1) and KSHV miRNAs may also play a similar role. It has been reported that the KSHV miRNA *miR-K12-11* attenuates host interferon signaling, a process that could be key to immune evasion (Liang et al. 2011). It has also been suggested that a cluster of EBV miRNAs at the BHRF1 locus facilitate persistence by reducing the recognition of infected cells by CD8[+] T cells (Feederle et al. 2011). This cluster has been reported to affect viral protein production, lowering the antigenic load of infected B cells, thus reducing chances of immune detection. The same cluster of EBV miRNAs has also been reported to inhibit apoptosis (Seto et al. 2010), which may also both encourage persistence and avoid immune detection.

In addition, KSHV miRNAs have been reported to be involved in the suppression of IE gene expression (Lu et al. 2010). A similar report suggested that two HSV-1 miRNAs expressed in latently infected sensory ganglia (*miR-H2-3p* and *miR-H6*) repress the IE proteins ICP0 and ICP4 (Umbach et al. 2008). It is encouraging that other herpesviruses have been shown not only to target latency, but via similar pathways to those targeted by HCMV, suggesting a level of functional redundancy among the herpesvirus miRNAs.

We have highlighted where HCMV miRNAs have been demonstrated to subtly affect gene expression of key cellular and viral pathways that may cooperate to contribute to the establishment and maintenance of latency; however, we would like to speculate further on this point, outlining the roles the miRNAs might yet be revealed to have. It has been demonstrated that IE gene expression can occur in differentiated hematopoietic stem cells without the reactivation of HCMV from latency (Taylor-Wiedeman et al. 1994), implying that other levels of regulation exist. We have outlined how *miR-UL112-1* has been shown to directly target viral IE gene expression, but this and other miRNAs may prove to have a more complex regulatory function, whereby they respond to external stimuli in reactivating cells, assisting in the silencing of IE expression until a threshold level of miRNA suppression is reached beyond which reactivation will occur. This could provide a mechanism for the subtle coupling of the cellular environment to viral gene expression, resulting in the induction of reactivation at a time when cellular conditions are optimal for viral replication.

The non-immunogenic nature of the viral miRNAs assists in the immune evasion required for viral persistence to occur, but may also work to overcome intrinsic intracellular antiviral mechanisms. Antiapoptotic functions have already been demonstrated for the BHRF1 cluster of EBV miRNAs (Seto et al. 2010), and similar pathways may be uncovered for the HCMV miRNAs targeting, for example, protein degradation pathways or the NFκB signaling pathway, which has been shown to be targeted by vaccinia (Stack et al. 2005). Such regulatory mechanisms targeting immune evasion may also contribute to the creation of a cellular environment in which viral persistence can occur in a robust and controlled manner. Further study alone will reveal the extent of the involvement of miRNAs in these processes.

As the number of targets identified for HCMV (and other herpesvirus) miRNAs increases, our level of understanding of the pathways controlling latency will deepen, and this brings about the possibility for new therapeutic studies. Since much of the health burden of HCMV arises from the reactivation of latent virus, the targeting of miRNAs that serve to preserve the latent state could prove to be an invaluable addition to clinical medicine. Following future study and clinical trials, we may find miRNA delivery to be in a position where patients experiencing immune suppression could potentially be protected from viral reactivation by the delivery of miRNAs targeting IE gene expression, such as *miR-UL112-1,* for example. With future study, such interventions could be a relevant strategy for the treatment of HCMV infection and could significantly reduce the health and associated financial burden of the virus.

REFERENCES

Dunn, W., C. Chou, H. Li, R. Hai, D. Patterson, V. Stolc, H. Zhu, and F. Liu. 2003. Functional profiling of a human cytomegalovirus genome. *Proc Natl Acad Sci U S A* **100**:14223–14228.

Dunn, W., P. Trang, Q. Zhong, E. Yang, C. van Belle, and F. Liu. 2005. Human cytomegalovirus expresses novel microRNAs during productive viral infection. *Cell Microbiol* **7**:1684–1695.

Emery, V.C. 2001. Investigation of CMV disease in immunocompromised patients. *J Clin Pathol* **54**:84–88.

Feederle, R., S.D. Linnstaedt, H. Bannert, H. Lips, M. Bencun, B.R. Cullen, and H.J. Delecluse. 2011. A viral microRNA cluster strongly potentiates the transforming properties of a human herpesvirus. *PLoS Pathog* **7**:e1001294.

REFERENCES

Fortunato, E.A., V. Sanchez, J.Y. Yen, and D.H. Spector. 2002. Infection of cells with human cytomegalovirus during S phase results in a blockade to immediate-early gene expression that can be overcome by inhibition of the proteasome. *J Virol* **76**:5369–5379.

Goodrum, F.D., C.T. Jordan, K. High, and T. Shenk. 2002. Human cytomegalovirus gene expression during infection of primary hematopoietic progenitor cells: a model for latency. *Proc Natl Acad Sci U S A* **99**:16255–16260.

Gottwein, E. 2012. Kaposi's sarcoma-associated herpesvirus microRNAs. *Front Microbiol* **3**:165.

Grey, F., A. Antoniewicz, E. Allen, J. Saugstad, A. McShea, J.C. Carrington, and J. Nelson. 2005. Identification and characterization of human cytomegalovirus-encoded microRNAs. *J Virol* **79**:12095–12099.

Grey, F., H. Meyers, E.A. White, D.H. Spector, and J. Nelson. 2007. A human cytomegalovirus-encoded microRNA regulates expression of multiple viral genes involved in replication. *PLoS Pathog* **3**:e163.

Grey, F., R. Tirabassi, H. Meyers, G. Wu, S. McWeeney, L. Hook, and J.A. Nelson. 2010. A viral microRNA down-regulates multiple cell cycle genes through mRNA 5'UTRs. *PLoS Pathog* **6**:e1000967.

Hafner, M., M. Landthaler, L. Burger, M. Khorshid, J. Hausser, P. Berninger, A. Rothballer, M. Ascano, Jr., A.C. Jungkamp, M. Munschauer, A. Ulrich, G.S. Wardle, et al. 2010. Transcriptome-wide identification of RNA-binding protein and microRNA target sites by PAR-CLIP. *Cell* **141**:129–141.

Kalejta, R.F. and T. Shenk. 2002. Manipulation of the cell cycle by human cytomegalovirus. *Front Biosci* **7**:d295–d306.

Kim, S., S. Lee, J. Shin, Y. Kim, I. Evnouchidou, D. Kim, Y.K. Kim, Y.E. Kim, J.H. Ahn, S.R. Riddell, E. Stratikos, V.N. Kim, et al. 2011. Human cytomegalovirus microRNA miR-US4-1 inhibits CD8(+) T cell responses by targeting the aminopeptidase ERAP1. *Nat Immunol* **12**:984–991.

Kim, Y., S. Lee, S. Kim, D. Kim, J.H. Ahn, and K. Ahn. 2012. Human cytomegalovirus clinical strain-specific microRNA miR-UL148D targets the human chemokine RANTES during infection. *PLoS Pathog* **8**:e1002577.

Lagos, D., G. Pollara, S. Henderson, F. Gratrix, M. Fabani, R.S. Milne, F. Gotch, and C. Boshoff. 2010. miR-132 regulates antiviral innate immunity through suppression of the p300 transcriptional co-activator. *Nat Cell Biol* **12**:513–519.

Liang, D., Y. Gao, X. Lin, Z. He, Q. Zhao, Q. Deng, and K. Lan. 2011. A human herpesvirus miRNA attenuates interferon signaling and contributes to maintenance of viral latency by targeting IKKepsilon. *Cell Res* **21**:793–806.

Lu, F., W. Stedman, M. Yousef, R. Renne, and P.M. Lieberman. 2010. Epigenetic regulation of Kaposi's sarcoma-associated herpesvirus latency by virus-encoded microRNAs that target Rta and the cellular Rbl2-DNMT pathway. *J Virol* **84**:2697–2706.

Murphy, E., J. Vanicek, H. Robins, T. Shenk, and A.J. Levine. 2008. Suppression of immediate-early viral gene expression by herpesvirus-coded microRNAs: implications for latency. *Proc Natl Acad Sci U S A* **105**:5453–5458.

Nachmani, D., D. Lankry, D.G. Wolf, and O. Mandelboim. 2010. The human cytomegalovirus microRNA miR-UL112 acts synergistically with a cellular microRNA to escape immune elimination. *Nat Immunol* **11**:806–813.

Nachmani, D., N. Stern-Ginossar, R. Sarid, and O. Mandelboim. 2009. Diverse herpesvirus microRNAs target the stress-induced immune ligand MICB to escape recognition by natural killer cells. *Cell Host Microbe* **5**:376–385.

Oduro, J.D., R. Uecker, C. Hagemeier, and L. Wiebusch. 2012. Inhibition of human cytomegalovirus immediate-early gene expression by cyclin A2-dependent kinase activity. *J Virol* **86**:9369–9383.

Payton, M. and S. Coats. 2002. Cyclin E2, the cycle continues. *Int J Biochem Cell Biol* **34**: 315–320.

Pfeffer, S., A. Sewer, M. Lagos-Quintana, R. Sheridan, C. Sander, F.A. Grasser, L.F. van Dyk, C.K. Ho, S. Shuman, M. Chien, J.J. Russo, J. Ju, et al. 2005. Identification of microRNAs of the herpesvirus family. *Nat Methods* **2**:269–276.

Poole, E., S.R. McGregor Dallas, J. Colston, R.S. Joseph, and J. Sinclair. 2011. Virally induced changes in cellular microRNAs maintain latency of human cytomegalovirus in CD34(+) progenitors. *J Gen Virol* **92**:1539–1549.

Santhakumar, D., T. Forster, N.N. Laqtom, R. Fragkoudis, P. Dickinson, C. Abreu-Goodger, S.A. Manakov, N.R. Choudhury, S.J. Griffiths, A. Vermeulen, A.J. Enright, B. Dutia, et al. 2010. Combined agonist-antagonist genome-wide functional screening identifies broadly active antiviral microRNAs. *Proc Natl Acad Sci U S A* **107**:13830–13835.

Seto, E., A. Moosmann, S. Gromminger, N. Walz, A. Grundhoff, and W. Hammerschmidt. 2010. Micro RNAs of Epstein-Barr virus promote cell cycle progression and prevent apoptosis of primary human B cells. *PLoS Pathog* **6**:e1001063.

Skalsky, R.L., D.L. Corcoran, E. Gottwein, C.L. Frank, D. Kang, M. Hafner, J.D. Nusbaum, R. Feederle, H.J. Delecluse, M.A. Luftig, T. Tuschl, U. Ohler, et al. 2012. The viral and cellular microRNA targetome in lymphoblastoid cell lines. *PLoS Pathog* **8**:e1002484.

Stack, J., I.R. Haga, M. Schroder, N.W. Bartlett, G. Maloney, P.C. Reading, K.A. Fitzgerald, G.L. Smith, and A.G. Bowie. 2005. Vaccinia virus protein A46R targets multiple Toll-like-interleukin-1 receptor adaptors and contributes to virulence. *J Exp Med* **201**:1007–1018.

Stark, T.J., J.D. Arnold, D.H. Spector, and G.W. Yeo. 2012. High-resolution profiling and analysis of viral and host small RNAs during human cytomegalovirus infection. *J Virol* **86**:226–235.

Stern-Ginossar, N., N. Elefant, A. Zimmermann, D.G. Wolf, N. Saleh, M. Biton, E. Horwitz, Z. Prokocimer, M. Prichard, G. Hahn, D. Goldman-Wohl, C. Greenfield, et al. 2007. Host immune system gene targeting by a viral miRNA. *Science* **317**:376–381.

Stern-Ginossar, N., C. Gur, M. Biton, E. Horwitz, M. Elboim, N. Stanietsky, M. Mandelboim, and O. Mandelboim. 2008. Human microRNAs regulate stress-induced immune responses mediated by the receptor NKG2D. *Nat Immunol* **9**:1065–1073.

Stern-Ginossar, N., N. Saleh, M.D. Goldberg, M. Prichard, D.G. Wolf, and O. Mandelboim. 2009. Analysis of human cytomegalovirus-encoded microRNA activity during infection. *J Virol* **83**: 10684–10693.

Taylor-Wiedeman, J., J.G. Sissons, L.K. Borysiewicz, and J.H. Sinclair. 1991. Monocytes are a major site of persistence of human cytomegalovirus in peripheral blood mononuclear cells. *J Gen Virol* **72** Pt 9:2059–2064.

Taylor-Wiedeman, J., P. Sissons, and J. Sinclair. 1994. Induction of endogenous human cytomegalovirus gene expression after differentiation of monocytes from healthy carriers. *J Virol* **68**: 1597–1604.

Tirabassi, R., L. Hook, I. Landais, F. Grey, H. Meyers, H. Hewitt, and J. Nelson. 2011. Human Cytomegalovirus US7 is regulated synergistically by two virally encoded microRNAs and by two distinct mechanisms. *J Virol* **85**:11938–11944.

Umbach, J.L., M.F. Kramer, I. Jurak, H.W. Karnowski, D.M. Coen, and B.R. Cullen. 2008. MicroRNAs expressed by herpes simplex virus 1 during latent infection regulate viral mRNAs. *Nature* **454**:780–783.

Wang, F.Z., F. Weber, C. Croce, C.G. Liu, X. Liao, and P.E. Pellett. 2008. Human cytomegalovirus infection alters the expression of cellular microRNA species that affect its replication. *J Virol* **82**:9065–9074.

Zydek, M., C. Hagemeier, and L. Wiebusch. 2010. Cyclin-dependent kinase activity controls the onset of the HCMV lytic cycle. *PLoS Pathog* **6**:e1001096.

11

INVOLVEMENT OF SMALL NON-CODING RNA IN HIV-1 INFECTION

Guihua Sun, John J. Rossi, and Daniela Castanotto

Department of Molecular and Cellular Biology, Beckman Research Institute of the City of Hope, Duarte, CA, USA

I.	Introduction	167
II.	miRNAs, siRNAs, and tdsmRNAs	168
III.	Regulation of miRNAs by HIV Infection	170
	A. Perturbation of the miRNA Biogenesis Pathway by HIV-1 Infection	170
	B. Perturbation of miRNA Expression Profiles by HIV-1 Infection	171
IV.	Regulation of HIV-1 Infection by Host miRNAs	172
V.	sncRNAs in Virally Infected Cells	173
	A. HIV-1 Naturally Encoded sncRNAs	174
	B. HIV-1 Artificially Encoded sncRNAs	175
VI.	tRNA-Derived sncRNAs and HIV-1	176
VII.	Concluding Remarks	177
	Acknowledgment	177
	References	177

ABBREVIATIONS

3′-UTR	3′-untranslated region
AATF	apoptosis-antagonizing transcription factor
Ago	Argonaute
AIDS	acquired immunodeficiency syndrome

MicroRNAs in Medicine, First Edition. Edited by Charles H. Lawrie.
© 2014 John Wiley & Sons, Inc. Published 2014 by John Wiley & Sons, Inc.

BART	Bam HI-A region rightward transcript
BLV	bovine leukemia virus
CD4+ T cells	CD4 protein-expressing T helper cells
CD8	cluster of differentiation 8 transmembrane glycoprotein
CEM	human lymphoid T-cell line
CRS	cis-acting repressive sequence
DGCR8	DiGeorge critical region 8
dsRNA	double-stranded RNA
EBV	Epstein–Barr virus
ENV	genomic region encoding the viral glycoprotein gp160
GAG	genomic region encoding the capsid proteins
GW182	multiple glycines (G)–tryptophan (W) repeats protein 182
HAART	highly active antiretroviral therapy
HIV-1	human immunodeficiency virus type 1
HTLV-1	human T-cell leukemia virus type 1
INS	inhibitory/instability RNA sequences
LSm-1	U6 snRNA-associated Sm-like protein
LTR	long terminal repeat
miRCURY LNA	microRNA array LNA-based system for expression profiling
miRNAs	microRNAs
mRNA	messenger RNA
ncRNAs	non-coding RNAs
NEF	multifunctional 27-kD myristoylated protein
P300/CBP	E1A binding protein P300/CREB-binding protein complex
PBMCs	peripheral blood mononucleocytes
P-bodies	processing bodies
PBS	primer-binding site
PCAF	P300/CBP-associated factor
PE	Psi element
PFV-1	primate foamy virus type 1
PITA	probability of interaction of target accessibility
PKR	protein kinase RNA activated
POL	genomic regions encoding the protease, the reverse transcriptase, and the integrase
PolIII	polymerase III
pre-miRNA	precursor microRNA
P-TEFb	positive transcription elongation factor b
RAKE	RNA-primed array-based Klenow enzyme
Rck/p54	Rck gene 54-kD product
REV	19-kD viral phosphoprotein
RISC	RNA-induced silencing complex
RNAi	RNA interference
RRE	Rev response element
RT	reverse transcriptase
siRNAs	small interfering RNAs
SLIP	slippery site (TTTTTT)
sncRNAs	small non-coding RNAs
SRS	suppressor of RNA silencing
TAR	trans-activation response element

TAS	transcript trans-activator factor S
TAT	trans-activator of HIV gene expression
TAX	trans-activating factor X
tdsmRNAs	transfer RNA-derived small RNAs
TRBP	TAR RNA-binding protein
tRNA	transfer RNA
VIF	viral infectivity factor protein
VPR	viral protein R
VPU	viral protein U
vsiRNA/vmiRNA	viral siRNA/miRNA
vsncRNAs	virally encoded siRNA/miRNA or microRNA-like small non-coding RNAs
WNV	West Nile virus
XRN-1	exoribonuclease-1

I. INTRODUCTION

Human immunodeficiency virus type 1 (HIV-1) is a lentivirus that is the primary cause of acquired immunodeficiency syndrome (AIDS) (Barre-Sinoussi et al. 1983; Chermann et al. 1983). HIV-1 primarily infects CD4+ T memory cells, and its infection can be roughly divided into two phases: an acute phase (early phase), which lasts about 1–2 months after the initial infection, and a chronic infection phase, which can last for upward of 10–20 years before the onset of AIDS (Li et al. 2005; Mattapallil et al. 2005). HIV-1 is a relatively small RNA virus; its 9.7-kilobase (kb) RNA genome consists of several secondary structures (LTR, TAR, RRE, PE, SLIP, CRS, INS) and nine genes (*gag, pol, env, tat, rev, nef, vif, vpr, vpu*) encoding 19 proteins (Figure 11.1). Despite numerous attempts by experts around the world, efforts to create an effective vaccine for

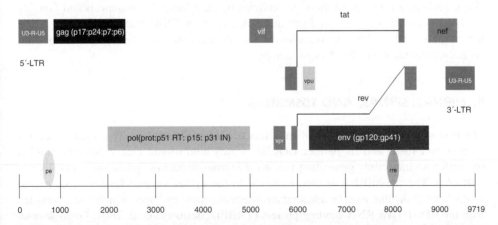

Figure 11.1. Diagram of HIV-1 NL4-3 genome. The HIV-1 genome contains two LTRs and nine major genes. The nine genes encode the structural or accessory proteins: Gag (p17, p24, p7, p6), Pol (RT, IN), Vif, Vpr, Vpu, Tat, Rev, Env (gp120, gp41), and Nef. *HIV-1-vmiRNA-H1* and *N367* are located in the U3 region of both LTRs. The TAR miRNA is also derived from the R region of both LTRs.

the treatment and prevention of HIV-1 infection are still far from reality. Thus, there is an urgency for more basic research into the host and viral mechanisms that support or ameliorate HIV-1 infection.

Technology advances in genome research have led to the discovery of many species of non-coding RNAs (ncRNAs), and great effort is being directed into elucidating their biological function. In the past decade, it has become apparent that ncRNAs play important roles in almost all aspects of biology (Sharp 2009). Among the small non-coding RNA (sncRNA) family, the function and mechanisms of action of siRNAs and microRNAs (miRNAs) are most clearly understood. Following the discovery of RNA interference (RNAi) in mammalian cells by Fire et al. (1998), numerous therapeutic strategies based on siRNA-triggered posttranscriptional silencing of homologous messenger RNAs (mRNAs), including HIV-1 RNA, have emerged in the field (Hannon and Rossi 2004). In the context of HIV-1, siRNAs, miRNAs, and transfer RNA (tRNA)-derived small RNAs (tdsmRNAs) are studied for their role in viral replication and pathogenesis and their potential use in gene therapy. tdsmRNAs have miRNA-like features and could have a viral regulatory role (see Section VII). They can be processed from the 3' or 5' end of the mature tRNA or from the trailer region of the precursor tRNA. miRNAs are a family of sncRNAs that can regulate gene expression primarily by binding to the 3'-UTR of targeted transcripts (Bartel 2009). In addition to their endogenous cellular targets, miRNAs could affect HIV-1 infection by directly targeting the 3'-UTR of the HIV-1 viral RNA genome. Since most of HIV transcripts use the Nef/3'-LTR as their 3'-UTR (Dennis 2002; Sun et al. 2012), miRNAs could possibly play a role in modulating HIV-1 replication and infection. Moreover, since HIV-1 uses host tRNA^{lys} as primer during its replication, functional tdsmRNAs could also be involved in the regulation of HIV-1 infection (Andreola et al. 1992).

A better understanding of the pathogenesis of HIV-1 infection and how it may affect ncRNAs is needed to provide new insights into the host miRNA pathway and improve our understanding of the mechanisms underlying HIV-1-mediated pathologies and T-lymphocyte depletion during the progression to AIDS. Particularly, it would be worthwhile to determine if in fact HIV-1 infection directly affects the miRNA pathways and if miRNAs directly or indirectly target HIV-1 to modulate infection. In the following sections, we discuss evidence for some of these possibilities by examining the biogenesis and function of HIV-1-encoded sncRNAs, cellular sncRNAs, and tdsmRNAs that participate in HIV-1 pathogenesis and replication. We also discuss the feasibility of using sncRNAs or anti-sncRNA molecules for HIV-1 gene therapy.

II. miRNAs, siRNAs, AND tdsmRNAs

The first step in the production of miRNAs occurs in the nucleus by a microprocessor formed by Drosha and its partner DGCR8, which bind to and cleave primary miRNA (pri-miRNA) transcripts generating precursor hairpin structures (precursor miRNA [pre-miRNA]). The pre-miRNA is then exported to the cytoplasm and further processed to miRNA/miRNA* duplexes (guide strand/passenger strand) by Dicer, which is in a complex with the HIV-1 TAR RNA-binding protein (TRBP), (Bernstein et al. 2001; Grishok et al. 2001; Hutvagner et al. 2001; Chendrimada et al. 2005; Haase et al. 2005; Lee et al. 2006). The selected "guide strand" is then incorporated into Ago2, another member of the RNA-induced silencing complex (RISC), and triggers sequence-specific degradation or an arrest in translation—often followed by degradation—of its RNA target (Doench and Sharp 2004). It has been proposed that RISC directs miRNA-repressed mRNAs to processing bodies (P-bodies: small cytoplasmic granules for RNA degradation) for deadenylation, decapping, and degradation, or alternatively for temporary storage and reuse of repressed transcripts (Behm-Ansmant et al. 2006; Giraldez et al. 2006; Wu et al. 2006; Eulalio et al. 2007). P-bodies are of interest in the context of HIV-1 because they could serve as a storage compartment for HIV-1 RNA and could contribute to latency (Figure 11.2).

siRNAs utilize the same basic cellular pathway as the miRNAs; gene suppression by cleavage or translational repression seems to rely solely on the extent of base-pairing interaction with the targeted RNA (Zeng et al. 2003).

The understanding of the biogenesis and mechanisms of action of tdsmRNA is still very limited (Pederson 2010). It has been proposed that tdsmRNAs are tRNA cleavage products generated under stress conditions, but their biological function is unclear (Kawaji et al. 2008; Fu et al. 2009; Lee et al. 2009; Yeung et al. 2009; Schopman et al. 2012).

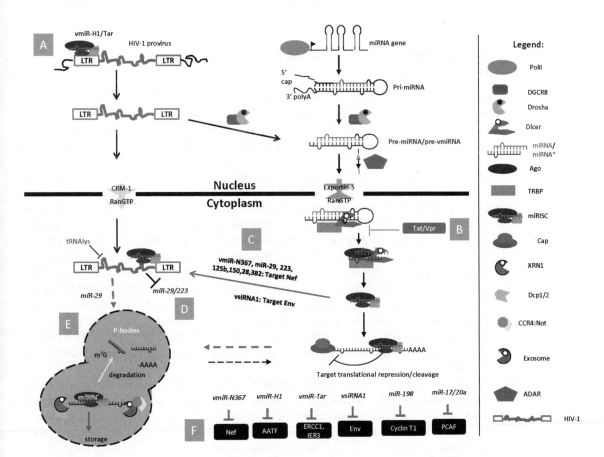

Figure 11.2. Schematic illustration of HIV-1-proposed interactions with the miRNA pathway. (A) vmiRNA-Tar may involve chromatin remodeling at the LTR region of the HIV-1 provirus. (B) Tat/Vpr may act as SRS by inhibiting Dicer activity. (C) Cellular miRNAs and viRNAs target HIV-1 genome. (D) A secondary structure in Nef/2'-LTR blocks *miR-29/223* RISC repression. (E) *miR-29* RISC bring HIV-1 virus to P-bodies, and P-bodies may act as HIV-1 viral RNA storage place and contribute to the latent phase of HIV infection. (F) vmiRNA and host miRNA target cellular factors or HIV-1 protein in HIV-1 infection. See color insert.

III. REGULATION OF miRNAs BY HIV INFECTION

An important concern is whether or not HIV-1 infection globally affects miRNA levels (affecting miRNA biogenesis through the miRNA processing pathway) or individually (affecting the biogenesis of individual miRNAs through transcription perturbation or effects on miRNA maturation). One way that HIV-1 could affect the miRNA pathway is through the production of a protein suppressor of RNA silencing (SRS). Virally encoded products often act as SRSs during viral infections to block any host response based on miRNA targeting of viral transcripts (Mallory et al. 2002; Chapman et al. 2004; Chao et al. 2005; Gatignol et al. 2005; Calabrese and Sharp 2006; Zhang et al. 2006; Zhou et al. 2006; Haasnoot et al. 2007; Hemmes et al. 2007; Abe et al. 2010). miRNA profiles comparing HIV-1-infected samples versus uninfected samples can reveal how HIV-1 infection affects the miRNA pathway. Deregulated miRNAs could also have prognostic and diagnostic value for HIV-1 therapies; unfortunately, controversial results have arisen from these studies as well.

A. Perturbation of the miRNA Biogenesis Pathway by HIV-1 Infection

Currently, there is conflicting evidence that Tat, Vpr, and the HIV TAR (cis-acting, trans-activation response) elements are used by HIV-1 as SRSs. TAR is an RNA structure that is used by the HIV-1 Tat protein to regulate HIV gene transcription. Functional TAR and Tat interaction, which requires the participation of cellular factors including TRBP and P-TEFb, promotes transcriptional activation from the stalled RNA polymerase complex on the LTR. Because TRBP is also a component of the RNAi dicing complex, the TAR structure could sequester TRBP and affect miRNA biogenesis (Chendrimada et al. 2005; Haase et al. 2005; Bennasser et al. 2006). Tat has also been shown to interact with Dicer and to act as an SRS (Bennasser et al. 2005; Bennasser and Jeang 2006). In contrast, a separate study showed that Tat, Tax (human T-cell leukemia virus type 1 [HTLV-1]), and Tas (PFV-1) failed to inhibit RNAi, and the stable expression of physiological levels of Tat did not globally inhibit miRNA production or expression in infected human cells (Lin and Cullen 2007). The latter finding was supported by a publication that demonstrated that both HIV-1 Tat and TAR expression do not reduce the efficacy of cellular RNA silencing (Sanghvi and Steel 2011b). However, there are additional studies that also support the role of Tat as an SRS (Qian et al. 2009; Hayes et al. 2011). Lastly, Coley et al. (2010) showed that HIV-1 Vpr can suppress Dicer expression during monocyte differentiated into macrophages.

The conflicting evidence concerning HIV-1-encoded SRSs may be attributable to the different systems and experimental conditions used for each study. Time course comparisons of infection, evaluations of different ratios of cells to virus, and comparisons of different subtypes of HIV-1 in carefully controlled experiments are necessary to resolve this issue.

A different but equally important and equally controversial question is whether or not the host miRNA pathway affects HIV-1 infection. There is evidence that knockdown of Dicer or Drosha boosts HIV-1 infection and produces infectious virus (Triboulet et al. 2007). Dicer and Drosha's suppressive role in HIV-1 infection is supported by work from Nathans et al. (2009), but contrary to the results mentioned, knockdown of Dicer or Drosha produced non-infectious virus. Yet another study found that TRBP contributes mainly to the enhancement of virus production and that Dicer does not mediate HIV-1 restriction by

RNAi (Christensen et al. 2007). TRBP's role in HIV-1 infection was further supported by the demonstration that it promotes viral infection through the suppression of PKR-mediated antiviral response and not through RNAi mechanisms (Sanghvi and Steel 2011a).

B. Perturbation of miRNA Expression Profiles by HIV-1 Infection

Yeung et al. (2005) first reported the RNA-primed array-based Klenow enzyme (RAKE) miRNA microarray data of miRNAs prepared from pNL4-3-transfected HeLa cells versus untransfected cells and concluded that HIV-1 down-regulates miRNAs (Yeung et al. 2005). Although several subsequent reports showed that HIV-1 infection up-regulates some host miRNAs, HIV-mediated miRNA down-regulation was also reported by Triboulet et al. (2007) using the same RAKE miRNA microarray platform and Northern blotting. The study showed down-regulation of the polycistronic miRNA cluster *miR-17/92* following HIV-1 NL4-3 infection in Jurkat cells. The authors concluded that down-regulation of the *miR-17/92* cluster would result in up-regulation of the target transcriptional coactivator P300/CBP-associated factor (PCAF), a histone acetyltransferase that can interact and activate Tat function for efficient viral replication. In contrast to the marked reduction of this miRNA cluster reported in this study, a more recent publication by Hayes et al. (2011) showed only a 50% HIV-1-mediated down-regulation of the same cluster. In our study using Jurkat cells, CEM cells, and PBMCs, we also observed a weak reduction of these miRNAs (Sun et al. 2012). The miRNAs that are part of the *miR-17/92* cluster can be considered as onco-miRs due to their up-regulation in tumors (He et al. 2005). These miRNAs have also been shown to be up-regulated in AIDS-related non-Hodgkin's lymphomas (Thapa et al. 2011), although one would expect their expression to be lowered if HIV-1 infection has in fact the ability to down-regulate them. Again, these results may be reflective of the different types of cells and methods used for these studies.

In addition to the *miR-17/92* cluster, other miRNAs have been reported to be affected by HIV-1 infection. Houzet et al. (2008) have proposed the interesting concept that various classes of HIV-1-positive individuals possess specific miRNA signatures. Thirty-six HIV-1 patients were categorized into four classes based on their CD4+ T-cell counts and viral loads. Through the RAKE miRNA microarray analysis of miRNA profiles in PBMCs, they observed the down-regulation of *miR-29a* and *miR-29b* in HIV-1 patients and infected PBMCs, and the down-regulation of *miR-29c*, *miR-26a*, and *miR-21* in HIV-1 patient samples. Part of their profiling data is consistent with data published by Hayes et al. and with our own miRCURY LNA™ microarray data from CEM, Jurkat, and PBMCs, which validates the above results (Hayes et al. 2011; Sun et al. 2012). Our data also provided some evidence that HIV-1 infection induced cell-cycle arrest and may contribute to the reduction of *miR-29* family members. One caveat for these types of profiling experiments is that although HIV-1 infection is more robust in PBMCs and more relevant than in cell lines, PBMCs are somewhat heterogeneic (usually a mixture of CD8− cells), may differentiate during infection, and die more readily. These factors could result in misleading profiling results. Nevertheless, if HIV-1 infection truly affects miRNA transcription and maturation in particular subsets of PBMCs, such knowledge could be used for the development of HIV-1 therapies, diagnosis, prognosis, and treatment response parameters.

Differences in sites of HIV-1 integration may also cause discrepancies in profiling results. Other factors, such as transfection, virus infection, differences in virus titers, duration of infection, choice of cell types for analyses, and differences in strains of HIV-1 used for such analyses, can further complicate the interpretation of the data.

IV. REGULATION OF HIV-1 INFECTION BY HOST miRNAs

Understanding the role of host miRNAs in HIV-1 infection is complicated by the likelihood that miRNAs could have an indirect role via targeting cellular factors that in turn suppress or activate HIV-1 replication. This scenario is further confounded by the possibility that HIV-1 encodes siRNAs/miRNAs during the infectious cycle. miRNAs involved in HIV-1 infection could thus be defined as HIV-1 or host encoded according to their source, and suppressors or activators based on their function. They could be further divided based on whether they directly or indirectly target HIV-1 or cellular transcripts or both. The latter would make it difficult to classify one miRNA as a suppressor or an activator of HIV-1 infection without extended knowledge of its targets. Nevertheless, it is of interest to determine if host miRNAs play a role in the regulation of HIV-1 gene expression, and whether or not any host miRNAs directly target the HIV-1 RNA genome.

Generally, miRNAs function as negative regulators of gene expression by binding to the 3'-UTR of the RNA target (Xie et al. 2005; Gu et al. 2009). Since the HIV-1 Nef sequence also serves as the 3'-UTR for most viral transcripts, miRNA-binding sites in this region will play a dual role by targeting the Nef coding sequencing and the 3'-UTR of other viral transcripts and thus could play a crucial role in HIV-1 infection (Schwartz et al. 1990) (Figure 11.1). Nef has been shown to play a positive role in viral replication and pathogenesis, and HIV-1 strains with Nef gene deletions result in slower progression to AIDS (reviewed in Foster and Garcia 2008). Thus, miRNAs targeting the Nef region have the potential to affect HIV-1 pathogenesis.

Several host miRNAs that target conserved regions of the HIV-1 genome were predicted *in silico* (Hariharan et al. 2005). Using the miRNA target predicting program probability of interaction by target accessibility (PITA) (Kertesz et al. 2007), we found thousands of potential cellular miRNA sites in the Nef/3'-LTR region (Sun et al. 2012). Although not every target site is functional, simultaneous targeting by several miRNAs can be synergistic in down-regulating gene expression. Based on a comparison of the miRNA expression patterns in activated CD4+ T lymphocytes with resting CD4+ T lymphocytes, it was concluded that several host miRNAs (*miR-125b*, *miR-150*, *miR-28*, *miR-223*, and *miR-382*), which are highly expressed in resting CD4+ T lymphocytes, may target the Nef/3'-LTR region and contribute to HIV-1 latency (miRNA interactions are summarized in Figure 11.2). Disappointingly, the most potent miRNA, *miR-223*, was then found to repress HIV-1 very weakly due to the structure in the targeted region. Thus, if *miR-223* has a bona fide anti-HIV-1 effect, it may be accomplished through the targeting of host factors (Sun et al. 2012). Furthermore, specific antagomirs (anti-miRNA antisense) against these miRNAs substantially increased HIV-1 protein translation in resting CD4 cells, but the results are complicated by the parallel up-regulation of many host factors (Huang et al. 2007). Nevertheless, it has been proposed that application of pooled antagomirs targeting this subset of miRNAs could possibly be used to purge latent HIV-1 reservoirs (Zhang 2009). The determination of enriched miRNAs in resting CD4 cells relative to activated CD4 cells is difficult to evaluate, however, because of their low expression levels (Landgraf et al. 2007; Betel et al. 2008). Despite this obstacle, profiling miRNAs in HIV-1-infected CD4 T cells versus resting CD4 T cells could provide valuable data supporting modulation of endogenous miRNAs for purging latently infected cells. It is worth pointing out that highly expressed miRNAs may be important for normal cellular homeostasis. Thus, the use of pooled antagomirs to block miRNA functions to activate HIV-1 in latently infected cells could result in toxicities in uninfected cells. Similar but somewhat controversial data have been reported for monocytes/macrophages (Wang et al. 2008; Swaminathan et al. 2009).

Ectopic expression of *miR-29a* has been shown to repress Nef expression and reduce viral levels (Ahluwalia et al. 2008; Nathans et al. 2009). Moreover, *miR-29a* seems to directly target HIV-1 transcripts to P-bodies, which could be a mechanism for maintaining HIV-1 in a latent state (Nathans et al. 2009). Other *miR-29* family members can also target HIV-1 (Hariharan et al. 2005). Interestingly, *miR-29* family members are observed to be down-regulated in both HIV-1 patients and infected PBMCs (Houzet et al. 2008; Hayes et al. 2011; Sun et al. 2012). Results showing that the *miR-29* family can target HIV-1 have been confirmed by several laboratories. Other supporting evidence for a role of the *miR-29* family in HIV infection is that Epstein–Barr virus (EBV) encodes miRNA-BART1-3p and -BART3-3p, which have a seed sequence similar to *miR-29* and that the bovine leukemia virus (BLV), also an RNA virus, encodes an miRNA that mimics the host *miR-29* and induces tumorigenesis (Kincaid et al. 2012; Skalsky et al. 2012). Unfortunately, our data show that a secondary structure formed in the *miR-29* and *miR-223* target regions can severely attenuate their potential repression of HIV-1 (Sun et al. 2012).

Finally, a study by Sung and Rice (2009) reported that host *miR-198* restricts HIV-1 replication in monocytes by repressing cyclin T1 expression, thereby supporting functional roles for miRNAs that do not directly target HIV-1 but indirectly contribute to HIV-1 infection by targeting host factors essential for HIV-1 pathogenesis, and a study by Chable-Bessia et al. (2009) showed that the major components of P-bodies, such as RCK/p54 (an RNA helicase), GW182 (a P-body marker protein that interacts with Argonautes), LSm-1 (an RNA-binding protein), and XRN-1 (a 5′ to 3′ exoribonuclease), negatively regulate HIV-1 gene expression by blocking viral mRNA association with polysomes, therefore supporting P-bodies as a place to sequester HIV-1 transcripts. Interestingly, their data also showed that knockdown of RCK/p54 or DGCR8 resulted in virus reactivation in PBMCs isolated from HIV-1-infected patients on highly active antiretroviral therapy (HAART). These results imply that HIV-1 RNAs can be sequestered in P-bodies and reenter translation upon P-body disruption, thus creating a type of latent infection different from the gene silencing-based latency.

Despite these findings concerning HIV-1 and host miRNA interactions, more carefully controlled experiments are needed to establish that host miRNAs are functionally active in HIV-1 replication.

V. sncRNAs IN VIRALLY INFECTED CELLS

It is currently accepted that several viruses utilize virally encoded siRNAs/miRNAs or miRNA-like sncRNAs (vsncRNAs) during their infection cycle (Grundhoff and Sullivan 2011; Hussain et al. 2012; Kincaid et al. 2012). In the case of the West Nile virus (WNV), the vsncRNA is encoded in a terminal stem loop located in the 3′-UTR (Hussain et al. 2012). However, if the RNA virus-encoded miRNAs or vsncRNAs follow the canonical miRNA biogenesis pathway, they could also target their own viral RNA genome and cause self-destruction. To avoid the Drosha processing of its own viral genome, BLV, for example, uses a PolIII promoter to generate a separate small transcript (Skalsky et al. 2012). Viruses with double-stranded DNA genomes such as EBV and Kaposi's sarcoma-associated herpesvirus have also been shown to encode several miRNAs, but their biological functions have not yet been entirely elucidated. The function of some viral non-coding small RNAs is documented as in the case of the adenovirus VA-1, which inactivates PKR (Kitajewski et al. 1986) or the seven small nuclear RNAs encoded by the herpesvirus saimiri, which up-regulate cellular genes involved in T-cell activation (Cook et al. 2005) and direct down-regulation of a mature host miRNA (Cazalla et al. 2010). Sullivan et al. (2005) also showed that SV40-encoded miRNAs down-regulate production of the T antigen to reduce susceptibility to the host cytotoxic T cells. However, functions and targets for the majority of viral sncRNAs are still unknown.

A. HIV-1 Naturally Encoded sncRNAs

It is still debatable whether or not HIV-1 encodes functional viral siRNAs/miRNAs (vsiRNAs/vmiRNAs) because of their low expression; reports are conflicting. Possible HIV-1-encoded small RNAs and their cellular targets were first described from *in silico* studies (Bennasser et al. 2004). Subsequently, there were several studies empirically addressing the existence and function of HIV-1-encoded sncRNAs (Omoto et al. 2004; Bennasser et al. 2005; Omoto and Fujii 2005; Kaul and Suman 2006; Klase et al. 2007, 2009; Kaul et al. 2008; Ouellet et al. 2008; Purzycka and Adamiak 2008; Yeung et al. 2009; Althaus et al. 2012; Schopman et al. 2012; Sun et al. 2012). There are currently three miRBase documented HIV-1-encoded vmiRNAs: *miR-H1-5p*, *miR-N367-3p*, and *miR-TAR-5p/3p* (Table 11.1 and Figure 11.2). HIV-1 *vmiRNA-N367* was first shown to be produced in HIV-1 persistently infected cells. *N367* was proposed to suppress HIV-1 infection by targeting both Nef and the LTR (Omoto et al. 2004; Omoto and Fujii 2005, 2006). The second vmiRNA, *miR-H1*, was found to be encoded in the LTR, and it seems to target the cellular HIV-1 suppression factor AATF to facilitate HIV-1 infection (Kaul and Suman 2006; Kaul et al. 2008). The third, a *vsiRNA1* derived from HIV-1 Rev response element (RRE) was first demonstrated in HIV-1-infected Jurkat cells, and it was easily detected 2 days postinfection (Bennasser et al. 2005). Finally, the putative TAR structure-processed vmiRNAs have been reported by two independent groups (Klase et al. 2007, 2009; Ouellet et al. 2008). But later, a systematic large-scale study showed that neither HIV-1 nor HTLV-1 expressed significant levels of either vsiRNAs or vmiRNAs

TABLE 11.1. Putative HIV-1 vsncRNAs and Proposed Functions

Name	Sequence (miRBase)	References	Function
miR-N367	acugaccuuuggauggugcuucaa	Omoto et al. (2004); Omoto and Fujii (2005)	Suppression of *nef*
miR-H1	ccagggaggcgugccugggc	Kaul and Suman (2006); Kaul et al. (2008)	Down-regulation of cellular *AATF*
TAR-5p	ucucucugguuagaccagaucuga	Klase et al. (2007, 2009); Ouellet et al. (2008); Purzycka and Adamiak (2008)	Inhibition of apoptosis by down-regulation of cellular apoptotic genes *ERCC1* and *IER3*; chromatin remodeling of the viral LTR
TAR-3p	ucucuggcuaacuagggaaccca	Klase et al. (2007, 2009); Ouellet et al. (2008); Purzycka and Adamiak (2008)	
vsiRNA1	uccuuggguucuuaggagc	Bennasser et al. (2005)	Rescues *Env* RNA expression

in persistently infected T cells (Lin and Cullen 2007), which supports the previous data by Pfeffer et al. (2005). In our own studies, we found that the proposed N367-encoded region overlaps with *miR-223* and *miR-29* target sites on HIV-1, and the sequence around this region forms a structure to block *miR-29* inhibition rather than encoding a vmiRNA (Sun et al. 2012). Consistent with these findings, we were not able to detect processed *N367*. The report of the vsiRNA1 is also questionable since the data could not be reproduced by a different group through both *in vitro* and *in vivo* experiments (Cullen 2006). Last, the expression levels of the putative TAR-miRNAs are extremely low. Only *in vivo* processed, ectopically expressed *TAR-3p* can be detected by Northern blotting (Klase et al. 2007, 2009; Ouellet et al. 2008; Purzycka and Adamiak 2008), and the cloned sequences from *TAR-3p* are only 17–18 nt long, thus shorter than the canonical 21–22 nt length of most Dicer-processed miRNAs (Klase et al. 2009). The findings discussed suggest that HIV-1-derived sncRNAs are expressed at very low levels, casting some doubts on their biological function (Mukherji et al. 2011). This is further corroborated by three separate attempts using deep sequencing technology to survey for low abundance sncRNAs. Pyro 454 sequencing of HIV-1-derived sncRNAs in infected MT4 cells (1 MOI for 2 days) by the same group that discovered *vsiRNA1* yielded only about 100 reads that mapped to the HIV-1 genome. They also failed to clone the HIV-1 *vsiRNA1*, which in their previous report was claimed to be highly expressed 2 days after infection, and were able to clone a sequence corresponding to only part of the proposed *vsiRNA1** strand (Yeung et al. 2009). SOLiD platform sequencing sncRNAs from HIV-1-infected SupT1 cells for 24 hours reported about 0.1 million reads corresponding to HIV-1, but 80% of them mapped to only two peaks located in the Nef-3′-LTR region, and the ratio of sense and antisense reads was about 6:5, which implies that they may be the sense and antisense transcripts from integrated provirus 3′-LTR and processed by Dicer (Lefebvre et al. 2011). However, using the same SOLiD plus deep sequence technology, Schopman et al. (2012) identified 26,000 HIV-1-derived sncRNA reads and discovered several "hot spots" for sncRNA production. Only *vmiRNA-TAR* with 53 reads is among the top five expressed vsncRNAs. Their data also support the existence of processed vsiRNA from the HIV-1 provirus 3′-UTR sense and antisense transcripts (Schopman et al. 2012). Finally, using a tailored enrichment strategy, 900 HIV-1 sncRNAs were captured and sequenced; 216 were found to be unique clones (87). Among these, about 16 were of the correct length to be Dicer products; six of them corresponded to *N367*; one to *TAR-3p*; and one to *miR-H1*. Since most of the literature concurs that *TAR-3p* is most abundant among the annotated HIV-1 sncRNAs, it is to be concluded that, at least using this strategy, the cloning frequency does not provide a faithful representation of the sncRNA level of expression (Althaus et al. 2012). Functional studies indicated that HIV-1 vsncRNAs mainly restrict their own HIV genome and do not target host cellular genes (Althaus et al. 2012). Targeting its own viral genome with HIV-1-derived vsncRNA could perhaps be a viral mechanism to introduce mutations and switch between different mutants.

B. HIV-1 Artificially Encoded sncRNAs

RNA viruses have also been engineered to produce functional miRNAs during viral infection without impairing their own replication (Rouha et al. 2010; Varble et al. 2010; Varble and tenOever 2011). One strategy that was explored was to insert the miRNA within an intron of a virus that undergoes splicing—the influenza A virus—which resulted in the correct processing of a functional miRNA without any alteration of splicing or viral fitness (Varble et al. 2010; Varble and tenOever 2011). Rouha et al. (2010) inserted a heterologous miRNA-precursor stem-loop sequence element into the RNA genome of the flavivirus tick-borne encephalitis virus, and they also achieved the production of a functional miRNA without affecting viral replication. Therefore, miRNA production can occur from cytoplasmic RNA viruses, and although the low abundance of the potential viral miRNAs makes their biological significance controversial, these findings show that RNA viruses have the ability to use the host cellular machinery for the production of functional miRNAs.

HIV-1 has also been used to express exogenous miRNAs. Klase et al. engineered HIV-1 to contain cis-embedded host miRNAs (*let-7a*, *miR-28*, *miR-29b*, *miR-138*, *miR-211*, *miR-326*, and *miR-329*) in Nef and concluded that these modified viruses are generally replication competent in T cells. However, they also observed that the expression of miRNAs with predicted target sequences in the HIV-1 genome inhibited viral replication when an inserted miRNA was particularly well processed by Drosha. An exception was *miR-29b*, which only weakly down-regulated replication. This is consistent with our findings that the putative *miR-29* HIV-1-targeted region forms a secondary structure that may prevent *miR-29* binding and suppression (Klase et al. 2011; Sun et al. 2012). Interestingly, it has been shown that HIV-1 is capable of escaping RNAi (Boden et al. 2003; Westerhout et al. 2005).

It is worth mentioning that because of the HIV-1 fast replication cycle coupled with the high mutation rate, one would expect that any HIV-1-encoded miRNA would rapidly change and lose any host target specificity. The exception may be sequences such as the TAR element, for which there is a strong selective pressure to maintain the sequence composition. It therefore remains a possibility that some sequences in the HIV-1 genome, such as the *TAR-3p* miRNA-like sequences, can play a functional physiological role in the HIV-1 life cycle, but the evidence to date is clearly less than compelling. The next-generation sequencing data also indicate that HIV-1-derived vsncRNAs are not a dominant sncRNA group, and their roles in HIV infection cycle are therefore still debatable.

VI. tRNA-DERIVED sncRNAs AND HIV-1

The involvement of tdsmRNAs in HIV-1 infection was discovered and studied by three independent groups (Yeung et al. 2009; Althaus et al. 2012; Schopman et al. 2012). Through the analysis of 454 pyrosequencing results, Yeung et al. (2009) first found a highly abundant small 18-nt RNA sequence that is the reverse complement of the HIV-1 primer-binding site (PBS), and this sequence is present only in HIV-1-infected cells. However, this sequence could potentially originate from the 3′ end of the human cellular tRNA^{lys}. During HIV-1 replication, reverse transcriptase (RT) uses tRNA^{lys} binding to its PBS as a primer for replication. Therefore, this sncRNA could be simply processed from the double-stranded RNA (dsRNA) hybrid formed between the PBS and the tRNA^{lys}. In contrast to Yeung's finding, deep sequencing data from Schopman et al. (2012) and our own SOLEXA sequencing data analyses (unpublished) show that tdsmRNAs are highly abundant in uninfected cells and most likely not induced by HIV-1. However, this does not rule out the possibility that a tdsmRNA could derive from a dsRNA hybrid formed between the HIV-1 PBS and the tRNA^{lys}. Interestingly, Schopman et al. also observed that tRNA^{lys} is dramatically reduced during HIV-1 infection. Several publications have addressed the processing of tRNAs to tdsmRNAs, which is amplified under stress conditions (Kawaji et al. 2008; Cole et al. 2009; Fu et al. 2009; Lee et al. 2009). To date, there is no published global profiling of tdsmRNAs in HIV-1-infected samples versus uninfected samples. It is possible that HIV-1 infection induces cell stress and results in tRNA cleavages. Only

Yeung et al. tested tdsmRNA^{lys} function and concluded that it may target the viral genome for suppression. The role of tdsmRNAs in HIV-1 infection certainly warrants further investigation.

VII. CONCLUDING REMARKS

In summary, despite many efforts and some progress in this field, the question of whether or not miRNAs (host or viral) play an essential role in HIV-1 pathogenesis remains largely disputed. The regulation of HIV-1 infection by host miRNAs that directly target the viral genome is possible, but it is unclear how effectively these cellular miRNAs are in regulating HIV-1 infection. The regulation of HIV-1 infection via miRNA targeting of cellular factors is also difficult to rationalize because they would affect the expression of genes that could facilitate or inhibit HIV infection. Finally, the question of whether or not HIV-1 can produce sufficient amounts of self-derived sncRNAs to regulate its own infectious cycle also remains unanswered. Given that cellular miRNAs are perturbed during HIV infection and that the virally encoded TAR miRNA-like sequences are present in productive infections, it is conceivable that sncRNAs could be used to monitor the pathogenic potential of viral isolates and could perhaps be used as prognostic markers for treatment response.

ACKNOWLEDGMENT

This work was supported by NIH grants AI29329, AI42552, and HL07470 to J.J.R.

REFERENCES

Abe, M., H. Suzuki, H. Nishitsuji, H. Shida, and H. Takaku. 2010. Interaction of human T-cell lymphotropic virus type I Rex protein with Dicer suppresses RNAi silencing. *FEBS Lett* **584**:4313–4318.

Ahluwalia, J.K., S.Z. Khan, K. Soni, P. Rawat, A. Gupta, M. Hariharan, V. Scaria, M. Lalwani, B. Pillai, D. Mitra, and S.K. Brahmachari. 2008. Human cellular microRNA hsa-miR-29a interferes with viral nef protein expression and HIV-1 replication. *Retrovirology* **5**:117.

Althaus, C.F., V. Vongrad, B. Niederost, B. Joos, F. Di Giallonardo, P. Rieder, J. Pavlovic, A. Trkola, H.F. Gunthard, K.J. Metzner, and M. Fischer. 2012. Tailored enrichment strategy detects low abundant small noncoding RNAs in HIV-1 infected cells. *Retrovirology* **9**:27.

Andreola, M.L., G.A. Nevinsky, P.J. Barr, L. Sarih-Cottin, B. Bordier, M. Fournier, S. Litvak, and L. Tarrago-Litvak. 1992. Interaction of tRNALys with the p66/p66 form of HIV-1 reverse transcriptase stimulates DNA polymerase and ribonuclease H activities. *J Biol Chem* **267**: 19356–19362.

Barre-Sinoussi, F., J.C. Chermann, F. Rey, M.T. Nugeyre, S. Chamaret, J. Gruest, C. Dauguet, C. Axler-Blin, F. Vezinet-Brun, C. Rouzioux, W. Rozenbaum, and L. Montagnier. 1983. Isolation of a T-lymphotropic retrovirus from a patient at risk for acquired immune deficiency syndrome (AIDS). *Science* **220**:868–871.

Bartel, D.P. 2009. MicroRNAs: target recognition and regulatory functions. *Cell* **136**:215–233.

Behm-Ansmant, I., J. Rehwinkel, T. Doerks, A. Stark, P. Bork, and E. Izaurralde. 2006. mRNA degradation by miRNAs and GW182 requires both CCR4:NOT deadenylase and DCP1:DCP2 decapping complexes. *Genes Dev* **20**:1885–1898.

Bennasser, Y. and K.T. Jeang. 2006. HIV-1 Tat interaction with Dicer: requirement for RNA. *Retrovirology* **3**:95.

Bennasser, Y., S.Y. Le, M.L. Yeung, and K.T. Jeang. 2004. HIV-1 encoded candidate micro-RNAs and their cellular targets. *Retrovirology* **1**:43.

Bennasser, Y., S.Y. Le, M. Benkirane, and K.T. Jeang. 2005. Evidence that HIV-1 encodes an siRNA and a suppressor of RNA silencing. *Immunity* **22**:607–619.

Bennasser, Y., M.L. Yeung, and K.T. Jeang. 2006. HIV-1 TAR RNA subverts RNA interference in transfected cells through sequestration of TAR RNA-binding protein, TRBP. *J Biol Chem* **281**:27674–27678.

Bernstein, E., A.A. Caudy, S.M. Hammond, and G.J. Hannon. 2001. Role for a bidentate ribonuclease in the initiation step of RNA interference. *Nature* **409**:363–366.

Betel, D., M. Wilson, A. Gabow, D.S. Marks, and C. Sander. 2008. The microRNA.org resource: targets and expression. *Nucleic Acids Res* **36**:D149–D153.

Boden, D., O. Pusch, F. Lee, L. Tucker, and B. Ramratnam. 2003. Human immunodeficiency virus type 1 escape from RNA interference. *J Virol* **77**:11531–11535.

Calabrese, J.M. and P.A. Sharp. 2006. Characterization of the short RNAs bound by the P19 suppressor of RNA silencing in mouse embryonic stem cells. *RNA* **12**:2092–2102.

Cazalla, D., T. Yario, and J.A. Steitz. 2010. Down-regulation of a host microRNA by a Herpesvirus saimiri noncoding RNA. *Science* **328**:1563–1566.

Chable-Bessia, C., O. Meziane, D. Latreille, R.T. Robinson, A. Zamborlini, A. Wagschal, J.M. Jacquet, J. Reynes, Y. Levy, A. Saib, Y. Bennasser, and M. Benkirane. 2009. Suppression of HIV-1 replication by microRNA effectors. *Retrovirology* **6**:26.

Chao, J.A., J.H. Lee, B.R. Chapados, E.W. Debler, A. Schneemann, and J.R. Williamson. 2005. Dual modes of RNA-silencing suppression by Flock House virus protein B2. *Nat Struct Mol Biol* **12**:952–957.

Chapman, E.J., A.I. Prokhnevsky, K. Gopinath, V.V. Dolja, and J.C. Carrington. 2004. Viral RNA silencing suppressors inhibit the microRNA pathway at an intermediate step. *Genes Dev* **18**:1179–1186.

Chendrimada, T.P., R.I. Gregory, E. Kumaraswamy, J. Norman, N. Cooch, K. Nishikura, and R. Shiekhattar. 2005. TRBP recruits the Dicer complex to Ago2 for microRNA processing and gene silencing. *Nature* **436**:740–744.

Chermann, J.C., F. Barre-Sinoussi, C. Dauguet, F. Brun-Vezinet, C. Rouzioux, W. Rozenbaum, and L. Montagnier. 1983. Isolation of a new retrovirus in a patient at risk for acquired immunodeficiency syndrome. *Antibiot Chemother* **32**:48–53.

Christensen, H.S., A. Daher, K.J. Soye, L.B. Frankel, M.R. Alexander, S. Laine, S. Bannwarth, C.L. Ong, S.W. Chung, S.M. Campbell, D.F. Purcell, and A. Gatignol. 2007. Small interfering RNAs against the TAR RNA binding protein, TRBP, a Dicer cofactor, inhibit human immunodeficiency virus type 1 long terminal repeat expression and viral production. *J Virol* **81**:5121–5131.

Cole, C., A. Sobala, C. Lu, S.R. Thatcher, A. Bowman, J.W. Brown, P.J. Green, G.J. Barton, and G. Hutvagner. 2009. Filtering of deep sequencing data reveals the existence of abundant Dicer-dependent small RNAs derived from tRNAs. *RNA* **15**:2147–2160.

Coley, W., R. Van Duyne, L. Carpio, I. Guendel, K. Kehn-Hall, S. Chevalier, A. Narayanan, T. Luu, N. Lee, Z. Klase, and F. Kashanchi. 2010. Absence of DICER in monocytes and its regulation by HIV-1. *J Biol Chem* **285**:31930–31943.

Cook, H.L., J.R. Lytle, H.E. Mischo, M.J. Li, J.J. Rossi, D.P. Silva, R.C. Desrosiers, and J.A. Steitz. 2005. Small nuclear RNAs encoded by Herpesvirus saimiri upregulate the expression of genes linked to T cell activation in virally transformed T cells. *Curr Biol* **15**:974–979.

Cullen, B.R. 2006. Is RNA interference involved in intrinsic antiviral immunity in mammals? *Nat Immunol* **7**:563–567.

Dennis, C. 2002. Small RNAs: the genome's guiding hand? *Nature* **420**:732.

REFERENCES

Doench, J.G. and P.A. Sharp. 2004. Specificity of microRNA target selection in translational repression. *Genes Dev* **18**:504–511.

Eulalio, A., J. Rehwinkel, M. Stricker, E. Huntzinger, S.F. Yang, T. Doerks, S. Dorner, P. Bork, M. Boutros, and E. Izaurralde. 2007. Target-specific requirements for enhancers of decapping in miRNA-mediated gene silencing. *Genes Dev* **21**:2558–2570.

Fire, A., S. Xu, M.K. Montgomery, S.A. Kostas, S.E. Driver, and C.C. Mello. 1998. Potent and specific genetic interference by double-stranded RNA in *Caenorhabditis elegans*. *Nature* **391**:806–811.

Foster, J.L. and J.V. Garcia. 2008. HIV-1 Nef: at the crossroads. *Retrovirology* **5**:84.

Fu, H., J. Feng, Q. Liu, F. Sun, Y. Tie, J. Zhu, R. Xing, Z. Sun, and X. Zheng. 2009. Stress induces tRNA cleavage by angiogenin in mammalian cells. *FEBS Lett* **583**:437–442.

Gatignol, A., S. Laine, and G. Clerzius. 2005. Dual role of TRBP in HIV replication and RNA interference: viral diversion of a cellular pathway or evasion from antiviral immunity? *Retrovirology* **2**:65.

Giraldez, A.J., Y. Mishima, J. Rihel, R.J. Grocock, S. Van Dongen, K. Inoue, A.J. Enright, and A.F. Schier. 2006. Zebrafish MiR-430 promotes deadenylation and clearance of maternal mRNAs. *Science* **312**:75–79.

Grishok, A., A.E. Pasquinelli, D. Conte, N. Li, S. Parrish, I. Ha, D.L. Baillie, A. Fire, G. Ruvkun, and C.C. Mello. 2001. Genes and mechanisms related to RNA interference regulate expression of the small temporal RNAs that control *C. elegans* developmental timing. *Cell* **106**:23–34.

Grundhoff, A. and C.S. Sullivan. 2011. Virus-encoded microRNAs. *Virology* **411**:325–343.

Gu, S., L. Jin, F. Zhang, P. Sarnow, and M.A. Kay. 2009. Biological basis for restriction of microRNA targets to the 3′ untranslated region in mammalian mRNAs. *Nat Struct Mol Biol* **16**:144–150.

Haase, A.D., L. Jaskiewicz, H. Zhang, S. Laine, R. Sack, A. Gatignol, and W. Filipowicz. 2005. TRBP, a regulator of cellular PKR and HIV-1 virus expression, interacts with Dicer and functions in RNA silencing. *EMBO Rep* **6**:961–967.

Haasnoot, J., W. de Vries, E.J. Geutjes, M. Prins, P. de Haan, and B. Berkhout. 2007. The Ebola virus VP35 protein is a suppressor of RNA silencing. *PLoS Pathog* **3**:e86.

Hannon, G.J. and J.J. Rossi. 2004. Unlocking the potential of the human genome with RNA interference. *Nature* **431**:371–378.

Hariharan, M., V. Scaria, B. Pillai, and S.K. Brahmachari. 2005. Targets for human encoded microRNAs in HIV genes. *Biochem Biophys Res Commun* **337**:1214–1218.

Hayes, A.M., S. Qian, L. Yu, and K. Boris-Lawrie. 2011. Tat RNA silencing suppressor activity contributes to perturbation of lymphocyte miRNA by HIV-1. *Retrovirology* **8**:36.

He, L., J.M. Thomson, M.T. Hemann, E. Hernando-Monge, D. Mu, S. Goodson, S. Powers, C. Cordon-Cardo, S.W. Lowe, G.J. Hannon, and S.M. Hammond. 2005. A microRNA polycistron as a potential human oncogene. *Nature* **435**:828–833.

Hemmes, H., L. Lakatos, R. Goldbach, J. Burgyan, and M. Prins. 2007. The NS3 protein of Rice hoja blanca tenuivirus suppresses RNA silencing in plant and insect hosts by efficiently binding both siRNAs and miRNAs. *RNA* **13**:1079–1089.

Houzet, L., M.L. Yeung, V. de Lame, D. Desai, S.M. Smith, and K.T. Jeang. 2008. MicroRNA profile changes in human immunodeficiency virus type 1 (HIV-1) seropositive individuals. *Retrovirology* **5**:118.

Huang, J., F. Wang, E. Argyris, K. Chen, Z. Liang, H. Tian, W. Huang, K. Squires, G. Verlinghieri, and H. Zhang. 2007. Cellular microRNAs contribute to HIV-1 latency in resting primary CD4+ T lymphocytes. *Nat Med* **13**:1241–1247.

Hussain, M., S. Torres, E. Schnettler, A. Funk, A. Grundhoff, G.P. Pijlman, A.A. Khromykh, and S. Asgari. 2012. West Nile virus encodes a microRNA-like small RNA in the 3′ untranslated region which up-regulates GATA4 mRNA and facilitates virus replication in mosquito cells. *Nucleic Acids Res* **40**:2210–2223.

Hutvagner, G., J. McLachlan, A.E. Pasquinelli, E. Balint, T. Tuschl, and P.D. Zamore. 2001. A cellular function for the RNA-interference enzyme Dicer in the maturation of the let-7 small temporal RNA. *Science* **293**:834–838.

Kaul, D. and K.A. Suman. 2006. Evidence and nature of a novel miRNA encoded by HIV-1. *Proc Indian Natn Sci Acad* **72**:91–95.

Kaul, D., A. Ahlawat, and S.D. Gupta. 2008. HIV-1 genome-encoded hiv1-mir-H1 impairs cellular responses to infection. *Mol Cell Biochem* **323**:143–148.

Kawaji, H., M. Nakamura, Y. Takahashi, A. Sandelin, S. Katayama, S. Fukuda, C.O. Daub, C. Kai, J. Kawai, J. Yasuda, P. Carninci, and Y. Hayashizaki. 2008. Hidden layers of human small RNAs. *BMC Genomics* **9**:157.

Kertesz, M., N. Iovino, U. Unnerstall, U. Gaul, and E. Segal. 2007. The role of site accessibility in microRNA target recognition. *Nat Genet* **39**:1278–1284.

Kincaid, R.P., J.M. Burke, and C.S. Sullivan. 2012. RNA virus microRNA that mimics a B-cell oncomiR. *Proc Natl Acad Sci U S A* **109**:3077–3082.

Kitajewski, J., R.J. Schneider, B. Safer, S.M. Munemitsu, C.E. Samuel, B. Thimmappaya, and T. Shenk. 1986. Adenovirus VAI RNA antagonizes the antiviral action of interferon by preventing activation of the interferon-induced eIF-2 alpha kinase. *Cell* **45**:195–200.

Klase, Z., P. Kale, R. Winograd, M.V. Gupta, M. Heydarian, R. Berro, T. McCaffrey, and F. Kashanchi. 2007. HIV-1 TAR element is processed by Dicer to yield a viral micro-RNA involved in chromatin remodeling of the viral LTR. *BMC Mol Biol* **8**:63.

Klase, Z., R. Winograd, J. Davis, L. Carpio, R. Hildreth, M. Heydarian, S. Fu, T. McCaffrey, E. Meiri, M. Ayash-Rashkovsky, S. Gilad, Z. Bentwich, et al. 2009. HIV-1 TAR miRNA protects against apoptosis by altering cellular gene expression. *Retrovirology* **6**:18.

Klase, Z., L. Houzet, and K.T. Jeang. 2011. Replication competent HIV-1 viruses that express intragenomic microRNA reveal discrete RNA-interference mechanisms that affect viral replication. *Cell Biosci* **1**:38.

Landgraf, P., M. Rusu, R. Sheridan, A. Sewer, N. Iovino, A. Aravin, S. Pfeffer, A. Rice, A.O. Kamphorst, M. Landthaler, C. Lin, N.D. Socci, et al. 2007. A mammalian microRNA expression atlas based on small RNA library sequencing. *Cell* **129**:1401–1414.

Lee, Y., I. Hur, S.Y. Park, Y.K. Kim, M.R. Suh, and V.N. Kim. 2006. The role of PACT in the RNA silencing pathway. *EMBO J* **25**:522–532.

Lee, Y.S., Y. Shibata, A. Malhotra, and A. Dutta. 2009. A novel class of small RNAs: tRNA-derived RNA fragments (tRFs). *Genes Dev* **23**:2639–2649.

Lefebvre, G., S. Desfarges, F. Uyttebroeck, M. Munoz, N. Beerenwinkel, J. Rougemont, A. Telenti, and A. Ciuffi. 2011. Analysis of HIV-1 expression level and sense of transcription by high-throughput sequencing of the infected cell. *J Virol* **85**:6205–6211.

Li, Q., L. Duan, J.D. Estes, Z.M. Ma, T. Rourke, Y. Wang, C. Reilly, J. Carlis, C.J. Miller, and A.T. Haase. 2005. Peak SIV replication in resting memory CD4+ T cells depletes gut lamina propria CD4+ T cells. *Nature* **434**:1148–1152.

Lin, J. and B.R. Cullen. 2007. Analysis of the interaction of primate retroviruses with the human RNA interference machinery. *J Virol* **81**:12218–12226.

Mallory, A.C., B.J. Reinhart, D. Bartel, V.B. Vance, and L.H. Bowman. 2002. A viral suppressor of RNA silencing differentially regulates the accumulation of short interfering RNAs and microRNAs in tobacco. *Proc Natl Acad Sci U S A* **99**:15228–15233.

Mattapallil, J.J., D.C. Douek, B. Hill, Y. Nishimura, M. Martin, and M. Roederer. 2005. Massive infection and loss of memory CD4+ T cells in multiple tissues during acute SIV infection. *Nature* **434**:1093–1097.

Mukherji, S., M.S. Ebert, G.X. Zheng, J.S. Tsang, P.A. Sharp, and A. van Oudenaarden. 2011. MicroRNAs can generate thresholds in target gene expression. *Nat Genet* **43**:854–859.

Nathans, R., C.Y. Chu, A.K. Serquina, C.C. Lu, H. Cao, and T.M. Rana. 2009. Cellular microRNA and P bodies modulate host-HIV-1 interactions. *Mol Cell* **34**:696–709.

Omoto, S. and Y.R. Fujii. 2005. Regulation of human immunodeficiency virus 1 transcription by nef microRNA. *J Gen Virol* **86**:751–755.

Omoto, S. and Y.R. Fujii. 2006. Cloning and detection of HIV-1-encoded microRNA. *Methods Mol Biol* **342**:255–265.

Omoto, S., M. Ito, Y. Tsutsumi, Y. Ichikawa, H. Okuyama, E.A. Brisibe, N.K. Saksena, and Y.R. Fujii. 2004. HIV-1 nef suppression by virally encoded microRNA. *Retrovirology* **1**:44.

Ouellet, D.L., I. Plante, P. Landry, C. Barat, M.E. Janelle, L. Flamand, M.J. Tremblay, and P. Provost. 2008. Identification of functional microRNAs released through asymmetrical processing of HIV-1 TAR element. *Nucleic Acids Res* **36**:2353–2365.

Pederson, T. 2010. Regulatory RNAs derived from transfer RNA? *RNA* **16**:1865–1869.

Pfeffer, S., A. Sewer, M. Lagos-Quintana, R. Sheridan, C. Sander, F.A. Grasser, L.F. van Dyk, C.K. Ho, S. Shuman, M. Chien, J.J. Russo, J. Ju, et al. 2005. Identification of microRNAs of the herpesvirus family. *Nat Methods* **2**:269–276.

Purzycka, K.J. and R.W. Adamiak. 2008. The HIV-2 TAR RNA domain as a potential source of viral-encoded miRNA. A reconnaissance study. *Nucleic Acids Symp Ser (Oxf)* **(52)**:511–512.

Qian, S., X. Zhong, L. Yu, B. Ding, P. de Haan, and K. Boris-Lawrie. 2009. HIV-1 Tat RNA silencing suppressor activity is conserved across kingdoms and counteracts translational repression of HIV-1. *Proc Natl Acad Sci U S A* **106**:605–610.

Rouha, H., C. Thurner, and C.W. Mandl. 2010. Functional microRNA generated from a cytoplasmic RNA virus. *Nucleic Acids Res* **38**:8328–8337.

Sanghvi, V.R. and L.F. Steel. 2011a. The cellular TAR RNA binding protein, TRBP, promotes HIV-1 replication primarily by inhibiting the activation of double-stranded RNA-dependent kinase PKR. *J Virol* **85**:12614–12621.

Sanghvi, V.R. and L.F. Steel. 2011b. A re-examination of global suppression of RNA interference by HIV-1. *PLoS ONE* **6**:e17246.

Schopman, N.C., M. Willemsen, Y.P. Liu, T. Bradley, A. van Kampen, F. Baas, B. Berkhout, and J. Haasnoot. 2012. Deep sequencing of virus-infected cells reveals HIV-encoded small RNAs. *Nucleic Acids Res* **40**:414–427.

Schwartz, S., B.K. Felber, D.M. Benko, E.M. Fenyo, and G.N. Pavlakis. 1990. Cloning and functional analysis of multiply spliced mRNA species of human immunodeficiency virus type 1. *J Virol* **64**:2519–2529.

Sharp, P.A. 2009. The centrality of RNA. *Cell* **136**:577–580.

Skalsky, R.L., D.L. Corcoran, E. Gottwein, C.L. Frank, D. Kang, M. Hafner, J.D. Nusbaum, R. Feederle, H.J. Delecluse, M.A. Luftig, T. Tuschl, U. Ohler, et al. 2012. The viral and cellular microRNA targetome in lymphoblastoid cell lines. *PLoS Pathog* **8**:e1002484.

Sullivan, C.S., A.T. Grundhoff, S. Tevethia, J.M. Pipas, and D. Ganem. 2005. SV40-encoded microRNAs regulate viral gene expression and reduce susceptibility to cytotoxic T cells. *Nature* **435**:682–686.

Sun, G., H. Li, X. Wu, M. Covarrubias, L. Scherer, K. Meinking, B. Luk, P. Chomchan, J. Alluin, A.F. Gombart, and J.J. Rossi. 2012. Interplay between HIV-1 infection and host microRNAs. *Nucleic Acids Res* **40**:2181–2196.

Sung, T.L. and A.P. Rice. 2009. miR-198 inhibits HIV-1 gene expression and replication in monocytes and its mechanism of action appears to involve repression of cyclin T1. *PLoS Pathog* **5**:e1000263.

Swaminathan, S., J. Zaunders, J. Wilkinson, K. Suzuki, and A.D. Kelleher. 2009. Does the presence of anti-HIV miRNAs in monocytes explain their resistance to HIV-1 infection? *Blood* **113**:5029–5030; author reply 5030–5021.

Thapa, D.R., X. Li, B.D. Jamieson, and O. Martinez-Maza. 2011. Overexpression of microRNAs from the miR-17-92 paralog clusters in AIDS-related non-Hodgkin's lymphomas. *PLoS ONE* **6**:e20781.

Triboulet, R., B. Mari, Y.L. Lin, C. Chable-Bessia, Y. Bennasser, K. Lebrigand, B. Cardinaud, T. Maurin, P. Barbry, V. Baillat, J. Reynes, P. Corbeau, et al. 2007. Suppression of microRNA-silencing pathway by HIV-1 during virus replication. *Science* **315**:1579–1582.

Varble, A. and B.R. tenOever. 2011. Implications of RNA virus-produced miRNAs. *RNA Biol* **8**:190–194.

Varble, A., M.A. Chua, J.T. Perez, B. Manicassamy, A. Garcia-Sastre, and B.R. tenOever. 2010. Engineered RNA viral synthesis of microRNAs. *Proc Natl Acad Sci U S A* **107**:11519–11524.

Wang, X., L. Ye, W. Hou, Y. Zhou, Y.J. Wang, D.S. Metzger, and W.Z. Ho. 2008. Cellular microRNA expression correlates with susceptibility of monocytes/macrophages to HIV-1 infection. *Blood* **113**:671–674.

Westerhout, E.M., M. Ooms, M. Vink, A.T. Das, and B. Berkhout. 2005. HIV-1 can escape from RNA interference by evolving an alternative structure in its RNA genome. *Nucleic Acids Res* **33**:796–804.

Wu, L., J. Fan, and J.G. Belasco. 2006. MicroRNAs direct rapid deadenylation of mRNA. *Proc Natl Acad Sci U S A* **103**:4034–4039.

Xie, X., J. Lu, E.J. Kulbokas, T.R. Golub, V. Mootha, K. Lindblad-Toh, E.S. Lander, and M. Kellis. 2005. Systematic discovery of regulatory motifs in human promoters and 3' UTRs by comparison of several mammals. *Nature* **434**:338–345.

Yeung, M.L., Y. Bennasser, T.G. Myers, G. Jiang, M. Benkirane, and K.T. Jeang. 2005. Changes in microRNA expression profiles in HIV-1-transfected human cells. *Retrovirology* **2**:81.

Yeung, M.L., Y. Bennasser, K. Watashi, S.Y. Le, L. Houzet, and K.T. Jeang. 2009. Pyrosequencing of small non-coding RNAs in HIV-1 infected cells: evidence for the processing of a viral-cellular double-stranded RNA hybrid. *Nucleic Acids Res* **37**:6575–6586.

Zeng, Y., R. Yi, and B.R. Cullen. 2003. MicroRNAs and small interfering RNAs can inhibit mRNA expression by similar mechanisms. *Proc Natl Acad Sci U S A* **100**:9779–9784.

Zhang, H. 2009. Reversal of HIV-1 latency with anti-microRNA inhibitors. *Int J Biochem Cell Biol* **41**:451–454.

Zhang, X., Y.R. Yuan, Y. Pei, S.S. Lin, T. Tuschl, D.J. Patel, and N.H. Chua. 2006. Cucumber mosaic virus-encoded 2b suppressor inhibits *Arabidopsis* Argonaute1 cleavage activity to counter plant defense. *Genes Dev* **20**:3255–3268.

Zhou, Z., M. Dell'Orco, P. Saldarelli, C. Turturo, A. Minafra, and G.P. Martelli. 2006. Identification of an RNA-silencing suppressor in the genome of Grapevine virus A. *J Gen Virol* **87**:2387–2395.

12

MicroRNA in Malaria

Panote Prapansilp and Gareth D.H. Turner

*Mahidol-Oxford Tropical Medicine Research Unit and Department of Tropical Pathology,
Faculty of Tropical Medicine, Mahidol University, Bangkok, Thailand
Centre for Tropical Medicine, Nuffield Department of Clinical Medicine,
Oxford University, Oxford, UK*

I. Introduction: Malaria as a Disease	184
II. The Pathogenesis of Severe Malaria	185
A. Sequestration and Cytoadherence	186
B. Reduction of Microvascular Flow	186
C. Endothelial Activation and Dysfunction	186
D. Cytokines and Host Immune Response	186
III. miRNAs and Malaria	187
A. miRNA in Malaria Parasites	187
B. miRNA in Mosquito Stages	188
C. miRNA in the Host Response to Malaria Infection	188
IV. Potential Roles for miRNAs in Regulating Host Response to Pathophysiological Mechanisms	189
A. Hypoxia	189
B. Apoptosis and Programmed Cell Death	191
C. Endothelial Activation	192
D. Central Nervous System (CNS) Injury in CM	192
V. Conclusions and Future Research Directions	193
References	193

MicroRNAs in Medicine, First Edition. Edited by Charles H. Lawrie.
© 2014 John Wiley & Sons, Inc. Published 2014 by John Wiley & Sons, Inc.

ABBREVIATIONS

BBB	blood–brain barrier
CM	cerebral malaria
EC	endothelial cell
ECM	experimental cerebral malaria
MARF	malaria-associated acute renal failure
miRNA	microRNA
mRNA	messenger RNA
PRBC	parasitized red blood cell
RNAi	RNA interference
siRNA	short interfering RNA
uRBC	uninfected red blood cell

I. INTRODUCTION: MALARIA AS A DISEASE

Malaria remains one of the world's major health burdens due to its toll of mortality and morbidity, the lack of an effective vaccine, and the difficulties of diagnosis and treatment. Approximately one-third of the world population, in most parts of Africa and South and Southeast Asia, is still at risk. During the period 1980–2010, there were an estimated 200–500 million new malaria cases and 0.5–2.5 million malaria-related deaths each year globally (Murray et al. 2012). Human malaria infection in nature is caused by one of five different species of the protozoan parasite *Plasmodium*, transmitted by the bite of female *Anopheles* mosquitoes. Most infections with malaria cause "mild" clinical disease, but severe disease and associated mortalities are predominantly linked with *Plasmodium falciparum* infection. *Plasmodium vivax*, which is now recognized as the most geographically widespread species, accounts for 50–90% of malaria episodes outside Africa (Carter and Mendis 2002) but typically displays a lower virulence. However, both *P. falciparum* and *P. vivax* cause a major burden of morbidity in pregnancy-associated malaria, and rare cases with severe complications or even fatal cases have been described in *P. vivax*- or *Plasmodium knowlesi*-infected individuals (Price et al. 2009; Lee et al. 2011). Other human malaria parasites include *Plasmodium malariae* and *Plasmodium ovale*, which cause less severe disease but can recur and cause chronic relapsing infection due to reemergence of hypnozoite stages from the liver.

Current control measures for severe malaria utilize efforts to prevent transmission and infection, such as environmental insecticide-spraying programs or insecticide-treated bed nets, and drug treatment in clinical infection or as a prophylactic measure for travelers. These approaches are constantly challenged by the ability of malaria parasites and their mosquito vectors to adapt and develop resistance. Reports of resistance to the latest and most effective antimalarial drugs, artemisinin and its derivatives, have been accumulating from Southeast Asian countries (e.g., Noedl et al. 2008; Dondorp et al. 2010). Mosquito resistance to pyrethroids, a commonly used insecticide, has now been reported in 41 countries (World Health Organization 2012). In addition, research into newer control measures such as new antimalarial drugs and insecticides, genetically engineered malaria-resistant mosquitoes, and the long-awaited malaria vaccine has been slow to progress. Despite over a century of active research, the pathophysiology of severe malaria is not well understood and a better understanding of these pathophysiological mechanisms remains a key to developing new interventions to fight this deadly disease.

One characteristic of malaria parasites is their ability to undergo large-scale changes in structure and function throughout their complex life cycle in both mosquito and human hosts. This represents a vast change in phenotype underpinned by the ability of the parasite to alter its transcriptome in cyclical fashion, which argues for a high level of genetic transcriptional control (Coulson et al. 2004). Interest in the mechanisms of parasite gene expression has been stimulated by the complete genome sequencing of the parasite in 2002 (Gardner et al. 2002). In addition, the host response to malaria infection shows evidence for strong selection pressures, which have molded human genetic development, as shown by the development of stable genetic polymorphisms in hemoglobinopathies in malaria-endemic areas. More specifically, genetic control of pathways involved in malaria pathogenesis has been investigated in murine models of disease, where immunological mechanisms have been proposed as important in the genesis of complications such as coma in cerebral malaria (CM) (Hunt et al. 2006). In several of these areas, the role of microRNAs (miRNAs) as regulators of host responses has begun to be investigated; although, interestingly the parasite itself seems to lack miRNA. This chapter will briefly review the pathophysiology of severe malaria infection and discuss evidence for the involvement of miRNA in disease pathways and the interaction between the human and mosquito hosts and *Plasmodium* parasite.

II. THE PATHOGENESIS OF SEVERE MALARIA

The outcome of an individual malaria infection is determined by multiple factors, including prior malaria exposure and resultant immunity, the genetics of host susceptibility and parasite virulence, coinfections such as HIV/AIDS and bacterial sepsis, timing of treatment, and socioeconomic factors such as nutrition and access to health care. Many malarial infections are clinically silent, showing the ability of the host's adaptive immune response to control the disease. Individuals living in high malaria transmission areas, such as sub-Saharan Africa, develop immunity against malaria due to repeated infection in childhood and usually have asymptomatic infection. In contrast, in non-immune individuals such as African children, travelers, and those living in low malaria transmission areas, infections are more clinically overt, with a spectrum of mild disease to the development of severe manifestations. Therefore, developing immunity may underlie some of the differences in the clinical spectrum of severe malaria between African children and adults in South and Southeast Asia. Severe clinical syndromes in both groups typically include CM (coma), metabolic acidosis, and severe malarial anemia. Adult disease is also associated with other complications, less in African children, such as renal failure, pulmonary edema, shock, coagulopathy, and jaundice.

Ultimately, all clinical manifestations of malaria are caused by parasitized red blood cells (PRBCs) circulating within the vasculature, which supplies every organ in the host, making malaria a multisystem disease. One fundamental pathophysiological question is how PRBCs, exclusively found in the vascular space, can cause functional and pathological changes that can ultimately lead to end-organ dysfunction. The infection of erythrocytes by malaria parasites results in progressive alterations to the erythrocyte's structure, biochemistry, and functions. These changes lead to severe complications and death if host immune response or drug therapy fails. Several potential mechanisms involved in the genesis of severe malaria have been proposed, some of which are summarized as follows.

A. Sequestration and Cytoadherence

Structural changes in the PRBC membrane, resulting in its increased rigidity and adhesiveness, can induce cytoadherence of infected erythrocytes to human endothelial cells (ECs) and other cells such as leukocytes, platelets, and uninfected red blood cell (uRBC). Cytoadherence leads to sequestration, which is a process whereby PRBCs infected with the late trophozoite and schizont parasite stages adhere to the endothelium in the deep microvasculature and disappear from the peripheral circulation. Sequestration of the PRBCs in deeper tissue of various organs provides the microaerophilic environment for malaria parasites and allows them to escape splenic clearance and to hide from the immune system (David et al. 1983). To the host, sequestration is a major contributor to the blockage of blood flow, which limits oxygen and nutrient supply to vital organs.

Cytoadherence of PRBCs to ECs is mediated via several parasite-derived proteins expressed at knob structures on the infected erythrocyte membrane, such as *P. falciparum* erythrocyte membrane protein 1 (PfEMP-1). These knobs bind to a number of potential ligands on human ECs and other cells. These ligands include CD36 (on the endothelium, leukocytes, and platelets), intercellular adhesion molecule 1 (ICAM-1, on the brain endothelium and leukocytes), and chondroitin sulfate A (on the placental endothelium) (reviewed in Miller et al. 2002). Their expression may be up-regulated by proinflammatory cytokines released in response to malaria infection, potentially increasing sequestration.

B. Reduction of Microvascular Flow

In addition to the sequestration, reduction in microvascular flow also results from decreased deformability of both PRBC and uRBC (Dondorp et al. 2000). Moreover, the phenomena of rosetting (the adherence of PRBC to uRBC), autoagglutination (the adherence of PRBC to PRBC), and platelet-mediated clumping could theoretically further impair flow in microvessels. Reduction in microvascular flow causes local tissue hypoxia, energy deprivation, lactic acid production, tissue injury, reduction in the supply of metabolic substrates, and cytokine stimulation, all of which can lead to organ dysfunction.

C. Endothelial Activation and Dysfunction

The endothelium represents the critical interface between the PRBCs in the blood circulation and the parenchyma of every organ system. Systemic activation of ECs (Turner et al. 1998) including the cerebral ECs of the blood–brain barrier (BBB) is observed during severe malarial infection and associated with increased permeability via the loosening of endothelial transmembrane junctions and up-regulation of various surface adhesion molecules (reviewed in Medana and Turner 2006; Nag et al. 2009). The consequences of activated endothelium in malaria include increased PRBC sequestration, immune cell recruitment, and leakage of the fluid to extravascular space, which can cause brain edema.

D. Cytokines and Host Immune Response

Although adaptive immune responses found in individuals living in areas of high malaria transmission can provide partial protection against malarial parasite replication, the imbalance of these complex immune responses to malaria also contributes to the genesis of severe complications and fatalities (reviewed in Angulo and Fresno 2002; Clark et al. 2006; Hunt et al. 2006). Cytokines of the proinflammatory cascade such as tumor necrosis factor (TNF), nitric oxide, interleukins, and interferon-gamma help in limiting the infection by inhibiting the replication of malaria parasites in low concentrations. However, excessive release of proinflammatory cytokines leads to immunopathology and undesirable complications through various mechanisms. These include reduced oxidative phosphorylation within the mitochondria, thus enhancing lactic acid production; increased cytoadherance by up-regulating the expression of surface molecules that worsens the degree of microvascular flow obstruction and leads to more tissue hypoxia and metabolic disturbances; and activation of blood coagulation cascades, causing thrombocytopenia and coagulopathy (Francischetti 2008). Moreover, host immune responses to malaria infection also contribute in part to severe malarial anemia (along with the rupture of erythrocytes due to parasite development and their premature destruction in the spleen) via suppression of erythropoiesis, induced apoptosis of erythroid cells in the bone marrow (Lamikanra et al. 2009), and induced hemolysis from the shearing effect on red blood cells (RBCs) by platelet-decorated ultralarge von Willebrand factor (VWF) strings (Lopez 2010). The appropriate regulation of immune responses to malaria is therefore important in the outcome of the infection.

III. miRNAs AND MALARIA

A. miRNA in Malaria Parasites

The malaria parasite *Plasmodium* is a complex eukaryotic organism with a unique repertoire of proteins expressed during multiple stages of life cycle in two hosts. This is made possible by its complex genome that helps the parasite invade host cells, evade immune responses, and pass through multiple phenotypically different stages of development (Florens et al. 2002). Its genome comprises more than 5000 genes (Gardner et al. 2002), and as such, it is reasonable to expect that the *Plasmodium* parasite would have complex mechanisms to control its expression of genes and proteins.

In other protozoa, such as *Tetrahymena thermophila* or *Trypanosoma brucei*, short interfering RNAs (siRNAs) of similar length to miRNAs have been found. RNA interference (RNAi), which is a phenomenon of double-stranded RNA (dsRNA)-mediated degradation of homologous mRNA, in *T. brucei*, is induced by either long dsRNA (Ngô et al. 1998) or siRNAs (Djikeng et al. 2001). In malaria parasites, several studies have independently reported the (RNAi) phenomenon in *P. falciparum* and *Plasmodium berghei* (Malhotra et al. 2002; McRobert and McConkey 2002; Mohmmed et al. 2003; Gissot et al. 2005; Crooke et al. 2006).

Data on the presence of miRNA in *Plasmodium* are scarce. So far, two separate investigations have failed to reveal specific parasite-encoded miRNA in *P. falciparum* (Rathjen et al. 2006; Xue et al. 2008). These two studies cloned and sequenced all short RNAs from a mixed stage of PRBCs, and further analyzed by bioinformatics prediction, which failed to match these cloned sequences with malaria genome. However, this approach was looking only for possible miRNA in the blood-stage *P. falciparum* parasites, which represents just a single point in the enzootic cycle of the parasite. To date, no study has reported the presence of RNAi-mediated sequences including siRNAs, miRNAs, repeat-associated small interfering RNAs (rasiRNAs), and PIWI-interacting RNAs (piRNAs) at any other stages of the parasite or in other *Plasmodium* species. The findings so far correspond with bioinformatics analysis, which showed that genes encoding Argonaute and dicer, important proteins in the biogenesis of miRNAs and siRNAs, are also missing from the *P. falciparum* genome (Coulson et al. 2004; Hall et al. 2005). These data imply that posttranscriptional gene silencing in *P. falciparum* must therefore utilize alternative mechanisms other than miRNAs.

Alternatively, the possibility remains that the RNAi effects observed in *Plasmodium* parasites might be a result of miRNAs or siRNAs from the host instead. Specific miRNAs from hosts could interfere the parasite gene transcription process, leading to either suppression or facilitation of parasite growth (as in the case of rhabdoviral vesicular stomatitis virus and hepatitis C virus [Jopling et al. 2006; Otsuka et al. 2007]). Host miRNAs could also control the expression of host factors, which could be either beneficial or harmful to the parasite.

This theory has recently gained acceptance as a result of a study by LaMonte et al. (2012) who found that human miRNAs that are enriched in sickle red cells, including *miR-451*, *let-7i*, and *miR-223*, could translocate into the *P. falciparum* parasite during its erythrocytic stage and integrate into parasite mRNAs, resulting in translational inhibition. These host-derived miRNAs inhibited the growth of the parasite, and their suppression resulted in two to three times increased parasite growth. These findings also represent a novel molecular mechanism that could contribute in part to the enhanced protection against malaria infection associated with the sickle-cell disease population in endemic malarial regions of Africa.

B. miRNA in Mosquito Stages

To date, there have been limited reports of the identification of miRNAs in mosquitoes. Of approximately 30–40 *Anopheles* species transmitting malaria in nature, miRNA sequences have so far been annotated for only two species: *Anopheles gambiae* (the main vector in Africa) and *Anopheles stephensi* (the chief vector in Asia) (Winter et al. 2007; Mead and Tu 2008). Over 67 miRNA sequences have been identified, and most of these have orthologues in other insects including *Drosophila*. The expression profile of four miRNAs in *A. gambiae* was demonstrated to be significantly affected by *Plasmodium* infection (Winter et al. 2007). This demonstrates the involvement of miRNAs in response to the infection of the malaria vector. Thus, miRNAs could also represent potential novel targets for the development of malarial control measures such as new insecticides or genetically engineered malaria-resistant mosquitoes.

C. miRNA in the Host Response to Malaria Infection

There have been a very limited number of studies on the role of miRNAs on the host response to malarial infection. Due to practical reasons, these studies have almost exclusively been conducted in the murine model of malaria disease. Two studies of miRNA expression in the liver and spleen of C57BL6 mice infected with *Plasmodium chabaudi* have identified hundreds of dysregulated miRNAs in response to the infection (Delić et al. 2010; Al-Quraishy et al. 2012). A single study of miRNA expression in the murine model of experimental CM (ECM) recently identified three miRNAs (*let-7i*, *miR-27a*, and *miR-150*) up-regulated in the brain of CBA mice infected with *P. berghei* ANKA (a model of ECM), in comparison with *P. berghei* K173 (a model of experimental non-CM) (El-Assaad et al. 2011). These studies have confirmed that malaria infection can alter the miRNA expression pattern in the host and that reprogramming of the expression of miRNAs may have a regulatory role in the pathogenesis of severe malaria. However, such animal studies are potentially limited in their usefulness for studying human malaria and, in particular, host response to malaria infection (Craig et al. 2012).

We have conducted miRNA expression studies in malaria-infected human tissues derived from postmortem studies in Asian adults (kidney) and African children (multiple organs). In the first study, miRNA microarray expression profiling and quantitative reverse transcription polymerase chain reaction (qRT-PCR) were used to examine miRNA changes in the kidneys of patients with and without malaria-associated acute renal failure (MARF). This was intended to identify human miRNA transcripts that could be a signature of malarial infection, and new biological pathways potentially contributing to the pathogenesis of acute kidney injury in MARF (Prapansilp 2012). These data showed that the miRNA expression profile of malaria patients is distinct from non-malaria controls (Figure 12.1). *In silico* target prediction algorithms and pathway analysis revealed that malarial infection altered the expression of miRNAs in the kidney involved in the repression of apoptosis/programmed cell death and catalytic activity, which could represent a protective response, either to parasite sequestration, host immune cell functions, or systemic complications of malaria such as shock.

IV. POTENTIAL ROLES FOR miRNAs IN REGULATING HOST RESPONSE TO PATHOPHYSIOLOGICAL MECHANISMS

A growing body of evidence has implicated the importance of miRNA to numerous molecular mechanisms involving developmental, physiological, and pathological changes of cells and tissues, such as cellular metabolism, proliferation, differentiation, apoptosis, infection, and tumorigenesis. Although very few miRNA studies have been performed in malarial disease to date, miRNAs are nevertheless important as negative regulators of several mechanisms associated with the genesis of malaria, and consequently likely to play an important role in the disease. Some of the pathophysiological mechanisms in malarial disease that have been shown to be regulated by miRNAs are discussed in the following sections, and the involvement of putative miRNAs is summarized in Figure 12.2.

A. Hypoxia

The reduction of microvascular blood flow by various causes discussed earlier would be expected to lead to some degree of local tissue ischemia and hypoxia. The relationship between hypoxia and susceptibility to ECM in murine model has recently been confirmed (Hempel et al. 2011), despite the fact that this model is characterized by a lack of cerebral sequestration of PRBCs, and such evidence remains more elusive in humans.

Since the first report of hypoxia-induced miRNAs by Kulshreshtha et al. (2007), a number of papers have implicated several miRNAs in regulating both upstream and downstream signaling of the hypoxia-inducible factor (HIF) pathway. miRNAs regulating HIF-1α under hypoxia include *miR-199a*, *miR-17-92* cluster, and *miR-20b* (Taguchi et al. 2008; Lei et al. 2009; Rane et al. 2009), whereas *miR-23*, *miR-24*, *miR-26*, *miR-107*, *miR-210*, and *miR-373* are induced by HIFs (Kulshreshtha et al. 2007; Crosby et al. 2009; Huang et al. 2009). Of these, *miR-210* is arguably the most consistently and stably induced miRNAs under hypoxic conditions (reviewed in Ivan et al. 2008; Huang et al. 2010). In comparison with the miRNA signature generated from the kidney of fatal human malaria (Prapansilp 2012), of these miRNAs, only miRNAs in *miR-17-92* cluster were found to be significantly up-regulated. However, the *miR-17-92* cluster is known to be involved in

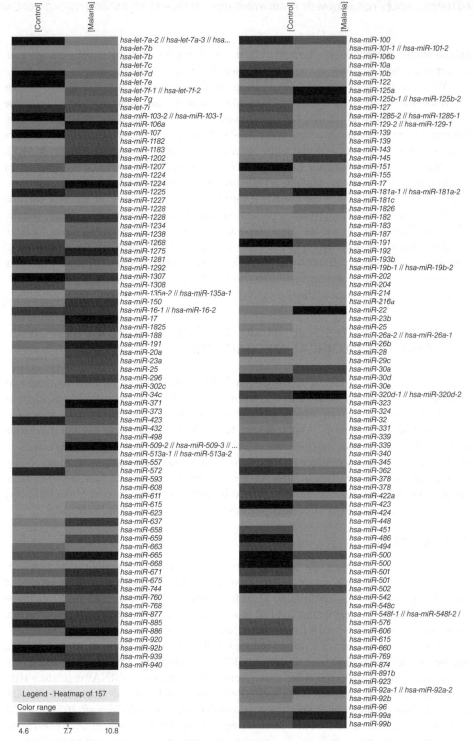

Figure 12.1. Heat map showing pooled data from miRNA array analysis of postmortem kidney tissues from control non-malaria patients (*n* = 9) compared with malaria-infected patients (*n* = 16) (derived from Affymetrix miRNA GeneChip version 1.0). It shows 157 differentially expressed miRNAs in malaria (using a cutoff of adjusted *P* < 0.05, *T*-test with Benjamini–Hochberg's control for false discovery rate (FDR), and a fold change cutoff of 2). See color insert.

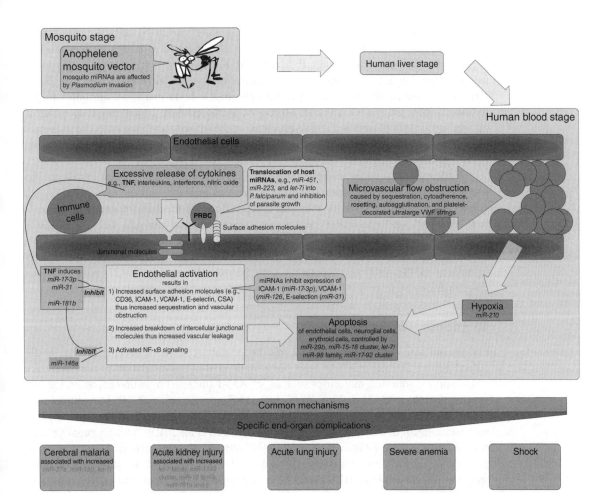

Figure 12.2. Diagram showing the relationship between different stages of malaria infection, relevant pathophysiological mechanisms at the infected red cell-endothelial interface, and sites where miRNA may play a role in regulation of these pathways, leading to specific complications of severe malaria. Most of these are hypothetical and have been identified in studies of other diseases. Confirmed data only currently exists for the role of miRNA in human malaria in acute kidney injury (shown in red). See color insert.

various mechanisms such as tumorigenesis, B-cell development, and immune response, and is thus not specific to hypoxia.

B. Apoptosis and Programmed Cell Death

Apoptosis have been observed in several pathophysiological studies of malaria. High levels of EC apoptosis via caspase activation have been reported using *in vitro* models of PRBC cocultures with human EC line (Pino et al. 2003). Tripathi et al. (2007, 2009) used an *in vitro* human BBB model to show that PRBC, but not uRBC, significantly decreased BBB integrity. Their model demonstrated that ICAM-1-mediated signaling following PRBC binding caused significant up-regulation of the mRNA transcripts associated with immune response and apoptosis, stimulating signaling via an NF-κB cascade. Soluble factors from *P. falciparum*-infected erythrocytes also induced apoptosis in ECs and neuroglial cells (Wilson et al. 2008). *P. berghei* ANKA-infected murine models also implicated a central role for apoptosis in the pathogenesis of ECM (Lovegrove et al. 2007). Hemozoin-induced apoptosis of erythroid cells in the bone marrow contributes in part to anemia in *P. falciparum* malaria. Moreover, increased Ang-2 expression, of which its circulating level in plasma is a strong predictor of the severity and fatal outcome in malaria (e.g., Yeo et al. 2008; Lovegrove et al. 2009), is associated with endothelial apoptosis and BBB breakdown (Nag et al. 2005).

Evidence from cancer studies implies that several miRNAs play roles in cell apoptosis. Some are limited to apoptosis pathways, such as *miR-29b* and the *miR-15-16* cluster, while others regulate both apoptosis and cell proliferation mechanisms, such as the *let-7/miR-98* family and the *miR-17-92* cluster (reviewed in Wang and Lee 2009). Some hypoxia-induced miRNAs such as *miR-210* also decrease proapoptotic signaling in a hypoxic environment (Kulshreshtha et al. 2007). These findings correspond with miRNA signatures in the kidneys of patients with MARF (Prapansilp 2012), which showed up-regulations of several miRNAs known to inhibit apoptotic signaling.

C. Endothelial Activation

Endothelial activation plays a crucial role in the pathogenesis of severe malaria, as discussed earlier. Several miRNAs have been implicated in the regulation of the expression of EC surface adhesion molecules such as VCAM-1 (targeted by *miR-126*, an endothelial specific miRNA), ICAM-1 (by *miR-17-3p*), and E-selectin (by *miR-31*). In addition, *miR-17-3p* and *miR-31* can be induced by the exposure of TNF-α, thereby acting as a negative feedback control of the endothelial activation (Suárez et al. 2010). Another negative feedback control is *miR-146a*, which inhibits NF-κB signal transduction through targeting gene encoding for signal transducer proteins in NF-κB pathway, including interleukin-1 receptor-associated kinase (IRK1) and TNF receptor-associated factor 6 (TRAF6) (Taganov et al. 2006).

Sun et al. (2012) have recently reported that *miR-181b* is rapidly induced in response to TNF-α and that *miR-181b* reduces NF-κB-mediated EC activation and vascular inflammation by inhibiting importin-a3, a regulator of NF-κB nuclear import. They have also demonstrated that systemic administration of *miR-181b* "mimic" decreases NF-κB downstream signaling and leukocyte influx to the endothelium, resulting in decreased lung injury and mortality in mouse. Interestingly, data from our miRNA microarray (Prapansilp 2012) in the kidneys of patients with malaria showed a significant down-regulation of *miR-181* family including *miR-181a* and *miR-181c*. Although each of the four miRNAs in *miR-181* family (*miR-181a* to *d*) is encoded from a different transcript, all of them have identical seed sequence, therefore potentially targeting the same mRNA transcripts. These findings imply that modulation of miRNAs might have a potential role in malaria adjunctive therapy.

D. Central Nervous System (CNS) Injury in CM

Malaria infection disturbs cerebral function by complex mechanisms involving both local effects of PRBC sequestration and systemic immunological responses of the host, leading to BBB leakage and dysfunction of astroglial cells and neurons. Numerous pathological studies have shown that common neuropathological features of human CM include the presence of PRBC sequestered in the capillaries of the brain, scattered microscopic hemorrhages and perivascular edema, activation of macrophages in Virchow–Robin space, some degree of leukocytic aggregation, vasculitis, and thrombus formation (reviewed in Medana and Turner 2006). Acute axonal injury, detected by an accumulation of beta-amyloid precursor protein (βAPP) has also been described as a final common pathway to neurological impairment in CM (Medana et al. 2002; Dorovini-Zis et al. 2011).

miR-107 has been shown to target beta-site amyloid precursor protein-cleaving enzyme 1 (BACE1), and the reduction in the expression of *miR-107* is associated with Alzheimer-type pathologies, such as the accumulation of amyloid beta, through increased activity of BACE1 (Wang et al. 2008). The decreased *miR-107* and thus increased BACE1 mRNA expression were also observed in the brain of mouse models of traumatic brain injury and brain-injured humans (Blasko et al. 2004; Uryu et al. 2007; Wang et al. 2010). These findings not only link control of APP expression in brain injury and Alzheimer's disease but also suggest the need to examine the potential roles for miRNAs such as *miR-107* in the pathogenesis of coma in CM.

V. CONCLUSIONS AND FUTURE RESEARCH DIRECTIONS

A better understanding of the pathophysiology of severe malaria and its various complications is a key to developing new control measures and treatments, as well as to achieve the ambitious goal of eradicating the disease. miRNA has emerged as a crucial regulator of gene expression in a range of physiological and pathological processes. In cancer research, where miRNA has been extensively studied, the miRNA expression profile is now being considered as a potentially diagnostic and prognostic marker, as well as a starting point in the development of therapeutic interventions (reviewed in Waldman and Terzic 2008). In malaria, miRNA research is still very limited. One reason for this is the difficulty of obtaining human tissue for use in research. Tissue from malaria-infected individuals can only be collected at postmortem after the disease progresses to death, despite treatment. Biopsy of tissues in living malaria patients, which is a common practice in cancer research, is ethically questionable because it is too invasive and does not provide clinical information for diagnosis or treatment of malaria. Most malarial deaths also occur in Africa, where facilities to conduct research may be limited and cultural acceptance of postmortem examination is low.

Despite the significance of miRNA in the pathogenesis of severe malaria, and its importance as a research tool, miRNA profiling has limited clinical implications at present. It currently has no role as clinical diagnostic tool because malaria is effectively diagnosed by screening the presence of PRBCs with a blood film, testing for malarial antigens using rapid immunochromatographic strip assay or detecting parasite DNA using polymerase chain reaction (PCR). However, in the future, when more data are available and the cost of miRNA examination is cheaper, a panel of miRNAs might be useful as a clinical prognostic factor of severe diseases or for monitoring response to therapeutic interventions.

REFERENCES

Al-Quraishy, S., M.A. Dkhil, D. Delić, A.A. Abdel-Baki, and F. Wunderlich. 2012. Organ-specific testosterone-insensitive response of miRNA expression of C57BL/6 mice to *Plasmodium chabaudi* malaria. *Parasitol Res* **111**:1093–1101.

Angulo, I. and M. Fresno. 2002. Cytokines in the pathogenesis of and protection against malaria. *Clin Diagn Lab Immunol* **9**:1145–1152.

Blasko, I., R. Beer, M. Bigl, J. Apelt, G. Franz, D. Rudzki, G. Ransmayr, A. Kampfl, and R. Schliebs. 2004. Experimental traumatic brain injury in rats stimulates the expression, production and activity of Alzheimer's disease beta-secretase (BACE-1). *J Neural Transm* **111**:523–536.

Carter, R. and K.N. Mendis. 2002. Evolutionary and historical aspects of the burden of malaria. *Clin Microbiol Rev* **15**:564–594.

Clark, I.A., A.C. Budd, L.M. Alleva, and W.B. Cowden. 2006. Human malarial disease: a consequence of inflammatory cytokine release. *Malar J* **5**:85.

Coulson, R.M.R., N. Hall, and C.A. Ouzounis. 2004. Comparative genomics of transcriptional control in the human malaria parasite *Plasmodium falciparum*. *Genome Res* **14**:1548–1554.

Craig, A.G., G.E.R. Grau, C.J. Janse, J.W. Kazura, D.A. Milner, J.W. Barnwell, G.D.H. Turner, J. Langhorne, and on behalf of the participants of the Hinxton Retreat meeting on "Animal Models for Research on Severe Malaria." 2012. The role of animal models for research on severe malaria. *PLoS Pathog* **8**:e1002401.

Crooke, A., A. Diez, P.J. Mason, and J.M. Bautista. 2006. Transient silencing of *Plasmodium falciparum* bifunctional glucose-6-phosphate dehydrogenase-6-phosphogluconolactonase. *FEBS J* **273**:1537–1546.

Crosby, M.E., R. Kulshreshtha, M. Ivan, and P.M. Glazer. 2009. MicroRNA regulation of DNA repair gene expression in hypoxic stress. *Cancer Res* **69**:1221–1229.

David, P.H., M. Hommel, L.H. Miller, I.J. Udeinya, and L.D. Oligino. 1983. Parasite sequestration in *Plasmodium falciparum* malaria: spleen and antibody modulation of cytoadherence of infected erythrocytes. *Proc Natl Acad Sci U S A* **80**:5075–5079.

Delić, D., M. Dkhil, S. Al-Quraishy, and F. Wunderlich. 2010. Hepatic miRNA expression reprogrammed by *Plasmodium chabaudi* malaria. *Parasitol Res* **108**:1111–1121.

Djikeng, A., H. Shi, C. Tschudi, and E. Ullu. 2001. RNA interference in *Trypanosoma brucei*: cloning of small interfering RNAs provides evidence for retroposon-derived 24-26-nucleotide RNAs. *RNA* **7**:1522–1530.

Dondorp, A.M., P.A. Kager, J. Vreeken, and N.J. White. 2000. Abnormal blood flow and red blood cell deformability in severe malaria. *Parasitol Today* **16**:228–232.

Dondorp, A.M., S. Yeung, L. White, C. Nguon, N.P.J. Day, D. Socheat, and L. von Seidlein. 2010. Artemisinin resistance: current status and scenarios for containment. *Nat Rev Microbiol* **8**: 272–280.

Dorovini-Zis, K., K. Schmidt, H. Huynh, W.J. Fu, R.O. Whitten, D.A. Milner, S.B. Kamiza, M.E. Molyneux, and T.E. Taylor. 2011. The neuropathology of fatal cerebral malaria in Malawian children. *Am J Pathol* **178**:2146–2158.

El-Assaad, F., C. Hempel, V. Combes, A.J. Mitchell, H.J. Ball, J.A.L. Kurtzhals, N.H. Hunt, J.-M. Mathys, and G.E.R. Grau. 2011. Differential microRNA expression in experimental cerebral and noncerebral malaria. *Infect Immun* **79**:2379–2384.

Florens, L., M.P. Washburn, J.D. Raine, R.M. Anthony, M. Grainger, J.D. Haynes, J.K. Moch, N. Muster, J.B. Sacci, D.L. Tabb, A.A. Witney, D. Wolters, et al. 2002. A proteomic view of the *Plasmodium falciparum* life cycle. *Nature* **419**:520–526.

Francischetti, I.M.B. 2008. Does activation of the blood coagulation cascade have a role in malaria pathogenesis? *Trends Parasitol* **24**:258–263.

Gardner, M.J., N. Hall, E. Fung, O. White, M. Berriman, R.W. Hyman, J.M. Carlton, A. Pain, K.E. Nelson, S. Bowman, I.T. Paulsen, K. James, et al. 2002. Genome sequence of the human malaria parasite *Plasmodium falciparum*. *Nature* **419**:498–511.

Gissot, M., S. Briquet, P. Refour, C. Boschet, and C. Vaquero. 2005. PfMyb1, a *Plasmodium falciparum* transcription factor, is required for intra-erythrocytic growth and controls key genes for cell cycle regulation. *J Mol Biol* **346**:29–42.

REFERENCES

Hall, N., M. Karras, J.D. Raine, J.M. Carlton, T.W.A. Kooij, M. Berriman, L. Florens, C.S. Janssen, A. Pain, G.K. Christophides, K. James, K. Rutherford, et al. 2005. A comprehensive survey of the *Plasmodium* life cycle by genomic, transcriptomic, and proteomic analyses. *Science* **307**: 82–86.

Hempel, C., V. Combes, N.H. Hunt, J.A.L. Kurtzhals, and G.E.R. Grau. 2011. CNS hypoxia is more pronounced in murine cerebral than noncerebral malaria and is reversed by erythropoietin. *Am J Pathol* **179**:1939–1950.

Huang, X., L. Ding, K.L. Bennewith, R.T. Tong, S.M. Welford, K.K. Ang, M. Story, Q.-T. Le, and A.J. Giaccia. 2009. Hypoxia-inducible *mir-210* regulates normoxic gene expression involved in tumor initiation. *Mol Cell* **35**:856–867.

Huang, X., Q.-T. Le, and A.J. Giaccia. 2010. MiR-210—micromanager of the hypoxia pathway. *Trends Mol Med* **16**:230–237.

Hunt, N.H., J. Golenser, T. Chan-Ling, S.B. Parekh, C. Rae, S. Potter, I.M. Medana, J. Miu, and H.J. Ball. 2006. Immunopathogenesis of cerebral malaria. *Int J Parasitol* **36**:569–582.

Ivan, M., A.L. Harris, F. Martelli, and R. Kulshreshtha. 2008. Hypoxia response and microRNAs: no longer two separate worlds. *J Cell Mol Med* **12**:1426–1431.

Jopling, C.L., K.L. Norman, and P. Sarnow. 2006. Positive and negative modulation of viral and cellular mRNAs by liver-specific microRNA *miR-122*. *Cold Spring Harb Symp Quant Biol* **71**:369–376.

Kulshreshtha, R., M. Ferracin, S.E. Wojcik, R. Garzon, H. Alder, F.J. Agosto-Perez, R. Davuluri, C.G. Liu, C.M. Croce, M. Negrini, G.A. Calin, and M. Ivan. 2007. A microRNA signature of hypoxia. *Mol Cell Biol* **27**:1859–1867.

Lamikanra, A.A., M. Theron, T.W.A. Kooij, and D.J. Roberts. 2009. Hemozoin (malarial pigment) directly promotes apoptosis of erythroid precursors. *PLoS ONE* **4**:e8446.

LaMonte, G., N. Philip, J. Reardon, J.R. Lacsina, W. Majoros, L. Chapman, C.D. Thornburg, M.J. Telen, U. Ohler, C.V. Nicchitta, T. Haystead, and J.T. Chi. 2012. Translocation of sickle cell erythrocyte microRNAs into *Plasmodium falciparum* inhibits parasite translation and contributes to malaria resistance. *Cell Host Microbe* **12**:187–199.

Lee, K.S., P.C.S. Divis, S.K. Zakaria, A. Matusop, R.A. Julin, D.J. Conway, J. Cox-Singh, and B. Singh. 2011. *Plasmodium knowlesi*: reservoir hosts and tracking the emergence in humans and macaques. *PLoS Pathog* **7**:e1002015.

Lei, Z., B. Li, Z. Yang, H. Fang, G.M. Zhang, Z.H. Feng, and B. Huang. 2009. Regulation of HIF-1alpha and VEGF by *miR-20b* tunes tumor cells to adapt to the alteration of oxygen concentration. *PLoS ONE* **4**:e7629.

Lopez, J.A. 2010. Malignant malaria and microangiopathies: merging mechanisms. *Blood* **115**: 1317–1318.

Lovegrove, F.E., S.A. Gharib, S.N. Patel, C.A. Hawkes, K.C. Kain, and W.C. Liles. 2007. Expression microarray analysis implicates apoptosis and interferon-responsive mechanisms in susceptibility to experimental cerebral malaria. *Am J Pathol* **171**:1894–1903.

Lovegrove, F.E., N. Tangpukdee, R.O. Opoka, E.I. Lafferty, N. Rajwans, M. Hawkes, S. Krudsood, S. Looareesuwan, C.C. John, W.C. Liles, and K.C. Kain. 2009. Serum angiopoietin-1 and -2 levels discriminate cerebral malaria from uncomplicated malaria and predict clinical outcome in African children. *PLoS ONE* **4**:e4912.

Malhotra, P., P.V.N. Dasaradhi, A. Kumar, A. Mohmmed, N. Agrawal, R.K. Bhatnagar, and V.S. Chauhan. 2002. Double-stranded RNA-mediated gene silencing of cysteine proteases (falcipain-1 and -2) of *Plasmodium falciparum*. *Mol Microbiol* **45**:1245–1254.

McRobert, L. and G.A. McConkey. 2002. RNA interference (RNAi) inhibits growth of *Plasmodium falciparum*. *Mol Biochem Parasitol* **119**:273–278.

Mead, E. and Z. Tu. 2008. Cloning, characterization, and expression of microRNAs from the Asian malaria mosquito, *Anopheles stephensi*. *BMC Genomics* **9**:244.

Medana, I.M. and G.D.H. Turner. 2006. Human cerebral malaria and the blood-brain barrier. *Int J Parasitol* **36**:555–568.

Medana, I.M., N.P.J. Day, T.T. Hien, N.T.H. Mai, D.B. Bethell, N.H. Phu, J. Farrar, M.M. Esiri, N.J. White, and G.D.H. Turner. 2002. Axonal injury in cerebral malaria. *Am J Pathol* **160**:655–666.

Miller, L.H., D.I. Baruch, K. Marsh, and O.K. Doumbo. 2002. The pathogenic basis of malaria. *Nature* **415**:673–679.

Mohmmed, A., P.V.N. Dasaradhi, R.K. Bhatnagar, V.S. Chauhan, and P. Malhotra. 2003. In vivo gene silencing in *Plasmodium berghei*—a mouse malaria model. *Biochem Biophys Res Commun* **309**:506–511.

Murray, C.J.L., L.C. Rosenfeld, S.S. Lim, K.G. Andrews, K.J. Foreman, D. Haring, N. Fullman, M. Naghavi, R. Lozano, and A.D. Lopez. 2012. Global malaria mortality between 1980 and 2010: a systematic analysis. *Lancet* **379**:413–431.

Nag, S., T. Papneja, R. Venugopalan, and D.J. Stewart. 2005. Increased angiopoietin2 expression is associated with endothelial apoptosis and blood-brain barrier breakdown. *Lab Invest* **85**:1189–1198.

Nag, S., J. Manias, and D.J. Stewart. 2009. Pathology and new players in the pathogenesis of brain edema. *Acta Neuropathol* **118**:197–217.

Ngô, H., C. Tschudi, K. Gull, and E. Ullu. 1998. Double-stranded RNA induces mRNA degradation in *Trypanosoma brucei*. *Proc Natl Acad Sci U S A* **95**:14687–14692.

Noedl, H., Y. Se, K. Schaecher, B.L. Smith, D. Socheat, M.M. Fukuda, and Artemisinin Resistance in Cambodia 1 (ARC1) Study Consortium. 2008. Evidence of artemisinin-resistant malaria in western Cambodia. *N Engl J Med* **359**:2619–2620.

Otsuka, M., Q. Jing, P. Georgel, L. New, J. Chen, J. Mols, Y.J. Kang, Z. Jiang, X. Du, R. Cook, S.C. Das, A.K. Pattnaik, et al. 2007. Hypersusceptibility to vesicular stomatitis virus infection in dicer1-deficient mice is due to impaired miR24 and miR93 expression. *Immunity* **27**:123–134.

Pino, P., I. Vouldoukis, J.P. Kolb, N. Mahmoudi, I. Desportes-Livage, F. Bricaire, M. Danis, B. Dugas, and D. Mazier. 2003. *Plasmodium falciparum*-infected erythrocyte adhesion induces caspase activation and apoptosis in human endothelial cells. *J Infect Dis* **187**:1283–1290.

Prapansilp, P. 2012. Molecular pathological investigation of the pathophysiology of fatal malaria. D.Phil. Thesis. University of Oxford. UK.

Price, R.N., N.M. Douglas, and N.M. Anstey. 2009. New developments in *Plasmodium vivax* malaria: severe disease and the rise of chloroquine resistance. *Curr Opin Infect Dis* **22**:430–435.

Rane, S., M. He, D. Sayed, H. Vashistha, A. Malhotra, J. Sadoshima, D.E. Vatner, S.F. Vatner, and M. Abdellatif. 2009. Downregulation of *miR-199a* derepresses hypoxia-inducible factor-1alpha and Sirtuin 1 and recapitulates hypoxia preconditioning in cardiac myocytes. *Circ Res* **104**:879–886.

Rathjen, T., C. Nicol, G. McConkey, and T. Dalmay. 2006. Analysis of short RNAs in the malaria parasite and its red blood cell host. *FEBS Lett* **580**:5185–5188.

Suárez, Y., C. Wang, T.D. Manes, and J.S. Pober. 2010. Cutting edge: TNF-induced microRNAs regulate TNF-induced expression of E-selectin and intercellular adhesion molecule-1 on human endothelial cells: feedback control of inflammation. *J Immunol* **184**:21–25.

Sun, X., B. Icli, A.K. Wara, N. Belkin, S. He, L. Kobzik, G.M. Hunninghake, M.P. Vera, T.S. Blackwell, R.M. Baron, and M.W. Feinberg. 2012. MicroRNA-181b regulates NF-κB-mediated vascular inflammation. *J C Invest* **122**:1973–1990.

Taganov, K.D., M.P. Boldin, K.J. Chang, and D. Baltimore. 2006. NF-kappaB-dependent induction of microRNA *miR-146*, an inhibitor targeted to signaling proteins of innate immune responses. *Proc Natl Acad Sci U S A* **103**:12481–12486.

REFERENCES

Taguchi, A., K. Yanagisawa, M. Tanaka, K. Cao, Y. Matsuyama, H. Goto, and T. Takahashi. 2008. Identification of hypoxia-inducible factor-1 alpha as a novel target for *miR-17-92* microRNA cluster. *Cancer Res* **68**:5540–5545.

Tripathi, A.K., D.J. Sullivan, and M.F. Stins. 2007. *Plasmodium falciparum*-infected erythrocytes decrease the integrity of human blood-brain barrier endothelial cell monolayers. *J Infect Dis* **195**:942–950.

Tripathi, A.K., W. Sha, V. Shulaev, M.F. Stins, and D.J. Sullivan. 2009. *Plasmodium falciparum*-infected erythrocytes induce NF-kappaB regulated inflammatory pathways in human cerebral endothelium. *Blood* **114**:4243–4252.

Turner, G.D.H., V.C. Ly, T.H. Nguyen, T.H. Tran, H.P. Nguyen, D.B. Bethell, S. Wyllie, K. Louwrier, S.B. Fox, K.C. Gatter, N.P.J. Day, T.H. Tran, et al. 1998. Systemic endothelial activation occurs in both mild and severe malaria. Correlating dermal microvascular endothelial cell phenotype and soluble cell adhesion molecules with disease severity. *Am J Pathol* **152**:1477–1487.

Uryu, K., X.-H. Chen, D. Martinez, K.D. Browne, V.E. Johnson, D.I. Graham, V.M.Y. Lee, J.Q. Trojanowski, and D.H. Smith. 2007. Multiple proteins implicated in neurodegenerative diseases accumulate in axons after brain trauma in humans. *Exp Neurol* **208**:185–192.

Waldman, S.A. and A. Terzic. 2008. MicroRNA signatures as diagnostic and therapeutic targets. *Clin Chem* **54**:943–944.

Wang, Y. and C.G.L. Lee. 2009. MicroRNA and cancer—focus on apoptosis. *J Cell Mol Med* **13**: 12–23.

Wang, W.-X., B.W. Rajeev, A.J. Stromberg, N. Ren, G. Tang, Q. Huang, I. Rigoutsos, and P.T. Nelson. 2008. The expression of microRNA *miR-107* decreases early in Alzheimer's disease and may accelerate disease progression through regulation of beta-bite amyloid precursor protein-cleaving enzyme 1. *J Neurosci* **28**:1213–1223.

Wang, W.-X., B.R. Wilfred, S.K. Madathil, G. Tang, Y. Hu, J. Dimayuga, A.J. Stromberg, Q. Huang, K.E. Saatman, and P.T. Nelson. 2010. *miR-107* regulates granulin/progranulin with implications for traumatic brain injury and neurodegenerative disease. *Am J Pathol* **177**:334–345.

Wilson, N.O., M.B. Huang, W. Anderson, V. Bond, M. Powell, W.E. Thompson, H.B. Armah, A.A. Adjei, R. Gyasi, Y. Tettey, and J.K. Stiles. 2008. Soluble factors from *Plasmodium falciparum*-infected erythrocytes induce apoptosis in human brain vascular endothelial and neuroglia cells. *Mol Biochem Parasitol* **162**:172–176.

Winter, F., S. Edaye, A. Huttenhofer, and C. Brunel. 2007. *Anopheles gambiae* miRNAs as actors of defence reaction against *Plasmodium* invasion. *Nucleic Acids Res* **35**:6953–6962.

World Health Organization. 2012. World Malaria Report 2011. World Health Organization, Geneva, Switzerland.

Xue, X., Q. Zhang, Y. Huang, L. Feng, and W. Pan. 2008. No miRNA were found in *Plasmodium* and the ones identified in erythrocytes could not be correlated with infection. *Malar J* **7**:47.

Yeo, T.W., D.A. Lampah, R. Gitawati, E. Tjitra, E. Kenangalem, K. Piera, R.N. Price, S.B. Duffull, D.S. Celermajer, and N.M. Anstey. 2008. Angiopoietin-2 is associated with decreased endothelial nitric oxide and poor clinical outcome in severe falciparum malaria. *Proc Natl Acad Sci U S A* **105**:17097–17102.

PART III

CANCER

13

THE MicroRNA DECALOGUE OF CANCER INVOLVEMENT

Tanja Kunej,[1] Irena Godnic,[1] Minja Zorc,[1] Simon Horvat,[1,2] and George A. Calin[3]

[1]*Department of Animal Science, Biotechnical Faculty, University of Ljubljana, Domzale, Slovenia*
[2]*Department of Biotechnology, National Institute of Chemistry, Ljubljana, Slovenia*
[3]*Experimental Therapeutics Department and Center for RNA Interference and Non-Coding RNA, The University of Texas MD Anderson Cancer Center, Houston, TX, USA*

I.	Introduction	202
II.	The Decalogue of Principles of miRNA Involvement in Human Cancers	203
	1. miRNAs are a class of ncRNAs that are estimated to regulate expression of two-thirds of the mammalian genome by binding to promoter, coding and untranslated regions (UTRs), proteins, or other ncRNAs	203
	2. Each miRNA can regulate the expression of numerous target genes, and conversely, multiple miRNAs can regulate the same target gene, and this interplay is involved in regulation of various physiological processes and pathophysiology of many diseases, including all analyzed types of human cancers	205
	3. About half of mammalian miRNAs are intragenic, predominantly intronic, and on the same strand as their host genes, and can be coordinately expressed and functionally linked with them	205
	4. miRNAs are frequently located within cancer-associated genomic regions (CAGRs) and can act as tumor suppressors or oncogenes	207
	5. Genetic variations in miRNA genes and their precursors, target sites, and genes encoding components of the processing machinery can affect phenotypic variation and disease susceptibility	208

MicroRNAs in Medicine, First Edition. Edited by Charles H. Lawrie.
© 2014 John Wiley & Sons, Inc. Published 2014 by John Wiley & Sons, Inc.

6. Three types of epigenetic concepts have been associated with miRNAs: (A) miRNA-mediated gene silencing is one of the epigenetic mechanisms, (B) miRNAs can also directly control the epigenetic machinery with a subclass of miRNAs (epi-miRNAs), and (C) miRNA expression can be down-regulated via promoter hypermethylation   210
7. Aberrant miRNA gene expression signatures characterize cancer cells, and miRNA profiling can be applied in diagnosis, prognosis, and treatment in cancer patients   211
8. Circulating miRNAs are potential non-invasive biomarkers in cancer   212
9. RNA inhibition using miRNAs is a potential treatment method for specific types of cancer   212
10. Interplay between miRNAs, other ncRNAs, and protein-coding genes forms a complex network of interactions in normal and disease tissues   213
III. Conclusion   214
References   214

ABBREVIATIONS

ceRNA	competitive endogenous RNA
lincRNA	long intergenic non-coding RNA
lncRNA	long non-coding RNA
miRNA	microRNA
MREs	miRNA response elements
ncRNA	non-coding RNA
pre-miRNA	miRNA precursor
pri-miRNA	miRNA primary transcript
RISC	RNA-induced silencing complex
UTR	untranslated region

I. INTRODUCTION

MicroRNAs (miRNAs) are non-coding RNAs (ncRNAs) with gene regulatory functions involved in a variety of molecular functions and biological processes in multicellular organisms (reviewed in Bartel 2004; Fabian et al. 2010). Initially transcribed by RNA polymerase II as long, capped, and polyadenylated primary transcripts (pri-miRNAs), they are processed by the microprocessor protein complex, which contains Drosha, an RNase III endonuclease, and *DGCR8* (DiGeorge syndrome critical region gene 8) (also known as Pasha) (Denli et al. 2004; Gregory et al. 2004). First, Drosha, in conjunction with its binding partner DGCR8, processes pri-miRNAs into hairpin RNAs of 70–100 nucleotides (nt) in length known as precursor miRNAs (pre-miRNAs). Translocated from the nucleus to the cytoplasm by exportin 5 (*XPO5*), pre-miRNAs are processed by an RNase III endonuclease Dicer and TAR RNA-binding protein (TRBP), in a 19- to 24-nt-long duplex. Finally, the duplex interacts with the RNA-induced silencing complex (RISC), which includes proteins of the Argonaute family (EIF2C 1–4 in humans) (Hammond et al. 2001). One strand of the miRNA duplex remains stably associated with RISC and becomes the mature miRNA, which guides the RISC to target messenger RNAs (mRNAs). Recent evidence shows that miRNAs are able to up- or down-regulate target gene expression by binding to its different regions (reviewed in Kunej et al. 2012a).

II. THE DECALOGUE OF PRINCIPLES OF miRNA INVOLVEMENT IN HUMAN CANCERS

The development of high-throughput methods to detect miRNA expression in human samples has provided invaluable tools to investigate the role of miRNAs both in physiological and pathological conditions. Exponentially accumulated data in the last years clearly show that perturbations of miRNA genes at a DNA, RNA, or expression level play a critical role in cancer initiation and progression. Determining the miRNome, the whole set of miRNAs, specific for cancer would further contribute to better diagnosis and prognosis of human diseases. Such findings are important not only for scientists in general but also for clinicians and oncologists, as the field of ncRNA touches every aspect of human oncology. The main principles for understanding miRNA involvement in human cancers could be summarized in the following way:

1. **miRNAs are a class of ncRNAs that are estimated to regulate expression of two-thirds of the mammalian genome by binding to promoter, coding and untranslated regions (UTRs), proteins, or other ncRNAs.**

The initial discovery and prevailing dogma that miRNAs target the 3'-UTRs of mRNAs and down-regulate the expression of protein-coding genes in cytoplasm has recently been expanded in response to the following additional observations: (A) miRNAs can be localized in the nucleus (Hwang et al. 2007); (B) in addition to 3'-UTR, miRNAs target other genic regions at a DNA or RNA level (5'-UTR, promoter regions, coding regions) (Lytle et al. 2007; Forman et al. 2008; Place et al. 2008; Tay et al. 2008) and even proteins (Eiring et al. 2010); (C) miRNAs can up-regulate as well as down-regulate translation (Vasudevan et al. 2007; Place et al. 2008); and (D) miRNAs interact with other ncRNAs and various types of RNA transcripts in a "competing endogenous RNA" (ceRNA) hypothesis (Salmena et al. 2011) (Figure 13.1):

(A) miRNAs have initially been considered to be located only in the cytoplasm. However, recent studies have reported that most miRNAs found in the cytoplasm might also localize to or function in the nucleus (Liao et al. 2010). Hwang et al. (2007) showed that human *miR-29b* was significantly enriched in the nucleus; the hexanucleotide terminal motif of *miR-29b* was found to be required for nuclear localization (Hwang et al. 2007).

(B) It has been shown that miRNAs can also, at a DNA level, affect transcription by direct binding to promoters. For example, human *miR-373* binds to the E-cadherin (*CDH1*) promoter, which induces gene expression (Place et al. 2008). miRNA-dependent mRNA repression also occurs through binding sites located in mRNA coding sequences, as shown for miRNAs targeting Nanog homeobox (*Nanog*), POU class 5 homeobox 1 (*Pou5f1*), sex determining region Y-box 2 (*Sox2*) (Tay et al. 2008), and *DICER* (Forman et al. 2008). In addition to miRNA-mediated gene silencing through base pairing with DNA or mRNA target sequences, miRNAs also interfere with the function of regulatory proteins (decoy activity). In particular, *miR-328* binds to poly-C-binding protein 2 (PCBP2), alternatively known as heterogeneous ribonucleoprotein (hnRNP) E2. This binding does not involve the miRNA's seed region and prevents its interaction with the target

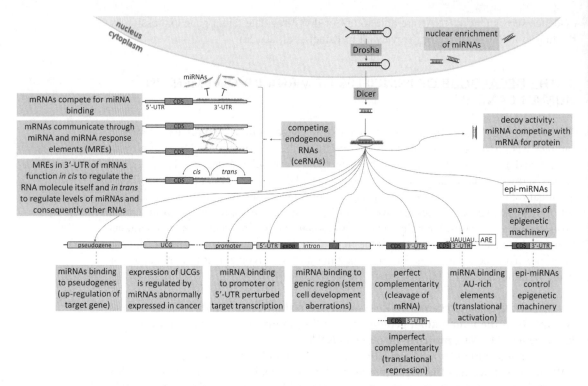

Figure 13.1. miRNAs can target and regulate different genic regions: promoters, 3'- and 5'-untranslated regions (UTRs), coding sequences (CDSs), AU-rich elements (AREs), other ncRNAs, including ultraconserved genes (UCGs), and proteins. Competing endogenous RNAs (ceRNAs) in this case represent an interacting activity between mRNAs and miRNAs.

mRNA. In chronic myelogenous leukemia, down-regulation of *miR-328* enables inhibition of myeloid differentiation with PCBP2, which leads to tumor progression (Eiring et al. 2010).

(C) It has also been shown that miRNAs up-regulate translation through binding to AU-rich elements (AREs) in cell-cycle arrested cells. For example, human *miR-369-3* targets AREs in the tumor necrosis factor (TNF) mRNA; during the cell cycle, while the cell was proliferating, the miRNA down-regulated translational activity, but upon cell-cycle arrest, it reversed its effect (Vasudevan et al. 2007). miRNAs can therefore both repress and activate gene expression at different points in the cell cycle.

(D) Other ncRNAs, such as non-coding ultraconserved genes (UCGs), have been found to be consistently altered at the genomic level in a high percentage of leukemias and carcinomas and may interact with miRNAs (Calin et al. 2007). Recently, the role of coding and non-coding RNAs has been emphasized and grouped in a unifying theory of ceRNAs that can regulate one another through their ability to compete for miRNA binding. The ceRNA hypothesis suggests that long non-coding RNAs (lncRNAs), transcribed pseudogenes, and mRNAs communicate with each other using miRNA response elements (MREs), which are sequences with partial complementarity to target mRNA transcripts (Salmena et al. 2011). ceRNAs may therefore be active partners in miRNA regulation, exerting effects on their expression levels, which may have important implications in pathological conditions, such as cancer.

2. Each miRNA can regulate the expression of numerous target genes, and conversely, multiple miRNAs can regulate the same target gene, and this interplay is involved in regulation of various physiological processes and pathophysiology of many diseases, including all analyzed types of human cancers.

It was estimated that about two-thirds of human protein-coding genes are considered as miRNA targets (Friedman et al. 2009b). Vertebrate miRNAs target about 200 transcripts each and more than one miRNA might coordinately regulate a single target, thereby providing a basis for complex networks (Krek et al. 2005). By specifically cleaving the homologous mRNAs, or by inhibiting protein synthesis, miRNAs are likely to target separate or multiple effectors of pathways involved in cell differentiation, proliferation, and survival. Several computational tools for miRNA target prediction have been developed (e.g., TargetScan, PicTar, miRanda); however, only a limited number of predicted targets were experimentally confirmed and are being collected in curated databases (e.g., miRecords, miRTarBase). Computational algorithms have been shown to predict miRNA targets with limited accuracy (Sethupathy et al. 2006), which represents a fundamental challenge in miRNA studies. The answer may lie in a systematic targetome evaluation, as described in a recent study where authors developed a transcriptome-wide identification of RNA-binding protein and miRNA target sites by photoactivatable ribonucleoside-enhanced cross-linking and immunoprecipitation (PAR-CLIP) (Hafner et al. 2010).

For several miRNAs, the involvement in essential biological processes has been demonstrated, such as B-cell lineage fate (*miR-181*) (Chen et al. 2004), B-cell survival (*miR-15a* and *miR-16-1*) (Calin et al. 2002), cell proliferation control (*miR-125B* and *let-7*) (Takamizawa et al. 2004; Lee et al. 2005), brain patterning (*miR-430*) (Giraldez et al. 2005), pancreatic cell insulin secretion (*miR-375*) (Poy et al. 2004), and adipocyte development (*miR-143*) (reviewed in Kunej et al. 2012b). miRNAs were found to be involved in the pathophysiology of all analyzed types of human tumors, including benign and malignant tumors. miRNAs differentially expressed between tumors and normal tissues have been identified in lymphoma, breast cancer, lung cancer, papillary thyroid carcinoma, glioblastoma, hepatocellular carcinoma, pancreatic tumors, pituitary adenomas, cervical cancer, brain tumors, prostate cancer, kidney and bladder cancers, and colorectal cancer (reviewed in Calin and Croce 2006a). Furthermore, miRNA alterations at the genic or expression levels were identified in many other human diseases, including schizophrenia and autoimmune or cardiac disorders (reviewed in Stahlhut Espinosa and Slack 2006; Ha 2011) (also see further chapters in this book).

3. About half of mammalian miRNAs are intragenic, predominantly intronic, and on the same strand as their host genes, and can be coordinately expressed and functionally linked with them.

Depending on the genomic position, miRNAs can be classified as intragenic and intergenic (Figure 13.2). It has been estimated that approximately half of vertebrate miRNAs are processed from introns of protein-coding and non-protein-coding transcripts (Erdmann et al. 2000; Rodriguez et al. 2004). A single host gene transcript can comprise multiple and overlapping resident miRNAs, called a cluster, which is processed from the same polycistronic primary transcript (Ambros 2004; Rodriguez et al. 2004). Intergenic miRNAs have their own transcriptional mechanisms, whereas intragenic, more specifically intronic miRNAs, can be cotranscribed with their host genes (Rodriguez et al. 2004;

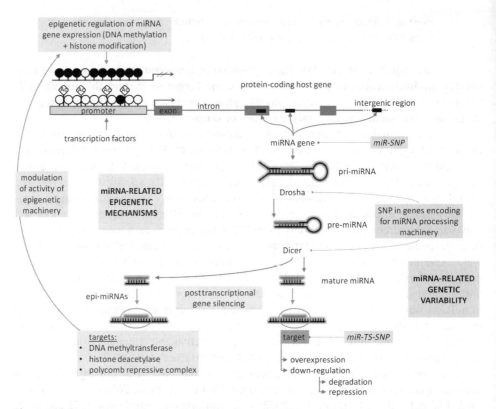

Figure 13.2. Interplay of mechanisms affecting miRNA biogenesis and function: genomic location of miRNA genes, polymorphisms, transcription factors, and epigenetic mechanisms. Ac, acetyl groups; *empty circles*, unmethylated CpG sites; *filled circles*, methylated CpG circles; SNP, single-nucleotide polymorphism; *miR-SNP*, SNP located within the miRNA gene; *miR-TS-SNP*, SNP located within miRNA target site.

Baskerville and Bartel 2005). However, there is also evidence that intronically encoded miRNAs are not just passively expressed as a part of host gene transcription, but RNA biogenesis can proceed before splicing catalysis, resulting in two separate mRNA and miRNA processing controls from a single primary transcript (Kim and Kim 2007). Recently, it was demonstrated that for conserved (evolutionary "older") intronic miRNAs the classical view of coexpression with the host gene holds true, but for non-conserved (phylogenetically "younger") intronic miRNAs, only rarely so, suggesting that we should be cautious in extrapolating host gene expression profiles for these "younger" intronic miRNAs (He et al. 2012). Because expression profiles of intronic miRNAs in many cases coincided with the transcription profiles of their host genes, this raised a question as to how these miRNAs were processed and coordinately regulated (Baskerville and Bartel 2005). Intronic miRNAs, like most ncRNAs, are released from the excised host introns in the postsplicing process (Kim and Kim 2007; Rearick et al. 2011), but it was later indicated that intronic miRNAs might also be processed from unspliced intronic regions prior to splicing catalysis (Kim and Kim 2007). A separate class of miRNA precursors named mirtrons is processed in an alternative pathway for miRNA biogenesis: Certain debranched introns mimic the structural features of pre-miRNAs to enter the miRNA processing pathway, however, without the Drosha-mediated cleavage (Ruby et al. 2007). While initially, mirtrons were determined in invertebrates, recent experiments provide evidence that mammalian miRNAs too can have mirtronic origin, presenting a biologically active class of splicing-dependent miRNAs (Sibley et al. 2012). Moreover, mirtrons appear to be very susceptible to epigenetic regulation, and hypermethylation was found to be commonly associated with urothelial cell carcinoma (Dudziec et al. 2011). It was observed that E2F transcription factor 1 (E2F1) regulates both an intronic miRNA cluster (*miR-106b-25*) and its host gene minichromosome maintenance complex component 7 (*MCM7*) and induces their accumulation in gastric primary tumors (Petrocca et al. 2008). Another study reported coordinate expression of *miR-218* and its host gene slit homologue 2 (*Drosophila*) (*SLIT2*), and between *miR-224* and its host gene *GABRE* (gamma-aminobutyric acid [GABA] A receptor, epsilon), in the clear cell renal cell carcinoma (White et al. 2011). miRNA genes located in closely linked clusters exhibit highly correlated expression patterns (Sempere et al. 2004; Baskerville and Bartel 2005). Cotranscription and correlated expression pattern of host genes and their resident miRNAs strongly support their transcriptional coregulation.

In addition to coexpression and proposed coregulation of miRNA and host genes, several studies have described a functional link between them in cells. Different roles have been attributed to intronic miRNAs, from providing a negative feedback regulatory mechanism with their host genes (Li et al. 2007), targeting genes that are functionally antagonistic to their host genes (Barik 2008), or having their host genes as targets of other miRNAs (Ballabio et al. 2010). For example, *miR-126* was found to regulate expression of its host gene EGF-like domain, multiple 7 (*EGFL7*) in a negative feedback loop (Fish et al. 2008). Intronically encoded *miR-342* was found to be down-regulated and to play a role in Sézary syndrome pathogenesis by inhibiting apoptosis and was regulated by binding of *miR-199a-1* to its host gene Enah/Vasp-like (*EVL*) (Ballabio et al. 2010). The association between introns and resident ncRNAs was also considered to have a synergistic effect with important implications for fine-tuning gene expression patterns in the genome (Rearick et al. 2011). Both *miR-126* and its host gene *EGFL7* are associated with vascular abnormalities (Wang et al. 2008) and are also epigenetically regulated in human cancer cells (Saito et al. 2009). Therefore, genomic location, especially of the intragenic miRNAs, strongly influences their coexpression regulation and often also functionally links miRNAs with their host genes.

4. miRNAs are frequently located within cancer-associated genomic regions (CAGRs) and can act as tumor suppressors or oncogenes.

The role of miRNAs in cancer was proposed early in the history of miRNA research by three important observations: miRNA genes are not randomly distributed in the genome, but are frequently located at fragile sites and CAGRs (Calin et al. 2004); miRNAs are involved in cell proliferation and apoptosis (Lee et al. 1993; Brennecke et al. 2003); and miRNA expression is deregulated in malignant tumors and tumor cell lines in comparison with normal tissues (Lu et al. 2005a; Calin and Croce 2006b; Gaur et al. 2007).

miRNAs can function as oncogenes, by activating malignant potential, or as tumor suppressors, by blocking the cell's malignant potential, and are therefore referred to as oncomiRs. Components required for miRNA biogenesis have also been associated with various cancers (reviewed in Esquela-Kerscher and Slack 2006). Direct evidence of miRNA oncogenic activity was reported when *miR-15a* and *miR-16-1* were found deleted or down-regulated in most chronic lymphocytic leukemia (CLL) patients (Calin et al.

2002). Follow-up studies reported that miRNAs can act by various mechanisms as oncogenes such as *miR-21* (Medina et al. 2010) or *miR-155* (Costinean et al. 2006), for which the transgenic mice models developed acute B-cell leukemias, or as tumor suppressors such as the *miR-15a/16-1* cluster, for which the knockout (KO) mice produced and developed CLL (reviewed in Croce 2009). In some instances, the same miRNA can act as an oncogene in one type of cells and as a suppressor in another due to different targets and mechanisms of action. For example, *miR-222* is overexpressed in liver cancers where it targets suppressor phosphatase and tensin homologue (*PTEN*) (Garofalo et al. 2009), while the same miRNA is down-regulated in erythroblastic leukemias where it targets oncogene v-kit Hardy–Zuckerman 4 feline sarcoma viral oncogene homologue (*KIT*) (Felli et al. 2005). Therefore, miRNAs can function both as oncogenes or tumor suppressors, but their mode of action in various cancers cannot be predicted as they function differentially depending on the cell or tissue type.

Modulation of miRNA biogenesis pathway can promote tumorigenesis through increased repression of tumor suppressors and/or through incomplete repression of oncogenes. It was indicated that the dysregulation of a single oncogenic miRNA can lead to the development of a malignant tumor. Such alterations can be a result of various mechanisms, such as deletions, amplifications, or mutations involving miRNA loci, and also epigenetic modifications and dysregulation of transcription factors (TFs), which target specific miRNAs (reviewed in Croce 2009) (Figure 13.2). In several cases, miRNAs have been shown to affect all hallmarks of malignant cells: (1) self-sufficiency in growth signals (*let-7* family), (2) insensitivity to antigrowth signals (*miR-17-92* cluster), (3) evasion from apoptosis (*miR-34a*), (4) limitless replicative potential (*miR-372/373* cluster), (5) angiogenesis (*miR-210*), and (6) invasion and metastases (*miR-10b*) (reviewed in Santarpia et al. 2010). Based on this, it is possible to propose that miRNAs are master regulators of tumor biology features, and their deregulation can therefore contribute to oncogenesis in various ways.

5. Genetic variations in miRNA genes and their precursors, target sites, and genes encoding components of the processing machinery can affect phenotypic variation and disease susceptibility.

Genetic variations have been shown to affect miRNA-mediated gene regulation. Genetic variations such as single-nucleotide polymorphisms (SNPs) and copy number variants (CNVs) within pri-miRNAs and genes encoding silencing components contribute to phenotypic variations, including disease susceptibility (reviewed in Georges et al. 2007). Germ line mutations or SNPs can occur in miRNA precursors, their target sites, and miRNA processing machinery (Figure 13.2) and have been found to affect miRNA-mediated regulatory functions, which can lead to phenotypic effects including activating cancerogenesis.

Germ line and somatic mutations in active mature, precursor, or primary miRNA molecules contribute to cancer predisposition and initiation, as was observed in *miR-15a/16* cluster mutations in rare families with CLL and breast cancer (Calin et al. 2005). Even though these mutations are rare, a similar mutation was observed in the New Zealand Black (NZB) strain of mice that were susceptible to the development of CLL late in life. Researchers identified a point mutation 6 nt downstream from the identical miRNA, murine *miR-16*, whose levels of expression were decreased in NZB lymphoid tissue. When delivered to an NZB malignant B-1 cell line, exogenous *miR-16* resulted in cell-cycle alterations and increased apoptosis (Raveche et al. 2007). Taken together, these two studies (one of human CLL and the other of a murine model of human indolent CLL) indicate that *miR-16* is the first miRNA proven to be involved in CLL predisposition and cancer predisposition in general.

Polymorphisms occurring in miRNA genes are referred to as *miR-SNPs*, and the term *miR-TS-SNP* is used for SNPs located within miRNA target-binding sites (Sun et al. 2009). Even though many miRNA sequence variations observed in cancer have altered the secondary structure with no demonstrated effects on miRNA processing (Diederichs and Haber 2006), several recent reports show that *miR-SNPs* can be associated with cancer susceptibility (Shen et al. 2008; Hu et al. 2009; Tian et al. 2009). It was observed that *miR-SNPs* affect function by modulating the miRNA precursor transcription, processing, and maturation (Zeng and Cullen 2005), or miRNA–mRNA interaction (Johnson et al. 2005). Sequence variations in the mature miRNA, especially in the seed region, may have an effect on miRNA target recognition (Sun et al. 2009) and can have an effect on a diverse array of traits (Zorc et al. 2012). A catalogue of genetic variations residing within miRNA seed region (*miR-seed-SNPs*) has been generated and will serve researchers as a starting point in testing more targeted hypothesis (Zorc et al. 2012). Because of the miRNA–target interaction, the *miR-SNPs* (including *miR-seed-SNPs*) and *miR-TS-SNPs* function in the same manner to create or destroy miRNA-binding sites.

Furthermore, polymorphisms in protein-coding mRNAs, which are targeted by miRNAs, can also influence cancer risk. For example, the *let-7* complementary SNP site in the *KRAS* 3′-UTR was found significantly associated with an increased risk for non-small-cell lung carcinoma among moderate smokers (Chin et al. 2008). Polymorphisms in miRNA-binding sites are also associated with other diseases; for example, a point mutation was identified in Tourette's syndrome patients in the 3′-UTR of *SLITRK1* (SLIT and NTRK-like family, member 1), disrupting the binding of *miR-189* (Abelson et al. 2005). Catalogues of SNPs residing within miRNA-binding regions of cancer genes have also been compiled (Landi et al. 2008; reviewed in Ryan et al. 2010). Even though *miR-TS-SNPs* were shown to influence susceptibility to tumorigenesis, additional association studies and follow-up functional experiments should still be applied to provide a clearer view on the interplay of these variations in disease development (Nicoloso et al. 2010). Indeed, there are cases of *miR-TS-SNPs* previously associated with the risk of cancer that could later not be confirmed in detailed functional studies, but a growing number of functionally supported cases (reviewed in Landi et al. 2012) corroborate the view that genetic variation in miRNA target-binding sites can affect cancer risk, treatment efficacy, and patient prognosis.

SNPs in the miRNA processing machinery are another class of polymorphisms with a potentially profound effect on the phenotype. They can have deleterious effects on the miRNome and global repression of miRNA maturation, shown to lead to tumorigenesis (Kumar et al. 2007). Several studies reported that such genetic polymorphisms affect cancer susceptibility: SNPs in gem (nuclear organelle)-associated protein 4 (*GEMIN4*) were significantly associated with altered renal cell carcinoma (Horikawa et al. 2008) and bladder cancer risk (Yang et al. 2008), whereas a SNP in the 3′-UTR of *DICER1* was associated with an increased risk of premalignant oral lesions in individuals with leukoplakia and/or erythroplakia (Clague et al. 2010). In a recent study, Sung et al. (2012) found SNPs located within eukaryotic translation initiation factor 2C, 2 (*EIF2C2*), *DICER1*, PIWI-like 1 (*Drosophila*) (*PIWIL1*), *DGCR8*, *DROSHA*, and *GEMIN4* that were associated with breast cancer survival.

The essential part in miRNA variation studies is identification of SNPs, which can be aided with bioinformatic tools by intercalating and cross-referencing data from the

Single Nucleotide Polymorphism Database (dbSNP). miRNA variation identification has improved the understanding of the disease complexity, and provided SNPs as genetic markers of increased cancer susceptibility as well as biomarkers of cancer type, outcome, and response to therapy (reviewed in Pelletier and Weidhaas 2010). Additionally, large-scale and high-throughput new generation sequencing (i.e., RNA-seq) will allow the identification of miRNA somatic mutations and expression alterations in a large number of patients and cancer types. This will eventually sort out which miRNA mutations (SNPs, insertions, deletions, amplifications) are most frequent in a particular cancer type or stage of cancerogenesis, and how that affects global miRNA and target gene expression profiles. Such genome-wide understanding of miRNA somatic mutations and expression changes they cause are likely to identify novel and previously unappreciated mechanisms in oncogenesis.

6. Three types of epigenetic concepts have been associated with miRNAs: (A) miRNA-mediated gene silencing is one of the epigenetic mechanisms, (B) miRNAs can also directly control the epigenetic machinery with a subclass of miRNAs (epi-miRNAs), and (C) miRNA expression can be down-regulated via promoter hypermethylation.

miRNAs have been found associated with several epigenetic mechanisms (Figure 13.2):

(A) miRNAs have the ability to regulate gene expression at a posttranscriptional level, and miRNA-mediated regulation is considered as one of the classes of epigenetic mechanisms (reviewed in Sharma et al. 2010). Together with other epigenetic mechanisms like promoter DNA methylation and histone modifications, miRNAs are involved in an interacting network of epigenetic regulation of gene expression (reviewed in Chuang and Jones 2007; Esteller 2008).

(B) A subclass of miRNAs (epi-miRNAs) directly controls the epigenetic machinery through a regulatory loop by targeting its regulating enzymes. The first epi-miRNAs identified were the *miR-29* family (*miR-29a*, *miR-29b*, and *miR-29c*), which are shown to directly target and down-regulate *de novo* DNA methyltransferases (*DNMT3A* and *DNMT3B*) and indirectly target *DNMT1* in lung cancer (Fabbri et al. 2007) and acute myeloid leukemia (Garzon et al. 2009). This led to demethylation of CpG islands in promoter regions of tumor suppressor genes, allowing their reactivation and a loss of the cell's tumorigenicity (Fabbri et al. 2007). It was also reported that *miR-449a* targets histone deacetylase 1 (*HDAC1*), which is frequently overexpressed in many types of cancer, and, for example, induces growth arrest in prostate cancer (Noonan et al. 2009). *miR-101* was shown to directly modulate the expression of enhancer of zeste homologue 2 (EZH2), a catalytic subunit of the polycomb repressive complex 2 (PRC2), which mediates epigenetic silencing of tumor suppressor genes in cancer (Friedman et al. 2009a).

(C) Expression of miRNA genes has also been found silenced in human tumors by epigenetic mechanisms, such as aberrant hypermethylation of CpG islands encompassing or in proximity of miRNA genes, and/or by histone acetylation (Weber et al. 2007). The first evidence of deregulated miRNA expression in cancer due to an altered methylation status was reported for *miR-127*, embedded within a CpG island promoter, which was silenced in several cancer cells, but strongly up-regulated after treatment with a hypomethylating agent (DNMT inhibitor 5-aza-2′-deoxycytidine) (Saito et al. 2006). A similar scenario was observed with *miR-124a*, whose function can be restored by erasing DNA methylation, and has functional consequences on cyclin D kinase 6 (CDK6) activity (Lujambio et al. 2007). On the other hand, Brueckner et al. (2007) observed that hypomethylation facilitated reactivation of *let-7a-3* and elevated its expression in human lung cancer cell lines, which resulted in enhanced tumor phenotypes. Compared with protein-coding genes, human oncomiRs were found to have an order of magnitude higher methylation frequency (Weber et al. 2007; Kunej et al. 2011). Epigenetically regulated miRNAs have been found to be associated with various types of cancer and also present cancer-specific biomarker potential (Kunej et al. 2011). Future studies of epigenetic regulation of miRNA expression together with downstream signaling pathways are likely to lead to development of novel drug targets in cancer therapy.

7. Aberrant miRNA gene expression signatures characterize cancer cells, and miRNA profiling can be applied in diagnosis, prognosis, and treatment in cancer patients.

Profiling of miRNA transcriptome is used to document typical expression signatures of a particular cell or tissue and to identify variability in disease states, such as cancer. The main mechanism of miRNome alterations in cancer cells is represented by aberrant gene expression, characterized by abnormal levels of expression for mature and/or pre-miRNA sequences in comparison with the corresponding normal tissues. It was shown that miRNome signatures and aberrations from a wild-type signature provide a more accurate diagnostic tool for cancer classification than the transcriptome of protein-coding genes (Lu et al. 2005b; reviewed in Calin and Croce 2006b). Deciphering the miRNome expression in normal and diseased states will be useful for the identification of miRNA targets, and alterations in miRNA expression patterns may disclose new pathogenic pathways in human tumorigenesis (Liu et al. 2004). Changes in miRNA expression pattern can be a consequence of mechanisms that can act independently or in combination, such as the location of miRNAs at CAGR, epigenetic regulation of miRNA expression, and abnormalities in miRNA processing genes and proteins, including mutations in *DICER1*, *TRBP*, or *XPO5*. In cancer, the loss of tumor suppressor miRNAs enhances the expression of target oncogenes, whereas increased expression of oncogenic miRNAs represses target tumor suppressor genes. Paired expression profiles of miRNAs and mRNAs can be used to identify functional miRNA–target relationships with high precision (Huang et al. 2007). The aberrant expression of miRNAs in cancer is characterized by abnormal levels of expression for mature and/or pre-miRNA transcripts in comparison with those in the corresponding normal tissue. Lu et al. (2005b) observed a general down-regulation of miRNAs in tumor samples compared with normal tissue samples. It was also found that miRNA expression profiles could be used to differentiate human cancers according to their developmental origin, with cancers of epithelial and hematopoietic origin having distinct miRNA profiles (Lu et al. 2005b). Therefore, determining the miRNA transcriptome profile in cancer cells or tissues and its deviations from expression signatures of normal cells or tissues is informative in identifying cancer subtypes and possibly even causal variability.

miRNA profiling achieved by various methods has allowed the identification of signatures associated with diagnosis, staging, progression, prognosis, and response to treatment of human tumors (reviewed in Ferdin et al. 2010). For example, the miRNA-based classifier is much better in establishing the correct diagnosis of metastatic cancer of unknown primary site than the classifier based on mRNAs of coding genes. As miRNA expression levels and tissue distribution pattern change with differentiation, the poorly differentiated tumors have lower global expression levels of miRNAs compared with well-differentiated tumors from control groups (Lu et al. 2005b). Because reduced expression levels of miRNAs present a hallmark in poorly differentiated tumors, miRNA profiling can therefore present an effective tool in the diagnosis of cancer of unknown primary site.

Profiling of miRNA expression correlates well with clinical and biological characteristics of tumors and has enabled the identification of signatures associated with diagnosis, staging, progression, prognosis, and response to treatment of human tumors (reviewed in Barbarotto et al. 2008). Profiling miRNA transcriptome of cancer cells or tissues therefore provides new insights of basic research interest as well as a novel fingerprinting tool to aid clinical oncology diagnostics.

8. Circulating miRNAs are potential non-invasive biomarkers in cancer.

miRNAs have been found to not only function within cells but can also act at neighboring cells and more distant sites within the body. The abundant circulating miRNAs have also been regarded having a potential biological role as extracellular messengers mediating short- and long-range cell–cell communication, as was observed with small RNAs in plants and *Caenorhabditis elegans* (reviewed in Mitchell et al. 2008). The measurement of miRNAs in body fluids, including plasma and serum, has been used to distinguish cancer patients from healthy subjects (Lawrie et al. 2008). Since deregulated miRNA expression is an early indicator in tumorigenesis, circulating miRNAs can be used for cancer detection and therefore represent a gold mine for non-invasive biomarkers in cancer (reviewed in Cortez et al. 2011). Biomarker potential of serum miRNAs relies mainly on their high stability and resistance to storage handling; they remain stable after being subjected to severe conditions that would normally degrade most RNAs, such as boiling, very low or high pH levels, extended storage, and 10 freeze–thaw cycles (Chen et al. 2008). Chen et al. (2008) identified expression patterns of serum miRNAs that were specific for lung and colorectal cancers, and diabetes, which provides evidence that serum miRNAs contain fingerprints for various diseases (Chen et al. 2008). Correlation between circulating miRNA levels and response to a given anticancer agent was also observed and may be useful in predicting patterns of resistance and sensitivity to drugs used in cancer treatment. This was shown in the case of serum *miR-21* levels that were higher in hormone refractory prostate cancer patients, whose disease was resistant to docetaxel-based chemotherapy, when compared with those with chemosensitive disease (Zhang et al. 2011). Additional and more detailed investigations of the types and levels of circulating miRNAs, and comparisons between cancer stages, types of cancer, and treatments may provide novel and more precise clinical laboratory cancer biomarkers with established reference intervals to allow appropriate clinical interpretation and therapy.

9. RNA inhibition using miRNAs is a potential treatment method for specific types of cancer.

RNA inhibition by using miRNAs, although not the "universal panacea" for any type of cancer, could represent valid options for the treatment of specific patients in the near future. These patients should have a concordant expression between a specific miRNA and the experimentally proven targets. RNA inhibition can be used to treat cancer patients in two ways: (1) by using RNA or DNA molecules as therapeutic drugs against mRNA of genes involved in the pathogenesis of cancers, and (2) by directly targeting ncRNAs that participate in cancer pathogenesis (reviewed in Spizzo et al. 2009). There are two advantages of using miRNAs: First, miRNAs are naturally occurring in human cells (by difference to chemotherapies or antisense oligonucleotides), and second, miRNAs target multiple genes from the same pathway and therefore the action occurs at multiple levels in the same pathway (e.g., *miR-16* targets both antiapoptotic genes B-cell CLL/lymphoma 2 [*BCL2*] [Cimmino et al. 2005] and myeloid cell leukemia sequence 1 [BCL2-related] [*MCL1*] [Calin et al. 2008]). On the other hand, approaches for sequence-specific inhibition with miRNAs in tumors address several difficulties, such as target specificity and delivery efficiency. miRNA-mediated therapy may lead to unwanted gene silencing (off-target effect). Delivery of the therapeutic miRNAs to the target tissue without compromising the integrity of the miRNA remains challenging (reviewed in Akhtar and Benter 2007; Castanotto and Rossi 2009; Whitehead et al. 2009). Current strategies for miRNA-based delivery use antisense oligonucleotides such as antagomirs, locked nucleic acid (LNA) anti-miR constructs, miRNA sponges, miR-masks, to block the oncogenic miRNAs, and synthetic miRNA mimics to restore miRNA expression (reviewed in Garzon et al. 2010). Appropriate target gene selection and therapeutic molecule design are crucial for efficient therapeutic design.

Virus-mediated delivery of *miR-26a*, which is normally expressed at high levels in diverse tissues but reduced in hepatocellular carcinoma cells, for liver cancer treatment in the mouse model resulted in inhibition of cancer cell proliferation, induction of tumor-specific apoptosis, and protection from disease progression (Kota et al. 2009). Exogenous delivery of synthetic *let-7* miRNA to established tumors significantly reduced tumor growth in mouse models of lung cancer (Trang et al. 2010). Therapeutic delivery of *miR-34a* mimic strongly inhibited cancer cell growth in mouse models for prostate (Liu et al. 2011) and lung cancers (Wiggins et al. 2010). Virus-mediated delivery of *miR-145* combined with 5-fluorouracil (5-FU) showed significant inhibition of tumor growth in breast tumor-bearing mice (Kim et al. 2011). Intravenous administration of antagomirs against *miR-16*, *miR-122*, *miR-192*, and *miR-194* effectively inhibited corresponding miRNA levels in the liver, lungs, kidneys, heart, intestine, fat, skin, bone marrow, muscle, ovaries, and adrenals (Krützfeldt et al. 2005). *miR-10b* silencing did not inhibit the growth of the primary tumor but drastically decreased the number of pulmonary metastases (Ma et al. 2010). The results of miRNA inhibition studies suggest great potential for miRNAs as a powerful tool for gene regulation research and therapeutic intervention.

10. Interplay between miRNAs, other ncRNAs, and protein-coding genes forms a complex network of interactions in normal and disease tissues.

Despite the leading role of miRNAs as cancer-related ncRNAs in published research, recently new categories of nontranslated RNAs have emerged. Other ncRNAs such as lncRNAs, including long intergenic non-coding RNAs (lincRNAs) (Gupta et al. 2010) and UCGs (Calin et al. 2007), were found to be abnormally expressed in cancer and involved in tumorigenic mechanisms. As the spectrum of ncRNAs is much larger than that of miRNAs (the estimates are as high as 1,000,000 ncRNA transcripts vs. as many as 10,000 potential miRNAs), it will have a profound impact on any aspect of basic and translational cancer research.

Recent analyses of miRNA genes, their targets, and genes encoding for processing machinery, genetic polymorphisms, and epigenetic modifications revealed that miRNA-mediated regulation in gene regulatory networks involves a far more complex system than initially expected. miRNA genes are linked with TFs in complex regulatory networks where they reciprocally regulate one another (Yu et al. 2008). It has been shown that tumor-associated transcribed ultraconserved regions (T-UCRs) in leukemias are negatively regulated by direct interaction with miRNAs (Calin et al. 2007). The ceRNA hypothesis asserts that RNA transcripts can indirectly regulate each other by competing for binding to miRNAs (Salmena et al. 2011). An example for ceRNA activity was described for tumor suppressor *PTEN* and its highly homologous transcript from phosphatase and tensin homologue pseudogene 1 (*PTENP1*) (Poliseno et al. 2010). The pseudogene transcript can compete with *PTEN* mRNA for miRNA binding and thereby modulate the expression of *PTEN*. Additionally, recent evidence implicates the interconnection of miRNAs and epigenetics. A subclass of miRNAs (epi-miRNAs) directly controls the epigenetic machinery through a regulatory loop by targeting key enzymes involved in establishing epigenetic memory (reviewed in Chuang and Jones 2007). Finally, the miRNA decoy functions have an effect on therapeutic approaches in human diseases, which include specific ways to overcome resistance to drug therapy and design of miRNA-based clinical trials in the future (reviewed in Almeida et al. 2012). The broken interactions within the complex network of miRNA regulation may lead to great disruption in the cell, possibly leading to cancerous phenotypes.

III. CONCLUSION

There can now be no more doubt that miRNAs are involved in the regulation of tumorigenic pathways involved in tumor initiation, development, progression, and dissemination. The important question of whether miRNAs represent the "dark side" of cancer predisposition has now started to be addressed by studies in large populations of cancer patients. miRNAs are identified as significant new diagnostic and prognostic tools for cancer patients, and as a consequence, miRNA-based cancer therapy should represent a bona fide option for medical oncologists in the not too distant future. A progressively increasing understanding of the implications of ncRNAs for the malignant phenotype represents the essential background necessary to achieve the ultimate goal of earlier detection and more effective treatment for cancer patients.

REFERENCES

Abelson, J.F., K.Y. Kwan, B.J. O'Roak, D.Y. Baek, A.A. Stillman, T.M. Morgan, C.A. Mathews, D.L. Pauls, M.-R. Rašin, M. Gunel, N.R. Davis, A.G. Ercan-Sencicek, et al. 2005. Sequence variants in SLITRK1 are associated with Tourette's syndrome. *Science* **310**:317–320.

Akhtar, S. and I.F. Benter. 2007. Nonviral delivery of synthetic siRNAs in vivo. *J Clin Invest* **117**:3623–3632.

Almeida, M.I., R.M. Reis, and G.A. Calin. 2012. Decoy activity through microRNAs: the therapeutic implications. *Expert Opin Biol Ther* **12**:1153–1159.

Ambros, V. 2004. The functions of animal microRNAs. *Nature* **431**:350–355.

Ballabio, E., T. Mitchell, M.S. van Kester, S. Taylor, H.M. Dunlop, J. Chi, I. Tosi, M.H. Vermeer, D. Tramonti, N.J. Saunders, J. Boultwood, J.S. Wainscoat, et al. 2010. MicroRNA expression in Sezary syndrome: identification, function, and diagnostic potential. *Blood* **116**:1105–1113.

Barbarotto, E., T.D. Schmittgen, and G.A. Calin. 2008. MicroRNAs and cancer: profile, profile, profile. *Int J Cancer* **122**:969–977.

REFERENCES

Barik, S. 2008. An intronic microRNA silences genes that are functionally antagonistic to its host gene. *Nucleic Acids Res* **36**:5232–5241.

Bartel, D.P. 2004. MicroRNAs: genomics, biogenesis, mechanism, and function. *Cell* **116**: 281–297.

Baskerville, S. and D. Bartel. 2005. Microarray profiling of microRNAs reveals frequent coexpression with neighboring miRNAs and host genes. *RNA* **11**:241–247.

Brennecke, J., D.R. Hipfner, A. Stark, R.B. Russell, and S.M. Cohen. 2003. bantam encodes a developmentally regulated microRNA that controls cell proliferation and regulates the proapoptotic gene hid in *Drosophila*. *Cell* **113**:25–36.

Brueckner, B., C. Stresemann, R. Kuner, C. Mund, T. Musch, M. Meister, H. Sültmann, and F. Lyko. 2007. The human let-7a-3 locus contains an epigenetically regulated microRNA gene with oncogenic function. *Cancer Res* **67**:1419–1423.

Calin, G.A. and C.M. Croce. 2006a. MicroRNA-cancer connection: the beginning of a new tale. *Cancer Res* **66**:7390–7394.

Calin, G.A. and C.M. Croce. 2006b. MicroRNA signatures in human cancers. *Nat Rev Cancer* **6**:857–866.

Calin, G.A., C.D. Dumitru, M. Shimizu, R. Bichi, S. Zupo, E. Noch, H. Aldler, S. Rattan, M. Keating, K. Rai, L. Rassenti, T. Kipps, et al. 2002. Frequent deletions and down-regulation of micro-RNA genes miR15 and miR16 at 13q14 in chronic lymphocytic leukemia. *Proc Natl Acad Sci U S A* **99**:15524–15529.

Calin, G.A., C. Sevignani, C.D. Dumitru, T. Hyslop, E. Noch, S. Yendamuri, M. Shimizu, S. Rattan, F. Bullrich, M. Negrini, and C.M. Croce. 2004. Human microRNA genes are frequently located at fragile sites and genomic regions involved in cancers. *Proc Natl Acad Sci U S A* **101**: 2999–3004.

Calin, G.A., M. Ferracin, A. Cimmino, G. Di Leva, M. Shimizu, S.E. Wojcik, M.V. Iorio, R. Visone, N.I. Sever, M. Fabbri, R. Iuliano, T. Palumbo, et al. 2005. A microRNA signature associated with prognosis and progression in chronic lymphocytic leukemia. *N Engl J Med* **353**:1793–1801.

Calin, G.A., C.G. Liu, M. Ferracin, T. Hyslop, R. Spizzo, C. Sevignani, M. Fabbri, A. Cimmino, E.J. Lee, S.E. Wojcik, M. Shimizu, E. Tili, et al. 2007. Ultraconserved regions encoding ncRNAs are altered in human leukemias and carcinomas. *Cancer Cell* **12**:215–229.

Calin, G.A., A. Cimmino, M. Fabbri, M. Ferracin, S.E. Wojcik, M. Shimizu, C. Taccioli, N. Zanesi, R. Garzon, R.I. Aqeilan, H. Alder, S. Volinia, et al. 2008. MiR-15a and miR-16-1 cluster functions in human leukemia. *Proc Natl Acad Sci U S A* **105**:5166–5171.

Castanotto, D. and J.J. Rossi. 2009. The promises and pitfalls of RNA-interference-based therapeutics. *Nature* **457**:426–433.

Chen, C.Z., L. Li, H.F. Lodish, and D.P. Bartel. 2004. MicroRNAs modulate hematopoietic lineage differentiation. *Science* **303**:83–86.

Chen, X., Y. Ba, L. Ma, X. Cai, Y. Yin, K. Wang, J. Guo, Y. Zhang, J. Chen, X. Guo, Q. Li, X. Li, et al. 2008. Characterization of microRNAs in serum: a novel class of biomarkers for diagnosis of cancer and other diseases. *Cell Res* **18**:997–1006.

Chin, L.J., E. Ratner, S. Leng, R. Zhai, S. Nallur, I. Babar, R.U. Muller, E. Straka, L. Su, E.A. Burki, R.E. Crowell, R. Patel, et al. 2008. A SNP in a let-7 microRNA complementary site in the KRAS 3′ untranslated region increases non-small cell lung cancer risk. *Cancer Res* **68**:8535–8540.

Chuang, J.C. and P.A. Jones. 2007. Epigenetics and microRNAs. *Pediatr Res* **61**:24R–29R.

Cimmino, A., G.A. Calin, M. Fabbri, M.V. Iorio, M. Ferracin, M. Shimizu, S.E. Wojcik, R.I. Aqeilan, S. Zupo, M. Dono, L. Rassenti, H. Alder, et al. 2005. miR-15 and miR-16 induce apoptosis by targeting BCL2. *Proc Natl Acad Sci U S A* **102**:13944–13949.

Clague, J., S.M. Lippman, H. Yang, M.A. Hildebrandt, Y. Ye, J.J. Lee, and X. Wu. 2010. Genetic variation in microRNA genes and risk of oral premalignant lesions. *Mol Carcinog* **49**:183–189.

Cortez, M.A., C. Bueso-Ramos, J. Ferdin, G. Lopez-Berestein, A.K. Sood, and G.A. Calin. 2011. MicroRNAs in body fluids—the mix of hormones and biomarkers. *Nat Rev Clin Oncol* **8**: 467–477.

Costinean, S., N. Zanesi, Y. Pekarsky, E. Tili, S. Volinia, N. Heerema, and C.M. Croce. 2006. Pre-B cell proliferation and lymphoblastic leukemia/high-grade lymphoma in E(mu)-miR155 transgenic mice. *Proc Natl Acad Sci U S A* **103**:7024–7029.

Croce, C.M. 2009. Causes and consequences of microRNA dysregulation in cancer. *Nat Rev Genet* **10**:704–714.

Denli, A.M., B.B. Tops, R.H. Plasterk, R.F. Ketting, and G.J. Hannon. 2004. Processing of primary microRNAs by the Microprocessor complex. *Nature* **432**:231–235.

Diederichs, S. and D.A. Haber. 2006. Sequence variations of microRNAs in human cancer: alterations in predicted secondary structure do not affect processing. *Cancer Res* **66**:6097–6104.

Dudziec, E., S. Miah, H.M. Choudhry, H.C. Owen, S. Blizard, M. Glover, F.C. Hamdy, and J.W. Catto. 2011. Hypermethylation of CpG islands and shores around specific microRNAs and mirtrons is associated with the phenotype and presence of bladder cancer. *Clin Cancer Res* **17**:1287–1296.

Eiring, A.M., J.G. Harb, P. Neviani, C. Garton, J.J. Oaks, R. Spizzo, S. Liu, S. Schwind, R. Santhanam, C.J. Hickey, H. Becker, J.C. Chandler, et al. 2010. miR-328 functions as an RNA decoy to modulate hnRNP E2 regulation of mRNA translation in leukemic blasts. *Cell* **140**: 652–665.

Erdmann, V.A., M. Szymanski, A. Hochberg, N. Groot, and J. Barciszewski. 2000. Non-coding, mRNA-like RNAs database Y2K. *Nucleic Acids Res* **28**:197–200.

Esquela-Kerscher, A. and F.J. Slack. 2006. Oncomirs—microRNAs with a role in cancer. *Nat Rev Cancer* **6**:259–269.

Esteller, M. 2008. Epigenetics in cancer. *N Engl J Med* **358**:1148–1159.

Fabbri, M., R. Garzon, A. Cimmino, Z. Liu, N. Zanesi, E. Callegari, S. Liu, H. Alder, S. Costinean, C. Fernandez-Cymering, S. Volinia, G. Guler, et al. 2007. MicroRNA-29 family reverts aberrant methylation in lung cancer by targeting DNA methyltransferases 3A and 3B. *Proc Natl Acad Sci U S A* **104**:15805–15810.

Fabian, M.R., N. Sonenberg, and W. Filipowicz. 2010. Regulation of mRNA translation and stability by microRNAs. *Annu Rev Biochem* **79**:351–379.

Felli, N., L. Fontana, E. Pelosi, R. Botta, D. Bonci, F. Facchiano, F. Liuzzi, V. Lulli, O. Morsilli, S. Santoro, M. Valtieri, G.A. Calin, et al. 2005. MicroRNAs 221 and 222 inhibit normal erythropoiesis and erythroleukemic cell growth via kit receptor down-modulation. *Proc Natl Acad Sci U S A* **102**:18081–18086.

Ferdin, J., T. Kunej, and G. Calin. 2010. Non-coding RNAs: identification of cancer-associated microRNAs by gene profiling. *Technol Cancer Res Treat* **9**:123–138.

Fish, J.E., M.M. Santoro, S.U. Morton, S. Yu, R.F. Yeh, J.D. Wythe, K.N. Ivey, B.G. Bruneau, D.Y. Stainier, and D. Srivastava. 2008. miR-126 regulates angiogenic signaling and vascular integrity. *Dev Cell* **15**:272–284.

Forman, J.J., A. Legesse-Miller, and H.A. Coller. 2008. A search for conserved sequences in coding regions reveals that the let-7 microRNA targets Dicer within its coding sequence. *Proc Natl Acad Sci U S A* **105**:14879–14884.

Friedman, J.M., G. Liang, C.C. Liu, E.M. Wolff, Y.C. Tsai, W. Ye, X. Zhou, and P.A. Jones. 2009a. The putative tumor suppressor microRNA-101 modulates the cancer epigenome by repressing the polycomb group protein EZH2. *Cancer Res* **69**:2623–2629.

REFERENCES

Friedman, R.C., K.K. Farh, C.B. Burge, and D.P. Bartel. 2009b. Most mammalian mRNAs are conserved targets of microRNAs. *Genome Res* **19**:92–105.

Garofalo, M., G. Di Leva, G. Romano, G. Nuovo, S.S. Suh, A. Ngankeu, C. Taccioli, F. Pichiorri, H. Alder, P. Secchiero, P. Gasparini, A. Gonelli, et al. 2009. miR-221&222 regulate TRAIL resistance and enhance tumorigenicity through PTEN and TIMP3 downregulation. *Cancer Cell* **16**:498–509.

Garzon, R., S.J. Liu, M. Fabbri, Z.F. Liu, C.E.A. Heaphy, E. Callegari, S. Schwind, J.X. Pang, J.H. Yu, N. Muthusamy, V. Havelange, S. Volinia, et al. 2009. MicroRNA-29b induces global DNA hypomethylation and tumor suppressor gene reexpression in acute myeloid leukemia by targeting directly DNMT3A and 3B and indirectly DNMT1. *Blood* **113**:6411–6418.

Garzon, R., G. Marcucci, and C.M. Croce. 2010. Targeting microRNAs in cancer: rationale, strategies and challenges. *Nat Rev Drug Discov* **9**:775–789.

Gaur, A., D.A. Jewell, Y. Liang, D. Ridzon, J.H. Moore, C. Chen, V.R. Ambros, and M.A. Israel. 2007. Characterization of microRNA expression levels and their biological correlates in human cancer cell lines. *Cancer Res* **67**:2456–2468.

Georges, M., W. Coppieters, and C. Charlier. 2007. Polymorphic miRNA-mediated gene regulation: contribution to phenotypic variation and disease. *Curr Opin Genet Dev* **17**:166–176.

Giraldez, A.J., R.M. Cinalli, M.E. Glasner, A.J. Enright, J.M. Thomson, S. Baskerville, S.M. Hammond, D.P. Bartel, and A.F. Schier. 2005. MicroRNAs regulate brain morphogenesis in zebrafish. *Science* **308**:833–838.

Gregory, R.I., K.P. Yan, G. Amuthan, T. Chendrimada, B. Doratotaj, N. Cooch, and R. Shiekhattar. 2004. The Microprocessor complex mediates the genesis of microRNAs. *Nature* **432**:235–240.

Gupta, R.A., N. Shah, K.C. Wang, J. Kim, H.M. Horlings, D.J. Wong, M.C. Tsai, T. Hung, P. Argani, J.L. Rinn, Y. Wang, P. Brzoska, et al. 2010. Long non-coding RNA HOTAIR reprograms chromatin state to promote cancer metastasis. *Nature* **464**:1071–1076.

Ha, T.Y. 2011. MicroRNAs in human diseases: from cancer to cardiovascular disease. *Immune Netw* **11**:135–154.

Hafner, M., M. Landthaler, L. Burger, M. Khorshid, J. Hausser, P. Berninger, A. Rothballer, M. Ascano, A.C. Jungkamp, M. Munschauer, A. Ulrich, G.S. Wardle, et al. 2010. Transcriptome-wide identification of RNA-binding protein and microRNA target sites by PAR-CLIP. *Cell* **141**:129–141.

Hammond, S.M., S. Boettcher, A.A. Caudy, R. Kobayashi, and G.J. Hannon. 2001. Argonaute2, a link between genetic and biochemical analyses of RNAi. *Science* **293**:1146–1150.

He, C., Z. Li, P. Chen, H. Huang, L.D. Hurst, J. Chen. 2012. Young intragenic miRNAs are less coexpressed with host genes than old ones: implications of miRNA-host gene coevolution. *Nucleic Acids Res* **40**:4002–4012.

Horikawa, Y., C.G. Wood, H. Yang, H. Zhao, Y. Ye, J. Gu, J. Lin, T. Habuchi, and X. Wu. 2008. Single nucleotide polymorphisms of microRNA machinery genes modify the risk of renal cell carcinoma. *Clin Cancer Res* **14**:7956–7962.

Hu, Z., J. Liang, Z. Wang, T. Tian, X. Zhou, J. Chen, R. Miao, Y. Wang, X. Wang, and H. Shen. 2009. Common genetic variants in pre-microRNAs were associated with increased risk of breast cancer in Chinese women. *Hum Mutat* **30**:79–84.

Huang, J.C., T. Babak, T.W. Corson, G. Chua, S. Khan, B.L. Gallie, T.R. Hughes, B.J. Blencowe, B.J. Frey, and Q.D. Morris. 2007. Using expression profiling data to identify human microRNA targets. *Nat Methods* **4**:1045–1049.

Hwang, H.W., E.A. Wentzel, and J.T. Mendell. 2007. A hexanucleotide element directs microRNA nuclear import. *Science* **315**:97–100.

Johnson, S.M., H. Grosshans, J. Shingara, M. Byrom, R. Jarvis, A. Cheng, E. Labourier, K.L. Reinert, D. Brown, and F.J. Slack. 2005. RAS is regulated by the let-7 microRNA family. *Cell* **120**:635–647.

Kim, Y.K. and V.N. Kim. 2007. Processing of intronic microRNAs. *EMBO J* **26**:775–783.

Kim, S.J., J.S. Oh, J.Y. Shin, K.D. Lee, K.W. Sung, S.J. Nam, and K.H. Chun. 2011. Development of microRNA-145 for therapeutic application in breast cancer. *J Control Release* **155**:427–434.

Kota, J., R.R. Chivukula, K.A. O'Donnell, E.A. Wentzel, C.L. Montgomery, H.W. Hwang, T.C. Chang, P. Vivekanandan, M. Torbenson, K.R. Clark, J.R. Mendell, and J.T. Mendell. 2009. Therapeutic microRNA delivery suppresses tumorigenesis in a murine liver cancer model. *Cell* **137**:1005–1017.

Krek, A., D. Grün, M.N. Poy, R. Wolf, L. Rosenberg, E.J. Epstein, P. MacMenamin, I. da Piedade, K.C. Gunsalus, M. Stoffel, and N. Rajewsky. 2005. Combinatorial microRNA target predictions. *Nat Genet* **37**:495–500.

Krützfeldt, J., N. Rajewsky, R. Braich, K.G. Rajeev, T. Tuschl, M. Manoharan, and M. Stoffel. 2005. Silencing of microRNAs in vivo with "antagomirs." *Nature* **438**:685–689.

Kumar, M.S., J. Lu, K.L. Mercer, T.R. Golub, and T. Jacks. 2007. Impaired microRNA processing enhances cellular transformation and tumorigenesis. *Nat Genet* **39**:673–677.

Kunej, T., I. Godnic, J. Ferdin, S. Horvat, P. Dovc, and G.A. Calin. 2011. Epigenetic regulation of microRNAs in cancer: an integrated review of literature. *Mutat Res* **717**:77–84.

Kunej, T., I. Godnic, S. Horvat, M. Zorc, and G.A. Calin. 2012a. Cross talk between microRNA and coding cancer genes. *Cancer J* **18**:223–231.

Kunej, T., D.J. Skok, M. Zorc, A. Ogrinc, J.J. Michal, M. Kovac, and Z. Jiang. 2012b. Obesity gene atlas in mammals. *J Genom* **1**:45–55.

Landi, D., F. Gemignani, R. Barale, and S. Landi. 2008. A catalog of polymorphisms falling in microRNA-binding regions of cancer genes. *DNA Cell Biol* **27**:35–43.

Landi, D., F. Gemignani, and S. Landi. 2012. Role of variations within microRNA-binding sites in cancer. *Mutagenesis* **27**:205–210.

Lawrie, C.H., S. Gal, H.M. Dunlop, B. Pushkaran, A.P. Liggins, K. Pulford, A.H. Banham, F. Pezzella, J. Boultwood, J.S. Wainscoat, C.S. Hatton, and A.L. Harris. 2008. Detection of elevated levels of tumour-associated microRNAs in serum of patients with diffuse large B-cell lymphoma. *Br J Haematol* **141**:672–675.

Lee, R.C., R.L. Feinbaum, and V. Ambros. 1993. The *C. elegans* heterochronic gene lin-4 encodes small RNAs with antisense complementarity to lin-14. *Cell* **75**:843–854.

Lee, Y.S., H.K. Kim, S. Chung, K.S. Kim, and A. Dutta. 2005. Depletion of human micro-RNA miR-125b reveals that it is critical for the proliferation of differentiated cells but not for the down-regulation of putative targets during differentiation. *J Biol Chem* **280**:16635–16641.

Li, S.C., P. Tang, and W.C. Lin. 2007. Intronic microRNA: discovery and biological implications. *DNA Cell Biol* **26**:195–207.

Liao, J.Y., L.M. Ma, Y.H. Guo, Y.C. Zhang, H. Zhou, P. Shao, Y.Q. Chen, and L.H. Qu. 2010. Deep sequencing of human nuclear and cytoplasmic small RNAs reveals an unexpectedly complex subcellular distribution of miRNAs and tRNA 3′ trailers. *PLoS ONE* **5**:e10563.

Liu, C.G., G.A. Calin, B. Meloon, N. Gamliel, C. Sevignani, M. Ferracin, C.D. Dumitru, M. Shimizu, S. Zupo, M. Dono, H. Alder, F. Bullrich, et al. 2004. An oligonucleotide microchip for genome-wide microRNA profiling in human and mouse tissues. *Proc Natl Acad Sci U S A* **101**:9740–9744.

Liu, C., K. Kelnar, B. Liu, X. Chen, T. Calhoun-Davis, H. Li, L. Patrawala, H. Yan, C. Jeter, S. Honorio, J.F. Wiggins, A.G. Bader, et al. 2011. The microRNA miR-34a inhibits prostate cancer stem cells and metastasis by directly repressing CD44. *Nat Med* **17**:211–215.

Lu, C., S.S. Tej, S. Luo, C.D. Haudenschild, B.C. Meyers, and P.J. Green. 2005a. Elucidation of the small RNA component of the transcriptome. *Science* **309**:1567–1569.

Lu, J., G. Getz, E.A. Miska, E. Alvarez-Saavedra, J. Lamb, D. Peck, A. Sweet-Cordero, B.L. Ebert, R.H. Mak, A.A. Ferrando, J.R. Downing, T. Jacks, et al. 2005b. MicroRNA expression profiles classify human cancers. *Nature* **435**:834–838.

REFERENCES

Lujambio, A., S. Ropero, E. Ballestar, M.F. Fraga, C. Cerrato, F. Setién, S. Casado, A. Suarez-Gauthier, M. Sanchez-Cespedes, A. Gitt, I. Spiteri, P.P. Das, et al. 2007. Genetic unmasking of an epigenetically silenced microRNA in human cancer cells. *Cancer Res* **67**:1424–1429.

Lytle, J.R., T.A. Yario, and J.A. Steitz. 2007. Target mRNAs are repressed as efficiently by microRNA-binding sites in the 5′ UTR as in the 3′ UTR. *Proc Natl Acad Sci U S A* **104**: 9667–9672.

Ma, L., F. Reinhardt, E. Pan, J. Soutschek, B. Bhat, E.G. Marcusson, J. Teruya-Feldstein, G.W. Bell, and R.A. Weinberg. 2010. Therapeutic silencing of miR-10b inhibits metastasis in a mouse mammary tumor model. *Nat Biotechnol* **28**:341–347.

Medina, P.P., M. Nolde, and F.J. Slack. 2010. OncomiR addiction in an in vivo model of microRNA-21-induced pre-B-cell lymphoma. *Nature* **467**:86–90.

Mitchell, P.S., R.K. Parkin, E.M. Kroh, B.R. Fritz, S.K. Wyman, E.L. Pogosova-Agadjanyan, A. Peterson, J. Noteboom, K.C. O'Briant, A. Allen, D.W. Lin, N. Urban, et al. 2008. Circulating microRNAs as stable blood-based markers for cancer detection. *Proc Natl Acad Sci U S A* **105**: 10513–10518.

Nicoloso, M.S., H. Sun, R. Spizzo, H. Kim, P. Wickramasinghe, M. Shimizu, S.E. Wojcik, J. Ferdin, T. Kunej, L. Xiao, S. Manoukian, G. Secreto, et al. 2010. Single-nucleotide polymorphisms inside microRNA target sites influence tumor susceptibility. *Cancer Res* **70**:2789–2798.

Noonan, E.J., R.F. Place, D. Pookot, S. Basak, J.M. Whitson, H. Hirata, C. Giardina, and R. Dahiya. 2009. miR-449a targets HDAC-1 and induces growth arrest in prostate cancer. *Oncogene* **28**:1714–1724.

Pelletier, C. and J.B. Weidhaas. 2010. MicroRNA binding site polymorphisms as biomarkers of cancer risk. *Expert Rev Mol Diagn* **10**:817–829.

Petrocca, F., R. Visone, M.R. Onelli, M.H. Shah, M.S. Nicoloso, I. de Martino, D. Iliopoulos, E. Pilozzi, C.G. Liu, M. Negrini, L. Cavazzini, S. Volinia, et al. 2008. E2F1-regulated microRNAs impair TGFβ-dependent cell-cycle arrest and apoptosis in gastric cancer. *Cancer Cell* **13**: 272–286.

Place, R.F., L.C. Li, D. Pookot, E.J. Noonan, and R. Dahiya. 2008. MicroRNA-373 induces expression of genes with complementary promoter sequences. *Proc Natl Acad Sci U S A* **105**: 1608–1613.

Poliseno, L., L. Salmena, J. Zhang, B. Carver, W.J. Haveman, and P.P. Pandolfi. 2010. A coding-independent function of gene and pseudogene mRNAs regulates tumour biology. *Nature* **465**:1033–1038.

Poy, M.N., L. Eliasson, J. Krutzfeldt, S. Kuwajima, X. Ma, P.E. Macdonald, S. Pfeffer, T. Tuschl, N. Rajewsky, P. Rorsman, and M. Stoffel. 2004. A pancreatic islet-specific microRNA regulates insulin secretion. *Nature* **432**:226–230.

Raveche, E.S., E. Salerno, B.J. Scaglione, V. Manohar, F. Abbasi, Y.C. Lin, T. Fredrickson, P. Landgraf, S. Ramachandra, K. Huppi, J.R. Toro, V.E. Zenger, et al. 2007. Abnormal microRNA-16 locus with synteny to human 13q14 linked to CLL in NZB mice. *Blood* **109**: 5079–5086.

Rearick, D., A. Prakash, A. McSweeny, S.S. Shepard, L. Fedorova, and A. Fedorov. 2011. Critical association of ncRNA with introns. *Nucleic Acids Res* **39**:2357–2366.

Rodriguez, A., S. Griffiths-Jones, J.L. Ashurst, and A. Bradley. 2004. Identification of mammalian microRNA host genes and transcription units. *Genome Res* **14**:1902–1910.

Ruby, J.G., C.H. Jan, and D.P. Bartel. 2007. Intronic microRNA precursors that bypass Drosha processing. *Nature* **448**:83–86.

Ryan, B.M., A.I. Robles, and C.C. Harris. 2010. Genetic variation in microRNA networks: the implications for cancer research. *Nat Rev Cancer* **10**:389–402.

Saito, Y., G. Liang, G. Egger, J.M. Friedman, J.C. Chuang, G.A. Coetzee, and P.A. Jones. 2006. Specific activation of microRNA-127 with downregulation of the proto-oncogene BCL6 by chromatin-modifying drugs in human cancer cells. *Cancer Cell* **9**:435–443.

Saito, Y., J.M. Friedman, Y. Chihara, G. Egger, J.C. Chuang, and G. Liang. 2009. Epigenetic therapy upregulates the tumor suppressor microRNA-126 and its host gene EGFL7 in human cancer cells. *Biochem Biophys Res Commun* **379**:726–731.

Salmena, L., L. Poliseno, Y. Tay, L. Kats, and P.P. Pandolfi. 2011. A ceRNA hypothesis: the Rosetta Stone of a hidden RNA language? *Cell* **146**:353–358.

Santarpia, L., M. Nicoloso, and G.A. Calin. 2010. MicroRNAs: a complex regulatory network drives the acquisition of malignant cell phenotype. *Endocr Relat Cancer* **17**:F51–F75.

Sempere, L.F., S. Freemantle, I. Pitha-Rowe, E. Moss, E. Dmitrovsky, and V. Ambros. 2004. Expression profiling of mammalian microRNAs uncovers a subset of brain-expressed microRNAs with possible roles in murine and human neuronal differentiation. *Genome Biol* **5**:R13.

Sethupathy, P., M. Megraw, and A.G. Hatzigeorgiou. 2006. A guide through present computational approaches for the identification of mammalian microRNA targets. *Nat Methods* **3**:881–886.

Sharma, S., T.K. Kelly, and P.A. Jones. 2010. Epigenetics in cancer. *Carcinogenesis* **31**:27–36.

Shen, J., C.B. Ambrosone, R.A. DiCioccio, K. Odunsi, S.B. Lele, and H. Zhao. 2008. A functional polymorphism in the miR-146a gene and age of familial breast/ovarian cancer diagnosis. *Carcinogenesis* **29**:1963–1966.

Sibley, C.R., Y. Seow, S. Saayman, K.K. Dijkstra, S. El Andaloussi, M.S. Weinberg, and M.J. Wood. 2012. The biogenesis and characterization of mammalian microRNAs of mirtron origin. *Nucleic Acids Res* **40**:438–448.

Spizzo, R., D. Rushworth, M. Guerrero, and G.A. Calin. 2009. RNA inhibition, microRNAs, and new therapeutic agents for cancer treatment. *Clin Lymphoma Myeloma* **9 Suppl 3**: S313–S318.

Stahlhut Espinosa, C.E. and F.J. Slack. 2006. The role of microRNAs in cancer. *Yale J Biol Med* **79**:131–140.

Sun, G., J. Yan, K. Noltner, J. Feng, H. Li, D.A. Sarkis, S.S. Sommer, and J.J. Rossi. 2009. SNPs in human miRNA genes affect biogenesis and function. *RNA* **15**:1640–1651.

Sung, H., S. Jeon, K.M. Lee, S. Han, M. Song, J.Y. Choi, S.K. Park, K.Y. Yoo, D.Y. Noh, S.H. Ahn, and D. Kang. 2012. Common genetic polymorphisms of microRNA biogenesis pathway genes and breast cancer survival. *BMC Cancer* **12**:195.

Takamizawa, J., H. Konishi, K. Yanagisawa, S. Tomida, H. Osada, H. Endoh, T. Harano, Y. Yatabe, M. Nagino, Y. Nimura, T. Mitsudomi, and T. Takahashi. 2004. Reduced expression of the let-7 microRNAs in human lung cancers in association with shortened postoperative survival. *Cancer Res* **64**:3753–3756.

Tay, Y., J. Zhang, A.M. Thomson, B. Lim, and I. Rigoutsos. 2008. MicroRNAs to Nanog, Oct4 and Sox2 coding regions modulate embryonic stem cell differentiation. *Nature* **455**:1124–1128.

Tian, T., Y. Shu, J. Chen, Z. Hu, L. Xu, G. Jin, J. Liang, P. Liu, X. Zhou, R. Miao, H. Ma, Y. Chen, et al. 2009. A functional genetic variant in microRNA-196a2 is associated with increased susceptibility of lung cancer in Chinese. *Cancer Epidemiol Biomark Prev* **18**:1183–1187.

Trang, P., P.P. Medina, J.F. Wiggins, L. Ruffino, K. Kelnar, M. Omotola, R. Homer, D. Brown, A.G. Bader, J.B. Weidhaas, and F.J. Slack. 2010. Regression of murine lung tumors by the let-7 microRNA. *Oncogene* **29**:1580–1587.

Vasudevan, S., Y. Tong, and J.A. Steitz. 2007. Switching from repression to activation: microRNAs can up-regulate translation. *Science* **318**:1931–1934.

Wang, S., A.B. Aurora, B.A. Johnson, X. Qi, J. McAnally, J.A. Hill, J.A. Richardson, R. Bassel-Duby, and E.N. Olson. 2008. The endothelial-specific microRNA miR-126 governs vascular integrity and angiogenesis. *Dev Cell* **15**:261–271.

Weber, B., C. Stresemann, B. Brueckner, and F. Lyko. 2007. Methylation of human microRNA genes in normal and neoplastic cells. *Cell Cycle* **6**:1001–1005.

White, N.M., T.T. Bao, J. Grigull, Y.M. Youssef, A. Girgis, M. Diamandis, E. Fatoohi, M. Metias, R.J. Honey, R. Stewart, K.T. Pace, G.A. Bjarnason, et al. 2011. miRNA profiling for clear cell renal cell carcinoma: biomarker discovery and identification of potential controls and consequences of miRNA dysregulation. *J Urol* **186**:1077–1083.

Whitehead, K.A., R. Langer, and D.G. Anderson. 2009. Knocking down barriers: advances in siRNA delivery. *Nat Rev Drug Discov* **8**:129–138.

Wiggins, J.F., L. Ruffino, K. Kelnar, M. Omotola, L. Patrawala, D. Brown, and A.G. Bader. 2010. Development of a lung cancer therapeutic based on the tumor suppressor microRNA-34. *Cancer Res* **70**:5923–5930.

Yang, H., C.P. Dinney, Y. Ye, Y. Zhu, H.B. Grossman, and X. Wu. 2008. Evaluation of genetic variants in microRNA-related genes and risk of bladder cancer. *Cancer Res* **68**:2530–2537.

Yu, X., J. Lin, D.J. Zack, J.T. Mendell, and J. Qian. 2008. Analysis of regulatory network topology reveals functionally distinct classes of microRNAs. *Nucleic Acids Res* **36**:6494–6503.

Zeng, Y. and B.R. Cullen. 2005. Efficient processing of primary microRNA hairpins by Drosha requires flanking nonstructured RNA sequences. *J Biol Chem* **280**:27595–27603.

Zhang, H.L., L.F. Yang, Y. Zhu, X.D. Yao, S.L. Zhang, B. Dai, Y.P. Zhu, Y.J. Shen, G.H. Shi, and D.W. Ye. 2011. Serum miRNA-21: elevated levels in patients with metastatic hormone-refractory prostate cancer and potential predictive factor for the efficacy of docetaxel-based chemotherapy. *Prostate* **71**:326–331.

Zorc, M., D. Jevsinek Skok, I. Godnic, G.A. Calin, S. Horvat, Z. Jiang, P. Dovc, and T. Kunej. 2012. Catalog of microRNA seed polymorphisms in vertebrates. *PLoS ONE* **7**:e30737.

14

MicroRNAs as Oncogenes and Tumor Suppressors

Eva E. Rufino-Palomares,[1] Fernando J. Reyes-Zurita,[1] Jose Antonio Lupiáñez,[1] and Pedro P. Medina[1,2]

[1]Department of Biochemistry and Molecular Biology, Faculty of Sciences, University of Granada, Granada, Spain
[2]Centre for Genomics and Oncological Research (GENYO), Granada, Spain

I.	Introduction	224
II.	miRNA Targets	224
III.	Role of miRNAs in Cancer	225
	A. miRNAs That Function as Oncogenes	227
	B. miRNAs That Function as Tumor Suppressors	230
IV.	miRNAs and Their Future Use in the Clinic: Diagnosis, Prognosis, and Therapy	233
	References	234

ABBREVIATIONS

miRNA	microRNA
mRNA	messenger RNA
ncRNAs	non-coding RNAs
UTR	untranslated region

MicroRNAs in Medicine, First Edition. Edited by Charles H. Lawrie.
© 2014 John Wiley & Sons, Inc. Published 2014 by John Wiley & Sons, Inc.

I. INTRODUCTION

Cancer is a complex disease where a group of abnormal cells grow without control and become able to invade adjacent tissues and colonize other organs. Such malignant behavior results in tissue dysfunction and ultimately organ failure. Several lines of evidence indicate that carcinogenesis is a multistep process where the malignant cells accumulate epigenetic and genetic alterations that drive the progressive transformation of normal cells into malignant ones. In this way, cancer cells select alterations in genes that promote cancer progression (oncogenes) or genes that prevent it (tumor suppressor genes) (Hanahan and Weinberg 2011). At the end of the transformation process, the malignant cells lose their cellular identity and acquire growth independence, resistance to apoptosis, senescence, and invasiveness.

Recent studies quantified that there are 20,684 human protein-coding genes. These genes comprise just 1.2% of the genome; however, 80% of the genome is functionally transcribed (Consortium et al. 2012). Therefore, it is evident that a significant contribution of the complexity underlying complex organisms is derived from non-protein-coding genes.

Until a few years ago, the relevance of non-protein-coding genes in mainstream biology was largely restricted to ribosomal RNAs (rRNAs), transfer RNAs (tRNAs), and some ribozymes. However, the scientific community have recently expanded the species of non-coding RNAs (ncRNAs) to include small nucleolar RNAs (snoRNAs), PIWI-interacting RNAs (piRNAs), large intergenic ncRNAs, transcribed ultraconserved regions (T-UCRs), microRNAs (miRNAs), and so on.

miRNAs are a class of small RNA molecules that regulate gene expression at the posttranscriptional level. Initially discovered in *Caenorhabditis elegans* (Lee et al. 1993; Wightman et al. 1993), they were considered an oddity of nematodes until it was realized that some of them were phylogenetically conserved in a wide variety of animals including humans (Pasquinelli et al. 2000; Reinhart et al. 2000). Today, miRNAs are increasingly seen as important regulators of gene expression, ushering in a renewed appreciation of the regulative capabilities of ncRNA. At the cellular level, miRNAs are significant in the establishment and maintenance of cell identity (Stadler and Ruohola-Baker 2008). Aberrant levels of miRNAs often result in loss of differentiation, a hallmark of cancer. Not surprisingly, therefore, dysfunctions of the miRNA pathway affect many cellular processes that are routinely altered in cancer, such as differentiation, proliferation, apoptosis, metastasis, and senescence (Sinkkonen et al. 2008).

II. miRNA TARGETS

It has been estimated that miRNAs can regulate the expression of between 30% and 60% of all human protein-coding genes (Lewis et al. 2005; Friedman et al. 2009). Plausibly, therefore, these regulators directly or indirectly affect most, if not all, cellular pathways. Interestingly, some genes involved in basic cellular processes avoid miRNA regulation due to short 3′-untranslated regions (3′-UTRs) that are specifically depleted of miRNA binding sites (Stark et al. 2005). The accurate prediction of the target genes is made difficult by the imperfect complementarity between the miRNA and the regulated mRNAs. To add further complexity to the issue, some genes can have alternative 3′-UTRs, which could be regulated by a different set of miRNAs. Several bioinformatics approaches for the prediction of miRNA targets have been developed (John et al. 2004; Krek et al. 2005; Grimson et al. 2007). Most search algorithms rely on the search for perfect complementarity in the so-called seed region, that is, the seven nucleotides between positions 2 and 8 of the miRNA (Lewis et al. 2005; Bartel 2009). However, when predictions based on such short complementary sequences are compared with experimental results from proteomic studies, the false-positive rate appears to be above 50% (Baek et al. 2008), highlighting the difficulty in deciphering the rules of miRNA target recognition. Other features of the target sequence, such as phylogenetic conservation, position within the 3'-UTR, absence of stable secondary structures, and new rules of binding, are also accounted to improve the accuracy of recent algorithms of prediction (Grimson et al. 2007; Chi et al. 2012). It has been estimated that a single miRNA family can regulate as many as 200 different genes (Lewis et al. 2005). Thus, the effects of the miRNAs are likely to be pleotropic, and their aberrant expression could feasibly unbalance the cell homeostasis, contributing to diseases, including cancer.

III. ROLE OF miRNAs IN CANCER

Since miRNAs can alter gene expression, they are plausible candidates to alter the physiological balance of the cell that contribute to human pathologies, including cancer (Figure 14.1). Pioneering studies showed a differential miRNA expression profile between normal

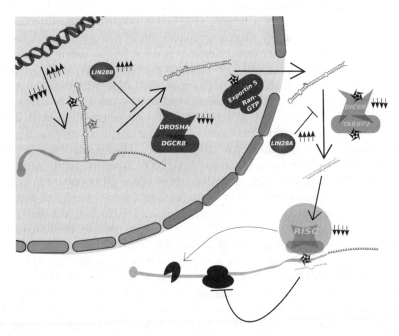

Figure 14.1. Mechanisms of microRNA (miRNA) dysfunction in cancer. Tumoral cells display overall a loss of miRNA expression. Although more specifically, there is a loss of miRNAs that suppresses oncogenic proteins/pathways and an accumulation of miRNAs that suppresses tumor suppressor proteins/pathways. Aberrations found in tumors contain changes of expression of miRNAs or proteins involved in their biogenesis (indicated by arrows) and/or mutations on these miRNAs, their binding sites, or the proteins involved in their biogenesis (indicated as stars). Aberrations in miRNA biogenesis modulators like LIN28, which regulates the biogenesis of *let-7* miRNA family that exemplifies this cartoon, have also been found.

tissue and tumors (Lu et al. 2005; Volinia et al. 2006; Yanaihara et al. 2006). However, many of these miRNAs are bystanders indirectly altered by the genomic, epigenomic, or physiological changes that arise during carcinogenesis and are not causative agents of tumor development.

Alterations in the components of the miRNA biogenesis machinery, and not only specific miRNAs (as it will be discussed later), are involved in the carcinogenesis process. Overall, it has been observed a general down-regulation of miRNAs in tumors compared with normal tissues (Lu et al. 2005). Experimental approaches have confirmed these observations, showing that global repression of miRNA maturation by mutation of *DROSHA*, *DGCR8*, and *DICER1* promotes cellular transformation and tumorigenesis (Kumar et al. 2007). Remarkably, conditional loss of *DICER* (Kanellopoulou et al. 2005; Murchison et al. 2005), or *DGCR8* (Wang et al. 2007), in murine embryonic stem cells (ESCs) results in impaired proliferation and differentiation.

Recently, mutations in some of the miRNA biogenesis machinery have been associated with cancer including *DICER1* (Hill et al. 2009), *TARBP2* (Melo et al. 2009), and Exportin 5 (XPO5) (Melo et al. 2010), strongly highlighting the relevance of the miRNA biogenesis in cellular transformation. Additionally, reduced expression of some of the miRNA machinery, like *DICER1*, correlates with short survival of non-small-cell lung cancer (Karube et al. 2005).

Other components of the miRNA machinery that have been implicated in cancer include the Argonaute family members hAgo1/EIF2C1 (eukaryotic translation–initiation factor 2 subunit 1), hAgo3/EIF2C3, hAgo4, and Hiwi. Argonaute proteins are the catalytic components of the RNA-induced silencing complex (RISC). hAgo1, hAgo3, and hAgo4 cluster on the 1p34-35 chromosomal region that is often lost in human cancers such as nephroblastoma, neuroblastoma, and carcinomas of the breast, liver, and colon (Koesters et al. 1999). On the other hand, overexpression of Hiwi (that belongs to the Argonaute protein family clade involved in germ line maintenance) (Aravin et al. 2006) has been related with germinal tumors such as seminomas, among others (Qiao et al. 2002; Liu et al. 2006; Taubert et al. 2007).

The posttranscriptional maturation of miRNAs is regulated in response to proliferative stimuli and cellular differentiation (Thomson et al. 2006) at several levels. Aberrations in the regulation of this miRNA maturation have been observed important in cancer. For example, it has been shown that, at least in the case of the *let-7* family of miRNAs, the RNA-binding protein LIN28 is necessary and sufficient for blocking the cleavage of the *pri-let-7* miRNAs (Viswanathan et al. 2008). In the human genome, there are two homologue LIN28 proteins: LIN28A and LIN28B. Both selectively block the expression of *let-7* miRNAs and function as oncogenes in a variety of human cancers. LIN28A recruits a TUTase (Zcchc11/TUT4) to *let-7* precursors to block processing by Dicer in the cell cytoplasm. LIN28B represses *let-7* processing in the nucleus by sequestering *pri-let-7* transcripts and inhibiting their processing by Drosha (Piskounova et al. 2011). Also, it has been observed that some RNA methyltransferases (such as BCDIN3D) can methylate the precursors of some miRNAs, regulating negatively the miRNA maturation by reducing the processing by Dicer (Xhemalce et al. 2012). Consistently, BCDIN3D depletion leads to lower levels of pre-miRNA and concomitantly increased mature miRNA. Additionally, it has been observed that the tumor suppressor breast cancer 1 (BRCA1) accelerates the processing of miRNA primary transcripts, increasing the expressions of both precursor and mature forms of at least the miRNAs: *let-7a-1*, *miR-16-1*, *miR-145*, and *miR-34a* (Kawai and Amano 2012).

In the following section, we will review, without claiming completeness given the rapid pace of the field, some of the better-characterized examples of miRNAs involved in tumor development. We have classified them from a classic point of view considering them as tumor suppressors when they impair tumor progression or as oncogenes when they promote it. However, since the effects of the miRNAs are intrinsically pleiotropic, this classification should be considered flexible. Actually, as it will be discussed ahead, it is conceivable that the same miRNA can act as tumor suppressor or as oncogene in different kinds of tumors, tissues, or physiological conditions (Table 14.1).

A. miRNAs That Function as Oncogenes

1. miR-17~92 Cluster. *miR-17~92* represents a cluster of miRNAs composed of *miR-17-5p*, *miR-17-3p*, *miR-18a*, *miR-19a*, *miR-20a*, *miR-19b-1*, and *miR-92a* that arises from the polycistronic transcript C13orf25 within the chromosomal region 13q31.3. Interestingly, this chromosomal region is often amplified in B-cell lymphomas and others malignancies, leading to overexpression of these miRNAs (Ota et al. 2004; He et al. 2005). Recent studies have shown that the *miR-17~92* cluster synergizes with the oncogenic properties of *c-myc* (He et al. 2005; O'Donnell et al. 2005; Dews et al. 2006). Overexpression of the *miR-17~92* cluster acted cooperatively with *c-myc* to accelerate tumor development in a mouse B-cell lymphoma model, increasing the tumor resistance to apoptosis (He et al. 2005). On the other hand, it was shown that *c-myc* binds directly to the *miR-17~92* cluster locus, enhancing its expression. Moreover, it was shown that E2F1 is negatively regulated by two miRNAs from the *miR-17~92* cluster: *miR-17-5p* and *miR-20a*. E2F1 is a member of the E2F transcription factor family that promotes the transition from G1 to S phase of the cell cycle. Interestingly, E2F1 is also a target of *c-myc*. Therefore, *c-myc* can regulate E2F1 expression by directly activating E2F1 transcription, but indirectly limiting its translation through the *miR-17-92-1* cluster, therefore achieving tight control of the proliferative signal. A further level of regulation seems to be in place as E2F1, E2F2, and E2F3 directly bind the promoter of the *miR-17~92* cluster and activate its transcription (O'Donnell et al. 2005; Sylvestre et al. 2007; Woods et al. 2007).

Although *c-myc* up-regulates the *miR-17~92* cluster, the predominant consequence of activation of *c-myc* is widespread repression of miRNA expression. Much of this repression is likely to be a direct result of *c-myc* binding to miRNA promoters. The reactivation of the miRNAs repressed by *c-myc* diminishes tumorigenicity in a lymphoma cell model (Chang et al. 2008b).

2. miR-21. Located in the chromosomal region 17q23.1, *miR-21* is overexpressed in many types of tumors, including neuroblastoma; glioblastoma; colorectal, lung, breast, liver, stomach, and pancreatic cancers; chronic lymphocytic leukemia (CLL); diffuse large B-cell lymphoma; acute myeloid leukemia (AML); and Hodgkin's lymphoma among others (Chan et al. 2005; Ciafre et al. 2005b; Volinia et al. 2006; Corsten et al. 2007; Fulci et al. 2007; Lawrie et al. 2007; Meng et al. 2007; Si et al. 2007; Chang et al. 2008a; Jongen-Lavrencic et al. 2008; Navarro et al. 2008; Cervigne et al. 2009; Gibcus et al. 2009; Li et al. 2009). Not surprisingly, increased expression of *miR-21* is among the most frequently associated with poor outcome in human cancer (Nair et al. 2012). *miR-21* has antiapoptotic abilities, since blocking it with antisense *miR-21* results in increased apoptosis (Chan et al. 2005; Si et al. 2007). Further studies have identified several tumor suppressors as *miR-21* targets, such as *PTEN*, *TPM1*, and *PDCD4* (Ciafre et al. 2005a; Meng

TABLE 14.1. Important MicroRNAs (miRNAs) in Cancer and Their Validated Targets

miRNA	Validated Targets	Dysregulation
miRNA as Oncogenes		
miR-17~92 cluster	MYC, E2F1, E2F2, E2F3, PTEN, CDKN1A, and PRKAA1	Lymphomas; breast, lung, colon, stomach, and pancreatic cancers
miR-21	PTEN, TPM1, and PDCD4	Neuroblastoma; glioblastoma; colorectal, lung, breast, liver, stomach, and pancreatic cancers; chronic lymphocytic leukemia; diffuse large B-cell lymphoma; acute myeloid leukemia; and Hodgkin's lymphoma
miR-155	TP53INP1, SHIP1	Chronic lymphocytic leukemia; B-cell lymphoma; lung, breast, and colon cancers
miR-221 and *miR-222*	p27(Kip1), CDKN1B, and PIK3R1	Chronic lymphocytic leukemia, thyroid carcinoma, and hepatocellular carcinoma
miR-372 and *miR-373*	LATS2, CDK2, and cyclin A1	Testicular tumors
miRNA as Tumor Suppressors		
let-7 family	RAS, MYC, CDK6, CDC25, and HMGA2	Lung, breast, colon, stomach, and ovarian cancers
miR15a and *miR-16-1*	BCL2, CCND1, CCND3, CCNE1, and CDK6	Chronic lymphocytic leukemia, prostate cancer, and multiple myelomas and pituitary adenomas
miR-29	TCL1, DNMT3A, and DNMT3B	Chronic lymphocytic leukemia, acute myeloid leukemia, lung and breast cancers, and cholangocarcinoma
miR-31	RhoA, NIK	Breast cancer
miR-34	E2F3, CDK4, CDK6, CCNE2, BIR3, DCR3, and BCL2	Neuroblastoma; melanoma; colon, lung, breast, kidney, bladder, pancreatic, and liver cancers
miR-126	IGFBP2, PITPNC1, and MERTK	Colorectal, gastric, lung, prostate, bladder, and breast cancers
miR-145	MYC, MUCIN1, JAM-A, FASCIN, OCT4, SOX2, and KLF4	Loss in breast and colon cancers
miR-203	P63, ABL1	Hepatocellular and pancreatic tumors

et al. 2006; Cervigne et al. 2009), which could help to explain its biological role in carcinogenesis. Overexpression of *miR-21* led to a pre-B malignant lymphoid-like phenotype *in vivo* (Medina et al. 2010). When *miR-21* was inactivated, the tumors regressed completely in a few days, partly as a result of apoptosis. Significantly, this was the first model for oncogene addiction in a non-protein-coding RNA. These results support efforts to treat human cancer through pharmacological inactivation of miRNAs such as *miR-21*.

Consistent with an oncogenic role, *miR-21* knockouts exhibit a decrease in the incidence of papillomas in skin carcinogenesis assays (Ma et al. 2011), and display a reduced susceptibility to lung tumorigenesis in response to K-RAS activation (Hatley et al. 2010). Importantly, measures of the level of *miR-21* could be useful in the diagnosis and prognosis of cancer (Schetter et al. 2008a), and high expression of *miR-21* predicts recurrence and unfavorable survival in non-small-cell lung cancer (Yang et al. 2012).

3. miR-155. *miR-155* is located in the chromosomal region 21q21.3 within a conserved region of the non-coding gene B-cell integration cluster (*BIC*). *BIC* was initially identified as a common integration site for avian leucosis virus, which induces B-cell lymphomas in collaboration with MYC (Clurman and Hayward 1989; Tam et al. 1997; Zhang et al. 2008a). Overexpression of *miR-155* has been observed in both hematological (Eis et al. 2005; Kluiver et al. 2005) and solid (Volinia et al. 2006; Yanaihara et al. 2006) tumors.

Experiments conducted in transgenic mice have shed light on the involvement of *miR-155* in cancer and the immune system. When *miR-155* was overexpressed under control of VH promoter-Ig heavy chain Eμ, it initially exhibited a preleukemic pre-B-cell proliferation, evident in spleen and bone marrow, followed by B-cell malignancy. These findings indicate that *miR-155* can induce polyclonal expansion, favoring the capture of secondary genetic changes necessary for full transformation (Costinean et al. 2006). Conversely, using genetic deletion and transgenic approaches, two groups independently showed that *miR-155* has an important role in the mammalian immune system (Rodriguez et al. 2007; Thai et al. 2007; Vigorito et al. 2007). Defective *miR-155* mice are immune deficient and have abnormalities in the maturation of B and T lymphocytes. Mouse transcriptome analysis identified a wide spectrum of *miR-155*-regulated genes, including cytokines, chemokines, and transcription factors. Another *miR-155* target, the tumor protein 53-induced nuclear protein 1 (TP53INP1), a proapoptotic factor, could explain, at least in part, the oncogenic role of the *miR-155* gene (Gironella et al. 2007). Interestingly, the *miR-155* expression level has a prognosis significance, and it is able to predict recurrence and survival in lung cancer and other tumors (Yang et al. 2012). The inhibition of *miR-155* may also have a therapeutic value as demonstrated in a recent report that showed that systemic delivery of antisense peptide nucleic acids encapsulated in nanoparticles slows the tumor growth in a mouse model lymphoma/leukemia driven by *miR-155* (Babar et al. 2012).

4. miR-221 *and* miR-222. These miRNAs are located in a cluster within the Xp11.3 region. *miR-221* and *miR-222* are frequently and consistently up-regulated in many tumors, including glioblastoma and pancreatic, kidney, bladder, colon, stomach, prostate, and thyroid cancers (le Sage et al. 2007; (Felicetti et al. 2008; (Garofalo et al. 2009). *miR-221* and *miR-222* were identified as potent regulators of dependent kinase inhibitor 1b (CDKN1B, p27/Kip1) and cyclin-dependent kinase inhibitor 1b (p57, Kip2), also known as CDKN1C, cell cycle inhibitors, and tumor suppressors (Fornari et al. 2008). Accordingly, high levels of *miR-221* and *miR-222* appear in glioblastomas and correlate with low levels of p27(Kip1) protein (le Sage et al. 2007). Another study found that *miR-221* and *miR-222* are overexpressed in aggressive non-small-cell lung cancer and hepatocarcinoma cells, as compared with less invasive and/or normal lung and liver cells (Garofalo et al. 2009). In melanoma, *miR-221* and *miR-222* control the tumor progression through downmodulation of CDKN1B/p27/Kip1 and c-KIT receptor (Felicetti et al. 2008). A transgenic model showed that *miR-221* deregulation in the liver produces spontaneous tumors and more sensitivity to diethylnitrosamine-induced tumors. Consequently, *in vivo* delivery of anti-*miR-221* oligonucleotides leads to a significant tumor reduction (Callegari et al. 2012). Recent results in a zebrafish model found *miR-221* to be important in cell migration, proliferation, and angiogenesis targeting the CDKN1B and phosphoinositide-3-kinase regulatory subunit 1 (PIK3R1) (Nicoli et al. 2012).

5. miR-372 and miR-373. These miRNAs are located in a cluster within the 19q13.42 region. A genetic screen for miRNAs that cooperate with oncogenes in cellular transformation identified oncogenic proprieties in *miR-372* and *miR-373* in human testicular germ cell tumors (Voorhoeve et al. 2007). These miRNAs promote tumorigenesis of primary human cells that harbor both oncogenic *RAS* and active wild-type p53 (Voorhoeve et al. 2007).

Functional assays determined that these miRNAs could disable the p53 pathway by inhibition of *CDK2*, possibly through direct inhibition of the expression of the large tumor suppressor homologue 2 (LATS2). Thus, these miRNAs render the cells insensitive to the suppression abilities of p53 and so act in overcoming senescence. However, other reports suggest that *miR-372* acts as a tumor suppressor gene in other tumors as *miR-372* is downregulated in cervical carcinoma tissues compared with adjacent normal cervical tissue. Growth curve and fluorescence-activated cell sorting (FACS) assays indicated that ectopic expression of *miR-372* suppressed cell growth and induced arrest in the S/G2 phases of cell cycle in HeLa cells (Tian et al. 2011). The antioncogenic role of *miR-372* may function through control of cell growth and cell cycle progression by down-regulating the cell cycle genes CDK2 and cyclin A1.

B. miRNAs That Function as Tumor Suppressors

1. The let-7 Family. *let-7* was the first miRNA identified in humans (Reinhart et al. 2000), and the *let-7* family consists of 12 very closely related sequences that encode for nine mature miRNAs (*let-7a, let-7b, let-7c, let-7d, let-7e, let-7f, let-7g, let-7i,* and *miR-98*). In *C. elegans, let-7* acts as a master temporal regulator of multiple genes required for cell cycle exit in seam cells, a stem cell-like population (Reinhart et al. 2000; Johnson et al. 2005). Many human *let-7* genes map to regions altered or deleted in human tumors (Calin et al. 2004), indicating that these genes may function as tumor suppressors. In fact, *let-7g* maps to 3p21, which has been implicated in the initiation of lung cancers (Yanaihara et al. 2006). Several studies have shown that *let-7* is expressed at lower levels in lung tumors than in normal tissue (Calin et al. 2004; Takamizawa et al. 2004; Johnson et al. 2005; Yanaihara et al. 2006). Furthermore, *let-7* levels have been correlated to lung cancer prognosis; patients with lower *let-7* expression survived for a shorter time than those with higher *let-7* expression in lung (Takamizawa et al. 2004; Yanaihara et al. 2006; Yu et al. 2008) and other tumors (Shell et al. 2007). Not surprisingly, decreased expression of *let-7* family members are among the miRNAs most frequently associated with poor outcome in human cancer (Nair et al. 2012).

A possible mechanistic explanation for the biological role of *let-7* as a tumor suppressor was provided by the discovery that the RAS oncogenes are *let-7* targets (Johnson et al. 2005). The RAS family encodes 21-kDa protein kinases that bind guanine nucleotides (guanosine triphosphate [GTP] and guanosine diphosphate [GDP]) and are implicated in signal transduction processes, including mitogenesis. Mutations in the *RAS* family of proto-oncogenes (comprising *H-RAS, N-RAS,* and *K-RAS*) are found in 20–30% of all human tumors. About one-third of human lung adenocarcinomas (LACs) carry *RAS* oncogenic mutations, mainly in codon 12 of the *KRAS* gene. Interestingly, the 3′-UTRs of human *RAS* genes contain multiple *let-7* complementary elements, and *let-7* represses the expression of *KRAS* and *NRAS* in tissue culture through their 3′-UTRs. Moreover, in lung squamous cell carcinoma, low *let-7* levels correlate with high RAS expression, consistent with *let-7*-mediated regulation of RAS protein levels *in vivo* (Johnson et al. 2005). Other oncogenes and proteins involved in cell cycle have been reported to be regulated by the *let-7* family, such as HGMA2 (Lee and Dutta 2007; Mayr et al. 2007), myc (Sampson et al. 2007), CDK6, and CDC25 (Johnson et al. 2007). Moreover, *let-7* expression also slows down cell growth in tissue culture (Johnson et al. 2007) and tumor growth both in xenograft and lung cancer mouse models (Kumar et al. 2008; Trang et al. 2009).

Several reports show that the tumor suppressor activity of *let-7* encompasses other tumor types such as gastric tumors (Zhang et al. 2008b), colon cancer (Akao et al. 2006), and Burkitt's lymphoma (Sampson et al. 2007). In breast cells, *let-7* represses self-renewal and tumorigenicity (Yu et al. 2007). Increased expression of *let-7* in self-renewing tumor-initiating cells (T-ICs) decreased their proliferation and metastatic capacity.

The regulation of *let-7* expression has become increasingly important in light of its involvement in tumor suppression. The expression of *let-7* miRNA family is regulated at various stages of its biogenesis, and the regulation processes involve numerous factors, for example, by pluripotency promoting factor LIN28 (Viswanathan et al. 2008), or is controlled by MYC binding to their promoters (Chang and Mendell 2007) in an autoregulatory loop with MYC.

2. miRNA-15a *and* miRNA-16-1. Pioneering studies of CLL revealed one of the first evidence that miRNAs can play a direct role in cancer. Frequent deletion of the 13q14 region in the majority of CLL cases suggested the presence of a tumor suppressor gene in that locus. For several years, the search for a candidate tumor suppressor in the 13q14 region had been frustrated by the absence of a suitable protein-coding gene. Deletion analyses showed a 30-kb region, which comprised a non-protein-coding gene called leukemia-associated gene 2 (*LEU2*), as the minimal common region lost in CLL patients (Bullrich et al. 2001). Calin and collaborators became aware that *LEU2* contains *miR-15a* and *miR-16-1* in its first intron. Reduced abundance of these two miRNAs was then documented in 68% of CLL cases analyzed (Calin et al. 2002). Accordingly, germ line mutations associated with the down-regulation of these miRNAs were found in CLL patients (Calin et al. 2005). Deletion of the *miR-15* and *miR-16* clusters in mice causes the development of B-cell lymphoproliferative disorders, recapitulating the spectrum of CLL-associated phenotypes observed in humans (Klein et al. 2010; Lia et al. 2012).

Further functional analysis described the antiapoptotic oncogene *BCL2* as one of targets regulated by *miR-15a* and *miR-16-1*. The levels of these miRNAs were inversely correlated to *BCL2* expression, and reporter assays determined that both miRNAs negatively regulate *BCL2* at a posttranscriptional level. Moreover, *BCL2* repression by these miRNAs induces apoptosis in a CLL cell line model (Cimmino et al. 2005).

Other studies provide increasing evidence that the *miR-15* family regulates the cell cycle through regulating multiple cell cycle genes including CyclinD1 (CCND1), CyclinD3 (CCND3), CyclinE1 (CCNE1), and CDK6 (Linsley et al. 2007; Liu et al. 2008).

3. miR-29. The *miR-29* family consists of four members: *miR-29a*, *miR-29b-1*, *miR-29b-2*, and *miR-29c*, which are encoded in two genetic clusters located in regions 7q32.3 and 1q32.2. Members of this family have been shown to be silenced or down-regulated in many different types of cancer and have subsequently been attributed tumor-suppressing properties, although exceptions have been described where *miR-29s* have tumor-promoting functions (Pekarsky and Croce 2010; Santanam et al. 2010). Experimental evidence suggests that putative targets of the *miR-29* family includes the antiapoptotic factor BCL-2 (Mott et al. 2007), T-cell leukemia/lymphoma protein 1A (TCL1A) oncogene, found to be disrupted in many T-cell leukemias (Pekarsky et al. 2006), and DNA methyltransferases DNMT3A and DNMT3B, which are frequently up-regulated in lung cancer (Fabbri et al. 2007). Restoration of *miR-29b* in AML cell lines and primary samples induces apoptosis, targeting myeloid leukemia cell differentiation protein (*Mcl-1*) and dramatically reduces tumorigenicity in a xenograft leukemia model (Mott et al. 2007; Garzon et al. 2009). A recent report found that the down-regulation of *miR-29a* may be used as an unfavorable prognostic marker in pediatric AML.

4. miR-31. Located in the chromosomal region 9p21.3, the expression of *miR-31* has been inversely correlated with metastasis in breast cancer patients (Valastyan et al. 2009). On the contrary, overexpression of *miR-31* in otherwise aggressive breast tumor cells can suppress metastasis; *miR-31*-mediated inhibition of several steps of metastasis, including local invasion, extravasation, or initial survival at a distant site; and metastatic colonization (Valastyan et al. 2009). Such antimetastatic effects are achieved via coordinate repression of a set of metastasis-promoting genes, including RhoA. Indeed, RhoA reexpression partially reverses *miR-31*-imposed metastasis suppression (Valastyan et al. 2009).

Recently, the loss of *miR-31* expression has been reported in adult T-cell leukemia (ATL) (Yamagishi et al. 2012). In ATL, there is a constitutive NF-κB activation that plays a causative role in the pathology. *miR-31* negatively regulates the noncanonical NF-κB pathway by targeting NF-κB-inducing kinase (NIK). Therefore, the loss of *miR-31* could contribute to trigger the oncogenic signaling of NF-κB (Yamagishi et al. 2012).

5. miR-34. In humans, the *miR-34* family is composed of three evolutionary conserved members (*miR-34a*, *miR-34b*, and *miR-34c*). *miR-34a* resides in the 1p36 chromosomal region, and *miR-34b* and *miR-34c* are derived from the same transcript in the 11q23.1 region. Initial links to cancer arose from the observation that low levels of *miR-34* are found in neuroblastomas (Welch et al. 2007). The reintroduction of *miR-34* into neuroblastoma cell lines caused a dramatic reduction in cell proliferation through the induction of a caspase-dependent apoptotic pathway. Shortly after this initial observation, independent studies documented the involvement of *miR-34* in the p53 pathway (Bommer et al. 2007; Chang and Mendell 2007; Chang et al. 2007; Corney et al. 2007; He et al. 2007; Tarasov et al. 2007). Interestingly, the expression of the *mir-34* family members is regulated directly by p53 (Hermeking 2010), and subsequently, the expression of the *mir-34* family reflects p53 activity. Accordingly, these miRNAs act as tumor suppressor genes, and their reintroduction in defective cells promotes cell cycle arrest and senescence or apoptosis depending on the genetic background. Experimental analysis and bioinformatics predictions have implicated the *miR-34* family in the regulation of important genes implicated in the control of cell cycle and apoptosis, such as *E2F3*, *CDK4*, *CDK6*, *CCNE2*, *BIR3*, *DCR3*, and *BCL2* (Bommer et al. 2007; He et al. 2007; Tazawa et al. 2007; Welch et al. 2007). Interestingly, the *miR-34* expression is negatively regulated by myc promoting high-grade transformation in B-cell lymphoma (Craig et al. 2011). On the other hand, during senescence, *miR-34* targets myc and thereby coordinately controls a set of cell cycle regulators (Christoffersen et al. 2010). Recently, it was reported that *in vivo* expression of *miR-34a* prevented tumor formation and progression in a therapeutically resistant K-RAS and p53-induced mouse model of LAC (Kasinski and Slack 2012), opening new possibilities for miRNA therapy based on this miRNA.

6. miR-126. Located in the chromosomal region 9q34.3, *miR-126* is silenced in a variety of cancers including colorectal, gastric, lung, prostate, bladder, and breast cancers (Liu et al. 2009; Feng et al. 2010; Li et al. 2011; Hamada et al. 2012). A recent report has shown that *miR-126* suppresses metastatic endothelial recruitment, metastatic angiogenesis, and metastatic colonization of breast cancer cells through coordinate targeting of proangiogenic genes and biomarkers of human metastasis including IGFBP2, PITPNC1, and MERTK (Png et al. 2012). Interestingly, reduction of the levels of *miR-126* in plasma correlated with poor prognosis in ATL, indicating their potential usefulness as a novel biomarker for the assessment of disease stage of this malignancy (Ishihara et al. 2012).

7. miR-145. *miR-145* is located in a cluster in the chromosomal region 5q32. Several reports have found that *miR-145* can promote an antistemness phenotype. *miR-145* expression is low in self-renewing human embryonic stem cells (hESCs) but it is highly up-regulated during differentiation (Xu et al. 2009). *miR-145* is able to negatively regulate the expression of OCT4, SOX2, and KLF4, important stem cell regulators. Increased *miR-145* expression inhibits hESC self-renewal, represses expression of pluripotency genes, and induces lineage-restricted differentiation (Gotte et al. 2010). On the contrary, loss of *miR-145* impairs differentiation and elevates the expression of OCT4, SOX2, and KLF4. Interestingly, *miR-145* is underexpressed in several tumor types including breast (Gotte et al. 2010) and colon cancers (Zhang et al. 2011). Additionally, it has been reported that *miR-145* is capable of inhibiting tumor cell growth and invasion by targeting genes such as c-Myc, JAM-A, and fascin (Sachdeva et al. 2009; Gotte et al. 2010).

8. miR-203. *miR-203* is encoded within the chromosomal region 14q32.33 within a fragile region that is lost is some tumors (Calin et al. 2004). Initially, *miR-203* was found to promote epidermal differentiation in keratinocytes by repressing "stemness" by targeting p63, an essential regulator of stem cell maintenance during epithelial development (Yi et al. 2008). The expression of *miR-203* genes has been found to be deregulated in several cancers, including hepatocellular and pancreatic tumors (Furuta et al. 2010; Greither et al. 2010; Ikenaga et al. 2010). *miR-203* is additionally hypermethylated in several hematopoietic tumors, including chronic myelogenous leukemias and some acute lymphoblastic leukemias (Bueno et al. 2008). A putative target of *miR-203*, ABL1, is specifically activated in these hematopoietic malignancies as the BCR-ABL1 fusion protein (Philadelphia chromosome). Overexpression of *miR-203* reduces ABL1 and BCR-ABL1 fusion protein levels and inhibits tumor cell proliferation in an ABL1-dependent manner (Bueno et al. 2008).

IV. miRNAs AND THEIR FUTURE USE IN THE CLINIC: DIAGNOSIS, PROGNOSIS, AND THERAPY

Although the miRNA era started only a few years ago, it has brought great promise for diagnosis, prognosis, and therapy of cancer. The quick development of powerful techniques such as miRNA microarrays, short-RNA deep sequencing, specific quantitative *polymerase chain* reaction (PCR) of miRNAs, and antisense technologies is expected to have a significant impact on clinical oncology in the next decade.

Since miRNAs are key factors that define cell identity, they could be used as a valuable tool in cancer diagnosis. Pioneering studies using miRNA microarray analysis and bead-based miRNA profiling identified statistically unique profiles, which could easily discriminate cancers from noncancerous tissues (Liu et al. 2004). Indeed, miRNA expression profiles seem to be more informative than traditional mRNA profiling for the classification of tumors with respect to their tissue of origin and differentiation (Lu et al. 2005). These findings demonstrate the effectiveness of miRNAs as biomarkers for tracing the tissue of origin of tumors of unknown primary origin, a major clinical problem (Gleave and Monia 2005).

In addition, miRNA expression profiles can also provide important information regarding the prognosis of cancer patients. For example, high expression of *miR-21* and *miR-155* predicts recurrence and unfavorable survival in non-small-cell lung cancer (Yang et al. 2012). The levels of these miRNAs could also help in predicting relapse of the cancer (Yu et al. 2008). Another study performed in colorectal cancer showed that high *miR-21* expression is associated with poor survival and poor therapeutic outcome (Schetter et al. 2008b).

Therapeutic strategies based on modulation of miRNA activity hold great promise due to the ability of these small RNAs to potently influence cellular functionality. The discovery of miRNAs acting as oncogenes suggests that antisense oligonucleotides could specifically block their pathogenic activity. These anti-miRs are oligonucleotides complementary to the miRNAs that have a chemical modification to improve their stability and/or the delivery. As in the case of many other drugs used in cancer therapies, problems related to the delivery, stability, or toxicity of these new compounds are to be expected as the experimentation moves more into the clinical phases. As each miRNA has multiple targets, its inhibition could cause side effects. Nevertheless, these possible pitfalls, the exploiting of miRNAs for clinical applications holds great promise, and the biomedicine community is embracing these regulators of gene expression with growing interest.

REFERENCES

Akao, Y., Y. Nakagawa, and T. Naoe. 2006. let-7 microRNA functions as a potential growth suppressor in human colon cancer cells. *Biol Pharm Bull* **29**:903–906.

Aravin, A., D. Gaidatzis, S. Pfeffer, M. Lagos-Quintana, P. Landgraf, N. Iovino, P. Morris, M.J. Brownstein, S. Kuramochi-Miyagawa, T. Nakano, M. Chien, J.J. Russo, et al. 2006. A novel class of small RNAs bind to MILI protein in mouse testes. *Nature* **442**:203–207.

Babar, I.A., C.J. Cheng, C.J. Booth, X. Liang, J.B. Weidhaas, W.M. Saltzman, and F.J. Slack. 2012. Nanoparticle-based therapy in an in vivo microRNA-155 (miR-155)-dependent mouse model of lymphoma. *Proc Natl Acad Sci U S A* **109**:E1695–E1704.

Baek, D., J. Villén, C. Shin, F.D. Camargo, S.P. Gygi, and D.P. Bartel. 2008. The impact of microRNAs on protein output. *Nature* **455**:64–71.

Bartel, D.P. 2009. MicroRNAs: target recognition and regulatory functions. *Cell* **136**:215–233.

Bommer, G.T., I. Gerin, Y. Feng, A.J. Kaczorowski, R. Kuick, R.E. Love, Y. Zhai, T.J. Giordano, Z.S. Qin, B.B. Moore, O.A. MacDougald, K.R. Cho, et al. 2007. p53-mediated activation of miRNA34 candidate tumor-suppressor genes. *Curr Biol* **17**:1298–1307.

Bueno, M.J., I. Perez de Castro, M. Gomez de Cedron, J. Santos, G.A. Calin, J.C. Cigudosa, C.M. Croce, J. Fernandez-Piqueras, and M. Malumbres. 2008. Genetic and epigenetic silencing of microRNA-203 enhances ABL1 and BCR-ABL1 oncogene expression. *Cancer Cell* **13**: 496–506.

Bullrich, F., H. Fujii, G. Calin, H. Mabuchi, M. Negrini, Y. Pekarsky, L. Rassenti, H. Alder, J.C. Reed, M.J. Keating, T.J. Kipps, and C.M. Croce. 2001. Characterization of the 13q14 tumor suppressor locus in CLL: identification of ALT1, an alternative splice variant of the LEU2 gene. *Cancer Res* **61**:6640–6648.

Calin, G.A., C.D. Dumitru, M. Shimizu, R. Bichi, S. Zupo, E. Noch, H. Aldler, S. Rattan, M. Keating, K. Rai, L. Rassenti, T. Kipps, et al. 2002. Frequent deletions and down-regulation of micro-RNA genes miR15 and miR16 at 13q14 in chronic lymphocytic leukemia. *Proc Natl Acad Sci U S A* **99**:15524–15529.

Calin, G.A., C. Sevignani, C.D. Dumitru, T. Hyslop, E. Noch, S. Yendamuri, M. Shimizu, S. Rattan, F. Bullrich, M. Negrini, and C.M. Croce. 2004. Human microRNA genes are frequently located at fragile sites and genomic regions involved in cancers. *Proc Natl Acad Sci U S A* **101**: 2999–3004.

Calin, G.A., M. Ferracin, A. Cimmino, G. Di Leva, M. Shimizu, S.E. Wojcik, M.V. Iorio, R. Visone, N.I. Sever, M. Fabbri, R. Iuliano, T. Palumbo, et al. 2005. A microRNA signature associated with prognosis and progression in chronic lymphocytic leukemia. *N Engl J Med* **353**:1793–1801.

Callegari, E., B.K. Elamin, F. Giannone, M. Milazzo, G. Altavilla, F. Fornari, L. Giacomelli, L. D'Abundo, M. Ferracin, C. Bassi, B. Zagatti, F. Corra, et al. 2012. Liver tumorigenicity promoted by microRNA-221 in a mouse transgenic model. *Hepatology* **56**:1025–1033.

Cervigne, N.K., P.P. Reis, J. Machado, B. Sadikovic, G. Bradley, N.N. Galloni, M. Pintilie, I. Jurisica, B. Perez-Ordonez, R. Gilbert, P. Gullane, J. Irish, et al. 2009. Identification of a microRNA signature associated with progression of leukoplakia to oral carcinoma. *Human Mol Genet* **18**:4818–4829.

Chan, J.A., A.M. Krichevsky, and K.S. Kosik. 2005. MicroRNA-21 is an antiapoptotic factor in human glioblastoma cells. *Cancer Res* **65**:6029–6033.

Chang, T.C. and J.T. Mendell. 2007. MicroRNAs in vertebrate physiology and human disease. *Annu Rev Genomics Hum Genet* **8**:215–239.

Chang, T.C., E.A. Wentzel, O.A. Kent, K. Ramachandran, M. Mullendore, K.H. Lee, G. Feldmann, M. Yamakuchi, M. Ferlito, C.J. Lowenstein, D.E. Arking, M.A. Beer, et al. 2007. Transactivation of miR-34a by p53 broadly influences gene expression and promotes apoptosis. *Mol Cell* **26**:745–752.

Chang, S.S., W.W. Jiang, I. Smith, L.M. Poeta, S. Begum, C. Glazer, S. Shan, W. Westra, D. Sidransky, and J.A. Califano. 2008a. MicroRNA alterations in head and neck squamous cell carcinoma. *Int J Cancer* **123**:2791–2797.

Chang, T.C., D. Yu, Y.S. Lee, E.A. Wentzel, D.E. Arking, K.M. West, C.V. Dang, A. Thomas-Tikhonenko, and J.T. Mendell. 2008b. Widespread microRNA repression by Myc contributes to tumorigenesis. *Nat Genet* **40**:43–50.

Chi, S.W., G.J. Hannon, and R.B. Darnell. 2012. An alternative mode of microRNA target recognition. *Nat Struct Mol Biol* **19**:321–327.

Christoffersen, N.R., R. Shalgi, L.B. Frankel, E. Leucci, M. Lees, M. Klausen, Y. Pilpel, F.C. Nielsen, M. Oren, and A.H. Lund. 2010. p53-independent upregulation of miR-34a during oncogene-induced senescence represses MYC. *Cell Death Differ* **17**:236–245.

Ciafre, S.A., S. Galardi, A. Mangiola, M. Ferracin, C.G. Liu, G. Sabatino, M. Negrini, G. Maira, C.M. Croce, and M.G. Farace. 2005a. Extensive modulation of a set of microRNAs in primary glioblastoma. *Biochem Biophys Res Commun* **334**:1351–1358.

Ciafre, S.A., S. Galardi, A. Mangiola, M. Ferracin, C.G. Liu, G. Sabatino, M. Negrini, G. Maira, C.M. Croce, and M.G. Farace. 2005b. Extensive modulation of a set of microRNAs in primary glioblastoma. *Biochem Biophys Res Commun* **334**:1351–1358.

Cimmino, A., G.A. Calin, M. Fabbri, M.V. Iorio, M. Ferracin, M. Shimizu, S.E. Wojcik, R.I. Aqeilan, S. Zupo, M. Dono, L. Rassenti, H. Alder, et al. 2005. miR-15 and miR-16 induce apoptosis by targeting BCL2. *Proc Natl Acad Sci U S A* **102**:13944–13949.

Clurman, B.E. and W.S. Hayward. 1989. Multiple proto-oncogene activations in avian leukosis virus-induced lymphomas: evidence for stage-specific events. *Mol Cell Biol* **9**:2657–2664.

Consortium, E.P., B.E. Bernstein, E. Birney, I. Dunham, E.D. Green, C. Gunter, and M. Snyder. 2012. An integrated encyclopedia of DNA elements in the human genome. *Nature* **489**:57–74.

Corney, D.C., A. Flesken-Nikitin, A.K. Godwin, W. Wang, and A.Y. Nikitin. 2007. MicroRNA-34b and microRNA-34c are targets of p53 and cooperate in control of cell proliferation and adhesion-independent growth. *Cancer Res* **67**:8433–8438.

Corsten, M.F., R. Miranda, R. Kasmieh, A.M. Krichevsky, R. Weissleder, and K. Shah. 2007. MicroRNA-21 knockdown disrupts glioma growth in vivo and displays synergistic cytotoxicity with neural precursor cell delivered S-TRAIL in human gliomas. *Cancer Res* **67**:8994–9000.

Costinean, S., N. Zanesi, Y. Pekarsky, E. Tili, S. Volinia, N. Heerema, and C.M. Croce. 2006. Pre-B cell proliferation and lymphoblastic leukemia/high-grade lymphoma in E(mu)-miR155 transgenic mice. *Proc Natl Acad Sci U S A* **103**:7024–7029.

Craig, V.J., S.B. Cogliatti, J. Imig, C. Renner, S. Neuenschwander, H. Rehrauer, R. Schlapbach, S. Dirnhofer, A. Tzankov, and A. Muller. 2011. Myc-mediated repression of microRNA-34a promotes high-grade transformation of B-cell lymphoma by dysregulation of FoxP1. *Blood* **117**:6227–6236.

Dews, M., A. Homayouni, D. Yu, D. Murphy, C. Sevignani, E. Wentzel, E.E. Furth, W.M. Lee, G.H. Enders, J.T. Mendell, and A. Thomas-Tikhonenko. 2006. Augmentation of tumor angiogenesis by a Myc-activated microRNA cluster. *Nat Genet* **38**:1060–1065.

Eis, P.S., W. Tam, L. Sun, A. Chadburn, Z. Li, M.F. Gomez, E. Lund, and J.E. Dahlberg. 2005. Accumulation of miR-155 and BIC RNA in human B cell lymphomas. *Proc Natl Acad Sci U S A* **102**:3627–3632.

Fabbri, M., R. Garzon, A. Cimmino, Z. Liu, N. Zanesi, E. Callegari, S. Liu, H. Alder, S. Costinean, C. Fernandez-Cymering, S. Volinia, G. Guler, et al. 2007. MicroRNA-29 family reverts aberrant methylation in lung cancer by targeting DNA methyltransferases 3A and 3B. *Proc Natl Acad Sci U S A* **104**:15805–15810.

Felicetti, F., M.C. Errico, L. Bottero, P. Segnalini, A. Stoppacciaro, M. Biffoni, N. Felli, G. Mattia, M. Petrini, M.P. Colombo, C. Peschle, and A. Care. 2008. The promyelocytic leukemia zinc finger-microRNA-221/-222 pathway controls melanoma progression through multiple oncogenic mechanisms. *Cancer Res* **68**:2745–2754.

Feng, R., X. Chen, Y. Yu, L. Su, B. Yu, J. Li, Q. Cai, M. Yan, B. Liu, and Z. Zhu. 2010. miR-126 functions as a tumour suppressor in human gastric cancer. *Cancer Lett* **298**:50–63.

Fornari, F., L. Gramantieri, M. Ferracin, A. Veronese, S. Sabbioni, G.A. Calin, G.L. Grazi, C. Giovannini, C.M. Croce, L. Bolondi, and M. Negrini. 2008. MiR-221 controls CDKN1C/p57 and CDKN1B/p27 expression in human hepatocellular carcinoma. *Oncogene* **27**:5651–5661.

Friedman, R.C., K.K.-H. Farh, C.B. Burge, and D.P. Bartel. 2009. Most mammalian mRNAs are conserved targets of microRNAs. *Genome Res* **19**:92–105.

Fulci, V., S. Chiaretti, M. Goldoni, G. Azzalin, N. Carucci, S. Tavolaro, L. Castellano, A. Magrelli, F. Citarella, M. Messina, R. Maggio, N. Peragine, et al. 2007. Quantitative technologies establish a novel microRNA profile of chronic lymphocytic leukemia. *Blood* **109**:4944–4951.

Furuta, M., K.I. Kozaki, S. Tanaka, S. Arii, I. Imoto, and J. Inazawa. 2010. miR-124 and miR-203 are epigenetically silenced tumor-suppressive microRNAs in hepatocellular carcinoma. *Carcinogenesis* **31**:766–776.

Garofalo, M., G. Di Leva, G. Romano, G. Nuovo, S.S. Suh, A. Ngankeu, C. Taccioli, F. Pichiorri, H. Alder, P. Secchiero, P. Gasparini, A. Gonelli, et al. 2009. miR-221&222 regulate TRAIL resistance and enhance tumorigenicity through PTEN and TIMP3 downregulation. *Cancer Cell* **16**:498–509.

REFERENCES

Garzon, R., C.E. Heaphy, V. Havelange, M. Fabbri, S. Volinia, T. Tsao, N. Zanesi, S.M. Kornblau, G. Marcucci, G.A. Calin, M. Andreeff, and C.M. Croce. 2009. MicroRNA 29b functions in acute myeloid leukemia. *Blood* **114**:5331–5341.

Gibcus, J.H., L.P. Tan, G. Harms, R.N. Schakel, D. de Jong, T. Blokzijl, P. Moller, S. Poppema, B.J. Kroesen, and A. van den Berg. 2009. Hodgkin lymphoma cell lines are characterized by a specific miRNA expression profile. *Neoplasia* **11**:167–176.

Gironella, M., M. Seux, M.J. Xie, C. Cano, R. Tomasini, J. Gommeaux, S. Garcia, J. Nowak, M.L. Yeung, K.T. Jeang, A. Chaix, L. Fazli, et al. 2007. Tumor protein 53-induced nuclear protein 1 expression is repressed by miR-155, and its restoration inhibits pancreatic tumor development. *Proc Natl Acad Sci U S A* **104**:16170–16175.

Gleave, M.E. and B.P. Monia. 2005. Antisense therapy for cancer. *Nat Rev Cancer* **5**:468–479.

Gotte, M., C. Mohr, C.Y. Koo, C. Stock, A.K. Vaske, M. Viola, S.A. Ibrahim, S. Peddibhotla, Y.H. Teng, J.Y. Low, K. Ebnet, L. Kiesel, et al. 2010. miR-145-dependent targeting of junctional adhesion molecule A and modulation of fascin expression are associated with reduced breast cancer cell motility and invasiveness. *Oncogene* **29**:6569–6580.

Greither, T., L.F. Grochola, A. Udelnow, C. Lautenschlager, P. Wurl, and H. Taubert. 2010. Elevated expression of microRNAs 155, 203, 210 and 222 in pancreatic tumors is associated with poorer survival. *Int J Cancer* **126**:73–80.

Grimson, A., K.K. Farh, W.K. Johnston, P. Garrett-Engele, L.P. Lim, and D.P. Bartel. 2007. MicroRNA targeting specificity in mammals: determinants beyond seed pairing. *Mol Cell* **27**:91–105.

Hamada, S., K. Satoh, W. Fujibuchi, M. Hirota, A. Kanno, J. Unno, A. Masamune, K. Kikuta, K. Kume, and T. Shimosegawa. 2012. MiR-126 acts as a tumor suppressor in pancreatic cancer cells via the regulation of ADAM9. *Mol Cancer Res* **10**:3–10.

Hanahan, D. and R.A. Weinberg. 2011. Hallmarks of cancer: the next generation. *Cell* **144**: 646–674.

Hatley, M.E., D.M. Patrick, M.R. Garcia, J.A. Richardson, R. Bassel-Duby, E.V. Rooij, and E.N. Olson. 2010. Modulation of K-ras-dependent lung tumorigenesis by microRNA-21. *Cancer Cell* **18**:282–293.

He, L., J.M. Thomson, M.T. Hemann, E. Hernando-Monge, D. Mu, S. Goodson, S. Powers, C. Cordon-Cardo, S.W. Lowe, G.J. Hannon, and S.M. Hammond. 2005. A microRNA polycistron as a potential human oncogene. *Nature* **435**:828–833.

He, L., X. He, L.P. Lim, E. de Stanchina, Z. Xuan, Y. Liang, W. Xue, L. Zender, J. Magnus, D. Ridzon, A.L. Jackson, P.S. Linsley, et al. 2007. A microRNA component of the p53 tumour suppressor network. *Nature* **447**:1130–1134.

Hermeking, H. 2010. The miR-34 family in cancer and apoptosis. *Cell Death Differ* **17**:193–199.

Hill, D.A., J. Ivanovich, J.R. Priest, C.A. Gurnett, L.P. Dehner, D. Desruisseau, J.A. Jarzembowski, K.A. Wikenheiser-Brokamp, B.K. Suarez, A.J. Whelan, G. Williams, D. Bracamontes, et al. 2009. DICER1 mutations in familial pleuropulmonary blastoma. *Science* **325**:965.

Ikenaga, N., K. Ohuchida, K. Mizumoto, J. Yu, T. Kayashima, H. Sakai, H. Fujita, K. Nakata, and M. Tanaka. 2010. MicroRNA-203 expression as a new prognostic marker of pancreatic adenocarcinoma. *Ann Surg Oncol* **17**:3120–3128.

Ishihara, K., D. Sasaki, K. Tsuruda, N. Inokuchi, K. Nagai, H. Hasegawa, K. Yanagihara, and S. Kamihira. 2012. Impact of miR-155 and miR-126 as novel biomarkers on the assessment of disease progression and prognosis in adult T-cell leukemia. *Cancer Epidemiol* **36**:560–565.

John, B., A.J. Enright, A. Aravin, T. Tuschl, C. Sander, and D.S. Marks. 2004. Human microRNA targets. *PLoS Biol* **2**:e363.

Johnson, S.M., H. Grosshans, J. Shingara, M. Byrom, R. Jarvis, A. Cheng, E. Labourier, K.L. Reinert, D. Brown, and F.J. Slack. 2005. RAS is regulated by the let-7 microRNA family. *Cell* **120**:635–647.

Johnson, C.D., A. Esquela-Kerscher, G. Stefani, M. Byrom, K. Kelnar, D. Ovcharenko, M. Wilson, X. Wang, J. Shelton, J. Shingara, L. Chin, D. Brown, et al. 2007. The let-7 microRNA represses cell proliferation pathways in human cells. *Cancer Res* **67**:7713–7722.

Jongen-Lavrencic, M., S.M. Sun, M.K. Dijkstra, P.J. Valk, and B. Lowenberg. 2008. MicroRNA expression profiling in relation to the genetic heterogeneity of acute myeloid leukemia. *Blood* **111**:5078–5085.

Kanellopoulou, C., S.A. Muljo, A.L. Kung, S. Ganesan, R. Drapkin, T. Jenuwein, D.M. Livingston, and K. Rajewsky. 2005. Dicer-deficient mouse embryonic stem cells are defective in differentiation and centromeric silencing. *Genes Dev* **19**:489–501.

Karube, Y., H. Tanaka, H. Osada, S. Tomida, Y. Tatematsu, K. Yanagisawa, Y. Yatabe, J. Takamizawa, S. Miyoshi, T. Mitsudomi, and T. Takahashi. 2005. Reduced expression of Dicer associated with poor prognosis in lung cancer patients. *Cancer Sci* **96**:111–115.

Kasinski, A.L. and F.J. Slack. 2012. miRNA-34 prevents cancer initiation and progression in a therapeutically resistant K-ras and p53-induced mouse model of lung adenocarcinoma. *Cancer Res* **72**:5576–5587.

Kawai, S. and A. Amano. 2012. BRCA1 regulates microRNA biogenesis via the DROSHA microprocessor complex. *J Cell Biol* **197**:201–208.

Klein, U., M. Lia, M. Crespo, R. Siegel, Q. Shen, T. Mo, A. Ambesi-Impiombato, A. Califano, A. Migliazza, G. Bhagat, and R. Dalla-Favera. 2010. The DLEU2/miR-15a/16-1 cluster controls B cell proliferation and its deletion leads to chronic lymphocytic leukemia. *Cancer Cell* **17**: 28–40.

Kluiver, J., S. Poppema, D. de Jong, T. Blokzijl, G. Harms, S. Jacobs, B.J. Kroesen, and A. van den Berg. 2005. BIC and miR-155 are highly expressed in Hodgkin, primary mediastinal and diffuse large B cell lymphomas. *J Pathol* **207**:243–249.

Koesters, R., V. Adams, D. Betts, R. Moos, M. Schmid, A. Siermann, S. Hassam, S. Weitz, P. Lichter, P.U. Heitz, M. von Knebel Doeberitz, and J. Briner. 1999. Human eukaryotic initiation factor EIF2C1 gene: cDNA sequence, genomic organization, localization to chromosomal bands 1p34-p35, and expression. *Genomics* **61**:210–218.

Krek, A., D. Grun, M.N. Poy, R. Wolf, L. Rosenberg, E.J. Epstein, P. MacMenamin, I. da Piedade, K.C. Gunsalus, M. Stoffel, and N. Rajewsky. 2005. Combinatorial microRNA target predictions. *Nat Genet* **37**:495–500.

Kumar, M.S., J. Lu, K.L. Mercer, T.R. Golub, and T. Jacks. 2007. Impaired microRNA processing enhances cellular transformation and tumorigenesis. *Nat Genet* **39**:673–677.

Kumar, M.S., S.J. Erkeland, R.E. Pester, C.Y. Chen, M.S. Ebert, P.A. Sharp, and T. Jacks. 2008. Suppression of non-small cell lung tumor development by the let-7 microRNA family. *Proc Natl Acad Sci U S A* **105**:3903–3908.

Lawrie, C.H., S. Soneji, T. Marafioti, C.D. Cooper, S. Palazzo, J.C. Paterson, H. Cattan, T. Enver, R. Mager, J. Boultwood, J.S. Wainscoat, and C.S. Hatton. 2007. MicroRNA expression distinguishes between germinal center B cell-like and activated B cell-like subtypes of diffuse large B cell lymphoma. *Int J Cancer* **121**:1156–1161.

Lee, Y.S. and A. Dutta. 2007. The tumor suppressor microRNA let-7 represses the HMGA2 oncogene. *Genes Dev* **21**:1025–1030.

Lee, R.C., R.L. Feinbaum, and V. Ambros. 1993. The *C. elegans* heterochronic gene lin-4 encodes small RNAs with antisense complementarity to lin-14. *Cell* **75**:843–854.

le Sage, C., R. Nagel, D.A. Egan, M. Schrier, E. Mesman, A. Mangiola, C. Anile, G. Maira, N. Mercatelli, S.A. Ciafre, M.G. Farace, and R. Agami. 2007. Regulation of the p27(Kip1) tumor suppressor by miR-221 and miR-222 promotes cancer cell proliferation. *EMBO J* **26**:3699–3708.

REFERENCES

Lewis, B.P., C.B. Burge, and D.P. Bartel. 2005. Conserved seed pairing, often flanked by adenosines, indicates that thousands of human genes are microRNA targets. *Cell* **120**:15–20.

Li, J., H. Huang, L. Sun, M. Yang, C. Pan, W. Chen, D. Wu, Z. Lin, C. Zeng, Y. Yao, P. Zhang, and E. Song. 2009. MiR-21 indicates poor prognosis in tongue squamous cell carcinomas as an apoptosis inhibitor. *Clin Cancer Res* **15**:3998–4008.

Li, X.M., A.M. Wang, J. Zhang, and H. Yi. 2011. Down-regulation of miR-126 expression in colorectal cancer and its clinical significance. *Med Oncol* **28**:1054–1057.

Lia, M., A. Carette, H. Tang, Q. Shen, T. Mo, G. Bhagat, R. Dalla-Favera, and U. Klein. 2012. Functional dissection of the chromosome 13q14 tumor-suppressor locus using transgenic mouse lines. *Blood* **119**:2981–2990.

Linsley, P.S., J. Schelter, J. Burchard, M. Kibukawa, M.M. Martin, S.R. Bartz, J.M. Johnson, J.M. Cummins, C.K. Raymond, H. Dai, N. Chau, M. Cleary, et al. 2007. Transcripts targeted by the microRNA-16 family cooperatively regulate cell cycle progression. *Mol Cell Biol* **27**:2240–2252.

Liu, C.G., G.A. Calin, B. Meloon, N. Gamliel, C. Sevignani, M. Ferracin, C.D. Dumitru, M. Shimizu, S. Zupo, M. Dono, H. Alder, F. Bullrich, et al. 2004. An oligonucleotide microchip for genome-wide microRNA profiling in human and mouse tissues. *Proc Natl Acad Sci U S A* **101**:9740–9744.

Liu, X., Y. Sun, J. Guo, H. Ma, J. Li, B. Dong, G. Jin, J. Zhang, J. Wu, L. Meng, and C. Shou. 2006. Expression of hiwi gene in human gastric cancer was associated with proliferation of cancer cells. *Int J Cancer* **118**:1922–1929.

Liu, Q., H. Fu, F. Sun, H. Zhang, Y. Tie, J. Zhu, R. Xing, Z. Sun, and X. Zheng. 2008. miR-16 family induces cell cycle arrest by regulating multiple cell cycle genes. *Nucleic Acids Res* **36**:5391–5404.

Liu, B., X.C. Peng, X.L. Zheng, J. Wang, and Y.W. Qin. 2009. MiR-126 restoration down-regulate VEGF and inhibit the growth of lung cancer cell lines in vitro and in vivo. *Lung Cancer* **66**:169–175.

Lu, J., G. Getz, E.A. Miska, E. Alvarez-Saavedra, J. Lamb, D. Peck, A. Sweet-Cordero, B.L. Ebert, R.H. Mak, A.A. Ferrando, J.R. Downing, T. Jacks, et al. 2005. MicroRNA expression profiles classify human cancers. *Nature* **435**:834–838.

Ma, X., M. Kumar, S.N. Choudhury, L.E. Becker Buscaglia, J.R. Barker, K. Kanakamedala, M.F. Liu, and Y. Li. 2011. Loss of the miR-21 allele elevates the expression of its target genes and reduces tumorigenesis. *Proc Natl Acad Sci U S A* **108**:10144–10149.

Mayr, C., M.T. Hemann, and D.P. Bartel. 2007. Disrupting the pairing between let-7 and Hmga2 enhances oncogenic transformation. *Science* **315**:1576–1579.

Medina, P.P., M. Nolde, and F.J. Slack. 2010. OncomiR addiction in an in vivo model of microRNA-21-induced pre-B-cell lymphoma. *Nature* **467**:86–90.

Melo, S.A., S. Ropero, C. Moutinho, L.A. Aaltonen, H. Yamamoto, G.A. Calin, S. Rossi, A.F. Fernandez, F. Carneiro, C. Oliveira, B. Ferreira, C.G. Liu, et al. 2009. A TARBP2 mutation in human cancer impairs microRNA processing and DICER1 function. *Nat Genet* **41**:365–370.

Melo, S.A., C. Moutinho, S. Ropero, G.A. Calin, S. Rossi, R. Spizzo, A.F. Fernandez, V. Davalos, A. Villanueva, G. Montoya, H. Yamamoto, S. Schwartz Jr., et al. 2010. A genetic defect in exportin-5 traps precursor microRNAs in the nucleus of cancer cells. *Cancer Cell* **18**:303–315.

Meng, F., R. Henson, M. Lang, H. Wehbe, S. Maheshwari, J.T. Mendell, J. Jiang, T.D. Schmittgen, and T. Patel. 2006. Involvement of human micro-RNA in growth and response to chemotherapy in human cholangiocarcinoma cell lines. *Gastroenterology* **130**:2113–2129.

Meng, F., R. Henson, H. Wehbe-Janek, K. Ghoshal, S.T. Jacob, and T. Patel. 2007. MicroRNA-21 regulates expression of the PTEN tumor suppressor gene in human hepatocellular cancer. *Gastroenterology* **133**:647–658.

Mott, J.L., S. Kobayashi, S.F. Bronk, and G.J. Gores. 2007. mir-29 regulates Mcl-1 protein expression and apoptosis. *Oncogene* **26**:6133–6140.

Murchison, E.P., J.F. Partridge, O.H. Tam, S. Cheloufi, and G.J. Hannon. 2005. Characterization of Dicer-deficient murine embryonic stem cells. *Proc Natl Acad Sci U S A* **102**:12135–12140.

Nair, V.S., L.S. Maeda, and J.P. Ioannidis. 2012. Clinical outcome prediction by microRNAs in human cancer: a systematic review. *J Natl Cancer Inst* **104**:528–540.

Navarro, A., A. Gaya, A. Martinez, A. Urbano-Ispizua, A. Pons, O. Balague, B. Gel, P. Abrisqueta, A. Lopez-Guillermo, R. Artells, E. Montserrat, and M. Monzo. 2008. MicroRNA expression profiling in classic Hodgkin lymphoma. *Blood* **111**:2825–2832.

Nicoli, S., C.P. Knyphausen, L.J. Zhu, A. Lakshmanan, and N.D. Lawson. 2012. miR-221 is required for endothelial tip cell behaviors during vascular development. *Dev Cell* **22**:418–429.

O'Donnell, K.A., E.A. Wentzel, K.I. Zeller, C.V. Dang, and J.T. Mendell. 2005. c-Myc-regulated microRNAs modulate E2F1 expression. *Nature* **435**:839–843.

Ota, A., H. Tagawa, S. Karnan, S. Tsuzuki, A. Karpas, S. Kira, Y. Yoshida, and M. Seto. 2004. Identification and characterization of a novel gene, C13orf25, as a target for 13q31-q32 amplification in malignant lymphoma. *Cancer Res* **64**:3087–3095.

Pasquinelli, A.E., B.J. Reinhart, F. Slack, M.Q. Martindale, M.I. Kuroda, B. Maller, D.C. Hayward, E.E. Ball, B. Degnan, P. Muller, J. Spring, A. Srinivasan, et al. 2000. Conservation of the sequence and temporal expression of let-7 heterochronic regulatory RNA. *Nature* **408**:86–89.

Pekarsky, Y. and C.M. Croce. 2010. Is miR-29 an oncogene or tumor suppressor in CLL? *Oncotarget* **1**:224–227.

Pekarsky, Y., U. Santanam, A. Cimmino, A. Palamarchuk, A. Efanov, V. Maximov, S. Volinia, H. Alder, C.G. Liu, L. Rassenti, G.A. Calin, J.P. Hagan, et al. 2006. Tcl1 expression in chronic lymphocytic leukemia is regulated by miR-29 and miR-181. *Cancer Res* **66**:11590–11593.

Piskounova, E., C. Polytarchou, J.E. Thornton, R.J. LaPierre, C. Pothoulakis, J.P. Hagan, D. Iliopoulos, and R.I. Gregory. 2011. Lin28A and Lin28B inhibit let-7 microRNA biogenesis by distinct mechanisms. *Cell* **147**:1066–1079.

Png, K.J., N. Halberg, M. Yoshida, and S.F. Tavazoie. 2012. A microRNA regulon that mediates endothelial recruitment and metastasis by cancer cells. *Nature* **481**:190–194.

Qiao, D., A.M. Zeeman, W. Deng, L.H. Looijenga, and H. Lin. 2002. Molecular characterization of hiwi, a human member of the piwi gene family whose overexpression is correlated to seminomas. *Oncogene* **21**:3988–3999.

Reinhart, B.J., F.J. Slack, M. Basson, A.E. Pasquinelli, J.C. Bettinger, A.E. Rougvie, H.R. Horvitz, and G. Ruvkun. 2000. The 21-nucleotide let-7 RNA regulates developmental timing in *Caenorhabditis elegans*. *Nature* **403**:901–906.

Rodriguez, A., E. Vigorito, S. Clare, M.V. Warren, P. Couttet, D.R. Soond, S. van Dongen, R.J. Grocock, P.P. Das, E.A. Miska, D. Vetrie, K. Okkenhaug, et al. 2007. Requirement of bic/microRNA-155 for normal immune function. *Science* **316**:608–611.

Sachdeva, M., S. Zhu, F. Wu, H. Wu, V. Walia, S. Kumar, R. Elble, K. Watabe, and Y.Y. Mo. 2009. p53 represses c-Myc through induction of the tumor suppressor miR-145. *Proc Natl Acad Sci U S A* **106**:3207–3212.

Sampson, V.B., N.H. Rong, J. Han, Q. Yang, V. Aris, P. Soteropoulos, N.J. Petrelli, S.P. Dunn, and L.J. Krueger. 2007. MicroRNA let-7a down-regulates MYC and reverts MYC-induced growth in Burkitt lymphoma cells. *Cancer Res* **67**:9762–9770.

Santanam, U., N. Zanesi, A. Efanov, S. Costinean, A. Palamarchuk, J.P. Hagan, S. Volinia, H. Alder, L. Rassenti, T. Kipps, C.M. Croce, and Y. Pekarsky. 2010. Chronic lymphocytic leukemia

REFERENCES

modeled in mouse by targeted miR-29 expression. *Proc Natl Acad Sci U S A* **107**:12210–12215.

Schetter, A.J., S.Y. Leung, J.J. Sohn, K.A. Zanetti, E.D. Bowman, N. Yanaihara, S.T. Yuen, T.L. Chan, D.L. Kwong, G.K. Au, C.G. Liu, G.A. Calin, et al. 2008a. MicroRNA expression profiles associated with prognosis and therapeutic outcome in colon adenocarcinoma. *JAMA* **299**: 425–436.

Schetter, A.J., S.Y. Leung, J.J. Sohn, K.A. Zanetti, E.D. Bowman, N. Yanaihara, S.T. Yuen, T.L. Chan, D.L. Kwong, G.K. Au, C.G. Liu, G.A. Calin, et al. 2008b. MicroRNA expression profiles associated with prognosis and therapeutic outcome in colon adenocarcinoma. *JAMA* **299**: 425–436.

Shell, S., S.M. Park, A.R. Radjabi, R. Schickel, E.O. Kistner, D.A. Jewell, C. Feig, E. Lengyel, and M.E. Peter. 2007. Let-7 expression defines two differentiation stages of cancer. *Proc Natl Acad Sci U S A* **104**:11400–11405.

Si, M.L., S. Zhu, H. Wu, Z. Lu, F. Wu, and Y.Y. Mo. 2007. miR-21-mediated tumor growth. *Oncogene* **26**:2799–2803.

Sinkkonen, L., T. Hugenschmidt, P. Berninger, D. Gaidatzis, F. Mohn, C.G. Artus-Revel, M. Zavolan, P. Svoboda, and W. Filipowicz. 2008. MicroRNAs control de novo DNA methylation through regulation of transcriptional repressors in mouse embryonic stem cells. *Nat Struct Mol Biol* **15**:259–267.

Stadler, B.M. and H. Ruohola-Baker. 2008. Small RNAs: keeping stem cells in line. *Cell* **132**: 563–566.

Stark, A., J. Brennecke, N. Bushati, R.B. Russell, and S.M. Cohen. 2005. Animal microRNAs confer robustness to gene expression and have a significant impact on 3'UTR evolution. *Cell* **123**:1133–1146.

Sylvestre, Y., V. De Guire, E. Querido, U.K. Mukhopadhyay, V. Bourdeau, F. Major, G. Ferbeyre, and P. Chartrand. 2007. An E2F/miR-20a autoregulatory feedback loop. *J Biol Chem* **282**: 2135–2143.

Takamizawa, J., H. Konishi, K. Yanagisawa, S. Tomida, H. Osada, H. Endoh, T. Harano, Y. Yatabe, M. Nagino, Y. Nimura, T. Mitsudomi, and T. Takahashi. 2004. Reduced expression of the let-7 microRNAs in human lung cancers in association with shortened postoperative survival. *Cancer Res* **64**:3753–3756.

Tam, W., D. Ben-Yehuda, and W.S. Hayward. 1997. bic, a novel gene activated by proviral insertions in avian leukosis virus-induced lymphomas, is likely to function through its noncoding RNA. *Mol Cell Biol* **17**:1490–1502.

Tarasov, V., P. Jung, B. Verdoodt, D. Lodygin, A. Epanchintsev, A. Menssen, G. Meister, and H. Hermeking. 2007. Differential regulation of microRNAs by p53 revealed by massively parallel sequencing: miR-34a is a p53 target that induces apoptosis and G1-arrest. *Cell Cycle* **6**: 1586–1593.

Taubert, H., T. Greither, D. Kaushal, P. Wurl, M. Bache, F. Bartel, A. Kehlen, C. Lautenschlager, L. Harris, K. Kraemer, A. Meye, M. Kappler, et al. 2007. Expression of the stem cell self-renewal gene Hiwi and risk of tumour-related death in patients with soft-tissue sarcoma. *Oncogene* **26**:1098–1100.

Tazawa, H., N. Tsuchiya, M. Izumiya, and H. Nakagama. 2007. Tumor-suppressive miR-34a induces senescence-like growth arrest through modulation of the E2F pathway in human colon cancer cells. *Proc Natl Acad Sci U S A* **104**:15472–15477.

Thai, T.H., D.P. Calado, S. Casola, K.M. Ansel, C. Xiao, Y. Xue, A. Murphy, D. Frendewey, D. Valenzuela, J.L. Kutok, M. Schmidt-Supprian, N. Rajewsky, et al. 2007. Regulation of the germinal center response by microRNA-155. *Science* **316**:604–608.

Thomson, J.M., M. Newman, J.S. Parker, E.M. Morin-Kensicki, T. Wright, and S.M. Hammond. 2006. Extensive post-transcriptional regulation of microRNAs and its implications for cancer. *Genes Dev* **20**:2202–2207.

Tian, R.Q., X.H. Wang, L.J. Hou, W.H. Jia, Q. Yang, Y.X. Li, M. Liu, X. Li, and H. Tang. 2011. MicroRNA-372 is down-regulated and targets cyclin-dependent kinase 2 (CDK2) and cyclin A1 in human cervical cancer, which may contribute to tumorigenesis. *J Biol Chem* **286**: 25556–25563.

Trang, P., P.P. Medina, J.F. Wiggins, L. Ruffino, K. Kelnar, M. Omotola, R. Homer, D. Brown, A.G. Bader, J.B. Weidhaas, and F.J. Slack. 2009. Regression of murine lung tumors by the let-7 microRNA. *Oncogene* **29**:1580–1587.

Valastyan, S., F. Reinhardt, N. Benaich, D. Calogrias, A.M. Szasz, Z.C. Wang, J.E. Brock, A.L. Richardson, and R.A. Weinberg. 2009. A pleiotropically acting microRNA, miR-31, inhibits breast cancer metastasis. *Cell* **137**:1032–1046.

Vigorito, E., K.L. Perks, C. Abreu-Goodger, S. Bunting, Z. Xiang, S. Kohlhaas, P.P. Das, E.A. Miska, A. Rodriguez, A. Bradley, K.G. Smith, C. Rada, et al. 2007. MicroRNA-155 regulates the generation of immunoglobulin class-switched plasma cells. *Immunity* **27**:847–859.

Viswanathan, S.R., G.Q. Daley, and R.I. Gregory. 2008. Selective blockade of microRNA processing by Lin28. *Science* **320**:97–100.

Volinia, S., G.A. Calin, C.G. Liu, S. Ambs, A. Cimmino, F. Petrocca, R. Visone, M. Iorio, C. Roldo, M. Ferracin, R.L. Prueitt, N. Yanaihara, et al. 2006. A microRNA expression signature of human solid tumors defines cancer gene targets. *Proc Natl Acad Sci U S A* **103**:2257–2261.

Voorhoeve, P.M., C. le Sage, M. Schrier, A.J. Gillis, H. Stoop, R. Nagel, Y.P. Liu, J. van Duijse, J. Drost, A. Griekspoor, E. Zlotorynski, N. Yabuta, et al. 2007. A genetic screen implicates miRNA-372 and miRNA-373 as oncogenes in testicular germ cell tumors. *Adv Exp Med Biol* **604**:17–46.

Wang, Y., R. Medvid, C. Melton, R. Jaenisch, and R. Blelloch. 2007. DGCR8 is essential for microRNA biogenesis and silencing of embryonic stem cell self-renewal. *Nat Genet* **39**: 380–385.

Welch, C., Y. Chen, and R.L. Stallings. 2007. MicroRNA-34a functions as a potential tumor suppressor by inducing apoptosis in neuroblastoma cells. *Oncogene* **26**:5017–5022.

Wightman, B., I. Ha, and G. Ruvkun. 1993. Posttranscriptional regulation of the heterochronic gene lin-14 by lin-4 mediates temporal pattern formation in *C. elegans*. *Cell* **75**:855–862.

Woods, K., J.M. Thomson, and S.M. Hammond. 2007. Direct regulation of an oncogenic micro-RNA cluster by E2F transcription factors. *J Biol Chem* **282**:2130–2134.

Xhemalce, B., S.C. Robson, and T. Kouzarides. 2012. Human RNA methyltransferase BCDIN3D regulates microRNA processing. *Cell* **151**:278–288.

Xu, N., T. Papagiannakopoulos, G. Pan, J.A. Thomson, and K.S. Kosik. 2009. MicroRNA-145 regulates OCT4, SOX2, and KLF4 and represses pluripotency in human embryonic stem cells. *Cell* **137**:647–658.

Yamagishi, M., K. Nakano, A. Miyake, T. Yamochi, Y. Kagami, A. Tsutsumi, Y. Matsuda, A. Sato-Otsubo, S. Muto, A. Utsunomiya, K. Yamaguchi, K. Uchimaru, et al. 2012. Polycomb-mediated loss of miR-31 activates NIK-dependent NF-kappaB pathway in adult T cell leukemia and other cancers. *Cancer Cell* **21**:121–135.

Yanaihara, N., N. Caplen, E. Bowman, M. Seike, K. Kumamoto, M. Yi, R.M. Stephens, A. Okamoto, J. Yokota, T. Tanaka, G.A. Calin, C.G. Liu, et al. 2006. Unique microRNA molecular profiles in lung cancer diagnosis and prognosis. *Cancer Cell* **9**:189–198.

Yang, M., H. Shen, C. Qiu, Y. Ni, L. Wang, W. Dong, Y. Liao, and J. Du. 2012. High expression of miR-21 and miR-155 predicts recurrence and unfavourable survival in non-small cell lung cancer. *Eur J Cancer* **49**:604–615.

Yi, R., M.N. Poy, M. Stoffel, and E. Fuchs. 2008. A skin microRNA promotes differentiation by repressing "stemness". *Nature* **452**:225–229.

REFERENCES

Yu, F., H. Yao, P. Zhu, X. Zhang, Q. Pan, C. Gong, Y. Huang, X. Hu, F. Su, J. Lieberman, and E. Song. 2007. let-7 regulates self renewal and tumorigenicity of breast cancer cells. *Cell* **131**:1109–1123.

Yu, S.L., H.Y. Chen, G.C. Chang, C.Y. Chen, H.W. Chen, S. Singh, C.L. Cheng, C.J. Yu, Y.C. Lee, H.S. Chen, T.J. Su, C.C. Chiang, et al. 2008. MicroRNA signature predicts survival and relapse in lung cancer. *Cancer Cell* **13**:48–57.

Zhang, T., K. Nie, and W. Tam. 2008a. BIC is processed efficiently to microRNA-155 in Burkitt lymphoma cells. *Leukemia* **22**:1795–1797.

Zhang, Z., Z. Li, C. Gao, P. Chen, J. Chen, W. Liu, S. Xiao, and H. Lu. 2008b. miR-21 plays a pivotal role in gastric cancer pathogenesis and progression. *Lab Invest* **88**:1358–1366.

Zhang, J., H. Guo, H. Zhang, H. Wang, G. Qian, X. Fan, A.R. Hoffman, J.F. Hu, and S. Ge. 2011. Putative tumor suppressor miR-145 inhibits colon cancer cell growth by targeting oncogene Friend leukemia virus integration 1 gene. *Cancer* **117**:86–95.

15

LONG NON-CODING RNAs AND THEIR ROLES IN CANCER

Yolanda Sánchez and Maite Huarte

Center for Applied Medical Research (CIMA), Division of Oncology, University of Navarra, Pamplona, Spain

I.	Introduction	246
II.	General Features of lncRNAs	246
III.	Classifications of lncRNAs	247
IV.	Functions and Mechanisms of lncRNAs	250
	A. lncRNAs and Transcriptional Regulation	251
	B. Epigenetic Regulation and lncRNAs	252
	C. lncRNAs in Posttranscriptional Regulation	253
V.	lncRNAs and Cancer	254
	A. lncRNAs with Abnormal Expression in Cancer: Tumor Suppressor and Oncogenic lncRNAs	254
	B. Genetic Alterations in lncRNAs Associated with Cancer	258
VI.	Future Perspectives	259
	Acknowledgments	259
	References	259

ABBREVIATIONS

AS ncRNA	antisense non-coding RNA
eRNAs	enhancer RNAs
lincRNA	large intergenic non-coding RNA

MicroRNAs in Medicine, First Edition. Edited by Charles H. Lawrie.
© 2014 John Wiley & Sons, Inc. Published 2014 by John Wiley & Sons, Inc.

lncRNAs	long non-coding RNAs
miRNA	microRNA
NAT	natural antisense transcript
ncRNAs	non-coding RNAs
ORF	open reading frame
PRC1	polycomb repressive complex 1
PRC2	polycomb repressive complex 2
snoRNA	small nucleolar RNA
T-UCRs	transcribed ultraconserved regions

I. INTRODUCTION

The latest transcriptomic and genomic technologies have brought into focus an amazing fact: at least 90% of the genome is transcribed, although protein-coding genes represent only <2% of the total genome sequence. The realization that a very significant number of the RNA species in the cell are non-coding has prompted researchers to make an effort toward understanding this intricate transcriptional landscape.

The attempts to classify the large diversity of ncRNAs have arbitrarily established two major groups, mostly based on practical considerations due to the separation of RNAs in common experimental protocols: (1) short ncRNAs, which are those shorter than 200 nt in length and include miRNAs, PIWI-interacting RNAs, small nucleolar RNAs (snoRNA), and transcription initiation RNAs (tiRNAs) (Beckerman and Prives 2010), and (2) long non-coding RNAs (lncRNAs), which are non-coding transcripts longer than 200 nt and sometimes as long as several thousand nucleotides. Data from the Encyclopedia of DNA Elements (ENCODE) project show over 9640 human genome loci transcribed solely into lncRNAs, often with multiple transcript isoforms (Derrien et al. 2012).

II. GENERAL FEATURES OF LNCRNAS

In contrast to miRNAs, which are processed into an ~22-nt active form and act through a relatively well-defined and uniform mechanism of posttranscriptional regulation, lncRNAs constitute a very heterogeneous group of long RNA molecules, which allows them to cover a broad spectrum of cellular functions by implementing different modes of action. Additionally, the number of lncRNAs encoded by mammalian genomes is much higher than that of miRNAs, although lncRNAs are generally less evolutionarily conserved. Even though miRNAs are different from lncRNAs in their biogenesis and modes of action, some lncRNA transcripts have a dual role both as lncRNAs and miRNAs, because they can be processed to produce an active miRNA.

By definition, lncRNAs lack functional open reading frames (ORFs) and do not have apparent coding capacity. However, determining whether a transcript is non-coding is challenging because a long non-coding transcript is likely to contain non-functional ORFs (Dinger et al. 2008). To determine the coding potential of lncRNAs, computational methods such as the "codon substitution frequency" (CSF) analysis have been applied. CSF determines if codons for amino acids are conserved through evolution, indicating the protein-coding potential of the transcript (Clamp et al. 2007; Lin et al. 2007a, 2008, 2011). Based on such analyses, lncRNAs do not show protein-coding capacity, although it cannot be excluded that some may codify for small non-conserved peptides. Additionally, experimental methods such as ribosome profiling have provided a strategy for identifying ribosome occupancy on RNA, which has been proposed as a way to distinguish between coding and non-coding transcripts (Galindo et al. 2007).

A common feature of many lncRNAs is that their expression levels are usually lower than those of protein-coding genes. However, lncRNAs present more cell type- and tissue-specific expression patterns (Pang et al. 2006; Mercer et al. 2008; Guttman et al. 2009, 2010; Cabili et al. 2011), suggesting that lncRNAs may have very specific regulatory roles and their correct expression and function could be a key to maintaining cellular homeostasis (Huarte and Rinn 2011).

As mentioned earlier, another characteristic of lncRNAs is their lower evolutionary conservation at the sequence level compared with miRNAs and protein-coding genes, which is consistent with a species-specific function. However, there are some exceptions, such as the ultraconserved lncRNAs, which are transcribed from ultraconserved regions (T-UCR) of the genome (Bejerano et al. 2004) (see later discussion).

The absence of coding capacity and low conservation of the nucleotide sequence of most lncRNAs suggests that the most important feature of lncRNAs could be their capacity to adopt a secondary or tertiary structure required for their function (Wan et al. 2011).

III. CLASSIFICATIONS OF LncRNAs

The poor understanding of the biology of the thousands of lncRNAs encoded by mammalian cells makes their classification difficult and sometimes confusing. Several classes have been defined that could in some cases overlap with each other.

In order to classify lncRNAs, a distinction can be made based on the RNA polymerase enzyme that transcribes the RNA. Possibly the most abundant lncRNAs are those transcribed by RNA polymerase II (PolII), which, like other PolII transcripts, present a cap structure in their 5' end and a 3' PolyA tail (Cheng et al. 2005; Wu et al. 2008). These transcripts are structurally identical to messenger RNAs (mRNAs), often have multiple exons and introns, and can undergo alternative splicing (Guttman et al. 2010). Furthermore, their promoters are susceptible to the same mechanisms of regulation by transcription factors and chromatin regulators as promoters of protein-coding genes.

Other RNA polymerases like PolIII and PolI exclusively produce non-coding RNAs. PolIII classic products include small RNAs such as transfer RNAs (tRNAs), 7SL RNA, MRP, and H1 RNAs, BC1, the 200-nucleotide-long BC200 RNA, or the short interspersed repeated DNA element (SINE)-encoded RNAs. Recent studies have expanded the set of known PolIII-synthesized ncRNAs, suggesting that gene-specific PolIII regulation is more common than previously appreciated (Dieci et al. 2007). It has been proposed that lncRNAs transcribed by RNA PolIII and PolI could be involved in cancer, as both polymerases are frequently deregulated in cancer, resulting in increasing activity (White 2004, 2008).

An alternative criterion for the classifications of lncRNAs is their genomic position relative to protein-coding genes (Table 15.1).

lncRNAs produced from intergenic regions, that is, lncRNAs encoded completely in the intergenic space between protein-coding loci, are denominated large intergenic non-coding RNAs (lincRNAs) (Guttman et al. 2009). lincRNAs were first defined based on the distinctive chromatin signature that marks genes actively transcribed by RNA PolII (Guttman et al. 2009). As all other genes transcribed by RNA PolII, lincRNAs are marked in their promoter region with histone H3 lysine 4 trimethylation (H3K4me3) *and* with histone H3 lysine 36 trimethylation (H3K36me3) along the entire transcribed region. It is

TABLE 15.1. Classifications of lncRNAs

Name	Location	Function	Examples
Intergenic (lincRNAs)	Encoded in the intergenic space between protein-coding loci	Regulate the expression of neighboring genes and distant genomic sequences by multiple mechanisms; epigenetic regulators in some cases	lincRNA-p21, HOTAIR, MALAT-1 (Huarte et al. 2010; Guttman et al. 2011)
Antisense (AS ncRNAs or NATs)	Transcribed in the opposite strand of a protein-coding gene	Regulate expression of sense mRNA	ANRIL, Zeb2/SipNAT, p21NAT/BX332409 (Cunnington et al. 2010; Magistri et al. 2012)
Intronic	Inside an intron of a protein-coding gene	Multiple functions	ncRNA DHFR (Martianov et al. 2007)
Pseudogenes	Genomic loci that have premature stop codons, deletions/insertions, and mutations that abrogate their translation into functional protein	Regulation of the wild-type gene expression/function by serving as a source of endogenous small interference RNA (siRNA), antisense transcript, or sequestering miRNA	PTENP1 (Poliseno et al. 2010)
Ultraconserved (T-UCR)	Transcribed from T-UCRs	Unknown but some of them bind to miRNAs	uc.29, uc.73, uc.111 (Calin et al. 2007)
Enhancer (eRNAs)	Transcribed from enhancers	Involved in gene activation	c-fos eRNA (Kim et al. 2010)
Imprinted (macroRNAs)	Transcribed at imprinted gene loci	Target chromatin regulators involved in gene silencing in *cis*	Xist, Air, H19, Kcnq1 (Nagano et al. 2008; Zhao et al. 2008)

estimated that human cells encode more than 5000 lincRNAs, which function by multiple mechanisms and have different biological roles (Khalil et al. 2009; Tsai et al. 2010; Guttman et al. 2011). Although relatively few lincRNAs have been functionally characterized, several are known to be involved in cancer progression, such as HOTAIR and metastasis-associated lung adenocarcinoma transcript 1 (MALAT-1) (see later discussion) (Gupta et al. 2010; Xu et al. 2011).

When lncRNAs are transcribed from the opposite DNA strand of a known protein-coding gene overlapping with it, they are named antisense non-coding RNAs (AS ncRNAs) or natural antisense transcripts (NATs). NATs initiate inside or 3′ of a protein-coding gene and are transcribed in the opposite direction, overlapping at least with one coding exon. Other lncRNAs are encoded within introns of protein-coding genes, therefore called intronic lncRNAs (Louro et al. 2009; Rearick et al. 2011). An intronic lncRNA initiates inside an intron of a protein-coding gene in either direction and terminates without overlapping with exons.

Multiple lncRNAs are transcribed from the promoter of a protein-coding gene, therefore named promoter-associated lncRNAs. They can be bidirectional and initiate the transcription in a divergent way from the promoter of a protein-coding gene. For instance, lncRNA CCND1 is transcribed from the 5′ regulatory region of the *cyclin D1* gene when cells are subjected to DNA damage, mediating *cyclin D1* transcriptional repression (Wang et al. 2008b) (Figure 15.1B).

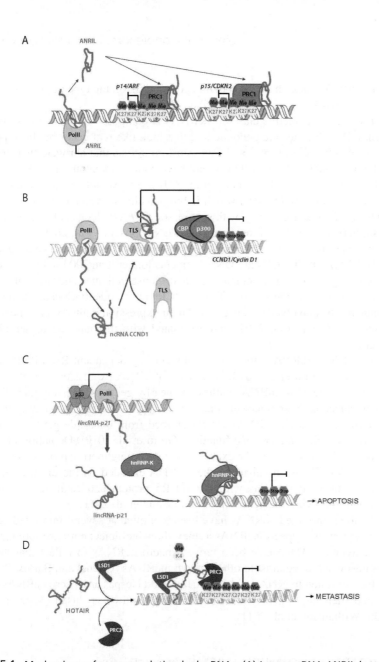

Figure 15.1. Mechanisms of gene regulation by lncRNAs. (A) Large ncRNA ANRIL is transcribed in antisense of the *p14/ARF* and *p15/CDKN2B* genes. ANRIL mediates gene silencing of the locus by interaction and recruitment of CBX7, a component of PRC1 histone 3–lysine 27 methyltransferase complex. (B) lncRNA CCND1 is transcribed from the 5′ of *cyclin D1/CCND1* gene. lncRNA CCND1 interacts with TLS protein, inducing a conformational change that allows its binding to the *cyclin D1* promoter and inhibiting *cyclin D1* gene expression by blocking CBP and p300 HAT activity. (C) Large intergenic non-coding RNA-p21 (lincRNA-p21) is induced by p53 and interacts with the protein hnRNP-K for binding to gene promoters for repression, which results in apoptosis induction. (D) lincRNA HOTAIR is transcribed from the HOXC locus. HOTAIR recruits PRC2 complex to gene promoters for histone 3–lysine 27 (H3K27) methylation. HOTAIR also binds LSD1–CoREST complex, which removes an active chromatin mark (H3K4 methylation), causing gene repression and leading to metastasis. See color insert.

Other lncRNA classes have been defined depending on the type of genomic sequence from which the RNA transcript is transcribed (Table 15.1).

For instance, a distinctive class of lncRNAs is composed by those transcribed from pseudogenes. Pseudogenes are genomic loci that look like real genes but have premature stop codons, deletions/insertions, and frameshift mutations that abrogate their translation into functional proteins. Because they do not have coding potential, they can be classified as lncRNAs (Zheng et al. 2007; Zhang et al. 2010). An example of this type of lncRNA is *PTEN pseudogene 1* (*PTENP1*), which is biologically active and can regulate the cellular levels of *PTEN* acting as a growth suppressor. *PTENP1* locus is lost in some human cancers, supporting its role as a regulator of PTEN (Poliseno et al. 2010).

Another class is composed of the lncRNAs produced from transcribed ultraconserved regions (T-UCRs). The T-UCRs are DNA segments longer than 200 bp that are conserved between orthologous regions in human, rat, and mouse genomes and most of them even in chicken and dog (Bejerano et al. 2004). T-UCRs are frequently located at fragile sites and genomic regions involved in cancers, and their expression is altered in some leukemias and carcinomas (Calin et al. 2007) and associated with the outcome in neuroblastoma (Scaruffi et al. 2009).

Enhancer RNAs (eRNAs) are produced by activity-dependent RNA PolII binding to specific enhancers (Kim et al. 2010). The level of eRNA expression at these enhancers correlates with the level of mRNA synthesis at nearby genes, suggesting that eRNA synthesis occurs specifically at enhancers that are actively involved in promoting the mRNA synthesis (Kim et al. 2010). eRNAs are transcribed from bona fide genomic enhancers, unlike lncRNAs with *enhancer-like* function (Orom et al. 2010). Depletion of the latter type also results in a decrease in the expression of their neighboring protein-coding genes. For instance, HOTTIP is an enhancer-like lncRNA encoded on the distal 5′ end of the HOXA gene cluster; it interacts with the WDR5 protein and catalyzes the activating H3K4me3 mark allowing the gene expression (Wang et al. 2011).

Finally, long imprinted ncRNAs have been identified at several imprinted gene clusters and are an unusual type as lncRNAs as they show inefficient cotranscriptional splicing and polyadenylation. They have been termed macro ncRNAs to reflect that their main product is unspliced, in contrast to other known lncRNAs (Guenzl and Barlow 2012). For instance, the imprinted lncRNAs Airn, Kcnq1ot1, and Nespas are macro ncRNAs required for imprinted silencing of genes in the igfr2, Kcnq1, and Gnas cluster (Mancini-Dinardo et al. 2006; Williamson et al. 2011).

IV. FUNCTIONS AND MECHANISMS OF LncRNAs

Although we know of the existence of thousands of lncRNAs transcribed by the mammalian genome, to this date, only a small number have been characterized in detail, revealing that lncRNAs are likely to participate in diverse biological processes through distinct mechanisms. lncRNAs have been implicated in gene regulatory roles virtually at every level, including chromosome dosage compensation, imprinting, epigenetic regulation, cell cycle control, nuclear and cytoplasmic trafficking, transcription, translation, splicing, cell differentiation, and other many processes (Mattick and Makunin 2006; Mattick 2009; Wilusz et al. 2009; Qureshi et al. 2010). To perform their functions, lncRNAs often form ribonucleoprotein complexes with different cellular factors, modulating their activities to regulate gene expression.

A. lncRNAs and Transcriptional Regulation

Eukaryotic RNA transcription is a tightly regulated process traditionally explained by the direct interaction of proteins with other proteins or with DNA to modulate the expression of a protein-coding gene. But the transcriptional process has an additional layer of complexity. Some lncRNAs can regulate different steps of this process, targeting transcriptional activators or repressors, components of the basal transcriptional machinery, including RNA PolII, and even the DNA duplex to regulate gene transcription and expression (Goodrich and Kugel 2006). Additionally, many lncRNAs can regulate gene transcription by interfacing with the epigenetic machinery and affecting chromatin structure (Koziol and Rinn 2010).

The expression of many lncRNAs is finely regulated in coordination with the downstream transcriptional programs in which they participate. Typically, the induction of a particular lncRNA is followed by the transcriptional regulation of its target genes, either *in cis* or *in trans*. For instance, DNA damage signals induce the expression of the lncRNA CCND1 associated with the *cyclin D1* gene promoter and mediating its transcriptional repression. This lncRNA interacts with the translocated in liposarcoma (TLS) protein, inducing a conformational change that allows the association of TLS to the *cyclin D1* promoter to silence it by impeding the histone acetyltransferase activities of CREB and p300 (Wang et al. 2008b) (Figure 15.1B). Similarly, when cells are stressed by DNA damage or oncogenic stress, the transcription factor p53, one of the more commonly mutated tumor suppressors in cancer, initiates a program that involves the induction of many genes, including lncRNAs. In particular, one lncRNA specifically induced by p53 named lincRNA-p21 mediates global gene repression to induce apoptosis by physically interacting with the protein hnRNP-K, allowing its localization to promoters of genes to be repressed in a p53-dependent manner (Huarte et al. 2010) (Figure 15.1C). These examples illustrate how the specific transcriptional regulation of lncRNAs may be critical for the activation of an adequate gene expression program in a tumor suppressor response.

Transcription of some lncRNAs is known to regulate the expression of genes in close genomic proximity (*cis*-acting regulation), functioning as transcriptional enhancers (Orom et al. 2010). For instance, the lncRNA *EVF2* recruits the binding and action of the transcription factor DLX2, which plays an important role in forebrain development by inducing the expression of the adjacent protein-coding genes Dlx5 and Dlx6 (Panganiban and Rubenstein 2002; Feng et al. 2006). Other *cis*-acting lncRNAs can regulate RNA PolII activity by interacting with the initiation complex. This is the case of an lncRNA transcribed from an upstream region of the dihydrofolate reductase (DHFR) locus, which forms a triplex between the single-stranded lncRNA and the double-stranded DHFR DNA at the major promoter of the gene to prevent the binding of the transcriptional cofactor TFIID (Martianov et al. 2007).

Additionally, some lncRNAs may promote the accessibility of RNA polymerases to promoter regions shared between the lncRNA and protein-coding genes. In yeast, glucose starvation results in the induction of the fbp1 gene and also of the several lncRNAs transcribed 5' upstream of the fbp1 (Hirota et al. 2008). When the lncRNAs are transcribed, the chromatin structure around the fbp1 promoter is accessible to the transcriptional machinery allowing fbp1 expression. This mechanism is not specific to yeast, as similar regulation has been observed in the human β-globin locus (Gribnau et al. 2000).

However, transcriptional regulation by lncRNAs occurs not only at proximal loci. Some lncRNAs can regulate distant genes (*trans*-acting lncRNAs) by targeting transcriptional activators or repressors to distant loci by different mechanisms, including functioning themselves as coregulators, modifying transcription factor activity, or regulating the association and activity of coregulators. For instance, human growth arrest-specific 5 (GAS5) lncRNA is induced under starvation and binds to the glucocorticoid receptor and prevents the receptor for binding to its correct regulatory elements (Kino et al. 2010).

B. Epigenetic Regulation and lncRNAs

Epigenetic modifications (DNA methylation and histone posttranslational modifications) regulate the expression of a large number of genes by remodeling the chromatin domains (Mikkelsen et al. 2007). Although RNA has long been considered as an integral component of chromatin (Rodriguez-Campos and Azorin 2007), nowadays, a number of studies suggest that lncRNAs are key components of the epigenetic regulatory processes, serving as a scaffold that directly or indirectly links chromatin complexes to DNA-binding proteins or transcription factors (Rinn et al. 2007; Khalil et al. 2009). It has been suggested that thousands of lncRNAs may mediate epigenetic changes by recruiting chromatin remodeling complexes to specific genomic loci, although the exact mechanism is not yet understood (Nagano et al. 2008; Pandey et al. 2008; Khalil et al. 2009; Ponting et al. 2009; Magistri et al. 2012).

The initial link between epigenetic regulation and lncRNAs was established by the discovery of the lncRNA X inactive specific transcript (Xist), involved in X-chromosome inactivation (XCI) and representing a classic and dramatic example of the epigenetic regulation by lncRNAs. Xist is expressed from one of the female X chromosomes and induces the transcriptional silencing of the entire chromosome (Borsani et al. 1991; Herzing et al. 1997). It has been shown that a segment of the Xist RNA (called RepA) is important for the targeting of Ezh2 (component of the polycomb repressive complex 2 [PRC2]) to the inactivated X chromosome, leading to H3K27me3 methylation and chromosome heterochromatin formation (Zhao et al. 2008).

The association between lncRNAs and epigenetic silencing has also been observed at imprinted gene loci. Many imprinted loci encode for at least one lncRNA that is involved in *cis*-silencing of the locus. Kcnq1ot1 and Air are two imprinting-associated lncRNAs paternally expressed and repressed on the maternal allele by DNA methylation on the promoter. lncRNA Air associates with the histone methyltransferase G9a and localizes to chromatin to silence three imprinted genes known as Slc22a3, Slc22a2, and Igf2r (Nagano et al. 2008). With a mechanism similar to Air, the AS ncRNA Kcnq1ot1 also regulates the expression of imprinted genes in a lineage-specific manner by recruiting repressive chromatin complexes such as G9a, PRC2, and the DNA methylase Dnmt1 to the Kcnq1 locus (Pandey et al. 2008; Mohammad et al. 2012).

The epigenetic regulation by lncRNAs expands beyond the above-described classical examples. Several lncRNAs have been found to function at this level, some of them with implications in cancer, where epigenetic changes are one of the determinants of the disease.

The lncRNA HOTAIR illustrates the relationship between epigenetic regulation by ncRNAs and cancer. HOTAIR is expressed from the HOXC locus and was initially discovered as a gene repressor of HOXD genes. HOTAIR acts as a molecular scaffold by interacting with PRC2 and the lysine-specific demethylase 1 (LSD1)–corepressor for element-1-silencing transcription factor (CoREST) complex to silence the HOXD loci (Tsai et al. 2010) (Figure 15.1D). HOTAIR is involved in cancer progression and poor prognostic of breast and colorectal cancer, phenotype associated with the PRC2-dependent gene repression induced by HOTAIR (Gupta et al. 2010; Kogo et al. 2011).

Another lncRNA involved in epigenetic regulation in cancer is the AS ncRNA ANRIL, transcribed in the antisense orientation of the INK4b-ARF-INK4a gene cluster, which controls expression in the INK4A/ARF locus. The INK4b/ARF locus encodes three tumor suppressor genes, INK4n/ARF/INK4α, p16/CDKN2A, and p15/CDKN2B, regulators of cell cycle progression and senescence. ANRIL was shown to interact with PRC2 as well as the polycomb/chromobox 7 (CBX7) protein, a member of the polycomb repressive complex 1 (PRC1), resulting in the targeting of both complexes to the chromatin and the establishment of repressive epigenetic marks (Yap et al. 2010) (Figure 15.1A). Altered ANRIL activity has been found in prostate cancer, and might be a cancer-initiating factor through silencing of the INK4b/ARF/INK4α locus (Yap et al. 2010).

ANRIL and HOTAIR act as scaffold RNA molecules by interacting with chromatin complexes, mechanism that has been suggested to be common to thousands of lncRNAs (Koziol and Rinn 2010; Guttman and Rinn 2012; Rinn and Chang 2012). In both cases, overexpression of the lncRNA causes changes to the chromatin landscape that can facilitate cancer initiation and/or progression, although the mechanism by which ANRIL and HOTAIR are altered in disease remains unclear.

C. lncRNAs in Posttranscriptional Regulation

lncRNAs can also modulate gene expression at the posttranscriptional level. The potential capacity of ncRNAs to recognize complementary sequences of the mRNA may allow for specific interactions necessary for the posttranscriptional processing of the mRNA, including the splicing, transport, editing, translation, and degradation. Some NATs or AS ncRNAs are involved in this type of posttranscriptional regulation. For instance, the BACE1-AS ncRNA controls the expression of BACE1 protein at posttranscriptional level (Faghihi et al. 2008). BACE1-AS ncRNA is up-regulated in the brain of Alzheimer's disease patients, increasing the stability of BACE1 protein and the presence of β-amyloid toxic plaques, characteristics of Alzheimer's disease.

However, NATs are not the only type of lncRNA involved in the regulation of protein expression. For instance, besides its role in regulating gene transcription, lincRNA-p21 has been found to associate with *JUNB* and *CTNNB1* mRNAs in HeLa cells lowering their translation in a selective manner (Yoon et al. 2012).

Some lncRNAs may function as molecular decoys for miRNAs, acting as natural sponges that compete for the binding of miRNAs with their mRNA targets (Karreth et al. 2011). For instance, miRNAs that target the tumor suppressor Pten mRNA can bind ncRNAs transcribed from pseudogenes (Karreth et al. 2011; Salmena et al. 2011). The competition for miRNA binding results in an additional level of regulation of gene expression where different classes of ncRNAs intervene and which is probably common to many tumor suppressor and oncogenes.

lncRNAs can also be involved in the regulation of splicing. Alternative splicing of mRNA allows the production of different protein isoforms with different cellular functions, presenting an opportunity for gene regulation. One example of an lncRNA involved at this level is a NAT whose expression is associated with human tumors with low E-cadherin expression. This NAT overlaps with an intronic 5′ splice site of the ZEB2 gene and prevents its splicing. The retained intron contains an internal ribosome entry site (IRES) necessary for the efficient translation of the ZEB2 protein, which functions as a transcriptional repressor of the E-cadherin, involved in the progression of different tumors (Beltran et al. 2008). But not just AS ncRNAs can regulate mRNA splicing. The lncRNA MALAT-1 plays a role in pre-mRNA alternative splicing regulation through its interaction with the serine/arginine-rich (SR) family of the nuclear phosphoprotcins that are involved in the splicing machinery (Bernard et al. 2010; Tripathi et al. 2010). MALAT-1 is up-regulated in a variety of human cancers (Ji et al. 2003; Lin et al. 2007b; Perez et al. 2008) such as non-small-cell lung cancer (NSCLC) metastasizing tumors, where it is overexpressed compared with the non-metastasizing tumors (Ji et al. 2003).

Besides the above described, there are additional steps where lncRNAs can modulate gene expression posttranscriptionally. For instance, lncRNAs have been involved in the process of protein trafficking between the nucleus and the cytoplasm of the cell, like the ncRNA NRON, which has been shown to regulate the trafficking of the transcription factor NFAT (nuclear factor of activated T cells) (Willingham et al. 2005).

Predictably, as our knowledge of the functions and biological roles of lncRNAs grows, the variety of mechanisms where lncRNAs intervene will expand, as well as our understanding of how lncRNA alterations affect cancer and other diseases.

V. LNCRNAS AND CANCER

Over the last few decades, studies have focused on the role of protein-coding genes in the pathogenesis of cancer. However, recent research points toward the need for an expanded view of cancer biology that includes ncRNAs. Our understanding of the contribution of lncRNAs to cancer is still poor, as less than 1% of the thousands of lncRNAs identified have been functionally characterized. Despite this relatively small number, lncRNAs have emerged as important regulatory players in several diseases including cancer. Mounting evidence shows that lncRNAs function as regulatory molecules in specific oncogenic and tumor suppressor pathways such as p53, MYC, and NF-κB (Khalil et al. 2009; Huarte et al. 2010; Hung et al. 2011). Additionally, a number of studies suggest that large ncRNAs interface with the epigenetic machinery (Guttman and Rinn 2012); epigenetic alterations are a major factor contributing to tumor transformation and cancer (Jones and Baylin 2007). Furthermore, many lncRNAs present altered expression in human cancers, and mutations linked with cancer lie in regions that encode for lncRNAs. It is not surprising, then, that lncRNAs are not just important for the maintenance of cellular homeostasis but are also key players in disease.

A. lncRNAs with Abnormal Expression in Cancer: Tumor Suppressor and Oncogenic lncRNAs

Many research groups are focusing their efforts to globally identify lncRNAs whose expression is altered in different cancer types, such as glioblastoma (Han et al. 2012), primary and metastatic pancreatic cancer (Tahira et al. 2011), or oral premalignant lesions (Gibb et al. 2011). As result of these studies, an increasing number or lncRNAs have been associated with cancer development, either as oncogenes (Table 15.2) or tumor suppressor gene candidates (Table 15.3).

One of the first lncRNAs found to be abnormally expressed in cancer was H19. H19 is an imprinted and maternally expressed lncRNA that is spliced, polyadenylated, and exported into the cytoplasm. H19 is an oncofetal gene that demonstrates maternal monoallelic expression in fetal tissues but can be reactivated during adult tissue regeneration and tumorigenesis. H19 has been implicated as a tumor suppressor or oncogene (Brannan et al. 1990; Ripoche et al. 1997; Matouk et al. 2007; Schoenfelder et al. 2007; Yoshimizu et al. 2008). It is known that H19 overexpression promotes tumorigenic properties of breast

TABLE 15.2. Examples of Oncogenic lncRNAs

ncRNA	Organism	Size	Genomic Location	Functional Characteristics	Mechanism	References
p21 NAT/ Bx332409	*Homo sapiens*	≥423 nt	Antisense *cdkn1a/p21*	Antisense of *Cdkn1a/p21*; negative regulation of *Cdkn1a/p21*	Requirement of Ago1 but not Ago2 for epigenetic silencing of *Cdkn1a/p21*	Morris et al. (2008)
HOTAIR	*Homo sapiens*	2200 nt	Intergenic *HoxC* locus	Gene silencing in *trans*; metastasis of breast and colon tumors	Interaction with PRC2 and LSD1 complexes and targeting to repressed genes	Rinn et al. (2007); Gupta et al. (2010); Kogo et al. (2011)
MALAT-1	*Homo sapiens*	8700 nt	Intergenic Chr11	Metastasis; overexpressed multiple tumor types	Induction of *GAGE6* proto-oncogene transcription by inhibition of protein-associated splicing factor (PSF) repressor	Lin et al. (2007b); Perez et al. (2008); Li et al. (2009)
VL30-1	*Mus musculus*	4900 nt	Retroelement non-coding RNA	Ras-mediated transformation of mouse fibroblasts	Induction of *Rab23* proto-oncogene transcription by inhibition of PSF repressor	Owen et al. (1990); Song et al. (2002, 2004)
ANRIL	*Homo sapiens*	2200 nt	Antisense of *INK4a/ARF/ INK4a* and *p15/CDKN2B*	Gene silencing of *INK4a/ARF/ INK4a* and *p15/CDKN2B*; up-regulated in prostate cancer	Interaction with CBX7 component of PRC1 complex	Yu et al. (2008); Cunnington et al. (2010); Yap et al. (2010); Pasmant et al. (2011)
H19	*Homo sapiens*	2300 nt	Imprinted *H19-Igf2* locus in Chr11	Control of imprinting; oncogenic or tumor suppressor; containing miRNA miR-675; overexpressed in multiple tumor types	Unknown	Brannan et al. (1990); Lottin et al. (2002b); Matouk et al. (2007)
CUDR	*Homo sapiens*	2200 nt	Intergenic Chr19	May regulate drug sensitivity and promote cellular transformation through resistance to apoptosis	Unknown	Tsang et al. (2007)
Zeb2/Sip1 NAT	*Homo sapiens*	680 nt	Antisense of *Zeb2/Sip1*	Overexpressed in human tumors with low E-cadherin expression; inhibition of E-cadherin through induction of Zeb2 protein levels	Inhibition of splicing of *Zeb2* first exon-containing IRES sequence	Beltran et al. (2008)

(*Continued*)

TABLE 15.2. (Continued)

ncRNA	Organism	Size	Genomic Location	Functional Characteristics	Mechanism	References
SRA-1	Homo sapiens	875 nt	Alternative splicing of *SRA* gene, loss of coding frame	Breast cancer; coactivator of steroid receptors and other transcription factors as MyoD; increased levels of non-coding isoform associated with metastasis	Interaction in ribonucleoprotein complexes with several positive regulators, including SRC-1, p68, and p72 and Pus1p and Pus3p, as well as negative regulators, such as Sharp and SLIRP to be recruited as promoters of regulated genes	Caretti et al. (2007); Colley et al. (2008); Colley and Leedman (2009); Cooper et al. (2009)
PCGEM1	Homo sapiens	1600 nt	Intergenic, Chr2	Inhibition of apoptosis, promotion of cell growth; up-regulated in prostate cancer in African American patients	Unknown	Petrovics et al. (2004); Fu et al. (2006); Ifere and Ananaba (2009)
UCA1	Homo sapiens	1400 nt	Intergenic, Chr19	Increases proliferation, migration, invasion, and drug resistance of human bladder cell line; associated with bladder cancer	Unknown	Wang et al. (2008a)
PCAT-1	Homo sapiens	7800 nt	Intergenic, Chr 8	Promotes cell proliferation; overexpressed in prostate tissue and prostate cancer cell lines	Interaction with PCR2	Prensner et al. (2011)
HULC	Homo sapiens	1639 nt	Chr6	Role in tumorigenesis; overexpressed in liver cancer	Acts as a sponge for miR-372, reducing its expression and activity	Du et al. (2012)
PANDA	Homo sapiens	1500 nt	Antisense, Chr 6	Induced by p53 upon DNA damage; inhibits the expression of the pro-apoptotic genes	Interacts with the transcription factor NF-YA	Hung et al. (2011)

TABLE 15.3. Examples of Tumor Suppressor lncRNAs

RNA	Organism	Size	Genomic Location	Functional Characteristics	Mechanism	References
GAS5	Homo sapiens, Mus musculus	Multiple splicing isoforms 600–1800 nt (human)	Intergenic; some introns encode snoRNAs	Induces growth arrest and apoptosis; down-regulated in breast cancer, up-regulated by induced growth arrest	Blocks transcriptional induction by glucocorticoid receptor (GR) by binding to GR DNA binding domain and competing with DNA glucocorticoid responsive element (GRE)	Coccia et al. (1992); Smith and Steitz (1998); Mourtada-Maarabouni et al. (2009); Kino et al. (2010)
lincRNA-p21	Mus musculus	3100 nt	Intergenic, upstream of p21/cdkn1a	Global gene repression in the p53 transcriptional response inducing cellular apoptosis	Physical interaction with hnRNP-K, targeting genes for transcriptional repression	Huarte et al. (2010)
MEG3	Homo sapiens	Several isoforms 1600–1800 nt	Intergenic, Chr 14	Activates p53 activity on specific genes by increasing p53 protein levels by suppressing MDM2 levels; also inhibits cell proliferation independently of p53; maternally expressed imprinted gene; hypermethylated in myeloid leukemia, multiple myeloma, and pituitary tumors	Unknown	Benetatos et al. (2008, 2011); Zhang et al. (2009)
ncRNA CCND1/cyclin D1	Homo sapiens	≥200–300 nt	Transcribed from the 5′ end of the cyclin D1 gene	Induces repression of the cyclin D1 gene; induced by DNA damage	Binding to TLS protein induces TLS allosteric change, allowing interaction with the cyclin D1 gene, inhibiting CBP/p300 HAT activity and resulting in cyclin D1 repression	Wang et al. (2008b)
PTENP1	Homo sapiens	3600 nt	Pseudogene, Chr9	Cell growth-suppressive pseudogene lost in some human cancer	Regulate cellular levels of PTEN	Poliseno et al. (2010)

cancer cells *in vivo* (Lottin et al. 2002a), and loss of imprinting at the H19 locus results in high H19 expression in cancers of the esophagus, colon, liver, bladder, and hepatic metastasis (Hibi et al. 1996; Fellig et al. 2005; Matouk et al. 2007). Additionally, H19 was shown to be directly activated by the oncogenic transcription factor c-MYC in colon cancer, suggesting that H19 can be an intermediate between c-Myc and downstream gene expression (Barsyte-Lovejoy et al. 2006).

lncRNA HOTAIR is also unequivocally associated with tumor malignancy and metastasis. HOTAIR is highly expressed in human breast cancer metastasis and in primary tumors predisposed to future metastasis (Gupta et al. 2010). Elevated HOTAIR levels are also predictive of metastasis or progression in colon and liver cancers (Kogo et al. 2011; Yang et al. 2011), and loss of HOTAIR inhibits cancer invasiveness, indicating a potentially direct role for lincRNAs in modulating cancer progression (Gupta et al. 2010).

MALAT-1 is another well-known oncogenic lncRNA. MALAT-1 is frequently over-expressed in many different tumor types (Ji et al. 2003; Lin et al. 2007b; Perez et al. 2008), including metastasis of early stage in NSCLC (Fu et al. 2006). MALAT-1 is an example of lncRNA that could be used as a tumor marker, although its cellular role is still unknown.

Although the expression of some lncRNAs is associated with tumor malignancy, others may function as tumor suppressors (Table 15.3). In fact, multiple lncRNAs are transcriptionally induced by p53, such as lincRNA-p21, and play a role in this tumor suppressor pathway (Huarte et al. 2010). Even though it is not induced by p53, MEG3 is an lncRNA involved in the p53 pathway, as it physically interacts with p53 protein and controls cell growth. MEG3 is expressed in several types of normal tissues, but its expression is decreased or absent by CpG methylation in malignant tissues (Benetatos et al. 2011).

The identification of lncRNAs associated with cancer opens up the possibility to their use as prognostic markers. This might encounter the difficulty of lncRNA low abundance and stability in plasma as large RNA molecules. However, there are already some promising candidates, such as lncRNA PCAT-1, which is expressed in prostate cancer and can predict poor-prognosis patients easily as it can be detected in urine (Prensner et al. 2011), or the highly expressed hepatocarcinoma-associated lncRNA HULC, which is detectable in the blood of hepatocarcinoma patients by conventional polymerase chain reaction (PCR) (Panzitt et al. 2007).

B. Genetic Alterations in lncRNAs Associated with Cancer

Genetic studies on lncRNA sequences reveal that there are large- and small-scale mutations in the lncRNA sequences that are correlated with disease. In fact, the most frequent mutations in the genome occur in the non-coding and intergenic regions (Halvorsen et al. 2010). These mutations can be transferred to lncRNAs transcribed from these regions affecting their function in the tumor cells. Much work still has to be done to be able to establish this direct association, although some studies are already pointing in this direction. This is the case of germ line and somatic mutations in lncRNAs that are possibly implicated in colorectal cancer and leukemias (Wojcik et al. 2010) or a human polymorphism at lncRNA associated with prostate cancer risk (Jin et al. 2011).

Genetic alterations such as whole-gene deletions or amplifications and chromosomal translocations also point toward the implication of lncRNAs in cancer pathogenesis. One of the most significant examples is the large deletion (403,231 bp) affecting the INK4/ARF locus and the ANRIL lncRNA that has been associated with hereditary cutaneous malignant melanoma (CMM) and neural system tumor (NST) syndrome (Pasmant et al. 2011). Based on this large chromosomal deletion, ANRIL was identified as a key player in the development of a hereditary and predisposition to cancer (Pasmant et al. 2011).

VI. FUTURE PERSPECTIVES

The role in tumorigenesis of the vast majority of lncRNAs with altered expression in cancer is still unclear. However, their use as markers for diagnosis is imminent, and their therapeutic applications may be possible in a more distant future. The progress in the use of gene silencing mediated by RNA interference (RNAi) or antisense oligonucleotides for the treatment of different diseases is encouraging and could be applied to silence selectively oncogenic lncRNAs. lncRNAs may be ideal targets for therapy due to their high turnover rate and their direct and specific regulatory functions. Predictably, therapeutic targeting of lncRNAs will carry fewer negative effects than that of protein-coding genes, given that they function by regulating specific facets of their protein interacting partners.

The latest genomic technologies are now widely used for the profiling of the whole genome and transcriptome of multiple cancers. The immense amount of data generated by these projects not only presents great possibilities for prognosis and therapeutics but also brings new challenges. This has called for the creation of the International Cancer Genome Consortium (ICGC), which coordinates the international effort to systematically study more than 25,000 cancer genomes at the genomic, epigenomic, and transcriptomic levels (Hudson et al. 2010). Predictably, in the next few years, a complete catalog of the large ncRNA expression as well as the genetic mutations, amplifications, and deletions in non-coding regions associated with different types of tumors will be available. Thanks to these collaborative projects, and together with the research efforts of many laboratories, our knowledge of the function of the long non-coding transcriptome in cancer will increase significantly in the near future.

ACKNOWLEDGMENTS

The authors are grateful to the members of Dr. Huarte's laboratory at the Center for Applied Medical Research (CIMA) for critical reading of the manuscript.

REFERENCES

Barsyte-Lovejoy, D., S.K. Lau, P.C. Boutros, F. Khosravi, I. Jurisica, I.L. Andrulis, M.S. Tsao, and L.Z. Penn. 2006. The c-Myc oncogene directly induces the H19 noncoding RNA by allele-specific binding to potentiate tumorigenesis. *Cancer Res* **66**:5330–5337.

Beckerman, R. and C. Prives. 2010. Transcriptional regulation by p53. *Cold Spring Harb Symp Quant Biol* **2**:a000935.

Bejerano, G., M. Pheasant, I. Makunin, S. Stephen, W.J. Kent, J.S. Mattick, and D. Haussler. 2004. Ultraconserved elements in the human genome. *Science* **304**:1321–1325.

Beltran, M., I. Puig, C. Pena, J.M. Garcia, A.B. Alvarez, R. Pena, F. Bonilla, and A.G. de Herreros. 2008. A natural antisense transcript regulates Zeb2/Sip1 gene expression during Snail1-induced epithelial-mesenchymal transition. *Genes Dev* **22**:756–769.

Benetatos, L., A. Dasoula, E. Hatzimichael, I. Georgiou, M. Syrrou, and K.L. Bourantas. 2008. Promoter hypermethylation of the MEG3 (DLK1/MEG3) imprinted gene in multiple myeloma. *Clin Lymphoma Myeloma* **8**:171–175.

Benetatos, L., G. Vartholomatos, and E. Hatzimichael. 2011. MEG3 imprinted gene contribution in tumorigenesis. *Int J Cancer* **129**:773–779.

Bernard, D., K.V. Prasanth, V. Tripathi, S. Colasse, T. Nakamura, Z. Xuan, M.Q. Zhang, F. Sedel, L. Jourdren, F. Coulpier, A. Triller, D.L. Spector, et al. 2010. A long nuclear-retained non-coding RNA regulates synaptogenesis by modulating gene expression. *EMBO J* **29**:3082–3093.

Borsani, G., R. Tonlorenzi, M.C. Simmler, L. Dandolo, D. Arnaud, V. Capra, M. Grompe, A. Pizzuti, D. Muzny, C. Lawrence, H.F. Willard, P. Avner, et al. 1991. Characterization of a murine gene expressed from the inactive X chromosome. *Nature* **351**:325–329.

Brannan, C.I., E.C. Dees, R.S. Ingram, and S.M. Tilghman. 1990. The product of the H19 gene may function as an RNA. *Mol Cell Biol* **10**:28–36.

Cabili, M.N., C. Trapnell, L. Goff, M. Koziol, B. Tazon-Vega, A. Regev, and J.L. Rinn. 2011. Integrative annotation of human large intergenic noncoding RNAs reveals global properties and specific subclasses. *Genes Dev* **25**:1915–1927.

Calin, G.A., C.G. Liu, M. Ferracin, T. Hyslop, R. Spizzo, C. Sevignani, M. Fabbri, A. Cimmino, E.J. Lee, S.E. Wojcik, M. Shimizu, E. Tili, et al. 2007. Ultraconserved regions encoding ncRNAs are altered in human leukemias and carcinomas. *Cancer Cell* **12**:215–229.

Caretti, G., E.P. Lei, and V. Sartorelli. 2007. The DEAD-box p68/p72 proteins and the noncoding RNA steroid receptor activator SRA: eclectic regulators of disparate biological functions. *Cell Cycle* **6**:1172–1176.

Cheng, J., P. Kapranov, J. Drenkow, S. Dike, S. Brubaker, S. Patel, J. Long, D. Stern, H. Tammana, G. Helt, V. Sementchenko, A. Piccolboni, et al. 2005. Transcriptional maps of 10 human chromosomes at 5-nucleotide resolution. *Science* **308**:1149–1154.

Clamp, M., B. Fry, M. Kamal, X. Xie, J. Cuff, M.F. Lin, M. Kellis, K. Lindblad-Toh, and E.S. Lander. 2007. Distinguishing protein-coding and noncoding genes in the human genome. *Proc Natl Acad Sci U S A* **104**:19428–19433.

Coccia, E.M., C. Cicala, A. Charlesworth, C. Ciccarelli, G.B. Rossi, L. Philipson, and V. Sorrentino. 1992. Regulation and expression of a growth arrest-specific gene (gas5) during growth, differentiation, and development. *Mol Cell Biol* **12**:3514–3521.

Colley, S.M. and P.J. Leedman. 2009. SRA and its binding partners: an expanding role for RNA-binding coregulators in nuclear receptor-mediated gene regulation. *Crit Rev Biochem Mol Biol* **44**:25–33.

Colley, S.M., K.R. Iyer, and P.J. Leedman. 2008. The RNA coregulator SRA, its binding proteins and nuclear receptor signaling activity. *IUBMB Life* **60**:159–164.

Cooper, C., J. Guo, Y. Yan, S. Chooniedass-Kothari, F. Hube, M.K. Hamedani, L.C. Murphy, Y. Myal, and E. Leygue. 2009. Increasing the relative expression of endogenous non-coding steroid receptor RNA activator (SRA) in human breast cancer cells using modified oligonucleotides. *Nucleic Acids Res* **37**:4518–4531.

Cunnington, M.S., M. Santibanez Koref, B.M. Mayosi, J. Burn, and B. Keavney. 2010. Chromosome 9p21 SNPs associated with multiple disease phenotypes correlate with ANRIL expression. *PLoS Genet* **6**:e1000899.

Derrien, T., R. Johnson, G. Bussotti, A. Tanzer, S. Djebali, H. Tilgner, G. Guernec, D. Martin, A. Merkel, D.G. Knowles, J. Lagarde, L. Veeravalli, et al. 2012. The GENCODE v7 catalog of human long noncoding RNAs: analysis of their gene structure, evolution, and expression. *Genome Res* **22**:1775–1789.

Dieci, G., G. Fiorino, M. Castelnuovo, M. Teichmann, and A. Pagano. 2007. The expanding RNA polymerase III transcriptome. *Trends Genet* **23**:614–622.

REFERENCES

Dinger, M.E., K.C. Pang, T.R. Mercer, and J.S. Mattick. 2008. Differentiating protein-coding and noncoding RNA: challenges and ambiguities. *PLoS Comput Biol* **4**:e1000176.

Du, Y., G. Kong, X. You, S. Zhang, T. Zhang, Y. Gao, L. Ye, and X. Zhang. 2012. Elevation of highly up-regulated in liver cancer (HULC) by hepatitis B virus X protein promotes hepatoma cell proliferation via down-regulating p18. *J Biol Chem* **287**:26302–26311.

Faghihi, M.A., F. Modarresi, A.M. Khalil, D.E. Wood, B.G. Sahagan, T.E. Morgan, C.E. Finch, G. St Laurent 3rd, P.J. Kenny, and C. Wahlestedt. 2008. Expression of a noncoding RNA is elevated in Alzheimer's disease and drives rapid feed-forward regulation of beta-secretase. *Nat Med* **14**:723–730.

Fellig, Y., I. Ariel, P. Ohana, P. Schachter, I. Sinelnikov, T. Birman, S. Ayesh, T. Schneider, N. de Groot, A. Czerniak, and A. Hochberg. 2005. H19 expression in hepatic metastases from a range of human carcinomas. *J Clin Pathol* **58**:1064–1068.

Feng, J., C. Bi, B.S. Clark, R. Mady, P. Shah, and J.D. Kohtz. 2006. The Evf-2 noncoding RNA is transcribed from the Dlx-5/6 ultraconserved region and functions as a Dlx-2 transcriptional coactivator. *Genes Dev* **20**:1470–1484.

Fu, X., L. Ravindranath, N. Tran, G. Petrovics, and S. Srivastava. 2006. Regulation of apoptosis by a prostate-specific and prostate cancer-associated noncoding gene, PCGEM1. *DNA Cell Biol* **25**:135–141.

Galindo, M.I., J.I. Pueyo, S. Fouix, S.A. Bishop, and J.P. Couso. 2007. Peptides encoded by short ORFs control development and define a new eukaryotic gene family. *PLoS Biol* **5**:e106.

Gibb, E.A., K.S. Enfield, G.L. Stewart, K.M. Lonergan, R. Chari, R.T. Ng, L. Zhang, C.E. MacAulay, M.P. Rosin, and W.L. Lam. 2011. Long non-coding RNAs are expressed in oral mucosa and altered in oral premalignant lesions. *Oral Oncol* **47**:1055–1061.

Goodrich, J.A. and J.F. Kugel. 2006. Non-coding-RNA regulators of RNA polymerase II transcription. *Nat Rev Mol Cell Biol* **7**:612–616.

Gribnau, J., K. Diderich, S. Pruzina, R. Calzolari, and P. Fraser. 2000. Intergenic transcription and developmental remodeling of chromatin subdomains in the human beta-globin locus. *Mol Cell* **5**:377–386.

Guenzl, P. and D. Barlow. 2012. Macro lncRNAs: a new layer of cis-regulatory information in the mammalian genome. *RNA Biol* **9**:731–741.

Gupta, R.A., N. Shah, K.C. Wang, J. Kim, H.M. Horlings, D.J. Wong, M.C. Tsai, T. Hung, P. Argani, J.L. Rinn, Y. Wang, P. Brzoska, et al. 2010. Long non-coding RNA HOTAIR reprograms chromatin state to promote cancer metastasis. *Nature* **464**:1071–1076.

Guttman, M. and J.L. Rinn. 2012. Modular regulatory principles of large non-coding RNAs. *Nature* **482**:339–346.

Guttman, M., I. Amit, M. Garber, C. French, M.F. Lin, D. Feldser, M. Huarte, O. Zuk, B.W. Carey, J.P. Cassady, M.N. Cabili, R. Jaenisch, et al. 2009. Chromatin signature reveals over a thousand highly conserved large non-coding RNAs in mammals. *Nature* **458**:223–227.

Guttman, M., M. Garber, J.Z. Levin, J. Donaghey, J. Robinson, X. Adiconis, L. Fan, M.J. Koziol, A. Gnirke, C. Nusbaum, J.L. Rinn, E.S. Lander, et al. 2010. Ab initio reconstruction of cell type-specific transcriptomes in mouse reveals the conserved multi-exonic structure of lincRNAs. *Nat Biotechnol* **28**:503–510.

Guttman, M., J. Donaghey, B.W. Carey, M. Garber, J.K. Grenier, G. Munson, G. Young, A.B. Lucas, R. Ach, L. Bruhn, X. Yang, I. Amit, et al. 2011. lincRNAs act in the circuitry controlling pluripotency and differentiation. *Nature* **477**:295–300.

Halvorsen, M., J.S. Martin, S. Broadaway, and A. Laederach. 2010. Disease-associated mutations that alter the RNA structural ensemble. *PLoS Genet* **6**:e1001074.

Han, L., K. Zhang, Z. Shi, J. Zhang, J. Zhu, S. Zhu, A. Zhang, Z. Jia, G. Wang, S. Yu, P. Pu, L. Dong, et al. 2012. LncRNA profile of glioblastoma reveals the potential role of lncRNAs in contributing to glioblastoma pathogenesis. *Int J Oncol* **40**:2004–2012.

Herzing, L.B., J.T. Romer, J.M. Horn, and A. Ashworth. 1997. Xist has properties of the X-chromosome inactivation centre. *Nature* **386**:272–275.

Hibi, K., H. Nakamura, A. Hirai, Y. Fujikake, Y. Kasai, S. Akiyama, K. Ito, and H. Takagi. 1996. Loss of H19 imprinting in esophageal cancer. *Cancer Res* **56**:480–482.

Hirota, K., T. Miyoshi, K. Kugou, C.S. Hoffman, T. Shibata, and K. Ohta. 2008. Stepwise chromatin remodelling by a cascade of transcription initiation of non-coding RNAs. *Nature* **456**:130–134.

Huarte, M. and J.L. Rinn. 2011. Large non-coding RNAs: missing links in cancer? *Hum Mol Genet* **19**:R152–R161.

Huarte, M., M. Guttman, D. Feldser, M. Garber, M.J. Koziol, D. Kenzelmann-Broz, A.M. Khalil, O. Zuk, I. Amit, M. Rabani, L.D. Attardi, A. Regev, et al. 2010. A large intergenic noncoding RNA induced by p53 mediates global gene repression in the p53 response. *Cell* **142**:409–419.

Hudson, T.J., W. Anderson, A. Artez, A.D. Barker, C. Bell, R.R. Bernabe, M.K. Bhan, F. Calvo, I. Eerola, D.S. Gerhard, A. Guttmacher, M. Guyer, et al. 2010. International network of cancer genome projects. *Nature* **464**:993–998.

Hung, T., Y. Wang, M.F. Lin, A.K. Koegel, Y. Kotake, G.D. Grant, H.M. Horlings, N. Shah, C. Umbricht, P. Wang, Y. Wang, B. Kong, et al. 2011. Extensive and coordinated transcription of noncoding RNAs within cell-cycle promoters. *Nat Genet* **43**:621–629.

Ifere, G.O. and G.A. Ananaba. 2009. Prostate cancer gene expression marker 1 (PCGEM1): a patented prostate-specific non-coding gene and regulator of prostate cancer progression. *Recent Pat DNA Gene Seq* **3**:151–163.

Ji, P., S. Diederichs, W. Wang, S. Boing, R. Metzger, P.M. Schneider, N. Tidow, B. Brandt, H. Buerger, E. Bulk, M. Thomas, W.E. Berdel, et al. 2003. MALAT-1, a novel noncoding RNA, and thymosin beta4 predict metastasis and survival in early-stage non-small cell lung cancer. *Oncogene* **22**:8031–8041.

Jin, G., J. Sun, S.D. Isaacs, K.E. Wiley, S.T. Kim, L.W. Chu, Z. Zhang, H. Zhao, S.L. Zheng, W.B. Isaacs, and J. Xu. 2011. Human polymorphisms at long non-coding RNAs (lncRNAs) and association with prostate cancer risk. *Carcinogenesis* **32**:1655–1659.

Jones, P.A. and S.B. Baylin. 2007. The epigenomics of cancer. *Cell* **128**:683–692.

Karreth, F.A., Y. Tay, D. Perna, U. Ala, S.M. Tan, A.G. Rust, G. DeNicola, K.A. Webster, D. Weiss, P.A. Perez-Mancera, M. Krauthammer, R. Halaban, et al. 2011. In vivo identification of tumor-suppressive PTEN ceRNAs in an oncogenic BRAF-induced mouse model of melanoma. *Cell* **147**:382–395.

Khalil, A.M., M. Guttman, M. Huarte, M. Garber, A. Raj, D. Rivea Morales, K. Thomas, A. Presser, B.E. Bernstein, A. van Oudenaarden, A. Regev, E.S. Lander, et al. 2009. Many human large intergenic noncoding RNAs associate with chromatin-modifying complexes and affect gene expression. *Proc Natl Acad Sci U S A* **106**:11667–11672.

Kim, T.K., M. Hemberg, J.M. Gray, A.M. Costa, D.M. Bear, J. Wu, D.A. Harmin, M. Laptewicz, K. Barbara-Haley, S. Kuersten, E. Markenscoff-Papadimitriou, D. Kuhl, et al. 2010. Widespread transcription at neuronal activity-regulated enhancers. *Nature* **465**:182–187.

Kino, T., D.E. Hurt, T. Ichijo, N. Nader, and G.P. Chrousos. 2010. Noncoding RNA gas5 is a growth arrest- and starvation-associated repressor of the glucocorticoid receptor. *Sci Signal* **3**:ra8.

Kogo, R., T. Shimamura, K. Mimori, K. Kawahara, S. Imoto, T. Sudo, F. Tanaka, K. Shibata, A. Suzuki, S. Komune, S. Miyano, and M. Mori. 2011. Long noncoding RNA HOTAIR regulates polycomb-dependent chromatin modification and is associated with poor prognosis in colorectal cancers. *Cancer Res* **71**:6320–6326.

Koziol, M.J. and J.L. Rinn. 2010. RNA traffic control of chromatin complexes. *Curr Opin Genet Dev* **20**:142–148.

Li, L., T. Feng, Y. Lian, G. Zhang, A. Garen, and X. Song. 2009. Role of human noncoding RNAs in the control of tumorigenesis. *Proc Natl Acad Sci U S A* **106**:12956–12961.

REFERENCES

Lin, M.F., J.W. Carlson, M.A. Crosby, B.B. Matthews, C. Yu, S. Park, K.H. Wan, A.J. Schroeder, L.S. Gramates, S.E. St Pierre, M. Roark, K.L. Wiley Jr., et al. 2007a. Revisiting the protein-coding gene catalog of *Drosophila melanogaster* using 12 fly genomes. *Genome Res* **17**: 1823–1836.

Lin, R., S. Maeda, C. Liu, M. Karin, and T.S. Edgington. 2007b. A large noncoding RNA is a marker for murine hepatocellular carcinomas and a spectrum of human carcinomas. *Oncogene* **26**:851–858.

Lin, M.F., A.N. Deoras, M.D. Rasmussen, and M. Kellis. 2008. Performance and scalability of discriminative metrics for comparative gene identification in 12 *Drosophila* genomes. *PLoS Comput Biol* **4**:e1000067.

Lin, M.F., I. Jungreis, and M. Kellis. 2011. PhyloCSF: a comparative genomics method to distinguish protein coding and non-coding regions. *Bioinformatics* **27**:i275–i282.

Lottin, S., E. Adriaenssens, T. Dupressoir, N. Berteaux, C. Montpellier, J. Coll, T. Dugimont, and J.J. Curgy. 2002a. Overexpression of an ectopic H19 gene enhances the tumorigenic properties of breast cancer cells. *Carcinogenesis* **23**:1885–1895.

Lottin, S., A.S. Vercoutter-Edouart, E. Adriaenssens, X. Czeszak, J. Lemoine, M. Roudbaraki, J. Coll, H. Hondermarck, T. Dugimont, and J.J. Curgy. 2002b. Thioredoxin post-transcriptional regulation by H19 provides a new function to mRNA-like non-coding RNA. *Oncogene* **21**:1625–1631.

Louro, R., A.S. Smirnova, and S. Verjovski-Almeida. 2009. Long intronic noncoding RNA transcription: expression noise or expression choice? *Genomics* **93**:291–298.

Magistri, M., M.A. Faghihi, G. St Laurent 3rd, and C. Wahlestedt. 2012. Regulation of chromatin structure by long noncoding RNAs: focus on natural antisense transcripts. *Trends Genet* **28**:389–396.

Mancini-Dinardo, D., S.J. Steele, J.M. Levorse, R.S. Ingram, and S.M. Tilghman. 2006. Elongation of the Kcnq1ot1 transcript is required for genomic imprinting of neighboring genes. *Genes Dev* **20**:1268–1282.

Martianov, I., A. Ramadass, A. Serra Barros, N. Chow, and A. Akoulitchev. 2007. Repression of the human dihydrofolate reductase gene by a non-coding interfering transcript. *Nature* **445**: 666–670.

Matouk, I.J., N. DeGroot, S. Mezan, S. Ayesh, R. Abu-lail, A. Hochberg, and E. Galun. 2007. The H19 non-coding RNA is essential for human tumor growth. *PLoS ONE* **2**:e845.

Mattick, J.S. 2009. The genetic signatures of noncoding RNAs. *PLoS Genet* **5**:e1000459.

Mattick, J.S. and I.V. Makunin. 2006. Non-coding RNA. *Hum Mol Genet* **15** Spec No. 1:R17–R29.

Mercer, T.R., M.E. Dinger, S.M. Sunkin, M.F. Mehler, and J.S. Mattick. 2008. Specific expression of long noncoding RNAs in the mouse brain. *Proc Natl Acad Sci U S A* **105**: 716–721.

Mikkelsen, T.S., M. Ku, D.B. Jaffe, B. Issac, E. Lieberman, G. Giannoukos, M. Alvarez, W. Brockman, T.K. Kim, R.P. Koche, W. Lee, E. Mendenhall, et al. 2007. Genome-wide maps of chromatin state in pluripotent and lineage-committed cells. *Nature* **448**:553–560.

Mohammad, F., T. Mondal, N. Guseva, G.K. Pandey, and C. Kanduri. 2012. Kcnq1ot1 noncoding RNA mediates transcriptional gene silencing by interacting with Dnmt1. *Development* **137**: 2493–2499.

Morris, K.V., S. Santoso, A.M. Turner, C. Pastori, and P.G. Hawkins. 2008. Bidirectional transcription directs both transcriptional gene activation and suppression in human cells. *PLoS Genet* **4**:e1000258.

Mourtada-Maarabouni, M., M.R. Pickard, V.L. Hedge, F. Farzaneh, and G.T. Williams. 2009. GAS5, a non-protein-coding RNA, controls apoptosis and is downregulated in breast cancer. *Oncogene* **28**:195–208.

Nagano, T., J.A. Mitchell, L.A. Sanz, F.M. Pauler, A.C. Ferguson-Smith, R. Feil, and P. Fraser. 2008. The Air noncoding RNA epigenetically silences transcription by targeting G9a to chromatin. *Science* **322**:1717–1720.

Orom, U.A., T. Derrien, R. Guigo, and R. Shiekhattar. 2010. Long noncoding RNAs as enhancers of gene expression. *Cold Spring Harb Symp Quant Biol* **75**:325–331.

Owen, R.D., D.M. Bortner, and M.C. Ostrowski. 1990. ras oncogene activation of a VL30 transcriptional element is linked to transformation. *Mol Cell Biol* **10**:1–9.

Pandey, R.R., T. Mondal, F. Mohammad, S. Enroth, L. Redrup, J. Komorowski, T. Nagano, D. Mancini-Dinardo, and C. Kanduri. 2008. Kcnq1ot1 antisense noncoding RNA mediates lineage-specific transcriptional silencing through chromatin-level regulation. *Mol Cell* **32**:232–246.

Pang, K.C., M.C. Frith, and J.S. Mattick. 2006. Rapid evolution of noncoding RNAs: lack of conservation does not mean lack of function. *Trends Genet* **22**:1–5.

Panganiban, G. and J.L. Rubenstein. 2002. Developmental functions of the Distal-less/Dlx homeobox genes. *Development* **129**:4371–4386.

Panzitt, K., M.M. Tschernatsch, C. Guelly, T. Moustafa, M. Stradner, H.M. Strohmaier, C.R. Buck, H. Denk, R. Schroeder, M. Trauner, and K. Zatloukal. 2007. Characterization of HULC, a novel gene with striking up-regulation in hepatocellular carcinoma, as noncoding RNA. *Gastroenterology* **132**:330–342.

Pasmant, E., A. Sabbagh, M. Vidaud, and I. Bieche. 2011. ANRIL, a long, noncoding RNA, is an unexpected major hotspot in GWAS. *FASEB J* **25**:444–448.

Perez, D.S., T.R. Hoage, J.R. Pritchett, A.L. Ducharme-Smith, M.L. Halling, S.C. Ganapathiraju, P.S. Streng, and D.I. Smith. 2008. Long, abundantly expressed non-coding transcripts are altered in cancer. *Hum Mol Genet* **17**:642–655.

Petrovics, G., W. Zhang, M. Makarem, J.P. Street, R. Connelly, L. Sun, I.A. Sesterhenn, V. Srikantan, J.W. Moul, and S. Srivastava. 2004. Elevated expression of PCGEM1, a prostate-specific gene with cell growth-promoting function, is associated with high-risk prostate cancer patients. *Oncogene* **23**:605–611.

Poliseno, L., L. Salmena, J. Zhang, B. Carver, W.J. Haveman, and P.P. Pandolfi. 2010. A coding-independent function of gene and pseudogene mRNAs regulates tumour biology. *Nature* **465**:1033–1038.

Ponting, C.P., P.L. Oliver, and W. Reik. 2009. Evolution and functions of long noncoding RNAs. *Cell* **136**:629–641.

Prensner, J.R., M.K. Iyer, O.A. Balbin, S.M. Dhanasekaran, Q. Cao, J.C. Brenner, B. Laxman, I.A. Asangani, C.S. Grasso, H.D. Kominsky, X. Cao, X. Jing, et al. 2011. Transcriptome sequencing across a prostate cancer cohort identifies PCAT-1, an unannotated lincRNA implicated in disease progression. *Nat Biotechnol* **29**:742–749.

Qureshi, I.A., J.S. Mattick, and M.F. Mehler. 2010. Long non-coding RNAs in nervous system function and disease. *Brain Res* **1338**:20–35.

Rearick, D., A. Prakash, A. McSweeny, S.S. Shepard, L. Fedorova, and A. Fedorov. 2011. Critical association of ncRNA with introns. *Nucleic Acids Res* **39**:2357–2366.

Rinn, J.L. and H.Y. Chang. 2012. Genome regulation by long noncoding RNAs. *Annu Rev Biochem* **81**:145–166.

Rinn, J.L., M. Kertesz, J.K. Wang, S.L. Squazzo, X. Xu, S.A. Brugmann, L.H. Goodnough, J.A. Helms, P.J. Farnham, E. Segal, and H.Y. Chang. 2007. Functional demarcation of active and silent chromatin domains in human HOX loci by noncoding RNAs. *Cell* **129**:1311–1323.

Ripoche, M.A., C. Kress, F. Poirier, and L. Dandolo. 1997. Deletion of the H19 transcription unit reveals the existence of a putative imprinting control element. *Genes Dev* **11**:1596–1604.

Rodriguez-Campos, A. and F. Azorin. 2007. RNA is an integral component of chromatin that contributes to its structural organization. *PLoS ONE* **2**:e1182.

Salmena, L., L. Poliseno, Y. Tay, L. Kats, and P.P. Pandolfi. 2011. A ceRNA hypothesis: the Rosetta Stone of a hidden RNA language? *Cell* **146**:353–358.

Scaruffi, P., S. Stigliani, S. Moretti, S. Coco, C. De Vecchi, F. Valdora, A. Garaventa, S. Bonassi, and G.P. Tonini. 2009. Transcribed-ultra conserved region expression is associated with outcome in high-risk neuroblastoma. *BMC Cancer* **9**:441.

Schoenfelder, S., G. Smits, P. Fraser, W. Reik, and R. Paro. 2007. Non-coding transcripts in the H19 imprinting control region mediate gene silencing in transgenic *Drosophila*. *EMBO Rep* **8**:1068–1073.

Smith, C.M. and J.A. Steitz. 1998. Classification of gas5 as a multi-small-nucleolar-RNA (snoRNA) host gene and a member of the 5′-terminal oligopyrimidine gene family reveals common features of snoRNA host genes. *Mol Cell Biol* **18**:6897–6909.

Song, X., B. Wang, M. Bromberg, Z. Hu, W. Konigsberg, and A. Garen. 2002. Retroviral-mediated transmission of a mouse VL30 RNA to human melanoma cells promotes metastasis in an immunodeficient mouse model. *Proc Natl Acad Sci U S A* **99**:6269–6273.

Song, X., A. Sui, and A. Garen. 2004. Binding of mouse VL30 retrotransposon RNA to PSF protein induces genes repressed by PSF: effects on steroidogenesis and oncogenesis. *Proc Natl Acad Sci U S A* **101**:621–626.

Tahira, A.C., M.S. Kubrusly, M.F. Faria, B. Dazzani, R.S. Fonseca, V. Maracaja-Coutinho, S. Verjovski-Almeida, M.C. Machado, and E.M. Reis. 2011. Long noncoding intronic RNAs are differentially expressed in primary and metastatic pancreatic cancer. *Mol Cancer* **10**:141.

Tripathi, V., J.D. Ellis, Z. Shen, D.Y. Song, Q. Pan, A.T. Watt, S.M. Freier, C.F. Bennett, A. Sharma, P.A. Bubulya, B.J. Blencowe, S.G. Prasanth, et al. 2010. The nuclear-retained noncoding RNA MALAT1 regulates alternative splicing by modulating SR splicing factor phosphorylation. *Mol Cell* **39**:925–938.

Tsai, M.C., O. Manor, Y. Wan, N. Mosammaparast, J.K. Wang, F. Lan, Y. Shi, E. Segal, and H.Y. Chang. 2010. Long noncoding RNA as modular scaffold of histone modification complexes. *Science* **329**:689–693.

Tsang, W.P., T.W. Wong, A.H. Cheung, C.N. Co, and T.T. Kwok. 2007. Induction of drug resistance and transformation in human cancer cells by the noncoding RNA CUDR. *RNA* **13**:890–898.

Wan, Y., M. Kertesz, R.C. Spitale, E. Segal, and H.Y. Chang. 2011. Understanding the transcriptome through RNA structure. *Nat Rev Genet* **12**:641–655.

Wang, F., X. Li, X. Xie, L. Zhao, and W. Chen. 2008a. UCA1, a non-protein-coding RNA up-regulated in bladder carcinoma and embryo, influencing cell growth and promoting invasion. *FEBS Lett* **582**:1919–1927.

Wang, X., S. Arai, X. Song, D. Reichart, K. Du, G. Pascual, P. Tempst, M.G. Rosenfeld, C.K. Glass, and R. Kurokawa. 2008b. Induced ncRNAs allosterically modify RNA-binding proteins in cis to inhibit transcription. *Nature* **454**:126–130.

Wang, K.C., Y.W. Yang, B. Liu, A. Sanyal, R. Corces-Zimmerman, Y. Chen, B.R. Lajoie, A. Protacio, R.A. Flynn, R.A. Gupta, J. Wysocka, M. Lei, et al. 2011. A long noncoding RNA maintains active chromatin to coordinate homeotic gene expression. *Nature* **472**:120–124.

White, R.J. 2004. RNA polymerase III transcription and cancer. *Oncogene* **23**:3208–3216.

White, R.J. 2008. RNA polymerases I and III, non-coding RNAs and cancer. *Trends Genet* **24**:622–629.

Williamson, C.M., S.T. Ball, C. Dawson, S. Mehta, C.V. Beechey, M. Fray, L. Teboul, T.N. Dear, G. Kelsey, and J. Peters. 2011. Uncoupling antisense-mediated silencing and DNA methylation in the imprinted Gnas cluster. *PLoS Genet* **7**:e1001347.

Willingham, A.T., A.P. Orth, S. Batalov, E.C. Peters, B.G. Wen, P. Aza-Blanc, J.B. Hogenesch, and P.G. Schultz. 2005. A strategy for probing the function of noncoding RNAs finds a repressor of NFAT. *Science* **309**:1570–1573.

Wilusz, J.E., H. Sunwoo, and D.L. Spector. 2009. Long noncoding RNAs: functional surprises from the RNA world. *Genes Dev* **23**:1494–1504.

Wojcik, S.E., S. Rossi, M. Shimizu, M.S. Nicoloso, A. Cimmino, H. Alder, V. Herlea, L.Z. Rassenti, K.R. Rai, T.J. Kipps, M.J. Keating, C.M. Croce, et al. 2010. Non-coding RNA sequence variations in human chronic lymphocytic leukemia and colorectal cancer. *Carcinogenesis* **31**: 208–215.

Wu, Q., Y.C. Kim, J. Lu, Z. Xuan, J. Chen, Y. Zheng, T. Zhou, M.Q. Zhang, C.I. Wu, and S.M. Wang. 2008. Poly A- transcripts expressed in HeLa cells. *PLoS ONE* **3**:e2803.

Xu, C., M. Yang, J. Tian, X. Wang, and Z. Li. 2011. MALAT-1: a long non-coding RNA and its important 3′ end functional motif in colorectal cancer metastasis. *Int J Oncol* **39**:169–175.

Yang, Z., L. Zhou, L.M. Wu, M.C. Lai, H.Y. Xie, F. Zhang, and S.S. Zheng. 2011. Overexpression of long non-coding RNA HOTAIR predicts tumor recurrence in hepatocellular carcinoma patients following liver transplantation. *Ann Surg Oncol* **18**:1243–1250.

Yap, K.L., S. Li, A.M. Munoz-Cabello, S. Raguz, L. Zeng, S. Mujtaba, J. Gil, M.J. Walsh, and M.M. Zhou. 2010. Molecular interplay of the noncoding RNA ANRIL and methylated histone H3 lysine 27 by polycomb CBX7 in transcriptional silencing of INK4a. *Mol Cell* **38**:662–674.

Yoon, J.H., K. Abdelmohsen, S. Srikantan, X. Yang, J.L. Martindale, S. De, M. Huarte, M. Zhan, K.G. Becker, and M. Gorospe. 2012. LincRNA-p21 suppresses target mRNA translation. *Mol Cell* **47**:648–655.

Yoshimizu, T., A. Miroglio, M.A. Ripoche, A. Gabory, M. Vernucci, A. Riccio, S. Colnot, C. Godard, B. Terris, H. Jammes, and L. Dandolo. 2008. The H19 locus acts in vivo as a tumor suppressor. *Proc Natl Acad Sci U S A* **105**:12417–12422.

Yu, W., D. Gius, P. Onyango, K. Muldoon-Jacobs, J. Karp, A.P. Feinberg, and H. Cui. 2008. Epigenetic silencing of tumour suppressor gene p15 by its antisense RNA. *Nature* **451**:202–206.

Zhang, X., K. Rice, Y. Wang, W. Chen, Y. Zhong, Y. Nakayama, Y. Zhou, and A. Klibanski. 2009. Maternally expressed gene 3 (MEG3) noncoding ribonucleic acid: isoform structure, expression, and functions. *Endocrinology* **151**:939–947.

Zhang, Z.D., A. Frankish, T. Hunt, J. Harrow, and M. Gerstein. 2010. Identification and analysis of unitary pseudogenes: historic and contemporary gene losses in humans and other primates. *Genome Biol* **11**:R26.

Zhao, J., B.K. Sun, J.A. Erwin, J.J. Song, and J.T. Lee. 2008. Polycomb proteins targeted by a short repeat RNA to the mouse X chromosome. *Science* **322**:750–756.

Zheng, D., A. Frankish, R. Baertsch, P. Kapranov, A. Reymond, S.W. Choo, Y. Lu, F. Denoeud, S.E. Antonarakis, M. Snyder, Y. Ruan, C.L. Wei, et al. 2007. Pseudogenes in the ENCODE regions: consensus annotation, analysis of transcription, and evolution. *Genome Res* **17**:839–851.

16

REGULATION OF HYPOXIA RESPONSES BY MicroRNA EXPRESSION

Carme Camps,[1] Adrian L. Harris,[2] and Jiannis Ragoussis[1]

[1]*Genomics Research Group, The Wellcome Trust Centre for Human Genetics, University of Oxford, Oxford, UK*
[2]*Growth Factor Group, Cancer Research UK, Molecular Oncology Laboratories, Weatherall Institute of Molecular Medicine, John Radcliffe Hospital, University of Oxford, Oxford, UK*

I.	Hypoxia in Solid Tumors	268
II.	Hypoxia-Inducible Factors (HIFs) and Transcriptional Regulation upon Hypoxia Response	269
III.	MicroRNA (miRNA) Regulation under Hypoxia	269
IV.	*miR-210* as Key Player in Hypoxia	270
	A. Discovery and Validation of *miR-210* Targets	271
	B. *miR-210* and Mitochondrial Functions	271
	C. *miR-210* and Angiogenesis	273
	D. *miR-210* and DNA Repair	273
	E. *miR-210* and the Cell Cycle	273
	F. *miR-210*, Apoptosis, and Cell Differentiation	274
	G. *miR-210* and Immunosuppression	275
	H. *miR-210* as a Biomarker	275
V.	miRNA-Mediated Regulation of HIF	275
	A. miRNAs Regulated by HIF	275
	B. miRNAs Regulated by Hypoxia in a HIF-Independent Manner	278
	C. miRNAs Not Regulated by Hypoxia	278
	D. Interaction between Different Levels of miRNA Regulation under Hypoxia	279
VI.	Conclusions	279
	References	280

MicroRNAs in Medicine, First Edition. Edited by Charles H. Lawrie.
© 2014 John Wiley & Sons, Inc. Published 2014 by John Wiley & Sons, Inc.

ABBREVIATIONS

AML	acute myeloid leukemia
ChIP	chromatin immunoprecipitation
ER	estrogen receptor
HCC	hepatocellular carcinoma
HGF	hepatocyte growth factor
HIF	hypoxia-inducible factor
HREs	hypoxia response elements
HUVEC	human umbilical vein endothelial cells
MDSs	myelodysplastic syndromes
mRNA	messenger ribonucleic acid
MSC	mesenchymal stem cells
PASMC	pulmonary artery smooth muscle cells
PHD	prolyl hydroxylase domain
pVHL	von Hippel–Lindau protein
RISC	RNA-induced silencing complex
UTR	untranslated region

I. HYPOXIA IN SOLID TUMORS

Oxygen plays an essential role in a broad range of biological processes, and for that reason, mammalian cells have developed mechanisms to face a decrease of oxygen levels and survive by adapting to this condition. Apart from a fall of atmospheric oxygen tension, hypoxia can arise at specific locations in the human body as a consequence of inflammation, tissue ischemia and injury, and solid tumor growth.

Hypoxia in solid tumors is generally associated with poor prognosis and resistance to conventional therapy. It arises as a result of a combination of factors including reduced oxygen diffusion due to tumor growth and lower oxygen perfusion from blood vessels because of poor and aberrant vascularization. The tumor- or therapy-related anemia, which is caused by reduced oxygen transport capacity, also contributes to the poor oxygenation of tumors. The capability of cancer cells to adapt to hypoxia contributes to their malignancy and aggressive phenotype. Indeed, hypoxia has been linked to several pathological features of cancer disease such as tumor progression, invasion, angiogenesis, changes in metabolism, and increased risk of metastasis (Wilson and Hay 2011).

Hypoxic tumors are generally more resistant to radiotherapy and chemotherapy, which has an impact in treatment outcome. For instance, the lack of oxidation of DNA free radicals by oxygen leads to resistance to ionizing radiation and antibiotics that induce DNA breaks. Other factors contributing to drug resistance in hypoxic tumors include cell cycle arrest, poor diffusion of the drug due to tumor growth and impaired vasculature, and extracellular acidification (review in Wilson and Hay 2011). However, conversely, hypoxia can provide targeted treatment opportunities as well. For instance, the hypoxia-activated prodrug PR610 is a proprietary drug currently under clinical trial (http://clinicaltrials.gov/ct2/show/study/NCT01631279). It is an irreversible multikinase inhibitor that is activated in areas of severe hypoxia. This drug inhibits cellular proliferation and differentiation of tumor cells overexpressing HER kinases, members of the epidermal growth factor receptor family of receptor kinases. Due to the hypoxia-specific activity, this drug has the potential to be more effective and less toxic. Therefore, it is important to identify hypoxia in cancer and use its characteristic biology to target therapies.

II. HYPOXIA-INDUCIBLE FACTORS (HIFs) AND TRANSCRIPTIONAL REGULATION UPON HYPOXIA RESPONSE

The family of transcription factors known as HIFs has a key role in the transcriptional regulation that takes place in response to hypoxia. HIFs are heterodimers composed of one alpha unit, which is regulated by oxygen levels as well as by growth factors, and one beta subunit, which is constitutively expressed. Under normoxia (normal oxygen load), the HIF alpha subunit is hydroxylated at conserved prolyl residues by prolyl hydroxylase domain (PHD) proteins. These "marks" are recognized by the von Hippel–Lindau protein (pVHL), which mediates the ubiquitination of HIF alpha subunits and following degradation via the proteasomal pathway. When the oxygen levels decrease, the HIF alpha subunit is no longer hydroxylated and can therefore form a stable complex with the beta subunit in the nucleus of the cell. It then binds to the hypoxia response elements (HREs) in the promoter regions of hypoxia-sensitive genes in order to induce its transcription (Ratcliffe et al. 1998; Semenza 1998). To date, there are three isoforms of the HIF alpha subunit that have been characterized, HIF-1α, HIF-2α, and HIF-3α, and one isoform of the HIF beta subunit, HIF-1β. Among the alpha subunits, HIF-1α is the one with the broadest pattern of expression in cells, whereas HIF-2α and HIF-3α are more tissue specific (Bertout et al. 2008).

HIF-1α plays the major role in the activation of transcriptional gene expression upon hypoxia (Elvidge et al. 2006). The genes regulated by HIF-1α are involved in many of the important processes of cancer biology such as glucose metabolism, cell proliferation and survival, pH regulation, migration, and angiogenesis. There is a certain overlap between HIF-1α and HIF-2α DNA binding sites (Mole et al. 2009; Schodel et al. 2011), and only a small number of genes seem to be regulated specifically by HIF-2α in the breast cancer cell line MCF-7. In contrast, HIF-3α acts in a dominant-negative fashion to inhibit HIF-1α and HIF-2α (Makino et al. 2001).

III. MicroRNA (miRNA) REGULATION UNDER HYPOXIA

The regulation of miRNA expression under hypoxia has been reported by several groups in a broad range of cell lines (for a review, see Kulshreshtha et al. 2008). Overall, more than 50 miRNAs were shown to be either up-regulated or down-regulated under hypoxia, although there was little overlap between the findings reported between different studies. This suggests that the regulation of miRNAs by hypoxia is cell type specific, although additionally, differences between technical platforms used and statistical analyses could also go some way toward explaining these discrepancies.

The direct regulation of *miR-26a-2* and *miR-210* by HIF-1α was demonstrated through chromatin immunoprecipitation (ChIP) (Kulshreshtha et al. 2007). In addition, it has been shown in human umbilical vein endothelial cells (HUVECs) exposed to chronic hypoxia (24 hours) that several of the components of the miRNA biosynthetic pathway including

DGCR8, EXPORTIN5, DICER, TRBP, AGO1, and AGO2 are significantly down-regulated at both protein and mRNA levels (Ho et al. 2012). Moreover, the decrease in DICER protein and mRNA levels has been confirmed *in vivo* in several tissues from mice exposed to chronic hypoxia, and also, low mRNA levels for DICER have been observed in rat models for pulmonary arterial hypertension (Caruso et al. 2010). On the other hand, it has been shown in human primary pulmonary artery smooth muscle cells (PASMCs) that AGO2 can be subjected to a posttranslational modification that increases its stability and translocation rate to the stress granules, which has an impact on miRNA–target interaction (Wu et al. 2011).

The first screenings for hypoxia-regulated miRNAs in breast cancer and other solid tumors suggested that hypoxia could be a key factor in miRNA modulation in cancer (Kulshreshtha et al. 2007). Moreover, the hypoxia-regulated miRNAs have been shown to have functional roles that are of great relevance in cancer. For instance, *miR-26*, *miR-107*, and *miR-210* have an antiapoptotic effect in hypoxia, leading to caspase inhibition (Kulshreshtha et al. 2007). Hypoxia-regulated miRNAs are also associated with cell invasion and metastasis. This is the case for *miR-373*, a HIF-1α-dependent miRNA, the expression of which has been correlated with increased proliferation and tumorigenesis (Voorhoeve et al. 2006). It is able to promote cancer cell migration and invasion *in vitro* and *in vivo* through suppression of CD44. Consistent with this observation, high levels of *miR-373* have been correlated with low levels of CD44 in breast cancer metastatic samples (Huang et al. 2008). Another mechanism of action for *miR-373* in promoting metastasis involves the direct suppression of mTOR and SIRT1, resulting in the activation of the Ras/Raf/MEK/Erk signaling pathway and NF-κB, in turn increasing the levels of metalloproteinase 9 (MMP9) and increasing cell migration and growth (Liu and Wilson 2012). However, some studies suggest a contrary effect of *miR-373* in metastasis. *miR-373* can directly transactivate E-cadherin gene expression by binding to complementary promoter regions, which inhibits invasion and metastasis (Place et al. 2008). Also, it seems to have a metastasis-suppressive role in ER negative breast cancer cell lines, in a process mediated by direct down-regulation of TGFBR2 by this miRNA (Keklikoglou et al. 2012). These publications indicate that the targets and biological effects of *miR-373* are context dependent, as for example, NF-κB signaling is constitutively activated in ER negative breast cancer (Keklikoglou et al. 2012).

miR-210 and *miR-495* have been related to the maintenance of cancer stem cell populations; for instance, increased *miR-495* has been found in a subset of a breast cancer stem cell population. It is able to promote cell invasion through suppression of E-cadherin expression. It can also down-regulate the expression of REDD1, which would result in an increase in cell proliferation in hypoxia through a posttranscriptional mechanism (Hwang-Verslues et al. 2011).

IV. *MIR-210* AS KEY PLAYER IN HYPOXIA

Among all miRNAs found to be modulated by hypoxia, *miR-210* has proved to be the most interesting due to its consistent up-regulation across a broad range of cell types. Its expression has been confirmed to be directly regulated by HIF-1α in a pVHL-dependent manner (Kulshreshtha et al. 2007; Camps et al. 2008). It has been found up-regulated in many cancers and also linked to hypoxia in cancer (for a review, see Chan and Loscalzo 2010). Indeed, *miR-210* levels have been correlated with a hypoxia gene expression signature in breast and head and neck cancers (Camps et al. 2008; Huang et al. 2009; Gee et al. 2010). These findings strongly support the idea that *miR-210* is also regulated by hypoxia *in vivo*. In addition, high levels of *miR-210* have been correlated with poor prognosis in breast, head and neck, and pancreatic cancers (Camps et al. 2008; Gee et al. 2010; Greither et al. 2010), and *miR-210* expression was found to be an independent prognostic factor in a study of 210 breast cancers (Camps et al. 2008).

A. Discovery and Validation of *miR-210* Targets

Studies to identify potential target genes of *miR-210* range from the characterization of particular predicted targets to large screenings involving immunoprecipitation of the RISC or the manipulation of *miR-210* levels in combination with gene expression microarrays, RNA sequencing, and/or proteomics (Fasanaro et al. 2009; Huang et al. 2009; Zhang et al. 2009; Puissegur et al. 2011). Another approach has been taken by Buffa et al. (2011) whereby expression levels of *miR-210* in breast cancer samples were correlated with expression levels of predefined hypoxia signature mRNAs and the patterns of predicted targets, leading to the discovery of novel prognostic mRNAs that are also predicted *miR-210* targets. This approach validates predicted targets through anticorrelation to miRNA and pathway gene expression and provides links between miRNAs, pathways, and miRNA targets (Frampton et al. 2012). Combining all the studies published up to date generates more than 600 putative target genes of *miR-210*. It should be noted however that just a few of them have been experimentally validated. Taking into account only those validated targets, *miR-210* is proposed to have a regulatory role in many important biological processes including mitochondrial function, DNA repair, cell cycle, apoptosis, stem cell differentiation, immunosuppression, angiogenesis, and hypoxia, many of which are discussed in the following sections (Table 16.1).

B. *miR-210* and Mitochondrial Functions

One important change that cells undergo for adapting to hypoxia involves the suppression of mitochondrial oxidative phosphorylation and the subsequent shift to glycolysis as main mechanism for obtaining energy. The shift to glycolysis is mainly directed by HIF, which is able to induce the expression of most of the genes involved in this pathway including pyruvate dehydrogenase kinase (*PDK1*) and lactate dehydrogenase A (*LDHA*). Several studies have shown the involvement of *miR-210* in this metabolic shift through its target iron–sulfur cluster scaffold homologue (*ISCU*) (Chan et al. 2009; Chen et al. 2010; Favaro et al. 2010). ISCU is a mitochondrial protein required for the assembly of iron–sulfur clusters that are incorporated into enzymes involved in the Krebs cycle, electron transport, and iron metabolism. Therefore, the down-regulation of *ISCU* by *miR-210* upon hypoxia would reduce the activity of these enzymes, which would contribute to the shift to glycolysis. In addition, it induces the production of reactive oxygen species (ROS), increases cell survival, and raises the iron uptake required for cell growth. Indeed, in an analysis of more than 900 patients with different tumor types, low levels of *ISCU* expression were correlated with a worse prognosis (Favaro et al. 2010). Moreover, some components of the aerobic respiratory chain have also been shown to be targets of *miR-210*. These include NADH dehydrogenase ubiquinone 1 subcomplex 4 (*NDUFA4*), succinate dehydrogenase complex subunit D (*SDHD*), and cytochrome c oxidase assembly homologue 10 (*COX10*), underlining its effect on mitochondrial functions.

TABLE 16.1. *miR-210* Validated Targets

Gene Symbol	Gene Name	Biological Process Affected upon *miR-210* Regulation	Reference
EFNA3	Ephrin-A3	Cell migration Angiogenesis	Fasanaro et al. (2008); Pulkkinen et al. (2008)
NPTX1	Neuronal pentraxin 1	Not determined	Pulkkinen et al. (2008)
E2F3	E2F transcription factor 3	Cell cycle	Giannakakis et al. (2008)
RAD52	Rad52 homologue	DNA repair	Crosby et al. (2009)
MNT	MAX-binding protein	Cell cycle	Zhang et al. (2009)
HOXA1	Homeobox A1	Tumor initiation Immune escape of tumor cells	Huang et al. (2009); Noman et al. (2012)
HOXA9	Homeobox A9	Not determined	Huang et al. (2009)
FGFRL1	Fibroblast growth factor-like 1	Tumor initiation	Huang et al. (2009); Tsuchiya et al. (2010)
CASP8AP2	Caspase-8-associated protein 2	Apoptosis	Kim et al. (2009)
ACVR1B	Activin receptor 1B	Cell differentiation	Mizuno et al. (2009)
BDNF	Brain-derived neurotrophic factor	Not determined	Fasanaro et al. (2009)
PTPN1	Tyrosine-protein phosphatase non-receptor type I	Not determined	Fasanaro et al. (2009)
P4HB	Protein disulfide isomerase	Not determined	Fasanaro et al. (2009)
GPD1L	Glycerol-3-phosphate dehydrogenase 1-like	Proteasome HIF-1α degradation	Fasanaro et al. (2009); Kelly et al. (2011)
ISCU	Iron–sulfur cluster scaffold homologue	Mitochondrial function ROS production Iron uptake Cell survival	Chan et al. (2009); Chen et al. (2010); Favaro et al. (2010)
COX10	Cytochrome c oxidase assembly homologue 10	Mitochondrial function	Chen et al. (2010)
SDHD	Succinate dehydrogenase complex subunit D	Mitochondrial function	Puissegur et al. (2011)
NDUFA4	NADH dehydrogenase ubiquinone 1 subcomplex 4	Mitochondrial function	Puissegur et al. (2011)
TCF7L2	Transcription factor 7-like 2	Cell differentiation	Qin et al. (2010)
AIFM3	Apoptosis-inducing factor, mitochondrion-associated 3	Apoptosis	Mutharasan et al. (2011); Yang et al. (2012)
HIF3A	Hypoxia-inducible factor 3	Not determined	Mutharasan et al. (2011)
VMP1	Vacuole membrane protein 1	Metastasis	Ying et al. (2011)
HSD17B1	Hydroxysteroid (17-beta) dehydrogenase 1	Preeclampsia	Ishibashi et al. (2012)
SHIP1	Phosphatidylinositol 3,4,5-triphosphate 5-phosphatase 1	Apoptosis	Lee et al. (2012)
FOXP3	Forkhead box P3	Regulatory T-cell function	Fayyad-Kazan et al. (2012)
NFKB1	Nuclear factor of kappa light polypeptide gene enhancer in B-cells 1	LPS-induced expression of proinflammatory cytokines	Qi et al. (2012)
PTBP3	Polypyrimidine tract binding protein 3	Apoptosis	Fasanaro et al. (2012)
PTPN1	Protein tyrosine phosphatase, non-receptor type 1	Immune escape of tumor cells	Noman et al. (2012)
TP53I11	Tumor protein p53 inducible protein 11	Apoptosis Cell survival Immune escape of tumor cells	Noman et al. (2012)

ROS, reactive oxygen species; LPS, lipopolysaccharides.

C. miR-210 and Angiogenesis

The ability of *miR-210* to regulate Ephrin-A3 (*EFNA3*) levels suggests an important role for this miRNA in angiogenesis (Fasanaro et al. 2008; Pulkkinen et al. 2008). EFNA3 is a negative modulator of vascular endothelial growth factor (VEGF)-dependent endothelial cell migration and vessel generation. In HUVEC cells, it has been shown that *miR-210* can control the formation of tubular structures through regulating EFNA3 expression (Fasanaro et al. 2008). In agreement with these findings, *miR-210* expression has been correlated to *VEGF* expression, hypoxia, and angiogenesis in breast cancer patients (Foekens et al. 2008).

D. miR-210 and DNA Repair

Hypoxia can affect the DNA repair capacity of tumor cells and increase their genetic instability. The mismatch repair and the homology-dependent repair pathways are inhibited in hypoxic cells due to the down-regulation of the *MLH1* and *MSH2* genes and the *BRCA1* and *RAD51* genes, respectively (Bindra et al. 2007). In addition, it has been shown that *miR-210* and *miR-373* regulate the levels of RAD52 and RAD23B, which are involved in the homology-dependent DNA repair pathway (Crosby et al. 2009). Indeed, forced expression of *miR-210* leads to a reduction of RAD52 protein levels whereas forced expression of *miR-373* results in a reduction of both RAD52 and RAD23B levels. These findings add a new level of regulation to this important pathway, although it is not clear whether the induction of these miRNAs alone is enough to cause DNA damage or genetic instability during hypoxia.

E. miR-210 and the Cell Cycle

The involvement of *miR-210* in the regulation of the cell cycle seems to be complex, and different outcomes have been reported depending on the cellular context. On the one hand, there are several mechanisms described for which hypoxia, through HIF-1α, leads to cell cycle arrest by inhibition of c-MYC function (Goda et al. 2003); induction by HIF-1α of the c-MYC antagonist MXI1; displacement of c-MYC by HIF-1α from target gene promoters and stimulation of its degradation; and competition of HIF-1α with c-MYC for the binding of the c-MYC functional partner MAX. In that context, it has been shown that *miR-210* is able to regulate the expression of MNT, an antagonist of c-MYC (Zhang et al. 2009). Therefore, the overexpression of *miR-210* in cancer cell lines circumvents the hypoxia-induced cell cycle arrest by down-regulating MNT. On the other hand, *miR-210* can lead to the opposite effect and decrease cell cycle division by repressing other targets. One of these is E2F3, a member of the E2F family of transcription factors that is often overexpressed in cancers and leads to an increase in the cell division rate. High expression levels of *E2F3* have been linked to low levels of *miR-210* in ovarian cancers, which are often caused by the loss of the genomic region encompassing *miR-210* genomic location on chromosome 11 (11p15.5) in these tumors (Giannakakis et al. 2008). Other targets for which repression by *miR-210* results in similar effects are HOXA1 and FGFRL1. When human pancreatic or head and neck cell lines overexpressing *miR-210* were implanted in nude mice, the initiation rate of xenograft growth was significantly lower than in controls. This phenotype was partially rescued by stable cotransfection of either *HOXA1* or *FGFRL1*, suggesting that these genes are implicated in the inhibition of tumor growth by *miR-210* but contribution of other targets would be necessary for full effect (Huang et al. 2009).

In hepatocellular carcinoma (HCC) cell lines, hypoxia slightly inhibits cell growth but increases cell migration and invasion. Several experiments have demonstrated that *miR-210* is mediating this effect through the direct regulation of vacuole membrane protein 1 (*VMP1*). In addition, VMP1 protein immunohistochemical staining was weak or non-detectable in a large proportion of HCC samples (30 out of 48), while it was found strong in all non-cancerous liver tissues analyzed. Moreover, the expression of VMP1 was found inversely correlated to *miR-210* levels in HCC samples (Ying et al. 2011). Therefore, these findings suggest that *miR-210* mediates hypoxia-induced metastasis in HCC through regulation of VMP1. In agreement with this role, *VMP1* has also been found down-regulated in metastatic breast cancer (Sauermann et al. 2008).

F. *miR-210*, Apoptosis, and Cell Differentiation

The hypoxic up-regulation of *miR-210* has also been related with the conservation and differentiation of stem cells. *miR-210* expression has been found to be increased in bone marrow-derived mesenchymal stem cells (MSCs) during ischemic preconditioning in anoxia (Kim et al. 2009). It is known that ischemic preconditioning increases cell survival in hypoxia/anoxia conditions. In agreement with that, increased expression of *miR-210* also correlates with improved survival of transplanted MSC in a rat model. This has been related to the ability of down-regulating caspase-8-associated protein 2 (*Casp8ap*) by *miR-210*. *Casp8ap* is a regulator of Fas-mediated apoptosis, and its expression promotes cellular death. Therefore, *miR-210* could exert a protective effect by targeting this gene (Kim et al. 2009). *miR-210* is also induced during the osteoblastic differentiation of mouse ST2 mesenchymal stem cells, where it targets the TGFβ/activin signaling pathway regulator activin A receptor type 1B (*AcvR1b*). The inhibition of TGFβ/activin signaling leads to osteoblastic differentiation, and for that reason, *miR-210* can promote this process by down-regulating *AcvR1b* (Mizuno et al. 2009). *miR-210* has also been shown to promote adipogenesis by repressing WNT signaling through targeting Tcf7l2 (Qin et al. 2010).

Myelodysplastic syndromes (MDSs) comprise a set of diseases affecting myeloid cells that normally evolve to acute myeloid leukemia (AML). They are characterized at an early stage of the disease by a high rate of apoptosis that is progressively lost toward its development to AML. There is evidence that *miR-210* could contribute to the progress from low-risk to high-risk MDS by regulating the expression of SHIP-1, a protein that dephosphorylates phosphoinositides and which is only expressed in hematopoietic cells (Geier et al. 1997). Indeed, *miR-210* has been found to be up-regulated in CD34+ MDS cells compared with control (Lee et al. 2012). In contrast, SHIP-1 expression has been found decreased in primary MDS-enriched bone marrow cells and also in high-risk MDS samples. This correlates with a higher PI 3′ kinase/Akt activity in these samples, which is known to promote cell survival. In agreement with this, it has been demonstrated that the loss of SHIP-1 decreases apoptosis in MDS-enriched bone marrow cells and increases the number of spontaneously growing colonies in primary AML cells (Lee et al. 2012). Fasanaro et al. (2012) have shown that *miR-210* antiapoptotic action is also mediated through suppression of ROD1 expression. This work demonstrated also for the first time that this miRNA can interact with its target without the "classical" seed sequence pairing.

G. miR-210 and Immunosuppression

Hypoxia is also implicated in the immune escape of tumor cells. Noman et al. (2012) have reported that *miR-210* plays a role in this process by targeting *PTPN1*, *HOXA1*, and *TP53I11*. *PTPN1* and *TP53I11* are known to play a role in apoptosis and cell survival. When the expression of these genes is suppressed in lung cancer or melanoma, then these have the potential to escape from cytotoxic T cells.

H. miR-210 as a Biomarker

The prognostic value of *miR-210* for breast and head and neck cancers has now been demonstrated in several studies (Camps et al. 2008; Foekens et al. 2008; Gee et al. 2010; Buffa et al. 2011; Rothe et al. 2011). It is encouraging that circulating levels of *miR-210* have been already measured in some series of cancer patients as shown in patients with diffuse large B-cell lymphoma compared with healthy controls (Lawrie et al. 2008), confirming its potential as cancer biomarker. Moreover, *miR-210* levels were also greater in the plasma of pancreatic cancer patients compared with healthy controls (Wang et al. 2009; Ho et al. 2010).

V. MiRNA-MEDIATED REGULATION OF HIF

Special attention has been spent on investigating the ability of miRNAs in regulating HIF function, possibly via a feedback loop. Several scenarios and mechanisms of regulation have been proposed including direct targeting of HIF and indirect regulation of HIF expression and protein stabilization, leading to both increase and decrease of HIF function (see references in the following sections). The miRNAs exerting these functions include ones regulated directly by HIF, miRNAs regulated by hypoxia but not through HIF, and miRNAs not regulated by hypoxia (Table 16.2).

A. miRNAs Regulated by HIF

Among the HIF-dependent miRNAs that in turn can regulate HIF are *miR-210* and *miR-20b*. There are two positive regulatory loops between *miR-210* and HIF-1α reported up to date. One of them is mediated by SDHD, one of the *miR-210* targets in mitochondrial function (Puissegur et al. 2011). Indeed, it has been demonstrated that the down-regulation of SDHD by *miR-210* leads to HIF-1α stabilization and maintenance of its activity (Puissegur et al. 2011). The second loop relating *miR-210* and HIF-1α involves the glycerol-3-phosphate dehydrogenase 1-like (*GPD1L*). This gene has been identified as a negative regulator of HIF since it is implicated in the pVHL pathway that leads to HIF degradation. It has also been found to be a target of *miR-210*. Therefore, *miR-210* is able to decrease GPD1L levels and contribute to the stabilization of HIF-1α protein (Kelly et al. 2011).

Concerning *miR-20b*, this has been shown to directly regulate HIF-1α and VEGF expression. In turn, overexpression of HIF-1α in normoxic hepatocarcinoma cells leads to the down-regulation of *miR-20b*. Low expression of *miR-20b* in hypoxia inhibits tumor cell growth but increases the resistance of tumor cells to apoptosis, suggesting that this mechanism could be involved in the higher resistance of hypoxic cells to chemo- and radiotherapy (Lei et al. 2009).

TABLE 16.2. miRNA-Mediated Regulation of HIF

miRNA	Regulation of miRNA Expression	Mechanism by which miRNA Regulates HIF-1α	Effect on HIF-1α Expression/Function	Situation Reported in Hypoxic Cells	Biological Effect/Clinical Relevance	References
miR-210	Hypoxia; HIF-1α dependent	Indirect: through targeting SDHD Indirect: through targeting GPD1L	HIF-1α proteasomal degradation HIF-1α proteasomal degradation	miR-210 up-regulated Increase in HIF-1α stabilization and function	High levels of miR-210 are associated with increased hazard in breast cancer and head and neck cancers Anticorrelation with predicted targets	Camps et al. (2008); Gee et al. (2010); Buffa et al. (2011); Kelly et al. (2011); Puissegur et al. (2011)
miR-20b	Hypoxia; HIF-1α dependent	Direct HIF-1α targeting	Regulation of HIF-1α protein levels	miR-20b down-regulated HIF-1α protein levels increased	Low levels of miR-20b in hepatocarcinoma cell lines inhibit tumor cell growth but increase resistance to apoptosis	Lei et al. (2009)
miR-199a	Hypoxia; AKT-dependent	Direct HIF-1α targeting	Regulation of HIF-1α protein levels	miR-199a down-regulated HIF-1α protein levels increased	Not determined	Rane et al. (2009, 2010)
miR-107	Hypoxia; P53 dependent	Indirect: through targeting HIF-1β	Regulation of HIF-1α transcriptional function	miR-107 increased HIF-1α transcriptional activity decreased	Overexpression of miR-107 in tumor cells leads to suppression of angiogenesis and tumor growth In human colon cancer, expression of miR-107 found inversely associated with HIF-1β expression	Yamakuchi et al. (2010)
miR-424	Hypoxia; PU.1 transactivation	Indirect: through targeting CUL2	HIF-1α proteasomal degradation	miR-424 up-regulated Increase in HIF-1α stabilization and function	miR-424 promote angiogenesis in vitro and in mice	Ghosh et al. (2010)
miR-130	Hypoxia	Indirect: through targeting DDX6	HIF-1α entry to P-bodies	miR-130 up-regulated HIF-1α translation increased	Not determined	Saito et al. (2011)

miRNAs Not Regulated by Hypoxia or HIF

miRNA	Regulation of miRNA Expression	Mechanism by which miRNA Regulates HIF-1α	Effect on HIF-1α Expression/Function	Situation Reported/Suggested in Cancer	Biological Effect/Clinical Relevance	References
miR-519c	Hepatocyte growth factor (HGF) in AKT-dependent manner	Direct HIF-1α targeting	Regulation of HIF-1α protein levels	HGF induces HIF-1α expression. HGF represses miR-519c expression, which would contribute to increase HIF-1α levels	miR-519c expression can suppress tumor angiogenesis, growth, and metastasis. Overexpression of miR-519c in cancer patients linked to better prognosis	Cha et al. (2010)
miR-22	c-MYC	Direct HIF-1α targeting	Regulation of HIF-1α protein levels	c-MYC represses miR-22 expression in cancer cell lines, which would increase HIF-1α protein levels	Overexpression of miR-22 in colon cancer cell lines decreases proliferation and migration. miR-22 found down-regulated in human colon cancers	Chang et al. (2008); Yamakuchi et al. (2011)
miR-17-92 cluster	c-MYC	Direct HIF-1α targeting	Regulation of HIF-1α protein levels	c-MYC induces miR-17-97 cluster expression, which would decrease HIF-1α protein levels	Not determined	Taguchi et al. (2008)
miR-31	Not determined	Indirect; through targeting FIH	Regulation of HIF-1α transcriptional function	An increase in miR-31 levels would increase HIF-1α transcriptional activity	Overexpression of miR-31 increases the oncogenic potential in HNSCC. Inhibition of miR-31 reduces the growth of tumor xenografts	Liu et al. (2010)
miR-145	Not determined	Indirect; through targeting p70S6K1	Regulation of HIF-1α expression	An increase in miR-145 levels would decrease HIF-1α expression	miR-145 found down-regulated in human colon cancer. In addition, miR-145 levels inversely correlated to p70S6K1 protein levels	Xu et al. (2012)

B. miRNAs Regulated by Hypoxia in a HIF-Independent Manner

Other miRNAs regulated by hypoxia in a HIF-independent manner but able to modulate HIF include *miR-199a*, *miR-107*, *miR-424*, and *miR-130*. The first one, *miR-199a*, has been found down-regulated under hypoxia in cardiac myocytes in an AKT-dependent manner. Since HIF-1α is a direct target of *miR-199a*, the down-regulation of this miRNA leads to an increase in HIF-1α levels (Rane et al. 2009, 2010).

The other three miRNAs are, in contrast, up-regulated under hypoxia. The hypoxic induction of *miR-107* is mediated by P53, and *miR-107* in turn can directly modulate HIF-1β, the partner of HIF-1α in the control of gene expression upon hypoxia. Therefore, it has been shown that the manipulation of *miR-107* levels results in the modulation of the hypoxia response mediated by HIF. In agreement with that, overexpression of *miR-107* in tumor cells leads to suppression of tumor angiogenesis, tumor growth, and tumor *VEGF* expression when injected in mice. In human colon cancer samples, expression of *miR-107* is inversely associated with HIF-1β expression (Yamakuchi et al. 2010).

Concerning the other two miRNAs, *miR-424* and *miR-130* modulate HIF function by indirect mechanisms. *miR-424* has been found up-regulated in endothelial cells via PU.1 transactivation. It targets Cullin 2 (*CUL2*), which is the scaffolding protein that assembles the ubiquitin ligase complex responsible for the degradation of HIF-1α isoform. Therefore, the up-regulation of *miR-424* by hypoxia leads to the stabilization of HIF-1α. In agreement with that, *miR-424* has been shown to promote angiogenesis *in vitro* and in mice (Ghosh et al. 2010). On the other hand, *miR-130* targets DDX6, a component of the P-bodies. The P-bodies are cytoplasmatic compartments containing repressed messenger ribonucleoprotein (mRNP) complexes. The down-regulation of DDX6 by *miR-130* in hypoxia would impair the entry of HIF-1α mRNA into the P-bodies, thus resulting in an increase in HIF-1α translation (Saito et al. 2011).

C. miRNAs Not Regulated by Hypoxia

Also of interest is the group of miRNAs able to control HIF expression although not directly regulated by hypoxia or HIF (Table 16.2). Some miRNAs such as *miR-519c*, *miR-22*, and the *miR-17-92* cluster can directly target HIF-1α. Most of these miRNAs have been found altered in cancer, and therefore, their relationship with HIF is intriguing. *miR-519c* has been shown to suppress tumor angiogenesis, growth, and metastasis through down-regulating HIF-1α levels. Overexpression of *miR-519c* in cancer patients has been linked to better prognosis (Cha et al. 2010). In addition, it has been shown that *miR-519c* and HIF-1α are commonly regulated by the hepatocyte growth factor (HGF). Interestingly, HGF induces HIF-1α expression but represses *miR-519c* expression through a posttranscriptional mechanism involving the Akt pathway (Cha et al. 2010).

Overexpression of *miR-22* in colon cancer cell lines was also shown to decrease the proliferation and migration stimulated by hypoxia. In agreement with this, *miR-22* down-regulation correlated with higher levels of VEGF expression in human colon cancer samples (Yamakuchi et al. 2011). It is known that c-MYC decreases the expression of this miRNA in cancer cell lines (Chang et al. 2008); therefore, tumors overexpressing c-MYC might be expected to have lower levels of *miR-22* and in turn higher levels of HIF-1α. In contrast, the expression of the miRNA cluster *miR-17-92* is induced by c-MYC, which leads to a reduction of HIF-1α (Taguchi et al. 2008). In conclusion, miRNA regulation by c-MYC can lead to opposite effects in terms of modulation of the HIF pathway.

Other hypoxia-independent miRNAs have been shown to regulate HIF-1α through indirect mechanisms. Two examples are *miR-31* and *miR-145*. *miR-31* targets the factor-inhibiting HIF (*FIH*), which is a regulatory factor that impairs HIF transcriptional activity. As a consequence, an increase in *miR-31* can lead to an increase in HIF-1α activity in normoxia, which has been seen to increase the oncogenic potential in head and neck squamous cell carcinoma (HNSCC) in cell culture or tumor xenografts. In agreement, inhibition of *miR-31* expression reduced the growth of tumor xenografts (Liu et al. 2010). In contrast, *miR-145* would decrease HIF-1α expression through its direct target p70S6K1, which suppresses tumor growth and angiogenesis. p70S6K1 is a component of the mTor signaling pathway that is involved in HIF-1α regulation. The level of *miR-145* is decreased in colon cancers, and low levels of this miRNA have been correlated to high levels of p70S6K1 protein in these cancers (Xu et al. 2012).

D. Interaction between Different Levels of miRNA Regulation under Hypoxia

All of these reports point to the notion that HIF-1α is regulated by miRNAs that in turn are regulated by HIF itself, other hypoxia factors, or by hypoxia-independent factors. In addition, components involved in the miRNA machinery are also affected by hypoxia (Ho et al. 2012). If miRNAs down-regulated due to these changes are once targeting HIF, as in the case of *miR-185*, then this generates a feedback loop (Ho et al. 2012).

All these mechanisms may be relevant for the treatment of cancer. For instance, metformin is a drug used to treat diabetes that seems to have anticancer properties. Diabetic patients treated with this drug have reduced incidence of cancer and also reduced mortality due to cancer (Muti et al. 2009). In addition, metformin affects the progression and relapse of breast, prostate, and lung cancer mouse xenografts when used in combination with standard chemotherapeutic drugs at suboptimal doses (Bae et al. 2007; Hirsch et al. 2009; Iliopoulos et al. 2011; Rattan et al. 2011). It has been shown that metformin induces the expression of *DICER* at transcriptional level, leading to the regulation of several miRNAs as well as other genes including *HIF-1α* (Blandino et al. 2012). The predicted targets for miRNAs induced by metformin not only are mostly involved in metabolism and insulin signaling but also include *miR-33a* that targets MYC (Blandino et al. 2012).

VI. CONCLUSIONS

The hypoxic regulation of miRNAs has been widely demonstrated over recent years. Genes involved in processes such as cell cycle, apoptosis, migration, and metastasis have been shown to be targets of hypoxia-regulated miRNAs. miRNA levels are also controlled during their posttranscriptional processing by hypoxia. In addition, there are miRNAs that can modulate the expression and activity of HIF, adding another layer of regulation. Others are hypoxia independent, but their levels are altered in certain cancers so their ability to regulate HIF can be relevant for understanding the processes involved in the disease and its progress (Figure 16.1).

One of the main players in hypoxia and cancer is *miR-210*, which has been found to have a regulatory role in important biological processes such as mitochondrial function, DNA repair, cell cycle, apoptosis, cell differentiation, immunosuppression, angiogenesis, and hypoxia. Research on *miR-210* is still very new; nevertheless, it is important to continue investigating the role of other miRNAs in hypoxia.

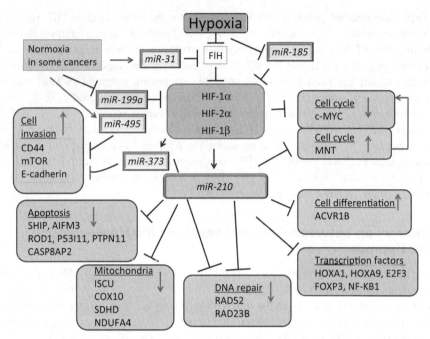

Figure 16.1. Actions of hypoxia-induced microRNAs. Green arrows and outlines indicate up-regulation and red ones indicate down-regulation. HIFs are regulated through *miR-185* and *miR-199a* directly or by *miR-31* through down-regulation of FIH. *miR-210* is the master hypoxamir with effects on mitochondrial functions, apoptosis, DNA repair, cell cycle, and differentiation through the down-regulation of targets with functions in these pathways. *miR-210* is up-regulated through direct binding of HIF-1α and HIF-2α to host transcript promoter elements. *miR-210* also indirectly regulates c-MYC, which is antagonized by HIF. The action of HIF and *miR-210* on c-MYC appears antagonistic. Furthermore, hypoxia pathways are regulated through *miR-373* and *miR-495*. See color insert.

REFERENCES

Bae, E.J., M.J. Cho, and S.G. Kim. 2007. Metformin prevents an adaptive increase in GSH and induces apoptosis under the conditions of GSH deficiency in H4IIE cells. *J Toxicol Environ Health A* **70**:1371–1380.

Bertout, J.A., S.A. Patel, and M.C. Simon. 2008. The impact of O2 availability on human cancer. *Nat Rev Cancer* **8**:967–975.

Bindra, R.S., M.E. Crosby, and P.M. Glazer. 2007. Regulation of DNA repair in hypoxic cancer cells. *Cancer Metastasis Rev* **26**:249–260.

Blandino, G., M. Valerio, M. Cioce, F. Mori, L. Casadei, C. Pulito, A. Sacconi, F. Biagioni, G. Cortese, S. Galanti, C. Manetti, G. Citro, et al. 2012. Metformin elicits anticancer effects through the sequential modulation of DICER and c-MYC. *Nat Commun* **3**:865.

Buffa, F.M., C. Camps, L. Winchester, C.E. Snell, H.E. Gee, H. Sheldon, M. Taylor, A.L. Harris, and J. Ragoussis. 2011. microRNA-associated progression pathways and potential therapeutic targets identified by integrated mRNA and microRNA expression profiling in breast cancer. *Cancer Res* **71**:5635–5645.

REFERENCES

Camps, C., F.M. Buffa, S. Colella, J. Moore, C. Sotiriou, H. Sheldon, A.L. Harris, J.M. Gleadle, and J. Ragoussis. 2008. hsa-*miR-210* is induced by hypoxia and is an independent prognostic factor in breast cancer. *Clin Cancer Res* **14**:1340–1348.

Caruso, P., M.R. MacLean, R. Khanin, J. McClure, E. Soon, M. Southgate, R.A. MacDonald, J.A. Greig, K.E. Robertson, R. Masson, L. Denby, Y. Dempsie, et al. 2010. Dynamic changes in lung microRNA profiles during the development of pulmonary hypertension due to chronic hypoxia and monocrotaline. *Arterioscler Thromb Vasc Biol* **30**:716–723.

Cha, S.T., P.S. Chen, G. Johansson, C.Y. Chu, M.Y. Wang, Y.M. Jeng, S.L. Yu, J.S. Chen, K.J. Chang, S.H. Jee, C.T. Tan, M.T. Lin, et al. 2010. MicroRNA-519c suppresses hypoxia-inducible factor-1alpha expression and tumor angiogenesis. *Cancer Res* **70**:2675–2685.

Chan, S.Y. and J. Loscalzo. 2010. MicroRNA-210: a unique and pleiotropic hypoxamir. *Cell Cycle* **9**:1072–1083.

Chan, S.Y., Y.Y. Zhang, C. Hemann, C.E. Mahoney, J.L. Zweier, and J. Loscalzo. 2009. MicroRNA-210 controls mitochondrial metabolism during hypoxia by repressing the iron-sulfur cluster assembly proteins ISCU1/2. *Cell Metab* **10**:273–284.

Chang, T.C., D. Yu, Y.S. Lee, E.A. Wentzel, D.E. Arking, K.M. West, C.V. Dang, A. Thomas-Tikhonenko, and J.T. Mendell. 2008. Widespread microRNA repression by Myc contributes to tumorigenesis. *Nat Genet* **40**:43–50.

Chen, Z., Y. Li, H. Zhang, P. Huang, and R. Luthra. 2010. Hypoxia-regulated microRNA-210 modulates mitochondrial function and decreases ISCU and COX10 expression. *Oncogene* **29**: 4362–4368.

Crosby, M.E., R. Kulshreshtha, M. Ivan, and P.M. Glazer. 2009. MicroRNA regulation of DNA repair gene expression in hypoxic stress. *Cancer Res* **69**:1221–1229.

Elvidge, G.P., L. Glenny, R.J. Appelhoff, P.J. Ratcliffe, J. Ragoussis, and J.M. Gleadle. 2006. Concordant regulation of gene expression by hypoxia and 2-oxoglutarate-dependent dioxygenase inhibition: the role of HIF-1alpha, HIF-2alpha, and other pathways. *J Biol Chem* **281**:15215–15226.

Fasanaro, P., Y. D'Alessandra, V. Di Stefano, R. Melchionna, S. Romani, G. Pompilio, M.C. Capogrossi, and F. Martelli. 2008. MicroRNA-210 modulates endothelial cell response to hypoxia and inhibits the receptor tyrosine kinase ligand Ephrin-A3. *J Biol Chem* **283**:15878–15883.

Fasanaro, P., S. Greco, M. Lorenzi, M. Pescatori, M. Brioschi, R. Kulshreshtha, C. Banfi, A. Stubbs, G.A. Calin, M. Ivan, M.C. Capogrossi, and F. Martelli. 2009. An integrated approach for experimental target identification of hypoxia-induced *miR-210*. *J Biol Chem* **284**:35134–35143.

Fasanaro, P., S. Romani, C. Voellenkle, B. Maimone, M.C. Capogrossi, and F. Martelli. 2012. ROD1 is a seedless target gene of hypoxia-induced *miR-210*. *PLoS ONE* **7**:e44651.

Favaro, E., A. Ramachandran, R. McCormick, H. Gee, C. Blancher, M. Crosby, C. Devlin, C. Blick, F. Buffa, J.L. Li, B. Vojnovic, R. Pires das Neves, et al. 2010. MicroRNA-210 regulates mitochondrial free radical response to hypoxia and Krebs cycle in cancer cells by targeting iron sulfur cluster protein ISCU. *PLoS ONE* **5**:e10345.

Fayyad-Kazan, H., R. Rouas, M. Fayyad-Kazan, R. Badran, N. El Zein, P. Lewalle, M. Najar, E. Hamade, F. Jebbawi, M. Merimi, P. Romero, A. Burny, et al. 2012. MicroRNA profile of circulating CD4-positive regulatory T cells in human adults and impact of differentially expressed microRNAs on expression of two genes essential to their function. *J Biol Chem* **287**: 9910–9922.

Foekens, J.A., A.M. Sieuwerts, M. Smid, M.P. Look, V. de Weerd, A.W. Boersma, J.G. Klijn, E.A. Wiemer, and J.W. Martens. 2008. Four miRNAs associated with aggressiveness of lymph node-negative, estrogen receptor-positive human breast cancer. *Proc Natl Acad Sci U S A* **105**: 13021–13026.

Frampton, A.E., J. Krell, L. Pellegrino, L. Roca-Alonso, L.R. Jiao, J. Stebbing, L. Castellano, and J. Jacob. 2012. Integrated analysis of miRNA and mRNA profiles enables target acquisition in human cancers. *Expert Rev Anticancer Ther* **12**:323–330.

Gee, H.E., C. Camps, F.M. Buffa, S. Patiar, S.C. Winter, G. Betts, J. Homer, R. Corbridge, G. Cox, C.M. West, J. Ragoussis, and A.L. Harris. 2010. hsa-mir-210 is a marker of tumor hypoxia and a prognostic factor in head and neck cancer. *Cancer* **116**:2148–2158.

Geier, S.J., P.A. Algate, K. Carlberg, D. Flowers, C. Friedman, B. Trask, and L.R. Rohrschneider. 1997. The human SHIP gene is differentially expressed in cell lineages of the bone marrow and blood. *Blood* **89**:1876–1885.

Ghosh, G., I.V. Subramanian, N. Adhikari, X. Zhang, H.P. Joshi, D. Basi, Y.S. Chandrashekhar, J.L. Hall, S. Roy, Y. Zeng, and S. Ramakrishnan. 2010. Hypoxia-induced microRNA-424 expression in human endothelial cells regulates HIF-alpha isoforms and promotes angiogenesis. *J Clin Invest* **120**:4141–4154.

Giannakakis, A., R. Sandaltzopoulos, J. Greshock, S. Liang, J. Huang, K. Hasegawa, C. Li, A. O'Brien-Jenkins, D. Katsaros, B.L. Weber, C. Simon, G. Coukos, et al. 2008. *miR-210* links hypoxia with cell cycle regulation and is deleted in human epithelial ovarian cancer. *Cancer Biol Ther* **7**:255–264.

Goda, N., H.E. Ryan, B. Khadivi, W. McNulty, R.C. Rickert, and R.S. Johnson. 2003. Hypoxia-inducible factor 1alpha is essential for cell cycle arrest during hypoxia. *Mol Cell Biol* **23**:359–369.

Greither, T., L.F. Grochola, A. Udelnow, C. Lautenschlager, P. Wurl, and H. Taubert. 2010. Elevated expression of microRNAs 155, 203, 210 and 222 in pancreatic tumors is associated with poorer survival. *Int J Cancer* **126**:73–80.

Hirsch, H.A., D. Iliopoulos, P.N. Tsichlis, and K. Struhl. 2009. Metformin selectively targets cancer stem cells, and acts together with chemotherapy to block tumor growth and prolong remission. *Cancer Res* **69**:7507–7511.

Ho, A.S., X. Huang, H. Cao, C. Christman-Skieller, K. Bennewith, Q.T. Le, and A.C. Koong. 2010. Circulating *miR-210* as a novel hypoxia marker in pancreatic cancer. *Transl Oncol* **3**:109–113.

Ho, J.J., J.L. Metcalf, M.S. Yan, P.J. Turgeon, J.J. Wang, M. Chalsev, T.N. Petruzziello-Pellegrini, A.K. Tsui, J.Z. He, H. Dhamko, H.S. Man, G.B. Robb, et al. 2012. Functional importance of dicer protein in the adaptive cellular response to hypoxia. *J Biol Chem* **287**:29003–29020.

Huang, Q., K. Gumireddy, M. Schrier, C. le Sage, R. Nagel, S. Nair, D.A. Egan, A. Li, G. Huang, A.J. Klein-Szanto, P.A. Gimotty, D. Katsaros, et al. 2008. The microRNAs *miR-373* and *miR-520c* promote tumour invasion and metastasis. *Nat Cell Biol* **10**:202–210.

Huang, X., L. Ding, K.L. Bennewith, R.T. Tong, S.M. Welford, K.K. Ang, M. Story, Q.T. Le, and A.J. Giaccia. 2009. Hypoxia-inducible mir-210 regulates normoxic gene expression involved in tumor initiation. *Mol Cell* **35**:856–867.

Hwang-Verslues, W.W., P.H. Chang, P.C. Wei, C.Y. Yang, C.K. Huang, W.H. Kuo, J.Y. Shew, K.J. Chang, E.Y. Lee, and W.H. Lee. 2011. *miR-495* is upregulated by E12/E47 in breast cancer stem cells, and promotes oncogenesis and hypoxia resistance via downregulation of E-cadherin and REDD1. *Oncogene* **30**:2463–2474.

Iliopoulos, D., H.A. Hirsch, and K. Struhl. 2011. Metformin decreases the dose of chemotherapy for prolonging tumor remission in mouse xenografts involving multiple cancer cell types. *Cancer Res* **71**:3196–3201.

Ishibashi, O., A. Ohkuchi, M.M. Ali, R. Kurashina, S.S. Luo, T. Ishikawa, T. Takizawa, C. Hirashima, K. Takahashi, M. Migita, G. Ishikawa, K. Yoneyama, et al. 2012. Hydroxysteroid (17-beta) dehydrogenase 1 is dysregulated by *miR-210* and *miR-518c* that are aberrantly expressed in preeclamptic placentas: a novel marker for predicting preeclampsia. *Hypertension* **59**:265–273.

Keklikoglou, I., C. Koerner, C. Schmidt, J.D. Zhang, D. Heckmann, A. Shavinskaya, H. Allgayer, B. Guckel, T. Fehm, A. Schneeweiss, O. Sahin, S. Wiemann, et al. 2012. MicroRNA-520/373 family functions as a tumor suppressor in estrogen receptor negative breast cancer by targeting NF-kappaB and TGF-beta signaling pathways. *Oncogene* **31**:4150–4163.

REFERENCES

Kelly, T.J., A.L. Souza, C.B. Clish, and P. Puigserver. 2011. A hypoxia-induced positive feedback loop promotes hypoxia-inducible factor 1alpha stability through *miR-210* suppression of glycerol-3-phosphate dehydrogenase 1-like. *Mol Cell Biol* **31**:2696–2706.

Kim, H.W., H.K. Haider, S. Jiang, and M. Ashraf. 2009. Ischemic preconditioning augments survival of stem cells via *miR-210* expression by targeting caspase-8-associated protein 2. *J Biol Chem* **284**:33161–33168.

Kulshreshtha, R., M. Ferracin, S.E. Wojcik, R. Garzon, H. Alder, F.J. Agosto-Perez, R. Davuluri, C.G. Liu, C.M. Croce, M. Negrini, G.A. Calin, and M. Ivan. 2007. A microRNA signature of hypoxia. *Mol Cell Biol* **27**:1859–1867.

Kulshreshtha, R., R.V. Davuluri, G.A. Calin, and M. Ivan. 2008. A microRNA component of the hypoxic response. *Cell Death Differ* **15**:667–671.

Lawrie, C.H., S. Gal, H.M. Dunlop, B. Pushkaran, A.P. Liggins, K. Pulford, A.H. Banham, F. Pezzella, J. Boultwood, J.S. Wainscoat, C.S. Hatton, and A.L. Harris. 2008. Detection of elevated levels of tumour-associated microRNAs in serum of patients with diffuse large B-cell lymphoma. *Br J Haematol* **141**:672–675.

Lee, D.W., M. Futami, M. Carroll, Y. Feng, Z. Wang, M. Fernandez, Z. Whichard, Y. Chen, S. Kornblau, E.J. Shpall, C.E. Bueso-Ramos, and S.J. Corey. 2012. Loss of SHIP-1 protein expression in high-risk myelodysplastic syndromes is associated with *miR-210* and *miR-155*. *Oncogene* **31**:4085–4094.

Lei, Z., B. Li, Z. Yang, H. Fang, G.M. Zhang, Z.H. Feng, and B. Huang. 2009. Regulation of HIF-1alpha and VEGF by *miR-20b* tunes tumor cells to adapt to the alteration of oxygen concentration. *PLoS ONE* **4**:e7629.

Liu, P. and M.J. Wilson. 2012. *miR-520c* and *miR-373* upregulate MMP9 expression by targeting mTOR and SIRT1, and activate the Ras/Raf/MEK/Erk signaling pathway and NF-kappaB factor in human fibrosarcoma cells. *J Cell Physiol* **227**:867–876.

Liu, C.J., M.M. Tsai, P.S. Hung, S.Y. Kao, T.Y. Liu, K.J. Wu, S.H. Chiou, S.C. Lin, and K.W. Chang. 2010. *miR-31* ablates expression of the HIF regulatory factor FIH to activate the HIF pathway in head and neck carcinoma. *Cancer Res* **70**:1635–1644.

Makino, Y., R. Cao, K. Svensson, G. Bertilsson, M. Asman, H. Tanaka, Y. Cao, A. Berkenstam, and L. Poellinger. 2001. Inhibitory PAS domain protein is a negative regulator of hypoxia-inducible gene expression. *Nature* **404**:550–554.

Mizuno, Y., Y. Tokuzawa, Y. Ninomiya, K. Yagi, Y. Yatsuka-Kanesaki, T. Suda, T. Fukuda, T. Katagiri, Y. Kondoh, T. Amemiya, H. Tashiro, and Y. Okazaki. 2009. *miR-210* promotes osteoblastic differentiation through inhibition of AcvR1b. *FEBS Lett* **583**:2263–2268.

Mole, D.R., C. Blancher, R.R. Copley, P.J. Pollard, J.M. Gleadle, J. Ragoussis, and P.J. Ratcliffe. 2009. Genome-wide association of hypoxia-inducible factor (HIF)-1alpha and HIF-2alpha DNA binding with expression profiling of hypoxia-inducible transcripts. *J Biol Chem* **284**: 16767–16775.

Mutharasan, R.K., V. Nagpal, Y. Ichikawa, and H. Ardehali. 2011. microRNA-210 is upregulated in hypoxic cardiomyocytes through Akt- and p53-dependent pathways and exerts cytoprotective effects. *Am J Physiol Heart Circ Physiol* **301**:H1519–H1530.

Muti, P., F. Berrino, V. Krogh, A. Villarini, M. Barba, S. Strano, and G. Blandino. 2009. Metformin, diet and breast cancer: an avenue for chemoprevention. *Cell Cycle* **8**:2661.

Noman, M.Z., S. Buart, P. Romero, S. Ketari, B. Janji, B. Mari, F. Mami-Chouaib, and S. Chouaib. 2012. Hypoxia-inducible *miR-210* regulates the susceptibility of tumor cells to lysis by cytotoxic T cells. *Cancer Res* **72**:4629–4641.

Place, R.F., L.C. Li, D. Pookot, E.J. Noonan, and R. Dahiya. 2008. MicroRNA-373 induces expression of genes with complementary promoter sequences. *Proc Natl Acad Sci U S A* **105**:1608–1613.

Puissegur, M.P., N.M. Mazure, T. Bertero, L. Pradelli, S. Grosso, K. Robbe-Sermesant, T. Maurin, K. Lebrigand, B. Cardinaud, V. Hofman, S. Fourre, V. Magnone, et al. 2011. *miR-210* is overexpressed in late stages of lung cancer and mediates mitochondrial alterations associated with modulation of HIF-1 activity. *Cell Death Differ* **18**:465–478.

Pulkkinen, K., T. Malm, M. Turunen, J. Koistinaho, and S. Yla-Herttuala. 2008. Hypoxia induces microRNA *miR-210* in vitro and in vivo ephrin-A3 and neuronal pentraxin 1 are potentially regulated by *miR-210*. *FEBS Lett* **582**:2397–2401.

Qi, J., Y. Qiao, P. Wang, S. Li, W. Zhao, and C. Gao. 2012. MicroRNA-210 negatively regulates LPS-induced production of proinflammatory cytokines by targeting NF-kappaB1 in murine macrophages. *FEBS Lett* **586**:1201–1207.

Qin, L., Y. Chen, Y. Niu, W. Chen, Q. Wang, S. Xiao, A. Li, Y. Xie, J. Li, X. Zhao, Z. He, and D. Mo. 2010. A deep investigation into the adipogenesis mechanism: profile of microRNAs regulating adipogenesis by modulating the canonical Wnt/beta-catenin signaling pathway. *BMC Genomics* **11**:320.

Rane, S., M. He, D. Sayed, H. Vashistha, A. Malhotra, J. Sadoshima, D.E. Vatner, S.F. Vatner, and M. Abdellatif. 2009. Downregulation of *miR-199a* derepresses hypoxia-inducible factor-1alpha and Sirtuin 1 and recapitulates hypoxia preconditioning in cardiac myocytes. *Circ Res* **104**:879–886.

Rane, S., M. He, D. Sayed, L. Yan, D. Vatner, and M. Abdellatif. 2010. An antagonism between the AKT and beta-adrenergic signaling pathways mediated through their reciprocal effects on *miR-199a-5p*. *Cell Signal* **22**:1054–1062.

Ratcliffe, P.J., J.F. O'Rourke, P.H. Maxwell, and C.W. Pugh. 1998. Oxygen sensing, hypoxia-inducible factor-1 and the regulation of mammalian gene expression. *J Exp Biol* **201**:1153–1162.

Rattan, R., S. Giri, L.C. Hartmann, and V. Shridhar. 2011. Metformin attenuates ovarian cancer cell growth in an AMP-kinase dispensable manner. *J Cell Mol Med* **15**:166–178.

Rothe, F., M. Ignatiadis, C. Chaboteaux, B. Haibe-Kains, N. Kheddoumi, S. Majjaj, B. Badran, H. Fayyad-Kazan, C. Desmedt, A.L. Harris, M. Piccart, and C. Sotiriou. 2011. Global microRNA expression profiling identifies *MiR-210* associated with tumor proliferation, invasion and poor clinical outcome in breast cancer. *PLoS ONE* **6**:e20980.

Saito, K., E. Kondo, and M. Matsushita. 2011. MicroRNA 130 family regulates the hypoxia response signal through the P-body protein DDX6. *Nucleic Acids Res* **39**:6086–6099.

Sauermann, M., O. Sahin, H. Sultmann, F. Hahne, S. Blaszkiewicz, M. Majety, K. Zatloukal, L. Fuzesi, A. Poustka, S. Wiemann, and D. Arlt. 2008. Reduced expression of vacuole membrane protein 1 affects the invasion capacity of tumor cells. *Oncogene* **27**:1320–1326.

Schodel, J., S. Oikonomopoulos, J. Ragoussis, C.W. Pugh, P.J. Ratcliffe, and D.R. Mole. 2011. High-resolution genome-wide mapping of HIF-binding sites by ChIP-seq. *Blood* **117**:e207–e217.

Semenza, G.L. 1998. Hypoxia-inducible factor 1: master regulator of O2 homeostasis. *Curr Opin Genet Dev* **8**:588–594.

Taguchi, A., K. Yanagisawa, M. Tanaka, K. Cao, Y. Matsuyama, H. Goto, and T. Takahashi. 2008. Identification of hypoxia-inducible factor-1 alpha as a novel target for *miR-17-92* microRNA cluster. *Cancer Res* **68**:5540–5545.

Tsuchiya, S., T. Fujiwara, F. Sato, Y. Shimada, E. Tanaka, Y. Sakai, K. Shimizu, and G. Tsujimoto. 2010. MicroRNA-210 regulates cancer cell proliferation through targeting fibroblast growth factor receptor-like 1 (FGFRL1). *J Biol Chem* **286**:420–428.

Voorhoeve, P.M., C. le Sage, M. Schrier, A.J. Gillis, H. Stoop, R. Nagel, Y.P. Liu, J. van Duijse, J. Drost, A. Griekspoor, E. Zlotorynski, N. Yabuta, et al. 2006. A genetic screen implicates miRNA-372 and miRNA-373 as oncogenes in testicular germ cell tumors. *Cell* **124**:1169–1181.

Wang, J., J. Chen, P. Chang, A. LeBlanc, D. Li, J.L. Abbruzzese, M.L. Frazier, A.M. Killary, and S. Sen. 2009. MicroRNAs in plasma of pancreatic ductal adenocarcinoma patients as novel blood-based biomarkers of disease. *Cancer Prev Res (Phila)* **2**:807–813.

REFERENCES

Wilson, W.R. and M.P. Hay. 2011. Targeting hypoxia in cancer therapy. *Nat Rev Cancer* **11**: 393–410.

Wu, C., J. So, B.N. Davis-Dusenbery, H.H. Qi, D.B. Bloch, Y. Shi, G. Lagna, and A. Hata. 2011. Hypoxia potentiates microRNA-mediated gene silencing through posttranslational modification of Argonaute2. *Mol Cell Biol* **31**:4760–4774.

Xu, Q., L.Z. Liu, X. Qian, Q. Chen, Y. Jiang, D. Li, L. Lai, and B.H. Jiang. 2012. *MiR-145* directly targets p70S6K1 in cancer cells to inhibit tumor growth and angiogenesis. *Nucleic Acids Res* **40**:761–774.

Yamakuchi, M., C.D. Lotterman, C. Bao, R.H. Hruban, B. Karim, J.T. Mendell, D. Huso, and C.J. Lowenstein. 2010. P53-induced microRNA-107 inhibits HIF-1 and tumor angiogenesis. *Proc Natl Acad Sci U S A* **107**:6334–6339.

Yamakuchi, M., S. Yagi, T. Ito, and C.J. Lowenstein. 2011. MicroRNA-22 regulates hypoxia signaling in colon cancer cells. *PLoS ONE* **6**:e20291.

Yang, W., T. Sun, J. Cao, F. Liu, Y. Tian, and W. Zhu. 2012. Downregulation of *miR-210* expression inhibits proliferation, induces apoptosis and enhances radiosensitivity in hypoxic human hepatoma cells in vitro. *Exp Cell Res* **318**:944–954.

Ying, Q., L. Liang, W. Guo, R. Zha, Q. Tian, S. Huang, J. Yao, J. Ding, M. Bao, C. Ge, M. Yao, J. Li, et al. 2011. Hypoxia-inducible microRNA-210 augments the metastatic potential of tumor cells by targeting vacuole membrane protein 1 in hepatocellular carcinoma. *Hepatology* **54**: 2064–2075.

Zhang, Z., H. Sun, H. Dai, R.M. Walsh, M. Imakura, J. Schelter, J. Burchard, X. Dai, A.N. Chang, R.L. Diaz, J.R. Marszalek, S.R. Bartz, et al. 2009. MicroRNA *miR-210* modulates cellular response to hypoxia through the MYC antagonist MNT. *Cell Cycle* **8**:2756–2768.

17

CONTROL OF RECEPTOR FUNCTION BY MicroRNAs IN BREAST CANCER

Claudia Piovan[1,2] and Marilena V. Iorio[2]

[1]*Department of Molecular Virology, Immunology and Medical Genetics and Comprehensive Cancer Center, Ohio State University, Columbus, OH, USA*
[2]*Start Up Unit, Department of Experimental Oncology, Fondazione IRCCS, Istituto Nazionale Tumori, Milano, Italy*

I.	Breast Cancer: Introduction	288
II.	MicroRNAs (miRNAs) in Breast Cancer: Diagnostic, Prognostic, and Predictive Biomarkers	289
	A. Aberrant Expression of miRNAs in Breast Cancer	289
	B. Use of miRNA Signatures to Predict Tumor Subtype	291
III.	ER and miRNAs	292
	A. The ER Family	292
	B. miRNA Regulation of ERs	293
	C. Estrogenic Regulation of miRNA Expression	294
	D. miRNAs and Endocrine Resistance in Breast Cancer	295
IV.	ErbB Receptor Signaling and miRNAs	297
	A. ErbB Signaling and Breast Cancer	297
	B. Regulation of EGFR Signaling by miRNAs	298
	C. Regulation of ErbB2 and ErbB3 Signaling by miRNAs	299
	D. The Influence of miRNA on HER2-Targeting Therapies	300
V.	Conclusions	302
	Acknowledgments	302
	References	303

MicroRNAs in Medicine, First Edition. Edited by Charles H. Lawrie.
© 2014 John Wiley & Sons, Inc. Published 2014 by John Wiley & Sons, Inc.

ABBREVIATIONS

EGF	epidermal growth factor
ER	estrogen receptor
miR	microRNA
miRNA	microRNA
PR	progesterone receptor
RTKs	receptor tyrosine kinases

I. BREAST CANCER: INTRODUCTION

Historically, breast cancer was perceived as one disease despite different histopathological features and varying responses to systemic treatment. Treatment decisions were exclusively based on clinicopathological variables such as tumor size, presence of lymph node metastasis, and histological grade as well as estrogen receptor (ER), progesterone receptor (PR), and HER2 expression as predictive markers of response to endocrine and HER2-targeting therapies. In light of the steady reduction in breast cancer mortality during the past three decades, these approaches have been successful; however, they are not sufficient for implementation of personalized therapy (Reis-Filho and Pusztai 2011). The practice of oncology continually faces the challenge of matching the right therapeutic regimen with the right patient, balancing relative benefit with risk to achieve the most successful outcome. The marginal success rate achieved in many types of cancer is likely a reflection of the enormous complexity of the disease process coupled with an inability to properly guide the use of available therapeutics.

The advent of high-throughput platforms for analysis of gene expression, such as microarrays, has strongly challenged the idea that breast cancer is a single disease. Data from these systems suggest that there exists a collection of different diseases that affect the same organ site and which originate from the same anatomical structure, but have different risk factors, clinical presentation, histopathological features, outcome, and response to systemic therapies. A highly important concept highlighted by genome-wide expression profiling studies was that response to treatment could be determined by intrinsic molecular characteristics of the tumors that can be probed with molecular methods.

The class discovery approach and subsequent hierarchical cluster analysis used by Perou et al. (2000) and by Sorlie et al. (2001) uncovered the existence of at least four different subtypes of breast cancer: luminal, HER2 enriched, basal-like, and normal breast-like.

Luminal tumors are a breast cancer subtype that are almost always positive for ER-alpha and which show expression of genes usually found in luminal epithelial cells of the normal mammary gland. Subsequent studies have demonstrated that luminal cancers could be subclassified into two groups (luminal A and luminal B), based on expression levels of proliferation-related genes. Luminal A tumors exhibit robust expression of ER, PR, and other markers of mature luminal epithelial cells including the transcription factor GATA3 and luminal cytokeratins (CK8 and CK18), and likely arise from malignant transformation of the mature luminal ductal or lobular epithelial cell (Perou et al. 2000; Visvader 2009). Luminal B tumors commonly express ER at a lower level than luminal A tumors and likely stem from the transformation of a cell with an intermediate degree of terminal luminal commitment. Accordingly, luminal B tumors usually exhibit lower expression of estrogen-related genes, higher mitotic indices and histological grade, and a significantly poorer prognosis compared with luminal A malignancies (Sorlie et al. 2001). In addition, as suggested from gene expression profiling, coexpression of HER2 and ER and/or PR can identify some luminal B tumors (i.e., the luminal–HER2-positive group). However, only approximately 30% of luminal B tumors are HER2 positive, indicating that this clinical marker alone is not sensitive enough to identify most luminal B breast cancers (Cheang et al. 2009).

The HER2-enriched group is composed of tumors that are preferentially ER negative by immunohistochemistry and gene expression profiling, and express genes mapping to the *HER2* amplicon. Overexpression of HER2 is observed in approximately 25–30% of human breast cancers, and is usually caused by amplification of the 17q12 locus (containing the *ERBB2* gene), which results in a strongly increased expression of wild-type HER2 RTK at the plasma membrane (Moasser 2007). Compared with luminal tumors, which lack ERBB2 amplification, HER2 cancers follow a more aggressive clinical course, with increased resistance to chemotherapeutic agents and an increased risk of distant metastasis (Sorlie et al. 2001).

Basal-like cancers are characterized by little to no expression of ER and ER-related genes (including PR), and frequent absence of HER2 overexpression (i.e., triple-negative phenotype), but exhibit high expression of proliferation-related genes and also of genes usually found in basal and myoepithelial cells of the breast: cytokeratins 5/6 and 17, and the epidermal growth factor receptor (EGFR) (Reis-Filho and Pusztai 2011).

Finally, the normal breast-like subtype tumors show remarkable similarities at the messenger RNA (mRNA) expression level with samples of normal breast and fibroadenomas (Reis-Filho and Pusztai 2011).

In the last 5 years, additional molecular subtypes of tumors that are preferentially ER negative have been proposed, including molecular apocrine and claudin-low. Molecular apocrine cancers are characterized by histological features suggestive of apocrine differentiation and the expression of androgen receptor and androgen receptor-associated genes (Farmer et al. 2005; Doane et al. 2006). The claudin-low subtype identifies a group of ER-negative tumors expressing low levels of genes involved in tight junctions and cell–cell adhesion, and luminal genes including potential GATA3 target genes (Herschkowitz et al. 2007). Interestingly, this subclass is also characterized by high levels of expression of genes involved in the epithelial-to-mesenchymal transition as well as expression of genes usually found in the so-called cancer stem cells (i.e., cells within a tumor bulk that have the ability to repopulate the tumor and recreate its original heterogeneity).

II. MicroRNAs (miRNAs) in Breast Cancer: Diagnostic, Prognostic, and Predictive Biomarkers

A. Aberrant Expression of miRNAs in Breast Cancer

To date, every type of breast tumor analyzed by miRNA profiling has shown significantly different miRNA profiles (for mature and/or precursor miRNAs) compared with normal cells from the same tissue. A review of published large-scale miRNA profiles reveals that several miRNAs are consistently deregulated in tumors from breast cancer patients. The first report describing the existence of an miRNA signature characterizing human breast cancer was published in 2005 (Iorio et al. 2005), suggesting the involvement of miRNAs in the pathogenesis of breast cancer. Iorio and colleagues described the first breast cancer-specific miRNA signature by performing a genome-wide miRNA expression analysis on a large set of normal and tumor breast tissues, which resulted in the identification of 29 differentially expressed miRNAs. This demonstrated that it was possible to distinguish tumors from normal tissues using miRNA profiling. Among the miRNAs differentially expressed, *miR-10b*, *miR-125b*, and *miR-145* were down-regulated, while *miR-21* and *miR-155* were up-regulated, suggesting that these miRNAs could exert a role as tumor suppressor genes or oncogenes, respectively. Subsequent studies have identified a large number of genes involved in breast cancer development and progression that are targets of miRNAs deregulated in breast cancer (Gusev et al. 2007). Commonly regulated miRNAs and their validated target genes in breast cancer are summarized in Table 17.1.

Most profiling studies have focused on miRNAs that are deregulated in primary breast cancer tissues or breast cancer cell lines. However, because of their resistance to degradation, miRNAs are stable in blood, and there is emerging interest in profiling circulating

TABLE 17.1. Commonly Deregulated MicroRNAs (miRNAs) and Their Targets in Breast Cancer

miRNA	Expression Levels in Breast Cancer	Functional Pathways	Known Targets	References
miR-21	↑	Apoptosis, invasion, metastasis	BCL2, TPM1, PDCD4, PTEN, MASPIN, RHOB, MMP3	Huang et al. (2008a,b); Zhu et al. (2008); Qi et al. (2009); Qian et al. (2009); Song et al. (2010)
miR-125b	↓	Proliferation, apoptosis, migration	HER2, HER3, BAK, MUC1, RTKN	Scott et al. (2007); Hofmann et al. (2009); Zhou et al. (2010)
miR-155	↑	TGF-beta signaling	FOXO3A, SOCS1, RHOA	Kong et al. (2008, 2010); Jiang et al. (2010)
miR-145	↓	Proliferation, apoptosis, invasion	ER-alpha, MUC1	Gotte et al. (2010); Sachdeva and Mo (2010); Spizzo et al. (2010)
miR-205	↓	Proliferation, invasion, senescence	HER3, VEGFA, E2F1, LAMC1, ZEB1	Gregory et al. (2008); Iorio et al. (2009); Wu et al. (2009); Piovan et al. (2012)
miR-200	↓ Basal-like ↑ Luminal A	EMT, TGF-beta signaling	ZEB1, ZEB2, TrKB, BMI1	Gregory et al. (2008); Dykxhoorn et al. (2009); Shimono et al. (2009)
miR-206	↑ ER-alpha positive	ER signaling	ER-alpha	Di Leva et al. (2010)
miR-221/222	↑ Basal-like ↓ Luminal A	EMT, ER signaling	ER-alpha, p27, p57, FOXO3A,	Di Leva et al. (2010)
miR-10b	↑	Metastasis	HOXD10	Ma et al. (2007)
miR-31	↓	Metastasis	FZD3, ITGA5, MMP16, RDX, RHOA	Valastyan et al. (2009)
miR-17-5p	↓	Proliferation	AIB1, CCND1, E2F1	Hossain et al. (2006); Yu et al. (2008)
miR-373/520c	↑	Metastasis	CD44	Huang et al. (2008a,b)
miR-210	↑	Hypoxia	MNT, RAD52	Camps et al. (2008); Zhang et al. (2009)

Several studies have identified altered miRNA expression profiles in breast cancer. The loss of tumor suppressor miRNAs and overexpression of oncogenic miRNAs leads to loss of regulation of several cellular functions that may be involved in breast cancer pathogenesis. TGF, transforming growth factor; EMT, epithelial-mesenchymal transition.

miRNAs as non-invasive surrogate markers. Circulating miRNAs have been found to be significantly elevated in the blood of cancer patients compared with healthy controls, and these levels are reflected in the primary tumors as well. Pilot studies by Roth et al. (2010) demonstrated that *miR-10b*, *miR-34a*, and *miR-155* discriminated patients with metastatic disease from healthy controls, and Heneghan et al. (2010) found that the expression of *miR-195* was significantly elevated in breast cancer patients. Other reports have described the correlation between circulating miRNAs and breast cancer: Wu et al. (2012) very recently reported that certain miRNAs can serve as potential blood-based biomarkers, and, in particular, *miR-122* prevalence in the circulation predicts breast cancer metastasis in early-stage patients; in another study, three of the four most deregulated miRNAs in breast cancer versus normal breast tissue were also found to be significantly altered in serum samples from cancer patients compared with healthy controls (van Schooneveld et al. 2012). Although circulating miRNAs have been found to be elevated in breast cancer patients, common breast cancer-specific miRNAs have yet to emerge, making it difficult to interpret the significance of individual circulating miRNAs.

Despite discrepancies in the results described by different studies that could be due to the use of different microarray platforms, techniques, or analytical tools, the available experimental evidence underscores the role of miRNAs in the biology of breast cancer and provides a starting point for investigating new molecular mechanisms of breast cancer initiation, progression, and metastasis.

B. Use of miRNA Signatures to Predict Tumor Subtype

Despite the complexities of understanding how miRNAs influence tumorigenesis, aberrantly expressed miRNAs have considerable potential for use as biomarkers for the detection, diagnosis, classification, and treatment of cancer. Considering that miRNA profiling seems to correlate with cell differentiation and development more accurately than gene profiling, it is very interesting and important to deeply investigate how miRNAs could be used as cancer biomarkers. Different tissue types have unique expression levels of individual miRNAs, from as low as a few copies to a million copies per cell, and thereby have unique miRNA "signatures." Likewise, each tumor type seems to have a unique miRNA signature, and such signatures are being exploited to identify the tissue of origin of metastatic tumors and to differentiate between different cancer subtypes.

Given the intrinsic heterogeneity of breast tumor samples, the identification of subtype-specific miRNA signatures (detected in a tumor specimen or from a patient's blood) might provide useful additional information for facilitating the accurate classification of a patient's tumor subtype, complementing current methods of classification and guiding treatment decisions (Andorfer et al. 2011). Several studies have attempted to profile miRNA expression in breast tumors as a function of breast cancer subtype. In one study, 309 miRNAs (out of greater than 900 human miRNA sequences) were quantified from samples of 93 breast tumors of known molecular tumor subtypes (Blenkiron et al. 2007). miRNAs were differentially expressed according to tumor subtype (luminal A, luminal B, basal-like, normal-like, and HER+), although miRNA signature did not predict tumor status. However, the miRNA signature could correctly classify basal versus luminal subtypes in an independent test set, suggesting that the expression patterns of miRNA differ among tumor subtypes and also highlighting the possibility of defining a predictive signature for hormone receptor status.

A recent analysis of 453 different miRNAs in 29 early-stage breast cancer tumors identified predictive signatures corresponding to subtype ER (*miR-342*, *miR-299*, *miR-217*, *miR-190*, *miR-135b*, and *miR-218*), PR (*miR-520g*, *miR-377*, *miR-527-518a*, and *miR-520f-520c*), or HER2 (*miR-520d, miR-181c, miR-302c, miR-376b,* and *miR-30e*) (Lowery et al. 2009). These signatures classified cases with 100% accuracy when compared with immunohistochemistry results.

Mattie et al. (2006) profiled 208 miRNAs in 20 breast cancer tumors to identify predictive signatures of ER (*miR-30d* and *miR-30e*), PR (*miR-106b, miR-19a, miR-29c,* and *miR-30a-5p*), HER2+ (*let-7f, let-7g, miR-107, miR-10b, miR-126, miR-154,* and *miR-159*), and ER/PR (*miR-142-5p, miR-200a, miR-205,* and *miR-25*).

In another study, miRNA expression patterns were profiled between normal and tumor breast tissue (Sempere et al. 2007). Although miRNAs were differentially expressed between the two tissue types, expression patterns did not correlate exclusively with ER or HER2 status. A particular strength of this study is that the authors were able to use *in situ* hybridization directly in breast tissue and thereby could examine the differential expressions of miRNAs in different cell types within a single tumor. This information helps to distinguish whether the up- or down-regulation of miRNAs is associated with changes in the tumor itself or in the surrounding stromal tissue that is likely to be included in typical tumor samples.

Taken together, these studies illustrate that the development of miRNA signatures for predicting tumor subtype is still in the early stages of development. A lack of profiling studies combined with apparent discrepancies among reported miRNA signatures highlights the need for (1) the validation of existing signatures in larger data sets, (2) the validation of tumor data sets with heterogeneous clinicopathological variables, and (3) validation with alternative platforms. An additional, and perhaps more important, use of miRNA profiling is the identification of patient subgroups within tumor subtypes that are more or less likely to respond to a particular therapy (Andorfer et al. 2011). This application of miRNAs is being actively researched and has indeed great potential for their use in clinics as "predictive biomarkers."

III. ER AND miRNAs

A. The ER Family

Estrogens participate in different physiological actions and have been recognized as central hormones in stimulating the growth and development of breast cancers. The activities of estrogens are mediated by the two main isoforms of intracellular ERs, ER-alpha and ER-beta, which are respectively encoded by the ESR1 and ESR2 genes. These cytoplasm/nuclear ERs have structural characteristics of the nuclear receptor superfamily and can form homo- or heterodimers when activated by their effective ligand 17-β-estradiol (E2). These dimers function as transcription factors to regulate gene expression, recognizing and stably binding to estrogen response elements (EREs). EREs were historically thought to be located in promoter regions but have been shown to be located at great distances from the transcription start site (Carroll et al. 2006). DNA binding increases ER interaction with basal transcription factor and coregulator proteins.

There are significant molecular differences between ER-alpha-positive and ER-alpha-negative breast cancers (Perou et al. 2000; Sorlie et al. 2003). Endocrine therapy has become the most important treatment option for women with ER-alpha-positive breast cancer, and approximately 70% of primary breast cancers express ER-alpha. ER-alpha is essential for estrogen-dependent growth, and its level of expression is a crucial determinant of response to endocrine therapy and prognosis in ER-alpha-positive breast cancer (Harvey et al. 1999; Yamashita et al. 2006; Dowsett et al. 2008). Multiple mechanisms involved in the regulation of ER-alpha expression in breast cancer have been identified, including mutations of the ER-alpha gene (Herynk and Fuqua 2004) and transcriptional silencing by DNA methylation within the ER-alpha promoter (Iwase et al. 1999; Giacinti et al. 2006; Gaudet et al. 2009). ER-alpha gene amplification has been recently reported in breast cancer (Holst et al. 2007; Tomita et al. 2009). However, breast cancer patients show a wide range of ER-alpha expression levels (Yamashita et al. 2011), and the levels of expression in individual patients change during disease progression and in response to systemic therapies (Yamashita et al. 2009). Therefore, other mechanisms may also regulate ER-alpha expression in breast cancer, such as miRNAs.

B. miRNA Regulation of ERs

To date, several miRNAs have been identified as bona fide ER-alpha regulators as reviewed by Klinge (2012): *miR-22* (Pandey and Picard 2009), *miR-206*, *miR-221* and *miR-222*, *miR-18a* (Castellano et al. 2009), *miR-18b*, *miR-193b*, *miR-302c* (Leivonen et al. 2009), *let-7*, and *miR-145*. miRNAs can influence estrogen-regulated gene expression by directly reducing ER-alpha mRNA stability or inhibiting translation. Although most of these miRNAs bind to the 3′-untranslated region (3′-UTR) of the ER-alpha transcript, *miR-145* affects the translation of ER-alpha by binding to a site in its coding region (Spizzo et al. 2010). Since estrogen signaling increases breast cell proliferation, miRNA-mediated ER repression generally causes a decrease in proliferation.

miR-206 has been shown to be inversely correlated with the expression of ER-alpha, but not ER-beta, in human breast tumors (Kondo et al. 2008). Estradiol (E2) and the ER-alpha-selective agonist PPT [4,4′,4″-(propyl-[1H]-pyrazole-1,3,5-triyl) trisphenol] decrease *miR-206* in MCF-7 cells and increase *miR-206* in ER-alpha-negative MDA-MB-231 cells (Adams et al. 2007; Di Leva et al. 2010). This finding offers a mechanism, in addition to ER-alpha promoter methylation, for reduced ER-alpha expression in MDA-MB-231 cells. Two predicted sites in the ER-alpha 3′-UTR were confirmed to be functional miR-binding sites, and overexpression of *miR-206* in MCF-7 or T47D cells represses ER-alpha mRNA and protein expression, inhibits cell growth, and suppresses oncogenic signaling in ER-positive cells (Di Leva et al. 2010). In a double-negative feedback loop, *miR-206* is suppressed by E2/ER-alpha signaling. This explains the correlation between high ER-alpha expression and low levels of *miR-206* in luminal A cancers (Guttilla et al. 2012).

miR-221 and *miR-222* are highly enriched in ER-alpha-negative breast cancer cells (Di Leva et al. 2010), as well as ER-alpha-negative breast tumors (Zhao et al. 2008). *miR-221* and *miR-222* each directly target ER-alpha mRNA at a conserved site. In addition, several studies have shown that *miR-221* and *miR-222* inhibit the translation of ER-alpha mRNA. Overexpression of *miR-221* and *miR-222* in ER-alpha-positive breast cancer tissues also decreases the expression of tumor suppressors such as CDKN1B, CDKN1C, BIM, PTEN, TIMP3, DNA-damage-inducible transcript 4, and FOXO3. Consequently, *miR-221* and *miR-222* promote high proliferation and estrogen-independent growth. As described for *miR-206*, *miR-221* and *miR-222* are repressed by ER-alpha signaling in a double-negative feedback loop. Utilizing *miR-221/222* gene promoter–luciferase constructs and chromatin immunoprecipitation, Di Leva et al. (2010) demonstrated that ER-alpha directly represses the *miR-221/222* gene promoter by recruiting corepressors NCoR and SMRT. Interestingly, these findings suggest that the activity of this regulatory loop may confer both a proliferative advantage and migratory activity to breast cancer cells while also promoting the transition from ER-positive to ER-negative tumors.

The repression of ER-alpha activity can be indirect, as is the case of *miR-27a*. ER-alpha can induce gene transcription at non-consensus EREs or ERE half sites through specificity protein (Sp) transcription factors. *miR-27a* indirectly regulates ER-alpha expression by targeting ZBTB10, an Sp repressor, that in turn reduces Sp1, Sp3, and Sp4 (Li et al. 2010). In addition, direct interaction of ER-alpha with Sp1, Sp3, and Sp4 bound to DNA GC-box motifs regulates transcription (Wu et al. 2009).

Recently, it has been reported that transient overexpression of *let-7a*, *let-7b*, and *let-7i* inhibits ERa expression in MCF-7 cells (Zhao et al. 2011).

To date, only one miRNA has been identified as regulating ER-beta: *miR-92* (Al-Nakhle et al. 2010). The role of *miR-92* has not been extensively explored, and future studies will provide more details regarding the effects of this or other miRNAs on the ER-beta receptor.

C. Estrogenic Regulation of miRNA Expression

Estrogenic regulation of miRNA expression has been studied by several groups, most of them focusing on how E2 treatment of breast cancer cell lines affects mature miRNA expression. The effect of E2 on miRNA expression has been examined in zebrafish (Cohen et al. 2008), August Copenhagen Irish rats (Kovalchuk et al. 2007), and mouse splenocytes (Dai et al. 2008), and has also been exhaustively reviewed (Klinge 2009, 2012). There are few studies in which miRNA regulation by E2 has been directly examined in human cell lines, and, notably, most of them were performed in MCF-7 cells (Adams et al. 2007; Bhat-Nakshatri et al. 2009; Castellano et al. 2009; Maillot et al. 2009; Wickramasinghe et al. 2009; Cicatiello et al. 2010; Di Leva et al. 2010; Hah et al. 2011; Rao et al. 2011). As noted by Klinge and colleagues, and by others, there is a lack of consistency of E2-regulated changes in miRNA expression even within the best-studied MCF-7 cell line. Several factors may have influenced these differing results, including timing and treatment conditions, methodology employed to "serum starve" the cells in order to see a response, E2 concentration, specific characteristics of MCF-7 cells between laboratories, choice of control gene used for miRNA normalization, and the assay method used to measure mature miRNA expression. Despite these factors, a majority of studies suggest that E2-up-regulated miRNAs have a tumor suppressor role by participating in a negative feedback loop to restrict E2 action.

The *miR-17-92* cluster and its paralogs are often overexpressed during breast carcinogenesis, and their expression is also increased with estradiol treatment (Li et al. 2011). This occurs indirectly when estradiol stimulates c-myc, which then binds to the promoter of the *miR-17-92* cluster (Castellano et al. 2009). E2-mediated activation of *miR-17-92* cluster functions as a tumor suppressor mechanism in breast cancer through downregulation of cyclin D1 and AIB1 by the *miR-17/20/106* family and also through the direct down-regulation of ER-alpha mediated by *miR-18* and *miR-19* (Castellano et al. 2009).

The *let-7* family of miRNAs is induced by estradiol and tends to be lower in breast cancers compared with normal breast tissue (Yu et al. 2007; Bhat-Nakshatri et al. 2009). Also, like *let-7*, the *miR-200* family is often associated with an ER-alpha-positive, epithelial phenotype (Park et al. 2008) and is also induced by estradiol (Bhat-Nakshatri et al. 2009).

Conversely, *miR-21*, a well-studied oncomiR that promotes cell proliferation and inhibits cell death (Lu et al. 2008), is repressed by estradiol. This causes an increase in the targets of *miR-21* (Pdcd4, PTEN, and Bcl-2), and this effect is blocked by ER-alpha antagonists (Wickramasinghe et al. 2009). ER-alpha directly suppresses pre-*miR-21* transcription by binding to its promoter region (Bhat-Nakshatri et al. 2009). Moreover, as previously reported in this chapter, Di Leva et al. (2010) have described a double-negative feedback loop involving the E2-suppressed miRNAs that target ER-alpha, such as *miR-206* and *miR-221/222*, which guarantees strong ER-alpha expression and low levels of the miRNAs in luminal A-type breast cancers.

In addition, global genome-binding studies were performed to identify E2-regulated miRNAs (as reviewed by Klinge 2012) by analyzing the position and orientation of all engaged RNA polymerases (RNA PolI, II, and III) across the genome of MCF-7 cells treated with 100 nM E2 for 10, 60, or 160 minutes (Hah et al. 2011). The authors in this study identified 322 expressed miRNA-containing transcripts and noted that 119 were regulated by E2 at least at one of the time points examined. Furthermore, using RefSeq, 2700 putative mRNA targets for these miRNAs were identified. Since the clinical course of ER-positive breast cancers may be dependent on the balance between E2-regulated tumor suppressor miRNAs and oncogenic miRNAs, further research is highly important for understanding estrogen-dependent breast cancer.

Yamagata et al. (2009) reported another important mechanism through which E2/ER-alpha may regulate miRNA expression: In addition to the direct effect of ER-alpha on miRNA transcription, the processing from pri-miRNA to pre-miRNA also appears to be hormonally regulated by ER-alpha. Activated ER-alpha interacts with and suppresses Drosha activity in mouse uterine epithelial cells and MCF-7 cells, and this interaction is enhanced by E2 and by direct binding to p68 and p72 RNA helicases in the Drosha complex. Estrogen-bound ER-alpha down-regulated the expression of a set of miRNAs in both animals and cultured cells by suppressing the Drosha-mediated processing of pri-miRNAs to pre-miRNAs. Thus, a steroid hormone achieves posttranscriptional control by regulating the maturation of miRNA.

miRNA regulation of ERa and subsequent ERa transcriptional activity are summarized in Figure 17.1.

D. miRNAs and Endocrine Resistance in Breast Cancer

The term *endocrine therapy* is given to those breast cancer treatments that target the ER. Endocrine therapies work by two different mechanisms to antagonize the growth-promoting activity of estrogen: lowering estrogen levels (by ovarian ablation in premenopausal, or by using aromatase inhibitors, AIs, in postmenopausal women) or modulating the ER directly, by using selective estrogen receptor modulators (SERMs), such as tamoxifen (Frasor et al. 2004; Jordan 2006). The SERMs bind to the ER on tumor cells, thus blocking the ability of the cells to proliferate (Goodsell 2002); the AIs, on the other hand, block the ability of the aromatase enzyme to convert androgen precursors in peripheral tissues to estrogen, which is the major source of circulating estrogen in postmenopausal women (Murphy 1998). Second-line endocrine therapies, using alternative ER-alpha inhibitory mechanisms, have also been developed (McDonnell 2006). One of these, the pure steroidal antiestrogen fulvestrant (Faslodex, ICI 182,780, Astra-Zeneca Corp., London, UK), completely suppresses ER-alpha activity, inactivating both ER-alpha-mediated genomic and non-genomic signaling, and has been approved for the treatment of breast cancer in postmenopausal women following failure of previous antiestrogen therapy (Howell 2006).

Endocrine therapy to block the ER pathway is highly effective, but its usefulness is limited by common intrinsic and acquired resistance. Multiple mechanisms responsible for endocrine resistance have been proposed, including deregulation of various components of the ER pathway, alterations in cell cycle and cell survival signaling molecules,

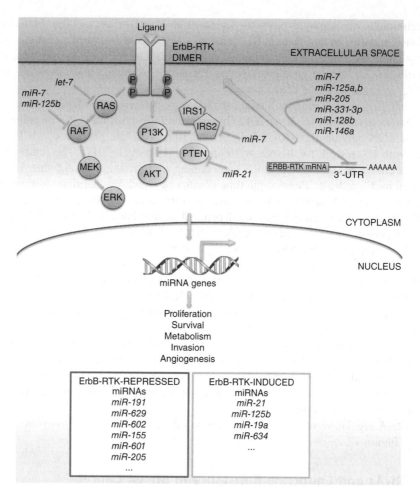

Figure 17.1. miRNA regulation of ER-alpha and subsequent ER-alpha transcriptional activity. The 3′-UTR of the ER-alpha transcript is directly targeted by *miR-145*, *miR-206*, *miR-221/222*, *miR-193b*, *miR-22*, *let-7*, *miR-18a,b*, and *miR-302c*. Estradiol (E2)-bound ER-alpha protein can regulate the transcription of pre-miRNAs either by directly binding to the miRNA promoter, by indirectly modulating the expression of other transcription factors that bind to the promoter of miRNAs (not shown), or by suppressing the Drosha-mediated processing of pri-miRNAs to pre-miRNAs. In turn, estrogen-repressed miRNAs can participate to a regulatory loop by modulating ER-alpha expression. See color insert.

and the activation of escape pathways that can provide tumors with alternative proliferative and survival stimuli (Osborne and Schiff 2011).

To date, only a few investigators have examined the role of miRNAs in endocrine resistance. A study by Miller et al. (2008) showed a significant up-regulation of eight miRNAs (*miR-221*, *miR-222*, *miR-181*, *miR-375*, *miR-32*, *miR-171*, *miR-213*, and *miR-203*) and down-regulation of seven miRNAs (*miR-342*, *miR-489*, *miR-21*, *miR-24*, *miR-27*, *miR-23*, and *miR-200*) in a tamoxifen-resistant cell line compared with a tamoxifen-sensitive cell line. They reported increased expression of *miR-221* and *miR-222* in HER2/neu-positive primary breast tumors, typically resistant to endocrine therapy, compared with

HER2/neu-negative tumors. They also suggested a relationship between tamoxifen resistance and reduced levels of the cell-cycle inhibitor p27 (Kip1) by augmenting *miR-221/222* expression. In a study by Rao et al. (2011), antiestrogen-sensitive and -resistant breast cancer cell lines were used as models to comprehensively investigate the role of *miR-221/222* in hormone-independent growth and acquired resistance to fulvestrant. In the study, *miR-221/222*-transfected cells were used to demonstrate that global gene expression changes associated with multiple (ER-independent) growth-promoting, oncogenic pathways are critical for acquired selective estrogen receptor downregulator (SERD) resistance. Pogribny et al. (2007) demonstrated that prolonged exposure of rats to tamoxifen was associated with altered expression of known tumor-associated miRNAs and their protein targets including *miR-16* (BCL2), *miR-17-5p* (E2F1), *miR-20* (E2F1), *miR-106a* (RB1), and *miR-34* (NOTCH1).

Overexpression of an oncogenic isoform of HER2 (HER2D16) causes tamoxifen resistance in MCF-7 cells by reducing *miR-15a* and *miR-16*, which normally suppress BCL-2 (Cittelly et al. 2010a). Overexpression of *miR-342* in HER2D16-MCF-7 cells sensitizes the cells to tamoxifen-induced apoptosis and reduces expression of BMP-7, GEMIN4, and SEMAD3, although further studies are required to determine the significance of these observations (Cittelly et al. 2010b). Microarrays identified 97 miRNAs differentially expressed in MCF-7 endocrine-sensitive versus -resistant LY2 breast cancer cells (Manavalan et al. 2011), among them are *miR-10a*, *miR-21*, *miR-22*, *miR-29a*, *miR-93*, *miR-125b*, *miR-181*, *miR-200a*, *miR-200b*, *miR-200c*, *miR-205*, and *miR-222*. A screen of miRNAs involved in estrogen resistance in MCF-7 cells identified up-regulation of *miR-101* as promoting estrogen-independent growth of MCF-7 cells, without affecting ER-alpha levels or activity (Sachdeva et al. 2011).

The majority of miRNA expression data in tamoxifen-resistant cells comes from studies on cell lines. However, Rodriguez-Gonzalez et al. (2011) validated three miRNAs (*miR-30a-3p*, *miR-30c*, and *miR-182*) that were originally identified (Foekens et al. 2008) for their abilities to predict the clinical benefit of tamoxifen in advanced breast cancer. The three miRNAs, when used together, were significantly associated with the benefit of tamoxifen and longer progression-free survival. Of these three miRNAs, only *miR-30c* independently predicted clinical benefit of tamoxifen in advanced breast cancer.

miRNA-mediated targeting of ER-alpha cofactors, which influence agonistic and antagonistic effects of ligand binding, has also been linked to impaired chemotherapeutic response. Indeed, one of the major coactivators for ER-alpha, amplified in breast cancer (AIB1), is a target for *miR-17* and *miR-106*, which are found to be dysregulated in breast cancer cells exhibiting drug resistance (Liang et al. 2010). Similarly, the transcriptional corepressor receptor interacting protein 140 (RIP140) was found to be targeted by *miR-346*, which is down-regulated in endocrine-resistant cell lines (Xin et al. 2009). These findings indicate an important role for miRNAs in regulating genes that determine responsiveness to endocrine therapies.

IV. ErbB RECEPTOR SIGNALING AND miRNAs

A. ErbB Signaling and Breast Cancer

The ErbB receptor family is composed of four members: EGFR (ErbB1), ErbB2, ErbB3, and ErbB4 (also known as HER1, HER2, HER3, and HER4, respectively). The basic ErbB signaling unit is a receptor dimer, which may be stimulated by ligand binding that promotes tyrosine kinase activity and recruitment of signaling molecules to phosphorylated docking sites at the C terminal end of the ErbB molecules (Burgess 2008). This results in the activation of a number of downstream pathways that include the MAPK and PI3K-Akt pathways, which promote cell growth and survival, as well as invasion and angiogenesis. ErbB receptors (especially EGFR and ErbB2) are often constitutively active in human cancers as a result of receptor overexpression or mutation, and and/or autocrine ligand production (Hynes and MacDonald 2009). Strikingly, up to 30% of breast carcinomas overexpress HER2, frequently as a consequence of genomic amplification of a region of the long arm of chromosome 17 (17q21) which includes the HER2 locus.

One of the most important steps in the management of breast cancer of the last decade has been the discovery of the HER2-targeting therapies. One of the approaches utilized to achieve ErbB2 inhibition is the use of monoclonal antibodies to block ligand binding and receptor tyrosine kinase activation, such as trastuzumab or pertuzumab. The "success story" of trastuzumab has been attributed to the identification of a subset of breast cancer tumors (HER2 positive) that show dependency on the target to which the therapy is directed and existence of a predictive biomarker (HER2-positive status of the primary tumor) for benefit from trastuzumab.

Beyond monoclonal antibodies, small molecule tyrosine kinase inhibitors (TKIs) were also generated. TKIs, such as the dual inhibitor lapatinib, targeting EGFR and HER2, compete with the tyrosine kinase for ATP binding and therefore block the phosphorylation of downstream intracellular signaling intermediates.

The benefit of anti-HER2 therapies demonstrated in clinical trials (increased rate of curability in early breast cancer and also overall survival in metastatic breast cancer) indicates that HER2 is, to date, one of the most promising molecules for targeted therapy. Nevertheless, since tumor cells utilizing alternative growth signaling pathways through transmembrane receptors as well as intracellular signaling transduction molecules can bypass HER2 blockade, a future ambitious aim is the successful combination of anti-HER2 strategies with drugs directed to molecules that contribute to anti-HER2 resistance (Tagliabue et al. 2010).

B. Regulation of EGFR Signaling by miRNAs

Several groups have examined the regulation of EGFR expression in human cancer cells by miRNAs. Two groups have focused on the control of EGFR expression and signaling by *miR-7* in a variety of cancer systems, including lungs, breast, prostate, and glioma (Kefas et al. 2008; Webster et al. 2009). The human EGFR mRNA 3'-UTR contains three target sites for *miR-7* that are poorly conserved among mammals. Overexpression in a variety of cancer cell types (lungs, glioma, breast, or prostate) of *miR-7* precursor leads to a significant reduction in both EGFR mRNA and protein, and results in non-apoptotic cell death in at least some cell lines.

EGFR overexpression in glioblastoma, often but not always caused by gene amplification, is a critical step in tumorigenesis (Toth 2009). A significant down-regulation of *miR-7* has been observed in glioblastoma cell lines and patient specimens (Kefas et al. 2008; Webster et al. 2009), highlighting the role of *miR-7* as a potential tumor suppressor in this system by its modulation of EGFR expression and signaling activity. *miR-7* also regulates expression of insulin receptor substrates 1 and 2 (IRS1/IRS2), which regulate PI3K/Akt signaling, RAF1, and other downstream effectors of EGFR signaling (Kefas et al. 2008; Webster et al. 2009). This demonstrates that *miR-7* causes a coordinated down-regulation of multiple members of the EGFR signaling cascade, suggesting that therapeutic up-regulation of this miRNA in tumors could block EGFR signaling at several levels (Barker et al. 2010).

Avraham et al. (2010) used a systems biology approach to elucidate how miRNAs are involved in the regulation of oncogenic signaling networks. The authors measured the coordinated regulation of miRNAs and mRNAs over time following epidermal growth factor (EGF) stimulation of MCF10A human mammary epithelial cells. They showed that EGF stimulation initiates a coordinated transcriptional program of miRNAs and transcription factors. The earliest event involves a decrease in the abundance of a subset of 23 miRNAs. This step permits rapid induction of oncogenic transcription factors, such as c-FOS, encoded by immediate early genes. In accordance with their role as suppressors of EGFR signaling, the abundance of the early subset of miRNAs is decreased in breast and in brain tumors driven by EGFR or the closely related HER2 (Avraham et al. 2010).

A very recent work by Uhlmann et al. (2012) combined a large-scale miRNA screening approach with a high-throughput proteomic readout and network-based data analysis to identify which miRNAs are involved in the EGFR-driven cell-cycle protein network and to uncover potential regulatory patterns. They validated three miRNAs (*miR-124*, *miR-147*, and *miR-193a-3p*) as novel tumor suppressors that cotarget EGFR-driven cell-cycle network proteins (e.g., AKT2, STAT3, p38, and JNK1) and inhibit cell-cycle progression and proliferation in breast cancer.

Notably, Garofalo et al. (2012) have recently identified a set of miRNAs modulated by both EGF and MET receptors in models of non-small-cell lung cancer (NSCLC). These miRNAs play important roles in gefitinib-induced apoptosis and the epithelial–mesenchymal transition *in vitro* and *in vivo* by inhibiting the expression of the genes encoding BCL2-like 11 (BIM), apoptotic peptidase activating factor 1 (APAF-1), protein kinase C ε (PKC-ε), and sarcoma viral oncogene homologue (SRC).

C. Regulation of ErbB2 and ErbB3 Signaling by miRNAs

The down-regulation of miRNAs that target HER2 might account for a subset of HER2-enriched tumors, and the therapeutic delivery of such miRNAs could hold great promise in treating HER2-positive tumors (Andorfer et al. 2011). The 3′-UTR of ErbB2 mRNA does not contain miRNA target sites that are well conserved among mammals. However, bioinformatics predictions suggest the presence of a number of poorly conserved putative target sites within the human ErbB2 mRNA 3′-UTR; thus, ErbB2 and ErbB3 expression is regulated by miRNAs. Such miRNAs include *miR-125a* and *miR-125b*, shown to directly regulate ErbB2 in breast cancer cells (Scott et al. 2007); *miR-205*, which regulates ErbB3 expression in breast cancer cells (Iorio et al. 2009); and *miR-331-3p*, which directly regulates ErbB2 by its specific binding to two separate target sites (Epis et al. 2009). Interestingly, the *miR-125a* and *miR-125b* target elements in the proximal ErbB2 3′-UTR are also common to the ErbB3 3′-UTR, and these miRNAs were shown to regulate both ErbB2 and ErbB3 expression and to suppress phosphorylation of ERK1/ERK2 and AKT. This is of particular significance as overexpressed ErbB2/ErbB3 heterodimers are potent activators of PI3K/Akt signaling, which promotes growth and disease progression in aggressive ErbB2-positive breast cancer cells. *miR-125b* has also been shown to regulate expression of Raf1 in breast cancer cell lines (Hofmann et al. 2009), which further highlights the capacity for a specific miRNA to regulate multiple members of a signaling pathway. *miR-205*, which is down-modulated in breast tumors compared with normal breast tissue, directly targets ErbB3 receptor and inhibits the activation of the downstream mediator Akt. Introducing *miR-205* into SKBr3 cells inhibits their clonogenic potential and increases responsiveness to TKIs gefitinib and lapatinib, abrogating HER3-mediated resistance and restoring potent proapoptotic activity (Iorio et al. 2009).

A study by Adachi et al. (2011) reports the down-regulation of *miR-205* by overexpressing HER2 in MCF10A breast cancer cells, thus suggesting the existence of a possible negative feedback loop between *miR-205* and HER signaling. Additional experimental evidence indicates that miRNA expression is regulated by members of the HER family, in particular, HER2: Ectopic expression of a clinically important oncogenic isoform of HER2, HER2D16, in MCF-7 cells induces tamoxifen resistance through suppression of *miR-15a-16* and consequent BCL2 up-regulation (Cittelly et al. 2010a).

Regulation of ErbB signaling in cancer cells by miRNAs is depicted in Figure 17.2.

D. The Influence of miRNA on HER2-Targeting Therapies

Overexpression of human epidermal growth factor (HER) family members occurs in ~20–25% of invasive breast cancer cases and correlates with poor patient prognosis (Murphy and Modi 2009). Indeed, the overexpression of HER2 through either gene amplification or transcriptional deregulation (Slamon et al. 1987) confers a worse prognosis, including a relative resistance to certain chemotherapeutic and hormonal agents (Serrano-Olvera et al. 2006). Two approved therapies targeting HER2, the monoclonal antibody trastuzumab and the TKI lapatinib, are clinically active against this type of breast cancer. However, a significant fraction of patients with HER2-positive breast cancer treated with these agents eventually relapse or develop progressive disease. This suggests that tumors acquire or possess intrinsic mechanisms of resistance that allow escape from HER2 inhibition.

Understanding the mechanisms of action and resistance to trastuzumab is therefore crucial for the development of new therapeutic strategies.

The identification of miRNA biomarker signatures that predict patient risk, disease outcome, and tolerability to trastuzumab therapy would greatly improve the personalized management of HER2-positive disease. Very recently, high levels of circulating *miR-210* have been associated with poor response to trastuzumab (Jung et al. 2012). Unfortunately, no other published studies have successfully identified a prognostic miRNA signature for trastuzumab response or toxicity, although several studies addressing the biological connection between miRNAs and the HER family of receptor tyrosine kinases were recently published.

It has been speculated that drug efficacy is reduced by the continuous overexpression of the drug target, a phenomenon that may be mediated by the down-regulation of *miR-125* and *miR-331* whose normal function is to suppress HER2 (Kovalchuk et al. 2008). Moreover, high expression of HER2 in cancer cells has been shown to induce ligand-independent constitutive activation of the receptor and its downstream MAPK and PI3K signaling pathways, thereby facilitating cross talk with the ER to induce multidrug resistance (Britton et al. 2006). miRNA-based suppression of EGFR-induced PI3K activation, including the repression of other EGFR members, such as HER3 by *miR-205*, may prove an important strategy for resensitizing breast cancer cells that have acquired cross-resistance to therapeutic agents.

Following stimulation of ErbB2 activity in breast cancer cells, Huang et al. (2009) identified *miR-21*, an established oncomiR, as an ERK1/ERK2-regulated miRNA. In turn, elevated expression of *miR-21* promotes tumor cell invasion at least in part by suppressing levels of programmed cell death 4 (PDCD4), a known metastasis suppressor. Similarly,

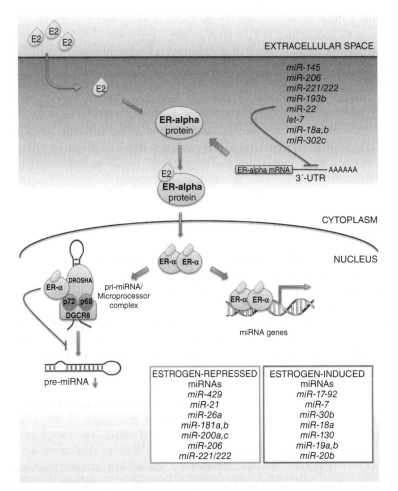

Figure 17.2. Regulation of ErbB signaling in cancer cells by miRNAs. ErbB receptor dimers are frequently overexpressed or aberrantly activated in human cancers, leading to activation of downstream ERK1/ERK2 and Akt signaling pathways. In turn, these modulate gene expression in order to promote cellular proliferation, apoptosis resistance, metabolism, migration and invasion, and angiogenesis. ErbB signaling represents an important therapeutic target. Recent research has identified miRNAs that modulate expression of a number of molecules belonging to these pathways. Specific examples reported in the text are included in this figure. Furthermore, ErbB signaling could regulate tumor development and progression via promoting expression of oncogenic miRNAs, or inhibiting tumor suppressor miRNAs. See color insert.

Seike et al. (2009) demonstrated that the EGFR signaling pathway up-regulates *miR-21* expression in lung cancer cells with a significant correlation existing between the levels of *miR-21* and phosphorylated EGFR (P-EGFR) in lung cancer cell lines.

The complexity of this system is further emphasized by the capacity of *miR-21* to regulate expression of molecules that inhibit ErbB signaling, such as the tumor suppressor PTEN, a key negative regulator of PI3K/Akt signaling (Zhou et al. 2010). Despite this, it appears that a positive regulatory loop between *miR-21* and ErbB signaling exists at least in some cancer systems, whereby *miR-21* promotes ErbB signaling (in part via suppression of PTEN), which in turn up-regulates expression of *miR-21*. Notably, PTEN was demonstrated to affect trastuzumab resistance (Wickramasinghe et al. 2009). Thus, *miR-21* represents an attractive target to augment the therapeutic inhibition of ErbB signaling in cancer.

As expected, trastuzumab treatment can modulate miRNA expression: Ichikawa et al. (2012) performed miRNA profiling after trastuzumab treatment in different breast cancer cell lines. Among the modulated miRNAs, *miR-26a* and *miR-30b* were able to mimic the effects of trastuzumab on cell-cycle arrest when ectopically overexpressed in breast cancer cells, thus revealing a probable role of miRNAs in the biological effect mediated by this drug.

V. CONCLUSIONS

The interest on miRNA field has increased extremely fast in the last few years, and it is now well recognized that these small RNA molecules represent an essential part of the encoding genome, finely tuning gene expression and thus exerting a critical role in all the most important processes and in different species.

miRNAs are major regulators of gene expression, with roles in nearly every area of cell behavior, development, and survival; therefore, it is not surprising that miRNAs are actively altered in all types of cancers, acting as oncogenes or tumor suppressor genes, and playing a role in the most important pathways driving proliferation and survival, as the ER and the HER-mediated networks in breast cancer.

miRNA profiling has been convincingly demonstrated to classify tumor and non-tumor samples more effectively than gene expression profiling and, more importantly, identifying breast cancer subtypes. The diagnostic role of miRNAs is crucial since different breast cancer subtypes exhibit different responsiveness to chemotherapeutic agents. In the future, predicting responses to a specific therapy will ideally involve a combination of gene expression data, miRNA profiling, and receptor status determination, with the goal of developing a personalized treatment strategy to improve patient outcome.

Referring to the use of miRNAs as the targeted therapy of the future is probably premature at this point, since many issues still need to be addressed (as the validation of the targets, the accurate prevision of the putative unwanted off target effects, and the development of efficient methods of a specific drug delivery); however, the number of discoveries, increasing so fast in the last few years, is certainly encouraging and promising.

Finally, drug resistance is widely accepted as the main cause of treatment failure. Recent studies has shown a possible link between miRNA deregulation and drug resistance, highlighting the role of miRNAs in regulating chemoresistance and focusing on potential therapeutic targets for reversing miRNA-mediated drug resistance. miRNA-based treatments, in combination with traditional chemotherapy, might be a new strategy for the clinical management of drug-resistant breast cancers.

ACKNOWLEDGMENTS

The authors are grateful to Dario Palmieri and Timothy Richmond for critically reading and revising the manuscript.

REFERENCES

Adachi, R., S. Horiuchi, Y. Sakurazawa, T. Hasegawa, K. Sato, and T. Sakamaki. 2011. ErbB2 down-regulates microRNA-205 in breast cancer. *Biochem Biophys Res Commun* **411**:804–808.

Adams, B.D., H. Furneaux, and B.A. White. 2007. The micro-ribonucleic acid (miRNA) *miR-206* targets the human estrogen receptor-alpha (ERalpha) and represses ERalpha messenger RNA and protein expression in breast cancer cell lines. *Mol Endocrinol* **21**:1132–1147.

Al-Nakhle, H., P.A. Burns, M. Cummings, A.M. Hanby, T.A. Hughes, S. Satheesha, A.M. Shaaban, L. Smith, and V. Speirs. 2010. Estrogen receptor {beta}1 expression is regulated by *miR-92* in breast cancer. *Cancer Res* **70**:4778–4784.

Andorfer, C.A., B.M. Necela, E.A. Thompson, and E.A. Perez. 2011. MicroRNA signatures: clinical biomarkers for the diagnosis and treatment of breast cancer. *Trends Mol Med* **17**:313–319.

Avraham, R., A. Sas-Chen, O. Manor, I. Steinfeld, R. Shalgi, G. Tarcic, N. Bossel, A. Zeisel, I. Amit, Y. Zwang, E. Enerly, H.G. Russnes, et al. 2010. EGF decreases the abundance of microRNAs that restrain oncogenic transcription factors. *Sci Signal* **3**:ra43.

Barker, A., K.M. Giles, M.R. Epis, P.M. Zhang, F. Kalinowski, and P.J. Leedman. 2010. Regulation of ErbB receptor signalling in cancer cells by microRNA. *Curr Opin Pharmacol* **10**:655–661.

Bhat-Nakshatri, P., G. Wang, N.R. Collins, M.J. Thomson, T.R. Geistlinger, J.S. Carroll, M. Brown, S. Hammond, E.F. Srour, Y. Liu, and H. Nakshatri. 2009. Estradiol-regulated microRNAs control estradiol response in breast cancer cells. *Nucleic Acids Res* **37**:4850–4861.

Blenkiron, C., L.D. Goldstein, N.P. Thorne, I. Spiteri, S.F. Chin, M.J. Dunning, N.L. Barbosa-Morais, A.E. Teschendorff, A.R. Green, I.O. Ellis, S. Tavare, C. Caldas, et al. 2007. MicroRNA expression profiling of human breast cancer identifies new markers of tumor subtype. *Genome Biol* **8**:R214.

Britton, D.J., I.R. Hutcheson, J.M. Knowlden, D. Barrow, M. Giles, R.A. McClelland, J.M. Gee, and R.I. Nicholson. 2006. Bidirectional cross talk between ERalpha and EGFR signalling pathways regulates tamoxifen-resistant growth. *Breast Cancer Res Treat* **96**:131–146.

Burgess, A.W. 2008. EGFR family: structure physiology signalling and therapeutic targets. *Growth Factors* **26**:263–274.

Camps, C., F.M. Buffa, S. Colella, J. Moore, C. Sotiriou, H. Sheldon, A.L. Harris, J.M. Gleadle, and J. Ragoussis. 2008. hsa-miR-210 Is induced by hypoxia and is an independent prognostic factor in breast cancer. *Clin Cancer Res* **14**:1340–1348.

Carroll, J.S., C.A. Meyer, J. Song, W. Li, T.R. Geistlinger, J. Eeckhoute, A.S. Brodsky, E.K. Keeton, K.C. Fertuck, G.F. Hall, Q. Wang, S. Bekiranov, et al. 2006. Genome-wide analysis of estrogen receptor binding sites. *Nat Genet* **38**:1289–1297.

Castellano, L., G. Giamas, J. Jacob, R.C. Coombes, W. Lucchesi, P. Thiruchelvam, G. Barton, L.R. Jiao, R. Wait, J. Waxman, G.J. Hannon, and J. Stebbing. 2009. The estrogen receptor-alpha-induced microRNA signature regulates itself and its transcriptional response. *Proc Natl Acad Sci U S A* **106**:15732–15737.

Cheang, M.C., S.K. Chia, D. Voduc, D. Gao, S. Leung, J. Snider, M. Watson, S. Davies, P.S. Bernard, J.S. Parker, C.M. Perou, M.J. Ellis, et al. 2009. Ki67 index, HER2 status, and prognosis of patients with luminal B breast cancer. *J Natl Cancer Inst* **101**:736–750.

Cicatiello, L., M. Mutarelli, O.M. Grober, O. Paris, L. Ferraro, M. Ravo, R. Tarallo, S. Luo, G.P. Schroth, M. Seifert, C. Zinser, M.L. Chiusano, et al. 2010. Estrogen receptor alpha controls a gene network in luminal-like breast cancer cells comprising multiple transcription factors and microRNAs. *Am J Pathol* **176**:2113–2130.

Cittelly, D.M., P.M. Das, V.A. Salvo, J.P. Fonseca, M.E. Burow, and F.E. Jones. 2010a. Oncogenic HER2{Delta}16 suppresses *miR-15a/16* and deregulates BCL-2 to promote endocrine resistance of breast tumors. *Carcinogenesis* **31**:2049–2057.

Cittelly, D.M., P.M. Das, N.S. Spoelstra, S.M. Edgerton, J.K. Richer, A.D. Thor, and F.E. Jones. 2010b. Downregulation of *miR-342* is associated with tamoxifen resistant breast tumors. *Mol Cancer* **9**:317.

Cohen, A., M. Shmoish, L. Levi, U. Cheruti, B. Levavi-Sivan, and E. Lubzens. 2008. Alterations in micro-ribonucleic acid expression profiles reveal a novel pathway for estrogen regulation. *Endocrinology* **149**:1687–1696.

Dai, R., R.A. Phillips, Y. Zhang, D. Khan, O. Crasta, and S.A. Ahmed. 2008. Suppression of LPS-induced interferon-gamma and nitric oxide in splenic lymphocytes by select estrogen-regulated microRNAs: a novel mechanism of immune modulation. *Blood* **112**:4591–4597.

Di Leva, G., P. Gasparini, C. Piovan, A. Ngankeu, M. Garofalo, C. Taccioli, M.V. Iorio, M. Li, S. Volinia, H. Alder, T. Nakamura, G. Nuovo, et al. 2010. MicroRNA cluster 221-222 and estrogen receptor alpha interactions in breast cancer. *J Natl Cancer Inst* **102**:706–721.

Doane, A.S., M. Danso, P. Lal, M. Donaton, L. Zhang, C. Hudis, and W.L. Gerald. 2006. An estrogen receptor-negative breast cancer subset characterized by a hormonally regulated transcriptional program and response to androgen. *Oncogene* **25**:3994–4008.

Dowsett, M., C. Allred, J. Knox, E. Quinn, J. Salter, C. Wale, J. Cuzick, J. Houghton, N. Williams, E. Mallon, H. Bishop, I. Ellis, et al. 2008. Relationship between quantitative estrogen and progesterone receptor expression and human epidermal growth factor receptor 2 (HER-2) status with recurrence in the Arimidex, Tamoxifen, Alone or in Combination trial. *J Clin Oncol* **26**:1059–1065.

Dykxhoorn, D.M., Y. Wu, H. Xie, F. Yu, A. Lal, F. Petrocca, D. Martinvalet, E. Song, B. Lim and J. Lieberman. 2009. miR-200 enhances mouse breast cancer cell colonization to form distant metastases. *PLoS One* **4**:e7181.

Epis, M.R., K.M. Giles, A. Barker, T.S. Kendrick, and P.J. Leedman. 2009. *miR-331-3p* regulates ERBB-2 expression and androgen receptor signaling in prostate cancer. *J Biol Chem* **284**: 24696–24704.

Farmer, P., H. Bonnefoi, V. Becette, M. Tubiana-Hulin, P. Fumoleau, D. Larsimont, G. Macgrogan, J. Bergh, D. Cameron, D. Goldstein, S. Duss, A.L. Nicoulaz, et al. 2005. Identification of molecular apocrine breast tumors by microarray analysis. *Oncogene* **24**:4660–4671.

Foekens, J.A., A.M. Sieuwerts, M. Smid, M.P. Look, V. de Weerd, A.W. Boersma, J.G. Klijn, E.A. Wiemer, and J.W. Martens. 2008. Four miRNAs associated with aggressiveness of lymph node-negative, estrogen receptor-positive human breast cancer. *Proc Natl Acad Sci U S A* **105**: 13021–13026.

Frasor, J., F. Stossi, J.M. Danes, B. Komm, C.R. Lyttle, and B.S. Katzenellenbogen. 2004. Selective estrogen receptor modulators: discrimination of agonistic versus antagonistic activities by gene expression profiling in breast cancer cells. *Cancer Res* **64**:1522–1533.

Garofalo, M., G. Romano, G. Di Leva, G. Nuovo, Y.J. Jeon, A. Ngankeu, J. Sun, F. Lovat, H. Alder, G. Condorelli, J.A. Engelman, M. Ono, et al. 2012. EGFR and MET receptor tyrosine kinase-altered microRNA expression induces tumorigenesis and gefitinib resistance in lung cancers. *Nat Med* **18**:74–82.

Gaudet, M.M., M. Campan, J.D. Figueroa, X.R. Yang, J. Lissowska, B. Peplonska, L.A. Brinton, D.L. Rimm, P.W. Laird, M. Garcia-Closas, and M.E. Sherman. 2009. DNA hypermethylation of ESR1 and PGR in breast cancer: pathologic and epidemiologic associations. *Cancer Epidemiol Biomarkers Prev* **18**:3036–3043.

Giacinti, L., P.P. Claudio, M. Lopez, and A. Giordano. 2006. Epigenetic information and estrogen receptor alpha expression in breast cancer. *Oncologist* **11**:1–8.

Goodsell, D.S. 2002. The molecular perspective: tamoxifen and the estrogen receptor. *Stem Cells* **20**:267–268.

Gotte, M., C. Mohr, C.Y. Koo, C. Stock, A.K. Vaske, M. Viola, S.A. Ibrahim, S. Peddibhotla, Y.H. Teng, J.Y. Low, K. Ebnet, L. Kiesel, et al. 2010. miR-145-dependent targeting of junctional

REFERENCES

adhesion molecule A and modulation of fascin expression are associated with reduced breast cancer cell motility and invasiveness. *Oncogene* **29**:6569–6580.

Gregory, P.A., A.G. Bert, E.L. Paterson, S.C. Barry, A. Tsykin, G. Farshid, M.A. Vadas, Y. Khew-Goodall, and G.J. Goodall. 2008. The miR-200 family and miR-205 regulate epithelial to mesenchymal transition by targeting ZEB1 and SIP1. *Nat Cell Biol* **10**:593–601.

Gusev, Y., T.D. Schmittgen, M. Lerner, R. Postier, and D. Brackett. 2007. Computational analysis of biological functions and pathways collectively targeted by co-expressed microRNAs in cancer. *BMC Bioinformatics* **8 Suppl 7**:S16.

Guttilla, I.K., B.D. Adams, and B.A. White. 2012. ERalpha, microRNAs, and the epithelial-mesenchymal transition in breast cancer. *Trends Endocrinol Metab* **23**:73–82.

Hah, N., C.G. Danko, L. Core, J.J. Waterfall, A. Siepel, J.T. Lis, and W.L. Kraus. 2011. A rapid, extensive, and transient transcriptional response to estrogen signaling in breast cancer cells. *Cell* **145**:622–634.

Harvey, J.M., G.M. Clark, C.K. Osborne, and D.C. Allred. 1999. Estrogen receptor status by immunohistochemistry is superior to the ligand-binding assay for predicting response to adjuvant endocrine therapy in breast cancer. *J Clin Oncol* **17**:1474–1481.

Heneghan, H.M., N. Miller, R. Kelly, J. Newell, and M.J. Kerin. 2010. Systemic miRNA-195 differentiates breast cancer from other malignancies and is a potential biomarker for detecting noninvasive and early stage disease. *Oncologist* **15**:673–682.

Herschkowitz, J.I., K. Simin, V.J. Weigman, I. Mikaelian, J. Usary, Z. Hu, K.E. Rasmussen, L.P. Jones, S. Assefnia, S. Chandrasekharan, M.G. Backlund, Y. Yin, et al. 2007. Identification of conserved gene expression features between murine mammary carcinoma models and human breast tumors. *Genome Biol* **8**:R76.

Herynk, M.H. and S.A. Fuqua. 2004. Estrogen receptor mutations in human disease. *Endocr Rev* **25**:869–898.

Hofmann, M.H., J. Heinrich, G. Radziwill, and K. Moelling. 2009. A short hairpin DNA analogous to *miR-125b* inhibits C-Raf expression, proliferation, and survival of breast cancer cells. *Mol Cancer Res* **7**:1635–1644.

Holst, F., P.R. Stahl, C. Ruiz, O. Hellwinkel, Z. Jehan, M. Wendland, A. Lebeau, L. Terracciano, K. Al-Kuraya, F. Janicke, G. Sauter, and R. Simon. 2007. Estrogen receptor alpha (ESR1) gene amplification is frequent in breast cancer. *Nat Genet* **39**:655–660.

Hossain, A., M.T. Kuo, and G.F. Saunders. 2006. Mir-17-5p regulates breast cancer cell proliferation by inhibiting translation of AIB1 mRNA. *Mol Cell Biol* **26**:8191–8201.

Howell, A. 2006. Is fulvestrant ("Faslodex") just another selective estrogen receptor modulator? *Int J Gynecol Cancer* **16 Suppl 2**:521–523.

Huang, G.L., X.H. Zhang, G.L. Guo, K.T. Huang, K.Y. Yang, and X.Q. Hu. 2008a. [Expression of microRNA-21 in invasive ductal carcinoma of the breast and its association with phosphatase and tensin homolog deleted from chromosome expression and clinicopathologic features]. *Zhonghua Yi Xue Za Zhi* **88**:2833–2837.

Huang, Q., K. Gumireddy, M. Schrier, C. le Sage, R. Nagel, S. Nair, D.A. Egan, A. Li, G. Huang, A.J. Klein-Szanto, P.A. Gimotty, D. Katsaros, et al. 2008b. The microRNAs miR-373 and miR-520c promote tumour invasion and metastasis. *Nat Cell Biol* **10**:202–210.

Huang, T.H., F. Wu, G.B. Loeb, R. Hsu, A. Heidersbach, A. Brincat, D. Horiuchi, R.J. Lebbink, Y.Y. Mo, A. Goga, and M.T. McManus. 2009. Up-regulation of *miR-21* by HER2/neu signaling promotes cell invasion. *J Biol Chem* **284**:18515–18524.

Hynes, N.E. and G. MacDonald. 2009. ErbB receptors and signaling pathways in cancer. *Curr Opin Cell Biol* **21**:177–184.

Ichikawa, T., F. Sato, K. Terasawa, S. Tsuchiya, M. Toi, G. Tsujimoto, and K. Shimizu. 2012. Trastuzumab produces therapeutic actions by upregulating *miR-26a* and *miR-30b* in breast cancer cells. *PLoS ONE* **7**:e31422.

Iorio, M.V., M. Ferracin, C.G. Liu, A. Veronese, R. Spizzo, S. Sabbioni, E. Magri, M. Pedriali, M. Fabbri, M. Campiglio, S. Menard, J.P. Palazzo, et al. 2005. MicroRNA gene expression deregulation in human breast cancer. *Cancer Res* **65**:7065–7070.

Iorio, M.V., P. Casalini, C. Piovan, G. Di Leva, A. Merlo, T. Triulzi, S. Menard, C.M. Croce, and E. Tagliabue. 2009. MicroRNA-205 regulates HER3 in human breast cancer. *Cancer Res* **69**:2195–2200.

Iwase, H., Y. Omoto, H. Iwata, T. Toyama, Y. Hara, Y. Ando, Y. Ito, Y. Fujii, and S. Kobayashi. 1999. DNA methylation analysis at distal and proximal promoter regions of the oestrogen receptor gene in breast cancers. *Br J Cancer* **80**:1982–1986.

Jiang, S., H.W. Zhang, M.H. Lu, X.H. He, Y. Li, H. Gu, M.F. Liu, and E.D. Wang. 2010. MicroRNA-155 functions as an OncomiR in breast cancer by targeting the suppressor of cytokine signaling 1 gene. *Cancer Res* **70**:3119–3127.

Jordan, V.C. 2006. The science of selective estrogen receptor modulators: concept to clinical practice. *Clin Cancer Res* **12**:5010–5013.

Jung, E.J., L. Santarpia, J. Kim, F.J. Esteva, E. Moretti, A.U. Buzdar, A. Di Leo, X.F. Le, R.C. Bast Jr., S.T. Park, L. Pusztai, and G.A. Calin. 2012. Plasma microRNA 210 levels correlate with sensitivity to trastuzumab and tumor presence in breast cancer patients. *Cancer* **118**: 2603–2614.

Kefas, B., J. Godlewski, L. Comeau, Y. Li, R. Abounader, M. Hawkinson, J. Lee, H. Fine, E.A. Chiocca, S. Lawler, and B. Purow. 2008. MicroRNA-7 inhibits the epidermal growth factor receptor and the Akt pathway and is down-regulated in glioblastoma. *Cancer Res* **68**: 3566–3572.

Klinge, C.M. 2009. Estrogen regulation of microRNA expression. *Curr Genomics* **10**:169–183.

Klinge, C.M. 2012. miRNAs and estrogen action. *Trends Endocrinol Metab* **23**:223–233.

Kondo, N., T. Toyama, H. Sugiura, Y. Fujii, and H. Yamashita. 2008. *miR-206* expression is down-regulated in estrogen receptor alpha-positive human breast cancer. *Cancer Res* **68**:5004–5008.

Kong, W., H. Yang, L. He, J.J. Zhao, D. Coppola, W.S. Dalton, and J.Q. Cheng. 2008. MicroRNA-155 is regulated by the transforming growth factor beta/Smad pathway and contributes to epithelial cell plasticity by targeting RhoA. *Mol Cell Biol* **28**:6773–6784.

Kong, W., L. He, M. Coppola, J. Guo, N.N. Esposito, D. Coppola, and J.Q. Cheng. 2010. MicroRNA-155 regulates cell survival, growth, and chemosensitivity by targeting FOXO3a in breast cancer. *J Biol Chem* **285**:17869–17879.

Kovalchuk, O., V.P. Tryndyak, B. Montgomery, A. Boyko, K. Kutanzi, F. Zemp, A.R. Warbritton, J.R. Latendresse, I. Kovalchuk, F.A. Beland, and I.P. Pogribny. 2007. Estrogen-induced rat breast carcinogenesis is characterized by alterations in DNA methylation, histone modifications and aberrant microRNA expression. *Cell Cycle* **6**:2010–2018.

Kovalchuk, O., J. Filkowski, J. Meservy, Y. Ilnytskyy, V.P. Tryndyak, V.F. Chekhun, and I.P. Pogribny. 2008. Involvement of microRNA-451 in resistance of the MCF-7 breast cancer cells to chemotherapeutic drug doxorubicin. *Mol Cancer Ther* **7**:2152–2159.

Leivonen, S.K., R. Makela, P. Ostling, P. Kohonen, S. Haapa-Paananen, K. Kleivi, E. Enerly, A. Aakula, K. Hellstrom, N. Sahlberg, V.N. Kristensen, A.L. Borresen-Dale, et al. 2009. Protein lysate microarray analysis to identify microRNAs regulating estrogen receptor signaling in breast cancer cell lines. *Oncogene* **28**:3926–3936.

Li, X., S.U. Mertens-Talcott, S. Zhang, K. Kim, J. Ball, and S. Safe. 2010. MicroRNA-27a indirectly regulates estrogen receptor {alpha} expression and hormone responsiveness in MCF-7 breast cancer cells. *Endocrinology* **151**:2462–2473.

Li, H., C. Bian, L. Liao, J. Li, and R.C. Zhao. 2011. *miR-17-5p* promotes human breast cancer cell migration and invasion through suppression of HBP1. *Breast Cancer Res Treat* **126**:565–575.

REFERENCES

Liang, Z., H. Wu, J. Xia, Y. Li, Y. Zhang, K. Huang, N. Wagar, Y. Yoon, H.T. Cho, S. Scala, and H. Shim. 2010. Involvement of *miR-326* in chemotherapy resistance of breast cancer through modulating expression of multidrug resistance-associated protein 1. *Biochem Pharmacol* **79**:817–824.

Lowery, A.J., N. Miller, A. Devaney, R.E. McNeill, P.A. Davoren, C. Lemetre, V. Benes, S. Schmidt, J. Blake, G. Ball, and M.J. Kerin. 2009. MicroRNA signatures predict oestrogen receptor, progesterone receptor and HER2/neu receptor status in breast cancer. *Breast Cancer Res* **11**:R27.

Lu, Z., M. Liu, V. Stribinskis, C.M. Klinge, K.S. Ramos, N.H. Colburn, and Y. Li. 2008. MicroRNA-21 promotes cell transformation by targeting the programmed cell death 4 gene. *Oncogene* **27**: 4373–4379.

Ma, L., J. Teruya-Feldstein, and R.A. Weinberg. 2007. Tumour invasion and metastasis initiated by microRNA-10b in breast cancer. *Nature* **449**:682–688.

Maillot, G., M. Lacroix-Triki, S. Pierredon, L. Gratadou, S. Schmidt, V. Benes, H. Roche, F. Dalenc, D. Auboeuf, S. Millevoi, and S. Vagner. 2009. Widespread estrogen-dependent repression of microRNAs involved in breast tumor cell growth. *Cancer Res* **69**:8332–8340.

Manavalan, T.T., Y. Teng, S.N. Appana, S. Datta, T.S. Kalbfleisch, Y. Li, and C.M. Klinge. 2011. Differential expression of microRNA expression in tamoxifen-sensitive MCF-7 versus tamoxifen-resistant LY2 human breast cancer cells. *Cancer Lett* **313**:26–43.

Mattie, M.D., C.C. Benz, J. Bowers, K. Sensinger, L. Wong, G.K. Scott, V. Fedele, D. Ginzinger, R. Getts, and C. Haqq. 2006. Optimized high-throughput microRNA expression profiling provides novel biomarker assessment of clinical prostate and breast cancer biopsies. *Mol Cancer* **5**:24.

McCafferty, M.P., R.E. McNeill, N. Miller, and M.J. Kerin. 2009. Interactions between the estrogen receptor, its cofactors and microRNAs in breast cancer. *Breast Cancer Res Treat* **116**:425–432.

McDonnell, D.P. 2006. Mechanism-based discovery as an approach to identify the next generation of estrogen receptor modulators. *FASEB J* **20**:2432–2434.

Miller, T.E., K. Ghoshal, B. Ramaswamy, S. Roy, J. Datta, C.L. Shapiro, S. Jacob, and S. Majumder. 2008. MicroRNA-221/222 confers tamoxifen resistance in breast cancer by targeting p27Kip1. *J Biol Chem* **283**:29897–29903.

Moasser, M.M. 2007. The oncogene HER2: its signaling and transforming functions and its role in human cancer pathogenesis. *Oncogene* **26**:6469–6487.

Murphy, M.J., Jr. 1998. Molecular action and clinical relevance of aromatase inhibitors. *Oncologist* **3**:129–130.

Murphy, C.G. and S. Modi. 2009. HER2 breast cancer therapies: a review. *Biologics* **3**:289–301.Osborne, C.K. and R. Schiff. 2011. Mechanisms of endocrine resistance in breast cancer. *Annu Rev Med* **62**:233–247.

Pandey, D.P. and D. Picard. 2009. *miR-22* inhibits estrogen signaling by directly targeting the estrogen receptor alpha mRNA. *Mol Cell Biol* **29**:3783–3790.

Park, S.M., A.B. Gaur, E. Lengyel, and M.E. Peter. 2008. The *miR-200* family determines the epithelial phenotype of cancer cells by targeting the E-cadherin repressors ZEB1 and ZEB2. *Genes Dev* **22**:894–907.

Perou, C.M., T. Sorlie, M.B. Eisen, M. van de Rijn, S.S. Jeffrey, C.A. Rees, J.R. Pollack, D.T. Ross, H. Johnsen, L.A. Akslen, O. Fluge, A. Pergamenschikov, et al. 2000. Molecular portraits of human breast tumors. *Nature* **406**:747–752.

Piovan, C., D. Palmieri, G. Di Leva, L. Braccioli, P. Casalini, G. Nuovo, M. Tortoreto, M. Sasso, I. Plantamura, T. Triulzi, C. Taccioli, E. Tagliabue, et al. 2012. Oncosuppressive role of p53-induced miR-205 in triple negative breast cancer. *Mol Oncol* **6**:458–472.

Pogribny, I.P., V.P. Tryndyak, A. Boyko, R. Rodriguez-Juarez, F.A. Beland, and O. Kovalchuk. 2007. Induction of microRNAome deregulation in rat liver by long-term tamoxifen exposure. *Mutat Res* **619**:30–37.

Qi, Y.T., Y. Zhao, Z. Zhang, D. Li, J.Y. Gu, X.N. Fu, X.N. Chen, C.G. Yuan, and L.H. Lai. 2009. [The expression pattern and possible function of microRNA-34a in neurons]. *Fen Zi Xi Bao Sheng Wu Xue Bao* **42**:217–223.

Qian, B., D. Katsaros, L. Lu, M. Preti, A. Durando, R. Arisio, L. Mu, and H. Yu. 2009. High miR-21 expression in breast cancer associated with poor disease-free survival in early stage disease and high TGF-beta1. *Breast Cancer Research and Treatment* **117**:131–140.

Rao, X., G. Di Leva, M. Li, F. Fang, C. Devlin, C. Hartman-Frey, M.E. Burow, M. Ivan, C.M. Croce, and K.P. Nephew. 2011. MicroRNA-221/222 confers breast cancer fulvestrant resistance by regulating multiple signaling pathways. *Oncogene* **30**:1082–1097.

Reis-Filho, J.S. and L. Pusztai. 2011. Gene expression profiling in breast cancer: classification, prognostication, and prediction. *Lancet* **378**:1812–1823.

Rodriguez-Gonzalez, F.G., A.M. Sieuwerts, M. Smid, M.P. Look, M.E. Meijer-van Gelder, V. de Weerd, S. Sleijfer, J.W. Martens, and J.A. Foekens. 2011. MicroRNA-30c expression level is an independent predictor of clinical benefit of endocrine therapy in advanced estrogen receptor positive breast cancer. *Breast Cancer Res Treat* **127**:43–51.

Roth, C., B. Rack, V. Muller, W. Janni, K. Pantel, and H. Schwarzenbach. 2010. Circulating microRNAs as blood-based markers for patients with primary and metastatic breast cancer. *Breast Cancer Res* **12**:R90.

Sachdeva, M. and Y.Y. Mo. 2010. miR-145-mediated suppression of cell growth, invasion and metastasis. *Am J Transl Res* **2**:170–180.

Sachdeva, M., H. Wu, P. Ru, L. Hwang, V. Trieu, and Y.Y. Mo. 2011. MicroRNA-101-mediated Akt activation and estrogen-independent growth. *Oncogene* **30**:822–831.

Scott, G.K., A. Goga, D. Bhaumik, C.E. Berger, C.S. Sullivan, and C.C. Benz. 2007. Coordinate suppression of ERBB2 and ERBB3 by enforced expression of micro-RNA *miR-125a* or *miR-125b*. *J Biol Chem* **282**:1479–1486.

Seike, M., A. Goto, T. Okano, E.D. Bowman, A.J. Schetter, I. Horikawa, E.A. Mathe, J. Jen, P. Yang, H. Sugimura, A. Gemma, S. Kudoh, et al. 2009. *MiR-21* is an EGFR-regulated anti-apoptotic factor in lung cancer in never-smokers. *Proc Natl Acad Sci U S A* **106**:12085–12090.

Sempere, L.F., M. Christensen, A. Silahtaroglu, M. Bak, C.V. Heath, G. Schwartz, W. Wells, S. Kauppinen, and C.N. Cole. 2007. Altered microRNA expression confined to specific epithelial cell subpopulations in breast cancer. *Cancer Res* **67**:11612–11620.

Serrano-Olvera, A., A. Duenas-Gonzalez, D. Gallardo-Rincon, M. Candelaria, and J. De la Garza-Salazar. 2006. Prognostic, predictive and therapeutic implications of HER2 in invasive epithelial ovarian cancer. *Cancer Treat Rev* **32**:180–190.

Shimono, Y., M. Zabala, R.W. Cho, N. Lobo, P. Dalerba, D. Qian, M. Diehn, H. Liu, S.P. Panula, E. Chiao, F.M. Dirbas, G. Somlo, et al. 2009. Downregulation of miRNA-200c links breast cancer stem cells with normal stem cells. *Cell* **138**:592–603.

Slamon, D.J., G.M. Clark, S.G. Wong, W.J. Levin, A. Ullrich, and W.L. McGuire. 1987. Human breast cancer: correlation of relapse and survival with amplification of the HER-2/neu oncogene. *Science* **235**:177–182.

Song, B., C. Wang, J. Liu, X. Wang, L. Lv, L. Wei, L. Xie, Y. Zheng, and X. Song. 2010. MicroRNA-21 regulates breast cancer invasion partly by targeting tissue inhibitor of metalloproteinase 3 expression. *J Exp Clin Cancer Res* **29**:29.

Sorlie, T., C.M. Perou, R. Tibshirani, T. Aas, S. Geisler, H. Johnsen, T. Hastie, M.B. Eisen, M. van de Rijn, S.S. Jeffrey, T. Thorsen, H. Quist, et al. 2001. Gene expression patterns of breast carcinomas distinguish tumor subclasses with clinical implications. *Proc Natl Acad Sci U S A* **98**:10869–10874.

Sorlie, T., R. Tibshirani, J. Parker, T. Hastie, J.S. Marron, A. Nobel, S. Deng, H. Johnsen, R. Pesich, S. Geisler, J. Demeter, C.M. Perou, et al. 2003. Repeated observation of breast tumor subtypes in independent gene expression data sets. *Proc Natl Acad Sci U S A* **100**:8418–8423.

Spizzo, R., M.S. Nicoloso, L. Lupini, Y. Lu, J. Fogarty, S. Rossi, B. Zagatti, M. Fabbri, A. Veronese, X. Liu, R. Davuluri, C.M. Croce, et al. 2010. *miR-145* participates with TP53 in a death-promoting regulatory loop and targets estrogen receptor-alpha in human breast cancer cells. *Cell Death Differ* **17**:246–254.

Tagliabue, E., A. Balsari, M. Campiglio, and S.M. Pupa. 2010. HER2 as a target for breast cancer therapy. *Expert Opin Biol Ther* **10**:711–724.

Tomita, S., Z. Zhang, M. Nakano, M. Ibusuki, T. Kawazoe, Y. Yamamoto, and H. Iwase. 2009. Estrogen receptor alpha gene ESR1 amplification may predict endocrine therapy responsiveness in breast cancer patients. *Cancer Sci* **100**:1012–1017.

Toth, J., K. Egervari, A. Klekner, L. Bognar, J. Szanto, Z. Nemes, and Z. Szollosi. 2009. Analysis of EGFR gene amplification, protein over-expression and tyrosine kinase domain mutation in recurrent glioblastoma. *Pathol Oncol Res* **15**:225–229.

Uhlmann, S., H. Mannsperger, J.D. Zhang, E.A. Horvat, C. Schmidt, M. Kublbeck, F. Henjes, A. Ward, U. Tschulena, K. Zweig, U. Korf, S. Wiemann, et al. 2012. Global microRNA level regulation of EGFR-driven cell-cycle protein network in breast cancer. *Mol Syst Biol* **8**:570.

Valastyan, S., F. Reinhardt, N. Benaich, D. Calogrias, A.M. Szasz, Z.C. Wang, J.E. Brock, A.L. Richardson, and R.A. Weinberg. 2009. A pleiotropically acting microRNA, miR-31, inhibits breast cancer metastasis. *Cell* **137**:1032–1046.

van Schooneveld, E., M.C. Wouters, I. Van der Auwera, D.J. Peeters, H. Wildiers, P.A. Van Dam, I. Vergote, P.B. Vermeulen, L.Y. Dirix, and S.J. Van Laere. 2012. Expression profiling of cancerous and normal breast tissues identifies microRNAs that are differentially expressed in serum from patients with (metastatic) breast cancer and healthy volunteers. *Breast Cancer Res* **14**:R34.

Visvader, J.E. 2009. Keeping abreast of the mammary epithelial hierarchy and breast tumorigenesis. *Genes Dev* **23**:2563–2577.

Webster, R.J., K.M. Giles, K.J. Price, P.M. Zhang, J.S. Mattick, and P.J. Leedman. 2009. Regulation of epidermal growth factor receptor signaling in human cancer cells by microRNA-7. *J Biol Chem* **284**:5731–5741.

Wickramasinghe, N.S., T.T. Manavalan, S.M. Dougherty, K.A. Riggs, Y. Li, and C.M. Klinge. 2009. Estradiol downregulates *miR-21* expression and increases *miR-21* target gene expression in MCF-7 breast cancer cells. *Nucleic Acids Res* **37**:2584–2595.

Wu, F., I. Ivanov, R. Xu, and S. Safe. 2009. Role of SP transcription factors in hormone-dependent modulation of genes in MCF-7 breast cancer cells: microarray and RNA interference studies. *J Mol Endocrinol* **42**:19–33.

Wu, X., G. Somlo, Y. Yu, M.R. Palomares, A.X. Li, W. Zhou, A. Chow, Y. Yen, J.J. Rossi, H. Gao, J. Wang, Y.C. Yuan, et al. 2012. De novo sequencing of circulating miRNAs identifies novel markers predicting clinical outcome of locally advanced breast cancer. *J Transl Med* **10**:42.

Xin, F., M. Li, C. Balch, M. Thomson, M. Fan, Y. Liu, S.M. Hammond, S. Kim, and K.P. Nephew. 2009. Computational analysis of microRNA profiles and their target genes suggests significant involvement in breast cancer antiestrogen resistance. *Bioinformatics* **25**:430–434.

Yamagata, K., S. Fujiyama, S. Ito, T. Ueda, T. Murata, M. Naitou, K. Takeyama, Y. Minami, B.W. O'Malley, and S. Kato. 2009. Maturation of microRNA is hormonally regulated by a nuclear receptor. *Mol Cell* **36**:340–347.

Yamashita, H., M. Nishio, Y. Ando, Z. Zhang, M. Hamaguchi, K. Mita, S. Kobayashi, Y. Fujii, and H. Iwase. 2006. Stat5 expression predicts response to endocrine therapy and improves survival in estrogen receptor-positive breast cancer. *Endocr Relat Cancer* **13**:885–893.

Yamashita, H., S. Takahashi, Y. Ito, T. Yamashita, Y. Ando, T. Toyama, H. Sugiura, N. Yoshimoto, S. Kobayashi, Y. Fujii, and H. Iwase. 2009. Predictors of response to exemestane as primary endocrine therapy in estrogen receptor-positive breast cancer. *Cancer Sci* **100**:2028–2033.

Yamashita, H., H. Iwase, T. Toyama, S. Takahashi, H. Sugiura, N. Yoshimoto, Y. Endo, Y. Fujii, and S. Kobayashi. 2011. Estrogen receptor-positive breast cancer in Japanese women: trends in incidence, characteristics, and prognosis. *Ann Oncol* **22**:1318–1325.

Yu, F., H. Yao, P. Zhu, X. Zhang, Q. Pan, C. Gong, Y. Huang, X. Hu, F. Su, J. Lieberman, and E. Song. 2007. *let-7* regulates self renewal and tumorigenicity of breast cancer cells. *Cell* **131**:1109–1123.

Yu, Z., C. Wang, M. Wang, Z. Li, M.C. Casimiro, M. Liu, K. Wu, J. Whittle, X. Ju, T. Hyslop, P. McCue, and R.G. Pestell. 2008. A cyclin D1/microRNA 17/20 regulatory feedback loop in control of breast cancer cell proliferation. *J Cell Biol* **182**:509–517.

Zhang, Z., H. Sun, H. Dai, R.M. Walsh, M. Imakura, J. Schelter, J. Burchard, X. Dai, A.N. Chang, R.L. Diaz, J.R. Marszalek, S.R. Bartz, et al. 2009. MicroRNA miR-210 modulates cellular response to hypoxia through the MYC antagonist MNT. *Cell Cycle* **8**:2756–2768.

Zhao, J.J., J. Lin, H. Yang, W. Kong, L. He, X. Ma, D. Coppola, and J.Q. Cheng. 2008. MicroRNA-221/222 negatively regulates estrogen receptor alpha and is associated with tamoxifen resistance in breast cancer. *J Biol Chem* **283**:31079–31086.

Zhao, Y., C. Deng, J. Wang, J. Xiao, Z. Gatalica, R.R. Recker, and G.G. Xiao. 2011. *Let-7* family miRNAs regulate estrogen receptor alpha signaling in estrogen receptor positive breast cancer. *Breast Cancer Res Treat* **127**:69–80.

Zhou, X., Y. Ren, L. Moore, M. Mei, Y. You, P. Xu, B. Wang, G. Wang, Z. Jia, P. Pu, W. Zhang, and C. Kang. 2010. Downregulation of *miR-21* inhibits EGFR pathway and suppresses the growth of human glioblastoma cells independent of PTEN status. *Lab Invest* **90**:144–155.

Zhu, S., H. Wu, F. Wu, D. Nie, S. Sheng, and Y.Y. Mo. 2008. MicroRNA-21 targets tumor suppressor genes in invasion and metastasis. *Cell Res* **18**:350–359.

18

MicroRNAs IN HUMAN PROSTATE CANCER: FROM PATHOGENESIS TO THERAPEUTIC IMPLICATIONS

Mustafa Ozen[1,2,3] and Omer Faruk Karatas[1,4]

[1]*Department of Medical Genetics, Istanbul University Cerrahpasa Medical School, Istanbul, Turkey*
[2]*Bezmialem Vakif University, Istanbul, Turkey*
[3]*Department of Pathology & Immunology, Baylor College of Medicine, Houston, TX, USA*
[4]*Molecular Biology and Genetics Department, Erzurum Technical University, Erzurum, Turkey*

I.	Introduction	312
II.	PCa Pathogenesis and miRNAs	313
III.	miRNA Profiling in PCa	313
	A. miRNAs as Oncomirs	314
	B. miRNAs as Tumor Suppressors	316
IV.	miRNAs and Their Targets in PCa	317
V.	miRNAs as Potential Markers for Diagnosis and Progression of PCa	318
VI.	miRNAs Modulating Tumor Progression and Metastasis in PCa	319
VII.	miRNAs that Play a Role in the Epigenetic Machinery of PCa Pathogenesis	320
VIII.	miRNAs as Implicated in PCa Therapy and Their Future Potential	321
IX.	Conclusion	321
	Acknowledgment	322
	References	322

ABBREVIATIONS

AIF apoptosis-inducing factor
AR androgen receptor

MicroRNAs in Medicine, First Edition. Edited by Charles H. Lawrie.
© 2014 John Wiley & Sons, Inc. Published 2014 by John Wiley & Sons, Inc.

ASAP1	ArfGAP with SH3 domain, ankyrin repeat, and PH domain 1
Bak1	BCL2-antogonist/killer 1
BIRC5	baculovirol IAP repeat containing 5
BNIP3	BCL2/adenovirus E1B 19-kDa interacting protein 3
BPH	benign prostatic hyperplasia
CD44+	CD44 positive
CKAP2	cytoskeleton-associated protein 2
DIM	3,3′-diindolylmethane
EIF4EBP1	eukaryotic translation initiation factor 4E binding protein 1
EMT	epithelial to mesenchymal transition
ERK5	extracellular signal-regulated kinase 5
EZH2	enhancer of zeste homologue 2
HDAC1	histone deacetylase 1
hK2	human glandular kallikrein 2
HMGA2	high-mobility group A2
IL-6	interleukin-6
IL-6R	interleukin-6 receptor
IL-24	interleukin-24
IL-32	interleukin-32
LASP1	LIM and SH3 protein 1
MAPK	mitogen-activated protein kinase
MARCKS	myristoylated alanine-rich protein kinase c substrate
miRNAs	microRNAs
mRNA	messenger RNA
PCa	prostate cancer
PDCD4	programmed cell death 4
PNP	purine nucleoside phosphorylase
PSA	prostate-specific antigen
PTEN	phosphatase and tensin homologue
RB1	retinoblastoma 1
RNA	ribonucleic acid
RUNX2	runt-related transcription factor 2
TGF-β1	transforming growth factor-beta 1
TSmiRs	tumor suppressor miRNAs
uPA	urokinase plasminogen activator
uPAR	urokinase plasminogen activator receptor
WASF1	WAS protein family, member 1

I. INTRODUCTION

MicroRNAs (miRNAs) are small, approximately 18- to 24-nucleotide-long, non-coding, and endogenously synthesized ribonucleic acids (RNAs) that have recently become a very popular subject in cancer research.

miRNAs are among the key regulators of posttranscriptional messenger RNA (mRNA) expression. Altered miRNA levels are observed in various tumor types, and their specific deregulation pattern in different tumors makes them potential therapeutic targets and diagnostic markers. In prostate cancer (PCa) profiling studies, a common deregulation pattern of miRNAs has been revealed through utilization of different cell lines, tumor xenografts, and clinical radical prostatectomy samples.

The mechanisms of how miRNAs target and inhibit their targets' expressions are not fully understood yet. To elucidate clearly how miRNAs function in PCa pathogenesis, more studies detailing miRNA functional mechanisms need to be performed.

II. PCA PATHOGENESIS AND miRNAs

PCa is the most commonly diagnosed malignant type of cancer, constituting 29% (240,890) of incident cancer cases in 2011 in the United States alone. It was also the second leading cause of cancer deaths among men, with 33,720 attributed cases in 2011 (Siegel et al. 2011). Of the PCa cases, primary tumors mostly stay locally inside the organ; however, some prostate carcinomas are known to metastasize to other organs (Damber and Aus 2008). At the beginning, circulating androgens play important roles in the growth of prostate tumors through their increased interaction with the androgen receptor (AR). Activation of AR leads to transactivation of its target genes and therefore PCa cell proliferation (Fletcher et al. 2012). Invasion and metastasis of PCa are accompanied by an important step, which is epithelial to mesenchymal transition (EMT). This process involves the substitution of epithelial cell adhesion molecules to mesenchymal-specific cytoskeletal components (Thiery et al. 2009; Iwatsuki et al. 2010). Recently, there have been enormous developments made in PCa therapy applications targeting early and non-metastasized PCa cases, although the need for therapeutic approaches against advanced PCa cases still remains (Viticchie et al. 2011). The most commonly utilized treatments are surgery, radiation, and hormone ablation therapy for early and localized PCas. Meanwhile, chemotherapy is the only choice for the advanced and metastasized tumors, although mostly it fails to result in positive clinical response (Bhatnagar et al. 2010).

miRNAs work in combination with each other rather than working individually. They operate in overlapping regulatory networks and are implied in fine-tuning of mRNA levels (Mallory and Vaucheret 2006). miRNAs are shown to play significant roles in regulation of almost every cellular process, such as development, proliferation, and apoptosis, through their spatial and temporal expression. The alterations in their expressions are implicated in pathogenesis of a variety of diseases, including human cancers, and they can act as potent tumor oncogenes or tumor suppressor genes (Filipowicz et al. 2008; Majid et al. 2010).

There are now several reports that show evidence for the participation of miRNAs in prostate carcinogenesis (Folini et al. 2010; Sevli et al. 2010; Galardi et al. 2011; Boll et al. 2012; Chen et al. 2012). Aberrant expression of miRNAs has been detected in PCa cell lines, xenografts, and clinical samples (Shi et al. 2008). These alterations may play significant roles in the PCa pathogenesis. A diagram of miRNAs and their validated targets is represented in Figure 18.1.

III. miRNA PROFILING IN PCA

Although there have been significant improvements in the detection and diagnosis of PCa in the last decades, it is still one of the leading causes of cancer deaths among men. Certain mechanisms playing significant roles in PCa pathogenesis have been elucidated; however, the mechanisms of the development and progression of PCa are still mostly unknown. Investigations of the possible roles that miRNAs can play in tumorigenesis help enlighten

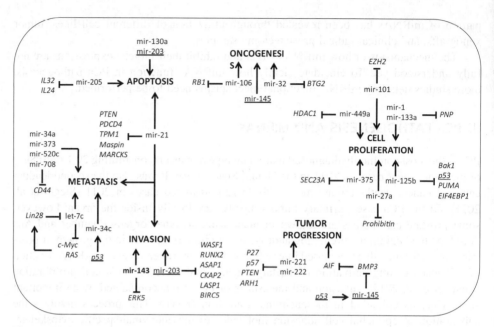

Figure 18.1. Schematic representation of key miRNAs and their validated targets as implicated in PCa pathogenesis. miRNAs and genes that are mentioned more than once in the figure are underlined.

both the genetic and epigenetic changes occurring during cancer initiation, progression, invasion, or metastasis.

A. miRNAs as Oncomirs

Oncomirs are miRNAs that are overexpressed in tumors. They act as oncogenes and play important roles in several kinds of cancer pathogenesis. *miR-125b*, *miR-221*, *miR-222*, *miR-21*, *miR-27a*, and *miR-106a* are among the key oncogenic miRNAs in PCa tumorigenesis.

miR-125b is a lin-4 homologue, which was the first identified miRNA (in *Caenorhabditis elegans*) (John et al. 2004), and it is necessary for cell proliferation (Lee et al. 2005). Its overexpression has been reported in several cancer types, including stomach cancer (Ueda et al. 2010), colon cancer (Baffa et al. 2009), pancreatic cancer (Bloomston et al. 2007), and ovarian cancer (Sorrentino et al. 2008). Additionally, deregulation of *miR-125b* has an important impact on PCa tumorigenesis, and it is up-regulated in androgen-independent PCa cells and in clinical tissues (Lee et al. 2005; Shi et al. 2007; Ozen et al. 2008). Ectopic expression of *miR-125b* in PCa cells resulted in androgen-independent cell growth *in vitro* and down-regulation of BCL2-antogonist/killer 1 (*Bak1*) (Shi et al. 2007). The *miR-125b*-mediated phenotype in PCa cells in terms of cell growth, however, cannot be mimicked by down-regulation of *Bak1* per se (Chiosea et al. 2006). In a recent study, in addition to *Bak1*, it was demonstrated that *p53* and *Puma*, which are key apoptotic genes, are also directly targeted by *miR-125b*, and their expression is significantly reduced in PCa cells (Shi et al. 2011). In the same study, *in vivo* experiments demonstrated that *miR-125b* can promote cell growth upon repression of these target genes in both intact and castrated male nude mice. Eukaryotic translation initiation factor 4E binding protein 1 (*EIF4EBP1*) was also shown to be one of the targets of *miR-125b* in PCa (Ozen et al. 2008).

miR-221 and *miR-222*, processed from a single transcript that is encoded from a genomic region on X chromosome, have been documented as being overexpressed in several tumors including PCa. In a recent study, it was reported that NF-κB and c-Jun, which are known to be important factors for the initiation and progression of cancer, directly modulate the transcription of these two miRNAs through binding their upstream region (Galardi et al. 2011). Upon their induction, *miR-221* and *miR-222* target and repress the expression of tumor suppressor genes such as *p27*, *p57*, and phosphatase and tensin homologue (*PTEN*) (Galardi et al. 2007; Garofalo et al. 2009; Pineau et al. 2010). Additionally, *miR-221* and *miR-222* target a maternally imprinted tumor suppressor gene, *ARH1*, in several cancers. *ARH1* is known to be expressed in numerous normal tissues; however, its expression is significantly reduced in cancer tissues through different mechanisms including loss of heterozygosity, epigenetic mechanisms, and binding of E2F to the *ARH1* promoter (Chen et al. 2011). In PCa, *ARH1* expression is reported to be down-regulated in comparison with the corresponding normal tissues, and *ARH1* repression in PCa is closely associated with the overexpression of *miR-221* and *miR-222* through their interaction with 3′-untranslated region (3′-UTR) of *ARH1* (Chen et al. 2011). In a recent study, it was proposed that *miR-221* controls the migration of androgen-independent PCa cells through targeting *DVL2* (Zheng et al. 2012). On the other hand, in stage T2a/b prostatectomy samples and androgen-dependent cell lines, *miR-221* and *miR-222* have been shown to be down-regulated (Tong et al. 2009). Interestingly, another study documented that *miR-221* is relatively down-regulated in a progressive manner in aggressive prostate tumors and that this down-regulation is associated with tumor progression, recurrence, and metastasis. Therefore, in PCa at least, the roles of *miR-221* and *miR-222* in tumorigenesis still need to be clarified.

Increased levels of *miR-21* have been observed in a number of tumor tissues including glioblastoma (Chan et al. 2005), breast cancer (Yan et al. 2008), gastric cancer (Zhang et al. 2008), and pancreatic cancer (Dillhoff et al. 2008) in comparison with corresponding normal tissues. Its overexpression leads to the suppression of tumor suppressor genes such as *PTEN* (Wang et al. 2008), programmed cell death 4 (*PDCD4*) (Lu et al. 2008), *tropomyosin 1* (Zhu et al. 2008), *maspin* (Zhu et al. 2008), and myristoylated alanine-rich protein kinase c substrate (*MARCKS*) (Li et al. 2009). In PCa, there is limited information about the possible functional role of *miR-21*. Although *miR-21* has no direct effect on the prostate tumor proliferation capacity, it is postulated that it could increase the invasion and motility potential of PCa and induce resistance to apoptosis (Li et al. 2009). In PCa cell line DU-145, *miR-21* directly represses *RECK*, a membrane-anchored glycoprotein that inhibits tumor cell invasion (Reis et al. 2012). In a recent study in PCa cell lines, however, it was demonstrated that the knockdown of *miR-21* is not enough on its own to affect either the proliferative and invasive potential or the chemo- and radiosensitivity capacity of PCa cells (Folini et al. 2010). In addition, in this same study, it was shown that *miR-21* lacks the ability to modulate the expression levels of *PTEN* and *PDCD4*. It is therefore suggested that, although it might have important roles in the tumorigenesis of PCa, *miR-21* is not necessarily a good candidate for therapeutic approaches in PCa (Folini et al. 2010).

miR-27a has been recently demonstrated to be an androgen-regulated oncomir in PCa. In addition to several cancers such as breast cancer (Mertens-Talcott et al. 2007), hepatocellular cancer (Huang et al. 2008b), and ovarian cancer (Nam et al. 2008), *miR-27a* is differentially expressed in PCa as well (Porkka et al. 2007). Its up-regulation results in the down-regulation of prohibitin, which is a tumor suppressor and AR corepressor, as well as increases PCa cell growth through elevated expression levels of AR target genes (Fletcher et al. 2012).

miR-106a shows elevated levels of expression in gastric tumor, lung tumor, and prostate tumor samples. Up-regulation of *miR-106b* with respect to the corresponding normal tissues in these tumors has been demonstrated to be correlated with the down-regulation of the putative target retinoblastoma 1 (*RB1*) (Volinia et al. 2006).

B. miRNAs as Tumor Suppressors

Tumor suppressor miRNAs (TSmiRs) are down-regulated in the PCa cells compared with the normal tissue in prostate. Their reduced expression results in the elevated expression of oncogenes in the cancer cells. Restoration of the normal expression patterns of these miRNAs leads to the repression of proliferative, invasive, clonogenic, and metastatic potential of the PCa cells both *in vivo* and *in vitro*. Among the most well-known TSmiRs are *miR-205*, *miR-145*, *miR-34c*, and *let-7c*. Recently, *miR-130a* and *miR-203* have also been shown to be down-regulated in PCa.

It has been previously shown that *miR-205* is significantly down-regulated in PCa cell lines with respect to control cells. The decrease in *miR-205* level is also reported in cancer tissues in comparison with benign prostatic hyperplasia (BPH) samples. In a recent study, it was demonstrated that *miR-205* acts as a TSmiR via specifically targeting the promoter of tumor suppressor *interleukin-24* (*IL-24*) and *interleukin-32* (*IL-32*) genes and up-regulating their expression. This study revealed that the effects of miRNAs are not restricted to only gene silencing (Majid et al. 2010).

In addition to *miR-205*, recently, it was shown that *miR-130a* and *miR-203* are also down-regulated in PCa, and they collectively repress AR and mitogen-activated protein kinase (MAPK) signaling pathways. Overexpression of these miRNAs in PCa cell lines resulted in inhibition of cell growth through apoptosis or cell-cycle arrest (Boll et al. 2012).

miR-145 is one of the crucial TSmiRs whose transcription is controlled by *p53*. It is down-regulated in many cancers such as colorectal, mammary, ovarian, and B-cell tumors (Chen et al. 2010). In addition, although its role is not clarified yet, microarray profiling showed that *miR-145* was also down-regulated in PCa (Porkka et al. 2007; Ozen et al. 2008). Bioinformatics analysis showed that BCL2/adenovirus E1B 19-kDa interacting protein 3 (*BNIP3*), which is overexpressed in many cancers including PCa, is a putative target of *miR-145*. Therefore, it is postulated that deficiency of *miR-145*, possibly stemming from dysfunction of its transcriptional regulator *p53*, causes the overexpression of *BNIP3*. Because *BNIP3* is known to act as a transcriptional corepressor of the apoptosis-inducing factor (*AIF*) gene, its overexpression corresponds with the down-regulation of *AIF*. This sequential aberrancy in the above-mentioned pathway has been proposed to be related to prostate tumor progression and its prognosis (Chen et al. 2010). *CCNA2* has been identified as another potential target for *miR-145* in LNCaP and VcaP cells (Wang et al. 2009).

miR-34c is another TSmiR whose expression is regulated by the transcription factor *p53* (He et al. 2007). *miR-34c*, along with *miR-34b*, is located at chromosome 11q23, and loss of heterozygosity within this region has been documented in different solid tumors such as breast cancer, lung cancer, and prostate adenocarcinoma (Rasio et al. 1995; Dahiya et al. 1997; Ellsworth et al. 2008). It is down-regulated in a variety of malignancies including neuroblastoma, lung cancer, colorectal cancer, and PCa (Cole et al. 2008; Liang 2008;

Toyota et al. 2008; Hagman et al. 2010). It has been reported that there is a reverse correlation between the *miR-34c* expression levels and aggressiveness of the tumor, World Health Organization (WHO) grade, prostate-specific antigen (PSA) levels, and occurrence of metastases in PCa (Hagman et al. 2010).

The *let-7* miRNA family is comprised of 13 homologous miRNAs, which flank genomic regions frequently deleted in human cancers (Calin et al. 2004). *let-7* members have been shown to play important roles in cell proliferation capacity, metastasis potential, resistance to radiation therapy, and postoperative survival rate of tumors (Lee and Dutta 2007; Yu et al. 2007; Nadiminty et al. 2012). It has also been demonstrated that they control the expression levels of oncogenes such as high-mobility group A2 (*HMGA2*) (Lee and Dutta 2007), *RAS* (Johnson et al. 2005), *Myc* (Kumar et al. 2007), and several other cell-cycle control genes. Down-regulation of *let-7* has been associated with pathogenesis of breast cancer (Yu et al. 2007), lung cancer (Landi et al. 2010), ovarian cancer (Helland et al. 2011), head and neck squamous cell carcinoma (Childs et al. 2009), and PCa (Nadiminty et al. 2012). *let-7* is transcriptionally regulated by a highly conserved RNA-binding protein, Lin28, whose overexpression in primary human tumors results in repression of *let-7* family miRNAs (Viswanathan et al. 2009). This causes up-regulation of *c-Myc* in many cancers, and it is known that c-Myc in turn transcriptionally activates *Lin28*. It is therefore speculated that this transcriptional activation loop might be important in miRNA cancer pathogenesis (Chang et al. 2009). In localized PCa, it has been shown that *let-7* is down-regulated compared with benign peripheral zone tissues (Ozen et al. 2008). In another study, it has been postulated that *let-7c*, one of the *let-7* family members, is down-regulated in clinical PCa specimens, and the overexpression of *let-7c* inhibited the growth of PCa cells. Moreover, reexpression of *let-7c* in xenografts of human PCa cells also significantly inhibited the tumor growth.

IV. MiRNAs AND THEIR TARGETS IN PCa

A single miRNA can target and modulate expression of many mRNAs posttranscriptionally, because there is only partial complementarity between the seed sequence of miRNAs and their target mRNA sequences. It is therefore not easy to find a specific target for a known miRNA. Microarray profiling experiments provide the opportunity to determine the differentially expressed miRNAs in a tissue or cell type. Then, typically various bioinformatics tools are used to find putative mRNA targets for these deregulated miRNAs. Functional validation of the target genes through investigating alterations in their transcriptional and translational levels upon ectopic miRNA expression points to the possible involvement of miRNAs in the expression control of the target mRNAs. Here, we give some examples of validated targets of miRNAs that are deregulated in PCa samples.

miR-375 is among the known oncomirs in PCa, and it has been demonstrated to target and repress *Sec23A*, which is suggested to play a role in the growth of PCa cells (Szczyrba et al. 2011). An androgen-regulated oncomir, *miR-32* has been found to be up-regulated in castration-resistant PCa. mRNA microarray analysis that utilized LNCaP cells transfected with pre-*miR-32*, and subsequent luciferase and immunostaining experiments revealed that overexpression of *miR-32* resulted in reduced levels of *BTG2* (Jalava et al. 2012).

miR-1 and *miR-133a* have been shown to be significantly down-regulated in PCa tissues, and they were both reported to directly target purine nucleoside phosphorylase (*PNP*) *in vitro*. Repression of *PNP*, which is a potential oncogene, resulted in decreased proliferation, migration, and invasion rates (Kojima et al. 2012). In addition to *miR-34a*, *miR-373*, and *miR-520c*, in a recent study, *miR-708* has also found to be directly targeting *CD44*. This same study also revealed that *miR-708* targets a serine/threonine kinase *AKT2*. Elevated levels of *CD44* and *AKT2* as a result of reduced *miR-708* have been proposed to result in PCa initiation, progression, and development (Saini et al. 2012). A known TSmiR, *miR-203*, which is important for PCa cell proliferation, invasion, and migration, has several validated targets such as cytoskeleton-associated protein 2 (*CKAP2*); LIM and SH3 protein 1 (*LASP1*); baculovirol IAP repeat containing 5 (*BIRC5*); WAS protein family, member 1 (*WASF1*); ArfGAP with SH3 domain, ankyrin repeat, and PH domain 1 (*ASAP1*); and runt-related transcription factor 2 (*RUNX2*) (Viticchie et al. 2011). *miR-143* acts as a TSmiR in PCa, and it is associated with cell proliferation and growth. The function of *miR-143* has been proposed to be partly mediated through inhibition of extracellular signal-regulated kinase 5 (*ERK5*), which is a direct target of *miR-143* (Clape et al. 2009). Additionally, its down-regulation has been associated with increased PCa cell proliferation, migration, and resistance to chemotherapeutics via activating KRAS and subsequent MAPK pathway (Xu et al. 2011).

Validated targets of known TSmiRs and oncomirs in PCa are summarized in Table 18.1.

V. miRNAs AS POTENTIAL MARKERS FOR DIAGNOSIS AND PROGRESSION OF PCa

Although PCa is highly prevalent among men, its early detection still presents significant problems as currently available diagnostic techniques lack high specificity and sensitivity. Current tests for early PCa detection include digital rectal examinations and PSA, which both have restricted diagnostic value. Moreover, recently several new markers including

TABLE 18.1. The List of MicroRNAs (miRNAs) and Their Biologically Verified Targets in PCa

miRNA	Target Gene
let-7	HMGA2, RAS, Myc
miR-1, miR-133a	PNP
miR-21	PTEN, PDCD4, TPM1, Maspin, MARCKS, RECK
miR-27a	Prohibitin
miR-32	BTG2
miR-34, miR-373, miR-520c, miR-708	CD44
miR-101	EZH2
miR-106a	RB1
miR-125b	Bak1, p53, PUMA, EIF4EBP1
miR-143	ERK5
miR-145	BNIP3, CCNA2
miR-203	CKAP2, LASP1, BIRC5, WASF1, ASAP1, RUNX2, ZEB2, Bmi
miR-205	IL-24, IL-32
miR-221/222	p27, p57, PTEN, ARH1
miR-375	Sec23A
miR-449a	HDAC1
miR-708	AKT2

human glandular kallikrein 2 (hK2), urokinase plasminogen activator (uPA) and its receptor (uPAR), transforming growth factor-beta 1 (TGF-β1), and interleukin-6 (IL-6) and its receptor (IL-6R) have been suggested as biomarkers alone or in combination with PSA for monitoring PCa (Srivastava et al. 2011). However, there is still need for a highly specific and sensitive biomarker that will diagnose patients experiencing PCa in its early stages.

miRNAs are promising diagnostic markers for PCa. A recent study revealed five circulating miRNAs that may differentiate PCa from BPH and healthy controls. Of these miRNAs, *let-7e*, *let-7c*, and *miR-30c* were down-regulated, while *miR-622* and *miR-1285* were up-regulated in the sera of PCa patients. The combination of these miRNAs has been suggested as a much more specific and sensitive diagnostic tool when they are used along with PSA (Chen et al. 2012). *miR-107* and *miR-574-3p* have also been shown to be a good predictive miRNA combination that are represented significantly in high concentrations in the urine of PCa patients compared with controls (Bryant et al. 2012). Elevated levels of *miR-16*, *miR-195*, and *miR-let-7i* have been detected in PCa patients' sera compared with that of controls, which gives the opportunity for non-invasive discrimination of PCa from BPH (Mahn et al. 2011). *miR-20a*, *miR-21*, *miR-145*, and *miR-221* have been shown to have diagnostic value in terms of distinguishing high-risk tumors from low-risk tumors in PCa. In the sera of patients with stage 3 tumors, *miR-20a* is up-regulated when compared with patients with stage 2 or below tumors. Elevated levels of *miR-20a* along with *miR-21* have been detected in patients with high-risk PCa (Shen et al. 2012). *miR-125b* and *miR-141* have also been shown to be up-regulated in the circulating sera of PCa patients with metastasis in comparison with those of healthy people (Pang et al. 2010). In a different study, various miRNAs including *miR-375*, *miR-9**, *miR-141*, *miR-200b*, and *miR-516a-3p* were detected as high in the sera of patients with metastasis. Of these miRNAs, *miR-141* and *miR-375* have shown to be prevalently pronounced among others for high-risk tumors (Brase et al. 2011).

VI. miRNAs MODULATING TUMOR PROGRESSION AND METASTASIS IN PCa

Along with their oncogenic and TSmiR properties, miRNAs are significantly involved in tumor progression and metastasis. Although the mechanisms are not fully clarified, some miRNAs have been identified as important effectors in metastatic pathways.

The CD44 positive (CD44+) subpopulation in PCa is enriched with cells that have cancer stem cell properties such as high clonogenicity (Collins et al. 2005) with tumor-initiating and metastatic capacity (Patrawala et al. 2006). In a recent study, it has been shown that *miR-34a*, which is transcriptionally activated by *p53* (He et al. 2007), directly targets *CD44* and represses its expression. It is down-regulated in CD44+ cells extracted from xenografts and primary tumors, resulting in an increased state of metastatic potential (Liu et al. 2011). *miR-373* and *miR-520c*, which are members of the same miRNA family (Huang et al. 2008a), have also been reported to suppress CD44 in protein level in PCa through preventing its translation. They have reduced expression levels in PCa cell lines and tissues, which correlate with increased tumor invasion and metastatic potential (Yang et al. 2009). In a recent study, *miR-708* was also reported to be underexpressed in CD44+ cells from PCa xenografts, and *CD44* has been validated as a direct target for *miR-708* (Saini et al. 2012).

Another important miRNA in tumor progression and metastasis is *miR-221*, which is underexpressed in metastatic tissue in comparison with their primary carcinoma tissue (Spahn et al. 2010). It has also been suggested as a prognostic marker for high-risk PCa (Spahn et al. 2010). Additionally, *miR-203* is significantly down-regulated miRNA in bone metastatic PCa. It progressively disappears in advanced metastatic PCa and is involved in the regulation of a number of prometastatic genes such as *ZEB2*, *Bmi*, and *Runx*, which is a master regulator of bone metastasis (Saini et al. 2011). In another study utilizing next-generation sequencing, several miRNAs including *miR-16*, *miR-34a*, *miR-126**, *miR-145*, and *miR-205*, which are already associated with PCa metastasis, have been identified as potential markers for the metastatic status of PCa (Watahiki et al. 2011). *let-7c*, *miR-100*, and *miR-218* have been found to be differentially expressed in high-grade localized PCa with respect to metastatic carcinoma (Leite et al. 2011).

VII. miRNAs THAT PLAY A ROLE IN THE EPIGENETIC MACHINERY OF PCa PATHOGENESIS

The modifications of DNA and related histone proteins, which do not affect the DNA sequence itself, are referred to as epigenetic changes. In addition to genetic changes, epigenetic alterations such as DNA hypermethylation, loss of imprinting, and altered histone modification patterns, are implicated in tumorigenesis in general and PCa in particular. Hypermethylation of the CpG islands in the promoter regions of several genes, which have mostly tumor suppressor properties in PCa, has been reported to promote tumorigenesis (Paone et al. 2011). In addition to genes, the promoter regions of miRNAs can be hypermethylated. *miR-145*, a significantly down-regulated miRNA in PCa, is silenced through DNA hypermethylation at its promoter region. Wild-type p53 can bind effectively to the p53 responsive element upstream of *miR-145* and up-regulate its expression, whereas the hypermethylation status or the existence of mutated p53 prevents the binding of p53 to the *miR-145* promoter. Therefore, it is postulated that in PCa, *miR-145* expression levels might be affected through hypermethylation of the promoter and p53 mutation pathways (Suh et al. 2011).

Additionally, histones on *miR-205* locus of PCa cells have been reported to be trimethylated on H3K27, while the methylation mark is lost on H3K4, leading to *miR-205* repression (Ke et al. 2009). Hypermethylation of *miR-205* promoter and acquisition of repressive chromatin modifications have also been reported in other cancer types including invasive bladder tumors and undifferentiated bladder cell lines (Wiklund et al. 2011).

Furthermore, deregulation of the regulators of epigenetic mechanisms is also associated with pathogenesis, proliferative capacity, and aggressiveness of PCa cells. *miR-29* family members in PCa cells were reported to target the DNA methyl transferases either directly or indirectly, which results in the reduction of global methylation pattern and the activation of epigenetically silenced tumor suppressor genes (Fabbri et al. 2007). Another example of deregulated miRNAs that disturb the epigenetic mechanisms in PCa is *miR-101*. It is frequently down-regulated in PCa (Lu et al. 2005) and has been shown to target and negatively regulate enhancer of zeste homologue 2 (*EZH2*) (Cao et al. 2010), which plays a role in gene silencing through incorporation into a multiprotein complex, polycomb repressive complex 2. Overexpression of *EZH2* in PCa has been demonstrated to increase cell proliferation, colony formation, and tumor invasion *in vitro* and *in vivo* (Varambally et al. 2008). Similarly, *miR-449a* is also down-regulated in PCa in parallel with the up-regulation of histone deacetylase 1 (*HDAC1*), which has been shown to be directly targeted by *miR-449a* (Noonan et al. 2009). *p27* is known to be targeted by *HDAC1*, and loss of tumorigenic properties upon ectopic *miR-449a* expression in PCa cell line PC3 is phenocopied in *p27* overexpressed cells (Noonan et al. 2009).

Despite the few reports that have been published suggesting the potential roles of deregulated miRNAs involved in the epigenetic machinery, it is evident that many more remain to be discovered with relevance to PCa pathogenesis.

VIII. miRNAs AS IMPLICATED IN PCa THERAPY AND THEIR FUTURE POTENTIAL

In the last decade, it has been noted that miRNAs are important players in cancer pathogenesis either as genetic or epigenetic regulators. Although they have a wide range of targets because of their partial complementarity, scientists and clinicians aim to use them in therapeutic applications against cancer. Clarifying the mechanism of actions, their true targets, and precise expression information is needed to develop efficient and practical therapies. Until now, there are various miRNAs that are suggested as potential therapeutic agents for PCa. For example, reconstitution of *let-7* expression has been proposed through using either lenti- or adenoviruses or transient transfection of *let-7* precursors, which may decrease the survival and proliferation of tumor cells (Barh et al. 2010; Nadiminty et al. 2012). In another report, a chemical, 3,3′-diindolylmethane (DIM), has been utilized in patients prior to radical prostatectomy to analyze its therapeutic role. The ongoing phase II clinical trial has shown that DIM intervention resulted in increased levels of *let-7* and down-regulation of *EZH2*, which is a direct target of *let-7* in PCa. DIM-introduced tumor samples had low self-renewal and clonogenic capacity (Kong et al. 2012). This result confirmed the potential benefits of therapeutic approaches targeting deregulated miRNAs in PCa. Moreover, *ARHI*, a tumor suppressor gene, is down-regulated in PCa as a result of overexpression of *miR-221* and *miR-222*. Genistein, a non-toxic chemotherapeutic agent has been shown to restore ARHI expression levels, and it has been suggested as a dietary therapeutic agent for PCa treatment (Chen et al. 2011). *miR-29b* (Steele et al. 2010), *miR-101* (Cao et al. 2010; Hao et al. 2011), *miR-145* (Ozen et al. 2008; Chen et al. 2010 and our unpublished data), and several other miRNAs have also been implicated as having therapeutic potential, although more efforts should be given to accomplish a successful therapy with miRNAs.

IX. CONCLUSION

Now more than 1000 miRNAs have been identified that are expressed in *Homo sapiens* (Kozomara and Griffiths-Jones 2011), and they are thought to target and modulate the expression of at least 60% of human genes. Their aberrant expression is involved in many diseases including PCa. miRNA expression profiling remains very important for enlightening the pathogenesis of PCa. miRNAs are also proposed as biomarkers for the early detection and diagnosis of PCa due to their differential expression in tumor tissues and stability in patient blood samples. Although most of the identified miRNA functions still remain to be resolved, they are promising candidates for novel therapeutic applications against PCa.

ACKNOWLEDGMENT

Part of the work presented in this chapter is supported by a grant (108S051) from The Scientific and Technological Research Council of Turkey (TUBITAK).

REFERENCES

Baffa, R., M. Fassan, S. Volinia, B. O'Hara, C.G. Liu, J.P. Palazzo, M. Gardiman, M. Rugge, L.G. Gomella, C.M. Croce, and A. Rosenberg. 2009. MicroRNA expression profiling of human metastatic cancers identifies cancer gene targets. *J Pathol* **219**:214–221.

Barh, D., R. Malhotra, B. Ravi, and P. Sindhurani. 2010. MicroRNA *let-7*: an emerging next-generation cancer therapeutic. *Curr Oncol* **17**:70–80.

Bhatnagar, N., X. Li, S.K. Padi, Q. Zhang, M.S. Tang, and B. Guo. 2010. Downregulation of *miR-205* and *miR-31* confers resistance to chemotherapy-induced apoptosis in prostate cancer cells. *Cell Death Dis* **1**:e105.

Bloomston, M., W.L. Frankel, F. Petrocca, S. Volinia, H. Alder, J.P. Hagan, C.G. Liu, D. Bhatt, C. Taccioli, and C.M. Croce. 2007. MicroRNA expression patterns to differentiate pancreatic adenocarcinoma from normal pancreas and chronic pancreatitis. *JAMA* **297**:1901–1908.

Boll, K., K. Reiche, K. Kasack, N. Morbt, A.K. Kretzschmar, J.M. Tomm, G. Verhaegh, J. Schalken, M. von Bergen, F. Horn, and J. Hackermuller. 2012. *MiR-130a*, *miR-203* and *miR-205* jointly repress key oncogenic pathways and are downregulated in prostate carcinoma. *Oncogene* **32**:277–285.

Brase, J.C., M. Johannes, T. Schlomm, M. Falth, A. Haese, T. Steuber, T. Beissbarth, R. Kuner, and H. Sultmann. 2011. Circulating miRNAs are correlated with tumor progression in prostate cancer. *Int J Cancer* **128**:608–616.

Bryant, R.J., T. Pawlowski, J.W. Catto, G. Marsden, R.L. Vessella, B. Rhees, C. Kuslich, T. Visakorpi, and F.C. Hamdy. 2012. Changes in circulating microRNA levels associated with prostate cancer. *Br J Cancer* **106**:768–774.

Calin, G.A., C. Sevignani, C.D. Dumitru, T. Hyslop, E. Noch, S. Yendamuri, M. Shimizu, S. Rattan, F. Bullrich, M. Negrini, and C.M. Croce. 2004. Human microRNA genes are frequently located at fragile sites and genomic regions involved in cancers. *Proc Natl Acad Sci U S A* **101**:2999–3004.

Cao, P., Z. Deng, M. Wan, W. Huang, S.D. Cramer, J. Xu, M. Lei, and G. Sui. 2010. MicroRNA-101 negatively regulates Ezh2 and its expression is modulated by androgen receptor and HIF-1alpha/HIF-1beta. *Mol Cancer* **9**:108.

Chan, J.A., A.M. Krichevsky, and K.S. Kosik. 2005. MicroRNA-21 is an antiapoptotic factor in human glioblastoma cells. *Cancer Res* **65**:6029–6033.

Chang, T.C., L.R. Zeitels, H.W. Hwang, R.R. Chivukula, E.A. Wentzel, M. Dews, J. Jung, P. Gao, C.V. Dang, M.A. Beer, A. Thomas-Tikhonenko, and J.T. Mendell. 2009. Lin-28B transactivation is necessary for Myc-mediated *let-7* repression and proliferation. *Proc Natl Acad Sci U S A* **106**:3384–3389.

Chen, X., J. Gong, H. Zeng, N. Chen, R. Huang, Y. Huang, L. Nie, M. Xu, J. Xia, F. Zhao, W. Meng, and Q. Zhou. 2010. MicroRNA145 targets BNIP3 and suppresses prostate cancer progression. *Cancer Res* **70**:2728–2738.

Chen, Y., M.S. Zaman, G. Deng, S. Majid, S. Saini, J. Liu, Y. Tanaka, and R. Dahiya. 2011. MicroRNAs 221/222 and genistein-mediated regulation of ARHI tumor suppressor gene in prostate cancer. *Cancer Prev Res (Phila)* **4**:76–86.

Chen, Z.H., G.L. Zhang, H.R. Li, J.D. Luo, Z.X. Li, G.M. Chen, and J. Yang. 2012. A panel of five circulating microRNAs as potential biomarkers for prostate cancer. *Prostate* **72**:1443–1452.

REFERENCES

Childs, G., M. Fazzari, G. Kung, N. Kawachi, M. Brandwein-Gensler, M. McLemore, Q. Chen, R.D. Burk, R.V. Smith, M.B. Prystowsky, T.J. Belbin, and N.F. Schlecht. 2009. Low-level expression of microRNAs *let-7d* and *miR-205* are prognostic markers of head and neck squamous cell carcinoma. *Am J Pathol* **174**:736–745.

Chiosea, S., E. Jelezcova, U. Chandran, M. Acquafondata, T. McHale, R.W. Sobol, and R. Dhir. 2006. Up-regulation of dicer, a component of the microRNA machinery, in prostate adenocarcinoma. *Am J Pathol* **169**:1812–1820.

Clape, C., V. Fritz, C. Henriquet, F. Apparailly, P.L. Fernandez, F. Iborra, C. Avances, M. Villalba, S. Culine, and L. Fajas. 2009. *miR-143* interferes with ERK5 signaling, and abrogates prostate cancer progression in mice. *PLoS ONE* **4**:e7542.

Cole, K.A., E.F. Attiyeh, Y.P. Mosse, M.J. Laquaglia, S.J. Diskin, G.M. Brodeur, and J.M. Maris. 2008. A functional screen identifies *miR-34a* as a candidate neuroblastoma tumor suppressor gene. *Mol Cancer Res* **6**:735–742.

Collins, A.T., P.A. Berry, C. Hyde, M.J. Stower, and N.J. Maitland. 2005. Prospective identification of tumorigenic prostate cancer stem cells. *Cancer Res* **65**:10946–10951.

Dahiya, R., J. McCarville, C. Lee, W. Hu, G. Kaur, P. Carroll, and G. Deng. 1997. Deletion of chromosome 11p15, p12, q22, q23-24 loci in human prostate cancer. *Int J Cancer* **72**:283–288.

Damber, J.E. and G. Aus. 2008. Prostate cancer. *Lancet* **371**:1710–1721.

Dillhoff, M., J. Liu, W. Frankel, C. Croce, and M. Bloomston. 2008. MicroRNA-21 is overexpressed in pancreatic cancer and a potential predictor of survival. *J Gastrointest Surg* **12**:2171–2176.

Ellsworth, R.E., A. Vertrees, B. Love, J.A. Hooke, D.L. Ellsworth, and C.D. Shriver. 2008. Chromosomal alterations associated with the transition from in situ to invasive breast cancer. *Ann Surg Oncol* **15**:2519–2525.

Fabbri, M., R. Garzon, A. Cimmino, Z. Liu, N. Zanesi, E. Callegari, S. Liu, H. Alder, S. Costinean, C. Fernandez-Cymering, S. Volinia, G. Guler, et al. 2007. MicroRNA-29 family reverts aberrant methylation in lung cancer by targeting DNA methyltransferases 3A and 3B. *Proc Natl Acad Sci U S A* **104**:15805–15810.

Filipowicz, W., S.N. Bhattacharyya, and N. Sonenberg. 2008. Mechanisms of post-transcriptional regulation by microRNAs: are the answers in sight? *Nat Rev Genet* **9**:102–114.

Fletcher, C.E., D.A. Dart, A. Sita-Lumsden, H. Cheng, P.S. Rennie, and C.L. Bevan. 2012. Androgen-regulated processing of the oncomir *MiR-27a*, which targets Prohibitin in prostate cancer. *Hum Mol Genet* **21**:3112–3127.

Folini, M., P. Gandellini, N. Longoni, V. Profumo, M. Callari, M. Pennati, M. Colecchia, R. Supino, S. Veneroni, R. Salvioni, R. Valdagni, M.G. Daidone, et al. 2010. *miR-21*: an oncomir on strike in prostate cancer. *Mol Cancer* **9**:12.

Galardi, S., N. Mercatelli, E. Giorda, S. Massalini, G.V. Frajese, S.A. Ciafre, and M.G. Farace. 2007. *miR-221* and *miR-222* expression affects the proliferation potential of human prostate carcinoma cell lines by targeting p27Kip1. *J Biol Chem* **282**:23716–23724.

Galardi, S., N. Mercatelli, M.G. Farace, and S.A. Ciafre. 2011. NF-kB and c-Jun induce the expression of the oncogenic *miR-221* and *miR-222* in prostate carcinoma and glioblastoma cells. *Nucleic Acids Res* **39**:3892–3902.

Garofalo, M., G. Di Leva, G. Romano, G. Nuovo, S.S. Suh, A. Ngankeu, C. Taccioli, F. Pichiorri, H. Alder, P. Secchiero, P. Gasparini, A. Gonelli, et al. 2009. *miR-221&222* regulate TRAIL resistance and enhance tumorigenicity through PTEN and TIMP3 downregulation. *Cancer Cell* **16**:498–509.

Hagman, Z., O. Larne, A. Edsjo, A. Bjartell, R.A. Ehrnstrom, D. Ulmert, H. Lilja, and Y. Ceder. 2010. *miR-34c* is downregulated in prostate cancer and exerts tumor suppressive functions. *Int J Cancer* **127**:2768–2776.

Hao, Y., X. Gu, Y. Zhao, S. Greene, W. Sha, D.T. Smoot, J. Califano, T.C. Wu, and X. Pang. 2011. Enforced expression of *miR-101* inhibits prostate cancer cell growth by modulating the COX-2 pathway in vivo. *Cancer Prev Res (Phila)* **4**:1073–1083.

He, L., X. He, L.P. Lim, E. de Stanchina, Z. Xuan, Y. Liang, W. Xue, L. Zender, J. Magnus, D. Ridzon, A.L. Jackson, P.S. Linsley, et al. 2007. A microRNA component of the p53 tumour suppressor network. *Nature* **447**:1130–1134.

Helland, A., M.S. Anglesio, J. George, P.A. Cowin, C.N. Johnstone, C.M. House, K.E. Sheppard, D. Etemadmoghadam, N. Melnyk, A.K. Rustgi, W.A. Phillips, H. Johnsen, et al. 2011. Deregulation of MYCN, LIN28B and LET7 in a molecular subtype of aggressive high-grade serous ovarian cancers. *PLoS ONE* **6**:e18064.

Huang, Q., K. Gumireddy, M. Schrier, C. le Sage, R. Nagel, S. Nair, D.A. Egan, A. Li, G. Huang, A.J. Klein-Szanto, P.A. Gimotty, D. Katsaros, et al. 2008a. The microRNAs *miR-373* and *miR-520c* promote tumour invasion and metastasis. *Nat Cell Biol* **10**:202–210.

Huang, S., X. He, J. Ding, L. Liang, Y. Zhao, Z. Zhang, X. Yao, Z. Pan, P. Zhang, J. Li, D. Wan, and J. Gu. 2008b. Upregulation of *miR-23a* approximately 27a approximately 24 decreases transforming growth factor-beta-induced tumor-suppressive activities in human hepatocellular carcinoma cells. *Int J Cancer* **123**:972–978.

Iwatsuki, M., K. Mimori, T. Yokobori, H. Ishi, T. Beppu, S. Nakamori, H. Baba, and M. Mori. 2010. Epithelial-mesenchymal transition in cancer development and its clinical significance. *Cancer Sci* **101**:293–299.

Jalava, S.E., A. Urbanucci, L. Latonen, K.K. Waltering, B. Sahu, O.A. Janne, J. Seppala, H. Lahdesmaki, T.L. Tammela, and T. Visakorpi. 2012. Androgen-regulated *miR-32* targets BTG2 and is overexpressed in castration-resistant prostate cancer. *Oncogene* **31**:4460–4471.

John, B., A.J. Enright, A. Aravin, T. Tuschl, C. Sander, and D.S. Marks. 2004. Human microRNA targets. *PLoS Biol* **2**:e363.

Johnson, S.M., H. Grosshans, J. Shingara, M. Byrom, R. Jarvis, A. Cheng, E. Labourier, K.L. Reinert, D. Brown, and F.J. Slack. 2005. RAS is regulated by the *let-7* microRNA family. *Cell* **120**:635–647.

Ke, X.S., Y. Qu, K. Rostad, W.C. Li, B. Lin, O.J. Halvorsen, S.A. Haukaas, I. Jonassen, K. Petersen, N. Goldfinger, V. Rotter, L.A. Akslen, et al. 2009. Genome-wide profiling of histone h3 lysine 4 and lysine 27 trimethylation reveals an epigenetic signature in prostate carcinogenesis. *PLoS ONE* **4**:e4687.

Kojima, S., T. Chiyomaru, K. Kawakami, H. Yoshino, H. Enokida, N. Nohata, M. Fuse, T. Ichikawa, Y. Naya, M. Nakagawa, and N. Seki. 2012. Tumour suppressors *miR-1* and *miR-133a* target the oncogenic function of purine nucleoside phosphorylase (PNP) in prostate cancer. *Br J Cancer* **106**:405–413.

Kong, D., E. Heath, W. Chen, M.L. Cher, I. Powell, L. Heilbrun, Y. Li, S. Ali, S. Sethi, O. Hassan, C. Hwang, N. Gupta, et al. 2012. Loss of *let-7* up-regulates EZH2 in prostate cancer consistent with the acquisition of cancer stem cell signatures that are attenuated by BR-DIM. *PLoS ONE* **7**:e33729.

Kozomara, A. and S. Griffiths-Jones. 2011. miRBase: integrating microRNA annotation and deep-sequencing data. *Nucleic Acids Res* **39**:D152–D157.

Kumar, M.S., J. Lu, K.L. Mercer, T.R. Golub, and T. Jacks. 2007. Impaired microRNA processing enhances cellular transformation and tumorigenesis. *Nat Genet* **39**:673–677.

Landi, M.T., Y. Zhao, M. Rotunno, J. Koshiol, H. Liu, A.W. Bergen, M. Rubagotti, A.M. Goldstein, I. Linnoila, F.M. Marincola, M.A. Tucker, P.A. Bertazzi, et al. 2010. MicroRNA expression differentiates histology and predicts survival of lung cancer. *Clin Cancer Res* **16**:430–441.

Lee, Y.S. and A. Dutta. 2007. The tumor suppressor microRNA *let-7* represses the HMGA2 oncogene. *Genes Dev* **21**:1025–1030.

Lee, Y.S., H.K. Kim, S. Chung, K.S. Kim, and A. Dutta. 2005. Depletion of human micro-RNA *miR-125b* reveals that it is critical for the proliferation of differentiated cells but not for the down-regulation of putative targets during differentiation. *J Biol Chem* **280**:16635–16641.

REFERENCES

Leite, K.R., J.M. Sousa-Canavez, S.T. Reis, A.H. Tomiyama, L.H. Camara-Lopes, A. Sanudo, A.A. Antunes, and M. Srougi. 2011. Change in expression of *miR-let7c*, *miR-100*, and *miR-218* from high grade localized prostate cancer to metastasis. *Urol Oncol* **29**:265–269.

Li, T., D. Li, J. Sha, P. Sun, and Y. Huang. 2009. MicroRNA-21 directly targets MARCKS and promotes apoptosis resistance and invasion in prostate cancer cells. *Biochem Biophys Res Commun* **383**:280–285.

Liang, Y. 2008. An expression meta-analysis of predicted microRNA targets identifies a diagnostic signature for lung cancer. *BMC Med Genomics* **1**:61.

Liu, C., K. Kelnar, B. Liu, X. Chen, T. Calhoun-Davis, H. Li, L. Patrawala, H. Yan, C. Jeter, S. Honorio, J.F. Wiggins, A.G. Bader, et al. 2011. The microRNA *miR-34a* inhibits prostate cancer stem cells and metastasis by directly repressing CD44. *Nat Med* **17**:211–215.

Lu, J., G. Getz, E.A. Miska, E. Alvarez-Saavedra, J. Lamb, D. Peck, A. Sweet-Cordero, B.L. Ebert, R.H. Mak, A.A. Ferrando, J.R. Downing, T. Jacks, et al. 2005. MicroRNA expression profiles classify human cancers. *Nature* **435**:834–838.

Lu, Z., M. Liu, V. Stribinskis, C.M. Klinge, K.S. Ramos, N.H. Colburn, and Y. Li. 2008. MicroRNA-21 promotes cell transformation by targeting the programmed cell death 4 gene. *Oncogene* **27**: 4373–4379.

Mahn, R., L.C. Heukamp, S. Rogenhofer, A. von Ruecker, S.C. Muller, and J. Ellinger. 2011. Circulating microRNAs (miRNA) in serum of patients with prostate cancer. *Urology* **77**:1265 e1269–1265 e1216.

Majid, S., A.A. Dar, S. Saini, S. Yamamura, H. Hirata, Y. Tanaka, G. Deng, and R. Dahiya. 2010. MicroRNA-205-directed transcriptional activation of tumor suppressor genes in prostate cancer. *Cancer* **116**:5637–5649.

Mallory, A.C. and H. Vaucheret. 2006. Functions of microRNAs and related small RNAs in plants. *Nat Genet* **38 Suppl**:S31–S36.

Mertens-Talcott, S.U., S. Chintharlapalli, X. Li, and S. Safe. 2007. The oncogenic microRNA-27a targets genes that regulate specificity protein transcription factors and the G2-M checkpoint in MDA-MB-231 breast cancer cells. *Cancer Res* **67**:11001–11011.

Nadiminty, N., R. Tummala, W. Lou, Y. Zhu, X.B. Shi, J.X. Zou, H. Chen, J. Zhang, X. Chen, J. Luo, R.W. deVere White, H.J. Kung, et al. 2012. MicroRNA *let-7c* is downregulated in prostate cancer and suppresses prostate cancer growth. *PLoS ONE* **7**:e32832.

Nam, E.J., H. Yoon, S.W. Kim, H. Kim, Y.T. Kim, J.H. Kim, J.W. Kim, and S. Kim. 2008. MicroRNA expression profiles in serous ovarian carcinoma. *Clin Cancer Res* **14**:2690–2695.

Noonan, E.J., R.F. Place, D. Pookot, S. Basak, J.M. Whitson, H. Hirata, C. Giardina, and R. Dahiya. 2009. *miR-449a* targets HDAC-1 and induces growth arrest in prostate cancer. *Oncogene* **28**:1714–1724.

Ozen, M., C.J. Creighton, M. Ozdemir, and M. Ittmann. 2008. Widespread deregulation of microRNA expression in human prostate cancer. *Oncogene* **27**:1788–1793.

Pang, Y., C.Y. Young, and H. Yuan. 2010. MicroRNAs and prostate cancer. *Acta Biochim Biophys Sin (Shanghai)* **42**:363–369.

Paone, A., R. Galli, and M. Fabbri. 2011. MicroRNAs as new characters in the plot between epigenetics and prostate cancer. *Front Genet* **2**:62.

Patrawala, L., T. Calhoun, R. Schneider-Broussard, H. Li, B. Bhatia, S. Tang, J.G. Reilly, D. Chandra, J. Zhou, K. Claypool, L. Coghlan, and D.G. Tang. 2006. Highly purified CD44+ prostate cancer cells from xenograft human tumors are enriched in tumorigenic and metastatic progenitor cells. *Oncogene* **25**:1696–1708.

Pineau, P., S. Volinia, K. McJunkin, A. Marchio, C. Battiston, B. Terris, V. Mazzaferro, S.W. Lowe, C.M. Croce, and A. Dejean. 2010. *miR-221* overexpression contributes to liver tumorigenesis. *Proc Natl Acad Sci U S A* **107**:264–269.

Porkka, K.P., M.J. Pfeiffer, K.K. Waltering, R.L. Vessella, T.L. Tammela, and T. Visakorpi. 2007. MicroRNA expression profiling in prostate cancer. *Cancer Res* **67**:6130–6135.

Rusio, D., M. Negrini, G. Manenti, T.A. Dragani, and C.M. Croce. 1995. Loss of heterozygosity at chromosome 11q in lung adenocarcinoma: identification of three independent regions. *Cancer Res* **55**:3988–3991.

Reis, S.T., J. Pontes-Junior, A.A. Antunesnes, M.F. Dall Oglio, N. Dip, C.C. Passerotti, G.A. Rossini, D.R. Morais, A.J. Nesrallah, C. Piantino, M. Srougi, and K.R. Leite. 2012. *miR-21* may acts as an oncomir by targeting RECK, a matrix metalloproteinase regulator, in prostate cancer. *BMC Urol* **12**:14.

Saini, S., S. Majid, S. Yamamura, L. Tabatabai, S.O. Suh, V. Shahryari, Y. Chen, G. Deng, Y. Tanaka, and R. Dahiya. 2011. Regulatory role of *miR-203* in prostate cancer progression and metastasis. *Clin Cancer Res* **17**:5287–5298.

Saini, S., S. Majid, V. Shahryari, S. Arora, S. Yamamura, I. Chang, M.S. Zaman, G. Deng, Y. Tanaka, and R. Dahiya. 2012. miRNA-708 control of CD44+ prostate cancer-initiating cells. *Cancer Res* **72**:3618–3630.

Sevli, S., A. Uzumcu, M. Solak, M. Ittmann, and M. Ozen. 2010. The function of microRNAs, small but potent molecules, in human prostate cancer. *Prostate Cancer Prostatic Dis* **13**:208–217.

Shen, J., G.W. Hruby, J.M. McKiernan, I. Gurvich, M.J. Lipsky, M.C. Benson, and R.M. Santella. 2012. Dysregulation of circulating microRNAs and prediction of aggressive prostate cancer. *Prostate* **72**:1469–1477.

Shi, X.B., L. Xue, J. Yang, A.H. Ma, J. Zhao, M. Xu, C.G. Tepper, C.P. Evans, H.J. Kung, and R.W. deVere White. 2007. An androgen-regulated miRNA suppresses Bak1 expression and induces androgen-independent growth of prostate cancer cells. *Proc Natl Acad Sci U S A* **104**:19983–19988.

Shi, X.B., C.G. Tepper, and R.W. White. 2008. MicroRNAs and prostate cancer. *J Cell Mol Med* **12**:1456–1465.

Shi, X.B., L. Xue, A.H. Ma, C.G. Tepper, H.J. Kung, and R.W. White. 2011. *miR-125b* promotes growth of prostate cancer xenograft tumor through targeting pro-apoptotic genes. *Prostate* **71**:538–549.

Siegel, R., E. Ward, O. Brawley, and A. Jemal. 2011. Cancer statistics, 2011: the impact of eliminating socioeconomic and racial disparities on premature cancer deaths. *CA Cancer J Clin* **61**:212–236.

Sorrentino, A., C.G. Liu, A. Addario, C. Peschle, G. Scambia, and C. Ferlini. 2008. Role of microR-NAs in drug-resistant ovarian cancer cells. *Gynecol Oncol* **111**:478–486.

Spahn, M., S. Kneitz, C.J. Scholz, N. Stenger, T. Rudiger, P. Strobel, H. Riedmiller, and B. Kneitz. 2010. Expression of microRNA-221 is progressively reduced in aggressive prostate cancer and metastasis and predicts clinical recurrence. *Int J Cancer* **127**:394–403.

Srivastava, A., S. Suy, S.P. Collins, and D. Kumar. 2011. Circulating microRNA as biomarkers: an update in prostate cancer. *Mol Cell Pharmacol* **3**:115–124.

Steele, R., J.L. Mott, and R.B. Ray. 2010. MBP-1 upregulates *miR-29b* that represses Mcl-1, collagens, and matrix-metalloproteinase-2 in prostate cancer cells. *Genes Cancer* **1**:381–387.

Suh, S.O., Y. Chen, M.S. Zaman, H. Hirata, S. Yamamura, V. Shahryari, J. Liu, Z.L. Tabatabai, S. Kakar, G. Deng, Y. Tanaka, and R. Dahiya. 2011. MicroRNA-145 is regulated by DNA methylation and p53 gene mutation in prostate cancer. *Carcinogenesis* **32**:772–778.

Szczyrba, J., E. Nolte, S. Wach, E. Kremmer, R. Stohr, A. Hartmann, W. Wieland, B. Wullich, and F.A. Grasser. 2011. Downregulation of Sec23A protein by miRNA-375 in prostate carcinoma. *Mol Cancer Res* **9**:791–800.

Thiery, J.P., H. Acloque, R.Y. Huang, and M.A. Nieto. 2009. Epithelial-mesenchymal transitions in development and disease. *Cell* **139**:871–890.

Tong, A.W., P. Fulgham, C. Jay, P. Chen, I. Khalil, S. Liu, N. Senzer, A.C. Eklund, J. Han, and J. Nemunaitis. 2009. MicroRNA profile analysis of human prostate cancers. *Cancer Gene Ther* **16**:206–216.

REFERENCES

Toyota, M., H. Suzuki, Y. Sasaki, R. Maruyama, K. Imai, Y. Shinomura, and T. Tokino. 2008. Epigenetic silencing of microRNA-34b/c and B-cell translocation gene 4 is associated with CpG island methylation in colorectal cancer. *Cancer Res* **68**:4123–4132.

Ueda, T., S. Volinia, H. Okumura, M. Shimizu, C. Taccioli, S. Rossi, H. Alder, C.G. Liu, N. Oue, W. Yasui, K. Yoshida, H. Sasaki, et al. 2010. Relation between microRNA expression and progression and prognosis of gastric cancer: a microRNA expression analysis. *Lancet Oncol* **11**:136–146.

Varambally, S., Q. Cao, R.S. Mani, S. Shankar, X. Wang, B. Ateeq, B. Laxman, X. Cao, X. Jing, K. Ramnarayanan, J.C. Brenner, J. Yu, et al. 2008. Genomic loss of microRNA-101 leads to overexpression of histone methyltransferase EZH2 in cancer. *Science* **322**:1695–1699.

Viswanathan, S.R., J.T. Powers, W. Einhorn, Y. Hoshida, T.L. Ng, S. Toffanin, M. O'Sullivan, J. Lu, L.A. Phillips, V.L. Lockhart, S.P. Shah, P.S. Tanwar, et al. 2009. Lin28 promotes transformation and is associated with advanced human malignancies. *Nat Genet* **41**:843–848.

Viticchie, G., A.M. Lena, A. Latina, A. Formosa, L.H. Gregersen, A.H. Lund, S. Bernardini, A. Mauriello, R. Miano, L.G. Spagnoli, R.A. Knight, E. Candi, et al. 2011. *MiR-203* controls proliferation, migration and invasive potential of prostate cancer cell lines. *Cell Cycle* **10**: 1121–1131.

Volinia, S., G.A. Calin, C.G. Liu, S. Ambs, A. Cimmino, F. Petrocca, R. Visone, M. Iorio, C. Roldo, M. Ferracin, R.L. Prueitt, N. Yanaihara, et al. 2006. A microRNA expression signature of human solid tumors defines cancer gene targets. *Proc Natl Acad Sci U S A* **103**:2257–2261.

Wang, Q., Y.C. Li, J. Wang, J. Kong, Y. Qi, R.J. Quigg, and X. Li. 2008. *miR-17-92* cluster accelerates adipocyte differentiation by negatively regulating tumor-suppressor Rb2/p130. *Proc Natl Acad Sci U S A* **105**:2889–2894.

Wang, L., H. Tang, V. Thayanithy, S. Subramanian, A.L. Oberg, J.M. Cunningham, J.R. Cerhan, C.J. Steer, and S.N. Thibodeau. 2009. Gene networks and microRNAs implicated in aggressive prostate cancer. *Cancer Res* **69**:9490–9497.

Watahiki, A., Y. Wang, J. Morris, K. Dennis, H.M. O'Dwyer, M. Gleave, P.W. Gout, and Y. Wang. 2011. MicroRNAs associated with metastatic prostate cancer. *PLoS ONE* **6**:e24950.

Wiklund, E.D., J.B. Bramsen, T. Hulf, L. Dyrskjot, R. Ramanathan, T.B. Hansen, S.B. Villadsen, S. Gao, M.S. Ostenfeld, M. Borre, M.E. Peter, T.F. Orntoft, et al. 2011. Coordinated epigenetic repression of the *miR-200* family and *miR-205* in invasive bladder cancer. *Int J Cancer* **128**:1327–1334.

Xu, B., X. Niu, X. Zhang, J. Tao, D. Wu, Z. Wang, P. Li, W. Zhang, H. Wu, N. Feng, Z. Wang, L. Hua, et al. 2011. *miR-143* decreases prostate cancer cells proliferation and migration and enhances their sensitivity to docetaxel through suppression of KRAS. *Mol Cell Biochem* **350**:207–213.

Yan, L.X., X.F. Huang, Q. Shao, M.Y. Huang, L. Deng, Q.L. Wu, Y.X. Zeng, and J.Y. Shao. 2008. MicroRNA *miR-21* overexpression in human breast cancer is associated with advanced clinical stage, lymph node metastasis and patient poor prognosis. *RNA* **14**:2348–2360.

Yang, K., A.M. Handorean, and K.A. Iczkowski. 2009. MicroRNAs 373 and 520c are downregulated in prostate cancer, suppress CD44 translation and enhance invasion of prostate cancer cells in vitro. *Int J Clin Exp Pathol* **2**:361–369.

Yu, F., H. Yao, P. Zhu, X. Zhang, Q. Pan, C. Gong, Y. Huang, X. Hu, F. Su, J. Lieberman, and E. Song. 2007. *let-7* regulates self renewal and tumorigenicity of breast cancer cells. *Cell* **131**:1109–1123.

Zhang, Z., Z. Li, C. Gao, P. Chen, J. Chen, W. Liu, S. Xiao, and H. Lu. 2008. *miR-21* plays a pivotal role in gastric cancer pathogenesis and progression. *Lab Invest* **88**:1358–1366.

Zheng, C., S. Yinghao, and J. Li. 2012. *MiR-221* expression affects invasion potential of human prostate carcinoma cell lines by targeting DVL2. *Med Oncol* **29**:815–822.

Zhu, S., H. Wu, F. Wu, D. Nie, S. Sheng, and Y.Y. Mo. 2008. MicroRNA-21 targets tumor suppressor genes in invasion and metastasis. *Cell Res* **18**:350–359.

MicroRNA SIGNATURES AS BIOMARKERS OF COLORECTAL CANCER

Katrin Pfütze,[1,2] Xiaoya Luo,[3] and Barbara Burwinkel[1,2]

[1]*Molecular Epidemiology (C080), German Cancer Research Center, Heidelberg, Germany*
[2]*Molecular Biology of Breast Cancer, Department of Obstetrics and Gynecology, University of Heidelberg, Heidelberg, Germany*
[3]*Division of Clinical Epidemiology and Aging Research (C070), German Cancer Research Center, Heidelberg, Germany*

I. Introduction	330
II. Circulating miRNAs in Plasma/Serum of CRC Patients	330
A. Future Directions for Circulating miRNA Analysis	331
III. miRNAs in CRC Tumor Tissue	332
A. Future Directions for miRNA Analysis in CRC Tumor Tissue	336
IV. Conclusion	336
References	337

ABBREVIATIONS

5-FU	5-fluorouracil
CRC	colorectal cancer
FFPE	formalin fixed and paraffin embedded
gFOBT	guaiac-based fecal occult blood test
iFOBT	immunochemical fecal occult blood test
miRNA	microRNA
MSI	microsatellite instable
MSS	microsatellite stable
TNM	tumor–node–metastasis

MicroRNAs in Medicine, First Edition. Edited by Charles H. Lawrie.
© 2014 John Wiley & Sons, Inc. Published 2014 by John Wiley & Sons, Inc.

I. INTRODUCTION

Colorectal cancer (CRC) is the third most common type of cancer in the Western world and the second most frequent cause of cancer-related death. The 5-year survival rate for CRC patients diagnosed at an early, localized stage is about 90% decreasing to 69% if lymph nodes or adjacent organs are affected. When the disease has already spread to distant organs, the 5-year survival is only 12%. However, only about 39% of CRCs are diagnosed in an early stage, in part due to the underuse of screening (Jemal et al. 2008). To date, the most important prognostic factor is pathological staging according to the tumor–node–metastasis (TNM) system (Greene et al. 2004), which designates the treatment of the patients. In contrast, little or no attention is given to additional predictive factors of treatment tolerance and effectiveness. The influence of such factors is supported by different studies, which show that a link exists between tumor aggressiveness, relapse potential, therapy response, and genetics as well as epigenetic alterations in tumor cells (Jankowski and Odze 2009; Walther et al. 2009).

To address these issues and establish new, less invasive screening methods, numerous studies have evaluated miRNAs on the one hand as early CRC detection markers in blood serum or plasma, and on the other hand as potential prognostic and predictive biomarkers in CRC tumor tissues.

Since miRNAs are involved in many cellular pathways, including cell proliferation, differentiation, apoptosis, and immune response, changes in their expression levels are considered as important factors involved in carcinogenesis (for review, see Bartels and Tsongalis 2009). The frequently observed expression alterations in different cancer types support the utilization of miRNAs as biomarker in CRC. To date, several miRNAs are considered as potential biomarkers in CRC, which will be discussed in the following chapter.

II. CIRCULATING miRNAs IN PLASMA/SERUM OF CRC PATIENTS

To date, colonoscopy is still the most reliable method for early CRC detection. Primarily due to its invasive nature and high cost, non-invasive screening tools have been developed as an alternative to colonoscopy. To date, the guaiac-based fecal occult blood test (gFOBT) and immunochemical fecal occult blood test (iFOBT) are the most common tests in use for CRC screening. The sensitivity of these tests for detecting advanced adenomas was 9% for gFOBT and ranged from 25% to 72% for iFOBTs with specificities of 96% and 70–97%, respectively (Whitlock et al. 2008). These are non-invasive and economic tests. However, given the large differences in diagnostic performance among these tests, careful evaluation of the different test variants is important (Hundt et al. 2009).

miRNA signatures may provide improved early detection abilities, since alterations in circulating miRNA signatures analyzed in blood serum or plasma could be observed in several cancer types (Zen and Zhang 2012; Cuk et al. 2013). The origin of circulating miRNAs is not yet fully understood. Although recent reports discuss whether circulating miRNAs results from dead/dying cells, or from secretion of tumor cells or from immunocytes, most likely the signatures reflect all of these aspects. For detailed discussion about these theories, please review Turchinovich et al. (2012).

Based on the minimal invasive accessibility of miRNAs detected in circulation, their stability based on the binding to the Ago2 protein or exosomes, as well as the reproducibility and (interindividual) consistence of miRNA signatures, circulating miRNAs are

TABLE 19.1. Circulating miRNAs as Potential Early Detection and Prognostic Biomarker in Plasma/Serum in CRC

miRNA	Expression[a]	Potential Detection Marker/ Comments	Association with Prognosis	References
miR-17-3p	↑	Reduced after surgery		Ng et al. (2009)
miR-29a	↑	Distinguish CRC patients and advanced adenomas from healthy individuals = early detection marker		Huang et al. (2010)
miR-92	↑	Distinguish CRC patients from healthy individuals/reduced after surgery		Ng et al. (2009); Huang et al. (2010)
miR-141	↑		Advanced stage, poor overall survival	Cheng et al. (2011)
miR-221	↑	Distinguish CRC patients from healthy individuals	Poor overall survival	Pu et al. (2010)

[a]Expression reported in plasma of patients compared with healthy controls: ↑ higher expression level in plasma of CRC patients.

promising biomarker candidates for CRC and other cancer types (Turchinovich et al. 2011).

In three plasma-based studies published, so far, *miR-29a*, *miR-92*, and *miR-221* have been detected as potential early detection markers of CRC (Ng et al. 2009; Huang et al. 2010; Pu et al. 2010). Of these, *miR-92* has been detected in two studies (Ng et al. 2009; Huang et al. 2010). All three miRNAs were significantly higher expressed in CRC patients compared with healthy controls with sensitivity and specificity, indicating a performance comparable with FOBT. Interestingly, the expression of circulating *miR-17-3p* and *miR-92* was lowered after surgery, and there was no association between elevated expression with inflammation and other gastrointestinal cancer (Ng et al. 2009). These observations suggest that these miRNAs are more than likely CRC related (Table 19.1).

Recently, Cheng et al. (2011) proposed circulating *miR-141* as an independent prognostic marker for advanced CRC. They observed a high correlation between its elevated expression and CRC stage and found that expression levels were capable of predicting poor overall survival. Interestingly, in an earlier study, plasma *miR-141* was also found to be a potential marker for metastatic prostate cancer (Mitchell et al. 2008).

A. Future Directions for Circulating miRNA Analysis

Circulating miRNAs offer great hope for the diagnosis and prognosis, and possibly prediction, of CRC. The number of studies assessing miRNA expression in cancer patients in general covering blood plasma and serum has rapidly increased in recent years; however, this field is still in its infancy. Unsurprisingly then, a lot of questions still need to be resolved.

So far, four studies assessing miRNA expression in plasma of CRC patients have been reported; however, only 96 miRNAs (from >2000 human miRNAs) were investigated in these studies (Ng et al. 2009; Huang et al. 2010; Pu et al. 2010; Cheng et al. 2011). In addition to the small fraction of analyzed miRNAs, the results of these studies are conflicting. There are many reasons why this might be the case, including inconsistent blood sample processing, which could be a major source of variability (Duttagupta et al. 2011; McDonald et al. 2011). For instance, blood cell contamination might produce artificially high miRNA concentrations. This can be minimized by a two-step centrifugation protocol consisting of a slow and a high-speed centrifugation step of whole blood or by filtration (McDonald et al. 2011). In future studies, a standardized sample processing and a two-step centrifugation procedure of plasma/serum are advisable.

Second, due to the lack of a standardized normalization strategy and reliable endogenous controls, studies are difficult to compare and the translation of results to the clinic is consequently impaired. For example, many studies use RNU6B as endogenous control. This small nuclear RNA is a component of the spliceosome, which is involved in splicing pre-messenger RNA (mRNA) and assumed to be constitutively expressed. Although RNU6B and miRNAs are similar in size, it is not clear whether purification efficiency is comparable and stable when using blood plasma/serum. Most importantly, RNU6B has different stability than the Ago-bound (and protected) circulating miRNAs, which is an important aspect in the RNase-rich environment of blood plasma/serum (Reddi and Holland 1976). In other studies, the use of *miR-16* as an endogenous normalization control has been suggested. However, due to its predominant expression in red blood cells, its qualification as proper control is questionable (Pritchard et al. 2012). Moreover, miRNAs, which are not predominantly expressed in blood cells, might turn out to be the most robust potential marker, which underlines the importance of knowledge about the origin of the measured miRNAs.

As preparation control, synthetic spiked-in miRNA such as miRNA originating from *Caenorhabditis elegans* has been investigated and used for normalization (Cortez and Calin 2009). So far, the combination of a preparation control and a standardized sample processing within a short time frame after blood collection followed by a two-step centrifugation protocol is probably the best, although far from optimal, way to counter the normalization problem. What is clear, however, is that more studies are necessary to achieve an accurate and reliable normalization strategy. Another limitation in these studies is the sensitivity of detection method itself. Although, quantitative real-time polymerase chain reaction (PCR) is the most frequently used, sensitive, and reproducible method to quantify gene expression, its accuracy is limited if miRNAs' expression is very low, which is often the case in serum/plasma samples. It will be interesting to evaluate to what extent next-generation sequencing approaches will contribute to the identification of novel miRNA/small RNAs not yet represented on array platforms.

To validate the usability of circulating miRNAs as early detection markers for CRC, compare them directly with gFOBT and iFOBT and in study cohorts including a large number of patients with preneoplastic lesions and adenomas. Analogously, to evaluate the prognostic value of circulating miRNAs in CRC, a direct comparison with existing prognostic markers such as circulating tumor cells (CTCs), a Food and Drug Administration (FDA)-approved test to estimate the prognosis of metastatic CRC cases, is needed.

III. miRNAs IN CRC TUMOR TISSUE

The expression of several miRNAs has been found to be altered in precancerous lesions, adenomas, and colorectal tumor tissue in comparison with normal colon tissue, and several miRNAs have been suggested as potential biomarkers in CRC tumor tissue (Table 19.2).

After surgery, the tumor tissue is often formalin fixed and paraffin embedded (FFPE) for the histological classification by the pathologist. This fixation allows long-term storage

TABLE 19.2. Prognostic and Predictive Biomarker in CRC Tumor Tissue

miRNA	Expression[a]	Association with Prognosis	Association with Response to Given Therapy	References
let-7a	↑	Metastasis		Vickers et al. (2012)
let-7g	↑		Clinical response to S-1	Nakajima et al. (2006); Hummel et al. (2010)
miR-10b	↑	Lymphatic invasion and shortened survival	Chemoresistance to 5-FU	Nishida et al. (2012)
miR-17-5p	↑	Shortened disease-free survival at early CRC stages (I and II)		Diaz et al. (2008)
miR-18a	↑	Poorer prognosis	Chemoresistance to 5-FU	Motoyama et al. (2009); Kurokawa et al. (2012)
miR-19b	↑		Chemoresistance to 5-FU	Kurokawa et al. (2012)
miR-20a	↑	Poorer prognosis	Chemoresistance to 5-FU, oxaliplatin, and teniposide	Schetter et al. (2008); Yantiss et al. (2009); Chai et al. (2010); Y.X. Wang et al. (2010)
miR-21	↑	CRC progression, stage and worse overall and progression-free survival	Poor therapeutic outcome (primarily FU based [5-FU, tegafur with uracil]) with or without generics	Brueckner et al. (2007); Slaby et al. (2007); Schetter et al. (2008); Liu et al. (2010); Shibuya et al. (2010); Valeri et al. (2010); Nielsen et al. (2011)
miR-29a/c	↑	Longer disease-free survival (DFS) in stage II		Weissmann-Brenner et al. (2012)
miR-31	↑	Advanced TNM stage and deeper invasion of tumors, poor prognosis	Chemoresistance to 5-FU	Motoyama et al. (2009); C.J. Wang et al. (2010)
miR-92a-1	↑	Poor prognosis		Motoyama et al. (2009)
miR-106a	↑	Progression and metastasis, poor survival		Volinia et al. (2006); Schetter et al. (2008); Arndt et al. (2009); Baffa et al. (2009); Xiao et al. (2009)
miR-125	↑	Advanced tumor size and tumor invasion, poorer prognosis		Nishida et al. (2011)
miR-155	↑	Lymph node metastases, worse overall, and DFS		Shibuya et al. (2010)
miR-181	↑	Poor survival		Schetter et al. (2008)
miR-181b-2	↑		Clinical response to S-1	Borralho et al. (2009); Hummel et al. (2010)
miR-183	↑	Poorer prognosis		Motoyama et al. (2009); Sarver et al. (2009)
miR-185	↑	Metastasis and poor survival		Tang et al. (2011)
miR-200c	↑	Poor survival		Xi et al. (2006)
miR-203	↑	Poor survival	Increased sensitivity to paclitaxel in the p53-mutated CRC	Schetter et al. (2008); Chiang et al. (2010); Li et al. (2011a)

(Continued)

TABLE 19.2. (*Continued*)

miRNA	Expression[a]	Association with Prognosis	Association with Response to Given Therapy	References
miR-224	↑	Tumor progression	Low expression associated with methotrexate resistance	Arndt et al. (2009); Mencia et al. (2011)
let-7b	↓		Chemoresistance to cetuximab	Ragusa et al. (2010)
miR-22	↓	Liver metastasis and poor overall survival	Overexpression in p53-mutated CRC cells led to increased chemosensitivity to paclitaxel	Li et al. (2011b); Zhang et al. (2012)
miR-34a	↓	Tumor-suppressive abilities	Chemoresistance to 5-FU	Tazawa et al. (2007); Akao et al. (2010)
miR-133b	↓	Metastasis and poor survival		Bandres et al. (2006); Tang et al. (2011)
miR-143	↓	Primary tumor size, more aggressive phenotype, and shorter DFS	Increases sensitivity to 5-FU	Slaby et al. (2007); Borralho et al. (2009); Chen et al. (2009); Motoyama et al. (2009); Ng et al. (2009); Kulda et al. (2010); Chang et al. (2011)
miR-145	↓	Inhibits tumor growth and angiogenesis		Shi et al. (2007); Chen et al. (2009); Motoyama et al. (2009); Kanwal and Gupta (2010); Chang et al. (2011); Liu et al. (2011)
miR-150	↓	Poor survival	Unfavorable response to 5-FU-based adjuvant chemotherapy in stage II and III patients	Ma et al. (2011)
miR-195	↓	Lymph node metastasis, advanced tumor stage, and poor overall survival		Liu et al. (2010); Wang et al. (2011)
miR-215	↓	Poor overall survival	Chemoresistance to methotrexate and tomudex	Braun et al. (2008); Boni et al. (2010); Song et al. (2010)
miR-365	↓	Cancer progression and poor survival	Increases sensitivity to 5-FU	Nie et al. (2012)
miR-451	↓	Poor prognosis	Reduced sensitivity to radiotherapy, chemoresistance to irinotecan	Bandres et al. (2009); Bitarte et al. (2011)
miR-498	↓	Worse progression-free survival in stage II MSS CRC		Schepeler et al. (2008)

[a]Expression reported in tumor compared with normal tissue: ↑ higher expression level in tumor compared with normal tissue; ↓ lower expression level in tumor compared with normal tissue.
5-FU, 5-flourouracil; TNM, tumor–node–metastasis; MSS, microsatellite stable.

of the material for retrospective studies. The fact that miRNAs are highly stable under these conditions creates the possibility to utilize them in the context of expression profiling, next-generation sequencing, and biomarker discovery from archived tumor samples (Xi et al. 2007; Bovell et al. 2012).

Most of the studies that have investigated miRNAs in the CRC patients' tissue samples use normal colon mucosa as a control (Bakirtzi et al. 2011; Balaguer et al. 2011; Chang et al. 2011; Huang et al. 2011; Luo et al. 2011, 2012; Song et al. 2011; Almeida et al. 2012; Migliore et al. 2012; Mosakhani et al. 2012; Nie et al. 2012; Reid et al. 2012; Vickers et al. 2012).

Overall, 234 miRNAs were reported to be differentially expressed when comparing normal colon tissue with tumor tissue. Some of these are consistently dysregulated in multiple studies such as *miR-143* and *miR-145*, were found to be lowered in CRC tissue in eight or more studies, while others, for example, *miR-20a* and *miR-31*, were found to be significantly higher expressed in CRC tissue compared with normal tissue in just two studies. Interestingly, *miR-203* was found to be highly expressed when comparing tumor tissue and controls in six studies, but low expressed in one study. These inconsistent results may be caused by different population origins, different environments, or diverse sample preparation methods. For a detailed review of miRNAs differentially expressed in CRC and between CRC stages, please see Luo et al. (2011) and Vickers et al. (2012).

The most intensely studied miRNAs regarding prognosis and prediction are *miR-21*, *miR-143*, and *miR-145*. *miR-21* was one of the first miRNAs described with a potential oncogenic function. While it inhibits the expression of various tumor suppressor genes, it also influences cell proliferation, apoptosis, invasion, and tumor progression. An increased expression of *miR-21* in CRC tumor tissue was correlated with an increased metastatic capacity of primary tumors (Brueckner et al. 2007; Hurst et al. 2009), leading to shortened disease-free (Balaguer et al. 2010; Kulda et al. 2010) and overall survival (Schetter et al. 2008). Due to the antiapoptotic effects of *miR-21*, its overexpression also influences therapy response, for example, by reducing the efficacy of 5-fluorouracil (5-FU) chemotherapy (Schetter et al. 2008; Huang et al. 2011).

In contrast, *miR-143* and *miR-145* are putative tumor suppressors. A reduced expression in tumor tissue was observed in several studies (Slaby et al. 2007; Chen et al. 2009; Motoyama et al. 2009), and the expression levels are negatively correlated to cell proliferation. Furthermore, reduced expression levels were found to be associated with increased tumor growth and angiogenesis, consequently leading to shorter disease-free survival (Akao et al. 2007; Kulda et al. 2010; Xu et al. 2012). In addition, the expression of *miR-143* directly correlates with the sensitivity of HCT116 CRC cells to 5-FU treatment (Borralho et al. 2009).

For a further promising CRC biomarker, *miR-215*, low expression levels in stage II and III tumors indicate good prognosis. This is a result of alterations in the cell cycle and in the colony formation, and cell adhesion abilities (Braun et al. 2008; Boni et al. 2010), leading to an overall decreased cell proliferation. Due to these effects, reduced *miR-215* expression levels could also predict resistance to chemotherapy with methotrexate and tomudex (Song et al. 2010).

Since CRC is a very heterogeneous disease, it is not surprising that several subgroups, for example, the microsatellite instable (MSI) phenotype, are described having distinct influences on the course and response to a given therapy. Patients with MSI tumors display a reduced disease-free survival compared with patients with microsatellite stable (MSS) ones (Popat and Houlston 2005), and furthermore often show reduced sensitivity to chemotherapeutic drugs that induce DNA damage, including 5-FU (Sinicrope and Sargent

2009). The first evidence for a correlation of miRNA expression profiles and MSS status was given in 2007, by identification of a panel of 14 miRNAs, which were differentially expressed between MSS and MSI tumors (Lanza et al. 2007). The expression of another panel of four miRNAs, *miR-142-3p*, *miR-212*, *miR-151*, and *miR-144*, was able to distinguish between MSS and MSI stage II CRC tumors with a sensitivity and specificity of 92% and 81%, respectively (Schepeler et al. 2008). Moreover, a panel of 17 miRNA expression profiles mastered to separate different MSS tumor subtypes according to their metastatic recurrence status in the patient. Out of this panel, the expression levels of *miR-320* and *miR-498* in the primary tumor could be correlated directly to the probability of recurrence-free survival (Schepeler et al. 2008).

A. Future Directions for miRNA Analysis in CRC Tumor Tissue

Among the investigated miRNAs in tumor tissue mentioned in all studies described here, 234 miRNAs were found to be significantly altered expressed in at least one study. Out of these, high levels of *miR-31* and low expression levels of *miR-145* were the most often reported to have a prognostic relevance. Furthermore, controversial results are reported in between studies (Luo et al. 2011).

Most of the studies reported chose RNU6B as endogenous control. As mentioned earlier, due to the different stability of RNU6B and Ago-bound (and protected) miRNA, it is questionable whether this is the optimal normalization strategy. In other studies, *miR-16* has been suggested to be used as an endogenous normalization control (Chang et al. 2010). Again, more studies are required to agree on a reliable normalization strategy. Another limitation of these studies is the sample size. Most studies reported to date analyzed less than 200 samples. More studies with large sample sizes will be needed before results can be translated into clinic.

Furthermore, many studies reported so far were candidate-based approaches meaning that miRNAs were selected based on previous reports. Even when microarrays were applied, just a subset of miRNAs known today has been covered. Thus, the conduction of unbiased, comprehensive miRNA studies is still essential.

IV. CONCLUSION

Due to the need of further minimal invasive screening methods with high sensitivity and specificity, evaluation of miRNAs as potential early detection and prognostic markers is a field of intensive research. Especially for the investigation of circulating miRNAs, problems of standardized sample processing and reliable normalization will have to be solved before this translates into clinical reality.

Despite many challenges to develop and implement a test with clinical effectiveness, the examples in this chapter illustrate that miRNAs have the potential as indicators of early-stage CRC, tumor recurrence, or metastasis, as well as predicting the response of a CRC patient to a given therapy. Their value will have to be judged in comparison and combination with existing markers. Eventually, miRNA signatures will improve the sensitivity and specificity of CRC screening and prognosis only in combination with other molecular markers. Due to their high stability in plasma and freshly frozen material as well as FFPE archival tissue and the evolving detection methods such as next-generation sequencing, it might be possible to detect and validate more tumor-related miRNAs in the near future. The increasing knowledge of miRNA signatures, their origin, and impact on early detection, prognosis, and therapy response will lead to an increased understanding of CRC and availability of miRNAs as biomarkers.

REFERENCES

Akao, Y., Y. Nakagawa, and T. Naoe. 2007. MicroRNA-143 and -145 in colon cancer. *DNA Cell Biol* **26**:311–320.

Akao, Y., S. Noguchi, A. Iio, K. Kojima, T. Takagi, and T. Naoe. 2010. Dysregulation of microRNA-34a expression causes drug-resistance to 5-FU in human colon cancer DLD-1 cells. *Cancer Lett* **300**:197–204.

Almeida, M.I., M.S. Nicoloso, L. Zeng, C. Ivan, R. Spizzo, R. Gafa, L. Xiao, X. Zhang, I. Vannini, F. Fanini, M. Fabbri, G. Lanza, et al. 2012. Strand-specific *miR-28-5p* and *miR-28-3p* have distinct effects in colorectal cancer cells. *Gastroenterology* **142**:886–896 e889.

Arndt, G.M., L. Dossey, L.M. Cullen, A. Lai, R. Druker, M. Eisbacher, C. Zhang, N. Tran, H. Fan, K. Retzlaff, A. Bittner, and M. Raponi. 2009. Characterization of global microRNA expression reveals oncogenic potential of miR-145 in metastatic colorectal cancer. *BMC Cancer* **9**:374.

Baffa, R., M. Fassan, S. Volinia, B. O'Hara, C.G. Liu, J.P. Palazzo, M. Gardiman, M. Rugge, L.G. Gomella, C.M. Croce, and A. Rosenberg. 2009. MicroRNA expression profiling of human metastatic cancers identifies cancer gene targets. *J Pathol* **219**:214–221.

Bakirtzi, K., M. Hatziapostolou, I. Karagiannides, C. Polytarchou, S. Jaeger, D. Iliopoulos, and C. Pothoulakis. 2011. Neurotensin signaling activates microRNAs-21 and -155 and Akt, promotes tumor growth in mice, and is increased in human colon tumors. *Gastroenterology* **141**:1749–1761 e1741.

Balaguer, F., A. Link, J.J. Lozano, M. Cuatrecasas, T. Nagasaka, C.R. Boland, and A. Goel. 2010. Epigenetic silencing of *miR-137* is an early event in colorectal carcinogenesis. *Cancer Res* **70**:6609–6618.

Balaguer, F., L. Moreira, J.J. Lozano, A. Link, G. Ramirez, Y. Shen, M. Cuatrecasas, M. Arnold, S.J. Meltzer, S. Syngal, E. Stoffel, R. Jover, et al. 2011. Colorectal cancers with microsatellite instability display unique miRNA profiles. *Clin Cancer Res* **17**:6239–6249.

Bandres, E., E. Cubedo, X. Agirre, R. Malumbres, R. Zarate, N. Ramirez, A. Abajo, A. Navarro, I. Moreno, M. Monzo, and J. Garcia-Foncillas. 2006. Identification by Real-time PCR of 13 mature microRNAs differentially expressed in colorectal cancer and non-tumoral tissues. *Mol Cancer* **5**:29.

Bandres, E., N. Bitarte, F. Arias, J. Agorreta, P. Fortes, X. Agirre, R. Zarate, J.A. Diaz-Gonzalez, N. Ramirez, J.J. Sola, P. Jimenez, J. Rodriguez, et al. 2009. microRNA-451 regulates macrophage migration inhibitory factor production and proliferation of gastrointestinal cancer cells. *Clin Cancer Res* **15**:2281–2290.

Bartels, C.L. and G.J. Tsongalis. 2009. MicroRNAs: novel biomarkers for human cancer. *Clin Chem* **55**:623–631.

Bitarte, N., E. Bandres, V. Boni, R. Zarate, J. Rodriguez, M. Gonzalez-Huarriz, I. Lopez, J. Javier Sola, M.M. Alonso, P. Fortes, and J. Garcia-Foncillas. 2011. MicroRNA-451 is involved in the self-renewal, tumorigenicity, and chemoresistance of colorectal cancer stem cells. *Stem Cells* **29**:1661–1671.

Boni, V., N. Bitarte, I. Cristobal, R. Zarate, J. Rodriguez, E. Maiello, J. Garcia-Foncillas, and E. Bandres. 2010. *miR-192/miR-215* influence 5-fluorouracil resistance through cell cycle-mediated mechanisms complementary to its post-transcriptional thymidylate synthase regulation. *Mol Cancer Ther* **9**:2265–2275.

Borralho, P.M., B.T. Kren, R.E. Castro, I.B. da Silva, C.J. Steer, and C.M. Rodrigues. 2009. MicroRNA-143 reduces viability and increases sensitivity to 5-fluorouracil in HCT116 human colorectal cancer cells. *FEBS J* **276**:6689–6700.

Bovell, L., C. Shanmugam, V.R. Katkoori, B. Zhang, E. Vogtmann, W.E. Grizzle, and U. Manne. 2012. miRNAs are stable in colorectal cancer archival tissue blocks. *Front Biosci (Elite Ed)* **4**:1937–1940.

Braun, C.J., X. Zhang, I. Savelyeva, S. Wolff, U.M. Moll, T. Schepeler, T.F. Orntoft, C.L. Andersen, and M. Dobbelstein. 2008. p53-Responsive microRNAs 192 and 215 are capable of inducing cell cycle arrest. *Cancer Res* **68**:10094–10104.

Brueckner, B., C. Stresemann, R. Kuner, C. Mund, T. Musch, M. Meister, H. Sultmann, and F. Lyko. 2007. The human *let-7a-3* locus contains an epigenetically regulated microRNA gene with oncogenic function. *Cancer Res* **67**:1419–1423.

Chai, H., M. Liu, R. Tian, X. Li, and H. Tang. 2010. miR-20a targets BNIP2 and contributes chemotherapeutic resistance in colorectal adenocarcinoma SW480 and SW620 cell lines. *Acta Biochim Biophys Sin (Shanghai)* **43**:217–225.

Chang, K.H., P. Mestdagh, J. Vandesompele, M.J. Kerin, and N. Miller. 2010. MicroRNA expression profiling to identify and validate reference genes for relative quantification in colorectal cancer. *BMC Cancer* **10**:173.

Chang, K.H., N. Miller, E.A. Kheirelseid, C. Lemetre, G.R. Ball, M.J. Smith, M. Regan, O.J. McAnena, and M.J. Kerin. 2011. MicroRNA signature analysis in colorectal cancer: identification of expression profiles in stage II tumors associated with aggressive disease. *Int J Colorectal Dis* **26**:1415–1422.

Chen, X., X. Guo, H. Zhang, Y. Xiang, J. Chen, Y. Yin, X. Cai, K. Wang, G. Wang, Y. Ba, L. Zhu, J. Wang, et al. 2009. Role of *miR-143* targeting KRAS in colorectal tumorigenesis. *Oncogene* **28**:1385–1392.

Cheng, H., L. Zhang, D.E. Cogdell, H. Zheng, A.J. Schetter, M. Nykter, C.C. Harris, K. Chen, S.R. Hamilton, and W. Zhang. 2011. Circulating plasma *MiR-141* is a novel biomarker for metastatic colon cancer and predicts poor prognosis. *PLoS ONE* **6**:e17745.

Chiang, Y., Y. Song, Z. Wang, Y. Chen, Z. Yue, H. Xu, C. Xing, and Z. Liu. 2010. Aberrant expression of miR-203 and its clinical significance in gastric and colorectal cancers. *J Gastrointest Surg* **15**:63–70.

Cortez, M.A. and G.A. Calin. 2009. MicroRNA identification in plasma and serum: a new tool to diagnose and monitor diseases. *Expert Opin Biol Ther* **9**:703–711.

Cuk, K., M. Zucknick, J. Heil, D. Madhavan, S. Schott, A. Turchinovich, D. Arlt, M. Rath, C. Sohn, A. Benner, H. Junkermann, A. Schneeweiss, et al. 2013. Circulating microRNAs in plasma as early detection markers for breast cancer. *Int J Cancer* **132**:1602–1612.

Diaz, R., J. Silva, J.M. Garcia, Y. Lorenzo, V. Garcia, C. Pena, R. Rodriguez, C. Munoz, F. Garcia, F. Bonilla, and G. Dominguez. 2008. Deregulated expression of miR-106a predicts survival in human colon cancer patients. *Genes Chromosomes Cancer* **47**:794–802.

Duttagupta, R., R. Jiang, J. Gollub, R.C. Getts, and K.W. Jones. 2011. Impact of cellular miRNAs on circulating miRNA biomarker signatures. *PLoS ONE* **6**:e20769.

Greene, F.L., A.K. Stewart, and H.J. Norton. 2004. New tumor-node-metastasis staging strategy for node-positive (stage III) rectal cancer: an analysis. *J Clin Oncol* **22**:1778–1784.

Huang, Z., D. Huang, S. Ni, Z. Peng, W. Sheng, and X. Du. 2010. Plasma microRNAs are promising novel biomarkers for early detection of colorectal cancer. *Int J Cancer* **127**:118–126.

Huang, Z., S. Huang, Q. Wang, L. Liang, S. Ni, L. Wang, W. Sheng, X. He, and X. Du. 2011. MicroRNA-95 promotes cell proliferation and targets sorting Nexin 1 in human colorectal carcinoma. *Cancer Res* **71**:2582–2589.

Hummel, R., D.J. Hussey, and J. Haier. 2010. MicroRNAs: predictors and modifiers of chemo- and radiotherapy in different tumour types. *Eur J Cancer* **46**:298–311.

Hundt, S., U. Haug, and H. Brenner. 2009. Comparative evaluation of immunochemical fecal occult blood tests for colorectal adenoma detection. *Ann Intern Med* **150**:162–169.

REFERENCES

Hurst, D.R., M.D. Edmonds, and D.R. Welch. 2009. Metastamir: the field of metastasis-regulatory microRNA is spreading. *Cancer Res* **69**:7495–7498.

Jankowski, J.A. and R.D. Odze. 2009. Biomarkers in gastroenterology: between hope and hype comes histopathology. *Am J Gastroenterol* **104**:1093–1096.

Jemal, A., R. Siegel, E. Ward, Y. Hao, J. Xu, T. Murray, and M.J. Thun. 2008. Cancer statistics, 2008. *CA Cancer J Clin* **58**:71–96.

Kanwal, R. and S. Gupta. 2010. Epigenetics and cancer. *J Appl Physiol* **109**:598–605.

Kulda, V., M. Pesta, O. Topolcan, V. Liska, V. Treska, A. Sutnar, K. Rupert, M. Ludvikova, V. Babuska, L. Holubec Jr., and R. Cerny. 2010. Relevance of *miR-21* and *miR-143* expression in tissue samples of colorectal carcinoma and its liver metastases. *Cancer Genet Cytogenet* **200**:154–160.

Kurokawa, K., T. Tanahashi, T. Iima, Y. Yamamoto, Y. Akaike, K. Nishida, K. Masuda, Y. Kuwano, Y. Murakami, M. Fukushima, and K. Rokutan. 2012. Role of miR-19b and its target mRNAs in 5-fluorouracil resistance in colon cancer cells. *J Gastroenterol* **47**:883–895.

Lanza, G., M. Ferracin, R. Gafa, A. Veronese, R. Spizzo, F. Pichiorri, C.G. Liu, G.A. Calin, C.M. Croce, and M. Negrini. 2007. mRNA/microRNA gene expression profile in microsatellite unstable colorectal cancer. *Mol Cancer* **6**:54.

Li, J., Y. Chen, J. Zhao, F. Kong, and Y. Zhang. 2011a. miR-203 reverses chemoresistance in p53-mutated colon cancer cells through downregulation of Akt2 expression. *Cancer Lett* **304**: 52–59.

Li, J., Y. Zhang, J. Zhao, F. Kong, and Y. Chen. 2011b. Overexpression of miR-22 reverses paclitaxel-induced chemoresistance through activation of PTEN signaling in p53-mutated colon cancer cells. *Mol Cell Biochem* **357**:31–38.

Liu, L., L. Chen, Y. Xu, R. Li, and X. Du. 2010. microRNA-195 promotes apoptosis and suppresses tumorigenicity of human colorectal cancer cells. *Biochem Biophys Res Commun* **400**:236–240.

Liu, M., N. Lang, M. Qiu, F. Xu, Q. Li, Q. Tang, J. Chen, X. Chen, S. Zhang, Z. Liu, J. Zhou, Y. Zhu, et al. 2011. miR-137 targets Cdc42 expression, induces cell cycle G1 arrest and inhibits invasion in colorectal cancer cells. *Int J Cancer* **128**:1269–1279.

Luo, X., B. Burwinkel, S. Tao, and H. Brenner. 2011. MicroRNA signatures: novel biomarker for colorectal cancer? *Cancer Epidemiol Biomarkers Prev* **20**:1272–1286.

Luo, H., J. Zou, Z. Dong, Q. Zeng, D. Wu, and L. Liu. 2012. Up-regulated *miR-17* promotes cell proliferation, tumour growth and cell cycle progression by targeting the RND3 tumour suppressor gene in colorectal carcinoma. *Biochem J* **442**:311–321.

Ma, Y., P. Zhang, F. Wang, H. Zhang, J. Yang, J. Peng, W. Liu, and H. Qin. 2011. miR-150 as a potential biomarker associated with prognosis and therapeutic outcome in colorectal cancer. *Gut* **61**(10):1447–1453.

McDonald, J.S., D. Milosevic, H.V. Reddi, S.K. Grebe, and A. Algeciras-Schimnich. 2011. Analysis of circulating microRNA: preanalytical and analytical challenges. *Clin Chem* **57**: 833–840.

Mencia, N., E. Selga, V. Noe, and C.J. Ciudad. 2011. Underexpression of miR-224 in methotrexate resistant human colon cancer cells. *Biochem Pharmacol* **82**:1572–1582.

Migliore, C., V. Martin, V.P. Leoni, A. Restivo, L. Atzori, A. Petrelli, C. Isella, L. Zorcolo, I. Sarotto, G. Casula, P.M. Comoglio, A. Columbano, et al. 2012. *MiR-1* downregulation cooperates with MACC1 in promoting MET overexpression in human colon cancer. *Clin Cancer Res* **18**:737–747.

Mitchell, P.S., R.K. Parkin, E.M. Kroh, B.R. Fritz, S.K. Wyman, E.L. Pogosova-Agadjanyan, A. Peterson, J. Noteboom, K.C. O'Briant, A. Allen, D.W. Lin, N. Urban, et al. 2008. Circulating microRNAs as stable blood-based markers for cancer detection. *Proc Natl Acad Sci U S A* **105**:10513–10518.

Mosakhani, N., V.K. Sarhadi, I. Borze, M.L. Karjalainen-Lindsberg, J. Sundstrom, R. Ristamaki, P. Osterlund, and S. Knuutila. 2012. MicroRNA profiling differentiates colorectal cancer according to KRAS status. *Genes Chromosomes Cancer* **51**:1–9.

Motoyama, K., H. Inoue, Y. Takatsuno, F. Tanaka, K. Mimori, H. Uetake, K. Sugihara, and M. Mori. 2009. Over- and under-expressed microRNAs in human colorectal cancer. *Int J Oncol* **34**:1069–1075.

Nakajima, G., K. Hayashi, Y. Xi, K. Kudo, K. Uchida, K. Takasaki, M. Yamamoto, and J. Ju. 2006. Non-coding MicroRNAs hsa-let-7g and hsa-miR-181b are Associated with Chemoresponse to S-1 in Colon Cancer. *Cancer Genomics Proteomics* **3**:317–324.

Ng, E.K., W.W. Chong, H. Jin, E.K. Lam, V.Y. Shin, J. Yu, T.C. Poon, S.S. Ng, and J.J. Sung. 2009. Differential expression of microRNAs in plasma of patients with colorectal cancer: a potential marker for colorectal cancer screening. *Gut* **58**:1375–1381.

Nie, J., L. Liu, W. Zheng, L. Chen, X. Wu, Y. Xu, X. Du, and W. Han. 2012. MicroRNA-365, down-regulated in colon cancer, inhibits cell cycle progression and promotes apoptosis of colon cancer cells by probably targeting Cyclin D1 and Bcl-2. *Carcinogenesis* **33**:220–225.

Nielsen, B.S., S. Jorgensen, J.U. Fog, R. Sokilde, I.J. Christensen, U. Hansen, N. Brunner, A. Baker, S. Moller, and H.J. Nielsen. 2011. High levels of microRNA-21 in the stroma of colorectal cancers predict short disease-free survival in stage II colon cancer patients. *Clin Exp Metastasis* **28**:27–38.

Nishida, N., T. Yokobori, K. Mimori, T. Sudo, F. Tanaka, K. Shibata, H. Ishii, Y. Doki, H. Kuwano, and M. Mori. 2011. MicroRNA miR-125b is a prognostic marker in human colorectal cancer. *Int J Oncol* **38**:1437–1443.

Nishida, N., S. Yamashita, K. Mimori, T. Sudo, F. Tanaka, K. Shibata, H. Yamamoto, H. Ishii, Y. Doki, and M. Mori. 2012. MicroRNA-10b is a Prognostic Indicator in Colorectal Cancer and Confers Resistance to the Chemotherapeutic Agent 5-Fluorouracil in Colorectal Cancer Cells. *Ann Surg Oncol* **19**(9):3065–3071.

Popat, S. and R.S. Houlston. 2005. A systematic review and meta-analysis of the relationship between chromosome 18q genotype, DCC status and colorectal cancer prognosis. *Eur J Cancer* **41**:2060–2070.

Pritchard, C.C., E. Kroh, B. Wood, J.D. Arroyo, K.J. Dougherty, M.M. Miyaji, J.F. Tait, and M. Tewari. 2012. Blood cell origin of circulating microRNAs: a cautionary note for cancer biomarker studies. *Cancer Prev Res (Phila)* **5**:492–497.

Pu, X.X., G.L. Huang, H.Q. Guo, C.C. Guo, H. Li, S. Ye, S. Ling, L. Jiang, Y. Tian, and T.Y. Lin. 2010. Circulating *miR-221* directly amplified from plasma is a potential diagnostic and prognostic marker of colorectal cancer and is correlated with p53 expression. *J Gastroenterol Hepatol* **25**:1674–1680.

Ragusa, M., A. Majorana, L. Statello, M. Maugeri, L. Salito, D. Barbagallo, M.R. Guglielmino, L.R. Duro, R. Angelica, R. Caltabiano, A. Biondi, M. Di Vita, et al. 2010. Specific alterations of microRNA transcriptome and global network structure in colorectal carcinoma after cetuximab treatment. *Mol Cancer Ther* **9**:3396–3409.

Reddi, K.K. and J.F. Holland. 1976. Elevated serum ribonuclease in patients with pancreatic cancer. *Proc Natl Acad Sci U S A* **73**:2308–2310.

Reid, J.F., V. Sokolova, E. Zoni, A. Lampis, S. Pizzamiglio, C. Bertan, S. Zanutto, F. Perrone, T. Camerini, G. Gallino, P. Verderio, E. Leo, et al. 2012. miRNA profiling in colorectal cancer highlights *miR-1* involvement in MET-dependent proliferation. *Mol Cancer Res* **10**:504–515.

Sarver, A.L., A.J. French, P.M. Borralho, V. Thayanithy, A.L. Oberg, K.A. Silverstein, B.W. Morlan, S.M. Riska, L.A. Boardman, J.M. Cunningham, S. Subramanian, L. Wang, et al. 2009. Human colon cancer profiles show differential microRNA expression depending on mismatch repair status and are characteristic of undifferentiated proliferative states. *BMC Cancer* **9**:401.

REFERENCES

Schepeler, T., J.T. Reinert, M.S. Ostenfeld, L.L. Christensen, A.N. Silahtaroglu, L. Dyrskjot, C. Wiuf, F.J. Sorensen, M. Kruhoffer, S. Laurberg, S. Kauppinen, T.F. Orntoft, et al. 2008. Diagnostic and prognostic microRNAs in stage II colon cancer. *Cancer Res* **68**:6416–6424.

Schetter, A.J., S.Y. Leung, J.J. Sohn, K.A. Zanetti, E.D. Bowman, N. Yanaihara, S.T. Yuen, T.L. Chan, D.L. Kwong, G.K. Au, C.G. Liu, G.A. Calin, et al. 2008. MicroRNA expression profiles associated with prognosis and therapeutic outcome in colon adenocarcinoma. *JAMA* **299**:425–436.

Shi, B., L. Sepp-Lorenzino, M. Prisco, P. Linsley, T. deAngelis, and R. Baserga. 2007. Micro RNA 145 targets the insulin receptor substrate-1 and inhibits the growth of colon cancer cells. *J Biol Chem* **282**:32582–32590.

Shibuya, H., H. Iinuma, R. Shimada, A. Horiuchi, and T. Watanabe. 2010. Clinicopathological and prognostic value of microRNA-21 and microRNA-155 in colorectal cancer. *Oncology* **79**: 313–320.

Sinicrope, F.A. and D.J. Sargent. 2009. Clinical implications of microsatellite instability in sporadic colon cancers. *Curr Opin Oncol* **21**:369–373.

Slaby, O., M. Svoboda, P. Fabian, T. Smerdova, D. Knoflickova, M. Bednarikova, R. Nenutil, and R. Vyzula. 2007. Altered expression of *miR-21*, *miR-31*, *miR-143* and *miR-145* is related to clinicopathologic features of colorectal cancer. *Oncology* **72**:397–402.

Song, B., Y. Wang, M.A. Titmus, G. Botchkina, A. Formentini, M. Kornmann, and J. Ju. 2010. Molecular mechanism of chemoresistance by *miR-215* in osteosarcoma and colon cancer cells. *Mol Cancer* **9**:96.

Song, Y., Y. Xu, Z. Wang, Y. Chen, Z. Yue, P. Gao, C. Xing, and H. Xu. 2011. MicroRNA-148b suppresses cell growth by targeting cholecystokinin-2 receptor in colorectal cancer. *Int J Cancer* **131**:1042–1051.

Tang, J.T., J.L. Wang, W. Du, J. Hong, S.L. Zhao, Y.C. Wang, H. Xiong, H.M. Chen, and J.Y. Fang. 2011. MicroRNA 345, a methylation-sensitive microRNA is involved in cell proliferation and invasion in human colorectal cancer. *Carcinogenesis* **32**:1207–1215.

Tazawa, H., N. Tsuchiya, M. Izumiya, and H. Nakagama. 2007. Tumor-suppressive miR-34a induces senescence-like growth arrest through modulation of the E2F pathway in human colon cancer cells. *Proc Natl Acad Sci U S A* **104**:15472–15477.

Turchinovich, A., L. Weiz, A. Langheinz, and B. Burwinkel. 2011. Characterization of extracellular circulating microRNA. *Nucleic Acids Res* **39**:7223–7233.

Turchinovich, A., L. Weiz, and B. Burwinkel. 2012. Extracellular miRNAs: the mystery of their origin and function? *Trends Biochem Sci* **37**:460–465.

Valeri, N., P. Gasparini, C. Braconi, A. Paone, F. Lovat, M. Fabbri, K.M. Sumani, H. Alder, D. Amadori, T. Patel, G.J. Nuovo, R. Fishel, et al. 2010. MicroRNA-21 induces resistance to 5-fluorouracil by down-regulating human DNA MutS homolog 2 (hMSH2). *Proc Natl Acad Sci U S A* **107**:21098–21103.

Vickers, M.M., J. Bar, I. Gorn-Hondermann, N. Yarom, M. Daneshmand, J.E. Hanson, C.L. Addison, T.R. Asmis, D.J. Jonker, J. Maroun, I.A. Lorimer, G.D. Goss, et al. 2012. Stage-dependent differential expression of microRNAs in colorectal cancer: potential role as markers of metastatic disease. *Clin Exp Metastasis* **29**:123–132.

Volinia, S., G.A. Calin, C.G. Liu, S. Ambs, A. Cimmino, F. Petrocca, R. Visone, M. Iorio, C. Roldo, M. Ferracin, R.L. Prueitt, N. Yanaihara, et al. 2006. A microRNA expression signature of human solid tumors defines cancer gene targets. *Proc Natl Acad Sci U S A* **103**:2257–2261.

Walther, A., E. Johnstone, C. Swanton, R. Midgley, I. Tomlinson, and D. Kerr. 2009. Genetic prognostic and predictive markers in colorectal cancer. *Nat Rev Cancer* **9**:489–499.

Wang, C.J., J. Stratmann, Z.G. Zhou, and X.F. Sun. 2010. Suppression of microRNA-31 increases sensitivity to 5-FU at an early stage, and affects cell migration and invasion in HCT-116 colon cancer cells. *BMC Cancer* **10**:616.

Wang, X., J. Wang, H. Ma, J. Zhang, and X. Zhou. 2011. Downregulation of miR-195 correlates with lymph node metastasis and poor prognosis in colorectal cancer. *Med Oncol* **29**:919–927.

Wang, Y.X., X.Y. Zhang, B.F. Zhang, C.Q. Yang, X.M. Chen, and H.J. Gao. 2010. Initial study of microRNA expression profiles of colonic cancer without lymph node metastasis. *J Dig Dis* **11**:50–54.

Weissmann-Brenner, A., M. Kushnir, G. Lithwick Yanai, R. Aharonov, H. Gibori, O. Purim, Y. Kundel, S. Morgenstern, M. Halperin, Y. Niv, and B. Brenner. 2012. Tumor microRNA-29a expression and the risk of recurrence in stage II colon cancer. *Int J Oncol* **40**:2097–2103.

Whitlock, E.P., J.S. Lin, E. Liles, T.L. Beil, and R. Fu. 2008. Screening for colorectal cancer: a targeted, updated systematic review for the U.S. Preventive Services Task Force. *Ann Intern Med* **149**:638–658.

Xi, Y., A. Formentini, M. Chien, D.B. Weir, J.J. Russo, J. Ju, and M. Kornmann. 2006. Prognostic Values of microRNAs in Colorectal Cancer. *Biomark Insights* **2**:113–121.

Xi, Y., G. Nakajima, E. Gavin, C.G. Morris, K. Kudo, K. Hayashi, and J. Ju. 2007. Systematic analysis of microRNA expression of RNA extracted from fresh frozen and formalin-fixed paraffin-embedded samples. *RNA* **13**:1668–1674.

Xiao, B., J. Guo, Y. Miao, Z. Jiang, R. Huan, Y. Zhang, D. Li, and J. Zhong. 2009. Detection of miR-106a in gastric carcinoma and its clinical significance. *Clin Chim Acta* **400**:97–102.

Xu, Q., L.Z. Liu, X. Qian, Q. Chen, Y. Jiang, D. Li, L. Lai, and B.H. Jiang. 2012. MiR-145 directly targets p70S6K1 in cancer cells to inhibit tumor growth and angiogenesis. *Nucleic Acids Res* **40**:761–774.

Yantiss, R.K., M. Goodarzi, X.K. Zhou, H. Rennert, E.C. Pirog, B.F. Banner, and Y.T. Chen. 2009. Clinical, pathologic, and molecular features of early-onset colorectal carcinoma. *Am J Surg Pathol* **33**:572–582.

Zen, K. and C.Y. Zhang. 2012. Circulating microRNAs: a novel class of biomarkers to diagnose and monitor human cancers. *Med Res Rev* **32**:326–348.

Zhang, G., S. Xia, H. Tian, Z. Liu, and T. Zhou. 2012. Clinical significance of miR-22 expression in patients with colorectal cancer. *Med Oncol* **29**(5):3108–3112.

20

GENETIC VARIATIONS IN MicroRNA-ENCODING SEQUENCES AND MicroRNA TARGET SITES ALTER LUNG CANCER SUSCEPTIBILITY AND SURVIVAL

Ming Yang[1] and Dongxin Lin[2]

[1]*College of Life Science and Technology, Beijing University of Chemical Technology, Beijing, China*
[2]*State Key Laboratory of Molecular Oncology and Beijing Key Laboratory of Carcinogenesis and Cancer Prevention, Cancer Institute and Hospital, Chinese Academy of Medical Sciences and Peking Union Medical College, Beijing, China*

I. Introduction	344
II. Genetic Variations in miR-Encoding Sequences and Lung Cancer	345
A. SNPs in Pre-miRNA Sequence	345
B. SNPs in Pre-miRNA Flanking Regions	346
III. Genetic Polymorphisms in miRNA Target Sites and Lung Cancer	347
A. A SNP in *KRAS* 3′-UTR Alters miRNA *let-7* Binding	347
B. Genetic Variation in an *miR-1827*-Binding Site of *MYCL1*	347
C. *REV3L* 3′-UTR 460 T > C SNP Influences Gene Regulation by *miR-25* and *miR-32*	348
D. A Functional Polymorphism at the *miR-629*-Binding Site of *NBS1*	349
IV. Conclusions	349
References	350

ABBREVIATIONS

3′-UTR	3′-untranslated region
CI	confidence interval
HR	hazard ratio
MAF	minor allele frequency

MicroRNAs in Medicine, First Edition. Edited by Charles H. Lawrie.
© 2014 John Wiley & Sons, Inc. Published 2014 by John Wiley & Sons, Inc.

miRNA	microRNA
NSCLC	non-small-cell lung cancer
OR	odds ratio
OS	overall survival
SCLC	small-cell lung cancer
SNP	single-nucleotide polymorphism
TNM	tumor–node–metastasis

I. INTRODUCTION

Lung cancer is one of the leading causes of cancer death all around the world, with more than 1,000,000 deaths each year (Parkin et al. 2005). About 80% of all lung cancer is classified as non-small-cell lung cancer (NSCLC) and the remaining 20% belongs to small-cell lung cancer (SCLC). SCLC is the most aggressive subtype of lung cancer, characterized by rapid doubling time, high growth fraction, and early development of widespread metastases (Jackman and Johnson 2005). It is well known that more than 75% of lung cancer is attributed to environmental carcinogen exposure, such as tobacco smoking (Parkin et al. 2005). However, not all exposed individuals develop lung cancer, suggesting that the genetic makeup is also important in the development of this malignancy.

For localized NSCLC, the 5-year survival rate is about 50%. However, over one-third of NSCLC cases are diagnosed at a locally advanced stage (Gandara et al. 2005), with a 3-year overall survival (OS) of 10–20%, even after multidisciplinary treatment, including a combination of radiotherapy and chemotherapy (Lee et al. 2006). The classic prognostic determinants for lung cancer include the tumor–node–metastasis (TNM) staging system, performance status, sex, and weight loss. Unfortunately, all these factors are far from sufficient to explain the patient-to-patient variability of prognosis. Several recent studies showed that some host genetic factors including single-nucleotide polymorphisms (SNPs) are associated with survival of NSCLC (Gurubhagavatula et al. 2004; Hu et al. 2008, 2011; Tibaldi et al. 2008; Yu et al. 2008; Bi et al. 2010), demonstrating that genetic factors are also good predictors for lung cancer survival and may have potential implication in therapeutic interventions.

MicroRNAs (miRNAs) are small RNAs that bind to target messenger RNAs (mRNAs) acting as posttranscriptional gene expression regulators (Ambros 2004). By pairing to near-complementary binding sites within the 3′-untranslated region (3′-UTR) of hundreds of target mRNAs, miRNAs impair their translation or promote their degradation, thus participating in the control of crucial cell processes. Previous studies have demonstrated the important roles of miRNAs in lung cancer development and progression (Yanaihara et al. 2006; Cho 2009; Lin et al. 2010). It has also been found that SNPs in the pre-miRNA genes may alter miRNA processing and expression and, therefore, contribute to lung cancer susceptibility and survival (Hu et al. 2008, 2011; Tian et al. 2009; Kim et al. 2010; Chu et al. 2011; Hong et al. 2011). In addition, it has been shown that SNPs in miRNAs target sites of oncogenes or tumor suppressor genes affect miRNA–mRNA interaction, and they are associated with lung cancer risk, alone and in combination with tobacco smoking (Chin et al. 2008; Nelson et al. 2010; Xiong et al. 2011; Zhang et al. 2013).

In this chapter, we review the current studies on literature (Table 20.1) and discuss the impact of genetic variants in miR-encoding sequences and miRNA target sites on lung cancer risk and survival.

TABLE 20.1. Studies of Single-Nucleotide Polymorphisms (SNPs) in miRNA Associated with Lung Cancer Susceptibility and Survival

No.	Studies	SNP	SNP Location	Study Design	No. of Cases	No. of Controls	Ethnicity	Cancer Type
1	Tian et al. (2009)	miR-196a2 rs11614913	Pre-miRNA genes	Case-control	1058	1035	Chinese	Lung cancer
2	Kim et al. (2010)	miR-196a2 rs11614913	Pre-miRNA gene	Case-control	654	640	Korean	Lung cancer
3	Hong et al. (2011)	miR-196a2 rs11614913	Pre-miRNA gene	Case-control	406	428	Korean	NSCLC
4	Hu et al. (2008)	miR-196a2 rs11614913	Pre-miRNA genes	Cohort	663		Chinese	NSCLC
5	Hu et al. (2011)	miR-30c-1 rs928508	Pre-miRNA flanking regions	Cohort	923		Chinese	NSCLC
6	Chin et al. (2008)	KRAS rs61764370	KRAS 3'-UTR altering miRNA let-7 binding	Case-control	2423	1822	Mainly Caucasian	NSCLC
7	Nelson et al. (2010)	KRAS rs61764370	KRAS 3'-UTR altering miRNA let-7 binding	Cohort	218		NA	NSCLC
8	Xiong et al. (2011)	MYCL1 rs3134615	MYCL1 3'-UTR altering miRNA miR-1827 binding	Case-control	666	758	Chinese	SCLC
9	Zhang et al. (2013)	REV3L rs465646	REV3L 3'-UTR altering miR-25 and miR-32 binding	Case-control	1072	1064	Chinese	Lung cancer
10	Yang et al. (2012)	NBS1 rs2735383	NBS1 3'-UTR altering miR-629 binding	Case-control	1559	1679	Chinese	Lung cancer

NSCLC, non-small-cell lung cancer; 3'-UTR, 3'-untranslated region; NA, not available; SCLC, small-cell lung cancer.

II. GENETIC VARIATIONS IN MiR-ENCODING SEQUENCES AND LUNG CANCER

A. SNPs in Pre-miRNA Sequence

Since polymorphisms in pre-miRNA sequence may alter miRNA processing and/or expression, several studies have investigated whether such genetic variants contribute to lung cancer susceptibility and patient survival (Hu et al. 2008; Tian et al. 2009; Kim et al. 2010; Hong et al. 2011). To study common genetic polymorphisms in pre-miRNA gene loci in the Chinese population, Hu et al. performed a systematical screening with the following criteria: (1) SNPs in pre-miRNA gene loci of 400 known human miRNAs (Griffiths-Jones 2004) and (2) common SNPs (minor allele frequency [MAF] >0.05) in the SNP databases (the NCBI dbSNP database, build 127; http://www.ncbi.nlm.nih.gov/projects/SNP/) (Hu et al. 2008). As a result, four candidate pre-miRNA SNPs were identified, including miR-146a rs2910164 C > G, miR-196a2 rs11614913 C > T, miR-499 rs3746444 G > A, and *miR-149* rs2292832 G > T. By genotyping these four SNPs in a case-control analysis including 1058 lung cancer patients and 1035 age-, sex-, and residential area-matched cancer-free controls, Tian et al. (2009) found that individuals with the CC genotype of *miR-196a2* rs11614913 had about 25% significantly increased risk for the development of lung cancer compared with individuals with the TT and TC genotype carriers (odds ratio [OR] = 1.25, 95% confidence interval [CI] = 1.01–1.54, $P = 0.038$). However, no significant associations were observed between the other three SNPs and lung cancer risk (Tian et al. 2009). In another two case-control studies in Korea, a similar significant effect of *miR-196a2* rs11614913 polymorphism on lung cancer risk was found (Kim et al. 2010; Hong et al. 2011). Kim et al. (2010) showed that the risk of lung cancer increased as the number of C alleles increased ($P_{trend} = 0.02$), and the CC genotype was associated with a significantly increased lung cancer risk compared with the TT genotype (OR = 1.45, 95% CI = 1.07–1.98, $P = 0.02$) in 654 Korean lung cancer cases and 640 Korean controls. Hong et al. (2011) also found that participants with TC/CC genotypes had higher risk for NSCLC compared with those with the TT genotype (OR = 1.42, 95% CI = 1.03–1.96, $P = 0.05$) in 406 Korean lung cancer patients and 428 Korean healthy controls. However, they reported no association of the CC genotype with lung cancer in the recessive genetic model, which is inconsistent with the results reported in the Chinese population. These aforementioned three studies suggest that *miR-196a2* rs11614913 polymorphism might be a common genetic factor in lung cancer etiology in East Asian populations.

The impact of the four SNPs (*miR-146a* rs2910164 C > G, *miR-196a2* rs11614913 C > T, *miR-499* rs3746444 G > A, and *miR-149* rs2292832 G > T) on lung cancer survival has also been examined in 663 Chinese patients (556 cases in test set and 107 in validation set) (Hu et al. 2008). It was found that only *miR-196a2* rs11614913 polymorphism was significantly associated with NSCLC survival. Significantly shorter median survival time was observed in individuals carrying the homozygous rs11614913 CC genotype (21.4 months) compared with individuals with the TT/CT genotype (27.3 months) (log-rank test, $P = 0.004$). Multivariate Cox proportional hazard regression analysis also showed that the rs11614913 CC genotype was an unfavorable prognostic factor (hazard ratio [HR] = 1.76, 95% CI = 1.34–2.33). Functional studies revealed that the rs11614913 CC genotype was associated with significantly increased levels of mature *miR-196a* expression but not with expression levels of pre-*miR-196a* in 23 human lung cancer tissue samples. This suggests that this SNP might enhance processing of the pre-*miR-196a* to its mature form. In addition, the rs11614913 polymorphism may influence binding of mature *miR-196a2-3p* to its target mRNA. Therefore, the *miR-196a2* rs11614913 C > T polymorphism might be a candidate biomarker for NSCLC survival.

B. SNPs in Pre-miRNA Flanking Regions

Genetic variants in the flanking regions (up to 450 bp in both sides) of pre-miRNA genes can also influence miRNA expression and, thus, might have impact on clinical phenotypes of NSCLC. Hu et al. evaluated the role of 85 SNPs in the flanking region of 400 known human pre-miRNA genes among 923 NSCLC patients in China and found that the *miR-30c-1* rs928508 A > G polymorphism was significantly associated with NSCLC survival. The association was more pronounced among stage I/II patients and those treated with surgery. Functional analyses demonstrated that rs928508 G allele was associated with a significantly decreased expression of precursor and mature *miR-30c*, but not with that of its primary miRNA. The authors suggest that *miR-30c-1* rs928508 SNP in the pre-miRNA flanking region is a potential prognostic biomarker of NSCLC (Hu et al. 2011).

III. GENETIC POLYMORPHISMS IN miRNA TARGET SITES AND LUNG CANCER

A. A SNP in *KRAS* 3′-UTR Alters miRNA *let-7* Binding

As one of the earliest identified human miRNAs, *let-7* functions as a regulator of expression of many important oncogenes including *KRAS*. Decreased expression of the *let-7* family of miRNAs has been observed in NSCLC (Esquela-Kerscher and Slack 2006). Therefore, the *let-7* family members may act as tumor suppressors. By sequencing the 3′-UTR of the *KRAS* gene in tissue DNA samples from 74 NSCLC patients, Chin et al. firstly identified a SNP (rs61764370) in a *let-7* complementary site in *KRAS* 3′-UTR. They investigated the association between the SNP and susceptibility to NSCLC in two independent case-control sets (New Mexico set and Boston set) (Chin et al. 2008). Although no association was seen between the allele and lung cancer risk in both case-control sets, data of the New Mexico case-control set did show that rs61764370 TG and GG carriers who smoked <41 pack-years had a 2.3-fold increased risk for developing NSCLC (95% CI = 1.1–4.6, $P = 0.02$) compared with the TT genotype carriers who smoked <41 pack-years. This association was validated in an independent case-control set in Boston (for those who smoked <40 pack-years: OR = 1.36, 95% CI = 1.07–1.73, $P = 0.01$). Functional analyses revealed that rs61764370 G allele (risk allele) leads to KRAS overexpression in lung cancer cells (Chin et al. 2008). In all, the functional *KRAS* 3′-UTR polymorphism alters *let-7* binding and increases KRAS expression, which, in turn, might result in elevated NSCLC risk among moderate smokers.

To further evaluate clinical utility of the *KRAS* 3′-UTR SNP in NSCLC, Nelson et al. examined its association with the *KRAS* codon 12 mutation in tumor tissues as well as lung cancer survival (Nelson et al. 2010) since the *KRAS* mutation is a strong predictor for NSCLC survival (Nelson et al. 1999). Unfortunately, they did not find any association between the *KRAS* rs61764370 polymorphism and *KRAS* mutation status or lung cancer survival in 218 NSCLC patients, suggesting that this polymorphism, if any, has limited clinical utility for NSCLC.

The allele frequency of *KRAS* 3′-UTR rs61764370 SNP was also examined in 2433 individuals from a global set of 46 populations (Chin et al. 2008). Interestingly, the MAF of the SNP is significantly different across geographic populations, with 7.6% of the chromosomes tested in European populations, <2.0% of chromosomes tested in African populations, and 0.4% of chromosomes tested in Asian and Native American populations. This suggests that there might be different lung cancer genetic determinants for different populations, such as SNPs in 3′-UTR of other oncogenes or tumor suppressor genes.

B. Genetic Variation in an *miR-1827*-Binding Site of *MYCL1*

To systematically evaluate how SNPs within miR-binding sites of target gene 3′-UTRs influence SCLC risk, we developed a novel *in silico* approach using public available profiling data (Xiong et al. 2011). First, we queried against the cancer-related microarray databases using Web-based NextBio software (http://www.nextbio.com) to find genes deregulated at the RNA level in SCLC and used the existing Patrocles database (http://www.patrocles.org) to identify putative SNPs within the miR-targeting site of the 3′-UTR of interest genes (Hiard et al. 2010; Kupershmidt et al. 2010). We identified 26 genes that are considerably deregulated in SCLC when the cutoff score is set above 85 (using NextBio, a higher score means that a gene is more substantially associated with a disease).

Among these genes, 17 had a total of 53 putative SNPs located in the 3′-UTR that might create or destroy miR-binding sites. We chose the SNPs for further analysis according to their MAF of 5% or above in the Han Chinese population (HapMap data release 21a), and with this criterion, only two SNPs, rs3134615 G > T in the 3′-UTR of *MYCL1* (*L-MYC*) and rs2291854 C > T in the 3′-UTR of mammalian achaete–scute complex homologue 1 (*MASH1* or *ASCL1*), were finally selected. We then investigated their associations with SCLC susceptibility in 666 SCLC patients and 758 controls. We found that the rs3134615 T allele was associated with a significantly increased risk of SCLC, with the OR for carrying the GT or TT genotype being 2.08 (95% CI = 1.39–3.21, P = 0.0004) compared with the GG genotype. However, no statistically significant association between *MASH1* rs2291854 C > T polymorphism and SCLC risk was observed. Stratification analyses showed that the rs3134615 GT/TT genotypes were associated with SCLC risk in both smokers (OR = 3.15, 95% CI = 1.86–5.32) and non-smokers (OR = 2.01, 95% CI = 1.03–3.92), indicating that rs3134615 polymorphism is a risk factor for the development of SCLC independent on smoking.

A set of functional assays *in vitro* and in cells showed that MYCL1 might be a target gene of *miR-1827*, which negatively regulates MYCL1 expression. The rs3134615 G-to-T change may inhibit the interaction of has-*miR-1827* with *MYCL1* mRNA 3′-UTR, resulting in higher constitutive expression of MYCL1. Since MYCL1 is one of MYC oncogene family members that play a critical part during malignant transformation, individuals carrying the rs3134615 T allele would be expected to have elevated risk for the development of SCLC. Because SCLC is one of the most aggressive cancers with very poor prognosis, clear understanding of genetic factors for the development of this cancer is imperative. Our results provide a new insight into SCLC carcinogenesis and have potential implications in early detection and targeted treatment of SCLC (Liu and Chen 2011; Xiong et al. 2011).

C. *REV3L* 3′-UTR 460 T > C SNP Influences Gene Regulation by *miR-25* and *miR-32*

REV3L, a catalytic subunit of DNA polymerase zeta, mainly participates in translesion DNA synthesis (translesion DNA synthesis and homologous DNA recombination are two major postreplicational DNA repair pathways). Recent evidence suggests that REV3L also plays an important role in the maintenance of genome stability despite its mutagenic characteristics. Colony formation assays indicate that REV3L can attenuate proliferation of cells, therefore potentially reducing the possibility of cancer development (Zhang et al. 2013). A significant reduction of REV3L expression in lung carcinomas compared with that of the normal adjacent tissues also suggests a tumor suppressor role for REV3L (Brondello et al. 2008; Zhang et al. 2013). This functional evidence suggests that *REV3L* might be a candidate susceptibility gene of lung cancer. To study association between *REV3L* polymorphisms and lung cancer risk, Zhang et al. first genotyped 15 common *REV3L* SNPs in a Chinese population and found that three SNPs (rs465646, rs459809, and rs1002481) were significantly associated with lung cancer risk. One of the strongest associations observed was for the 3′-UTR 460 T > C polymorphism (rs465646). The *REV3L* 3′-UTR 460TC/CC genotypes had a decreased lung cancer risk compared with the 460TT genotype (OR = 0.69, 95% CI = 0.53–0.90, P = 0.008). This result was then successfully validated in a subsequent replication study (OR = 0.72, 95% CI = 0.55–0.94, P = 0.016). A stronger association was found in combined data from the two studies (1072 lung cancer patients vs. 1064 cancer-free controls, OR = 0.71, 95% CI = 0.59–0.85, $P = 3.04 \times 10^{-4}$). To assess the *in silico* prediction that this 3'-UTR T-to-C change may influence the binding of *miR-25* and *miR-32*, surface plasmon resonance analysis and luciferase reporter assays were done. It was found that the T allele has a stronger binding affinity for *miR-25* and *miR-32* and significantly decreased reporter gene expression level compared with the C allele. All these data suggest that the *REV3L* rs465646 polymorphism affects lung cancer susceptibility by modifying *miR-25/-32*-mediated posttranscriptional gene regulation.

D. A Functional Polymorphism at the *miR-629*-Binding Site of *NBS1*

It has been reported that several genetic polymorphisms in *NBS1* are associated with increased risk of different cancers (Lu et al. 2009; Park et al. 2010; Zheng et al. 2011). Since genetic variants in the 3'-UTR region of *NBS1* might affect miR-mediated expression regulation of the gene, Yang et al. (2012) examined whether genetic polymorphisms in 3'-UTR of the *NBS1* gene influence gene expression and, thus, lung cancer susceptibility. In two independent case-control studies conducted in Southern and Eastern Chinese, three NBS1 tag SNPs (rs14448, rs13312986, and rs2735383) were genotyped. The authors reported that rs2735383 but not rs13312986 or rs14448 variant was significantly associated with lung cancer susceptibility. Under a recessive genetic model, the rs2735383 CC genotype showed a significantly increased risk of lung cancer in a total of 1559 patients versus 1679 controls (OR = 1.40, 95% CI = 1.18–1.66, $P = 0.0001$) compared with the GG or GC genotype. Results of quantitative polymerase chain reaction and Western blotting assays revealed that individuals carrying the rs2735383 CC genotype had lower mRNA and protein expression levels in tumor tissues than those with the other genotypes. Luciferase reporter gene assays indicated that the rs2735383 C allele had a lower transcription activity than the G allele. *miR-629* had substantial impact on the expression of *NBS1 in vitro*. These observations demonstrated that the *NBS1* rs2735383 G > C variation contributes to an increased risk of lung cancer by diminishing gene's expression probably through blocking *miR-629* bound to the 3'-UTR of *NBS1*.

IV. CONCLUSIONS

miRNAs are a class of small non-coding RNA molecules that regulate gene expression through binding to the 3'-UTR of target mRNAs, resulting in mRNA cleavage or translation repression (Bartel 2004). As a result, miRNAs act as oncogenes or tumor suppressor genes by regulating tumor suppressor genes or oncogenes and are involved in carcinogenesis and cancer progression (Esquela-Kerscher and Slack 2006). It has been shown that SNPs located in miR-encoding sequences or the 3'-UTR of target mRNA might affect gene regulation and, thus, contribute to individual differences in susceptibility to lung cancer or disease progression (Hu et al. 2008, 2011; Ryan et al. 2010). We could speculate that with more and more new miRNAs discovered and their biological function identified, more miR-related genetic polymorphisms and their impact on lung cancer (and other cancers) risk will be revealed in the future, which would enable us to better understand the etiology and mechanism for the development of human lung cancer.

REFERENCES

Ambros, V. 2004. The functions of animal microRNAs. *Nature* **431**:350–355.

Bartel, D.P. 2004. MicroRNAs: genomics, biogenesis, mechanism, and function. *Cell* **116**: 281–297.

Bi, N., M. Yang, L. Zhang, X. Chen, W. Ji, G. Ou, D. Lin, and L. Wang. 2010. Cyclooxygenase-2 genetic variants are associated with survival in unresectable locally advanced non-small cell lung cancer. *Clin Cancer Res* **16**:2383–2390.

Brondello, J.M., M.J. Pillaire, C. Rodriguez, P.A. Gourraud, J. Selves, C. Cazaux, and J. Piette. 2008. Novel evidences for a tumor suppressor role of Rev3, the catalytic subunit of Pol zeta. *Oncogene* **27**:6093–6101.

Chin, L.J., E. Ratner, S. Leng, R. Zhai, S. Nallur, I. Babar, R.U. Muller, E. Straka, L. Su, E.A. Burki, R.E. Crowell, R. Patel, et al. 2008. A SNP in a *let-7* microRNA complementary site in the KRAS 3′ untranslated region increases non-small cell lung cancer risk. *Cancer Res* **68**:8535–8540.

Cho, W.C. 2009. Role of miRNAs in lung cancer. *Expert Rev Mol Diagn* **9**:773–776.

Chu, H., M. Wang, D. Shi, L. Ma, Z. Zhang, N. Tong, X. Huo, W. Wang, D. Luo, Y. Gao, and Z. Zhang. 2011. *MiR-196a2* Rs11614913 polymorphism contributes to cancer susceptibility: evidence from 15 case-control studies. *PLoS ONE* **6**:e18108.

Esquela-Kerscher, A. and F.J. Slack. 2006. Oncomirs-microRNAs with a role in cancer. *Nat Rev Cancer* **6**:259–269.

Gandara, D., S. Narayan, P.N. Lara Jr., Z. Goldberg, A. Davies, D.H. Lau, P. Mack, P. Gumerlock, and S. Vijayakumar. 2005. Integration of novel therapeutics into combined modality therapy of locally advanced non-small cell lung cancer. *Clin Cancer Res* **11**:5057s–5062s.

Griffiths-Jones, S. 2004. The microRNA registry. *Nucleic Acids Res* **32**:D109–D111.

Gurubhagavatula, S., G. Liu, S. Park, W. Zhou, L. Su, J.C. Wain, T.J. Lynch, D.S. Neuberg, and D.C. Christiani. 2004. XPD and XRCC1 genetic polymorphisms are prognostic factors in advanced non-small-cell lung cancer patients treated with platinum chemotherapy. *J Clin Oncol* **22**:2594–2601.

Hiard, S., C. Charlier, W. Coppieters, M. Georges, and D. Baurain. 2010. Patrocles: a database of polymorphic *miR-mediated* gene regulation in vertebrates. *Nucleic Acids Res* **38**:D640–D651.

Hong, Y.S., H.J. Kang, J.Y. Kwak, B.L. Park, C.H. You, Y.M. Kim, and H. Kim. 2011. Association between microRNA196a2 rs11614913 genotypes and the risk of non-small cell lung cancer in Korean population. *J Prev Med Public Health* **44**:125–130.

Hu, Z., J. Chen, T. Tian, X. Zhou, H. Gu, L. Xu, Y. Zeng, R. Miao, G. Jin, H. Ma, Y. Chen, and H. Shen. 2008. Genetic variants of miRNA sequences and non-small cell lung cancer survival. *J Clin Invest* **118**:2600–2608.

Hu, Z., Y. Shu, Y. Chen, J. Chen, J. Dong, Y. Liu, S. Pan, L. Xu, J. Xu, Y. Wang, J. Dai, H. Ma, et al. 2011. Genetic polymorphisms in the precursor microRNA flanking region and non-small cell lung cancer survival. *Am J Respir Crit Care Med* **183**:641–648.

Jackman, D.M. and B.E. Johnson. 2005. Small-cell lung cancer. *Lancet* **366**:1385–1396.

Kim, M.J., S.S. Yoo, Y.Y. Choi, and J.Y. Park. 2010. A functional polymorphism in the pre-microRNA-196a2 and the risk of lung cancer in a Korean population. *Lung Cancer* **69**: 127–129.

Kupershmidt, I., Q.J. Su, A. Grewal, S. Sundaresh, I. Halperin, J. Flynn, M. Shekar, H. Wang, J. Park, W. Cui, G.D. Wall, R. Wisotzkey, et al. 2010. Ontology-based meta-analysis of global collections of high-throughput public data. *PLoS ONE* **5**:e13066.

Lee, C.B., T.E. Stinchcombe, J.G. Rosenman, and M.A. Socinski. 2006. Therapeutic advances in local-regional therapy for stage III non-small-cell lung cancer: evolving role of dose-escalated conformal (3-dimensional) radiation therapy. *Clin Lung Cancer* **8**:195–202.

REFERENCES

Lin, P.Y., S.L. Yu, and P.C. Yang. 2010. MicroRNA in lung cancer. *Br J Cancer* **103**:1144–1148.

Liu, H.Y. and J. Chen. 2011. Polymorphisms in miRNA binding site: new insight into small cell lung cancer susceptibility. *Acta Pharmacol Sin* **32**:1191–1192.

Lu, M., J. Lu, X. Yang, M. Yang, H. Tan, B. Yun, and L. Shi. 2009. Association between the NBS1 E185Q polymorphism and cancer risk: a meta-analysis. *BMC Cancer* **9**:124.

Nelson, H.H., B.C. Christensen, S.L. Plaza, J.K. Wiencke, C.J. Marsit, and K.T. Kelsey. 2010. KRAS mutation, KRAS-LCS6 polymorphism, and non-small cell lung cancer. *Lung Cancer* **69**:51–53.

Nelson, H.H., E.J. Mark, J.K. Wiencke, J.C. Wain, and K.T. Kelsey. 1999. Implications and prognostic value of K-ras mutation for early-stage lung cancer in women. *J Natl Cancer Inst* **91**:2032–2038.

Park, S.L., D. Bastani, B.Y. Goldstein, S.C. Chang, W. Cozen, L. Cai, C. Cordon-Cardo, B. Ding, S. Greenland, N. He, S.K. Hussain, Q. Jiang, et al. 2010. Associations between NBS1 polymorphisms, haplotypes and smoking-related cancers. *Carcinogenesis* **31**:1264–1271.

Parkin, D.M., F. Bray, J. Ferlay, and P. Pisani. 2005. Global cancer statistics, 2002. *CA Cancer J Clin* **55**:74–108.

Ryan, B.M., A.I. Robles, and C.C. Harris. 2010. Genetic variation in microRNA networks: the implications for cancer research. *Nat Rev Cancer* **10**:389–402.

Tian, T., Y. Shu, J. Chen, Z. Hu, L. Xu, G. Jin, J. Liang, P. Liu, X. Zhou, R. Miao, H. Ma, Y. Chen, et al. 2009. A functional genetic variant in microRNA-196a2 is associated with increased susceptibility of lung cancer in Chinese. *Cancer Epidemiol Biomarkers Prev* **18**:1183–1187.

Tibaldi, C., E. Giovannetti, E. Vasile, V. Mey, A.C. Laan, S. Nannizzi, R. Di Marsico, A. Antonuzzo, C. Orlandini, S. Ricciardi, M. Del Tacca, G.J. Peters, et al. 2008. Correlation of CDA, ERCC1, and XPD polymorphisms with response and survival in gemcitabine/cisplatin-treated advanced non-small cell lung cancer patients. *Clin Cancer Res* **14**:1797–1803.

Xiong, F., C. Wu, J. Chang, D. Yu, B. Xu, P. Yuan, K. Zhai, J. Xu, W. Tan, and D. Lin. 2011. Genetic variation in an *miR-1827* binding site in MYCL1 alters susceptibility to small-cell lung cancer. *Cancer Res* **71**:5175–5181.

Yanaihara, N., N. Caplen, E. Bowman, M. Seike, K. Kumamoto, M. Yi, R.M. Stephens, A. Okamoto, J. Yokota, T. Tanaka, G.A. Calin, C.G. Liu, et al. 2006. Unique microRNA molecular profiles in lung cancer diagnosis and prognosis. *Cancer Cell* **9**:189–198.

Yang, L., Y. Li, M. Cheng, D. Huang, J. Zheng, B. Liu, X. Ling, Q. Li, X. Zhang, W. Ji, Y. Zhou, and J. Lu. 2012. A functional polymorphism at microRNA-629-binding site in the 3′-untranslated region of NBS1 gene confers an increased risk of lung cancer in Southern and Eastern Chinese population. *Carcinogenesis* **33**:338–347.

Yu, D., X. Zhang, J. Liu, P. Yuan, W. Tan, Y. Guo, T. Sun, D. Zhao, M. Yang, J. Liu, B. Xu, and D. Lin. 2008. Characterization of functional excision repair cross-complementation group 1 variants and their association with lung cancer risk and prognosis. *Clin Cancer Res* **14**:2878–2886.

Zhang, S., H. Chen, X. Zhao, J. Cao, J. Tong, J. Lu, W. Wu, H. Shen, Q. Wei, and D. Lu. 2013. REV3L 3′UTR 460 T>C polymorphism in microRNA target sites contributes to lung cancer susceptibility. *Oncogene* **32**:242–250.

Zheng, J., C. Zhang, L. Jiang, Y. You, Y. Liu, J. Lu, and Y. Zhou. 2011. Functional NBS1 polymorphism is associated with occurrence and advanced disease status of nasopharyngeal carcinoma. *Mol Carcinog* **50**:689–696.

21

MicroRNA IN MYELOPOIESIS AND MYELOID DISORDERS

Sara E. Meyer[1] and H. Leighton Grimes[1,2]

[1]*Division of Cellular and Molecular Immunology, Cincinnati Children's Hospital Medical Center, Cincinnati, OH, USA*
[2]*Division of Experimental Hematology and Cancer Biology, Cincinnati Children's Hospital Medical Center, Cincinnati, OH, USA*

I.	Introduction	355
II.	Significance of Myeloid Biology in the Earliest Mammalian miRNA Discoveries	356
III.	Critical Requirements of miRNA Signaling to Instruct Proper Myeloid Cell Fate	357
	A. miRNA Expression Patterns in Normal Human Stem/Progenitor and Differentiated Cells of the Myeloid Lineage	357
	B. miRNA and Transcription Factor Coregulation Is Required to Control Normal Myelopoiesis and Are Commonly Deregulated in Myeloid Diseases	358
IV.	Diagnostic, Prognostic, Predictive, and Therapeutic Applications of miRNA in Myeloid Disorders and AML	360
	A. miRNA Expression Has Diagnostic, Prognostic, and Predictive Significance in Myeloid Disorders and Malignancies	360
	B. The Future of miRNA in Myeloid Disorders and Leukemias: RNA Therapeutics Move toward the Clinic	363
Acknowledgments		364
References		365

MicroRNAs in Medicine, First Edition. Edited by Charles H. Lawrie.
© 2014 John Wiley & Sons, Inc. Published 2014 by John Wiley & Sons, Inc.

ABBREVIATIONS

ALL	acute lymphoblastic leukemia
AML	acute myelogenous leukemia
AML1-ETO	fusion oncoprotein created by chromosomal translocation between AML1 gene (chromosome 21) and ETO gene (chromosome 8)
AMO	anti-miRNA oligonucleotide
APL	acute promyelocytic leukemia
ATRA	all-trans retinoic acid
Baso	basophil
BM	bone marrow
CBFβ	core binding factor-beta
CD	cluster of differentiation
C/EBPα	CCAAT/enhancer binding protein-alpha
CML	chronic myelogenous leukemia
CMP	common myeloid progenitor
CN-AML	cytogenetically normal acute myeloid leukemia
ELANE	neutrophil elastase
Eos	eosinophil
EP	erythroid progenitor
Er	erythrocyte
G-CSF	granulocyte colony-stimulating factor
GFI1	growth factor independent 1
GMP	granulocyte–monocyte progenitor
Gr1	myeloid differentiation antigen Ly-6G a cell surface marker of granulocytes and neutrophils
GTP	guanosine-5′-triphosphate
HSC	hematopoietic stem cell
HSPC	hematopoietic stem and progenitor cell
Lin–	cells negative for lineage markers
LNA	locked nucleic acid
Mac	macrophage
Mac1	CD11b antigen a cell surface marker for monocytes
M-CSFR	macrophage colony-stimulating factor receptor
mDC	myeloid-derived dendritic cell
MDS	myelodysplastic syndrome
MeK	megakaryocyte
MEP	megakaryocyte–erythrocyte progenitor
miRNA*	microRNA star species/strand
miRNA/miR	microRNA
MLL	mixed-lineage leukemia
MLL-AF9	fusion oncoprotein created by chromosomal translocation between MLL (chromosome 11q23) and AF9 (chromosome 9)
MPN/MPD	myeloproliferative neoplasm/disorder
MPP	multipotent progenitor
Neut	neutrophil
NPM1	nucleophosmin 1
nt	nucleotide
PB	peripheral blood

Plat	platelet
PV	polycythemia vera
RISC	RNA-induced silencing complex
RNAi	RNA interference
RUNX1/AML1	runt-related transcription factor 1

I. INTRODUCTION

Adult hematopoiesis is a multilineage differentiation cascade of hierarchical cell divisions originating from adult multipotent hematopoietic stem and progenitor cells (HSPCs) in the bone marrow (BM) leading to the generation of all mature blood cell types of the innate and adaptive immune system. The focus of this chapter is the production of all mature myeloid lineage cell types including granulocytes (neutrophils, eosinophils, and basophils; i.e., granulopoiesis), macrophages (i.e., monopoiesis), megakaryocytes (platelets) (i.e., megakaryopoiesis), erythrocytes (i.e., erythropoiesis), and myeloid–dendritic cells. It is broadly accepted that during adult human hematopoiesis, HSCs in the BM give rise to multipotent progenitor (MPP) cells that in turn give rise to subtypes of hematopoietic progenitor cells that become successively more lineage committed. Human common myeloid progenitors (CMPs), granulocyte–monocyte progenitors (GMPs), and megakaryocyte–erythrocyte progenitors (MEPs) have been phenotypically defined by *in vivo* xenograft models (Doulatov et al. 2012). CMPs generate GMP and MEP as progeny progenitors. GMPs give rise to neutrophils, eosinophils, basophils, and monocytes/macrophages, while MEPs differentiate into megakaryocytes, platelets, and erythrocytes. The precise function of the carefully controlled intra- and extracellular signaling networks that determine myeloid cell fate must be intact to produce a normal myeloid system and immune function. One consequence of this hierarchical developmental is that disruption of a singular signaling entity, such as mutations in key myelopoietic transcription factors early on in the differentiation cascade, can have a dramatic impact on the myeloid system. In addition, signaling from cell surface receptors toward transcription factors is known to be central to regulate the proper development and maintenance of all blood cell types. For example, disruption of normal myeloid differentiation can result in myeloproliferative syndromes (too many neutrophils), or neutropenias (too few neutrophils). Additional aberrations such as mutations or chromosomal translocations involving cytokine/growth factor receptors or epigenetic regulators are causal in the development of myeloid malignancies, including myelodysplastic syndrome (MDS, wherein not enough mature myeloid cells are made), myeloproliferative neoplasms/disorders (MPNs/MPDs, which occur upon disruption of myeloid progenitor maintenance resulting in the accumulation of myeloid cells), and acute myelogenous leukemia (AML).

It is now evident that microRNAs (miRNAs) have an essential role as downstream intermediates and/or effectors of intracellular signaling pathways that control normal myelopoiesis and are deregulated in myeloid malignancies (Stoffers et al. 2012).

miRNAs demonstrate distinct expression profiles associated with a wide variety of normal and malignant tissues. For example, *miR-142*, *miR-144*, *miR-155*, *miR-150*, and *miR-223* are significantly expressed in the hematopoietic system compared with all other human tissues (Landgraf et al. 2007). The expression patterns of miRNAs throughout normal and malignant myelopoiesis suggest that miRNAs control myeloid differentiation, proliferation, and survival; the disruption of which leads to myeloid diseases.

The characterization of miRNAs in normal human myelopoiesis could, however, be considered incomplete in part due to the obscurity of normal human myeloid cell populations that can be purified from normal donors' peripheral blood (PB) and BM. As a supplement, studies of miRNAs in human cell lines of myeloid origin and comparative normal murine myelopoiesis have been critical to our understanding on this topic. One advantage of myelopoiesis as a model is the power of mouse genetics and the application of mouse models of human disease, including humanized murine models that closely mimic normal and defective human myelopoiesis and myeloid malignancies. Such models allow experimental *in vivo* validation of miRNA function in many aspects of myeloid biology (Zuber et al. 2009; Doulatov et al. 2012). On the other hand, cells are more readily attainable from patients with myeloid disorders and malignancies such as MDS and AML, allowing the identification of many critical miRNAs involved in myeloid diseases. Importantly, some studies have used anti-miRNA and miRNA mimic technologies to correct deregulated miRNA expression and associated disease pathology *in vitro*. Very little is known about the efficacy of such miRNA therapeutic technologies on myeloid disease *in vivo*. Furthermore, the limited understanding of the mechanisms governing miRNA signaling presents significant challenges to any field studying miRNA biology. Highlighted herein are several important miRNAs, some of which laid the foundation for investigations of miRNAs in myeloid biology that are deregulated in myeloid disease with significant roles during normal myelopoiesis. Lastly, the possible future clinical utility of miRNAs as disease markers, diagnostic tools, prognostic indicators, RNA interference (RNAi) therapeutic targets, and gene therapy modulators will be addressed.

II. SIGNIFICANCE OF MYELOID BIOLOGY IN THE EARLIEST MAMMALIAN miRNA DISCOVERIES

lin-4, the first miRNA described in 1993, was shown to regulate the translation of *lin-14* through complimentary binding to the 3′-untranslated region (3′-UTR) of *lin-14* (Lee et al. 1993). In *Caenorhabditis elegans*, *lin-4* regulates developmental patterning at all larval stages (Reinhart et al. 2000). Moving from *C. elegans* to a mammalian animal model, Lagos-Quintana et al. (2002) identified 34 novel miRNAs, in addition to those previously identified in worms, through cloning and sequencing from mouse tissues. Calin et al. (2002) provided the first evidence of deregulated miRNA expression in human leukemia patients. Then, Chen et al. (2004) were the first to show differential expression of miRNAs in hematopoietic tissues and to experimentally demonstrate the miRNA function in mammals. *miR-142s* was expressed in the BM, spleen, and thymus in both lymphoid and myeloid cell types, whereas *miR-181* expression was found preferentially in HSPCs of the BM and *miR-223* expression was restricted to myeloid lineages (Gr1+ and Mac1+) within the BM compartment. Transplantation of murine HSPCs ectopically expressing each of these miRNAs into irradiated mice in an *in vivo* murine hematopoietic differentiation assay resulted in altered hematopoietic lineage outputs (Chen et al. 2004). Together, these seminal research studies on miRNA differential expression patterns in the mammalian hematopoietic system provided the very first evidence that miRNAs might have important roles in normal hematopoietic differentiation and the pathogenesis of defective malignant hematopoiesis.

III. CRITICAL REQUIREMENTS OF miRNA SIGNALING TO INSTRUCT PROPER MYELOID CELL FATE

A. miRNA Expression Patterns in Normal Human Stem/Progenitor and Differentiated Cells of the Myeloid Lineage

Over the last decade, researchers have identified temporal, lineage-specific, cell type-specific patterns of miRNA expression profiles in purified human and murine hematopoietic cell populations. This section highlights miRNAs discovered in progenitor and some differentiated myeloid cell types purified from normal human patient samples; however, these studies and others including immortalized human cell lines and murine hematopoietic cells have been comprehensively reviewed elsewhere (Garzon and Croce 2008; Pelosi et al. 2009; Petrocca and Lieberman 2009a; Navarro and Lieberman 2010; O'Connell et al. 2011; Palma et al. 2012). The CD34 protein is a cell surface antigen routinely used to distinguish myeloid and lymphoid stem and progenitor cells (CD34+) from (CD34−) lineage-committed and terminally differentiated cells, which gain other surface marker expression. In the hematopoietic stem/progenitor compartment, Georgantas et al. (2007) identified 35 miRNAs expressed in PB and 78 in BM, 33 of which were expressed in CD34+ HSPCs from both compartments. Next, using miRNA–messenger RNA (mRNA) target prediction algorithms integrated with mRNA expression profiles derived from purified lineage-committed hematopoietic cells, the authors postulated the functional or regulatory roles the miRNAs may have in the differentiation of specific myeloid cell types. In particular, miRNAs *miR-17*, *miR-24a*, and *miR-155* (myeloid restricted), and miRNAs *miR-128a* and *miR-181a* (myeloid and lymphoid) exhibited the strongest association with regulators of early cell fate decisions of the myeloid lineage transition from MPP to CMP. *miR-16a* and *miR-107* showed significant correlation with genes involved in the CMP to GMP transition. *miR-221* and *miR-222* predicted targets are involved in erythroid differentiation from MEP, and *miR-223* predicted targets are involved in granulocytic differentiation from GMPs. Allantaz et al. (2012) performed miRNA expression profiling on nine human mature myeloid blood cell types including neutrophils, eosinophils, myeloid dendritic cells, and monocytes isolated from normal donors each finding six cell type-specific miRNAs including *miR-143* in neutrophils, *miR-500* in monocytes, and *miR-652* and *miR-223* in myeloid cell types. Ramkissoon et al. (2006) investigated the expression of a select set of known hematopoietic-associated miRNAs in a panel of malignant hematopoietic cell lines and normal human granulocytes and monocytes isolated from normal donor PB. In agreement with the above-mentioned studies, the authors found *miR-223* specifically expressed in primary human myeloid cells (granulocytes and monocytes compared with low–no expression in T and B cells). Although studies on primary normal human hematopoietic cells are limited, these analyses suggest that miRNAs may be master regulators of human myelopoiesis. In Figure 21.1, a general outline of normal human myelopoiesis including miRNAs expressed (white letters) in purified human myeloid populations is shown. Although separable human progenitor populations have been characterized and are illustrated in Figure 21.1 (e.g., CMP, GMP) (Doulatov et al. 2012), only inferences have been made as to which miRNAs may function in each human progenitor population. Experimentation is required to resolve the role of differential miRNA expression in the different human myeloid progenitor populations.

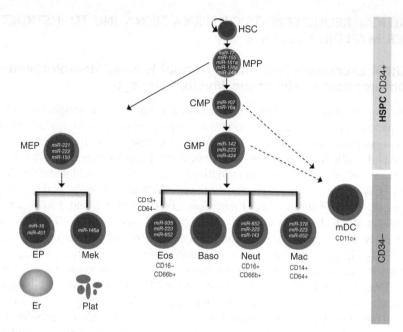

Figure 21.1. MicroRNAs expressed in purified human myeloid cells that function during normal myelopoiesis. The hierarchy of normal human myelopoiesis is illustrated. Depicted is each cell type (purple with black letters) during the myeloid differentiation cascade (denoted by arrows). mDCs are thought to derive from either CMP or GMP; this relationship is not yet firmly understood (dashed arrows). The microRNAs illustrated (white letters) were chosen based on two criteria. First, these microRNAs were identified in purified myeloid cell populations from normal human donors without *in vitro* manipulations. The microRNAs were further restricted to those with experimental functional evidence in *in vitro* or *in vivo* human and murine models. CD markers used to purify myeloid subsets are indicated. See color insert.

B. miRNA and Transcription Factor Coregulation Is Required to Control Normal Myelopoiesis and Are Commonly Deregulated in Myeloid Diseases

Moving beyond basic analyses of miRNA expression in unmanipulated human progenitor and differentiated myeloid cell populations, the field employs *in vitro* human cell and *in vivo* murine models to test the hypothesis that differential expression of miRNA has a functional role in myelopoiesis and myeloid diseases. There are several well-established factors that signal to specify myeloid lineage cell fate. Importantly, not only do these factors have key roles during normal myelopoiesis but disruption of these signaling pathways is also frequently the basis of myeloid disorders and malignancies. Notably, most of these factors also directly or indirectly regulate miRNA expression and are subject to miRNA regulation themselves. Surprisingly, only neutropenia, MDS, MPN, and AML have been characterized for miRNA deregulation in primary unmanipulated human patient samples diagnosed with myeloid diseases; therefore, this section focuses on the miRNAs found to be central to these diseases, which are also enforced by other *in vitro* or *in vivo* models.

The transcription factor growth factor independent 1 (*GFI1*) is critical for neutrophil formation and myeloid progenitor maintenance (Horman et al. 2009). *GFI1*, or other genes such as neutrophil elastase (*ELANE*), is found mutated in neutropenia, and disruption of GFI1 is sufficient to produce neutropenia in mice. GFI1 directly represses *miR-21* and *miR-196b* (Horman et al. 2009; Velu et al. 2009). Overexpression of *miR-196b* decreases granulopoiesis, whereas *miR-21* overexpression increases monopoiesis in Lin− murine BM cells (Velu et al. 2009). *miR-196b* and *miR-21* are deregulated in human neutropenia patients with GFI1 inactivating mutations (Velu et al. 2009). Conversely, the histone methyltransferase mixed-lineage leukemia (MLL) can positively regulate expression of the *miR-196b* locus, and leukemias harboring chromosomal translocations involving *MLL* (11q23) significantly overexpress *miR-196b* compared with other subtypes of AML (Jongen-Lavrencic et al. 2008; Popovic et al. 2009). Importantly, *miR-196b* inhibition in murine MLL-AF9-transformed cells significantly reduces the colony-forming (transformation assay) and replating ability (immortalization) (Popovic et al. 2009).

miR-223, one of the first miRNAs described in mammals and one of the most well-studied miRNAs in myeloid biology, is a transcriptional target of CCAAT/enhancer binding protein, alpha (C/EBPα). C/EBPα is another essential factor in myelopoiesis. *miR-223* expression is significantly increased during normal murine and human granulopoiesis and upon all-trans retinoic acid (ATRA)-induced myeloid differentiation of acute promyelocytic leukemia (APL) cells (Fazi et al. 2005; Fukao et al. 2007; Saumet et al. 2009; Pulikkan et al. 2010a). Seemingly contradictory to this, *miR-223* null mice have increased granulocyte production (Johnnidis et al. 2008). This may be explained at least in part by a lack of cell-cycle exit that cells normally undergo during terminal differentiation. Indeed, *miR-223* was shown to target cell-cycle regulator *E2F1* blocking cell-cycle progression (Pulikkan et al. 2010a). Dicer cleavage of the miRNA hairpin stem loop results in a 21- to 24-nt miRNA duplex consisting of the mature strand that is incorporated into RNA-induced silencing complex (RISC) and the star (miRNA*) strand that is often degraded. In myeloid cells, *miR-223** and *miR-126** strands are readily detectable (Li et al. 2008; Kuchenbauer et al. 2011). Thus, another possibility is that *miR-223** functions during normal granulopoiesis. Expression of *miR-223** in *miR-223* null mice reduces granulopoiesis but leads to accumulation of myeloid progenitors (Kuchenbauer et al. 2011). C/EBPα also positively regulates *miR-34a* expression, which is specifically decreased in association with *C/EBPα* mutant leukemias (Pulikkan et al. 2010b; Marcucci et al. 2011). Consistent with C/EBPα regulation, *miR-34a* expression increases during granulopoiesis and may induce megakaryopoiesis through c-Myb inhibition (Navarro et al. 2009; Pulikkan et al. 2010b).

The tumor suppressor *miR-29b* increases during myeloid differentiation. *miR-29b* is transcriptionally increased by C/EBPα or transcriptionally repressed by c-Myc and nuclear factor kappa-light chain enhancer of activated B cells (NF-κB) in AML compared with normal BM (Eyholzer et al. 2010; Liu et al. 2010).

Runt-related transcription factor 1 (RUNX1/AML1) is part of a heterodimeric monocyte-macrophage differentiation regulatory core binding factor (CBF) transcription factor complex with core binding factor-beta (CBFβ). The RUNX1/CBFβ transcription factor complex is required for normal monopoiesis. AML1 negatively transcriptionally regulates the expression of the miRNA clusters *miR-17-92* and *miR-106a-92*. These miRNA clusters can in turn repress RUNX1 protein expression, leading to reduced macrophage colony-stimulating factor receptor (M-CSFR) expression required for differentiation, thereby resulting in increased blast cell proliferation and decreased monopoiesis (Fontana et al. 2007). Coordinately, the *miR-17-92* cluster is significantly decreased in AML, harboring chromosomal rearrangements in members of the CBF complex *AML1* and *CBFβ* (Li et al. 2008). In addition, *miR-223* is epigenetically silenced in AML1-ETO, t(8;21) leukemia cells; however, whether RUNX1 controls *miR-223* expression during normal myelopoiesis is not known (Fazi et al. 2007). Conversely, c-Myc drives increased expression of the oncogenic miRNA cluster *miR-17-92* found in MDS and contributing to myeloid leukemias associated with 11q23 translocations (He et al. 2005; Aguda et al. 2008; Li et al. 2008; Pons et al. 2009).

miR-451 expression increases in human CD34+ cells induced to undergo multistage erythroid differentiation, and its expression is further detected in polycythemia vera (PV), an MPD wherein too many red blood cells are produced (Zhan et al. 2007; Bruchova et al. 2008). The *miR-144/451* cluster is under control of the transcription factor GATA-1, a critical regulator of erythropoiesis (Dore et al. 2008). The transcription factor PU.1 positively regulates *miR-155* expression and is involved in neutrophil–monocyte cell fate (Ghani et al. 2011). The balance between PU.1 and GATA-1 is important for megakaryocyte–erythroid cell fates, as these factors function antagonistically to balance lineage cell fates (Zhang et al. 1999). The decline in *miR-155* expression promotes megakaryocytic differentiation of CD34+ cells *in vitro*, and forced expression of *miR-155* decreases erythromegakaryocyte differentiation (Georgantas et al. 2007; Romania et al. 2008). *miR-155* is overexpressed in MPN/MPD and AML patients with activating mutations in the tyrosine kinase receptor *FLT3*, including FLT3 internal tandem duplication (FLT3-ITD). Importantly, *miR-155* overexpression alone is sufficient to cause a MPD in mice (O'Connell et al. 2008). Chronic myelogenous leukemia (CML) is the result of a chromosomal translocation (t(9;22) or the Philadelphia chromosome) involving the tyrosine kinase *ABL* and the *BCR* gene, generating the BCR-ABL fusion oncoprotein. CML is associated with decreased expression of *miR-150* and increased expression of several miRNAs in the *miR-17-92* cluster, possibly through c-Myc, a known transcriptional regulator of the miRNA cluster that is activated downstream of BCR-ABL activity (Xie et al. 2002; O'Donnell et al. 2005; Machova Polakova et al. 2011). In sum, miRNAs are regulated by central hematopoietic differentiation transcription factors, are deregulated during leukemogenesis, and have been demonstrated to play functionally important roles in both normal hematopoiesis and leukemogenesis.

IV. DIAGNOSTIC, PROGNOSTIC, PREDICTIVE, AND THERAPEUTIC APPLICATIONS OF miRNA IN MYELOID DISORDERS AND AML

A. miRNA Expression Has Diagnostic, Prognostic, and Predictive Significance in Myeloid Disorders and Malignancies

miRNA expression profiling in normal and malignant hematopoietic tissues has uncovered tremendous opportunities for the possible use of miRNAs in disease diagnosis, prognosis, and prediction of outcome. Some studies suggest that the use of miRNA as molecular biomarkers may be even more accurate and reliable than clinically established protein-coding genes. While not yet implemented in the clinic, the use of miRNAs as biomarkers could be of significant usefulness to medicine in the near future.

miRNA expression signatures, and in some cases a single miRNA, can distinguish between normal and disease hematopoietic cells, and between hematologic disease types. The following are examples of miRNAs that may have diagnostic value in myeloid malignancies. There are seven miRNAs that have been shown in at least two independent studies to be differentially expressed in AML compared with normal BM: up-regulated *miR-21*, *miR-155*, *miR-222*, *miR-223*, *let-7b*, and *miR-181a* and down-regulated *miR-29b* (Dixon-McIver et al. 2008; Isken et al. 2008; Jongen-Lavrencic et al. 2008; Li et al. 2008; O'Connell et al. 2008; Zhu et al. 2011). *miR-10a*, *miR-10b*, and *miR-126* expressions are increased in patients with MDS compared with normal controls, and *miR-145* is encoded within the deleted region on chromosome 5q and decreased in association with 5q– MDS found in several independent studies (Pons et al. 2009; Starczynowski et al. 2010; Kumar et al. 2011; Votavova et al. 2011). *miR-378*, also encoded within the 5q deleted region, is the most discriminatory miRNA in 5q– MDS compared with normal controls (Dostalova Merkerova et al. 2011; Erdogan et al. 2011; Votavova et al. 2011). The expressions of *miR-181a*, *miR-223*, *miR-128a*, and *miR-128b* are the most discriminatory between AML and acute lymphoblastic leukemia (ALL) patients wherein *miR-181a* and *miR-223* are highly expressed in AML and *miR-128a/b* in ALL (Mi et al. 2007; Zhu et al. 2011). miRNA expression profiles can discriminate between the French–American–British (FAB) classification of different cytogenetic subtypes of AML (Lu et al. 2005; Jongen-Lavrencic et al. 2008; Li et al. 2008).

In managing treatment strategies for patients with hematologic malignancies, it can be helpful to stratify patients based on known risks and outcomes associated with their specific diseases. Different miRNA expression patterns correlate with overall survival, prognostic favorable or unfavorable cytogenetic subgroups, and with disease/outcome risk groups (Dixon-McIver et al. 2008; Jongen-Lavrencic et al. 2008; Marcucci et al. 2009, 2011). In the future, miRNAs may be used as prognostic biomarkers to provide information about overall patient outcome. Much less is known about the utility of miRNAs as predictive biomarkers as there are very few attempts to correlate miRNA expression with the effects of a therapeutic intervention in myeloid malignancies, some of which have even failed to find a connection (Lopotova et al. 2011; Scholl et al. 2012). Table 21.1 illustrates the most promising miRNAs as putative diagnostic markers. *miR-451* and *miR-181a* (bold letters) have reported prognostic value in myeloid disease. CML patients with low *miR-451* expression have a greater risk of relapse with frontline tyrosine kinase inhibitor (imatinib/dasatinib) treatment (Lopotova et al. 2011; Scholl et al. 2012). Elevated *miR-181a* expression in patients with cytogenetically normal AML (CN-AML), independent of known gene mutation status, is associated with increased chance of complete remission and longer overall survival (Marcucci et al. 2008a, 2008b; Schwind et al. 2010; Havelange et al. 2011).

NPM1c mutations are the most common (50–60%) mutation in CN-AML conferring a favorable prognosis associated with increased expression of *miR-10a/b*, *miR-196a/b*, and *let-7b* (Jongen-Lavrencic et al. 2008; Becker et al. 2010; Cammarata et al. 2010; Russ et al. 2011; Bryant et al. 2012; Danen-van Oorschot et al. 2012). *NPM1c* mutants have a better prognosis than wild-type *NPM1* CN-AML; however, *NPM1* wild-type patients with higher expression of *miR-181* are associated with a better outcome (Marcucci et al. 2008b; Schwind et al. 2010). Expression of *miR-181a* is strongly associated with CN-AML, in particular, those harboring *C/EBPα* mutations and is indicative of a favorable prognosis (Marcucci et al. 2005, 2008a; Jongen-Lavrencic et al. 2008; Havelange et al. 2011). *C/EBPα* mutant CN-AML also demonstrates reduced expression of *miR-196b* and *miR-196a*. This is congruent with CBF chromosomal rearrangements in AML that also have low *miR-196b* expression associated with a low-risk subgroup; however, in contrast, AML harboring 11q23 translocations are an intermediate-risk subgroup and have high expression of *miR-196b*.

TABLE 21.1. Potential Diagnostic and Prognostic MicroRNA (miRNA) Biomarkers in Myeloid Diseases

Myeloid Disease	miRNA Up-Regulated	miRNA Down-Regulated	Genetic Abnormality	Prognosis	Prevalence (%)	References
CN-AML					40–50	Garzon et al. (2008a)
	miR-155		FLT3-ITD	Poor	25	Garzon et al. (2008b); Jongen-Lavrencic et al. (2008); Cammarata et al. (2010); Foran (2010); Havelange et al. (2011)
	miR-181a	miR-34a	C/EBPα	Favorable	15	Jongen-Lavrencic et al. (2008); Marcucci et al. (2008a); Pulikkan et al. (2010b); Havelange et al. (2011)
	miR-10a/b miR-196a/b		NPM1c	Favorable	50–60	Dixon-McIver et al. (2008); Garzon et al. (2008a); Foran (2010)
	miR-181a		Variable	Favorable		Marcucci et al. (2008a, 2008b); Schwind et al. (2010); Havelange et al. (2011)
AML-abnormal karyotype					25	Garzon et al. (2008a)
	miR-17-92 miR-196a/b		t(9;11), 11q23	Intermediate–poor		Jongen-Lavrencic et al. (2008); Li et al. (2008); Marcucci et al. (2009); Foran (2010); Mi et al. (2010); Danen-van Oorschot et al. (2012)
	miR-126/126*	miR-223	t(8;21), inv(16)	Favorable		Fazi et al. (2007); Jongen-Lavrencic et al. (2008); Li et al. (2008); Foran (2010)
	miR-127 miR-224		t(15;17)	Favorable		Dixon-McIver et al. (2008); Jongen-Lavrencic et al. (2008); Li et al. (2008); Foran (2010)
CML		**miR-451**	BCR-ABL	Poor		Lopotova et al. (2011); Scholl et al. (2012)
PV	miR-16	miR-150	JAK2 (V617F)		90	Bruchova et al. (2007, 2008); Guglielmelli et al. (2011)

Note: Prevalence is noted as percentage within myeloid disease type; for CN-AML and AML-abnormal karyotype prevalence is noted as percentage of total AML. miRNAs in bold letters are prognostic and plain texts are diagnostic.

In spite of all of these important clinical findings relating to the potential use of miRNAs as biomarkers, one major issue that has arisen is an apparent lack of consensus by and large across expression studies in myeloid malignancies and disorders as to which miRNAs are deregulated (Marcucci et al. 2011). All of the aforementioned miRNAs in this section were handpicked as representative of findings in the field because they were found in at least two independent studies; however, given the abundance of miRNA expression studies, particularly on AML patients, the majority of differentially expressed miRNAs found in one study do not translate to the next. This challenge to the field can be a result of many variables such as which population is appropriate to normalize miRNA expression: normal whole PB, BM, mononuclear (MNC), or CD34+ cells? Each of these choices, although derived from normal patients, consists of vastly different cell populations. Based on the changing miRNA expression patterns throughout myelopoiesis as discussed earlier, a comparison of miRNA expression in AML patients to PB, which contains many differentiated cell types of the immune system versus a comparison made to CD34+ HSPCs, is likely to give completely different miRNA expression results. Other possibilities for this discrepancy are the method for isolation and purification of miRNA, differences in technological global miRNA quantification assay platforms, and statistical analyses.

B. The Future of miRNA in Myeloid Disorders and Leukemias: RNA Therapeutics Move toward the Clinic

The disruption of normal human myelopoiesis has a range of consequences from reduced immune function to more severe diseases like leukemia. Although substantial progress has been made in the field finding better treatments for patients with myeloid disorders, not all patients respond to the currently available therapies. Thus, miRNAs may represent a way to personalize therapeutics, which is postulated to increase the chance for individual response. Severe neutropenias can be treated with granulocyte colony-stimulating factor (G-CSF) cytokine therapy; however, some patients do not respond to this treatment (Dale et al. 2003; Donini et al. 2007). The 5-year overall survival rate of patients with AML is approximately 15–20% in the United States (http://seer.cancer.gov) (Howlader et al. 2012). The loss and gain of miRNA function contribute to the pathogenesis of these and many other diseases, and represent plausible targets for therapeutic intervention. Such potential molecular therapies include miRNA mimics to restore loss of function and anti-miRNA oligonucleotides (AMOs) to inhibit gain-of-function miRNAs in myeloid diseases. Although delivery, stability, specificity, toxicity, and efficacy remain major obstacles to the implementation of each miRNA-targeted therapeutic in the clinic, there is encouraging research demonstrating the effectiveness of these technologies *in vitro*, and in more recent studies *in vivo*. The topic of miRNA in cancer therapeutics has been expertly reviewed in recent years (Petrocca and Lieberman 2009b; Phalon et al. 2010; Wahid et al. 2010; Garofalo and Croce 2011; Toscano et al. 2011). Alas, to date, myeloid diseases seem to be severely underrepresented in even the preclinical trial phases (United States) in the miRNA therapeutic category (http://clinicaltrials.gov 2012). It is apparent, however, that basic research is currently paving the way toward this end goal.

In 2005, the "antagomir," an RNA analogue with $2'$-O-methyl ($2'$-OMe) modified on every ribonucleotide, partially phosphorothioated backbone linkage (for stability) with a $3'$ cholesterol conjugated (for cell entry) complimentary to *miR-122* was the first type of anti-miRNA successfully used in an *in vivo* animal model to block miRNA function (Krutzfeldt et al. 2005). Furthermore, Krutzfeldt et al. (2005) demonstrated that antagomirs could target miRNA expressed in multiple murine tissues including BM. Since this landmark study, several different AMO chemistries have been developed. One such example is a fully phosphorothioated locked nucleic acid (LNA) DNA chemistry as short as eight nucleotides in length complimentary to the seed region of an miRNA, or even family of conserved miRNAs, that sufficiently inhibited miRNA function *in vivo* (Obad et al. 2011).

Presented here are highlighted examples of possible miRNA therapeutics that show promise for translation to future clinical use for myeloid malignancies. An attractive target for miRNA inhibitory therapeutics is the widely studied *miR-21*. *miR-21* is overexpressed in the majority of AML (Jongen-Lavrencic et al. 2008). Several groups are putting forth efforts to investigate the optimal AMO design to specifically inhibit *miR-21* activity *in vivo* in other cancer types; however, it is conceivable that the approval of miRNA-targeted therapy for use in one disease could be applied to a multitude of others (Obad et al. 2011; Munoz-Alarcon et al. 2012). Another instance where myeloid malignancies may benefit from the studies on miRNA therapeutics in other diseases is the development of antagomir against *miR-17-5p* that was shown to inhibit the growth and therapy-resistant neuroblastoma *in vivo* (Fontana et al. 2008). *miR-17-5p* is part of the oncogenic *miR-17-92* cluster highly up-regulated in subtypes of AML (Li et al. 2008; Mi et al. 2010). *miR-155* expression is elevated in many cancers including in specific association with AML harboring FLT3 mutations (Garzon et al. 2008b). *miR-155* was shown to be targetable in hematopoietic tissues as nanoparticle-based delivery of anti-*miR-155* inhibited *miR-155* to slow the growth of B-cell lymphomas in mice (Babar et al. 2012). An LNA against *miR-10a* in an NPM1c mutant human AML cell line induced cell death *in vitro* (Bryant et al. 2012). Antagomir directed against *miR-196b* significantly reduced the immortalization and transforming ability of the fusion oncoprotein MLL-AF9, generated by the t(9;11) in human AML, in murine Lin– BM cells *in vitro* (Popovic et al. 2009). Conversely, successful use of lentiviral gene transfer to increase *miR-29a* or *miR-142-3p* in AML patient blasts resulted in a gain of miRNA function that promoted myeloid differentiation *in vitro* (Wang et al. 2012).

The most recent and perhaps novel use of miRNAs is in gene therapeutics. miRNA target site sequences included in the next-generation gene therapy vectors can be used to better restrict transgene expression to target cells. The very first report on this concept was in a hematopoietic system in mice (Brown et al. 2006; Gentner et al. 2010). Because this next-generation gene therapy is so new, it is difficult to predict whether miRNA regulation will be the key to sustained gene transfer *in vivo*.

The use of miRNA technologies in the clinic may be in the distant future for myeloid diseases given that the majority of studies are still in initial target validation phases *in vitro*. Perhaps many of these studies in the recent past have stalled prior to preclinical examination since most gain- and loss-of-function miRNA studies aim to merely understand the general miRNA biology or mechanism of action in myeloid biology. Thus, the tools for *in vivo* targeting and inhibition miRNA function exist; however, the future direction of this research should focus to implement these technologies *in vivo*.

ACKNOWLEDGMENTS

The authors would like to acknowledge funding from The Ladies Auxiliary to the Veterans of Foreign Wars Postdoctoral Cancer Research Fellowship (S.E.M.) and NIH/NCI R01-CA159845 (H.L.G.).

REFERENCES

Aguda, B.D., Y. Kim, M.G. Piper-Hunter, A. Friedman, and C.B. Marsh. 2008. MicroRNA regulation of a cancer network: consequences of the feedback loops involving *miR-17-92*, E2F, and Myc. *Proceedings of the National Academy of Sciences of the United States of America* **105**: 19678–19683.

Allantaz, F., D.T. Cheng, T. Bergauer, P. Ravindran, M.F. Rossier, M. Ebeling, L. Badi, B. Reis, H. Bitter, M. D'Asaro, A. Chiappe, S. Sridhar, et al. 2012. Expression profiling of human immune cell subsets identifies miRNA-mRNA regulatory relationships correlated with cell type specific expression. *PLoS ONE* **7**:e29979.

Babar, I.A., C.J. Cheng, C.J. Booth, X. Liang, J.B. Weidhaas, W.M. Saltzman, and F.J. Slack. 2012. Nanoparticle-based therapy in an in vivo microRNA-155 (*miR-155*)-*dependent* mouse model of lymphoma. *Proceedings of the National Academy of Sciences of the United States of America* **109**:E1695–E1704.

Becker, H., G. Marcucci, K. Maharry, M.D. Radmacher, K. Mrozek, D. Margeson, S.P. Whitman, Y.Z. Wu, S. Schwind, P. Paschka, B.L. Powell, T.H. Carter, et al. 2010. Favorable prognostic impact of NPM1 mutations in older patients with cytogenetically normal de novo acute myeloid leukemia and associated gene- and microRNA-expression signatures: a Cancer and Leukemia Group B study. *Journal of Clinical Oncology: Official Journal of the American Society of Clinical Oncology* **28**:596–604.

Brown, B.D., M.A. Venneri, A. Zingale, L. Sergi Sergi, and L. Naldini. 2006. Endogenous microRNA regulation suppresses transgene expression in hematopoietic lineages and enables stable gene transfer. *Nature Medicine* **12**:585–591.

Bruchova, H., D. Yoon, A.M. Agarwal, J. Mendell, and J.T. Prchal. 2007. Regulated expression of microRNAs in normal and polycythemia vera erythropoiesis. *Experimental Hematology* **35**:1657–1667.

Bruchova, H., M. Merkerova, and J.T. Prchal. 2008. Aberrant expression of microRNA in polycythemia vera. *Haematologica* **93**:1009–1016.

Bryant, A., C.A. Palma, V. Jayaswal, Y.W. Yang, M. Lutherborrow, and D.D. Ma. 2012. *miR-10a* is aberrantly overexpressed in Nucleophosmin1 mutated acute myeloid leukaemia and its suppression induces cell death. *Molecular Cancer* **11**:8.

Calin, G.A., C.D. Dumitru, M. Shimizu, R. Bichi, S. Zupo, E. Noch, H. Aldler, S. Rattan, M. Keating, K. Rai, L. Rassenti, T. Kipps, et al. 2002. Frequent deletions and down-regulation of micro-RNA genes miR15 and miR16 at 13q14 in chronic lymphocytic leukemia. *Proceedings of the National Academy of Sciences of the United States of America* **99**:15524–15529.

Cammarata, G., L. Augugliaro, D. Salemi, C. Agueli, M. La Rosa, L. Dagnino, G. Civiletto, F. Messana, A. Marfia, M.G. Bica, L. Cascio, P.M. Floridia, et al. 2010. Differential expression of specific microRNA and their targets in acute myeloid leukemia. *American Journal of Hematology* **85**:331–339.

Chen, C.Z., L. Li, H.F. Lodish, and D.P. Bartel. 2004. MicroRNAs modulate hematopoietic lineage differentiation. *Science* **303**:83–86.

Dale, D.C., T.E. Cottle, C.J. Fier, A.A. Bolyard, M.A. Bonilla, L.A. Boxer, B. Cham, M.H. Freedman, G. Kannourakis, S.E. Kinsey, R. Davis, D. Scarlata, et al. 2003. Severe chronic neutropenia: treatment and follow-up of patients in the Severe Chronic Neutropenia International Registry. *American Journal of Hematology* **72**:82–93.

Danen-van Oorschot, A.A., J.E. Kuipers, S. Arentsen-Peters, D. Schotte, V. de Haas, J. Trka, A. Baruchel, D. Reinhardt, R. Pieters, C.M. Zwaan, and M.M. van den Heuvel-Eibrink. 2012. Differentially expressed miRNAs in cytogenetic and molecular subtypes of pediatric acute myeloid leukemia. *Pediatric Blood & Cancer* **58**:715–721.

Dixon-McIver, A., P. East, C.A. Mein, J.B. Cazier, G. Molloy, T. Chaplin, T. Andrew Lister, B.D. Young, and S. Debernardi. 2008. Distinctive patterns of microRNA expression associated with karyotype in acute myeloid leukaemia. *PLoS ONE* **3**:e2141.

Donini, M., S. Fontana, G. Savoldi, W. Vermi, L. Tassone, F. Gentili, E. Zenaro, D. Ferrari, L.D. Notarangelo, F. Porta, F. Facchetti, L.D. Notarangelo, et al. 2007. G-CSF treatment of severe congenital neutropenia reverses neutropenia but does not correct the underlying functional deficiency of the neutrophil in defending against microorganisms. *Blood* **109**:4716–4723.

Dore, L.C., J.D. Amigo, C.O. Dos Santos, Z. Zhang, X. Gai, J.W. Tobias, D. Yu, A.M. Klein, C. Dorman, W. Wu, R.C. Hardison, B.H. Paw, et al. 2008. A GATA-1-regulated microRNA locus essential for erythropoiesis. *Proceedings of the National Academy of Sciences of the United States of America* **105**:3333–3338.

Dostalova Merkerova, M., Z. Krejcik, H. Votavova, M. Belickova, A. Vasikova, and J. Cermak. 2011. Distinctive microRNA expression profiles in CD34+ bone marrow cells from patients with myelodysplastic syndrome. *European Journal of Human Genetics: EJHG* **19**:313–319.

Doulatov, S., F. Notta, E. Laurenti, and J.E. Dick. 2012. Hematopoiesis: a human perspective. *Cell Stem Cell* **10**:120–136.

Erdogan, B., C. Facey, J. Qualtieri, J. Tedesco, E. Rinker, R.B. Isett, J. Tobias, D.A. Baldwin, J.E. Thompson, M. Carroll, and A.S. Kim. 2011. Diagnostic microRNAs in myelodysplastic syndrome. *Experimental Hematology* **39**:915–926 e912.

Eyholzer, M., S. Schmid, L. Wilkens, B.U. Mueller, and T. Pabst. 2010. The tumour-suppressive *miR-29a/b1* cluster is regulated by CEBPA and blocked in human AML. *British Journal of Cancer* **103**:275–284.

Fazi, F., A. Rosa, A. Fatica, V. Gelmetti, M.L. De Marchis, C. Nervi, and I. Bozzoni. 2005. A minicircuitry comprised of microRNA-223 and transcription factors NFI-A and C/EBPα regulates human granulopoiesis. *Cell* **123**:819–831.

Fazi, F., S. Racanicchi, G. Zardo, L.M. Starnes, M. Mancini, L. Travaglini, D. Diverio, E. Ammatuna, G. Cimino, F. Lo-Coco, F. Grignani, and C. Nervi. 2007. Epigenetic silencing of the myelopoiesis regulator microRNA-223 by the AML1/ETO oncoprotein. *Cancer Cell* **12**:457–466.

Fontana, L., E. Pelosi, P. Greco, S. Racanicchi, U. Testa, F. Liuzzi, C.M. Croce, E. Brunetti, F. Grignani, and C. Peschle. 2007. MicroRNAs 17-5p-20a-106a control monocytopoiesis through AML1 targeting and M-CSF receptor upregulation. *Nature Cell Biology* **9**:775–787.

Fontana, L., M.E. Fiori, S. Albini, L. Cifaldi, S. Giovinazzi, M. Forloni, R. Boldrini, A. Donfrancesco, V. Federici, P. Giacomini, C. Peschle, and D. Fruci. 2008. Antago*mir-17-5p* abolishes the growth of therapy-resistant neuroblastoma through p21 and BIM. *PLoS ONE* **3**:e2236.

Foran, J.M. 2010. New prognostic markers in acute myeloid leukemia: perspective from the clinic. *Hematology/the Education Program of the American Society of Hematology. American Society of Hematology. Education Program* **2010**:47–55.

Fukao, T., Y. Fukuda, K. Kiga, J. Sharif, K. Hino, Y. Enomoto, A. Kawamura, K. Nakamura, T. Takeuchi, and M. Tanabe. 2007. An evolutionarily conserved mechanism for microRNA-223 expression revealed by microRNA gene profiling. *Cell* **129**:617–631.

Garofalo, M. and C.M. Croce. 2011. MicroRNAs: master regulators as potential therapeutics in cancer. *Annual Review of Pharmacology and Toxicology* **51**:25–43.

Garzon, R. and C.M. Croce. 2008. MicroRNAs in normal and malignant hematopoiesis. *Current Opinion in Hematology* **15**:352–358.

Garzon, R., M. Garofalo, M.P. Martelli, R. Briesewitz, L. Wang, C. Fernandez-Cymering, S. Volinia, C.G. Liu, S. Schnittger, T. Haferlach, A. Liso, D. Diverio, et al. 2008a. Distinctive microRNA signature of acute myeloid leukemia bearing cytoplasmic mutated nucleophosmin. *Proceedings of the National Academy of Sciences of the United States of America* **105**:3945–3950.

Garzon, R., S. Volinia, C.G. Liu, C. Fernandez-Cymering, T. Palumbo, F. Pichiorri, M. Fabbri, K. Coombes, H. Alder, T. Nakamura, N. Flomenberg, G. Marcucci, et al. 2008b. MicroRNA signatures associated with cytogenetics and prognosis in acute myeloid leukemia. *Blood* **111**:3183–3189.

Gentner, B., I. Visigalli, H. Hiramatsu, E. Lechman, S. Ungari, A. Giustacchini, G. Schira, M. Amendola, A. Quattrini, S. Martino, A. Orlacchio, J.E. Dick, et al. 2010. Identification of hematopoietic stem cell-specific miRNAs enables gene therapy of globoid cell leukodystrophy. *Science Translational Medicine* **2**:58ra84.

Georgantas, R.W., 3rd, R. Hildreth, S. Morisot, J. Alder, C.G. Liu, S. Heimfeld, G.A. Calin, C.M. Croce, and C.I. Civin. 2007. CD34+ hematopoietic stem-progenitor cell microRNA expression and function: a circuit diagram of differentiation control. *Proceedings of the National Academy of Sciences of the United States of America* **104**:2750–2755.

Ghani, S., P. Riemke, J. Schonheit, D. Lenze, J. Stumm, M. Hoogenkamp, A. Lagendijk, S. Heinz, C. Bonifer, J. Bakkers, S. Abdelilah-Seyfried, M. Hummel, et al. 2011. Macrophage development from HSCs requires PU.1-coordinated microRNA expression. *Blood* **118**:2275–2284.

Guglielmelli, P., L. Tozzi, C. Bogani, I. Iacobucci, V. Ponziani, G. Martinelli, A. Bosi, and A.M. Vannucchi. 2011. Overexpression of microRNA-16-2 contributes to the abnormal erythropoiesis in polycythemia vera. *Blood* **117**:6923–6927.

Havelange, V., N. Stauffer, C.C. Heaphy, S. Volinia, M. Andreeff, G. Marcucci, C.M. Croce, and R. Garzon. 2011. Functional implications of microRNAs in acute myeloid leukemia by integrating microRNA and messenger RNA expression profiling. *Cancer* **117**:4696–4706.

He, L., J.M. Thomson, M.T. Hemann, E. Hernando-Monge, D. Mu, S. Goodson, S. Powers, C. Cordon-Cardo, S.W. Lowe, G.J. Hannon, and S.M. Hammond. 2005. A microRNA polycistron as a potential human oncogene. *Nature* **435**:828–833.

Horman, S.R., C.S. Velu, A. Chaubey, T. Bourdeau, J. Zhu, W.E. Paul, B. Gebelein, and H.L. Grimes. 2009. Gfi1 integrates progenitor versus granulocytic transcriptional programming. *Blood* **113**:5466–5475.

Howlader, N., A.M. Noone, M. Krapcho, N. Neyman, R. Aminou, S.F. Altekruse, C.L. Kosary, J. Ruhl, Z. Tatalovich, H. Cho, A. Mariotto, M.P. Eisner, et al., Eds. 2012. SEER Cancer Statistics Review, 1975-2009 (Vintage 2009 Populations).

http://clinicaltrials.gov. 2012. ClinicalTrials.gov a service provided by the U.S. National Institutes of Health.

Isken, F., B. Steffen, S. Merk, M. Dugas, B. Markus, N. Tidow, M. Zuhlsdorf, T. Illmer, C. Thiede, W.E. Berdel, H. Serve, and C. Muller-Tidow. 2008. Identification of acute myeloid leukaemia associated microRNA expression patterns. *British Journal of Haematology* **140**:153–161.

Johnnidis, J.B., M.H. Harris, R.T. Wheeler, S. Stehling-Sun, M.H. Lam, O. Kirak, T.R. Brummelkamp, M.D. Fleming, and F.D. Camargo. 2008. Regulation of progenitor cell proliferation and granulocyte function by microRNA-223. *Nature* **451**:1125–1129.

Jongen-Lavrencic, M., S.M. Sun, M.K. Dijkstra, P.J. Valk, and B. Lowenberg. 2008. MicroRNA expression profiling in relation to the genetic heterogeneity of acute myeloid leukemia. *Blood* **111**:5078–5085.

Krutzfeldt, J., N. Rajewsky, R. Braich, K.G. Rajeev, T. Tuschl, M. Manoharan, and M. Stoffel. 2005. Silencing of microRNAs in vivo with "antagomirs". *Nature* **438**:685–689.

Kuchenbauer, F., S.M. Mah, M. Heuser, A. McPherson, J. Ruschmann, A. Rouhi, T. Berg, L. Bullinger, B. Argiropoulos, R.D. Morin, D. Lai, D.T. Starczynowski, et al. 2011. Comprehensive analysis of mammalian miRNA* species and their role in myeloid cells. *Blood* **118**:3350–3358.

Kumar, M.S., A. Narla, A. Nonami, A. Mullally, N. Dimitrova, B. Ball, J.R. McAuley, L. Poveromo, J.L. Kutok, N. Galili, A. Raza, E. Attar, et al. 2011. Coordinate loss of a microRNA and protein-coding gene cooperate in the pathogenesis of 5q- syndrome. *Blood* **118**:4666–4673.

Lagos-Quintana, M., R. Rauhut, A. Yalcin, J. Meyer, W. Lendeckel, and T. Tuschl. 2002. Identification of tissue-specific microRNAs from mouse. *Current Biology: CB* **12**:735–739.

Landgraf, P., M. Rusu, R. Sheridan, A. Sewer, N. Iovino, A. Aravin, S. Pfeffer, A. Rice, A.O. Kamphorst, M. Landthaler, C. Lin, N.D. Socci, et al. 2007. A mammalian microRNA expression atlas based on small RNA library sequencing. *Cell* **129**:1401–1414.

Lee, R.C., R.L. Feinbaum, and V. Ambros. 1993. The *C. elegans* heterochronic gene lin-4 encodes small RNAs with antisense complementarity to lin-14. *Cell* **75**:843–854.

Li, Z., J. Lu, M. Sun, S. Mi, H. Zhang, R.T. Luo, P. Chen, Y. Wang, M. Yan, Z. Qian, M.B. Neilly, J. Jin, et al. 2008. Distinct microRNA expression profiles in acute myeloid leukemia with common translocations. *Proceedings of the National Academy of Sciences of the United States of America* **105**:15535–15540.

Liu, S., L.C. Wu, J. Pang, R. Santhanam, S. Schwind, Y.Z. Wu, C.J. Hickey, J. Yu, H. Becker, K. Maharry, M.D. Radmacher, C. Li, et al. 2010. Sp1/NFkappaB/HDAC/*miR-29b* regulatory network in KIT-driven myeloid leukemia. *Cancer Cell* **17**:333–347.

Lopotova, T., M. Zackova, H. Klamova, and J. Moravcova. 2011. MicroRNA-451 in chronic myeloid leukemia: *miR-451-BCR-ABL* regulatory loop? *Leukemia Research* **35**:974–977.

Lu, J., G. Getz, E.A. Miska, E. Alvarez-Saavedra, J. Lamb, D. Peck, A. Sweet-Cordero, B.L. Ebert, R.H. Mak, A.A. Ferrando, J.R. Downing, T. Jacks, et al. 2005. MicroRNA expression profiles classify human cancers. *Nature* **435**:834–838.

Machova Polakova, K., T. Lopotova, H. Klamova, P. Burda, M. Trneny, T. Stopka, and J. Moravcova. 2011. Expression patterns of microRNAs associated with CML phases and their disease related targets. *Molecular Cancer* **10**:41.

Marcucci, G., K. Mrozek, and C.D. Bloomfield. 2005. Molecular heterogeneity and prognostic biomarkers in adults with acute myeloid leukemia and normal cytogenetics. *Current Opinion in Hematology* **12**:68–75.

Marcucci, G., K. Maharry, M.D. Radmacher, K. Mrozek, T. Vukosavljevic, P. Paschka, S.P. Whitman, C. Langer, C.D. Baldus, C.G. Liu, A.S. Ruppert, B.L. Powell, et al. 2008a. Prognostic significance of, and gene and microRNA expression signatures associated with, CEBPA mutations in cytogenetically normal acute myeloid leukemia with high-risk molecular features: a Cancer and Leukemia Group B Study. *Journal of Clinical Oncology: Official Journal of the American Society of Clinical Oncology* **26**:5078–5087.

Marcucci, G., M.D. Radmacher, K. Maharry, K. Mrozek, A.S. Ruppert, P. Paschka, T. Vukosavljevic, S.P. Whitman, C.D. Baldus, C. Langer, C.G. Liu, A.J. Carroll, et al. 2008b. MicroRNA expression in cytogenetically normal acute myeloid leukemia. *The New England Journal of Medicine* **358**:1919–1928.

Marcucci, G., K. Mrozek, M.D. Radmacher, C.D. Bloomfield, and C.M. Croce. 2009. MicroRNA expression profiling in acute myeloid and chronic lymphocytic leukaemias. *Best Practice & Research. Clinical Haematology* **22**:239–248.

Marcucci, G., K. Mrozek, M.D. Radmacher, R. Garzon, and C.D. Bloomfield. 2011. The prognostic and functional role of microRNAs in acute myeloid leukemia. *Blood* **117**:1121–1129.

Mi, S., J. Lu, M. Sun, Z. Li, H. Zhang, M.B. Neilly, Y. Wang, Z. Qian, J. Jin, Y. Zhang, S.K. Bohlander, M.M. Le Beau, et al. 2007. MicroRNA expression signatures accurately discriminate acute lymphoblastic leukemia from acute myeloid leukemia. *Proceedings of the National Academy of Sciences of the United States of America* **104**:19971–19976.

Mi, S., Z. Li, P. Chen, C. He, D. Cao, A. Elkahloun, J. Lu, L.A. Pelloso, M. Wunderlich, H. Huang, R.T. Luo, M. Sun, et al. 2010. Aberrant overexpression and function of the *miR-17-92* cluster in MLL-rearranged acute leukemia. *Proceedings of the National Academy of Sciences of the United States of America* **107**:3710–3715.

Munoz-Alarcon, A., P. Guterstam, C. Romero, M.A. Behlke, K.A. Lennox, J. Wengel, S. El Andaloussi, and U. Langel. 2012. Modulating anti-microRNA-21 activity and specificity using oligonucleotide derivatives and length optimization. *ISRN Pharmaceutics* **2012**:407154.

Navarro, F. and J. Lieberman. 2010. Small RNAs guide hematopoietic cell differentiation and function. *Journal of Immunology* **184**:5939–5947.

Navarro, F., D. Gutman, E. Meire, M. Caceres, I. Rigoutsos, Z. Bentwich, and J. Lieberman. 2009. *miR-34a* contributes to megakaryocytic differentiation of K562 cells independently of p53. *Blood* **114**:2181–2192.Obad, S., C.O. dos Santos, A. Petri, M. Heidenblad, O. Broom, C. Ruse, C. Fu,

REFERENCES

M. Lindow, J. Stenvang, E.M. Straarup, H.F. Hansen, T. Koch, et al. 2011. Silencing of microRNA families by seed-targeting tiny LNAs. *Nature Genetics* **43**:371–378.

O'Connell, R.M., D.S. Rao, A.A. Chaudhuri, M.P. Boldin, K.D. Taganov, J. Nicoll, R.L. Paquette, and D. Baltimore. 2008. Sustained expression of microRNA-155 in hematopoietic stem cells causes a myeloproliferative disorder. *The Journal of Experimental Medicine* **205**:585–594.

O'Connell, R.M., J.L. Zhao, and D.S. Rao. 2011. MicroRNA function in myeloid biology. *Blood* **118**:2960–2969.

O'Donnell, K.A., E.A. Wentzel, K.I. Zeller, C.V. Dang, and J.T. Mendell. 2005. c-Myc-regulated microRNAs modulate E2F1 expression. *Nature* **435**:839–843.

Palma, C.A., E.J. Tonna, D.F. Ma, and M.A. Lutherborrow. 2012. MicroRNA control of myelopoiesis and the differentiation block in acute myeloid leukaemia. *Journal of Cellular and Molecular Medicine* **16**:978–987.

Pelosi, E., C. Labbaye, and U. Testa. 2009. MicroRNAs in normal and malignant myelopoiesis. *Leukemia Research* **33**:1584–1593.

Petrocca, F. and J. Lieberman. 2009a. Micromanagers of immune cell fate and function. *Advances in Immunology* **102**:227–244.

Petrocca, F. and J. Lieberman. 2009b. Micromanipulating cancer: microRNA-based therapeutics? *RNA Biology* **6**:335–340.

Phalon, C., D.D. Rao, and J. Nemunaitis. 2010. Potential use of RNA interference in cancer therapy. *Expert Reviews in Molecular Medicine* **12**:e26.

Pons, A., B. Nomdedeu, A. Navarro, A. Gaya, B. Gel, T. Diaz, S. Valera, M. Rozman, M. Belkaid, E. Montserrat, and M. Monzo. 2009. Hematopoiesis-related microRNA expression in myelodysplastic syndromes. *Leukemia & Lymphoma* **50**:1854–1859.

Popovic, R., L.E. Riesbeck, C.S. Velu, A. Chaubey, J. Zhang, N.J. Achille, F.E. Erfurth, K. Eaton, J. Lu, H.L. Grimes, J. Chen, J.D. Rowley, et al. 2009. Regulation of *mir-196b* by MLL and its overexpression by MLL fusions contributes to immortalization. *Blood* **113**:3314–3322.

Pulikkan, J.A., V. Dengler, P.S. Peramangalam, A.A. Peer Zada, C. Muller-Tidow, S.K. Bohlander, D.G. Tenen, and G. Behre. 2010a. Cell-cycle regulator E2F1 and microRNA-223 comprise an autoregulatory negative feedback loop in acute myeloid leukemia. *Blood* **115**:1768–1778.

Pulikkan, J.A., P.S. Peramangalam, V. Dengler, P.A. Ho, C. Preudhomme, S. Meshinchi, M. Christopeit, O. Nibourel, C. Muller-Tidow, S.K. Bohlander, D.G. Tenen, and G. Behre. 2010b. C/EBPα regulated microRNA-34a targets E2F3 during granulopoiesis and is down-regulated in AML with CEBPA mutations. *Blood* **116**:5638–5649.

Ramkissoon, S.H., L.A. Mainwaring, Y. Ogasawara, K. Keyvanfar, J.P. McCoy Jr., E.M. Sloand, S. Kajigaya, and N.S. Young. 2006. Hematopoietic-specific microRNA expression in human cells. *Leukemia Research* **30**:643–647.

Reinhart, B.J., F.J. Slack, M. Basson, A.E. Pasquinelli, J.C. Bettinger, A.E. Rougvie, H.R. Horvitz, and G. Ruvkun. 2000. The 21-nucleotide *let-7* RNA regulates developmental timing in *Caenorhabditis elegans*. *Nature* **403**:901–906.

Romania, P., V. Lulli, E. Pelosi, M. Biffoni, C. Peschle, and G. Marziali. 2008. MicroRNA 155 modulates megakaryopoiesis at progenitor and precursor level by targeting Ets-1 and Meis1 transcription factors. *British Journal of Haematology* **143**:570–580.

Russ, A.C., S. Sander, S.C. Luck, K.M. Lang, M. Bauer, F.G. Rucker, H.A. Kestler, R.F. Schlenk, H. Dohner, K. Holzmann, K. Dohner, and L. Bullinger. 2011. Integrative nucleophosmin mutation-associated microRNA and gene expression pattern analysis identifies novel microRNA—target gene interactions in acute myeloid leukemia. *Haematologica* **96**:1783–1791.

Saumet, A., G. Vetter, M. Bouttier, E. Portales-Casamar, W.W. Wasserman, T. Maurin, B. Mari, P. Barbry, L. Vallar, E. Friederich, K. Arar, B. Cassinat, et al. 2009. Transcriptional repression of microRNA genes by PML-RARA increases expression of key cancer proteins in acute promyelocytic leukemia. *Blood* **113**:412–421.

Scholl, V., R. Hassan, and I.R. Zalcberg. 2012. miRNA-451: a putative predictor marker of Imatinib therapy response in chronic myeloid leukemia. *Leukemia Research* **36**:119–121.

Schwind, S., K. Maharry, M.D. Radmacher, K. Mrozek, K.B. Holland, D. Margeson, S.P. Whitman, C. Hickey, H. Becker, K.H. Metzeler, P. Paschka, C.D. Baldus, et al. 2010. Prognostic significance of expression of a single microRNA, *miR-181a*, in cytogenetically normal acute myeloid leukemia: a Cancer and Leukemia Group B study. *Journal of Clinical Oncology: Official Journal of the American Society of Clinical Oncology* **28**:5257–5264.

Starczynowski, D.T., F. Kuchenbauer, B. Argiropoulos, S. Sung, R. Morin, A. Muranyi, M. Hirst, D. Hogge, M. Marra, R.A. Wells, R. Buckstein, W. Lam, et al. 2010. Identification of *miR-145* and *miR-146a* as mediators of the 5q- syndrome phenotype. *Nature Medicine* **16**:49–58.

Stoffers, S.L., S.E. Meyer, and H.L. Grimes. 2012. MicroRNAs in the midst of myeloid signal transduction. *Journal of Cellular Physiology* **227**:525–533.

Toscano, M.G., Z. Romero, P. Munoz, M. Cobo, K. Benabdellah, and F. Martin. 2011. Physiological and tissue-specific vectors for treatment of inherited diseases. *Gene Therapy* **18**:117–127.

Velu, C.S., A.M. Baktula, and H.L. Grimes. 2009. Gfi1 regulates *miR-21* and *miR-196b* to control myelopoiesis. *Blood* **113**:4720–4728.

Votavova, H., M. Grmanova, M. Dostalova Merkerova, M. Belickova, A. Vasikova, R. Neuwirtova, and J. Cermak. 2011. Differential expression of microRNAs in CD34+ cells of 5q- syndrome. *Journal of Hematology & Oncology* **4**:1.

Wahid, F., A. Shehzad, T. Khan, and Y.Y. Kim. 2010. MicroRNAs: synthesis, mechanism, function, and recent clinical trials. *Biochimica et Biophysica Acta* **1803**:1231–1243.

Wang, X.S., J.N. Gong, J. Yu, F. Wang, X.H. Zhang, X.L. Yin, Z.Q. Tan, Z.M. Luo, G.H. Yang, C. Shen, and J.W. Zhang. 2012. MicroRNA-29a and microRNA-142-3p are regulators of myeloid differentiation and acute myeloid leukemia. *Blood* **119**:4992–5004.

Xie, S., H. Lin, T. Sun, and R.B. Arlinghaus. 2002. Jak2 is involved in c-Myc induction by Bcr-Abl. *Oncogene* **21**:7137–7146.

Zhan, M., C.P. Miller, T. Papayannopoulou, G. Stamatoyannopoulos, and C.Z. Song. 2007. MicroRNA expression dynamics during murine and human erythroid differentiation. *Experimental Hematology* **35**:1015–1025.

Zhang, P., G. Behre, J. Pan, A. Iwama, N. Wara-Aswapati, H.S. Radomska, P.E. Auron, D.G. Tenen, and Z. Sun. 1999. Negative cross-talk between hematopoietic regulators: GATA proteins repress PU.1. *Proceedings of the National Academy of Sciences of the United States of America* **96**:8705–8710.

Zhu, Y.D., L. Wang, C. Sun, L. Fan, D.X. Zhu, C. Fang, Y.H. Wang, Z.J. Zou, S.J. Zhang, J.Y. Li, and W. Xu. 2011. Distinctive microRNA signature is associated with the diagnosis and prognosis of acute leukemia. *Medical Oncology (Northwood, London, England)* **29**:2323–2331.

Zuber, J., I. Radtke, T.S. Pardee, Z. Zhao, A.R. Rappaport, W. Luo, M.E. McCurrach, M.M. Yang, M.E. Dolan, S.C. Kogan, J.R. Downing, and S.W. Lowe. 2009. Mouse models of human AML accurately predict chemotherapy response. *Genes & Development* **23**:877–889.

22

MicroRNA DEREGULATION BY ABERRANT DNA METHYLATION IN ACUTE LYMPHOBLASTIC LEUKEMIA

Xabier Agirre[1] and Felipe Prósper[1,2]

[1]*Oncology Division, Foundation for Applied Medical Research,*
University of Navarra, Pamplona, Spain
[2]*Hematology Service and Area of Cell Therapy, Clínica Universidad de Navarra,*
University of Navarra, Pamplona, Spain

I.	Introduction	372
	A. DNA Methylation in Human Cancer	372
	B. Acute Lymphoblastic Leukemia (ALL)	373
II.	miRNA Deregulation by DNA Methylation in ALL	374
	A. Aberrant DNA Methylation of miRNAs as a Prognostic Factor for ALL	374
	B. Functional Involvement of Aberrantly Methylated miRNAs in ALL	376
	C. Chromosomal Translocations and Aberrantly Methylated miRNAs in ALL	377
III.	Conclusion	379
	References	379

ABBREVIATIONS

ABL1	Abelson murine leukemia viral oncogene homologue 1
AF4	ALL1-fused gene from chromosome 4 gene
ALL	acute lymphoblastic leukemia
BCR	breakpoint cluster region
CDK6	cyclin-dependent kinase 6
CDKs	cyclin-dependent kinases
CpG	cytosine–phosphate–guanine

MicroRNAs in Medicine, First Edition. Edited by Charles H. Lawrie.
© 2014 John Wiley & Sons, Inc. Published 2014 by John Wiley & Sons, Inc.

DNA	deoxyribonucleic acid
DNMTs	DNA methyltransferases
ETV6	ets variant 6
HOXA	homeobox A
miRNA	microRNA
MLL	myeloid/lymphoid or mixed-lineage leukemia
PBX1	pre-B-cell leukemia homeobox 1
RB	retinoblastoma
RNA	ribonucleic acid
RUNX1	runt-related transcription factor 1
TCF3	*Homo sapiens* transcription factor 3
TP53	tumor protein p53 gene
UTR	untranslated region
ZEB2	zinc finger E-box-binding homeobox 2

I. INTRODUCTION

Many studies have shown that expression of microRNAs (miRNAs) is deregulated in a range of human cancers. This alteration in miRNA expression plays a very important role in the development, progression, and metastasis of these human neoplasias. miRNA expression studies also indicate that most miRNAs have lower expression levels in tumors than in normal tissues (Lu et al. 2005), suggesting that some of these miRNAs may act as putative tumor suppressor genes, as has subsequently been shown in some cases (Agirre et al. 2009b). In contrast, it has been observed that a smaller number of miRNAs are more strongly expressed in human cancers, including those responsible for oncogenic activity (Medina et al. 2010).

While hundreds of miRNAs have been discovered and a variety of functions of miRNAs have been intensively studied, the mechanisms controlling the expression of most of them are largely unknown. However, it seems likely that the same mechanisms that control the expression of classic or coding genes are also involved in regulating the expression of miRNAs. These include genetic mechanisms such as point mutations, deletions, amplifications (Lv et al. 2012), loss of heterozygosity (Agueli et al. 2010), translocations (Bousquet et al. 2008), and, as demonstrated more recently, epigenetic mechanisms such as DNA methylation (Lujambio et al. 2008; Bandres et al. 2009). This chapter focuses on this latter group of epigenetic mechanisms and, in particular, their association with the pathogenesis of acute lymphoblastic leukemia (ALL).

A. DNA Methylation in Human Cancer

Epigenetic mechanisms are those heritable changes in gene expression patterns that occur in a cell without alterations in the normal sequence of nucleotides comprising the genomic sequence. Unlike genetic alterations, which are permanent changes in the genome, epigenetic modifications are reversible. Epigenetic mechanisms encompass DNA methylation, modification of the histone code, and positioning of nucleosomes. The most thoroughly studied and best characterized of these in cancer are alterations in DNA methylation.

DNA methylation occurs mainly in the cytosines of CpG dinucleotides and involves the addition of a methyl group at the carbon in position 5 of cytosine by the action of DNA methyltransferases (DNMT1, DNMT3A, and DNMT3B). These CpGs are underrepresented

INTRODUCTION

in the genome, except for small regions known as CpG islands (Bird 2002). Most CpG islands are located in the proximal promoter regions of almost half of the human genes and are usually unmethylated in normal cells. By contrast, in tumor cells, the most common epigenetic aberration is the hypermethylation of promoter regions, which is associated with inappropriate transcriptional silencing (Esteller 2008). Tumor cells may have abnormal DNA methylation of the promoter regions of the genes, associated with transcriptional silencing and thereby with the loss of function of these genes. This ultimately leads to loss of cell-cycle control, regulation of transcription factors, the mechanisms of DNA repair, apoptosis, angiogenesis and metastasis, and invasion, creating a genomically unstable phenotype. The pattern of gene methylation provides important information about the neoplastic cell, so that each type of tumor has a specific profile of DNA methylation (Fernandez et al. 2012). Moreover, each patient has a specific DNA methylation profile that differs from that of other patients with the same type of tumor. These specific profiles could be used as biomarkers to determine the diagnosis and prognosis of these patients (Agirre et al. 2009a). However, besides the inappropriate DNA hypermethylation occurring in the promoter regions of genes associated with the presence of CpG islands, in some malignancies, there is an overall hypomethylation of the genome that leads to genomic instability and oncogene activation, which directly contribute to cell tumorigenesis (Brueckner et al. 2007).

B. Acute Lymphoblastic Leukemia (ALL)

ALL is a biologically and clinically heterogeneous disease caused by the uncontrolled proliferation of immature B or T lymphoid precursors. This hematological malignancy is the commonest neoplasia and the leading cause of cancer mortality in childhood. The new treatment algorithms have produced a cure for most children with ALL, but although there has also been a significant improvement in adult patients, long-term rates of disease-free survival in this group of patients are less than 40%. This motivates us to learn more about the biology and course of the disease in order to improve the survival and quality of life of these patients (Faderl et al. 2010). Various genetic alterations promoting cell proliferation, differentiation, apoptosis, and gene transcription have been found in ALL cells. Among these, chromosomal translocations are undoubtedly one of the most important steps in the oncogenic development of ALL (Look 1997). The most common chromosomal translocations associated with ALL are t(9;22)(q32;q11), which involves the fusion of the *ABL1* and *BCR* genes; t(12;21), which leads to the fusion of the *ETV6* and *RUNX1* genes; and translocations involving the MLL gene located in the 11q23 chromosome region, which is associated with several genes, the most common being *AF4* t(4;11)(q21;q23) and t(1;19)(q23;p13), which give rise to the fusion of the *TCF3* and *PBX1* genes. These chromosomal translocations serve as biomarkers in diagnosis and prognosis, and help to monitor minimal residual disease in patients with ALL. Unfortunately, they are present at low percentages in certain morphological subtypes of ALL (Faderl et al. 1998). Interestingly, in addition to these genetic changes, epigenetic silencing has recently been identified as one of the basic changes in the development of cancer (Baylin and Ohm 2006). Aberrant epigenetic regulation, particularly DNA hypermethylation of gene promoters, is a frequent mechanism of gene silencing that has been associated with the development, diagnosis and prognosis of the disease, and the response to therapy in patients with ALL (Davidsson et al. 2009; Stumpel et al. 2009; Milani et al. 2010). This aberrant DNA methylation includes the regulation of non-coding RNAs such as miRNAs (Schotte et al. 2009; Vilas-Zornoza et al. 2011).

II. miRNA DEREGULATION BY DNA METHYLATION IN ALL

Extensive analysis of genomic sequences of miRNA genes has shown around half of them to be associated with CpG islands, suggesting that they could be subject to this regulation by DNA methylation (Weber et al. 2007). The regulation of expression of various miRNAs due to inappropriate methylation of DNA has been demonstrated in patients with ALL (Table 22.1). Furthermore, this aberrant methylation of DNA plays an important role as a biomarker in the diagnosis and prognosis of patients with ALL. Furthermore, it is involved in the regulation of important signaling pathways and could even be a useful therapeutic target (Figure 22.1).

A. Aberrant DNA Methylation of miRNAs as a Prognostic Factor for ALL

In the case of ALL, two studies have demonstrated that aberrant DNA methylation affects two families of miRNAs: the *miR-124a* family (comprising *miR-124a-1*, *miR-124a-2*,

TABLE 22.1. miRNAs Deregulated by Aberrant DNA Methylation in ALL

miRNA	Genomic Location	DNA Methylation	Regulation by Genes on Chromosome Translocations	Target in ALL
miR-9-1	1q22	Hyper		CDK6
miR-9-2	5q14.3			FGFR1
miR-9-3	15q26.1			
miR-10a	17q21.32	Hyper		HOXA3
miR-10b	2q31.1	Hyper	MLL	
miR-34b	11q23.1	Hyper		CDK6
miR-34c				
miR-124a1	8p23.1	Hyper		CDK6
miR-124a2	8q12.3			
miR-124a3	20q13.33			
miR-132	17p13.3	Hyper		
miR-143	5q32	Hyper		MLL–AF4
miR-152	17q21.32	Hyper	MLL	MLL DNMT1 AF4–MLL
miR-196b	7p15.2	Hyper Hypo in MLL rearranged ALL		MYC
miR-203	14q32.33	Hyper		ABL1 BCR–ABL1
hsa-*miR-200a*	1p36.33	Hyper	MLL	ZEB2
miR-200b				
miR-212	17p13.3	Hyper		
miR-429	1p36.33	Hyper	MLL	ZEB2
miR-432	14q32.2	Hyper	MLL	
miR-503	Xq26.3	Hyper	MLL	
miR-let-7b	22q13.31	Hyper	MLL	

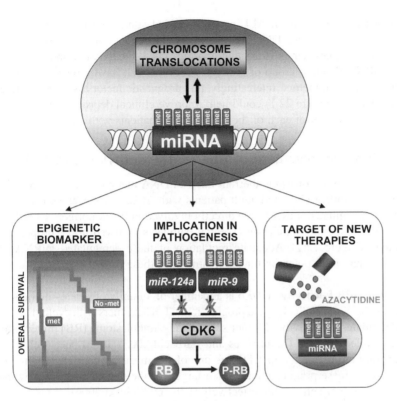

Figure 22.1. Regulation and involvement of miRNAs deregulated by aberrant DNA methylation in ALL. See color insert.

and *miR-124a-3*) and the *miR-9* family (comprising *miR-9-1*, *miR-9-2*, and *miR-9-3*). The miRNAs of each of these family members are located in a different chromosomal region (e.g., *miR-124a-1* is located in chromosome region 8p23.1, *miR-124a-2* in 8q12.3, and *miR-124a-3* in 20q13.33). Each has a specific pri- and pre-miRNA sequence, but the three miRNA members give rise to the same mature miRNA. This indicates the importance of aberrant DNA methylation in the regulation of miRNAs in ALL because having three different genes that encode the same mature miRNA product, the inappropriate methylation of DNA is present in all members comprising the miRNA family. Considering the three members of each family, aberrant DNA methylation of *miR-124a* and *miR-9* families was observed in 59% and 54% of ALL patients, respectively. Furthermore, hypermethylation of the *miR-124a* and *miR-9* families was associated with higher relapse and mortality rates, each of them being an independent prognostic factor for disease-free survival and overall survival in ALL patients (Agirre et al. 2009b; Rodriguez-Otero et al. 2011).

In addition to these six miRNAs, seven other miRNAs (*miR-10b*, *miR-34b*, *miR-34c*, *miR-132*, *miR-196b*, *miR-203*, and *miR-212*) are known to be regulated by inappropriate DNA methylation and histone modifications associated with a close chromatin conformation (Agirre et al. 2009a). In this case, DNA hypermethylation of at least 1 of the 13 miRNAs was found in 65% of cases. Patients with at least 1 methylated miRNA of the 13 had significantly higher relapse and mortality rates; this methylation was an independent adverse prognostic factor for disease-free survival and overall survival. The separate analysis of children and adult ALL patients yielded similar results, suggesting that the prognostic value of aberrant DNA methylation is independent of age in ALL (Agirre et al. 2009a). These studies demonstrate that aberrant DNA methylation and expression of a specific miRNA define a group of patients with ALL that have a poorer outcome, independent of other risk factors. Interestingly, this prognostic factor associated with aberrant DNA methylation (Figure 22.1) could help improve clinical decisions and develop therapies tailored to the risk of each of the subgroups of patients with ALL.

B. Functional Involvement of Aberrantly Methylated miRNAs in ALL

Besides the possibility of using aberrant miRNA DNA methylation as a biomarker or as a prognostic factor for child and adult patients with ALL, it has been shown that several epigenetically regulated miRNAs are directly involved in the tumorigenesis of this hematological malignancy by regulating essential pathways involved in the proper control of cell biology (Figure 22.1). Adequate regulation of these deregulated miRNAs or their target genes may have potential for the treatment of these patients.

1. CDK6-RB Pathway. CDK6 is a member of a family of serine–threonine kinases involved in the control of cell-cycle progression. CDK6 partners with cyclin D phosphorylate and regulates the activity of tumor suppressor retinoblastoma (RB) protein during the G1-S cell-cycle transition (Malumbres and Barbacid 2005). It has been well established that inappropriate regulation of CDKs is one of the most frequent alterations in human cancer, making these proteins an attractive target for the development of new inhibitors that eliminate tumor cells highly effectively (Lee and Sicinski 2006).

One mechanism that leads to the inappropriate regulation of CDKs, specifically of CDK6, is the loss of expression of miRNAs that bind specifically to the 3′-UTR of the gene. In the case of ALL, the inappropriate DNA methylation of the three members of the *miR-124a* and *miR-9* families, and the two members of the *miR-34* family of miRNAs (*miR-34b* and *miR-34c*), is associated with a reduction in the expression of their mature miRNAs and up-regulation of their target CDK6. The forced reexpression of mature miRNAs of *miR-124a* and *miR-9* reduces the expression of CDK6 and the level of phosphorylation of RB, and is associated with inhibition of cell proliferation and, in some cases, increased apoptosis of ALL cells (Figure 22.1). However, the CDK6 inhibitor (PD-0332991) induces a significant inhibition of cell growth and proliferation of ALL cell lines (Agirre et al. 2009a, 2009b; Rodriguez-Otero et al. 2011; Vilas-Zornoza et al. 2011). These results highlight the fact that CDK6-RB1 pathway may be a preferred target of aberrantly methylated miRNAs in ALL, establishing the pathway as an attractive therapeutic target in these hematological malignancies.

2. TP53 Pathway. TP53 is a tumor suppressor gene *par excellence* that regulates a range of cellular processes such as cell-cycle arrest, apoptosis, gene transcription, DNA synthesis, DNA repair, and senescence. It is one of the most frequently mutated genes in human cancers, with inactivating mutations (associated with tumor progression and genetic instability) present in more than 50% of patients with solid tumors. This percentage is much lower in hematological malignancies, with fewer than 10% of patients with ALL bearing mutations of *TP53*, despite ALL cells being abnormally resistant to apoptosis, which is a hallmark of deregulation of the p53 pathway. This suggests that other mechanisms must be involved in the inability of p53 to carry out its biological functions (Agirre et al. 2003).

Recent studies have shown that members of the *miR-34* family, *miR-34a*, *miR-34b*, and *miR-34c*, are direct transcriptional targets of the tumor suppressor gene *TP53*. These three miRNAs are also essential elements for the correct functioning of the p53 pathway, with respect to its level of cell-cycle control and to the induction of apoptosis (Corney et al. 2007; He et al. 2007). Aberrant DNA methylation of *miR-34b* and *miR-34c* has been found in 35% of ALL patient samples, along with 13 genes that are involved in p53 regulation and p53-dependent cell cycle, and the induction of apoptosis. These results indicate that aberrant DNA methylation of p53-dependent genes, including the three members of the *miR-34* family, is an important mechanism associated with the deregulated function of the *TP53* pathway in ALL. Interestingly, the use of compounds such as 5-aza-2-deoxycytidine, curcumin, and Nutlin 3 leads, directly or indirectly, to an increase in *TP53*-dependent apoptosis in ALL cells. These results are important because they open up the possibility of new treatment strategies for patients with ALL (Vilas-Zornoza et al. 2011).

C. Chromosomal Translocations and Aberrantly Methylated miRNAs in ALL

As explained earlier, the most common cytogenetic abnormalities in ALL are chromosomal translocations that usually activate and deregulate transcription factors involved in the control of cell differentiation, proliferation, and apoptosis. Several studies have demonstrated that genes involved in these chromosomal translocations that occur in ALL can regulate the expression of miRNAs and can, in turn, undergo regulation mediated by these small non-coding RNAs (Figure 22.1). In addition, some of these miRNAs are deregulated by aberrant DNA methylation in ALL.

1. ABL1 and the BCR–ABL1 Fusion Oncoprotein.
The *ABL1* proto-oncogene encodes a cytoplasmic and a nuclear protein tyrosine kinase that has been implicated in processes of cell differentiation, division and adhesion, and stress response. The translocation between the long arms of chromosomes 9 and 22, t(9;22)(q34;q11), known as the Philadelphia (Ph) chromosome, gives rise to the fusion of the *BCR* (chromosome 22) and *ABL1* (chromosome 9) genes, generating the BCR–ABL1 fusion oncogene, which has abnormal and constitutive ABL1 tyrosine kinase activity. This translocation is mainly associated with chronic myeloid leukemia and B-cell ALL. However, its positivity is a very poor prognostic factor for patients with ALL, although the recent use of tyrosine kinase inhibitors in the treatment of these ALL patients has yielded very encouraging improvements (San José-Enériz et al. 2008; Hunger et al. 2011).

ABL1 and the BCR–ABL1 oncoprotein are both relevant targets for the miRNA *miR-203* (Bueno et al. 2008). The expression of this miRNA is weaker in ALL cell lines and patient samples due to aberrant DNA methylation (Agirre et al. 2009a; Chim et al. 2011). The treatment of ALL cell lines with the combination of the demethylating agent 5′-azacytidine and histone deacetylase inhibitor 4-phenylbutyrate, results in the efficient demethylation of *miR-203*, the restoration of its expression, and a significant decrease in ABL1 and BCR–ABL1 oncoprotein levels. These results suggest that inactivation of *miR-203* may provide a proliferative advantage in BCR–ABL1-positive ALL and that restoration of its expression, probably by treatment with epigenetic drugs, could be a novel treatment option for these patients.

2. The MLL Gene and Its Rearrangements.
The *MLL* gene, located on chromosome 11q23, encodes a protein with a histone methyltransferase activity that participates

in development and hematopoiesis. It is considered one of the most promiscuous oncogenes because it has been involved in rearrangements with almost 70 partners. The vast majority of these rearrangements yield a chimeric oncoprotein that fuses the amino terminal portion of the MLL gene with a carboxy terminal portion of the associated gene (Harper and Aplan 2008). Childhood ALL patients with *MLL* gene rearrangements constitute a high-risk subgroup characterized by a complex biology, with a very bad prognosis in the case of infant patients (Bueno et al. 2011).

Recent studies have demonstrated that distinct aberrant DNA methylation patterns distinguish several genetic subtypes of MLL-rearranged ALL in infants, and that these can influence their survival to varying degrees (Milani et al. 2010). For example, seven miRNAs whose regulation is affected by aberrant DNA methylation (*miR-10a*, *miR-152*, *hsa-miR-200a*, *miR-200b*, *miR-429*, *miR-432*, and *miR-503*) have been noted in infant ALL patients with the most frequent MLL translocation, t(4;11), which results in a fusion of *MLL* with the *AF4* gene. Interestingly, patients with a high level of methylation of one of these miRNAs, *miR-152*, are associated with a greater risk of relapse and shorter overall survival. The expression of putative targets of these miRNAs, such as DNMT1 in the case of *miR-152*, HOXA3 for *miR-10a*, and ZEB2 in the case of the miRNA cluster *miR-220a/200b/429*, all of which are strongly expressed in ALL cells with the MLL–AF4 fusion, is decreased after up-regulation of miRNA expression by demethylating agents (Stumpel et al. 2011). Like these miRNAs, the expression of *let-7b* is also recovered following treatment with demethylating agents in ALL cell lines with MLL rearrangements. MLL fusion proteins regulate the expression of *let-7b*, and this miRNA is hypermethylated in ALL cases, probably as a consequence of oncogenic MLL fusion proteins (Nishi et al. 2013). These results indicate a role for the MLL and MLL fusion oncoproteins in deregulating miRNAs, making them interesting potential therapeutic targets for the treatment of patients with ALL.

Unlike previous miRNAs, the expression of *miR-196b*, regulated by MLL and MLL fusion products, is higher in ALL cells with MLL rearrangements. This miRNA is encoded in the *HOXA* cluster, and its expression is strongly associated with the expression level of *HOXA* family genes due, at least in part, to less DNA methylation at their promoter regions. Interestingly, the up-regulated expression of *miR-196b* causes an increase in the proliferation and differentiation block of bone marrow progenitor cells, and so directly contributes to the development of leukemia (Popovic et al. 2009; Schotte et al. 2010).

MLL and MLL fusion proteins, in addition to directly regulating the expression of miRNAs, may also be regulated by these small non-coding RNAs. Stumpel et al. showed that *miR-152*, a member of the *miR-148/152* miRNA family, is hypermethylated and downregulated in ALL patients who possess the MLL–AF4 fusion protein. They identified the wild-type MLL as a potential target of *miR-152* in these patients. Therefore, *miR-148/152* would not be able to target the MLL–AF4 fusion protein due to a loss of the C-terminus of MLL in this fusion. However, *in silico*, *miR-152* could regulate the expression of the reciprocal AF4–MLL fusion product detected in the majority of ALL patients that are positive to t(4;11). The AF4–MLL fusion contributes to leukemogenic transformation, and *miR-152* could aid this process. These results illustrate the mutual regulation between MLL and miRNAs, particularly in the case *of hsa-miR-152* (Stumpel et al. 2011).

In a similar vein, Dou et al. have shown that *miR-143* is hypermethylated in MLL–AF4-positive ALL cells and its reexpression induces apoptosis and inhibits proliferation of these leukemic cells. However, in addition to the possible regulation of *miR-143* by the MLL–AF4 fusion, *miR-143* functions as a tumor suppressor gene regulating the expression of the MLL–AF4 fusion oncoprotein in B-cell ALL. These results suggest that inhibition of *miR-143* expression due to inappropriate methylation confers a proliferative advantage on MLL–AF4 positive ALL cells, and restoration of this miRNA may be a beneficial treatment for these patients (Dou et al. 2012).

III. CONCLUSION

This chapter describes why, as happens with classic or coding genes, the inappropriate methylation of DNA is one of the mechanisms that deregulate the expression of miRNAs in human tumors. In the specific case of ALL, various aberrantly methylated miRNAs are of direct application as biomarkers in the prognosis of these patients. In addition, some of the miRNAs that are improperly regulated in this hematological malignancy regulate key pathways involved in the development of human tumors, such as the CDK6-RB or *TP53* pathways. Along with this, several of these miRNAs mutually regulate each other, with common translocations occurring in ALL. This clearly indicates that these small non-coding RNAs play an important role in the leukemogenesis of ALL. Interestingly, all this knowledge regarding the involvement of DNA methylation in the regulation of miRNAs makes them excellent therapeutic targets for the treatment of patients with ALL, either directly, through their reexpression, or indirectly, through the appropriate regulation of the genes or pathways that they affect (Figure 22.1).

REFERENCES

Agirre, X., F.J. Novo, M.J. Calasanz, M.J. Larrayoz, I. Lahortiga, M. Valgañón, et al. 2003. TP53 is frequently altered by methylation, mutation, and/or deletion in acute lymphoblastic leukaemia. *Mol Carcinog* **38**:201–208.

Agirre, X., J. Roman-Gomez, A. Jimenez-Velasco, V. Arqueros, A. Vilas-Zornoza, P. Rodriguez-Otero, et al. 2009a. Epigenetic regulation of microRNAs in acute lymphoblastic leukemia. *J Clin Oncol* **27**:1316–1322.

Agirre, X., A. Vilas-Zornoza, A. Jimenez-Velasco, J.I. Martn-Subero, L. Cordeu, L. Garate, et al. 2009b. Epigenetic silencing of the tumor suppressor microRNA *Hsa-miR-124a* regulates CDK6 expression and confers a poor prognosis in acute lymphoblastic leukemia. *Cancer Res* **69**: 4443–4453.

Agueli, C., G. Cammarata, D. Salemi, L. Dagnino, R. Nicoletti, M. La Rosa, et al. 2010. 14q32/miRNA clusters loss of heterozygosity in acute lymphoblastic leukemia is associated with up-regulation of BCL11a. *Am J Hematol* **85**:575–578.

Bandres, E., X. Agirre, N. Bitarte, N. Ramirez, R. Zarate, J. Roman-Gomez, et al. 2009. Epigenetic regulation of microRNA expression in colorectal cancer. *Int J Cancer* **125**:2737–2743.

Baylin, S.B. and J.E. Ohm. 2006. Epigenetic gene silencing in cancer—a mechanism for early oncogenic pathway addiction? *Nat Rev Cancer* **6**:107–116.

Bird, A. 2002. DNA methylation patterns and epigenetic memory. *Genes Dev* **16**:6–21.

Bousquet, M., C. Quelen, R. Rosati, V. Mansat-De Mas, R. La Starza, C. Bastard, et al. 2008. Myeloid cell differentiation arrest by miR-125b-1 in myelodysplastic syndrome and acute myeloid leukemia with the t(2;11)(p21;q23) translocation. *J Exp Med* **205**:2499–2506.

Brueckner, B., C. Stresemann, R. Kuner, C. Mund, T. Musch, M. Meister, H. Sültmann, and F. Lyko. 2007. The human let-7a-3 locus contains an epigenetically regulated microRNA gene with oncogenic function. *Cancer Res* **67**:1419–1423.

Bueno, M.J., I. Perez de Castro, M. Gomez de Cedron, J. Santos, G.A. Calin, J.C. Cigudosa, et al. 2008. Genetic and epigenetic silencing of microRNA-203 enhances ABL1 and BCR-ABL1 oncogene expression. *Cancer Cell* **13** 6:496–506.

Bueno, C., R. Montes, P. Catalina, R. Rodríguez, and P. Menendez. 2011. Insights into the cellular origin and etiology of the infant pro-B acute lymphoblastic leukemia with MLL-AF4 rearrangement. *Leukemia* **25**:400–410.

Chim, C.S., K.Y. Wong, C.Y. Leung, L.P. Chung, P.K. Hui, S.Y. Chan, et al. 2011. Epigenetic inactivation of the *miR-203* in haematological malignancies. *J Cell Mol Med* **15**:2760–2767.

Corney, D.C., A. Flesken-Nikitin, A.K. Godwin, W. Wang, and A.Y. Nikitin. 2007. MicroRNA-34b and microRNA-34c are targets of p53 and cooperate in control of cell proliferation and adhesion-independent growth. *Cancer Res* **67**:8433–8438.

Davidsson, J., H. Lilljebjorn, A. Andersson, S. Veerla, J. Heldrup, M. Behrendtz, et al. 2009. The DNA methylome of pediatric acute lymphoblastic leukemia. *Hum Mol Genet* **18**:4054–4065.

Dou, L., D. Zheng, J. Li, Y. Li, L. Gao, L. Wang, and L. Yu. 2012. Methylation-mediated repression of microRNA-143 enhances MLL-AF4 oncogene expression. *Oncogene* **31**:507 517.

Esteller, M. 2008. Epigenetics in cancer. *N Engl J Med* **358**:1148–1159.

Faderl, S., H.M. Kantarjian, M. Talpaz, and Z. Estrov. 1998. Clinical significance of cytogenetic abnormalities in adult acute lymphoblastic leukemia. *Blood* **91**:3995–4019.

Faderl, S., S. O'Brien, C.H. Pui, W. Stock, M. Wetzler, D. Hoelzer, and H.M. Kantarjian. 2010. Adult acute lymphoblastic leukemia: concepts and strategies. *Cancer* **116**:1165–1176.

Fernandez, A.F., Y. Assenov, J. Martin-Subero, B. Balint, R. Siebert, H. Taniguchi, H. Yamamoto, M. Hidalgo, A.C. Tan, O. Galm, I. Ferrer, M. Sanchez-Cespedes, A. Villanueva, J. Carmona, J.V. Sanchez-Mut, M. Berdasco, V. Moreno, G. Capella, D. Monk, E. Ballestar, S. Ropero, R. Martinez, M. Sanchez-Carbayo, F. Prosper, X. Agirre, M.F. Fraga, O. Graña, L. Perez-Jurado, J. Mora, S. Puig, J. Prat, L. Badimon, A.A. Puca, S.J. Meltzer, T. Lengauer, J. Bridgewater, C. Bock, and M. Esteller. 2012. A DNA methylation fingerprint of 1628 human samples. *Genome Res* **22**:407–419.

Harper, D.P. and P.D. Aplan. 2008. Chromosomal rearrangements leading to MLL gene fusions: clinical and biological aspects. *Cancer Res* **68**:10024–10027.

He, L., X. He, L.P. Lim, E. de Stanchina, Z. Xuan, Y. Liang, et al. 2007. A microRNA component of the p53 tumour suppressor network. *Nature* **447**:1130–1134.

Hunger, S.P., E.A. Raetz, M.L. Loh, and C.G. Mullighan. 2011. Improving outcomes for high-risk ALL: translating new discoveries into clinical care. *Pediatr Blood Cancer* **56**:984–993.

Lee, Y.M. and P. Sicinski. 2006. Targeting cyclins and cyclin-dependent kinases in cancer: lessons from mice, hopes for therapeutic applications in human. *Cell Cycle* **5**:2110–2114.

Look, A.T. 1997. Oncogenic transcription factors in the human acute leukemias. *Science* **278**: 1059–1064.

Lu, J., G. Getz, E.A. Miska, E. Alvarez-Saavedra, J. Lamb, D. Peck, et al. 2005. MicroRNA expression profiles classify human cancers. *Nature* **435**:834–838.

Lujambio, A., G.A. Calin, A. Villanueva, S. Ropero, M. Sanchez-Cespedes, D. Blanco, et al. 2008. A microRNA DNA methylation signature for human cancer metastasis. *Proc Natl Acad Sci U S A* **105**:13556–13561.

Lv, S.Q., Y.H. Kim, F. Giulio, T. Shalaby, S. Nobusawa, H. Yang, et al. 2012. Genetic alterations in microRNAs in medulloblastomas. *Brain Pathol* **22**:230–239.

Malumbres, M. and M. Barbacid. 2005. Mammalian cyclin-dependent kinases. *Trends Biochem Sci* **30**:630–641.

Medina, P.P., M. Nolde, and F.J. Slack. 2010. OncomiR addiction in an in vivo model of microRNA-21-induced pre-B-cell lymphoma. *Nature* **467**:86–90.

REFERENCES

Milani, L., A. Lundmark, A. Kiialainen, J. Nordlund, T. Flaegstad, E. Forestier, et al. 2010. DNA methylation for subtype classification and prediction of treatment outcome in patients with childhood acute lymphoblastic leukemia. *Blood* **115**:1214–1225.

Nishi, M., M. Eguchi-Ishimae, Z. Wu, W. Gao, H. Iwabuki, S. Kawakami, H. Tauchi, T. Inukai, K. Sugita, Y. Hamasaki, E. Ishii, and M. Eguchi. 2013. Suppression of the let-7b microRNA pathway by DNA hypermethylation in infant acute lymphoblastic leukemia with MLL gene rearrangements. *Leukemia* **27**:389–397.

Popovic, R., L.E. Riesbeck, C.S. Velu, A. Chaubey, J. Zhang, N.J. Achille, F.E. Erfurth, K. Eaton, J. Lu, H.L. Grimes, J. Chen, J.D. Rowley, and N.J. Zeleznik-Le. 2009. Regulation of mir-196b by MLL and its overexpression by MLL fusions contributes to immortalization. *Blood* **113**: 3314–3322.

Rodriguez-Otero, P., J. Roman-Gomez, A. Vilas-Zornoza, E.S. Jose-Eneriz, V. Martin-Palanco, J. Rifon, et al. 2011. Deregulation of FGFR1 and CDK6 oncogenic pathways in acute lymphoblastic leukaemia harbouring epigenetic modifications of the MIR9 family. *Br J Haematol* **155**:73–83.

San José-Enériz, E., J. Román-Gómez, L. Cordeu, E. Ballestar, L. Garate, E.J. Andreu, I. Isidro, E. Guruceaga, A. Jiménez-Velasco, A. Heiniger, A. Torres, M.J. Calasanz, M. Esteller, N.C. Gutierrez, A. Rubio, I. Pérez-Roger, F. Prosper, and X. Agirre. 2008. BCR-ABL1—induced expression of HSPA8 promotes cell survival in chronic myeloid leukemia. *Br J Haematol* **142**:571–582.

Schotte, D., J.C. Chau, G. Sylvester, G. Liu, C. Chen, V.H. van der Velden, et al. 2009. Identification of new microRNA genes and aberrant microRNA profiles in childhood acute lymphoblastic leukemia. *Leukemia* **23**:313–322.

Schotte, D., E.A. Lange-Turenhout, D.J. Stumpel, R.W. Stam, J.G. Buijs-Gladdines, J.P. Meijerink, et al. 2010. Expression of miR-196b is not exclusively MLL-driven but is especially linked to activation of HOXA genes in pediatric acute lymphoblastic leukemia. *Haematologica* **95** **10**:1675–1682.

Stumpel, D.J., P. Schneider, E.H. van Roon, J.M. Boer, P. de Lorenzo, M.G. Valsecchi, et al. 2009. Specific promoter methylation identifies different subgroups of MLL-rearranged infant acute lymphoblastic leukemia, influences clinical outcome, and provides therapeutic options. *Blood* **114**:5490–5498.

Stumpel, D.J., D. Schotte, E.A. Lange-Turenhout, P. Schneider, L. Seslija, R.X. de Menezes, et al. 2011. Hypermethylation of specific microRNA genes in MLL-rearranged infant acute lymphoblastic leukemia: major matters at a micro scale. *Leukemia* **25**:429–439.

Vilas-Zornoza, A., X. Agirre, V. Martin-Palanco, J.I. Martin-Subero, E. San Jose-Eneriz, L. Garate, et al. 2011. Frequent and simultaneous epigenetic inactivation of TP53 pathway genes in acute lymphoblastic leukemia. *PLoS ONE* **6**:e17012.

Weber, B., C. Stresemann, B. Brueckner, and F. Lyko. 2007. Methylation of human microRNA genes in normal and neoplastic cells. *Cell Cycle* **6**:1001–1005.

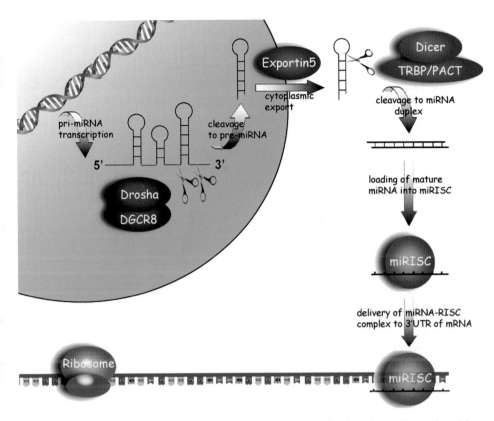

Figure 1.1. Schematic diagram of the canonical miRNA biosynthetic pathway. Reproduced from Lawrie, C.H. (2007) *Br J Haematol* **136** (6):503–512.

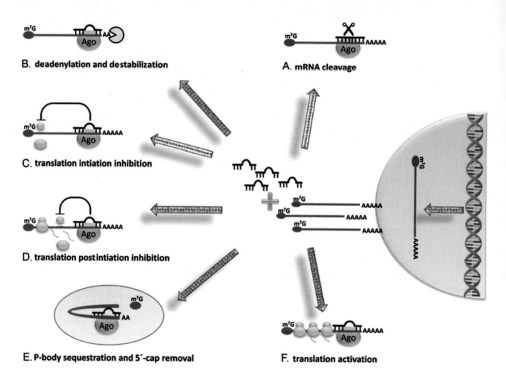

Figure 1.2. Schematic diagram of proposed mechanisms for miRNA function. (A) Ago-mediated cleavage of mRNA can occur when the miRNA sequence is complementary to the target gene-binding site. (B) Removal of poly(A) tail by deadenylases causes destabilization and degradation of mRNA. (C) Translation initiation inhibited by miRISC interactions with eukaryotic translation initiation factors (eIFs). (D) Inhibition of translation postinitiation. (E) Sequestration of mRNA in P-bodies. (F) miRNA-mediated translational activation.

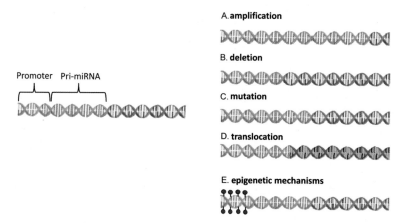

Figure 1.3. The various genetic and epigenetic alterations that can result in aberrant expression of miRNAs. (A) Amplification of miRNA-encoding regions. (B) Deletion of miRNA-encoding regions. (C) Mutations in the miRNA sequence (including SNPs). (D) Translocation occurring between distal, usually gene promoter regions, and miRNA-encoding regions. (E) Epigenetic mechanisms, such as histone modification and methylation of promoter regions of miRNAs, can silence miRNA expression.

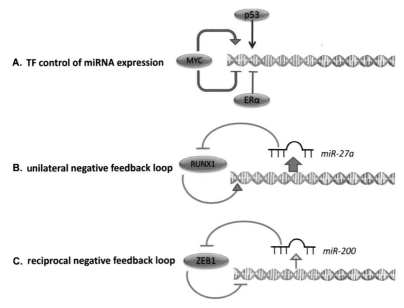

Figure 1.4. Examples of simple regulatory loop motifs involving miRNAs and transcription factors (TFs). (A) TFs can bind directly to the promoter region of miRNAs, either up-regulating or down-regulating expression, or a single TF (e.g., MYC) can up-regulate expression of one miRNA but down-regulate another. (B) When the TF that regulates the miRNA is itself regulated by that miRNA, a simple loop motif results. Up-regulation of the miRNA by the TF decreases its own expression in a unilateral negative feedback loop. (C) When the TF down-regulates the miRNA, this can increase TF expression in a reciprocal (positive) feedback loop.

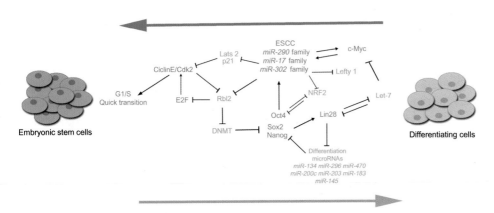

Figure 2.1. The pluripotency/differentiation regulatory network. Propluripotency elements are in blue and prodifferentiation in orange. There is a balance between ESc state and differentiation. ESCC miRNAs target cell cycle inhibitors and early differentiation markers, promoting self-renewal and the maintenance of the pluripotent state. On the other hand, differentiation miRNAs directly target the core pluripotency transcription factors, destabilizing the network and promoting differentiation. The final choice will depend on external signals, which will up-regulate the expression of ones, causing the repression of the others.

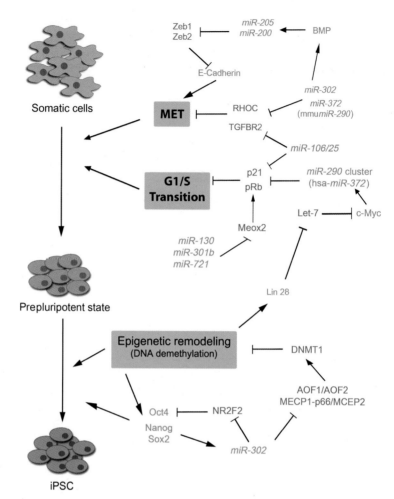

Figure 2.2. MiRNAs involved in reprogramming: In the figure are represented the two phases in the reprogramming process. Embryonic-specific miRNAs (*miR-290* cluster and *miR-302* family) target inhibitors of MET (RHOC and TGFBR2) and G1/S transition (p21 and pRb). Indirectly can up-regulate MET promoting factors, such as BMP. In the reprogramming process, some endogenous miRNAs are up-regulated helping in overcoming MET (*miR-205* and *miR-200* target MET inhibitors Zeb1 and 2) or promoting G1/S transition (*miR-130, miR-301b,* and *miR-721* inhibit Meox2, a G1/S inhibitor). *miR-302* is also involved in DNA demethylation needed for the epigenetic remodeling in the cell, as well as in the consolidation of the pluripotency network, together with the endogenous pluripotency markers (Oct4, Nanog, Sox2, and Lin-28). Proreprogramming players are colored in green; antireprogramming ones are colored in red.

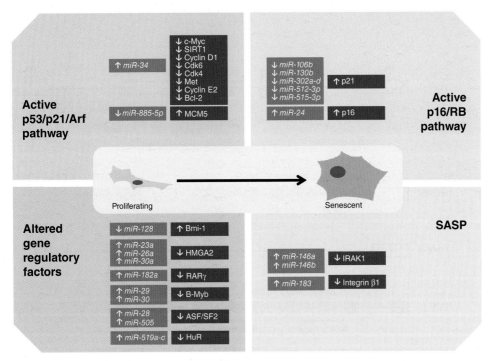

Figure 4.1. miRNAs that influence senescence-relevant pathways. Proliferating cells progress to senescence by acquiring several prominent phenotypes: an active p53/p21/Arf pathway (green), an active pRB/p16 pathway (blue), a senescence-associated secretory phenotype (SASP, yellow), and changes in senescence-associated gene regulatory factors (pink). Dark green, the main senescence-associated (SA)miRNAs identified to date as mediating each senescence phenotype. Dark blue, the principal target proteins influenced by SA-miRNAs. The major downstream consequences of activation of p53/p21/Arf and p16/RB and gene regulatory factors are the inhibition of the cell division machinery and the implementation of senescence-associated gene expression patterns. The main consequences of SASP is the secretion of factors that cause inflammation and compromise the integrity of the ECM.

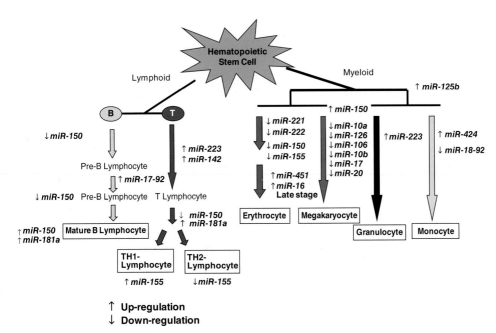

Figure 6.1. Expression levels of miRNAs during hematopoiesis.

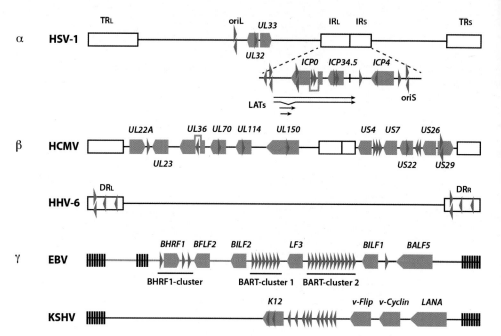

Figure 8.1. Genomic location of miRNAs encoded by human herpesviruses. Almost all human herpesviruses code for miRNA genes with the exception of VZV and HHV-7. For simplicity, only HSV-1 genome was represented, HSV-2 being very similar. The genomes were not drawn to scale, and only ORF close to or embedding miRNA genes (in green) were pictured. White boxes and successive black bars schematize repeats. TR, IR, and DR stand for terminal, internal, and direct repeats, respectively; $_L$ for long and $_S$ for short; oriL for lytic origin of replication; LAT for latency-associated transcript.

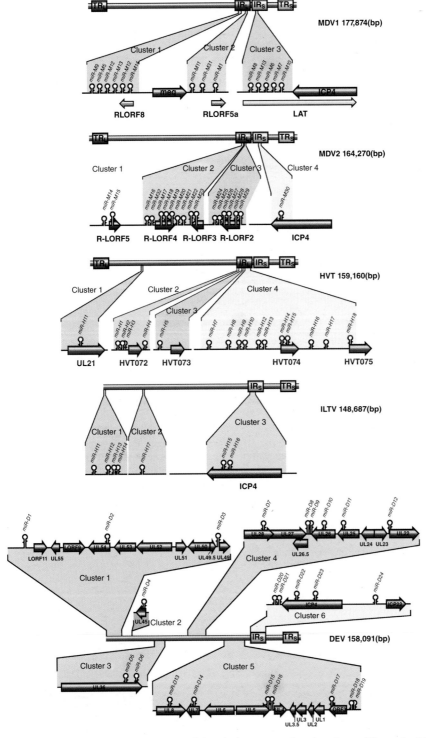

Figure 9.1. Diagrammatic representation of the viral genomes showing the positions of miRNAs. Position and orientation of selected transcripts are shown. The genome size (base pairs) of viruses (MDV1, MDV2, HVT, ILTV, and DEV) is shown on the right.

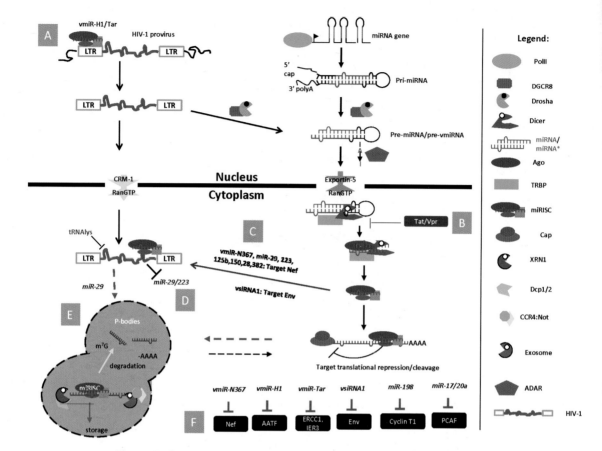

Figure 11.2. Schematic illustration of HIV-1-proposed interactions with the miRNA pathway. (A) vmiRNA-Tar may involve chromatin remodeling at the LTR region of the HIV-1 provirus. (B) Tat/Vpr may act as SRS by inhibiting Dicer activity. (C) Cellular miRNAs and viRNAs target HIV-1 genome. (D) A secondary structure in Nef/2'-LTR blocks *miR-29/223* RISC repression. (E) *miR-29* RISC bring HIV-1 virus to P-bodies, and P-bodies may act as HIV-1 viral RNA storage place and contribute to the latent phase of HIV infection. (F) vmiRNA and host miRNA target cellular factors or HIV-1 protein in HIV-1 infection.

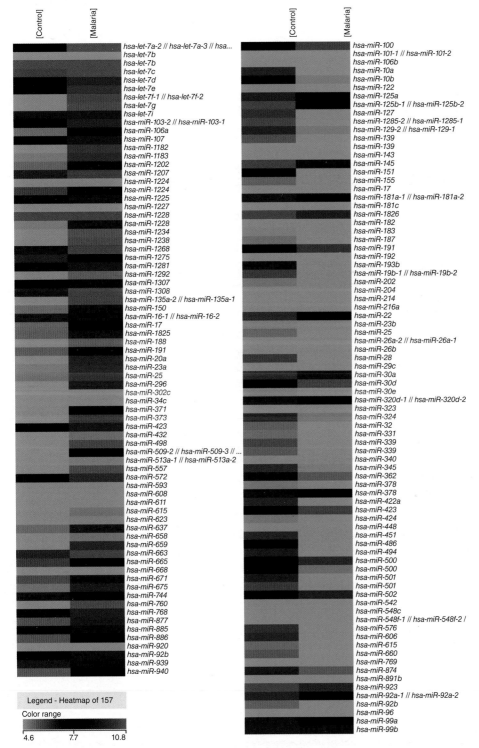

Figure 12.1. Heat map showing pooled data from miRNA array analysis of postmortem kidney tissues from control non-malaria patients ($n = 9$) compared with malaria-infected patients ($n = 16$) (derived from Affymetrix miRNA GeneChip version 1.0). It shows 157 differentially expressed miRNAs in malaria (using a cutoff of adjusted $P < 0.05$, T-test with Benjamini–Hochberg's control for false discovery rate (FDR), and a fold change cutoff of 2).

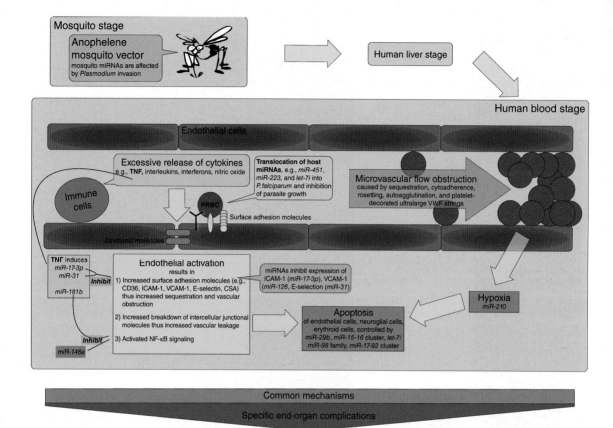

Figure 12.2. Diagram showing the relationship between different stages of malaria infection, relevant pathophysiological mechanisms at the infected red cell-endothelial interface, and sites where miRNA may play a role in regulation of these pathways, leading to specific complications of severe malaria. Most of these are hypothetical and have been identified in studies of other diseases. Confirmed data only currently exists for the role of miRNA in human malaria in acute kidney injury (shown in red).

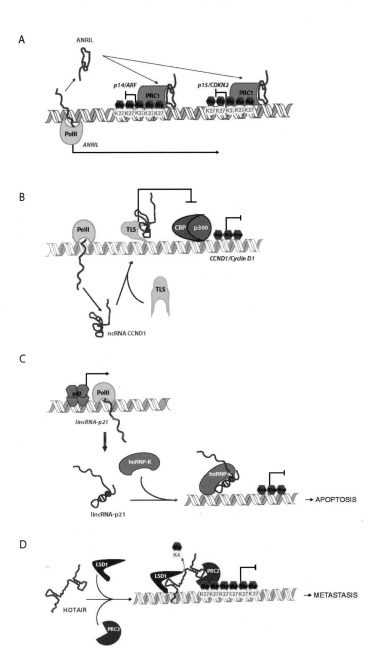

Figure 15.1. Mechanisms of gene regulation by lncRNAs. (A) Large ncRNA ANRIL is transcribed in antisense of the *p14/ARF* and *p15/CDKN2B* genes. ANRIL mediates gene silencing of the locus by interaction and recruitment of CBX7, a component of PRC1 histone 3–lysine 27 methyltransferase complex. (B) lncRNA CCND1 is transcribed from the 5′ of *cyclin D1/CCDN1* gene. lncRNA CCND1 interacts with TLS protein, inducing a conformational change that allows its binding to the *cyclin D1* promoter and inhibiting *cyclin D1* gene expression by blocking CBP and p300 HAT activity. (C) Large intergenic non-coding RNA-p21 (lincRNA-p21) is induced by p53 and interacts with the protein hnRNP-K for binding to gene promoters for repression, which results in apoptosis induction. (D) lincRNA HOTAIR is transcribed from the HOXC locus. HOTAIR recruits PRC2 complex to gene promoters for histone 3–lysine 27 (H3K27) methylation. HOTAIR also binds LSD1–CoREST complex, which removes an active chromatin mark (H3K4 methylation), causing gene repression and leading to metastasis.

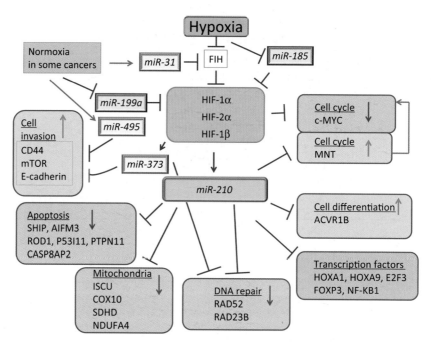

Figure 16.1. Actions of hypoxia-induced microRNAs. Green arrows and outlines indicate up-regulation and red ones indicate down-regulation. HIFs are regulated through *miR-185* and *miR-199a* directly or by *miR-31* through down-regulation of FIH. *miR-210* is the master hypoxamir with effects on mitochondrial functions, apoptosis, DNA repair, cell cycle, and differentiation through the down-regulation of targets with functions in these pathways. *miR-210* is up-regulated through direct binding of HIF-1α and HIF-2α to host transcript promoter elements. *miR-210* also indirectly regulates c-MYC, which is antagonized by HIF. The action of HIF and *miR-210* on c-MYC appears antagonistic. Furthermore, hypoxia pathways are regulated through *miR-373* and *miR-495*.

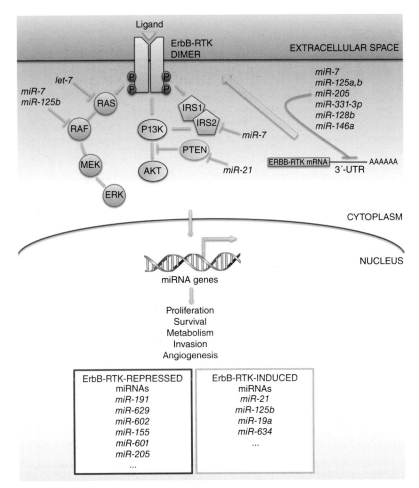

Figure 17.1. miRNA regulation of ER-alpha and subsequent ER-alpha transcriptional activity. The 3′-UTR of the ER-alpha transcript is directly targeted by *miR-145, miR-206, miR-221/222, miR-193b, miR-22, let-7, miR-18a,b,* and *miR-302c*. Estradiol (E2)-bound ER-alpha protein can regulate the transcription of pre-miRNAs either by directly binding to the miRNA promoter, by indirectly modulating the expression of other transcription factors that bind to the promoter of miRNAs (not shown), or by suppressing the Drosha-mediated processing of pri-miRNAs to pre-miRNAs. In turn, estrogen-repressed miRNAs can participate to a regulatory loop by modulating ER-alpha expression.

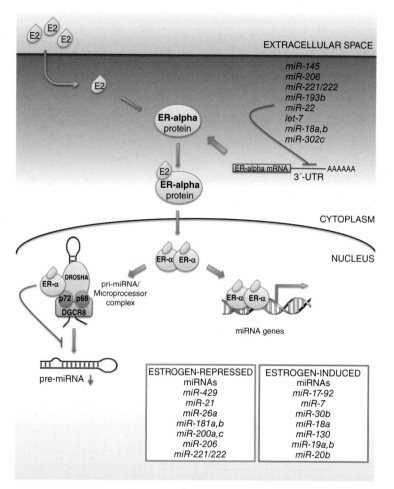

Figure 17.2. Regulation of ErbB signaling in cancer cells by miRNAs. ErbB receptor dimers are frequently overexpressed or aberrantly activated in human cancers, leading to activation of downstream ERK1/ERK2 and Akt signaling pathways. In turn, these modulate gene expression in order to promote cellular proliferation, apoptosis resistance, metabolism, migration and invasion, and angiogenesis. ErbB signaling represents an important therapeutic target. Recent research has identified miRNAs that modulate expression of a number of molecules belonging to these pathways. Specific examples reported in the text are included in this figure. Furthermore, ErbB signaling could regulate tumor development and progression via promoting expression of oncogenic miRNAs, or inhibiting tumor suppressor miRNAs.

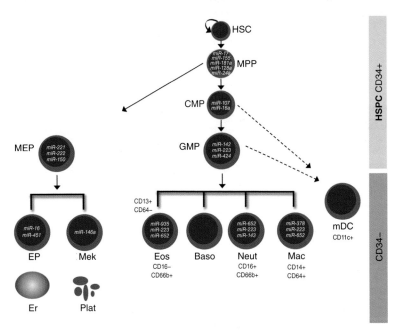

Figure 21.1. MicroRNAs expressed in purified human myeloid cells that function during normal myelopoiesis. The hierarchy of normal human myelopoiesis is illustrated. Depicted is each cell type (purple with black letters) during the myeloid differentiation cascade (denoted by arrows). mDCs are thought to derive from either CMP or GMP; this relationship is not yet firmly understood (dashed arrows). The microRNAs illustrated (white letters) were chosen based on two criteria. First, these microRNAs were identified in purified myeloid cell populations from normal human donors without *in vitro* manipulations. The microRNAs were further restricted to those with experimental functional evidence in *in vitro* or *in vivo* human and murine models. CD markers used to purify myeloid subsets are indicated.

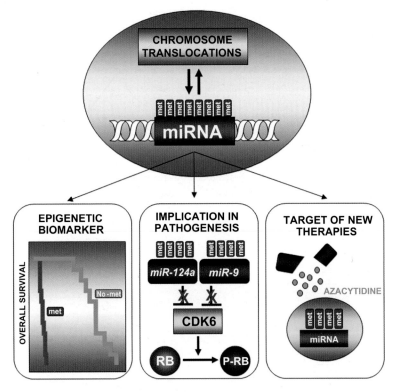

Figure 22.1. Regulation and involvement of miRNAs deregulated by aberrant DNA methylation in ALL.

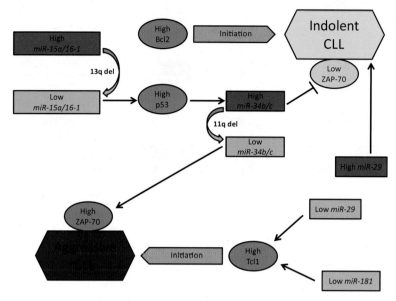

Figure 23.1. MicroRNAs in molecular mechanisms of CLL.

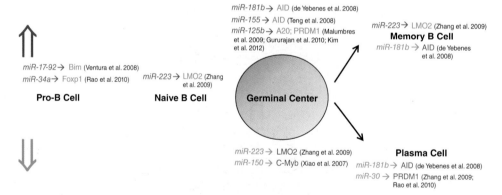

Figure 24.1. Master miRNAs that regulate B-cell development. Red (miRNA or genes) denotes high expression, while green (miRNA or genes) denotes low expression.

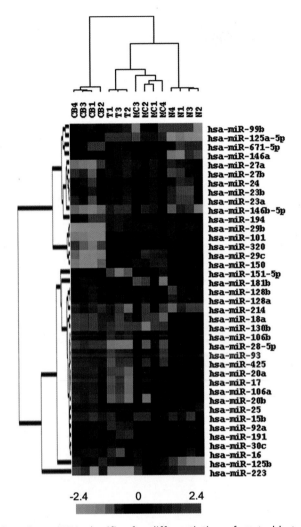

Figure 25.1. Thirty-nine miRNA classifier for differentiation of centroblasts (CB), naive B cell (N), memory B cells (MC), and T cells (T). Mean centered log ratios for each miRNA are represented. Missing values are in gray.

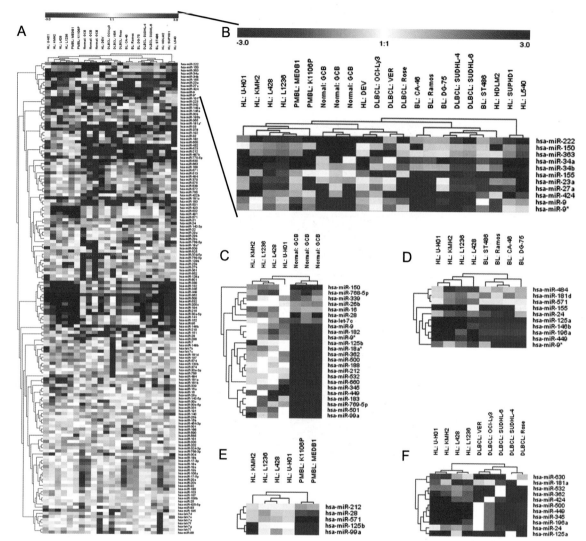

Figure 26.1. Heatmaps of miRNA profiling study in HL, non-Hodgkin's lymphoma-derived cell lines and normal GC B cells. Eight HL and 11 non-HL cell lines were analyzed on an Agilent miRNA microarray platform (version V1, Agilent, Santa Clara, CA). GC B cells (GCB) purified from three individuals (CD19+, IgD−, and CD38+) were analyzed as their normal counterparts. (A) Heat map of unsupervised clustering of the 151 miRNAs flagged present in all samples of at least one subtype. (B) Enlargement of the clustering pedigree from part A. The B-cell classical (c)HL cell lines cluster together with the PMBL cell lines, whereas the two T-cell cHL cell lines (HDLM2 and L540) and one cHL of uncertain origin (SUPHD1) cluster separately. The NLPHL cell line DEV clustered together with the normal GC B cells. (C) Heat map of significantly differentially expressed miRNAs between B-cell cHL and GC B cells. (D) Heat map of significantly differentially expressed miRNAs between B-cell cHL and BL. (E) Heat map of significantly differentially expressed miRNAs between B-cell cHL and PMBL. (F) Heat map of significantly differentially expressed miRNAs between B-cell cHL and DLBCL (unpaired t-test, miRNAs with P-value of <0.01 are shown).

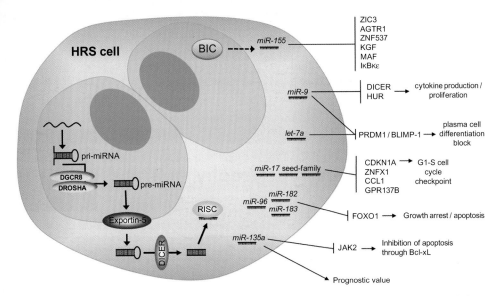

Figure 26.2. Schematic presentation of the currently proven miRNA targets in HL and their pathogenetic relevance.

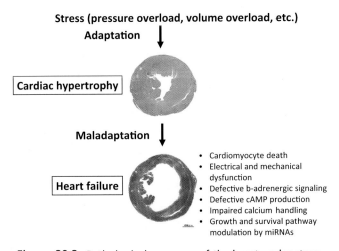

Figure 29.2. Pathological processes of the heart under stress.

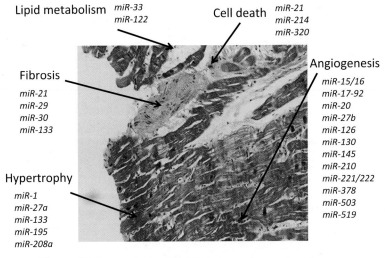

Figure 29.3. Dysregulated miRNAs in cardiovascular diseases.

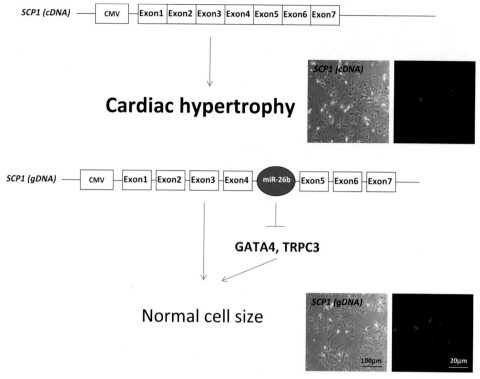

Figure 29.4. Scheme of the function of the intronic *miR-26b*.

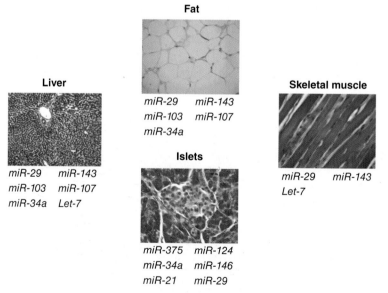

Figure 30.1. Selected miRNAs up-regulated in insulin-secreting cells and in insulin target tissues under diabetes conditions. See text for the description of the role and mode of action of each of these non-coding RNAs.

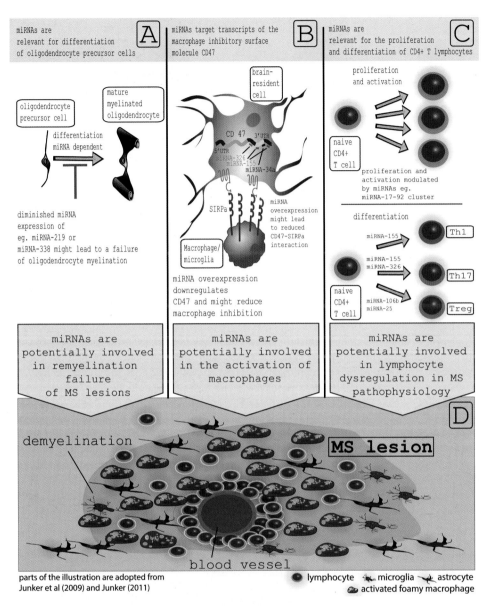

Figure 32.1. MiRNAs participate in diverse pathophysiological processes in MS. The differentiation of oligodendrocyte precursor cells is miRNA dependent. The same miRNAs are downregulated in inactive MS lesions (Junker et al. 2009), and thus might lead to a remyelination failure of the lesions (A). Transcripts of the immunomodulatory surface molecule CD47 are targeted by miRNAs, for example, *miR-155, miR-338*, and *miR-34a*, which are up-regulated in active MS lesions. A reduced interaction of CD47 with its ligand SIRPα on phagocytes might enhance their phagocytic activity (B). MiRNAs participate in the proliferation, activation, and differentiation of diverse lymphocyte subsets, such as CD4+ T cells. An involvement of Th17 cells or Treg cells in MS pathophysiology is suggested by several studies (see text) (C). These diverse miRNA interactions probably participate in several pathophysiological events in MS, such as lesion formation (D), immune cell dysregulation, and tissue destruction and regeneration.

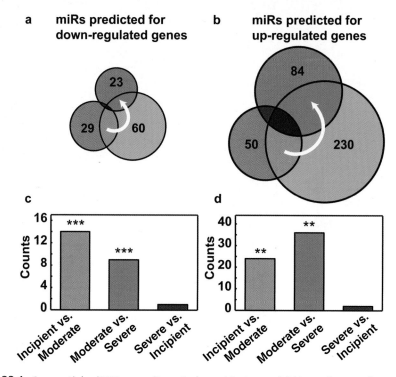

Figure 33.1. Sequential miRNAs recruitment along AD stages. (a) Venn diagram for predicted microRNAs targeted to AD down-regulated transcripts. Arrow indicates direction of disease progression. (b) Same as (a) for up-regulated genes. (c) Numbers of shared predicted miRNAs between pairs of stages for miRNAs predicted to be down-regulated. Two or three asterisks indicate Fisher's exact test P-values of $P < 0.01$ or $P < 0.001$, respectively. (d) Same as (c) for miRNAs predicted to be up-regulated. Courtesy of Current Alzheimer Research, reprinted from Barbash and Soreq (2012).

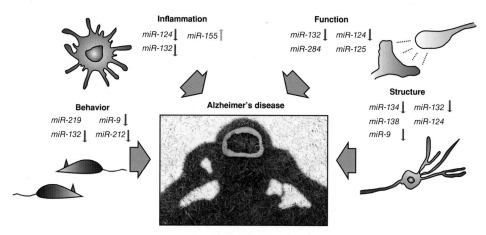

Figure 33.2. miRNAs regulate processes underlying the structural, functional, inflammation-related, and behavioral impairments characteristic of AD. Each miRNA is shown under the relevant process. References for these miRNAs can be found in Table 33.1. For some of the miRNAs, altered regulation was shown to occur in AD. For these genes, colored arrows indicate the direction of regulation: Red downward arrow indicates down-regulation of this gene, while green upward arrow indicates up-regulation of this gene. Central photograph: creative demonstration of AD by a patient participant in a therapeutic workshop (courtesy of Mrs. Tema Barbash).

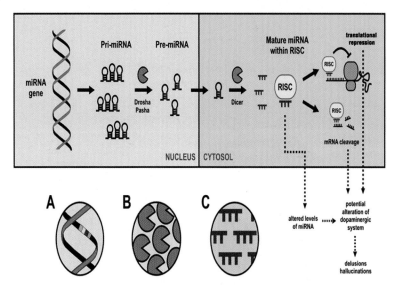

Figure 34.1. Biosynthesis of miRNA and potential impact on dopamine leading to psychosis The mature miRNA is much smaller than the gene involved in encoding the molecule. (See text for full caption).

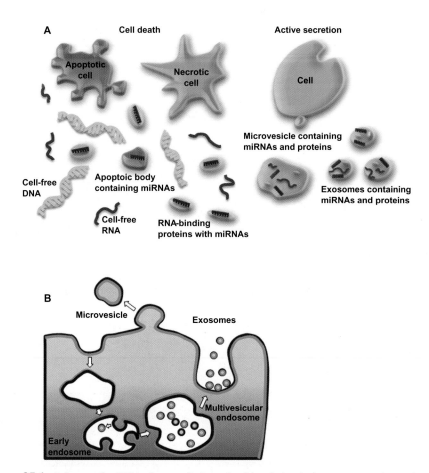

Figure 35.1. Release of miRNAs from cells into the blood circulation can occur by various cell physiological events, such as apoptosis, necrosis, and active secretion. (See text for full caption.)

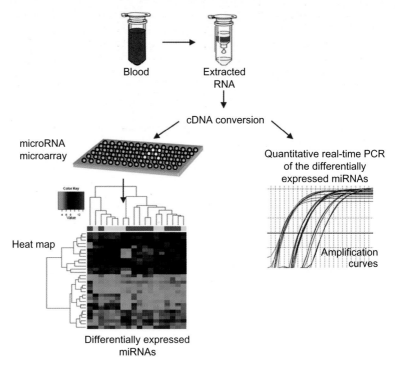

Figure 35.2. Experimental flowchart of quantification of circulating miRNAs in blood.

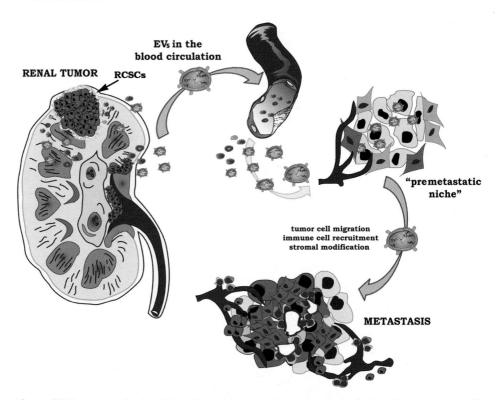

Figure 37.1. Potential role of EVs released from renal cancer stem cells. Renal cancer stem cells (RCSCs) expressing the CD105 marker may release EVs that enter the circulation, reach distant sites, and modify the microenvironment by triggering angiogenesis and formation of a "premetastatic niche" (i.e., condition favoring recruitment and implant of circulating tumor cells). miRNAs carried by EVs not only may sustain the ability of RCSCs to modify the tumor microenvironment, but they can also be used for tumor staging and as prognostic markers.

23

ROLE OF miRNAs IN THE PATHOGENESIS OF CHRONIC LYMPHOCYTIC LEUKEMIA

Veronica Balatti, Yuri Pekarsky, Lara Rizzotto, and Carlo M. Croce

Department of Molecular Virology, Immunology and Medical Genetics, Comprehensive Cancer Center and the Wexner Medical Center, The Ohio State University, Columbus, OH, USA

I. Introduction	384
A. Chronic Lymphocytic Leukemia (CLL) Characteristics and Clinical Outcomes	384
B. Association with MicroRNAs (miRNAs)	385
C. miRNA Signatures in CLLs	385
D. Functional Role of miRNAs in CLL	386
II. *miR-15a/16-1*	386
A. *miR-15a/16-1* at 13q14: First Example of Tumor Suppressor miRNA	386
B. *miR-15a/16-1* and *DLEU7*	389
C. *BCL2-TRAF2* Transgenic Mice: Cooperation of *miR-15a/16-1* and *DLEU7*	390
III. *miR-29*	391
A. Expression of *miR-29* in B Cells and in CLL	391
B. Role of *miR-29* and *miR-181* in Aggressive CLLs: Up-Regulation of *TCL1*	391
C. *TCL1* Transgenic Mice	392
D. Role of *miR-29* in Indolent CLL: *miR-29* Transgenic Mice Model	392
IV. *miR-34b/c*	393
A. Molecular Association between Chromosomal Abnormalities and Severity of CLLs	393
B. *miR-15a/16-1*, *miR-34b/c*, *TP53*, and ZAP-70 Circuit: Molecular Interactions in Patients with 17p–13q–11q Deletions in CLL	393

MicroRNAs in Medicine, First Edition. Edited by Charles H. Lawrie.
© 2014 John Wiley & Sons, Inc. Published 2014 by John Wiley & Sons, Inc.

V. miRNAs, Single-Nucleotide Polymorphisms (SNPs), and DNA Methylation	394
VI. Conclusions	395
Acknowledgments	396
References	396

ABBREVIATIONS

APRIL	a proliferation-inducing ligand
BCL2	B-cell lymphoma 2
BCMA	B-cell maturation antigen
BrdU	bromodeoxyuridine
CLL	chronic lymphocytic leukemia
HDAC	histone deacetylase
IgH V_H	immunoglobulin heavy-chain variable-region gene
KO	knockout
Mdr	minimal deleted region
NFATs	nuclear factor of activated T cells
NF-κB	nuclear factor kappa-light-chain enhancer of activated B cells
NZB	New Zealand Black
NZW	New Zealand White
pRb	phosphorylated retinoblastoma
SNPs	single-nucleotide polymorphisms
TNF	tumor necrosis factor
TRAF2	TNF receptor-associated factor 2
UTR	untranslated region
WT	wild type
ZAP-70	70-kD zeta-associated protein

I. INTRODUCTION

A. Chronic Lymphocytic Leukemia (CLL) Characteristics and Clinical Outcomes

CLL is the most common human leukemia, accounting for ~30% of all cases of adult leukemia. In the United States, almost 15,000 new cases are observed each year (Jemal et al. 2010). CLL is mostly a disease of elderly people, with the incidence increasing linearly with each decade (Bullrich and Croce 2001). This disease occurs in two forms, aggressive and indolent, both characterized by the clonal expansion of CD5 positive B cells (Bullrich and Croce 2001). More than 90% of the leukemic cells are non-dividing and are at the G0/G1 phase of the cell cycle (Bullrich and Croce 2001). However, several reports showed that high lymphocyte count in CLL patients is caused also by proliferating cells from the bone morrow, spleen, or lymph nodes (Messmer et al. 2005; Chiorazzi 2007; Sieklucka et al. 2008). CLL cells are also quite resistant to apoptosis (Bullrich and Croce 2001).

The clinical course of CLL can be defined by several predictive factors such as mutational status of the immunoglobulin heavy-chain variable-region gene (IgH V_H), expression levels of the 70-kD zeta-associated protein (ZAP-70), and the presence of different

INTRODUCTION

chromosomal alterations (Orchard et al. 2004; Rassenti et al. 2004). CLLs with unmutated IgH V_H and high expression of the ZAP-70 tend to have an aggressive course, whereas patients with mutated V_H clones and low ZAP-70 expression have an indolent course (Chiorazzi et al. 2005). Genomic alterations in CLL are important independent predictors of disease progression and survival (Dohner et al. 2000); however, the molecular basis of these associations was largely unknown until recently. About 80% of patients have cytogenetic abnormalities that can be detected by fluorescence *in situ* hybridization (FISH). The incidence of the most relevant genetic abnormalities range from 14% to 40% for deletion of 13q14, from 10% to 32% for deletion of 11q23, from 11% to 18% for trisomy 12, from 3% to 27% for deletion of 17p, and from 2% to 9% for deletion of 6q (Neilson et al. 1997; Stilgenbauer et al. 1999). Prognosis is worst in patients with 17p deletion, followed by 11q deletion, trisomy 12, and normal karyotype (negative FISH panel), while patients with deletion of 13q as the only abnormality have the best prognosis (Neilson et al. 1997; Zenz et al. 2008). Cytogenetic abnormalities can be used to identify subsets of patients with different clinical forms, time to progression, and survival rates. According to recent studies, three risk groups can be differentiated: (1) low risk: patients with a normal karyotype or isolated 13q deletion; (2) intermediate risk: subjects with del11q deletion, trisomy 12, or 6q deletion; and (3) high risk: patients with 17p deletion or a complex karyotype (Moreno and Montserrat 2010).

The clinical course of CLL is highly variable, and approximately one-third of patients never require treatment; in another third, the initial indolent phase is followed by progression of the disease, and the remaining third has aggressive disease at the outset and needs immediate treatment (Dighiero and Binet 2000).

B. Association with MicroRNAs (miRNAs)

Since the first association of miRNAs with cancer by Calin et al. (2002), it was clear that these genes could play a role in the clinical management of cancer patients. miRNAs are endogenous non-coding RNAs, 19–25 nucleotides in size (Bartel 2004). Recent studies have shown that miRNAs are involved in various cellular processes, including DNA methylation (Fabbri et al. 2007), cellular growth, differentiation, and apoptosis (Zhang and Chen 2009). Moreover, miRNAs can modulate gene expression in a tissue-specific manner and are able to bind target messenger RNAs (mRNAs) either inhibiting their translation or promoting their degradation (Iorio and Croce 2012).

C. miRNA Signatures in CLLs

Numerous reports demonstrated that, as with protein-coding genes, miRNAs are differentially expressed in cancers, indicating that miRNA deregulation could play tumor suppressor or oncogenic roles in cancer pathogenesis (Volinia et al. 2006). Moreover, it has been demonstrated that miRNA expression profiles can be used to distinguish normal B cells from malignant CLL cells and, more importantly, that miRNA signatures are associated with prognosis, progression, and drug resistance of CLL (Ferracin et al. 2010). Specifically, a signature profile was reported, describing 13 miRNAs that differentiate aggressive and indolent CLL (Calin et al. 2004). Another report showed that the expression profile of 32 miRNAs is able to discriminate between cytogenetic subgroups (Visone et al. 2009). For example, expression level of *miR-21* can distinguish between good-prognosis and poor-prognosis CLL. Indeed, patients with high levels of *miR-21* had a higher risk of death compared with patients with low expression levels (Rossi et al. 2010). Moreover, high expression of *miR-155* was reported in the aggressive form of CLL (Calin et al. 2007).

Recently, it has also been found that miRNA signature can be used to predict CLLs that are refractory to fludarabine treatment (Ferracin et al. 2010). To elucidate whether miRNAs are involved in the development of fludarabine resistance, Ferracin et al. analyzed the expression of miRNAs before and after therapy in patients classified as responder or refractory and identified an miRNA signature able to distinguish between these two classes. Expression levels of several miRNAs were also able to predict fludarabine resistance in an independent test cohort. Among these miRNAs, *miR-148a*, *miR-222*, and *miR-21* exhibited a significantly higher expression in non-responders either before or after treatment. Recently, Zenz et al. (2009) found that fludarabine refractory CLLs are frequently characterized by lower levels of *miR-34a*, and low expression of *miR-34a* was associated with fludarabine resistance even in the absence of p53 aberrations (Zenz et al. 2009).

Intriguingly, we recently found that expression levels of *miR-181b* not only can distinguish between indolent and aggressive cohorts of patients but can also be used as a time-to-treatment biomarker of the disease progression. We studied serial time points from the same patients and found that expression of *miR-181b* decreases with the severity of the disease. This new discovery highlights the importance of this miRNA in clinic, suggesting that expression levels of miRNAs can be used not only to classify patients in a cohort but also to keep track of the disease course (Visone et al. 2011).

Thus, miRNA expression levels can distinguish normal B cells from CLL, discriminate indolent between aggressive CLL forms, and separate responder and refractory cohorts of patients. Moreover, our latest findings provide a new role for miRNAs as potential monitors for patients who might require therapy (Visone et al. 2011).

D. Functional Role of miRNAs in CLL

Besides using miRNA expression levels as tools to discriminate different CLL forms or to keep track of disease progression, researchers have recently focused on the molecular impact of miRNA deregulation in CLL. Interestingly, the *miR-15/16* cluster, *miR-29*, *miR-181* family members, and *miRs-34b/c* were found as the most deregulated miRNAs in CLL. The same miRNAs were found to regulate gene expression patterns. These findings can clarify molecular steps that lead to the onset of the disease or drive its progression (Table 23.1).

II. *MIR-15A/16-1*

A. *miR-15a/16-1* at 13q14: First Example of Tumor Suppressor miRNA

As mentioned earlier, genomic aberrations are detected in over 80% of CLL cases and include 13q, 11q, 17p, and 6q deletions, and trisomy 12 (Dohner et al. 2000). The most frequently deleted genomic region in CLL occurs at chromosome 13q14.3 (in about 50% of cases) and is associated with the longest treatment-free interval (Dohner et al. 2000). To identify tumor suppressor genes at 13q14, several laboratories used positional cloning and sequencing of a region of more than 1 Mb (Bullrich et al. 2001; Migliazza et al. 2001).

TABLE 23.1. MicroRNAs in CLL Pathogenesis

	Initiation	Progression	Marker	References
miR-15a/16-1	Down-regulation in 13q del patients			Calin et al. (2002); Pekarsky et al. (2005)
miR-29	Up-regulation in indolent versus normal B cells	Down-regulation in aggressive versus indolent form		Pekarsky and Croce (2010); Santanam et al. (2010)
miR-34b/c		Down-regulation in 11q del patients		Rassenti et al. (2004); Fabbri et al. (2011)
miR-181			Expression decreases during progression	Visone et al. (2011)

However, none of the known genes in this region were found to be down-regulated in CLL by deletions or mutations (Bullrich et al. 2001; Migliazza et al. 2001; Rondeau et al. 2001; Mertens et al. 2002). In 2001, we generated somatic cell hybrids using mouse and CLL cells carrying 13q14 deletion and translocation. Using these hybrids, we identified a 30-kb region of deletion between exons 2 and 5 of the *LEU2* gene (Calin et al. 2002; Pekarsky et al. 2005). Interestingly, the translocation breakpoint was mapped to the same region (Calin et al. 2002; Pekarsky et al. 2005). Since *LEU2* had previously been sequenced and excluded as a candidate tumor suppressor gene in 13q14 (Bullrich et al. 2001; Migliazza et al. 2001; Wolf et al. 2001; Mertens et al. 2002), we continued to investigate the region and finally discovered a cluster of two non-coding miRNA genes, *miR-15a* and *miR-16-1*, located exactly within the deleted region and near the translocation breakpoint (Calin et al. 2002). Importantly, the *miR-15a/16-1* cluster was found deleted or its expression down-regulated in ~66% of CLL cases (Calin et al. 2002). In contrast, expression of the rest of the genes in the region (*DLEU1*, *DLEU2*, and *RFP5*) was not affected by the 13q14 deletions (Bullrich et al. 2001; Migliazza et al. 2001; Pekarsky et al. 2005).

The importance of the *miR-15a/16-1* cluster in CLL was confirmed in a recent study of CLL development in New Zealand Black (NZB) mice, the only mouse strain naturally susceptible to CLL (Raveche et al. 2007). In NZB mice, CLL develops late in life, with an autoimmune phenotype and B-cell hyperproliferation followed by slow progression to late-onset CLL (Raveche 1990; Zanesi et al. 2010). Older NZB animals show a clonal expansion of the subpopulation of B-1 B cells similar to that found in human CLL (Raveche 1990; Zanesi et al. 2010). Linkage analysis has found that the mouse genomic region homologous to 13q14 is one of the *loci* associated with CLL development. Subsequent DNA sequencing resulted in the identification of a point mutation in *miR-15a/16-1* precursor, causing a decrease of *miR-16-1* expression in NZB lymphoid tissues, accompanied by elevated levels of Bcl2 (Raveche et al. 2007). Other strains of mice including the closest relative of NZB, the NZW strain, did not show this mutation. Accordingly, lymphoid tissues from NZB mice were analyzed for the levels of mature *miR-16-1* and showed reduced expression of this miRNA. Finally, delivery of exogenous *miR-16-1* to an NZB malignant cell line led to cell-cycle alterations such as decrease in S phase cells and G1 arrest (Raveche et al. 2007).

The first genetic manipulation in mice, which confirmed the importance of *miR-15a/16-1* deletion in CLL, was carried out by Dr. Dalla-Favera and colleagues (Klein et al. 2010). These authors designed a model with conditional alleles that either resembled the loss of the minimal deleted region (*Mdr*), already characterized in human CLL (Migliazza et al. 2001) and spanning entirely the *DLEU2* gene, or, without altering Dleu2 expression, contained only the specific deletion of the *miR-15a/16-1* cluster (Klein et al. 2010). Both *Mdr* knockout (KO) and *miR-15a/16-1* KO strains at 1 year of age presented approximately 50% of CD5+ B220+ B cells among mononuclear cells in the peritoneum versus 15% in control animals. In total, mice with CLL were 27% of *Mdr* KO and 21% *miR-15a/16-1* KO, while some type of clonal B-cell proliferation affected 42% of *Mdr* KO and 26% of *miR-15a/16-1* KO mice between 15 and 18 months of age. *Mdr* KO animals lived less than wild-type (WT) siblings and eventually succumbed to leukemias, while the differential survival between *miR-15a/16-1* and their WT littermates was not statistically significant, providing evidence that the latter were affected by a phenotype milder than the former. Because of the more aggressive disease shown by *Mdr* KO mice, it is likely that other elements included in the *Mdr* locus, like the *DLEU2* gene itself, may participate in CLL tumor suppression (Klein et al. 2010). Mechanisms leading to B-cell proliferations were investigated with different approaches. By quantifying active DNA synthesis through BrdU incorporation experiments, *miR-15a/16-1* KO B cells were shown to begin DNA synthesis earlier than WT B cells (Klein et al. 2010). The authors also analyzed levels of phosphorylated retinoblastoma (pRb) protein, an indicator of entry into the cell cycle, in mitogen-stimulated B cells isolated from *miR-15a/16-1* KO or *Mdr* KO and WT animals. Thus, pRb was produced in both KO B cells at earlier time points than in WT B cells. Individual contributions of *miR-15a/16-1* cluster versus *DLEU2* gene to the lymphoproliferation were dissected by investigators generating an inducible system where these two genetic elements underwent separate *in vitro* reexpression in a human cell line derived from a 13q14 KO CLL. These findings demonstrated that impaired proliferation occurred in *miR-15a/16-1*-expressing cells, with higher fraction of cells in G0/G1 phase, but not in those expressing Dleu2, thus suggesting a possible control of the inhibition of G0/G1 phase transition by *miR-15a/16-1* (Klein et al. 2010).

B-cell lymphoma 2 (*BCL2*) gene is a central player in the genetic program of eukaryotic cells promoting survival by inhibiting cell death (Cory and Adams 2002). Overexpression of Bcl2 protein has been reported in many types of human cancers, including leukemias, lymphomas, and carcinomas (Sanchez-Beato et al. 2003). In follicular lymphomas and in a fraction of diffuse large B-cell lymphomas, *BCL2* is activated due to the translocation t(14,18)(q32;q21), which places the *BCL2* gene under the control of Ig heavy-chain enhancers, resulting in the overexpression of the gene (Tsujimoto et al. 1984, 1985). In CLL, malignant B cells overexpress Bcl2 (Kitada et al. 1998). However, with the exception of less than 5% of cases, in which the *BCL2* gene is juxtaposed to Ig loci (Adachi et al. 1990), no mechanism has been discovered to explain *BCL2* up-regulation in CLL. It has been shown that *miR-15a* and *miR-16-1* expression is inversely correlated to Bcl2 expression in CLL and that these miRNAs negatively regulate *BCL2* at posttranscriptional level (Cimmino et al. 2005). Since *BCL2* is a predicted target of both *miR-15a* and *miR-16-1*, the down-regulation of these miRNAs in a leukemic cell line resulted in an increase of Bcl2 expression with consequent inhibition of apoptosis (Cimmino et al. 2005). Interestingly, *miR-15a/16-1* expression also resulted in growth inhibition of tumor engraftments of leukemic cells in nude mice, confirming the tumor suppression properties of these miRNAs (Calin et al. 2008). In summary, Bcl2 overexpression driven by down-regulation of *miR-15a* and *miR-16-1* seems to be a regulatory mechanism involved in the

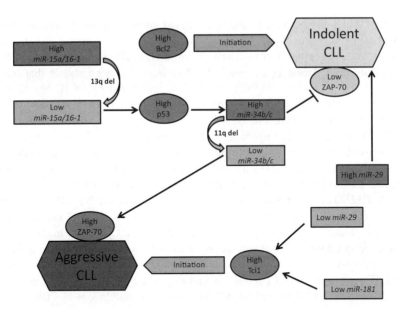

Figure 23.1. MicroRNAs in molecular mechanisms of CLL. See color insert.

pathogenesis of the major fraction of human CLL. These studies determined that the *miR-15a/16-1* cluster functions as a tumor suppressor in CLL by inhibiting Bcl2, and deletions at 13q14 represent an initializing step in CLL development (Cimmino et al. 2005).

Recently, an *miR-15a/16-1-TP53* feedback circuitry was also reported (Fabbri et al. 2011). This study showed that p53 directly transactivates *miR-15a/16-1* promoter, while the *miR-15a/16-1* cluster targets *TP53* expression (Fabbri et al. 2011) (Figure 23.1).

B. *miR-15a/16-1* and *DLEU7*

Recently, Ouillette et al. (2008) used microarray technology to determine the *Mdr* at 13q14 in CLL. In addition to *miR-15a/16-1*, this region contained *DLEU7* gene located telometic to *miR-15a/16-1*. *DLEU7* was previously identified as a candidate tumor suppressor gene at 13q14 (Hammarsund et al. 2004). Since *DLEU7* is the only protein-coding gene located within reported *Mdr* at 13q14, we investigated whether *DLEU7* can cooperate with *miR-15a/16-1* (Palamarchuk et al. 2010). Sequencing of *DLEU7* coding exons failed to find mutations in CLL samples, although a previous study reported hypermethylation of *DLEU7* promoter, with consequent silencing of this gene in 61% of CLL cases (Hammarsund et al. 2004). Real-time polymerase chain reaction (RT-PCR) experiments confirmed that expression of *DLEU7* in CLL samples is decreased when compared with normal CD19+ B cells. *miR-15a/16-1* was also found down-regulated in these CLL samples (Palamarchuk et al. 2010).

Since recent studies demonstrated a significant role for the NF-κB pathway in the pathogenesis of CLL (Pekarsky et al. 2007), we investigated whether Dleu7 might function as an inhibitor of NF-κB. In the inactive state, NF-κB proteins are bound to IκB proteins in the cytoplasm. After stimulation, IκB is degraded and NF-κB translocates the nucleus (Brockman et al. 1995; Chen et al. 1995; Ghosh et al. 1998). Induction of NF-κB can be driven by a variety of stimuli including exposure to members of the tumor necrosis factor

(TNF) superfamily, chemotherapy, and ionizing radiation (Beg and Baltimore 1996; Van Antwerp et al. 1996; Wang et al. 1996). Activation of NF-κB prevents B cells from undergoing apoptosis and regulates growth and differentiation (Beg and Baltimore 1996; Van Antwerp et al. 1996; Wang et al. 1996). In B cells, it has been shown that transgenic expression of the TNF ligand APRIL resulted in an expansion of B220+ CD5+ cells (Planelles et al. 2004). APRIL binds B-cell maturation antigen (BCMA) and TACI (Haiat et al. 2006), which stimulate the NF-κB pathway, suggesting that NF-κB activation through TACI and BCMA is important in the pathogenesis of CLL (Palamarchuk et al. 2010). Moreover, nuclear factor of activated T cells (NFATs) can also be activated by TACI and BCMA (Mackay et al. 2003), and NFAT was previously reported as a hallmark of unstimulated CLL cells (Schuh et al. 1996; Berland and Wortis 1998).

Since *DLEU7* is located within the 13q14 deleted region and NF-κB/NFAT activation can be critical in CLL pathogenesis, we investigated whether Dleu7 expression has an effect on NF-κB and NFAT activation by TACI and BCMA. Our experiments showed that Dleu7 expression inhibits NF-κB activation by BCMA over fivefold, while activation by TACI was inhibited over fourfold (Palamarchuk et al. 2010). Also, Dleu7 expression can inhibit NFAT activation by TACI and BCMA approximately eightfold. Thus, we concluded that Dleu7 functions as NFAT and NF-κB inhibitor (Palamarchuk et al. 2010).

C. *BCL2-TRAF2* Transgenic Mice: Cooperation of *miR-15a/16-1* and *DLEU7*

Since the indolent form of the disease is often characterized by 13q14 deletions, it is likely that up-regulation of *BCL2* plays a major role in this subset of CLL. Evidence for this hypothesis came from Dr. Reed and colleagues whose research used two previously described mouse models, one with Bcl2 overexpression in the lymphoid system (Katsumata et al. 1992), and the second with up-regulation of an isoform of *TRAF2* in B and T cells (Lee et al. 1997). TRAF2, or TNF receptor-associated factor 2, can bind to TNF receptor family and mediate the activation of NF-κB by TNF proteins (Chung et al. 2002). TNF-mediated signaling increased lymphocyte proliferation and survival (Haiat et al. 2006).

TRAF2 transgenic mice failed to develop a frank leukemia but showed an increased number of B cells accompanied by lymphadenopathy and splenomegaly (Lee et al. 1997). *BCL2* transgenic animals, which were designed with a construct mimicking t(14;18) translocation juxtaposing *BCL2* gene with the immunoglobulin heavy-chain locus at 14q32 as reported in human follicular lymphomas, did not develop malignancies either, presenting only prolonged *in vitro* B-cell survival and *in vivo* polyclonal B-cell expansions (Katsumata et al. 1992).

TRAF2DN-BCL2 double transgenic mice, on the other hand, displayed severe splenomegaly, and most animals were affected by a CLL-like disease with high B-cell blood count (Zapata et al. 2004). While single transgenics showed a normal lifespan, the double ones survived only between 6 and 14 months. Because of their complex features, it was not clear whether *TRAF2DN-BCL2* transgenics were a model of indolent or aggressive CLL (Pekarsky et al. 2007).

Our report showed that 13q14 region deleted in indolent CLL contains two cooperating tumor suppressors, indicating a collaboration of coding and non-coding genes, *miR-15a/16-1* and *DLEU7*. Both cooperating tumor suppressors, *miR-15a/16-1* and *DLEU7*, are inactivated by the 13q14 deletions. *DLEU7* deletions result in the induction of TNF signaling through TRAFs, while *miR-15a/16-1* deletions cause a constitutive increase of Bcl2 expression. Thus, 13q14 deletions cause CLL development by a molecular mechanism resembling the oncogenic events in *TRAF2DN/BCL2* transgenics (Palamarchuk et al. 2010).

III. *MIR-29*

A. Expression of *miR-29* in B Cells and in CLL

The role of *miR-29* in development/progression of CLLs is still ambiguous. In both indolent and aggressive CLLs, *miR-29* is overexpressed compared with normal B cells. In addition, expression levels of *miR-29* are higher in indolent than in aggressive CLLs (Calin et al. 2005; Pekarsky et al. 2006; Santanam et al. 2010). This led the research to evaluate the role of this miRNA in CLL.

Recently, we reported that *miR-29* is up-regulated in indolent CLL compared with normal B cells (Santanam et al. 2010), which implies an oncogenic function for *miR-29*, initiating or at least significantly contributing to the pathogenesis of CLL. On the other hand, we showed that *miR-29* expression is down-regulated in aggressive CLLs versus indolent CLLs (Calin et al. 2005; Pekarsky et al. 2006), that expression levels of *TCL1* and *miR-29* are inversely correlated and that *miR-29* targets *TCL1* expression (Pekarsky et al. 2006). These data suggest a possible tumor suppressor function for *miR-29* in aggressive CLL.

B. Role of *miR-29* and *miR-181* in Aggressive CLLs: Up-Regulation of *TCL1*

As mentioned earlier, an miRNA signature was published with 13 miRNAs that differentiate aggressive and indolent CLL (Calin et al. 2004). Intriguingly, of the four down-regulated miRNAs in aggressive CLL, three are different isoforms of *miR-29* (*miR-29a-2*, *miR-29b-2*, and *miR-29c*) (Calin et al. 2004), strongly suggesting that deregulation of *miR-29* can play a role in the pathogenesis of aggressive CLLs. Furthermore, expression of members of *miR-29* family could discriminate between CLL samples with good and bad prognosis (Calin et al. 2005).

In aggressive CLLs, down-regulation of *miR-29* appears to be involved in Tcl1 overexpression, along with *miR-181* (Pekarsky et al. 2006). Activation of the *TCL1* oncogene is an important initiating event in the pathogenesis of aggressive CLL. T-cell leukemia/lymphoma 1 (*TCL1*) was identified as a target of translocations and inversions at 14q32.1, in T-cell prolymphocytic leukemias (T-PLLs) (Virgilio et al. 1994). High Tcl1 expression in human CLL correlates with aggressive phenotype (Herling et al. 2006). Tcl1 functions as an activator of PI3K–Akt(PKB) oncogenic pathway (Laine et al. 2000; Pekarsky et al. 2000). Tcl1 activates Akt, driving its nuclear translocation and leading to an increased proliferation, inhibition of apoptosis, and transformation (Pekarsky et al. 2000). At the same time, Tcl1 activates NF-κB, inhibits AP-1 (Pekarsky et al. 2008), and inhibits *DNMT3a* (Palamarchuk et al. 2012), which is involved in the epigenetic deregulation of gene expression. This leads to defects in cell death, increased survival, and CLL pathogenesis.

Recently, we investigated whether *TCL1* expression in CLL is regulated by miRNAs (Pekarsky et al. 2006). *miR-29b* and also *miR-181b* are down-regulated in aggressive CLLs with 11q deletions and predicted to target Tcl1 (Pekarsky et al. 2006). Interestingly, *miR-181* is differentially expressed in B cells, and *TCL1* is mostly B-cell-specific gene (Ramkissoon et al. 2006). This suggests that Tcl1 might be a target of *miR-181* not only in CLL cells but also in normal B lymphocytes. We thus proceeded to verify if these miRNAs target Tcl1 expression. Our experiments revealed that coexpression of Tcl1 with *miR-29* and *miR-181* significantly decreased Tcl1 expression (Pekarsky et al. 2006). We therefore concluded that *miR-29b* and *miR-181b* target *TCL1* expression on mRNA and protein levels (Pekarsky et al. 2006). Concordantly, we found inverse correlation between *miR-29b* and *miR-181b* expression and Tcl1 protein expression in CLL samples (Pekarsky et al. 2006). These results suggest that Tcl1 expression in CLL is, at least in part, regulated by *miR-29* and *miR-181* (Pekarsky et al. 2006) (Figure 23.1).

C. *TCL1* Transgenic Mice

Since *TCL1* expression is regulated by miRNAs, like *miR-29* and *miR-181*, that target the 3′-untranslated region (3′-UTR) of the gene, we generated transgenic mice of Eµ-*TCL1* full length (Eµ-*TCL1* FL) containing both the 3′- and 5′-UTRs of *TCL1* under a B-cell-specific promoter (Efanov et al. 2010). These animals showed a CLL-like leukemia between 16 and 20 months of age. A population of CD5+ CD23+ B cells accumulated in spleens and lymph nodes of these mice. Immunological abnormalities like hypoimmunoglobulinemia, impaired immune response, and abnormal levels of cytokines were also found in Eµ-*TCL1* FL animals and were similar to those observed in human CLL (Efanov et al. 2010). In conclusion, both classical Eµ-*TCL1* and Eµ-*TCL1* FL transgenic mouse models of CLL displayed important biological similarities with their human counterpart that went beyond the simple resemblance between the two leukemias. Our study demonstrated that *TCL1* up-regulation in mouse B cells results in aggressive CLL (Bichi et al. 2002).

D. Role of *miR-29* in Indolent CLL: *miR-29* Transgenic Mice Model

Since *miR-29* is overexpressed in indolent human CLL compared with aggressive CLL and normal B cells, we designed a transgenic mouse characterized by overexpression of *miR-29* in B cells to evaluate the role of this miRNA in B-cell leukemias. We reported an increase in CD5+ CD19+ IgM+ B-cell populations, a hallmark of CLL, in splenocytes from these transgenics (Santanam et al. 2010). Eighty-five percent of *miR-29* animals showed a marked expansion of CD5+ B cells, between 12 and 14 months of age, representing up to 50% of the total B cells. Only 20% of the transgenics between 24 and 26 months of age died of leukemia. These data led us to conclude that *miR-29* mice mimicked the indolent form of CLL. In fact, the percentage of leukemic cells increased with age, from 20% of all B cells in mice below 15 months of age to more than 65% in mice above 20 months of age, indicating a gradual progression of indolent CLL (Santanam et al. 2010). Using BrdU incorporation experiments to measure the proliferative capacity of leukemic cells, we demonstrated a significant proliferation in *miR-29* transgenic B cells compared with WT CD19+ cells where no proliferation was found. Thus, *miR-29* overexpression seems to play a role in promoting B-cell proliferation. Furthermore, since immune incompetence and progressive hypogammaglobulinemia are typical features of human CLL, immune response to sheep red blood cell (SRBC) antigen and serum levels of immunoglobulins were analyzed in *miR-29* mice and their WT littermates. Both parameters were significantly decreased in transgenic animals, confirming the mimicking of indolent human CLL in *miR-29* transgenics (Santanam et al. 2010).

The current idea of the role of *miR-29* in CLLs is associated with its effect on Tcl1 expression pattern in both indolent and aggressive forms. Since *TCL1* is generally not expressed in indolent CLL (Pekarsky et al. 2006), it likely does not play an important function in indolent CLL. Thus, its down-regulation due to *miR-29* overexpression does not slow indolent CLL development. Up-regulation of *miR-29* expression is not sufficient to cause aggressive CLL. In contrast, up-regulation of Tcl1 is absolutely required for the initiation of the aggressive form of CLL. Down-regulation of *miR-29* expression in aggressive CLL (compared with the indolent form) contributes to up-regulation of Tcl1 and development of aggressive CLL (Pekarsky and Croce 2010).

IV. MIR-34B/C

A. Molecular Association between Chromosomal Abnormalities and Severity of CLLs

At present, it is not known how the 13q, 11q, and 17p deletions contribute to CLL pathogenesis and progression (Dohner et al. 2000). To determine the possible existence of molecular interactions among 13q, 17p, and 11q deletions that can affect the outcome of patients, we investigated if the *miR-15a/16-1* cluster (located at 13q), tumor protein p53 (located at 17p), and the *miR-34b/c* cluster (located at 11q) are linked in a molecular pathway that could explain the prognostic implications (indolent vs. aggressive form) of 13q, 17p, and 11q deletions in CLL (Fabbri et al. 2011). The loss of the long arm of chromosome 11 involves the 11q23.1 region where the *miR-34b/c* cluster is located (Auer et al. 2007). Deletion of 17p leads to abrogation of the p53 tumor suppressor (Merkel et al. 2010), while 13q deletion, as explained later, involves *miR-15a/16-1* down-regulation.

B. *miR-15a/16-1*, *miR-34b/c*, *TP53*, and ZAP-70 Circuit: Molecular Interactions in Patients with 17p–13q–11q Deletions in CLL

Several *TP53*-binding sites were found upstream of the *miR-15a/16-1* on chromosome 13 and *miR-34b/c* on chromosome 11. Chromatin immunoprecipitation analysis revealed that *TP53* directly binds to its predicted binding sites on both chromosomes 13 and 11. Thus, *TP53* can induce the expression of both these miRNAs (Fabbri et al. 2011). On the other hand, *miR-15a/16-1* targets *TP53*, while a binding site for the *miR-34* family was predicted in ZAP-70 mRNA (Fabbri et al. 2011). Consequences of these interactions may depend on chromosomal alterations and lead to different outcomes via feedback circuits involving protein-coding genes and miRNAs (Fabbri et al. 2011). In this model, *TP53* (on chromosome 17p) represents the molecular connection between *miR-15a/16-1* (on chromosome 13q) and *miR-34b/c* (on chromosome 11q) (Fabbri et al. 2011).

In 13q-deleted patients, the loss of *miR-15a/16-1* expression shifts the balance not only toward higher levels of antiapoptotic proteins Bcl2 (Cimmino et al. 2005; Calin et al. 2008) but also toward higher levels of the tumor suppressor protein p53. Consequently, in 13q patients, while the number of apoptotic cells may decrease because of the increased levels of Bcl2, the p53 tumor suppressor pathway remains intact, thus keeping the increase in tumor burden relatively low. This finding could explain how 13q deletions are associated with the indolent form of CLL. Moreover, increased p53 levels in patients with 13q deletions are associated with transactivation of *miR-34b/c* and with reduced levels of ZAP-70, positively correlating with survival of CLL patients (Rassenti et al. 2004), and explaining the indolent course of CLLs carrying 13q deletions.

CLL patients with 11q deletion, instead, express significantly lower levels of *miR-34b/c* and significantly higher levels of ZAP-70, both at mRNA and protein levels. These patients show poorer overall survival than patients with normal cytogenetic profiles and lower levels of ZAP-70. In these patients, TP53 is not up-regulated because *miR-15a/16-1* is not deleted. This condition is associated with lower control on apoptosis. TP53-driven transactivation of *miR-34b/c* is indeed ineffective, since the genomic locus of *miR-34b/c* is deleted, leading to a higher expression of ZAP-70, which correlates with poor prognosis (Fabbri et al. 2011).

In conclusion, we found that an miRNA/TP53 feedback circuitry is associated with pathogenesis and prognosis of CLL. These results also demonstrate that restoring expression of *miR-15a/16-1* indirectly affects expression of *miR-34* family by modulating levels of TP53 expression. Moreover, *miR-34* family is a downstream target of p53, and its overexpression can cause p53-like effects on apoptosis or cell-cycle arrest (Fabbri et al. 2011) (Figure 23.1).

V. miRNAs, SINGLE-NUCLEOTIDE POLYMORPHISMS (SNPs), AND DNA METHYLATION

The complexity of the pathways involving miRNAs in CLL development/progression was found to extend beyond their ability to directly regulate gene expression. miRNA expression can be modulated by trasactivator factors and respond to the presence of SNPs (Asslaber et al. 2010). Moreover, deregulation of epigenetic processes can also modify miRNA expression, leading to a differing progression of the disease and a different prognosis (Sampath et al. 2012).

A good example of SNPs being involved in altered miRNA expression is offered by *miR-34a* (Asslaber et al. 2010). *miR-34a* has been implicated in the CLL response to DNA damage through a p53-mediated induction (Dijkstra et al. 2009; Mraz et al. 2009; Zenz et al. 2009). TP53 protein transactivates *miR-34a* on chromosome 1p36, inducing tumor suppressor effects and enhancing apoptosis and cycle arrest (Bommer et al. 2007; Chang et al. 2007; He et al. 2007; Tarasov et al. 2007). The presence of a SNP (SNP 309) in the intronic region of the promoter of ubiquitin ligase *MDM2* leads to increased expression of *MDM2*, which binds p53, inhibiting its transactivation effects on *miR-34a* (Asslaber et al. 2010). In patients with intact p53, it has been reported that the presence of this SNP at the promoter of *MDM2* gene can induce down-regulation of *miR-34a* (Asslaber et al. 2010). In many types of cancer, this SNP has been associated with accelerated tumor formation and poor prognosis (Menin et al. 2006; Ohmiya et al. 2006; Gryshchenko et al. 2008). Asslaber et al. (2010) have shown that the GG genotype of *MDM2* SNP 309 leads to reduced overall survival and treatment-free survival in CLL. CLL cells of patients with the GG genotype had a significantly lower mean expression of *miR-34a* as compared with the TT genotype, suggesting the attenuation of the p53 pathway by the SNP 309. *miR-34a* levels in cells with the heterozygous GT genotype were found between those with the GG and the TT genotypes. Thus, the presence of this SNP restrains p53 activity on the *miR-34a* expression in CLL patients without p53 deletion/mutation.

miRNAs can also be involved in epigenetic gene regulation with positive and negative feedback circuits (Sampath et al. 2012). The histone deacetylases (HDACs) are chromatin-modulating enzymes that catalyze the removal of acetyl groups on specific lysines around gene promoters (van der Vlag and Otte 1999). Thus, they can trigger the demethylation of lysine 4 on histones (H3K4me2/3) promoting chromatin compaction and leading to epigenetic gene silencing (van der Vlag and Otte 1999). Recent data established that HDACs can also silence miRNAs. In particular, it has been observed that *miR-15a/16-1* are silenced by epigenetic mechanisms in 30–35% of CLL samples cooperating with 13q14 deletion to account for the low expression levels of these miRNAs in CLL (Sampath et al. 2012). Indeed, it has been found that HDAC1-3 is overexpressed in CLL but not normal lymphocytes identifying an independent mechanism for the silencing of *miR-15a/16-1* (Sampath et al. 2012).

In samples with monoallelic 13q14 deletion, it has been observed that the HDACs repressed *miR-15a/16-1* expression on the residual allele providing an example of functional cooperation between a genetic and an epigenetic mechanism to achieve gene repression. Induction of *miR-15a/16-1* in response to HDAC inhibition is associated with activation of cell death. Future prospective trials should evaluate the specific impact of epigenetic silencing of *miR-15a/16-1* on disease behavior and progression. Therefore, this offers a therapeutic strategy that antagonizes an important survival mechanism in cells, and CLL patients who exhibit such epigenetic silencing may represent a group that may benefit from HDAC inhibitor-based therapy (Sampath et al. 2012).

VI. CONCLUSIONS

CLL is a heterogeneous disease, and karyotypic aberrations are strongly predictive of survival. In order of highest to lowest risk, the genomic categories are 17p deletion, 11q deletion, trisomy 12, normal FISH, and 13q deletion. Patients with 17p deletion respond poorly to treatment, while patients with 11q deletion CLL, even if progresses early, show a better response to treatment. Indolent and aggressive forms are characterized by a different IgH V_H mutational status and ZAP-70 expression. Unmutated IgH V_H/ZAP-70-positive patients have increased rates of progression and reduced remission durations. miRNA expression is a new important tool in the management of the disease.

miRNAs differentially expressed in cancers and their deregulation could play tumor suppressor or oncogenic roles in cancer pathogenesis. miRNA expression profiles have been found as useful tools to distinguish normal B cells from malignant CLL cells and can be associated with the prognosis, progression, and drug resistance of CLL. miRNAs modify gene expression, and their deregulation involves downstream effects on cell cycle and proliferation. Deregulation of miRNAs can happen as a consequence of chromosomal alteration, epigenetic modulation, or interaction with other genes. Deletion of *miR-15a/16-1* has been correlated to Bcl2 up-regulation in indolent CLL, while down-regulation of *miR-29* and *miR-181* has been correlated to Tcl1 up-regulation in aggressive CLL. On the other hand, overexpression of *miR-29* in B cells results in the development of indolent CLL. *miR-34* family members are involved in a fine regulated feedback circuit with p53 and *miR-15a/16-1* in 13q-deleted CLL, suggesting that the interplay between miRNAs and genes is bidirectional. Finally, miRNAs can also be epigenetically silenced, suggesting a new cooperating system of abnormal regulation of these molecules. The study of these mechanisms can clarify the role of miRNAs in the development and progression of CLL and detect new targets for therapy.

ACKNOWLEDGMENTS

This work was supported by the ACS Research Scholar Award and Swan Family Award (to Yuri Pekarsky).

REFERENCES

Adachi, M., A. Tefferi, P.R. Greipp, T.J. Kipps, and Y. Tsujimoto. 1990. Preferential linkage of bcl-2 to immunoglobulin light chain gene in chronic lymphocytic leukemia. *J Exp Med* **171**:559–564.

Asslaber, D., J.D. Pinon, I. Seyfried, P. Desch, M. Stocher, I. Tinhofer, A. Egle, O. Merkel, and R. Greil. 2010. MicroRNA-34a expression correlates with MDM2 SNP309 polymorphism and treatment-free survival in chronic lymphocytic leukemia. *Blood* **115**:4191–4197.

Auer, R.L., S. Riaz, and F.E. Cotter. 2007. The 13q and 11q B-cell chronic lymphocytic leukaemia-associated regions derive from a common ancestral region in the zebrafish. *Br J Haematol* **137**:443–453.

Bartel, D.P. 2004. MicroRNAs: genomics, biogenesis, mechanism, and function. *Cell* **116**: 281–297.

Beg, A.A. and D. Baltimore. 1996. An essential role for NF-kappaB in preventing TNF-alpha-induced cell death. *Science* **274**:782–784.

Berland, R. and H.H. Wortis. 1998. An NFAT-dependent enhancer is necessary for anti-IgM-mediated induction of murine CD5 expression in primary splenic B cells. *J Immunol* **161**:277–285.

Bichi, R., S.A. Shinton, E.S. Martin, A. Koval, G.A. Calin, R. Cesari, G. Russo, R.R. Hardy, and C.M. Croce. 2002. Human chronic lymphocytic leukemia modeled in mouse by targeted TCL1 expression. *Proc Natl Acad Sci U S A* **99**:6955–6960.

Bommer, G.T., I. Gerin, Y. Feng, A.J. Kaczorowski, R. Kuick, R.E. Love, Y. Zhai, T.J. Giordano, Z.S. Qin, B.B. Moore, O.A. MacDougald, K.R. Cho, et al. 2007. p53-mediated activation of miRNA34 candidate tumor-suppressor genes. *Curr Biol* **17**:1298–1307.

Brockman, J.A., D.C. Scherer, T.A. McKinsey, S.M. Hall, X. Qi, W.Y. Lee, and D.W. Ballard. 1995. Coupling of a signal response domain in I kappa B alpha to multiple pathways for NF-kappa B activation. *Mol Cell Biol* **15**:2809–2818.

Bullrich, F. and C. Croce. 2001. Molecular biology of chronic lymphocytic leukemia. In *Chronic lymphocytic leukemias*. B. Cheson, Ed. Marcel Dekker, New York. pp. 9–32.

Bullrich, F., H. Fujii, G. Calin, H. Mabuchi, M. Negrini, Y. Pekarsky, L. Rassenti, H. Alder, J.C. Reed, M.J. Keating, T.J. Kipps, and C.M. Croce. 2001. Characterization of the 13q14 tumor suppressor locus in CLL: identification of ALT1, an alternative splice variant of the LEU2 gene. *Cancer Res* **61**:6640–6648.

Calin, G.A., C.D. Dumitru, M. Shimizu, R. Bichi, S. Zupo, E. Noch, H. Aldler, S. Rattan, M. Keating, K. Rai, L. Rassenti, T. Kipps, et al. 2002. Frequent deletions and down-regulation of micro-RNA genes miR15 and miR16 at 13q14 in chronic lymphocytic leukemia. *Proc Natl Acad Sci U S A* **99**:15524–15529.

Calin, G.A., C.G. Liu, C. Sevignani, M. Ferracin, N. Felli, C.D. Dumitru, M. Shimizu, A. Cimmino, S. Zupo, M. Dono, M.L. Dell'Aquila, H. Alder, et al. 2004. MicroRNA profiling reveals distinct signatures in B cell chronic lymphocytic leukemias. *Proc Natl Acad Sci U S A* **101**:11755–11760.

Calin, G.A., M. Ferracin, A. Cimmino, G. Di Leva, M. Shimizu, S.E. Wojcik, M.V. Iorio, R. Visone, N.I. Sever, M. Fabbri, R. Iuliano, T. Palumbo, et al. 2005. A microRNA signature associated with prognosis and progression in chronic lymphocytic leukemia. *N Engl J Med* **353**:1793–1801.

REFERENCES

Calin, G.A., Y. Pekarsky, and C.M. Croce. 2007. The role of microRNA and other non-coding RNA in the pathogenesis of chronic lymphocytic leukemia. *Best Pract Res Clin Haematol* **20**: 425–437.

Calin, G.A., A. Cimmino, M. Fabbri, M. Ferracin, S.E. Wojcik, M. Shimizu, C. Taccioli, N. Zanesi, R. Garzon, R.I. Aqeilan, H. Alder, S. Volinia, et al. 2008. MiR-15a and *miR-16-1* cluster functions in human leukemia. *Proc Natl Acad Sci U S A* **105**:5166–5171.

Chang, T.C., E.A. Wentzel, O.A. Kent, K. Ramachandran, M. Mullendore, K.H. Lee, G. Feldmann, M. Yamakuchi, M. Ferlito, C.J. Lowenstein, D.E. Arking, M.A. Beer, et al. 2007. Transactivation of *miR-34a* by p53 broadly influences gene expression and promotes apoptosis. *Mol Cell* **26**:745–752.

Chen, Z., J. Hagler, V.J. Palombella, F. Melandri, D. Scherer, D. Ballard, and T. Maniatis. 1995. Signal-induced site-specific phosphorylation targets I kappa B alpha to the ubiquitin-proteasome pathway. *Genes Dev* **9**:1586–1597.

Chiorazzi, N. 2007. Cell proliferation and death: forgotten features of chronic lymphocytic leukemia B cells. *Best Pract Res Clin Haematol* **20**:399–413.

Chiorazzi, N., K.R. Rai, and M. Ferrarini. 2005. Chronic lymphocytic leukemia. *N Engl J Med* **352**:804–815.

Chung, J.Y., Y.C. Park, H. Ye, and H. Wu. 2002. All TRAFs are not created equal: common and distinct molecular mechanisms of TRAF-mediated signal transduction. *J Cell Sci* **115**:679–688.

Cimmino, A., G.A. Calin, M. Fabbri, M.V. Iorio, M. Ferracin, M. Shimizu, S.E. Wojcik, R.I. Aqeilan, S. Zupo, M. Dono, L. Rassenti, H. Alder, et al. 2005. *miR-15* and *miR-16* induce apoptosis by targeting BCL2. *Proc Natl Acad Sci U S A* **102**:13944–13949.

Cory, S. and J.M. Adams. 2002. The Bcl2 family: regulators of the cellular life-or-death switch. *Nat Rev Cancer* **2**:647–656.

Dighiero, G. and J.L. Binet. 2000. When and how to treat chronic lymphocytic leukemia. *N Engl J Med* **343**:1799–1801.

Dijkstra, M.K., K. van Lom, D. Tielemans, F. Elstrodt, A.W. Langerak, M.B. van 't Veer, and M. Jongen-Lavrencic. 2009. 17p13/TP53 deletion in B-CLL patients is associated with microRNA-34a downregulation. *Leukemia* **23**:625–627.

Dohner, H., S. Stilgenbauer, A. Benner, E. Leupolt, A. Krober, L. Bullinger, K. Dohner, M. Bentz, and P. Lichter. 2000. Genomic aberrations and survival in chronic lymphocytic leukemia. *N Engl J Med* **343**:1910–1916.

Efanov, A., N. Zanesi, N. Nazaryan, U. Santanam, A. Palamarchuk, C.M. Croce, and Y. Pekarsky. 2010. CD5+CD23+ leukemic cell populations in TCL1 transgenic mice show significantly increased proliferation and Akt phosphorylation. *Leukemia* **24**:970–975.

Fabbri, M., M. Ivan, A. Cimmino, M. Negrini, and G.A. Calin. 2007. Regulatory mechanisms of microRNAs involvement in cancer. *Expert Opin Biol Ther* **7**:1009–1019.

Fabbri, M., A. Bottoni, M. Shimizu, R. Spizzo, M.S. Nicoloso, S. Rossi, E. Barbarotto, A. Cimmino, B. Adair, S.E. Wojcik, N. Valeri, F. Calore, et al. 2011. Association of a microRNA/TP53 feedback circuitry with pathogenesis and outcome of B-cell chronic lymphocytic leukemia. *JAMA* **305**:59–67.

Ferracin, M., B. Zagatti, L. Rizzotto, F. Cavazzini, A. Veronese, M. Ciccone, E. Saccenti, L. Lupini, A. Grilli, C. De Angeli, M. Negrini, and A. Cuneo. 2010. MicroRNAs involvement in fludarabine refractory chronic lymphocytic leukemia. *Mol Cancer* **9**:123.

Ghosh, S., M.J. May, and E.B. Kopp. 1998. NF-kappa B and Rel proteins: evolutionarily conserved mediators of immune responses. *Annu Rev Immunol* **16**:225–260.

Gryshchenko, I., S. Hofbauer, M. Stoecher, P.T. Daniel, M. Steurer, A. Gaiger, K. Eigenberger, R. Greil, and I. Tinhofer. 2008. MDM2 SNP309 is associated with poor outcome in B-cell chronic lymphocytic leukemia. *J Clin Oncol* **26**:2252–2257.

Haiat, S., C. Billard, C. Quiney, F. Ajchenbaum-Cymbalista, and J.P. Kolb. 2006. Role of BAFF and APRIL in human B-cell chronic lymphocytic leukaemia. *Immunology* **118**:281–292.

Hammarsund, M., M.M. Corcoran, W. Wilson, C. Zhu, S. Einhorn, O. Sangfelt, and D. Grander. 2004. Characterization of a novel B-CLL candidate gene—DLEU7—located in the 13q14 tumor suppressor locus. *FEBS Lett* **556**:75–80.

He, L., X. He, L.P. Lim, E. de Stanchina, Z. Xuan, Y. Liang, W. Xue, L. Zender, J. Magnus, D. Ridzon, A.L. Jackson, P.S. Linsley, et al. 2007. A microRNA component of the p53 tumour suppressor network. *Nature* **447**:1130–1134.

Herling, M., K.A. Patel, J. Khalili, E. Schlette, R. Kobayashi, L.J. Medeiros, and D. Jones. 2006. TCL1 shows a regulated expression pattern in chronic lymphocytic leukemia that correlates with molecular subtypes and proliferative state. *Leukemia* **20**:280–285.

Iorio, M.V. and C.M. Croce. 2012. MicroRNA involvement in human cancer. *Carcinogenesis* **33**:1126–1133.

Jemal, A., R. Siegel, J. Xu, and E. Ward. 2010. Cancer statistics, 2010. *CA Cancer J Clin* **60**:277–300.

Katsumata, M., R.M. Siegel, D.C. Louie, T. Miyashita, Y. Tsujimoto, P.C. Nowell, M.I. Greene, and J.C. Reed. 1992. Differential effects of Bcl-2 on T and B cells in transgenic mice. *Proc Natl Acad Sci U S A* **89**:11376–11380.

Kitada, S., J. Andersen, S. Akar, J.M. Zapata, S. Takayama, S. Krajewski, H.G. Wang, X. Zhang, F. Bullrich, C.M. Croce, K. Rai, J. Hines, et al. 1998. Expression of apoptosis-regulating proteins in chronic lymphocytic leukemia: correlations with in vitro and in vivo chemoresponses. *Blood* **91**:3379–3389.

Klein, U., M. Lia, M. Crespo, R. Siegel, Q. Shen, T. Mo, A. Ambesi-Impiombato, A. Califano, A. Migliazza, G. Bhagat, and R. Dalla-Favera. 2010. The DLEU2/*miR-15a/16-1* cluster controls B cell proliferation and its deletion leads to chronic lymphocytic leukemia. *Cancer Cell* **17**:28–40.

Laine, J., G. Kunstle, T. Obata, M. Sha, and M. Noguchi. 2000. The protooncogene TCL1 is an Akt kinase coactivator. *Mol Cell* **6**:395–407.

Lee, S.Y., A. Reichlin, A. Santana, K.A. Sokol, M.C. Nussenzweig, and Y. Choi. 1997. TRAF2 is essential for JNK but not NF-kappaB activation and regulates lymphocyte proliferation and survival. *Immunity* **7**:703–713.

Mackay, F., P. Schneider, P. Rennert, and J. Browning. 2003. BAFF AND APRIL: a tutorial on B cell survival. *Annu Rev Immunol* **21**:231–264.

Menin, C., M.C. Scaini, G.L. De Salvo, M. Biscuola, M. Quaggio, G. Esposito, C. Belluco, M. Montagna, S. Agata, E. D'Andrea, D. Nitti, A. Amadori, et al. 2006. Association between MDM2-SNP309 and age at colorectal cancer diagnosis according to p53 mutation status. *J Natl Cancer Inst* **98**:285–288.

Merkel, O., D. Asslaber, J.D. Pinon, A. Egle, and R. Greil. 2010. Interdependent regulation of p53 and *miR-34a* in chronic lymphocytic leukemia. *Cell Cycle* **9**:2764–2768.

Mertens, D., S. Wolf, P. Schroeter, C. Schaffner, H. Dohner, S. Stilgenbauer, and P. Lichter. 2002. Down-regulation of candidate tumor suppressor genes within chromosome band 13q14.3 is independent of the DNA methylation pattern in B-cell chronic lymphocytic leukemia. *Blood* **99**:4116–4121.

Messmer, B.T., D. Messmer, S.L. Allen, J.E. Kolitz, P. Kudalkar, D. Cesar, E.J. Murphy, P. Koduru, M. Ferrarini, S. Zupo, G. Cutrona, R.N. Damle, et al. 2005. In vivo measurements document the dynamic cellular kinetics of chronic lymphocytic leukemia B cells. *J Clin Invest* **115**:755–764.

Migliazza, A., F. Bosch, H. Komatsu, E. Cayanis, S. Martinotti, E. Toniato, E. Guccione, X. Qu, M. Chien, V.V. Murty, G. Gaidano, G. Inghirami, et al. 2001. Nucleotide sequence, transcription map, and mutation analysis of the 13q14 chromosomal region deleted in B-cell chronic lymphocytic leukemia. *Blood* **97**:2098–2104.

REFERENCES

Moreno, C. and E. Montserrat. 2010. Genetic lesions in chronic lymphocytic leukemia: what's ready for prime time use? *Haematologica* **95**:12–15.

Mraz, M., K. Malinova, J. Kotaskova, S. Pavlova, B. Tichy, J. Malcikova, K. Stano Kozubik, J. Smardova, Y. Brychtova, M. Doubek, M. Trbusek, J. Mayer, et al. 2009. *miR-34a, miR-29c* and *miR-17-5p* are downregulated in CLL patients with TP53 abnormalities. *Leukemia* **23**: 1159–1163.

Neilson, J.R., R. Auer, D. White, N. Bienz, J.J. Waters, J.A. Whittaker, D.W. Milligan, and C.D. Fegan. 1997. Deletions at 11q identify a subset of patients with typical CLL who show consistent disease progression and reduced survival. *Leukemia* **11**:1929–1932.

Ohmiya, N., A. Taguchi, N. Mabuchi, A. Itoh, Y. Hirooka, Y. Niwa, and H. Goto. 2006. MDM2 promoter polymorphism is associated with both an increased susceptibility to gastric carcinoma and poor prognosis. *J Clin Oncol* **24**:4434–4440.

Orchard, J.A., R.E. Ibbotson, Z. Davis, A. Wiestner, A. Rosenwald, P.W. Thomas, T.J. Hamblin, L.M. Staudt, and D.G. Oscier. 2004. ZAP-70 expression and prognosis in chronic lymphocytic leukaemia. *Lancet* **363**:105–111.

Ouillette, P., H. Erba, L. Kujawski, M. Kaminski, K. Shedden, and S.N. Malek. 2008. Integrated genomic profiling of chronic lymphocytic leukemia identifies subtypes of deletion 13q14. *Cancer Res* **68**:1012–1021.

Palamarchuk, A., A. Efanov, N. Nazaryan, U. Santanam, H. Alder, L. Rassenti, T. Kipps, C.M. Croce, and Y. Pekarsky. 2010. 13q14 deletions in CLL involve cooperating tumor suppressors. *Blood* **115**:3916–3922.

Palamarchuk, A., P.S. Yan, N. Zanesi, L. Wang, B. Rodrigues, M. Murphy, V. Balatti, A. Bottoni, N. Nazaryan, H. Alder, L. Rassenti, T.J. Kipps, et al. 2012. Tcl1 protein functions as an inhibitor of de novo DNA methylation in B-cell chronic lymphocytic leukemia (CLL). *Proc Natl Acad Sci U S A* **109**:2555–2560.

Pekarsky, Y. and C.M. Croce. 2010. Is *miR-29* an oncogene or tumor suppressor in CLL? *Oncotarget* **1**:224–227.

Pekarsky, Y., A. Koval, C. Hallas, R. Bichi, M. Tresini, S. Malstrom, G. Russo, P. Tsichlis, and C.M. Croce. 2000. Tcl1 enhances Akt kinase activity and mediates its nuclear translocation. *Proc Natl Acad Sci U S A* **97**:3028–3033.

Pekarsky, Y., G.A. Calin, and R. Aqeilan. 2005. Chronic lymphocytic leukemia: molecular genetics and animal models. *Curr Top Microbiol Immunol* **294**:51–70.

Pekarsky, Y., U. Santanam, A. Cimmino, A. Palamarchuk, A. Efanov, V. Maximov, S. Volinia, H. Alder, C.G. Liu, L. Rassenti, G.A. Calin, J.P. Hagan, et al. 2006. Tcl1 expression in chronic lymphocytic leukemia is regulated by *miR-29* and *miR-181*. *Cancer Res* **66**:11590–11593.

Pekarsky, Y., N. Zanesi, R.I. Aqeilan, and C.M. Croce. 2007. Animal models for chronic lymphocytic leukemia. *J Cell Biochem* **100**:1109–1118.

Pekarsky, Y., A. Palamarchuk, V. Maximov, A. Efanov, N. Nazaryan, U. Santanam, L. Rassenti, T. Kipps, and C.M. Croce. 2008. Tcl1 functions as a transcriptional regulator and is directly involved in the pathogenesis of CLL. *Proc Natl Acad Sci U S A* **105**:19643–19648.

Planelles, L., C.E. Carvalho-Pinto, G. Hardenberg, S. Smaniotto, W. Savino, R. Gomez-Caro, M. Alvarez-Mon, J. de Jong, E. Eldering, A.C. Martinez, J.P. Medema, and M. Hahne. 2004. APRIL promotes B-1 cell-associated neoplasm. *Cancer Cell* **6**:399–408.

Ramkissoon, S.H., L.A. Mainwaring, Y. Ogasawara, K. Keyvanfar, J.P. McCoy Jr., E.M. Sloand, S. Kajigaya, and N.S. Young. 2006. Hematopoietic-specific microRNA expression in human cells. *Leuk Res* **30**:643–647.

Rassenti, L.Z., L. Huynh, T.L. Toy, L. Chen, M.J. Keating, J.G. Gribben, D.S. Neuberg, I.W. Flinn, K.R. Rai, J.C. Byrd, N.E. Kay, A. Greaves, et al. 2004. ZAP-70 compared with immunoglobulin heavy-chain gene mutation status as a predictor of disease progression in chronic lymphocytic leukemia. *N Engl J Med* **351**:893–901.

Raveche, E.S. 1990. Possible immunoregulatory role for CD5 + B cells. *Clin Immunol Immunopathol* **56**:135–150.

Raveche, E.S., E. Salerno, B.J. Scaglione, V. Manohar, F. Abbasi, Y.C. Lin, T. Fredrickson, P. Landgraf, S. Ramachandra, K. Huppi, J.R. Toro, V.E. Zenger, et al. 2007. Abnormal microRNA-16 locus with synteny to human 13q14 linked to CLL in NZB mice. *Blood* **109**: 5079–5086.

Rondeau, G., I. Moreau, S. Bezieau, J.L. Petit, R. Heilig, S. Fernandez, E. Pennarun, J.S. Myers, M.A. Batzer, J.P. Moisan, and M.C. Devilder. 2001. Comprehensive analysis of a large genomic sequence at the putative B-cell chronic lymphocytic leukaemia (B-CLL) tumour suppresser gene locus. *Mutat Res* **458**:55–70.

Rossi, S., M. Shimizu, E. Barbarotto, M.S. Nicoloso, F. Dimitri, D. Sampath, M. Fabbri, S. Lerner, L.L. Barron, L.Z. Rassenti, L. Jiang, L. Xiao, et al. 2010. MicroRNA fingerprinting of CLL patients with chromosome 17p deletion identify a *miR-21* score that stratifies early survival. *Blood* **116**:945–952.

Sampath, D., C. Liu, K. Vasan, M. Sulda, V.K. Puduvalli, W.G. Wierda, and M.J. Keating. 2012. Histone deacetylases mediate the silencing of *miR-15a*, *miR-16*, and *miR-29b* in chronic lymphocytic leukemia. *Blood* **119**:1162–1172.

Sanchez-Beato, M., A. Sanchez-Aguilera, and M.A. Piris. 2003. Cell cycle deregulation in B-cell lymphomas. *Blood* **101**:1220–1235.

Santanam, U., N. Zanesi, A. Efanov, S. Costinean, A. Palamarchuk, J.P. Hagan, S. Volinia, H. Alder, L. Rassenti, T. Kipps, C.M. Croce, and Y. Pekarsky. 2010. Chronic lymphocytic leukemia modeled in mouse by targeted *miR-29* expression. *Proc Natl Acad Sci U S A* **107**:12210–12215.

Schuh, K., A. Avots, H.P. Tony, E. Serfling, and C. Kneitz. 1996. Nuclear NF-ATp is a hallmark of unstimulated B cells from B-CLL patients. *Leuk Lymphoma* **23**:583–592.

Sieklucka, M., P. Pozarowski, A. Bojarska-Junak, I. Hus, A. Dmoszynska, and J. Rolinski. 2008. Apoptosis in B-CLL: the relationship between higher ex vivo spontaneous apoptosis before treatment in III-IV Rai stage patients and poor outcome. *Oncol Rep* **19**:1611–1620.

Stilgenbauer, S., L. Bullinger, A. Benner, K. Wildenberger, M. Bentz, K. Dohner, A.D. Ho, P. Lichter, and H. Dohner. 1999. Incidence and clinical significance of 6q deletions in B cell chronic lymphocytic leukemia. *Leukemia* **13**:1331–1334.

Tarasov, V., P. Jung, B. Verdoodt, D. Lodygin, A. Epanchintsev, A. Menssen, G. Meister, and H. Hermeking. 2007. Differential regulation of microRNAs by p53 revealed by massively parallel sequencing: *miR-34a* is a p53 target that induces apoptosis and G1-arrest. *Cell Cycle* **6**: 1586–1593.

Tsujimoto, Y., L.R. Finger, J. Yunis, P.C. Nowell, and C.M. Croce. 1984. Cloning of the chromosome breakpoint of neoplastic B cells with the t(14;18) chromosome translocation. *Science* **226**: 1097–1099.

Tsujimoto, Y., J. Cossman, E. Jaffe, and C.M. Croce. 1985. Involvement of the bcl-2 gene in human follicular lymphoma. *Science* **228**:1440–1443.

Van Antwerp, D.J., S.J. Martin, T. Kafri, D.R. Green, and I.M. Verma. 1996. Suppression of TNF-alpha-induced apoptosis by NF-kappaB. *Science* **274**:787–789.

van der Vlag, J. and A.P. Otte. 1999. Transcriptional repression mediated by the human polycomb-group protein EED involves histone deacetylation. *Nat Genet* **23**:474–478.

Virgilio, L., M.G. Narducci, M. Isobe, L.G. Billips, M.D. Cooper, C.M. Croce, and G. Russo. 1994. Identification of the TCL1 gene involved in T-cell malignancies. *Proc Natl Acad Sci U S A* **91**:12530–12534.

Visone, R., L.Z. Rassenti, A. Veronese, C. Taccioli, S. Costinean, B.D. Aguda, S. Volinia, M. Ferracin, J. Palatini, V. Balatti, H. Alder, M. Negrini, et al. 2009. Karyotype-specific microRNA signature in chronic lymphocytic leukemia. *Blood* **114**:3872–3879.

REFERENCES

Visone, R., A. Veronese, L.Z. Rassenti, V. Balatti, D.K. Pearl, M. Acunzo, S. Volinia, C. Taccioli, T.J. Kipps, and C.M. Croce. 2011. *miR-181b* is a biomarker of disease progression in chronic lymphocytic leukemia. *Blood* **118**:3072–3079.

Volinia, S., G.A. Calin, C.G. Liu, S. Ambs, A. Cimmino, F. Petrocca, R. Visone, M. Iorio, C. Roldo, M. Ferracin, R.L. Prueitt, N. Yanaihara, et al. 2006. A microRNA expression signature of human solid tumors defines cancer gene targets. *Proc Natl Acad Sci U S A* **103**:2257–2261.

Wang, C.Y., M.W. Mayo, and A.S. Baldwin Jr. 1996. TNF- and cancer therapy-induced apoptosis: potentiation by inhibition of NF-kappaB. *Science* **274**:784–787.

Wolf, S., D. Mertens, C. Schaffner, C. Korz, H. Dohner, S. Stilgenbauer, and P. Lichter. 2001. B-cell neoplasia associated gene with multiple splicing (BCMS): the candidate B-CLL gene on 13q14 comprises more than 560 kb covering all critical regions. *Hum Mol Genet* **10**:1275–1285.

Zanesi, N., Y. Pekarsky, F. Trapasso, G. Calin, and C.M. Croce. 2010. MicroRNAs in mouse models of lymphoid malignancies. *J Nucleic Acids Investig* **1**:36–40.

Zapata, J.M., M. Krajewska, H.C. Morse 3rd, Y. Choi, and J.C. Reed. 2004. TNF receptor-associated factor (TRAF) domain and Bcl-2 cooperate to induce small B cell lymphoma/chronic lymphocytic leukemia in transgenic mice. *Proc Natl Acad Sci U S A* **101**:16600–16605.

Zenz, T., D. Mertens, H. Dohner, and S. Stilgenbauer. 2008. Molecular diagnostics in chronic lymphocytic leukemia—pathogenetic and clinical implications. *Leuk Lymphoma* **49**:864–873.

Zenz, T., J. Mohr, E. Eldering, A.P. Kater, A. Buhler, D. Kienle, D. Winkler, J. Durig, M.H. van Oers, D. Mertens, H. Dohner, and S. Stilgenbauer. 2009. *miR-34a* as part of the resistance network in chronic lymphocytic leukemia. *Blood* **113**:3801–3808.

Zhang, H. and Y. Chen. 2009. New insight into the role of miRNAs in leukemia. *Sci China C Life Sci* **52**:224–231.

24

MicroRNA in B-Cell Non-Hodgkin's Lymphoma: Diagnostic Markers and Therapeutic Targets

Nerea Martínez,[1] Lorena Di Lisio,[1] and Miguel Angel Piris[1,2]

[1]*Cancer Genomics Laboratory, IFIMAV, Santander, Spain*
[2]*Department of Pathology, Hospital U. Marqués de Valdecilla, Santander, Spain*

I. Introduction	404
II. The Process of B-Cell Differentiation	405
A. The Role of miRNA in B-Cell Differentiation	405
III. B-Cell Lymphomas and miRNAs	406
A. Burkitt's Lymphoma (BL)	406
B. Chronic Lymphocytic Leukemia (CLL)	409
C. Diffuse Large B-Cell Lymphoma (DLBCL)	410
D. Follicular Lymphoma (FL)	410
E. Mantle Cell Lymphoma (MCL)	411
F. Marginal Zone Lymphoma: MALT Type	411
G. Nodal Marginal Zone Lymphoma (NMZL)	412
H. Splenic Marginal Zone Lymphoma (SMZL)	412
IV. miRNA-Targeted Therapy in B-Cell Lymphomas	412
V. Summary	413
References	413

ABBREVIATIONS

BL	Burkitt's lymphoma
CLL	chronic lymphocytic leukemia

MicroRNAs in Medicine, First Edition. Edited by Charles H. Lawrie.
© 2014 John Wiley & Sons, Inc. Published 2014 by John Wiley & Sons, Inc.

DLBCL	diffuse large B-cell lymphoma
FL	follicular lymphoma
HL	Hodgkin's lymphoma
MALT	mucosa-associated lymphoid tissue
MCL	mantle cell lymphoma
miRNA	microRNA
NHL	non-Hodgkin's lymphoma
NMZL	nodal marginal zone lymphoma
nt	nucleotide
qRT-PCR	quantitative real-time polymerase chain reaction
SMZL	splenic marginal zone lymphoma
WHO	World Health Organization

I. INTRODUCTION

Neoplasms derived from B cells are recognized in the WHO classification as a family that comprises at least 26 members, specific clinicopathological disorders that can be recognized through an integrative analysis of clinical, genomic, morphologic, and immunohistochemical features (Swerdlow et al. 2008).

The recognition and diagnosis of the different B-cell malignancies have evolved over time: first using data derived from immunohistochemical or gene expression studies and, more recently, through the development and application of next-generation sequencing studies to the knowledge of lymphoproliferative disorders. This molecular dissection has been paralleled by functional studies, allowing a better understanding of lymphoma pathogenesis. The ultimate aim of the continuous reelaboration of the lymphoma classification system is to assign effective therapy with minimal side effects.

Lymphoma classification is largely based on the assumption that B-lymphoma cells recapitulate the features of specific B-cell populations, thus making it possible to correlate the phenotype of particular lymphoma types with that of their normal B-cell counterparts. Nevertheless, there is a growing realization that lymphoma cells express additional, surrogate markers that provide information about specific molecular events playing a role in lymphoma pathogenesis. Therefore, lymphoma diagnosis uses a dynamic combination of markers for the identification of specific B-cell subsets (e.g., Tdt, CD10, BCL6) and others that recognize discrete oncogenic events such as cyclin D1 expression or *MYD88* mutations.

Understandably, hematologists and pathologists in charge of lymphoma diagnosis received with high expectations the news that a new class of molecules, microRNAs (miRNAs), playing an essential role in B-cell differentiation, seems to offer a robust tool for the recognition of the different lymphoma types (Fabbri et al. 2009; Di Lisio et al. 2012). miRNAs are a recently identified class of non-coding 21- to 23-nt RNAs that function as posttranscriptional regulators of gene expression, by targeting their corresponding messenger RNAs for degradation or translational repression. The small size of miRNAs makes it feasible to study their levels in paraffin-embedded tissue or serum by qRT-PCR, thus adding to the panel of potential diagnostic markers an arsenal of more than a thousand new molecules (Di Lisio et al. 2012).

Calin et al. (2002) provided the first indication that miRNAs might be useful in lymphoma diagnosis and involved in lymphomagenesis in 2002. Since then, a number of studies have established a role for miRNAs in B-cell differentiation and lymphoma pathogenesis, together with their potential as diagnostic, prognostic, and predictor markers.

More recently, it has also been shown that miRNAs can provide novel targets for therapy, an issue that is attracting great interest (Babar et al. 2012; Zhang et al. 2012).

II. THE PROCESS OF B-CELL DIFFERENTIATION

B-cell differentiation is a highly regulated process in which uncommitted hematopoietic precursors differentiate into antibody-secreting plasma cells or memory B cells by a multistep process, with multiple checkpoints where a subtle balance between multiple transcription factors determine cell differentiation progress and ultimately cell fate. This balance is at least partially determined by the posttranscriptional regulatory capacity of a myriad of miRNAs concertedly acting at these multiple decision points.

An essential part in B-cell maturation is performed in the germinal centers, where naive B cells undergo somatic hypermutation (SHM) of the variable regions of the immunoglobulin (Ig) genes and Ig class switch recombination (CSR). Then, the specific antigen-reactive B cells differentiate into the major effector B cells of the adaptive immune system: memory and plasma cells. Alternative processes of B-cell maturation are also possible, since germinal center-independent memory B cells have also been identified, and exposure to the germinal center environment may be consecutively repeated for some B-cell subpopulations (Berkowska et al. 2011).

A. The Role of miRNA in B-Cell Differentiation

B-cell lineage differentiation decision seems to be taken as a consequence of delicate changes in a choreography where the balance among *miR-181*, *miR-23a* cluster, *miR-34a*, *miR-125b*, *miR-150*, and *miR-17-92* cluster plays a determining role (Zhou et al. 2007; Ventura et al. 2008; Kong et al. 2010; Chaudhuri et al. 2012). Chen et al. (2004) demonstrated that *miR-181* was preferentially expressed in the B-lymphoid cells of mouse bone marrow, and its ectopic expression in hematopoietic stem/progenitor cells led to an increased fraction of B cells in both *in vivo* and *in vitro* models. In contrast, an increased expression of *miR-150* in hematopoietic progenitors has been shown to reduce the normal quantity of mature B cells, suggesting a blockage in the maturation process at the pro-B-cell stage (Zhou et al. 2007), at least in part through regulation of the expression of C-MYB, a transcription factor highly expressed in lymphocyte progenitors (Xiao et al. 2007). Additionally, the absence of *miR-17-92* has been demonstrated to lead to an increased level of the proapoptotic protein Bim, inhibiting B-cell development at the pro-B to pre-B transition (Ventura et al. 2008). Finally, *miR-34a*, which is expressed at low levels in early B cells, regulates B-cell differentiation from pro-B cell to pre-B cell. Experiments in which the expression of *miR-34a* has been forcibly introduced confirmed the existence of a blockage at the pro-B-cell stage as a consequence of Foxp1 down-regulation, a transcriptional regulator required for B-cell differentiation, which is a direct target of *miR-34a* (Rao et al. 2010).

Germinal center differentiation is also finely regulated by several miRNAs, in particular, by *miR-181b* and *miR-155*, chief controllers of germinal center reaction that target the activation-induced cytidine deaminase (AID) (de Yebenes et al. 2008; Teng et al. 2008). AID expression is required during the B-cell SHM process and class switching of Ig genes (Muramatsu et al. 2000), the two main phenomena that take place in the germinal center.

Additionally to AID, other genes expressed by germinal center cells, such as CD10 and PU1, have been shown to be targeted by *miR-181b* and *miR-155* (Vigorito et al. 2007; Thompson et al. 2011).

Germinal center B cells are distinguished by the expression of specific transcription factors such as BCL6 and LMO2, and the absence of plasma cell markers PRDM1/BLIMP1 and XBP1. The expression of all these transcription factors has been demonstrated to be regulated by miRNAs. Thus, BCL6 and PRDM1 are regulated by the *miR-30* family, *miR-9*, and *let-7a* (Lin et al. 2011), while *miR-125b* down-regulates the expression of IRF4 and PRDM1 (Gururajan et al. 2010), and the memory B-cell-enriched *miR-223* inhibits the expression of LMO2 (Malumbres et al. 2009).

Activated B-cells seem to display some specific miRNA markers. Thus, increased *miR-155* expression seems to play a role, through NF-κB regulation, in silencing PU1 and CD10 (Thompson et al. 2011), while TNFAIP3 (also known as A20) is inhibited by *miR-125a/b*, leading to enhanced NF-κB activation (Kim et al. 2012). Marginal zone B cells are distinguished by the expression of *miR-223* (Zhang et al. 2012), an miRNA whose expression is characteristic of memory B cells.

The main miRNAs that have been demonstrated so far to be involved in the different stages of B-cell differentiation are shown in Figure 24.1.

III. B-CELL LYMPHOMAS AND miRNAs

Some of the main validated observations are summarized here (see also Table 24.1 and Table 24.2). Additionally, a summary of the main findings concerning prognostic value of miRNA expression is included in Table 24.2, and specific comments are described for each lymphoma type.

A. Burkitt's Lymphoma (BL)

BL is characterized by the deregulation of the MYC oncogene, in most cases as a consequence of translocations involving the *MYC* locus (8q24) and Ig genes (14q32). MYC regulates and is itself regulated by a large set of miRNAs, leading to a complex regulatory loop that may have a particular importance in BL. Some of the miRNAs involved in this autoregulatory loop are *let-7a*, *e*, and *f*; *miR-26* (*a* and *b*); the *miR-17-92* cluster; *miR-34b*;

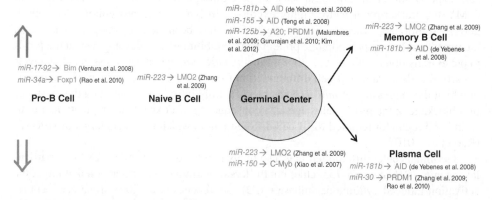

Figure 24.1. Master miRNAs that regulate B-cell development. Red (miRNA or genes) denotes high expression, while green (miRNA or genes) denotes low expression. See color insert.

TABLE 24.1. miRNA Expression in B-Cell Non-Hodgkin's Lymphoma

Lymphoma Type and Frequency	Cell Morphology	Main Alterations	Immunophenotype	Deregulated miRNA Associated with Relevant Target		References
BL 1%		*MYC-Ig* translocations: t(8;14)(q24;q32) or t(8;22)(q24;q11) or t(8;2)(q24;p12)	Positive for CD19, CD20, CD22, CD10, BCL6, CD38, CD77, CD43, IgM, Ki67 Negative for BCL2, TdT	miRNA DOWN *miR-98, miR-331, miR-363, miR-34b, let-7a, let-7e, let-7f* *miR-155*	Target UP Myc AID	Sampson et al. (2007); Leucci et al. (2008); Bueno et al. (2011) Dorsett et al. (2008)
CLL 12%		13q14.3del, trisomy 12 is frequent; 11q22-23del (*ATM*), 17p13del (*TP53*), 6q21del, *NOTCH1, S3F1* mutations	Positive for IgM, IgD, CD20, CD22, CD5, CD19, CD79a, CD23, CD43 Negative for CD10 and cyclin D1	miRNA DOWN *miR-29, miR-181* *miR-15a/16-1*	Target UP Tcl1 Bcl2	Pekarsky et al. (2006) Cimmino et al. (2005)
DLBCL 27%		ABC subtype: gain 3q, 18q21-q22, loss 6q21-q22 GCB subtype: gain 12q12, BCL2, BCL6, and MYC rearrangement	Positive for CD19, CD20, CD22, CD79a, Ki67, IgM > IgG > IgA Note: CD10+ up to 60%, BCL6+ up to 80%, IRF4/MUM1+ up to 65%, p53+ up to 60%	miRNA DOWN *miR-34a* miRNA UP (in ABC type) *miR-155*	Target UP Foxp1 Target DOWN CD10	Craig et al. (2011a) Thompson et al. (2011)
FL 15%		t(14;18)(q32;q21), *BCL2* rearrangements 5–15%: 3q27 and/or *BCL6* rearrangements	Positive for SIg, CD19, CD20, CD22, CD79a, BCL2, BCL6, CD10, GCET1 Negative for CD5, CD43, IRF4/MUM1 Note: (IgM+/−, IgD, IgG, or rarely IgA)	miRNA UP *miR-20a/b,* *miR-194*	Target DOWN CHECK1, CDKN1A, SOCS2	Wang et al. (2012)

(*Continued*)

TABLE 24.1. (Continued)

Lymphoma Type and Frequency	Cell Morphology	Main Alterations	Immunophenotype	Deregulated miRNA Associated with Relevant Target		References
MCL 6%		CCND1 translocation t(11;14)(q13;q32)	Positive for IgM/IgD, CD5, CD43, cyclin D1 Negative for CD10, CD23, BCL6	miRNA *miR-16-1*, *miR-17-92*	Target Cyclin D1	Chen et al. (2008); Deshpande et al. (2009)
MALT 6%		t(11;18)(q21;q21) (*API2-MALT1*) t(1;14)(p22;q32) (*BCL10-IGH*) t(14;18)(q32;q21) (*IGH-MALT1*) t(3;14)(p14.1;q32) (*FOXP1-IGH*)	Positive for IgM (less frequently for IgA or IgG), CD20, CD79a, CD21, and CD35 Negative for CD5, CD10, CD23 Note: CD43+/− and CD11c+/− (weak)	miRNA DOWN *miR-29a/b/c* *miR-26a* *miR-92b/96* miRNA UP *miR-200a, b, c*	Target UP CDK6 NF-κB PRMT5 Target DOWN Cyclin E2	Zhao et al. (2010) Di Lisio et al. (2010) Pal et al. (2007) Cai et al. (2012)
NMZL <1%		Trisomies 3, 18, and 7	Positive for B-cell markers, MNDA, BCL2, CD43 50% Negative for CD5, CD23, CD10, BCL6, cyclin D1 Note: IgD+ 40%	miRNA UP *miR-221*, *miR-223*, and *let-7f*	Target DOWN LMO2	Arribas et al. (2012)

TABLE 24.2. Prognostic Value of miRNAs in Lymphomas

Malignancy	miRNA Proposed for Patients' Stratification		Reference
	Correlation	miRNA	
CLL[a]	OS and PFS	↓ miR-223: worse outcome	Stamatopoulos et al. (2009, 2010); Li et al. (2011a, 2011b); Zhou et al. (2012)
	OS and PFS	↓ miR-29c: worse outcome	Calin et al. (2005); Stamatopoulos et al. (2009, 2010); Li et al. (2011b)
	TFS; response to therapy	↓ miR-34a: worse outcome	Zenz et al. (2009b); Asslaber et al. (2010)
	Progression to active disease	↓ miR-181b: progression	Visone et al. (2011, 2012)
DLBCL[a]	OS and PFS	↑ miR-222: worse outcome	Malumbres et al. (2009); Alencar et al. (2011); Montes-Moreno et al. (2011)
	OS and RFS	↓ miR-21: worse outcome	Lawrie et al. (2007, 2008); Roehle et al. (2008)
MCL	OS	↓ miR-29a/b/c: worse outcome	Zhao et al. (2010)
	OS	↑ miR-20b: worse outcome	Di Lisio et al. (2010)
	OS	↑ miR-17-5p and miR-20a: worse outcome	Navarro et al. (2009)
	OS	↑ cluster miR-17-92: worse outcome	Rao et al. (2012)

[a]Only miRNAs confirmed in a second series of cases are reported.
OS, overall survival; TFS, treatment-free survival; PFS, progression-free survival; RFS, relapse-free Survival.

miR-98; miR-145; miR-155; miR-331; and miR-363 (O'Donnell et al. 2005; Sampson et al. 2007; Leucci et al. 2008; Bueno et al. 2011).

One of the most recurrent features of miRNA expression profile in BL is the loss of miR-155 (Kluiver et al. 2006, 2007). miR-155 is a chief regulator of B-cell development that targets AID, an enzyme that is essential for the development of the MYC-IGH translocations (Dorsett et al. 2008). Additionally, the increased expression of miR-155 in diffuse large B-cell lymphoma (DLBCL) (more pronounced in the ABC type) can be used for the differential between BL and DLBCL (Lawrie et al. 2007; Clarkson et al. 2009; Di Lisio et al. 2012). The use of miRNA signatures may also be useful for a more accurate delineation of the group of B-cell lymphomas with intermediate features between DLBCL and BL (Leucci et al. 2008).

Interestingly, the different epidemiologic subtypes of BL share a homogenous miRNA profile distinct from DLBCL, as shown by Lenze et al. (2011).

B. Chronic Lymphocytic Leukemia (CLL)

CLL has been a disease model for the study of the pathogenic relevance of miRNAs, where it was possible to learn about CLL pathogenesis and also some general principles on the role of miRNAs in cancer pathogenesis (Calin et al. 2002).

The first study of miRNAs in CLL described the loss of *miR-15* and *miR-16* in association with the loss of chromosomal region 13q14 (present in about 50% of CLL cases) (Calin et al. 2002). miRNAs included in this cluster were demonstrated to target the expression of BCL2 antiapoptotic protein, and the loss of these two miRNAs was correlated with high expression of BCL2 and diminished apoptosis of CLL neoplastic cells (Kitada et al. 1998; Cimmino et al. 2005). Studies performed also in other genes contained in the 13q14 deleted area showed that the loss of *DLEU7* and *DLEU2* may cooperate with the loss of *miR-15/16* in the pathogenesis of CLL (Klein et al. 2010; Santanam et al. 2010).

Other miRNAs identified for having a role in the pathogenesis of CLL are *miR-29* and *miR-181*, whose loss in CLL cases favors the overexpression of TCL1 (Pekarsky et al. 2006). Increased expression of *miR-29* has been related with an indolent form of CLL in transgenic mice (Santanam et al. 2010), while a loss of *miR-29*, with parallel increased TCL1 expression, has been found to distinguish aggressive CLL cases (Pekarsky et al. 2006). Other relevant miRNAs identified in CLL are *miR-34a*, which regulates E2F1 and C-MYB (Zauli et al. 2011); *miR-221/222*, which regulates p27 (Frenquelli et al. 2010); and *miR-106b*, which regulates p73 (Sampath et al. 2009).

CLL studies have also shown increased expression of *miR-155*, here activated by MYB. Thus, increased expression of MYB (v-myb myeloblastosis viral oncogene homologue), an oncogene overexpressed in a subset of B-CLL patients, regulates the expression of *miR-155* through its binding to *miR-155* promoter, coinciding with the hypermethylated histone H3K4 residue and hyperacetylation of H3K9 at the MIR155HG promoter (Vargova et al. 2011).

The expression of some specific miRNAs has been proposed also to be a solid marker for patient stratification. For instance, dynamic changes in *miR-181b* expression values are associated with increased risk of disease progression in CLL and adverse clinical outcome (Visone et al. 2011).

A low expression of *miR-34a* has been found to be associated with p53 inactivation, chemotherapy-refractory disease, shorter time to treatment, impaired DNA damage response, and apoptosis resistance, irrespective of 17p deletion/*TP53* mutation (Zenz et al. 2009a, 2009b; Asslaber et al. 2010).

Finally, the use of plasma miRNA expression levels has been suggested as a novel parameter for CLL prognosis and prediction (Moussay et al. 2011).

C. Diffuse Large B-Cell Lymphoma (DLBCL)

The diagnostic and prognostic relevance of miRNAs in DLBCL is largely covered in Chapter 25, so it will not be mentioned extensively here, although some relevant miRNAs are given in Table 24.1 and Table 24.2.

D. Follicular Lymphoma (FL)

Translocation (and consequent overexpression) of BCL2 locus t(14;18)(q32;q21) is considered the hallmark of this disease even if in a small percentage of cases, no translocation can be identified. The miRNA signature of FL compared with non-tumor controls includes overexpression of *miR-20a/b* and *miR-194*, which target CDKN1A and SOCS2, a feature that contributes to tumor cell proliferation and survival (Wang et al. 2012).

A differential miRNA expression profile between t(14;18) positive and negative cases, including *miR-16*, *miR-26a*, *miR-101*, *miR-29c*, and *miR-138*, has been described, and the negative cases have been associated with an increased proliferative capacity and a "late" germinal center B-cell phenotype (Roehle et al. 2008; Lawrie et al. 2009; Leich et al. 2011).

Interestingly, some of the miRNAs that characterize the FL signature (*let-7*, *miR-30*) and regulate the expression of BCL6 and BLIMP1/PRDM1 are induced after follicular dendritic cell (FDC) contact (Lin et al. 2011), thus illustrating an interesting interrelationship between the tumor and the stroma.

E. Mantle Cell Lymphoma (MCL)

MCL is a neoplasm characterized by the overexpression of cyclin D1 as a consequence of t(11;14) translocation. DNA coding and miRNA gene signatures have been identified by different groups for this lymphoma type (Navarro et al. 2009; Di Lisio et al. 2010; Zhao et al. 2010; Iqbal et al. 2012). Functional and clinical studies have already established the relevance of some of these findings.

Chen and Deshpande demonstrated that the loss of miRNA target sites by the *miR-15/16* family and the *miR-17-92* cluster, at the CCND1 3′-untranslated region (3′-UTR), contributes to the pathogenic overexpression of the cyclin D1 protein (Chen et al. 2008; Deshpande et al. 2009). Zhao et al. (2010) identified *miR-29* down-regulation as a mechanism involved in the activation of CDK4/CDK6, being also a prognostic marker for this condition. Navarro et al. (2009) identified overexpression of *miR-17-5p/miR-20a* as associated with high MYC mRNA levels in tumors with a more aggressive behavior, while our own group demonstrated that miRNAs deregulated in MCL targeted essential pathways for lymphoma survival such as CD40, mitogen-activated protein kinase, and NF-κB (Di Lisio et al. 2010). Finally, Iqbal et al. (2012) identified a cluster characterized by high expression of miRNAs from the *miR-17-92* cluster and its paralogues, *miR-106a-363* and *miR-106b-25*, and associated with high proliferation gene signature in MCL. The relevance of the *miR-17-92* cluster was also confirmed by Chaudhuri et al. (2012), who demonstrated that overexpression of *miR-17-92* activates the PI3K/AKT pathway and inhibits chemotherapy-induced apoptosis in MCL cell lines.

Multiple miRNAs have been proposed to prognosticate clinical outcome in MCL (Navarro et al. 2009; Di Lisio et al. 2010; Zhao et al. 2010; Iqbal et al. 2012), but none of them has been validated in independent studies. Some of these studies have been carried out in paraffin-embedded tissue and anticipate that miRNAs could be used in the future for identifying MCL patients whose disease is indolent or aggressive.

F. Marginal Zone Lymphoma: MALT Type

Multiple studies highlighted the different phases of MALT lymphoma development. A signature of 27 deregulated miRNAs, transcriptionally repressed by MYC, has been found to characterize gastric DLBCL, when compared with MALT lymphoma (Craig et al. 2011a). The same researchers have also described how transformation from gastritis to MALT lymphoma is epigenetically regulated by *miR-203* promoter methylation, identifying ABL1 as a potential target for the treatment of this malignancy (Craig et al. 2011b). miRNA signatures were recently identified by RT-PCR in 68 gastric biopsy samples representing normal mucosa, gastritis, suspicious lymphoid infiltrates, and overt MALT lymphoma. The study identified a set of five miRNAs (*miR-150*, *miR-550*, *miR-124a*, *miR-518b*, and *miR-539*) differentially expressed in gastritis as opposed to MALT lymphoma (Thorns et al. 2012). A study of gastric MALT lymphoma revealed that a high level of expression of *miR-223*, a marker of memory B cells, correlates with an increased E2A expression in gastric MALT lymphomas (Liu et al. 2010).

Other studies conducted in different organs have generated organ-specific signatures for MALT lymphomas. Thus, a study comparing tumors diagnosed in the conjunctiva with adjacent normal conjunctiva tissue revealed up-regulation of *miR-150* and *miR-155* and down-regulation of *miR-184*, *miR-200a*, *b*, and *c*, and *miR-205* (Cai et al. 2012). Further *in vitro* experiments with *miR-200a*, *b*, and *c* demonstrated their capacity to down-regulate cyclin E2, alone or in combination.

G. Nodal Marginal Zone Lymphoma (NMZL)

Studies performed in NMZL show a signature that can be used for the differential diagnosis with FL. Thus, NMZL cases show increased expression of *miR-221*, *miR-223*, and *let-7f*, a signature that closely mimics that exhibited by memory B cells and cells isolated from the normal marginal zone. Up-regulation of *miR-223* and *miR-221*, which target the germinal center (GC)-related genes LMO2 and CD10, could be partially responsible for the expression of a marginal zone gene signature (Arribas et al. 2012).

H. Splenic Marginal Zone Lymphoma (SMZL)

SMZL miRNA signatures reflect the splenic origin of the analyzed samples in most cases, marginal zone differentiation, presence of hepatitis C virus (HCV), or common genetic alterations such as 7q32 deletion (Bouteloup et al. 2012).

An miRNA signature has been identified in SMZL, including *miR-21*, *miR-155*, and *miR-146a*, that are overexpressed, while seven miRNAs, including *miR-139*, *miR-345*, *miR-125a*, and *miR-126*, had a reduced expression. Some of these miRNAs were confirmed in an independent study that found increased expression of *miR-21* to be associated with an adverse outcome (Bouteloup et al. 2012; Peveling-Oberhag et al. 2012). The 7q32 chromosomal region contains a cluster of miRNAs (including *miR-29a* and *miR-29b-1*), commonly lost in these tumors (Ruiz-Ballesteros et al. 2007; Bouteloup et al. 2012). Furthermore, *miR-26b* (a tumor suppressor miRNA) was found significantly down-regulated in SMZLs arising in HCV-positive patients (Peveling-Oberhag et al. 2012).

IV. miRNA-TARGETED THERAPY IN B-CELL LYMPHOMAS

Despite the fact that the potential utility or miRNA/anti-miRNA delivery therapy is generally recognized, the miRNA/anti-miRNA treatment is still a challenging issue (Garzon et al. 2010; Kota and Balasubramanian 2010), where the efficient targeting of miRNAs in tumor cells *in vivo* is the main difficulty.

Nevertheless, promising findings in lymphoma therapy have been achieved at the preclinical stage. Medina et al. (2010) have shown that inactivation of *miR-21* leads to regression of *miR-21*-induced murine lymphoma in a few days, partly as a result of apoptosis, demonstrating that tumors can become "addicted" to oncomiRs. Babar and coworkers have identified potential ways to inhibit oncogenic miRNAs using *miR-155*-induced lymphoma as a model; they proved that systemic delivery of antisense peptide nucleic acids encapsulated in unique polymer nanoparticles inhibits *miR-155* and slows the growth of these "addicted" pre-B-cell tumors *in vivo* (Babar et al. 2012; Zhang et al. 2012).

Salerno et al. (2009) have also shown that addition of exogenous *miR-15a* and *miR-16* led to an accumulation of quiescent CLL cells and augments apoptosis inducted by nutlin, a mouse double minute 2 (MDM2) antagonist, and genistein, a tyrosine kinase inhibitor.

Other researchers have proposed that systemically administered locked nucleic acid (LNA) antimiRs could be used for exploring miRNA function in rodents and primates. Their findings support the potential of these compounds as a new class of therapeutics for disease-associated miRNAs. This strategy has been eventually applied to models of Waldestrom macroglobulinemia (Elmen et al. 2008; Zhang et al. 2012).

V. SUMMARY

We are progressively discovering the complexity of the posttranscriptional mechanisms that regulate the expression of the main oncogenic genes and pathways involved in B-cell lymphoma pathogenesis. A subtle choreography links the stroma and the neoplastic cells and contributes to explain the survival of the neoplastic cells and their resistance to chemotherapy. Along the way, potential diagnostic markers and therapeutic targets for specific lymphoma types are emerging. miRNAs are preserved in paraffin-embedded tissue and present in the serum of lymphoma patients, thus providing opportunities for better diagnosis and cure.

REFERENCES

Alencar, A.J., R. Malumbres, G.A. Kozloski, R. Advani, N. Talreja, S. Chinichian, J. Briones, Y. Natkunam, L.H. Sehn, R.D. Gascoyne, R. Tibshirani, and I.S. Lossos. 2011. MicroRNAs are independent predictors of outcome in diffuse large B-cell lymphoma patients treated with R-CHOP. *Clin Cancer Res* **17**:4125–4135.

Arribas, A.J., Y. Campos-Martin, C. Gomez-Abad, P. Algara, M. Sanchez-Beato, M.S. Rodriguez-Pinilla, S. Montes-Moreno, N. Martinez, J. Alves-Ferreira, M.A. Piris, and M. Mollejo. 2012. Nodal marginal zone lymphoma: gene expression and miRNA profiling identify diagnostic markers and potential therapeutic targets. *Blood* **119**:e9–e21.

Asslaber, D., J.D. Pinon, I. Seyfried, P. Desch, M. Stocher, I. Tinhofer, A. Egle, O. Merkel, and R. Greil. 2010. MicroRNA-34a expression correlates with MDM2 SNP309 polymorphism and treatment-free survival in chronic lymphocytic leukemia. *Blood* **115**:4191–4197.

Babar, I.A., C.J. Cheng, C.J. Booth, X. Liang, J.B. Weidhaas, W.M. Saltzman, and F.J. Slack. 2012. Nanoparticle-based therapy in an in vivo microRNA-155 (miR-155)-dependent mouse model of lymphoma. *Proc Natl Acad Sci U S A* **109**:E1695–E1704.

Berkowska, M.A., G.J. Driessen, V. Bikos, C. Grosserichter-Wagener, K. Stamatopoulos, A. Cerutti, B. He, K. Biermann, J.F. Lange, M. van der Burg, J.J. van Dongen, and M.C. van Zelm. 2011. Human memory B cells originate from three distinct germinal center-dependent and -independent maturation pathways. *Blood* **118**:2150–2158.

Bouteloup, M., A. Verney, N. Rachinel, E. Callet-Bauchu, M. Ffrench, B. Coiffier, J.P. Magaud, F. Berger, G.A. Salles, and A. Traverse-Glehen. 2012. MicroRNA expression profile in splenic marginal zone lymphoma. *Br J Haematol* **156**:279–281.

Bueno, M.J., M. Gomez de Cedron, G. Gomez-Lopez, I. Perez de Castro, L. Di Lisio, S. Montes-Moreno, N. Martinez, M. Guerrero, R. Sanchez-Martinez, J. Santos, D.G. Pisano, M.A. Piris, et al. 2011. Combinatorial effects of microRNAs to suppress the Myc oncogenic pathway. *Blood* **117**:6255–6266.

Cai, J., X. Liu, J. Cheng, Y. Li, X. Huang, X. Ma, H. Yu, H. Liu, and R. Wei. 2012. MicroRNA-200 is commonly repressed in conjunctival MALT lymphoma, and targets cyclin E2. *Graefes Arch Clin Exp Ophthalmol* **250**:523–531.

Calin, G.A., C.D. Dumitru, M. Shimizu, R. Bichi, S. Zupo, E. Noch, H. Aldler, S. Rattan, M. Keating, K. Rai, L. Rassenti, T. Kipps, et al. 2002. Frequent deletions and down-regulation of micro-RNA genes miR15 and miR16 at 13q14 in chronic lymphocytic leukemia. *Proc Natl Acad Sci U S A* **99**:15524–15529.

Calin, G.A., M. Ferracin, A. Cimmino, G. Di Leva, M. Shimizu, S.E. Wojcik, M.V. Iorio, R. Visone, N.I. Sever, M. Fabbri, R. Iuliano, T. Palumbo, et al. 2005. A microRNA signature associated with prognosis and progression in chronic lymphocytic leukemia. *N Engl J Med* **353**:1793–1801.

Chaudhuri, A.A., A.Y. So, A. Mehta, A. Minisandram, N. Sinha, V.D. Jonsson, D.S. Rao, R.M. O'Connell, and D. Baltimore. 2012. Oncomir miR-125b regulates hematopoiesis by targeting the gene Lin28A. *Proc Natl Acad Sci U S A* **109**:4233–4238.

Chen, C.Z., L. Li, H.F. Lodish, and D.P. Bartel. 2004. MicroRNAs modulate hematopoietic lineage differentiation. *Science* **303**:83–86.

Chen, R.W., L.T. Bemis, C.M. Amato, H. Myint, H. Tran, D.K. Birks, S.G. Eckhardt, and W.A. Robinson. 2008. Truncation in CCND1 mRNA alters miR-16-1 regulation in mantle cell lymphoma. *Blood* **112**:822–829.

Cimmino, A., G.A. Calin, M. Fabbri, M.V. Iorio, M. Ferracin, M. Shimizu, S.E. Wojcik, R.I. Aqeilan, S. Zupo, M. Dono, L. Rassenti, H. Alder, et al. 2005. miR-15 and miR-16 induce apoptosis by targeting BCL2. *Proc Natl Acad Sci U S A* **102**:13944–13949.

Clarkson, J.H., J.J. Kirkpatrick, and R.S. Lawrie. 2009. "Gearing to a time table"; the evolution of earlier surgical eschar excision in massive burns by British burns surgeons at the battles of Cassino, 1944: an example of real-time audit. *Burns* **35**:221–231.

Craig, V.J., S.B. Cogliatti, J. Imig, C. Renner, S. Neuenschwander, H. Rehrauer, R. Schlapbach, S. Dirnhofer, A. Tzankov, and A. Muller. 2011a. Myc-mediated repression of microRNA-34a promotes high-grade transformation of B-cell lymphoma by dysregulation of FoxP1. *Blood* **117**:6227–6236.

Craig, V.J., S.B. Cogliatti, H. Rehrauer, T. Wundisch, and A. Muller. 2011b. Epigenetic silencing of microRNA-203 dysregulates ABL1 expression and drives *Helicobacter*-associated gastric lymphomagenesis. *Cancer Res* **71**:3616–3624.

Deshpande, A., A. Pastore, A.J. Deshpande, Y. Zimmermann, G. Hutter, M. Weinkauf, C. Buske, W. Hiddemann, and M. Dreyling. 2009. 3'UTR mediated regulation of the cyclin D1 proto-oncogene. *Cell Cycle* **8**:3584–3592.

de Yebenes, V.G., L. Belver, D.G. Pisano, S. Gonzalez, A. Villasante, C. Croce, L. He, and A.R. Ramiro. 2008. miR-181b negatively regulates activation-induced cytidine deaminase in B cells. *J Exp Med* **205**:2199–2206.

Di Lisio, L., G. Gomez-Lopez, M. Sanchez-Beato, C. Gomez-Abad, M.E. Rodriguez, R. Villuendas, B.I. Ferreira, A. Carro, D. Rico, M. Mollejo, M.A. Martinez, J. Menarguez, et al. 2010. Mantle cell lymphoma: transcriptional regulation by microRNAs. *Leukemia* **24**:1335–1342.

Di Lisio, L., M. Sánchez-Beato, G. Gómez-López, M.E. Rodríguez, S. Montes-Moreno, M. Mollejo, J. Menárguez, M.A. Martínez, J. Alves, D.G. Pisano, M.A. Piris, and N. Martínez. 2012. MicroRNA signatures in B-cell lymphomas. *Blood Cancer J* **2**:e57.

Dorsett, Y., K.M. McBride, M. Jankovic, A. Gazumyan, T.H. Thai, D.F. Robbiani, M. Di Virgilio, B. Reina San-Martin, G. Heidkamp, T.A. Schwickert, T. Eisenreich, K. Rajewsky, et al. 2008. MicroRNA-155 suppresses activation-induced cytidine deaminase-mediated Myc-Igh translocation. *Immunity* **28**:630–638.

Elmen, J., M. Lindow, S. Schutz, M. Lawrence, A. Petri, S. Obad, M. Lindholm, M. Hedtjarn, H.F. Hansen, U. Berger, S. Gullans, P. Kearney, et al. 2008. LNA-mediated microRNA silencing in non-human primates. *Nature* **452**:896–899.

REFERENCES

Fabbri, M., C.M. Croce, and G.A. Calin. 2009. MicroRNAs in the ontogeny of leukemias and lymphomas. *Leuk Lymphoma* **50**:160–170.

Frenquelli, M., M. Muzio, C. Scielzo, C. Fazi, L. Scarfo, C. Rossi, G. Ferrari, P. Ghia, and F. Caligaris-Cappio. 2010. MicroRNA and proliferation control in chronic lymphocytic leukemia: functional relationship between miR-221/222 cluster and p27. *Blood* **115**:3949–3959.

Garzon, R., G. Marcucci, and C.M. Croce. 2010. Targeting microRNAs in cancer: rationale, strategies and challenges. *Nat Rev Drug Discov* **9**:775–789.

Gururajan, M., C.L. Haga, S. Das, C.M. Leu, D. Hodson, S. Josson, M. Turner, and M.D. Cooper. 2010. MicroRNA 125b inhibition of B cell differentiation in germinal centers. *Int Immunol* **22**:583–592.

Iqbal, J., Y. Shen, Y. Liu, K. Fu, E.S. Jaffe, C. Liu, Z. Liu, C.M. Lachel, K. Deffenbacher, T.C. Greiner, J.M. Vose, S. Bhagavathi, et al. 2012. Genome-wide miRNA profiling of mantle cell lymphoma reveals a distinct subgroup with poor prognosis. *Blood* **119**:4939–4948.

Kim, S.W., K. Ramasamy, H. Bouamar, A.P. Lin, D. Jiang, and R.C. Aguiar. 2012. MicroRNAs miR-125a and miR-125b constitutively activate the NF-kappaB pathway by targeting the tumor necrosis factor alpha-induced protein 3 (TNFAIP3, A20). *Proc Natl Acad Sci U S A* **109**:7865–7870.

Kitada, S., J. Andersen, S. Akar, J.M. Zapata, S. Takayama, S. Krajewski, H.G. Wang, X. Zhang, F. Bullrich, C.M. Croce, K. Rai, J. Hines, et al. 1998. Expression of apoptosis-regulating proteins in chronic lymphocytic leukemia: correlations with in vitro and in vivo chemoresponses. *Blood* **91**:3379–3389.

Klein, U., M. Lia, M. Crespo, R. Siegel, Q. Shen, T. Mo, A. Ambesi-Impiombato, A. Califano, A. Migliazza, G. Bhagat, and R. Dalla-Favera. 2010. The DLEU2/miR-15a/16-1 cluster controls B cell proliferation and its deletion leads to chronic lymphocytic leukemia. *Cancer Cell* **17**:28–40.

Kluiver, J., E. Haralambieva, D. de Jong, T. Blokzijl, S. Jacobs, B.J. Kroesen, S. Poppema, and A. van den Berg. 2006. Lack of BIC and microRNA miR-155 expression in primary cases of Burkitt lymphoma. *Genes Chromosomes Cancer* **45**:147–153.

Kluiver, J., A. van den Berg, D. de Jong, T. Blokzijl, G. Harms, E. Bouwman, S. Jacobs, S. Poppema, and B.J. Kroesen. 2007. Regulation of pri-microRNA BIC transcription and processing in Burkitt lymphoma. *Oncogene* **26**:3769–3776.

Kong, K.Y., K.S. Owens, J.H. Rogers, J. Mullenix, C.S. Velu, H.L. Grimes, and R. Dahl. 2010. MIR-23A microRNA cluster inhibits B-cell development. *Exp Hematol* **38**:629–640 e621.

Kota, S.K. and S. Balasubramanian. 2010. Cancer therapy via modulation of micro RNA levels: a promising future. *Drug Discov Today* **15**:733–740.

Lawrie, C.H., S. Soneji, T. Marafioti, C.D. Cooper, S. Palazzo, J.C. Paterson, H. Cattan, T. Enver, R. Mager, J. Boultwood, J.S. Wainscoat, and C.S. Hatton. 2007. MicroRNA expression distinguishes between germinal center B cell-like and activated B cell-like subtypes of diffuse large B cell lymphoma. *Int J Cancer* **121**:1156–1161.

Lawrie, C.H., S. Gal, H.M. Dunlop, B. Pushkaran, A.P. Liggins, K. Pulford, A.H. Banham, F. Pezzella, J. Boultwood, J.S. Wainscoat, C.S. Hatton, and A.L. Harris. 2008. Detection of elevated levels of tumour-associated microRNAs in serum of patients with diffuse large B-cell lymphoma. *Br J Haematol* **141**:672–675.

Lawrie, C.H., J. Chi, S. Taylor, D. Tramonti, E. Ballabio, S. Palazzo, N.J. Saunders, F. Pezzella, J. Boultwood, J.S. Wainscoat, and C.S. Hatton. 2009. Expression of microRNAs in diffuse large B cell lymphoma is associated with immunophenotype, survival and transformation from follicular lymphoma. *J Cell Mol Med* **13**:1248–1260.

Leich, E., A. Zamo, H. Horn, E. Haralambieva, B. Puppe, R.D. Gascoyne, W.C. Chan, R.M. Braziel, L.M. Rimsza, D.D. Weisenburger, J. Delabie, E.S. Jaffe, et al. 2011. MicroRNA profiles of t(14;18)-negative follicular lymphoma support a late germinal center B-cell phenotype. *Blood* **118**:5550–5558.

Lenze, D., L. Leoncini, M. Hummel, S. Volinia, C.G. Liu, T. Amato, G. De Falco, J. Githanga, H. Horn, J. Nyagol, G. Ott, J. Palatini, et al. 2011. The different epidemiologic subtypes of Burkitt lymphoma share a homogenous micro RNA profile distinct from diffuse large B-cell lymphoma. *Leukemia* **25**:1869–1876.

Leucci, E., M. Cocco, A. Onnis, G. De Falco, P. van Cleef, C. Bellan, A. van Rijk, J. Nyagol, B. Byakika, S. Lazzi, P. Tosi, H. van Krieken, et al. 2008. MYC translocation-negative classical Burkitt lymphoma cases: an alternative pathogenetic mechanism involving miRNA deregulation. *J Pathol* **216**:440–450.

Li, S., Z. Li, F. Guo, X. Qin, B. Liu, Z. Lei, Z. Song, L. Sun, H.T. Zhang, J. You, and Q. Zhou. 2011a. miR-223 regulates migration and invasion by targeting Artemin in human esophageal carcinoma. *J Biomed Sci* **18**:24.

Li, S., H.F. Moffett, J. Lu, L. Werner, H. Zhang, J. Ritz, D. Neuberg, K.W. Wucherpfennig, J.R. Brown, and C.D. Novina. 2011b. MicroRNA expression profiling identifies activated B cell status in chronic lymphocytic leukemia cells. *PLoS ONE* **6**:e16956.

Lin, J., T. Lwin, J.J. Zhao, W. Tam, Y.S. Choi, L.C. Moscinski, W.S. Dalton, E.M. Sotomayor, K.L. Wright, and J. Tao. 2011. Follicular dendritic cell-induced microRNA-mediated upregulation of PRDM1 and downregulation of BCL-6 in non-Hodgkin's B-cell lymphomas. *Leukemia* **25**:145–152.

Liu, T.Y., S.U. Chen, S.H. Kuo, A.L. Cheng, and C.W. Lin. 2010. E2A-positive gastric MALT lymphoma has weaker plasmacytoid infiltrates and stronger expression of the memory B-cell-associated miR-223: possible correlation with stage and treatment response. *Mod Pathol* **23**:1507–1517.

Malumbres, R., K.A. Sarosiek, E. Cubedo, J.W. Ruiz, X. Jiang, R.D. Gascoyne, R. Tibshirani, and I.S. Lossos. 2009. Differentiation stage-specific expression of microRNAs in B lymphocytes and diffuse large B-cell lymphomas. *Blood* **113**:3754–3764.

Medina, P.P., M. Nolde, and F.J. Slack. 2010. OncomiR addiction in an in vivo model of microRNA-21-induced pre-B-cell lymphoma. *Nature* **467**:86–90.

Montes-Moreno, S., N. Martinez, B. Sanchez-Espiridion, R. Diaz Uriarte, M.E. Rodriguez, A. Saez, C. Montalban, G. Gomez, D.G. Pisano, J.F. Garcia, E. Conde, E. Gonzalez-Barca, et al. 2011. miRNA expression in diffuse large B-cell lymphoma treated with chemoimmunotherapy. *Blood* **118**:1034–1040.

Moussay, E., K. Wang, J.H. Cho, K. van Moer, S. Pierson, J. Paggetti, P.V. Nazarov, V. Palissot, L.E. Hood, G. Berchem, and D.J. Galas. 2011. MicroRNA as biomarkers and regulators in B-cell chronic lymphocytic leukemia. *Proc Natl Acad Sci U S A* **108**:6573–6578.

Muramatsu, M., K. Kinoshita, S. Fagarasan, S. Yamada, Y. Shinkai, and T. Honjo. 2000. Class switch recombination and hypermutation require activation-induced cytidine deaminase (AID), a potential RNA editing enzyme. *Cell* **102**:553–563.

Navarro, A., S. Bea, V. Fernandez, M. Prieto, I. Salaverria, P. Jares, E. Hartmann, A. Mozos, A. Lopez-Guillermo, N. Villamor, D. Colomer, X. Puig, et al. 2009. MicroRNA expression, chromosomal alterations, and immunoglobulin variable heavy chain hypermutations in mantle cell lymphomas. *Cancer Res* **69**:7071–7078.

O'Donnell, K.A., E.A. Wentzel, K.I. Zeller, C.V. Dang, and J.T. Mendell. 2005. c-Myc-regulated microRNAs modulate E2F1 expression. *Nature* **435**:839–843.

Pal, S., R.A. Baiocchi, J.C. Byrd, M.R. Grever, S.T. Jacob, and S. Sif. 2007. Low levels of miR-92b/96 induce PRMT5 translation and H3R8/H4R3 methylation in mantle cell lymphoma. *EMBO J* **26**(15):3558–3569.

Pekarsky, Y., U. Santanam, A. Cimmino, A. Palamarchuk, A. Efanov, V. Maximov, S. Volinia, H. Alder, C.G. Liu, L. Rassenti, G.A. Calin, J.P. Hagan, et al. 2006. Tcl1 expression in chronic lymphocytic leukemia is regulated by miR-29 and miR-181. *Cancer Res* **66**:11590–11593.

Peveling-Oberhag, J., G. Crisman, A. Schmidt, C. Doring, M. Lucioni, L. Arcaini, S. Rattotti, S. Hartmann, A. Piiper, W.P. Hofmann, M. Paulli, R. Kuppers, et al. 2012. Dysregulation of global microRNA expression in splenic marginal zone lymphoma and influence of chronic hepatitis C virus infection. *Leukemia* **26**:1654–1662.

Rao, D.S., R.M. O'Connell, A.A. Chaudhuri, Y. Garcia-Flores, T.L. Geiger, and D. Baltimore. 2010. MicroRNA-34a perturbs B lymphocyte development by repressing the forkhead box transcription factor Foxp1. *Immunity* **33**:48–59.

Rao, E., C. Jiang, M. Ji, X. Huang, J. Iqbal, G. Lenz, G. Wright, L.M. Staudt, Y. Zhao, T.W. McKeithan, W.C. Chan, and K. Fu. 2012. The miRNA-17 approximately 92 cluster mediates chemoresistance and enhances tumor growth in mantle cell lymphoma via PI3K/AKT pathway activation. *Leukemia* **26**:1064–1072.

Roehle, A., K.P. Hoefig, D. Repsilber, C. Thorns, M. Ziepert, K.O. Wesche, M. Thiere, M. Loeffler, W. Klapper, M. Pfreundschuh, A. Matolcsy, H.W. Bernd, et al. 2008. MicroRNA signatures characterize diffuse large B-cell lymphomas and follicular lymphomas. *Br J Haematol* **142**:732–744.

Ruiz-Ballesteros, E., M. Mollejo, M. Mateo, P. Algara, P. Martinez, and M.A. Piris. 2007. MicroRNA losses in the frequently deleted region of 7q in SMZL. *Leukemia* **21**:2547–2549.

Salerno, E., B.J. Scaglione, F.D. Coffman, B.D. Brown, A. Baccarini, H. Fernandes, G. Marti, and E.S. Raveche. 2009. Correcting miR-15a/16 genetic defect in New Zealand Black mouse model of CLL enhances drug sensitivity. *Mol Cancer Ther* **8**:2684–2692.

Sampath, D., G.A. Calin, V.K. Puduvalli, G. Gopisetty, C. Taccioli, C.G. Liu, B. Ewald, C. Liu, M.J. Keating, and W. Plunkett. 2009. Specific activation of microRNA106b enables the p73 apoptotic response in chronic lymphocytic leukemia by targeting the ubiquitin ligase Itch for degradation. *Blood* **113**:3744–3753.

Sampson, V.B., N.H. Rong, J. Han, Q. Yang, V. Aris, P. Soteropoulos, N.J. Petrelli, S.P. Dunn, and L.J. Krueger. 2007. MicroRNA let-7a down-regulates MYC and reverts MYC-induced growth in Burkitt lymphoma cells. *Cancer Res* **67**:9762–9770.

Santanam, U., N. Zanesi, A. Efanov, S. Costinean, A. Palamarchuk, J.P. Hagan, S. Volinia, H. Alder, L. Rassenti, T. Kipps, C.M. Croce, and Y. Pekarsky. 2010. Chronic lymphocytic leukemia modeled in mouse by targeted miR-29 expression. *Proc Natl Acad Sci U S A* **107**:12210–12215.

Stamatopoulos, B., N. Meuleman, B. Haibe-Kains, P. Saussoy, E. Van Den Neste, L. Michaux, P. Heimann, P. Martiat, D. Bron, and L. Lagneaux. 2009. MicroRNA-29c and microRNA-223 down-regulation has in vivo significance in chronic lymphocytic leukemia and improves disease risk stratification. *Blood* **113**:5237–5245.

Stamatopoulos, B., N. Meuleman, C. De Bruyn, K. Pieters, G. Anthoine, P. Mineur, D. Bron, and L. Lagneaux. 2010. A molecular score by quantitative PCR as a new prognostic tool at diagnosis for chronic lymphocytic leukemia patients. *PLoS ONE* **5**:e12780.

Swerdlow, S.H., E. Campo, N.L. Harris, E.S. Jaffe, S.A. Pileri, H. Stein, J. Thiele, and J.W. Vardiman. 2008. *WHO classification of tumours of haematopoietic and lymphoid tissues*. IARC Press, Lyon.

Teng, G., P. Hakimpour, P. Landgraf, A. Rice, T. Tuschl, R. Casellas, and F.N. Papavasiliou. 2008. MicroRNA-155 is a negative regulator of activation-induced cytidine deaminase. *Immunity* **28**:621–629.

Thompson, R.C., M. Herscovitch, I. Zhao, T.J. Ford, and T.D. Gilmore. 2011. NF-kappaB down-regulates expression of the B-lymphoma marker CD10 through a miR-155/PU.1 pathway. *J Biol Chem* **286**:1675–1682.

Thorns, C., J. Kuba, V. Bernard, A. Senft, S. Szymczak, A.C. Feller, and H.W. Bernd. 2012. Deregulation of a distinct set of microRNAs is associated with transformation of gastritis into MALT lymphoma. *Virchows Arch* **460**:371–377.

Vargova, K., N. Curik, P. Burda, P. Basova, V. Kulvait, V. Pospisil, F. Savvulidi, J. Kokavec, E. Necas, A. Berkova, P. Obrtlikova, J. Karban, et al. 2011. MYB transcriptionally regulates the miR-155 host gene in chronic lymphocytic leukemia. *Blood* **117**:3816–3825.

Ventura, A., A.G. Young, M.M. Winslow, L. Lintault, A. Meissner, S.J. Erkeland, J. Newman, R.T. Bronson, D. Crowley, J.R. Stone, R. Jaenisch, P.A. Sharp, et al. 2008. Targeted deletion reveals essential and overlapping functions of the miR-17 through 92 family of miRNA clusters. *Cell* **132**:875–886.

Vigorito, E., K.L. Perks, C. Abreu-Goodger, S. Bunting, Z. Xiang, S. Kohlhaas, P.P. Das, E.A. Miska, A. Rodriguez, A. Bradley, K.G. Smith, C. Rada, et al. 2007. MicroRNA-155 regulates the generation of immunoglobulin class-switched plasma cells. *Immunity* **27**:847–859.

Visone, R., A. Veronese, L.Z. Rassenti, V. Balatti, D.K. Pearl, M. Acunzo, S. Volinia, C. Taccioli, T.J. Kipps, and C.M. Croce. 2011. miR-181b is a biomarker of disease progression in chronic lymphocytic leukemia. *Blood* **118**:3072–3079.

Visone, R., A. Veronese, V. Balatti, and C.M. Croce. 2012. MiR-181b: new perspective to evaluate disease progression in chronic lymphocytic leukemia. *Oncotarget* **3**:195–202.

Wang, W., M. Corrigan-Cummins, J. Hudson, I. Maric, O. Simakova, S.S. Neelapu, L.W. Kwak, J.E. Janik, B. Gause, E.S. Jaffe, and K.R. Calvo. 2012. MicroRNA profiling of follicular lymphoma identifies microRNAs related to cell proliferation and tumor response. *Haematologica* **97**:586–594.

Xiao, C., D.P. Calado, G. Galler, T.H. Thai, H.C. Patterson, J. Wang, N. Rajewsky, T.P. Bender, and K. Rajewsky. 2007. MiR-150 controls B cell differentiation by targeting the transcription factor c-Myb. *Cell* **131**:146–159.

Zauli, G., R. Voltan, M.G. di Iasio, R. Bosco, E. Melloni, M.E. Sana, and P. Secchiero. 2011. miR-34a induces the downregulation of both E2F1 and B-Myb oncogenes in leukemic cells. *Clin Cancer Res* **17**:2712–2724.

Zenz, T., S. Habe, T. Denzel, J. Mohr, D. Winkler, A. Buhler, A. Sarno, S. Groner, D. Mertens, R. Busch, M. Hallek, H. Dohner, et al. 2009a. Detailed analysis of p53 pathway defects in fludarabine-refractory chronic lymphocytic leukemia (CLL): dissecting the contribution of 17p deletion, TP53 mutation, p53-p21 dysfunction, and miR34a in a prospective clinical trial. *Blood* **114**: 2589–2597.

Zenz, T., J. Mohr, E. Eldering, A.P. Kater, A. Buhler, D. Kienle, D. Winkler, J. Durig, M.H. van Oers, D. Mertens, H. Dohner, and S. Stilgenbauer. 2009b. miR-34a as part of the resistance network in chronic lymphocytic leukemia. *Blood* **113**:3801–3808.

Zhang, J., D.D. Jima, C. Jacobs, R. Fischer, E. Gottwein, G. Huang, P.L. Lugar, A.S. Lagoo, D.A. Rizzieri, D.R. Friedman, J.B. Weinberg, P.E. Lipsky, and S.S. Dave. 2009. Patterns of microRNA expression characterize stages of human B-cell differentiation. *Blood* **113**(19):4586–4594.

Zhang, Y., A.M. Roccaro, C. Rombaoa, L. Flores, S. Obad, S.M. Fernandes, A. Sacco, Y. Liu, H. Ngo, P. Quang, A.K. Azab, F. Azab, et al. 2012. LNA-mediated anti-microRNA-155 silencing in low-grade B cell lymphomas. *Blood* **120**:1678–1686.

Zhao, J.J., J. Lin, T. Lwin, H. Yang, J. Guo, W. Kong, S. Dessureault, L.C. Moscinski, D. Rezania, W.S. Dalton, E. Sotomayor, J. Tao, et al. 2010. MicroRNA expression profile and identification of miR-29 as a prognostic marker and pathogenetic factor by targeting CDK6 in mantle cell lymphoma. *Blood* **115**:2630–2639.

Zhou, B., S. Wang, C. Mayr, D.P. Bartel, and H.F. Lodish. 2007. miR-150, a microRNA expressed in mature B and T cells, blocks early B cell development when expressed prematurely. *Proc Natl Acad Sci U S A* **104**:7080–7085.

Zhou, K., S. Yi, Z. Yu, Z. Li, Y. Wang, D. Zou, J. Qi, Y. Zhao, and L. Qiu. 2012. MicroRNA-223 expression is uniformly down-regulated in B cell lymphoproliferative disorders and is associated with poor survival in patients with chronic lymphocytic leukemia. *Leuk Lymphoma* **53**: 1155–1161.

25

MICRORNAS IN DIFFUSE LARGE B-CELL LYMPHOMA

Izidore S. Lossos and Alvaro J. Alencar

Department of Medicine, Division of Hematology-Oncology and Molecular and Cellular Pharmacology, Sylvester Comprehensive Cancer Center, University of Miami, Miami, FL, USA

I.	Introduction	420
II.	Mature B-Cell Differentiation Stage-Specific MicroRNA Expression Patterns	420
III.	MicroRNA Expression in DLBCL	422
IV.	Potential Roles of MicroRNAs in Lymphomagenesis	425
	A. *miR-155* Involvement in Lymphoid Malignancies	425
	B. *miR-17-92* Cluster Involvement in Lymphoid Malignancies	426
V.	MicroRNAs as Potential Biomarkers in DLBCL	427
VI.	Conclusions	429
	References	430

ABBREVIATIONS

ABC	activated B-cell like
CHOP	cyclophosphamide, doxorubicin, vincristine, prednisone
COO	cell of origin
DLBCL	diffuse large B-cell lymphoma
EFS	event-free survival
GCB	germinal center B-cell like
IHC	immunohistochemistry
MiRNA	microRNA
NHL	non-Hodgkin's lymphoma
OS	overall survival

MicroRNAs in Medicine, First Edition. Edited by Charles H. Lawrie.
© 2014 John Wiley & Sons, Inc. Published 2014 by John Wiley & Sons, Inc.

I. INTRODUCTION

Diffuse large B-cell lymphoma (DLBCL), the most common subtype of non-Hodgkin's lymphomas (NHL), is a genetically and clinically heterogeneous disease (Lossos 2005) (Morton et al. 2006). DLBCL is characterized by an aggressive clinical course, and only about 50–60% of patients are cured with the current standard treatment, which includes the anti-CD20 B-cell antibody rituximab in addition to CHOP (R-CHOP-cyclophosphamide, doxorubicin, vincristine, prednisone) (Coiffier et al. 2002; Habermann et al. 2006). Patients treated with chemotherapy regimens without rituximab demonstrate lower remission rates and shorter survival. The pathogenesis of DLBCL represents a complex multistep process involving a collaboration between biological programs of normal B cells hijacked by the transformed malignant cells and multiple acquired molecular and genetic aberrations (Lossos 2005; Lenz and Staudt 2010). Several attempts to divide DLBCL into subtypes according to morphology, genetic abnormalities, or association with viral infections have been made (Swerdlow et al. 2008) but resulted in poor reproducibility or did not explain the clinical heterogeneity and/or correlate with the clinical outcome of patients. Currently, the most clinically meaningful classification is based on gene expression profiling, which divides DLBCL into three major subtypes—primary mediastinal large B-cell lymphoma, activated B-cell (ABC) like, and germinal center B-cell (GCB) like. Primary mediastinal large B-cell lymphoma is an uncommon subtype of DLBCL that demonstrates a characteristic gene signature partially overlapping with the gene expression signature of Hodgkin's lymphoma (Rosenwald et al. 2003). GCB-like DLBCL is characterized by a gene expression signature typical of germinal center (GC) B-cells and is associated with better survival than the ABC-like DLBCL, which is characterized by a gene expression signature similar to stimulated and activated peripheral B cells (Alizadeh et al. 2000). The cell of origin (COO) of the ABC-like DLBCL is still controversial, with data supporting the origin from either late GC cells or early plasmablasts. This classification of DLBCL tumors has also led to the appreciation that while some genetic and molecular lesions with known or predicted oncogenic potential occur in both GCB-like and ABC-like DLBCL subtypes, many oncogenic pathways are exclusively or predominantly used by only one subtype (e.g., predominant BCL-6 expression in the GCB-like and NF-κB pathway activation and inactivation of the tumor suppressor BLIMP1 in the ABC-like DLBCL) (Davis et al. 2001; Lossos et al. 2001; Pasqualucci et al. 2006; Lenz et al. 2008a, 2008b; Mandelbaum et al. 2010). However, the identified pathobiological events do not account for all the pathogenic mechanisms underlying DLBCL.

The recent discovery of microRNAs (miRNAs) that regulate gene expression at the posttranscriptional level, potentially affecting multiple signaling pathways, led to an extensive evaluation of their role in normal B cell differentiation, DLBCL pathogenesis, and applicability as potential prognostic biomarkers. In this chapter, we will present the current state of knowledge of miRNAs in DLBCL.

II. MATURE B-CELL DIFFERENTIATION STAGE-SPECIFIC MicroRNA EXPRESSION PATTERNS

To secure proper immune responses and protection from foreign pathogens, normal lymphocytes undergo a complex process of differentiation and maturation starting from lymphoid precursors in the bone marrow to terminally differentiated and immunologically active plasma cells and memory B cells. This differentiation process is associated with and regulated by orchestrated changes in RNA expression patterns between the successive stages of B-cell differentiation. Since lymphomas, including DLBCL, originate from normal B lymphocytes, they are characterized by the same transcriptional programs present in the specific ontogenic stages of B cell differentiation from which the tumors originate. Similarly, normal B-cell differentiation should be characterized by global changes in miRNA expression, and thus lymphoma cells should express miRNA programs characteristic of the specific ontogenic stages of B cell differentiation from which they originate. Consequently, to start understanding the role of miRNAs in DLBCL, it was important to establish miRNA expression patterns specific to the ontogenic stages of B-cell differentiation. To this end, we and two other groups examined changes in miRNA expression during peripheral B-cell differentiation (Basso et al. 2009; Malumbres et al. 2009; Zhang et al. 2009). These three studies demonstrated remarkable changes in miRNA expression patterns between the successive stages of peripheral B cell differentiation, with excellent congruency for specific miRNA expression changes during differentiation across the studies (Table 25.1). In addition, Jima et al. and Basso et al., using deep sequencing and a combination of cloning and computational analyses of short-RNA libraries, identified new miRNAs specifically or predominantly expressed in peripheral B lymphocytes (Basso et al. 2009; Jima et al. 2010). In our study, we found a distinguishable pattern of miRNA expression in centroblasts compared with other lymphocyte subpopulations, with specific expression increases of 51 miRNAs and down-regulation of various miRNAs in the GC

TABLE 25.1. Differential Expression of miRNAs during B-Cell Peripheral Differentiation

Stage	Naïve B Cell	Centroblast	Memory B Cell
MiRNAs	miR-223	miR-146a	miR-223
	miR-320	miR-99b	miR-320
	miR-29b	miR-125a-5p	miR-29b
	miR-101	miR-92a	miR-101
	miR-29c	miR-214	miR-29c
	miR-150	miR-125b	miR-150
	miR-151-5p	miR-191	miR-194
		miR-128b	miR-146b-5p
		miR-20b	miR-23a
		miR-18a	miR-24
		miR-130b	miR-23b
		miR-106b	miR-27b
		miR-425	miR-27a
		miR-93	miR-671-5p
		miR-28-5p	miR-146a
		miR-17	
		miR-106a	
		miR-25	
		miR-20a	
		miR-15b	
		miR-30c	
		miR-16	
		miR-181b	
		miR-151-5p	

lymphocytes (Malumbres et al. 2009). Although naive and memory B cells could be distinguished based on their miRNA expression patterns, the observed differences were less striking compared with the GC lymphocytes. Remarkable changes in expression upon B-cell maturation were observed for specific miRNAs. For example, the expressions of *miR-18a* and *miR-28* were 15- and 10-fold higher, respectively, in centroblasts compared to memory B cells, whereas a higher expression (10- to 20-fold) in memory B cells compared with centroblasts was observed for *miR-101c, miR-150,* and *miR-29c.* Similarly, Zhang et al. identified 32 miRNAs that were differentially expressed between naive and GC lymphocytes, 33 miRNAs differentially expressed between GC lymphocytes and plasma cells, and 27 miRNAs differentially expressed between GC memory lymphocytes (Zhang et al. 2009). These findings demonstrate that similar to mRNAs, unique miRNA expression patterns are established in GC cells. The specificity of miRNA expression in the GC cells allowed us to find a signature of 39 miRNAs whose expression could differentiate GC cells from naïve and memory B cells (Figure 25.1).

Furthermore, miRNAs specifically expressed in B cell subpopulations repressed expression of genes with known roles in B cell differentiation. *miR-223,* whose expression was down-regulated in the GC cells, directly repressed expression of LMO2 protein, which is specifically up-regulated in the GC lymphocytes (Malumbres et al. 2009). *miR-125b,* whose expression was down-regulated in memory B cells, directly repressed expression of IRF4 and BLIMP1 proteins, necessary for post-GC differentiation (Malumbres et al. 2009). Similarly, the *miR-30* family and *miR-9* also repressed expression of the BLIMP1 protein (Zhang et al. 2009), while *miR-181a,* whose expression is up-regulated in the GC lymphocytes, specifically repressed expression of the FOXP1 protein, which encodes a transcription factor whose expression was associated with outcome in DLBCL patients (Alencar et al. 2011).

III. MicroRNA Expression in DLBCL

miRNAs are implicated in carcinogenesis and can function as oncogenes or tumor suppressors in a context-dependent manner. To better understand the potential roles of miRNAs in lymphomas, several studies analyzed global miRNome or the expression of specific miRNAs in DLBCL with specific attention to changes in miRNA expression compared with normal B-cell counterparts. Additionally, differential miRNA expression between the GCB-like and ABC-like DLBCL was explored. While these studies generated valuable data sets important for deciphering the roles and significance of miRNAs in DLBCL, they also produced perplexing and contradictory results, whose confirmation and interpretation will require additional independent studies.

We analyzed miRNA expression profiles in eight cell lines representing the GCB-like and ABC-like DLBCL (Malumbres et al. 2009). Using unsupervised hierarchical clustering of 217 miRNAs for which data were available in at least 50% of samples, GCB-like and ABC-like DLBCL cell lines were perfectly segregated based on global similarities in miRNA expression patterns. Furthermore, we have observed that both GCB-like and ABC-like cell lines exhibit expression signatures more similar to centroblasts than memory and naïve cells, suggesting that they arise from different stages of lymphocytes in GCs. To identify a miRNA signature sufficient to separate GCB-like and ABC-like DLBCL cell lines, we performed a statistical analysis of microarrays (SAM analysis) using these 217 miRNA probes. This approach identified nine miRNAs (*miR-146b-5p, miR-146a, miR-21, miR-155, miR-500, miR-222, miR-363, miR-574-3p,* and *miR-574-5p*) differen-

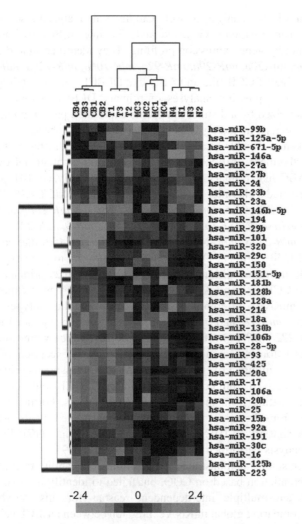

Figure 25.1. Thirty-nine miRNA classifier for differentiation of centroblasts (CB), naive B cell (N), memory B cells (MC), and T cells (T). Mean centered log ratios for each miRNA are represented. Missing values are in gray. See color insert.

tially expressed between the DLBCL subtypes, at a false discovery rate (FDR) of 10%. These nine miRNAs were expressed at higher levels in the ABC-like DLBCL, and only two (*miR-146b-5p* and *miR-146a*) were included in the COO classifier constructed to differentiate distinct normal B-cell subpopulations (Figure 25.1). This finding suggests that differences between the GCB-like and ABC-like DLBCL subtypes are not based solely on cellular origin, but most probably also reflects the different biology of these tumors. Lawrie et al. also demonstrated higher expression of *miR-155*, *miR-221*, *miR-222*, *miR-21*, *miR-363*, and *miR-518a* and lower expression of *miR-181a*, *miR-590*, *miR-421*, and *miR-324* in ABC-like compared with GCB-like DLBCL cell lines, but did not attempt to construct a specific classifier (Lawrie et al. 2008b). While not incorporated in our classifier, many of these latter miRNAs were differentially expressed between the GCB-like and

ABC-like cell lines in our analysis as well. Culpin et al. analyzed a set of eight different DLBCL cell lines subclassified to GCB-like and ABC-like DLBCL by immunohistochemistry (IHC) and not by gene expression profiling. They identified nine different miRNAs (*miR-17, miR-19b,* miR20a, *miR-29a, miR-92a, miR-106a, miR-720a, miR1260,* and *miR-1280*) that discriminated GCB-like and ABC-like DLBCL cell lines (Culpin et al. 2010).

Global miRNA expression analyses of primary DLBCL demonstrated that these tumors are characterized by a distinct miRNome expression pattern compared with other subtypes of lymphomas, including follicular and Burkitt lymphomas, as well as chronic lymphocytic leukemia (Roehle et al. 2008; Lawrie et al. 2009; Zhang et al. 2009; Lenze et al. 2011). Furthermore, miRNA expression could differentiate DLBCL tumors into GCB-like and ABC-like primary DLBCL. Jima et al. analyzed 101 primary DLBCL tumors and constructed a classifier based on 25 miRNAs (*miR-128, miR-129-3p, miR-152, miR-155, miR-185, miR-193-5p, miR-196b, miR-199b-3p, miR-20b, miR-23a, miR-27a, miR-28-5p, miR-301a, miR-331-3p, miR-365, miR-625, miR-9, miR-2282, miR-2287, miR-2290, miR-2311, miR-2348, miR-2412, miR-2432,* and *miR-2450*) that could differentiate GCB-like and ABC-like primary DLBCL with equal efficacy to gene expression profiling (Jima et al. 2010). In contrast, Montes Moreno et al. analyzed 29 primary DLBCL tumors subclassified into GCB-like and ABC-like subtypes based on gene expression profiling, identifying eight miRNAs differentially expressed between the subtypes. These included *miR-331, miR-151, miR-28,* and *miR-454-3p*, which were up-regulated in the GCB-like tumors, and *miR-222, miR-144, miR-451,* and *miR-221*, which were up-regulated in the ABC-like tumors (Montes-Moreno et al. 2011). With the exception of a few specific miRNAs, there was a minimal overlap between the COO classifiers derived from cell line studies, classifiers derived from cell line and primary DLBCL studies and even between the classifiers originating from the primary DLBCL studies. Furthermore, eight of the miRNAs derived from the nine miRNA classifier constructed by us (Malumbres et al. 2009) failed to subclassify 53 IHC-defined primary DLBCL into GCB-like and ABC-like subtypes in an analysis performed by Li et al. (Li et al. 2009).

Overall, these studies demonstrated the potential applicability of miRNA expression for DLBCL differentiation based on COO, but failed to identify a universal and verified classifier. There are multiple interdependent reasons for this variability and non-reproducibility. Comparing global miRNA expression between DLBCL cell line and whole primary tumors demonstrated marked differences and segregated cell lines separately from primary tumors (Lawrie et al. 2009). The differences in miRNA expression between the cell lines and primary tumors might be attributed to: (1) different source of the miRNA—viable growing cell lines versus paraffin-embedded primary tumors analyzed simultaneously without normalization for differences in the source of the RNA; (2) specific changes in miRNA expression induced by prolonged *in vitro* growth of cell lines in culture media; (3) presence of non-tumor cells in primary tumors that may have served as an additional source of miRNA expression. Indeed, in contrast to genes, some of which are specifically and exclusively expressed in GC B cells and tumors derived from them, the currently known miRNAs usually do not demonstrate exclusive expression patterns in only one type of cell and are commonly expressed in both malignant B-lymphocytes and non-malignant cells comprising the tumor microenvironment, thus potentially preventing the application of cell line-derived classifiers to non-selected whole primary DLBCL tumors. The lack of consensus between the various cell line classifiers and COO predictors derived from primary DLBCL studies most likely results from the use of gene expression versus IHC for subtype definition, analysis of relatively small number of cases, use of different platforms for analyzing miRNA expression, and examining a different number of miRNAs, reflecting the ongoing advances in miRNA identification and cloning. Further large-scale studies that will address these potential confounding factors and verify the classifiers in independent cohorts will hopefully resolve the current conflicting results and lead to the identification of a clinically useful classifier.

IV. POTENTIAL ROLES OF MicroRNAs IN LYMPHOMAGENESIS

miRNAs play important roles in controlling many biological processes, including cell differentiation, stem cell maintenance, metabolism, proliferation, and apoptosis. Multiple studies demonstrated that miRNAs may regulate expression of genes implicated in carcinogenesis, and numerous miRNAs are deregulated in cancers. Further, a major role for miRNAs in carcinogenesis was suggested by the observation that many miRNAs are located in genomic regions involved in chromosomal deletions and amplifications and in which presumed tumor suppressor genes or oncogenes, respectively, failed to be discovered despite extensive investigations (Calin et al. 2004). Several miRNAs were also implicated in lymphoma and DLBCL pathogenesis (Figure 25.2).

A. *miR-155* Involvement in Lymphoid Malignancies

miR-155 is located at chromosome 21q21.3 inside the *B-cell integration cluster* (BIC) gene (Eis et al. 2005), a common site for insertion of proviral DNA in avian leukosis virus-induced lymphomas (Clurman and Hayward 1989; Tam et al. 1997). This is a unique miRNA not included in any miRNA cluster and characterized by a distinctive seed region that is not shared by other miRNAs in the human genome (Griffiths-Jones 2004; Griffiths-Jones et al. 2006, 2008; Sanger-Institute 2009). Expression of BIC and *miR-155* is induced by B-cell receptor (BCR) cross-linking through the extracellular signaling-regulated kinase (ERK) and c-Jun N-terminal kinase (JNK) pathways (Yin et al. 2008).

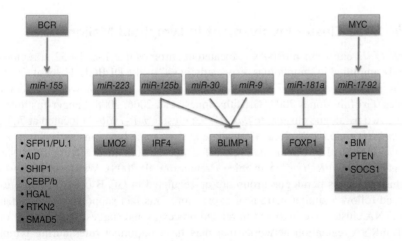

Figure 25.2. Summary of miRNAs and their targets implicated in the pathogenesis of the DLBCL.

miR-155 was demonstrated to be a key factor in the regulation of GC responses (Thai et al. 2007). In *miR-155* knockout (KO) mice, the number of GCs in the spleen after chicken gamma globulin immunization was decreased compared with wild-type mice. Furthermore, the number of B cells in the GCs was also reduced, as well as the titers of specific IgG_1 antibodies. Accordingly, in mice with B cells transgenic for *miR-155*, all these indicators of GC response were increased. The reduction in specific IgG_1 production in *miR-155*-deficient mice might be attributed to targeting of SFPI1/PU.1 and AID by this miRNA in B lymphocytes (Muramatsu et al. 2000; Vigorito et al. 2007). Overexpression of the SFPI1/PU.1 transcription factor in B cells inhibits Ig class switching to IgG_1, while AID is one of the key mediators of the Ig class switching and somatic hypermutation in the GC (Dorsett et al. 2008; Teng et al. 2008). *miR-155* can also affect the GC reaction by regulating cytokine production by B and T cells. INPPD5 (SHIP) and NF-IL6 (CEBP) have been recently described as targets of *miR-155*, suggesting the role of this miRNA in IL-6 signaling during B cell maturation (Costinean et al. 2009). Furthermore, in *miR-155* KO mice, the production of TNF-α and lymphotoxin-β by B cells was greatly reduced, and T-cell differentiation seemed to be biased toward a T_H2 phenotype, with increased IL-4, IL-5, and IL-10 secretion in detriment of IFNγ production *in vitro* (Rodriguez et al. 2007). This cytokine profile would contribute to the impaired immune function observed in the *miR-155*-deficient mice. Furthermore, we have recently demonstrated that *miR-155* increases GC lymphocyte and GC-derived lymphoma cell motility by directly suppressing expression of GC-specific HGAL protein and Rhotekin 2 (RTKN2) (Dagan et al. 2012).

miR-155 overexpression in murine Pro-B cells under the control of the immunoglobulin heavy chain Eμ enhancer induced increased Pre-B-cell proliferation and led to a leukemic malignancy (Costinean et al. 2006), suggesting that it may play a role in lymphomagenesis. *miR-155* is expressed at high levels in primary mediastinal B-cell lymphoma (Kluiver et al. 2005) and is particularly overexpressed in the ABC-like DLBCL (Eis et al. 2005; Lawrie et al. 2007; Malumbres et al. 2009). In DLBCL, overexpression of *miR-155* leads to SMAD5 down-regulation and ablates the *in vitro* growth-inhibitory effects of TGF-β1 and BMP2/4 (Rai et al. 2010). In xenograft models of human DLBCL, *miR-155* overexpression led to the development of larger and more widespread tumors (Rai et al. 2010).

B. *miR-17-92* Cluster Involvement in Lymphoid Malignancies

The *miR-17-92* cluster of miRNAs is located at chromosome 13q31-q32, a region that is frequently amplified in lymphomas, especially in GCB-like DLBCL (He et al. 2005; Lenz et al. 2008b). It is composed of six different miRNAs, some of which harbor related sequences (Griffiths-Jones 2004; Griffiths-Jones et al. 2006, 2008; Sanger-Institute 2009). There are two paralogue clusters to *miR-17-92*: cluster *miR-106b-25* located at 7q22.1 and cluster *miR-106a-363* at Xq26.2. They share at least some functions with cluster *miR-17-92*, as was suggested by the more severe effects of the combined deletion of cluster *miR-17-92* and cluster *miR-106b-25* in mice (Ventura et al. 2008). We have shown that the three clusters of this paralogue group are up-regulated in GC B cells (Malumbres et al. 2009), and follow a similar pattern of expression. This fact supports the hypothesis that these miRNA clusters are involved in related processes and suggest the presence of coordinated miRNA regulation networks that may have important roles during lymphocyte differentiation.

The *miR-17-92* cluster is implicated in B-cell development. Targeted deletion of this cluster in mice induces an increase in expression of the proapoptotic protein BIM in B cells, leading to a decrease in cell survival and blocking the transition from the pro- to pre-B cell stage (Ventura et al. 2008). Selective over-expression of *miR-17-92* cluster in mouse lymphocytes induces increased proliferation and decreased activation-induced cell death, leading to lymphoproliferative disease and autoimmunity (Xiao et al. 2008). These effects could be attributed to direct down-regulation of the antiproliferative protein PTEN and the proapoptotic protein BIM by this miRNA cluster (Xiao et al. 2008). In addition, a regulatory loop between MYC, cluster *miR-17-92*, and E2F1 may also contribute to this phenotype. Indeed, overexpression of the *miR-17-92* cluster in Eμ-myc mice overexpressing MYC from B-cell precursor stages accelerated development of lymphoid malignancies, exhibiting a synergistic effect and increasing the tumor aggressiveness (He et al. 2005) (Tagawa et al. 2007). Myc was shown to directly up-regulate the *miR-17-92* polycistron by binding upstream of its locus. It was also demonstrated that MYC can induce E2F1, which contributes to its effects on the enhancement of transcription of the *miR-17-92* cluster. It is well known that MYC can have either oncogenic or proapoptotic functions depending on the cell context; when MYC induces E2F1 function, the proapoptotic features of E2F1 can shift the balance to apoptosis, but if *miR-17-92* cluster is also induced, proliferation is enhanced and apoptosis diminished.

V. MicroRNAs AS POTENTIAL BIOMARKERS IN DLBCL

DLBCL is characterized by a very heterogeneous clinical course and outcome. Thus multiple studies have tried to identify biomarkers that might be useful to predict patients' survival and response to therapy (Lossos and Morgensztern 2006). Since expression of specific miRNAs changes during B cell differentiation, potentially reflecting the COO of DLBCL tumors, and because there is emerging evidence pointing to dysfunctional expression of miRNAs in tumors, including DLBCL, and suggesting that they may play an important role in the pathogenesis of these malignancies, numerous investigators hypothesized that miRNAs may also serve as reliable biomarkers. Furthermore, since miRNAs are very small RNA molecules comprising of only 18–24 nucleotides, they are less prone to degradation compared with mRNA, leading to their better preservation and stable expression in archival tissues, such as routinely prepared formalin-fixed paraffin-embedded (FFPE) tissue and biological fluids, such as serum/plasma and so on.

Several investigators tried to use quantification of serum and cerebrospinal fluid (CSF) miRNAs for diagnosis of DLBCL and diagnosis of its dissemination to the central nervous system (CNS), respectively (Lawrie et al. 2008a; Baraniskin et al. 2011; Fang et al. 2012). While these studies demonstrated changes in specific miRNA expression patterns in the serum of DLBCL patients and in CSF of patients with CNS involvement, these studies were based on very small cohorts of patients and need to be verified in independent, larger studies before any recommendation on their clinical value and use can be made.

Simultaneously, measurement of miRNA expression in primary DLBCL tumor tissues was assessed as potential biomarkers to predict patients' response to therapy and outcome. Roehle et al. analyzed expression of 157 miRNAs in 58 DLBCL patients treated with CHOP-like chemotherapy without rituximab and examined the association between patients' overall survival (OS) and event-free survival (EFS), with expression of each individual miRNA dichotomized based on the median expression value of the analyzed cohort (Roehle et al. 2008). In a multivariate analysis, patients with low expression of *miR-127* had shorter OS and EFS. Patients with low expression of *miR-21, miR-23a, miR-27a*, and *mir-34a* had only shorter OS. Low expression of *miR-19a* was associated with shorter EFS, while low expression of *miR-195* and let7g was associated with longer EFS. Expression of *miR-155* was not associated with survival in this study. Jung et al. also did not observe correlation between expression of *miR-155* and outcome in 129 DLBCL patients; however, they observed higher *miR-155* expression levels in the ABC-like DLBCL (Jung and Aguiar 2009). When the authors examined the effect of *miR-155* expression exclusively within the ABC-like subtype, they found a marked trend toward better survival rates for patients expressing high levels of *miR-155*.

Lawrie et al. demonstrated that low expression of *miR-21* was associated with shorter relapse-free survival (RFS) based on analysis of 35 *de novo* DLBCL cases, 28 of which were treated with R-CHOP chemotherapy, while the others were not treated at all or treated with chemotherapy regimens without rituximab (Lawrie et al. 2007). In another study, the same group examined miRNA expression in 64 de novo DLBCL patients heterogeneously treated with and without rituximab-containing chemotherapy regimens and used median expression of specific miRNAs as a cutoff for definition of DLBCL subgroups (Lawrie et al. 2009). The authors showed that high expressions of *miR-637, miR-608*, and *miR-302* were associated with shorter EFS, while high expressions of *miR-330, miR-30e, miR-425, miR-27a, miR-24, miR-23a, miR-199b, miR-199a*, and *miR-100* were associated with longer EFS. In analyses limited only to patients treated with R-CHOP, it was shown that expression of *miR-302, miR-330, miR-425, miR-27a, miR-199b, miR-142, miR-519*, and *miR-222* were associated with EFS.

Montes-Moreno et al. used 258 DLBCL cases, 243 of which were treated with rituximab containing regimens to construct a nine-miRNA classifier (*miR-221, miR-222, miR-331, miR-451, miR-28, miR151, miR148, miR-93*, and *miR-491*) based on miRNAs consisting of the COO signature and identifying additional miRNAs whose individual expression correlated with OS (Montes-Moreno et al. 2011). This classifier could subdivide DLBCL patients with significantly different OS and progression-free survival (PFS).

We used 176 specimens of DLBCL patients treated with R-CHOP to examine the correlation between OS and PFS and expression of 11 miRNAs selected based on their ability to distinguish GCB- and ABC-like DLBCL cell lines or exhibiting highly variable expression in DLBCL patients (*miR-21, miR-146a, miR-146b-5p, miR-155, miR-222, miR363, miR-500, miR-574-3p, miR-18a, miR-140-3p*, and *miR-181a*) (Alencar et al. 2011). Expression of three (*miR-18a, miR-181a*, and *miR-222*) of the 11 analyzed miRNAs was individually associated with survival of DLBCL patients in a univariate analysis. Expression of *miR-18a*, analyzed as a continuous variable, was statistically correlated with OS, and increased expression of this miRNA was associated with a shorter OS. Expression of *miR-18a* was not associated with PFS. Expression of *miR-181a*, analyzed as a continuous variable, was statistically correlated with PFS but not with OS, and increased expression of this miRNA was associated with longer PFS. Expression of *miR-222*, analyzed as a dichotomous variable (above and below median expression) was statistically correlated with PFS, but not with OS. Higher expression of *miR-222* was associated with shorter PFS. Multivariate analysis incorporating clinical and laboratory variables associated with DLBCL patients' survival confirmed the independent prognostic value of these three miRNAs.

Careful analysis of the published data presented herein undoubtedly demonstrates that miRNAs have tremendous potential to be used as biomarkers for estimating the prognosis of DLBCL patients; however, at the same time, it also shows that current knowledge does not allow the selection of individual miRNAs that can be used now for this purpose. The latter results from marked controversies between published studies that may stem from: (1) analysis of small cohorts of DLBCL patients, which may lead to misleading conclusions, since change in outcome of only few patients may alter proposed conclusions; and (2) incorporation of heterogeneously treated patients, as it is well known that rituximab improves DLBCL patients' outcomes. Presently, only our study was exclusively limited to patients treated with rituximab-containing regimens. Incorporation of suboptimally treated patients may bias proposed study conclusions; (3) In some of the published studies, the expression cutoffs for increased and decreased expression of the analyzed miRNAs were not rigorously determined and might be preselected and optimized for the reported cohorts. This may lead to reduced reproducibility of the results. Demonstrating that a biomarker is associated with an outcome as a continuous variable in addition to unbiased cutoff determination usually eliminates these concerns; (4) Validation of the findings in independent cohorts of patients. Use of retrospective cohort of patients may be associated with inherent biases related to uncontrolled selection of retrospectively available specimens that may lead to overfitted models in the tested study group. Therefore, there is a need for formal external validation in independent groups of patients. Although internal validation on subsets of the original patient cohort is performed frequently as an essential part of prognostic model development and refinement, this process is not sufficient. Moreover, careful analysis of the presented miRNA studies shows that even internal validation was not performed in most of the studies.

The ultimate goal for every prognostic biomarker is validation in an independent, well-designed, large prospective study, since this will confirm that the suggested methodologies to assess the expression of prognostic markers are robust, reproducible, and are independent of potential variability in sample handling or applied methodology. In addition, these prospective studies will also serve as an external validation of method reproducibility in different laboratories.

Despite all these caveats, the currently available data already suggest that at least several miRNAs (e.g., *miR-222*) shown to be associated with outcome in independent reports should be further evaluated in prospective studies.

VI. CONCLUSIONS

miRNA discovery and analyses of miRNA expression and function in lymphocytes and DLBCL have already provided valuable information on the mechanisms regulating peripheral B cell differentiation and DLBCL pathogenesis hinting at the potential applicability of these small molecules as molecular biomarkers. Highly reproducible studies that assessed changes in miRNA expression upon peripheral B cell differentiation and examined miRNA involvement in regulating proteins controlling these processes and contributing to lymphomagenesis have markedly advanced our knowledge. In contrast, studies aimed to identify tumor miRNAs representing distinct COO signatures and to discover specific miRNAs that can be useful as biomarkers resulted in valuable but frequently controversial findings due to pitfalls recognized and discussed in this chapter. Future studies taking into account the lessons learned from these studies will resolve the current controversies and will broaden our knowledge on the roles of miRNAs in DLBCL and their potential use in daily clinical practice.

REFERENCES

Alencar, A.J., R. Malumbres, G.A. Kozloski, R. Advani, N. Talreja, S. Chinichian, J. Briones, Y. Natkunam, L.H. Sehn, R.D. Gascoyne, R. Tibshirani, and I.S. Lossos. 2011. MicroRNAs are independent predictors of outcome in diffuse large B-cell lymphoma patients treated with R-CHOP. *Clin Cancer Res* **17**:4125–4135.

Alizadeh, A.A., M.B. Eisen, R.E. Davis, C. Ma, I.S. Lossos, A. Rosenwald, J.C. Boldrick, H. Sabet, T. Tran, X. Yu, J.I. Powell, L. Yang, et al. 2000. Distinct types of diffuse large B-cell lymphoma identified by gene expression profiling. *Nature* **403**:503–511.

Baraniskin, A., J. Kuhnhenn, U. Schlegel, A. Chan, M. Deckert, R. Gold, A. Maghnouj, H. Zollner, A. Reinacher-Schick, W. Schmiegel, S.A. Hahn, and R. Schroers. 2011. Identification of microRNAs in the cerebrospinal fluid as marker for primary diffuse large B-cell lymphoma of the central nervous system. *Blood* **117**:3140–3146.

Basso, K., P. Sumazin, P. Morozov, C. Schneider, R.L. Maute, Y. Kitagawa, J. Mandelbaum, J. Haddad, Jr., C.Z. Chen, A. Califano, and R. Dalla-Favera. 2009. Identification of the human mature B cell miRNome. *Immunity* **30**:744–752.

Calin, G.A., C. Sevignani, C.D. Dumitru, T. Hyslop, E. Noch, S. Yendamuri, M. Shimizu, S. Rattan, F. Bullrich, M. Negrini, and C.M. Croce. 2004. Human microRNA genes are frequently located at fragile sites and genomic regions involved in cancers. *Proc Natl Acad Sci U S A* **101**:2999–3004.

Clurman, B.E. and W.S. Hayward. 1989. Multiple proto-oncogene activations in avian leukosis virus-induced lymphomas: evidence for stage-specific events. *Mol Cell Biol* **9**:2657–2664.

Coiffier, B., E. Lepage, J. Briere, R. Herbrecht, H. Tilly, R. Bouabdallah, P. Morel, E. Van Den Neste, G. Salles, P. Gaulard, F. Reyes, P. Lederlin, et al. 2002. CHOP chemotherapy plus rituximab compared with CHOP alone in elderly patients with diffuse large-B-cell lymphoma. *N Engl J Med* **346**:235–242.

Costinean, S., S.K. Sandhu, I.M. Pedersen, E. Tili, R. Trotta, D. Perrotti, D. Ciarlariello, P. Neviani, J. Harb, L.R. Kauffman, A. Shidham, and C.M. Croce. 2009. Ship and C/ebp{beta} are targeted by *miR-155* in B cells of E{micro}-*miR-155* transgenic mice. *Blood* **114**:1374–1382.

Costinean, S., N. Zanesi, Y. Pekarsky, E. Tili, S. Volinia, N. Heerema, and C.M. Croce. 2006. Pre-B cell proliferation and lymphoblastic leukemia/high-grade lymphoma in E(mu)-miR155 transgenic mice. *Proc Natl Acad Sci U S A* **103**:7024–7029.

Culpin, R.E., S.J. Proctor, B. Angus, S. Crosier, J.J. Anderson, and T. Mainou-Fowler. 2010. A 9 series microRNA signature differentiates between germinal centre and activated B-cell-like diffuse large B-cell lymphoma cell lines. *Int J Oncol* **37**:367–376.

Dagan, L.N., X. Jiang, S. Bhatt, E. Cubedo, K. Rajewsky, and I.S. Lossos. 2012. *miR-155* regulates HGAL expression and increases lymphoma cell motility. *Blood* **119**:513–520.

Davis, R.E., K.D. Brown, U. Siebenlist, and L.M. Staudt. 2001. Constitutive nuclear factor kappaB activity is required for survival of activated B cell-like diffuse large B cell lymphoma cells. *J Exp Med* **194**:1861–1874.

Dorsett, Y., K.M. McBride, M. Jankovic, A. Gazumyan, T.H. Thai, D.F. Robbiani, M. Di Virgilio, B.R. San-Martin, G. Heidkamp, T.A. Schwickert, T. Eisenreich, K. Rajewsky, et al. 2008. MicroRNA-155 suppresses activation-induced cytidine deaminase-mediated Myc-Igh translocation. *Immunity* **28**:630–638.

Eis, P.S., W. Tam, L. Sun, A. Chadburn, Z. Li, M.F. Gomez, E. Lund, and J.E. Dahlberg. 2005. Accumulation of *miR-155* and BIC RNA in human B cell lymphomas. *Proc Natl Acad Sci U S A* **102**:3627–3632.

Fang, C., D.X. Zhu, H.J. Dong, Z.J. Zhou, Y.H. Wang, L. Liu, L. Fan, K.R. Miao, P. Liu, W. Xu, and J.Y. Li. 2012. Serum microRNAs are promising novel biomarkers for diffuse large B cell lymphoma. *Ann Hematol* **91**:553–559.

REFERENCES

Griffiths-Jones, S. 2004. The microRNA registry. *Nucleic Acids Res* **32**:D109–D111.

Griffiths-Jones, S., R.J. Grocock, S. van Dongen, A. Bateman, and A.J. Enright. 2006. miRBase: microRNA sequences, targets and gene nomenclature. *Nucleic Acids Res* **34**:D140–D144.

Griffiths-Jones, S., H.K. Saini, S. van Dongen, and A.J. Enright. 2008. miRBase: tools for microRNA genomics. *Nucleic Acids Res* **36**:D154–D158.

Habermann, T.M., E.A. Weller, V.A. Morrison, R.D. Gascoyne, P.A. Cassileth, J.B. Cohn, S.R. Dakhil, B. Woda, R.I. Fisher, B.A. Peterson, and S.J. Horning. 2006. Rituximab-CHOP versus CHOP alone or with maintenance rituximab in older patients with diffuse large B-cell lymphoma. *J Clin Oncol* **24**:3121–3127.

He, L., J.M. Thomson, M.T. Hemann, E. Hernando-Monge, D. Mu, S. Goodson, S. Powers, C. Cordon-Cardo, S.W. Lowe, G.J. Hannon, and S.M. Hammond. 2005. A microRNA polycistron as a potential human oncogene. *Nature* **435**:828–833.

Jima, D.D., J. Zhang, C. Jacobs, K.L. Richards, C.H. Dunphy, W.W. Choi, W.Y. Au, G. Srivastava, M.B. Czader, D.A. Rizzieri, A.S. Lagoo, P.L. Lugar, et al. 2010. Deep sequencing of the small RNA transcriptome of normal and malignant human B cells identifies hundreds of novel micro-RNAs. *Blood* **116**:e118–e127.

Jung, I. and R.C. Aguiar. 2009. MicroRNA-155 expression and outcome in diffuse large B-cell lymphoma. *Br J Haematol* **144**:138–140.

Kluiver, J., S. Poppema, D. de Jong, T. Blokzijl, G. Harms, S. Jacobs, B.J. Kroesen, and A. van den Berg. 2005. BIC and *miR-155* are highly expressed in Hodgkin, primary mediastinal and diffuse large B cell lymphomas. *J Pathol* **207**:243–249.

Lawrie, C.H., S. Soneji, T. Marafioti, C.D. Cooper, S. Palazzo, J. Paterson, H. Cattan, T. Enver, R. Mager, J. Boultwood, J. Wainscoat, and C.S. Hatton. 2007. MicroRNA expression distinguishes between germinal center B cell-like and activated B cell-like subtypes of diffuse large B cell lymphoma. *Int J Cancer* **121**:1156–1161.

Lawrie, C.H., S. Gal, H. Dunlop, B. Pushkaran, A. Liggins, K. Pulford, A.H. Banham, F. Pezzella, J. Boultwood, J. Wainscoat, C.S. Hatton, and A. Harris. 2008a. Detection of elevated levels of tumour-associated microRNAs in serum of patients with diffuse large B-cell lymphoma. *Br J Haematol* **141**:672–675.

Lawrie, C.H., N.J. Saunders, S. Soneji, S. Palazzo, H.M. Dunlop, C.D. Cooper, P.J. Brown, X. Troussard, H. Mossafa, T. Enver, F. Pezzella, J. Boultwood, et al. 2008b. MicroRNA expression in lymphocyte development and malignancy. *Leukemia* **22**:1440–1446.

Lawrie, C.H., J. Chi, S. Taylor, D. Tramonti, E. Ballabio, S. Palazzo, N.J. Saunders, F. Pezzella, J. Boultwood, J.S. Wainscoat, and C.S. Hatton. 2009. Expression of microRNAs in diffuse large B cell lymphoma is associated with immunophenotype, survival and transformation from follicular lymphoma. *J Cell Mol Med* **13**:1248–1260.

Lenz, G., R.E. Davis, V.N. Ngo, L. Lam, T.C. George, G.W. Wright, S.S. Dave, H. Zhao, W. Xu, A. Rosenwald, G. Ott, H.K. Muller-Hermelink, et al. 2008a. Oncogenic CARD11 mutations in human diffuse large B cell lymphoma. *Science* **319**:1676–1679.

Lenz, G. and L.M. Staudt. 2010. Aggressive lymphomas. *N Engl J Med* **362**:1417–1429.

Lenz, G., G.W. Wright, N.C. Emre, H. Kohlhammer, S.S. Dave, R.E. Davis, S. Carty, L.T. Lam, A.L. Shaffer, W. Xiao, J. Powell, A. Rosenwald, et al. 2008b. Molecular subtypes of diffuse large B-cell lymphoma arise by distinct genetic pathways. *Proc Natl Acad Sci U S A* **105**: 13520–13525.

Lenze, D., L. Leoncini, M. Hummel, S. Volinia, C.G. Liu, T. Amato, G. De Falco, J. Githanga, H. Horn, J. Nyagol, G. Ott, J. Palatini, et al. 2011. The different epidemiologic subtypes of Burkitt lymphoma share a homogenous micro RNA profile distinct from diffuse large B-cell lymphoma. *Leukemia* **25**:1869–1876.

Li, C., S.W. Kim, D. Rai, A.R. Bolla, S. Adhvaryu, M.C. Kinney, R.S. Robetorye, and R.C. Aguiar. 2009. Copy number abnormalities, MYC activity, and the genetic fingerprint of normal B cells mechanistically define the microRNA profile of diffuse large B-cell lymphoma. *Blood* **113**: 6681–6690.

Lossos, I.S. 2005. Molecular pathogenesis of diffuse large B-cell lymphoma. *J Clin Oncol* **23**: 6351–6357.

Lossos, I.S., C.D. Jones, R. Warnke, Y. Natkunam, H. Kaizer, J.L. Zehnder, R. Tibshirani, and R. Levy. 2001. Expression of a single gene, BCL-6, strongly predicts survival in patients with diffuse large B-cell lymphoma. *Blood* **98**:945–951.

Lossos, I.S. and D. Morgensztern. 2006. Prognostic biomarkers in diffuse large B-cell lymphoma. *J Clin Oncol* **24**:995–1007.

Malumbres, R., K.A. Sarosiek, E. Cubedo, J.W. Ruiz, X. Jiang, R.D. Gascoyne, R. Tibshirani, and I.S. Lossos. 2009. Differentiation stage-specific expression of microRNAs in B lymphocytes and diffuse large B-cell lymphomas. *Blood* **113**:3754–3764.

Mandelbaum, J., G. Bhagat, H. Tang, T. Mo, M. Brahmachary, Q. Shen, A. Chadburn, K. Rajewsky, A. Tarakhovsky, L. Pasqualucci, and R. Dalla-Favera. 2010. BLIMP1 is a tumor suppressor gene frequently disrupted in activated B cell-like diffuse large B cell lymphoma. *Cancer Cell* **18**.568–579.

Montes-Moreno, S., N. Martinez, B. Sanchez-Espiridion, R. Diaz Uriarte, M.E. Rodriguez, A. Saez, C. Montalban, G. Gomez, D.G. Pisano, J.F. Garcia, E. Conde, E. Gonzalez-Barca, et al. 2011. miRNA expression in diffuse large B-cell lymphoma treated with chemoimmunotherapy. *Blood* **118**:1034–1040.

Morton, L.M., S.S. Wang, S.S. Devesa, P. Hartge, D.D. Weisenburger, and M.S. Linet. 2006. Lymphoma incidence patterns by WHO subtype in the United States, 1992–2001. *Blood* **107**: 265–276.

Muramatsu, M., K. Kinoshita, S. Fagarasan, S. Yamada, Y. Shinkai, and T. Honjo. 2000. Class switch recombination and hypermutation require activation-induced cytidine deaminase (AID), a potential RNA editing enzyme. *Cell* **102**:553–563.

Pasqualucci, L., M. Compagno, J. Houldsworth, S. Monti, A. Grunn, S.V. Nandula, J.C. Aster, V.V. Murty, M.A. Shipp, and R. Dalla-Favera. 2006. Inactivation of the PRDM1/BLIMP1 gene in diffuse large B cell lymphoma. *J Exp Med* **203**:311–317.

Rai, D., S.W. Kim, M.R. McKeller, P.L. Dahia, and R.C. Aguiar. 2010. Targeting of SMAD5 links microRNA-155 to the TGF-beta pathway and lymphomagenesis. *Proc Natl Acad Sci U S A* **107**:3111–3116.

Rodriguez, A., E. Vigorito, S. Clare, M.V. Warren, P. Couttet, D.R. Soond, S. van Dongen, R.J. Grocock, P.P. Das, E.A. Miska, D. Vetrie, K. Okkenhaug, et al. 2007. Requirement of bic/microRNA-155 for normal immune function. *Science* **316**:608–611.

Roehle, A., K. Hoefig, D. Repsilber, C. Thorns, M. Ziepert, K. Wesche, M. Thiere, M. Loeffler, W. Klapper, M. Pfreundschuh, A. Matolcsy, H. Bernd, et al. 2008. MicroRNA signatures characterize diffuse large B-cell lymphomas and follicular lymphomas. *Br J Haematol* **142**:732–744.

Rosenwald, A., G. Wright, K. Leroy, X. Yu, P. Gaulard, R.D. Gascoyne, W.C. Chan, T. Zhao, C. Haioun, T.C. Greiner, D.D. Weisenburger, J.C. Lynch, et al. 2003. Molecular diagnosis of primary mediastinal B cell lymphoma identifies a clinically favorable subgroup of diffuse large B cell lymphoma related to Hodgkin lymphoma. *J Exp Med* **198**:851–862.

Sanger-Institute. 2009. miRBase relase 13.0.

Swerdlow, S.H., E. Campo, N.L. Harris, E.S. Jaffe, S.A. Pileri, H. Stein, J. Thiele, and J.W. Vardiman. 2008. *WHO classification of tumours of haematopoietic and lymphoid tissues*, Fourth Edition. World Health Organization, Geneva, Switzerland.

Tagawa, H., K. Karube, S. Tsuzuki, K. Ohshima, and M. Seto. 2007. Synergistic action of the microRNA-17 polycistron and Myc in aggressive cancer development. *Cancer Sci* **98**:1482–1490.

REFERENCES

Tam, W., D. Ben-Yehuda, and W.S. Hayward. 1997. bic, a novel gene activated by proviral insertions in avian leukosis virus-induced lymphomas, is likely to function through its noncoding RNA. *Mol Cell Biol* **17**:1490–1502.

Teng, G., P. Hakimpour, P. Landgraf, A. Rice, T. Tuschl, R. Casellas, and F.N. Papavasiliou. 2008. MicroRNA-155 is a negative regulator of activation-induced cytidine deaminase. *Immunity* **28**:621–629.

Thai, T.H., D.P. Calado, S. Casola, K.M. Ansel, C. Xiao, Y. Xue, A. Murphy, D. Frendewey, D. Valenzuela, J.L. Kutok, M. Schmidt-Supprian, N. Rajewsky, et al. 2007. Regulation of the germinal center response by microRNA-155. *Science* **316**:604–608.

Ventura, A., A.G. Young, M.M. Winslow, L. Lintault, A. Meissner, S.J. Erkeland, J. Newman, R.T. Bronson, D. Crowley, J.R. Stone, R. Jaenisch, P.A. Sharp, et al. 2008. Targeted deletion reveals essential and overlapping functions of the *miR-17* through 92 family of miRNA clusters. *Cell* **132**:875–886.

Vigorito, E., K.L. Perks, C. Abreu-Goodger, S. Bunting, Z. Xiang, S. Kohlhaas, P.P. Das, E.A. Miska, A. Rodriguez, A. Bradley, K.G. Smith, C. Rada, et al. 2007. microRNA-155 regulates the generation of immunoglobulin class-switched plasma cells. *Immunity* **27**:847–859.

Xiao, C., L. Srinivasan, D.P. Calado, H.C. Patterson, B. Zhang, J. Wang, J.M. Henderson, J.L. Kutok, and K. Rajewsky. 2008. Lymphoproliferative disease and autoimmunity in mice with increased *miR-17-92* expression in lymphocytes. *Nat Immunol* **9**:405–414.

Yin, Q., X. Wang, J. McBride, C. Fewell, and E. Flemington. 2008. B-cell receptor activation induces BIC/*miR-155* expression through a conserved AP-1 element. *J Biol Chem* **283**:2654–2662.

Zhang, J., D.D. Jima, C. Jacobs, R. Fischer, E. Gottwein, G. Huang, P.L. Lugar, A.S. Lagoo, D.A. Rizzieri, D.R. Friedman, J.B. Weinberg, P.E. Lipsky, et al. 2009. Patterns of microRNA expression characterize stages of human B-cell differentiation. *Blood* **113**:4586–4594.

26

THE ROLE OF MicroRNAs IN HODGKIN'S LYMPHOMA

Wouter Plattel, Joost Kluiver, Arjan Diepstra, Lydia Visser, and Anke van den Berg

Department of Pathology and Medical Biology, University of Groningen, University Medical Center Groningen, Groningen, The Netherlands

I. Introduction	436
II. miRNAs in Hodgkin's Lymphoma	437
III. miRNA Profiling Studies in Hodgkin's Lymphoma	438
A. HL Cell Lines	438
B. Microdissected HRS Cells	441
C. Total Tissue Samples	441
IV. Functional miRNA Studies in HL	441
V. Clinical Value of miRNAs in HL	443
VI. Concluding Remarks and Future Perspectives	443
References	444

ABBREVIATIONS

cHL	classical Hodgkin's lymphoma
EBV	Epstein–Barr virus
GC B cells	germinal center B cells
HL	Hodgkin's lymphoma
HRS cells	Hodgkin Reed Sternberg cells
LP cells	lymphocyte predominant cells
miRNAs	microRNAs

MicroRNAs in Medicine, First Edition. Edited by Charles H. Lawrie.
© 2014 John Wiley & Sons, Inc. Published 2014 by John Wiley & Sons, Inc.

NLPHL   nodular lymphocyte predominant Hodgkin's lymphoma
PMBL   primary mediastinal B-cell lymphoma

I. INTRODUCTION

Hodgkin's lymphoma (HL) is the second most common lymphoma subtype with an incidence of about 9,100 and 18,000 new HL patients per year in the United States and Europe respectively. About 50% of the HL patients are diagnosed between 15 and 34 years of age, but the disease can occur at any age. In Western countries, a bimodal age-incidence distribution is observed, but there is a considerable variation in incidence and age distribution in different parts of the world.

The tumor cells of HL are in almost all cases of B-cell origin, although some rare cases of T-cell type HL have been reported. HL is a very peculiar lymphoma subtype since it consists of a minority of neoplastic cells that generally comprise less than 1% of the total cell population in a background of reactive cells (Poppema 1996). The vast majority of the HL patients present with the so-called classical (c)HL subtype, which includes nodular sclerosis, lymphocyte rich, mixed cellularity, and lymphocyte-depleted subtypes. The cellular background consists of varying amounts of T and B cells, eosinophils, mast cells, macrophages, and plasma cells and this composition determines the cHL subtype. Nodular lymphocyte predominant HL (NLPHL) is considered to be a different entity based on pathological and clinical features and accounts for only 5% of all HL cases. The tumor cells of cHL are referred to as Hodgkin Reed Sternberg (HRS) cells, and the tumor cells in NLPHL are called lymphocyte predominant (LP) cells.

Approximately 85% of cHL patients are cured by risk-adapted chemotherapy with or without radiotherapy, but treatment-related late toxic effects like cardiovascular disease, solid malignancies, and secondary leukemias can occur. NLPHL is a much more indolent disease than cHL, and is therefore only treated with more intensive chemotherapeutic regimens in patients with locally advanced or advanced-stage disease. Deaths in NLPHL are usually caused by treatment toxicity and occasional transformation to non-Hodgkin's lymphoma.

The origin of the HRS cells in cHL has been controversial for a long time because the immunophenotype of these cells is strikingly different from other hematopoietic cells (Hugh and Poppema 1992; Poppema et al. 1992). Detection of clonal immunoglobulin gene rearrangements, despite the lack of immunoglobulin protein production, confirmed a B-cell origin (Küppers et al. 1994). HRS cells frequently have a high number of immunoglobulin (Ig) gene mutations leading to stop codons and other alterations that disable the gene in a proportion of the cases (Kanzler et al. 1996). These studies indicated that HRS cells originate from germinal center B cells that lost the capacity to be selected by antigens. Such alterations are lethal to normal B cells, but HRS precursor cells are able to survive. At the time of diagnosis, HRS cells have virtually lost their B-cell identity since they show no or strongly reduced expression of many common B-cell markers (sIg, CD19, CD20, CD22, and CD79a) and B-cell transcription factors (Bob-1, Oct-2, PU.1, and Pax5) (Hertel et al. 2002; Schwering et al. 2003). LP cells are also derived from germinal B cells, but these cells do have a functional immunoglobulin gene rearrangement and also express most common B-cell markers (Atayar and Poppema 2011).

Infection with Epstein–Barr virus (EBV) is found in 20–40% of cHL patients in the Western world (Jarrett et al. 2005). This infection is clonal and considered to be an early event that is essential for the malignant transformation of HRS precursor cells. In EBV−cHL the causative tumor initiating event is still unknown. Many candidate genes have been screened and mutations in known oncogenes, including *TP53*, *FAS*, and *IkBa*, have been found in a variable number of cases in EBV+ and EBV− cHL patients. More recently, somatic mutations in the NF-κB suppressor gene, *TNFAIP3*, have been reported in 14 out of 20 (70%) of the EBV− cHL and in only 2 out of 16 (12.5%) of the EBV+ cHL (Schmitz et al. 2009). The high mutation frequency in EBV− cHL might indicate that inactivation of NF-κB plays an important role in cHL pathogenesis. EBV infection and loss of TNFAIP3 are two pathogenetic mechanisms that lead to constitutional activation of NF-κB, which is a hallmark of cHL.

II. MiRNAs IN HODGKIN'S LYMPHOMA

Deregulation of miRNAs in hematopoietic cells has been linked directly to the development of several lymphoma subtypes. Both oncogenic, for example, *miR-17~92* and *miR-21*, and tumor suppressor miRNAs, for example, *miR-15a~16-1* and *miR-150*, have been reported in various hematological malignancies (Calin et al. 2002; He et al. 2005; Xiao et al. 2007, 2008; Zhou et al. 2007; Mu et al. 2009; Olive et al. 2009; Klein et al. 2010; Medina et al. 2010).

Retrospectively, the first report on a miRNA in HL was published in 2003, when a high expression of the B cell integration cluster (*BIC*) gene was observed in HL cell lines by qRT-PCR (van den Berg et al. 2003). A tumor cell-specific expression pattern was confirmed by RNA-ISH in the HRS cells of primary HL cases. Expression of *BIC* was also shown in part of the germinal center B cells, and the transcripts were usually located in the nucleus in both HRS cells and germinal center B cells. The *BIC* gene locus was originally described as a commonly targeted region in an avian leukosis virus-induced chicken model for B-cell lymphoma (Tam et al. 1997). Overexpression of the *BIC* gene in chicken embryos resulted in the induction of B-cell lymphoma and enhanced the oncogenic potential of MYC (Tam et al. 2002). The *BIC* transcript does not have a functional open reading frame, but does contain a highly conserved stem loop-like region. Cloning of small RNAs from mouse hematopoietic cells resulted in the identification of *miR-155*, which was derived from the conserved stem-loop region that was located in the third exon of the human *BIC* gene (Lagos-Quintana et al. 2002).

miR-155 plays a crucial role in the development, function, and regulation of immune cells, including B cells, T cells, and dendritic cells (Rodriguez et al. 2007; Thai et al. 2007; Kohlhaas et al. 2009; Leng et al. 2011; Lu et al. 2011). Aberrant expression of *miR-155* is involved in the pathogenesis of several autoimmune diseases, for example, rheumatoid arthritis, multiple sclerosis, and systemic lupus erythematosus (Dai et al. 2010; Leng et al. 2011). The oncogenic potential of *BIC/miR-155* is supported by its high expression in many B-cell lymphoma subtypes (Eis et al. 2005; Kluiver et al. 2005) and by the development of cancer in transgenic mice (Costinean et al. 2006; O'Connell et al. 2008). Interestingly, *miR-155* may also function as a tumor suppressor by regulating the expression of activation-induced cytidine deaminase (AID), and thereby limiting or even preventing the occurrence of *MYC-IgH* translocations in activated B cells (Dorsett et al. 2008). In line with this finding, *miR-155* is expressed at very low levels in Burkitt lymphoma, which is characterized by *MYC* translocations (Kluiver et al. 2006).

III. miRNA PROFILING STUDIES IN HODGKIN'S LYMPHOMA

In HL, both miRNA profiling and functional studies have been hampered due to the scarcity of the tumor cells in the affected tissue samples and the lack of appropriate animal models. Studies on HL pathogenesis frequently depend on the use of HL cell lines or total tissue samples. The total number of HL cell lines is limited, but includes the main cHL subtypes of both B- and T-cell origin and one NLP HL cell line. These cell lines are usually derived from end-stage and progressive or relapsed HLs and are thus not entirely representative of HRS and LP cells at the time of diagnosis. Currently published miRNA profiling studies have been performed using these HL cell lines, microdissected HRS cells, and total tissue sections (Landgraf et al. 2007; Lawrie et al. 2008; Navarro et al. 2008; Gibcus et al. 2009; Van Vlierberghe et al. 2009).

A. HL Cell Lines

Landgraf et al. sequenced 250 small RNA libraries, including four cHL cell lines and various normal B-cell subsets, as well as other hematological cell lines (Landgraf et al. 2007). Three of the four cHL cell lines that were EBV− clustered together and were found to most closely resemble the plasmacytoma cell lines. The fourth HL cell line was EBV+ and clustered together with other EBV immortalized lymphoma derived cell lines. A very high expression level was observed specifically for *miR-16, miR-21, miR-29b, miR-142*, and *miR-155* in the three EBV− cHL-derived cell lines.

Lawrie et al. also analyzed miRNA expression in four HL cell lines and compared the expression pattern profile of B cells of different developmental stages (Lawrie et al. 2008). Two of the four HL cell lines (L428 and L1236) clustered closely together. These two cell lines also clustered closely together with a primary mediastinal B-cell lymphoma (PMBL) cell line, a lymphoma subtype that was shown to have many similarities with HL in gene expression studies. KM-H2 clustered separately, and the T-cell derived L540 HL cell line clustered closely together with ALK+ anaplastic large cell lymphoma. Three of the four cell lines were also included in the study by Landgraf et al., and a direct comparison by Lawrie et al. showed a comparable expression pattern.

Gibcus et al. subsequently profiled three B-cell derived cHL cell lines and one B-cell derived NLP HL cell line. A high expression was observed for the oncogenic *miR-17-92* cluster, *miR-16, miR-21, miR-24*, and *miR-155* in the three cHL cell lines. Several miRNAs were differentially expressed in cHL as compared with either EBV transformed lymphoblastoid cell lines, PMBL, and Burkitt lymphoma cell lines (Gibcus et al. 2009). The most pronounced differences were observed between cHL and BL cell lines with 16 differentially expressed miRNAs, including up-regulation of *miR-155* and *miR-9* and down-regulation of *miR-150*. Down-regulation of *miR-150* was observed also in comparison to lymphoblastoid and PMBL cell lines. The overlap between the most abundantly expressed miRNAs in this study and the most abundantly expressed miRNAs in the small RNA sequencing libraries of Landgraf et al. consisted of 18 miRNAs (listed in Table 26.1).

In addition to the cell lines reported in the study by Gibcus et al., we now also profiled five additional lymphoma cell lines and purified primary GC B cells (Figure 26.1). Unsupervised clustering revealed one cluster that contained the cHL cell lines with a confirmed B-cell origin based on immunoglobulin rearrangements and PMBL cell lines, a clustering pattern similar to that observed by Lawrie et al. (2008). A second cluster contained the NLP HL cell line DEV, normal GC B cells, and part of the diffuse large B-cell lymphoma

TABLE 26.1. Overview of the Most Abundantly Expressed miRNAs Observed in Four HL Profiling Studies

miRNA	cHL Cell Lines (Top 30) Landgraf	cHL Cell Lines (Top 27) Gibcus	HRS Cells and cHL Cell Lines (Top 6) Vlierberghe	Total Tissue Samples (Top 45) Navarro
let-7a	+	+	−	−
let-7b	+	−	−	−
let-7f	+	+	−	−
let-7g	−	+	−	−
let-7i	+	−	−	+
miR-103	−	+	−	−
miR-106a	−	+	−	+
miR-106b	+	+	−	−
miR-140	+	−	−	−
miR-142-3p	+	+	−	−
miR-142-5p	+	+	−	+
miR-155	+	+	+	−
miR-15a	+	+	−	−
miR-15b	−	+	−	−
miR-16	+	+	+	−
miR-17	+	+	−	−
miR-186	+	−	−	−
miR-18a	+	−	−	−
miR-191	+	+	−	−
miR-195	+	−	−	−
miR-19a	+	+	−	−
miR-19b	+	+	−	−
miR-20a	+	+	+	−
miR-20b	−	+	−	−
miR-21	+	+	+	+
miR-24	+	−	−	−
miR-25	−	+	−	−
miR-27a	+	−	−	+
miR-27b	+	−	−	−
miR-29a	+	+	−	−
miR-29b	+	+	−	−
miR-29c	−	+	−	−
miR-30b	−	+	+	−
miR-30d	+	−	−	−
miR-30e	+	−	−	−
miR-425	+	−	−	−
miR-565	−	+	−	−
miR-9	+	−	+	+
miR-92a	+	+	−	−
miR-93	+	+	−	−
	−	−	−	39 non-consistent miRNAs

+, present in top list; −, not present in top list.

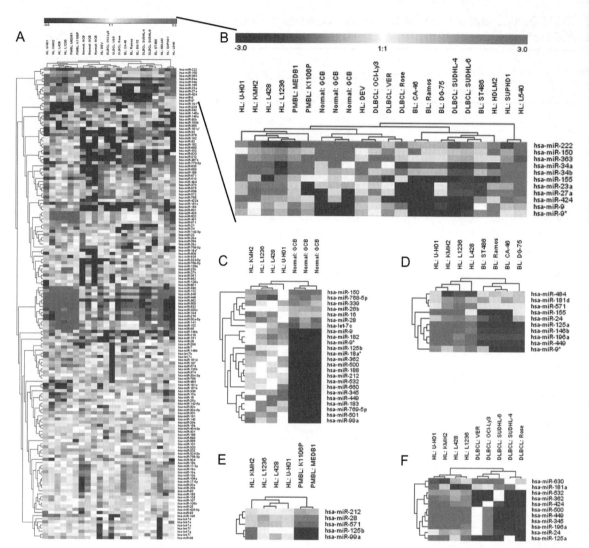

Figure 26.1. Heatmaps of miRNA profiling study in HL, non-Hodgkin's lymphoma-derived cell lines and normal GC B cells. Eight HL and 11 non-HL cell lines were analyzed on an Agilent miRNA microarray platform (version V1, Agilent, Santa Clara, CA). GC B cells (GCB) purified from three individuals (CD19+, IgD−, and CD38+) were analyzed as their normal counterparts. (A) Heat map of unsupervised clustering of the 151 miRNAs flagged present in all samples of at least one subtype. (B) Enlargement of the clustering pedigree from part A. The B-cell classical (c)HL cell lines cluster together with the PMBL cell lines, whereas the two T-cell cHL cell lines (HDLM2 and L540) and one cHL of uncertain origin (SUPHD1) cluster separately. The NLPHL cell line DEV clustered together with the normal GC B cells. (C) Heat map of significantly differentially expressed miRNAs between B-cell cHL and GC B cells. (D) Heat map of significantly differentially expressed miRNAs between B-cell cHL and BL. (E) Heat map of significantly differentially expressed miRNAs between B-cell cHL and PMBL. (F) Heat map of significantly differentially expressed miRNAs between B-cell cHL and DLBCL (unpaired t-test, miRNAs with P-value of <0.01 are shown). See color insert.

B. Microdissected HRS Cells

There is one miRNA profiling study that used laser microdissection to isolate primary HRS cells from patient material. Thirty microdissected primary HRS cells per patient were pooled, and RNA was isolated to generate miRNA signatures for nine cHL patients (Van Vlierberghe et al. 2009). In comparison with CD77+ B cells, 41 miRNAs were differentially expressed. Fifteen of these miRNAs were also differentially expressed in cHL cell lines in comparison to CD77+ B cells; 12 miRNAs were up-regulated, and three miRNAs were down-regulated. The overlap with our analysis (as shown in Figure 26.1) comprises three miRNAs, that is, *miR-9, miR-16,* and *miR-18a*. The relatively small overlap between both studies might be explained by different strategies applied to sort GC B cells. Six of the 12 up-regulated differentially expressed miRNAs were also among the most abundantly expressed miRNAs in the cell lines, that is, *miR-9, miR-16, miR-20a, miR-21, miR-30b,* and *miR-155* (Table 26.1). Consistent findings with respect to the most abundantly expressed miRNAs observed in the studies by Landgraf et al. and Gibcus et al. included *miR-16, miR-20a, miR-21,* and *miR-155*.

C. Total Tissue Samples

The second study, using primary cases, analyzed total tissue samples of 49 cHL patients and 10 reactive lymph nodes (Navarro et al. 2008). A signature of 25 miRNAs could differentiate cHL from reactive lymph node tissue, whereas 36 miRNAs were differentially expressed between nodular sclerosis and mixed cellularity subtypes. For four miRNAs, that is *miR-21, miR-134, miR-138,* and *miR-155, in situ* hybridization confirmed expression in the HRS cells. Twenty of the 25 miRNAs differentially expressed between cHL and lymph node were also highly expressed in cHL cell lines, suggesting a tumor cell-specific expression pattern. Comparing the top 45 most abundantly expressed miRNAs in total tissue samples of this study with the most abundant miRNAs observed in primary HRS cells or cell lines in the previously discussed studies revealed only a marginal overlap (Table 26.1). This difference is most likely explained by the analysis of total tissue, including only a minority of tumor cells and a vast majority of inflammatory cells. Navarro et al. (2008) also compared EBV+ and EBV− cHL tissue samples and found ten differentially expressed miRNAs. Three miRNAs, that is, *miR-96, miR-128a,* and *miR-128b*, showed low levels specifically in nodular sclerosis EBV+ cHL as compared with nodular sclerosis EBV− cHL. These differences might reflect putative differences in microenvironment or indirect effects caused by EBV proteins or miRNAs. No specific information was given on the expression of EBV-derived miRNAs.

IV. FUNCTIONAL MIRNA STUDIES IN HL

A substantial number (>25) of *miR-155* target genes have been identified in different normal and malignant cell types in the past few years. Some of these proven targets, that is *AID, FOXO3a, IL13Ra1, PU.1, SHIP1,* and *SOCS1* (Teng et al. 2008; O'Connell et al.

2009; Kong et al. 2010; Martinez-Nunez et al. 2011; Thompson et al. 2011; Zhang et al. 2011), have been shown to be involved in the pathogenesis of HL. However, to date, there are no studies that have addressed the functional consequences of the high *miR-155* levels in HL on these proven target genes.

Gibcus et al. (2009) tested the 3′-UTR sequence of 11 previously experimentally validated target genes by luciferase reporter assays in three HL cell lines and observed a *miR-155*-dependent targeting for six genes, that is *ZIC3, AGTR1, ZNF537, KGF, MAF*, and *IkBkE*, in one or more of the cell lines. No further validation at the protein level of these target genes has been done. A recent study by Dagan et al. (2012) showed that *HGAL* is also a direct target of *miR-155* in diffuse large B cell lymphoma. HGAL causes decreased lymphocyte and lymphoma cell motility by activating the RhoA signaling cascade. In HL, *miR-155* levels are high, and HGAL levels are typically low (Natkunam et al. 2007), which indicates that *HGAL* might be a pathogenically relevant target gene of *miR-155* in HL.

In a ribonucleoprotein immunoprecipitation Chip (RIP-Chip) experiment using antibodies against Ago2, a comprehensive overview of miRNA targets was generated for two HL cell lines. A significant overrepresentation of genes involved in proliferation, apoptosis, and the p53 pathway were identified in the Ago2 IP-fraction, indicating that these processes are regulated by miRNAs in HL (Tan et al. 2009). A high proportion of the miRNA target genes were regulated by the *miR-17* seed family. Validation of 11 of the *miR-17* identified targets by luciferase reporter assay indicated a *miR-17*-dependent regulation for nine genes, for example, *ZNFX1, CCL1*, and *GPR137B*. In a follow-up study, Gibcus et al. (2011) showed that *CDKN1A* encoding the p21 protein—one of the target genes identified by RIP-ChIP analysis—was a valid *miR-17* seed family target gene in HL cell lines using luciferase reporter assay and Western blotting. Inhibition of the *miR-17* seed family using antisense oligonucleotides resulted in increased p21 levels and a block in the G1-S cell cycle transition.

A second functional study in HL focused on *miR-9* and *let-7a*, targeting *PRDM1*, a master regulator of terminal B-cell differentiation (Nie et al. 2008). High levels of *miR-9* and *let-7a* correlated with low levels of PRDM1 in HL cell lines. The majority of HRS cells in primary HL cases also showed weak or no PRDM1 expression. Inhibition of *miR-9* or *let-7a* in HL cell lines using antisense oligonucleotides resulted in reduced PRDM1 levels. It can be speculated that high *miR-9* and *let-7a* levels prevent plasma cell differentiation of HRS precursor cells and thereby contribute to the pathogenesis and phenotype of HRS cells. In a very recent paper, Leucci et al. (2012) also studied the role of *miR-9* in HL. Inhibition of *miR-9* resulted in increased mRNA levels of *DICER1* and *HuR*. Subsequent luciferase and Western blot analysis confirmed targeting of these two genes by *miR-9*. Besides induction of these two genes, the authors also observed significant differences of genes that have a consensus HuR-binding motif. HuR can bind to AU rich transcripts and prevent degradation via the AU-mediated decay pathway. These binding motifs are found in the transcripts of many cytokines and chemokines, and the authors showed a HuR-dependent effect of *miR-9* inhibition on the chemoattracting potential of HL cell line culture supernatant. Moreover, effects of *miR-9* inhibition were observed on TNFa, CCL5, IL-5, and IL-6 expression. In a xenograft mouse model, reduced tumor outgrowth was observed upon subcutaneous delivery of *miR-9* inhibitors (Leucci et al. 2012). It is unclear whether these effects are caused by targeting of *PRDM1*, *HuR*, or one or more of the other *miR-9* target genes.

Based on the prognostic value of *miR-135a* in cHL (see later in the chapter), Navarro et al. (2009) studied *miR-135a* target genes in HL. Overexpression of *miR-135a* in HL cell lines resulted in increased caspase levels and decreased growth of the cells. The cytoplasmic tyrosine kinase *JAK2* contained a *miR-135a*-binding site in the 3′-UTR and a direct regulation was shown using a luciferase reporter assay and by Western blot. Overexpression of *miR-135a* induced down-regulation of the JAK2 protein levels and as a consequence Bcl-xL was also down-regulated in HL. This suggests a role for Bcl-xL in *miR-135a*-mediated apoptosis via JAK2 (Navarro et al. 2009).

In a recent study, Xie et al. (2012) showed a consistent down-regulation of FOXO1 in both cHL and NLPHL. The *FOXO1* locus at 13q14 was frequently deleted in HL cell lines and HRS cells purified from primary HL tissue samples. A second factor that contributed to the down-regulation of *FOXO1* was the constitutively activated PI3K/AKT and ERK pathways. A third factor associated with the low FOXO1 levels was the high expression of *miR-96, miR-182,* and *miR-183*, three miRNAs previously shown to target *FOXO1* (Myatt et al. 2010). A marked induction of FOXO1 protein levels was shown upon inhibition of these miRNAs in HL cell lines, confirming an effective targeting. Reintroduction of FOXO1 induced a growth arrest and apoptosis in all five cHL cell lines studied and indicated a functional role of FOXO1 down-regulation in the pathogenesis of HL.

V. CLINICAL VALUE OF miRNAs IN HL

To date, potential clinical applications of miRNAs include their use as prognostic markers, biomarkers, or as therapeutic targets. Many studies already report on the value of miRNAs as prognostic, tissue, or serum biomarkers in different malignancies (Lujambio and Lowe 2012). There is currently only one study that analyzed the prognostic value of miRNAs for HL patients (Navarro et al. 2009). In this study, a group of 89 cHL patients was analyzed; 76 patients had a complete remission, 5 a partial remission, and 8 patients were chemoresistant. Of 25 miRNAs tested, a significant difference in survival was observed only for *miR-135a* by comparing patients with a high level ($n = 42$) of this miRNA to patients with a low level ($n = 22$). Patients with a low *miR-135a* level had a shorter disease free survival ($P = 0.02$) and showed a higher frequency of relapse ($P = 0.04$), consistent with its proposed biological role in inhibiting apoptosis by modulating JAK2 (Navarro et al. 2009). In multivariate analysis, *miR-135a* remained a prognostic factor for disease-free survival. Given the relatively small patient cohort, it is essential to validate these findings in other larger cohorts to establish the true prognostic value of *miR-135a*.

VI. CONCLUDING REMARKS AND FUTURE PERSPECTIVES

It is obvious that miRNAs are involved in the pathogenesis of HL and can have an effect on different cellular pathways (Figure 26.2). Only a few studies show functional effects upon miRNA induction or inhibition that directly affect cell growth and apoptosis. These miRNAs, that is, *miR-9*, the *miR-17* seed family, and *miR-135a*, might thus be considered as potential therapeutic targets for the treatment of HL. For the highly expressed *miR-155*, the direct pathobiological consequences remain unknown, and additional studies are required. In many malignancies, the use of circulating miRNAs as possible prognostic or disease markers has been studied, but the value of circulating miRNAs in HL is still unknown. Currently, the potential for clinical applications of miRNA in HL is limited, and more studies are needed to fully explore the strength and breadth of opportunities.

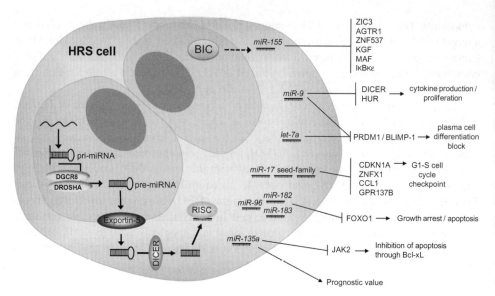

Figure 26.2. Schematic presentation of the currently proven miRNA targets in HL and their pathogenetic relevance. See color insert.

REFERENCES

Atayar, C. and S. Poppema. 2011. Nodular lymphocyte predominance type of Hodgkin lymphoma. In *Hematopathology*. E.S. Jaffe, N.L. Harris, J.W. Vardiman, E. Campo, and D.A. Arber, Eds. Saunders, Elsevier, Philadelphia, pp. 436–453.

Calin, G.A., C.D. Dumitru, M. Shimizu, R. Bichi, S. Zupo, E. Noch, H. Aldler, S. Rattan, M. Keating, K. Rai, L. Rassenti, T. Kipps, M. Negrini, F. Bullrich, and C.M. Croce. 2002. Frequent deletions and down-regulation of micro- RNA genes miR15 and miR16 at 13q14 in chronic lymphocytic leukemia. *Proc Natl Acad Sci U S A* **99 24**:15524–15529.

Costinean, S., N. Zanesi, Y. Pekarsky, E. Tili, S. Volinia, N. Heerema, and C.M. Croce. 2006. Pre-B cell proliferation and lymphoblastic leukemia/high-grade lymphoma in E(mu)-miR155 transgenic mice. *Proc Natl Acad Sci U S A* **103 18**:7024–7029.

Dagan, L.N., X. Jiang, S. Bhatt, E. Cubedo, K. Rajewsky, and I.S. Lossos. 2012. *miR-155* regulates HGAL expression and increases lymphoma cell motility. *Blood* **119 2**:513–520.

Dai, R., Y. Zhang, D. Khan, B. Heid, D. Caudell, O. Crasta, and S.A. Ahmed. 2010. Identification of a common lupus disease-associated microRNA expression pattern in three different murine models of lupus. *PLoS ONE* **5 12**:e14302.

Dorsett, Y., K.M. McBride, M. Jankovic, A. Gazumyan, T.H. Thai, D.F. Robbiani, M. Di Virgilio, B. Reina San-Martin, G. Heidkamp, T.A. Schwickert, T. Eisenreich, K. Rajewsky, and M.C. Nussenzweig. 2008. MicroRNA-155 suppresses activation-induced cytidine deaminase-mediated Myc-Igh translocation. *Immunity* **28 5**:630–638.

Eis, P.S., W. Tam, L. Sun, A. Chadburn, Z. Li, M.F. Gomez, E. Lund, and J.E. Dahlberg. 2005. Accumulation of *miR-155* and BIC RNA in human B cell lymphomas. *Proc Natl Acad Sci U S A* **102 10**:3627–3632.

Gibcus, J.H., B.J. Kroesen, R. Koster, N. Halsema, D. de Jong, S. de Jong, S. Poppema, J. Kluiver, A. Diepstra, and A. van den Berg. 2011. *MiR-17/106b* seed family regulates p21 in Hodgkin's lymphoma. *J Pathol* **225 4**:609–617.

REFERENCES

Gibcus, J.H., L.P. Tan, G. Harms, R.N. Schakel, D. de Jong, T. Blokzijl, P. Möller, S. Poppema, B.J. Kroesen, and A. van den Berg. 2009. Hodgkin lymphoma cell lines are characterized by a specific miRNA expression profile. *Neoplasia* **11** 2:167–176.

He, L., J.M. Thomson, M.T. Hemann, E. Hernando-Monge, D. Mu, S. Goodson, S. Powers, C. Cordon-Cardo, S.W. Lowe, G.J. Hannon, and S.M. Hammond. 2005. A microRNA polycistron as a potential human oncogene. *Nature* **435** 7043:828–833.

Hertel, C.B., X.G. Zhou, S.J. Hamilton-Dutoit, and S. Junker. 2002. Loss of B cell identity correlates with loss of B cell-specific transcription factors in Hodgkin/Reed-Sternberg cells of classical Hodgkin lymphoma. *Oncogene* **21** 32:4908–4920.

Hugh, J. and S. Poppema. 1992. Immunophenotype of Reed-Sternberg cells. *Int Rev Exp Pathol* **33**:81–114.

Jarrett, R.F., G.L. Stark, J. White, B. Angus, F.E. Alexander, A.S. Krajewski, J. Freeland, G.M. Taylor, P.R. Taylor, and Scotland and Newcastle Epidemiology of Hodgkin Disease Study Group. 2005. Impact of tumor Epstein-Barr virus status on presenting features and outcome in age-defined subgroups of patients with classic Hodgkin lymphoma: a population-based study. *Blood* **106** 7:2444–2451.

Kanzler, H., R. Küppers, M.L. Hansmann, and K. Rajewsky. 1996. Hodgkin and Reed-Sternberg cells in Hodgkin's disease represent the outgrowth of a dominant tumor clone derived from (crippled) germinal center B cells. *J Exp Med* **184** 4:1495–1505.

Klein, U., M. Lia, M. Crespo, R. Siegel, Q. Shen, T. Mo, A. Ambesi-Impiombato, A. Califano, A. Migliazza, G. Bhagat, and R. Dalla-Favera. 2010. The DLEU2/*miR-15a/16-1* cluster controls B cell proliferation and its deletion leads to chronic lymphocytic leukemia. *Cancer Cell* **17** 1:28–40.

Kluiver, J., E. Haralambieva, D. de Jong, T. Blokzijl, S. Jacobs, B.J. Kroesen, S. Poppema, and A. van den Berg. 2006. Lack of BIC and microRNA *miR-155* expression in primary cases of Burkitt lymphoma. *Genes Chromosomes Cancer* **45** 2:147–153.

Kluiver, J., S. Poppema, D. de Jong, T. Blokzijl, G. Harms, S. Jacobs, B.J. Kroesen, and A. van den Berg. 2005. BIC and *miR-155* are highly expressed in Hodgkin, primary mediastinal and diffuse large B cell lymphomas. *J Pathol* **207** 2:243–249.

Kohlhaas, S., O.A. Garden, C. Scudamore, M. Turner, K. Okkenhaug, and E. Vigorito. 2009. Cutting edge: the Foxp3 target *miR-155* contributes to the development of regulatory T cells. *J Immunol* **182** 5:2578–2582.

Kong, W., L. He, M. Coppola, J. Guo, N.N. Esposito, D. Coppola, and J.Q. Cheng. 2010. MicroRNA-155 regulates cell survival, growth, and chemosensitivity by targeting FOXO3a in breast cancer. *J Biol Chem* **285** 23:17869–17879.

Küppers, R., K. Rajewsky, M. Zhao, G. Simons, R. Laumann, R. Fischer, and M.L. Hansmann. 1994. Hodgkin disease: Hodgkin and Reed-Sternberg cells picked from histological sections show clonal immunoglobulin gene rearrangements and appear to be derived from B cells at various stages of development. *Proc Natl Acad Sci U S A* **91** 23:10962–10966.

Lagos-Quintana, M., R. Rauhut, A. Yalcin, J. Meyer, W. Lendeckel, and T. Tuschl. 2002. Identification of tissue-specific microRNAs from mouse. *Curr Biol* **12** 9:735–739.

Landgraf, P., M. Rusu, R. Sheridan, A. Sewer, N. Iovino, A. Aravin, S. Pfeffer, A. Rice, A.O. Kamphorst, M. Landthaler, C. Lin, N.D. Socci, L. Hermida, V. Fulci, S. Chiaretti, R. Foà, J. Schliwka, U. Fuchs, A. Novosel, R.U. Müller, B. Schermer, U. Bissels, et al. 2007. A mammalian microRNA expression atlas based on small RNA library sequencing. *Cell* **129** 7:1401–1414.

Lawrie, C.H., N.J. Saunders, S. Soneji, S. Palazzo, H.M. Dunlop, C.D. Cooper, P.J. Brown, X. Troussard, H. Mossafa, T. Enver, F. Pezzella, J. Boultwood, J.S. Wainscoat, and C.S. Hatton. 2008. MicroRNA expression in lymphocyte development and malignancy. *Leukemia* **22** 7:1440–1446.

Leng, R.X., H.F. Pan, W.Z. Qin, G.M. Chen, and D.Q. Ye. 2011. Role of microRNA-155 in autoimmunity. *Cytokine Growth Factor Rev* **22** 3:141–147.

Leucci, E., A. Zriwil, L.H. Gregersen, K.T. Jensen, S. Obad, C. Bellan, L. Leoncini, S. Kauppinen, and A.H. Lund. 2012. Inhibition of *miR-9* de-represses HuR and DICER1 and impairs Hodgkin lymphoma tumour outgrowth in vivo. *Oncogene* **31**(49):5081–5089.

Lu, C., X. Huang, X. Zhang, K. Roensch, Q. Cao, K.I. Nakayama, B.R. Blazar, Y. Zeng, and X. Zhou. 2011. *miR-221* and *miR-155* regulate human dendritic cell development, apoptosis, and IL-12 production through targeting of p27kip1, KPC1, and SOCS-1. *Blood* **117 16**:4293–4303.

Lujambio, A. and S.W. Lowe. 2012. The microcosmos of cancer. *Nature* **482 7385**:347–355.

Martinez-Nunez, R.T., F. Louafi, and T. Sanchez-Elsner. 2011. The interleukin 13 (IL-13) pathway in human macrophages is modulated by microRNA-155 via direct targeting of interleukin 13 receptor alpha1 (IL13Ralpha1). *J Biol Chem* **286 3**:1786–1794.

Medina, P.P., M. Nolde, and F.J. Slack. 2010. OncomiR addiction in an in vivo model of microRNA-21-induced pre-B-cell lymphoma. *Nature* **467 7311**:86–90.

Mu, P., Y.C. Han, D. Betel, E. Yao, M. Squatrito, P. Ogrodowski, E. de Stanchina, A. D'Andrea, C. Sander, and A. Ventura. 2009. Genetic dissection of the *miR-17~92* cluster of microRNAs in Myc-induced B-cell lymphomas. *Genes Dev* **23 24**:2806–2811.

Myatt, S.S., J. Wang, L.J. Monteiro, M. Christian, K.K. Ho, L. Fusi, R.E. Dina, J.J. Brosens, S. Ghaem-Maghami, and E.W. Lam. 2010. Definition of microRNAs that repress expression of the tumor suppressor gene FOXO1 in endometrial cancer. *Cancer Res* **70 1**:367–377.

Natkunam, Y., E.D. Hsi, P. Aoun, S. Zhao, P. Elson, B. Pohlman, H. Naushad, M. Bast, R. Levy, and I.S. Lossos. 2007. Expression of the human germinal center-associated lymphoma (HGAL) protein identifies a subset of classic Hodgkin lymphoma of germinal center derivation and improved survival. *Blood* **109 1**:298–305.

Navarro, A., T. Diaz, A. Martinez, A. Gaya, A. Pons, B. Gel, C. Codony, G. Ferrer, C. Martinez, E. Montserrat, and M. Monzo. 2009. Regulation of JAK2 by *miR-135a:* prognostic impact in classic Hodgkin lymphoma. *Blood* **114 14**:2945–2951.

Navarro, A., A. Gaya, A. Martinez, A. Urbano-Ispizua, A. Pons, O. Balagué, B. Gel, P. Abrisqueta, A. Lopez-Guillermo, R. Artells, E. Montserrat, and M. Monzo. 2008. MicroRNA expression profiling in classic Hodgkin lymphoma. *Blood* **111 5**:2825–2832.

Nie, K., M. Gomez, P. Landgraf, J.F. Garcia, Y. Liu, L.H. Tan, A. Chadburn, T. Tuschl, D.M. Knowles, and W. Tam. 2008. MicroRNA-mediated down-regulation of PRDM1/Blimp-1 in Hodgkin/Reed-Sternberg cells: a potential pathogenetic lesion in Hodgkin lymphomas. *Am J Pathol* **173 1**:242–252.

O'Connell, R.M., A.A. Chaudhuri, D.S. Rao, and D. Baltimore. 2009. Inositol phosphatase SHIP1 is a primary target of *miR-155*. *Proc Natl Acad Sci U S A* **106 17**:7113–7118.

O'Connell, R.M., D.S. Rao, A.A. Chaudhuri, M.P. Boldin, K.D. Taganov, J. Nicoll, R.L. Paquette, and D. Baltimore. 2008. Sustained expression of microRNA-155 in hematopoietic stem cells causes a myeloproliferative disorder. *J Exp Med* **205 3**:585–594.

Olive, V., M.J. Bennett, J.C. Walker, C. Ma, I. Jiang, C. Cordon-Cardo, Q.J. Li, S.W. Lowe, G.J. Hannon, and L. He. 2009. *miR-19* is a key oncogenic component of mir-17-92. *Genes Dev* **23** 24:2839–2849.

Poppema, S. 1996. Immunology of Hodgkin's disease. *Baillieres Clin Haematol* **9 3**:447–457.

Poppema, S., J. Kaleta, B. Hepperle, and L. Visser. 1992. Biology of Hodgkin's disease. *Ann Oncol* **3 Suppl 4**:5–8.

Rodriguez, A., E. Vigorito, S. Clare, M.V. Warren, P. Couttet, D.R. Soond, S. van Dongen, R.J. Grocock, P.P. Das, E.A. Miska, D. Vetrie, K. Okkenhaug, A.J. Enright, G. Dougan, M. Turner, and A. Bradley. 2007. Requirement of bic/microRNA-155 for normal immune function. *Science* **316 5824**:608–611.

Schmitz, R., M.L. Hansmann, V. Bohle, J.I. Martin-Subero, S. Hartmann, G. Mechtersheimer, W. Klapper, I. Vater, M. Giefing, S. Gesk, J. Stanelle, R. Siebert, and R. Küppers. 2009. TNFAIP3

(A20) is a tumor suppressor gene in Hodgkin lymphoma and primary mediastinal B cell lymphoma. *J Exp Med* **206** 5:981–989.

Schwering, I., A. Bräuninger, U. Klein, B. Jungnickel, M. Tinguely, V. Diehl, M.L. Hansmann, R. Dalla-Favera, K. Rajewsky, and R. Küppers. 2003. Loss of the B-lineage-specific gene expression program in Hodgkin and Reed-Sternberg cells of Hodgkin lymphoma. *Blood* **101** 4:1505–1512.

Tam, W., D. Ben-Yehuda, and W.S. Hayward. 1997. bic, a novel gene activated by proviral insertions in avian leukosis virus-induced lymphomas, is likely to function through its noncoding RNA. *Mol Cell Biol* **17** 3:1490–1502.

Tam, W., S.H. Hughes, W.S. Hayward, and P. Besmer. 2002. Avian bic, a gene isolated from a common retroviral site in avian leukosis virus-induced lymphomas that encodes a noncoding RNA, cooperates with c-myc in lymphomagenesis and erythroleukemogenesis. *J Virol* **76** 9:4275–4286.

Tan, L.P., E. Seinen, G. Duns, D. de Jong, O.C. Sibon, S. Poppema, B.J. Kroesen, K. Kok, and A. van den Berg. 2009. A high throughput experimental approach to identify miRNA targets in human cells. *Nucleic Acids Res* **37** 20:e137.

Teng, G., P. Hakimpour, P. Landgraf, A. Rice, T. Tuschl, R. Casellas, and F.N. Papavasiliou. 2008. MicroRNA-155 is a negative regulator of activation-induced cytidine deaminase. *Immunity* **28** 5:621–629.

Thai, T.H., D.P. Calado, S. Casola, K.M. Ansel, C. Xiao, Y. Xue, A. Murphy, D. Frendewey, D. Valenzuela, J.L. Kutok, M. Schmidt-Supprian, N. Rajewsky, G. Yancopoulos, A. Rao, and K. Rajewsky. 2007. Regulation of the germinal center response by microRNA-155. *Science* **316** 5824:604–608.

Thompson, R.C., M. Herscovitch, I. Zhao, T.J. Ford, and T.D. Gilmore. 2011. NF-kappaB downregulates expression of the B-lymphoma marker CD10 through a *miR-155/PU.1* pathway. *J Biol Chem* **286** 3:1675–1682.

van den Berg, A., B.J. Kroesen, K. Kooistra, D. de Jong, J. Briggs, T. Blokzijl, S. Jacobs, J. Kluiver, A. Diepstra, E. Maggio, and S. Poppema. 2003. High expression of B-cell receptor inducible gene BIC in all subtypes of Hodgkin lymphoma. *Genes Chromosomes Cancer* **37** 1:20–28.

Van Vlierberghe, P., A. De Weer, P. Mestdagh, T. Feys, K. De Preter, P. De Paepe, K. Lambein, J. Vandesompele, N. Van Roy, B. Verhasselt, B. Poppe, and F. Speleman. 2009. Comparison of miRNA profiles of microdissected Hodgkin/Reed-Sternberg cells and Hodgkin cell lines versus CD77+ B-cells reveals a distinct subset of differentially expressed miRNAs. *Br J Haematol* **147** 5:686–690.

Xiao, C., D.P. Calado, G. Galler, T.H. Thai, H.C. Patterson, J. Wang, N. Rajewsky, T.P. Bender, and K. Rajewsky. 2007. *MiR-150* controls B cell differentiation by targeting the transcription factor c-Myb. *Cell* **131** 1:146–159.

Xiao, C., L. Srinivasan, D.P. Calado, H.C. Patterson, B. Zhang, J. Wang, J.M. Henderson, J.L. Kutok, and K. Rajewsky. 2008. Lymphoproliferative disease and autoimmunity in mice with increased *miR-17-92* expression in lymphocytes. *Nat Immunol* **9** 4:405–414.

Xie, L., A. Ushmorov, F. Leithäuser, H. Guan, C. Steidl, J. Färbinger, C. Pelzer, M.J. Vogel, H.J. Maier, R.D. Gascoyne, P. Möller, and T. Wirth. 2012. FOXO1 is a tumor suppressor in classical Hodgkin lymphoma. *Blood* **119** 15:3503–3511.

Zhang, M., Q. Zhang, F. Liu, L. Yin, B. Yu, and J. Wu. 2011. MicroRNA-155 may affect allograft survival by regulating the expression of suppressor of cytokine signaling 1. *Med Hypotheses* **77** 4:682–684.

Zhou, B., S. Wang, C. Mayr, D.P. Bartel, and H.F. Lodish. 2007. *miR-150*, a microRNA expressed in mature B and T cells, blocks early B cell development when expressed prematurely. *Proc Natl Acad Sci U S A* **104** 17:7080–7085.

27

MicroRNA EXPRESSION IN CUTANEOUS T-CELL LYMPHOMAS

Cornelis P. Tensen

Department of Dermatology, Leiden University Medical Center, Leiden, The Netherlands

I.	Introduction	450
II.	The Pathology of Cutaneous Lymphoma	450
	A. Mycosis Fungoides	450
	B. Sézary Syndrome	450
	C. Primary Cutaneous Anaplastic Large-Cell Lymphoma	451
III.	Aberrant Expression of Components of the MicroRNA Biogenesis Machinery in Cutaneous T-Cell Lymphoma	452
	A. The Microprocessor Complex	452
	B. Modulators of miRNA Processing	452
	C. DICER and the RISC Complex	453
IV.	miRNA Expression Profiling in Cutaneous T-Cell Lymphoma	453
V.	Functional Consequences of (Aberrant) miRNA Expression in Cutaneous T-Cell Lymphoma	457
VI.	Conclusions and Perspectives	457
	Acknowledgments	458
	References	458

ABBREVIATIONS

CTCL   cutaneous T-cell lymphoma
MF   mycosis fungoides

MicroRNAs in Medicine, First Edition. Edited by Charles H. Lawrie.
© 2014 John Wiley & Sons, Inc. Published 2014 by John Wiley & Sons, Inc.

MiRNA   microRNA
Sz   Sézary syndrome

I. INTRODUCTION

Although the majority of lymphomas arise in the lymph node, a considerable proportion involve extranodal sites. After the gastrointestinal tract, the skin is the second most common site of extranodal non-Hodgkin lymphoma, with an estimated annual incidence of 1:100,000 (Groves et al. 2000). Primary cutaneous lymphomas are a group of lymphoproliferative disorders of neoplastic lymphocytes presenting in the skin with no evidence of extracutaneous disease at the time of diagnosis. In the Western world, primary cutaneous lymphomas are more often of T-cell origin (75%) than of B-cell origin (25%) (Willemze et al. 2005).

Several different classification systems were in use for classifying cutaneous lymphomas before a consensus was reached with the implementation of the so-called World Health Organization-European Organization for Research and Treatment of Cancer (WHO-EORTC) classification (see Table 27.1; Willemze et al. 2005). Proper classification is not only an essential prerequisite for determining prognosis and treatment, but also for defining groups for molecular studies (including those focusing on miRNAs). The pathology of primary cutaneous T-cell lymphoma (CTCL) entities that are best described and most often studied on the molecular level are listed and explained in some more detail in the next section.

II. THE PATHOLOGY OF CUTANEOUS LYMPHOMA

A. Mycosis Fungoides

Mycosis fungoides (MF) is the most common type of CTCL, with an estimated annual incidence of 1/200,000 (Trautinger et al. 2006). MF generally has an indolent course, with slow progression from patches to more infiltrated plaques and eventually tumors over years and sometimes decades (Willemze et al. 2005). Early patch/plaque stage MF is characterized by the presence atypical T cells with hyperchromatic, cerebriform nuclei, which preferentially infiltrate into the epidermis (epidermotropism). With progression to tumor stage (MF-T), epidermotropism may be lost and dermal infiltrates become more diffuse with increasing number of blast cells, and transformation to large-cell lymphoma can occur (Benner et al. 2009a, 2009b, 2012b). The neoplastic T-cells have a $CD3^+$, $CD4^+$, $CD8^-$, and $CD45RO^+$ memory T-cell phenotype. The prognosis is dependent on the stage of disease, ranging from a 10-year disease-specific survival of 98% in patients with limited plaque stage disease to 42% in patients with tumor stage disease (van Doorn et al. 2000; Kim et al. 2003).

B. Sézary Syndrome

Sézary syndrome (Sz) is a malignancy of skin-homing CD4+ T cells characterized by a triad of erythroderma, generalized lymphadenopathy, and the presence of neoplastic T cells in the skin, lymph nodes, and peripheral blood (Sézary and Bouvrain 1938; Fink-Puches et al. 2002; Willemze et al. 2005).

In addition to the clinical presentation diagnostic criteria are an identical T-cell clone in both the peripheral blood and skin as determined by molecular or cytogenetic tests in

TABLE 27.1. WHO-EORTC Classification of Primary Cutaneous Lymphoma

WHO-EORTC Classification	Frequency, %[a]	Disease-Specific 5-Year Survival, %
Cutaneous T-cell lymphoma		
Mycosis fungoides	44	88
Variants of Mycosis fungoides		
Folliculotropic MF	4	80
Pagetoid reticulosis	<1	100
Granulomatous slack skin	<1	100
Sezary syndrome	3	24
Primary cutaneous CD30-positive lymphoproliferative disorders		
Primary cutaneous anaplastic large cell lymphoma	8	95
Lymphomatoid papulosis	12	100
Subcutaneous panniculitis-like T-cell lymphoma	1	82
Primary cutaneous NK/T-cell lymphoma, nasal-type	<1	NR
Primary cutaneous peripheral T-cell lymphoma, unspecified	2	16
Primary cutaneous peripheral T-cell lymphoma, rare subtypes		
Primary cutaneous gamma/delta-T-cell lymphoma	<1	NR
Primary cutaneous aggressive CD8+ T-cell lymphoma	<1	18
Primary cutaneous CD4+ small/medium pleomorphic T-cell lymphoma	2	75
Cutaneous B-cell lymphoma		
Primary cutaneous marginal zone B-cell lymphoma	7	99
Primary cutaneous follicle center lymphoma	11	95
Primary cutaneous diffuse large B-cell lymphoma, leg type	4	55

[a]Lawrie et al. (2008).
NR indicates not reached.
Adapted from Willemze et al. (2005).

combination with one or more of the following criteria: Sézary cell count > 1000 cells/mm^3, a CD4/CD8 ratio of >10 caused by an expanding population of CD4$^+$ T cells, or loss of T-cell markers CD2, CD3, CD4, and/or CD5 (Olsen et al. 2011). Sézary patients have a poor prognosis, with a disease specific 5-year survival of 24% (Willemze et al. 2005). Sz is often considered as a leukemic phase or variant of MF, but in the recent classification, MF and Sz are included as separate disease entities based on their distinctive clinical features and disease behavior (see Table 27.1). This division is supported by clear genomic differences of tumor cells between MF and Sz (van Doorn et al. 2009), which suggest that the molecular pathogenesis, but also therapeutic requirements of these CTCL, may be distinct.

C. Primary Cutaneous Anaplastic Large-Cell Lymphoma

Primary cutaneous anaplastic large-cell lymphoma (cALCL) is a CTCL composed of large cells with an anaplastic, pleomorphic, or immunoblastic cytomorphology that shows expression of the CD30 receptor in more than 75% of the neoplastic cells (Willemze et al. 2005). The neoplastic cells have an activated CD4+ T-cell phenotype with variable loss of T-cell markers and frequent expression of cytotoxic proteins (Kaudewitz et al. 1989; Paulli et al. 1995; Kummer et al. 1997; Bekkenk et al. 2003). cALCL has an indolent clinical behavior, rarely shows extracutaneous dissemination and patients have a good prognosis with a 5-year survival exceeding 90% (Bekkenk et al. 2003). cALCL, together with lymphomatoid papulosis (LyP), constitutes a spectrum of primary cutaneous CD30+ lymphoproliferative disorders (Willemze and Beljaards 1993). Lymphomatoid papulosis is a chronic, recurrent, self-healing papulonecrotic or papulonodular skin disease with histologic features suggestive of cALCL (Willemze et al. 2005).

III. ABERRANT EXPRESSION OF COMPONENTS OF THE MicroRNA BIOGENESIS MACHINERY IN CUTANEOUS T-CELL LYMPHOMA

A. The Microprocessor Complex

The generation of microRNAs (miRNAs) requires a cascade of synthesis and processing steps (see Chapter 1 for an overview). In brief, it starts with transcription of primary miRNA transcripts (pri-miRNAs) by RNA polymerase II, which are cleaved into 60- to 70-nt stem-loop hairpin pre-miRNA structures by the microprocessor complex in the nucleus. This complex consists of DGCR8 (encoded by *DGCR8*) and DROSHA (encoded by *RNASEN*). *In vivo* studies have provided evidence that these molecules might function as haploinsufficient tumor suppressors, that is, partial depletion of Drosha accelerates cellular transformation and tumorigenesis in mouse models (Kumar et al. 2007). Consistent with the potential relevance of these mechanisms, reduced DROSHA mRNA levels have been associated with poor prognosis in several human cancers (Sugito et al. 2006; Merritt et al. 2008). Previous work by Lawrie et al. (Lawrie et al. 2009) demonstrated that levels of *DGCR8* were down-regulated in T cell lines (including the CTCL cell line HUT78 [Bunn and Foss 1996]) compared with control counterparts. In contrast, levels of *RNASEN* were up-regulated in T- and B-cell lines, which was confirmed for malignant B-cells in biopsy samples from DLBCL patients (Lawrie et al. 2009).

Array-based analysis of 20 MF tumor skin biopsies did not reveal significant expression differences of *DGCR8* between tumor stage MF (the most common type of CTCL) and controls (normal skin, inflamed skin, and T-cell subsets [van Kester et al. 2012]). To investigate *RNASEN* expression in CTCL patient material, we used RT-Q-PCR and quantified mRNA levels in MF skin biopsies and relevant controls (van Kester et al. 2012). We observed that the expression of *RNASEN* in tumor stage MF (MF-T) is higher compared to early stage MF and benign dermatoses (van Kester et al. 2012). This is in line with the cell line data and suggests that *RNASEN* does not function as a tumor suppressor in MF.

B. Modulators of miRNA Processing

In addition to the core machinery, modulators of miRNA processing can also function as haploinsufficient tumor suppressors. For example, after cleavage by DROSHA, pre-miRNAs are exported from the nucleus, which is facilitated by Exportin-5 (encoded by *XPO5*). Point mutations that impair *XPO5* function are correlated with sporadic and hereditary carcinomas (Melo et al. 2010). XPO5 mRNA levels were found to be higher in both B- and T-cell lines (including CTCL cell lines) compared with their respective control samples (Lawrie et al. 2009); however, these findings were not confirmed (yet) using patient material.

Cytoplasmic pre-miRNA is further cleaved to the mature miRNA by DICER (see next section) along with PRKRA (encoded by *PACT*) and TARBP2P (encoded by *TRBP*). Since mutations in these genes affect miRNA processing, Lawrie et al. (Lawrie et al. 2009) determined expression levels of *PACT* and *TRBP* in T-cell lines, including HuT78, and found that levels of PRKRA mRNA were up-regulated in CTCL T-cell lines, while TARBP2P mRNA levels were not significantly different compared with controls. To date, follow up on these observations using clinical material are not performed.

C. DICER and the RISC Complex

Recently, DICER was recognized as a bona fide tumor suppressor gene; deletion of a single *Dicer1* allele in lung epithelia (using conditional mouse models) promotes *Kras*-driven lung adenocarcinomas (Kumar et al. 2007). In addition, aberrant expression levels of DICER have been linked with prognosis and clinical course in several cancers (Karube et al. 2005; Chiosea et al. 2006; Merritt et al. 2008) (Shu et al. 2012; Wu et al. 2012), though both overexpression, as well as reduced expression, has been reported as a prognostic factor. Initial studies showed that DICER1 was down-regulated in the CTCL-derived T-cell line HUT78 (Lawrie et al. 2009). Array-based gene expressing analysis of clinical samples (MF-T skin biopsies) also indicated down-regulation of DICER mRNA in patients compared with normal skin, inflamed skin, and peripheral blood-derived T-cells (van Kester et al. 2012, supplementary table S4). However, this result could not be confirmed using RT-Q-PCR (van Kester et al. 2012).

DICER protein expression was studied in a cohort of 50 patients with primary CTCL (Valencak et al. 2011). High DICER expression levels were found to be associated with a significant negative prognostic impact on disease-specific survival in patients with certain subtypes of CTCL. As stated by the authors, these data have to be interpreted with caution since a mixed collection of patients with primary cutaneous T-cell lymphomas were included in this study with small sample sizes of certain subtypes.

The information on expression of other components of the RISC complex in CTCL is very limited. One study used arrays and RT-Q-PCR to measure expression of *EIF2C1*, *EIF2C2*, *EIF2C3*, and *EIF2C4* in MF-T in comparison with normal skin, inflamed skin, and peripheral blood-derived T-cells (van Kester et al. 2012). According to these data, it turned out that EIF2C3 and EIF2C4 were expressed at similar levels across all tested samples. A more refined analysis showed decreased expression of *EIF2C2* (Argonaute 2), an essential component of the miRNA machinery (Esquela-Kerscher and Slack 2006) in MF-T in comparison with healthy CD4+ T cells and a benign dermatose (chronic eczematous dermatitis, CED), but not in comparison with normal skin, early-stage MF or another type of benign skin disease, chronic discoid lupus erythematosus (CDLE). *EIF2C1* expression in MF-T is not different from control CD4+ T cells, normal skin, early-stage MF, but higher compared with CED, but lower expressed compared with CDLE.

IV. miRNA EXPRESSION PROFILING IN CUTANEOUS T-CELL LYMPHOMAS

The first study describing a genome-wide analysis of miRNA expression in CTCL was performed on peripheral blood-derived malignant T cells from Sézary patients using miRNA microarrays (Ballabio et al. 2010). This analysis identified >100 differentially expressed miRNAs compared with normal peripheral blood-derived CD4$^+$ T cells. According to this study, the majority of Sz-associated miRNAs were down-regulated compared with normal T cells. Most prominent up-regulated miRs identified were *miR-145, miR-574-5p, miR200c, miR-199a*, miR-143*, and *miR-214*, while *miR-342, miR-223*, and *miR-150* were among the most down-regulated (see Table 27.2). The majority of observed changes in miR expression levels correlated with previously reported genomic copy number abnormalities (Vermeer et al. 2008), which is in line with the observation that miRNAs exhibit high frequency genomic alterations in human cancers (Zhang et al. 2006; Croce 2009). However, this correlation was not found for all, suggesting that in addition to the copy number effect, miRNA expression in Sz is also regulated by other mechanisms. Narducci et al. (Narducci et al. 2011) also identified *miR-214* and *miR-199a** up-regulation in Sz using a commercial array-based miR expression detection platform. In this study, they also confirmed up-regulation of *miR-7* and decreased expression of *miR-342, miR-223, miR-92, miR-181a* and *miR-191* in Sz.

A recent study using deep sequence technology (Qin et al. 2012) confirmed the down-regulation of *miR-181a* and up-regulation of *miR-214* and *miR-199a** and *miR-486* in Sézary cells compared with peripheral blood-derived CD4+ T cells from healthy controls. In addition, it was demonstrated that expression of these miRs is also increased when compared with (*in vivo* activated) CD4+ T cells isolated from patients with erythroderma secondary to atopic dermatitis, indicating up-regulation of those miRNAs is not simply a consequence of T-cell activation. In combination with a Q-PCR analysis of *miR precursors*, it was shown that the *miR-199a2/214* cluster within the DNM3os transcript represents the vast majority of aberrantly expressed miRNAs in Sézary syndrome. The deep sequence analysis also permitted the identification of sequence variants of individual miRs. More specifically, it showed that the length of the predominant form of *miR-21* consists of 23 nucleotides (nts) instead of the 22 nts mainly reported in literature.

Profiling of miRNAs in tumor-stage MF was performed by van Kester et al. by comparison of miRNA expression pattern of MF tumors skin biopsies with benign inflammatory dermatoses using miRNA microarrays (van Kester et al. 2011). Identified differences were validated in biopsies of an independent group of patients and controls by Q-PCR. In contrast to Sz, it was found that in MF-T, the majority (30/49) of the differentially expressed miRNAs is up-regulated compared with benign controls. For most of the identified dysregulated miRNAs, aberrant expression in cancer is described, and several up-regulated miRNAs (*miR-93, miR-155*, and *miR-17-92*) have been validated functionally as oncomirs (Esquela-Kerscher and Slack 2006; O'Connell et al. 2008, 2010).

An identical approach using the same controls and miRNA detection platform was used to unravel the miRNome of cALCL (Benner et al. 2012a). Thirteen miRNAs were found that are differentially expressed between cALCL and benign controls. Of these miRNAs, the up-regulation of *miR-155, miR-27b, miR-30c*, and *miR-29b* in cALCL was validated by miRNA-Q-PCR on independent study groups (cALCL biopsies and controls).

Although characterized by a completely different clinical behavior, the μRNA microarray analysis revealed no statistically significantly differences between cALCL and MF-T. However, miRNA-Q-PCR analysis identified *miR-155, miR-27b, miR-93, miR-29b*, and *miR-92a* as being statistically significantly differentially expressed between cALCL and tumor stage MF. Unfortunately, direct comparison between the miRNome of Sz and MF-T/cALCL was not possible, since different reference RNAs, as well as different

TABLE 27.2. Overview of Profiling Studies Aimed at the Identification of Differentially Expressed miRNAs in Distinct CTCL Entities

CTCL Entity	Sz	Sz	Sz	Sz	MF	cALCL
Publication	Ballabio et al. (2010)	Narducci et al. (2011)	Qin et al. (2012)	Qin et al. (2012)	van Kester et al. (2011)	Benner et al. (2012a)
Platform	μRNA microarray[a]	miRNA array[b]	NGS	NGS	μRNA microarray[a]	μRNA microarray[a]
Common reference	Tonsil miRNA	None	None	None	Synthetic miRNA	Synthetic miRNA
Comparison	CD3+ T-cells healthy donors	CD3+ T-cells healthy donors	CD4+ T-cells healthy donors	CD4+ T-cells benign erythroderma	Benign inflammatory skin disease (biopsies)	Benign inflammatory skin disease (biopsies)

Ranking (Fold Change)

Up	hsa-miR-145	**hsa-miR-214**	hsa-miR-214*	**hsa-miR-214**	hsa-miR-93	hsa-miR-425-5p
	hsa-miR-574-5p	**hsa-miR-199a***	**hsa-miR-214**	hsa-miR-199a	hsa-miR-425-5p	hsa-miR-155
	hsa-miR-200c	hsa-miR-199a	hsa-miR-199a	hsa-miR-214*	hsa-miR-155	hsa-miR-29b
	hsa-miR-199a*	hsa-miR-142-3p	**hsa-miR-199a***	**hsa-miR-199a***	hsa-miR-21	hsa-miR-30c
	hsa-miR-143	hsa-miR-486	hsa-miR-486-5p	hsa-miR-486-5p	hsa-miR-92a	hsa-miR-92b
	hsa-miR-214	hsa-miR-29b	hsa-miR-181b	let-7d	hsa-miR-142-3p	hsa-miR-27b
	hsa-miR-98	hsa-miR-146a			hsa-miR-92	hsa-miR-17-5p
	hsa-miR-518a-3p	hsa-miR-34a			hsa-miR-146a	hsa-miR-26b
	has-miR-7	hsa-miR-18a			hsa-miR-30b	hsa-miR-146b
	hsa-miR-152	hsa-miR-21			hsa-miR-16	hsa-miR-342-3p
Down	hsa-miR-342	let-7b	hsa-miR-31	hsa-miR-126*	hsa-miR-620	hsa-miR-197
	hsa-miR-223	let-7c	hsa-miR-99a	hsa-miR-193b	hsa-miR-302d	
	hsa-miR-150	hsa-miR-223	hsa-miR-142-3p	hsa-miR-125b	hsa-miR-483	
	hsa-miR-189(24*)	hsa-miR-125b	hsa-miR-126	hsa-miR-326	hsa-miR-204	
	hsa-miR-186	hsa-miR-145	hsa-miR-342-5p	hsa-miR-126	hsa-miR-323b-5p	
	hsa-miR-423-3p	hsa-miR-193b	hsa-miR-30e	hsa-miR-100	hsa-miR-380-5p	
	hsa-miR-92	hsa-miR-31	hsa-miR-140-3p	hsa-miR-99a	hsa-miR-383	
	hsa-miR-181a	hsa-miR-342	hsa-miR-146b-5p		hsa-miR-211	
	hsa-miR-191	hsa-miR-197	hsa-miR-192		hsa-miR-363	
	hsa-miR-376a	hsa-miR-361	hsa-miR-29b		hsa-miR-133b	

In bold are miRs identified in all Sz studies and comparisons.
[a]Lawrie et al. (2008).
[b]Commercial platform of Agilent.
NGS, next-generation sequencing; Sz, Sézary syndrome; MF, mycosis fungoides; cALCL, primary cutaneous anaplastic large cell lymphoma.

miRNA sources (CD4+ T cells in Sz vs. skin biopsies in the MF-T and cALCL study) were used.

Nevertheless, the combination of data from all profiling studies in Sz, MF-T, and cALCL suggests that distinctive miRs are involved in the pathogenesis of the different disease entities. *MiR-214, -199a*, -486,* and *-181a* in Sz, *miR-93* in MF, and *miR-155* in cALCL (and to a lesser extent, also in MF; see Table 27.2). It was therefore unexpected that Ralfkiaer et al., studying a heterogeneous group of CTCL, were able to identify general miR classifiers distinguishing CTCL from benign dermatoses (Ralfkiaer et al. 2011).

Using microarrays, Ralkiaer et al. showed that the most induced (*miR-326, miR-663b,* and *miR-711*) and repressed (*miR-203* and *miR-205*) miRNAs distinguish CTCL from benign skin diseases (Ralfkiaer et al. 2011). A subsequently developed Q-PCR-based classifier consisting of *miR-155, miR-203*, and *miR-205* was able to distinguish CTCL (including Sz) from benign skin disorders.

This apparent discrepancies with previous miRNA profiling studies, prompted us to reevaluate the deep sequence data of Sz and control T-cells (Qin et al. 2012; see Figure 27.1) for expression of miRs belonging to this classifier).

According to this analysis *miR-203* and *205* are indeed lowly expressed in Sz, but absent in healthy T cells. It is therefore not likely that these miRs are down-regulated in tumor T-cells as suggested by Ralfkiaer et al. The observed down-regulation of *miR-203* and *-205* in CTCL skin biopsies is still noteworthy, since these miRs were previously described as hallmarks of the skin (*miR-203*; Yi et al. 2008) or down-regulated during epithelial to mesenchymal transition (*miR-205*; Gregory et al. 2008). This suggests that (loss of) expression of these miRs in CTCL skin biopsies merely reflects the amount of skin tissue over T cells (high in benign controls and lower in tumor skin biopsies) rather than down-regulation of these miRs in tumor T cells.

The data presented in Figure 27.1 also raises some questions about the ability to use up-regulation of *miR-155* and *miR-326* (no difference between Sz and controls) or

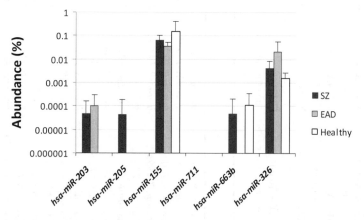

Figure 27.1. Expression of microRNAs belonging to a proposed diagnostic CTCL classifier (Ralfkiaer et al. 2011). Data are from the study of Qin et al., (Qin et al. 2012) using deep sequence analysis for the quantification of miRNA expression in peripheral blood derived CD4+ T cells from Sézary (Sz) syndrome ($n = 12$), erythroderma secondary to atopic dermatitis (EAD) patients ($n = 4$) and healthy controls ($n = 4$). The expression of each miR was normalized within samples, and miRs per group are given as percentage (% of mean ± SD).

miR-663b (no difference between Sz and healthy cells) as markers for Sz as proposed by Ralfkiaer et al. For MF and cALCL only the up-regulation of *miR-155* (as determined by Q-PCR and array analysis) is in accordance with the proposed classifier (van Kester et al. 2011; Benner et al. 2012a) and recently also confirmed for MF by Maj et al. (Maj et al. 2012). Close re-inspection of our own microarray data on MF and cALCL did not provide any support for the aberrant expression of the other miRs belonging to the proposed classifier.

V. FUNCTIONAL CONSEQUENCES OF (ABERRANT) miRNA EXPRESSION IN CUTANEOUS T-CELL LYMPHOMA

A very limited number of studies are published that determine the functional relevance of miRs in CTCL and nearly all of them used cell lines. Ballabio et al. (Ballabio et al. 2010) studied the functional consequences of down-regulation of *miR-17-5p*, part of the *miR-17-92* cluster, and *miR-342* in Sz. They showed that reintroduction of the miRs into the Sz cell line Seax (Kaltoft et al. 1987) resulted in increased apoptosis, implying a tumor suppressive role for these miRs. van der Fits et al. (van der Fits et al. 2011) demonstrated that *miR-21* is a direct target of STAT3, a transcription factor constitutively activated in many cancers, including Sézary syndrome. Stimulation of patient-derived Sézary- or healthy CD4+ T-cells with IL-21 resulted in a strong activation of STAT3, and subsequent up-regulation of (pri)*miR-21* expression. Silencing of *miR-21* in Seax resulted in increased apoptosis, suggesting a functional role for *miR-21* in the leukomogenic process.

The potential effect on apoptosis and cell survival of *miR-21*, but also of *miR-214* and *miR-486*, were investigated by Narducci et al. (Narducci et al. 2011) using the CTCL cell line HUT78. Since this cell line shows endogenous expression of *miR-21*, whereas it does not express *miR-214* nor *miR-486*, an *miR-21* loss-of-function and *miR-214/-486* gain-of-function approaches were applied. *miR-21* knockdown significantly enhanced the apoptosis, while an increase of cell viability in *miR-214* and *miR-486* transfected cells was seen. These findings indicate that *miR-21*, *miR-214*, and *miR-486* contribute to the apoptotic resistance of the CTCL cell line.

Finally Manfe et al. studied the function of *miR-122* in CTCL and showed, using Seax and MyLa (a cell line from a MF patient [Kaltoft et al. 1992]), that *miR-122* overexpression decreased the sensitivity to chemotherapy-induced apoptosis via a signaling circuit involving the activation of Akt and inhibition of p53 (Manfe et al. 2012). The described Q-PCR results on *miR-122* expression in CTCL in this study are somewhat difficult to interpret. On the one hand, the authors report that *miR-122* was not detectable in quiescent T cells, but on the other hand, they state that the data on *miR-122* expression levels in CTCL patients were normalized to its relative expression in quiescent T-cells (which is zero; Manfe et al. 2012).

VI. CONCLUSIONS AND PERSPECTIVES

In recent years, major steps have been made in the molecular characterization of several types of (clinically characterized) cutaneous lymphoma entities. This progress has also been made in the identification of miRNAs, which might be pathogenetically involved in the genesis and progression of several types of CTCL. Although most knowledge has been gathered from profiling studies, a complete picture is still lacking due to the use of different platforms (several types of [commercial] arrays, next generation deep sequencing and miRNA-Q-PCR kits from several vendors), as well as the diversity of used controls. An improved description of the miRNA landscape in CTCL can be expected from additional deep sequencing efforts (having major advances in robustness, resolution, and interlab portability and no need for reference miRs [t Hoen et al. 2008]) Such studies should not solely aim at increasing the number of clinically well defined CTCL samples, but also of relevant controls.

Little progress has been made in the understanding of the functional consequences of (aberrant) miRNA expression in CTCL. This might, in part, be explained by the limited number of cell lines available for these studies and the known difficulties with obtaining and culturing tumor cells from CTCL patients, despite recent improvements in the isolation of tumor cells from CTCL skin lesions (Campbell et al. 2010).

The recently described novel mouse model for Sézary syndrome, permitting long-term systemic repopulation of the RAG2–/– γc–/– mice with Sz cell lines or primary patient-derived tumor cells (van der Fits et al. 2012), will offer novel possibilities to study the functional role of miRNAs in Sézary syndrome. However, additional suitable (animal) models representing other types of cutaneous lymphoma, including different CTCL entities, are warranted.

ACKNOWLEDGMENTS

The author is grateful to Dr. L. van der Fits and Prof. Dr. R. Willemze (LUMC) for critically reading and help to improve this manuscript.

REFERENCES

Ballabio, E., T. Mitchell, M.S. van Kester, S. Taylor, H.M. Dunlop, J. Chi, I. Tosi, M.H. Vermeer, D. Tramonti, N.J. Saunders, J. Boultwood, J.S. Wainscoat, et al. 2010. MicroRNA expression in Sezary syndrome: identification, function, and diagnostic potential. *Blood* **116**:1105–1113.

Bekkenk, M.W., M.H. Vermeer, P.M. Jansen, A.M. van Marion, M.R. Canninga-van Dijk, P.M. Kluin, M.L. Geerts, C.J. Meijer, and R. Willemze. 2003. Peripheral T-cell lymphomas unspecified presenting in the skin: analysis of prognostic factors in a group of 82 patients. *Blood* **102**: 2213–2219.

Benner, M.F., E. Ballabio, M.S. van Kester, N.J. Saunders, M.H. Vermeer, R. Willemze, C.H. Lawrie, and C.P. Tensen. 2012a. Primary cutaneous anaplastic large cell lymphoma shows a distinct miRNA expression profile and reveals differences from tumor stage mycosis fungoides. *Exp Dermatol* **21**:632–634.

Benner, M.F., P.M. Jansen, C.J.L.M. Meijer, and R. Willemze. 2009a. Diagnostic and prognostic evaluation of phenotypic markers TRAF1, MUM1, BCL2 and CD15 in cutaneous CD30-positive lymphoproliferative disorders. *Br J Dermatol* **161**:121–127.

Benner, M.F., P.M. Jansen, C.J.L.M. Meijer, and R. Willemze. 2009b. Diagnostic and prognostic evaluation of phenotypic markers TRAF1, MUM1, bcl-2 and CD15 in cutaneous CD30-positive lymphoproliferations. *Br J Dermatol* **161**:121–127.

Benner, M.F., P.M. Jansen, M.H. Vermeer, and R. Willemze. 2012b. Prognostic factors in transformed mycosis fungoides: a retrospective analysis of 100 cases. *Blood* **119**:1643–1649.

Bunn, P.A. and F.M. Foss. 1996. T-cell lymphoma cell lines (HUT102 and HUT78) established at the National Cancer Institute: history and importance to understanding the biology, clinical

REFERENCES

features, and therapy of cutaneous T-cell lymphomas (CTCL) and adult T-cell leukemia-lymphomas (ATLL). *J Cell Biochem* **24**:12–23.

Campbell, J.J., R.A. Clark, R. Watanabe, and T.S. Kupper. 2010. Sezary syndrome and mycosis fungoides arise from distinct T-cell subsets: a biologic rationale for their distinct clinical behaviors. *Blood* **116**:767–771.

Chiosea, S., E. Jelezcova, U. Chandran, M. Acquafondata, T. McHale, R.W. Sobol, and R. Dhir. 2006. Up-regulation of dicer, a component of the microrna machinery, in prostate adenocarcinoma. *Am J Pathol* **169**:1812–1820.

Croce, C.M. 2009. Causes and consequences of microRNA dysregulation in cancer. *Nat Rev Genet* **10**:704–714.

Esquela-Kerscher, A. and F.J. Slack. 2006. Oncomirs—microRNAs with a role in cancer. *Nat Rev Cancer* **6**:259–269.

Fink-Puches, R., P. Zenahlik, B. Back, J. Smolle, H. Kerl, and L. Cerroni. 2002. Primary cutaneous lymphomas: applicability of current classification schemes (European Organization for Research and Treatment of Cancer, World Health Organization) based on clinicopathologic features observed in a large group of patients. *Blood* **99**:800–805.

Gregory, P.A., A.G. Bert, E.L. Paterson, S.C. Barry, A. Tsykin, G. Farshid, M.A. Vadas, Y. Khew-Goodall, and G.J. Goodall. 2008. The *miR-200* family and *miR-205* regulate epithelial to mesenchymal transition by targeting ZEB1 and SIP1. *Nat Cell Biol* **10**:593–601.

Groves, F.D., M.S. Linet, L.B. Travis, and S.S. Devesa. 2000. Cancer surveillance series: non-Hodgkin's lymphoma incidence by histologic subtype in the United States from 1978 through 1995. *J Natl Cancer Inst* **92**:1240–1251.

Kaltoft, K., S. Bisballe, H.F. Rasmussen, K. Thestruppedersen, K. Thomsen, and W. Sterry. 1987. A continuous T-cell line from a patient with Sezary-syndrome. *Arch Dermatol Res* **279**:293–298.

Kaltoft, K., S. Bisballe, T. Dyrberg, E. Boel, P.B. Rasmussen, and K. Thestrup-Pedersen. 1992. Establishment of two continuous T-cell strains from a single plaque of a patient with mycosis fungoides. *In Vitro Cell Dev Biol* **28A**:161–167.

Karube, Y., H. Tanaka, H. Osada, S. Tomida, Y. Tatematsu, K. Yanagisawa, Y. Yatabe, J. Takamizawa, S. Miyoshi, T. Mitsudomi, and T. Takahashi. 2005. Reduced expression of Dicer associated with poor prognosis in lung cancer patients. *Cancer Sci* **96**:111–115.

Kaudewitz, P., H. Stein, F. Dallenbach, F. Eckert, K. Bieber, G. Burg, and O. Braun-Falco. 1989. Primary and secondary cutaneous Ki-1+ (CD30+) anaplastic large cell lymphomas. Morphologic, immunohistologic, and clinical-characteristics. *Am J Pathol* **135**:359–367.

Kim, Y.H., H.L. Liu, S. Mraz-Gernhard, A. Varghese, and R.T. Hoppe. 2003. Long-term outcome of 525 patients with mycosis fungoides and Sezary syndrome: clinical prognostic factors and risk for disease progression. *Arch Dermatol* **139**:857–866.

Kumar, M.S., J. Lu, K.L. Mercer, T.R. Golub, and T. Jacks. 2007. Impaired microRNA processing enhances cellular transformation and tumorigenesis. *Nat Genet* **39**:673–677.

Kummer, J.A., M.H. Vermeer, D. Dukers, C.J. Meijer, and R. Willemze. 1997. Most primary cutaneous CD30-positive lymphoproliferative disorders have a CD4-positive cytotoxic T-cell phenotype. *J Invest Dermatol* **109**:636–640.

Lawrie, C.H., N.J. Saunders, S. Soneji, S. Palazzo, H.M. Dunlop, C.D.O. Cooper, P.J. Brown, X. Troussard, H. Mossafa, T. Enver, F. Pezzella, J. Boultwood, J.S. Wainscoat, and C.S.R. Hatton. 2008. MicroRNA expression in lymphocyte development and malignancy. *Leukemia* **22**: 1440–1446.

Lawrie, C.H., C.D. Cooper, E. Ballabio, J. Chi, D. Tramonti, and C.S. Hatton. 2009. Aberrant expression of microRNA biosynthetic pathway components is a common feature of haematological malignancy. *Br J Haematol* **145**:545–548.

Maj, J., A. Jankowska-Konsur, A. Sadakierska-Chudy, L. Noga, and A. Reich. 2012. Altered microRNA expression in mycosis fungoides. *Br J Dermatol* **166**:331–336.

Manfe, V., E. Biskup, A. Rosbjerg, M. Kamstrup, A.G. Skov, C.M. Lerche, B.T. Lauenborg, N. Odum, and R. Gniadecki. 2012. *miR-122* regulates p53/Akt signalling and the chemotherapy-induced apoptosis in cutaneous T-cell lymphoma. *PLoS ONE* **7**:e29541.

Melo, S.A., C. Moutinho, S. Ropero, G.A. Calin, S. Rossi, R. Spizzo, A.F. Fernandez, V. Davalos, A. Villanueva, G. Montoya, H. Yamamoto, S. Schwartz, et al. 2010. A genetic defect in exportin-5 traps precursor microRNAs in the nucleus of cancer cells. *Cancer Cell* **18**: 303–315.

Merritt, W.M., Y.G. Lin, L.Y. Han, A.A. Kamat, W.A. Spannuth, R. Schmandt, D. Urbauer, L.A. Pennacchio, J.F. Cheng, A.M. Nick, M.T. Deavers, A. Mourad-Zeidan, et al. 2008. Dicer, Drosha, and outcomes in patients with ovarian cancer. *N Engl J Med* **359**:2641–2650.

Narducci, M.G., D. Arcelli, M.C. Picchio, C. Lazzeri, E. Pagani, F. Sampogna, E. Scala, P. Fadda, C. Cristofoletti, A. Facchiano, M. Frontani, A. Monopoli, et al. 2011. MicroRNA profiling reveals that *miR-21*, miR486 and *miR-214* are upregulated and involved in cell survival in Sezary syndrome. *Cell Death Dis* **2**:e151.

O'Connell, R.M., D.S. Rao, A.A. Chaudhuri, and D. Baltimore. 2010. Physiological and pathological roles for microRNAs in the immune system. *Nat Rev Immunol* **10**:111–122.

O'Connell, R.M., D.S. Rao, A.A. Chaudhuri, M.P. Boldin, K.D. Taganov, J. Nicoll, R.L. Paquette, and D. Baltimore. 2008. Sustained expression of microRNA-155 in hematopoietic stem cells causes a myeloproliferative disorder. *J Exp Med* **205**:585–594.

Olsen, E.A., A.H. Rook, J. Zic, Y. Kim, P. Porcu, C. Querfeld, G. Wood, M.F. Demierre, M. Pittelkow, L.D. Wilson, L. Pinter-Brown, R. Advani, et al. 2011. Sezary syndrome: immunopathogenesis, literature review of therapeutic options, and recommendations for therapy by the United States Cutaneous Lymphoma Consortium (USCLC). *J Am Acad Dermatol* **64**:352–404.

Paulli, M., E. Berti, R. Rosso, E. Boveri, S. Kindl, C. Klersy, M. Lazzarino, G. Borroni, F. Menestrina, and M. Santucci. 1995. CD30/Ki-1-positive lymphoproliferative disorders of the skin–clinicopathologic correlation and statistical analysis of 86 cases: a multicentric study from the European Organization for Research and Treatment of Cancer Cutaneous Lymphoma Project Group. *J Clin Oncol* **13**:1343–1354.

Qin, Y., H.P. Buermans, M.S. van Kester, L. van der Fits, J.J. Out-Luiting, S. Osanto, R. Willemze, M.H. Vermeer, and C.P. Tensen. 2012. Deep-sequencing analysis reveals that the *miR-199a2/214* cluster within DNM3os represents the vast majority of aberrantly expressed microRNAs in Sezary syndrome. *J Invest Dermatol* **132**:1520–1522.

Ralfkiaer, U., P.H. Hagedorn, N. Bangsgaard, M.B. Lovendorf, C.B. Ahler, L. Svensson, K.L. Kopp, M.T. Vennegaard, B. Lauenborg, J.R. Zibert, T. Krejsgaard, C.M. Bonefeld, et al. 2011. Diagnostic microRNA profiling in cutaneous T-cell lymphoma (CTCL). *Blood* **118**:5891–5900.

Sézary, A.E. and Y. Bouvrain. 1938. Erythrodermie avec présence de cellules monstrueuses dans derme et dans sang circulant. *Bull Soc Fr Dermatol Syphiligr* **45**:254–260.

Shu, G.S., Z.L. Yang, and D.C. Liu. 2012. Immunohistochemical study of Dicer and Drosha expression in the benign and malignant lesions of gallbladder and their clinicopathological significances. *Pathol Res Pract* **208**:392–397.

Sugito, N., H. Ishiguro, Y. Kuwabara, M. Kimura, A. Mitsui, H. Kurehara, T. Ando, R. Mori, N. Takashima, R. Ogawa, and Y. Fujii. 2006. RNASEN regulates cell proliferation and affects survival in esophageal cancer patients. *Clin Cancer Res* **12**:7322–7328.

t Hoen, P.A., Y. Ariyurek, H.H. Thygesen, E. Vreugdenhil, R.H. Vossen, R.X. de Menezes, J.M. Boer, G.J. van Ommen, and J.T. den Dunnen. 2008. Deep sequencing-based expression analysis shows major advances in robustness, resolution and inter-lab portability over five microarray platforms. *Nucleic Acids Res* **36**:e141.

Trautinger, F., R. Knobler, R. Willemze, K. Peris, R. Stadler, L. Laroche, M. D'Incan, A. Ranki, N. Pimpinelli, P. Ortiz-Romero, R. Dummer, T. Estrach, et al. 2006. EORTC consensus recommendations for the treatment of mycosis fungoides/Sezary syndrome. *Eur J Cancer* **42**:1014–1030.

REFERENCES

Valencak, J., K. Schmid, F. Trautinger, W. Wallnöfer, L. Muellauer, A. Soleiman, R. Knobler, A. Haitel, H. Pehamberger, and M. Raderer. 2011. High expression of Dicer reveals a negative prognostic influence in certain subtypes of primary cutaneous T cell lymphomas. *J Dermatol Sci* **64**:185–190.

van der Fits, L., M.S. van Kester, Y.J. Qin, J.J. Out-Luiting, F. Smit, W.H. Zoutman, R. Willemze, C.P. Tensen, and M.H. Vermeer. 2011. MicroRNA-21 expression in CD4+T cells is regulated by STAT3 and is pathologically involved in Sezary syndrome. *J Invest Dermatol* **131**:762–768.

van der Fits, L., H.G. Rebel, J.J. Out-Luiting, S.M. Pouw, F. Smit, K.G. Vermeer, L. van Zijl, C.P. Tensen, K. Weijer, and M.H. Vermeer. 2012. A novel mouse model for Sézary syndrome using xenotransplantation of Sézary cells into immunodeficient RAG2(–/–) γc(–/–) mice. *Exp Dermatol* **21**(9):706–709.

van Doorn, R., M.S. van Kester, R. Dijkman, M.H. Vermeer, A.A. Mulder, K. Szuhai, J. Knijnenburg, J.M. Boer, R. Willemze, and C.P. Tensen. 2009. Oncogenomic analysis of mycosis fungoides reveals major differences with Sezary syndrome. *Blood* **113**:127–136.

van Doorn, R., C.W. Van Haselen, P.C. van Voost Vader, M.L. Geerts, F. Heule, M. de Rie, P.M. Steijlen, S.K. Dekker, W.A. van Vloten, and R. Willemze. 2000. Mycosis fungoides: disease evolution and prognosis of 309 Dutch patients. *Arch Dermatol* **136**:504–510.

van Kester, M.S., E. Ballabio, M.F. Benner, X.H. Chen, N.J. Saunders, L. van der Fits, R. van Doorn, M.H. Vermeer, R. Willemze, C.P. Tensen, and C.H. Lawrie. 2011. miRNA expression profiling of mycosis fungoides. *Mol Oncol* **5**:273–280.

van Kester, M.S., M.K. Borg, W.H. Zoutman, J.J. Out-Luiting, P.M. Jansen, E.J. Dreef, M.H. Vermeer, R. van Doorn, R. Willemze, and C.P. Tensen. 2012. A meta-analysis of gene expression data identifies a molecular signature characteristic for tumor-stage mycosis fungoides. *J Invest Dermatol* **132**:2050–2059.

Vermeer, M.H., R. van Doorn, R. Dijkman, X. Mao, S. Whittaker, V. van Voorst, M.J. Gerritsen, M.L. Geerts, S. Gellrich, O. Soderberg, K.J. Leuchowius, U. Landegren, et al. 2008. Novel and highly recurrent chromosomal alterations in Sezary syndrome. *Cancer Res* **68**:2689–2698.

Willemze, R. and R.C. Beljaards. 1993. Spectrum of primary cutaneous CD30 (Ki-1)-positive lymphoproliferative disorders. A proposal for classification and guidelines for management and treatment. *J Am Acad Dermatol* **28**:973–980.

Willemze, R., E.S. Jaffe, G. Burg, L. Cerroni, E. Berti, S.H. Swerdlow, E. Ralfkiaer, S. Chimenti, J.L. Diaz-Perez, L.M. Duncan, F. Grange, N.L. Harris, et al. 2005. WHO-EORTC classification for cutaneous lymphomas. *Blood* **105**:3768–3785.

Wu, D.Y., J. Tao, B. Xu, P.C. Li, Q. Lu, and W. Zhang. 2012. Downregulation of Dicer, a component of the microRNA machinery, in bladder cancer. *Mol Med Rep* **5**:695–699.

Yi, R., M.N. Poy, M. Stoffel, and E. Fuchs. 2008. A skin microRNA promotes differentiation by repressing "stemness." *Nature* **452**:225–229.

Zhang, L., J. Huang, N. Yang, J. Greshock, M.S. Megraw, A. Giannakakis, S. Liang, T.L. Naylor, A. Barchetti, M.R. Ward, G. Yao, A. Medina, et al. 2006. microRNAs exhibit high frequency genomic alterations in human cancer. *Proc Natl Acad Sci U S A* **103**:9136–9141.

PART IV

HEREDITARY AND OTHER NON-INFECTIOUS DISEASES

PART IV

HEREDITARY AND OTHER NON-INFECTIOUS DISEASES

28

MICRORNAs AND HEREDITARY DISORDERS

Matías Morín and Miguel A. Moreno-Pelayo

Unidad de Genética Molecular, Ramón y Cajal Institute of Health Research (IRYCIS) and Biomedical Network Research Centre on Rare Diseases (CIBERER), Madrid, Spain

I.	Introduction	465
II.	MicroRNAs and Hereditary Disorders	466
	A. Mutations Affecting the 3′-UTR of mRNAs	467
	B. Mutations in Genes Involved in miRNA Processing and Function	468
	C. Genomic Rearrangements Affecting the miRNAs Sequence	468
	D. Point Mutations in the miRNA Mature Sequence Resulting in Monogenic-Based Disorders: *miR-96* and *miR-184*	469
III.	Conclusion	473
	References	473

I. INTRODUCTION

For the last decade, all eyes have focused on microRNAs (miRNAs) as a new paradigm in the field of regulation of gene expression. These molecules are small non-coding RNAs that modulate the expression of hundreds of genes involved in a variety of biological processes (Nilsen 2007). miRNAs undergo sequential processing by RNase III from a primary transcript (pri-miRNA) to generate a precursor molecule of about 70 bp (pre-miRNA) with a stem-loop structure (Figure 28.1) that is exported to the cytoplasm where it is processed in a second step to obtain the functional mature sequence of about 20–25 bp. The miRNA functions as an adapter that allows the silencing inducer complex (miRISC) to recognize a particular mRNA and block translation. The sequence that determines the

---

MicroRNAs in Medicine, First Edition. Edited by Charles H. Lawrie.
© 2014 John Wiley & Sons, Inc. Published 2014 by John Wiley & Sons, Inc.

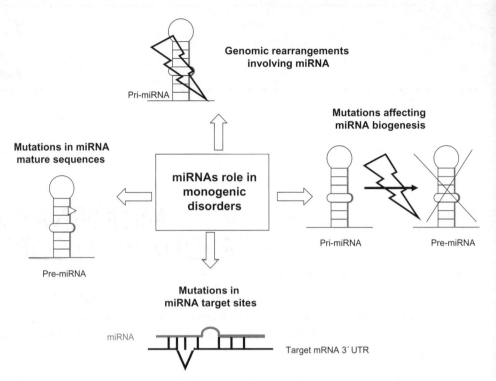

Figure 28.1. Schematic drawing indicating the potential pathogenic mechanisms of the main types of microRNA mutations associated with human genetic disorders.

specificity of the interaction between miRNA and its targets is called the seed region. This region encompasses seven bases from position 2 to 8 in the 5′ end of the mature sequence. The target sequences that recognize miRNAs are characterized by perfect complementarity with the seed region and are usually in the 3′-untranslated region (3′-UTR) of the mRNAs (Bartel 2009). The mechanisms of posttranscriptional repression mediated by miRNAs include direct inhibition of translation initiation, the nascent peptide degradation, or premature separation of the ribosome and the mRNA. It appears that the intensity of regulation will depend on the number of targets present in the mRNA (Kloosterman and Plasterk 2006). For further details on miRNA form and function, see Chapter 1 of this book.

II. MicroRNAs AND HEREDITARY DISORDERS

MiRNAs are implicated in a wide range of basic biological processes, including development, differentiation, apoptosis, and proliferation (Bartel 2004; Harfe 2005). Since the discovery of the strong impact of miRNAs on biological processes, it has been hypothesized that mutations affecting miRNA function may have a pathogenic role in human diseases. Previous studies have already shown that aberrant miRNA expression is implicated in most forms of human cancer (Bonci et al. 2008; Deng et al. 2008; Visone and Croce 2009), but fewer studies have established a clear link between miRNAs and human genetic disorders. Several arguments historically have risen against the hypothesis of miRNAs as genes responsible for human genetic diseases. The first one is the implication of the miRNAs in the regulation of basic cellular processes; hence, a significant alteration of their function would not be compatible with cell survival. The second argument emerged from the natural redundancy in miRNA action, so that the functional alteration of a single miRNA might not result in a significant perturbation of biological processes that ultimately lead to a diseased phenotype. In this review, we provide an overview of the evidence available to date that supports a pathogenic role for miRNAs in human genetic diseases, with a particular focus on monogenic disorders. For this aim, we have structured the chapter by identifying the main types of mutational mechanisms affecting miRNA function with a pathogenic role in human mendelian disorders (Figure 28.1).

A. Mutations Affecting the 3′-UTR of mRNAs

These mutations can lead to the removal or to the *de novo* generation of a target recognition site for a specific miRNA. In animal cells, most miRNAs form imperfect hybrids with sequences in the 3′-UTR, with the miRNA 5′-proximal "seed" region (positions 2–8) providing most of the pairing specificity (Eulalio et al. 2008; Filipowicz et al. 2008). It is conceivable that some sequence variations falling within the 3′ UTR of mRNA may alter miRNA recognition sites, either by altering functional miRNA target sites or by creating aberrant miRNA target sites.

One of the first animal disorders with a Mendelian transmission reported to be caused by dysregulation of a specific miRNA–mRNA target pair was the Texel sheep model. The Texel sheep phenotype is characterized by an inherited muscular hypertrophy that is more pronounced in the hindquarters of sheep (Clop et al. 2006). These authors demonstrated that the myostatin (GDF8) gene of Texel sheep is characterized by a G-to-A transition in the 3′-UTR that creates a target site for *miR-1* and *miR-206*, which are highly expressed in the skeletal muscle. This sequence change leads to a translational inhibition of the myostatin gene and, hence, is responsible for the muscular hypertrophy of Texel sheep.

There are now some examples of sequence variations in the 3′-UTR of mRNAs altering miRNA recognition sites that have been suggested to have a pathogenic role in human genetic diseases. The first was reported by Abelson et al. in 2005 (Abelson et al. 2005), who identified two independent occurrences of the identical sequence variant in the binding site for the miRNA *miR-189* (now termed *miR-24**) in the 3′-UTR of the *SLITRK1* mRNA in familial cases of Tourette's syndrome, a developmental neuropsychiatric disorder characterized by chronic vocal and motor tics. This 3′-UTR sequence variation in *SLITRK1* was proposed in order to determine an increased extent of repression of this gene by *miR-189* (*miR-24**). However, the lack of association between *SLITRK1* and Tourette syndrome has also been reported in other studies (Scharf et al. 2008).

The second example is represented by two different point mutations in the 3′-UTR of the *REEP1* gene, which have been associated with an autosomal dominant form of hereditary spastic paraplegia (SPG31) (Zuchner et al. 2006; Beetz et al. 2008). These mutations, which alter the sequence of a predicted target site for *miR-140*, were found to segregate with the disease phenotype and were not detected in a large set of human controls. These data strongly suggest the pathogenic role of the impaired *miR-140-REEP1* binding in some SPG31 families, although so far no functional data have been provided to consolidate this hypothesis.

Other studies have focused on deciphering the potential implications of sequence variations in the 3′-UTR of mRNAs in the pathogenesis of human diseases. Sequence variations creating or destroying putative miRNA target sites are abundant in the human genome and might be important effectors of phenotypic variation (Georges et al. 2007). A list of additional sequence variations altering putative miRNA recognition sites and with a potential role in human disease was revised, evidencing many cases lacking of true association between the presence of polymorphisms/mutations in miRNA target sites (poly-miRTSs) and human diseases (Sethupathy and Collins 2008).

B. Mutations in Genes Involved in miRNA Processing and Function

A number of different proteins are involved in the processing of miRNAs, hence mutations altering the function of these proteins are predicted to result in a global alteration of miRNA function. Complete loss-of-function mutations of certain key members of the miRNA processing pathway (such as Drosha and Dicer) are expected to be incompatible with life and, therefore, are not believed to play a role in the pathogenesis of human monogenic disorders. However, two human diseases characterized by mutations in genes involved in miRNA processing/activity are classically thought to be included in this category. The first example is represented by the Fragile X syndrome. The product of the *FRM1* gene, whose loss of function is responsible for this condition, is a selective RNA-binding protein. It has been proposed that the FMRP1 protein may function as a translational repressor of its mRNA targets at synapses by recruiting the RISC complex along with miRNAs and by facilitating the recognition between miRNAs and a specific subset of their mRNA targets. This interaction is suggested to be important in the process of synaptic plasticity, which, instead, is largely compromised in Fragile X syndrome patients (Jin et al. 2004; Li et al. 2008). The other example is related to the *DGCR8* gene, which maps to chromosomal region 22q11.2 and is commonly deleted in DiGeorge syndrome (Shiohama et al. 2003). This syndrome is characterized by cardiovascular defects, craniofacial defects, immunodeficiency, and neurobehavioral alterations. DGCR8 is a component of the Drosha complex, and its haploinsufficiency in DiGeorge syndrome patients might have a potential impact on miRNA processing (Wang et al. 2007). Neither of these studies provided experimental evidence to support a direct role of altered global miRNA processing in human hereditary disorders; however, a recent study has shed light on this matter.

A miRNA expression profile of human tumors has been characterized by an overall miRNA down-regulation. Explanations include a failure of miRNA posttranscriptional regulation, transcriptional silencing associated with hypermethylation of CpG island promoters, and miRNA transcriptional repression by oncogenic factors. Another possibility is that the enzymes and cofactors involved in miRNA processing pathways may themselves be targets of genetic disruption, further enhancing cellular transformation. Melo and colleagues reported truncating mutations in *TARBP2* (TAR RNA-binding protein 2), encoding an integral component of a DICER1-containing complex, in sporadic and hereditary carcinomas with microsatellite instability. The presence of *TARBP2* frameshift mutations causes diminished TRBP protein expression and a defect in the processing of miRNAs. The reintroduction of TRBP in the deficient cells restores the efficient production of miRNAs and inhibits tumor growth. Most important, the TRBP impairment is associated with a destabilization of the DICER1 protein. These results provide, for a subset of human tumors, an explanation for the observed defects in the expression of mature miRNAs (Melo et al. 2009).

C. Genomic Rearrangements Affecting the miRNA Sequence

Similar to protein-coding loci, miRNA loci can themselves be subjected to chromosomal rearrangements, leading to large mutations, such as deletions, insertions, translocations, or duplications. To date, however, there are no examples of such mutations that are clearly associated with human Mendelian diseases. However, a careful analysis of the genomic organization of miRNAs reveals that a number of intragenic miRNAs are localized within host genes whose mutations are responsible for human genetic disorders.

A comprehensive analysis of the mutation spectrum in reported disease genes in the Human Gene Mutation Database (HGMD) (Stenson et al. 2009) showed that some mutations do, indeed, significantly affect one or more miRNAs. This is the case, for instance, in certain intragenic deletions responsible for Duchenne muscular dystrophy (Den Dunnen et al. 1989; Winnard et al. 1993; Arikawa-Hirasawa et al. 1995), choroideraemia (van Bokhoven et al. 1994; Fujiki et al. 1999), and epidermolysis bullosa (Scheffer et al. 1997; Huber et al. 2002), among others (for detailed information, see Meola et al. 2009), which are also predicted to encompass some miRNAs. It is of crucial importance to confirm these predictions and to determine whether or not the deletion of these miRNAs is able to play a role in the phenotype observed. Furthermore, several miRNAs loci are also either deleted or duplicated in some well-known human aneuploidy syndromes, and there is initial evidence of their contribution to the pathogenic mechanisms of the complex manifestations of these disorders (Sethupathy et al. 2007).

D. Point Mutations in the miRNA Mature Sequence Resulting in Monogenic-Based Disorders: *miR-96* and *miR-184*

In contrast to the larger genomic rearrangements discussed earlier, there are clear examples of point mutations in the mature sequence of miRNAs that lead to distinct forms of monogenic diseases. Duan et al. (2007) described a single-nucleotide polymorphism (SNP) within the seed region of *miR-125a* that in addition to reducing miRNA-mediated translational suppression, significantly altered the processing from pri-miRNA to pre-miRNA. However, this SNP has not been associated with a disease status, although it is suggested that SNPs that reside within miRNA genes may, indeed, impair miRNA biogenesis and alter target selection (Duan et al. 2007).

Two years later, the first example of point mutations in the mature sequence of a miRNA with an aetiopathogenic role in a human Mendelian disease was reported as occurring in the autosomal dominant form of deafness, namely, DFNA50 (Mencia et al. 2009). Hereditary hearing impairment, especially the nonsyndromic forms in which the hearing deficit is not accompanied by other clinical signs, are the most frequent sensorineural traits that are characterized by a vast genetic and clinical heterogeneity. The autosomal dominant nonsyndromic hearing loss (ADNSHL) forms represent around 10–20% of the hereditary cases. To date more than 50 loci (DFNA) have been mapped associated with ADNSHL. DFNA50 was mapped to the 7q32 region after studying a family who had a progressive deafness of postlingual manifestation (it appears after acquiring language). In the critical interval, the *miR-96-182-183* cluster was identified. This family of miRNAs comprises a sensory tissue-specific miRNA cluster with an exceptional conservation of expression in ciliated neurosensory organs among both vertebrate and invertebrate organisms (Wienholds et al. 2005; Kloosterman et al. 2006; Weston et al. 2006; Xu et al. 2007; Pierce et al. 2008). In mouse, the *miR96-182-183* family was expressed in sensory hair cells and neurons of both the vestibule and the cochlea and are evolutionarily highly conserved (Weston et al. 2006), being detected in sensory cells of the eye, olfactory epithelium, neuromastomas, and ears in zebrafish (Wienholds et al. 2005). These characteristics made the miRNAs of the *miR-96-182-183* cluster excellent candidates for DFNA50-associated deafness. Sequencing of this cluster identified two different nucleotide substitutions in the seed region of the human *miR-96* in two Spanish families affected by DFNA50 hearing loss. In particular, both the mutations, *miR-96* (+13G > A) and (+14C > A), which were not present in several unrelated normal-hearing Spanish controls, were segregated in both of the families with a hearing impairment (Figure 28.2A). *miR-96*, together with *miR-182* and *miR-183*, is transcribed as a single polycistronic transcript and is reported to be expressed in the inner ear. For this reason, the authors also carried out a mutation screening of *miR-182* and *miR-183* in the same cohort of patients, tested for *miR-96*. However, they did not find any potential mutation. The fact that both of the earlier-mentioned families manifested the hearing loss postlingually indicated that probably neither of the two *miR-96* mutations resulted in impaired development of the inner ear. Instead, they could have had an impact on the regulatory role that *miR-96* plays in the hair cells of the adult cochlea, which maintain the gene expression profiles required for its normal function. *In vitro* experiments showed that both mutations impaired, but did not abrogate, the processing of *miR-96* to its mature form. Furthermore, a luciferase reporter assay confirmed that both mutations were able to affect the targeting of a subset of selected *miR-96* target genes, mostly expressed in the inner ear (*AQP5, CELSR2, MYRIP, ODF2,* and *RYK*). In contrast, no significant gain of function was associated with these two mutations, at least for the potentially new acquired *miR-96* targets investigated. In addition, after an ophthalmologic revision, no ocular phenotype was observed in individuals carrying mutations in *miR-96* (age range between 2 and 66 years), suggesting that its specific targets in the retina, a site in which *miR-96* is also strongly expressed, were not critical for its function or that the translation of these targets was not markedly affected (Mencia et al. 2009).

The finding of a single base change (A > T) in the seed region of *miR-96* in a mouse mutant (diminuendo) with a progressive hearing loss phenotype provided additional support to the finding that a single base change in *miR-96* is the causative mutation behind the hearing loss phenotype in both man and mouse (Lewis et al. 2009). In particular, the diminuendo mutant showed progressive hearing impairment in heterozygotes and profound deafness in homozygotes associated with hair cell defects. Lewis and colleagues suggested that the degeneration observed in homozygotes could be a consequence of a prior dysfunction of the hair cells. Bioinformatic analysis indicated that the mutation has a direct effect on the expression of many genes, including transcription factor genes, that are directly required for hair cell development and survival. The large number of genes whose expression is affected by *miR-96* suggests that the mechanism that explains the effects of the mutation may not be simple but, rather, may be the result of a combination of different small effects that act in concert to cause hair cell dysfunction (Lewis et al. 2009).

At this point, it was clear that the simple modification of a base in the region in *miR-96* seed was enough to cause deafness. The results obtained in human and mouse taken together indicate that the mutant *miR-96* molecules are less efficient silencing their direct targets and are able to silence new targets with complementary sites to the mutated *miR-96* seed regions. Therefore, an attractive hypothesis is that the existing phenotypic differences (age of onset and hearing profile) between the two families with DFNA50 deafness could be due to a different repertoire of targets deregulated, whose decipherment sheds light on the biological processes that are affected and helps in designing specific therapies to alleviate the progressive deterioration of hearing in individuals carrying these mutations.

More recently, another group identified one putative novel mutation within the *miR-96* gene in a family with autosomal dominant nonsyndromic hearing loss after screening 882 hearing-impaired patients and 836 normal-hearing Italian controls (Soldà et al. 2012). Interestingly, although located outside the mature *miR-96* sequence (Figure 28.2A), the

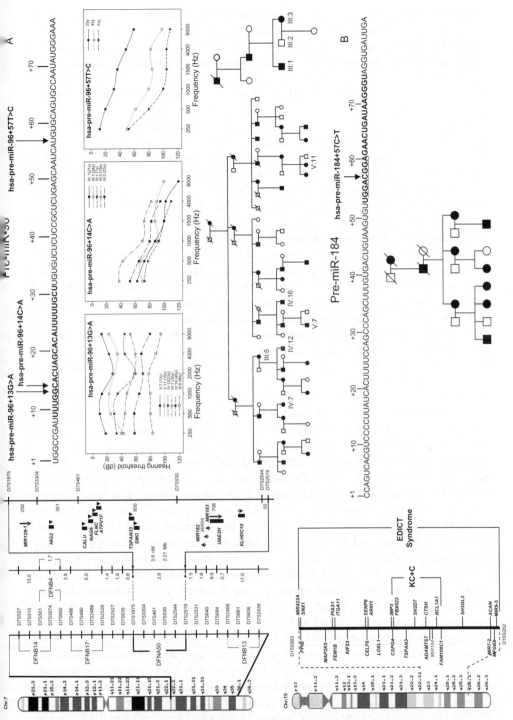

Figure 28.2. Monogenic disorders associated with mutations in the mature sequence of miRNAs. (A) Chromosomal localization of *miR-96* within the critical region of DFNA50 locus at 7q32. Description of the Spanish (*miR-96*+13G > A and +14C > A) and Italian (*miR-96*+57T > C) mutations in the precursor sequence of *miR96* and the corresponding audiograms. Pedigrees of the Spanish families are shown indicating the affected members from which the air conduction values were obtained. (B) Chromosomal localization of *miR-184* associated with the EDICT syndrome at 15q24-25. The mutation localization in the precursor sequence of *miR-184* and the genealogical tree are shown.

detected variant replaces a highly conserved nucleotide within the companion *miR-96**, and is predicted to reduce the stability of the pre-miRNA hairpin. The impact on *miR-96/mir-96** biogenesis caused by the mutation was evaluated by transient expression in mammalian cells and the amounts of mature *miR-96* gathered by real-time reverse-transcription polymerase chain reaction (PCR). These experiments revealed that both *miR-96* and *miR-96** levels were significantly reduced in the mutant, whereas the precursor levels were unaffected. Moreover, *miR-96* and *miR-96** expression levels could be restored by a compensatory mutation that reconstitutes the secondary structure of the pre-*miR-96* hairpin, demonstrating that the mutation hinders precursor processing, probably interfering with Dicer cleavage. In fact, cotransfection of the mutant pre-*miR-96* hairpin, in which the mature *miR-96* sequence is not altered, revealed that the identified mutation significantly impacts *miR-96* regulation of selected targets (*MYRIP*, *ACVR2B*, and *CACNB4*) as a result of reduced levels of *miR-96*. This work has provided definitively further evidence of the involvement of *miR-96* mutations in human deafness, demonstrating in this case that a quantitative defect of this miRNA may contribute to hearing loss.

The miRNA *miR-184* has also been recently implicated in two additional monogenic disorders: keratoconus associated with early-onset anterior polar cataracts and EDICT syndromes. Keratoconus is a noninflammatory ectatic corneal disorder characterized by progressive stromal thinning with an unaffected endothelium and Descemet membrane. EDICT syndrome is an autosomal dominant syndromal anterior segment dysgenesis characterized by endothelial dystrophy, iris hypoplasia, congenital cataract, and stromal thinning, but does not demonstrate a keratoconus phenotype (Akpek et al. 2002; Jun et al. 2002). The stromal thinning noted in the EDICT syndrome, however, is uniform and nonectatic. An autosomal dominant syndrome of anterior polar cataract and endothelial dystrophy consistent with EDICT syndrome was first described in a large Swedish family in 1951 (Dohlman 1951). The affected individuals developed anterior polar cataracts in early childhood and corneal guttae as early as age 10, with corneal disease progressing with age. The severity of the disease varied between family members: while the majority of affected individuals developed both corneal guttae and cataracts, one individual had guttae and no cataract. A similar disease phenotype was later reported in an American family of Scandinavian descent. Through at least five generations, this family inherited autosomal dominant anterior polar cataract with corneal guttae (Traboulsi and Weinberg 1989). Histologic examination of the cornea of the affected proband revealed epithelial edema, normal stroma, and thickening of the Descemet membrane with focal excrescences. A third instance of a dominant syndrome of cataract and corneal dystrophy was reported in a large Northern Irish family in 2003 (Hughes et al. 2003). Similar to the previous reports, affected members of this family developed anterior polar cataract in childhood; however, in contrast to previously described cases, the corneal dystrophy in this family was described as keratoconus rather than as cornea guttata.

Hughes and colleagues reported in 2011 a causal mutation (+57 C > T) in the seed region of *miR-184* implicated in a syndrome of keratoconus and early-onset anterior polar cataracts (Hughes et al. 2011). The locus for keratoconus with cataract was mapped to a 5.5-Mb region within the previous linkage interval for EDICT, suggesting a possible common causative mutation (Jun et al. 2002; Hughes et al. 2003; Dash et al. 2006) (Figure 28.2B). The common linkage interval harbors 75 annotated genes and 4 miRNAs. Interestingly, the same mutation has been recently implicated in a family with EDICT syndrome. In this family, the mutation segregates completely with the disease phenotype in the family, and the variant is absent in 28 nonhuman vertebrates and has not been identified in 1130 control chromosomes and in the 1000 Genomes database. *In silico* analyses suggest that the *miR-184*(+57 C > T) substitution alters the stability of pre-*miR-184* and thus may interfere with Dicer binding or cleavage. Computational analyses suggest several potential mechanisms for pathogenesis by the *miR-184*(+57 C > T) mutation. The substitution is predicted to strengthen binding at the 5′ end of the mature miRNA, a region that is critical for Dicer binding and RISC assembly (Khvorova et al. 2003; Krol et al. 2004). This prediction suggests that *miR-184*(+57 C > T) could reduce the expression of mature *miR-184* by interfering with Dicer cleavage, or that it could reduce the activity of mature *miR-184* by preventing assembly of the RISC. The 5′ binding affinity also plays an integral role in selection of pre-microRNA cleavage sites by Dicer, raising the possibility that *miR-184*(+57 C > T) alters Dicer cleavage, resulting in production of a dramatically different mature miRNA (Schwarz et al. 2003).

In silico structure modeling offers conflicting data regarding the thermodynamic stability of the *miR-184*(+57 C>T) variant. Although calculations of secondary structure suggest that *miR-184*(+57 C>T) stabilizes pre-*miR-184*, predictions of the tertiary structure show little or no difference in free energy. Thus, it seems unlikely that a change in overall thermodynamic stability of pre-*miR-184* is the main causal factor in the pathogenesis of EDICT syndrome.

In vitro, *miR-184* has been shown to directly inhibit Akt2 in cultured human glioma and neuroblastoma cell lines, and overexpression of *miR-184* suppresses cell viability and proliferation in those cell types (Chen and Stallings 2007; Foley et al. 2010; Malzkorn et al. 2010). The Akt pathway is involved in epithelial–mesenchymal transition, and defects in this process have been associated with Fuchs corneal dystrophy and posterior polymorphous dystrophy (Larue and Bellacosa 2005).

The findings that *miR-184* modulates the Akt signaling pathway and that the *miR-184*(+57 C > T) variant is less effective at regulating the upstream Akt regulator SHIP2 offer a promising possibility for the pathogenesis of the corneal disease in EDICT syndrome. Defects in epithelial–mesenchymal transition could represent a common disease mechanism among Fuchs corneal dystrophy, posterior polymorphous dystrophy, and the endothelial pathology observed in EDICT syndrome.

III. CONCLUSION

MiRNAs have emerged as a new paradigm in posttranscriptional gene regulation due to their ability to finely tune gene dosage. Their implication in the regulation of many cellular processes and their role in the proper differentiation and function of tissues and organs are progressively being uncovered, as well as their role in different pathologies, including those of genetic origin. Of key interest is the identification of monogenic forms associated with point mutations in the miRNAs mature sequences (i.e., *mi-R96* and *mi-R184*) or in the counterpart sites at the 3′-UTR regions of gene targets. This fact will likely be the base to initiate novel research areas focused on the development of promising and accurate therapies based on miRNAs in the oncoming years.

REFERENCES

Abelson, J.F., K.Y. Kwan, B.J. O'Roak, D.Y. Baek, A.A. Stillman, T.M. Morgan, C.A. Mathews, D.L. Pauls, M.R. Rasin, M. Gunel, et al. 2005. Sequence variants in SLITRK1 are associated with Tourette's syndrome. *Science* **310**:317–320.

Akpek, E.K., A.S. Jun, D.F. Goodman, W.R. Green, and J.D. Gottsch. 2002. Clinical and ultrastructural features of a novel hereditary anterior segment dysgenesis. *Ophthalmology* **109**: 513–519.

Arikawa-Hirasawa, E., R. Koga, T. Tsukahara, I. Nonaka, A. Mitsudome, K. Goto, A.H. Beggs, and K. Arahata. 1995. A severe muscular dystrophy patient with an internally deleted very short (110 kD) dystrophin: presence of the binding site for dystrophin-associated glycoprotein (DAG) may not be enough for physiological function of dystrophin. *Neuromuscul Disord* **5**(5):429–438.

Bartel, D.P. 2004. MicroRNAs: genomics, biogenesis, mechanism,and function. *Cell* **116**:281–297.

Bartel, D.P. 2009. MicroRNAs: target recognition and regulatory functions. *Cell* **136** 2:215–233.

Beetz, C., R. Schule, T. Deconinck, K.N. Tran-Viet, H. Zhu, B.P. Kremer, S.G. Frints, W.A. van Zelst-Stams, P. Byrne, S. Otto, et al. 2008. REEP1 mutation spectrum and genotype/phenotype correlation in hereditary spastic paraplegia type 31. *Brain* **131**:1078–1086.

Bonci, D., V. Coppola, M. Musumeci, A. Addario, R. Giuffrida, L. Memeo, L. D'Urso, A. Pagliuca, M. Biffoni, C. Labbaye, et al. 2008. The *miR-15a-miR-16-1* cluster controls prostate cancer by targeting multiple oncogenic activities. *Nat Med* **14**:1271–1277.

Casula, L., S. Murru, M. Pecorara, M.S. Ristaldi, G. Restagno, G. Mancuso, M. Morfini, R. De Biasi, F. Baudo, A. Carbonara, et al. 1990. Recurrent mutations and three novel rearrangements in the factor VIII gene of hemophilia A patients of Italian descent. *Blood* **75**:662–670.

Chen, Y. and R.L. Stallings. 2007. Differential patterns of microRNA expression in neuroblastoma are correlated with prognosis, differentiation, and apoptosis. *Cancer Res* **67**:976–983.

Clop, A., F. Marcq, H. Takeda, D. Pirottin, X. Tordoir, B. Bibe, J. Bouix, F. Caiment, J.M. Elsen, F. Eychenne, et al. 2006. A mutation creating a potential illegitimate microRNA target site in the myostatin gene affects muscularity in sheep. *Nat Genet* **38**:813–818.

Dash, D.P., G. Silvestri, and A.E. Hughes. 2006. Fine mapping of the keratoconus with cataract locus on chromosome 15q and candidate gene analysis. *Mol Vis* **12**:499–505.

Den Dunnen, J.T., P.M. Grootscholten, E. Bakker, L.A. Blonden, H.B. Ginjaar, M.C. Wapenaar, H.M. van Paassen, C. van Broeckhoven, P.L. Pearson, and G.J. van Ommen. 1989. Topography of the Duchenne muscular dystrophy (DMD) gene: FIGE and cDNA analysis of 194 cases reveals 115 deletions and 13 duplications. *Am J Hum Genet* **45**:835–847.

Deng, S., G.A. Calin, C.M. Croce, G. Coukos, and L. Zhang. 2008. Mechanisms of microRNA deregulation in human cancer. *Cell Cycle* **7**:2643–2646.

Dohlman, C.H. 1951. Familial congenital cornea guttata in association with anterior polar cataract. *Acta Ophthalmol* **29**:445–473.

Duan, R., C. Pak, and P. Jin. 2007. Single nucleotide polymorphism associated with mature *miR-125a* alters the processing of primiRNA. *Hum Mol Genet* **16**:1124–1131.

Eulalio, A., E. Huntzinger, and E. Izaurralde. 2008. Getting to the root of miRNA-mediated gene silencing. *Cell* **132**:9–14.

Filipowicz, W., S.N. Bhattacharyya, and N. Sonenberg. 2008. Mechanisms of post-transcriptional regulation by microRNAs: are the answers in sight? *Nat Rev Genet* **9**:102–114.

Foley, N., I. Bray, A. Tivnan, et al. 2010. MicroRNA-184 inhibits neuroblastoma cell survival through targeting the serine/threonine kinase AKT2. *Mol Cancer* **9**:83.

Fujiki, K., Y. Hotta, M. Hayakawa, A. Saito, Y. Mashima, M. Mori, M. Yoshii, A. Murakami, M. Matsumoto, S. Hayasaka, et al. 1999. REP-1 gene mutations in Japanese patients with choroideremia. *Graefes Arch Clin Exp Ophthalmol* **237**:735–740.

Georges, M., W. Coppieters, and C. Charlier. 2007. Polymorphic miRNAmediated gene regulation: contribution to phenotypic variation and disease. *Curr Opin Genet Dev* **17**:166–176.

Harfe, B.D. 2005. MicroRNAs in vertebrate development. *Curr Opin Genet Dev* **15**:410–415.

REFERENCES

Huber, M., M. Floeth, L. Borradori, H. Schacke, E.L. Rugg, E.B. Lane, E. Frenk, D. Hohl, and L. Bruckner-Tuderman. 2002. Deletion of the cytoplasmatic domain of BP180/collagen XVII causes a phenotype with predominant features of epidermolysis bullosa simplex. *J Invest Dermatol* **118**:185–192.

Hughes, A., D. Bradley, M. Campbell, et al. 2011. Mutation altering the *miR-184* seed region causes familial keratoconus with cataract. *Am J Hum Genet* **89**:628–633.

Hughes, A.E., D.P. Dash, A.J. Jackson, D.G. Frazer, and G. Silvestri. 2003. Familial keratoconus with cataract: linkage to the long arm of chromosome 15 and exclusion of candidate genes. *Invest Ophthalmol Vis Sci* **44**:5063–5066.

Jin, P., D.C. Zarnescu, S. Ceman, M. Nakamoto, J. Mowrey, T.A. Jongens, D.L. Nelson, K. Moses, and S.T. Warren. 2004. Biochemical and genetic interaction between the fragile X mental retardation protein and the microRNA pathway. *Nat Neurosci* **7**:113–117.

Jun, A.S., K.W. Broman, D.V. Do, E.K. Akpek, W.J. Stark, and J.D. Gottsch. 2002. Endothelial dystrophy, iris hypoplasia, congenital cataract, and stromal thinning (EDICT) syndrome maps to chromosome 15q22.1-q25.3. *Am J Ophthalmol* **134**:172–176.

Khvorova, A., A. Reynolds, and S.D. Jayasena. 2003. Functional siRNAs and miRNAs exhibit strand bias. *Cell* **115**:209–216.

Kloosterman, W.P. and R.H. Plasterk. 2006. The diverse functions of microRNAs in animal development and disease. *Dev Cell* **11** 4:441–450.

Kloosterman, W.P., E. Wienholds, E. de Bruijn, S. Kauppinen, and R.H. Plasterk. 2006. In situ detection of miRNAs in animal embryos using LNA-modified oligonucleotide probes. *Nat Methods* **3**:27–29.

Krol, J., K. Sobczak, U. Wilczynska, et al. 2004. Structural Features of MicroRNA (miRNA) precursors and their relevance to miRNA biogenesis and small interfering RNA/short hairpin RNA design. *J Biol Chem* **279**:42230–42239.

Larue, L. and A. Bellacosa. 2005. Epithelial-mesenchymal transition in development and cancer: role of phosphatidylinositol 3_ kinase/AKT pathways. *Oncogene* **24**:7443–7454.

Lewis, M.A., E. Quint, A.M. Glazier, H. Fuchs, M.H. De Angelis, C. Langford, S. van Dongen, C. Abreu-Goodger, M. Piipari, N. Redshaw, et al. 2009. An ENU-induced mutation of *miR-96* associated with progressive hearing loss in mice. *Nat Genet* **41**:614–618.

Li, Y., L. Lin, and P. Jin. 2008. The microRNA pathway and fragile X mental retardation protein. *Biochim Biophys Acta* **1779**:702–705.

Malzkorn, B., M. Wolter, F. Liesenberg, et al. 2010. Identification and functional characterization of microRNAs involved in the malignant progression of gliomas. *Brain Pathol* **20**: 539–550.

Melo, S.A., S. Ropero, C. Moutinho, L.A. Aaltonen, H. Yamamoto, G.A. Calin, S. Rossi, A.F. Fernandez, F. Carneiro, C. Oliveira, B. Ferreira, C.G. Liu, A. Villanueva, G. Capella, S. Schwartz, Jr., R. Shiekhattar, and M. Esteller. 2009. A TARBP2 mutation in human cancer impairs microRNA processing and DICER1 function. *Nat Genet* **41** 3:365–370. Erratum in: Nat Genet. 2010 42(5):464.

Mencia, A., S. Modamio-Hoybjor, N. Redshaw, M. Morin, F. Mayo-Merino, L. Olavarrieta, L.A. Aguirre, I. del Castillo, K.P. Steel, T. Dalmay, et al. 2009. Mutations in the seed region of human *miR-96* are responsable for nonsyndromic progressive hearing loss. *Nat Genet* **41**:609–613.

Meola, N., V.A. Gennarino, and S. Banfi. 2009. microRNAs and genetic diseases. *Pathogenetics* **2** 1:7.

Nilsen, T.W. 2007. Mechanisms of microRNA-mediated gene regulation in animal cells. *Trends Genet* **23** 5:243–249.

Pierce, M.L., et al. 2008. MicroRNA-183 family conservation and ciliated neurosensory organ expression. *Evol Dev* **10**:106–113.

Scharf, J.M., P. Moorjani, J. Fagerness, J.V. Platko, C. Illmann, B. Galloway, E. Jenike, S.E. Stewart, and D.L. Pauls. 2008. Lack of association between SLITRK1var321 and Tourette syndrome in a large family based sample. *Neurology* **70**:1495–1496.

Scheffer, H., R.P. Stulp, E. Verlind, M. van der Meulen, L. Bruckner-Tuderman, T. Gedde-Dahl, Jr., G.J. te Meerman, A. Sonnenberg, C.H. Buys, and M.F. Jonkman. 1997. Implications of intragenic marker homozygosity and haplotype sharing in a rare autosomal recessive disorder: the example of the collagen type XVII (COL17A1) locus in generalised atrophic benign epidermolysis bullosa. *Hum Genet* **100**:230–235.

Schwarz, D.S., G.R. Hutva'gner, T. Du, Z. Xu, N. Aronin, and P.D. Zamore. 2003. Asymmetry in the assembly of the RNAi enzyme complex. *Cell* **115**:199–208.

Sethupathy, P. and F.S. Collins. 2008. MicroRNA target site polymorphisms and human disease. *Trends Genet* **24**:489–497.

Sethupathy, P., C. Borel, M. Gagnebin, G.R. Grant, S. Deutsch, T.S. Elton, A.G. Hatzigeorgiou, and S.E. Antonarakis. 2007. Human microRNA-155 on chromosome 21 differentially interacts with its polymorphic target in the AGTR1 3′ untranslated region: a mechanism for functional single-nucleotide polymorphisms related to phenotypes. *Am J Hum Genet* **81**:405–413.

Shiohama, A., T. Sasaki, S. Noda, S. Minoshima, and N. Shimizu. 2003. Molecular cloning and expression analysis of a novel gene DGCR8 located in the DiGeorge syndrome chromosomal region. *Biochem Biophys Res Commun* **304**:184–190.

Soldà, G., M. Robusto, P. Primignani, P. Castorina, E. Benzoni, A. Cesarani, U. Ambrosetti, R. Asselta, and S. Duga. 2012. A novel mutation within the MIR96 gene causes non-syndromic inherited hearing loss in an Italian family by altering pre-miRNA processing. *Hum Mol Genet* **21** 3:577–585.

Stenson, P.D., M. Mort, E.V. Ball, K. Howells, A.D. Phillips, N.S. Thomas, and D.N. Cooper. 2009. The Human Gene Mutation Database: 2008 update. *Genome Med* **1**:13.

Traboulsi, E.I. and R.J. Weinberg. 1989. Familial congenital cornea guttata with anterior polar cataracts. *Am J Ophthalmol* **108**:123–125.

van Bokhoven, H., M. Schwartz, S. Andreasson, J.A. van den Hurk, L. Bogerd, M. Jay, K. Ruther, B. Jay, I.H. Pawlowitzki, E.M. Sankila, et al. 1994. Mutation spectrum in the CHM gene of Danish and Swedish choroideremia patients. *Hum Mol Genet* **3**:1047–1051.

Visone, R. and C.M. Croce. 2009. MiRNAs and cancer. *Am J Pathol* **174**:1131–1138.

Wang, Y., R. Medvid, C. Melton, R. Jaenisch, and R. Blelloch. 2007. DGCR8 is essential for microRNA biogenesis and silencing of embryonic stem cell self-renewal. *Nat Genet* **39**:380–385.

Weston, M.D., M.L. Pierce, S. Rocha-Sanchez, K.W. Beisel, and G.A. Soukup. 2006. MicroRNA gene expression in the mouse inner ear. *Brain Res* **1111**:95–104.

Wienholds, E., et al. 2005. MicroRNA expression in zebrafish embryonic development. *Science* **309**:310–311.

Winnard, A.V., C.J. Klein, D.D. Coovert, T. Prior, A. Papp, P. Snyder, D.E. Bulman, P.N. Ray, P. McAndrew, W. King, et al. 1993. Characterization of translational frame exception patients in Duchenne/Becker muscular dystrophy. *Hum Mol Genet* **2**:737–744.

Xu, S., P.D. Witmer, S. Lumayag, B. Kovacs, and D. Valle. 2007. MicroRNA (miRNA) transcriptome of mouse retina and identification of a sensory organ-specific miRNA cluster. *J Biol Chem* **282**:25053–25066.

Zuchner, S., G. Wang, K.N. Tran-Viet, M.A. Nance, P.C. Gaskell, J.M. Vance, A.E. Ashley-Koch, and M.A. Pericak-Vance. 2006. Mutations in the novel mitochondrial protein REEP1 cause hereditary spastic paraplegia type 31. *Am J Hum Genet* **79**:365–369.

29

MicroRNAs and Cardiovascular Diseases

Koh Ono

*Department of Cardiovascular Medicine, Graduate School of Medicine,
Kyoto University, Kyoto, Japan*

I.	Introduction	478
II.	Cardiac Hypertrophy	480
III.	Myocardial Ischemia and Cell Death	481
IV.	Cardiac Fibrosis	483
V.	Arrhythmia	484
VI.	Angiogenesis and Vascular Disease	484
VII.	Heart Failure	485
VIII.	Lipid Metabolism	486
IX.	Conclusion	487
	Acknowledgments	487
	References	487

ABBREVIATIONS

3′-UTR	3′-untranslated region
ABCA1	ATP-binding cassette transporter A1
AP1	activator protein 1
CAD	coronary artery disease
CHF	congestive heart failure
CTGF	connective tissue growth factor

MicroRNAs in Medicine, First Edition. Edited by Charles H. Lawrie.
© 2014 John Wiley & Sons, Inc. Published 2014 by John Wiley & Sons, Inc.

Cx43	connexin43
ECM	extracellular matrix
ERK-MAPK	extracellular signal-regulated kinase/mitogen-activated protein kinase
GJA1	gap junction protein a1
GLUT4	glucose transporter 4
HDL	high density lipoprotein
HERG	human ether-à-go-go-related gene
HIF-1α	hypoxia-inducible factor-1α
HSP20	heat shock protein 20
I/R	ischemia and reperfusion
IGF-1	insulin-like growth factor-1
KCNE1	potassium voltage-gated channel, Isk-related family, member
KCNH2	potassium voltage-gated channel, subfamily H, member 2
KCNJ2	potassium inwardly rectifying channel, subfamily J, member 2
KCNQ1	potassium voltage-gated channel, KQT-like subfamily, member 1
KLF15	Kruppel-like factor-15
MHC	myosin heavy chain
MI	myocardial infarction
MicroRNA	miRNA
MMP	matrix metalloprotease
Ncx1	sodium/calcium exchanger 1
PDCD4	programmed cell death 4
PE	phenylephrine
PTEN	phosphatase and tensin homologue
SCP1	small CTD phaphatase 1
SPRY1	Sprouty homologue 1
SREBP	sterol regulatory element binding protein
TAC	transverse aortic constriction
THRAP 1	thyroid hormone receptor-associated protein 1
TNF	tumor necrosis factor
TRPC3	transient receptor potential canonical 3
VCAM-1	vascular cell adhesion molecule 1
VLDL	very low density lipoprotein
VSMC	vascular smooth muscle cell

I. INTRODUCTION

MicroRNAs (miRNAs) are endogenous, single-stranded, small, ~22-nucleotide noncoding RNAs. miRNAs are generally regarded as negative regulators of gene expression by inhibiting translation and/or promoting mRNA degradation by base pairing to complementary sequences within the 3'-untranslated region (3'-UTR) of protein-coding mRNA transcripts (Bagga et al. 2005; Humphreys et al. 2005; Kiriakidou et al. 2007). However, there is evidence that some miRNA can stimulate gene expression in specific conditions.

Cardiovascular disease is the leading cause of morbidity and mortality in developed countries (Figure 29.1). Generally, pathological processes of the heart are associated with an altered expression profile of genes that are important for cardiac function (Figure 29.2) (Kairouz et al. 2012). The regulation of cardiac gene expression is complex, with

INTRODUCTION

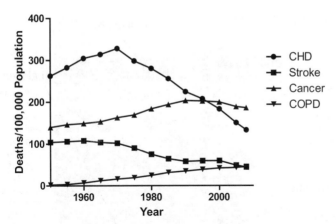

Figure 29.1. Unadjusted death rates for selected causes, United States, 1950–2008. Modified from the data from National Institute of Health (http://www.nhlbi.nih.gov/resources/docs/2012_ChartBook_508.pdf).

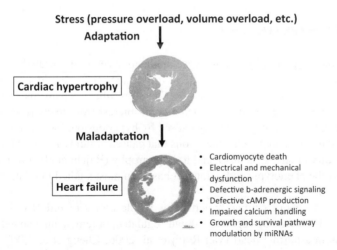

Figure 29.2. Pathological processes of the heart under stress. See color insert.

individual genes controlled by multiple transcription factors associate with their regulatory enhancer/promoter sequences to activate gene expression (Olson 2006). miRNAs have reshaped our view of how gene expression is regulated by adding another layer of regulation at the posttranscriptional level. Cardiovascular diseases encompass many pathologies, including cardiac hypertrophy, myocardial ischemia, cardiac fibrosis, arrhythmia, vascular diseases, and lipid metabolism, which will be discussed in more detail in the following sections (Figure 29.3).

The implications of miRNAs in cardiovascular pathology have been recognized only very recently, and research on miRNAs in relation to such diseases has now become a rapidly evolving field. In this review, we will summarize the current understanding of miRNA function in the pathogenesis of cardiovascular diseases.

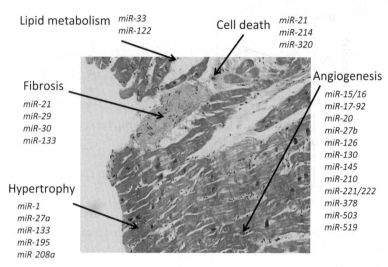

Figure 29.3. Dysregulated miRNAs in cardiovascular diseases. See color insert.

II. CARDIAC HYPERTROPHY

Because cardiac hypertrophy, an increase in heart size, is associated with nearly all forms of heart failure, it is of great clinical importance that we understand the mechanisms responsible for cardiac hypertrophy. Therefore, the mechanisms of regulation of hypertrophy-associated genes have attracted great interest from many researchers.

Pathological hypertrophy is mainly caused by hypertension, loss of myocytes following ischemic damage, and genetic alterations that cause cardiomyopathy. Moreover, metabolic abnormality or stress can also lead to hypertrophy (Rajabi et al. 2007). Pathological hypertrophy is the phenotypic end point that has been most studied in relation to miRNAs of the heart to date.

In animal models of cardiac hypertrophy, whole arrays of miRNAs have indicated that separate miRNAs are up-regulated, down-regulated, or remain unchanged with respect to their levels in a normal heart (van Rooij et al. 2006; Cheng et al. 2007; Ikeda et al. 2007; Sayed et al. 2007; Tatsuguchi et al. 2007; Thum et al. 2007; Chen et al. 2008; Chen and Gorski 2008). In these studies, some miRNAs have been more frequently reported as differentially expressed in the same direction than others, indicating the possibility that these miRNAs might have common roles in hypertrophy pathogenesis.

miR-1 was reported to target a cytoskeletal regulatory protein, twinfilin 1 (Twf1), that binds to actin monomers, preventing their assembly into filaments (Li et al. 2010). Down-regulation of *miR-1* induced by hypertrophic stimuli, such as transverse aortic constriction (TAC) or a-adrenergic stimulation with phenylephrine (PE), results in increased Twf1 expression, and overexpression of Twf1 is sufficient to induce cardiac hypertrophy. Another target of *miR-1* is insulin-like growth factor (IGF-1). Repression of *miR-1* and up-regulation of IGF-1 was also demonstrated in models of cardiac hypertrophy (Elia et al. 2009). It is also well known that acromegalic patients, in whom IGF-1 is induced, display cardiac hypertrophy.

miR-1 is encoded by two bicistronic clusters: *miR-133a-1/miR-1-2* and *miR-133a-2/miR-1-1*. Like *miR-1, miR-133* also has the potential to attenuate agonist-induced hypertrophy (Care et al. 2007; Matkovich et al. 2010), whereas repression of *miR-133* sensitizes the myocardium to excessive cardiac growth. Therefore, these clusters generate antagonizing effects on the stimulation of cardiac hypertrophy.

In contrast, *miR-195* was sufficient to drive pathological cardiac growth when overexpressed in neonatal cardiac myocytes and in transgenic mice (van Rooij et al. 2006). These results suggest that *miR-195* is a prohypertrophic factor that actively participates in the hypertrophic process; however, no direct targets of *miR-195* have been reported in the context of cardiac hypertrophy.

Interestingly, a family of so-called myomiRs was discovered that are encoded within the introns of the separate myosin heavy chain genes. *miR-208a*, *miR-208b*, and *miR-499* are located within the *Myh6*, *Myh7*, and *Myh7b* genes, respectively. It was reported that targets of *miR-208a* include thyroid hormone receptor (THR)-associated protein 1 (THRAP1) (van Rooij et al. 2007; Callis et al. 2009), suggesting that *miR-208a* initiates cardiomyocyte hypertrophy by regulating triiodothyronine-dependent repression of β-myosin heavy chain (MHC) expression. *miR-27a* also regulates β-MHC gene expression by targeting TRβ1 in cardiomyocytes (Nishi et al. 2011). Overexpression of *miR-208a* was sufficient to up-regulate Myh7 and to elicit cardiac hypertrophy, resulting in systolic dysfunction (Callis et al. 2009). Although *miR-208a* is required for cardiac hypertrophy, the role of the cotranscribed *miR-208b* in these pathological conditions remains to be elucidated. *miR-499* is encoded in an intron of the *myh7* gene and is considered likely to play a role in myosin gene regulation (van Rooij et al. 2009; Bell et al. 2010).

Intronic miRNAs have been shown to be critically involved in the regulation of their host genes as well as other genes, and thus important in regulating cellular events and pathological processes (Barik 2008; Ronchetti et al. 2008). Recently, it was shown that intronic *miR-26b* derived from intron 4 functioned as a negative regulator of its host gene small CTD phaphatase 1 (SCP1) in cardiomyocytes (Figure 29.4) (Sowa et al. 2012).

III. MYOCARDIAL ISCHEMIA AND CELL DEATH

A rapidly increasing number of studies have shown that cardiac and circulating miRNAs are markedly altered in myocardial ischemia or infarction (MI). These novel findings shed new light on the mechanisms that lead to MI complications, post-MI ventricular remodeling, and cardiac repair. Furthermore, recent studies show that circulating miRNAs may represent novel and sensitive biomarkers of MI and, possibly, also function as an intercellular signaling mechanism.

Cardiomyocyte death/apoptosis is a key cellular event in ischemic hearts. It was found that *miR-320* expression was consistently dysregulated in ischemic hearts (Ren et al. 2009). Ren et al. identified heat-shock protein 20 (HSP20), a known cardioprotective protein, as a target of for *miR-320*. Knockdown of endogenous *miR-320* provided protection against ischemia/reperfusion (I/R)-induced cardiomyocyte death and apoptosis through the up-regulation of HSP20. The miRNA expression signature in rat hearts at 6 hours after MI revealed that *miR-21* expression was significantly down-regulated in infracted areas but was up-regulated in boarder areas (Dong et al. 2009). Adenoviral transfer of *miR-21 in vivo* decreased cell apoptosis in the border and infracted areas through its target gene, programmed cell death 4 (PDCD4), and activator protein 1 (AP1) pathway.

Early reperfusion of ischemic heart remains the most effective intervention for improving clinical outcome following myocardial infarction. However, abnormal increases

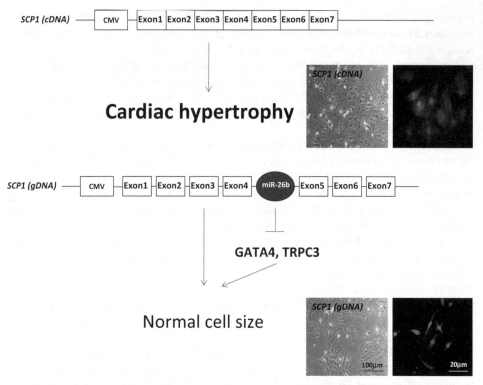

Figure 29.4. Scheme of the function of the intronic *miR-26b*. See color insert.

in intracellular Ca^{2+} during myocardial reperfusion can cause cardiomyocyte death, known as I/R injury. Cardiac I/R is also accompanied by dynamic changes in the expression of miRNAs; for example, *miR-214* is up-regulated during ischemic injury. Genetic deletion of *miR-214* in mice caused a loss of cardiac contractility, increased apoptosis, and excessive fibrosis in response to I/R injury (Aurora et al. 2012). The cardioprotective roles of *miR-214* during I/R injury were attributed to repression of the mRNA encoding sodium/calcium exchanger 1 (Ncx1), a key regulator of Ca^{2+} influx; and to repression of several downstream effectors of Ca^{2+} signaling that mediate cell death. These results suggest a pivotal role for *miR-214* as a regulator of cardiomyocyte Ca^{2+} homeostasis and survival during cardiac injury.

Recent studies have shown that some miRNAs are present in circulating blood and that they are included in exosomes and microparticles (Valadi et al. 2007; Hunter et al. 2008). Recently, circulating miRNAs have been reported in patients with MI (Wang et al. 2010; Kuwabara et al. 2011). From these results, it has been hypothesized that miRNAs in the systemic circulation may reflect tissue damage, and for this reason, they can be used as a biomarker of MI (Ai et al. 2010; Cheng et al. 2010; D'Alessandra et al. 2010). Moreover, plasma levels of endothelial cell-enriched miRNAs, such as *miR-126, miR-17*, and *miR-92a*; inflammation associated *miR-155*; and smooth muscle-enriched *miR-145* were reported to be significantly reduced in coronary artery disease (CAD) patients compared with healthy controls. These results also indicate that they can be used as biomarker candidates for CAD (Fichtlscherer et al. 2010). Therefore, the source and the mechanism for the change determine the set of miRNAs that can be used for myocardial ischemia/infarction.

IV. CARDIAC FIBROSIS

Cardiac fibrosis is an important contributor to the development of cardiac dysfunction in diverse pathological conditions, such as MI, ischemic, dilated, and hypertrophic cardiomyopathies, and heart failure, and can be defined as excessive accumulation of extracellular matrix (ECM) proteins in the heart (Rossi 1998; Swynghedauw 1999; Manabe et al. 2002; Brown et al. 2005; Khan and Sheppard 2006; Martos et al. 2007). Cardiac fibrosis leads to an increased mechanical stiffness, initially causing diastolic dysfunction, and eventually resulting in systolic dysfunction and overt heart failure. In addition, fibrosis can also disturb the electrical continuity between cardiomyocytes, leading to conduction slowing and hence an increase in the chance of arrhythmias. It is also possible that the enhanced diffusion distance for cardiac substrates and oxygen to the cardiac myocytes, caused by fibrosis, negatively influenced the myocardial balance between energy demand and supply (Manabe et al. 2002; Brown et al. 2005).

The *miR-29* family is composed of three members, *miR-29a, b,* and *c*, which differ only by one to two nucleotides. It was shown that the *miR-29* family, which is highly expressed in fibroblasts, targets mRNAs encoding a multitude of ECM-related proteins involved in fibrosis, including col1a1, col3a1, elastin, and fibrillin (van Rooij et al. 2008). *miR-29* is dramatically repressed in the border zone flanking the infracted area in a mouse model of MI. Down-regulation of *miR-29* would be predicted to counter the repression of these mRNAs and enhance the fibrotic responses. Therefore, it is tempting to speculate that up-regulation of *miR-29* may be a therapeutic option for MI.

miR-21 is expressed in all cell types of the cardiovascular system; its expression is most prominent in cardiac fibroblasts and rather weak in cadiomyocytes. Furthermore, *miR-21* is among the most strongly up-regulated miRNAs in response to a variety of forms of cardiac stress (van Rooij et al. 2006, 2008; Ichimura 2011). Thum et al. showed that *miR-21* is up-regulated in cardiac fibroblasts in the failing heart, where it represses the expression of Sprouty homolog1 (SPRY1), a negative regulator of the extracellular signal-regulated kinase/mitogen-activated protein kinase (ERK-MAPK) signaling pathway (Thum et al. 2008). Up-regulation of *miR-21* in response to cardiac injury was shown to enhance ERK-MAPK signaling, leading to fibroblast proliferation and fibrosis. Phosphatase and tensin homologue (PTEN) has also been demonstrated to be a direct target of *miR-21* in cardiac fibroblasts (Roy et al. 2009). Previous reports characterize PTEN as a suppressor of matrix metalloprotease-2 (MMP-2) expression (Park et al. 2002; Zheng et al. 2006). I/R in the heart induced *miR-21* in cardiac fibroblasts in the infracted region. Thus, I/R-induced *miR-21* limits PTEN function and causes activation of the Akt pathway and increased MMP-2 expression in cardiac fibroblasts.

Connective tissue growth factor (CTGF), a key molecule involved in fibrosis, was shown to be regulated by two miRNAs, *miR-133* and *miR-30*, which are both consistently down-regulated in several models of pathological hypertrophy and heart failure. (Duisters et al. 2009). The authors indicated that *miR-133* and *miR-30* are down-regulated during cardiac disease, which inversely correlated with the up-regulation of CTGF. *In vitro* experiments designed to overexpress or inhibit these miRNAs can effectively repress CTGF expression by interacting directly with the 3′-UTR region of the CTGF mRNA.

Together, these data indicate that miRNAs are important regulators of cardiac fibrosis and are involved in structural heart disease.

V. ARRHYTHMIA

Recently, it was established that gap junction protein a1 (GJA1), encoding connexin43 (Cx43), and potassium inwardly rectifying channel, subfamily J, member 2 (KCNJ2) (encoding the K+ channel subunit Kir2.1) are target genes for *miR-1* (Yang et al. 2007). Cx43 is critical for intercell conductance of excitation (Jongsma 2000; Lerner et al. 2000; Saffitz et al. 2000), and Kir2.1 governs cardiac membrane potential (Wang et al. 1998; Diaz et al. 2004), both of which are important determinants for cardiac excitability.

To date, the cardiac ion channel genes that have been confirmed experimentally to be targets of *miR-1* or *miR-133* include GJA1/Cx43/I_J (Yang et al. 2007), KCNJ2/Kir2.1/I_{K1} (Yang et al. 2007), potassium voltage-gated channel, subfamily H (eag-related) member 2 (KCNH2)/human ether-à-go-go-related gene (HERG)/I_{Kr} (Xiao et al. 2007), potassium voltage-gated channel, KQT-like subfamily, member 1 (KCNQ1)/KvLQT1/I_{Ks} (Luo et al. 2007), and potassium voltage-gated channel, Isk-related family, member 1 (KCNE1)/mink/I_{Ks} (Luo et al. 2007). The fact that altered expression of miRNAs can deregulate expression of cardiac ion channels provided novel insight into the molecular understanding of cardiac excitability.

It is also reported that *miR-328* was up-regulated in atria from dogs with induced atrial fibrillation and targets the L-type calcium channel (Lu et al. 2010). Strikingly, inhibition of *miR-328* levels with an antagomir reversed the conditions. The fact that genetic knockdown of endogenous *miR-328* reduced atrial fibrillation vulnerability also suggests the potential of *miR-328* as a target for atrial fibrillation treatment.

VI. ANGIOGENESIS AND VASCULAR DISEASES

Recently, a few specific miRNAs that regulate endothelial cell functions and angiogenesis have been described. Pro-angiogenic miRNAs include *let7f* and *miR-27b* (Kuehbacher et al. 2007), *miR-17-92* cluster (Dews et al. 2006), *miR-126* (Fish et al. 2008; Wang et al. 2008), *miR-130a* (Chen and Gorski 2008), *miR-210*, and *miR-378*, (Lee et al. 2007; Fasanaro et al. 2008). MiRNAs that exert antiangiogenic effects include *miR-15/16* (Cimmino et al. 2005; Hua et al. 2006), *miR-20a/b* (Hua et al. 2006), *miR-92a* (Bonauer et al. 2009), and *miR-221/222* (Poliseno et al. 2006; Suarez et al. 2007).

The Dimmeler group showed that the *miR-17-92* cluster is highly expressed in human endothelial cells and that *miR-92a* controlled the growth of new blood vessels (angiogenesis) (Bonauer et al. 2009). Forced overexpression of *miR-92a* in endothelial cells blocked angiogenesis, and systemic administration of an antagomir to inhibit *miR-92a* led to enhanced blood vessel growth and functional recovery of damaged tissue in mouse models of limb ischemia and MI. Therefore, *miR-92a* may serve as a valuable therapeutic target in the setting of ischemic disease. *miR-503* expression is enriched in endothelial cells isolated from ischemic hind limbs, especially in a diabetic setting (Caporali et al. 2011). Postischemic foot blood flow recovery was impaired in diabetic mice, and local injections of Ad decoy-*miR-503* completely normalized blood flow recovery in diabetic mice. *miR-126* is an endothelial cell-enriched abundant miRNA, and mechano-sensitive zinc finger transcription factor klf2a was shown to induce *miR-126* expression to activate vascular endothelial growth factor signaling (Nicoli et al. 2010). This work described a novel genetic mechanism in which a miRNA facilitates integration of a physiological stimulus with growth factor signaling in endothelial cells to guide angiogenesis. On the other hand, transfection of endothelial cells with an oligonucleotide that decreases *miR-126* permitted an increase in tumor necrosis factor (TNF)-α, stimulated vascular adhesion molecule 1 (VCAM-1) expression, and increased leukocyte adherence to endothelial cells (Harris et al. 2008). It is known that hypoxia-inducible factor-1α (HIF-1α), which is regulated by *miR-519c*, controls angiogenic events. Antagomir-mediated inhibition of *miR-519c* increased HIF-1a protein and enhanced angiogenic activities in tumors, and it may be utilized during myocardial ischemia (Cha et al. 2010).

Recently, Ji et al. revealed miRNAs that are aberrantly expressed in the vascular walls after balloon injury (Ji et al. 2007). Modulating an aberrantly overexpressed *miR-21* via antisense-mediated depletion had a significant negative effect on neointimal lesion formation. They also demonstrated that PTEN and BCL2 were involved in *miR-21-mediated* cellular effects. The same group also revealed that *miR-221* and *miR-222* expression levels were elevated in rat carotid arteries after angioplasty (Liu et al. 2009). Moreover, they found that p27 (Kip1) and p57 (Kip2) were target genes involved in *miR-221-* and *miR-222-mediated* effects on vascular smooth muscle cell (VSMC) growth. Knockdown of *miR-221* and *miR-222* resulted in decreased VSMC proliferation both *in vitro* and *in vivo*. *miR-145* is selectively expressed in VSMCs of the vascular wall, and its expression is significantly down-regulated in the vascular walls with neointimal lesion formation. The target of *miR-145* is KLF5 and its downstream signaling molecule, myocardin. Restoration of *miR-145* in balloon-injured arteries via Ad-*miR-145* inhibited neointimal growth and can be used as therapy of a variety of proliferative vascular diseases.

Aortic aneurysms are a common clinical condition that can cause death due to aortic dissection or rupture. The association between aortic aneurysm pathogenesis and altered TGF-β signaling, inflammation, and apoptosis has been the subjects of numerous investigations. Recently, a TGF-β-responsive *miR-29* (Maegdefessel et al. 2012b; Merk et al. 2012) and *miR-21* (Maegdefessel et al. 2012a), whose targets include PTEN, SPRY1, PDCD4, and BCL2 have been identified to play roles in cellular phenotypic modulation during aortic development. It was demonstrated that decreasing the levels of *miR-29b* or increasing the levels of *miR-21* in the aortic wall could attenuate aortic aneurysm progression in the porcine pancreatic elastase infusion and angiotensin II infusion model of abdominal aortic aneurysms in mice (Maegdefessel et al. 2012a, 2012b).

VII. HEART FAILURE

Because all of the previously described pathologies that is, cardiac hypertrophy, fibrosis, arrhythmia, and CAD, can cause heart failure, all of the miRNAs discussed so far are also relevant to this disease entity.

It is well known that heart failure is characterized by left ventricular remodeling and dilatation associated with activation of a fetal gene program triggering pathological changes in the myocardium associated with progressive dysfunction. Consistent with the reactivation of the fetal gene program during heart failure, an impressive similarity has been found between the miRNA expression pattern occurring in human failing hearts and that seen in the hearts of 12- to 14-week-old fetuses (Thum et al. 2007). Indeed, more than about 80% of the induced and repressed miRNAs were regulated in the same direction in fetal and failing heart tissue compared with the healthy adult control left ventricle tissue.

The most consistent changes were up-regulation of *miR-21, miR-29b, miR-129, miR-210, miR-211, miR-212*, and *miR-423*, with down-regulation of *miR-30, miR-182*, and *miR-526*. Interestingly, gene expression analysis revealed that most of the up-regulated genes were characterized by the presence of a significant number of the predicted binding sites for down-regulated miRNAs and *vice versa*.

Recently, many profiling studies have been conducted and revealed a large number of miRNAs that are differentially expressed in heart failure, pointing to the new mode of regulation of cardiovascular diseases (Bagga et al. 2005; Cheng et al. 2007; Ikeda et al. 2007, 2009; Tatsuguchi et al. 2007; Duisters et al. 2009). Horie et al. indicated that *miR-133* may fine-tune glucose transporter 4 (GLUT4) via targeting of kruppel-like factor-15 (KLF15) in heart failure (Horie et al. 2009), and there may be many other miRNA functions in specific disease settings. Nishi et al. suggested that four different miRNAs, which have the same seed sequence, regulate mitochondrial membrane potential during the transition from cardiac hypertrophy to failure (Nishi et al. 2010).

It has been proposed that miRNAs can exert their roles in response to treatment with chemotherapeutic agents. For example, it is suggested that the up-regulation of *miR-146a* after Dox treatment is involved in acute Dox-induced cardiotoxicity by targeting ErbB4 (Horie et al. 2010b). Inhibition of both ErbB2 and ErbB4 signaling may be one of the reasons why those patients who receive concurrent therapy with Dox and trastuzumab suffer from congestive heart failure (CHF).

VIII. LIPID METABOLISM

miR-122 is highly expressed in the liver, and it is estimated to account for approximately 70% of all liver miRNA (Orom et al. 2008). Silencing of *miR-122* in mice resulted in a sustained reduction in total plasma cholesterol, observed in both the LDL and high-density lipoprotein (HDL) fractions (Elmen et al. 2008b). Furthermore, *miR-122* antagonism in mice fed a high fat diet showed a significant improvement in liver steatosis, as evidenced by reductions in liver triglyceride content, and an increase in the rate of fatty acid β oxidation (Esau et al. 2006). Similar to the observation in mice, silencing *miR-122* in African green monkeys (Elmen et al. 2008a) and chimpanzees (Lanford et al. 2010) resulted in substantial reductions in total plasma cholesterol, ranging from 20% to 30% without apparent toxicities.

Recent reports have indicated that *miR-33* controls cholesterol homeostasis based on knockdown experiments using antisense technology (Marquart et al. 2010; Najafi-Shoushtari et al. 2010; Rayner et al. 2010). *miR-33-deficient* mice were generated and the critical role of *miR-33* in the regulation of ATP-binding cassette transporter A1 (ABCA1) expression, and HDL biosynthesis was confirmed *in vivo* (Horie et al. 2010a). It has already been reported that silencing of *miR-33a* resulted in a reduction of atherosclerosis in mice (Rayner et al. 2011b).

In humans, sterol regulatory element binding protein 1 (*SREBP1*) and *SREBP2* encode *miR-33b* and *miR-33a*, respectively (Najafi-Shoushtari et al. 2010). It is well known that hypertriglycemia in metabolic syndrome is caused by the insulin-induced increase in *SREBP1c* mRNA and protein levels (Kim et al. 1998; Chen et al. 2004). Low HDL often accompanies this situation and it is possible that the reduction in HDL is caused by a decrease in ABCA1, because of the increased production of *miR-33b* from the insulin-induced induction of *SREBP1c*. Although it is impossible to prove this in rodent models that lack *miR-33b*, antagonizing *miR-33a* and *-33b* could be a promising way to raise HDL

TABLE 29.1. Identified miRNAs of Therapeutic Interest

Target Diseases/Conditions	miRNAs to be Modulated
Cardiac hypertrophy	*miR-1*↑, *miR-133*↑, *miR-195*↓, *miR-208a*↓, *miR-26b*↑
Cell death	*miR-320*↓, *miR-214*↑
Fibrosis	*miR-29*↑, *miR-21*↓, *miR-133*↑, *miR-30*↑
Arrythmia	*miR-328*↓
Angiogenesis	*miR-92a*↓, *miR-503*↓, *miR-126*↑, *miR-519c*↓
Neointimal formation	*miR-21*↓, *miR-145*↑, *miR-221/222*↓
Aortic aneurysm	*miR-29b*↓, *miR-21*↑
Lipid metabolism	*miR-122*↓, *miR-33*↓

levels when the transcription of both *SREBPs* is up-regulated. Recently, it was shown that inhibition of *miR-33a* and *-33b* resulted in a rise in plasma HDL and reduction of very low-density lipoprotein (VLDL) triglycerides in non-human primates (Rayner et al. 2011a). Thus, a combination of silencing of endogenous *miR-33* and statins may be a useful therapeutic strategy for raising HDL and lowering LDL levels especially for metabolic syndrome patients.

IX. CONCLUSIONS

The biology of miRNAs in cardiovascular disease is a young research area and an emerging field. Identifying the gene targets and signaling pathways responsible for their cardiovascular effects is critical for future studies. Currently identified miRNAs of therapeutic interest are summarized in Table 29.1. Taken together, these recent pieces of evidence show that miRNAs play powerful roles in the cardiovascular system.

ACKNOWLEDGMENTS

This work was supported in part by a Grant-in-Aid for Scientific Research (23390211) from the Ministry of Education, Culture, Sports, Science and Technology of Japan and by Grant-in-Aid for Scientific Research on Innovative Areas (3307) from the Ministry of Education, Culture, Sports, Science and Technology of Japan to K. Ono.

REFERENCES

Ai, J., R. Zhang, Y. Li, J. Pu, Y. Lu, J. Jiao, K. Li, B. Yu, Z. Li, R. Wang, L. Wang, Q. Li, et al. 2010. Circulating microRNA-1 as a potential novel biomarker for acute myocardial infarction. *Biochem Biophys Res Commun* **391**:73–77.

Aurora, A.B., A.I. Mahmoud, X. Luo, B.A. Johnson, E. van Rooij, S. Matsuzaki, K.M. Humphries, J.A. Hill, R. Bassel-Duby, H.A. Sadek, and E.N. Olson. 2012. MicroRNA-214 protects the mouse heart from ischemic injury by controlling Ca(2)(+) overload and cell death. *J Clin Invest* **122**:1222–1232.

Bagga, S., J. Bracht, S. Hunter, K. Massirer, J. Holtz, R. Eachus, and A.E. Pasquinelli. 2005. Regulation by *let-7* and lin-4 miRNAs results in target mRNA degradation. *Cell* **122**:553–563.

Barik, S. 2008. An intronic microRNA silences genes that are functionally antagonistic to its host gene. *Nucleic Acids Res* **36**:5232–5241.

Bell, M.L., M. Buvoli, and L.A. Leinwand. 2010. Uncoupling of expression of an intronic microRNA and its myosin host gene by exon skipping. *Mol Cell Biol* **30**:1937–1945.

Bonauer, A., G. Carmona, M. Iwasaki, M. Mione, M. Koyanagi, A. Fischer, J. Burchfield, H. Fox, C. Doebele, K. Ohtani, E. Chavakis, M. Potente, et al. 2009. MicroRNA-92a controls angiogenesis and functional recovery of ischemic tissues in mice. *Science* **324**:1710–1713.

Brown, R.D., S.K. Ambler, M.D. Mitchell, and C.S. Long. 2005. The cardiac fibroblast: therapeutic target in myocardial remodeling and failure. *Annu Rev Pharmacol Toxicol* **45**:657–687.

Callis, T.E., K. Pandya, H.Y. Seok, R.H. Tang, M. Tatsuguchi, Z.P. Huang, J.F. Chen, Z. Deng, B. Gunn, J. Shumate, M.S. Willis, C.H. Selzman, et al. 2009. MicroRNA-208a is a regulator of cardiac hypertrophy and conduction in mice. *J Clin Invest* **119**:2772–2786.

Caporali, A., M. Meloni, C. Vollenkle, D. Bonci, G.B. Sala-Newby, R. Addis, G. Spinetti, S. Losa, R. Masson, A.H. Baker, R. Agami, C. le Sage, et al. 2011. Deregulation of microRNA-503 contributes to diabetes mellitus-induced impairment of endothelial function and reparative angiogenesis after limb ischemia. *Circulation* **123**:282–291.

Care, A., D. Catalucci, F. Felicetti, D. Bonci, A. Addario, P. Gallo, M.L. Bang, P. Segnalini, Y. Gu, N.D. Dalton, L. Elia, M.V. Latronico, et al. 2007. MicroRNA-133 controls cardiac hypertrophy. *Nat Med* **13**:613–618.

Cha, S.T., P.S. Chen, G. Johansson, C.Y. Chu, M.Y. Wang, Y.M. Jeng, S.L. Yu, J.S. Chen, K.J. Chang, S.H. Jee, C.T. Tan, M.T. Lin, et al. 2010. MicroRNA-519c suppresses hypoxia-inducible factor-1alpha expression and tumor angiogenesis. *Cancer Res* **70**:2675–2685.

Chen, G., G. Liang, J. Ou, J.L. Goldstein, and M.S. Brown. 2004. Central role for liver X receptor in insulin-mediated activation of Srebp-1c transcription and stimulation of fatty acid synthesis in liver. *Proc Natl Acad Sci U S A* **101**:11245–11250.

Chen, J.F., E.P. Murchison, R. Tang, T.E. Callis, M. Tatsuguchi, Z. Deng, M. Rojas, S.M. Hammond, M.D. Schneider, C.H. Selzman, G. Meissner, C. Patterson, et al. 2008. Targeted deletion of Dicer in the heart leads to dilated cardiomyopathy and heart failure. *Proc Natl Acad Sci U S A* **105**:2111–2116.

Chen, Y. and D.H. Gorski. 2008. Regulation of angiogenesis through a microRNA (*miR-130a*) that down-regulates antiangiogenic homeobox genes GAX and HOXA5. *Blood* **111**:1217–1226.

Cheng, Y., R. Ji, J. Yue, J. Yang, X. Liu, H. Chen, D.B. Dean, and C. Zhang. 2007. MicroRNAs are aberrantly expressed in hypertrophic heart: do they play a role in cardiac hypertrophy? *Am J Pathol* **170**:1831–1840.

Cheng, Y., N. Tan, J. Yang, X. Liu, X. Cao, P. He, X. Dong, S. Qin, and C. Zhang. 2010. A translational study of circulating cell-free microRNA-1 in acute myocardial infarction. *Clin Sci (Lond)* **119**:87–95.

Cimmino, A., G.A. Calin, M. Fabbri, M.V. Iorio, M. Ferracin, M. Shimizu, S.E. Wojcik, R.I. Aqeilan, S. Zupo, M. Dono, L. Rassenti, H. Alder, et al. 2005. *miR-15* and *miR-16* induce apoptosis by targeting BCL2. *Proc Natl Acad Sci U S A* **102**:13944–13949.

D'Alessandra, Y., P. Devanna, F. Limana, S. Straino, A. Di Carlo, P.G. Brambilla, M. Rubino, M.C. Carena, L. Spazzafumo, M. De Simone, B. Micheli, P. Biglioli, et al. 2010. Circulating microRNAs are new and sensitive biomarkers of myocardial infarction. *Eur Heart J* **31**:2765–2773.

Dews, M., A. Homayouni, D. Yu, D. Murphy, C. Sevignani, E. Wentzel, E.E. Furth, W.M. Lee, G.H. Enders, J.T. Mendell, and A. Thomas-Tikhonenko. 2006. Augmentation of tumor angiogenesis by a Myc-activated microRNA cluster. *Nat Genet* **38**:1060–1065.

Diaz, R.J., C. Zobel, H.C. Cho, M. Batthish, A. Hinek, P.H. Backx, and G.J. Wilson. 2004. Selective inhibition of inward rectifier K+ channels (Kir2.1 or Kir2.2) abolishes protection by ischemic preconditioning in rabbit ventricular cardiomyocytes. *Circ Res* **95**:325–332.

REFERENCES

Dong, S., Y. Cheng, J. Yang, J. Li, X. Liu, X. Wang, D. Wang, T.J. Krall, E.S. Delphin, and C. Zhang. 2009. MicroRNA expression signature and the role of microRNA-21 in the early phase of acute myocardial infarction. *J Biol Chem* **284**:29514–29525.

Duisters, R.F., A.J. Tijsen, B. Schroen, J.J. Leenders, V. Lentink, I. van der Made, V. Herias, R.E. van Leeuwen, M.W. Schellings, P. Barenbrug, J.G. Maessen, S. Heymans, et al. 2009. *miR-133* and *miR-30* regulate connective tissue growth factor: implications for a role of microRNAs in myocardial matrix remodeling. *Circ Res* **104**:170–178, 176p following 178.

Elia, L., R. Contu, M. Quintavalle, F. Varrone, C. Chimenti, M.A. Russo, V. Cimino, L. De Marinis, A. Frustaci, D. Catalucci, and G. Condorelli. 2009. Reciprocal regulation of microRNA-1 and insulin-like growth factor-1 signal transduction cascade in cardiac and skeletal muscle in physiological and pathological conditions. *Circulation* **120**:2377–2385.

Elmen, J., M. Lindow, S. Schutz, M. Lawrence, A. Petri, S. Obad, M. Lindholm, M. Hedtjarn, H.F. Hansen, U. Berger, S. Gullans, P. Kearney, et al. 2008a. LNA-mediated microRNA silencing in non-human primates. *Nature* **452**:896–899.

Elmen, J., M. Lindow, A. Silahtaroglu, M. Bak, M. Christensen, A. Lind-Thomsen, M. Hedtjarn, J.B. Hansen, H.F. Hansen, E.M. Straarup, K. McCullagh, P. Kearney, et al. 2008b. Antagonism of microRNA-122 in mice by systemically administered LNA-antimiR leads to up-regulation of a large set of predicted target mRNAs in the liver. *Nucleic Acids Res* **36**:1153–1162.

Esau, C., S. Davis, S.F. Murray, X.X. Yu, S.K. Pandey, M. Pear, L. Watts, S.L. Booten, M. Graham, R. McKay, A. Subramaniam, S. Propp, et al. 2006. *miR-122* regulation of lipid metabolism revealed by in vivo antisense targeting. *Cell Metab* **3**:87–98.

Fasanaro, P., Y. D'Alessandra, V. Di Stefano, R. Melchionna, S. Romani, G. Pompilio, M.C. Capogrossi, and F. Martelli. 2008. MicroRNA-210 modulates endothelial cell response to hypoxia and inhibits the receptor tyrosine kinase ligand Ephrin-A3. *J Biol Chem* **283**:15878–15883.

Fichtlscherer, S., S. De Rosa, H. Fox, T. Schwietz, A. Fischer, C. Liebetrau, M. Weber, C.W. Hamm, T. Roxe, M. Muller-Ardogan, A. Bonauer, A.M. Zeiher, et al. 2010. Circulating microRNAs in patients with coronary artery disease. *Circ Res* **107**:677–684.

Fish, J.E., M.M. Santoro, S.U. Morton, S. Yu, R.F. Yeh, J.D. Wythe, K.N. Ivey, B.G. Bruneau, D.Y. Stainier, and D. Srivastava. 2008. *miR-126* regulates angiogenic signaling and vascular integrity. *Dev Cell* **15**:272–284.

Harris, T.A., M. Yamakuchi, M. Ferlito, J.T. Mendell, and C.J. Lowenstein. 2008. MicroRNA-126 regulates endothelial expression of vascular cell adhesion molecule 1. *Proc Natl Acad Sci U S A* **105**:1516–1521.

Horie, T., K. Ono, M. Horiguchi, H. Nishi, T. Nakamura, K. Nagao, M. Kinoshita, Y. Kuwabara, H. Marusawa, Y. Iwanaga, K. Hasegawa, M. Yokode, et al. 2010a. MicroRNA-33 encoded by an intron of sterol regulatory element-binding protein 2 (Srebp2) regulates HDL in vivo. *Proc Natl Acad Sci U S A* **107**:17321–17326.

Horie, T., K. Ono, H. Nishi, Y. Iwanaga, K. Nagao, M. Kinoshita, Y. Kuwabara, R. Takanabe, K. Hasegawa, T. Kita, and T. Kimura. 2009. MicroRNA-133 regulates the expression of GLUT4 by targeting KLF15 and is involved in metabolic control in cardiac myocytes. *Biochem Biophys Res Commun* **389**:315–320.

Horie, T., K. Ono, H. Nishi, K. Nagao, M. Kinoshita, S. Watanabe, Y. Kuwabara, Y. Nakashima, R. Takanabe-Mori, E. Nishi, K. Hasegawa, T. Kita, et al. 2010b. Acute doxorubicin cardiotoxicity is associated with *miR-146a-induced* inhibition of the neuregulin-ErbB pathway. *Cardiovasc Res* **87**:656–664.

Hua, Z., Q. Lv, W. Ye, C.K. Wong, G. Cai, D. Gu, Y. Ji, C. Zhao, J. Wang, B.B. Yang, and Y. Zhang. 2006. MiRNA-directed regulation of VEGF and other angiogenic factors under hypoxia. *PLoS ONE* **1**:e116.

Humphreys, D.T., B.J. Westman, D.I. Martin, and T. Preiss. 2005. MicroRNAs control translation initiation by inhibiting eukaryotic initiation factor 4E/cap and poly(A) tail function. *Proc Natl Acad Sci U S A* **102**:16961–16966.

Hunter, M.P., N. Ismail, X. Zhang, B.D. Aguda, E.J. Lee, L. Yu, T. Xiao, J. Schafer, M.L. Lee, T.D. Schmittgen, S.P. Nana-Sinkam, D. Jarjoura, et al. 2008. Detection of microRNA expression in human peripheral blood microvesicles. *PLoS ONE* **3**:e3694.

Ichimura, A. 2011. miRNAs and regulation of cell signaling. *FEBS J* **278**:1610–1618.

Ikeda, S., A. He, S.W. Kong, J. Lu, R. Bejar, N. Bodyak, K.H. Lee, Q. Ma, P.M. Kang, T.R. Golub, and W.T. Pu. 2009. MicroRNA-1 negatively regulates expression of the hypertrophy-associated calmodulin and Mef2a genes. *Mol Cell Biol* **29**:2193–2204.

Ikeda, S., S.W. Kong, J. Lu, E. Bisping, H. Zhang, P.D. Allen, T.R. Golub, B. Pieske, and W.T. Pu. 2007. Altered microRNA expression in human heart disease. *Physiol Genomics* **31**:367–373.

Ji, R., Y. Cheng, J. Yue, J. Yang, X. Liu, H. Chen, D.B. Dean, and C. Zhang. 2007. MicroRNA expression signature and antisense-mediated depletion reveal an essential role of microRNA in vascular neointimal lesion formation. *Circ Res* **100**:1579–1588.

Jongsma, H.J. 2000. Diversity of gap junctional proteins: does it play a role in cardiac excitation? *J Cardiovasc Electrophysiol* **11**:228–230.

Kairouz, V., L. Lipskaia, R.J. Hajjar, and E.R. Chemaly. 2012. Molecular targets in heart failure gene therapy: current controversies and translational perspectives. *Ann N Y Acad Sci* **1254**: 42–50.

Khan, R. and R. Sheppard. 2006. Fibrosis in heart disease: understanding the role of transforming growth factor-beta in cardiomyopathy, valvular disease and arrhythmia. *Immunology* **118**: 10–24.

Kim, J.B., P. Sarraf, M. Wright, K.M. Yao, E. Mueller, G. Solanes, B.B. Lowell, and B.M. Spiegelman. 1998. Nutritional and insulin regulation of fatty acid synthetase and leptin gene expression through ADD1/SREBP1. *J Clin Invest* **101**:1–9.

Kiriakidou, M., G.S. Tan, S. Lamprinaki, M. De Planell-Saguer, P.T. Nelson, and Z. Mourelatos. 2007. An mRNA m7G cap binding-like motif within human Ago2 represses translation. *Cell* **129**:1141–1151.

Kuehbacher, A., C. Urbich, A.M. Zeiher, and S. Dimmeler. 2007. Role of Dicer and Drosha for endothelial microRNA expression and angiogenesis. *Circ Res* **101**:59–68.

Kuwabara, Y., K. Ono, T. Horie, H. Nishi, K. Nagao, M. Kinoshita, S. Watanabe, O. Baba, Y. Kojima, S. Shizuta, M. Imai, T. Tamura, et al. 2011. Increased microRNA-1 and microRNA-133a levels in serum of patients with cardiovascular disease indicate myocardial damage. *Circ Cardiovasc Genet* **4**:446–454.

Lanford, R.E., E.S. Hildebrandt-Eriksen, A. Petri, R. Persson, M. Lindow, M.E. Munk, S. Kauppinen, and H. Orum. 2010. Therapeutic silencing of microRNA-122 in primates with chronic hepatitis C virus infection. *Science* **327**:198–201.

Lee, D.Y., Z. Deng, C.H. Wang, and B.B. Yang. 2007. MicroRNA-378 promotes cell survival, tumor growth, and angiogenesis by targeting SuFu and Fus-1 expression. *Proc Natl Acad Sci U S A* **104**:20350–20355.

Lerner, D.L., K.A. Yamada, R.B. Schuessler, and J.E. Saffitz. 2000. Accelerated onset and increased incidence of ventricular arrhythmias induced by ischemia in Cx43-deficient mice. *Circulation* **101**:547–552.

Li, Q., X.W. Song, J. Zou, G.K. Wang, E. Kremneva, X.Q. Li, N. Zhu, T. Sun, P. Lappalainen, W.J. Yuan, Y.W. Qin, and Q. Jing. 2010. Attenuation of microRNA-1 derepresses the cytoskeleton regulatory protein twinfilin-1 to provoke cardiac hypertrophy. *J Cell Sci* **123**:2444–2452.

Liu, X., Y. Cheng, S. Zhang, Y. Lin, J. Yang, and C. Zhang. 2009. A necessary role of *miR-221* and *miR-222* in vascular smooth muscle cell proliferation and neointimal hyperplasia. *Circ Res* **104**:476–487.

Lu, Y., Y. Zhang, N. Wang, Z. Pan, X. Gao, F. Zhang, H. Shan, X. Luo, Y. Bai, L. Sun, W. Song, C. Xu, et al. 2010. MicroRNA-328 contributes to adverse electrical remodeling in atrial fibrillation. *Circulation* **122**:2378–2387.

Luo, X., J. Xiao, H. Lin, B. Li, Y. Lu, B. Yang, and Z. Wang. 2007. Transcriptional activation by stimulating protein 1 and post-transcriptional repression by muscle-specific microRNAs of IKs-encoding genes and potential implications in regional heterogeneity of their expressions. *J Cell Physiol* **212**:358–367.

Maegdefessel, L., J. Azuma, R. Toh, A. Deng, D.R. Merk, A. Raiesdana, N.J. Leeper, U. Raaz, A.M. Schoelmerich, M.V. McConnell, R.L. Dalman, J.M. Spin, et al. 2012a. MicroRNA-21 blocks abdominal aortic aneurysm development and nicotine-augmented expansion. *Sci Transl Med* **4**:122ra122.

Maegdefessel, L., J. Azuma, R. Toh, D.R. Merk, A. Deng, J.T. Chin, U. Raaz, A.M. Schoelmerich, A. Raiesdana, N.J. Leeper, M.V. McConnell, R.L. Dalman, et al. 2012b. Inhibition of microRNA-29b reduces murine abdominal aortic aneurysm development. *J Clin Invest* **122**:497–506.

Manabe, I., T. Shindo, and R. Nagai. 2002. Gene expression in fibroblasts and fibrosis: involvement in cardiac hypertrophy. *Circ Res* **91**:1103–1113.

Marquart, T.J., R.M. Allen, D.S. Ory, and A. Baldan. 2010. *miR-33* links SREBP-2 induction to repression of sterol transporters. *Proc Natl Acad Sci U S A* **107**:12228–12232.

Martos, R., J. Baugh, M. Ledwidge, C. O'Loughlin, C. Conlon, A. Patle, S.C. Donnelly, and K. McDonald. 2007. Diastolic heart failure: evidence of increased myocardial collagen turnover linked to diastolic dysfunction. *Circulation* **115**:888–895.

Matkovich, S.J., W. Wang, Y. Tu, W.H. Eschenbacher, L.E. Dorn, G. Condorelli, A. Diwan, J.M. Nerbonne, and G.W. Dorn, 2nd. 2010. MicroRNA-133a protects against myocardial fibrosis and modulates electrical repolarization without affecting hypertrophy in pressure-overloaded adult hearts. *Circ Res* **106**:166–175.

Merk, D.R., J.T. Chin, B.A. Dake, L. Maegdefessel, M.O. Miller, N. Kimura, P.S. Tsao, C. Iosef, G.J. Berry, F.W. Mohr, J.M. Spin, C.M. Alvira, et al. 2012. *miR-29b* participates in early aneurysm development in Marfan syndrome. *Circ Res* **110**:312–324.

Najafi-Shoushtari, S.H., F. Kristo, Y. Li, T. Shioda, D.E. Cohen, R.E. Gerszten, and A.M. Naar. 2010. MicroRNA-33 and the SREBP host genes cooperate to control cholesterol homeostasis. *Science* **328**:1566–1569.

Nicoli, S., C. Standley, P. Walker, A. Hurlstone, K.E. Fogarty, and N.D. Lawson. 2010. MicroRNA-mediated integration of haemodynamics and Vegf signalling during angiogenesis. *Nature* **464**:1196–1200.

Nishi, H., K. Ono, T. Horie, K. Nagao, M. Kinoshita, Y. Kuwabara, S. Watanabe, T. Takaya, Y. Tamaki, R. Takanabe-Mori, H. Wada, K. Hasegawa, et al. 2011. MicroRNA-27a regulates beta cardiac myosin heavy chain gene expression by targeting thyroid hormone receptor {beta}1 in neonatal rat ventricular myocytes. *Mol Cell Biol* **31**:744–755.

Nishi, H., K. Ono, Y. Iwanaga, T. Horie, K. Nagao, G. Takemura, M. Kinoshita, Y. Kuwabara, R.T. Mori, K. Hasegawa, T. Kita, and T. Kimura. 2010. MicroRNA-15b modulates cellular ATP levels and degenerates mitochondria via Arl2 in neonatal rat cardiac myocytes. *J Biol Chem* **285**:4920–4930.

Olson, E.N. 2006. Gene regulatory networks in the evolution and development of the heart. *Science* **313**:1922–1927.

Orom, U.A., F.C. Nielsen, and A.H. Lund. 2008. MicroRNA-10a binds the 5′UTR of ribosomal protein mRNAs and enhances their translation. *Mol Cell* **30**:460–471.

Park, M.J., M.S. Kim, I.C. Park, H.S. Kang, H. Yoo, S.H. Park, C.H. Rhee, S.I. Hong, and S.H. Lee. 2002. PTEN suppresses hyaluronic acid-induced matrix metalloproteinase-9 expression in U87MG glioblastoma cells through focal adhesion kinase dephosphorylation. *Cancer Res* **62**:6318–6322.

Poliseno, L., A. Tuccoli, L. Mariani, M. Evangelista, L. Citti, K. Woods, A. Mercatanti, S. Hammond, and G. Rainaldi. 2006. MicroRNAs modulate the angiogenic properties of HUVECs. *Blood* **108**:3068–3071.

Rajabi, M., C. Kassiotis, P. Razeghi, and H. Taegtmeyer. 2007. Return to the fetal gene program protects the stressed heart: a strong hypothesis. *Heart Fail Rev* **12**:331–343.

Rayner, K.J., C.C. Esau, F.N. Hussain, A.L. McDaniel, S.M. Marshall, J.M. van Gils, T.D. Ray, F.J. Sheedy, L. Goedeke, X. Liu, O.G. Khatsenko, V. Kaimal, et al. 2011a. Inhibition of *miR-33a/b* in non-human primates raises plasma HDL and lowers VLDL triglycerides. *Nature* **478**: 404–407.

Rayner, K.J., F.J. Sheedy, C.C. Esau, F.N. Hussain, R.E. Temel, S. Parathath, J.M. van Gils, A.J. Rayner, A.N. Chang, Y. Suarez, C. Fernandez-Hernando, E.A. Fisher, et al. 2011b. Antagonism of *miR-33* in mice promotes reverse cholesterol transport and regression of atherosclerosis. *J Clin Invest* **121**:2921–2931.

Rayner, K.J., Y. Suarez, A. Davalos, S. Parathath, M.L. Fitzgerald, N. Tamehiro, E.A. Fisher, K.J. Moore, and C. Fernandez-Hernando. 2010. *MiR-33* contributes to the regulation of cholesterol homeostasis. *Science* **328**:1570–1573.

Ren, X.P., J. Wu, X. Wang, M.A. Sartor, J. Qian, K. Jones, P. Nicolaou, T.J. Pritchard, and G.C. Fan. 2009. MicroRNA-320 is involved in the regulation of cardiac ischemia/reperfusion injury by targeting heat-shock protein 20. *Circulation* **119**:2357–2366.

Ronchetti, D., M. Lionetti, L. Mosca, L. Agnelli, A. Andronache, S. Fabris, G.L. Deliliers, and A. Neri. 2008. An integrative genomic approach reveals coordinated expression of intronic *miR-335*, *miR-342*, and *miR-561* with deregulated host genes in multiple myeloma. *BMC Med Genomics* **1**:37.

Rossi, M.A. 1998. Pathologic fibrosis and connective tissue matrix in left ventricular hypertrophy due to chronic arterial hypertension in humans. *J Hypertens* **16**:1031–1041.

Roy, S., S. Khanna, S.R. Hussain, S. Biswas, A. Azad, C. Rink, S. Gnyawali, S. Shilo, G.J. Nuovo, and C.K. Sen. 2009. MicroRNA expression in response to murine myocardial infarction: *miR-21* regulates fibroblast metalloprotease-2 via phosphatase and tensin homologue. *Cardiovasc Res* **82**:21–29.

Saffitz, J.E., J.G. Laing, and K.A. Yamada. 2000. Connexin expression and turnover: implications for cardiac excitability. *Circ Res* **86**:723–728.

Sayed, D., C. Hong, I.Y. Chen, J. Lypowy, and M. Abdellatif. 2007. MicroRNAs play an essential role in the development of cardiac hypertrophy. *Circ Res* **100**:416–424.

Sowa, N., T. Horie, Y. Kuwabara, O. Baba, S. Watanabe, H. Nishi, M. Kinoshita, R. Takanabe-Mori, H. Wada, A. Shimatsu, K. Hasegawa, T. Kimura, et al. 2012. MicroRNA 26b encoded by the intron of small CTD phosphatase (SCP) 1 has an antagonistic effect on its host gene. *J Cell Biochem* **113**:3455–3465.

Suarez, Y., C. Fernandez-Hernando, J.S. Pober, and W.C. Sessa. 2007. Dicer dependent microRNAs regulate gene expression and functions in human endothelial cells. *Circ Res* **100**:1164–1173.

Swynghedauw, B. 1999. Molecular mechanisms of myocardial remodeling. *Physiol Rev* **79**: 215–262.

Tatsuguchi, M., H.Y. Seok, T.E. Callis, J.M. Thomson, J.F. Chen, M. Newman, M. Rojas, S.M. Hammond, and D.Z. Wang. 2007. Expression of microRNAs is dynamically regulated during cardiomyocyte hypertrophy. *J Mol Cell Cardiol* **42**:1137–1141.

Thum, T., P. Galuppo, C. Wolf, J. Fiedler, S. Kneitz, L.W. van Laake, P.A. Doevendans, C.L. Mummery, J. Borlak, A. Haverich, C. Gross, S. Engelhardt, et al. 2007. MicroRNAs in the human heart: a clue to fetal gene reprogramming in heart failure. *Circulation* **116**:258–267.

Thum, T., C. Gross, J. Fiedler, T. Fischer, S. Kissler, M. Bussen, P. Galuppo, S. Just, W. Rottbauer, S. Frantz, M. Castoldi, J. Soutschek, et al. 2008. MicroRNA-21 contributes to myocardial disease by stimulating MAP kinase signalling in fibroblasts. *Nature* **456**:980–984.

Valadi, H., K. Ekstrom, A. Bossios, M. Sjostrand, J.J. Lee, and J.O. Lotvall. 2007. Exosome-mediated transfer of mRNAs and microRNAs is a novel mechanism of genetic exchange between cells. *Nat Cell Biol* **9**:654–659.

REFERENCES

van Rooij, E., D. Quiat, B.A. Johnson, L.B. Sutherland, X. Qi, J.A. Richardson, R.J. Kelm, Jr., and E.N. Olson. 2009. A family of microRNAs encoded by myosin genes governs myosin expression and muscle performance. *Dev Cell* **17**:662–673.

van Rooij, E., L.B. Sutherland, N. Liu, A.H. Williams, J. McAnally, R.D. Gerard, J.A. Richardson, and E.N. Olson. 2006. A signature pattern of stress-responsive microRNAs that can evoke cardiac hypertrophy and heart failure. *Proc Natl Acad Sci U S A* **103**:18255–18260.

van Rooij, E., L.B. Sutherland, X. Qi, J.A. Richardson, J. Hill, and E.N. Olson. 2007. Control of stress-dependent cardiac growth and gene expression by a microRNA. *Science* **316**:575–579.

van Rooij, E., L.B. Sutherland, J.E. Thatcher, J.M. DiMaio, R.H. Naseem, W.S. Marshall, J.A. Hill, and E.N. Olson. 2008. Dysregulation of microRNAs after myocardial infarction reveals a role of *miR-29* in cardiac fibrosis. *Proc Natl Acad Sci U S A* **105**:13027–13032.

Wang, G.K., J.Q. Zhu, J.T. Zhang, Q. Li, Y. Li, J. He, Y.W. Qin, and Q. Jing. 2010. Circulating microRNA: a novel potential biomarker for early diagnosis of acute myocardial infarction in humans. *Eur Heart J* **31**:659–666.

Wang, S., A.B. Aurora, B.A. Johnson, X. Qi, J. McAnally, J.A. Hill, J.A. Richardson, R. Bassel-Duby, and E.N. Olson. 2008. The endothelial-specific microRNA *miR-126* governs vascular integrity and angiogenesis. *Dev Cell* **15**:261–271.

Wang, Z., L. Yue, M. White, G. Pelletier, and S. Nattel. 1998. Differential distribution of inward rectifier potassium channel transcripts in human atrium versus ventricle. *Circulation* **98**:2422–2428.

Xiao, J., X. Luo, H. Lin, Y. Zhang, Y. Lu, N. Wang, B. Yang, and Z. Wang. 2007. MicroRNA *miR-133* represses HERG K+ channel expression contributing to QT prolongation in diabetic hearts. *J Biol Chem* **282**:12363–12367.

Yang, B., H. Lin, J. Xiao, Y. Lu, X. Luo, B. Li, Y. Zhang, C. Xu, Y. Bai, H. Wang, G. Chen, and Z. Wang. 2007. The muscle-specific microRNA *miR-1* regulates cardiac arrhythmogenic potential by targeting GJA1 and KCNJ2. *Nat Med* **13**:486–491.

Zheng, H., H. Takahashi, Y. Murai, Z. Cui, K. Nomoto, H. Niwa, K. Tsuneyama, and Y. Takano. 2006. Expressions of MMP-2, MMP-9 and VEGF are closely linked to growth, invasion, metastasis and angiogenesis of gastric carcinoma. *Anticancer Res* **26**:3579–3583.

# 30

MICRORNAS AND DIABETES

Romano Regazzi

Department of Fundamental Neurosciences, University of Lausanne, Lausanne, Switzerland

I.	Introduction	496
II.	Diabetes Mellitus Is Associated with Changes in Gene Expression in Many Organs	496
III.	Role of miRNAs in the Regulation of β-Cell Functions	497
IV.	Role of miRNAs in Insulin Target Tissues	499
V.	Contribution of miRNAs to Diabetes Complications	502
VI.	Circulating miRNAs as Diabetes Biomarkers	504
VII.	Conclusion	504
	References	504

ABBREVIATIONS

3′-UTR	3′-untranslated region
DIO mice	diet-induced obese mice
GK rats	Goto-Kakizaki rats
miRNAs	microRNAs
PI3K	phosphoinositide-3-kinase
PTEN	phosphatase and tensin homologue
T1D	type 1 diabetes
T2D	type 2 diabetes
TGFβ	transforming growth factor beta1
VEGF	vascular endothelial growth factor

MicroRNAs in Medicine, First Edition. Edited by Charles H. Lawrie.
© 2014 John Wiley & Sons, Inc. Published 2014 by John Wiley & Sons, Inc.

I. INTRODUCTION

Diabetes mellitus is a very common metabolic disorder characterized by chronically elevated blood glucose levels. According to the International Diabetes Federation, more than 350 million peoples worldwide were estimated to live with diabetes in 2011, and, in view of the current trends, about 500 million affected people are expected by 2030 (http://www.idf.org). Individuals with diabetes have an increased risk of developing micro- and macrovascular complications, potentially leading to blindness, lower limb amputations, stroke, and heart or kidney failure. Consequently, this chronic metabolic disease has a significant impact on life expectancy and is a major public health concern. Diabetes can have different etiologies, but is invariably linked to a release of insufficient amounts of insulin to match the organism's needs. Indeed, insulin secretion from β cells, located within the pancreatic islets of Langerhans, plays a central role in blood glucose homeostasis, and dysfunction or loss of these cells can have devastating effects on body metabolism. Type 1 diabetes (T1D) (about 5–10% of diabetes cases) develops as a result of an autoimmune destruction of β cells (Eizirik et al. 2009). During the initial phases of the disease, immune cells infiltrate the islets of Langerhans, chronically exposing β cells to elevated concentrations of proinflammatory cytokines. These inflammatory mediators cause a progressive decline of the insulin secretory activities and contribute to the loss of β-cells by apoptosis during the autoimmune reaction. T1D is characterized by a >75% reduction in β-cell mass and consequent inability of these cells to ensure proper control of blood glucose levels. Type 2 diabetes (T2D) is the most common form of the disease (about 90% of cases) and is initiated by a diminished insulin sensitivity of peripheral tissues (Nolan et al. 2011). This insulin-resistant state, often linked to obesity, can normally be compensated by β-cell mass expansion and by an increase in insulin secretory activity. However, in genetically predisposed individuals, this compensatory mechanism fails, resulting in profound perturbations in blood glucose homeostasis. Chronic exposure of β-cells to elevated glucose concentrations, in particular if combined with high circulating free fatty acid levels (a condition commonly observed in obese individuals), has a deleterious impact on β-cells, leading to impairment in insulin biosynthesis and secretion and, eventually, to apoptosis. Beside T1D and T2D, less common forms of diabetes mellitus exist but will not be discussed in this chapter.

II. DIABETES MELLITUS IS ASSOCIATED WITH CHANGES IN GENE EXPRESSION IN MANY ORGANS

Under prediabetic and diabetic conditions, a variety of cells in the body are chronically exposed to abnormal concentrations of glucose, insulin, fatty acids, inflammatory mediators, and others. This has a major impact on gene expression, causing cellular dysfunction in several organs. In the last two decades, a large number of studies have scrutinized the changes in gene expression occurring in tissues isolated from different diabetes animal models and in cultured conditions mimicking the diabetic milieu. This highlighted the importance of a number of signaling cascades, culminating in the activation of key transcription factors. However, at the beginning of this century, the discovery of microRNAs (miRNAs) revolutionized the understanding of the mechanisms controlling gene expression (Bushati and Cohen 2007) and subsequently also opened new perspectives for the diabetes field. This chapter summarizes the results of experiments carried out in the last few years in the miRNA field in an attempt to elucidate the contribution of these

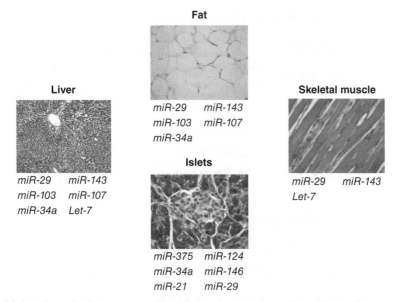

Figure 30.1. Selected miRNAs up-regulated in insulin-secreting cells and in insulin target tissues under diabetes conditions. See text for the description of the role and mode of action of each of these non-coding RNAs. See color insert.

non-coding RNA molecules in the development of diabetes and its long-term complications. The most important miRNAs up-regulated in insulin-secreting cells and in insulin target tissues under diabetes conditions are summarized in Figure 30.1.

III. ROLE OF miRNAs IN THE REGULATION OF β-CELL FUNCTIONS

Pancreatic β cells are highly specialized cells that produce and secrete insulin in response to a rise in circulating concentrations of nutrients (in particular, glucose), hormones, and neurotransmitters. Although short-term exposure of β cells to nutrients triggers their secretory activity, chronic exposure to elevated concentrations of glucose and free fatty acids has a deleterious impact on β cells and results in dysfunction and apoptosis (Prentki and Nolan 2006). The studies carried out within the last few years focused either on the identification of the miRNAs contributing to β-cell differentiation and in the accomplishment of β cell-specific tasks or on the analysis of the changes in miRNA levels occurring in prediabetic and diabetic conditions. Figure 30.2 summarizes the identified targets and the functional impact of the miRNAs described later in the text. The pioneering work of Poy et al. led to the compilation of the first catalogue of miRNAs expressed in pancreatic islets and to the identification of *miR-375*, one of the miRNAs highly enriched in β cells, as a negative regulator of insulin secretion (Poy et al. 2004). Subsequent studies revealed that this non-coding RNA is essential for proper differentiation of pancreatic islets and is required for β-cell mass expansion under insulin-resistant conditions. Indeed, *miR-375* knockout mice display reduced β-cell proliferation, and genetic ablation of this miRNA in obese leptin-deficient *ob/ob* mice, a model of insulin resistance, resulted in a severe diabetic state (Poy et al. 2009). *miR-375* directly targets *PDK1*, a key component of the

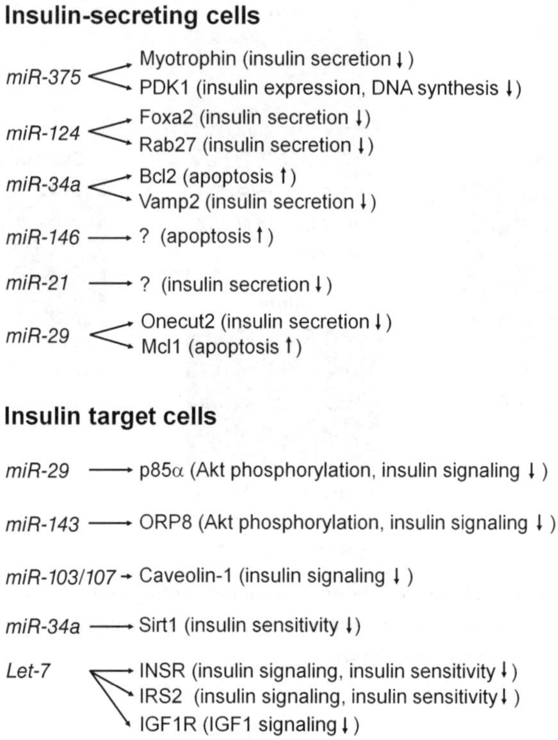

Insulin-secreting cells

miR-375 — Myotrophin (insulin secretion ↓)
         PDK1 (insulin expression, DNA synthesis ↓)

miR-124 — Foxa2 (insulin secretion ↓)
         Rab27 (insulin secretion ↓)

miR-34a — Bcl2 (apoptosis ↑)
         Vamp2 (insulin secretion ↓)

miR-146 ⟶ ? (apoptosis ↑)

miR-21 ⟶ ? (insulin secretion ↓)

miR-29 — Onecut2 (insulin secretion ↓)
        Mcl1 (apoptosis ↑)

Insulin target cells

miR-29 ⟶ p85α (Akt phosphorylation, insulin signaling ↓)

miR-143 ⟶ ORP8 (Akt phosphorylation, insulin signaling ↓)

miR-103/107 ⟶ Caveolin-1 (insulin signaling ↓)

miR-34a ⟶ Sirt1 (insulin sensitivity ↓)

Let-7 — INSR (insulin signaling, insulin sensitivity ↓)
       IRS2 (insulin signaling, insulin sensitivity ↓)
       IGF1R (IGF1 signaling ↓)

Figure 30.2. Mode of action and functional impact of the miRNAs up-regulated in insulin-secreting cells and insulin target tissues under diabetes conditions. The figure shows the experimentally validated target genes and, in brackets, the consequences for the cellular activities. The direction of the arrows indicates whether the cellular function is increased or reduced.

phosphoinositide-3-kinase (PI3K) signaling pathway, resulting in decreased glucose-induced insulin expression and DNA synthesis (El Ouaamari et al. 2008). The precise mechanisms regulating the expression of this miRNA are not completely understood, but *miR-375* levels have been reported to be reduced by glucose and cAMP-raising agents (Keller et al. 2012). A small increase (~30–40%) in *miR-375* was observed in pancreatic islets of *ob/ob* mice (Poy et al. 2009), while a decrease in *miR-375* precursor was detected in islets of Goto-Kakizaki (GK) rats, a non-obese T2D model (El Ouaamari et al. 2008). However, in another study, the level of mature *miR-375* in the islets of diabetic GK rats was not significantly different from controls (Esguerra et al. 2011).

A second miRNA that has attracted the interest of islet specialists is *miR-124*. This miRNA is highly enriched in neuronal cells, but is present at lower levels in β cells. The expression of *miR-124* increases during the embryonic development of the pancreas concomitantly with the differentiation of β cells, suggesting a role in this process (Baroukh et al. 2007). This non-coding RNA modulates the level of the transcription factor Foxa2, a master regulator of β-cell development governing the expression of several genes involved in glucose metabolism and insulin secretion (Baroukh et al. 2007). Moreover, its up-regulation can inhibit the expression of several components of the exocytotic machinery (Lovis et al. 2008a). Thus, it is likely that the level of *miR-124* has to be tightly controlled to permit proper differentiation of β cells without interfering with the capacity of accomplishing specialized insulin secretory activities. Indeed, forced expression of *miR-124* in insulin-secreting cell lines leads to alterations in intracellular free Ca^{2+} concentrations and causes impairment in insulin release (Baroukh et al. 2007; Lovis et al. 2008a).

Systematic investigation of changes in miRNA expression occurring in β-cells under physiopathological conditions and in diabetes animal models led to the identification of an additional group of miRNAs potentially involved in the development of diabetes. Global profiling of islets cells in control Wistar rats compared with those of GK rats, a non-obese T2D model, highlighted 30 miRNAs displaying differential expression (Esguerra et al. 2011). These miRNAs included *miR-124*, which was up-regulated in diabetic animals. Some of these non-coding RNAs, for example, *miR-132*, were found to be expressed in a glucose-dependent manner and their predicted targets enriched in known exocytotic genes. These observations led to the suggestion that the secretory defects observed in diabetic GK rats may be linked to differential islet miRNA expression with a consequent reduction in the level of key components necessary for insulin exocytosis.

Two other miRNAs, *miR-34* and *miR-146*, are up-regulated upon prolonged exposure of β cells to palmitate, a free fatty acid with a deleterious impact on insulin secretion and cell survival (Lovis et al. 2008b). Elevated levels of *miR-34* and *miR-146* were also observed in *db/db* mice displaying T2D diabetes. In contrast to GK rats, *db/db* mice do not express the leptin receptor and become severely obese before developing T2D. Forced expression of *miR-34* and *miR-146* in β-cell lines led to defective glucose-induced insulin secretion and sensitization toward apoptosis (Lovis et al. 2008b), suggesting that these two regulatory RNAs may contribute to β-cell dysfunction observed in *db/db* mice and to the development of T2D. These two miRNAs and *miR-21* were also up-regulated upon prolonged exposure of β cells to proinflammatory cytokines and in prediabetic NOD mice, a spontaneous model of T1D (Roggli et al. 2010). Introduction of anti-miR oligonucleotides specifically blocking the activity of these miRNAs permitted the attenuation of the deleterious impact of IL-1β on glucose-induced insulin secretion and cell survival. Systematic inspection of the changes occurring in islets of NOD mice during the prediabetic phases highlighted variations in several additional miRNAs and, in particular, a strong up-regulation of the *miR-29* family members *miR-29a/b/c* (Roggli et al. 2012). An increase in the level of *miR-29a/b/c* in β cells resulted in reduced capacity to release insulin in response to glucose. This could be explained at least in part by direct binding of *miR-29* family members to the 3′-UTR of the Onecut2 mRNA. Indeed, the consequent reduction in the level of this transcription factor results in an inappropriate expression of granuphilin, a major component of the machinery of exocytosis and a potent inhibitor of insulin release. As is the case with other cell types, in β cells, a rise in *miR-29* expression to levels comparable with those observed in prediabetic NOD mice triggered apoptosis. This deleterious effect was prevented by introducing oligonucleotides specifically masking the *miR-29* binding site present in the 3′-UTR of the mRNA coding for Mcl1. A reduction in the level of this antiapoptotic protein, a member of the Bcl2 family, has been proposed to contribute to cytokine-induced β-cell death (Allagnat et al. 2011). Consistent with this hypothesis, the oligonucleotides designed to bind the Mcl1 mRNA and to mask the *miR-29* target site not only restored the normal levels of this antiapoptotic protein, but also prevented insulin-secreting cells from cytokine-mediated death.

IV. ROLE OF miRNAs IN INSULIN TARGET TISSUES

Changes in miRNA levels in diabetes animal models have been reported in different insulin target tissues with consequent impairment in insulin signaling and glucose homeostasis.

Microarray analysis of miRNAs expressed in skeletal muscles from control Wistar rats compared with GK rats revealed changes in several miRNAs, including an up-regulation of *miR-29* family members (He et al. 2007). Elevated expression of the three *miR-29* family members was also detected in two other target tissues of insulin action, liver and fat. As mentioned earlier, under certain conditions, this particular miRNA family is also up-regulated in pancreatic β cells (Roggli et al. 2012). However, in insulin-secreting cells, this occurs during the early phases of T1D and has not been reported in animal models of T2D. Adenovirus-mediated overexpression of *miR-29a/b/c* in 3T3-L1 adipocytes caused reduction of Akt activation and impairment in insulin-stimulated glucose uptake (He et al. 2007). *miR-29a* levels were also significantly elevated in the liver of diabetic *db/db* mice (Pandey et al. 2011). As was the case for 3T3-L1 adipocytes, overexpression of *miR-29a* in HepG2 cells resulted in impairment in insulin-mediated Akt phosphorylation. This effect was attributed to direct inhibition of the expression of the p85α subunit of PI3K.

miR-143, another miRNA up-regulated in the liver and fat of *db/db* and diet-induced obese (DIO) mice (Takanabe et al. 2008), was also found to affect insulin-stimulated Akt phosphorylation and glucose sensitivity. Indeed, overexpression of *miR-143* in the liver resulted in impaired glucose tolerance and reduced Akt activation. This effect was attributed to down-regulation of the oxysterol-binding protein-related protein ORP8, a direct target of *miR-143*. In agreement with these observations, mice lacking the *miR-143* and *miR-145* cluster did not develop obesity-associated insulin resistance (Jordan et al. 2011).

Global profiling of skeletal muscle in humans led to the identification of more than 60 differentially expressed miRNAs in T2D patients, including an up-regulation of *miR-143* and down-regulation of two muscle-specific miRNAs, *miR-206* and *miR-133a* (Gallagher et al. 2010). Interestingly, approximately 15% of these miRNAs was already modified in individuals with impaired glucose tolerance, suggesting an involvement in the early phases of the disease process.

Two highly related miRNAs that differ by only one nucleotide, *miR-103* and *miR-107*, are significantly up-regulated in the liver of obese, leptin-deficient *ob/ob* mice and in DIO mice (Li et al. 2009; Trajkovski et al. 2011). Moreover, *miR-103* is increased in the liver of diabetic GK rats (Herrera et al. 2010). Injection of adenoviral constructs leading to *miR-103/107* overexpression in mice led to impaired insulin sensitivity and increased hepatic glucose production (Trajkovski et al. 2011). In contrast, silencing of *miR-103/107* with anti-miRNA oligonucleotides improved glucose tolerance in *ob/ob* and DIO mice by favoring glucose uptake in the liver and fat, but not in skeletal muscle (Trajkovski et al. 2011). This beneficial effect was associated with reduced levels of subcutaneous and visceral fat and smaller adipocyte size. Although the mechanisms underlying these phenomena remain to be precisely defined, at least part of the effect of *miR-103/107* on glucose homeostasis was proposed to be mediated by targeting the mRNA of caveolin-1, an essential component of plasma membrane caveolae and a key regulator of insulin receptor signaling (Trajkovski et al. 2011). Interestingly, continuous administration of plant-derived polyphenols that are able to prevent diet-induced fatty liver disease was found to prevent *miR-103/107* induction (Joven et al. 2012), further confirming the central role played by these non-coding RNAs in the control of glucose homeostasis.

Another miRNA potentially contributing to insulin resistance is *miR-34a*. This non-coding RNA is strongly up-regulated in liver and adipose tissue of *ob/ob* mice (Li et al. 2009; Zhao et al. 2009; Trajkovski et al. 2011), DIO mice (Lee et al. 2010), and in the liver of mice in which T1D diabetes was induced by streptozotocin injection (Li et al. 2009). *mir-34a* has been shown to directly target the mRNA of SIRT1 (Lee et al. 2010), an NAD^+-dependent deacetylase that exerts protective effects against metabolic diseases, including diabetes. Indeed, mice overexpressing SIRT1 or treated with SIRT1 activators are protected from diet-induced obesity and insulin resistance (Banks et al. 2008; Feige et al. 2008). Thus, reduction of SIRT1 expression in response to *miR-34a* up-regulation is likely to contribute to the diminished insulin sensitivity of liver and fat observed in T2D models.

Recently, *Let-7* family members have been discovered to play a central and unexpected role in glucose metabolism in many organs. Several members of this family were previously reported to be up-regulated in the liver of *ob/ob* and DIO mice (Trajkovski et al. 2011). However, strong evidence for the contribution of these non-coding RNAs to the regulation of glucose metabolism was provided by the phenotype of *Let-7* overexpressing mice. Indeed, these mice display impaired glucose tolerance, reduced glucose-induced insulin secretion, and a decrease in fat mass (Frost and Olson 2011). Blockade of *Let-7* family members with anti-miRNA oligonucleotides improved insulin sensitivity in liver and muscle and prevented impaired glucose tolerance in mice with diet-induced obesity (Frost and Olson 2011). In agreement with these findings, overexpression of Lin28a/b, two RNA-binding proteins that inhibit *Let-7* biogenesis, led to improvement in insulin sensitivity (Zhu et al. 2011). These phenomena were demonstrated to be at least in part linked to the capacity of *Let-7* to reduce the level of several key components of the insulin-PI3K-mTOR signaling pathway, including the receptors for insulin and IGF1 and the insulin-receptor substrate IRS2. The central position occupied by *Let-7* family in the complex signaling network controlling glucose metabolism is also inferred by the enrichment of the list of predicted targets of these non-coding RNAs for genes that in genome-wide association studies are linked to the control of fasting glucose levels and T2D (Zhu et al. 2011).

Disturbance of fatty acid and cholesterol homeostasis are crucial risk factors for metabolic disorders. Indeed, insulin resistance and T2D are often associated with abnormalities in circulating cholesterol and lipid profiles. A better understanding of the mechanisms controlling the plasmatic levels of cholesterol and fatty acids could open new perspectives for the treatment of insulin resistance. Although their expression is not significantly affected in the liver of diabetes animal models, a series of recent studies have demonstrated that *miR-122* and *miR-33* play a central role in the regulation of plasma cholesterol levels. Indeed, blockade of *miR-122*, which is primarily expressed in the liver, using an antisense-based strategy reduced the expression of several genes involved in cholesterol biosynthesis and led to lowering of plasma cholesterol levels in both mouse (Krutzfeldt et al. 2005) and African green monkeys (Elmen et al. 2008). However, blockade of *miR-122* in mice and nonhuman primates reduces both low- (LDL) and high- (HDL) density lipoproteins. Thus, therapeutic strategies aiming at preventing cholesterol-related diseases through *miR-122* antagonism may not yield the expected cardiometabolic benefit. *MiR-33a* and *miR-33b* are two closely related miRNAs that are embedded in the intronic sequences of the genes encoding the transcription factors SREBP1 and SREBP2 (Gerin et al. 2010; Horie et al. 2010). These transcription factors regulate the expression of numerous genes involved in cholesterol and fatty acid biosynthesis. *MiR-33a/b* were found to cooperate with their hosting genes in the control of cholesterol homeostasis by reducing the expression of the ATP-binding cassette transporter ABCA1, a pump playing a crucial role in the efflux of cholesterol from the cells and in the formation of HDL particles. In fact, *miR-33*-deficient mice and mice treated with anti-*miR-33* molecules display higher liver ABCA1 levels and increased ApoAI-dependent cholesterol efflux (Horie et al. 2010; Marquart et al. 2010; Najafi-Shoushtari et al. 2010; Rayner et al. 2010, 2011b). Conversely, ABCA1 and plasma HDL levels decline after overexpression of *miR-33* in the liver (Marquart et al. 2010;

Rayner et al. 2010). In addition to their effect on cholesterol homeostasis, *miR-33a/b* also inhibit translation of key genes involved in fatty acid metabolism and insulin signaling (Gerin et al. 2010; Davalos et al. 2011). Interestingly, hepatic SREBP1 and *miR-33b* levels are increased in nonhuman primates in response to a high-carbohydrate diet and high circulating insulin levels (Rayner et al. 2011a). Under these conditions, anti-*miR-33* treatment permitted the levels of plasma triglyceride to be lowered and those of HDL to be raised.

V. CONTRIBUTION OF miRNAs TO DIABETES COMPLICATIONS

Chronic elevation of blood glucose levels can lead to micro- and macrovascular damage, resulting in dysfunction and failure of numerous organs, including kidney, retina, peripheral nerves, and heart. In the last few years, a large number of studies highlighted a potential role for miRNAs in diabetic complications and, in particular, in diabetic nephropathy. This condition is characterized by glomerular fibrosis and a progressive decline in glomerular filtration rate, leading to kidney failure (Brosius et al. 2010). At least part of this degenerative process is attributed to the activation of the transforming growth factor beta 1 (TGFβ 1) signaling pathway in the kidney and the consequent accumulation in the glomerulus of extracellular matrix proteins (Sharma and Ziyadeh 1995). Indeed, TGFβ, a key regulator of the expression of extracellular matrix proteins, is increased in mesangial cells located in the glomerulus of animals displaying diabetic nephropathy (Kato et al. 2007). TGFβ induces its own expression, leading to an autocrine feedforward loop that causes renal dysfunction and failure. TGFβ increases the expression of extracellular matrix proteins, such as collagens, by reducing the levels of the E-box repressors deltaEF1 and Smad-interacting protein 1. The latter effect is at least in part due to translational repression mediated by *miR-192*, a miRNA that is up-regulated in response to TGFβ and is strongly increased in glomeruli isolated from streptozotocin-treated diabetic mice, as well as diabetic *db/db* mice. Treatment of mesangial cells with TGFβ results also in the rise of *miR-200b/c* (Kato et al. 2011). Up-regulation of these miRNAs, which are among those increased in glomeruli from T1D and T2D mice, occurs downstream to the induction of *miR-192*. In fact, inhibitors of *miR-192* attenuate the expression of *miR-200b/c*, collagen 1α, and TGFβ (Kato et al. 2011). TGFβ and *miR-192* are also involved in the activation of Akt, a protein kinase playing a central role in fibrosis and survival of glomerular mesangial cells (Figure 30.3). This occurs indirectly through the induction of two other miRNAs, *miR-216a* and *miR-217*, which target the mRNA of the phosphatase and tensin homologue (PTEN), an inhibitor of Akt activation (Kato et al. 2009). Another miRNA that is likely to contribute to diabetic nephropathy is *miR-377*. This non-coding RNA is induced upon incubation of mesangial cells at high concentrations of glucose and TGFβ and is up-regulated in mouse diabetic nephropathy models (Wang et al. 2008). Forced expression of *miR-377* led to enhanced production of fibronectin, an extracellular matrix protein that accumulates in excess during diabetic nephropathy. An additional miRNA that is present in excessive amounts in glomeruli from *db/db* mice and in kidney endothelial cells and podocytes treated with high glucose is *miR-29c* (Long et al. 2011). Overexpression of *miR-29c* resulted in accumulation of extracellular matrix proteins and podocyte apoptosis, while blockade of this miRNA protected against high glucose-induced cell death. This effect appears to be mediated by translational repression of Sprouty homologue 1 with a consequent activation of Rho kinase (Long et al. 2011). Interestingly, anti-*miR-29c* treatment of *db/db* mice permitted a reduction in

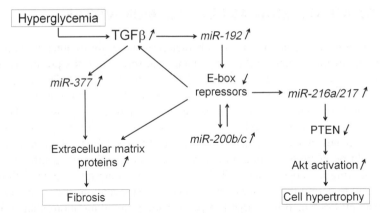

Figure 30.3. Signaling network elicited by diabetic conditions in kidney glomerular mesangial cells. Prolonged exposure to hyperglycemic conditions triggers an increase in TGFβ expression in renal cells. This results in the induction of *miR-192* and *miR-377* and in the accumulation of extracellular matrix proteins causing glomerular fibrosis. Most of these effects are due to translational inhibition of the E-box repressors Zeb1/2 that are direct targets of *miR-192*. Inhibition of E-box repressors leads also to the induction of *miR-216/217* and to the activation of a cascade causing hypertrophy of mesangial cells.

albuminuria and kidney mesangial matrix accumulation, confirming the role of *miR-29c* in diabetic nephropathy (Long et al. 2011).

Diabetic retinopathy is another common complication of diabetes mellitus, involving a damage of the retina microvasculature that can eventually lead to blindness. Sustained hyperglycemia causes metabolic alterations in endothelial cells and augmented expression of several angiogenic molecules, including the vascular endothelial growth factor (VEGF), resulting in perturbation in blood vessel function (Brownlee 2001). During the initial stages of the disease, fluids and lipids leaking out from damaged retinal capillaries cause macular edema and blurred vision. As the disease progresses, new blood vessels form in attempt to compensate for the lack of oxygen in the retina. Without timely treatment, these new fragile capillaries can bleed, perturbing the vision and destroying the retina. Microarray analysis permitted the identification of several miRNAs displaying expression changes in the retina of streptozotocin-induced diabetic rats, including down-regulation of *miR-200b* (McArthur et al. 2011). Decreased *miR-200b* levels were observed also by *in situ* hybridization in human retinas from diabetic patients. This miRNA targets the mRNA of VEGF, and its reduction is capable of mimicking the increase in endothelial permeability and angiogenesis triggered by glucose (McArthur et al. 2011). In agreement with these observations, injection of *miR-200b* oligonucleotides in the vitreous cavity of the eye of diabetic rats decreased VEGF expression and prevented the diabetes-induced increase in retinal vascular permeability (McArthur et al. 2011). Global miRNA profiling of the retina and of retinal endothelial cells was also reported by others. Kovacs and collaborators observed differential expression of a large number of miRNAs in the retina of rats 3 months after injection of streptozotocin and in retinal endothelial cells (Kovacs et al. 2011). In particular, they detected an up-regulation of several miRNAs that are controlled by VEGF, p53, and NFκB. Another study observed also abnormal miRNA expression profiles in the retina of streptozotocin-injected rats with several miRNAs displaying changes that parallel the course of diabetic retinopathy (Wu et al. 2012).

VI. CIRCULATING miRNAs AS DIABETES BIOMARKERS

Diabetic and prediabetic conditions are characterized not only by alterations in miRNA expression in several organs, but also by changes in serum miRNA profiles (Chen et al. 2008; Zampetaki et al. 2010; Karolina et al. 2011; Kong et al. 2011). Characteristic modifications in the miRNA profile have been shown to be useful tools to signalize the presence of diseases, in particular, some types of cancer. Since diabetes diagnosis can be achieved with simple, noninvasive methods, such as measurements of blood glucose levels, the analysis of changes in serum miRNAs to distinguish diabetic patients from healthy individuals would be of limited value. In contrast, the identification of specific miRNA signatures predicting the appearance of T1D, T2D, or their long-term complications would be very useful to permit the instauration of appropriate prevention strategies. The analysis of blood samples obtained in a large prospective study involving more than 800 persons led to the identification of characteristic miRNA expression changes that precede by several years the manifestation of T2D and of some of its vascular complications (Zampetaki et al. 2010). In view of these encouraging findings, it can be anticipated that in the coming years, many more studies will scrutinize the available collections of blood samples obtained from diabetic and prediabetic individuals for changes in miRNA levels.

VII. CONCLUSION

The discovery of miRNAs has revolutionized the understanding of the mechanisms governing gene expression and has elicited a new wave of studies in many fields, including diabetes research. There is now no doubt that miRNAs are central players in the control of metabolism and are involved in the processes underlying the appearance of diabetes and its long-term complications. The exciting findings gathered in the last few years promise to open new avenues in the development of therapeutic tools to prevent and treat this metabolic disease. This is good news, considering the current socioeconomic burden of this disease and the extraordinary public health challenge that diabetes mellitus will represent for the coming generations.

REFERENCES

Allagnat, F., D. Cunha, F. Moore, J.M. Vanderwinden, D.L. Eizirik, and A.K. Cardozo. 2011. Mcl-1 downregulation by pro-inflammatory cytokines and palmitate is an early event contributing to beta-cell apoptosis. *Cell Death Differ* **18**:328–337.

Banks, A.S., N. Kon, C. Knight, M. Matsumoto, R. Gutierrez-Juarez, L. Rossetti, W. Gu, and D. Accili. 2008. SirT1 gain of function increases energy efficiency and prevents diabetes in mice. *Cell Metab* **8**:333–341.

Baroukh, N., M.A. Ravier, M.K. Loder, E.V. Hill, A. Bounacer, R. Scharfmann, G.A. Rutter, and E. Van Obberghen. 2007. MicroRNA-124a regulates Foxa2 expression and intracellular signaling in pancreatic beta-cells lines. *J Biol Chem* **282**:19575–19588.

Brosius, F.C., C.C. Khoury, C.L. Buller, and S. Chen. 2010. Abnormalities in signaling pathways in diabetic nephropathy. *Expert Rev Endocrinol Metab* **5**:51–64.

Brownlee, M. 2001. Biochemistry and molecular cell biology of diabetic complications. *Nature* **414**:813–820.

Bushati, N. and S.M. Cohen. 2007. microRNA functions. *Annu Rev Cell Dev Biol* **23**:175–205.

REFERENCES

Chen, X., Y. Ba, L. Ma, X. Cai, Y. Yin, K. Wang, J. Guo, Y. Zhang, J. Chen, X. Guo, Q. Li, X. Li, et al. 2008. Characterization of microRNAs in serum: a novel class of biomarkers for diagnosis of cancer and other diseases. *Cell Res* **18**:997–1006.

Davalos, A., L. Goedeke, P. Smibert, C.M. Ramirez, N.P. Warrier, U. Andreo, D. Cirera-Salinas, K. Rayner, U. Suresh, J.C. Pastor-Pareja, E. Esplugues, E.A. Fisher, et al. 2011. *miR-33a/b* contribute to the regulation of fatty acid metabolism and insulin signaling. *Proc Natl Acad Sci U S A* **108**:9232–9237.

Eizirik, D.L., M.L. Colli, and F. Ortis. 2009. The role of inflammation in insulitis and beta-cell loss in type 1 diabetes. *Nat Rev Endocrinol* **5**:219–226.

El Ouaamari, A., N. Baroukh, G.A. Martens, P. Lebrun, D. Pipeleers, and E. van Obberghen. 2008. *miR-375* targets 3′-phosphoinositide-dependent protein kinase-1 and regulates glucose-induced biological responses in pancreatic beta-cells. *Diabetes* **57**:2708–2717.

Elmen, J., M. Lindow, S. Schutz, M. Lawrence, A. Petri, S. Obad, M. Lindholm, M. Hedtjarn, H.F. Hansen, U. Berger, S. Gullans, P. Kearney, et al. 2008. LNA-mediated microRNA silencing in non-human primates. *Nature* **452**:896–899.

Esguerra, J.L., C. Bolmeson, C.M. Cilio, and L. Eliasson. 2011. Differential glucose-regulation of microRNAs in pancreatic islets of non-obese type 2 diabetes model Goto-Kakizaki rat. *PLoS ONE* **6**:e18613.

Feige, J.N., M. Lagouge, C. Canto, A. Strehle, S.M. Houten, J.C. Milne, P.D. Lambert, C. Mataki, P.J. Elliott, and J. Auwerx. 2008. Specific SIRT1 activation mimics low energy levels and protects against diet-induced metabolic disorders by enhancing fat oxidation. *Cell Metab* **8**:347–358.

Frost, R.J. and E.N. Olson. 2011. Control of glucose homeostasis and insulin sensitivity by the *Let-7* family of microRNAs. *Proc Natl Acad Sci U S A* **108**:21075–21080.

Gallagher, I.J., C. Scheele, P. Keller, A.R. Nielsen, J. Remenyi, C.P. Fischer, K. Roder, J. Babraj, C. Wahlestedt, G. Hutvagner, B.K. Pedersen, and J.A. Timmons. 2010. Integration of microRNA changes in vivo identifies novel molecular features of muscle insulin resistance in type 2 diabetes. *Genome Med* **2**:9.

Gerin, I., L.A. Clerbaux, O. Haumont, N. Lanthier, A.K. Das, C.F. Burant, I.A. Leclercq, O.A. MacDougald, and G.T. Bommer. 2010. Expression of *miR-33* from an SREBP2 intron inhibits cholesterol export and fatty acid oxidation. *J Biol Chem* **285**:33652–33661.

He, A., L. Zhu, N. Gupta, Y. Chang, and F. Fang. 2007. Overexpression of micro ribonucleic acid 29, highly up-regulated in diabetic rats, leads to insulin resistance in 3T3-L1 adipocytes. *Mol Endocrinol* **21**:2785–2794.

Herrera, B.M., H.E. Lockstone, J.M. Taylor, M. Ria, A. Barrett, S. Collins, P. Kaisaki, K. Argoud, C. Fernandez, M.E. Travers, J.P. Grew, J.C. Randall, et al. 2010. Global microRNA expression profiles in insulin target tissues in a spontaneous rat model of type 2 diabetes. *Diabetologia* **53**:1099–1109.

Horie, T., K. Ono, M. Horiguchi, H. Nishi, T. Nakamura, K. Nagao, M. Kinoshita, Y. Kuwabara, H. Marusawa, Y. Iwanaga, K. Hasegawa, M. Yokode, et al. 2010. MicroRNA-33 encoded by an intron of sterol regulatory element-binding protein 2 (Srebp2) regulates HDL in vivo. *Proc Natl Acad Sci U S A* **107**:17321–17326.

Jordan, S.D., M. Kruger, D.M. Willmes, N. Redemann, F.T. Wunderlich, H.S. Bronneke, C. Merkwirth, H. Kashkar, V.M. Olkkonen, T. Bottger, T. Braun, J. Seibler, et al. 2011. Obesity-induced overexpression of miRNA-143 inhibits insulin-stimulated AKT activation and impairs glucose metabolism. *Nat Cell Biol* **13**:434–446.

Joven, J., E. Espinel, A. Rull, G. Aragones, E. Rodriguez-Gallego, J. Camps, V. Micol, M. Herranz-Lopez, J.A. Menendez, I. Borras, A. Segura-Carretero, C. Alonso-Villaverde, et al. 2012. Plant-derived polyphenols regulate expression of miRNA paralogs *miR-103/107* and *miR-122* and prevent diet-induced fatty liver disease in hyperlipidemic mice. *Biochim Biophys Acta* **1820**:894–899.

Karolina, D.S., A. Armugam, S. Tavintharan, M.T. Wong, S.C. Lim, C.F. Sum, and K. Jeyaseelan. 2011. MicroRNA 144 impairs insulin signaling by inhibiting the expression of insulin receptor substrate 1 in type 2 diabetes mellitus. *PLoS ONE* **6**:e22839.

Kato, M., J. Zhang, M. Wang, L. Lanting, H. Yuan, J.J. Rossi, and R. Natarajan. 2007. MicroRNA-192 in diabetic kidney glomeruli and its function in TGF-beta-induced collagen expression via inhibition of E-box repressors. *Proc Natl Acad Sci U S A* **104**:3432–3437.

Kato, M., S. Putta, M. Wang, H. Yuan, L. Lanting, I. Nair, A. Gunn, Y. Nakagawa, H. Shimano, I. Todorov, J.J. Rossi, and R. Natarajan. 2009. TGF-beta activates Akt kinase through a microRNA-dependent amplifying circuit targeting PTEN. *Nat Cell Biol* **11**:881–889.

Kato, M., L. Arce, M. Wang, S. Putta, L. Lanting, and R. Natarajan. 2011. A microRNA circuit mediates transforming growth factor-beta1 autoregulation in renal glomerular mesangial cells. *Kidney Int* **80**:358–368.

Keller, D.M., E.A. Clark, and R.H. Goodman. 2012. Regulation of microRNA-375 by cAMP in pancreatic beta-cells. *Mol Endocrinol* **26**:989–999.

Kong, L., J. Zhu, W. Han, X. Jiang, M. Xu, Y. Zhao, Q. Dong, Z. Pang, Q. Guan, L. Gao, J. Zhao, and L. Zhao. 2011. Significance of serum microRNAs in pre-diabetes and newly diagnosed type 2 diabetes: a clinical study. *Acta Diabetol* **48**:61–69.

Kovacs, B., S. Lumayag, C. Cowan, and S. Xu. 2011. MicroRNAs in early diabetic retinopathy in streptozotocin-induced diabetic rats. *Invest Ophthalmol Vis Sci* **52**:4402–4409.

Krutzfeldt, J., N. Rajewsky, R. Braich, K.G. Rajeev, T. Tuschl, M. Manoharan, and M. Stoffel. 2005. Silencing of microRNAs in vivo with "antagomirs". *Nature* **438**:685–689.

Lee, J., A. Padhye, A. Sharma, G. Song, J. Miao, Y.Y. Mo, L. Wang, and J.K. Kemper. 2010. A pathway involving farnesoid X receptor and small heterodimer partner positively regulates hepatic sirtuin 1 levels via microRNA-34a inhibition. *J Biol Chem* **285**:12604–12611.

Li, S., X. Chen, H. Zhang, X. Liang, Y. Xiang, C. Yu, K. Zen, Y. Li, and C.Y. Zhang. 2009. Differential expression of microRNAs in mouse liver under aberrant energy metabolic status. *J Lipid Res* **50**:1756–1765.

Long, J., Y. Wang, W. Wang, B.H. Chang, and F.R. Danesh. 2011. MicroRNA-29c is a signature microRNA under high glucose conditions that targets Sprouty homolog 1, and its in vivo knockdown prevents progression of diabetic nephropathy. *J Biol Chem* **286**:11837–11848.

Lovis, P., S. Gattesco, and R. Regazzi. 2008a. Regulation of the expression of components of the machinery of exocytosis of insulin-secreting cells by microRNAs. *Biol Chem* **389**:305–312.

Lovis, P., E. Roggli, D.R. Laybutt, S. Gattesco, J.Y. Yang, C. Widmann, A. Abderrahmani, and R. Regazzi. 2008b. Alterations in microRNA expression contribute to fatty acid-induced pancreatic beta-cell dysfunction. *Diabetes* **57**:2728–2736.

Marquart, T.J., R.M. Allen, D.S. Ory, and A. Baldan. 2010. *miR-33* links SREBP-2 induction to repression of sterol transporters. *Proc Natl Acad Sci U S A* **107**:12228–12232.

McArthur, K., B. Feng, Y. Wu, S. Chen, and S. Chakrabarti. 2011. MicroRNA-200b regulates vascular endothelial growth factor-mediated alterations in diabetic retinopathy. *Diabetes* **60**:1314–1323.

Najafi-Shoushtari, S.H., F. Kristo, Y. Li, T. Shioda, D.E. Cohen, R.E. Gerszten, and A.M. Naar. 2010. MicroRNA-33 and the SREBP host genes cooperate to control cholesterol homeostasis. *Science* **328**:1566–1569.

Nolan, C.J., P. Damm, and M. Prentki. 2011. Type 2 diabetes across generations: from pathophysiology to prevention and management. *Lancet* **378**:169–181.

Pandey, A.K., G. Verma, S. Vig, S. Srivastava, A.K. Srivastava, and M. Datta. 2011. *miR-29a* levels are elevated in the db/db mice liver and its overexpression leads to attenuation of insulin action on PEPCK gene expression in HepG2 cells. *Mol Cell Endocrinol* **332**:125–133.

REFERENCES

Poy, M.N., L. Eliasson, J. Krutzfeldt, S. Kuwajima, X. Ma, P.E. Macdonald, S. Pfeffer, T. Tuschl, N. Rajewsky, P. Rorsman, and M. Stoffel. 2004. A pancreatic is*let-specific* microRNA regulates insulin secretion. *Nature* **432**:226–230.

Poy, M.N., J. Hausser, M. Trajkovski, M. Braun, S. Collins, P. Rorsman, M. Zavolan, and M. Stoffel. 2009. *miR-375* maintains normal pancreatic alpha- and beta-cell mass. *Proc Natl Acad Sci U S A* **106**:5813–5818.

Prentki, M. and C.J. Nolan. 2006. Islet beta cell failure in type 2 diabetes. *J Clin Invest* **116**: 1802–1812.

Rayner, K.J., Y. Suarez, A. Davalos, S. Parathath, M.L. Fitzgerald, N. Tamehiro, E.A. Fisher, K.J. Moore, and C. Fernandez-Hernando. 2010. *MiR-33* contributes to the regulation of cholesterol homeostasis. *Science* **328**:1570–1573.

Rayner, K.J., C.C. Esau, F.N. Hussain, A.L. McDaniel, S.M. Marshall, J.M. van Gils, T.D. Ray, F.J. Sheedy, L. Goedeke, X. Liu, O.G. Khatsenko, V. Kaimal, et al. 2011a. Inhibition of *miR-33a/b* in non-human primates raises plasma HDL and lowers VLDL triglycerides. *Nature* **478**: 404–407.

Rayner, K.J., F.J. Sheedy, C.C. Esau, F.N. Hussain, R.E. Temel, S. Parathath, J.M. van Gils, A.J. Rayner, A.N. Chang, Y. Suarez, C. Fernandez-Hernando, E.A. Fisher, et al. 2011b. Antagonism of *miR-33* in mice promotes reverse cholesterol transport and regression of atherosclerosis. *J Clin Invest* **121**:2921–2931.

Roggli, E., A. Britan, S. Gattesco, N. Lin-Marq, A. Abderrahmani, P. Meda, and R. Regazzi. 2010. Involvement of microRNAs in the cytotoxic effects exerted by proinflammatory cytokines on pancreatic beta-cells. *Diabetes* **59**:978–986.

Roggli, E., S. Gattesco, D. Caille, C. Briet, C. Boitard, P. Meda, and R. Regazzi. 2012. Changes in microRNA expression contribute to pancreatic beta-cell dysfunction in prediabetic NOD mice. *Diabetes* **61**:1742–1751.

Sharma, K. and F.N. Ziyadeh. 1995. Hyperglycemia and diabetic kidney disease. The case for transforming growth factor-beta as a key mediator. *Diabetes* **44**:1139–1146.

Takanabe, R., K. Ono, Y. Abe, T. Takaya, T. Horie, H. Wada, T. Kita, N. Satoh, A. Shimatsu, and K. Hasegawa. 2008. Up-regulated expression of microRNA-143 in association with obesity in adipose tissue of mice fed high-fat diet. *Biochem Biophys Res Commun* **376**:728–732.

Trajkovski, M., J. Hausser, J. Soutschek, B. Bhat, A. Akin, M. Zavolan, M.H. Heim, and M. Stoffel. 2011. MicroRNAs 103 and 107 regulate insulin sensitivity. *Nature* **474**:649–653.

Wang, Q., Y. Wang, A.W. Minto, J. Wang, Q. Shi, X. Li, and R.J. Quigg. 2008. MicroRNA-377 is up-regulated and can lead to increased fibronectin production in diabetic nephropathy. *FASEB J* **22**:4126–4135.

Wu, J.H., Y. Gao, A.J. Ren, S.H. Zhao, M. Zhong, Y.J. Peng, W. Shen, M. Jing, and L. Liu. 2012. Altered MicroRNA Expression Profiles in Retinas with Diabetic Retinopathy. *Ophthalmic Res* **47**:195–201.

Zampetaki, A., S. Kiechl, I. Drozdov, P. Willeit, U. Mayr, M. Prokopi, A. Mayr, S. Weger, F. Oberhollenzer, E. Bonora, A. Shah, J. Willeit, et al. 2010. Plasma microRNA profiling reveals loss of endothelial *miR-126* and other microRNAs in type 2 diabetes. *Circ Res* **107**:810–817.

Zhao, E., M.P. Keller, M.E. Rabaglia, A.T. Oler, D.S. Stapleton, K.L. Schueler, E.C. Neto, J.Y. Moon, P. Wang, I.M. Wang, P.Y. Lum, I. Ivanovska, et al. 2009. Obesity and genetics regulate microRNAs in islets, liver, and adipose of diabetic mice. *Mamm Genome* **20**:476–485.

Zhu, H., N. Shyh-Chang, A.V. Segre, G. Shinoda, S.P. Shah, W.S. Einhorn, A. Takeuchi, J.M. Engreitz, J.P. Hagan, M.G. Kharas, A. Urbach, J.E. Thornton, et al. 2011. The Lin28/*let-7* axis regulates glucose metabolism. *Cell* **147**:81–94.

31

MicroRNAs in Liver Diseases

Patricia Munoz-Garrido,[1] Marco Marzioni,[2] Elizabeth Hijona,[1] Luis Bujanda,[1] and Jesus M. Banales[1,3]

[1]*Division of Hepatology and Gastroenterology, Biodonostia Research Institute, San Sebastián, Spain*
[2]*Department of Gastroenterology, "Università Politecnica delle Marche," Ancona, Italy*
[3]*IKERBASQUE, Basque Foundation of Science, Bilbao, Spain*

I.	Introduction	510
II.	Viral Hepatitis	511
III.	Alcoholic Liver Disease (ALD)	512
IV.	Non-Alcoholic Fatty Liver Disease (NAFLD)	513
V.	Drug-Induced Liver Injury (DILI)	513
VI.	Hepatocellular Carcinoma (HCC)	514
VII.	Biliary Diseases	515
VIII.	Fibrosis	516
IX.	miRNA-Based Therapeutic Approaches to Liver Disease	516
X.	Conclusions and Future Directions	517
	Acknowledgments	517
	References	518

ABBREVIATIONS

AE2	anion exchanger 2
ALD	alcoholic liver disease
APAP	acetaminophen
CCA	cholangiocarcinoma

MicroRNAs in Medicine, First Edition. Edited by Charles H. Lawrie.
© 2014 John Wiley & Sons, Inc. Published 2014 by John Wiley & Sons, Inc.

CCl₄	carbon tetrachloride
Cdc25A	cell division cycle 25A
DILI	drug-induced liver injury
ET-1	endothelin-1
HBV	hepatitis B virus
HCC	hepatocellular carcinoma
HCV	hepatitis C virus
HIF-1α	hipoxia-induced factor-1α
IL6	interleukin 6
IL8	interleukin 8
INFβ	interferon β
LNA	locked nucleic acid
LSECs	liver sinusoidal endothelial cells
miRNA	microRNA
NAFLD	non-alcoholic fatty liver disease
NASH	non-alcoholic steatohepatitis
NF-κB	nuclear factor-kappa B
NSC	neural stem cell
2'-OMe	2'-O-methyl
PBC	primary biliary cirrhosis
PCLDs	polycystic liver diseases
PPAR-α	peroxisome proliferator-activated receptor-α
TGF-β	transforming growth factor-β
TNF-α	tumor necrosis factor-α

I. INTRODUCTION

The liver, the largest organ in the human body, is composed of two types of epithelial cells: hepatocytes and cholangiocytes. Hepatocytes represent 70–80% of total liver mass and participate in fundamental physiological processes, such as drug detoxification, synthesis of cholesterol and bile salts, and primary bile generation. Cholangiocytes, the epithelial cells lining the bile ducts, are also key players in normal liver physiology, modulating the composition and flow of primary bile generated at the canaliculi of hepatocytes upstream of the biliary tree (Banales et al. 2006b). Although cholangiocytes represent only 3–5% of total liver cell population, they may account for up to 30% of total bile flow, mediating the fluidization and alkalinization of the bile (Banales et al. 2006b).

Both hepatocytes and cholangiocytes may directly or indirectly be affected by many types of diseases, including viral, genetic, neoplastic, immune-associated, idiopathic, infectious, vascular, or drug-induced diseases. Important advances are being achieved in understanding the molecular mechanisms involved in the development and progression of those disorders. Increasing evidence strongly suggests a role for microRNAs (miRNAs) in the etiopathogenesis of liver diseases, and consequently their potential role as targets for diagnosis and therapy (Wang et al. 2012) (Figure 31.1).

miRNAs are small non-coding endogenously transcribed RNAs (~20–22 nucleotides) able to regulate the expression of multiple genes by binding to complementary sites of targeted mRNAs, causing translational repression (imperfect target duplexes) or degradation (perfect matches) (Bartel 2009; Garzon et al. 2010). They participate in the regulation of multiple cell types under physiological and pathological conditions and are fundamental

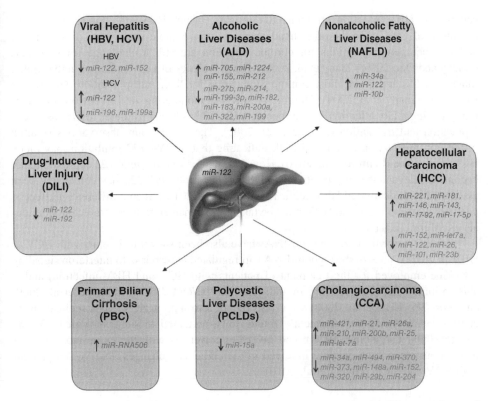

Figure 31.1. Figure summarizing the major miRNA alterations associated with various forms of liver diseases.

in different cellular processes, such as development, proliferation, apoptosis, metabolism, morphogenesis, and in disease. The human miRNA family includes 1733 mature miRNAs, encoded by 1424 precursors (Wang et al. 2012). Some miRNAs may be liver specific, such as *miR-122*, which in experimental down-regulation in mice results in blockage of cholesterol and lipid-metabolizing enzymes (Krutzfeldt et al. 2005).

In the liver, increasing evidence demonstrates the significant role of miRNAs in the development and/or progression of chronic liver diseases, such as viral hepatitis (i.e., hepatitis B or C viruses [HBV or HCV]), alcoholic liver disease (ALD), non-alcoholic fatty liver disease (NAFLD), drug-induced liver injury (DILI), and biliary diseases. Those chronic liver diseases represent a perpetuating injury, which can lead to the development of cirrhosis and primary liver cancers, such as hepatocellular carcinoma (HCC) and/or cholangiocarcinoma (CCA).

This chapter highlights some of the major advances that have been made toward understanding the role of miRNAs in the development and/or progression of liver diseases, and also discusses their potential use as biomarkers and therapy for liver injury.

II. VIRAL HEPATITIS

The exposure to HBV or HCV leads to chronic infection in the majority of subjects and is an important risk factor for the development of cirrhosis and HCC. Although vaccination can prevent HBV infection, strategies to eliminate chronic HBV infection are ineffective (Chang et al. 1997). On the other hand, there are no available vaccines to prevent HCV, and the therapeutical options for chronic infection are still a challenge for clinicians (Ghany and Doo 2009; Ghany et al. 2009). The prevalence of HCV infection in the United States is 1.6% and is the most common indication for liver transplantation (Ghany and Doo 2009; Ghany et al. 2009). The role of miRNAs modulating viral liver tropism is currently getting major attention. Liver-specific *miR-122* has been postulated to increase HCV replication and translation (Jopling et al. 2005). These important observations resulted in the first miRNA-based clinical trial, indicating that *miRNA-122* inhibition may block HCV replication (Lanford et al. 2010). However, while targeting *miR-122* expression may become a relevant strategy to attenuate HCV replication, *miR-122* has conversely been associated with decreased HBV replication (Qiu et al. 2010). Such a differential effect of *miR-122* between HCV and HBV must be taken into consideration in coinfected individuals (Kerr et al. 2011).

Asides from *miR-122*, other miRNAs have also been shown to interact with HCV or HBV. *MiR-196* is one of the eight miRNAs up-regulated in response to interferon signaling (cytokine employed for the treatment of patients with HCV and HBV infection) and is able to inhibit HCV expression. In addition, both *miRNA-199a* (Murakami et al. 2009) and *miR-196* (Hritz et al. 2008) were also reported to inhibit HCV RNA replication. In contrast, *miRNA-152* was frequently found down-regulated in subjects with HBV and hepatocellular carcinoma (HCC), and inversely correlates with the expression of DNA methyltransferase I, an enzyme involved in the epigenetic changes observed in patients with HBV and HCC (Huang et al. 2010).

III. ALCOHOLIC LIVER DISEASE (ALD)

Alcoholic liver disease (ALD) is the hepatic manifestation of alcohol abuse and one of the major global causes of chronic liver disease (Gao and Bataller 2011). ALD mainly results from the direct action of ethanol on the liver. However, ethanol and/or its metabolites (i.e., acetaldehyde) are also postulated to induce intestinal hyperpermeability, which leads to penetration of bacterial products (i.e., endotoxin) from the lumen into the systemic circulation and results in inflammatory processes in the liver (Hill et al. 2000; Hritz et al. 2008; Rao 2008; Keshavarzian et al. 2009; Bala and Szabo 2012; Szabo and Bala 2010). Chronic alcohol consumption alters intracellular signaling pathways in different cell types within the liver (i.e., hepatocytes, hepatic stem cells, stromal cells, and inflammatory cells), resulting in secretion of proinflammatory cytokines (i.e., tumor necrosis factor-α [TNF-α], interleukin-6 [IL6], and interleukin-8 [IL8]), oxidative stress, lipid peroxidation, and acetaldehyde toxicity, which cause inflammation and fatty liver, and in some cases, lead to fibrosis, cirrhosis, and hepatocellular carcinoma. Recent data indicate that ethanol may provoke some of these effects via alteration of miRNA levels and miRNA-regulated pathways. Indeed, miRNAs have been postulated as mediators of ethanol-induced tolerance, neural stem cell (NSC) proliferation and differentiation, gut leakiness, alcoholic liver disease, hepatocellular carcinoma, and other gastrointestinal cancers (Miranda et al. 2010; Bala and Szabo 2012; Wang et al. 2012).

Alterations in the hepatic miRNA profile have been reported in an experimental animal model of ALD (Dolganiuc et al. 2009), where 1% of known miRNAs were found to be up-regulated (e.g., *miR-705* and *miR-1224*), and another 1% were down-regulated (e.g., *miR-27b, miR-214, miR-199a-3p, miR-182, miR-183, miR-200a,* and *miR-322*).

Moreover, inflammation-related miRNAs deserve special attention in ALD, since activation of the innate immune system is a hallmark of alcohol steatohepatitis (Baltimore et al. 2008). Thus, alcohol induces *miR-155* expression in Kupffer cells, leading to the increase of TNF-α production (Bala et al. 2011). Moreover, increased *miR-155* was found in isolated hepatocytes of alcohol-fed mice (Bala et al. 2011). Alcohol was also reported to down-regulate *miR-199* in rat liver sinusoidal endothelial cells (LSECs) and human endothelial cells, increasing the endothelin-1 (ET-1) mRNA expression and the hypoxia-induced factor-1α (HIF-1α), and likely contributing to inflammation in patients with cirrhosis (Yeligar et al. 2009).

On the other hand, miRNAs may deregulate the gut permeability and support the development of ALD (Bala and Szabo 2012). For instance, ethanol increases the intestinal expression of *miR-212*, leading to gut leakiness by down-regulation of ZO-1 expression, a key enteric tight junction scaffolding protein (Tang et al. 2008). The resulting endotoxemia initiates hepatic damage and may lead indirectly to the development of ALD (Tang et al. 2008).

IV. NON-ALCOHOLIC FATTY LIVER DISEASE (NAFLD)

Non-alcoholic fatty liver disease (NAFLD) is a common chronic liver disease in the majority of developed countries, affecting 20-30% of adults (Vernon et al. 2011). NAFLD is caused by lipid deposition in the liver (i.e., steatosis) and is associated with insulin resistance and metabolic syndrome. NAFLD may progress to non-alcoholic steatohepatitis (NASH), a more aggressive phenotype that causes liver cirrhosis (Vernon et al. 2011). Levels of circulating miRNAs, such as *miR-34a* and *miR-122*, were found up-regulated in patients with NAFLD (Cermelli et al. 2011), suggesting their potential use as markers for diagnosis. In addition, culture of hepatoma cells with a high concentration of free fatty acids (experimental *in vitro* model of NAFLD) resulted in deregulation of the miRNAs profile (17 up-regulated and 15 down-regulated) (Zheng et al. 2010); in particular, increased miRNA-10b expression in steatotic hepatocytes was demonstrated to directly target the peroxisome proliferator-activated receptor-α (PPAR-α), a nuclear receptor that improves steatosis (Zheng et al. 2010). PPAR-α is a lipid sensor in the liver that regulate the transcriptional expression of genes involved in hepatic fat deposition and inflammation (Lefebvre et al. 2006).

V. DRUG-INDUCED LIVER INJURY (DILI)

Drug-induced liver injury (DILI) is a frequent side effect derived from the consumption of different drugs and represents a serious clinical and economical health problem worldwide. Acetaminophen (APAP) is one of the most widely employed painkillers in the United States due to its prescription-free status, and abuse of this drug can result in acute liver failure due to massive centrilobular hepatocyte death and inflammation (Lee 2004; Larson et al. 2005). In experimental animal models of APAP-induced DILI, liver injury was correlated with increased plasma levels of *miR-122* and *miR-192* and with down-regulation of these miRNAs in the liver (Starkey Lewis et al. 2011; Bala and Szabo 2012; Bala et al. 2012). In addition, the concomitant hepatic inflammation was associated with a modest plasma increase of inflammatory miRNAs (i.e., *miR-155, -146a,* and *-125b*).

VI. HEPATOCELLULAR CARCINOMA (HCC)

Hepatocellular carcinoma (HCC) is the most common primary liver cancer, and the fifth most frequent type of cancer worldwide. HCC represents the third deadliest type of cancer, with treatment options and prognosis being dependent on tumor size and staging (El-Serag 2011). HCC pathogenesis depends on the etiology and is mostly caused by viral hepatitis infection (HBV or HCV) or cirrhosis.

The role of miRNAs in hepatic carcinogenesis is complex due to the differential etiology of HCC, tumor stage, and ethnic background. Thus, only a few miRNAs have been commonly found to be dysregulated in HCC (down-regulated: *let-7*, *miR-122*, *miR-26*, and *miR-101*; up-regulated: *miR-221*, *miR-181*, and *miR-17-92*), indicating the heterogeneity of this cancer and suggesting the potential use of these miRNAs as biomarkers or therapeutic targets (Wang et al. 2012).

MiRNA expression is commonly down-regulated in most cancer types, an event associated with down-regulation of Dicer1, a key enzyme of the miRNA biosynthetic pathway (Nelson and Weiss 2008). Conditional Dicer1-knockout mice were found to develop prominent steatosis (i.e., abnormal retention of lipids within a cell), depletion of glycogen storage in hepatocytes, increased hepatocyte proliferation, and vast apoptosis (Sekine et al. 2009). In addition, approximately 60% of Dicer1-mutant mice spontaneously developed HCC at 1 year of age (Sekine et al. 2009). These results strongly indicate that miRNA processing through Dicer1 is essential for hepatocyte survival, metabolism, and hepatic tumoral suppression.

miRNAs can function as onco-miRNAs or tumor suppressors in the neoplastic transformation of the liver. A stem cell-like miRNA profile, that is, overexpression of the *miR-371-3* cluster, was found in undifferentiated, high-grade human HCC samples with poor prognosis (Cairo et al. 2010). The *miR-371-3* cluster is typically overexpressed in embryonic stem cells and is down-regulated during differentiation *in vivo* (Cairo et al. 2010). In addition, *miR-146* was found to promote hepatocyte proliferation and colony formation during hepatocarcinogenesis (Albulescu et al. 2011). In HCC cell lines, *miR-29a* was found to induce a synergic effect, with arsenic trioxide (a chemotherapeutic agent) inhibiting cell growth and inducing apoptosis (Meng et al. 2011). Likewise, HCC is associated with overexpression of *miR-221*, which induces hepatocyte proliferation and apoptotic alterations (Sharma et al. 2011), as well as with down-regulation of the tumor suppressor *miR-23b* (Nagata et al. 2009; Yuan et al. 2011).

MiRNAs have also been associated with the metastatic features of HCC. Thus, changes in the expression pattern of 20 miRNAs were observed in metastatic human HCC (16 of them down-regulated, including *miR-122*) (Budhu et al. 2008). *MiR-122* down-regulation promoted cell growth and migration of HCC cells in mice (Gramantieri et al. 2007; Coulouarn et al. 2009). On the other hand, overexpression of *miR-17-5p*, *miR-143*, and *miR-181* in HCC was associated with tumor cell growth and metastasis (Ji et al. 2009b; Wang et al. 2010; Yang et al. 2010).

Men show a higher incidence of HCC than women (between two- and sixfold) (El-Serag and Rudolph 2007). Likewise, women with HCC survive longer than men, indicating a sex-related mechanism that prevents HCC formation or slows its progression (El-Serag and Rudolph 2007). In this regard, differences in the expression of specific miRNAs were found in HCC samples between men and women (five miRNAs overexpressed and three down-regulated in women), suggesting their potential role in the sex-related differences observed in HCC (Ji et al. 2009a; Wang et al. 2012).

VII. BILIARY DISEASES

Cholangiocytes, the epithelial cells of bile ducts, are essential for normal liver physiological regulating the bile flow and composition (Banales et al. 2006b). Therefore, liver diseases targeting bile ducts are called cholangiopathies, and their etiopathogenesis may be of diverse origin (e.g., neoplastic, genetic, and immune-mediated) (Marzioni et al. 2010). Increasing evidence strongly indicates that miRNAs are important players in biliary pathophysiological processes, such as cholangiocarcinoma (CCA), polycystic liver diseases (PCLDs), and primary biliary cirrhosis (PBC).

CCA is a rare and aggressive malignancy affecting the bile ducts that shows an increasing incidence worldwide (Blechacz and Gores 2008; Blechacz et al. 2011). The role of miRNAs modulating the neoplastic transformation of cholangiocytes is receiving much attention. Several reports have shown a differential miRNA profile between CCA cells and normal cholangiocytes (Meng et al. 2006; Stutes et al. 2007; Chen et al. 2009; Selaru et al. 2009; Olaru et al. 2011). CCAs are associated with up-regulation of *miR-421* (Zhong et al. 2012), *miR-21* (He et al. 2011), *miR-26a* (Zhang et al. 2012), and *miR-210* (Yang et al. 2011), and down-regulation of *miR-34a* (Yang et al. 2011) and *miR-494* (Olaru et al. 2011), leading to cell proliferation and migration. In addition, several reports have revealed that CCA also promotes cell survival and proliferation through up-regulation of antiapoptotic miRNAs (i.e., *miR-25* [Razumilava et al. 2012] and *miR-21* [Selaru et al. 2009]) and down-regulation of proapoptotic miRNAs (i.e., *miR-320* [Chen et al. 2009], *miR-29b* [Mott et al. 2007; Stutes et al. 2007], and *miR-204* [Chen et al. 2009]). Although little is known about epigenetic changes associated with CCA, some studies have indicated that down-regulation of *miR-370* (Meng et al. 2006), *miR-373* (Chen et al. 2011), *miR-148a*, and *miR-152* (Braconi et al. 2010) in CCA result in hypermethylation and subsequent silencing of tumor suppressor genes. Additionally, up-regulation of *miR-21, miR-200b,* and *miR-let-7a* contribute to CCA chemoresistance by modulating the chemotherapy-induced apoptosis (Meng et al. 2006, 2007).

Polycystic liver diseases (PCLDs) are genetic disorders characterized by bile-duct dilatation and/or cyst development (Drenth et al. 2010). Polycystic cholangiocytes isolated from an animal model of hepatorenal cystogenesis (i.e., the PCK rat, which possesses a mutation in the human orthologue *PKHD1* gene) showed alteration in the miRNA expression profile compared with normal rat cholangiocytes, as most of the miRNAs were down-regulated (Lee et al. 2008). Among those highly down-regulated miRNAs, *miR-15a* was shown to target the cell-cycle regulator "cell division cycle 25A" (Cdc25A), resulting in cell proliferation and cystogenesis (Lee et al. 2008). In addition, *miR-15a* down-regulation and Cdc25A overexpression was also found in the bile ducts of PCLD patients. Importantly, preclinical studies to target this cdc25A overexpression with vitamin K (a natural inhibitor of Cdc25a) have shown the inhibition of hepatorenal cystogenesis in rodent models of polycystic kidney and liver disease (Masyuk et al. 2012).

Primary biliary cirrhosis (PBC) is a chronic cholestatic liver disease with an unknown etiopathogenesis and is associated with autoimmunity (Poupon 2010). A recent report demonstrated that PBC livers show alterations in the miRNA expression profile compared with healthy human livers (Padgett et al. 2009). Among those miRNAs that were up-regulated, *miR-506* was also found to be overexpressed in the bile ducts of PBC patients, resulting in inhibition of the protein expression and activity of anion exchanger 2 (AE2), a Cl^-/HCO_3^- exchanger located in the apical membrane of the hepatobiliary tract that mediates the alkalinization and fluidization of the bile, as well as control the intracellular pH (Banales et al. 2006a, 2006b, 2012). This observation is very important, since down-regulation of AE2 has been suggested to have an important etiopathogenic role in the development and progression of PBC (Salas et al. 2008). So, *miR-506* could represent a potential therapeutic target for PBC.

VIII. FIBROSIS

Liver fibrosis results from chronic liver injury and a tightly organized compensatory process of proliferation and repair. The complex process of fibrogenesis, which includes the deposition of extracellular matrix proteins, is mediated by hepatic stellate cells (HSCs) and inflammatory cells, such as Kupffer cells (Bataller and Brenner 2005). HSCs are located in the space of Disse (i.e., perisinusoidal space, located in the liver between the hepatocyte and sinusoid); they acquire an activated (i.e., profibrogenic) phenotype via a broad array of mediators, growth factors, and cytokines, the most important of which is transforming growth factor-β (TGF-β) (Bataller and Brenner 2005). All three *miR-29* family members (*miR-29a, b,* and *c*) have been reported to be down-regulated in the liver of human patients with advanced liver fibrosis, as well as in experimental animal models of fibrosis mediated by carbon tetrachloride (CCl_4) (Roderburg et al. 2011). *miR-29* is able to inhibit liver fibrosis through the targeting of TGF-β and nuclear factor kappa-B (NF-κB) signaling cascades in HSCs (Roderburg et al. 2011). In addition, *miR-29* is also involved in the down-regulation of different extracellular matrix genes, suggesting its potential therapeutic use for the treatment of liver fibrosis (Roderburg et al. 2011). Another miRNA involved in liver fibrosis is *miR-195*, which was reported to target cyclin E1 and to account for interferon-β (IFNβ)-induced inhibition of HSC proliferation (Sekiya et al. 2011). Finally, a recent study revealed that *miRNA-199* and *miR-200* families are overexpressed in human patients with liver fibrosis, as well as experimental animal models of fibrosis (Murakami et al. 2011).

IX. MIRNA-BASED THERAPEUTIC APPROACHES TO LIVER DISEASE

There are two different available strategies to modulate the levels of specific miRNAs: miRNA supplementation and miRNA inhibition.

Administration of short RNA duplexes mimicking miRNAs could be an effective approach for those diseases characterized by a specific down-regulation of a key miRNA. To avoid potential side effects in healthy cells, this strategy is mainly focused on those miRNAs that are widely expressed in normal tissues but exclusively down-regulated in pathological cells. Thus, systemic administration of *miR-124* is able to prevent and suppress hepatocarcinogenesis in experimental animal models (Hatziapostolou et al. 2011). Likewise, *miR-let-7* and *miR-34* are in preclinical development for cancer therapy (Wang et al. 2012). However, technology to safely and effectively deliver miRNAs or siRNAs to specific target cells is proving to be an important challenge.

On the other hand, miRNA inhibitors are a new class of chemically engineered single-stranded oligonucleotides that antagonize the overexpression of specific miRNAs by either degradation (i.e., antagomiRs) or sequestration (i.e., antimiRs) of the mature miRNA.

AntagomiRs possess a perfect complementary sequence to the entire mature miRNA and are based on a medium-affinity chemistry termed 2'-O-methyl (2'-OMe), that make it more resistant to degradation (Krutzfeldt et al. 2005). AntagomiRs are also phosphorylated and may be conjugated with cholesterol. A preclinical proof of efficacy for cholesterol-conjugated anti-*miR-221* (chol-anti-*miR-221*) was tested in an orthotopic mouse model of HCC (Park et al. 2011). Intravenous administration of chol-anti-*miR-221* resulted in the targeting of oncogenic *miR-221* in HCC xenograft tumors, and subsequent inhibition of tumor cell proliferation, increased levels of markers of apoptosis, cell-cycle arrest, and increased mouse survival. These findings strongly suggest that this strategy may be beneficial for the treatment of HCC patients.

AntimiRs are maybe complementary to only part of the mature miRNA target and are generally developed using locked nucleic acid (LNA) chemistry, which increases the hybridization properties (melting temperature) of oligonucleotides. Thus, systemic administration of eight nucleotide-long naked LNA-antimiRs showed a potent dose-dependent inhibition of an miRNA family (based on the 7–8 seed sequence) in different organs, such as the liver, as well as in xenograft tumors (Wang et al. 2012). However, in some cases, these tiny-LNA-antimiRs may cause unspecific binding and subsequent side effects.

There is currently an ongoing clinical trial using Miravirsen, an antimiR targeting *miR-122*, which is being tested for the therapeutic treatment of chronic HCV infection. *miR-122*, a liver-specific abundant miRNA, is involved in lipid and cholesterol metabolism and in HCV replication (Jopling et al. 2005). In preclinical studies using animal models (Elmen et al. 2008a, 2008b), intravenous administration of naked Miravirsen potently inhibited *miR-122* in a dose-dependent manner and resulted in up-regulation of predicted mRNA targets. In chimpanzees chronically infected with HCV subtype 1a or 1b, intravenous administration of Miravisen once a week for 12 weeks reduced the serological viral titers in a dose-dependent manner (Lanford et al. 2010). Importantly, viral suppression was maintained 3 months after the last dose, with no evidence of viral resistance or side effects in any treated animal. In 2008, Miravirsen became the first miRNA-based therapy entering in a clinical trial, and showed a significant reduction of HCV RNA in patients. Moreover, this miRNA inhibitor was found safe and nontoxic at any dose employed.

X. CONCLUSIONS AND FUTURE DIRECTIONS

miRNAs represent a new research area in the field of liver pathophysiology. As stated in this chapter, they may regulate different features in the liver (e.g., apoptosis, proliferation, migration, and metabolism) and can be regulated by several mechanisms and agents (e.g., virus, alcohol, drugs, lipids, epigenetics, hormones, and cytokines) (Figure 31.1). Although limited information is available regarding the role of miRNAs in the development and progression of liver diseases, future investigations will increase our understanding of the miRNA-related molecular mechanisms involved in liver pathophysiology and their potential use for diagnosis and therapy. *miR-122*-based clinical trials for the treatment of HCV infection represent a new era in the field and strongly support the potential therapeutic use of miRNAs for the treatment of liver diseases.

ACKNOWLEDGMENTS

J.M.B. and his research are funded by IKERBASQUE, the Basque Foundation for Science, and the Asociación Española Contra el Cáncer.

REFERENCES

Albulescu, R., M. Neagu, L. Albulescu, and C. Tanase. 2011. Tissular and soluble miRNAs for diagnostic and therapy improvement in digestive tract cancers. *Expert Review of Molecular Diagnostics* **11**:101–120.

Bala, S., M. Marcos, K. Kodys, T. Csak, D. Catalano, P. Mandrekar, and G. Szabo. 2011. Up-regulation of microRNA-155 in macrophages contributes to increased tumor necrosis factor {alpha} (TNF{alpha}) production via increased mRNA half-life in alcoholic liver disease. *The Journal of Biological Chemistry* **286**:1436–1444.

Bala, S., J. Petrasek, S. Mundkur, D. Catalano, I. Levin, J. Ward, H. Alao, K. Kodys, and G. Szabo. 2012. Circulating microRNAs in exosomes indicate hepatocyte injury and inflammation in alcoholic, drug-induced, and inflammatory liver diseases. *Hepatology* **56**:1946–1957.

Bala, S. and G. Szabo. 2012. MicroRNA signature in alcoholic liver disease. *International Journal of Hepatology* **2012**:498232.

Baltimore, D., M.P. Boldin, R.M. O'Connell, D.S. Rao, and K.D. Taganov. 2008. MicroRNAs: new regulators of immune cell development and function. *Nature Immunology* **9**:839–845.

Banales, J.M., F. Arenas, C.M. Rodriguez-Ortigosa, E. Saez, I. Uriarte, R.B. Doctor, J. Prieto, and J.F. Medina. 2006a. Bicarbonate-rich choleresis induced by secretin in normal rat is taurocholate-dependent and involves AE2 anion exchanger. *Hepatology* **43**:266–275.

Banales, J.M., J. Prieto, and J.F. Medina. 2006b. Cholangiocyte anion exchange and biliary bicarbonate excretion. *World Journal of Gastroenterology: WJG* **12**:3496–3511.

Banales, J.M., E. Saez, M. Uriz, S. Sarvide, A.D. Urribarri, P. Splinter, P.S. Tietz Bogert, L. Bujanda, J. Prieto, J.F. Medina, and N.F. Larusso. 2012. Up-regulation of microRNA 506 leads to decreased Cl(-) /HCO(3) (-) anion exchanger 2 expression in biliary epithelium of patients with primary biliary cirrhosis. *Hepatology* **56**:687–697.

Bartel, D.P. 2009. MicroRNAs: target recognition and regulatory functions. *Cell* **136**:215–233.

Bataller, R. and D.A. Brenner. 2005. Liver fibrosis. *The Journal of Clinical Investigation* **115**:209–218.

Blechacz, B. and G.J. Gores. 2008. Cholangiocarcinoma: advances in pathogenesis, diagnosis, and treatment. *Hepatology* **48**:308–321.

Blechacz, B., M. Komuta, T. Roskams, and G.J. Gores. 2011. Clinical diagnosis and staging of cholangiocarcinoma. *Nature Reviews. Gastroenterology & Hepatology* **8**:512–522.

Braconi, C., N. Huang, and T. Patel. 2010. MicroRNA-dependent regulation of DNA methyltransferase-1 and tumor suppressor gene expression by interleukin-6 in human malignant cholangiocytes. *Hepatology* **51**:881–890.

Budhu, A., H.L. Jia, M. Forgues, C.G. Liu, D. Goldstein, A. Lam, K.A. Zanetti, Q.H. Ye, L.X. Qin, C.M. Croce, Z.Y. Tang, and X.W. Wang. 2008. Identification of metastasis-related microRNAs in hepatocellular carcinoma. *Hepatology* **47**:897–907.

Cairo, S., Y. Wang, A. de Reynies, K. Duroure, J. Dahan, M.J. Redon, M. Fabre, M. McClelland, X.W. Wang, C.M. Croce, and M.A. Buendia. 2010. Stem cell-like micro-RNA signature driven by Myc in aggressive liver cancer. *Proceedings of the National Academy of Sciences of the United States of America* **107**:20471–20476.

Cermelli, S., A. Ruggieri, J.A. Marrero, G.N. Ioannou, and L. Beretta. 2011. Circulating microRNAs in patients with chronic hepatitis C and non-alcoholic fatty liver disease. *PloS ONE* **6**:e23937.

Chang, M.H., C.J. Chen, M.S. Lai, H.M. Hsu, T.C. Wu, M.S. Kong, D.C. Liang, W.Y. Shau, and D.S. Chen. 1997. Universal hepatitis B vaccination in Taiwan and the incidence of hepatocellular carcinoma in children. Taiwan Childhood Hepatoma Study Group. *The New England Journal of Medicine* **336**:1855–1859.

REFERENCES

Chen, L., H.X. Yan, W. Yang, L. Hu, L.X. Yu, Q. Liu, L. Li, D.D. Huang, J. Ding, F. Shen, W.P. Zhou, M.C. Wu, et al. 2009. The role of microRNA expression pattern in human intrahepatic cholangiocarcinoma. *Journal of Hepatology* **50**:358–369.

Chen, Y., W. Gao, J. Luo, R. Tian, H. Sun, and S. Zou. 2011. Methyl-CpG binding protein MBD2 is implicated in methylation-mediated suppression of *miR-373* in hilar cholangiocarcinoma. *Oncology Reports* **25**:443–451.

Coulouarn, C., V., M. Factor, J.B. Andersen, M.E. Durkin, and S.S. Thorgeirsson. 2009. Loss of *miR-122* expression in liver cancer correlates with suppression of the hepatic phenotype and gain of metastatic properties. *Oncogene* **28**:3526–3536.

Dolganiuc, A., J. Petrasek, K. Kodys, D. Catalano, P. Mandrekar, A. Velayudham, and G. Szabo. 2009. MicroRNA expression profile in Lieber-DeCarli diet-induced alcoholic and methionine choline deficient diet-induced nonalcoholic steatohepatitis models in mice. *Alcoholism, Clinical and Experimental Research* **33**:1704–1710.

Drenth, J.P., M. Chrispijn, D.M. Nagorney, P.S. Kamath, and V.E. Torres. 2010. Medical and surgical treatment options for polycystic liver disease. *Hepatology* **52**:2223–2230.

El-Serag, H.B. 2011. Hepatocellular carcinoma. *The New England Journal of Medicine* **365**: 1118–1127.

El-Serag, H.B. and K.L. Rudolph. 2007. Hepatocellular carcinoma: epidemiology and molecular carcinogenesis. *Gastroenterology* **132**:2557–2576.

Elmen, J., M. Lindow, M. Schutz, M. Lawrence, A. Petri, S. Obad, M. Lindholm, M. Hedtjarn, H.F. Hansen, U. Berger, S. Gullans, P. Kearney, et al. 2008a. LNA-mediated microRNA silencing in non-human primates. *Nature* **452**:896–899.

Elmen, J., M. Lindow, A. Silahtaroglu, M. Bak, M. Christensen, A. Lind-Thomsen, M. Hedtjarn, J.B. Hansen, H.F. Hansen, E.M. Straarup, K. McCullagh, P. Kearney, et al. 2008b. Antagonism of microRNA-122 in mice by systemically administered LNA-antimiR leads to up-regulation of a large set of predicted target mRNAs in the liver. *Nucleic Acids Research* **36**:1153–1162.

Gao, B. and R. Bataller. 2011. Alcoholic liver disease: pathogenesis and new therapeutic targets. *Gastroenterology* **141**:1572–1585.

Garzon, R., G. Marcucci, and C.M. Croce. 2010. Targeting microRNAs in cancer: rationale, strategies and challenges. *Nature Reviews. Drug Discovery* **9**:775–789.

Ghany, M.G. and E.C. Doo. 2009. Antiviral resistance and hepatitis B therapy. *Hepatology* **49**:S174–S184.

Ghany, M.G., D.B. Strader, D.L. Thomas, and L.B. Seeff. 2009. Diagnosis, management, and treatment of hepatitis C: an update. *Hepatology* **49**:1335–1374.

Gramantieri, L., M. Ferracin, F. Fornari, A. Veronese, S. Sabbioni, C.G. Liu, G.A. Calin, C. Giovannini, E. Ferrazzi, G.L. Grazi, C.M. Croce, L. Bolondi, et al. 2007. Cyclin G1 is a target of *miR-122a*, a microRNA frequently down-regulated in human hepatocellular carcinoma. *Cancer Research* **67**:6092–6099.

Hatziapostolou, M., C. Polytarchou, E. Aggelidou, A. Drakaki, G.A. Poultsides, S.A. Jaeger, H. Ogata, M. Karin, K. Struhl, M. Hadzopoulou-Cladaras, and D. Iliopoulos. 2011. An HNF4alpha-miRNA inflammatory feedback circuit regulates hepatocellular oncogenesis. *Cell* **147**:1233–1247.

He, Q., L. Cai, L. Shuai, D. Li, C. Wang, Y. Liu, X. Li, Z. Li, and S. Wang. 2011. Ars2 is overexpressed in human cholangiocarcinomas and its depletion increases PTEN and PDCD4 by decreasing microRNA-21. *Molecular Carcinogenesis* **52**:286–296.

Hill, D.B., S. Barve, S. Joshi-Barve, and C. McClain. 2000. Increased monocyte nuclear factor-kappaB activation and tumor necrosis factor production in alcoholic hepatitis. *The Journal of Laboratory and Clinical Medicine* **135**:387–395.

Hritz, I., P. Mandrekar, A. Velayudham, D. Catalano, A. Dolganiuc, K. Kodys, E. Kurt-Jones, and G. Szabo. 2008. The critical role of toll-like receptor (TLR) 4 in alcoholic liver disease is independent of the common TLR adapter MyD88. *Hepatology* **48**:1224–1231.

Huang, J., Y. Wang, Y. Guo, and S. Sun. 2010. Down-regulated microRNA-152 induces aberrant DNA methylation in hepatitis B virus-related hepatocellular carcinoma by targeting DNA methyltransferase 1. *Hepatology* **52**:60–70.

Ji, J., J. Shi, A. Budhu, Z. Yu, M. Forgues, S. Roessler, S. Ambs, Y. Chen, P.S. Meltzer, C.M. Croce, L.X. Qin, K. Man, et al. 2009a. MicroRNA expression, survival, and response to interferon in liver cancer. *The New England Journal of Medicine* **361**:1437–1447.

Ji, J., T. Yamashita, A. Budhu, M. Forgues, H.L. Jia, C. Li, C. Deng, E. Wauthier, L.M. Reid, Q.H. Ye, L.X. Qin, W. Yang, et al. 2009b. Identification of microRNA-181 by genome-wide screening as a critical player in EpCAM-positive hepatic cancer stem cells. *Hepatology* **50**:472–480.

Jopling, C.L., M. Yi, A.M. Lancaster, S.M. Lemon, and P. Sarnow. 2005. Modulation of hepatitis C virus RNA abundance by a liver-specific MicroRNA. *Science* **309**:1577–1581.

Kerr, T.A., K.M. Korenblat, and N.O. Davidson. 2011. MicroRNAs and liver disease. *Translational Research: The Journal of Laboratory and Clinical Medicine* **157**:241–252.

Keshavarzian, A., A. Farhadi, C.B. Forsyth, J. Rangan, S. Jakate, M. Shaikh, A. Banan, and J.Z. Fields. 2009. Evidence that chronic alcohol exposure promotes intestinal oxidative stress, intestinal hyperpermeability and endotoxemia prior to development of alcoholic steatohepatitis in rats. *Journal of Hepatology* **50**:538–547.

Krutzfeldt, J., N. Rajewsky, R. Braich, K.G. Rajeev, T. Tuschl, M. Manoharan, and M. Stoffel. 2005. Silencing of microRNAs in vivo with "antagomirs". *Nature* **438**:685–689.

Lanford, R.E., E.S. Hildebrandt-Eriksen, A. Petri, R. Persson, M. Lindow, M.E. Munk, S. Kauppinen, and H. Orum. 2010. Therapeutic silencing of microRNA-122 in primates with chronic hepatitis C virus infection. *Science* **327**:198–201.

Larson, A.M., J. Polson, R.J. Fontana, T.J. Davern, E. Lalani, L.S. Hynan, J.S. Reisch, F.V. Schiodt, G. Ostapowicz, A.O. Shakil, and W.M. Lee. 2005. Acetaminophen-induced acute liver failure: results of a United States multicenter, prospective study. *Hepatology* **42**:1364–1372.

Lee, S.O., T. Masyuk, P. Splinter, J.M. Banales, A. Masyuk, A. Stroope, and N. Larusso. 2008. MicroRNA15a modulates expression of the cell-cycle regulator Cdc25A and affects hepatic cystogenesis in a rat model of polycystic kidney disease. *The Journal of Clinical Investigation* **118**:3714–3724.

Lee, W.M. 2004. Acetaminophen and the U.S. Acute Liver Failure Study Group: lowering the risks of hepatic failure. *Hepatology* **40**:6–9.

Lefebvre, P., G. Chinetti, J.C. Fruchart, and B. Staels. 2006. Sorting out the roles of PPAR alpha in energy metabolism and vascular homeostasis. *The Journal of Clinical Investigation* **116**:571–580.

Marzioni, M., S. Saccomanno, C. Candelaresi, C. Rychlicki, L. Agostinelli, L. Trozzi, S. De Minicis, and A. Benedetti. 2010. Clinical implications of novel aspects of biliary pathophysiology. *Digestive and Liver Disease: Official Journal of the Italian Society of Gastroenterology and the Italian Association for the Study of the Liver* **42**:238–244.

Masyuk, T.V., B.N. Radtke, A.J. Stroope, J.M. Banales, A.I. Masyuk, S.A. Gradilone, G.B. Gajdos, N. Chandok, J.L. Bakeberg, C.J. Ward, E.L. Ritman, H. Kiyokawa, et al. 2012. Inhibition of Cdc25A suppresses hepato-renal cystogenesis in rodent models of polycystic kidney and liver disease. *Gastroenterology* **142**:622–633 e624.

Meng, F., R. Henson, M. Lang, H. Wehbe, S. Maheshwari, J.T. Mendell, J. Jiang, T.D. Schmittgen, and T. Patel. 2006. Involvement of human micro-RNA in growth and response to chemotherapy in human cholangiocarcinoma cell lines. *Gastroenterology* **130**:2113–2129.

REFERENCES

Meng, F., R. Henson, H. Wehbe-Janek, H. Smith, Y. Ueno, and T. Patel. 2007. The MicroRNA *let-7a* modulates interleukin-6-dependent STAT-3 survival signaling in malignant human cholangiocytes. *The Journal of Biological Chemistry* **282**:8256–8264.

Meng, X.Z., T.S. Zheng, X. Chen, J.B. Wang, W.H. Zhang, S.H. Pan, H.C. Jiang, and L.X. Liu. 2011. microRNA expression alteration after arsenic trioxide treatment in HepG-2 cells. *Journal of Gastroenterology and Hepatology* **26**:186–193.

Miranda, R.C., A.Z. Pietrzykowski, Y. Tang, P. Sathyan, D. Mayfield, A. Keshavarzian, W. Sampson, and D. Hereld. 2010. MicroRNAs: master regulators of ethanol abuse and toxicity? *Alcoholism, Clinical and Experimental Research* **34**:575–587.

Mott, J.L., S. Kobayashi, S.F. Bronk, and G.J. Gores. 2007. mir-29 regulates Mcl-1 protein expression and apoptosis. *Oncogene* **26**:6133–6140.

Murakami, Y., H.H. Aly, A. Tajima, I. Inoue, and K. Shimotohno. 2009. Regulation of the hepatitis C virus genome replication by *miR-199a*. *Journal of Hepatology* **50**:453–460.

Murakami, Y., H. Toyoda, M. Tanaka, M. Kuroda, Y. Harada, F. Matsuda, A. Tajima, N. Kosaka, T. Ochiya, and K. Shimotohno. 2011. The progression of liver fibrosis is related with overexpression of the *miR-199* and 200 families. *PloS ONE* **6**:e16081.

Nagata, H., E. Hatano, M. Tada, M. Murata, K. Kitamura, H. Asechi, M. Narita, A. Yanagida, N. Tamaki, S. Yagi, I. Ikai, K. Matsuzaki, et al. 2009. Inhibition of c-Jun NH2-terminal kinase switches Smad3 signaling from oncogenesis to tumor- suppression in rat hepatocellular carcinoma. *Hepatology* **49**:1944–1953.

Nelson, K.M. and G.J. Weiss. 2008. MicroRNAs and cancer: past, present, and potential future. *Molecular Cancer Therapeutics* **7**:3655–3660.

Olaru, A.V., G. Ghiaur, S. Yamanaka, D. Luvsanjav, F. An, I. Popescu, S. Alexandrescu, S. Allen, T.M. Pawlik, M. Torbenson, C. Georgiades, L.R. Roberts, et al. 2011. MicroRNA down-regulated in human cholangiocarcinoma control cell cycle through multiple targets involved in the G1/S checkpoint. *Hepatology* **54**:2089–2098.

Padgett, K.A., R.Y. Lan, P.C. Leung, A. Lleo, K. Dawson, J. Pfeiff, T.K. Mao, R.L. Coppel, A.A. Ansari, and M.E. Gershwin. 2009. Primary biliary cirrhosis is associated with altered hepatic microRNA expression. *Journal of Autoimmunity* **32**:246–253.

Park, J.K., T. Kogure, G.J. Nuovo, J. Jiang, L. He, J.H. Kim, M.A. Phelps, T.L. Papenfuss, C.M. Croce, T. Patel, and T.D. Schmittgen. 2011. *miR-221* silencing blocks hepatocellular carcinoma and promotes survival. *Cancer Research* **71**:7608–7616.

Poupon, R. 2010. Primary biliary cirrhosis: a 2010 update. *Journal of Hepatology* **52**:745–758.

Qiu, L., H. Fan, W. Jin, B. Zhao, Y. Wang, Y. Ju, L. Chen, Y. Chen, Z. Duan, and S. Meng. 2010. *miR-122-induced* down-regulation of HO-1 negatively affects *miR-122-mediated* suppression of HBV. *Biochemical and Biophysical Research Communications* **398**:771–777.

Rao, R.K. 2008. Acetaldehyde-induced barrier disruption and paracellular permeability in Caco-2 cell monolayer. *Methods in Molecular Biology* **447**:171–183.

Razumilava, N., S.F. Bronk, R.L. Smoot, C.D. Fingas, N.W. Werneburg, L.R. Roberts, and J.L. Mott. 2012. *miR-25* targets TNF-related apoptosis inducing ligand (TRAIL) death receptor-4 and promotes apoptosis resistance in cholangiocarcinoma. *Hepatology* **55**:465–475.

Roderburg, C., G.W. Urban, K. Bettermann, M. Vucur, H. Zimmermann, S. Schmidt, J. Janssen, C. Koppe, P. Knolle, M. Castoldi, F. Tacke, C. Trautwein, et al. 2011. Micro-RNA profiling reveals a role for *miR-29* in human and murine liver fibrosis. *Hepatology* **53**:209–218.

Salas, J.T., J.M. Banales, S. Sarvide, S. Recalde, A. Ferrer, I. Uriarte, R.P. Oude Elferink, J. Prieto, and J.F. Medina. 2008. Ae2a,b-deficient mice develop antimitochondrial antibodies and other features resembling primary biliary cirrhosis. *Gastroenterology* **134**:1482–1493.

Sekine, S., R. Ogawa, R. Ito, N. Hiraoka, M.T. McManus, Y. Kanai, and M. Hebrok. 2009. Disruption of Dicer1 induces dysregulated fetal gene expression and promotes hepatocarcinogenesis. *Gastroenterology* **136**:2304–2315 e2301-2304.

Sekiya, Y., T. Ogawa, M. Iizuka, K. Yoshizato, K. Ikeda, and N. Kawada. 2011. Down-regulation of cyclin E1 expression by microRNA-195 accounts for interferon-beta-induced inhibition of hepatic stellate cell proliferation. *Journal of Cellular Physiology* **226**:2535–2542.

Selaru, F.M., A.V. Olaru, T. Kan, S. David, Y. Cheng, Y. Mori, J. Yang, B. Paun, Z. Jin, R. Agarwal, J.P. Hamilton, J. Abraham, et al. 2009. MicroRNA-21 is overexpressed in human cholangiocarcinoma and regulates programmed cell death 4 and tissue inhibitor of metalloproteinase 3. *Hepatology* **49**:1595–1601.

Sharma, A.D., N. Narain, E.M. Handel, M. Iken, N. Singhal, T. Cathomen, M.P. Manns, H.R. Scholer, M. Ott, and T. Cantz. 2011. MicroRNA-221 regulates FAS-induced fulminant liver failure. *Hepatology* **53**:1651–1661.

Starkey Lewis, P.J., J. Dear, V. Platt, K.J. Simpson, D.G. Craig, D.J. Antoine, N.S. French, N. Dhaun, D.J. Webb, E.M. Costello, J.P. Neoptolemos, J. Moggs, et al. 2011. Circulating microRNAs as potential markers of human drug-induced liver injury. *Hepatology* **54**:1767–1776.

Stutes, M., S. Tran, and S. DeMorrow. 2007. Genetic and epigenetic changes associated with cholangiocarcinoma: from DNA methylation to microRNAs. *World Journal of Gastroenterology: WJG* **13**:6465–6469.

Szabo, G. and S. Bala. 2010. Alcoholic liver disease and the gut-liver axis. *World Journal of Gastroenterology: WJG* **16**:1321–1329.

Tang, Y., A. Banan, C.B. Forsyth, J.Z. Fields, C.K. Lau, L.J. Zhang, and A. Keshavarzian. 2008. Effect of alcohol on *miR-212* expression in intestinal epithelial cells and its potential role in alcoholic liver disease. *Alcoholism, Clinical and Experimental Research* **32**:355–364.

Vernon, G., A. Baranova, and Z.M. Younossi. 2011. Systematic review: the epidemiology and natural history of non-alcoholic fatty liver disease and non-alcoholic steatohepatitis in adults. *Alimentary Pharmacology & Therapeutics* **34**:274–285.

Wang, B., S.H. Hsu, S. Majumder, H. Kutay, W. Huang, S.T. Jacob, and K. Ghoshal. 2010. TGFbeta-mediated upregulation of hepatic *miR-181b* promotes hepatocarcinogenesis by targeting TIMP3. *Oncogene* **29**:1787–1797.

Wang, X.W., N.H. Heegaard, and H. Orum. 2012. MicroRNAs in liver disease. *Gastroenterology* **142**:1431–1443.

Yang, F., Y. Yin, F. Wang, Y. Wang, L. Zhang, Y. Tang, and S. Sun. 2010. *miR-17-5p* Promotes migration of human hepatocellular carcinoma cells through the p38 mitogen-activated protein kinase-heat shock protein 27 pathway. *Hepatology* **51**:1614–1623.

Yang, H., T.W. Li, J. Peng, X. Tang, K.S. Ko, M. Xia, and M.A. Aller. 2011. A mouse model of cholestasis-associated cholangiocarcinoma and transcription factors involved in progression. *Gastroenterology* **141**:378–388, 388 e371–374.

Yeligar, S., H. Tsukamoto, and V.K. Kalra. 2009. Ethanol-induced expression of ET-1 and ET-BR in liver sinusoidal endothelial cells and human endothelial cells involves hypoxia-inducible factor-1alpha and microrNA-199. *Journal of Immunology* **183**:5232–5243.

Yuan, B., R. Dong, D. Shi, Y. Zhou, Y. Zhao, M. Miao, and B. Jiao. 2011. Down-regulation of *miR-23b* may contribute to activation of the TGF-beta1/Smad3 signalling pathway during the termination stage of liver regeneration. *FEBS Letters* **585**:927–934.

Zhang, J., C. Han, and T. Wu. 2012. MicroRNA-26a promotes cholangiocarcinoma growth by activating beta-catenin. *Gastroenterology* **143**:246–256 e248.

Zheng, L., G.C. Lv, J. Sheng, and Y.D. Yang. 2010. Effect of miRNA-10b in regulating cellular steatosis level by targeting PPAR-alpha expression, a novel mechanism for the pathogenesis of NAFLD. *Journal of Gastroenterology and Hepatology* **25**:156–163.

Zhong, X.Y., J.H. Yu, W.G. Zhang, Z.D. Wang, Q. Dong, S. Tai, Y.F. Cui, and H. Li. 2012. MicroRNA-421 functions as an oncogenic miRNA in biliary tract cancer through down-regulating farnesoid X receptor expression. *Gene* **493**:44–51.

32

MicroRNA REGULATION IN MULTIPLE SCLEROSIS

Andreas Junker

*Department of Neuropathology, University Medical Center Goettingen,
Georg-August University, Goettingen, Germany*

I. Introduction	524
A. miRNAs in MS	525
II. miRNA Profiles in MS Patients	525
III. Involvement of miRNAs in the Pathophysiology of MS	529
A. Myelination	529
B. Immune System	529
C. Degenerative Component	532
D. miRNAs in EAE	532
IV. Concluding Remarks and Open Questions	533
Acknowledgments	533
References	533

ABBREVIATIONS

AKR1	Aldo-keto reductase family 1 member
BCL2L11	Bcl-2-like protein 11
C/EBP-α	CCAAT/enhancer-binding protein-alpha
CDKN1A	cyclin-dependent kinase inhibitor 1
CNS	central nervous system
EAE	experimental autoimmune encephalomyelitis

MicroRNAs in Medicine, First Edition. Edited by Charles H. Lawrie.
© 2014 John Wiley & Sons, Inc. Published 2014 by John Wiley & Sons, Inc.

ELOVL7	elongation of very long chain fatty acids protein 7
FoxP3	forkhead box P3
miRNA	microRNA
MOG	myelin oligodendrocyte glycoprotein
MRI	magnetic resonance imaging
MS	multiple sclerosis
NAWM	normal-appearing white matter
NMO	neuromyelitis optica
OL	oligodendrocyte lineage
PBMC	peripheral blood mononuclear cells
PLP	proteolipid protein
PPMS	primary progressive multiple sclerosis disease course
RRMS	relapsing-remitting multiple sclerosis disease course
Sirp-α	signal-regulatory protein alpha
SPMS	secondary progressive multiple sclerosis disease course
TGF-β	transforming growth factor beta
TH-17 cells	T-helper 17 cells
Treg cells	CD4+ CD25+ regulatory T cells

I. INTRODUCTION

Multiple sclerosis (MS) is an immune-mediated disorder of the central nervous system in which genetic susceptibility may play a substantial role (Hohlfeld 1997). The multifarious appearance of the disease, with its large degree of heterogeneity of clinical, genetic, magnetic resonance imaging (MRI) and pathological findings, indicates that more than one pathogenetic mechanism might contribute to tissue injury (Lassmann et al. 2007). The course of MS is not predictable. There are 250,000–350,000 patients with MS in the United States (Anderson et al. 1992), and more than 500,000 affected patients in Europe (Flachenecker et al. 2010). Twenty-one percent of all patients will need walking aids within 15 years of disease onset (Hurwitz 2011). Eighty percent of MS patients follow a relapsing-remitting disease course (RRMS). Symptoms and signs typically evolve over a period of several days, stabilize, and then often improve, spontaneously or in response to corticosteroids, within weeks (Noseworthy et al. 2000). RRMS typically begins in the second or third decade of life and has a female predominance of approximately 2:1 (Noseworthy et al. 2000). Persistent signs of central nervous system dysfunction may develop after a relapse, and the disease may progress between relapses (secondary progressive multiple sclerosis—SPMS) (Noseworthy et al. 2000). Primary progressive multiple sclerosis (PPMS), which accounts for 20% of MS patients, is characterized by a gradually progressive clinical course and a similar incidence among men and women (Noseworthy et al. 2000).

A complex interaction between invading immune cells and brain-resident cells leads to the development of inflammatory demyelinating brain lesions in the gray and white matter, but it is still unclear how the disease process is initiated or how it becomes chronic (Fugger et al. 2009; Goverman 2009; Steinman 2009). The pathogenic role of brain-invading autoreactive T lymphocytes has been firmly established in experimental autoimmune encephalomyelitis, an animal model of MS, laying the foundation for current concepts of the pathogenesis of MS (reviewed by Hohlfeld and Wekerle 2004). The T-cell infiltration of the lesions consists of CD8$^+$ T cells and CD4$^+$ T cells. Frequently, the cytotoxic CD8$^+$ T cells outnumber CD4$^+$ T cells, and clonal expansions of T cells in MS lesions have been mainly detected in the CD8⁺ compartment (reviewed by (Saxena et al. 2011)). Therefore, a pathophysiological role of CD8$^+$ T cells for MS lesion development has been postulated. Other components of the immune system, such as complement (Lucchinetti et al. 2000) and antibodies (Meinl et al. 2006; Hauser et al. 2008), mediate immune processes, at least in a lesion subset. Besides T cells, monocytes and microglia cells are abundant in active MS lesions. Microglia and macrophages can phagocytose and produce abundant cytokines and chemokines. Their activity level is jointly responsible for the destructiveness of the lesions (Benveniste 1997; Sriram and Rodriguez 1997; Platten and Steinman 2005; Breij et al. 2008). Astrocytes have modulatory functions during lesion formation. They produce diverse chemokines and cytokines, and they form the characteristic gliotic scar that is the prominent feature of the chronic inactive MS plaques that comprise one of the final stages of a lesion (Farina et al. 2007; Williams et al. 2007).

A. miRNAs in MS

Numerous miRNAs are expressed differentially in the peripheral blood of MS patients compared with controls (Table 32.1). This might be of relevance due to their regulatory influence on immune cells, such as T cells or B cells, which are associated with the disease pathophysiology of MS. Differences in the expression level of miRNAs are also detectable in lesions and in the normal-appearing white matter (NAWM) of the central nervous system (CNS) of patients as compared with appropriate control tissue (Table 32.1). miRNAs, which are differentially expressed in MS lesions and NAWM, were linked to glia cells and infiltrating immune cells. Some of those conspicuous miRNAs could be associated with the regulation of immunological functions or the control of degenerative or regenerative CNS mechanisms. This implies that miRNAs might regulate all aspects of MS pathology, such as inflammation, demyelination, and axonal damage. In addition, there is evidence that miRNAs influence remyelination, which often remains incomplete, and is restricted to the lesion edge of MS lesions (Prineas and Connell 1979; Barkhof et al. 2003; Bruck et al. 2003). Remyelination is understood to exert an axon-protective function (Irvine and Blakemore 2008; Piaton et al. 2010).

The next section addresses in detail the current understanding of miRNAs in MS lesion pathology. An improved understanding of the complex regulatory mechanisms of miRNAs in MS pathogenesis should help to define novel therapeutic targets and biomarkers for the disease.

II. miRNA PROFILES IN MS PATIENTS

In all studies which examined miRNA profiles in the blood or brain tissue of MS patients (Table 32.1), a difference in miRNA expression was detected between the patient groups and those of healthy individuals. However, there was little agreement between the studies regarding this difference in miRNA expression. Several reasons, such as differences in defining the patient cohorts or differences between the sources of the studied material, may be responsible for this. Despite clear differences between studies, a few commonly dysregulated miRNAs were described in more than one MS miRNA expression study. Namely, *miR-146a* (Fenoglio et al. 2011; Waschbisch et al. 2011), *miR-155* (Paraboschi et al. 2011; Waschbisch et al. 2011), and *miR-326* (Du et al. 2009; Waschbisch et al. 2011), which have been described as being up-regulated in more than one study in the peripheral blood mononuclear cells (PBMCs) of patients with RRMS. Additionally, *miR-22* was found to be up-regulated in the plasma of MS patients (Siegel et al. 2012), as well as

TABLE 32.1. Overview of Regulated miRNAs in MS

Tissue or Cell Type	miRNAs Most Significantly Up-Regulated versus Control	miRNAs Most Significantly Down-Regulated versus Control	Reference
All leucocytes from peripheral blood	MS patients during relapse: miR-18b-5p, miR-493, miR-599		Otaegui et al. (2009)
Peripheral blood leukocytes	MS patients with RRMS: miR-326		Du et al. (2009)
Whole peripheral blood	MS patients with RRMS: miR-145, miR-186, miR-664, miR-422a, miRNA-142-3p, miR-584, miR-223, miR-1275, miR-491-5p	MS patients with RRMS: miR-20b	Keller et al. (2009)
Purified CD4+ and CD8+ T cells and B cells	CD4+ T cells from MS patients with RRMS: miR-485-3p, miR-376a, miR-1, miR-497, miR-193a, miR-200b, miR-126, miR-486, miR-17-5p CD8+ T cells from MS patients with RRMS: miR-629 B cells from MS patients with RRMS: miR-497	CD4+ T cells from MS patients with RRMS: miR-34a CD8+ T cells: miR-30a-3p, miR-149, miR-497 B cells from MS patients with RRMS: miR-92, miR-135b, miR-153, miR-189, miR-422a	Lindberg et al. (2010)
Whole peripheral blood	MS patients with PPMS, SPMS and RRMS: miR-768-3p	MS patients with PPMS, SPMS and RRMS: hsa-let-7d, hsa-let-7f, hsa-let-7g, hsa-let-7i, miR-106a, miR-126, miR-126*, miR-140-5p, miR-15a, miR-15b, miR-16, miR-17, miR-20a, miR-20b, miR-211, miR-27a, miR-27b, miR-374a, miR-454, miR-510, miR-579, miR-623, miR-624*, miR-93, miR-98	Cox et al. (2010)
Active and inactive lesions in the brains of MS patients	Active MS lesions: miR-650, miR-155, miR-326, miR-142-3p, miR-146a, miR-146b, miR-34a, miR-21, miR-23a, miR-199a, miR-27a, miR-142-5p, miR-193a, miR-15a, miR-200c, miR-130a, miR-223, miR-22, miR-320, miR-214 Inactive MS lesions: miR-629, miR-148a, miR-23a, miR-28, miR-195, miR-497, miR-214, miR-130a, miR-135a, miR-204, miR-200c, miR-660, miR-152, miR-30a-5p, miR-30a-3p, miR-365, miR-532, miR-126, hsa-let7c, miR-20b, miR-30d, miR-9	Active MS lesions: miR-656, miR-184, miR-139, miR-23b, miR-328, miR-487b, miR-181c, miR-340 Inactive MS lesions: miR-219, miR-338, miR-642, miR-181b, miR-18a, miR-340, miR-190, miR-213, miR-330, miR-181d, miR-151, miR-23b, miR-140	Junker et al. (2009)

Normal-appearing white matter in the brains of MS patients	NAWM: miR-25, miR-505*, miR-320b, miR-320a, miR-338-3p, miR-181b, miR-125b1, miR-92a, miR-155, miR-584, miR-219-2-3p, miR-338-5p, miR-219-5p, miR-142-5p	Noorbakhsh et al. (2011)
Purified Treg cells	CD4+CD25high Treg cells from MS patients with RRMS: miR-29c, miR-107, miR-210, hsa-let-7i, miR-15a, miR-19a, miR-19b, miR-301a, miR-22, miR-106b, miR-29a, miR-93, miR-148a, miR-590-5p, miR-223, miR-221	De Santis et al. (2010)
Peripheral blood mononuclear cells (PBMC)	MS patients with RRMS: miR-21, miR-146a, miR-146b	Fenoglio et al. (2011)
PBMC, purified CD4+ T cells, purified CD8+ T cells		Lorenzi et al. (2012)
Plasma samples	Blood plasma from MS patients: miR-614, miR-572, miR-648, miR-1826, miR-422a, miR-22	Siegel et al. (2012)
PBMC	MS patients with RRMS: miR-155	Paraboschi et al. (2011)
PBMC	MS patients with PPMS, SPMS and RRMS: miR-524-3p, miR-223*, miR-550*	Martinelli-Boneschi et al. (2012)
	NAWM: miR-7, miR-299-5p, miR-135a, miR-218, miR-129-3p, miR-9, miR-128, miR-130A, miR-126, miR-335, miR-98	
	CD4+CD25high Treg cells from MS patients with RRMS: miR-138-2*, miR-324-3p, miR-338-5p, miR-512-3p, miR-564, miR-886-3p, miR-489	
	CD4+ T cells from MS patients with RRMS: miR-15a-5p, miR-16	
	Blood plasma from MS patients: miR-1979	
	MS patients with PPMS, SPMS and RRMS: miR-363, miR-31*, miR-876-3p, hsa-let7g, miR-181c, miR-374a*, miR-150	

(Continued)

TABLE 32.1. (*Continued*)

Tissue or Cell Type	miRNAs Most Significantly *Up-Regulated* versus Control	miRNAs Most Significantly *Down-Regulated* versus Control	Reference
Purified naive CD4+CD45RA+ T cells and purified memory CD4+CD45RO+ T cells	Purified naive CD4+CD45RA+ T cells of MS patients with PPMS, SPMS and RRMS: *miR-128* Purified naive CD4+CD45RA+ T cells of MS patients with SPMS and RRMS: *miR-27b* Purified memory CD4+CD45RO+ T cells of MS patients with SPMS and RRMS: *miR-340*		Guerau-de-Arellano et al. (2011)
PBMC	MS patients with RRMS: *miR-326, miR-155, miR-146a, miR-142-3p*		Waschbisch et al. (2011)
B cells	Purified B cells of untreated vs. natalizumab treated MS patients with RRMS: *miR-551a, miR-19b, miR-191, miR-598, miR-150, EBV-miR-BART3-5p, miR-142-5p, EBV-miR-BART11-5p, miR-383, miR-106b*	Purified B cells of MS patients with RRMS: *miR-515, miR-411*, miR-25, miR-16, miR-297a, miR-329, miR-299-5p, miR-520g, miR-486-5p, miR-363, miR-221, miR-19b, miR-644, EBV-miR-BART7, miR-616*, miR-551a, miR-106b, miR-200a, miR-191, miR-152, miR-520A-3p, miR-let-7i, miR-203, EBV-miR-BART19-5b, miR-7-1*, miR-649, miR-15a, miR-28-5p, miR-204, miR-582-5p, miR-624*, miR-181a, miR-599, miR-548C-5p, EBV-miR-BART11-5p, miR-103-2, miR-93, miR-107, miR-655, miR-151-5p, miR-340*, miR-337-3p, miR-140-5p, miR-200b, miR-320b, miR-369-5p, miR-130b, miR-585, miR-218*	Sievers et al. (2012)

purified T-reg cells (De Santis et al. 2010). *miR-223* appeared to be up-regulated in patients' whole blood (Keller et al. 2009), as well as in the purified T-reg cells from MS patients (De Santis et al. 2010). Interestingly, *all* miRNAs that were found to be up-regulated in more than one study in the peripheral blood of patients were identified as up-regulated in the active brain lesions of MS patients (Junker et al. 2009). In contrast, *miR-16* was commonly down-regulated in $CD4^+$ T cells (Lorenzi et al. 2012), B cells (Sievers et al. 2012), and whole blood from MS patients (Cox et al. 2010).

III. INVOLVEMENT OF miRNAs IN THE PATHOPHYSIOLOGY OF MS

A. Myelination

The intact myelin sheath around axons is a necessary prerequisite for proper axonal function and axon protection (Nave and Trapp 2008). Demyelination in MS is one element in the pathophysiology of the disease that is jointly responsible for axonal destruction and consecutive disability of the patients (Kornek et al. 2000). Whether regulation of remyelination in MS necessitates equal miRNA regulation, such as the processes during primary (embryonic) myelination, still remains to be elaborated in detail.

The development and maintenance of the myelin sheath around axons require the regulatory function of miRNAs, as has been demonstrated in animal studies (Shin et al. 2009; Dugas et al. 2010; Zhao et al. 2010). Both were observed in animals in which Dicer-mediated miRNA processing was cell-specifically inactivated (using Cre recombinase under the control of selected promoters), either in early stages of oligodendrocyte development (Dugas et al. 2010; Zhao et al. 2010) or in mature oligodendrocytes (Shin et al. 2009). Dicer-mutant mice, in which Dicer was deleted in proteolipid-protein (PLP)-expressing oligodendrocytes, show neurodegenerative symptoms in terms of demyelination, oxidative damage, astrocytosis, and microglia activation—and possibly in combination with neuronal degeneration, a shorter lifespan. The main molecular components responsible for these effects were the *miR-219* and its target ELOVL7 (elongation of very long chain fatty acids protein 7) (Shin et al. 2009). Compared with controls, the oligodendrocyte lineage (OL)-specific Dicer mutant shows the down-regulation of several miRNAs, such as *miR-32*, *miR-144*, and *miR-219*, whereas *miR-7a*, *miR-7b*, *miR-181-1*, and *miR-592* appeared to be up-regulated (Shin et al. 2009). *miR-219* proved to be the most abundant miRNA in mature OL cells (Lau et al. 2008). The sound axonal condition and the generation of an intact myelin sheath require proper miRNA expression.

Moreover, specific miRNAs, such as *miR-219* (Dugas et al. 2010; Zhao et al. 2010) and *miR-338* (Zhao et al. 2010), are capable of initiating oligodendrocyte differentiation. Maturation of oligodendrocytes requires both miRNAs, probably due to the suppression of oligodendrocyte differentiation inhibitors (Figure 32.1).

Several other miRNAs, for example, *miR-23a*, might also participate in the physiology of myelination, as was recently shown (Lin and Fu 2009). *miR-23* modulates Lamin B1, which is an important factor for demyelination in adult-onset autosomal-dominant leukodystrophy.

B. Immune System

Diverse components of the innate and adaptive immune system participate in the pathophysiology of MS (for review, see Frohman et al. 2006). Some miRNAs that appeared to

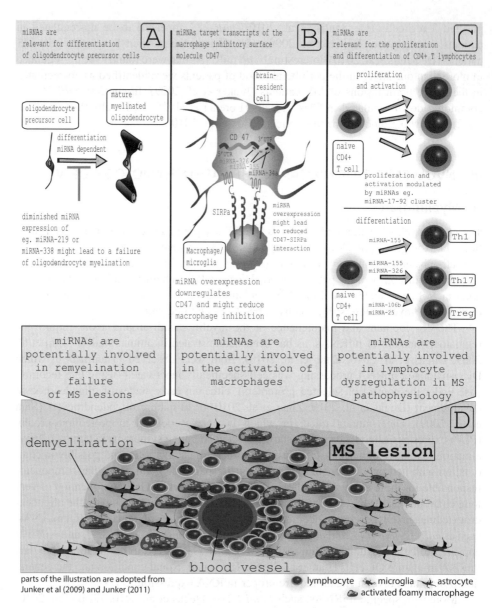

Figure 32.1. MiRNAs participate in diverse pathophysiological processes in MS. The differentiation of oligodendrocyte precursor cells is miRNA dependent. The same miRNAs are downregulated in inactive MS lesions (Junker et al. 2009), and thus might lead to a remyelination failure of the lesions (A). Transcripts of the immunomodulatory surface molecule CD47 are targeted by miRNAs, for example, *miR-155, miR-338*, and *miR-34a*, which are up-regulated in active MS lesions. A reduced interaction of CD47 with its ligand SIRPα on phagocytes might enhance their phagocytic activity (B). MiRNAs participate in the proliferation, activation, and differentiation of diverse lymphocyte subsets, such as CD4+ T cells. An involvement of Th17 cells or Treg cells in MS pathophysiology is suggested by several studies (see text) (C). These diverse miRNA interactions probably participate in several pathophysiological events in MS, such as lesion formation (D), immune cell dysregulation, and tissue destruction and regeneration. See color insert.

be regulated in MS lesions or MS patient blood were recently described as having regulatory functions in the immune system (O'Connell et al. 2010b) or in other autoimmune disorders (Pauley et al. 2009).

1. Acute Inflammatory Reaction. In the most frequent disease course of MS, the relapsing-remitting disease course, relapses are associated with CNS lesions with an acute inflammatory reaction and concomitant myelin destruction involving numerous macrophages with a sparse T cell infiltration (Noseworthy et al. 2000).

Several miRNAs appear to be up-regulated in these active demyelinating MS CNS lesions. *miR-155* was among the most up-regulated of miRNAs in active MS lesions compared with control brain specimens (Junker et al. 2009). *miR-155* is a miRNA with many different functions in the immune system. Notably, the B cell response, the Th2 helper T cell response, and the differentiation and activation of microglia and macrophages are affected (Rodriguez et al. 2007; Thai et al. 2007; Ponomarev et al. 2012).

Compared with naive T cells, $CD4^+$ $CD25^+$ regulatory T cells (Treg cells) display a distinct miRNA profile with up-regulation of miRNAs, such as *miR-21, miR-146a, miR-223, miR-214, miR-125a,* and *miR-155,* and down-regulation of *miR-150* and *miR-142-5p* (Cobb et al. 2006). Overexpression of the transcription factor forkhead box P3 (FoxP3) was shown to regulate the expression of most of these miRNAs (Cobb et al. 2006). Interestingly, this miRNA expression pattern resembles that of activated $CD4^+$ T helper cells (Cobb et al. 2006). Treg cells play an important role in the suppression of autoimmunity and are potentially relevant in the pathophysiology of MS (For review see (Costantino et al. 2008)). Notably, *miR-21, miR-146a, miR-223, miR-215,* and *miR-155* are all up-regulated in active MS lesions (Junker et al. 2009). Differences between MS patient blood and control subjects with regard to miRNAs expressed in Treg cells were investigated in a study by De Santis and colleagues (De Santis et al. 2010). In this study, miRNAs of the *miR-106b-25* cluster were, among others, identified as being down-regulated in T reg cells from MS patients compared with controls. *miR-106b* and *miR-25* from this cluster were shown to modulate the transforming growth factor beta (TGF-β) signaling pathway through their action on cyclin-dependent kinase inhibitor 1 (CDKN1A) and Bcl-2-like protein 11 (BCL2L11) (Petrocca et al. 2008). TGF-β signaling is relevant for the differentiation and maturation of Treg cells (Yamagiwa et al. 2001; Zheng et al. 2002; Bettelli et al. 2006; Davidson et al. 2007). It might well be that Treg cell activity is modulated by the dysregulation of miRNAs of the *miR-106b-25* cluster in the disease course of MS (Figure 32.1). Some of the miRNAs described by De Santis and colleagues have previously been found by other groups describing miRNA expression in MS (Table 32.1).

Other T cell subsets, such as T helper 17 (Th-17) cells, are also regulated through miRNAs (Du et al. 2009). Th-17 cells are thought to play important roles in the pathophysiology of MS (Steinman 2008; Tzartos et al. 2008) (Figure 32.1). *miR-326* might be critically involved in the modulation of these cells and is up-regulated in active human MS lesions (Junker et al. 2009), as well as in CNS lesions of experimental autoimmune encephalomyelitis (EAE) of myelin oligodendrocyte glycoprotein $(MOG)_{35-55}$ immunized mice (Lescher et al. 2012).

In the blood of MS patients, *miR-326* was found to be up-regulated during relapses, and the inhibition of *miR-326* was shown to reduce symptoms in a murine EAE model by preventing Th17 cell differentiation through a T cell-intrinsic mechanism (Du et al. 2009). Its action could be modulated by targeting Ets-1 (Du et al. 2009), which is a known negative regulator of Th-17 cells.

2. Chronic Inflammatory Reaction. One prominent feature of MS is a chronic inflammatory component in the CNS of patients (Frischer et al. 2009). Activated microglia cells, macrophages, activated astrocytes, and a sparse infiltration of T cells provide an inflammatory and oxidative milieu that is responsible for axonal damage and the consecutive progression of disability in affected people (Haider et al. 2011). Interestingly, active MS lesions, chronic inactive MS lesions, as well as the NAWM that surrounds the lesion area have considerably different miRNA profiles compared with control tissue (Junker et al. 2009; Noorbakhsh et al. 2011). miRNAs that appear to be up- or down-regulated in these tissue areas could be linked to the infiltrating immune cells, as well as to the brain-resident cells, such as microglia and astrocytes.

One relevant target of miRNAs regulated in active MS lesions is the immunomodulatory surface molecule CD47. The phagocytic activity of macrophages is inhibited by the interaction of CD47 with its ligand, signal-regulatory protein alpha (Sirp-α) (Oldenborg et al. 2001; Yamao et al. 2002; Ishikawa-Sekigami et al. 2006). This interaction also prevents the phenotypic and functional maturation of immature dendritic cells, as well as the cytokine production by mature dendritic cells (Latour et al. 2001). Due to its inhibitory effect on macrophages, CD47 has been regarded as a "don't eat me signal" and as a kind of "marker of self" (Oldenborg et al. 2000; Yamao et al. 2002). *miR-155, miR-34a*, and *miR-326* target the transcript of CD47, which is down-regulated in active and inactive MS lesions (Koning et al. 2007; Junker et al. 2009), and they might therefore have a direct effect on the activity of macrophages in active MS lesions. (Junker et al. 2009) (Figure 32.1).

The activation of microglia cells is modulated by *miR-124* (Ponomarev et al. 2011). *miR-124* is expressed in microglia and inhibits the transcription factor CCAAT/enhancer-binding protein alpha (C/EBP-α) and its downstream target, the transcription factor PU.1 (Ponomarev et al. 2011). This promotes microglia quiescence, and the *in vivo* application of *miR-124* in EAE mice caused marked suppression of disease (Ponomarev et al. 2011).

C. Degenerative Component

It was shown that miRNAs might affect inflammatory and degenerative processes in MS brains by modulating the neurosteroid biosynthesis (Noorbakhsh et al. 2011). Levels of important neurosteroids, including allopregnanolone, are suppressed in the white matter of patients with MS. Moreover, transcripts of the enzymes aldo-keto reductase family 1 member (AKR1) C1 and AKR1 C2, which are essential for neurosteroid biosynthesis in the brain and which are suppressed in MS brain white matter, are targeted by the miRNAs *miR-338, miR-155*, and *miR-491* (Noorbakhsh et al. 2011). miRNA-mediated differences in neurosteroidogenic mechanisms might therefore support inflammation, demyelination, and axonal injury, which is also reinforced by the finding that allopregnanolone treatment suppresses neuroinflammation, demyelination, and axonal injury in mice affected by EAE (Noorbakhsh et al. 2011).

D. miRNAs in EAE

Several miRNAs, among them *miR-155*, which are highly up-regulated in active MS brain white matter lesions, can also be found to be up-regulated in EAE models of the disease (Lescher et al. 2012). Notably, *miR-155* knockout mice are highly resistant to EAE, which demonstrates the fundamental influence of this miRNA on the disease pathology (O'Connell et al. 2010a).

Diverse miRNA functions may be studied in animal models, also due to the fact that miRNA expression in EAE models resembles that of human MS.

IV. CONCLUDING REMARKS AND OPEN QUESTIONS

Alterations in the profiles of miRNAs in blood cells or body fluids might serve as MS biomarkers for assessing disease activity or disease subtype allocation. It might also well be possible that miRNAs serve as markers for successful therapy in MS. Only recently, it was demonstrated that the immunomodulatory drug glatirameracetate is able to normalize dysregulated miRNA expression in relapsing remitting MS (Waschbisch et al. 2011). Targeting of miRNAs represents a fascinatingly novel therapeutic strategy for regulating pathogenic gene expression (Czech 2006; Elmen et al. 2008). Discovering whether this is also applicable for MS will be a matter of exciting future studies.

ACKNOWLEDGMENTS

The author thanks Christine Crozier and Stefan Nessler for critical reading of the manuscript.

REFERENCES

Anderson, D.W., J.H. Ellenberg, C.M. Leventhal, S.C. Reingold, M. Rodriguez, and D.H. Silberberg. 1992. Revised estimate of the prevalence of multiple sclerosis in the United States. *Ann Neurol* **31**:333–336.

Barkhof, F., W. Bruck, C.J. De Groot, E. Bergers, S. Hulshof, J. Geurts, C.H. Polman, and P. van der Valk. 2003. Remyelinated lesions in multiple sclerosis: magnetic resonance image appearance. *Arch Neurol* **60**:1073–1081.

Benveniste, E.N. 1997. Role of macrophages/microglia in multiple sclerosis and experimental allergic encephalomyelitis. *J Mol Med (Berl)* **75**:165–173.

Bettelli, E., Y. Carrier, W. Gao, T. Korn, T.B. Strom, M. Oukka, H.L. Weiner, and V.K. Kuchroo. 2006. Reciprocal developmental pathways for the generation of pathogenic effector TH17 and regulatory T cells. *Nature* **441**:235–238.

Breij, E.C., B.P. Brink, R. Veerhuis, C. van den Berg, R. Vloet, R. Yan, C.D. Dijkstra, P. van der Valk, and L. Bo. 2008. Homogeneity of active demyelinating lesions in established multiple sclerosis. *Ann Neurol* **63**:16–25.

Bruck, W., T. Kuhlmann, and C. Stadelmann. 2003. Remyelination in multiple sclerosis. *J Neurol Sci* **206**:181–185.

Cobb, B.S., A. Hertweck, J. Smith, E. O'Connor, D. Graf, T. Cook, S.T. Smale, S. Sakaguchi, F.J. Livesey, A.G. Fisher, and M. Merkenschlager. 2006. A role for Dicer in immune regulation. *J Exp Med* **203**:2519–2527.

Costantino, C.M., C. Baecher-Allan, and D.A. Hafler. 2008. Multiple sclerosis and regulatory T cells. *J Clin Immunol* **28**:697–706.

Cox, M.B., M.J. Cairns, K.S. Gandhi, A.P. Carroll, S. Moscovis, G.J. Stewart, S. Broadley, R.J. Scott, D.R. Booth, and J. Lechner-Scott. 2010. MicroRNAs *miR-17* and *miR-20a* inhibit T cell activation genes and are under-expressed in MS whole blood. *PLoS ONE* **5**:e12132.

Czech, M.P. 2006. MicroRNAs as therapeutic targets. *N Engl J Med* **354**:1194–1195.

Davidson, T.S., R.J. DiPaolo, J. Andersson, and E.M. Shevach. 2007. Cutting Edge: IL-2 is essential for TGF-beta-mediated induction of Foxp3+ T regulatory cells. *J Immunol* **178**:4022–4026.

De Santis, G., M. Ferracin, A. Biondani, L. Caniatti, T.M. Rosaria, M. Castellazzi, B. Zagatti, L. Battistini, G. Borsellino, E. Fainardi, R. Gavioli, M. Negrini, R. Furlan, and E. Granieri. 2010. Altered miRNA expression in T regulatory cells in course of multiple sclerosis. *J Neuroimmunol* **226**:165–171.

Du, C., C. Liu, J. Kang, G. Zhao, Z. Ye, S. Huang, Z. Li, Z. Wu, and G. Pei. 2009. MicroRNA *miR-326* regulates TH-17 differentiation and is associated with the pathogenesis of multiple sclerosis. *Nat Immunol* **10**:1252–1259.

Dugas, J.C., T.L. Cuellar, A. Scholze, B. Ason, A. Ibrahim, B. Emery, J.L. Zamanian, L.C. Foo, M.T. McManus, and B.A. Barres. 2010. Dicer1 and *miR-219* are required for normal oligodendrocyte differentiation and myelination. *Neuron* **65**:597–611.

Elmen, J., M. Lindow, S. Schutz, M. Lawrence, A. Petri, S. Obad, M. Lindholm, M. Hedtjarn, H.F. Hansen, U. Berger, S. Gullans, P. Kearney, P. Sarnow, E.M. Straarup, and S. Kauppinen. 2008. LNA-mediated microRNA silencing in non-human primates. *Nature* **452**:896–899.

Farina, C., F. Aloisi, and E. Meinl. 2007. Astrocytes are active players in cerebral innate immunity. *Trends Immunol* **28**:138–145.

Fenoglio, C., C. Cantoni, M. De Riz, E. Ridolfi, F. Cortini, M. Serpente, C. Villa, C. Comi, F. Monaco, L. Mellesi, S. Valzelli, N. Bresolin, D. Galimberti, and E. Scarpini. 2011. Expression and genetic analysis of miRNAs involved in CD4+ cell activation in patients with multiple sclerosis. *Neurosci Lett* **504**:9–12.

Flachenecker, P., L. Khil, S. Bergmann, M. Kowalewski, I. Pascu, F. Perez-Miralles, J. Sastre-Garriga, and T. Zwingers. 2010. Development and pilot phase of a European MS register. *J Neurol* **257**:1620–1627.

Frischer, J.M., S. Bramow, A. Dal-Bianco, C.F. Lucchinetti, H. Rauschka, M. Schmidbauer, H. Laursen, P.S. Sorensen, and H. Lassmann. 2009. The relation between inflammation and neurodegeneration in multiple sclerosis brains. *Brain* **132**:1175–1189.

Frohman, E.M., M.K. Racke, and C.S. Raine. 2006. Multiple sclerosis–the plaque and its pathogenesis. *N Engl J Med* **354**:942–955.

Fugger, L., M.A. Friese, and J.I. Bell. 2009. From genes to function: the next challenge to understanding multiple sclerosis. *Nat Rev Immunol* **9**:408–417.

Goverman, J. 2009. Autoimmune T cell responses in the central nervous system. *Nat Rev Immunol* **9**:393–407.

Guerau-de-Arellano, M., K.M. Smith, J. Godlewski, Y. Liu, R. Winger, S.E. Lawler, C.C. Whitacre, M.K. Racke, and A.E. Lovett-Racke. 2011. Micro-RNA dysregulation in multiple sclerosis favours pro-inflammatory T-cell-mediated autoimmunity. *Brain* **134**:3578–3589.

Haider, L., M.T. Fischer, J.M. Frischer, J. Bauer, R. Hoftberger, G. Botond, H. Esterbauer, C.J. Binder, J.L. Witztum, and H. Lassmann. 2011. Oxidative damage in multiple sclerosis lesions. *Brain* **134**:1914–1924.

Hauser, S.L., E. Waubant, D.L. Arnold, T. Vollmer, J. Antel, R.J. Fox, A. Bar-Or, M. Panzara, N. Sarkar, S. Agarwal, A. Langer-Gould, and C.H. Smith. 2008. B-cell depletion with rituximab in relapsing-remitting multiple sclerosis. *N Engl J Med* **358**:676–688.

Hohlfeld, R. 1997. Biotechnological agents for the immunotherapy of multiple sclerosis. Principles, problems and perspectives. *Brain* **120**:865–916.

Hohlfeld, R. and H. Wekerle. 2004. Autoimmune concepts of multiple sclerosis as a basis for selective immunotherapy: from pipe dreams to (therapeutic) pipelines. *Proc Natl Acad Sci U S A* **101 Suppl 2**:14599–14606.

Hurwitz, B.J. 2011. Analysis of current multiple sclerosis registries. *Neurology* **76**:S7–S13.

Irvine, K.A. and W.F. Blakemore. 2008. Remyelination protects axons from demyelination-associated axon degeneration. *Brain* **131**:1464–1477.

REFERENCES

Ishikawa-Sekigami, T., Y. Kaneko, Y. Saito, Y. Murata, H. Okazawa, H. Ohnishi, P.A. Oldenborg, Y. Nojima, and T. Matozaki. 2006. Enhanced phagocytosis of CD47-deficient red blood cells by splenic macrophages requires SHPS-1. *Biochem Biophys Res Commun* **343**:1197–1200.

Junker, A., M. Krumbholz, S. Eisele, H. Mohan, F. Augstein, R. Bittner, H. Lassmann, H. Wekerle, R. Hohlfeld, and E. Meinl. 2009. MicroRNA profiling of multiple sclerosis lesions identifies modulators of the regulatory protein CD47. *Brain* **132**:3342–3352.

Keller, A., P. Leidinger, J. Lange, A. Borries, H. Schroers, M. Scheffler, H.P. Lenhof, K. Ruprecht, and E. Meese. 2009. Multiple sclerosis: microRNA expression profiles accurately differentiate patients with relapsing-remitting disease from healthy controls. *PLoS ONE* **4**:e7440.

Koning, N., L. Bo, R.M. Hoek, and I. Huitinga. 2007. Downregulation of macrophage inhibitory molecules in multiple sclerosis lesions. *Ann Neurol* **62**:504–514.

Kornek, B., M.K. Storch, R. Weissert, E. Wallstroem, A. Stefferl, T. Olsson, C. Linington, M. Schmidbauer, and H. Lassmann. 2000. Multiple sclerosis and chronic autoimmune encephalomyelitis: a comparative quantitative study of axonal injury in active, inactive, and remyelinated lesions. *Am J Pathol* **157**:267–276.

Lassmann, H., W. Bruck, and C.F. Lucchinetti. 2007. The immunopathology of multiple sclerosis: an overview. *Brain Pathol* **17**:210–218.

Latour, S., H. Tanaka, C. Demeure, V. Mateo, M. Rubio, E.J. Brown, C. Maliszewski, F.P. Lindberg, A. Oldenborg, A. Ullrich, G. Delespesse, and M. Sarfati. 2001. Bidirectional negative regulation of human T and dendritic cells by CD47 and its cognate receptor signal-regulator protein-alpha: down-regulation of IL-12 responsiveness and inhibition of dendritic cell activation. *J Immunol* **167**:2547–2554.

Lau, P., J.D. Verrier, J.A. Nielsen, K.R. Johnson, L. Notterpek, and L.D. Hudson. 2008. Identification of dynamically regulated microRNA and mRNA networks in developing oligodendrocytes. *J Neurosci* **28**:11720–11730.

Lescher, J., F. Paap, V. Schultz, L. Redenbach, U. Scheidt, H. Rosewich, S. Nessler, E. Fuchs, J. Gartner, W. Bruck, and A. Junker. 2012. MicroRNA regulation in experimental autoimmune encephalomyelitis in mice and marmosets resembles regulation in human multiple sclerosis lesions. *J Neuroimmunol* **246**:27–33.

Lin, S.T. and Y.H. Fu. 2009. *miR-23* regulation of lamin B1 is crucial for oligodendrocyte development and myelination. *Dis Model Mech* **2**:178–188.

Lindberg, R.L., F. Hoffmann, M. Mehling, J. Kuhle, and L. Kappos. 2010. Altered expression of *miR-17-5p* in CD4+ lymphocytes of relapsing-remitting multiple sclerosis patients. *Eur J Immunol* **40**:888–898.

Lorenzi, J.C., D.G. Brum, D.L. Zanette, S.A. de Paula Alves, F.G. Barbuzano, A.C. Dos Santos, A.A. Barreira, and W.A. Silva Jr. 2012. *miR-15a* and 16-1 are downregulated in CD4(+) T cells of multiple sclerosis relapsing patients. *Int J Neurosci* **122**(8):466–471.

Lucchinetti, C., W. Bruck, J. Parisi, B. Scheithauer, M. Rodriguez, and H. Lassmann. 2000. Heterogeneity of multiple sclerosis lesions: implications for the pathogenesis of demyelination. *Ann Neurol* **47**:707–717.

Martinelli-Boneschi, F., C. Fenoglio, P. Brambilla, M. Sorosina, G. Giacalone, F. Esposito, M. Serpente, C. Cantoni, E. Ridolfi, M. Rodegher, L. Moiola, B. Colombo, M. De Riz, V. Martinelli, E. Scarpini, G. Comi, and D. Galimberti. 2012. MicroRNA and mRNA expression profile screening in multiple sclerosis patients to unravel novel pathogenic steps and identify potential biomarkers. *Neurosci Lett* **508**:4–8.

Meinl, E., M. Krumbholz, and R. Hohlfeld. 2006. B lineage cells in the inflammatory central nervous system environment: migration, maintenance, local antibody production, and therapeutic modulation. *Ann Neurol* **59**:880–892.

Nave, K.A. and B.D. Trapp. 2008. Axon-glial signaling and the glial support of axon function. *Annu Rev Neurosci* **31**:535–561.

Noorbakhsh, F., K.K. Ellestad, F. Maingat, K.G. Warren, M.H. Han, L. Steinman, G.B. Baker, and C. Power. 2011. Impaired neurosteroid synthesis in multiple sclerosis. *Brain* **134**:2703–2721.

Noseworthy, J.H., C. Lucchinetti, M. Rodriguez, and B.G. Weinshenker. 2000. Multiple sclerosis. *N Engl J Med* **343**:938–952.

O'Connell, R.M., D. Kahn, W.S. Gibson, J.L. Round, R.L. Scholz, A.A. Chaudhuri, M.E. Kahn, D.S. Rao, and D. Baltimore. 2010a. MicroRNA-155 promotes autoimmune inflammation by enhancing inflammatory T cell development. *Immunity* **33**:607–619.

O'Connell, R.M., D.S. Rao, A.A. Chaudhuri, and D. Baltimore. 2010b. Physiological and pathological roles for microRNAs in the immune system. *Nat Rev Immunol* **10**:111–122.

Oldenborg, P.A., H.D. Gresham, and F.P. Lindberg. 2001. CD47-signal regulatory protein alpha (SIRPalpha) regulates Fcgamma and complement receptor-mediated phagocytosis. *J Exp Med* **193**:855–862.

Oldenborg, P.A., A. Zheleznyak, Y.F. Fang, C.F. Lagenaur, H.D. Gresham, and F.P. Lindberg. 2000. Role of CD47 as a marker of self on red blood cells. *Science* **288**:2051–2054.

Otaegui, D., S.E. Baranzini, R. Armananzas, B. Calvo, M. Munoz-Culla, P. Khankhanian, I. Inza, J.A. Lozano, T. Castillo-Trivino, A. Asensio, J. Olaskoaga, and A. López de Munain. 2009. Differential micro RNA expression in PBMC from multiple sclerosis patients. *PLoS ONE* **4**:e6309.

Paraboschi, E.M., G. Solda, D. Gemmati, E. Orioli, G. Zeri, M.D. Benedetti, A. Salviati, N. Barizzone, M. Leone, S. Duga, and R. Asselta. 2011. Genetic association and altered gene expression of mir-155 in multiple sclerosis patients. *Int J Mol Sci* **12**:8695–8712.

Pauley, K.M., S. Cha, and E.K. Chan. 2009. MicroRNA in autoimmunity and autoimmune diseases. *J Autoimmun* **32**:189–194.

Petrocca, F., A. Vecchione, and C.M. Croce. 2008. Emerging role of *miR-106b-25/miR-17-92* clusters in the control of transforming growth factor beta signaling. *Cancer Res* **68**:8191–8194.

Piaton, G., R.M. Gould, and C. Lubetzki. 2010. Axon-oligodendrocyte interactions during developmental myelination, demyelination and repair. *J Neurochem* **114**:1243–1260.

Platten, M. and L. Steinman. 2005. Multiple sclerosis: trapped in deadly glue. *Nat Med* **11**:252–253.

Ponomarev, E.D., T. Veremeyko, N. Barteneva, A.M. Krichevsky, and H.L. Weiner. 2011. MicroRNA-124 promotes microglia quiescence and suppresses EAE by deactivating macrophages via the C/EBP-alpha-PU.1 pathway. *Nat Med* **17**:64–70.

Ponomarev, E.D., T. Veremeyko, and H.L. Weiner. 2012. MicroRNAs are universal regulators of differentiation, activation, and polarization of microglia and macrophages in normal and diseased CNS. *Glia* **61**(1):91–103.

Prineas, J.W. and F. Connell. 1979. Remyelination in multiple sclerosis. *Ann Neurol* **5**:22–31.

Rodriguez, A., E. Vigorito, S. Clare, M.V. Warren, P. Couttet, D.R. Soond, S. van Dongen, R.J. Grocock, P.P. Das, E.A. Miska, D. Vetrie, K. Okkenhaug, A.J. Enright, G. Dougan, M. Turner, and A. Bradley. 2007. Requirement of bic/microRNA-155 for normal immune function. *Science* **316**:608–611.

Saxena, A., G. Martin-Blondel, L.T. Mars, and R.S. Liblau. 2011. Role of CD8 T cell subsets in the pathogenesis of multiple sclerosis. *FEBS Lett* **585**:3758–3763.

Shin, D., J.Y. Shin, M.T. McManus, L.J. Ptacek, and Y.H. Fu. 2009. Dicer ablation in oligodendrocytes provokes neuronal impairment in mice. *Ann Neurol* **66**:843–857.

Siegel, S.R., J. Mackenzie, G. Chaplin, N.G. Jablonski, and L. Griffiths. 2012. Circulating microRNAs involved in multiple sclerosis. *Mol Biol Rep* **39**:6219–6225.

Sievers, C., M. Meira, F. Hoffmann, P. Fontoura, L. Kappos, and R.L. Lindberg. 2012. Altered microRNA expression in B lymphocytes in multiple sclerosis: towards a better understanding of treatment effects. *Clin Immunol* **144**:70–79.

Sriram, S. and M. Rodriguez. 1997. Indictment of the microglia as the villain in multiple sclerosis. *Neurology* **48**:464–470.

REFERENCES

Steinman, L. 2008. A rush to judgment on Th17. *J Exp Med* **205**:1517–1522.

Steinman, L. 2009. A molecular trio in relapse and remission in multiple sclerosis. *Nat Rev Immunol* **9**:440–447.

Thai, T.H., D.P. Calado, S. Casola, K.M. Ansel, C. Xiao, Y. Xue, A. Murphy, D. Frendewey, D. Valenzuela, J.L. Kutok, M. Schmidt-Supprian, N. Rajewsky, G. Yancopoulos, A. Rao, and K. Rajewsky. 2007. Regulation of the germinal center response by microRNA-155. *Science* **316**: 604–608.

Tzartos, J.S., M.A. Friese, M.J. Craner, J. Palace, J. Newcombe, M.M. Esiri, and L. Fugger. 2008. Interleukin-17 production in central nervous system-infiltrating T cells and glial cells is associated with active disease in multiple sclerosis. *Am J Pathol* **172**:146–155.

Waschbisch, A., M. Atiya, R.A. Linker, S. Potapov, S. Schwab, and T. Derfuss. 2011. Glatiramer acetate treatment normalizes deregulated microRNA expression in relapsing remitting multiple sclerosis. *PLoS ONE* **6**:e24604.

Williams, A., G. Piaton, and C. Lubetzki. 2007. Astrocytes–friends or foes in multiple sclerosis? *Glia* **55**:1300–1312.

Yamagiwa, S., J.D. Gray, S. Hashimoto, and D.A. Horwitz. 2001. A role for TGF-beta in the generation and expansion of CD4+CD25+ regulatory T cells from human peripheral blood. *J Immunol* **166**:7282–7289.

Yamao, T., T. Noguchi, O. Takeuchi, U. Nishiyama, H. Morita, T. Hagiwara, H. Akahori, T. Kato, K. Inagaki, H. Okazawa, Y. Hayashi, T. Matozaki, K. Takeda, S. Akira, and M. Kasuga. 2002. Negative regulation of platelet clearance and of the macrophage phagocytic response by the transmembrane glycoprotein SHPS-1. *J Biol Chem* **277**:39833–39839.

Zhao, X., X. He, X. Han, Y. Yu, F. Ye, Y. Chen, T. Hoang, X. Xu, Q.S. Mi, M. Xin, F. Wang, B. Appel, and Q.R. Lu. 2010. MicroRNA-mediated control of oligodendrocyte differentiation. *Neuron* **65**:612–626.

Zheng, S.G., J.D. Gray, K. Ohtsuka, S. Yamagiwa, and D.A. Horwitz. 2002. Generation ex vivo of TGF-beta-producing regulatory T cells from CD4+CD25- precursors. *J Immunol* **169**:4183–4189.

33

THE ROLE OF MicroRNAs IN ALZHEIMER'S DISEASE

Shahar Barbash and Hermona Soreq

Department of Biological Chemistry and The Edmond and Lily Safra Center for Brain Sciences, The Hebrew University of Jerusalem, Jerusalem, Israel

I.	Introduction	540
II.	miRNAs in Alzheimer's Disease and Other Neurodegenerative Diseases	540
	A. One in Eight	540
	B. A Quest for More Effective Treatment	541
	C. miRNA Regulation in Neurodegenerative Diseases	542
III.	Neuronal miRNAs	542
	A. miRNAs and Memory	542
	B. Functions of miRNAs at the Synapse	543
	C. miRNAs as Controllers of Neuritic Structure, Synapse Formation and Maintenance	544
	D. miRNAs and Behavior	545
IV.	miRNAs in Inflammatory Processes and Their Link to the Nervous System	545
	A. Overreactive Immune System	545
	B. miRNAs at the Neuronal–Immune Interface	546
V.	miRNAs and AD Therapy	547
	A. Can miRNAs Serve as Diagnostic Biomarkers?	547
	B. Different Approaches for Therapeutic Use of miRNA Manipulations	547
	C. Potential Limitations of miRNA Manipulation Efforts	548
VI.	Future Prospects and Concluding Remarks	549
	Acknowledgments	549
	References	549

MicroRNAs in Medicine, First Edition. Edited by Charles H. Lawrie.
© 2014 John Wiley & Sons, Inc. Published 2014 by John Wiley & Sons, Inc.

ABBREVIATIONS

AChE	acetylcholinesterase
AD	Alzheimer's disease
ALS	amyotrophic lateral sclerosis
FGF20	fibroblast growth factor 20
fMRI	functional magnetic resonance imaging
LIMK1	Lim-domain-containing protein kinase 1
LTP	long-term potentiation
miRNA	microRNA
NMJ	neuromuscular junctions
PD	Parkinson's disease
RISK	RNA-induced silencing complex

I. INTRODUCTION

"Where is he? He said he'll be here at 12:00. . . . Did I ever tell you about her? The State of Israel would be officially declared in a couple of months. . . . The doctor said eat oranges, only oranges." This was a typical monologue of the grandfather of S.B. in his last years (around 2007; the State of Israel was declared in 1948), and this is the cognitive manifestation of Alzheimer's disease (AD) in a nutshell. Patients mix past and present, fail to identify even those people who are closest to them, or fail to maintain a single line of thought for longer than a minute (Welsh et al. 1991; Collie and Maruff 2000). The last stages of the disease entail living in constant confusion, constant disconnection from one's environment. But the molecular mechanisms leading to these consequences, especially in non-familial AD patients, are still incompletely understood (Benilova et al. 2012). This chapter describes the role of microRNAs (miRNAs) in AD while conveying the message of miRNA control of AD progression, covering harmful as well as protective *miR*-mediated processes in AD and describing neuronal and immune miRNAs as potential targets for therapeutic intervention. MiRNAs notably regulate the expression of a major fraction of the human genes in health and disease (He and Hannon 2004), and play essential roles in the development and maintenance of the nervous system (Fineberg et al. 2009). Correspondingly, terminally differentiated neurons may be particularly sensitive to aberrant miRNA regulation. Significant clinical implications and therapeutic value have recently been attributed to miRNAs, and new ways to employ therapeutic tools for manipulating their levels are rapidly emerging, especially in the field of cancer (Calin and Croce 2006; Esquela-Kerscher and Slack 2006). Given this progress, we predict parallel development in the near future in the realm of neurodegenerative disease.

II. miRNAs IN ALZHEIMER'S DISEASE AND OTHER NEURODEGENERATIVE DISEASES

A. One in Eight

It has been more than a century since Alois Alzheimer, a German physician, discovered AD, and its prevalence in Western societies increases steadily. According to the 2011

Alzheimer's Disease Facts and Figures report, which addresses the scope of the disease in the United States (http://www.alz.org/alzheimers_disease_facts_and_figures.asp), one in eight individuals aged 65 and older and half of those aged 85 and older is an AD patient. The corresponding health care, long-term assistance, and hospitalization costs are projected to increase from $183 billion in 2011 to $1.1 trillion in 2050. Despite the overwhelming scope of the disease and immense research efforts, no disease-changing treatment is yet available, and current treatment is limited to palliative effects (Hardy and Selkoe 2002), further highlighting the unmet need for new therapeutic directions.

The first signs of the disease involve difficulties in forming new memories and acquiring new information. These cognitive abilities presumably break down first due to damage to nervous system communication and loss of neurons that likely starts at brain regions engaged in memory formation, such as the hippocampus. By the time an individual encounters sufficient cognitive problems to be identified as an AD patient, he or she already lacks a major fraction of brain neurons. When the brain subsequently consumes all of its compensating power, it loses the ability to perform the same task with fewer nerve cells and to switch a task from one brain region to another; consequently, access to remote memories becomes difficult or is entirely lost. Therapeutic intervention at this late time point involves the use of inhibitors of acetylcholinesterase (AChE) for limiting the degradation of the neurotransmitter acetylcholine, the levels of which decline in the AD brain (Schliebs and Arendt 2006). However, this treatment is palliative at best; it may minimize the memory impairment and attenuate the functional deterioration, but fails to delay or reverse further loss of neurons. The amazing flexibility and redundancy of the brain is therefore a major reason for the relatively late diagnosis of the disease, which is one of the obstacles confounding effective treatment. Importantly, this flexibility also offers hope for novel therapeutic intervention tools aimed at changing regulatory processes, rather than the end point of cholinergic neurotransmission.

B. A Quest for More Effective Treatment

Alois Alzheimer first observed the pathological hallmarks accompanying the disease. These include extracellular deposits ("plaques") of the Aβ beta amyloid protein fragment, and the intracellular twisted strands of the cyto-skeletal protein tau ("tangles"). It is still argued if these pathological hallmarks are causally responsible for the brain malfunction or if they are the outcome of another "causative" molecular process, of an effort of the diseased brain to protect itself from the neurotoxicity of the amyloid peptide, or a combination of these possibilities. Nonetheless, most of the drugs that are currently being developed target these AD hallmarks. Some aim to avoid novel aggregation, others attempt to disintegrate existing aggregates, and yet others are immune system agents that recognize these protein aggregates as harmful entities and aim at inducing their effective clearance. The initial success in attenuating the aggregation phenotype was accompanied by encephalitis in some of the tested volunteers (Nicoll et al. 2003). This enforced a temporary halt of these clinical trials, and alternative ways are currently sought for avoiding these harmful consequences. Recent attention is focused on the soluble species of those Aβ fragments that build the plaques and that exert almost immediate toxic effects on synapse function and morphology (Benilova et al. 2012). Given that neurons that receive aberrant, excessive, or insufficient inputs degenerate and eventually die, the Aβ toxicity is thought to occur early in the disease process, long before the later events that include neuronal death and cognitive impairments. Also, interference with these early events is more likely to offer disease-changing effects. For all of these reasons, researchers are increasingly interested in investigating the early dysregulation processes in which miRNAs are predicted to be involved.

C. miRNA Regulation in Neurodegenerative Diseases

The brain holds trillions of synapses, the connection points between neurons. Synapse formation and plasticity depend on sensory input and are essential for information storage. In the adult brain, dynamic changes in synapse structure and function allow long-lasting information storage and depend on corresponding alterations in neuronal gene expression (Ho et al. 2011); recent reports demonstrate corresponding changes in the levels of particular miRNAs, as well as active involvement of certain miRNAs in controlling neuronal activity. This may involve direct regulation of neuronal gene expression, suggesting relevance for the difficulties in information storage and learning during the early stages of AD. Other examples include changes in the miRNA production pathway itself; thus, engineered depletion of Dicer, a major component of the miRNA machinery, impairs the integrity of several different neuronal types, including Purkinje cells in the cerebellum (Schaefer et al. 2007) and in cortical and hippocampal neurons (Davis et al. 2008). This induces a relatively rapid neuronal loss, accompanied by behavioral alterations.

Several miRNAs are already known to be causally involved in neurological diseases. In the context of Parkinson's disease (PD), a mutation in the PD-associated LRRK2 gene has been shown to interfere with miRNA regulation (Gehrke et al. 2010). This highlights the importance of miRNA regulation for the maintenance of dopaminergic neurons and might explain why inherited variants of LRRK2 are associated with increased risk of PD. In amyotrophic lateral sclerosis (ALS), *miR-206* was shown to target histone deacetylase 4, thereby inducing pronounced reinnervation by potentiating epigenetic control over neuromuscular innervation (Williams et al. 2009) and directly demonstrating therapeutic potential of a specific miRNA in a neurodegeneration process. In AD, several miRNAs were shown to regulate Aβ biogenesis. For example, the BACE1-targeted *miR-29a/b-1* cluster is prominently decreased in AD patients. Those patients display abnormally high levels of the BACE1 enzyme that cleaves the amyloid precursor protein APP to form the neurotoxic Aβ peptide (Hebert et al. 2008). AD progression is further accompanied by sequential and gradual alterations in the miRNA population, so that early and moderate stage AD patients share a large fraction of altered miRNAs; patients at the severe stage of AD show a considerable overlap of changes with those observed at the moderate stage more than the early stage. This sequential overlap effect covers miRNA targeting both up- and down-regulated genes, suggesting contribution of miRNA dysregulation to the AD processes already at early stages of the disease (Barbash and Soreq 2012; Berson et al. 2012) (Figure 33.1 and Figure 33.2). Lastly, ample evidence from the field of cancer research highlights a major role for miRNAs in cell cycle control and apoptosis, two processes that are also an integral part of neurodegeneration.

III. NEURONAL miRNAs

A. miRNAs and Memory

Intriguingly, several miRNAs have been physically localized at synaptic structures, pointing to the possibility of miRNA regulation of synaptic protein synthesis in response to

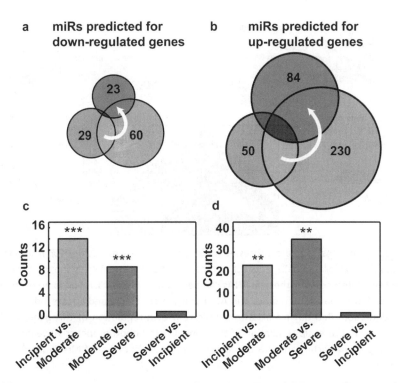

Figure 33.1. Sequential miRNAs recruitment along AD stages. (a) Venn diagram for predicted microRNAs targeted to AD down-regulated transcripts. Arrow indicates direction of disease progression. (b) Same as (a) for up-regulated genes. (c) Numbers of shared predicted miRNAs between pairs of stages for miRNAs predicted to be down-regulated. Two or three asterisks indicate Fisher's exact test P-values of $P < 0.01$ or $P < 0.001$, respectively. (d) Same as (c) for miRNAs predicted to be up-regulated. Courtesy of Current Alzheimer Research, reprinted from Barbash and Soreq (2012). See color insert.

extracellular signals. Supporting this notion, proteins that are required for synthesis and regulation of miRNAs are found in dendrites of mature neurons. These include the endoribonuclease Dicer, which cleaves pre-miRNAs into short double-stranded RNA fragments, and Argonaute, the catalytic component of the RNA-induced silencing complex (RISC) (Lugli et al. 2005; Barbee et al. 2006). Additionally, some miRNAs are found in the nuclear P-bodies, granule-shaped structures where key processes in mRNA regulation and synthesis take place, such as decapping, deadenylation, and exonucleation. That these granules respond to neuronal activation (Cougot et al. 2008) suggests corresponding occurrence of miRNA manipulations. An example of a dendritically localized miRNA with functional effects on protein synthesis and dendritic extension in hippocampal neurons is *miR-134*, which targets the Limk1 (Lim-domain-containing protein kinase 1) known to be involved in spine development (Schratt et al. 2006), further indicating that miRNA alterations may be a driving force in neurodegenerative events.

B. Functions of miRNAs at the Synapse

Numerous neuronal activity events are coupled to specific steps in the pathway of gene expression. Several of these processes are activated and regulated by miRNAs, suggesting

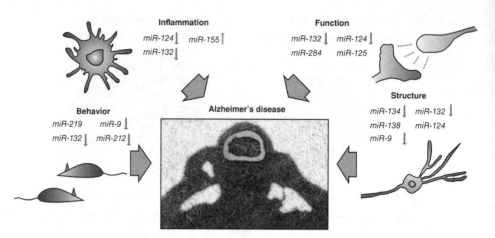

Figure 33.2. miRNAs regulate processes underlying the structural, functional, inflammation-related, and behavioral impairments characteristic of AD. Each miRNA is shown under the relevant process. References for these miRNAs can be found in Table 33.1. For some of the miRNAs, altered regulation was shown to occur in AD. For these genes, colored arrows indicate the direction of regulation: Red downward arrow indicates down-regulation of this gene, while green upward arrow indicates up-regulation of this gene. Central photograph: creative demonstration of AD by a patient participant in a therapeutic workshop (courtesy of Mrs. Tema Barbash). See color insert.

that these miRNAs may initiate and control molecular responses to specific external signals. These include neuronal activity-dependent miRNAs whose promoters harbor binding sites for early-immediate transcription activators, such as MEF2 and CREB. Rapid changes in calcium concentration drive the transcriptional activation of these activity-dependent miRNAs, initiating specific sets of neuronal genes (Vo et al. 2005; Fiore et al. 2009). A well-documented example for such CREB activation is that of *miR-132*, shown to potentiate the growth of dendrites and synapses and control stress-induced damages in cognitive performance, as well as inflammation (see later in the text). Additionally, miRNAs may be coupled to neuronal activity via posttranscriptional regulation of their synthesis, such as stimulation of Dicer activity through neuronal activation that may activate other miRNAs at the synapse (Lugli et al. 2005). Further down the cascade of mRNA targeting by miRNAs is the neuronal activity-dependent inactivation of Armitage, a major component of RNA-induced silencing complex (RISC) (Siegel et al. 2009) by miRNA-mediated translational inhibition. The aberrant neuronal activity in AD, which is an early event in the disease (Selkoe 2002), could hence deteriorate brain performance via the induction of miRNA deficiencies through multiple pathways.

C. miRNAs as Controllers of Neuritic Structure, Synapse Formation, and Maintenance

Morphological changes triggered by aberrant assembly of the cytoskeleton occur in many animal models of neurodegenerative diseases as well as in neuronal cultures exposed to Aβ aggregates. Dystrophic neurites are seen in AD (Selkoe 2002), and AD animal models present damaged maintenance of brain synapses. Several miRNAs exert their functional effect via targeting the cytoskeleton and may be relevant to this phenotype. MiRNA functions were also validated in the dendritic spines, postsynaptic neurite structures where the protein machinery that is essential for long-lasting plasticity resides. Two major examples are *miR-134* and *miR-138,* targeted at the depalmitoylation enzyme APT1, which is essential for sustaining the postsynaptic spine structure (Siegel et al. 2009). The remodeling and maintenance of mature synapses also require *miR-1,* which targets the nicotinic acetylcholine receptor and modifies the gain of cholinergic transmission in neuromuscular junctions (NMJ) of *Caenorhabditis elegans* (Simon et al. 2008). In glutamatergic brain synapses, *miR-284* can simultaneously regulate the amounts of several subunits of the glutamate receptor that shapes the postsynaptic signal (Karr et al. 2009), supporting the theory of rheostat-like function of miRNAs to dim an entire pathway by targeting several key transcripts (Chen et al. 2004). Although synapse formation and synapse maintenance may overlap, it is clear that miRNAs act in both of these domains. Thus, the major components of miRNA synthesis, including Dicer and Argonaute, have a functional role in synaptogenesis (Jin et al. 2004).

D. miRNAs and Behavior

Behavioral effects of interference with the synthesis of miRNAs, and especially multitargeted miRNAs, have been observed in several experimental systems. Examples include stress-associated neuronal excitation, which among other effects induces the neuronal modulator *miR-132*. Studying hippocampal stress-inducible induction of *miR-132* and suppression of its GTPase activator p250GAP target and the acetylcholine hydrolyzing enzyme acetylcholinesterase (AChE), we discovered predator scent stress-inducible hippocampal increases in *miR-132,* with inverse decreases in the levels and variability of both its p250GAP and AChE targets (Shaked et al. 2009). Furthermore, hippocampal AChE knockdown prevented *miR-132* induction, p250GAP suppression, and stress-inducible learning and memory deficits in a foot shock model, supporting cholinergic relevance, and engineered excess of human AChE inversely led to excessive *miR-132,* cosuppressed host AChE and p250GAP, and prevented cholinergic hyper-reactivity, anxiogenic-like phenotype, and impaired locomotion and memory. Given that *miR-132* levels are reduced in the AD hippocampus (Cogswell et al. 2008), this predicts therapeutic value for avoiding the *miR-132* decline. In mice, antisense inhibition of *miR-132* and the clock gene-targeted *miR-219* perturbed the circadian rhythm (Cheng et al. 2007). Another striking example at the behavioral level is *miR-9,* which targets several splice variants of K channels and by that plays a role in adaptation to alcohol (Pietrzykowski et al. 2008) by inducing synaptic alterations. A summary of the relevant miRNAs appears in Table 33.1.

IV. miRNAs IN INFLAMMATORY PROCESSES AND THEIR LINK TO THE NERVOUS SYSTEM

A. Overreactive Immune System

Inflammation is the body's innate immune reaction to invading pathogens. In healthy mammals, the duration of inflammation is terminated by several checks and balances, among them signaling from the central nervous system. The brain constantly monitors the peripheral levels of inflammation-related molecules, such as proinflammatory cytokines, and in turn controls peripheral inflammation via transduction molecules, such as hormones

TABLE 33.1. Relevant miRNAs for AD Phenotype

miRNA	Validated Target	Physiological Role	References
miR-9	Heavy neurofilament subunit, REST	Axonal cytoskeleton assembly, neuronal gene expression	Packer et al. (2008) and Haramati et al. (2010)
miR-124	C/EBP-a	Neuronal maturation, hippocampal LTP.	Chandrasekar and Dreyer (2009) and Cheng et al. (2009)
miR-132	MeCP2, AChE, p250GAP	Inhibition of inflammation, LTP.	Shaked et al. (2009), Wibrand et al. (2010), and Shaltiel et al. (2012)
miR-206	Histone deacetylase 4	Neuromuscular innervation	Williams et al. (2009)
miR-138	?	Postsynaptic spine structure	Siegel et al. (2009)
miR-134	Limk1, APT1	Postsynaptic spine structure development	Schratt et al. (2006)
miR-1	Nicotinic acetylcholine receptor	Cholinergic transmission	Simon et al. (2008)
miR-219		Circadian rhythm	Cheng et al. (2007)
miR-155	MeCP2, CD47, SHIP1	Neurite growth	Le et al. (2009) and Ooi et al. (2010)

and neurotransmitters. Furthermore, reactive agents of peripheral inflammation may change brain physiology (Glass et al. 2010). Several miRNAs were identified as being involved in regulating the molecular processes underlying this link between the brain and the immune system (Soreq and Wolf 2011). Importantly, these checks and balances are perturbed in neurodegenerative diseases, which involve impaired communication between the nervous and the immune system, and consequent loss of the body's capacity to localize and inhibit inflammation (Tracey 2010). When that happens the immune system may become overreactive in a harmful manner, exposing the body to infections while impairing cognitive processes (Meisel and Meisel 2011).

B. miRNAs at the Neuronal–Immune Interface

There are several examples of miRNAs that demonstrate involvement in both neuronal and immune functions. At the neuronal domain, *miR-124* potentiates neuronal maturation (Cheng et al. 2009) and stabilizes long-term potentiation (LTP) in sensory-motor (Rajasethupathy et al. 2009) and hippocampal neurons (Chandrasekar and Dreyer 2009). In addition, *miR-124* is expressed in the brain's immune cells: CNS macrophages and microglia, where it induces a phenotype switch from an inflammatory to a quiescent microglial state (Ponomarev et al. 2011). Also, *miR-124* targets several glucocorticoid receptors that mediate stress responses known to accompany and exacerbate the symptoms in various neurodegenerative diseases (Vreugdenhil et al. 2009). Another prominent example for a miRNA shared by the nervous and the immune systems is *miR-9,* the neuronal functions of which were described earlier. This miRNA targets the transcriptional activator NFκB1 in myeloid cells, and its levels increase under exposure to proinflammatory cytokines and following the activation of several toll-like receptors, the peripheral biosensors of the innate immune system. Yet another miRNA that regulates components of both the nervous and the immune system is *miR-155*, which robustly modulates neurite growth (Le et al. 2009) and, in parallel, regulates inflammation by targeting the proinflammatory cytokine tumor necrosis factor TNF-α and regulating lymphoid cell differentiation (Ooi et al. 2010). Last, and perhaps most interesting, *miR-132,* of which the neurological implications were discussed earlier, influences the intensity and resolution of inflammation by targeting AChE and inhibiting its production, which increases the anti-inflammatory activity of acetylcholine (Shaked et al. 2009). The *miR-132/AChE* interaction was validated in both loss- and gain-of-function experiments. These abilities to counteract chronic inflammation and, in parallel, to stabilize postsynaptic structures, thereby sustaining synaptic plasticity, make *miR-132* a fascinating therapeutic candidate for neurological diseases in general and AD in particular.

V. miRNAs AND AD THERAPY

A. Can miRNAs Serve as Diagnostic Biomarkers?

An important aim in AD research is to develop an early, minimally invasive approach for accurate diagnosis using "signature" biomarkers. Many believe that RNA molecules from white blood cells can offer higher sensitivity and specificity than protein biomarkers, and current technologies enable simultaneous detection of many thousand transcripts from minute blood cell samples by microarrays or next-generation sequencing. These data are then used in the search for the most informative transcripts for each disease: those that would provide the best and earliest discrimination between healthy individuals and patients. This strategy, together with cognitive measures and functional magnetic resonance imaging (fMRI) scans enables the best possible diagnostics today. Of note, potential miRNA-binding sites can be found in several key genes involved in neurodegenerative disease (e.g., APP for AD and α-synuclein for Parkinson's disease). Therefore, disease-related miRNAs would predictably show an even earlier change in patients compared with protein-coding mRNAs, because they represent a regulatory level that is higher up in the hierarchy of gene expression and is hence likely to affect the expression of many other genes at an earlier time than the current protein markers. This further raises the question if mutations in particular miRNAs or in *miR*-binding sites may increase the risk of disease, in which case the patients' DNA may be directly tested. One such example is a mutation in a miRNA-binding site on the mRNA for fibroblast growth factor 20 (FGF20), which associates with an increased risk in developing Parkinson's disease (Wang et al. 2008). We speculate that in the near future more studies addressing this question will be performed.

B. Different Approaches for Therapeutic Use of miRNA Manipulations

That miRNAs function in both harmful and protective aspects of AD and that each miRNA regulates several target genes that are often functionally connected in a given cell type highlights their advantages for therapeutic purposes. Thus, manipulating the activity of a particular miRNA would predictably lead to a strong effect on neuronal function, because by pushing one button, we are actually changing the state of several genes in a pathway. This would require far smaller doses and might involve fewer side effects than small molecule drugs targeted to a particular protein; also, targeting miRNAs can be the right

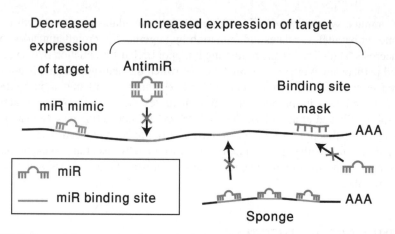

Figure 33.3. Potential approaches for manipulating miRNA levels. Shown is a schematic representation for the four major techniques for miRNA control. miRNA mimics can decrease the expression level of the target for a particular miRNA. AntimiRNA, Sponge or masking of the miRNA binding site, would inversely increase the expression of the miRNA target. For thorough reviews on these techniques, see Ebert et al. (2007) and Medina and Slack (2009).

regulatory level between intervening with the expression of a single gene, which might yield too small an effect, and intervening with transcription factors, which may lead to too-wide and nonspecific consequences. Intervention with miRNAs can be achieved via introduction of an artificial miRNA mimic to yield a gain of function or, inversely, by treatment with a complementary sequence (antisense oligonucleotide) to the modulated miRNA, which will inhibit its function by hybridizing with this miRNA. In addition, miRNA inhibition can be achieved using a reporter gene carrying several miRNA target decoy sequences in its 3′-untranslated region (these are called "sponges"). When vectors encoding these sponges are transfected into cultured cells, they strongly derepress miRNA targets. Finally, masking the miRNA-binding site using a synthetic oligonucleotide may prevent its activity. Figure 33.3 presents these alternative approaches schematically. For thorough reviews on these techniques, see Ebert et al. (2007) and Medina and Slack (2009). At the functional level, targeting each regulatory miRNA may add robustness to the neuronal network by stabilizing neuronal structures and connections and potentiating their functioning. Such a reinforced neuronal network would potentially cope better with the toxicity induced by Aβ. Nevertheless, while the investigation of the therapeutic potential of miRNA-regulating oligonucleotides is still in its infancy, the immune potential of injected oligonucleotides has already been validated (Kole et al. 2012). Developing miRNA therapeutics would hence be confronted with all of the caveats known for other biologics.

C. Potential Limitations of miRNA Manipulation Efforts

The multiple targets of a single miRNA, which is the strength of these posttranscriptional regulators, might be their weakness as well, since it would likely yield pronounced off-target effects. Thus, changing the concentration of a specific miRNA is likely to affect both the intended target and additional target genes that are expressed in the same host cell (Shaltiel et al. 2012). On the bright side, *miR*-regulating tools would predictably be more selective than small molecule drugs that often affect many different proteins. Also, the earlier the treatment is given, the better; using miRNA manipulations to limit amyloid fibril formation and thus strengthen the neuronal network has better prospects to succeed when the network is still viable and functioning. In AD, current diagnosis still happens relatively late, when neuronal death and other harmful processes have already been initiated. Nevertheless, stabilizing the remaining neuronal network might be more effective than trying to halt aggregation. Lastly, introducing a novel class of drugs would require tedious examinations of biocompatibility and toxicity. These putative agents would have to penetrate the blood–brain barrier, and thus the special requirements of drugs to be delivered to the brain must be kept in mind.

VI. FUTURE PROSPECTS AND CONCLUDING REMARKS

Throughout this chapter, we attempted to set the boundaries of knowledge existing today about miRNA involvement in the pathogenesis of AD at the level of those neurotoxic molecules involved in AD pathology (Aβ), the functional integrity of the neuronal network, and the aberrant inflammation that is part of the disease. We further covered the translational prospects and expected limitations in taking this knowledge a step forward and into the clinic. While challenging and complex, we believe that the rapidly accumulating knowledge on miRNAs in AD might offer new hope to those who badly need new and effective treatment modalities for this devastating disease.

ACKNOWLEDGMENTS

This study has been supported by The Legacy Heritage Biomedical Science Partnership Program of the Israel Science Foundation (Grant No. 378/11, to H.S.), The Rosetrees Foundation, and the German Israeli Foundation for Scientific Research and Development (G.I.F) (Grant No. 1093-32.2/2010, to H.S.). S.B. is an incumbent of a pre- doctoral fellowship by The Edmond and Lily Safra Center for Brain Sciences.

REFERENCES

Barbash, S. and H. Soreq. 2012. Threshold-independent meta-analysis of Alzheimer's disease transcriptomes shows progressive changes in hippocampal functions, epigenetics and microRNA regulation. *Curr Alzheimer Res* **9** **4**:425–435.

Barbee, S.A., P.S. Estes, A.M. Cziko, J. Hillebrand, R.A. Luedeman, J.M. Coller, N. Johnson, I.C. Howlett, C. Geng, R. Ueda, A.H. Brand, S.F. Newbury, et al. 2006. Staufen- and FMRP-containing neuronal RNPs are structurally and functionally related to somatic P bodies. *Neuron* **52**:997–1009.

Benilova, I., E. Karran, and B. De Strooper. 2012. The toxic Abeta oligomer and Alzheimer's disease: an emperor in need of clothes. *Nat Neurosci* **15**:349–357.

Berson, A., S. Barbash, G. Shaltiel, Y. Goll, G. Hanin, D.S. Greenberg, M. Ketzef, A.J. Becker, A. Friedman, and H. Soreq. 2012. Cholinergic-associated loss of hnRNP-A/B in Alzheimer's disease impairs cortical splicing and cognitive function in mice. *EMBO Mol Med* **4**:730–742.

Calin, G.A. and C.M. Croce. 2006. MicroRNA signatures in human cancers. *Nat Rev Cancer* **6**:857–866.

Chandrasekar, V. and J.L. Dreyer. 2009. microRNAs *miR-124, let-7d* and *miR-181a* regulate cocaine-induced plasticity. *Mol Cell Neurosci* **42**:350–362.

Chen, C.Z., L. Li, H.F. Lodish, and D.P. Bartel. 2004. MicroRNAs modulate hematopoietic lineage differentiation. *Science* **303**:83–86.

Cheng, H.Y., J.W. Papp, O. Varlamova, H. Dziema, B. Russell, J.P. Curfman, T. Nakazawa, K. Shimizu, H. Okamura, S. Impey, and K. Obrietan. 2007. microRNA modulation of circadian-clock period and entrainment. *Neuron* **54**:813–829.

Cheng, L.C., E. Pastrana, M. Tavazoie, and F. Doetsch. 2009. *miR-124* regulates adult neurogenesis in the subventricular zone stem cell niche. *Nat Neurosci* **12**:399–408.

Cogswell, J.P., J. Ward, I.A. Taylor, M. Waters, Y. Shi, B. Cannon, K. Kelnar, J. Kemppainen, D. Brown, C. Chen, R.K. Prinjha, J.C. Richardson, et al. 2008. Identification of miRNA changes in Alzheimer's disease brain and CSF yields putative biomarkers and insights into disease pathways. *J Alzheimers Dis* **14**:27–41.

Collie, A. and P. Maruff. 2000. The neuropsychology of preclinical Alzheimer's disease and mild cognitive impairment. *Neurosci Biobehav Rev* **24**:365–374.

Cougot, N., S.N. Bhattacharyya, L. Tapia-Arancibia, R. Bordonne, W. Filipowicz, E. Bertrand, and F. Rage. 2008. Dendrites of mammalian neurons contain specialized P-body-like structures that respond to neuronal activation. *J Neurosci* **28**:13793–13804.

Davis, T.H., T.L. Cuellar, S.M. Koch, A.J. Barker, B.D. Harfe, M.T. McManus, and E.M. Ullian. 2008. Conditional loss of Dicer disrupts cellular and tissue morphogenesis in the cortex and hippocampus. *J Neurosci* **28**:4322–4330.

Ebert, M.S., J.R. Neilson, and P.A. Sharp. 2007. MicroRNA sponges: competitive inhibitors of small RNAs in mammalian cells. *Nat Methods* **4**:721–726.

Esquela-Kercher, A. and F.J. Slack. 2006. Oncomirs - microRNAs with a role in cancer. *Nat Rev Cancer* **6**:259–269.

Fineberg, S.K., K.S. Kosik, and B.L. Davidson. 2009. MicroRNAs potentiate neural development. *Neuron* **64**:303–309.

Fiore, R., S. Khudayberdiev, M. Christensen, G. Siegel, S.W. Flavell, T.K. Kim, M.E. Greenberg, and G. Schratt. 2009. Mef2-mediated transcription of the miR379-410 cluster regulates activity-dependent dendritogenesis by fine-tuning Pumilio2 protein levels. *EMBO J* **28**:697–710.

Gehrke, S., Y. Imai, N. Sokol, and B. Lu. 2010. Pathogenic LRRK2 negatively regulates microRNA-mediated translational repression. *Nature* **466**:637–641.

Glass, C.K., K. Saijo, B. Winner, M.C. Marchetto, and F.H. Gage. 2010. Mechanisms underlying inflammation in neurodegeneration. *Cell* **140**:918–934.

Haramati, S., E. Chapnik, Y. Sztainberg, R. Eilam, R. Zwang, N. Gershoni, E. McGlinn, P.W. Heiser, A.M. Wills, I. Wirguin, L.L. Rubin, H. Misawa, et al. 2010. miRNA malfunction causes spinal motor neuron disease. *Proc Natl Acad Sci U S A* **107**:13111–13116.

Hardy, J. and D.J. Selkoe. 2002. The amyloid hypothesis of Alzheimer's disease: progress and problems on the road to therapeutics. *Science* **297**:353–356.

He, L. and G.J. Hannon. 2004. MicroRNAs: small RNAs with a big role in gene regulation. *Nat Rev Genet* **5**:522–531.

Hebert, S.S., K. Horre, L. Nicolai, A.S. Papadopoulou, W. Mandemakers, A.N. Silahtaroglu, S. Kauppinen, A. Delacourte, and B. De Strooper. 2008. Loss of microRNA cluster *miR-29a/b-1* in sporadic Alzheimer's disease correlates with increased BACE1/beta-secretase expression. *Proc Natl Acad Sci U S A* **105**:6415–6420.

Ho, V.M., J.A. Lee, and K.C. Martin. 2011. The cell biology of synaptic plasticity. *Science* **334**: 623–628.

Jin, P., D.C. Zarnescu, S. Ceman, M. Nakamoto, J. Mowrey, T.A. Jongens, D.L. Nelson, K. Moses, and S.T. Warren. 2004. Biochemical and genetic interaction between the fragile X mental retardation protein and the microRNA pathway. *Nat Neurosci* **7**:113–117.

Karr, J., V. Vagin, K. Chen, S. Ganesan, O. Olenkina, V. Gvozdev, and D.E. Featherstone. 2009. Regulation of glutamate receptor subunit availability by microRNAs. *J Cell Biol* **185**:685–697.

Kole, R., A.R. Krainer, and S. Altman. 2012. RNA therapeutics: beyond RNA interference and antisense oligonucleotides. *Nat Rev Drug Discov* **11**:125–140.

Le, M.T., H. Xie, B. Zhou, P.H. Chia, P. Rizk, M. Um, G. Udolph, H. Yang, B. Lim, and H.F. Lodish. 2009. MicroRNA-125b promotes neuronal differentiation in human cells by repressing multiple targets. *Mol Cell Biol* **29**:5290–5305.

Lugli, G., J. Larson, M.E. Martone, Y. Jones, and N.R. Smalheiser. 2005. Dicer and eIF2c are enriched at postsynaptic densities in adult mouse brain and are modified by neuronal activity in a calpain-dependent manner. *J Neurochem* **94**:896–905.

Medina, P.P. and F.J. Slack. 2009. Inhibiting microRNA function in vivo. *Nat Methods* **6**:37–38.

Meisel, C. and A. Meisel. 2011. Suppressing immunosuppression after stroke. *N Engl J Med* **365**:2134–2136.

Nicoll, J.A., D. Wilkinson, C. Holmes, P. Steart, H. Markham, and R.O. Weller. 2003. Neuropathology of human Alzheimer disease after immunization with amyloid-beta peptide: a case report. *Nat Med* **9**:448–452.

Ooi, A.G., D. Sahoo, M. Adorno, Y. Wang, I.L. Weissman, and C.Y. Park. 2010. MicroRNA-125b expands hematopoietic stem cells and enriches for the lymphoid-balanced and lymphoid-biased subsets. *Proc Natl Acad Sci U S A* **107**:21505–21510.

Packer, A.N., Y. Xing, S.Q. Harper, L. Jones, and B.L. Davidson. 2008. The bifunctional microRNA miR-9/miR-9* regulates REST and CoREST and is downregulated in Huntington's disease. *J Neurosci* **28**:14341–14346.

Pietrzykowski, A.Z., R.M. Friesen, G.E. Martin, S.I. Puig, C.L. Nowak, P.M. Wynne, H.T. Siegelmann, and S.N. Treistman. 2008. Posttranscriptional regulation of BK channel splice variant stability by *miR-9* underlies neuroadaptation to alcohol. *Neuron* **59**:274–287.

Ponomarev, E.D., T. Veremeyko, N. Barteneva, A.M. Krichevsky, and H.L. Weiner. 2011. MicroRNA-124 promotes microglia quiescence and suppresses EAE by deactivating macrophages via the C/EBP-alpha-PU.1 pathway. *Nat Med* **17**:64–70.

Rajasethupathy, P., F. Fiumara, R. Sheridan, D. Betel, S.V. Puthanveettil, J.J. Russo, C. Sander, T. Tuschl, and E. Kandel. 2009. Characterization of small RNAs in Aplysia reveals a role for *miR-124* in constraining synaptic plasticity through CREB. *Neuron* **63**:803–817.

Schaefer, A., D. O'Carroll, C.L. Tan, D. Hillman, M. Sugimori, R. Llinas, and P. Greengard. 2007. Cerebellar neurodegeneration in the absence of microRNAs. *J Exp Med* **204**:1553–1558.

Schliebs, R. and T. Arendt. 2006. The significance of the cholinergic system in the brain during aging and in Alzheimer's disease. *J Neural Transm* **113**:1625–1644.

Schratt, G.M., F. Tuebing, E.A. Nigh, C.G. Kane, M.E. Sabatini, M. Kiebler, and M.E. Greenberg. 2006. A brain-specific microRNA regulates dendritic spine development. *Nature* **439**:283–289.

Selkoe, D.J. 2002. Alzheimer's disease is a synaptic failure. *Science* **298**:789–791.

Shaked, I., A. Meerson, Y. Wolf, R. Avni, D. Greenberg, A. Gilboa-Geffen, and H. Soreq. 2009. MicroRNA-132 potentiates cholinergic anti-inflammatory signaling by targeting acetylcholinesterase. *Immunity* **31**:965–973.

Shaltiel, G., M. Hanan, Y. Wolf, S. Barbash, E. Kovalev, S. Shoham, and H. Soreq. 2012. Hippocampal microRNA-132 mediates stress-inducible cognitive deficits through its acetylcholinesterase target. *Brain Struct Funct* **218**:59–72.

Siegel, G., G. Obernosterer, R. Fiore, M. Oehmen, S. Bicker, M. Christensen, S. Khudayberdiev, P.F. Leuschner, C.J. Busch, C. Kane, K. Hubel, F. Dekker, et al. 2009. A functional screen implicates microRNA-138-dependent regulation of the depalmitoylation enzyme APT1 in dendritic spine morphogenesis. *Nat Cell Biol* **11**:705–716.

Simon, D.J., J.M. Madison, A.L. Conery, K.L. Thompson-Peer, M. Soskis, G.B. Ruvkun, J.M. Kaplan, and J.K. Kim. 2008. The microRNA *miR-1* regulates a MEF-2-dependent retrograde signal at neuromuscular junctions. *Cell* **133**:903–915.

Soreq, H. and Y. Wolf. 2011. NeurimmiRs: microRNAs in the neuroimmune interface. *Trends Mol Med* **17**:548–555.

Tracey, K.J. 2010. Understanding immunity requires more than immunology. *Nat Immunol* **11**: 561–564.

Vo, N., M.E. Klein, O. Varlamova, D.M. Keller, T. Yamamoto, R.H. Goodman, and S. Impey. 2005. A cAMP-response element binding protein-induced microRNA regulates neuronal morphogenesis. *Proc Natl Acad Sci U S A* **102**:16426–16431.

Vreugdenhil, E., C.S. Verissimo, R. Mariman, J.T. Kamphorst, J.S. Barbosa, T. Zweers, D.L. Champagne, T. Schouten, O.C. Meijer, E.R. de Kloet, and C.P. Fitzsimons. 2009. MicroRNA 18 and 124a down-regulate the glucocorticoid receptor: implications for glucocorticoid responsiveness in the brain. *Endocrinology* **150**:2220–2228.

Wang, G., J.M. van der Walt, G. Mayhew, Y.J. Li, S. Zuchner, W.K. Scott, E.R. Martin, and J.M. Vance. 2008. Variation in the miRNA-433 binding site of FGF20 confers risk for Parkinson disease by overexpression of alpha-synuclein. *Am J Hum Genet* **82**:283–289.

Welsh, K., N. Butters, J. Hughes, R. Mohs, and A. Heyman. 1991. Detection of abnormal memory decline in mild cases of Alzheimer's disease using CERAD neuropsychological measures. *Arch Neurol* **48**:278–281.

Wibrand, K., D. Panja, A. Tiron, M.L. Ofte, K.O. Skaftnesmo, C.S. Lee, J.T. Pena, T. Tuschl, and C.R. Bramham. 2010. Differential regulation of mature and precursor microRNA expression by NMDA and metabotropic glutamate receptor activation during LTP in the adult dentate gyrus in vivo. *Eur J Neurosci* **31**:636–645.

Williams, A.H., G. Valdez, V. Moresi, X. Qi, J. McAnally, J.L. Elliott, R. Bassel-Duby, J.R. Sanes, and E.N. Olson. 2009. MicroRNA-206 delays ALS progression and promotes regeneration of neuromuscular synapses in mice. *Science* **326**:1549–1554.

34

CURRENT VIEWS ON THE ROLE OF MicroRNAs IN PSYCHOSIS

Aoife Kearney, Javier A. Bravo, and Timothy G. Dinan

Department of Psychiatry, University College Cork, Cork, Ireland

I.	Introduction	554
II.	Phenotypic Expression of Schizophrenia: The Role of Neurotransmitters	554
	A. Dopamine	555
	B. Glutamate	555
	C. Serotonin	555
	D. Treatment Limitations	555
III.	Etiology of Psychosis	556
IV.	MicroRNA and Schizophrenia	557
V.	Bipolar Affective Disorder	559
VI.	miRNAs as Drug Targets for Treatment of Psychosis	560
VII.	Future MicroRNA-Based Therapies	561
VIII.	Conclusions	563
Acknowledgments		563
References		563

ABBREVIATIONS

ATF2	activating transcription factor 2
BMAL	brain and muscle aryl hydrocarbon receptor nuclear translocator (ARNT)-like
BDNF	brain-derived neurotrophic factor
CATIE	Clinical Antipsychotic Trials of Intervention Effectiveness

MicroRNAs in Medicine, First Edition. Edited by Charles H. Lawrie.
© 2014 John Wiley & Sons, Inc. Published 2014 by John Wiley & Sons, Inc.

CLOCK	circadian locomotor output cycles kaput
CNV	copy number variation
D2	dopamine 2 receptor
DISC1	disrupted-in-schizophrenia-1
GABA	gamma amino butyric acid
GAD	glutamate decarboxylase
JUN	transcription factor
LSD	lysergic acid diethylamide
MECP2	methyl CpG-binding protein 2
miRNA	microRNA
NMDA	n-methy-d-asparte
SLITRK1	Slit and Trk-like family member 1
TAF1	gene for transcription factor IID (TFIID)

I. INTRODUCTION

Psychiatric disease, including schizophrenia, bipolar affective disorder, and major depression, exerts an extensive burden on patients and society. The World Health Organization (WHO) estimates that worldwide disorders of mental health conditions account for over 25% of the total burden of disease (WHO 2011). This chapter outlines the definite limitations in the current gold standard therapeutic options, their biological origins, and the potential role miRNA could play in future novel treatments. Modern psychiatry endeavours to consolidate its body of research on the potential genetic and cell signaling deficits that result in the clinical phenotype. It is hoped that this methodology will prove effective in the development of more effective and streamlined treatment plans that have a robust and systematic evidence base. The current research into the role of miRNAs is of eminent importance in the path to meeting this unified and very timely therapeutic goal. Studies to date in miRNA have been most abundant in the field of schizophrenia as opposed to that of bipolar affective disorder, and the view that non-protein coding genes have an important regulatory role with implications for genetic liability to psychosis is a view that is gaining greater credence (Dinan 2010). Thus, the role of miRNAs in serious mental illness could provide a more precise and fitting explanation of its etiology and could serve as a novel and potentially very valuable treatment resource.

There are increasing data emerging to support the involvement of miRNA not just in human brain development but also in brain disease (Fiore et al. 2008; Pushparaj et al. 2008). Indeed it is a neuropsychiatric disorder, Tourette syndrome, that is the best example to date of a specific brain disorder induced by miRNA dysregulation Abelson et al. (2005). identified Slit and Trk-like family member 1 (SLITRK1), which is the leucine rich transmembrane protein, involved in Tourette syndrome. The identification of a 3'-UTR mutation at the *miR-189*-binding site was demonstrated to enhance miRNA target binding and postulated to be causative in this disorder.

II. PHENOTYPIC EXPRESSION OF SCHIZOPHRENIA: THE ROLE OF NEUROTRANSMITTERS

Psychotic illness is primarily categorized as schizophrenia and bipolar affective disorder. Of all major psychiatric illness, schizophrenia continues to present the major diagnostic challenges in clinical practice. The major classification systems (ICD-10 and DSM-IV) serve to provide clarity in terms of what fulfils specific diagnostic criteria from a phenomenological perspective. Patients with schizophrenia can display a variety of symptoms, which can be grouped into positive and negative subtypes. Positive symptoms (e.g., delusions, hallucinations, and disorganized speech) are more effectively treated with antipsychotic medication than negative symptoms (avolition, apathy, and affective flattening)—which constitutes a well-documented "deficit syndrome" (Carpenter et al. 1988). Five predominant subtypes of schizophrenia are recognized. The paranoid subtype is the most common, followed by hebephrenic, catatonic, simple and undifferentiated. The various subtypes are descriptive in nature rather than delineating a specific biological variability. Liddle instead proposed three overlapping clinical syndromes: reality distortion, disorganization, and psychomotor poverty (Liddle 1987).

A. Dopamine

To date, the most universally accepted hypothesis of schizophrenia is that of dopamine excess. Overactivity of dopamine neurotransmission in the mesolimbic pathway may underlie the positive symptoms of schizophrenia (Gurevich et al. 1997). The exact mechanisms have not been fully elucidated, but theories converge on a genetic component, and this is strongly supported by way of the fact that all antipsychotic medication are dopamine receptor antagonists. The potency of these various compounds is linked with the degree of affinity for the D2 receptor. Positron emission tomography (PET) studies suggest that an antipsychotic effect is obtained when 60% of D2 receptors are occupied (Farde et al. 1988). However, because receptors of the D2 subfamily are found in both limbic and striatal regions and the generally poor selectivity of antipsychotic agents, their blockade is responsible for both their antipsychotic effect and the Parkinson-like side effects seen with traditional drugs (Seeman and Van Tol 1994).

B. Glutamate

There is also strong evidence for the participation of other neurotransmitter systems, for example, drugs that act on the glutamatergic system (e.g., phencyclidine, ketamine, dizocilpine [MK-801]) can induce transient psychotomimetic effects (Halberstadt 1995; Moghaddam 2003). The involvement of NMDA receptors is supported by the fact that coagonists of the NMDA receptor or drugs acting on the metabotropic glutamate receptors 2/3 can ameliorate psychotic symptomatology (Javitt 2006; Patil et al. 2007). A role for the inhibitory GABAergic system is supported by postmortem studies indicating alterations in its functioning (Akbarian and Huang 2006).

C. Serotonin

The participation of the serotonergic system in the emergence of psychotic illness is also suggested due to the greater affinity that atypical antipsychotics have for 5-HT2A receptors than to D2 receptors. Moreover, the hallucinogenic LSD, which exerts its effects on the serotonergic system, can also induce psychotomimetic effects, which are generally of a transient nature (Jones et al. 2008).

D. Treatment Limitations

The pitfalls of current psychopharmacological agents lies not only in their suboptimal performance, in terms of response rates, but in oftentimes the emergence of severe side effects both metabolic and neurological. The most recent body of evidence looking at effectiveness of antipsychotic treatments, side effect profiling, and compliance is the CATIE study, a large multicentered pragmatic randomized control trial that enrolled almost 1500 individuals with chronic schizophrenia. This revealed a high rate of treatment discontinuation (up to 74%) among patients over the 18-month period of the trial. The most controversial finding of the CATIE study was the lack of significant differences in effectiveness between most of the second-generation antipsychotics and perphenazine (a typical antipsychotic), serving to illustrate the lack of progress in terms of psychotherapeutic options since the 1950s. A further limitation of atypical antipsychotics is that they, too, have limited efficacy in treating the negative symptoms found in the disorder.

Thus the current treatment regimens used have marked deficits, coupled with a significant side effect profile. To confound matters undoubtedly, there is also variation in the effectiveness at individual patient level, and there is no way to predict whether or not a patient will respond. Consequently, treatments for schizophrenia must be individualized, and pharmacogenetic predictors of response remain elusive.

III. ETIOLOGY OF PSYCHOSIS

The widely accepted etiology of schizophreniform illness is that of a complex interplay between genetic loading and environmental risk factors. The vastly diverse clinical phenotype of the illness represents a reciprocal complexity in terms of therapeutic challenges intensified by the absence of a complete elucidation of its complex biological framework. Of all major psychiatric illness, it remains the most difficult to define. The heritability of schizophrenia is estimated at circa 80%, supported by twin studies that show the monozygotic twin concordance rates of 48% and an estimated 12-fold increased risk of illness development in a first-degree relative (Farmer et al. 1987). In terms of genetic studies to date, the most promising data have arisen from small family-based linkage studies of which the identification of the DISC1 (disrupted-in-schizophrenia-1) locus was noted to cosegregate in a Scottish pedigree in whom high rates of psychotic illness were present (St Clair et al. 1990) The estimation prevails, however, that less than 10% of psychiatric illness is caused by rare, highly penetrant genes, such as DISC1, or by copy number variation (CNV), which, in turn supports the theory that the majority of cases are the result of the complex interaction of many genes, each exerting a small effect. It is well established, however, that 22q11.2 microdeletions account for 1–2% of sporadic cases of schizophrenia, and that 30% of children with this microdeletion, whose phenotypic expressions include diGeorge syndrome and velocardiofacial syndrome, will develop schizophrenia in adolescence or early adulthood. It also represents the only recurrent copy variant responsible for inducing new cases of schizophrenia within the population (Karayiorgou et al. 1995; Stefansson et al. 2008).

Research is now converging on an excess of *de novo* copy number variants in schizophrenia (Xu et al. 2008). Genome-wide association study results conceptualize both schizophrenia and bipolar disorder as being highly polygenic and it is the presence of a significant number of single nucleotide polymorphisms that is associated with phenotypic expression (Purcell et al. 2009). No specific locus has yet been reported for schizophrenia that reaches genome-wide levels of significance in any single or combined study (Dudbridge and Gusnanto 2008).

Given this biological complexity, it is possible that miRNAs exert an integral role at a pathophysiological level.

The neurodevelopmental model of schizophrenia (a model that has long been accepted for other neuropsychiatric disorders that present in childhood) puts forward the view that the manifestation of symptoms is as a result of preexisting abnormal neurodevelopmental processes, for example, synaptic pruning. Indeed, the peak onset of psychotic illness occurs during adolescence and overlaps with cortical dendritic pruning, supporting this illness model (Feinberg 1982). It is also further supported by the occurrence of coexisting deficits in cognitive and motor performances in children who later proceed to develop psychotic illness (Rapoport et al. 2012); however, early disorders of cognition, language, and motor performance are both general risks for adult psychopathology and too common to be effective in predicting schizophrenia.

IV. MicroRNA AND SCHIZOPHRENIA

The study of microRNAs (miRNAs) in psychiatric conditions remains, relatively speaking, in its infancy. The specific molecular mechanisms through which altered miRNA activity may cause psychiatric phenotypes are still poorly understood. miRNAs that are expressed in the brain accounts for more than 50% of miRNAs identified to date (Dinan 2010). The diverse biological action of miRNA to alter gene expression within multiple signaling pathways may be a part of disease pathogenesis in schizophrenia (Coyle 2009). In support of this is the accumulating evidence that miRNAs are involved in the control of neurotransmitter release. Substance P synthesis is inhibited by *miR-130a* and *miR-206*. In turn, the expression of these miRNAs is reduced by interleukin-1α (Greco and Rameshwar 2007). Hunsberger et al. (Hunsberger et al. 2009) postulate that miRNAs may serve as a unifying link between the diverse findings observed in schizophrenia, for example structural developmental anomalies, neurotransmitter alterations and response to treatment. They conclude that a key advantage of miRNAs is their capacity to target hundreds of genes that may be involved in the varied aspects of the disease process.

The role of miRNA in brain function is emerging, with accumulating evidence demonstrating its involvement in neuronal differentiation and neurogenesis (*miR-124* and *miR-9*). The role of *miR-134* in dendritic spine sizes has been established, which via its action on LIMK1 (a protein kinase regulating actin filaments), plays a key role in long-term potentiation and thus memory formation (Schratt et al. 2006). *miR-134* inhibits LIMK1 translation, but this is removed by exposure to brain-derived neurotrophic factor (BDNF). *miR30a-5p*, in the prefrontal cortex, is a posttranscriptional inhibitor of BDNF (Mellios et al. 2008).

In schizophrenia, miRNA profiling—predominately within the cortex—has now been completed in several patient populations. An early study involving miRNA was performed by Burmsitrova et al., where the role of methyl CpG-binding protein 2 gene was studied. This gene (MECP2) provides the instruction for the MECP2 protein that is vital for normal brain development and is known to be present in high levels in mature nerve cells. The primary roles of this protein are in gene silencing and synapse formation. Mutations of MECP2 are associated with autistic spectrum disorders, such as Rett syndrome, and also are postulated to play a role in development of psychotic illness (Piton et al. 2011). *miR-130b*, which potentially targets MECP2, is located in a well-documented susceptibility locus for schizophrenia. Using a comparative analysis, Burmsitrova et al. (Burmistrova et al. 2007) examined the expression of *miR-130b* in the neocortex of 24 schizophrenic and non-schizophrenic controls. No difference in the expression was observed, however, and genetic association analysis did not reveal an association of any of the *miR-130b* allelic variants with schizophrenia.

Alterations in the Dgcr8 gene, which encodes for an important subunit of the microprocessor complex that mediates the biogenesis of miRNAs from the primary miRNA transcript (see Chapter 1) were shown by Stark et al. (2008) with a mouse model carrying a 1.3-Mb chromosomal deficiency, to cause impairments in spatial working memory and sensorimotor gating but with no impairment in associative memory. This study was the primary study to demonstrate the association between abnormal miRNA biogenesis and cognitive performance in mice. This is strong evidence for a direct biological link between miRNA biogenesis and schizophrenia.

Perkins et al. (Perkins et al. 2007), in a postmortem study, compared the expression of 264 miRNAs using postmortem prefrontal cortex of individuals with schizophrenia (thirteen) and without any background psychiatric disorder (21). Differential expression was identified for 16 miRNAs of subjects versus controls, 15 of which were expressed at lower-fold levels and one at a higher-fold level. This was screened for using a custom-made miRNA microarray. Real-time PCR was subsequently employed to determine the expression level of 12 selected miRNAs. *miR-26b*, *miR-30b*, *miR-29b*, and *miR-106b* showed the greatest fold change. The study concluded that the miRNAs, which were shown in the study to have altered expression in psychotic illness, most probably played a role in synaptic plasticity at the level of the dendritic spines.

Another postmortem study was completed by Beveridge et al. (Beveridge et al. 2010), which evaluated miRNA expression in the superior temporal gyrus and dorsolateral prefrontal cortex. This study revealed an up-regulation of *miR-181b* via the usage of a high-throughput microarray system. This miRNA was found to have two genomic loci—chromosome 1 and chromosome 9. *miR-181b* is preferentially expressed in B-lymphoid cells of bone marrow and muscle. Chromosome 9 was shown to be a more active locus than chromosome 1 with the identification of the genes GR1A2 (ionotropic glutamate receptor) and VSNL1 (calcium sensor binding protein). Both of these genes were suppressed in the same tissue, and thus changes in the local miRNA environment may be influential. Of note, four miRNAs from this study overlapped with the findings of Perkins et al. (Perkins et al. 2007) (i.e., *miR-24, miR-26b, miR-29c,* and *miR-7*), but an inverse correlation was found, all four being up-regulated in the Beveridge study and down-regulated in Perkins et al. These contrary findings can be rationalized in part by the fact that the studies were conducted on postmortem specimens that are highly susceptible to confounding variables. Conclusions derived from these two studies offer strong support to the hypothesis that dysregulation of cortical gene expression in schizophrenia could be altered by miRNA levels.

Given the pivotal role that BDNF plays in maintaining neuronal structure and function it is not surprising that it has been an important focus for research into the pathophysiology of psychosis. This is highlighted by the miRNA studies of Mellios et al. (Mellios et al. 2008), which used postmortem samples from the prefrontal cortex. They demonstrated that miRNAs that were differentially expressed in schizophrenia could act as post-transcriptional inhibitors of BDNF. A further study which involved 20 cases and 20 controls examined BDNF and two regulatory miRNAS *miR-195* and *miR-30a-5p*. This suggested that the disease-related variability of BDNF protein levels is attributable to *miR-195* levels. *miR-195* was also shown to contribute to the regulation and variability of BDNF protein levels, which in turn produces alterations in neuropeptide Y and somatostatin mRNAs. This study suggests that up to 50% of disease-related variability in BDNF levels can be attributable to *miR-195*. It is also postulated that a significant portion of the variability in disease related changes in neuropeptide Y and somatostatin can also be attributed to *miR-195*.

Dysregulation of the GABAergic system via deficits in gene expression was also examined and the transcript for neuropeptide Y, somatostatin, and parvalbumin. These data serve to indicate that a regulatory cascade is involved in the dysregulated GABAergic system in the prefrontal cortex of people with schizophreniform illness. The view that *miR-195* is an important target for novel therapeutic strategies is supported by this study and further research conducted by Guo et al. (Guo et al. 2010) showing *miR-195* to be involved in a complex regulatory cascade reported to be implicated in the pathogenesis of this disease.

miRNAs that target specific schizophrenia-associated genes were identified using quantitative real-time PCR by Zhu et al. (Zhu et al. 2009). Brain RNA samples from the Stanley Array Collection (Stanley Medical Research Institute) of 35 patients with schizophrenia and 34 controls were used to identify expression levels of *miR-346* and GRID1. Expression of both were lower in the schizophrenia sample. This study suggests that decreased *miR-346* may result in alterations in mRNA targets in psychotic illness. GRID1 is a glutamate receptor subunit gene that is down-regulated in schizophrenia, and *miR-346* is located within an intron of this gene.

A Scandinavian study (Hansen et al. 2007) used a case-control design in three variable sample populations (Danish, Norwegian, and Swedish). Eighteen known SNPs were studied located within or near brain-expressed miRNA, two of which showed nominal significant allelic association to schizophrenia in both the Danish and Norwegian samples. Eight to 15 genes predicted to be regulated by both *miR-198* and *miR-206* were revealed to be related to the transcriptional factors JUN, ATF2, and TAF1.

A further case-control study conducted in a Chinese population by Zhu et al. (Zhu et al. 2009) reported a potential functional variant of *miR-30e*. The miRNA samples in this study were from peripheral blood, unlike those in Perkins et al. study. Both report an increase in expression of *miR-30e* in the prefrontal cortex of schizophrenia patient samples. Thus, *miR-30e* is being touted as a key player in potentially novel treatment strategies.

V. BIPOLAR AFFECTIVE DISORDER

Modern psychiatry does not recognize unipolar mania as a diagnostic entity. Such patients are viewed as genetically bipolar, which in turn is subdivided into bipolar I and bipolar II. Dopamine again appears to be of critical importance in the manifestation of manic symptoms, which can be provoked by dopamine agonists, such as bromocriptine (Silverstone 1984). Psychostimulants, which increase presynaptic dopamine release, can also provoke mania-like symptoms. Antipsychotic medication is used in the acute manic phase of the illness, as well as having a mood stabilizing role. Indeed, schizophrenia and bipolar disorder are now regarded as representing different expressions of a unifying continuum in which multiple signaling pathways have become dysregulated. Bipolar disorder again mirrors schizophrenia in its lack of fully effective treatment options resulting in significant morbidity, patient suffering, and economic burden. It is accepted that bipolar disorder, such as schizophrenia, is polygenic in etiology, and the clinical phenotype is the result of multiple single-nucleotide polymorphisms (Purcell et al. 2009). This again supports the postulated role of miRNA via its ability to regulate a large network of protein coding targets. The current mood-stabilizing medications that are prescribed as the gold standard treatment of bipolar disorder are only partially effective and have a myriad of unwanted side effects. Lithium remains at the forefront of most widely prescribed medication in the management with valproate, an anticonvulsant used as a second-line agent, or, in instances where there is a specific contraindication—such as in renal impairment. The specific extent to which these agents alter cellular plasticity cascades involved in the specific pathophysiology of bipolar disorder is, as of yet, poorly understood (Schloesser et al. 2008). Lithium has a multitude of roles at cellular level that are putatively thought to be involved in exerting its biological effect. It has been shown to be involved in decreasing levels of protein kinase C and increasing levels of BDNF and also glial cell neurotrophic factor (Fukumoto et al. 2001; Angelucci et al. 2003; Jacobsen and Mork 2004). Lithium also has the ability to enhance neurogenesis (Chen et al. 2000; Kim et al. 2004) and long-term potentiation (Son et al. 2003). Lithium has been shown to reduce hippocampal levels of BDNF in an animal model of mania (Frey et al. 2006).

The effects of chronic mood stabilizers and hippocampal miRNA expression have been examined by Zhou et al., whereby Wister rats were chronically treated with lithium or sodium valproate. Potential targets were screened using miRNA microarray. The hippocampal expression levels of 37 miRNAS were altered in the lithium-treated group and in 31 of the sodium valproate treated group wheareas only nine miRNAS were regulated by both drugs in a similar manner. A significant reduction in levels of the miRNAs *let-7b*, *let-7c*, *miR-128a*, *miR-24a*, *miR-30c*, *miR-30a*, and *miR-221* were observed with chronic treatment with either lithium or valproate. *miR-144* was, however, noted to be significantly up-regulated. Certain miRNA target genes, GRM7, DPP10, and THRB, are potential genetic risk factors for bipolar disorder. GRM7 has been pinpointed following whole genome association studies as being a candidate gene for both bipolar disorder and schizophrenia (Shibata et al. 2009). Down-regulation of *miR-34a* would lead to an up-regulation of its effector GRM7 (Zhou et al. 2009). Mood stabilizers are thus postulated as targeting specific miRNAs, and the data from Zhou et al. (2009) are the first data in support of this.

The CLOCK gene has also been implicated as a susceptibility gene for bipolar disorder in studies in mice models where this gene has been knocked out and produces mania-like behaviors (Roybal et al. 2007). The transcriptional factors BMAL1 and CLOCK and Period and Cryptochrome proteins are involved in the regulation of circadian rhythms. The activity of BMAL1 and CLOCK is counterbalanced by period and cryptochrome proteins to regulate time of day-dependent gene expression. The CLOCK-BMAL1 complex binds to the enhancer region of *miR-219* and regulates the circadian rhythmic expression of pre-*miR-219* (Cheng et al. 2007). The circadian period is increased by inhibition of *miR-219* expression in the suprachiasmatic nuclei by infusion of an antagomir in the lateral ventricle. As *miR-219* involvement has been explored both in the behavioral expression associated with NMDA-receptor hypofunction and in circadian rhythm posttranslational feedback loops, it is possible it has a key role in both schizophrenia and bipolar disorder (Coyle 2009).

VI. miRNAs AS DRUG TARGETS FOR TREATMENT OF PSYCHOSIS

As outlined, the current treatment options for psychotic illness are only partially effective, with many limitations in predicting their response rate, patient variability and overall poor efficacy at targeting negative symptomatology. The clinical phenotype thus has a chronic and enduring course punctuated with, oftentimes, frequent relapses. The pharmaceutical industry since the 1950s has focused extensively, if not exclusively, on the manipulation of the dopaminergic system, which has yielded, thus far, suboptimal results. It is postulated that modulation of certain miRNAs involved in schizophrenia could occur through the use

TABLE 34.1. Current Pharmacology and Some Implicated miRNAs

Disease	Current Drug Therapies	Implicated miRNAs
Schizophrenia	*Typical antipsychotics:* chlorpromazine trifluoperazine, haloperidol *Atypical antipsychotics*: olanzapine, risperidone, quetiapine	*miR-195, miR-134* *miR-181b, miR-346* *miR-24, 26b, 29c*
Bipolar affective disorder	Lithium *Antipsychotics*: olanzapine, *Anticonvulsants*: sodium valproate	*miR-144* *miR-34a, miR-219* *miR-let7b*, 7c *miR-221*

of conventional small molecules. Alcohol, for example, has been shown to up-regulate *miR-9* in the mammalian brain (Pietrzykowski et al. 2008), and *miR-140* is significantly up-regulated by nicotine (Huang and Li 2009).This is one of many mechanisms through which nicotine regulates gene expression at a posttranscriptional level (Table 34.1, Figure 34.1).

Antipsychotic medications primarily target the dopaminergic system, and miRNA effects on receptor expression could generate revolutionary treatment options. D1 receptor expression is up-regulated by *miR-504* through direct binding at the 3′-UTR, which leads to differential allele specific expression of the receptor, the opposite occurs with inhibition. This was illustrated by (Huang and Li 2009).

Perkins et al. (Perkins et al. 2007) also looked at the effects of the typical antipsychotic haloperidol on the expression of 179 miRNAs in haloperidol treated and haloperidol naïve rats. Three miRNAs were shown to be expressed at higher levels in the haloperidol-treated rats (*miR-199a, miR-128a*, and *miR-128b*). The effect of atypical antipsychotics on miRNA is an area yet to be explored. Kochera et al. (Kocerha et al. 2009) explored the use of the phencyclidine-like NMDA receptor antagonist dizocilpine on rodents. After a single dose of dizocilpine, within 15 minutes, there was a robust reduction of *miR-219* in the prefrontal cortex in rodents in which this agent usually produces hyperlocomotion and stereotypy. This study further explored the effect of pretreatment, with either haloperidol or clozapine on dizocilpine-mediation regulation of *miR-219*. It was noted that both drugs attenuated the locomotion induced by dizocilpine and *miR-219* concentration in the prefrontal cortex was not reduced. The data provide support for an integral role for *miR-219* in the behavioral expression associated with NMDA receptor hypofunction.

VII. FUTURE MicroRNA-BASED THERAPIES

Psychotic disorders are almost certainly polygenic in origin, and given the fact that miRNAs can regulate many genes, they are an especially appropriate target. Experimental evidence demonstrates that correction of specific miRNA alterations using miRNA mimics or antagomirs can normalize aberrant gene regulatory networks and signaling pathways. Antagomirs are a new class of modified oligonucleotides that were pioneered by Krutzfeldt et al. (Krutzfeldt et al. 2005) and were found to specifically and effectively silence miRNA expression. These cholesterol-conjugated single stranded RNA molecules are 21–23 nt in length and are complimentary to the mature target miRNA. It has been shown in rodent

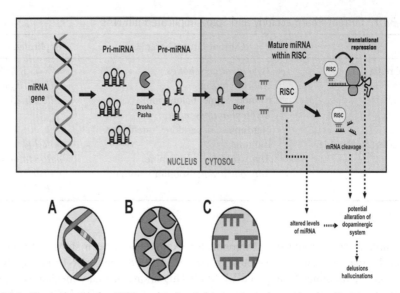

Figure 34.1. Biosynthesis of miRNA and potential impact on dopamine leading to psychosis The mature miRNA is much smaller than the gene involved in encoding the molecule. The primary transcripts (pri-mRNA) are hairpin structures with a poly(A) tail and a cap. This molecule is processed by the nuclease Drosha and the RNA-binding protein DGCR8 (Pasha in invertebrates), and the resulting molecule is called pre-mirNA. The smaller pre-miRNA passes from the nucleus into the cytosol and is processed by the endonuclease Dicer. Mature miRNA forms part of the RNA-induced silencing complex (RISC). This complex gives miRNAs its gene-silencing capacity. Current findings have suggested that alterations in miRNA biology could be the base for psychiatric conditions, such as schizophrenia. These alterations could be at the level of (A) single nucleotide polymorphisms, (B) miRNA biosynthesis, and/or (C) the amount of miRNA being expressed (up- or down-regulation). See color insert.

models that when antagomirs against *miR-122* and *miR-16* are administered intravenously, a marked reduction in miRNA levels in the liver, kidney, lung, heart, and skeletal muscle is produced. However, to effectively produce a down-regulation of targeted miRNAs in the CNS it is necessary to deliver them directly into the brain. This is one of the major challenges impeding their clinical use. Krutzfeldt et al. (2005) demonstrated that administrating these antagomirs peripherally was quite unlikely to prove effective in the CNS. Moreover, it has not yet been demonstrated that antagomirs administered centrally are capable of altering a specific behavioral phenotype. It has already been demonstrated that small molecules can impact on miRNA activity, including the postulated role of *miR-9*-dependent mechanisms contributing to alcohol dependence and the up-regulation of *miR-140* by nicotine (Huang and Li 2009). Thus the specific modulation of miRNAs in psychotic illness might take place via the manipulation of conventional small molecules. However, the blood–brain barrier could be a major hurdle in restricting the delivery of antagomirs/miRNA mimics to the CNS as it is only permeable to lipophilic molecules of less than 400 Da (Brasnjevic et al. 2009). It is unknown whether miRNAs can cross the blood–brain barrier or not. The future development of therapeutic targets thus seems likely to center on small molecules for miRNA manipulation or smart technologies for administration of miRNA mimics or antagomirs. Prior to these future novel psychopharmacological developments, it is of paramount importance to specifically demonstrate that the centrally administered technology is capable of producing behavioral changes in animals. This will be a most welcome and timely development in improving outcome measures in patients with psychosis.

VIII. CONCLUSIONS

Much has been learned regarding the epidemiology of psychotic illness, but biological understanding still remains rather rudimentary, and current treatments have significant limitations. miRNA studies in this field are in their infancy. However, they offer enormous potential to increase our understanding of complex brain disorders and provide a potential target for effective pharmacological intervention.

ACKNOWLEDGMENTS

The authors would like to acknowledge Dr. Marcela Julio-Pieper for her help in designing the figure of this review. T.G.D. is supported in part by Science Foundation Ireland in the form of a center grant (Alimentary Pharmabiotic Centre), by the Health Research Board (HRB) of Ireland and the Higher Education Authority (HEA) of Ireland.

REFERENCES

Abelson, J.F., K.Y. Kwan, B.J. O'Roak, D.Y. Baek, A.A. Stillman, T.M. Morgan, C.A. Mathews, D.L. Pauls, M.-R. Rasin, M. Gunel, N.R. Davis, A.G. Ercan-Sencicek, et al. 2005. Sequence variants in SLITRK1 are associated with Tourette's syndrome. *Science* **310**:317–320.

Akbarian, S. and H.-S. Huang. 2006. Molecular and cellular mechanisms of altered GAD1/GAD67 expression in schizophrenia and related disorders. *Brain Res Rev* **52**:293–304.

Angelucci, F., L. Aloe, P. Jimenez-Vasquez, and A.A. Mathe. 2003. Lithium treatment alters brain concentrations of nerve growth factor, brain-derived neurotrophic factor and glial cell line-derived neurotrophic factor in a rat model of depression. *Int J Neuropsychopharmacol* **6**:225–231.

Beveridge, N.J., E. Gardiner, A.P. Carroll, P.A. Tooney, and M.J. Cairns. 2010. Schizophrenia is associated with an increase in cortical microRNA biogenesis. *Mol Psychiatry* **15**:1176–1189.

Brasnjevic, I., H.W.M. Steinbusch, C. Schmitz, and P. Martinez-Martinez. 2009. Delivery of peptide and protein drugs over the blood-brain barrier. *Prog Neurobiol* **87**:212–251.

Burmistrova, O.A., A.Y. Goltsov, L.I. Abramova, V.G. Kaleda, V.A. Orlova, and E.I. Rogaev. 2007. MicroRNA in schizophrenia: genetic and expression analysis of *miR-130b* (22q11). *Biochemistry (Mosc)* **72**:578–582.

Carpenter, W.T., Jr., D.W. Heinrichs, and A.M. Wagman. 1988. Deficit and nondeficit forms of schizophrenia: the concept. *Am J Psychiatry* **145**:578–583.

Chen, G., G. Rajkowska, F. Du, N. Seraji-Bozorgzad, and H.K. Manji. 2000. Enhancement of hippocampal neurogenesis by lithium. *J Neurochem* **75**:1729–1734.

Cheng, H.Y., J.W. Papp, O. Varlamova, H. Dziema, B. Russell, J.P. Curfman, T. Nakazawa, K. Shimizu, H. Okamura, S. Impey, and K. Obrietan. 2007. microRNA modulation of circadian-clock period and entrainment. *Neuron* **54**:813–829.

Coyle, J.T. 2009. MicroRNAs suggest a new mechanism for altered brain gene expression in schizophrenia. *Proc Natl Acad Sci U S A* **106**:2975–2976.

Dinan, T.G. 2010. MicroRNAs as a target for novel antipsychotics: a systematic review of an emerging field. *Int J Neuropsychopharmacol* **13**:395–404.

Dudbridge, F. and A. Gusnanto. 2008. Estimation of significance thresholds for genomewide association scans. *Genet Epidemiol* **32**:227–234.

Farde, L., F.A. Wiesel, C. Halldin, and G. Sedvall. 1988. Central D2-dopamine receptor occupancy in schizophrenic patients treated with antipsychotic drugs. *Arch Gen Psychiatry* **45**:71–76.

Farmer, A.E., P. McGuffin, and I.I. Gottesman. 1987. Twin concordance for DSM-III schizophrenia. Scrutinizing the validity of the definition. *Arch Gen Psychiatry* **44**:634–641.

Feinberg, I. 1982. Schizophrenia: caused by a fault in programmed synaptic elimination during adolescence? *J Psychiatr Res* **17**:319–334.

Fiore, R., G. Siegel, and G. Schratt. 2008. MicroRNA function in neuronal development, plasticity and disease. *Biochim Biophys Acta* **1779**:471–478.

Frey, B.N., A.C. Andreazza, K.M. Cereser, M.R. Martins, F.C. Petronilho, D.F. de Souza, F. Tramontina, C.A. Goncalves, J. Quevedo, and F. Kapczinski. 2006. Evidence of astrogliosis in rat hippocampus after d-amphetamine exposure. *Prog Neuropsychopharmacol Biol Psychiatry* **30**:1231–1234.

Fukumoto, T., S. Morinobu, Y. Okamoto, A. Kagaya, and S. Yamawaki. 2001. Chronic lithium treatment increases the expression of brain-derived neurotrophic factor in the rat brain. *Psychopharmacology (Berl)* **158**:100–106.

Greco, S.J. and P. Rameshwar. 2007. MicroRNAs regulate synthesis of the neurotransmitter substance P in human mesenchymal stem cell-derived neuronal cells. *Proc Natl Acad Sci U S A* **104**:15484–15489.

Guo, A.-Y., J. Sun, P. Jia, and Z. Zhao. 2010. A novel microRNA and transcription factor mediated regulatory network in schizophrenia. *BMC Syst Biol* **4**:10.

Gurevich, E.V., Y. Bordelon, R.M. Shapiro, S.E. Arnold, R.E. Gur, and J.N. Joyce. 1997. Mesolimbic dopamine D3 receptors and use of antipsychotics in patients with schizophrenia. A postmortem study. *Arch Gen Psychiatry* **54**:225–232.

Halberstadt, A.L. 1995. The phencyclidine-glutamate model of schizophrenia. *Clin Neuropharmacol* **18**:237–249.

Hansen, T., L. Olsen, M. Lindow, K.D. Jakobsen, H. Ullum, E. Jonsson, O.A. Andreassen, S. Djurovic, I. Melle, I. Agartz, H. Hall, S. Timm, et al. 2007. Brain expressed microRNAs implicated in schizophrenia etiology. *PLoS ONE* **2**(9):e873.

Huang, W. and M.D. Li. 2009. Nicotine modulates expression of *miR-140**, which targets the 3'-untranslated region of dynamin 1 gene (Dnm1). *Int J Neuropsychopharmacol* **12**:537–546.

Hunsberger, J.G., D.R. Austin, G. Chen, and H.K. Manji. 2009. MicroRNAs in mental health: from biological underpinnings to potential therapies. *Neuromolecular Med* **11**:173–182.

Jacobsen, J.P. and A. Mork. 2004. The effect of escitalopram, desipramine, electroconvulsive seizures and lithium on brain-derived neurotrophic factor mRNA and protein expression in the rat brain and the correlation to 5-HT and 5-HIAA levels. *Brain Res* **1024**:183–192.

Javitt, D.C. 2006. Is the glycine site half saturated or half unsaturated? Effects of glutamatergic drugs in schizophrenia patients. *Curr Opin Psychiatry* **19**:151–157.

Jones, D.N., J.E. Gartlon, A. Minassian, W. Perry, and M.A. Geyer. 2008. In *Animal and translational models for CNS drug discovery*. R.A. McArthur and F. Borsini, Eds. Elsevier, Amsterdam. pp. 199–261.

Karayiorgou, M., M.A. Morris, B. Morrow, R.J. Shprintzen, R. Goldberg, J. Borrow, A. Gos, G. Nestadt, P.S. Wolyniec, and V.K. Lasseter. 1995. Schizophrenia susceptibility associated with interstitial deletions of chromosome 22q11. *Proc Natl Acad Sci U S A* **92**:7612–7616.

Kim, J.S., M.-Y. Chang, I.T. Yu, J.H. Kim, S.-H. Lee, Y.-S. Lee, and H. Son. 2004. Lithium selectively increases neuronal differentiation of hippocampal neural progenitor cells both in vitro and in vivo. *J Neurochem* **89**:324–336.

REFERENCES

Kocerha, J., M.A. Faghihi, M.A. Lopez-Toledano, J. Huang, A.J. Ramsey, M.G. Caron, N. Sales, D. Willoughby, J. Elmen, H.F. Hansen, H. Orum, S. Kauppinen, et al. 2009. MicroRNA-219 modulates NMDA receptor-mediated neurobehavioral dysfunction. *Proc Natl Acad Sci U S A* **106**:3507–3512.

Krutzfeldt, J., N. Rajewsky, R. Braich, K.G. Rajeev, T. Tuschl, M. Manoharan, and M. Stoffel. 2005. Silencing of microRNAs in vivo with "antagomirs". *Nature* **438**:685–689.

Liddle, P.F. 1987. Schizophrenic syndromes, cognitive performance and neurological dysfunction. *Psychol Med* **17**:49–57.

Mellios, N., H.-S. Huang, A. Grigorenko, E. Rogaev, and S. Akbarian. 2008. A set of differentially expressed miRNAs, including *miR-30a-5p,* act as post-transcriptional inhibitors of BDNF in prefrontal cortex. *Hum Mol Genet* **17**:3030–3042.

Moghaddam, B. 2003. Bringing order to the glutamate chaos in schizophrenia. *Neuron* **40**: 881–884.

Patil, S.T., L. Zhang, F. Martenyi, S.L. Lowe, K.A. Jackson, B.V. Andreev, A.S. Avedisova, L.M. Bardenstein, I.Y. Gurovich, M.A. Morozova, S.N. Mosolov, N.G. Neznanov, et al. 2007. Activation of mGlu2/3 receptors as a new approach to treat schizophrenia: a randomized Phase 2 clinical trial. *Nat Med* **13**:1102–1107.

Perkins, D.O., C.D. Jeffries, L.F. Jarskog, J.M. Thomson, K. Woods, M.A. Newman, J.S. Parker, J. Jin, and S.M. Hammond. 2007. microRNA expression in the prefrontal cortex of individuals with schizophrenia and schizoaffective disorder. *Genome Biol* **8**:R27.

Pietrzykowski, A.Z., R.M. Friesen, G.E. Martin, S.I. Puig, C.L. Nowak, P.M. Wynne, H.T. Siegelmann, and S.N. Treistman. 2008. Posttranscriptional regulation of BK channel splice variant stability by *miR-9* underlies neuroadaptation to alcohol. *Neuron* **59**:274–287.

Piton, A., J. Gauthier, F.F. Hamdan, R.G. Lafreniere, Y. Yang, E. Henrion, S. Laurent, A. Noreau, P. Thibodeau, L. Karemera, D. Spiegelman, F. Kuku, et al. 2011. Systematic resequencing of X-chromosome synaptic genes in autism spectrum disorder and schizophrenia. *Mol Psychiatry* **16**:867–880.

Purcell, S.M., N.R. Wray, J.L. Stone, P.M. Visscher, M.C. O'Donovan, P.F. Sullivan, and P. Sklar. 2009. Common polygenic variation contributes to risk of schizophrenia and bipolar disorder. *Nature* **460**:748–752.

Pushparaj, P.N., J.J. Aarthi, J. Manikandan, and S.D. Kumar. 2008. siRNA, miRNA, and shRNA: in vivo applications. *J Dent Res* **87**:992–1003.

Rapoport, J.L., J.N. Giedd, and N. Gogtay. 2012. Neurodevelopmental model of schizophrenia: update 2012. *Mol Psychiatry* **17**(12):1228–1238.

Roybal, K., D. Theobold, A. Graham, J.A. DiNieri, S.J. Russo, V. Krishnan, S. Chakravarty, J. Peevey, N. Oehrlein, S. Birnbaum, M.H. Vitaterna, P. Orsulak, et al. 2007. Mania-like behavior induced by disruption of CLOCK. *Proc Natl Acad Sci U S A* **104**:6406–6411.

Schloesser, R.J., J. Huang, P.S. Klein, and H.K. Manji. 2008. Cellular plasticity cascades in the pathophysiology and treatment of bipolar disorder. *Neuropsychopharmacology* **33**:110–133.

Schratt, G.M., F. Tuebing, E.A. Nigh, C.G. Kane, M.E. Sabatini, M. Kiebler, and M.E. Greenberg. 2006. A brain-specific microRNA regulates dendritic spine development. *Nature* **439**:283–289.

Seeman, P. and H.H. Van Tol. 1994. Dopamine receptor pharmacology. *Trends Pharmacol Sci* **15**:264–270.

Shibata, H., A. Tani, T. Chikuhara, R. Kikuta, M. Sakai, H. Ninomiya, N. Tashiro, N. Iwata, N. Ozaki, and Y. Fukumaki. 2009. Association study of polymorphisms in the group III metabotropic glutamate receptor genes, GRM4 and GRM7, with schizophrenia. *Psychiatry Res* **167**:88–96.

Silverstone, T. 1984. Response to bromocriptine distinguishes bipolar from unipolar depression. *Lancet* **1**:903–904.

Son, H., I.T. Yu, S.-J. Hwang, J.S. Kim, S.-H. Lee, Y.-S. Lee, B.-K. Kaang, and S.-H. Lee. 2003. Lithium enhances long-term potentiation independently of hippocampal neurogenesis in the rat dentate gyrus. *J Neurochem* **85**:872–881.

St Clair, D., D. Blackwood, W. Muir, A. Carothers, M. Walker, G. Spowart, C. Gosden, and H.J. Evans. 1990. Association within a family of a balanced autosomal translocation with major mental illness. *Lancet* **336**:13–16.

Stark, K.L., B. Xu, A. Bagchi, W.S. Lai, H. Liu, R. Hsu, X. Wan, P. Pavlidis, A.A. Mills, M. Karayiorgou, and J.A. Gogos. 2008. Altered brain microRNA biogenesis contributes to phenotypic deficits in a22q11-deletion mouse model. *Nat Genet* **40**(6):751–760.

Stefansson, H., D. Rujescu, S. Cichon, O.P.H. Pietilainen, A. Ingason, S. Steinberg, R. Fossdal, E. Sigurdsson, T. Sigmundsson, J.E. Buizer-Voskamp, T. Hansen, K.D. Jakobsen, et al. 2008. Large recurrent microdeletions associated with schizophrenia. *Nature* **455**:232–236.

World Health Organization (WHO). 2011. *Mental health atlas 2011*. WHO, Geneva.

Xu, B., J.L. Roos, S. Levy, E.J. van Rensburg, J.A. Gogos, and M. Karayiorgou. 2008. Strong association of de novo copy number mutations with sporadic schizophrenia. *Nat Genet* **40**: 880–885.

Zhou, R., P. Yuan, Y. Wang, J.G. Hunsberger, A. Elkahloun, Y. Wei, P. Damschroder-Williams, J. Du, G. Chen, and H.K. Manji. 2009. Evidence for selective microRNAs and their effectors as common long-term targets for the actions of mood stabilizers. *Neuropsychopharmacology* **34**:1395–1405.

Zhu, Y., T. Kalbfleisch, M.D. Brennan, and Y. Li. 2009. A MicroRNA gene is hosted in an intron of a schizophrenia-susceptibility gene. *Schizophr Res* **109**:86–89.

Part V

CIRCULATING MicroRNAs AS CELLULAR MESSENGERS AND NOVEL BIOMARKERS

35

CIRCULATING MicroRNAs AS NON-INVASIVE BIOMARKERS

Heidi Schwarzenbach and Klaus Pantel

Department of Tumor Biology, University Medical Center Hamburg-Eppendorf, Hamburg, Germany

I.	Introduction	570
II.	Characteristics of MicroRNAs	570
III.	Biological Role of Circulating MicroRNAs	570
IV.	Quantification of Circulating MicroRNAs	572
V.	MicroRNAs in Cancer	576
VI.	Circulating MicroRNAs in Cancer	576
	A. Circulating MicroRNAs in Breast and Ovarian Cancers	577
	B. Circulating MicroRNAs in Gastrointestinal Cancers	578
	C. Circulating MicroRNAs in Lung Cancer	578
	D. Circulating MicroRNAs in Prostate Cancer	579
	E. Circulating MicroRNAs in Other Cancer Types	579
VII.	Circulating MicroRNAs in Pregnancy and Benign Diseases	580
	A. Circulating MicroRNAs in Pregnancy	580
	B. Circulating MicroRNAs in Bowel Diseases and Diabetes	580
	C. Circulating MicroRNAs in Liver and Kidney Diseases	580
	D. Circulating MicroRNAs in Heart Disease	581
VIII.	Conclusion	581
References		582

ABBREVIATIONS

3′-UTR   3′-untranslated region
BPH   benign prostatic hyperplasia

MicroRNAs in Medicine, First Edition. Edited by Charles H. Lawrie.
© 2014 John Wiley & Sons, Inc. Published 2014 by John Wiley & Sons, Inc.

EMT	epithelial–mesenchymal transition
miRNA	microRNA
NGS	next-generation sequencing
NSCLC	non-small cell lung cancer

I. INTRODUCTION

Efficient management of patients relies on early diagnosis and monitoring of treatment. In this respect, significant efforts have been put into finding informative, blood-based biomarkers. These endeavours are reflected in a large number of research articles published on circulating nucleic acids in plasma or serum of patients with various diseases. With the exception of non-invasive prenatal diagnostic tests (Lo 2012), the approaches on circulating DNA and RNA applicable for clinical practice remain elusive. Based on their particular characteristics, circulating microRNAs (miRNAs) have a high level of sensitivity and specificity. MiRNA clusters are often expressed in a tissue-, development- or disease-specific manner and modify the expression of numerous genes. In the blood circulation, miRNAs are highly stable. Therefore, and to avoid biopsies by invasive methods, cell-free miRNAs in plasma or serum could serve as a "liquid biopsy," useful for diagnostic and prognostic application. The minimally invasive procedure of blood withdrawal delivers the possibility of taking repeated blood samples, consequently allowing the changes in the concentrations of miRNAs to be traced during the natural course of the disease or during treatment. These features suggest that aberrantly expressed miRNAs in the blood circulation might be attractive candidates for putative non-invasive biomarkers and could help to understand the pathogenesis of a disease. However, their clinical utility has to be validated in large prospective multicenter studies to reach the high level of evidence required for their introduction into clinical practice. In addition, studies are required to verify the reproducibility of data and validate the optimal plasma/serum miRNA cut-off levels for diagnosis and prognosis.

II. CHARACTERISTICS OF MicroRNAs

MiRNAs are a class of naturally occurring small non-coding RNA molecules. Mature miRNAs consisting of 19–25 nucleotides are single-stranded and derived from hairpin precursor molecules of 70–100 nucleotides. As one of the largest gene families, miRNAs accounts for approximately 1% of the human genome and are highly conserved in nearly all organisms (Kim 2005). In mammals, they are believed to regulate approximately 50% of all protein-coding genes (Krol et al. 2010). Mostly, they function as posttranscriptional regulators, which sequence-specifically bind to near complementary sequences in the 3′-untranslational region (UTR) of their target mRNAs (and occasionally 5′-UTR or coding sequences). In this way, the protein expression is posttranscriptionally repressed either by inhibiting the translation or degrading their target mRNA.

III. BIOLOGICAL ROLE OF CIRCULATING MicroRNAs

As a result of apoptotic and necrotic cell death, nucleic acids, such as DNA, mRNA and miRNAs, are released into the blood circulation. In addition, active secretion has also been

BIOLOGICAL ROLE OF CIRCULATING MicroRNAs

suggested as a potential source of circulating nucleic acids (Stroun et al. 2001) (Figure 35.1A). Increased levels of circulating nucleic acids have been detected in different normal and pathological circumstances. Despite accumulating evidence for the presence of miRNAs in the blood circulation, their origin, function, and half-life remain poorly understood.

The concentrations of these cell-free nucleic acids have been found to reflect disease development and progression, tumor load, and malignant progression toward metastatic relapse (Schwarzenbach et al. 2011). In blood, miRNAs circulate in a highly stable form and are protected against RNase digestion, presumably because most of them are included in apoptotic bodies, microvesicles, exosomes, and/or in complexes with RNA-binding

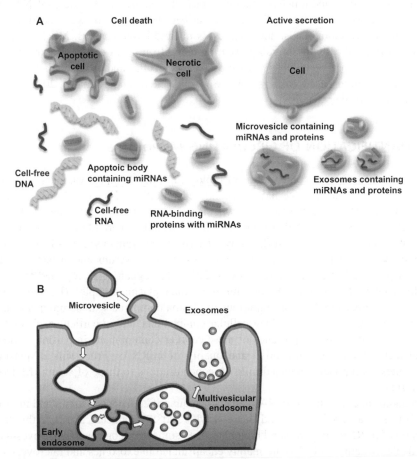

Figure 35.1. Release of miRNAs from cells into the blood circulation can occur by various cell physiological events, such as apoptosis, necrosis, and active secretion. In blood, miRNAs circulate in various secreted extracellular vesicles, such as apoptotic bodies, microvesicles, exosomes, and complexes with RNA-binding proteins (A). Mechanism of the secretion of exosomes (50–100 nm) and microvesicles (100–1000 nm). Within the endosomal system exosomes are built by inward budding of the limiting cell membrane of the multivesicular body, a late endosomal compartment. The fusion of the multivesicular body with the plasma membrane leads to the secretion of exosomes. Microvesicles can be formed at the plasma membrane by budding into the extracellular space (B). See color insert.

proteins (Kosaka et al. 2010; Asaga et al. 2011) (Figure 35.1A). As the genetically programmed cell death, apoptosis leads from cellular condensation to DNA fragmentation and results in apoptotic bodies containing nuclear fragments or remnants of the cytoplasm. Microvesicles are relatively large (~100 nm to 1 μm) vesicles released from cells through blebbing, whereas exosomes are small vesicles (~30–100 nm) released when endosomally derived multivesicular bodies fuse with the plasma membrane (Cocucci et al. 2009; Kosaka et al. 2010) (Figure 35.1B).

It has also been proposed that extracellular miRNAs can act as mediators of cell–cell communication and immune regulation (Valadi et al. 2007; Pegtel et al. 2010). Thus, miRNAs, which are secreted from cells, can be functionally transferred through exosomes and delivered to recipient cells. By endocytotic uptake, membrane fusion, or scavenger receptors, the transported miRNAs may then mediate repression of critical mRNA targets in the recipient cells. Recent studies have described the transfer between dendritic cells, hepatocellular carcinoma cells, and adipocytes in lipid-based carriers. MiRNAs are also transferred from T cells to antigen-presenting cells, from stem cells to endothelial cells and fibroblasts, from macrophages to breast cancer cells, and from epithelial cells to hepatocytes in lipid-based carriers (reviewed in Vickers and Remaley 2012; see also Chapter 36 for further details).

IV. QUANTIFICATION OF CIRCULATING MicroRNAs

Although present in many other forms of biological fluids (e.g., saliva, tears, and urine), the majority of studies to date have concentrated on either plasma or serum, which are easily obtained. However, the choice of whether to use serum or plasma, and within these materials, whether to use whole serum/plasma or purified/prepared fractions, is an important issue that has not been sufficiently discussed and remains unresolved. Whereas Mitchell et al. found no significant differences between serum and plasma levels of miRNAs (Mitchell et al. 2008; Kroh et al. 2010), Heneghan et al. reported that serum samples contain lower miRNA concentrations than plasma samples (Heneghan et al. 2010a). The higher miRNA concentrations in plasma compared with serum could mainly be due to the presence of subcellular/cellular components, and in particular, platelets and erythrocytes. Likewise, the presence of these cellular contaminants in plasma could be a critical factor for the quantification and quality of miRNAs. Processing of plasma to remove these components reduced miRNA concentrations to those of serum (McDonald et al. 2011).

In addition, circulating miRNAs can also be quantified from exosomes and microvesicles (Hunter et al. 2008; Taylor and Gercel-Taylor 2008; Bryant et al. 2012). In this regard, the origin of miRNAs in peripheral blood has been somewhat controversially investigated, in particular, whether miRNAs are mainly encapsulated in exosomes and microvesicles or also circulate cell-freely (Arroyo et al. 2011; Turchinovich et al. 2011). To date, the question has also been open whether the miRNA expression profile in the whole plasma is similar to the pattern in exosomes and microvesicles. Comparing the detection of miRNAs extracted from serum and isolated exosomes showed that the isolation of circulating exosomes improved the sensitivity of amplification of lowly expressed miRNAs from human biological fluids. Consequently, it was suggested that the extraction of exosomal miRNAs should be the starting point for early biomarker studies to reduce the probability of false negative results involving low abundance miRNAs that may be missed by using unfractionated serum. Once potential biomarkers are identified in exosomes, methods can be optimized to detect these specific miRNAs in whole serum for use in larger studies (Gallo et al. 2012). Furthermore, a recent study compared the transcript levels of two miRNAs in serum with the levels in exosomes and microvesicles, and showed that the association of cell-free serum miRNAs with metastatic prostate cancer was confirmed using serum-derived exosomes and microvesicles (Bryant et al. 2012).

An interesting aspect to consider when quantifying miRNAs in exosomes and microvesicles is that they are believed to be involved in intercellular communication, to stimulate cellular signaling and regulate metabolic function and homeostasis of hematopoietic cells (Howcroft et al. 2011; Lee et al. 2011). It was reported that macrophages regulate the invasiveness of breast cancer cells through exosome-mediated delivery of oncogenic miRNAs (Yang et al. 2011), and that microvesicles contribute to trigger the angiogenic switch and coordinate metastatic diffusion during tumor progression (Grange et al. 2011). However, with the exception of EpCAM (Taylor and Gercel-Taylor 2008), marker proteins that allow enrichment of tumor-derived exosomes over normal exosomes, which also circulate in peripheral blood of tumor patients, are less well defined.

Currently, there are several different platforms that can be used for quantifying extracted miRNAs (reviewed in Chugh and Dittmer 2012). Quantitative real-time PCR is considered the gold standard for quantifying miRNAs and is relatively cheap, easy to carry out, sensitive and hence a popularly used technique. Consequently, several companies offer quantitative PCR-based assays for the detection of specific miRNAs, including those specifically developed with circulating miRNAs in mind. A further sensitive method is the relative quantification of low abundance circulating miRNAs by stem-loop RT-PCR (Hurley et al. 2012). The use of PCR-based techniques is particularly suitable for measuring single miRs, whereas miRNA arrays containing hundreds of miRNAs (such as Taqman low density arrays [TLDA]) are the most frequently used technique to profile sera/plasma samples. However, the high costs of miRNA arrays make them unsuitable for a large study. More rarely, Northern blots and RNAse protection assays have been used to measure miRNA expression. These techniques are time-consuming and not suitable for high-throughput screenings, and also require a significant amount of RNA, something that is hard to achieve in sera-based studies. The most commonly used alternative to PCR-based platforms are microarray-based technologies that rely on hybridization to specific probes, which can cover more than 1000 mature human miRNAs sequences listed in the miRNA database (Sanger miRNABase). They can be challenging to optimize probes and hybridization conditions, and the results of these measurements can vary considerably between different manufactures. Next-generation sequencing (NGS) appears to be a very promising technique for identifying the expression levels of miRNAs in a sample. In contrast to microarrays, NGS is largely sequence-independent and does not rely on the design of primers or probes specific to each miRNA. In this method, adapter-ligated sample RNA and cDNA libraries are amplified by PCR and sequenced. The output delivers sequencing reads of varying lengths corresponding to miRNAs, which are then aligned to the reference sequence of choice. Despite these advantages, NGS is labor intensive and expensive (reviewed in Chugh and Dittmer 2012). Finally, the nanotechnology-based assay is a new method which uses nanopores that can detect the position and conformation of the target molecule within the pore (Y. Wang et al. 2011).

To date, in most publications, miRNA expression microarrays have been applied for the screening of deregulated transcript levels of miRNAs. Once aberrant transcript levels

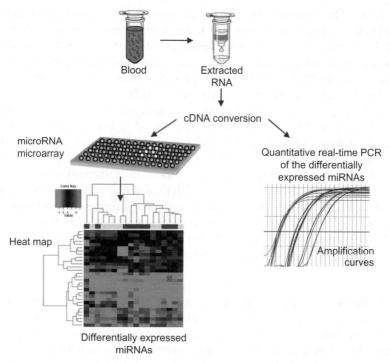

Figure 35.2. Experimental flowchart of quantification of circulating miRNAs in blood. See color insert.

of the miRNAs are deduced from the array data, validation studies can be carried out by quantitative real-time PCR in single miRNA assays by the TaqMan method (see Figure 35.2).

In different studies, the application of the different techniques has shown conflicting quantitative data on the deregulation of circulating miRNAs from the same disease type (reviewed in Kosaka et al. 2010). These discrepancies are probably due to differences between technical platforms and statistical analyses used. Besides, the expression profiles of miRNAs may change in respect to the established risk factors of the patients and whether the blood samples were drawn prior to or after treatment, surgery, or chemotherapy (Kosaka et al. 2010). In addition, a suitable endogenous miRNA control, to normalize miRNA levels, is lacking. The choice of reference gene for normalization of quantitative PCR data analysis has a great effect on the study outcome that is particularly problematic for plasma/sera studies due to wide variation in the levels of the snRNAs generally used for tissue miRNA studies (e.g., *RNU6B*) (Song et al. 2012). Therefore, it is necessary to choose a suitable reference for reliable expression data. For plasma/sera miRNA analyses the algorithms should reveal stably expressed reference genes across all patients and healthy controls. For example, *miR-16* or the nucleolar *RNU6-2* were frequently used as reference genes (Schaefer et al. 2010).

Nevertheless, based on their biological role and involvement in transforming cells, circulating miRNAs may have potential as diagnostic, prognostic, and predictive biomarkers and may also be considered as potential future therapeutic targets (Table 35.1 and Table 35.2).

TABLE 35.1. Dysregulated Levels of Circulating miRNAs in Plasma/Serum of Patients with Different Tumor Entities ($n \geq 40$) in Relationship to Their Diagnosis and Prognosis

Cancer	MicroRNAs	Diagnosis	Prognosis	Reference
Breast	miR-122, -375		X	Wu et al. (2012)
	miRNA-210	X		Jung et al. (2012)
	miR-215, -299-5P, -411, -452	X		van Schooneveld et al. (2012)
	miR-20a, -21, -214	X		Schwarzenbach et al. (2012)
	miR-10b, -34a, -141, -155	X		Roth et al. (2010)
	miRNA-21	X	X	Asaga et al. (2011)
	miR-10b, -21, -145, -155, -195, -16, let-7a	X		Heneghan et al. (2010b)
	miRNA-21, -106a, -126, -155, -199a, -335	X		F. Wang et al. (2010)
Cervical	miRNA-218	X		Yu et al. (2012)
Colorectal	miRNA-29a	X		Wang and Gu (2012)
	miRNA-141	X	X	Cheng et al. (2011)
	miR-29a, -92a	X		Huang et al. (2010)
	miR-17-3p, -92	X		Ng et al. (2009)
Gastric	miRNA-196a	X	X	Tsai et al. (2012)
	miR-17-5p, -20a	X	X	Wang and Gu (2012) and Wang et al. (2012)
	miRNA-378	X		H. Liu et al. (2012)
	miR-17-5p, -21, -106a, -106b, let-7a	X		Tsujiura et al. (2010)
	miR-106a, -17	X		Zhou et al. (2010)
Hepatocellular carcinoma	miRNA-122	X		Qi et al. (2011)
	miRNA-21	X		Tomimaru et al. (2012)
	miR-16, -199a	X		Qu et al. (2011)
	miR-15b, -130b	X		A.M. Liu et al. (2012)
Lung	miR-155, -197, -182	X		Zheng et al. (2011)
	miR-21, miR-210, and miRNA-486-5p	X		Shen et al. (2011)
	miRNA-21	X		Wei et al. (2011)
	let-7f, miRNA-30e-3p	X	X	Silva et al. (2011)
	miR-486, -30d, -1, -499		X	Hu et al. (2010)
	miR-25, -223	X		Chen et al. (2008)
Leukemia	miR-92a, -638	X		Tanaka et al. (2009)
Lymphoma	miRNA-221	X	X	Guo et al. (2010)
	miRNA-21	X	X	Lawrie et al. (2008)
Melanoma	miRNA-221	X	X	Kanemaru et al. (2011)
Myelodysplasis	let-7a, miRNA-16		X	Zuo et al. (2011)
Esophageal	miR-21, -375	X		Komatsu et al. (2011)
Prostate	miR-107, -574-3p, -141 -375, -200b	X		Bryant et al. (2012)
	let-7e, let-7c, miR-30c, -622, -1285	X		Chen et al. (2012)
	miR-20a, -21, -145, -221	X		Shen et al. (2012)
	miR-375, -141	X		Brase et al. (2011)
	miRNA-21	X		Zhang et al. (2011)
	miR-21, -141, -221	X		Yaman Agaoglu et al. (2011)

This table is not comprehensive and is based on our review of studies deemed as significant clinical translational events.

TABLE 35.2. Dysregulated Levels of Circulating miRNAs in Plasma/Serum of Patients with Different Diseases ($n \geq 40$) in Relationship to Their Diagnosis and Prognosis

Cancer	miRNAs	Diagnosis	Prognosis	Reference
Inflammatory bowel	MiR-16, -23a, -29a, -106a, -107, -126, -191, -199a-5p, -200c, -362-3p, -532-3p	X		Paraskevi et al. (2012)
	miRNAs-199a-5p, -362-3p, -340*, -532-3p, miRNAplus-1271	X		Wu et al. (2011)
Diabetes	miR-15a, -29b, -126, -223, -28-3p	X		Zampetaki et al. (2010)
Ectopic pregnancy	miRNA-323-3p	X		Zhao et al. (2012)
Hepatitis C	miR-34a, -122	X		Cermelli et al. (2011)
Kidney injury	miRNA-210	X	X	Lorenzen et al. (2011)
Liver injury	miR-21, -122, -223	X		Xu et al. (2011)
	miR-122, -192	X		Starkey Lewis et al. (2011)
	miRNA-122	X		Zhang et al. (2010)
Myocardial disease	miRNA-499	X		Emilian et al. (2012)
	miRNA-499-5p	X		Olivieri et al. (2013)
	miRNA-208b, -499	X		Devaux et al. (2012)
	miR-133, -328	X		R. Wang et al. (2011)
	miR-133a, -208b	X	X	Widera et al. (2011)
	miRNA-1	X		Ai et al. (2010)
Pulmonary tuberculosis	miRNA-29a	X		Fu et al. (2011)
Sepsis	miR-146a, -223	X		J.F. Wang et al. (2010)
Systemic lupus erythematosus	miR-200a, -200b, -200c, -429, -205, -192,	X		G. Wang et al. (2011)
Systemic sclerosis	miRNA-142-3p	X		Makino et al. (2011)

This table is not comprehensive and is based on our review of studies deemed as significant clinical translational events.

V. MICRORNAS IN CANCER

As half of human miRNAs are localized in fragile chromosomal regions, which may exhibit DNA amplifications, deletions, or translocations during tumor development, their expression is frequently deregulated in cancer (Croce 2009). MiRNAs have, therefore, important roles in repression of protein expression in tumorigenesis (Bartel 2009). To date, studies on solid cancers (ovarian, lung, breast, and colorectal cancer) have reported that miRNAs are involved in the regulation of different cellular processes, such as apoptosis, cell proliferation, epithelial to mesenchymal transition, and metastases (Heneghan et al. 2009). During tumorigenesis, miRNAs can act as tumor suppressor genes or oncogenes. A decrease in tumor suppressive miRNAs may reduce the translation of oncogenes leading to an increase in tumor-specific proteins. If there is an overexpression of oncogenic miRNAs, tumor suppressive genes are more strongly repressed than under physiological conditions (Esquela-Kerscher and Slack 2006).

VI. CIRCULATING MICRORNAS IN CANCER

So far, the concern whether miRNAs detected in the blood circulation of tumor patients mainly originate from tumor cells has not clearly been defined. As the expression of miRNAs seems to reflect aspects of the human physiological state, they may originate from different sources. Along with tumor cells, a proportion of these miRNAs may also be derived from blood cells and other organs. Cancer-associated miRNAs may also be released into the blood circulation by immunocytes in the tumor microenvironment or from other affected organs that mediate cellular responses to tumor burden and inflammatory reactions (Okada et al. 2010). Thus, blood may form a pool of a variety of circulating miRNAs discharged from different sources. Dependent on their origin, circulating miRNAs may have specific roles and serve as therapeutic targets (Cho 2012). The release of specific miRNAs by cancer cells may be involved in tumor progression, immune-suppressive, or angiogenic processes. Conversely, cells surrounding the primary tumor may secrete tumor-suppressive miRNAs, which block tumor growth and propagation (Ha 2011; Toffanin et al. 2012). Microvesicles derived from human melanomas and colorectal carcinomas promote tumor growth and immune escape by distorting monocyte differentiation (Valenti et al. 2006). In contrast, exosomal miRNAs discharged by dendritic cells and B lymphocytes can deliver signals for T cell activation (Kim et al. 2005).

In 2008, the presence of miRNAs in serum was first described for patients with diffuse large B-cell lymphoma. This was the first evidence showing the feasibility of quantifying circulating miRNAs from blood. In this study, the association of high expression levels of *miR-21* with relapse-free survival was reported, suggesting that miRNAs have potential as minimally invasive diagnostic markers (Lawrie et al. 2008). Subsequently, many further studies have been published, suggesting that miRNA expression is related to tumor classification, diagnosis, disease progression, and prognosis (Table 35.1). We will review the evidence in differing cancer types below.

A. Circulating MicroRNAs in Breast and Ovarian Cancers

In patients with early-stage breast cancer, the prevalence of *miR-122, miR-10b, miR-34a* and *miR-155* in the blood circulation correlated with the presence of overt metastasis (Roth et al. 2010; Wu et al. 2012). The serum concentrations of *miR-10b, miR-34a*, and *miR-155* were also significantly elevated in patients with ovarian as well as with lung cancer (Roth et al. 2011a, 2011b). *miR-155* was one of the first miRNAs which were quantified and extracted from the blood (Lawrie et al. 2008). Kong et al. detected that the knock down of *miR-155* can prevent the process of TGF-ß-induced EMT (epithelial mesenchymal transition). It is therefore likely that *miR-155* is involved in the invasive character of breast carcinoma (Kong et al. 2008). Furthermore, *miR-10b* seems to play a crucial role in metastases. In this regard, Ma et al. showed that in metastatic cell lines *miR-10b* was highly expressed and could affect both migration of cells and their invasive features (Ma et al. 2007). Heneghan et al. found an association of high serum level of *miR-10b* with the estrogen receptor status of breast cancer patients (Heneghan et al. 2010b). In patients with lung carcinoma and lymph node metastases particularly high concentrations of this miRNA, which correlated with high concentrations of the tumor marker TPA, were also reported (Roth et al. 2011b). TPA, a cytokeratin antigen and a degradation product of the cytoskeleton, is released by proliferating cells into the blood circulation (Buccheri and Ferrigno 1988). Thus, a combination of both markers (TPA and *miR-10b)* could improve the screening of lung carcinomas. Not only in early-stage breast cancer patients, but also in ovarian cancer patients, high levels of *miR-34a* seem to be involved in metastases and tumor progression. Ovarian cancer patients with lymph node metastases had significantly high *miR-34a* values (Roth et al. 2011a). In the literature, *miR-34a* has been discussed as p53-inducable tumor suppressor, which is down-regulated in lung carcinoma tissues (Bommer et al. 2007).

In addition, the concentrations of *miR-141* were significantly increased in the blood serum of patients with ovarian as well as with lung cancer (Roth et al. 2011a, 2011b). *MiR-141* has been described as tumor suppressor, which stabilizes the epithelial phenotype of cell lines and is repressed by the EMT-inducing ZEB1 factor (Burk et al. 2008). In the blood of patients with advanced prostate cancer, the concentrations of circulating *miR-141* were increased (Mitchell et al. 2008). Moreover, the elevated levels of *miR-141* were associated with high uPA concentrations in the serum of lung carcinoma patients (Roth et al. 2011b). UPA, a serine protease, is involved in the degradation of the extracellular matrix. In this process, uPA activates the protease plasmin, which in turn also degrades components of the extracellular matrix and activates matrix metalloproteinases (Duffy 1996). Thus, the degradation of the extracellular matrix contributes to the tumor progression (Borgono and Diamandis 2004).

It has been suggested that plasma *miR-210* could be used for monitoring the response of breast cancer patients to therapies that contain trastuzumab (Jung et al. 2012). As an indicator of malignant disease and metastatic spread to regional lymph nodes circulating *miR-214*, which was predicted to target the tumor suppressor gene phosphatase and tensin homologue deleted (PTEN), had diagnostic potential in breast cancer patients (Schwarzenbach et al. 2012). An assay for quantifying circulating *miR-21* was shown to have utility in detecting progression of early-stage breast cancer (Asaga et al. 2011). The expression levels of *miR-21*, *miR-126*, *miR-155*, *miR-199a*, and *miR-335* were associated with clinicopathologic features of breast cancer, such as histological tumor grades and sex hormone receptor expression (F. Wang et al. 2010). In most cervical cancer patients, *miR-218* was deregulated and associated with tumor invasion (Yu et al. 2012) (Table 35.1).

B. Circulating MicroRNAs in Gastrointestinal Cancers

In respect of colorectal cancer, serum *miR-29a* had strong potential as a minimally invasive biomarker for early detection of this cancer entity with liver metastasis (Wang and Gu 2012). Plasma *miR-141* was also associated with distant metastasis and poor prognosis (Cheng et al. 2011). The prevalence of *miR-92* in plasma was able to differentiate colorectal cancer from gastric cancer patients (Ng et al. 2009). In gastric cancer, circulating *miR-17-5p* and *miR-20a* may be a promising minimally invasive molecular marker for pathological progression, prediction of prognosis, and monitoring of chemotherapeutic effects (Wang and Gu 2012; Wang et al. 2012). Moreover, the detection of *miR-106a* and *miR-17* in peripheral blood of patients with gastric cancers may be a novel tool for monitoring circulating tumor cells (Zhou et al. 2010). An oncogenic role for *miR-196a*, that ectopic expression promoted EMT, migration, and invasion capability of transfected cells, was described in gastric cancer. Elevated levels of circulating *miR-196a* in serum were associated with gastric cancer disease status and relapse (Tsai et al. 2012) (Table 35.1).

C. Circulating MicroRNAs in Lung Cancer

As a definitive preoperative diagnosis of solitary pulmonary nodules found by CT has been a clinical challenge, the combined panel of *miR-21* and *miR-486-5p* could facilitate the diagnosis to distinguish lung tumors from benign solitary pulmonary nodules (Shen et al. 2011). The screening of *miR-21* in plasma of non-small cell lung cancer (NSCLC) patients could also serve for the assessment of the sensitivity to platinum-based chemotherapy (Wei et al. 2011). Moreover, in NSCLC patients and healthy individuals, who differ in vesicle-related miRNAs in plasma, the levels of *let-7f* and *miR-30e-3p* were associated with poor outcome. Thus, plasma vesicle-related miRNAs obtained by non-invasive methods could serve as circulating tumor biomarkers of discriminating and prognostic value (Silva et al. 2011). Also, the signature of *miR-486, miR-30d, miR-1* and *miR-499* in the serum may serve as a minimally invasive predictor for the overall survival of NSCLC (Hu et al. 2010) (Table 35.1).

D. Circulating MicroRNAs in Prostate Cancer

In prostate cancer patients, an association of the amounts of *miR-141* and *miR-375* with metastatic disease was described (Mitchell et al. 2008) and confirmed using serum-derived exosomes and microvesicles from a separate cohort of patients with recurrent or non-recurrent disease following radical prostatectomy (Bryant et al. 2012). A panel of five circulating miRNAs, *let-7e, let-7c, miR-30c, miR-622,* and *miR-1285,* could discriminate prostate cancer from benign prostatic hyperplasia (BPH) with high sensitivity and specificity, and therefore, combined with the routine PSA test, these five cancer-specific miRNAs might help to improve prostate cancer diagnosis in clinical applications (Chen et al. 2012). The altered plasma levels of *miR-20a, miR-21, miR-145,* and *miR-221* could be useful predictors to distinguish prostate cancer patients with varied aggressiveness (Shen et al. 2012). In addition, circulating *miR-375* and *miR-141* turned out to be the most pronounced markers for high-risk tumors. Their levels also correlated with high Gleason score and lymph-node positive status. These observations suggest that the release of *miR-375* and *miR-141* into the blood circulation is associated with advanced prostate cancer disease (Brase et al. 2011). The quantification of *miR-21* targeting the tumor suppressor gene PTEN and programmed cell death 4 (PDCD4) has been reported to be a useful biomarker for prostate cancer patients during disease progression. Patients with hormone-refractory prostate cancer expressed higher serum levels of *miR-21* than those with androgen-dependent and localized prostate cancer. Androgen-dependent prostate cancer patients with low serum PSA levels had serum levels of *miR-21* similar to those of patients with localized prostate cancer or BPH. The highest serum levels of *miR-21* were found in hormone-refractory prostate cancer patients who were resistant to docetaxel-based chemotherapy when compared with patients sensitive to chemotherapy. These findings suggest that *miR-21* is an indicator of the transformation to hormone refractory disease and a potential predictor for the efficacy of docetaxel-based chemotherapy (Zhang et al. 2011). The quantification of *miR-21,* together with *miR-141* and *miR-221,* revealed varying patterns in blood of clinical subgroups. In patients diagnosed with metastatic prostate cancer, levels of all three miRNAs were significantly higher than in patients with localized and local advanced disease (Yaman Agaoglu et al. 2011) (Table 35.1).

E. Circulating MicroRNAs in Other Cancer Types

Extra-nodal natural killer T-cell (NK/T-cell) lymphoma is a progressive cancer type with poor prognosis due to the lack of disease-specific treatment. To develop specific therapeutic strategies, it is essential to identify tumor markers for this cancer type. Univariate and multivariate analyses revealed that plasma levels of *miR-221* may be a diagnostic and prognostic marker for NK/T-cell lymphoma (Guo et al. 2010). Among patients with malignant melanoma, *miR-221* levels were significantly increased in patients with stage I–IV compared with those with *in situ* tumors, and additionally, correlated with tumor thickness (Kanemaru et al. 2011). In patients with myelodysplastic syndrome, the circulating levels of *let-7a* and *miR-16* were significantly associated with their progression-free and overall survival. This association persisted even after patients were stratified according to the International Prognostic Scoring System. These findings suggest that the plasma levels of both miRNAs can serve as minimally invasive prognostic markers for this disease (Zuo et al. 2011). Furthermore, esophageal cancer patients with a high plasma level of *miR-21* tended to have larger vascular invasion and to show a high correlation with tumor recurrence (Komatsu et al. 2011) (Table 35.1).

In summary, the findings discussed earlier highlight the potential clinical utility of circulating miRNA profiling in cancer diagnosis and prognosis. Conspicuously, in these studies, circulating, cell-free *miR-21* seems to be a biomarker with clinical value for screening of several tumor entities, such as breast, gastric, hepatocellular, lung, lymphoma, esophageal, and prostate cancer (Table 35.1).

VII. CIRCULATING MicroRNAs IN PREGNANCY AND BENIGN DISEASES

Apart from the clinical relevance of circulating miRNAs in the blood of cancer patients, deregulated levels of miRNAs may also play a role in benign diseases and altered physiological states (Table 35.2).

A. Circulating MicroRNAs in Pregnancy

The presence of specific pregnancy-associated miRNAs in the maternal circulation has also been investigated. As the application of serum human chorionic gonadotropin (hCG) and progesterone to identify patients with ectopic pregnancy has been shown to have poor clinical utility, pregnancy-associated circulating miRNAs have been suggested as potential biomarkers for the diagnosis of pregnancy associated complications. In particular, *miR-323-3p*, together with hCG and progesterone, demonstrated significant diagnostic accuracy for the diagnosis of ectopic pregnancy (Zhao et al. 2012) (Table 35.2).

B. Circulating MicroRNAs in Bowel Diseases and Diabetes

In a recent study the putative role of miRNAs as contributors to inflammatory bowel disease pathogenesis was described. *MiR-16, miR-23a, miR-29a, miR-106a, miR-107, miR-126, miR-191, miR-199a-5p, miR-200c, miR-362-3p*, and *miR-532-3p* were expressed at significantly high levels in the blood from patients with Crohn's disease (Paraskevi et al. 2012).

MiRNAs have also been implicated in the epigenetic regulation of key metabolic, inflammatory, and antiangiogenic pathways in type 2 diabetes. Reduced *miR-15a, miR-29b, miR-126, miR-223* levels, and elevated *miR-28-3p* levels antedated the manifestation of this disease. For endothelial *miR-126*, the reduction was confined to circulating vesicles in plasma. Investigation on hyperglycemic Lep(ob) mice showed that high glucose concentrations reduced the *miR-126* content of endothelial apoptotic bodies. These findings might explain the impaired peripheral angiogenic signaling in patients with type 2 diabetes (Zampetaki et al. 2010) (Table 35.2).

C. Circulating MicroRNAs in Liver and Kidney Diseases

In patients with chronic hepatitis C infection and non-alcoholic fatty liver disease, *miR-122* and *miR-34a* levels positively correlated with histological disease severity from simple steatosis to steatohepatitis. The serum levels of both miRNAs correlated with liver enzymes levels, fibrosis stage and inflammation activity (Cermelli et al. 2011). Concerning acute kidney injury, *miR-210* predicted mortality in this patient cohort and may serve as a novel biomarker reflecting pathophysiological changes on a cellular level (Lorenzen et al. 2011). Liver insult is frequently caused by viral infection, alcohol abuse, or toxic chemical exposure. The potential use of circulating *miR-21, miR-122, miR-223*, and *miR-192* as a novel, predictive, and reliable blood marker panel for viral-, alcohol-, or chemical-induced liver injury was described (Zhang et al. 2010; Starkey Lewis et al. 2011; Xu et al. 2011) (Table 35.2).

D. Circulating MicroRNAs in Heart Disease

Specific miRNAs have also been linked to heart disease. Diagnosis of acute myocardial injury by using biomarkers is difficult in patients with advanced renal failure. In these patients, increased levels of both troponins and *miR-499* were observed. However, whereas the levels of cardiac troponins are unaffected by hemodialysis, it is not the case for the amounts of *miR-499*. Therefore, these observations alleviate the potential of *miR-499* as a marker of myocardial injury (Emilian et al. 2012). In addition, geriatric patients with acute non-ST elevation myocardial infarction have frequently atypical symptoms and non-diagnostic electrocardiogram. The detection of modest elevation of troponins is challenging for physicians needing to routinely triage these patients. Unfortunately, non-coronary diseases, such as acute heart failure, may also cause this elevation. Interestingly, circulating *miR-499* and *miR-499-5p* were comparable with cardiac troponins in discriminating myocardial infarction from acute heart failure. As sensitive biomarkers, they may exhibit diagnostic accuracy for patients with modest elevation of troponins at initial contact (Devaux et al. 2012; Olivieri et al. 2013). Furthermore, patients with myocardial infarction may harbor higher levels of *miR-1, miR-133a*, and *miR-208b* than patients with unstable angina, but these miRNA levels also showed a large overlap between these patient cohorts. In univariate and age- and gender-adjusted analyses, the levels of *miR-133a* and *miR-208b* were significantly associated with the risk of death of the patients, but both miRNAs lost their independent association with the clinical outcome upon further adjustment for high-sensitivity troponin (Widera et al. 2011). Finally, it was reported that the significantly elevated levels of *miR-1* in plasma from these patients dropped to normal levels following medication (Ai et al. 2010) (Table 35.2).

To sum up, the findings discussed earlier highlight the potential clinical value of circulating miRNAs in diverse benign diseases. Deregulated levels of circulating miRNAs were also detected in the blood of patients with other benign diseases, for example, pulmonary tuberculosis, sepsis, systemic lupus erythematosus or sclerosis (Table 35.2).

VIII. CONCLUSION

Currently, efficient management of patients relies on early diagnosis and monitoring of treatment. Assays that allow the repetitive monitoring of diseases using blood samples may be efficient in assessing disease progression in patients. In the future, minimally invasive blood analyses of circulating miRNAs may have the potential to complement the existing biomarkers. In this respect, circulating miRNAs may be promising blood-based cancer biomarkers, as they are informative and disease-specifically modulated. This deregulation qualifies them as potential diagnostic biomarkers. MiRNAs prevent the translation of many different proteins, which can be involved in disease-relevant signal pathways of malignant and benign diseases, as well as altered physiological states. The study of these signal pathways offers additionally potential therapeutic targets in the treatment of these diseases. Due to the promising results regarding the expression profiles of miRNAs in the patient's blood in previous studies, the influence of miRNAs on its potential target mRNAs may reveal interesting aspects of the diseases. However, single miRNAs may bind to several mRNAs, which exacerbate the search of specific mRNA target molecules. Nevertheless, the bioinformatical prediction of potential mRNA targets for miRNAs through databases and protein-based microarray analyses may help delivering data on target molecules, which are essential in a particular disease.

In the continued development of biomarkers, several crucial issues should be addressed, such as the different analytical technical platforms, the use of serum or plasma samples, the extraction of miRNA from whole serum/plasma or exosomes/microvesicles isolated by specific disease-associated markers and the sample collection, storage, and processing. Depending on the time point of blood sample collection (before, during or after surgery or treatment) and the patient's treatment (surgery, chemotherapy, immunotherapy, and/or medication), the expression profile of circulating miRNAs may certainly change, and these parameters should therefore be considered. Especially critical to developing useful biomarkers is the establishment of an endogenous miRNA control to normalize the levels of circulating miRNAs. For prognostic evaluations, the acquisition of clinicopathological data of the patients may also be a critical matter. Finally, insufficient patient numbers and inappropriate statistical analyses are evidence for lacking diagnostic specificity. These common deficiencies in performing miRNA analyses may contribute to lacking comparability of data among the studies and exacerbate reproducibility of the studies. The standardization of the approaches will be a major task that will require international cooperation among the scientists to obtain a consensus on miRNA assays and reporting results. If these problems could be solved, blood-based miRNAs may become valuable biomarkers for the management of patients with a variety of different diseases. In particular, some identified miRNAs appear repeatedly significant for different malignancies and show promising prognostic associations with disease outcomes, suggesting that *miR-coordinated* regulatory pathways are common to many diseases.

REFERENCES

Ai, J., R. Zhang, Y. Li, J. Pu, Y. Lu, J. Jiao, K. Li, B. Yu, Z. Li, R. Wang, L. Wang, Q. Li, et al. 2010. Circulating microRNA-1 as a potential novel biomarker for acute myocardial infarction. *Biochem Biophys Res Commun* **391**:73–77.

Arroyo, J.D., J.R. Chevillet, E.M. Kroh, I.K. Ruf, C.C. Pritchard, D.F. Gibson, P.S. Mitchell, C.F. Bennett, E.L. Pogosova-Agadjanyan, D.L. Stirewalt, J.F. Tait, and M. Tewari. 2011. Argonaute2 complexes carry a population of circulating microRNAs independent of vesicles in human plasma. *Proc Natl Acad Sci U S A* **108**:5003–5008.

Asaga, S., C. Kuo, T. Nguyen, M. Terpenning, A.E. Giuliano, and D.S. Hoon. 2011. Direct serum assay for microRNA-21 concentrations in early and advanced breast cancer. *Clin Chem* **57**:84–91.

Bartel, D.P. 2009. MicroRNAs: target recognition and regulatory functions. *Cell* **136**:215–233.

Bommer, G.T., I. Gerin, Y. Feng, A.J. Kaczorowski, R. Kuick, R.E. Love, Y. Zhai, T.J. Giordano, Z.S. Qin, B.B. Moore, O.A. MacDougald, K.R. Cho, et al. 2007. p53-mediated activation of miRNA34 candidate tumor-suppressor genes. *Curr Biol* **17**:1298–1307.

Borgono, C.A. and E.P. Diamandis. 2004. The emerging roles of human tissue kallikreins in cancer. *Nat Rev Cancer* **4**:876–890.

Brase, J.C., M. Johannes, T. Schlomm, M. Falth, A. Haese, T. Steuber, T. Beissbarth, R. Kuner, and H. Sultmann. 2011. Circulating miRNAs are correlated with tumor progression in prostate cancer. *Int J Cancer* **128**:608–616.

Bryant, R.J., T. Pawlowski, J.W. Catto, G. Marsden, R.L. Vessella, B. Rhees, C. Kuslich, T. Visakorpi, and F.C. Hamdy. 2012. Changes in circulating microRNA levels associated with prostate cancer. *Br J Cancer* **106**:768–774.

Buccheri, G. and D. Ferrigno. 1988. Usefulness of tissue polypeptide antigen in staging, monitoring, and prognosis of lung cancer. *Chest* **93**:565–570.

Burk, U., J. Schubert, U. Wellner, O. Schmalhofer, E. Vincan, S. Spaderna, and T. Brabletz. 2008. A reciprocal repression between ZEB1 and members of the miR-200 family promotes EMT and invasion in cancer cells. *EMBO Rep* **9**:582–589.

Cermelli, S., A. Ruggieri, J.A. Marrero, G.N. Ioannou, and L. Beretta. 2011. Circulating microRNAs in patients with chronic hepatitis C and non-alcoholic fatty liver disease. *PLoS ONE* **6**:e23937.

Chen, X., Y. Ba, L. Ma, X. Cai, Y. Yin, K. Wang, J. Guo, Y. Zhang, J. Chen, X. Guo, Q. Li, X. Li, et al. 2008. Characterization of microRNAs in serum: a novel class of biomarkers for diagnosis of cancer and other diseases. *Cell Res* **18**:997–1006.

Chen, Z.H., G.L. Zhang, H.R. Li, J.D. Luo, Z.X. Li, G.M. Chen, and J. Yang. 2012. A panel of five circulating microRNAs as potential biomarkers for prostate cancer. *Prostate* **72**:1443–1452.

Cheng, H., L. Zhang, D.E. Cogdell, H. Zheng, A.J. Schetter, M. Nykter, C.C. Harris, K. Chen, S.R. Hamilton, and W. Zhang. 2011. Circulating plasma MiR-141 is a novel biomarker for metastatic colon cancer and predicts poor prognosis. *PLoS ONE* **6**:e17745.

Cho, W.C. 2012. MicroRNAs as therapeutic targets and their potential applications in cancer therapy. *Expert Opin Ther Targets* **16**:747–759.

Chugh, P. and D.P. Dittmer. 2012. Potential pitfalls in microRNA profiling. *Wiley Interdiscip Rev RNA* **3**:601–616.

Cocucci, E., G. Racchetti, and J. Meldolesi. 2009. Shedding microvesicles: artefacts no more. *Trends Cell Biol* **19**:43–51.

Croce, C.M. 2009. Causes and consequences of microRNA dysregulation in cancer. *Nat Rev Genet* **10**:704–714.

Devaux, Y., M. Vausort, E. Goretti, P.V. Nazarov, F. Azuaje, G. Gilson, M.F. Corsten, B. Schroen, M.L. Lair, S. Heymans, and D.R. Wagner. 2012. Use of circulating microRNAs to diagnose acute myocardial infarction. *Clin Chem* **58**:559–567.

Duffy, M.J. 1996. Proteases as prognostic markers in cancer. *Clin Cancer Res* **2**:613–618.

Emilian, C., E. Goretti, F. Prospert, D. Pouthier, P. Duhoux, G. Gilson, Y. Devaux, and D.R. Wagner. 2012. MicroRNAs in patients on chronic hemodialysis (MINOS study). *Clin J Am Soc Nephrol* **7**:619–623.

Esquela-Kerscher, A. and F.J. Slack. 2006. Oncomirs - microRNAs with a role in cancer. *Nat Rev Cancer* **6**:259–269.

Fu, Y., Z. Yi, X. Wu, J. Li, and F. Xu. 2011. Circulating microRNAs in patients with active pulmonary tuberculosis. *J Clin Microbiol* **49**:4246–4251.

Gallo, A., M. Tandon, I. Alevizos, and G.G. Illei. 2012. The majority of microRNAs detectable in serum and saliva is concentrated in exosomes. *PLoS ONE* **7**:e30679.

Grange, C., M. Tapparo, F. Collino, L. Vitillo, C. Damasco, M.C. Deregibus, C. Tetta, B. Bussolati, and G. Camussi. 2011. Microvesicles released from human renal cancer stem cells stimulate angiogenesis and formation of lung premetastatic niche. *Cancer Res* **71**:5346–5356.

Guo, H.Q., G.L. Huang, C.C. Guo, X.X. Pu, and T.Y. Lin. 2010. Diagnostic and prognostic value of circulating miR-221 for extranodal natural killer/T-cell lymphoma. *Dis Markers* **29**:251–258.

Ha, T.Y. 2011. The role of microRNAs in regulatory T cells and in the immune response. *Immune Netw* **11**:11–41.

Heneghan, H.M., N. Miller, and M.J. Kerin. 2010a. Circulating miRNA signatures: promising prognostic tools for cancer. *J Clin Oncol* **28**:e573–e574; author reply e575–e576.

Heneghan, H.M., N. Miller, A.J. Lowery, K.J. Sweeney, and M.J. Kerin. 2009. MicroRNAs as novel biomarkers for breast cancer. *J Oncol* **2009**:950201.

Heneghan, H.M., N. Miller, A.J. Lowery, K.J. Sweeney, J. Newell, and M.J. Kerin. 2010b. Circulating microRNAs as novel minimally invasive biomarkers for breast cancer. *Ann Surg* **251**:499–505.

Howcroft, T.K., H.G. Zhang, M. Dhodapkar, and S. Mohla. 2011. Vesicle transfer and cell fusion: emerging concepts of cell-cell communication in the tumor microenvironment. *Cancer Biol Ther* **12**:159–164.

Hu, Z., X. Chen, Y. Zhao, T. Tian, G. Jin, Y. Shu, Y. Chen, L. Xu, K. Zen, C. Zhang, and H. Shen. 2010. Serum microRNA signatures identified in a genome-wide serum microRNA expression profiling predict survival of non-small-cell lung cancer. *J Clin Oncol* **28**:1721–1726.

Huang, Z., D. Huang, S. Ni, Z. Peng, W. Sheng, and X. Du. 2010. Plasma microRNAs are promising novel biomarkers for early detection of colorectal cancer. *Int J Cancer* **127**:118–126.

Hunter, M.P., N. Ismail, X. Zhang, B.D. Aguda, E.J. Lee, L. Yu, T. Xiao, J. Schafer, M.L. Lee, T.D. Schmittgen, S.P. Nana-Sinkam, D. Jarjoura, et al. 2008. Detection of microRNA expression in human peripheral blood microvesicles. *PLoS ONE* **3**:e3694.

Hurley, J., D. Roberts, A. Bond, D. Keys, and C. Chen. 2012. Stem-loop RT-qPCR for microRNA expression profiling. *Methods Mol Biol* **822**:33–52.

Jung, E.J., L. Santarpia, J. Kim, F.J. Esteva, E. Moretti, A.U. Buzdar, A. Di Leo, X.F. Le, R.C. Bast Jr., S.T. Park, L. Pusztai, and G.A. Calin. 2012. Plasma microRNA 210 levels correlate with sensitivity to trastuzumab and tumor presence in breast cancer patients. *Cancer* **118**:2603–2614.

Kanemaru, H., S. Fukushima, J. Yamashita, N. Honda, R. Oyama, A. Kakimoto, S. Masuguchi, T. Ishihara, Y. Inoue, M. Jinnin, and H. Ihn. 2011. The circulating microRNA-221 level in patients with malignant melanoma as a new tumor marker. *J Dermatol Sci* **61**:187–193.

Kim, S.H., E.R. Lechman, N. Bianco, R. Menon, A. Keravala, J. Nash, Z. Mi, S.C. Watkins, A. Gambotto, and P.D. Robbins. 2005. Exosomes derived from IL-10-treated dendritic cells can suppress inflammation and collagen-induced arthritis. *J Immunol* **174**:6440–6448.

Kim, V.N. 2005. MicroRNA biogenesis: coordinated cropping and dicing. *Nat Rev Mol Cell Biol* **6**:376–385.

Komatsu, S., D. Ichikawa, H. Takeshita, M. Tsujiura, R. Morimura, H. Nagata, T. Kosuga, T. Iitaka, H. Konishi, A. Shiozaki, H. Fujiwara, K. Okamoto, et al. 2011. Circulating microRNAs in plasma of patients with oesophageal squamous cell carcinoma. *Br J Cancer* **105**:104–111.

Kong, W., H. Yang, L. He, J.J. Zhao, D. Coppola, W.S. Dalton, and J.Q. Cheng. 2008. MicroRNA-155 is regulated by the transforming growth factor beta/Smad pathway and contributes to epithelial cell plasticity by targeting RhoA. *Mol Cell Biol* **28**:6773–6784.

Kosaka, N., H. Iguchi, and T. Ochiya. 2010. Circulating microRNA in body fluid: a new potential biomarker for cancer diagnosis and prognosis. *Cancer Sci* **101**:2087–2092.

Kroh, E.M., R.K. Parkin, P.S. Mitchell, and M. Tewari. 2010. Analysis of circulating microRNA biomarkers in plasma and serum using quantitative reverse transcription-PCR (qRT-PCR). *Methods* **50**:298–301.

Krol, J., I. Loedige, and W. Filipowicz. 2010. The widespread regulation of microRNA biogenesis, function and decay. *Nat Rev Genet* **11**:597–610.

Lawrie, C.H., S. Gal, H.M. Dunlop, B. Pushkaran, A.P. Liggins, K. Pulford, A.H. Banham, F. Pezzella, J. Boultwood, J.S. Wainscoat, C.S. Hatton, and A.L. Harris. 2008. Detection of elevated levels of tumour-associated microRNAs in serum of patients with diffuse large B-cell lymphoma. *Br J Haematol* **141**:672–675.

Lee, T.H., E. D'Asti, N. Magnus, K. Al-Nedawi, B. Meehan, and J. Rak. 2011. Microvesicles as mediators of intercellular communication in cancer–the emerging science of cellular "debris". *Semin Immunopathol* **33**:455–467.

Liu, A.M., T.J. Yao, W. Wang, K.F. Wong, N.P. Lee, S.T. Fan, R.T. Poon, C. Gao, and J.M. Luk. 2012. Circulating miR-15b and miR-130b in serum as potential markers for detecting hepatocellular carcinoma: a retrospective cohort study. *BMJ Open* **2**:e000825.

Liu, H., L. Zhu, B. Liu, L. Yang, X. Meng, W. Zhang, Y. Ma, and H. Xiao. 2012. Genome-wide microRNA profiles identify miR-378 as a serum biomarker for early detection of gastric cancer. *Cancer Lett* **316**:196–203.

Lo, Y.M. 2012. Fetal nucleic acids in maternal blood: the promises. *Clin Chem Lab Med* **50**: 995–998.

Lorenzen, J.M., J.T. Kielstein, C. Hafer, S.K. Gupta, P. Kumpers, R. Faulhaber-Walter, H. Haller, D. Fliser, and T. Thum. 2011. Circulating miR-210 predicts survival in critically ill patients with acute kidney injury. *Clin J Am Soc Nephrol* **6**:1540–1546.

Ma, L., J. Teruya-Feldstein, and R.A. Weinberg. 2007. Tumour invasion and metastasis initiated by microRNA-10b in breast cancer. *Nature* **449**:682–688.

Makino, K., M. Jinnin, I. Kajihara, N. Honda, K. Sakai, S. Masuguchi, S. Fukushima, Y. Inoue, and H. Ihn. 2011. Circulating miR-142-3p levels in patients with systemic sclerosis. *Clin Exp Dermatol* **37**:34–39.

McDonald, J.S., D. Milosevic, H.V. Reddi, S.K. Grebe, and A. Algeciras-Schimnich. 2011. Analysis of circulating microRNA: preanalytical and analytical challenges. *Clin Chem* **57**:833–840.

Mitchell, P.S., R.K. Parkin, E.M. Kroh, B.R. Fritz, S.K. Wyman, E.L. Pogosova-Agadjanyan, A. Peterson, J. Noteboom, K.C. O'Briant, A. Allen, D.W. Lin, N. Urban, et al. 2008. Circulating microRNAs as stable blood-based markers for cancer detection. *Proc Natl Acad Sci U S A* **105**:10513–10518.

Ng, E.K., W.W. Chong, H. Jin, E.K. Lam, V.Y. Shin, J. Yu, T.C. Poon, S.S. Ng, and J.J. Sung. 2009. Differential expression of microRNAs in plasma of patients with colorectal cancer: a potential marker for colorectal cancer screening. *Gut* **58**:1375–1381.

Okada, H., G. Kohanbash, and M.T. Lotze. 2010. MicroRNAs in immune regulation–opportunities for cancer immunotherapy. *Int J Biochem Cell Biol* **42**:1256–1261.

Olivieri, F., R. Antonicelli, M. Lorenzi, Y. D'Alessandra, R. Lazzarini, G. Santini, L. Spazzafumo, R. Lisa, L. La Sala, R. Galeazzi, R. Recchioni, R. Testa, et al. 2013. Diagnostic potential of circulating miR-499-5p in elderly patients with acute non ST-elevation myocardial infarction. *Int J Cardiol* **167**:531–536.

Paraskevi, A., G. Theodoropoulos, I. Papaconstantinou, G. Mantzaris, N. Nikiteas, and M. Gazouli. 2012. Circulating MicroRNA in inflammatory bowel disease. *J Crohns Colitis* **6**:900–904.

Pegtel, D.M., K. Cosmopoulos, D.A. Thorley-Lawson, M.A. van Eijndhoven, E.S. Hopmans, J.L. Lindenberg, T.D. de Gruijl, T. Wurdinger, and J.M. Middeldorp. 2010. Functional delivery of viral miRNAs via exosomes. *Proc Natl Acad Sci U S A* **107**:6328–6333.

Qi, P., S.Q. Cheng, H. Wang, N. Li, Y.F. Chen, and C.F. Gao. 2011. Serum microRNAs as biomarkers for hepatocellular carcinoma in Chinese patients with chronic hepatitis B virus infection. *PLoS ONE* **6**:e28486.

Qu, K.Z., K. Zhang, H. Li, N.H. Afdhal, and M. Albitar. 2011. Circulating microRNAs as biomarkers for hepatocellular carcinoma. *J Clin Gastroenterol* **45**:355–360.

Roth, C., B. Rack, V. Muller, W. Janni, K. Pantel, and H. Schwarzenbach. 2010. Circulating microRNAs as blood-based markers for patients with primary and metastatic breast cancer. *Breast Cancer Res* **12**:R90.

Roth, C., S. Kasimir-Bauer, M. Heubner, K. Pantel, and H. Schwarzenbach. 2011a. Increase in circulating microRNA levels in blood of ovarian cancer patients. In *Circulating nucleic acids in plasma and serum*. P.B. Gahan, Ed. Springer.

Roth, C., S. Kasimir-Bauer, K. Pantel, and H. Schwarzenbach. 2011b. Screening for circulating nucleic acids and caspase activity in the peripheral blood as potential diagnostic tools in lung cancer. *Mol Oncol* **5**:281–291.

Schaefer, A., M. Jung, K. Miller, M. Lein, G. Kristiansen, A. Erbersdobler, and K. Jung. 2010. Suitable reference genes for relative quantification of miRNA expression in prostate cancer. *Exp Mol Med* **42**:749–758.

Schwarzenbach, H., D.S. Hoon, and K. Pantel. 2011. Cell-free nucleic acids as biomarkers in cancer patients. *Nat Rev Cancer* **11**:426–437.

Schwarzenbach, H., K. Milde-Langosch, B. Steinbach, V. Muller, and K. Pantel. 2012. Diagnostic potential of PTEN-targeting miR-214 in the blood of breast cancer patients. *Breast Cancer Res Treat* **134**:933–941.

Shen, J., G.W. Hruby, J.M. McKiernan, I. Gurvich, M.J. Lipsky, M.C. Benson, and R.M. Santella. 2012. Dysregulation of circulating microRNAs and prediction of aggressive prostate cancer. *Prostate* **72**:1469–1477.

Shen, J., Z. Liu, N.W. Todd, H. Zhang, J. Liao, L. Yu, M.A. Guarnera, R. Li, L. Cai, M. Zhan, and F. Jiang. 2011. Diagnosis of lung cancer in individuals with solitary pulmonary nodules by plasma microRNA biomarkers. *BMC Cancer* **11**:374.

Silva, J., V. Garcia, A. Zaballos, M. Provencio, L. Lombardia, L. Almonacid, J.M. Garcia, G. Dominguez, C. Pena, R. Diaz, M. Herrera, A. Varela, et al. 2011. Vesicle-related microRNAs in plasma of NSCLC patients and correlation with survival. *Eur Respir J* **37**:617–623.

Song, J., Z. Bai, W. Han, J. Zhang, H. Meng, J. Bi, X. Ma, S. Han, and Z. Zhang. 2012. Identification of suitable reference genes for qPCR analysis of serum microRNA in gastric cancer patients. *Dig Dis Sci* **57**:897–904.

Starkey Lewis, P.J., J. Dear, V. Platt, K.J. Simpson, D.G. Craig, D.J. Antoine, N.S. French, N. Dhaun, D.J. Webb, E.M. Costello, J.P. Neoptolemos, J. Moggs, et al. 2011. Circulating microRNAs as potential markers of human drug-induced liver injury. *Hepatology* **54**:1767–1776.

Stroun, M., J. Lyautey, C. Lederrey, A. Olson-Sand, and P. Anker. 2001. About the possible origin and mechanism of circulating DNA apoptosis and active DNA release. *Clin Chim Acta* **313**:139–142.

Tanaka, M., K. Oikawa, M. Takanashi, M. Kudo, J. Ohyashiki, K. Ohyashiki, and M. Kuroda. 2009. Down-regulation of miR-92 in human plasma is a novel marker for acute leukemia patients. *PLoS ONE* **4**:e5532.

Taylor, D.D. and C. Gercel-Taylor. 2008. MicroRNA signatures of tumor-derived exosomes as diagnostic biomarkers of ovarian cancer. *Gynecol Oncol* **110**:13–21.

Toffanin, S., D. Sia, and A. Villanueva. 2012. microRNAs: new ways to block tumor angiogenesis? *J Hepatol* **57**:490–491.

Tomimaru, Y., H. Eguchi, H. Nagano, H. Wada, S. Kobayashi, S. Marubashi, M. Tanemura, A. Tomokuni, I. Takemasa, K. Umeshita, T. Kanto, Y. Doki, et al. 2012. Circulating microRNA-21 as a novel biomarker for hepatocellular carcinoma. *J Hepatol* **56**:167–175.

Tsai, K.W., Y.L. Liao, C.W. Wu, L.Y. Hu, S.C. Li, W.C. Chan, M.R. Ho, C.H. Lai, H.W. Kao, W.L. Fang, K.H. Huang, and W.C. Lin. 2012. Aberrant expression of miR-196a in gastric cancers and correlation with recurrence. *Genes Chromosomes Cancer* **51**:394–401.

Tsujiura, M., D. Ichikawa, S. Komatsu, A. Shiozaki, H. Takeshita, T. Kosuga, H. Konishi, R. Morimura, K. Deguchi, H. Fujiwara, K. Okamoto, and E. Otsuji. 2010. Circulating microRNAs in plasma of patients with gastric cancers. *Br J Cancer* **102**:1174–1179.

Turchinovich, A., L. Weiz, A. Langheinz, and B. Burwinkel. 2011. Characterization of extracellular circulating microRNA. *Nucleic Acids Res* **39**:7223–7233.

REFERENCES

Valadi, H., K. Ekstrom, A. Bossios, M. Sjostrand, J.J. Lee, and J.O. Lotvall. 2007. Exosome-mediated transfer of mRNAs and microRNAs is a novel mechanism of genetic exchange between cells. *Nat Cell Biol* **9**:654–659.

Valenti, R., V. Huber, P. Filipazzi, L. Pilla, G. Sovena, A. Villa, A. Corbelli, S. Fais, G. Parmiani, and L. Rivoltini. 2006. Human tumor-released microvesicles promote the differentiation of myeloid cells with transforming growth factor-beta-mediated suppressive activity on T lymphocytes. *Cancer Res* **66**:9290–9298.

van Schooneveld, E., M.C. Wouters, I. Van der Auwera, D.J. Peeters, H. Wildiers, P.A. Van Dam, I. Vergote, P.B. Vermeulen, L.Y. Dirix, and S.J. Van Laere. 2012. Expression profiling of cancerous and normal breast tissues identifies microRNAs that are differentially expressed in serum from patients with (metastatic) breast cancer and healthy volunteers. *Breast Cancer Res* **14**:R34.

Vickers, K.C. and A.T. Remaley. 2012. Lipid-based carriers of microRNAs and intercellular communication. *Curr Opin Lipidol* **23**:91–97.

Wang, F., Z. Zheng, J. Guo, and X. Ding. 2010. Correlation and quantitation of microRNA aberrant expression in tissues and sera from patients with breast tumor. *Gynecol Oncol* **119**: 586–593.

Wang, G., L.S. Tam, E.K. Li, B.C. Kwan, K.M. Chow, C.C. Luk, P.K. Li, and C.C. Szeto. 2011. Serum and urinary free microRNA level in patients with systemic lupus erythematosus. *Lupus* **20**:493–500.

Wang, J.F., M.L. Yu, G. Yu, J.J. Bian, X.M. Deng, X.J. Wan, and K.M. Zhu. 2010. Serum miR-146a and miR-223 as potential new biomarkers for sepsis. *Biochem Biophys Res Commun* **394**: 184–188.

Wang, L.G. and J. Gu. 2012. Serum microRNA-29a is a promising novel marker for early detection of colorectal liver metastasis. *Cancer Epidemiol* **36**:e61–e67.

Wang, M., H. Gu, S. Wang, H. Qian, W. Zhu, L. Zhang, C. Zhao, Y. Tao, and W. Xu. 2012. Circulating miR-17-5p and miR-20a: molecular markers for gastric cancer. *Mol Med Rep* **5**:1514–1520.

Wang, R., N. Li, Y. Zhang, Y. Ran, and J. Pu. 2011. Circulating MicroRNAs are promising novel biomarkers of acute myocardial infarction. *Intern Med* **50**:1789–1795.

Wang, Y., D. Zheng, Q. Tan, M.X. Wang, and L.Q. Gu. 2011. Nanopore-based detection of circulating microRNAs in lung cancer patients. *Nat Nanotechnol* **6**:668–674.

Wei, J., W. Gao, C.J. Zhu, Y.Q. Liu, Z. Mei, T. Cheng, and Y.Q. Shu. 2011. Identification of plasma microRNA-21 as a biomarker for early detection and chemosensitivity of non-small cell lung cancer. *Chin J Cancer* **30**:407–414.

Widera, C., S.K. Gupta, J.M. Lorenzen, C. Bang, J. Bauersachs, K. Bethmann, T. Kempf, K.C. Wollert, and T. Thum. 2011. Diagnostic and prognostic impact of six circulating microRNAs in acute coronary syndrome. *J Mol Cell Cardiol* **51**:872–875.

Wu, F., N.J. Guo, H. Tian, M. Marohn, S. Gearhart, T.M. Bayless, S.R. Brant, and J.H. Kwon. 2011. Peripheral blood microRNAs distinguish active ulcerative colitis and Crohn's disease. *Inflamm Bowel Dis* **17**:241–250.

Wu, X., G. Somlo, Y. Yu, M.R. Palomares, A.X. Li, W. Zhou, A. Chow, Y. Yen, J.J. Rossi, H. Gao, J. Wang, Y.C. Yuan, et al. 2012. De novo sequencing of circulating miRNAs identifies novel markers predicting clinical outcome of locally advanced breast cancer. *J Transl Med* **10**:42.

Xu, J., C. Wu, X. Che, L. Wang, D. Yu, T. Zhang, L. Huang, H. Li, W. Tan, C. Wang, and D. Lin. 2011. Circulating microRNAs, miR-21, miR-122, and miR-223, in patients with hepatocellular carcinoma or chronic hepatitis. *Mol Carcinog* **50**:136–142.

Yaman Agaoglu, F., M. Kovancilar, Y. Dizdar, E. Darendeliler, S. Holdenrieder, N. Dalay, and U. Gezer. 2011. Investigation of miR-21, miR-141, and miR-221 in blood circulation of patients with prostate cancer. *Tumour Biol* **32**:583–588.

Yang, M., J. Chen, F. Su, B. Yu, L. Lin, Y. Liu, J.D. Huang, and E. Song. 2011. Microvesicles secreted by macrophages shuttle invasion-potentiating microRNAs into breast cancer cells. *Mol Cancer* **10**:117.

Yu, J., Y. Wang, R. Dong, X. Huang, S. Ding, and H. Qiu. 2012. Circulating microRNA-218 was reduced in cervical cancer and correlated with tumor invasion. *J Cancer Res Clin Oncol* **138**:671–674.

Zampetaki, A., S. Kiechl, I. Drozdov, P. Willeit, U. Mayr, M. Prokopi, A. Mayr, S. Weger, F. Oberhollenzer, E. Bonora, A. Shah, J. Willeit, et al. 2010. Plasma microRNA profiling reveals loss of endothelial miR-126 and other microRNAs in type 2 diabetes. *Circ Res* **107**:810–817.

Zhang, H.L., L.F. Yang, Y. Zhu, X.D. Yao, S.L. Zhang, B. Dai, Y.P. Zhu, Y.J. Shen, G.H. Shi, and D.W. Ye. 2011. Serum miRNA-21: Elevated levels in patients with metastatic hormone-refractory prostate cancer and potential predictive factor for the efficacy of docetaxel-based chemotherapy. *Prostate* **71**:326–331.

Zhang, Y., Y. Jia, R. Zheng, Y. Guo, Y. Wang, H. Guo, M. Fei, and S. Sun. 2010. Plasma microRNA-122 as a biomarker for viral-, alcohol-, and chemical-related hepatic diseases. *Clin Chem* **56**: 1830–1838.

Zhao, Z., Q. Zhao, J. Warrick, C.M. Lockwood, A. Woodworth, K.H. Moley, and A.M. Gronowski. 2012. Circulating microRNA miR-323-3p as a biomarker of ectopic pregnancy. *Clin Chem* **58**:896–905.

Zheng, D., S. Haddadin, Y. Wang, L.Q. Gu, M.C. Perry, C.E. Freter, and M.X. Wang. 2011. Plasma microRNAs as novel biomarkers for early detection of lung cancer. *Int J Clin Exp Pathol* **4**:575–586.

Zhou, H., J.M. Guo, Y.R. Lou, X.J. Zhang, F.D. Zhong, Z. Jiang, J. Cheng, and B.X. Xiao. 2010. Detection of circulating tumor cells in peripheral blood from patients with gastric cancer using microRNA as a marker. *J Mol Med (Berl)* **88**:709–717.

Zuo, Z., G.A. Calin, H.M. de Paula, L.J. Medeiros, M.H. Fernandez, M. Shimizu, G. Garcia-Manero, and C.E. Bueso-Ramos. 2011. Circulating microRNAs let-7a and miR-16 predict progression-free survival and overall survival in patients with myelodysplastic syndrome. *Blood* **118**: 413–415.

36

CIRCULATING MicroRNAs AS CELLULAR MESSENGERS

Kasey C. Vickers

*Division of Cardiovascular Medicine, Department of Medicine,
Vanderbilt University School of Medicine, Nashville, TN, USA*

I.	Introduction	590
II.	Extracellular miRNA	590
III.	Intercellular Communication	592
IV.	Cellular Export of Extracellular miRNA	594
V.	Lipid-Based miRNA Carriers	594
VI.	Protein miRNA Carriers	595
VII.	Delivery of Circulating miRNA to Recipient Cells	596
VIII.	Functional Roles of Extracellular miRNA	596
IX.	Summary	599
	References	601

ABBREVIATIONS

3′-UTR	3′-untranslated region
Apoe,	Apolipoprotein E
CXCL12	chemokine (C-X-C motif) ligand 12
CXCR4	C-X-C chemokine receptor type
FPLC	fast-protein liquid chromatography
GI	gastrointestinal
JAK	Janus kinase
KLF2	Krueppel-like factor 2.

MicroRNAs in Medicine, First Edition. Edited by Charles H. Lawrie.
© 2014 John Wiley & Sons, Inc. Published 2014 by John Wiley & Sons, Inc.

miRNA	microRNA
miRNP	miRNA ribonucleoprotein complex
MP	microparticle
MVB	multivesicular bodies
RISC	RNA induced silencing complex
SOCS5	suppressor of cytokine signaling 5
STAT	signal transducer and activator of transcription

I. INTRODUCTION

Cellular microRNAs (miRNA) are powerful regulators of metabolism and physiology and mediate key responses to cellular stress (Baek et al. 2008; Friedman et al. 2009; Mendell and Olson 2012). Primary miRNAs (pri-miRNA) are transcribed from the genome and rapidly processed into precursor hairpins in the nucleus (Bartel 2004). After nuclear export, miRNA hairpins (pre-miRNA) are processed into mature forms that are ultimately loaded into the miRNA ribonuleoprotein complex (miRNP) or RNA-induced silencing complex (RISC) (Preall and Sontheimer 2005; Bartel 2009). Some miRNAs present in plasma are found in their single-stranded form, thus sometime during processing, miRNAs are likely shuttled from the RISC complex, prepared for export, and secreted from cells into the extracellular environment (Vickers et al. 2011). miRNAs, thus, are not confined to cells and have been found in most biological fluids analyzed (Weber et al. 2010). Human blood (Lawrie et al. 2008; Mitchell et al. 2008), plasma/sera (Corsten et al. 2010; Zampetaki et al. 2010), breast milk (Kosaka et al. 2010c), urine (Hanke et al. 2010), cerebral spinal fluid (Cogswell et al. 2008; Baraniskin et al. 2011), semen (Wang et al. 2011; Li et al. 2012), and saliva (Hanson et al. 2009) all contain extracellular miRNAs. Low molecular weight RNAs, likely miRNAs, were observed in blood as early as 2004 (El-Hefnawy et al. 2004); however, miRNAs were first identified in human blood/serum in 2008 by Lawrie et al. (Lawrie et al. 2008) (Figure 36.1). Due to their stability in blood, urine, and saliva, circulating miRNAs hold tremendous potential as disease biomarkers. Differential abundances of specific miRNAs in distinct compartments have been reported for many pathophysiologies (Etheridge et al. 2011), including cardiovascular disease (Fichtlscherer et al. 2011; McManus and Ambros 2011), cancer (Kosaka et al. 2010a; Moussay et al. 2011), fatty liver disease (Cheung et al. 2008), diabetes (Zampetaki et al. 2010) and inflammatory diseases (J.F. Wang et al. 2010). As such, many researchers and *in vivo* diagnostic companies are very interested in the use of extracellular miRNAs for early detection, prevention, and treatment of diseases. Finally and most importantly, extracellular miRNAs are taken up by cells where they control gene expression and phenotype (Zhang et al. 2010; Vickers et al. 2011). Therefore, miRNAs in this regard are very much like soluble factors, hormones, and neurotransmitters in that they provide cell-to-cell communication in paracrine, endocrine, and exocrine networks. Here, we highlight the current biology of miRNA-based intercellular communication and address the physiological outcomes of miRNA transfer.

II. EXTRACELLULAR MiRNA

Extracellular miRNAs were first observed to be functional and act as systemic signals in plants as early as 1996 (Baulcombe 1996; Voinnet and Baulcombe 1997). It was not until

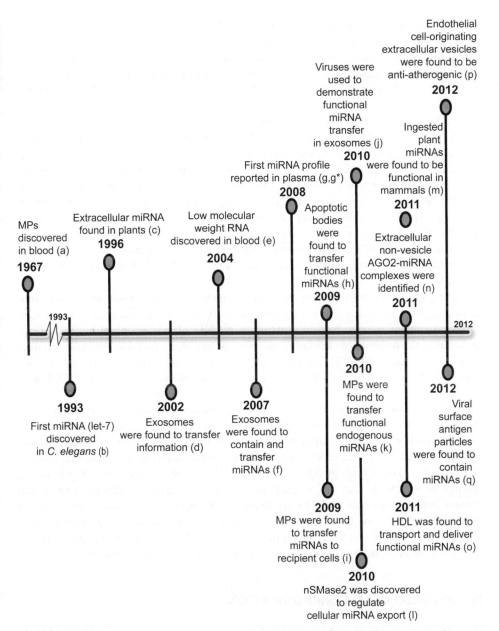

Figure 36.1. Chronological timeline of key discoveries in miRNA intercellular communication. Key references: (a) Wolf (1967); (b) Lee et al. (1993); (c) Baulcombe (1996); (d) Thery et al. (2002a); (e) El-Hefnawy et al. (2004); (f) Valadi et al. (2007); (g) Lawrie et al. (2008); (g*) Mitchell et al. (2008); (h) Zernecke et al. (2009); (i) Yuan et al. (2009); (j) Pegtel et al. (2010); (k) Zhang et al. (2010); (l) Kosaka et al. (2010b); (m) Zhang et al. (2012); (n) Arroyo et al. (2011); (o) Vickers et al. (2011); (p) Hergenreider et al. (2012); (q) Novellino et al. (2012).

a decade later before researchers truly recognized the potential of extracellular miRNAs in mammalian blood. A previously held concept, and one that most likely held back research into circulating miRNAs, was that extracellular RNA is rapidly degraded by circulating ribonucleases (RNases) (Weickmann and Glitz 1982). Given the presence of high levels of RNases in blood and the lability of RNA, it came as a surprise to many that circulating miRNAs in plasma were stable, resistant to RNases, and functional (Chen et al. 2008; Mitchell et al. 2008). Through interaction with lipids and proteins, or likely both together, miRNAs are protected from RNase degradation. Multiple studies have used proteases, detergents, and sonication to render extracellular miRNAs sensitive to RNases (Chen et al. 2008; Mitchell et al. 2008; Kosaka et al. 2010b; Zhang et al. 2010; Muller et al. 2011; Turchinovich et al. 2011). These studies suggest that extracellular miRNAs are not inherently protected from RNases though any chemical modification, but are simply protected by their carriers (Arroyo et al. 2011). Most interesting, circulating miRNAs were found to not be removed by conventional dialysis (Martino et al. 2012). Only trace amounts of extracellular miRNAs were found in the dialysate, indicating that a majority of circulating miRNAs are associated with lipid-based carriers or protein complexes.

The global pool of circulating miRNAs is made up of carrier subclasses, including exosomes, microparticles (MPs), lipoproteins, and other ribonucleoprotein complexes. Although somewhat controversial (Arroyo et al. 2011; Turchinovich et al. 2011; Gallo et al. 2012), the majority of circulating miRNAs are not likely found in membrane-derived vesicles (exosomes and MPs), but associated with circulating proteins or lipoproteins. One study found that only 10% of circulating miRNAs are associated with membrane-derived vesicles (Arroyo et al. 2011) (Figure 36.2). Nevertheless, some evidence does suggest that specific serum and saliva miRNAs are found in greater abundance in vesicle fractions (Gallo et al. 2012). Due to the physical manipulation of vesicles during high-speed ultracentrifugation or ultrafractionation, it remains to be determined if methods used to isolate the carrier subclasses alter miRNA associations. Furthermore, lipoproteins and extracellular vesicles likely exchange lipids and proteins in circulation; however, it is unknown whether they have the capacity to exchange or laterally transfer miRNAs. If so, it may explain in part how some miRNAs are found in all carriers. Nevertheless, each class of lipid-based carriers has been reported to possess their own miRNA signature (Vickers et al. 2011). As such, it is likely that miRNAs associated with specific carriers are part of different communication networks with different sources and recipient cells.

III. INTERCELLULAR COMMUNICATION

One of the most fascinating observations of extracellular RNAs is that circulating miRNAs are biological active and regulate gene expression in targeted cells. As such, a new cell-to-cell communication pathway has emerged. Classically, secreted communication between organs and cells was limited to soluble factors. In a way, extracellular miRNAs act very much like hormones and steroids, influencing global gene regulation and contributing to systemic homeostasis. However, whereas soluble factors act on specific receptors, extracellular vesicles and lipoprotein particles have the potential to deliver a cassette of miRNAs and generate a more sophisticated and complex gene regulatory response.

The three key criteria for miRNA intercellular communication include the following: (1) selective packaging and miRNA export, (2) protected transport, and (3) cellular delivery with functional targeting and altered gene expression. At this time, little is known

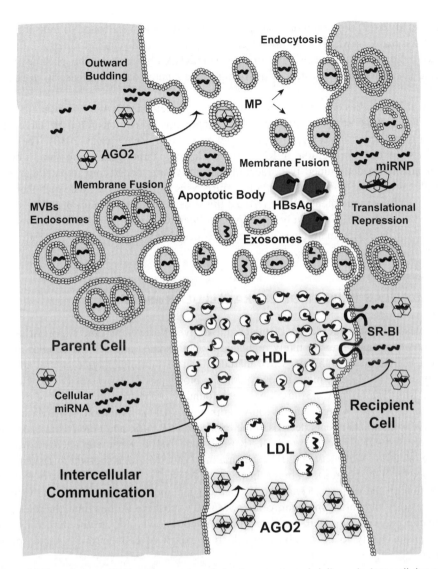

Figure 36.2. Cellular miRNA export, extracellular transport, and delivery in intercellular communication. MP, microparticle; AGO2, Argonaute 2; miRNP, miRNA ribonucleoprotein complex (RNA-induced silencing complex, RISC); HBsAg, hepatitis B viral surface antigen particle; MVBs, multivesicular bodies; SR-BI, scavenger receptor B1; miRNA, microRNA; HDL, high-density lipoproteins; LDL, low-density lipoproteins.

about how cells actively select miRNAs to be retained or exported, but a few critical observations have been made. Evidence suggests that exosomal contents, namely proteins and RNA, are not randomly transported, but specifically packaged and secreted through highly regulated mechanisms (Thery et al. 2002b; K. Wang et al. 2010; Zhang et al. 2010; Thery 2011). Multiple studies have found the export signature to be distinct from that of the parent cell, which strongly supports the selective export hypothesis (Valadi et al. 2007; K. Wang et al. 2010). Tumor cells selectively export specific miRNAs, and it is now apparent that some miRNAs are transcribed only to be exported (Ohshima et al. 2010;

Pigati et al. 2010). For example, exosomes from mast cells were found to transport ~120 miRNAs, many of which were not observed in parent cells (Valadi et al. 2007). Furthermore, biological fluids each have their own miRNA profiles, and the plasma miRNA profile is remarkably consistent, which both argue against passive release from necrotic cells.

Multiple studies have found that transferred extracellular RNA is functional in recipient cells, and the strongest evidence for functionality is demonstrated by target gene luciferase reporters in recipient cells (Pegtel et al. 2010; Zhang et al. 2010). These data clearly show that extracellular miRNAs repress specific target gene expression; however, it remains to be determined if extracellular miRNAs are loaded into intracellular RISC complexes or if the amount of circulating miRNAs is sufficient to regulate cell phenotype and physiology in recipient cells *in vivo*. At this time, fundamental *in vitro* studies suggest that extracellular miRNAs likely participate in intercellular communication; however, as pointed out by others, many questions remain concerning their impact *in vivo* (Turchinovich et al. 2012).

IV. CELLULAR EXPORT OF EXTRACELLULAR MiRNA

The release of different classes of membrane-derived vesicles and the export of cellular miRNAs to lipoproteins are likely regulated by both shared and distinct cellular mechanisms. The ceramide pathway, which includes the rate-limiting enzyme neutral sphingomyelinase 2 (nSMase2), plays a key role in the biogenesis of exosomes (Trajkovic et al. 2008). Most importantly, nSMase2 has been found to orchestrate the release of extracellular miRNAs (K. Wang et al. 2010; Vickers et al. 2011). Inhibition of nSMase2 with siRNAs or chemical agents decreased the export of *miR-146a* and *miR-16* via exosomes, but increased *miR-223* export to HDL (Kosaka et al. 2010b; Kogure et al. 2011; Mittelbrunn et al. 2011; Vickers et al. 2011). The majority of nSMase2 is thought to reside on the inner leaflet of the plasma membrane; however, nSMases2 is a critical regulator of exosome formation on the surface of endosomes. Cellular membranes, lipoproteins, and extracellular vesicles all contain sphingomyelin, nSMase2's substrate. At this time, the role sphingomyelin plays in the cellular export process is not known, nor are the numerous other mechanisms likely to be involved in selective export of cellular miRNAs to specific carriers.

V. LIPID-BASED MiRNA CARRIERS

Circulating miRNAs are transported by a diverse group of lipid-based carriers—lipoproteins, exosomes, MPs, and apoptotic bodies. Exosomes are membrane-derived vesicles that originate from the inward budding of the plasma membrane resulting in the formation of multivesicular bodies (MVB) within endosomes (Thery et al. 2002b; Kosaka et al. 2010b; Ohshima et al. 2010; Thery 2011). Inside the cell, exosomes are observed to have a uniform diameter of ~100 nm; however, upon secretion, the actual size may change, as extracellular exosomes have been observed to be as small as 40 nm in diameter (Mause and Weber 2010). Secretion occurs through the fusion of MVB-containing endosomes with the plasma membrane and this process is the defining characteristic of exosomes compared to other membrane-derived vesicles. Similar to intact cells, exosomes contain non-random sets of (1) cytoplasmic and transmembrane proteins, (2) long and small RNAs, and (3) soluble factors and small molecules. Exosomal nucleic acids, mRNAs and miRNAs, are often referred to as exosomal shuttled RNA or esRNA (Valadi et al. 2007). Exosomes are released from most cell-types *in vitro*, including inflammatory cells, muscle cells, mast cells, neurons, epithelial cells, stem cells, and tumor cells (Valadi et al. 2007; Skog et al. 2008; Rosell et al. 2009; Lachenal et al. 2010; Vrijsen et al. 2010; Kuwabara et al. 2011). Human chronic villi have even been found to release miRNA-containing exosomes into maternal circulation (Luo et al. 2009). As early as 2002, exosomes were identified as carriers of information (antigens) between cells (Thery et al. 2002a); however, exosomes were first reported to contain and transfer miRNAs in 2007 (Valadi et al. 2007). Since then many studies have profiled exosomes for differential miRNAs in various cancers, including breast, lung, ovarian, and gastric cancers (Taylor and Gercel-Taylor 2008; Rabinowits et al. 2009; Friel et al. 2010; Ohshima et al. 2010).

In addition to exosomes, membrane-derived microparticles (MPs), also known as microvesicles, also contain specific miRNAs. Until recently, the nomenclature of exosomes and MPs was loosely defined, but MPs are distinct from exosomes in that they are formed by the outward budding or blebbing of the plasma membrane. Although MPs previously have been observed with a wide-range of sizes (50–4000 nm), MPs are now characterized as extracellular vesicles >100 nm in diameter, whereas vesicles <100 nm are considered exosomes. Although this delineation is somewhat arbitrary, larger MPs are proposed to sediment at much lower centrifugal speeds than smaller exosomes, and current purification strategies now include step-wise isolations to differentiate exosomes and MPs. Nevertheless, future studies will be needed to determine the accuracy, sensitivity, and efficiency in differential centrifugation for subclass purification.

First observed in 1967, MPs are heterogeneous in size and composition and released by multiple cell types, including platelets, which account for the majority of MPs in plasma (Wolf 1967; VanWijk et al. 2003). During apoptosis, cells release larger MPs known as apoptotic bodies, which also contain functional miRNAs. MP secretion is the result of plasma membrane instability, but may be part of an early apoptotic mechanism serving as a stress signals to neighboring cells. Similar to other cell-to-cell networks, MPs likely provide communication within both microenvironments and across distant tissues. The most abundant class of lipid-based miRNA carriers in circulation is lipoproteins, namely low-density lipoproteins (LDL) and high-density lipoproteins (HDL). Lipoproteins are comprised of a single layer of phospholipid with a hydrophobic core compared with exosomes and MPs, which have a bilayer phospholipid shell and hydrohillic core. HDL has been found to transport specific miRNA signatures in health and cardiovascular disease in both humans and mice (Vickers et al. 2011). Less is known about LDL-miRNAs; however, LDL does contain miRNAs, and its profile was found to be distinct from that of HDL and exosomes (Vickers et al. 2011). At this time, chylomicrons and larger very low-density lipoproteins (VLDL) have not been reported to contain miRNAs; however, it would not be unexpected for them to contain small RNAs.

VI. PROTEIN miRNA CARRIERS

Argonaute 2 (AGO2) is the main structure–function protein of the cellular miRNP complex (Meister et al. 2004). Multiple studies have found evidence of extracellular AGO2-miRNA complexes in both vesicles and non-vesicle fractions (Collino et al. 2010; Arroyo et al. 2011; Turchinovich et al. 2011). Currently, it is unknown how AGO2-miRNA complexes are actively or passively secreted from cells or if they require membrane-derived vesicles.

While AGO2-bound miRNAs are likely present outside of vesicles in plasma, it remains to be determined (1) what cells release them, (2) if these complexes contain lipid, and (3) if they have the capacity to deliver functional miRNAs to recipient cells. Many extracellular miRNAs are not associated with AGO2, but most miRNA complexes are retained in 100-kDa ultrafiltrates (Arroyo et al. 2011). Evidence suggests that miRNAs are associated with protein complexes of 50–300 kDa. Nucleophosmin 1 (NPM1), another secreted ribonucleoprotein, has also been found in complex with specific extracellular miRNAs (K. Wang et al. 2010). HDL, which could be classified as a circulating protein scaffold, was not found to contain AGO2 or NPM1 (Vickers et al. 2011). Therefore, it is very likely that there are other unidentified ribonucleoproteins in addition to AGO2, HDL, and NPM1 that associate with, protect, and transport extracellular miRNAs.

In addition to protein complexes, viral surface antigen particles have also been reported to contain miRNAs (Novellino et al. 2012). Hepatitis B virus uses hepatocytes for the assembly and production of its double-shelled virions (42 nm) and subviral particles (20 nm). Hepatitis B surface antigen particles (HBsAg) have been found to contain miRNAs in association with AGO2 (Novellino et al. 2012 antigen particles). The most abundant miRNAs found in HBsAgs after immunoprecipitation were *miR-188-5p, miR-760, miR-17, miR-135a**, and *miR-138-1**. Common hepatic miRNAs, *miR-122, miR-27a*, and *miR-30b*, were also highly abundant. These particles are not thought to be associated with lipid; however, AGO2 was detected by enzyme-linked immunosorbent assay (ELISA) after particles were treated with detergents.

VII. DELIVERY OF CIRCULATING miRNA TO RECIPIENT CELLS

The role of circulating miRNAs as cellular messengers and their bioactivity is attributed to their uptake and ability to repress gene expression. Extracellular vesicles likely deliver miRNAs through one of two routes, endocytosis (Morelli et al. 2004; Tian et al. 2010) or membrane fusion (Parolini et al. 2009; Montecalvo et al. 2011). The actual transfer mechanism likely depends on the cell-type and surface proteins on the membranes of both the cell and vesicle. However, identification of cell-specific receptors and transmembrane proteins that facilitate cellular uptake remains to be determined. The prevailing hypotheses is that extracellular vesicles bind to surface proteins of recipient cells, which triggers signaling cascades that ultimately results in phagocytosis (Feng et al. 2010), macropinocytosis (Fitzner et al. 2011), or fusion events (Montecalvo et al. 2011). The transfer of HDL-miRNAs to recipient cells is dependent upon scavenger receptor B1 (SR-BI); however, HDL-miRNAs may be taken up by other receptors on cells that do not express HDL's receptor SRBI.

VIII. FUNCTIONAL ROLES OF EXTRACELLULAR miRNA

In 2007, Valadi et al. demonstrated that exosomes transfer miRNAs between mast cells, and in 2009, Yuan et al. found that embryonic stem cells transferred miRNAs to recipient fibroblasts. However, both studies did not experimentally determine whether the delivered miRNAs mediated gene regulation (Valadi et al. 2007). The first study to demonstrate that transferred miRNAs alter gene expression in recipient cells was a 2009 report that found that apoptotic bodies from endothelial cells transferred *miR-126* to recipient endothelial cells with target gene expression changes (Zernecke et al. 2009). Since then, only a limited number of studies have demonstrated altered gene expression and cellular physiology associated with intercellular communication. In 2010, viruses were used to demonstrate that infected cells utilize exosomes to transfer viral miRNAs between cells (Meckes et al. 2010; Pegtel et al. 2010; Meckes and Raab-Traub 2011). Epstein–Barr-infected B cells were found to release exosomes containing EBV-miRNAs (BART, BHRF1) that were taken up by monocyte-derived dendritic cells (DC) (Pegtel et al. 2010). Recipient DCs were found to have elevated EBV-miRNAs and significant repression of EBV-miRNA target genes, as determined by gene reporter luciferase assays (Pegtel et al. 2010). This was the first study to provide evidence that exosomes can transfer functional miRNAs to recipient cells, as EBV-miRNA-mediated targeting was differentiated from recipient cell miRNA-mediated repression. Similarly, EBV-infected nasopharygeal carcinoma cells (NPC) were also found to transfer exosomes containing EBV-miRNAs to endothelial cells; however, evidence of gene regulation was not presented (Meckes et al. 2010). In 2010, MPs from human monocytes and cultured THP-1 macrophages were found to transfer endogenous *miR-150* to endothelial cells (Zhang et al. 2010). This was the first study to demonstrate extracellular vesicles transfer functional miRNAs between different cell-types in a communicative pathway, as determined by target gene (luciferase) reporters to demonstrate direct targeting of 3'-UTRs. In this study, delivered *miR-150* repressed c-Myb expression and enhanced endothelial cell migration (Zhang et al. 2010). Furthermore, this study also found increased *miR-150* in the arterial wall of mice injected with MPs and observed that MPs from subjects with cardiovascular disease contained increased abundances of *miR-150* (Zhang et al. 2010).

Extracellular miRNAs likely serve as cellular messengers in multiple routes of communication within inflammation processes. During immune synapsis, exosomes were found to shuttle functional miRNAs to antigen-presenting cells (APCs) in an antigen-specific manner (Mittelbrunn et al. 2011). This study also found that exosomes released from T cells, B cells, and DCs each have their own distinct miRNA profile from their parent cell-type (Mittelbrunn et al. 2011). Bone-marrow derived DCs were also found to transfer functional miRNAs from immature DCs cells to mature DC2.4 cells (Montecalvo et al. 2011). Furthermore, tumor-associated macrophages (IL-4 activated) have been found to transfer specific miRNAs to breast cancer cells (Yang et al. 2011). Macrophage-originating miRNAs increased breast cancer cell invasiveness, which established a communication link between inflammation and cancer that leads to cell invasiveness.

Hepatocellular carcinoma cells (Hep3B) were found to regulate gene expression in neighboring Hep3B cells by miRNA transfer in extracellular vesicles (Kogure et al. 2011). This supports the hypothesis that extracellular miRNAs serve in some capacity as communicative messengers that promote tumorigenesis. Likewise, neoplastic cells were found to communicate with endothelial cells through transfer of extracellular *miR-92*, which repressed integrin α5 and enhanced tube formation and endothelial cell migration (Umezu et al. 2013). Kidney tumor cells (CD105+) were also found to transfer miRNAs (MPs) to endothelial cells (Grange et al. 2011). In addition, proangiogenic miRNAs (*miR-9*) from tumor cells were found to activate the JAK-STAT pathway through the targeting and down-regulation of SOCS5 (Zhuang et al. 2012). Further details regarding stimulation of angiogenesis and metastasis by miRNA-containing vesicles in renal carcinoma can be found in Chapter 37 of this book. Most interestingly, miRNAs released from cancer cells (exosomes) were found to activate Toll-like receptors in endosomes of recipient cells

(Fabbri et al. 2012). This study demonstrated that not only can exosomal miRNA activate Toll-like receptors, but again proposes that extracellular miRNAs promote tumor growth (Fabbri et al. 2012). Collectively, these studies suggest the tumor cells utilize extracellular miRNAs to promote tumorogenesis and malignancy through miRNA-mediated gene regulation.

Although multiple studies have found that endothelial cells act as recipient cells for intercellular communication, endothelial cells have also been found to secrete extracellular miRNAs. Extracellular vesicles were found to transport *miR-143/miR-145* from KLF2-inducted or shear-stress-stimulated endothelial cells to recipient smooth muscle cells and induce an atheroprotective phenotype (Hergenreider et al. 2012). Furthermore, extracellular vesicles secreted from KLF2-induced endothelial cells were able to reduce atherosclerosis in $Apoe^{-/-}$ mice after injection. These observations support earlier studies that also observed cross-species delivery of functional miRNAs and suggest that specific components within the intercellular communication pathway are likely conserved among mammals (Zhang et al. 2010). miRNA-based intercellular communication likely occurs between close or distant cells, providing both environmental and systemic gene regulation. During atherosclerosis, *miR-126* packaged in apoptotic bodies likely serves as anti-inflammatory vascular signals communicating with neighboring cells protecting against atherosclerosis (Zernecke et al. 2009). Injection of apoptotic bodies enriched with *miR-126* into atherogenic mouse models increased endothelial progenitor cells numbers and stabilized plaques (Zernecke et al. 2009).

Fundamental *in vitro* studies suggest that multiple biological pathways are influenced by miRNA intercellular communication. MPs have been found to transfer specific miRNAs from adipocytes to macrophages, which established a novel network between systemic fat storage and inflammation (Ogawa et al. 2010). Large adipocytes have also been found to transfer miRNAs to small adipocytes, thus stimulating lipid storage in a paracrine manner, but this pathway may very well be involved in systemic endocrine signaling (Muller et al. 2011). Even adult human bone marrow have been found to communicate with stem cells through miRNA signaling (Collino et al. 2010). Likewise, extracellular vesicles were found to transfer miRNAs (*miR-146a*) from kidney cells (HEK293) to prostate cancer cells and repress proliferation through targeting of ROCK1 (Kosaka et al. 2010b). ROCK1 is a critical regulator of apoptosis and apoptotic body formation. Although ROCK1 is activated by caspases during apoptosis, extracellular miRNA delivery and targeting of ROCK1 may antagonize apoptosis or even repress MP or apoptotic body biogenesis. Although other signaling networks suggest that extracellular miRNAs promote tumorogenesis, the *miR-146a-ROCK1* network demonstrates that specific miRNAs may also serve as extracellular tumor suppressors. In the brain, extracellular miRNAs may serve as cellular messengers between mesenchymal stromal and parenchymal cells (neurons and astrocytes), as transferred exosomal miRNAs increased neurite length and branch number (Xin et al. 2012). Lastly, human HDL was found to transfer functional miRNAs to hepatoma cells (Huh7) and SR-BI-expressing kidney cells (BHK).

In 2011, ingested plant miRNAs were found in mammalian circulation and livers, thus providing evidence of cross-kingdom miRNA communication and gene regulation. Most importantly, this study demonstrated miRNAs found in food may retain biological activity in mammalian tissues and regulate gene expression. This study proposed that the epithelial MPs may transport and deliver functional miRNAs to circulation and the liver. To our knowledge, this concept was first proposed in 2010, as breast milk was found to contain exosomes and miRNAs; however, functional ingested miRNAs was first demonstrated by Zhang et al. in 2012 (Iguchi et al. 2010; Kosaka et al. 2010c; Zhang et al. 2012). To fully understand the complexity and applicability of extracellular miRNAs after digestion, significant work is needed to determine how miRNAs are protected in the GI tract and how they are shuttled to the hepatic circulatory system. However, if cleverly designed, it is entirely possible that someday, there will be a new class of customized miRNA foods that may be useful in promoting health and countering disease.

IX. SUMMARY

Although studies outlined here demonstrate that intercellular communication likely occurs *in vivo* even with cross-species delivery, outstanding questions remain on the functionality of extracellular miRNAs. From afar, many global questions remain, including the following: What is the systemic physiological significance of extracellular miRNAs in health and disease communication? More specifically, how does miRNA signaling maintain homeostasis in health, and are disease related changes to miRNA communication adaptive or maladaptive? Most importantly, can we intercept and modify the miRNA signal for therapeutic gains?

Each of the lipid-based carrier subclasses appears to only transport limited numbers and small concentrations of miRNAs in plasma. To achieve gene regulation, a multitude of particles or vesicles could deliver small amounts of miRNAs over time; however, it remains to be determined exactly what the necessary payload is required to alter a biological process. Many studies demonstrated the functionality of transferred miRNAs by target gene reporters (luciferase) containing target sites or 3′-UTRs for delivered miRNAs. This is generally considered an appropriate measure of miRNA activity; however, it only suggests that exogenously transferred miRNAs are loaded into the miRNP complex. Direct evidence of extracellular miRNAs in complex with cellular AGO2 also remains to be demonstrated. Another limitation to studying the impact or physiological relevance of transferred miRNAs is that each subclass contains a diversity of bioactive lipids, cytokines, surface factors, dozens of proteins, and in the case of vesicles, translatable mRNA molecules (Valadi et al. 2007). Therefore, attributing gene expression changes solely to delivered miRNAs, and posttranscriptional regulation remains a significant challenge. A greater challenge is accurately characterizing extracellular miRNA communicative benefit to health, as it requires both pathway focused and systemic analyses. Multiple studies investigating the role of extracellular miRNAs released from tumor cells found that delivered miRNAs promote cancer growth and invasiveness in the recipient cells. Nevertheless, not all signaling appears to be maladaptive as miRNAs secreted from kidney cells were found to inhibit cell growth in prostate cancer cells (Kosaka et al. 2010b). Another outstanding and unresolved mechanism pertains to selective miRNA delivery and transfer to recipient cells. Multiple studies have reported that only specific miRNAs are transferred to recipient cells from lipid-based carriers (Collino et al. 2010; Zhang et al. 2010). Similar to selective loading and cellular export, miRNA delivery may also be specific to certain miRNAs, for example, *miR-150* delivery from MPs to endothelial cells. An alternative to this hypotheses is that all miRNAs are transferred, but the recipient cells selectively designates miRNAs for degradation (Yuan et al. 2009). Cellular miRNAs have been found to have unique and prolonged half-lives, the stability and half-life of exogenously delivered miRNA remains to be quantified (Gantier et al. 2011).

The biggest issue with the intercellular communication hypothesis is the low amount of miRNAs found in serum/blood. Opponents argue that there is not enough circulating miRNAs to achieve gene regulatory effects in recipient cells even though *in vitro* studies clearly suggest that predicted target genes are being altered by transferred miRNAs. The argument centers on the concept that the impact of miRNAs is attributed specifically to their concentration; however, recipient cells may have increased sensitivity for extracellular miRNAs, or we may as a field underestimate the impact of low abundant miRNAs in gene regulation. The amount of miRNAs per particle or vesicle may be low; however, cells and tissues rapidly interact with extreme numbers of particles and vesicles in plasma *in vivo*, which likely providing sustained levels of miRNAs and communicative signals. To date, the functional impact of circulating miRNAs *in vivo* remains to be determined, but the numerous *in vitro* studies and limited *in vivo* investigation support the intercellular communication hypothesis and the hypothesis should not be dismissed simply due to low abundance of circulating miRNAs.

Some have proposed that intercellular communication through miRNAs is particularly adept for microenvironments or space-confined processes (Zomer et al. 2010). For example, miRNAs likely contribute to tumor microenviroments and secondary or peripheral immune responses. The subendothelium and the microenvironment of the atherosclerotic plaque, that is, space between the endothelium and the internal elastic lamina, is well suited for miRNA intercellular communication. Hergenreider et al. found this environment to be particularly important for extracellular vesicle transport of miRNAs from endothelial cells to vascular smooth muscle cells to protect against atherosclerosis-associated mechanisms (Hergenreider et al. 2012). Distinct miRNA signatures are observed in most if not all biological fluids, including plasma, therefore, lipid-based miRNA carriers most likely also work in distant environments and provide systemic gene regulation. Microenvironments likely play a central role in shaping the composition and functionality of extracellular vesicles by influencing adjacent cell signaling, transcription, selection, and export of miRNAs. Secreted carriers then are available to control systemic gene expression and homeostasis (Mause and Weber 2010). As an example, atherosclerosis and the microenvironment of a plaque may influence adjacent smooth muscle cells, endothelial cells, and inflammatory cells, all of which can produce lipid-based carriers of miRNAs. Upon secretion, these carriers and their miRNA signals may travel to high-impact organs, for example, the liver, where delivered miRNAs could influence systemic lipid and cholesterol metabolism to adapt to dyslipidemia.

Although lipid-based carriers and miRNAs in intercellular communication present enormous complexities that require further investigation, they also offer unique therapeutic strategies and novel preventative and treatment approaches. For example, MPs enriched with *miR-126* were found to antagonize diet-induced atherosclerosis in mice (Zernecke et al. 2009). Likewise, extracellular vesicles enriched with endogenous *miR-143/miR-145* may also be used as preventative or therapeutic strategies against atherosclerosis. As such, future approaches will likely take advantage of naturally present miRNA signals or functionalize lipid-based carriers to prevent/treat cardiovascular disease. The hypothesis that extracellular miRNAs are protected in the gut lumen is very interesting and raises many potential strategies to deliver miRNAs to humans in part of therapeutic approaches. Genetically modified food could be engineered to overexpress specific miRNAs, or miRNAs could simply be given orally. Delivery of miRNAs for therapy could also be blended with blood transfusions. All are feasible and hold enormous potential, and future investigation into these approaches is certainly warranted. Although studies into extracellular miRNAs have only recently emerged, this entirely novel field of study has advanced rapidly and provides tremendous amounts of future discovery and high impact on medicine.

REFERENCES

Arroyo, J.D., J.R. Chevillet, E.M. Kroh, I.K. Ruf, C.C. Pritchard, D.F. Gibson, P.S. Mitchell, C.F. Bennett, E.L. Pogosova-Agadjanyan, D.L. Stirewalt, J.F. Tait, and M. Tewari. 2011. Argonaute2 complexes carry a population of circulating microRNAs independent of vesicles in human plasma. *Proc Natl Acad Sci U S A* **108**:5003–5008.

Baek, D., J. Villen, C. Shin, F.D. Camargo, S.P. Gygi, and D.P. Bartel. 2008. The impact of microRNAs on protein output. *Nature* **455**:64–71.

Baraniskin, A., J. Kuhnhenn, U. Schlegel, A. Chan, M. Deckert, R. Gold, A. Maghnouj, H. Zollner, A. Reinacher-Schick, W. Schmiegel, S.A. Hahn, and R. Schroers. 2011. Identification of microRNAs in the cerebrospinal fluid as marker for primary diffuse large B-cell lymphoma of the central nervous system. *Blood* **117**:3140–3146.

Bartel, D.P. 2004. MicroRNAs: genomics, biogenesis, mechanism, and function. *Cell* **116**:281–297.

Bartel, D.P. 2009. MicroRNAs: target recognition and regulatory functions. *Cell* **136**:215–233.

Baulcombe, D.C. 1996. RNA as a target and an initiator of post-transcriptional gene silencing in transgenic plants. *Plant Mol Biol* **32**:79–88.

Chen, X., Y. Ba, L. Ma, X. Cai, Y. Yin, K. Wang, J. Guo, Y. Zhang, J. Chen, X. Guo, Q. Li, X. Li, et al. 2008. Characterization of microRNAs in serum: a novel class of biomarkers for diagnosis of cancer and other diseases. *Cell Res* **18**:997–1006.

Cheung, O., P. Puri, C. Eicken, M.J. Contos, F. Mirshahi, J.W. Maher, J.M. Kellum, H. Min, V.A. Luketic, and A.J. Sanyal. 2008. Nonalcoholic steatohepatitis is associated with altered hepatic MicroRNA expression. *Hepatology* **48**:1810–1820.

Cogswell, J.P., J. Ward, I.A. Taylor, M. Waters, Y. Shi, B. Cannon, K. Kelnar, J. Kemppainen, D. Brown, C. Chen, R.K. Prinjha, J.C. Richardson, et al. 2008. Identification of miRNA changes in Alzheimer's disease brain and CSF yields putative biomarkers and insights into disease pathways. *J Alzheimers Dis* **14**:27–41.

Collino, F., M.C. Deregibus, S. Bruno, L. Sterpone, G. Aghemo, L. Viltono, C. Tetta, and G. Camussi. 2010. Microvesicles derived from adult human bone marrow and tissue specific mesenchymal stem cells shuttle selected pattern of miRNAs. *PLoS ONE* **5**:e11803.

Corsten, M.F., R. Dennert, S. Jochems, T. Kuznetsova, Y. Devaux, L. Hofstra, D.R. Wagner, J.A. Staessen, S. Heymans, and B. Schroen. 2010. Circulating MicroRNA-208b and MicroRNA-499 reflect myocardial damage in cardiovascular disease. *Circ Cardiovasc Genet* **3**:499–506.

El-Hefnawy, T., S. Raja, L. Kelly, W.L. Bigbee, J.M. Kirkwood, J.D. Luketich, and T.E. Godfrey. 2004. Characterization of amplifiable, circulating RNA in plasma and its potential as a tool for cancer diagnostics. *Clin Chem* **50**:564–573.

Etheridge, A., I. Lee, L. Hood, D. Galas, and K. Wang. 2011. Extracellular microRNA: a new source of biomarkers. *Mutat Res* **717**:85–90.

Fabbri, M., A. Paone, F. Calore, R. Galli, E. Gaudio, R. Santhanam, F. Lovat, P. Fadda, C. Mao, G.J. Nuovo, N. Zanesi, M. Crawford, et al. 2012. MicroRNAs bind to Toll-like receptors to induce prometastatic inflammatory response. *Proc Natl Acad Sci U S A* **109**:e2110–2116.

Feng, D., W.L. Zhao, Y.Y. Ye, X.C. Bai, R.Q. Liu, L.F. Chang, Q. Zhou, and S.F. Sui. 2010. Cellular internalization of exosomes occurs through phagocytosis. *Traffic* **11**:675–687.

Fichtlscherer, S., A.M. Zeiher, and S. Dimmeler. 2011. Circulating microRNAs: biomarkers or mediators of cardiovascular diseases? *Arterioscler Thromb Vasc Biol* **31**:2383–2390.

Fitzner, D., M. Schnaars, D. van Rossum, G. Krishnamoorthy, P. Dibaj, M. Bakhti, T. Regen, U.K. Hanisch, and M. Simons. 2011. Selective transfer of exosomes from oligodendrocytes to microglia by macropinocytosis. *J Cell Sci* **124**:447–458.

Friedman, R.C., K.K. Farh, C.B. Burge, and D.P. Bartel. 2009. Most mammalian mRNAs are conserved targets of microRNAs. *Genome Res* **19**:92–105.

Friel, A.M., C. Corcoran, J. Crown, and L. O'Driscoll. 2010. Relevance of circulating tumor cells, extracellular nucleic acids, and exosomes in breast cancer. *Breast Cancer Res Treat* **123**: 613–625.

Gallo, A., M. Tandon, I. Alevizos, and G.G. Illei. 2012. The majority of microRNAs detectable in serum and saliva is concentrated in exosomes. *PLoS ONE* **7**:e30679.

Gantier, M.P., C.E. McCoy, I. Rusinova, D. Saulep, D. Wang, D. Xu, A.T. Irving, M.A. Behlke, P.J. Hertzog, F. Mackay, and B.R. Williams. 2011. Analysis of microRNA turnover in mammalian cells following Dicer1 ablation. *Nucleic Acids Res* **39**:5692–5703.

Grange, C., M. Tapparo, F. Collino, L. Vitillo, C. Damasco, M.C. Deregibus, C. Tetta, B. Bussolati, and G. Camussi. 2011. Microvesicles released from human renal cancer stem cells stimulate angiogenesis and formation of lung premetastatic niche. *Cancer Res* **71**:5346–5356.

Hanke, M., K. Hoefig, H. Merz, A.C. Feller, I. Kausch, D. Jocham, J.M. Warnecke, and G. Sczakiel. 2010. A robust methodology to study urine microRNA as tumor marker: microRNA-126 and microRNA-182 are related to urinary bladder cancer. *Urol Oncol* **28**:655–661.

Hanson, E.K., H. Lubenow, and J. Ballantyne. 2009. Identification of forensically relevant body fluids using a panel of differentially expressed microRNAs. *Anal Biochem* **387**:303–314.

Hergenreider, E., S. Heydt, K. Treguer, T. Boettger, A.J. Horrevoets, A.M. Zeiher, M.P. Scheffer, A.S. Frangakis, X. Yin, M. Mayr, T. Braun, C. Urbich, et al. 2012. Atheroprotective communication between endothelial cells and smooth muscle cells through miRNAs. *Nat Cell Biol* **14**:249–256.

Iguchi, H., N. Kosaka, and T. Ochiya. 2010. Secretory microRNAs as a versatile communication tool. *Commun Integr Biol* **3**:478–481.

Kogure, T., W.L. Lin, I.K. Yan, C. Braconi, and T. Patel. 2011. Intercellular nanovesicle-mediated microRNA transfer: a mechanism of environmental modulation of hepatocellular cancer cell growth. *Hepatology* **54**:1237–1248.

Kosaka, N., H. Iguchi, and T. Ochiya. 2010a. Circulating microRNA in body fluid: a new potential biomarker for cancer diagnosis and prognosis. *Cancer Sci* **101**:2087–2092.

Kosaka, N., H. Iguchi, Y. Yoshioka, F. Takeshita, Y. Matsuki, and T. Ochiya. 2010b. Secretory mechanisms and intercellular transfer of microRNAs in living cells. *J Biol Chem* **285**: 17442–17452.

Kosaka, N., H. Izumi, K. Sekine, and T. Ochiya. 2010c. microRNA as a new immune-regulatory agent in breast milk. *Silence* **1**:7.

Kuwabara, Y., K. Ono, T. Horie, H. Nishi, K. Nagao, M. Kinoshita, S. Watanabe, O. Baba, Y. Kojima, S. Shizuta, M. Imai, T. Tamura, et al. 2011. Increased microRNA-1 and microRNA-133a levels in serum of patients with cardiovascular disease indicate myocardial damage. *Circ Cardiovasc Genet* **4**:446–454.

Lachenal, G., K. Pernet-Gallay, M. Chivet, F.J. Hemming, A. Belly, G. Bodon, B. Blot, G. Haase, Y. Goldberg, and R. Sadoul. 2010. Release of exosomes from differentiated neurons and its regulation by synaptic glutamatergic activity. *Mol Cell Neurosci* **46**:409–418.

Lawrie, C.H., S. Gal, H.M. Dunlop, B. Pushkaran, A.P. Liggins, K. Pulford, A.H. Banham, F. Pezzella, J. Boultwood, J.S. Wainscoat, C.S. Hatton, and A.L. Harris. 2008. Detection of elevated levels of tumour-associated microRNAs in serum of patients with diffuse large B-cell lymphoma. *Br J Haematol* **141**:672–675.

Lee, R.C., R.L. Feinbaum, and V. Ambros. 1993. The C. elegans heterochronic gene lin-4 encodes small RNAs with antisense complementarity to lin-14. *Cell* **75**:843–854.

Li, H., S. Huang, C. Guo, H. Guan, and C. Xiong. 2012. Cell-free seminal mRNA and microRNA exist in different forms. *PLoS ONE* **7**:e34566.

Luo, S.S., O. Ishibashi, G. Ishikawa, T. Ishikawa, A. Katayama, T. Mishima, T. Takizawa, T. Shigihara, T. Goto, A. Izumi, A. Ohkuchi, S. Matsubara, et al. 2009. Human villous trophoblasts express and secrete placenta-specific microRNAs into maternal circulation via exosomes. *Biol Reprod* **81**:717–729.

Martino, F., J. Lorenzen, J. Schmidt, M. Schmidt, M. Broll, Y. Gorzig, J.T. Kielstein, and T. Thum. 2012. Circulating MicroRNAs Are Not Eliminated by Hemodialysis. *PLoS ONE* **7**:e38269.

Mause, S.F. and C. Weber. 2010. Microparticles: protagonists of a novel communication network for intercellular information exchange. *Circ Res* **107**:1047–1057.

McManus, D.D. and V. Ambros. 2011. Circulating microRNAs in cardiovascular disease. *Circulation* **124**:1908–1910.

Meckes, D.G., Jr. and N. Raab-Traub. 2011. Microvesicles and viral infection. *J Virol* **85**: 12844–12854.

Meckes, D.G., Jr., K.H. Shair, A.R. Marquitz, C.P. Kung, R.H. Edwards, and N. Raab-Traub. 2010. Human tumor virus utilizes exosomes for intercellular communication. *Proc Natl Acad Sci U S A* **107**:20370–20375.

Meister, G., M. Landthaler, A. Patkaniowska, Y. Dorsett, G. Teng, and T. Tuschl. 2004. Human Argonaute2 mediates RNA cleavage targeted by miRNAs and siRNAs. *Mol Cell* **15**:185–197.

Mendell, J.T. and E.N. Olson. 2012. MicroRNAs in stress signaling and human disease. *Cell* **148**: 1172–1187.

Mitchell, P.S., R.K. Parkin, E.M. Kroh, B.R. Fritz, S.K. Wyman, E.L. Pogosova-Agadjanyan, A. Peterson, J. Noteboom, K.C. O'Briant, A. Allen, D.W. Lin, N. Urban, et al. 2008. Circulating microRNAs as stable blood-based markers for cancer detection. *Proc Natl Acad Sci U S A* **105**:10513–10518.

Mittelbrunn, M., C. Gutierrez-Vazquez, C. Villarroya-Beltri, S. Gonzalez, F. Sanchez-Cabo, M.A. Gonzalez, A. Bernad, and F. Sanchez-Madrid. 2011. Unidirectional transfer of microRNA-loaded exosomes from T cells to antigen-presenting cells. *Nat Commun* **2**:282.

Montecalvo, A., A.T. Larregina, W.J. Shufesky, D. Beer Stolz, M.L. Sullivan, J.M. Karlsson, C.J. Baty, G.A. Gibson, G. Erdos, Z. Wang, J. Milosevic, O.A. Tkacheva, et al. 2011. Mechanism of transfer of functional microRNAs between mouse dendritic cells via exosomes. *Blood* **119**:756–766.

Morelli, A.E., A.T. Larregina, W.J. Shufesky, M.L. Sullivan, D.B. Stolz, G.D. Papworth, A.F. Zahorchak, A.J. Logar, Z. Wang, S.C. Watkins, L.D. Falo Jr., and A.W. Thomson. 2004. Endocytosis, intracellular sorting, and processing of exosomes by dendritic cells. *Blood* **104**:3257–3266.

Moussay, E., K. Wang, J.H. Cho, K. van Moer, S. Pierson, J. Paggetti, P.V. Nazarov, V. Palissot, L.E. Hood, G. Berchem, and D.J. Galas. 2011. MicroRNA as biomarkers and regulators in B-cell chronic lymphocytic leukemia. *Proc Natl Acad Sci U S A* **108**:6573–6578.

Muller, G., M. Schneider, G. Biemer-Daub, and S. Wied. 2011. Microvesicles released from rat adipocytes and harboring glycosylphosphatidylinositol-anchored proteins transfer RNA stimulating lipid synthesis. *Cell Signal* **23**:1207–1223.

Novellino, L., R.L. Rossi, F. Bonino, D. Cavallone, S. Abrignani, M. Pagani, and M.R. Brunetto. 2012. Circulating hepatitis B surface antigen particles carry hepatocellular microRNAs. *PLoS ONE* **7**:e31952.

Ogawa, R., C. Tanaka, M. Sato, H. Nagasaki, K. Sugimura, K. Okumura, Y. Nakagawa, and N. Aoki. 2010. Adipocyte-derived microvesicles contain RNA that is transported into macrophages and might be secreted into blood circulation. *Biochem Biophys Res Commun* **398**:723–729.

Ohshima, K., K. Inoue, A. Fujiwara, K. Hatakeyama, K. Kanto, Y. Watanabe, K. Muramatsu, Y. Fukuda, S. Ogura, K. Yamaguchi, and T. Mochizuki. 2010. *Let-7* microRNA family is selectively secreted into the extracellular environment via exosomes in a metastatic gastric cancer cell line. *PLoS ONE* **5**:e13247.

Parolini, I., C. Federici, C. Raggi, L. Lugini, S. Palleschi, A. De Milito, C. Coscia, E. Iessi, M. Logozzi, A. Molinari, M. Colone, M. Tatti, et al. 2009. Microenvironmental pH is a key factor for exosome traffic in tumor cells. *J Biol Chem* **284**:34211–34222.

Pegtel, D.M., K. Cosmopoulos, D.A. Thorley-Lawson, M.A. van Eijndhoven, E.S. Hopmans, J.L. Lindenberg, T.D. de Gruijl, T. Wurdinger, and J.M. Middeldorp. 2010. Functional delivery of viral miRNAs via exosomes. *Proc Natl Acad Sci U S A* **107**:6328–6333.

Pigati, L., S.C. Yaddanapudi, R. Iyengar, D.J. Kim, S.A. Hearn, D. Danforth, M.L. Hastings, and D.M. Duelli. 2010. Selective release of microRNA species from normal and malignant mammary epithelial cells. *PLoS ONE* **5**:e13515.

Preall, J.B. and E.J. Sontheimer. 2005. RNAi: RISC gets loaded. *Cell* **123**:543–545.

Rabinowits, G., C. Gercel-Taylor, J.M. Day, D.D. Taylor, and G.H. Kloecker. 2009. Exosomal microRNA: a diagnostic marker for lung cancer. *Clin Lung Cancer* **10**:42–46.

Rosell, R., J. Wei, and M. Taron. 2009. Circulating microRNA signatures of tumor-derived exosomes for early diagnosis of non-small-cell lung cancer. *Clin Lung Cancer* **10**:8–9.

Skog, J., T. Wurdinger, S. van Rijn, D.H. Meijer, L. Gainche, M. Sena-Esteves, W.T. Curry Jr., B.S. Carter, A.M. Krichevsky, and X.O. Breakefield. 2008. Glioblastoma microvesicles transport RNA and proteins that promote tumour growth and provide diagnostic biomarkers. *Nat Cell Biol* **10**:1470–1476.

Taylor, D.D. and C. Gercel-Taylor. 2008. MicroRNA signatures of tumor-derived exosomes as diagnostic biomarkers of ovarian cancer. *Gynecol Oncol* **110**:13–21.

Thery, C. 2011. Exosomes: secreted vesicles and intercellular communications. *F1000 Biol Rep* **3**:15.

Thery, C., L. Duban, E. Segura, P. Veron, O. Lantz, and S. Amigorena. 2002a. Indirect activation of naive CD4+ T cells by dendritic cell-derived exosomes. *Nat Immunol* **3**:1156–1162.

Thery, C., L. Zitvogel, and S. Amigorena. 2002b. Exosomes: composition, biogenesis and function. *Nat Rev Immunol* **2**:569–579.

Tian, T., Y. Wang, H. Wang, Z. Zhu, and Z. Xiao. 2010. Visualizing of the cellular uptake and intracellular trafficking of exosomes by live-cell microscopy. *J Cell Biochem* **111**:488–496.

Trajkovic, K., C. Hsu, S. Chiantia, L. Rajendran, D. Wenzel, F. Wieland, P. Schwille, B. Brugger, and M. Simons. 2008. Ceramide triggers budding of exosome vesicles into multivesicular endosomes. *Science* **319**:1244–1247.

Turchinovich, A., L. Weiz, and B. Burwinkel. 2012. Extracellular miRNAs: the mystery of their origin and function. *Trends Biochem Sci* **37**:460–465.

Turchinovich, A., L. Weiz, A. Langheinz, and B. Burwinkel. 2013. Characterization of extracellular circulating microRNA. *Nucleic Acids Res* **39**:7223–7233.

Umezu, T., K. Ohyashiki, M. Kuroda, and J.H. Ohyashiki. 2013. Leukemia cell to endothelial cell communication via exosomal miRNAs. *Oncogene* **32**:2747–2755.

Valadi, H., K. Ekstrom, A. Bossios, M. Sjostrand, J.J. Lee, and J.O. Lotvall. 2007. Exosome-mediated transfer of mRNAs and microRNAs is a novel mechanism of genetic exchange between cells. *Nat Cell Biol* **9**:654–659.

VanWijk, M.J., E. VanBavel, A. Sturk, and R. Nieuwland. 2003. Microparticles in cardiovascular diseases. *Cardiovasc Res* **59**:277–287.

Vickers, K.C., B.T. Palmisano, B.M. Shoucri, R.D. Shamburek, and A.T. Remaley. 2011. MicroRNAs are transported in plasma and delivered to recipient cells by high-density lipoproteins. *Nat Cell Biol* **13**:423–433.

Voinnet, O. and D.C. Baulcombe. 1997. Systemic signalling in gene silencing. *Nature* **389**:553.

Vrijsen, K.R., J.P. Sluijter, M.W. Schuchardt, B.W. van Balkom, W.A. Noort, S.A. Chamuleau, and P.A. Doevendans. 2010. Cardiomyocyte progenitor cell-derived exosomes stimulate migration of endothelial cells. *J Cell Mol Med* **14**:1064–1070.

Wang, C., C. Yang, X. Chen, B. Yao, C. Zhu, L. Li, J. Wang, X. Li, Y. Shao, Y. Liu, J. Ji, J. Zhang, et al. 2011. Altered profile of seminal plasma microRNAs in the molecular diagnosis of male infertility. *Clin Chem* **57**:1722–1731.

Wang, J.F., M.L. Yu, G. Yu, J.J. Bian, X.M. Deng, X.J. Wan, and K.M. Zhu. 2010. Serum *miR-146a* and *miR-223* as potential new biomarkers for sepsis. *Biochem Biophys Res Commun* **394**: 184–188.

Wang, K., S. Zhang, J. Weber, D. Baxter, and D.J. Galas. 2010. Export of microRNAs and microRNA-protective protein by mammalian cells. *Nucleic Acids Res* **38**:7248–7259.

Weber, J.A., D.H. Baxter, S. Zhang, D.Y. Huang, K.H. Huang, M.J. Lee, D.J. Galas, and K. Wang. 2010. The microRNA spectrum in 12 body fluids. *Clin Chem* **56**:1733–1741.

Weickmann, J.L. and D.G. Glitz. 1982. Human ribonucleases. Quantitation of pancreatic-like enzymes in serum, urine, and organ preparations. *J Biol Chem* **257**:8705–8710.

Wolf, P. 1967. The nature and significance of platelet products in human plasma. *Br J Haematol* **13**:269–288.

Xin, H., Y. Li, B. Buller, M. Katakowski, Y. Zhang, X. Wang, X. Shang, Z.G. Zhang, and M. Chopp. 2012. Exosome-mediated transfer of *miR-133b* from multipotent mesenchymal stromal cells to neural cells contributes to neurite outgrowth. *Stem Cells* **30**:1556–1564.

Yang, M., J. Chen, F. Su, B. Yu, L. Lin, Y. Liu, J.D. Huang, and E. Song. 2011. Microvesicles secreted by macrophages shuttle invasion-potentiating microRNAs into breast cancer cells. *Mol Cancer* **10**:117.

Yuan, A., E.L. Farber, A.L. Rapoport, D. Tejada, R. Deniskin, N.B. Akhmedov, and D.B. Farber. 2009. Transfer of microRNAs by embryonic stem cell microvesicles. *PLoS ONE* **4**:e4722.

Zampetaki, A., S. Kiechl, I. Drozdov, P. Willeit, U. Mayr, M. Prokopi, A. Mayr, S. Weger, F. Oberhollenzer, E. Bonora, A. Shah, J. Willeit, et al. 2010. Plasma microRNA profiling reveals loss of endothelial *miR-126* and other microRNAs in type 2 diabetes. *Circ Res* **107**:810–817.

Zernecke, A., K. Bidzhekov, H. Noels, E. Shagdarsuren, L. Gan, B. Denecke, M. Hristov, T. Koppel, M.N. Jahantigh, E. Lutgens, S. Wang, E.N. Olson, et al. 2009. Delivery of microRNA-126 by apoptotic bodies induces CXCL12-dependent vascular protection. *Sci Signal* **2**:ra81.

Zhang, L., D. Hou, X. Chen, D. Li, L. Zhu, Y. Zhang, J. Li, Z. Bian, X. Liang, X. Cai, Y. Yin, C. Wang, et al. 2012. Exogenous plant MIR168a specifically targets mammalian LDLRAP1: evidence of cross-kingdom regulation by microRNA. *Cell Res* **22**:107–126.

Zhang, Y., D. Liu, X. Chen, J. Li, L. Li, Z. Bian, F. Sun, J. Lu, Y. Yin, X. Cai, Q. Sun, K. Wang, et al. 2010. Secreted monocytic *miR-150* enhances targeted endothelial cell migration. *Mol Cell* **39**:133–144.

Zhuang, G., X. Wu, Z. Jiang, I. Kasman, J. Yao, Y. Guan, J. Oeh, Z. Modrusan, C. Bais, D. Sampath, and N. Ferrara. 2012. Tumour-secreted *miR-9* promotes endothelial cell migration and angiogenesis by activating the JAK-STAT pathway. *EMBO J* **31**:3513–3523.

Zomer, A., T. Vendrig, E.S. Hopmans, M. van Eijndhoven, J.M. Middeldorp, and D.M. Pegtel. 2010. Exosomes: fit to deliver small RNA. *Commun Integr Biol* **3**:447–450.

37

RELEASE OF MicroRNA-CONTAINING VESICLES CAN STIMULATE ANGIOGENESIS AND METASTASIS IN RENAL CARCINOMA

Federica Collino, Cristina Grange, and Giovanni Camussi

Department of Internal Medicine, Molecular Biotechnology Center (MBC) and Centre for Research in Experimental Medicine (CeRMS), Torino, Italy

I.	Introduction	608
II.	Extracellular Vesicles in Cell Communication	608
III.	Interplay between Tumor and Their Surrounding Cells	609
IV.	Renal Carcinomas	609
	A. miRNAs and Renal Cell Carcinomas	610
	B. Circulating miRNAs and RCC	611
V.	Tumor Angiogenesis and miRNAs	612
VI.	Tumor Metastasis and miRNAs	613
VII.	Cancer Stem Cells	615
	A. Renal CSCs and EVs	616
VIII.	Conclusions	617
	References	617

ABBREVIATIONS

CSCs	cancer stem cells
CTLs	cytotoxic T lymphocytes
EC	endothelial cells
ECM	extracellular matrix
EMT	epithelial mesenchymal transition

MicroRNAs in Medicine, First Edition. Edited by Charles H. Lawrie.
© 2014 John Wiley & Sons, Inc. Published 2014 by John Wiley & Sons, Inc.

EVs	extracellular vesicles
HIF1/HIF2	hypoxia-inducible factors
HMECs	normal mammary epithelial cells
MAPK	mitogen-activated protein kinase
MARCKS	myristoylated alanine-rich protein kinase C substrate
miRNAs	microRNAs
PDCD4	tumor suppressor programmed cell death 4
RCC	renal cell carcinoma
RCSCs	renal CSC
ss-DNA	single-stranded DNA
TMV	tumor EVs
Tsp1	thrombospondin-1
VHL	Von Hippel–Lindau

I. INTRODUCTION

MicroRNAs (miRNAs) are a class of small non-coding RNAs able to modify the translational profile of cells through direct degradation of their target mRNAs or by blocking their translation into functional proteins. In recent years, miRNAs have been unveiled as biomarkers of physiological and pathological states. In fact, the miRNA signature for different tumors at different stages has been proposed as a new efficient biomarker for use in the clinical setting (Ludwig and Weinstein 2005). Moreover, miRNAs have been described as being released by their cell of origin and they function as secretory molecules that are able to induce epigenetic modification in target cells. A requirement for the messenger action of miRNAs is their protection from extracellular-degrading enzymes. Extracellular vesicles (EVs) have been proposed as a major protective mechanism for extracellular RNA. Indeed, EVs may vehicle nucleic acids, including miRNAs.

II. EXTRACELLULAR VESICLES IN CELL COMMUNICATION

The tissue microenvironment is governed by a complex network of signals generated by crosstalk between different cells. Cells may communicate with and influence each other not only through the secretion of soluble factors (Peinado et al. 2011), but also by direct cell-to-cell contact (Sherer and Mothes 2008) and intercellular trafficking of EVs. EVs behave as carriers of information between their cell of origin and neighboring cells. Interestingly, they also have the ability to enter the circulation and other biological fluids and, in doing so, are able to influence sites that are distant from the secretion site.

EVs are a heterogeneous population of vesicles with a spherical shape composed of a lipid bilayer and hydrophilic proteins. They comprise vesicles that differ in composition, size, and releasing mechanisms. EVs include shedding vesicles (also called ectosomes, microparticles, or exovesicles) formed by blebbing of the cell plasma membrane, and exosomes that are generated from the endosomal membrane compartment by exocytosis (Van Dommelen et al. 2012). Shedding vesicles are a heterogeneous population, with sizes ranging from 100 to 1000 nm, whereas exosomes tend to be more homogeneous, with sizes ranging from 30 to 120 nm. EVs contain membrane proteins from the cells of origin, but additionally contain a unique pattern of functional macromolecules, bioactive lipids, and genetic material. EVs are enriched in nucleic acids, including mRNAs, miRNAs, single-stranded DNA (ss-DNA), and mitochondrial DNA (Van Dommelen et al. 2012) that can be transferred from one cell to another, inducing epigenetic reprogramming of the recipient cell. Differentiation of exosomes from shedding vesicles is based not only on the biogenesis and size, but also on their content. Whereas exosome content is characterized by the presence of high levels of tetraspanins (CD9, CD63, and CD81), annexins, heat-shock proteins (such as Hsp60, Hsp70, and Hsp90), and low amounts of phosphatidylserines, shedding vesicles, expose high amounts of phosphatidylserine, cholesterol, sphingomyelin, and ceramide, and contain proteins associated with lipid rafts (György et al. 2011).

III. INTERPLAY BETWEEN TUMOR AND THEIR SURROUNDING CELLS

Cancer involves a complex machinery that is responsible for deregulation of intrinsic processes of tumor cells (e.g. proliferation, survival, and migration potential) and modification of the cell microenvironment, in order to create favorable conditions for cancer cell growth. In this scenario, normal cells may be able to influence surrounding tumor cells to suppress their growth (Velpula et al. 2011; Leung and Brugge 2012), whereas tumor cells could generate a favorable environment for their spread and dissemination.

Tumor-adjacent epithelial, stromal, or endothelial cells (EC) have been shown to acquire genetic tumor-related modifications. Hill et al. demonstrated that oncogenic stress generated in prostate epithelium is able to induce a mitogenic signal to the surrounding mesenchymal tissue, including induction of a p53 response able to suppress stromal fibroblast proliferation. This in turn generates a preferential selection against p53 in the stromal compartment, leading to prevalence of a p53-null fibroblast population able to support cancer growth (Hill et al. 2005). Akino et al. reported the presence of cytogenetic abnormalities in tumor endothelial cells that may play a significant role in modifying tumor–stromal interactions (Akino et al. 2009). Interestingly, reactive tumor–stroma and tumor endothelial cells are able to generate oncogenic signals that facilitate tumorigenesis. In this context, EVs have been shown to generate a cancer-selective microenvironment able to sustain tumor growth and progression. The ability of EVs to act as vehicles that spread cancer molecules has been demonstrated for several different cancers (D'Souza-Schorey and Clancy 2012). Al Nedawi et al. demonstrated that EVs released by aggressive glioma cells contained the oncogenic form of the epidermal growth factor receptor, EGFRvIII, that was subsequently transferred via EVs to non-aggressive tumor cells (Al-Nedawi et al. 2008). The oncogene transfer led to activation of molecules associated with the mitogen-activated protein kinase (MAPK) and Akt signaling in recipient cells (Al-Nedawi et al. 2008). Antonyak et al. showed that EVs released by breast carcinoma and glioma cells may release functional proteins, such as transglutaminase and fibronectin to fibroblasts and epithelial cells, respectively (Antonyak et al. 2011). Different studies have also reported the role cancer cell-released EVs in evading the immune system by means of the stimulation of apoptosis of NK cells and cytotoxic T lymphocytes (CTLs) (Kim et al. 2005).

IV. RENAL CARCINOMAS

Renal cell carcinoma (RCC) is a common form of urological tumor, representing 3% of total human malignancies, with a high metastatic index at diagnosis and a high rate of relapse. The incidence of RCC has increased in the past 30 years worldwide, with a mortality rate of over 40%, with RCC being the seventh most common cancer in men in the United States (Rathmell and Godley 2010). RCC includes a heterogeneous group of subtypes that differ in histopathological features and clinical behavior (namely: clear cell or conventional, papillary, chromophobe, and collecting duct [Lopez-Beltran et al. 2009]). Clear cell RCC, the most common subtype (75–80% of renal tumors), is a very vascularized tumor with constitutive activation of the angiogenic pathway, predominantly due to mutation or hypermethylation of the oncosuppressor gene, Von Hippel–Lindau (VHL), involved in degradation of hypoxia-inducible factors (HIF1/HIF2) (Rathmell and Godley 2010). It is the RCC subtype that determines the prognosis and response to therapy, and therefore correct diagnosis is critical. The classical approach to diagnosis is based on immuno-histochemical profiling. However, a remarkable number of needle biopsies used routinely for diagnosis are non-informative using the immunohistochemical technique (Fridman et al. 2010). This observation, associated with the absence of biomarkers for early detection and follow-up of the disease, underline the necessity to develop new types of molecular markers for RCC classification (Slaby et al. 2012).

A modern and innovative approach for molecular characterization of tumors is based on the miRNA expression profile (Slaby et al. 2012). Aberrant levels of miRNAs are present in many cancers compared with their normal tissue counterparts, and in general, there is believed to be a global down-regulation of miRNAs associated with malignancy. Through the generation of a miRNA marker algorithm to classify RCCs, Fridman et al. showed that this method reached an accuracy of 90% in tumor diagnosis compared with the histological method. This approach can provide additional information for characterization and differentiation of RCC, in conjunction with existing diagnostic tools (Fridman et al. 2010).

A. miRNAs and Renal Cell Carcinomas

Different groups have compared the levels of miRNAs expressed in RCC with normal tissue. These studies indicate that the miRNA profile may facilitate the ability to distinguish tumor tissue from normal parenchyma (Table 37.1). Most of these studies utilized hybridization microarray platforms, with subsequent validation of expression changes by qRT-PCR (Slaby et al. 2012). The alteration in miRNA patterns includes up-regulation of certain miRNAs and down-regulation of others.

Among the up-regulated miRNAs in RCC, the following miRNAs have been observed: *miR-210*, which is induced by HIF-α overexpression in hypoxic conditions (Valera et al. 2011); *miR-155* (Juan et al. 2010), expression of which has been correlated with the RCC size (White et al. 2011); and *miR-21*, which has been correlated with a general poor prognosis (Juan et al. 2010). The oncogenetic *miR-106b*, detected at high levels in primary RCC tumors, is reduced in metastatic patients with respect to patients in remission (Slaby et al. 2010). This particular work supports the possible role of *miR-106b* as a potential predictive marker of early metastasis in RCC patients (Slaby et al. 2010).

Among miRNAs that are down-regulated in RCC, *miR-9* was shown by Hildebrandt et al. to correlate with a risk of metastatic recurrence. There was a consequent decrease in the expression of *miR-9* following hypermethylation of the chromosomal region encoding for this miRNA (Hildebrandt et al. 2010). Other down-regulated miRNAs in RCC include *miR-141* and *miR-200c*, which are both members of the *miR-200* family, associated with the epithelial mesenchymal transition (EMT) (Neves et al. 2010). The EMT is characterized by metastatic behavior of cancer cells, and protein inducers of EMT directly suppress transcription of *miR-141* and *-200c* (Slaby et al. 2010). Recently, Slaby et al.

TABLE 37.1. miRNAs in RCC

miRNA	Function	Reference
miR-210	The hypoxia-regulated miR-210 is up-regulated in clear cell tumors compared with tumor of non-clear cell histology.	Valera et al. (2011)
miR-155	miR-155 is overexpressed in RCC samples and correlates with tumor size.	Juan et al. (2010) and White et al. (2011)
miR-21	miR-21 is overexpressed in RCC samples and correlates with a poor prognosis.	Juan et al. (2010)
miR-106b	miR-106b has been identified as potential predictive marker of early metastasis in RCC patients.	Slaby et al. (2010)
miR-9	miR-9 down-regulation correlates with risk of metastatic recurrence.	Hildebrandt et al. (2010)
miR-141 miR-200c	miR-200 family is down-regulated in RCC samples, and their expression is associated with EMT transition.	Slaby et al. (2010) Neves et al. (2010)
miR-127-3p miR-145 miR-126	These miRNAs significantly correlate with relapse-free survival of non-metastatic RCC.	Slaby et al. (2012)
miR-1233	Identification as potential circulating biomarker for RCC.	Wulfken et al. (2011)
miR-378 miR-451	miR-378 is increased and miR-451 is decreased in the serum of RCC patients.	Redova et al. (2012)

generated a miRNA signature associated with early relapse after nephrectomy in RCC patients (Slaby et al. 2012). The authors identified a relapse-associated signature consisting of 64 miRNAs. By validation assays, they confirmed the correlation between *miR-127-3p, -145*, and *miR-126* overexpression and relapse-free survival of non-metastatic RCC patients. This study supports the possibility of identifying RCC patients at high risk of early relapse after nephrectomy on the basis of their miRNA expression profile in clinical practice (Slaby et al. 2012).

B. Circulating miRNAs and RCC

Circulating miRNAs are present and stable in human plasma and could be used to distinguish between patients with tumors and healthy subjects (Lawrie et al. 2008). The expression profiles of circulating miRNAs have been extensively studied in plasma of patients with different malignancies. For example, in prostate cancer serum, *miR-141* levels were found to distinguish metastatic prostate patients from age-matched controls (Mitchell et al. 2008). In breast cancer, *miR-10b*, *miR-34a*, and *miR-155* correlated with the presence of overt metastases (Roth et al. 2010). There are currently only two studies that have been carried on miRNA levels in RCC. One study suggests that circulating *miR-1233* is a tumor biomarker (Wulfken et al. 2011), while the second study identified 30 miRNAs as being differentially expressed between the serum of RCC patients and healthy controls: 19 miRNAs were up-regulated and 11 miRNAs were down-regulated (Redova et al. 2012). Two of these have been successfully validated as potential biomarkers, namely *miR-378*, which is increased, and *miR-451*, which is decreased in the serum of RCC patients. Combination of the two miRNAs could be used as a predictable biomarker of RCC (Redova et al. 2012).

V. TUMOR ANGIOGENESIS AND miRNAs

Angiogenesis is the physiological process that involves growth of new blood vessels from preexisting vessels and is crucial for solid tumor growth and invasion because the vasculature provides metabolic support and access to the circulation. During tumorigenesis, the vascular system provides tumor cells with oxygen, growth factors, metabolites, and hormones (Bussolati et al. 2011). The process of angiogenesis is an attractive therapeutic target, and recently, different antiangiogenic agents have in fact been approved and introduced in clinical therapies while others are undergoing trials. Recent evidence indicates that tumor-derived endothelial cells (EC) possess a distinct and unique phenotype that differs from normal endothelial cells at the molecular and functional levels (Bussolati et al. 2010).

The importance of miRNAs in EC function has been clearly demonstrated *in vitro* and *in vivo* using mice deficient for Dicer or Drosha, two key enzymes in miRNA processing. Dicer deficiency in EC leads to a decrease of the angiogenic response to vascular endothelial growth factor (VEGF), as well as impaired proliferation and sprout formation, with a concomitant up-regulation of thrombospondin-1 (Tsp1), a potent antiangiogenic factor (Suárez et al. 2008). The generation of a EC-miRNA profile has identified approximately 30 specific miRNAs that are known as "Angio-miRNAs" (Caporali and Emanueli 2011). Within this group, *miR-126* has been described as being essential for vascular stability and integrity; in particular, it has been demonstrated to enhance the action of VEGF and FGF by repressing the expression of Spread-1, a negative regulator of VEGF signaling (S. Wang et al. 2008). Moreover, Urbich et al. described the potential of *miR-23*, *-27*, and *-130a* to enhance angiogenesis by down-regulating different anti-angiogenic factors (Urbich et al. 2008). Among the highly expressed miRNAs in EC, *miR-221* and *miR-222* have been associated with an anti-angiogenic function by directly targeting c-kit and eNOS (Urbich et al. 2008). Another group of miRNAs associated with EC function are those regulated by angiogenic stimuli (Caporali and Emanueli 2011). Recent data have demonstrated the potential of VEGF and epidermal growth factor (EGF) to induce *miR-296* and *miR-132* expression, and of the hypoxia to increase *miR-210* and *miR-424* expression levels (Caporali and Emanueli 2011).

During tumor angiogenesis, the physiological regulation of the angiogenic process is disrupted, and levels of miRNAs involved in angiogenesis are also altered, both in EC and tumor cells. Ablation of Dicer from the endothelium has shown a reduction in tumor growth, also suggesting a fundamental role of miRNAs in tumor angiogenesis (Suárez et al. 2008). The first family of miRNAs shown to be associated with tumor angiogenesis, as well as being relevant in physiological conditions, is the *miRNA-17-92* cluster, including *miR-17*, *-18a*, *-19a/b*, *-20a*, and *-92a*. These specific miRNAs modulate different targets that are involved in angiogenesis, including Tsp1 (Doebele et al. 2010). In tumor cells, Tsp1 has been also documented to be modulated by *miR-194*, which acts by sustaining the surrounding EC angiogenesis (Sundaram et al. 2011). During tumor progression, miRNAs that are modulated by VEGF, such as *miR-296*, *-130a*, and *-132* are usually overexpressed in EC. In particular, *miR-296*, *-130a*, and *-210* are considered as being hallmarks of tumor progression (Collet et al. 2012). Moreover, miRNAs involved in the regulation of oxygen levels, such as *miR-20* directly, and *miR-21* indirectly, are significantly up-regulated in tumor cells and regulate HIF-1α levels. In particular, *miR-21* has been defined as an "oncomiRNA," because it targets and represses PTEN (Liu et al. 2011).

Tumor angiogenesis is regulated not only by posttranscriptional modifications that occur in EC or cancer cells, but also by factors that are present as an integral part of the tumor microenvironment. In particular, tumor EVs (TMV) have a pivotal role in the modulation of tumor angiogenesis. Several studies have suggested that TMV-induced endothelial cell activation depends on the expression of sphingomyelins or tetraspanins by EVs. Tetraspanin 8, a constitutive component of exosomes, is enriched in TMVs and contributes to selective recruitment of proteins and mRNAs into exosomes. It has been suggested that tetraspanins have an important function not only in endothelial cell activation, but also in the premetastatic niche preparation (Nazarenko et al. 2010). In addition, Al-Nedawi et al. described that TMVs activate autocrine release of VEGF, by means of oncogenic EGFR transfer, which induces the angiogenic switch (Al-Nedawi et al. 2008). Moreover, activated endothelial cells may communicate over long distances by transferring the Dll4 Notch ligand via TMVs, thus propagating the angiogenic signal (Sheldon et al. 2010). Another mechanism used by ovarian cancer cells to induce endothelial cell activation and angiogenesis is mediated by the extracellular MMP inducer CD147 shed by TMVs (Millimaggi et al. 2007). Skog et al. demonstrated that EVs released by glioblastoma tumor cells contain selected patterns of mRNAs, miRNAs and angiogenic proteins. The RNAs carried inside EVs are able to be transferred to normal endothelial cells and are subsequently translated into functional proteins (Skog et al. 2008). Zhang et al., who demonstrated transfer of angiogenic *miR-150* to endothelial cells, via EVs derived from human monocyte/macrophage cell lines (Y. Zhang et al. 2010), showed that increasing concentrations of *miR-150* to endothelial cells resulted in modulation of c-Myb expression and enhanced migration. Recently, Cantaluppi et al. demonstrated that endothelial progenitor cell-derived EVs promote angiogenesis by delivering *miR-126* and *miR-296* (Cantaluppi et al. 2012). Currently, no studies on the angiogenic miRNA delivery through tumor-released EVs are available.

VI. TUMOR METASTASIS AND miRNAs

Metastatization is a multistep process characterized by tumor cell invasion of the extracellular matrix (ECM), followed by entry into the blood fluid and extravasation to distant sites where secondary tumors are established. All these steps are associated with modifications of the transcription profile of tumor cells that can alter the cells in such a way that the end result is successful completion of the metastatization process. Recently, miRNAs that play roles in various steps of the metastatic process have been characterized.

Cell invasion is the first step involved in cancer progression, characterized by migration of tumor cells to neighboring tissues through destruction of ECM proteins. Different miRNAs have been shown to either positively or negatively correlate with the migration potential of tumor cells. *miR-21*, in particular, has been described to regulate the invasion potential of tumor cells in different cancer models. In colorectal and breast cancer, *miR-21* has been shown to down-regulate expression of tumor suppressor programmed cell death 4 (PDCD4) at a posttranscriptional level, thus stimulating tumor cell invasion and metastatization (Huang et al. 2009; Reis et al. 2010). Recently, inactivation of *miR-21* by antisense oligonucleotides has been associated with the induction of sensitivity to apoptosis and inhibition of cell motility and invasion in prostate cancer cell lines (Li et al.

2009). In addition, the myristoylated alanine-rich protein kinase c substrate (MARCKS), a key protein involved in cell motility, has been identified as a new target of *miR-21* in prostate cancer cells (Li et al. 2009). *miR-21* has been also described as a modulator of expression of metalloproteinase inhibitors, RECK and TIMP3, resulting in increased glioma cell migration and invasion (Gabriely et al. 2008). The comparison of miRNA profiles in normal and malignant glioblastoma specimens revealed the role of *miR-146b* as an important regulator of glioblastoma U373 cell migration (Xia et al. 2009). Overexpression of *miR-146b* significantly reduced the invasion of glioblastoma U373 cells by targeting the matrix metalloproteinase 16 (MMP16) (Xia et al. 2009). Another miRNA, *miR-205*, has been described as being negatively correlated with the proliferative ability and clonogenic capacity of prostate cancer cells. Moreover, its overexpression has been demonstrated to promote E-cadherin expression associated with the reduction of prostate cancer cell migration and invasion (Gandellini et al. 2009). The inhibitory function on metastasis of miRNAs has also been demonstrated by *miR-1*, proposed to play a role in the inhibition of migration of A549 lung cancer cells, through the blocking of oncogenes or the activation of genes involved in apoptosis (Nasser et al. 2008). Conversely, *miR-182* over-expression has been correlated with the potential of melanoma cells to migrate and metastasize both *in vitro* and *in vivo*. *miR-182* targeted microphthalmia-associated transcription factor-M and FOXO3 and its overexpression has been related with progression from primary to metastatic melanoma (Segura et al. 2009). In addition, *miR-126* and *miR-183* overexpression was observed to be involved in the metastatization potential of lung cancer (Crawford et al. 2008; G. Wang et al. 2008).

Metastasis formation is correlated with the potential of cancer cells to acquire a mesenchymal phenotype through the induction of EMT. Numerous miRNAs have been demonstrated as regulators of this process. The *miR-200* family has emerged as a key regulator of the EMT process through the direct repression of ZEB transcription factors (Gregory et al. 2011). miRNA screening between metastatic breast cancer cells and normal mammary epithelial cells (HMECs) has also enlightened the important role of *miR-10b* in breast cancer metastasis (Ma et al. 2007): ectopic expression of *miR-10b* in HMECs enhanced their invasive and migration potential (Ma et al. 2007). Other studies demonstrated the important role of *miR-335*, *-126*, and *-206* as metastasis suppressors in breast cancer cells and their derivate bone and lung metastatic cells (Tavazoie et al. 2008).

EV production has been correlated with the invasive stage of cancer both *in vitro* and *in vivo*. The first demonstration of the role of EVs in metastasis progression was revealed by experiments showing transfer of information by exosomes derived from highly metastatic B16 melanoma cells to poorly metastatic melanoma cells, enabling them to acquire a more aggressive phenotype (Poste and Nicolson 1980). EVs have the potential to promote tumor cell invasion by means of numerous proteases contained within them, such as MMP2, MMP9, and MT1-MMP (Hakulinen et al. 2008). In addition, it has been suggested that EVs may contribute to the formation of the so-called premetastatic niche by enhancing recruitment of tumor cells in sites distant from the primary tumor, and thus supporting the generation of secondary tumors. Hood et al. observed the potential of melanoma exosomes to remodel the ECM, leading to the creation of an environment for metastasis development (Hood et al. 2011). Moreover, Jung et al. reported the same effects of exosomes from rat pancreatic adenocarcinoma cells in remodeling the lung microenvironment to generate secondary tumors (Jung et al. 2009). Castellana et al. observed that TMV released by highly metastatic prostate cancer cells can induce activation of different pathways in normal fibroblasts, supporting the increase in motility and resistance to apoptosis of recipient cells. In turn, activated fibroblasts are able to increase migration and invasion of highly metastatic PC3 cells (Castellana et al. 2009). Recently, the potential of EVs released by highly metastatic melanoma BL6-10 cells to modify poorly metastatic F1 melanoma cells has been highlighted, as molecules that are responsible for the metastatic phenotype are transferred (Hao et al. 2006). Taken together, these studies support the role of EVs in reprogramming recipient cells toward a more aggressive phenotype. The role of miRNA delivery in this context has only been addressed in a few studies. Ohshima et al. found an enrichment of the *let-7* miRNA family, in exosomes released by the highly metastatic gastric cancer cell line, AZ-P7a. The authors hypothesize that secretion of *let-7* miRNA is instrumental in maintaining the oncogenic and pro-metastatic properties of this gastric cancer cell line (Ohshima et al. 2010). The *Let-7* miRNA family, which is considered a tumor-suppressive family, can less frequently play an oncogenic function, through the direct targeting of caspase-3 mRNA (Tsang and Kwok 2008); it can therefore also be hypothesized that highly metastatic cancer cells can release these miRNAs to transfer their oncogenic potential to surrounding cells.

VII. CANCER STEM CELLS

Within the tumor cell population, cancer cells have shown a heterogeneous phenotype. Intrinsic and extrinsic mechanisms are involved in the cancer cell heterogeneity; on one hand, genetic or epigenetic modifications (Nowell 1976; Baylin 2011; You and Jones 2012) within cancer cells could determine the generation of cells with different tumorigenic potential; on the other hand, the different microenvironmental stimuli within a tumor could contribute to the phenotypic and functional changes upon cancer cells. Finally, although the majority of tumor cells are determined to differentiate and ultimately to stop dividing, a small population of cells has been identified in different tumors, defined as cancer stem cells (CSCs) or tumor-initiating cells, that possess self-renewal capability and from which non-tumorigenic cancer cells can be generated, creating a hierarchical organization (Reya et al. 2001). CSCs have been identified in several solid tumors, and although specific markers shared by CSCs derived from different tumors are lacking, they have shown an overall common property: the ability to generate serially transplantable tumors.

CSCs were first identified in acute myeloid leukemia as a rare subset of cells that when transplanted into immune-deficient mice were able to induce leukemia (Wang and Dick 2005). The presence of CSCs in solid tumors was first demonstrated in the breast cancer (Al-Hajj et al. 2003). The existence of a population of CSCs, showing a wide and variable range of markers, was also established in malignant pleural effusions of lung cancer (Basak et al. 2009).

Several studies have demonstrated CD44 as being an important marker of CSCs in different solid tumors, such as prostate, ovarian, colon, hepatocellular, bladder, gastric, and pancreatic cancers (Dimov et al. 2010; Gedye et al. 2010; Vries et al. 2010; Almhanna and Philip 2011). Other groups have identified CD133 as a marker of CSCs in brain tumors, including glioblastoma multiforme and medulloblastoma, as well as colorectal, hepatocellular, pancreatic, lung, endometrial, ovarian, and prostate cancers (O'Brien et al. 2009). CD147 and CD90 were also described as possible CSC markers in oral squamous cell carcinomas and hepatocellular carcinoma, respectively (Yao and Mishra 2009; Richard and Pillai 2010).

Recently, we identified a subset of CSCs expressing the mesenchymal stem cell marker CD105 in human renal cell carcinoma, with clonogenic ability, lack of epithelial differentiation markers, and expression of Nestin, Nanog, and Oct3-4 stem cell markers

(Bussolati et al. 2008). Unlike their negative counterpart, the CD105+ renal CSC (CD105+ RCSCs), showed the ability to grow as spheres and to differentiate into epithelial or endothelial cells both *in vitro and in vivo*. Moreover, the CD105+ population demonstrated the ability to generate serially transplantable tumors *in vivo* when a low number of cells were injected into immune-compromised mice. The tumors generated showed an epithelial carcinoma phenotype, resembling the tumor of origin (Bussolati et al. 2008).

A. Renal CSCs and EVs

Tumor cells release large amount of EVs, and the number of circulating EVs has been shown to correlate with poor prognosis. This may depend on the pleiotrophic effect of EVs. We previously demonstrated that normal stem cells are an abundant source of EVs, which have a fundamental role in cell-to-cell communication (Deregibus et al. 2007; Bruno et al. 2009). We recently found that CD105+ RCSCs are able to shed vesicles, and that these EVs may modify the tumor microenvironment by triggering angiogenesis and favor the formation of a lung "premetastatic niche" (Figure 37.1). The CD105− population showed also the potential to release EVs with the same morphology and size of EVs

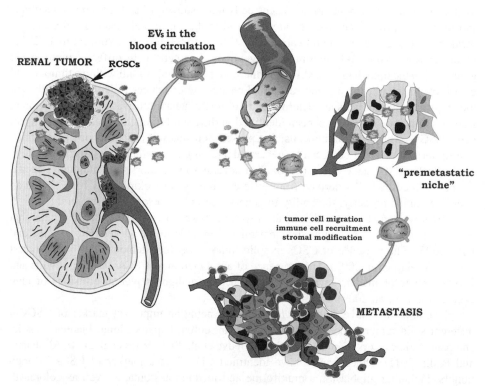

Figure 37.1. Potential role of EVs released from renal cancer stem cells. Renal cancer stem cells (RCSCs) expressing the CD105 marker may release EVs that enter the circulation, reach distant sites, and modify the microenvironment by triggering angiogenesis and formation of a "premetastatic niche" (i.e., condition favoring recruitment and implant of circulating tumor cells). miRNAs carried by EVs not only may sustain the ability of RCSCs to modify the tumor microenvironment, but they can also be used for tumor staging and as prognostic markers. See color insert.

derived from CD105+ RCSCs, but lacking the proangiogenic potential (Grange et al. 2011). EVs released by CD105⁺ RCSCs also expressed the CD105 molecule on their surface, contrary to EVs released by the CD105− population. Comparison of the RNA content in the EVs released by both populations showed the enrichment of transcripts coding for several proangiogenic proteins, such as VEGF, FGF, angiopoietin1, ephrin A3, and MMP2, MMP9 in the CD105⁺ EVs. These transcripts are absent in CD105− tumor EVs. The comparison of miRNA content in the two populations of released EVs showed a greater enrichment of small RNAs of the size of miRNAs (42.3 ± 2.5%) in CD105+ EVs compared with CD105− EVs (20.2 ± 1.7%). The miRNA expression profile of EVs released from CD105+ and CD105− cells revealed the presence of 82 and 87 miRNAs, respectively. Twenty-four miRNAs were significantly up-regulated in CD105+ EVs with respect to CD105− EVs whereas 33 miRNAs were significantly down regulated. Gene ontology analysis of the predicted genes modulated by the miRNAs that were up-regulated in CD105+ EVs showed a strong over-representation in terms of crucial biological processes like transcription, metabolic process, nucleic acid binding, cell adhesion molecules and regulation of cell proliferation. Among the miRNAs shuttled by CD105+ EVs, we detected *miR-200c, miR-92*, and *miR-141* that were described as being significantly up regulated in patients with ovarian, colorectal and prostate cancer, respectively (Grange et al. 2011). In addition, different miRNAs identified in CD105+ EVs, such as *miR-29a, miR-650*, and *miR-151*, were associated with tumor invasion and metastases (Gebeshuber et al. 2009; Luedde 2010; X. Zhang et al. 2010). Some miRNAs that were significantly enriched in CD105+ EVs, such as *miR-19b, miR-29c*, and *miR-151* were directly associated with renal carcinoma, whereas they resulted undetectable in normal renal tissue (Chow et al. 2010).

In conclusion, the presence of selected patterns of oncogenic miRNAs may represent not only the signature of CSCs, but also sustain their ability to modify the tumor microenvironment.

VIII. CONCLUSIONS

Characterization of the miRNA content of tumor-derived EVs may provide crucial diagnostic and prognostic information. The role of tumor EVs in delivering miRNAs capable of reprogramming recipient cells toward a more aggressive phenotype has only been addressed in a few studies. However, understanding the mechanisms involved in the crosstalk between normal and tumor cells may provide information useful for the design of new therapeutic strategies. Removal from the circulation of tumor-derived EVs or inhibition of their release could limit tumor spread. On the other hand, EVs could be exploited to deliver antioncogenic miRNAs to tumor cells in order to reprogram them to a more benign phenotype. This may be achieved using EVs derived from normal stem cells (Fonsato et al. 2012) or EVs specifically engineered to carry tumor suppressive molecules.

REFERENCES

Akino, T., K. Hida, Y. Hida, K. Tsuchiya, D. Freedman, C. Muraki, N. Ohga, K. Matsuda, K. Akiyama, T. Harabayashi, N. Shinohara, K. Nonomura, M. Klagsbrun, and M. Shindoh. 2009. Cytogenetic abnormalities of tumor-associated endothelial cells in human malignant tumors. *Am J Pathol* **175**:2657–2667.

Al-Hajj, M., M.S. Wicha, A. Benito-Hernandez, S.J. Morrison, and M.F. Clarke. 2003. Prospective identification of tumorigenic breast cancer cells. *Proc Natl Acad Sci U S A* **100**:3983–3988.

Almhanna, K. and P.A. Philip. 2011. Defining new paradigms for the treatment of pancreatic cancer. *Curr Treat Options Oncol* **12**:111–125.

Al-Nedawi, K., B. Meehan, J. Micallef, V. Lhotak, L. May, A. Guha, and J. Rak. 2008. Intercellular transfer of the oncogenic receptor EGFRvIII by microvesicles derived from tumour cells. *Nat Cell Biol* **10**:619–624.

Antonyak, M.A., B. Li, L.K. Boroughs, J.L. Johnson, J.E. Druso, K.L. Bryant, D.A. Holowka, and R.A. Cerione. 2011. Cancer cell-derived microvesicles induce transformation by transferring tissue transglutaminase and fibronectin to recipient cells. *Proc Natl Acad Sci U S A* **108**: 4852–4857.

Basak, S., M.S. Veena, S. Oh, G. Huang, E.S. Srivatsan, M. Huang, S. Sharma, and R.K. Batra. 2009. The malignant pleural effusion as a model to investigate intratumoral heterogeneity in lung cancer. *PLoS ONE* **4**:e5884.

Baylin, S.B. 2011. Resistance, epigenetics and the cancer ecosystem. *Nat Med* **17**:288–289.

Bruno, S., C. Grange, M.C. Deregibus, R.A. Calogero, S. Saviozzi, F. Collino, L. Morando, A. Busca, M. Falda, B. Bussolati, C. Tetta, and G. Camussi. 2009. Mesenchymal stem cell-derived microvesicles protect against acute tubular injury. *J Am Soc Nephrol* **20**:1053–1067.

Bussolati, B., S. Bruno, C. Grange, U. Ferrando, and G. Camussi. 2008. Identification of a tumor-initiating stem cell population in human renal carcinomas. *FASEB J* **22**:3696–3705.

Bussolati, B., M.C. Deregibus, and G. Camussi. 2010. Characterization of molecular and functional alterations of tumor endothelial cells to design anti-angiogenic strategies. *Curr Vasc Pharmacol* **8**:220–232.

Bussolati, B., C. Grange, and G. Camussi. 2011. Tumor exploits alternative strategies to achieve vascularization. *FASEB J* **25**:2874–2882.

Cantaluppi, V., L. Biancone, F. Figliolini, S. Beltramo, D. Medica, M.C. Deregibus, F. Galimi, R. Romagnoli, M. Salizzoni, C. Tetta, G.P. Segoloni, and G. Camussi. 2012. Microvesicles derived from endothelial progenitor cells enhance neoangiogenesis of human pancreatic islets. *Cell Transplant* **21**:1305–1320.

Caporali, A. and C. Emanueli. 2011. MicroRNA regulation in angiogenesis. *Vascul Pharmacol* **55**:79–86.

Castellana, D., F. Zobairi, M.C. Martinez, M.A. Panaro, V. Mitolo, J.M. Freyssinet, and C. Kunzelmann. 2009. Membrane microvesicles as actors in the establishment of a favorable prostatic tumoral niche: a role for activated fibroblasts and CX3CL1-CX3CR1 axis. *Cancer Res* **69**:69785–69793.

Chow, T.F., Y.M. Youssef, E. Lianidou, A.D. Romaschin, R.J. Honey, R. Stewart, K.T. Pace, and G.M. Yousef. 2010. Differential expression profiling of microRNAs and their potential involvement in renal cell carcinoma pathogenesis. *Clin Biochem* **43**:150–158.

Collet, G., K. Skrzypek, C. Grillon, A. Matejuk, B. El Hafni-Rahbi, N. Lamerant-Fayel, and C. Kieda. 2012. Hypoxia control to normalize pathologic angiogenesis: potential role for endothelial precursor cells and miRNAs regulation. *Vascul Pharmacol* **56**:252–261.

Crawford, M., E. Brawner, K. Batte, L. Yu, M.G. Hunter, G.A. Otterson, G. Nuovo, C.B. Marsh, and S.P. Nana-Sinkam. 2008. MicroRNA-126 inhibits invasion in non-small cell lung carcinoma cell lines. *Biochem Biophys Res Commun* **373**:607–612.

Deregibus, M.C., V. Cantaluppi, R. Calogero, M. Lo Iacono, C. Tetta, L. Biancone, S. Bruno, B. Bussolati, and G. Camussi. 2007. Endothelial progenitor cell derived microvesicles activate an angiogenic program in endothelial cells by a horizontal transfer of mRNA. *Blood* **110**: 2440–2448.

Dimov, I., M. Visnjic, and V. Stefanovic. 2010. Urothelial cancer stem cells. *ScientificWorldJournal* **10**:1400–1415.

Doebele, C., A. Bonauer, A. Fischer, A. Scholz, Y. Reiss, C. Urbich, W.K. Hofmann, A.M. Zeiher, and S. Dimmeler. 2010. Members of the microRNA-17-92 cluster exhibit a cell-intrinsic antiangiogenic function in endothelial cells. *Blood* **115**:4944–4950.

D'Souza-Schorey, C. and J.W. Clancy. 2012. Tumor-derived microvesicles: shedding light on novel microenvironment modulators and prospective cancer biomarkers. *Genes Dev* **26**:1287–1299.

Fonsato, V., F. Collino, M.B. Herrera, C. Cavallari, M.C. Deregibus, B. Cisterna, S. Bruno, R. Romagnoli, M. Salizzoni, C. Tetta, and G. Camussi. 2012. Human liver stem cell-derived microvesicles inhibit hepatoma growth in SCID mice by delivering antitumor microRNAs. *Stem Cells* **30**:1985–1998.

Fridman, E., Z. Dotan, I. Barshack, M.B. David, A. Dov, S. Tabak, O. Zion, S. Benjamin, H. Benjamin, H. Kuker, C. Avivi, K. Rosenblatt, S. Polak-Charcon, J. Ramon, N. Rosenfeld, and Y. Spector. 2010. Accurate molecular classification of renal tumors using microRNA expression. *J Mol Diagn* **12**:687–696.

Gabriely, G., T. Wurdinger, S. Kesari, C.C. Esau, J. Burchard, P.S. Linsley, and A.M. Krichevsky. 2008. MicroRNA 21 promotes glioma invasion by targeting matrix metalloproteinase regulators. *Mol Cell Biol* **28**:5369–5380.

Gandellini, P., M. Folini, N. Longoni, M. Pennati, M. Binda, M. Colecchia, R. Salvioni, R. Supino, R. Moretti, P. Limonta, R. Valdagni, M.G. Daidone, and N. Zaffaroni. 2009. *miR-205* Exerts tumor-suppressive functions in human prostate through down-regulation of protein kinase Cepsilon. *Cancer Res* **69**:2287–2295.

Gebeshuber, C.A., K. Zatloukal, and J. Martinez. 2009. *miR-29a* suppresses tristetraprolin, which is a regulator of epithelial polarity and metastasis. *EMBO Rep* **10**:400–405.

Gedye, C., A.J. Davidson, M.R. Elmes, J. Cebon, D. Bolton, and I.D. David. 2010. Cancer stem cells in urologic cancers. *Urol Oncol* **28**:585–590.

Grange, C., M. Tapparo, F. Collino, L. Vitillo, C. Damasco, M.C. Deregibus, C. Tetta, B. Bussolati, and G. Camussi. 2011. Microvesicles released from human renal cancer stem cells stimulate angiogenesis and formation of lung premetastatic niche. *Cancer Res* **71**:5346–5356.

Gregory, P.A., C.P. Bracken, E. Smith, A.G. Bert, J.A. Wright, S. Roslan, M. Morris, L. Wyatt, G. Farshid, Y.Y. Lim, G.J. Lindeman, M.F. Shannon, P.A. Drew, Y. Khew-Goodall, and G.J. Goodall. 2011. An autocrine TGF-beta/ZEB/*miR-200* signaling network regulates establishment and maintenance of epithelial-mesenchymal transition. *Mol Biol Cell* **22**:1686–1698.

György, B., T.G. Szabó, M. Pásztói, Z. Pál, P. Misják, B. Aradi, V. László, E. Pállinger, E. Pap, A. Kittel, G. Nagy, A. Falus, and E.I. Buzás. 2011. Membrane vesicles, current state-of-the-art: emerging role of extracellular vesicles. *Cell Mol Life Sci* **68**:2667–2688.

Hakulinen, J., L. Sankkila, N. Sugiyama, K. Lehti, and J. Keski-Oja. 2008. Secretion of active membrane type 1 matrix metalloproteinase (MMP-14) into extracellular space in microvesicular exosomes. *J Cell Biochem* **105**:1211–1218.

Hao, S., Z. Ye, F. Li, Q. Meng, M. Qureshi, J. Yang, and J. Xiang. 2006. Epigenetic transfer of metastatic activity by uptake of highly metastatic B16 melanoma cell-released exosomes. *Exp Oncol* **28**:126–131.

Hildebrandt, M.A., J. Gu, J. Lin, Y. Ye, W. Tan, P. Tamboli, C.G. Wood, and X. Wu. 2010. Hsa-*miR-9* methylation status is associated with cancer development and metastatic recurrence in patients with clear cell renal cell carcinoma. *Oncogene* **29**:5724–5728.

Hill, R., Y. Song, R.D. Cardiff, and T. Van Dyke. 2005. Selective evolution of stromal mesenchyme with p53 loss in response to epithelial tumorigenesis. *Cell* **123**:1001–1111.

Hood, J.L., R.S. San, and S.A. Wickline. 2011. Exosomes released by melanoma cells prepare sentinel lymph nodes for tumor metastasis. *Cancer Res* **71**:3792–3801.

Huang, T.H., F. Wu, G.B. Loeb, R. Hsu, A. Heidersbach, A. Brincat, D. Horiuchi, R.J. Lebbink, Y.Y. Mo, A. Goga, and M.T. McManus. 2009. Up-regulation of *miR-21* by HER2/neu signaling promotes cell invasion. *J Biol Chem* **284**:18515–18524.

Juan, D., G. Alexe, T. Antes, H. Liu, A. Madabhushi, C. Delisi, S. Ganesan, G. Bhanot, and L.S. Liou. 2010. Identification of a microRNA panel for clear-cell kidney cancer. *Urology* **75**: 835–841.

Jung, T., D. Castellana, P. Klingbeil, I. Cuesta Hernández, M. Vitacolonna, D.J. Orlicky, S.R. Roffler, P. Brodt, and M. Zöller. 2009. CD44v6 dependence of premetastatic niche preparation by exosomes. *Neoplasia* **11**:1093–1105.

Kim, J.W., E. Wieckowski, D.D. Taylor, T.E. Reichert, S. Watkins, and T.L. Whiteside. 2005. Fas ligand-positive membranous vesicles isolated from sera of patients with oral cancer induce apoptosis of activated T lymphocytes. *Clin Cancer Res* **11**:1010–1020.

Lawrie, C.H., S. Gal, H.M. Dunlop, B. Pushkaran, A.P. Liggins, K. Pulford, A.H. Banham, F. Pezzella, J. Boultwood, J.S. Wainscoat, C.S. Hatton, and A.L. Harris. 2008. Detection of elevated levels of tumour-associated microRNAs in serum of patients with diffuse large B-cell lymphoma. *Br J Haematol* **141**:672–675.

Leung, C.T. and J.S. Brugge. 2012. Outgrowth of single oncogene-expressing cells from suppressive epithelial environments. *Nature* **482**:410–413.

Li, T., D. Li, J. Sha, P. Sun, and Y. Huang. 2009. MicroRNA-21 directly targets MARCKS and promotes apoptosis resistance and invasion in prostate cancer cells. *Biochem Biophys Res Commun* **383**:280–285.

Liu, J., D.P. Lei, T. Jin, X.N. Zhao, G. Li, and X.L. Pan. 2011. Altered expression of *miR-21* and PTEN in human laryngeal and hypopharyngeal squamous cell carcinomas. *Asian Pac J Cancer Prev* **12**:2653–2657.

Lopez-Beltran, A., J.C. Carrasco, L. Cheng, M. Scarpelli, Z. Kirkali, and R. Montironi. 2009. 2009 update on the classification of renal epithelial tumors in adults. *Int J Urol* **16**:432–443.

Ludwig, J.A. and J.N. Weinstein. 2005. Biomarkers in cancer staging, prognosis and treatment selection. *Nat Rev Cancer* **5**:845–856.

Luedde, T. 2010. MicroRNA-151 and its hosting gene FAK (focal adhesion kinase) regulate tumor cell migration and spreading of hepatocellular carcinoma. *Hepatology* **52**:1164–1166.

Ma, L., J. Teruya-Feldstein, and R.A. Weinberg. 2007. Tumour invasion and metastasis initiated by microRNA-10b in breast cancer. *Nature* **449**:682–688.

Millimaggi, D., M. Mari, S. D'Ascenzo, E. Carosa, E.A. Jannini, S. Zucker, G. Carta, A. Pavan, and V. Dolo. 2007. Tumor vesicle-associated CD147 modulates the angiogenic capability of endothelial cells. *Neoplasia* **9**:349–357.

Mitchell, P.S., R.K. Parkin, E.M. Kroh, B.R. Fritz, S.K. Wyman, E.L. Pogosova-Agadjanyan, A. Peterson, J. Noteboom, K.C. O'Briant, A. Allen, D.W. Lin, N. Urban, C.W. Drescher, B.S. Knudsen, D.L. Stirewalt, R. Gentleman, R.L. Vessella, P.S. Nelson, D.B. Martin, and M. Tewari. 2008. Circulating microRNAs as stable blood-based markers for cancer detection. *Proc Natl Acad Sci U S A* **105**:10513–10518.

Nasser, M.W., J. Datta, G. Nuovo, H. Kutay, T. Motiwala, S. Majumder, B. Wang, S. Suster, S.T. Jacob, and K. Ghoshal. 2008. Down-regulation of micro-RNA-1 (*miR-1*) in lung cancer. Suppression of tumorigenic property of lung cancer cells and their sensitization to doxorubicin-induced apoptosis by *miR-1*. *J Biol Chem* **283**:33394–33405.

Nazarenko, I., S. Rana, A. Baumann, J. McAlear, A. Hellwig, M. Trendelenburg, G. Lochnit, K.T. Preissner, and M. Zöller. 2010. Cell surface tetraspanin Tspan8 contributes to molecular pathways of exosome-induced endothelial cell activation. *Cancer Res* **70**:1668–1678.

Neves, R., C. Scheel, S. Weinhold, E. Honisch, K.M. Iwaniuk, H.I. Trompeter, D. Niederacher, P. Wernet, S. Santourlidis, and M. Uhrberg. 2010. Role of DNA methylation in *miR-200c/141* cluster silencing in invasive breast cancer cells. *BMC Res Notes* **3**:219–226.

Nowell, P.C. 1976. The clonal evolution of tumor cell populations. *Science* **194**:23–28.

O'Brien, C.A., A. Kreso, and J. Dick. 2009. Cancer stem cells in solid tumors: an overview. *Semin Radiat Oncol* **19**:71–77.

REFERENCES

Ohshima, K., K. Inoue, A. Fujiwara, K. Hatakeyama, K. Kanto, Y. Watanabe, K. Muramatsu, Y. Fukuda, S. Ogura, K. Yamaguchi, and T. Mochizuki. 2010. *Let-7* microRNA family is selectively secreted into the extracellular environment via exosomes in a metastatic gastric cancer cell line. *PLoS ONE* **5**:e13247.

Peinado, H., S. Lavotshkin, and D. Lyden. 2011. The secreted factors responsible for pre-metastatic niche formation: old sayings and new thoughts. *Semin Cancer Biol* **21**:139–146.

Poste, G. and G.L. Nicolson. 1980. Arrest and metastasis of blood-borne tumor cells are modified by fusion of plasma membrane vesicles from highly metastatic cells. *Proc Natl Acad Sci U S A* **77**:399–403.

Rathmell, W.K. and P.A. Godley. 2010. Recent updates in renal cell carcinoma. *Curr Opin Oncol* **22**:250–256.

Redova, M., A. Poprach, J. Nekvindova, R. Iliev, L. Radova, R. Lakomy, M. Svoboda, R. Vyzula, and O. Slaby. 2012. Circulating *miR-378* and *miR-451* in serum are potential biomarkers for renal cell carcinoma. *J Transl Med* **10**:55–63.

Reis, P.P., M. Tomenson, N.K. Cervigne, J. Machado, I. Jurisica, M. Pintilie, M.A. Sukhai, B. Perez-Ordonez, R. Grénman, R.W. Gilbert, P.J. Gullane, J.C. Irish, and S. Kamel-Reid. 2010. Programmed cell death 4 loss increases tumor cell invasion and is regulated by *miR-21* in oral squamous cell carcinoma. *Mol Cancer* **9**:238–251.

Reya, T., S.J. Morrison, M.F. Clarke, and I.L. Weissman. 2001. Stem cells, cancer, and cancer stem cells. *Nature* **414**:105–111.

Richard, V. and M.R. Pillai. 2010. The stem cell code in oral epithelial tumorigenesis: "the cancer stem cell shift hypothesis". *Biochim Biophys Acta* **1806**:146–162.

Roth, C., B. Rack, V. Müller, W. Janni, K. Pantel, and H. Schwarzenbach. 2010. Circulating microRNAs as blood-based markers for patients with primary and metastatic breast cancer. *Breast Cancer Res* **12**:90–98.

Segura, M.F., D. Hanniford, S. Menendez, L. Reavie, X. Zou, S. Alvarez-Diaz, J. Zakrzewski, E. Blochin, A. Rose, D. Bogunovic, D. Polsky, J. Wei, P. Lee, I. Belitskaya-Levy, N. Bhardwaj, I. Osman, and E. Hernando. 2009. Aberrant *miR-182* expression promotes melanoma metastasis by repressing FOXO3 and microphthalmia-associated transcription factor. *Proc Natl Acad Sci U S A* **106**:1814–1819.

Sheldon, H., E. Heikamp, H. Turley, R. Dragovic, P. Thomas, C.E. Oon, R. Leek, M. Edelmann, B. Kessler, R.C. Sainson, I. Sargent, J.L. Li, and A.L. Harris. 2010. New mechanism for notch signaling to endothelium at a distance by Delta-like 4 incorporation into exosomes. *Blood* **116**:2385–2394.

Sherer, N.M. and W. Mothes. 2008. Cytonemes and tunneling nanotubules in cell-cell communication and viral pathogenesis. *Trends Cell Biol* **18**:414–420.

Skog, J., T. Würdinger, S. van Rijn, D.H. Meijer, L. Gainche, M. Sena-Esteves, W.T. Curry Jr., B.S. Carter, A.M. Krichevsky, and X.O. Breakefield. 2008. Glioblastoma microvesicles transport RNA and proteins that promote tumour growth and provide diagnostic biomarkers. *Nat Cell Biol* **10**:1470–1476.

Slaby, O., J. Jancovicova, R. Lakomy, M. Svoboda, A. Poprach, P. Fabian, L. Kren, J. Michalek, and R. Vyzula. 2010. Expression of miRNA-106b in conventional renal cell carcinoma is a potential marker for prediction of early metastasis after nephrectomy. *J Exp Clin Cancer Res* **29**:105–112.

Slaby, O., M. Redova, A. Poprach, J. Nekvindova, R. Iliev, L. Radova, R. Lakomy, M. Svoboda, and R. Vyzula. 2012. Identification of MicroRNAs associated with early relapse after nephrectomy in renal cell carcinoma patients. *Genes Chromosomes Cancer* **51** 7:707–716.

Suárez, Y., C. Fernández-Hernando, J. Yu, S.A. Gerber, K.D. Harrison, J.S. Pober, M.L. Iruela-Arispe, M. Merkenschlager, and W.C. Sessa. 2008. Dicer-dependent endothelial microRNAs are necessary for postnatal angiogenesis. *Proc Natl Acad Sci U S A* **105**:14082–14087.

Sundaram, P., S. Hultine, L.M. Smith, M. Dews, J.L. Fox, D. Biyashev, J.M. Schelter, Q. Huang, M.A. Cleary, O.V. Volpert, and A. Thomas-Tikhonenko. 2011. p53-responsive *miR-194* inhibits thrombospondin-1 and promotes angiogenesis in colon cancers. *Cancer Res* **71**:7490–7501.

Tavazoie, S.F., C. Alarcón, T. Oskarsson, D. Padua, Q. Wang, P.D. Bos, W.L. Gerald, and J. Massagué. 2008. Endogenous human microRNAs that suppress breast cancer metastasis. *Nature* **51**:147–152.

Tsang, W.P. and T.T. Kwok. 2008. *Let-7a* microRNA suppresses therapeutics-induced cancer cell death by targeting caspase-3. *Apoptosis* **13**:1215–1222.

Urbich, C., A. Kuehbacher, and S. Dimmeler. 2008. Role of microRNAs in vascular diseases, inflammation, and angiogenesis. *Cardiovasc Res* **79**:581–588.

Valera, V.A., B.A. Walter, W.M. Linehan, and M.J. Merino. 2011. Regulatory effects of microRNA-92 (*miR-92*) on VHL gene expression and the hypoxic activation of *miR-210* in clear cell renal cell carcinoma. *J Cancer* **2**:515–526.

Van Dommelen, S.M., P. Vader, S. Lakhal, S.A. Kooijmans, W.W. van Solinge, M.J. Wood, and R.M. Schiffelers. 2012. Microvesicles and exosomes: opportunities for cell-derived membrane vesicles in drug delivery. *J Control Release* **161**:635–644.

Velpula, K.K., V.R. Dasari, A.J. Tsung, C.S. Gondi, J.D. Klopfenstein, S. Mohanam, and J.S. Rao. 2011. Regulation of glioblastoma progression by cord blood stem cells is mediated by downregulation of cyclin D1. *PLoS ONE* **6**:e18017.

Vries, R.G.J., M. Huch, and H. Clevers. 2010. Stem cells and cancer of the stomach and intestine. *Mol Oncol* **4**:373–384.

Wang, G., W. Mao, and S. Zheng. 2008. MicroRNA-183 regulates Ezrin expression in lung cancer cells. *FEBS Lett* **582**:3663–3668.

Wang, J.C. and J.E. Dick. 2005. Cancer stem cells: lessons from leukemia. *Trends Cell Biol* **15**:494–501.

Wang, S., A.B. Aurora, B.A. Johnson, X. Qi, J. McAnally, J.A. Hill, J.A. Richardson, R. Bassel-Duby, and E.N. Olson. 2008. The endothelial-specific microRNA *miR-126* governs vascular integrity and angiogenesis. *Dev Cell* **15**:261–271.

White, N.M., T.T. Bao, J. Grigull, Y.M. Youssef, A. Girgis, E. Fatoohi, M. Metias, R.J. Honey, R. Stewart, K.T. Pace, G.A. Bjarnason, and G.M. Yousef. 2011. miRNA profiling for clear cell renal cell carcinoma: biomarker discovery and identification of potential controls and consequences of miRNA dysregulation. *J Urol* **186**:1077–1083.

Wulfken, L.M., R. Moritz, C. Ohlmann, S. Holdenrieder, V. Jung, F. Becker, E. Herrmann, G. Walgenbach-Brünagel, A. von Ruecker, S.C. Müller, and J. Ellinger. 2011. MicroRNAs in renal cell carcinoma: diagnostic implications of serum *miR-1233* levels. *PLoS ONE* **6**:e25787.

Xia, H., Y. Qi, S.S. Ng, X. Chen, D. Li, S. Chen, R. Ge, S. Jiang, G. Li, Y. Chen, M.L. He, H.F. Kung, L. Lai, and M.C. Lin. 2009. microRNA-146b inhibits glioma cell migration and invasion by targeting MMPs. *Brain Res* **1269**:158–165.

Yao, Z. and L. Mishra. 2009. Cancer stem cells and hepatocellular carcinoma. *Cancer Biol Ther* **8**:1691–1698.

You, J.S. and P.A. Jones. 2012. Cancer genetics and epigenetics: two sides of the same coin? *Cancer Cell* **22**:9–20.

Zhang, X., W. Zhu, J. Zhang, S. Huo, L. Zhou, Z. Gu, and M. Zhang. 2010. MicroRNA-650 targets ING4 to promote gastric cancer tumorigenicity. *Biochem Biophys Res Commun* **395**:275–280.

Zhang, Y., D. Liu, X. Chen, J. Li, L. Li, Z. Bian, F. Sun, J. Lu, Y. Yin, X. Cai, Q. Sun, K. Wang, Y. Ba, Q. Wang, D. Wang, J. Yang, P. Liu, T. Xu, Q. Yan, J. Zhang, K. Zen, and C.Y. Zhang. 2010. Secreted monocytic *miR-150* enhances targeted endothelial cell migration. *Mol Cell* **39**:133–144.

PART VI

THERAPEUTIC USES OF MicroRNAs: CURRENT PERSPECTIVES AND FUTURE DIRECTIONS

38

MicroRNA REGULATION OF CANCER STEM CELLS AND MicroRNAs AS POTENTIAL CANCER STEM CELL THERAPEUTICS

Can Liu and Dean G. Tang

Department of Molecular Carcinogenesis, The University of Texas MD Anderson Cancer Center, Smithville, TX, USA

I. Introduction 626
II. miRNA Regulation of Cancer Stem Cells 626
    A. Breast CSCs 627
    B. Glioma and Brain CSCs 630
    C. Prostate CSCs (PCSCs) 631
    D. Other CSCs 632
III. miRNAs in Cancer Diagnosis, Prognosis, and Therapy 633
IV. Conclusions and Perspectives 634
Acknowledgments 634
References 634

ABBREVIATIONS

ALDH	aldehyde dehydrogenase
ATM	ataxia telangiectasia mutated
BCSC	breast cancer stem cell
C. elegans	*Caenorhabditis elegans*
CDX2	caudal-type homeobox transcription factor 2
COX-2	cyclooxygenase-2
CSCs	cancer stem cells
Dll1	delta-like 1

MicroRNAs in Medicine, First Edition. Edited by Charles H. Lawrie.
© 2014 John Wiley & Sons, Inc. Published 2014 by John Wiley & Sons, Inc.

EMT	epithelial-mesenchymal transition
ESC	embryonic stem cell
GBM	glioblastoma multiforme
HDAC	histone deacetylase
MB	medulloblastoma
MIF	macrophage migration inhibitory factor
miRNA	microRNA
NLK	nemo-like kinase
NPC	neural precursor cells
NSCLC	non-small cell lung cancer
PCa	prostate cancer
PCSC	prostate cancer stem cell
PRC2	polycomb repressor complex 2
SP	side population
TGF-β	transforming growth factor-β
T-ICs	tumor-initiating cells
ZEB1	Zinc finger E-box-binding homeobox 1

I. INTRODUCTION

Research in the past decade suggests that tumor cells are heterogeneous and a subpopulation of cells within cancers called cancer stem cells (CSCs) or tumor-initiating cells (T-ICs) is the driving force for tumor initiation, progression, recurrence, and metastasis (Reya et al. 2001). CSCs, like their normal counterpart, possess significant proliferative, self-renewal, and differentiation abilities. More importantly, CSCs demonstrate high tumor-initiating capacity compared with the bulk or marker-negative cells, and regenerate tumors that can be serially transplanted in immune-deficient mice. Regardless of their origins, many human cancers are now considered to contain CSCs (Visvader and Linderman 2008). Characterization of the regulatory mechanisms of CSCs is expected to not only improve our understanding of tumor biology, but also help us in designing novel therapeutics that targets the roots of cancer.

microRNAs (miRNAs), a recently discovered category of non-coding RNA, have emerged as important regulators of cancer in general and CSCs in particular. By modulating gene expression through homologous recognition to their target mRNAs and inducing mRNA degradation and translation inhibition, miRNAs play powerful roles in most biological processes, including development, proliferation, and apoptosis (Bartel 2004). Initially discovered in regulating in embryonic stem cells (ESCs) in *Caenorhabditis elegans* development, miRNAs are now shown to be involved in various aspects of oncogenesis and CSC properties (Calin and Croce 2006).

II. MIRNA REGULATION OF CANCER STEM CELLS

Differential expression of miRNAs has been observed in various cancers when compared with the corresponding benign tissues and metastatic tumors (Calin and Croce 2006), and some of these miRNAs are identified to control tumor development by acting as either oncogenes (oncomiR) or tumor suppressors. Recent evidence also suggests that abnormal expression of miRNAs is important in maintaining CSC properties, such as self-renewal

A. Breast CSCs

miRNA analysis in SK-3rd BCSCs, which had been enriched by serially passaging breast cancer cell line SKBR3 in mice treated with chemotherapy, revealed altered expression of a number of miRNAs, including *let-7, miR-16, miR-107, miR-128a*, and *miR-20b*, providing the first link of miRNAs and CSCs (Yu et al. 2007).The authors further unraveled the important role of *let-7* in maintaining the key properties of BCSCs. Overexpression of *let-7* inhibited cell proliferation, mammosphere formation, tumor formation, and metastasis in NOD/SCID mice. H-RAS and HMGA2 were two direct downstream targets of *let-7* that mediated the inhibitory effects (Yu et al. 2007).

Interestingly, a recent study from the same group suggested a synergistic inhibitory effect of two underexpressed miRNAs, *let-7* and *miR-30*, on BCSCs and tumor formation and progression (Yu et al. 2010). Overexpression of *miR-30* alone in BCSCs, similar to *let-7*, diminished their self-renewal ability *in vitro* and tumor regeneration *in vivo*. When two miRNAs were introduced at the same time, a more complete blockage of BCSC "stemness" was observed. These observations suggest that multiple miRNAs may cooperatively regulate CSC properties (Table 38.1).

Another miRNA family frequently down-regulated in BCSC is the *miR-200* family, which is composed of five members, *miR-200a, miR-200b, miR-200c, miR-141*, and *miR-429*. In $CD44^+CD24^{-/lo}$ BCSCs, 37 miRNAs, including all five *miR-200* members, were found to be differentially expressed, compared with non-tumorigenic cancer cells (Shimono et al. 2009).Three miRNA clusters, *miR-200c-141, miR-200b-200a-429*, and *miR-183-96-182*, were also down-regulated in normal mammary stem/progenitor cells (Shimono et al. 2009). Overexpression of *miR-200c* reduced the clonal expansion and tumor-initiation activities in BCSCs, and interestingly, also suppressed formation of mammary ducts by normal mammary stem cells via directly targeting stem cell factor BMI-1. Additional evidence suggests that inhibition of *miR-200* family is required for formation and maintenance of BCSCs by targeting stem cell gene Suz12, a subunit of a polycomb repressor complex (PRC2) (Iliopoulos et al. 2010).

miR-200 family is also involved in a key biological process called epithelial-mesenchymal transition (EMT), an initial step of tumor metastasis. All five members of the *miR-200* family were significantly down-regulated in cells that undergo EMT in response to transforming growth factor-β (TGF-β), as well as in invasive breast cancer cell lines with mesenchymal characteristics (Gregory et al. 2008). *miR-200* overexpression prevented TGF-β-induced EMT by negatively regulating the expression of E-cadherin transcription repressor ZEB1 (zinc finger E-box-binding homeobox 1, also known as TCF8) and ZEB2 (also known as ZFXH1B and SMAD interacting protein 1 or SIP1). Intriguingly, ZEB1 and ZEB2 can also transcriptionally repress the expression of *miR-200* miRNAs by binding to their promoter regions, leading to strong activation of EMT (Burk et al. 2008; Park et al. 2008). Taken together, these data on miRNA-200 family provide links between EMT, cancer cell invasion and CSCs characteristics.

miR-205 is another miRNA implicated in regulating both normal and cancer stem cells. miRNA expression profiling in ALDH (aldehyde dehydrogenase) positive and Sca-1 positive mouse mammary epithelial cells revealed that *miR-205* and *miR-22* were highly expressed, whereas *let-7* family members and *miR-93* were depleted in stem cell-enriched populations (Ibarra et al. 2007). Interestingly, although *miR-205* was most abundant in

TABLE 38.1. miRNA Dysregulation in Cancer Stem Cells

microRNA	Cancercell Types	Expression	Observed Phenotypes	Target(s)	References
			Oncogenes		
***miR-181* family**	Mammospheres	Up-regulated	Induces sphere formation	ATM	Wang et al. (2011)
	EpCAM+AFP+ hepatic CSC	Up-regulated	Promotes proliferation and tumor initiation	CDX2, GATA6, NLK	J. Ji et al. (2009)
			Tumor Suppressors		
***Let-7* family**	BCSC	Down-regulated	Suppresses proliferation, mammosphere formation and tumor development	H-RAS and HMGA2	Yu et al. (2007)
miR-30	PCSC	Down-regulated	Inhibits proliferation, clonal expansion and tumor development	Ras and Myc	Liu et al. (2012)
	BCSC	Down-regulated	Inhibits self-renewal, induces apoptosis, and reducestumorigenesis and lung metastasis	Ubc9 and ITGB3	Yu et al. (2010)
miR-34	Glioblastoma, medulloblastom, glioma stem cell	Down-regulated	Induces cell cycle arrest, apoptosis, and inhibits xenograft growth	c-Met, Notch-1, Notch-2, CDK6; Dll1	Li et al. (2009); Guessous et al. (2010); de Antonellis et al. (2011)
	PCSC	Down-regulated	Inhibits cell proliferation, induce apoptosis, suppresses clonal expansion and tumor development	CD44	Liu et al. (2011)
	Gastric CSC	Down-regulated	Inhibits cell growth and induces chemosensitization and apoptosis, inhibits tumorsphere formation	Bcl-2, Notch and HMGA2	Ji et al. (2008)
	Pancreatic CSC	Down-regulated	Inhibits clonogenic cell growth and invasion, induces apoptosis and cell cycle arrest, and sensitizes the cells to chemotherapy	Bcl-2 and Notch1/2	Q. Ji et al. (2009)

miRNA	Cancer type	Regulation	Function	Targets	References
miR-200 family	BCSC	Down-regulated	Inhibits clonal expansion and tumor initiation	BMI-1, Suz2	Shimono et al. (2009); Iliopoulos et al. (2010)
	Breast cancer cells undergoing EMT and metastatic breast cancer cells	Down-regulated	Prevents TGF-β-induced EMT	ZEB1 and ZEB2	Gregory et al. (2008)
	Prostate cancer cells undergoing EMT	Down-regulated	Inhibits prostasphere formation	Notch1 and Lin28b	Kong et al. (2010)
	Metastatic prostate cancer cells	Down-regulated	Impedes proliferation	JAGGED1	Vallejo et al. (2011)
miR-124	Glioma and medulloblastoma	Down-regulated	Inhibits proliferation, induces neuronal-like differentiation, inhibits stem cell traits and tumor initiation and invasiveness	SNAI2	Silber et al. (2008); Xia et al. (2012)
miR-128	Glioma	Downregulated	Inhibits proliferation and xenograft growth	BMI-1 ARP-5	Godlewski et al. (2008); Cui et al. (2010)
miR-143 and miR-145	Bone metastatic PCa	Down-regulated	Suppress migration and invasion and tumor development and metastasis		Peng et al. (2011)
miR-199b-5p	Medulloblastoma	Down-regulated	Inhibits proliferation and anchorage-independent growth, and tumor development	HES-1	Garzia et al. (2009)
miR-451	Glioblastoma	Elevated in CD133− cells	Inhibits neurosphere formation		Gal et al. (2008)
	Colonosphere	Down-regulated	Inhibits self-renewal, tumorigenecity and chemoresistance to irinotecan.	MIF	Bitarte et al. (2011)
miR-93	Colon CSCs	Down-regulated	Inhibits proliferation and colony formation	HDAC8 and TLE4	Yu et al. (2011)

ALDH-positive normal mammary progenitor cells, it displayed a heterogeneous expression pattern in different subtypes or stages of breast cancer (Greene et al. 2010). Loss of *miR-205* was observed in metastatic breast cancer cells and clinical samples, consistent with previous findings that *miR-205* (together with *miR-200*) negatively regulates EMT (Gregory et al. 2008; Park et al. 2008).

Finally, *miR-181* family members (*miR-181a* and *miR-181b*) are shown to mediate the TGF-β-induced increases in BCSCs. TGF-β directly up-regulates *miR-181* miRNAs in mammospheres grown in differentiation-inhibitory conditions. Expression of *miR-181a/b*, or depletion of ATM, a direct target of *miR-181*, is sufficient to induce sphere formation in cancer cells (Wang et al. 2011).

B. Glioma and Brain CSCs

Several studies have reported miRNA dysregulation in glioblastoma multiforme (GBM), the most malignant form of glioma, and other brain CSCs. miRNA expression profiling in glioma has uncovered a distinct expression pattern of 71 miRNAs when compared with normal adult brain (Lavon et al. 2010). Of great interest, this profile is remarkably reminiscent of that of neural precursor cells (NPC) and ESCs, supporting the idea that cancer (or CSCs) and normal stem cells may share a common miRNA regulatory network (Lavon et al. 2010). Notably, about half of the 71 miRNAs were clustered in seven genomic regions that were previously linked to cancer development and/or stem cells: *miR-17-92*, *miR-106b-25*, *miR-106a-363*, *miR-183-182*, *miR-367-302*, *miR-371-373*, and the miRNA cluster in the Dlk1-Dio3 region (Lavon et al. 2010).

miR-124, one of the most abundant miRs in the adult brain, is frequently down-regulated in glioma and medulloblastoma (MB), suggesting that it may suppress brain tumor development by promoting differentiation. In support of this suggestion, Silber et al. have found that *miR-124* (and also *miR-137*) inhibited GBM cell proliferation and induced neuronal-like differentiation of GBM-derived stem cells (Silber et al. 2008). Another recent study also shows that loss of *miR-124* enhances stem cell traits and invasiveness of glioma cells (Xia et al. 2012). By targeting SNAI2, *miR-124* inhibits neurosphere formation, CD133$^+$ cell subpopulation, and stem cell marker expression (BMI-1, Nanog, and Nestin) *in vitro*, and tumorigencity and invasiveness *in vivo* (Xia et al. 2012).

Another interesting miRNA that regulates glioma and other brain CSCs is *miR-128*. Several studies have reported underexpression of *miR-128* in human GBM specimens compared with normal brain tissues (Ciafrè et al. 2005; Godlewski et al. 2008; Zhang et al. 2009). *miR-128* inhibited glioma stem cell proliferation *in vitro* and glioma xenograft growth *in vivo*. Moreover, *miR-128* significantly blocked glioma CSC self-renewal by directly targeting BMI-1 (Godlewski et al. 2008), an important stem cell regulator. Recently, Cui et al. have reported that *miR-128* regulates the proliferation of glioma and GBM cell by targeting ARP5 (angiopoietin-related growth factor protein 5; ANGPTL6), a transcriptional suppressor that promotes stem cell renewal (Cui et al. 2010).

A comparison of miRNA expression profiles in CD133$^+$ and CD133$^-$ glioblastoma cells showed that *miR-451*, together with several other miRNAs, were elevated in the non-stem CD133$^-$ population (Gal et al. 2008). Functional assays revealed that overexpression of *miR-451* in GBM cells inhibited neurosphere formation (Gal et al. 2008). A recent study showed that a tumor suppressive miRNA, *miR-145*, inversely correlated with the levels of Oct-4 and Sox-2 in CD133$^+$ glioblastoma cells and negatively regulated the tumorigenesis of GBM cells (Yang et al. 2012).

Finally, several miRNAs have recently been found to regulate the NOTCH signaling pathway, which is critical in stem cell differentiation and often dysregulated in human cancers. *miR-199b-5p* was down-regulated in MB and its overexpression inhibited proliferation and anchorage-independent growth of MB cells by targeting HES-1 (Garzia et al. 2009). Strikingly, overexpression of *miR-199-5p* decreased the CD133$^+$ subpopulation of MB cells and inhibited tumor regeneration (Garzia et al. 2009). *miR-34a* was recently found to be down-regulated in human GBM (Li et al. 2009). Transfection of *miR-34a* into bulk GBM, MB, and other glioma stem cells caused cell-cycle arrest, apoptosis, and also inhibited glioma xenograft growth, mediated by down-regulation of multiple oncogenic targets, including c-MET, Notch-1, Notch-2, and CDK6 (Li et al. 2009). Forced expression of c-Met or Notch-1/Notch-2 partially rescued the effects of *miR-34a* on cell death in glioma stem cells (Li et al. 2009; Guessous et al. 2010). In MB cells, *miR-34a* expression also diminishes the CD133$^+$/CD15$^+$ tumor-propagating cells, and negatively regulates cell proliferation and induces apoptosis and neural differentiation by targeting Notch ligand Delta-like 1 (Dll1) (de Antonellis et al. 2011).

C. Prostate CSCs (PCSCs)

CSCs with high tumor-initiating and metastatic potential were also identified in prostate cancer (PCa), by using cell surface marker CD44, CD133, α2β1, and side population techniques (Collins et al. 2005; Patrawala et al. 2005, 2006, 2007). Through an unbiased miRNA expression profiling in six PCa stem/progenitor cell populations purified from prostate cancer xenografts, including three CD44$^+$ populations from LAPC9, LAPC4 and Du145 tumors, CD133$^+$ from LAPC4, α2β1$^+$ from Du145, and side population (SP) from LAPC9, our group for the first time uncovered prostate CSC (PCSC)-specific miRNA expression profiles (Liu et al. 2011, 2012). Forty miRNAs were differentially expressed in the three CD44$^+$ PCa cell populations compared with corresponding CD44$^-$ PCa cell populations (Liu et al. 2012). Four underexpressed miRNAs, *miR-34a, let-7b, miR-106a*, and *miR-141*, and two overexpressed miRNAs, *miR-301* and *miR-452*, were identified in the five marker-positive cell populations compared with isogenic marker-negative cell populations. Most notably, *miR-34a* was the only miRNA that was commonly underexpressed in all six marker-positive cell populations (Liu et al. 2011). Further validation in CD44$^+$ PCa cells purified from ~20 patient prostate tumors confirmed the underexpression pattern of *miR-34a*. Overexpression of *miR-34a* in bulk or purified CD44$^+$PCa cells exerted a pronounced inhibition in tumor growth and metastasis *in vivo*. Strikingly, systemic delivery of *miR-34a* oligos through tail vein extended the survival of animals bearing orthotopic human PCa by inhibiting metastasis to the lung and other organs, indicating the therapeutic potential of this miRNA. Of significance, we demonstrated that CD44 itself represented a direct and relevant downstream target of *miR-34a* (Liu et al. 2011).

We recently demonstrated similar tumor inhibitory functions of another underexpressed miRNA identified in our profiling experiments, *let-7* (Liu et al. 2012). Overexpression of *let-7* by lenti-viral vectors suppressed PCa regeneration *in vivo* and clonal expansion *in vitro* by targeting RAS and MYC. Nevertheless, we observed differential effects between *let-7* and *miR-34a* on PCa and PCSC cells, in that *miR-34a* prominently induces G1 cell-cycle arrest followed by cell senescence, whereas *let-7* mainly causes G2-M arrest without inducing senescence (Liu et al. 2012). Taken together, our observations indicate that key tumor suppressive miRNAs may distinctively and coordinately regulate (prostate) tumor progression by targeting different aspect of CSC activities (Figure 38.1; Liu and Tang 2011).

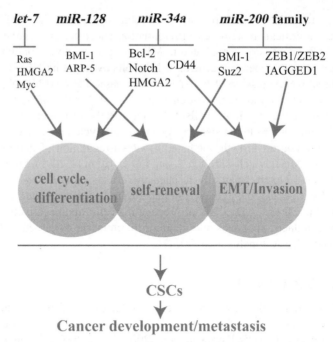

Figure 38.1. The emerging concept of major miRNAs distinctively and concertedly regulating key aspects of CSCs. CSCs are coordinately regulated by *let-7, miR-128, miR-200* family, and *miR-34a* miRNAs. These miRNAs, via targeting critical downstream signaling molecules, regulate several fundamental properties of CSCs, including cell-cycle exit and differentiation, self-renewal, morphological plasticity (EMT), migration, and invasion (represented by three shaded circles), which, in turn, contribute to cancer development and metastasis.

Several key miRNAs implicated in EMT and tumor metastasis in other cancer types also showed altered expression in PCa. For example, the *miR-200* and *let-7* family members were down-regulated in PCa cells undergoing EMT, which appeared to display a stem cell-like phenotype with increased expression of stem cell genes, such as Sox2, Nanog, Oct-4, Lin28b, and Notch, as well as enhanced sphere forming potential and tumor-initiating ability in mice (Kong et al. 2010). Furthermore, *miR-200c* and *miR-141* inhibited proliferation of human metastatic PCa cells by directly targeting JAGGED1, and overexpression of a JAGGED1 cDNA lacking its 3'-UTR restored PCa cell proliferation (Vallejo et al. 2011).

miRNA microarray analysis in primary and bone metastatic PCa samples revealed five miRNAs that were significantly down-regulated in metastasis samples, including *miR-143, miR-145, miR-508-5p, miR-33,* and *miR-100* (Peng et al. 2011). Overexpression of *miR-143* and *miR-145* reduced migration and invasion abilities of PCa cell *in vitro* and tumor development and bone metastasis *in vivo*.

D. Other CSCs

Interestingly, *miR-34* inhibits not only the GBM CSCs (Li et al. 2009) and PCSCs (Liu et al. 2011), but also pancreatic and gastric CSCs (Q. Ji et al. 2008, 2009). Restoration of *miR-34* expression in these latter CSCs inhibits sphere formation *in vitro* and tumor regeneration *in vivo* via modulating downstream targets, such as Bcl-2, NOTCH, and HMGA2 (Q. Ji et al. 2008, 2009).

In hepatic CSCs identified by EpCAM$^+$AFP$^+$ profile, Ji et al. uncovered a unique miRNA signature in which the *miR-181* family and several *miR-17-92* cluster members were up-regulated in the CSC population (J. Ji et al. 2009). Inhibition of *miR-181* led to a reduction in the number of EpCAM$^+$ hepatocellular carcinoma cells and in the tumor-initiating ability *in vivo*, whereas overexpression of *miR-181* increased the EpCAM$^+$ cells (J. Ji et al. 2009). The authors also found that *miR-181* directly targeted CDX2 (caudal type homeobox transcription factor 2), GATA6, and NLK (nemo-like kinase), a Wnt/beta-catenin pathway inhibitor.

In Type 1/CD44$^+$ ovarian CSCs, *miR-199a* and *miR-214*, a cluster located within the human *DNM3os* gene, were significantly down-regulated (Yin et al. 2010). Twist 1 was identified as an upstream regulator of *miR-199a* and *miR-214* cluster (Yin et al. 2010).

A comparison of miRNA expression in human colon CSC populations enriched by sphere formation versus the parental SW1116 cells has resulted in 35 up-regulated and 11 down-regulated miRNAs, with *miR-93* being the most strongly down-regulated (Yu et al. 2011). *miR-93* overexpression inhibited cell proliferation and colony formation of SW1116 CSCs cells by targeting histone deacetylase (HDAC8) and the transcription factor transducing-like enhancer protein 4 (TLE4). *miR-451* was also shown to be down-regulated in colonospheres derived from colon cancer cell lines (Bitarte et al. 2011). Ectopic expression of *miR-451* inhibited self-renewal, tumorigenicity, and chemoresistance to irinotecan. *miR-451* modulated cyclooxygenase (COX-2) expression by directly targeting macrophage migration inhibitory factor (MIF) gene. In turn, COX-2 allows Wnt/β-catenin activation, which is essential for CSC growth. Moreover, *miR-451* restoration decreases the expression of the multidrug transporter ABCB1, resulting in irinotecan sensitization (Bitarte et al. 2011).

III. miRNAs IN CANCER DIAGNOSIS, PROGNOSIS, AND THERAPY

Given that many CSC activities and tumor development per se are regulated by miRNAs, novel therapeutics that are based on miRNA have the possibility to become very promising anticancer therapeutics in the near future. First, miRNA expression profiles in CSCs and at various stages or in subtypes of cancer will be informative for diagnosing cancer origin and predicting prognosis for cancer patients. Second, functional studies of specific or groups of miRNAs that play a role in regulating CSCs will guide development of new anticancer therapeutics. Third, miRNAs have been shown to sensitize tumor cells to chemotherapy by targeting critical signaling pathways. Recent *in vivo* tumor experiments have established the proof of principle for the therapeutic efficacy of miRNAs in cancer by either replacing the tumor-suppressive miRNAs or targeting oncogenic miRNAs.

Systemic delivery of *miR-34a* through tail vein injection into mice bearing preformed LAPC9 PCa xenografts extended animal survival by inhibiting tumor progression and metastasis (Liu et al. 2011). Systemic or intratumoral injection of *miR-34* also impaired tumor development in non-small cell lung cancer (NSCLC) xenografts (Wiggins et al. 2010; Trang et al. 2011). "Replacement therapy" by delivering the tumor suppressive miRNA *let-7* into a NSCLC mouse model through different routes has revealed the therapeutic benefits of this miRNA. For instance, intranasal delivery of *let-7* viral particles or systemic delivery of *let-7* oligo mimics into a Kras$^{G12D/+}$ autochthonous NSCLC mouse model retarded tumor development (Esquela-Kerscher et al. 2008; Kumar et al. 2008;

Trang et al. 2010). Intratumoral injection of *let-7* oligo mimics similarly reduced tumor burden of a preestablished H460 lung cancer xenograft (Trang et al. 2011).

Meanwhile, oncogenic miRNAs are also potential therapeutic targets by using the antagomir that acts like a neutralizer to reduce the levels of endogenous target miRNAs. A prime example is that systemic intravenous delivery of the antagomir against *miR-10b* to tumor-bearing mice suppressed metastasis of 4T1 breast cancer cells to the lung without affecting primary tumor growth (Ma et al. 2010).

Although current studies offer us hope in developing miRNA-based therapeutics, the effective delivery of miRNA and managing toxicity of miRNA-formulated particles remain daunting challenges. Further studies are required for developing safer delivery vehicles, better specificity of miRNA distribution to the target tumor sites, and more controlled release of miRNAs.

IV. CONCLUSIONS AND PERSPECTIVES

miRNAs have been intimately implicated in tumor development and may regulate cancer development via modulating CSC properties. Studies reviewed here suggest an emerging perspective that several major tumor-suppressive miRNAs may distinctively and concertedly regulate key biological properties of CSCs (Figure 38.1). In this regard, *let-7* miRNAs appear to control the cell-cycle and differentiation properties of BCSCs (Yu et al. 2007), *miR-200c* may modulate the self-renewal of BCSCs by targeting Bmi-1 (Shimono et al. 2009) and regulate metastasis by targeting ZEB1 and ZEB2 (Burk et al. 2008; Gregory et al. 2008; Park et al. 2008), *miR-128* may inhibit GBM CSC self-renewal by targeting Bmi-1 and ARP-5 (Godlewski et al. 2008; Cui et al. 2010), and *miR-34a* may impede the migratory and invasive properties of PCSCs by directly repressing CD44 (Figure 38.1; Liu et al. 2011). Therefore, eradication of CSCs and cancer proper may require replacement of multiple tumor-suppressive miRNAs.

ACKNOWLEDGMENTS

We thank the other Tang Lab members for their support and assistance in actual experiments. miRNA-related research in our lab was supported in part by grants from NIH (R01-CA155693-01A), Department of Defense (W81XWH-11-1-0331), CPRIT funding (RP120380), and the MD Anderson Cancer Center the Center for Cancer Epigenetics ,and Laura and John Arnold Foundation RNA Center pilot grant (to D.G. Tang), and by two Center Grants (CCSG-5 P30 CA016672 and ES007784). C. Liu was supported in part by a predoctoral fellowship from the Department of Defense (W81XWH-10-1-0194). We apologize to the colleagues whose work could not be cited due to space constraints.

REFERENCES

Bartel, D.P. 2004. microRNA: genomics, biogenesis, mechanism, and function. *Cell* **116**:281–297.

Bitarte, N., E. Bandres, V. Boni, R. Zarate, J. Rodriguez, M. Gonzalez-Huarriz, I. Lopez, J. Javier Sola, M.M. Alonso, P. Fortes, and J. Garcia-Foncillas. 2011. MicroRNA-451 is involved in the self-renewal, tumorigenicity, and chemoresistance of colorectal cancer stem cells. *Stem Cells* **29**:1661–1671.

REFERENCES

Burk, U., J. Schubert, U. Wellner, O. Schmalhofer, E. Vincan, S. Spaderna, et al. 2008. A reciprocal repression between ZEB1 and members of the *miR-200* family promotes EMT and invasion in cancer cells. *EMBO Rep* **9**:582–589.

Calin, G.A. and C.M. Croce. 2006. MicroRNA signatures in human cancers. *Nat Rev Cancer* **6**:857–866.

Ciafrè, S.A., S. Galardi, A. Mangiola, M. Ferracin, C.G. Liu, G. Sabatino, M. Negrini, G. Maira, C.M. Croce, and M.G. Farace. 2005. Extensive modulation of a set of microRNAs in primary glioblastoma. *Biochem Biophys Res Commun* **334**:1351–1358.

Collins, A.T., P.A. Berry, C. Hyde, M.J. Stower, and N.J. Maitland. 2005. Prospective identification of tumorigenic prostate cancer stem cells. *Cancer Res* **65**:10946–10951.

Cui, J.G., Y. Zhao, P. Sethi, Y.Y. Li, A. Mahta, F. Culicchia, and W.J. Lukiw. 2010. Micro-RNA-128 (miRNA-128) down-regulation in glioblastomatargets ARP5 (ANGPTL6), Bmi-1 and E2F-3a, key regulators ofbrain cell proliferation. *J Neurooncol* **98**:297–304.

de Antonellis, P., C. Medaglia, E. Cusanelli, I. Andolfo, L. Liguori, G. De Vita, M. Carotenuto, A. Bello, F. Formiggini, A. Galeone, G. De Rosa, A. Virgilio, I. Scognamiglio, M. Sciro, G. Basso, J.H. Schulte, G. Cinalli, A. Iolascon, and M. Zollo. 2011. *MiR-34a* targeting of Notch ligand delta-like 1 impairs CD15+/CD133+ tumor-propagating cells and supports neural differentiation in medulloblastoma. *PLoS ONE* **6**:e24584.

Esquela-Kerscher, A., P. Trang, J.F. Wiggins, L. Patrawala, A. Cheng, L. Ford, J.B. Weidhaas, D. Brown, A.G. Bader, and F.J. Slack. 2008. The *let-7* microRNA reduces tumor growth in mouse models of lung cancer. *Cell Cycle* **7**:759–764.

Gal, H., G. Pandi, A.A. Kanner, Z. Ram, G. Lithwick-Yanai, N. Amariglio, G. Rechavi, and D. Givol. 2008. MIR-451 and Imatinibmesylate inhibit tumor growth of glioblastoma stem cells. *Biochem Biophys Res Commun* **376**:86–90.

Garzia, L., I. Andolfo, E. Cusanelli, N. Marino, G. Petrosino, D. De Martino, V. Esposito, A. Galeone, L. Navas, S. Esposito, S. Gargiulo, S. Fattet, V. Donofrio, G. Cinalli, A. Brunetti, L.D. Vecchio, P.A. Northcott, O. Delattre, M.D. Taylor, A. Iolascon, and M. Zollo. 2009. MicroRNA-199b-5p impairs cancer stem cells through negative regulation of HES1 in medulloblastoma. *PLoS ONE* **4**:e4998.

Godlewski, J., M.O. Nowicki, A. Bronisz, S. Williams, A. Otsuki, G. Nuovo, A. Raychaudhury, H.B. Newton, E.A. Chiocca, and S. Lawler. 2008. Targeting of the Bmi-1 oncogene/stem cell renewal factor by microRNA-128 inhibits glioma proliferation and self-renewal. *Cancer Res* **68**: 9125–9130.

Greene, S.B., J.I. Herschkowitz, and J.M. Rosen. 2010. The ups and downs of *miR-205:* identifying the roles of *miR-205* in mammary gland development and breast cancer. *RNA Biol* **7**:300–304.

Gregory, P.A., A.G. Bert, E.L. Paterson, S.C. Barry, A. Tsykin, G. Farshid, M.A. Vadas, Y. Khew-Goodall, and G.J. Goodall. 2008. The *miR-200* family and *miR-205* regulate epithelial to mesenchymal transition by targeting ZEB1 and SIP1. *Nat Cell Biol* **10**:593–601.

Guessous, F., Y. Zhang, A. Kofman, A. Catania, Y. Li, D. Schiff, B. Purow, and R. Abounader. 2010. microRNA-34a is tumor suppressive in brain tumors and glioma stem cells. *Cell Cycle* **9**:1031–1036.

Ibarra, I., Y. Erlich, S.K. Muthuswamy, R. Sachidanandam, and G.J. Hannon. 2007. A role for microRNAs in maintenance of mouse mammary epithelial progenitor cells. *Genes Dev* **21**: 3238–3243.

Iliopoulos, D., M. Lindahl-Allen, C. Polytarchou, H.A. Hirsch, P.N. Tsichlis, and K. Struhl. 2010. Loss of *miR-200* inhibition of Suz12 leads to polycomb-mediated repression required for the formation and maintenance of cancer stem cells. *Mol Cell* **39**:761–772.

Ji, J., T. Yamashita, A. Budhu, M. Forgues, H.L. Jia, C. Li, C. Deng, E. Wauthier, L.M. Reid, Q.H. Ye, L.X. Qin, W. Yang, H.Y. Wang, Z.Y. Tang, C.M. Croce, and X.W. Wang. 2009. Identification of microRNA-181 by genome-wide screening as a critical player in EpCAM-positive hepatic cancer stem cells. *Hepatology* **50**:472–480.

Ji, Q., X. Hao, Y. Meng, M. Zhang, J. Desano, D. Fan, and L. Xu. 2008. Restoration of tumor suppressor *miR-34* inhibits human p53-mutant gastric cancer tumorspheres. *BMC Cancer* **8**:266.

Ji, Q., X. Hao, M. Zhang, W. Tang, M. Yang, L. Li, D. Xiang, J.T. Desano, G.T. Bommer, D. Fan, E.R. Fearon, T.S. Lawrence, and L. Xu. 2009. MicroRNA *miR-34* inhibits human pancreatic cancer tumor-initiating cells. *PLoS ONE* **4**:e6816.

Kong, D., S. Banerjee, A. Ahmad, Y. Li, Z. Wang, S. Sethi, and F.H. Sarkar. 2010. Epithelial to mesenchymal transition is mechanistically linked with stem cell signatures in prostate cancer cells. *PLoS ONE* **5**:e12445.

Kumar, M.S., S.J. Erkeland, R.E. Pester, C.Y. Chen, M.S. Ebert, P.A. Sharp, and T. Jacks. 2008. Suppression of non-small cell lung tumor development by the *let-7* microRNA family. *Proc Natl Acad Sci U S A* **105**:3903–3908.

Lavon, I., D. Zrihan, A. Granit, O. Einstein, N. Fainstein, M.A. Cohen, M.A. Cohen, B. Zelikovitch, Y. Shoshan, S. Spektor, B.E. Reubinoff, Y. Felig, O. Gerlitz, T. Ben-Hur, Y. Smith, and T. Siegal. 2010. Gliomas display a microRNA expression profilereminiscent of neural precursor cells. *Neuro Oncol* **12**:422–433.

Li, Y., F. Guessous, Y. Zhang, C. Dipierro, B. Kefas, E. Johnson, L. Marcinkiewicz, J. Jiang, Y. Yang, T.D. Schmittgen, B. Lopes, D. Schiff, B. Purow, and R. Abounader. 2009. MicroRNA-34a inhibits glioblastoma growth by targeting multiple oncogenes. *Cancer Res* **69**:7569–7576.

Liu, C., K. Kelnar, B. Liu, X. Chen, T. Calhoun-Davis, H. Li, L. Patrawala, H. Yan, C. Jeter, S. Honorio, J.F. Wiggins, A.G. Bader, R. Fagin, D. Brown, and D.G. Tang. 2011. The microRNA *miR-34a* inhibits prostate cancer stem cells and metastasis by directly repressing CD44. *Nat Med* **17**:211–215.

Liu, C., K. Kelnar, A.V. Vlassov, D. Brown, J. Wang, and D.G. Tang. 2012. Distinct microRNA expression profiles in prostate cancer stem/progenitor cells and tumor-suppressive functions of *let-7*. *Cancer Res* **72**:3393–3404.

Liu, C. and D.G. Tang. 2011. MicroRNA regulation of cancer stem cells. *Cancer Res* **71**: 5950–5954.

Ma, L., F. Reinhardt, E. Pan, J. Soutschek, B. Bhat, E.G. Marcusson, J. Teruya-Feldstein, G.W. Bell, and R.A. Weinberg. 2010. Therapeutic silencing of *miR-10b* inhibits metastasis in a mouse mammary tumor model. *Nat Biotechnol* **28**:341–347.

Park, S.M., A.B. Gaur, E. Lengyel, and M.E. Peter. 2008. The *miR-200* family determines the epithelial phenotype of cancer cells by targeting the E-cadherin repressors ZEB1 and ZEB2. *Genes Dev* **22**:894–907.

Patrawala, L., T. Calhoun, R. Schneider-Broussard, H. Li, B. Bhatia, and S. Tang. 2006. Highly purified CD44+ prostate cancer cells from xenograft human tumors are enriched in tumorigenic and metastatic progenitor cells. *Oncogene* **25**:1696–1708.

Patrawala, L., T. Calhoun, R. Schneider-Broussard, J. Zhou, K. Claypool, and D.G. Tang. 2005. Side population is enriched in tumorigenic, stem-like cancer cells, whereas ABCG2+ and ABCG2- cancer cells are similarly tumorigenic. *Cancer Res* **65**:6207–6219.

Patrawala, L., T. Calhoun-Davis, R. Schneider-Broussard, and D.G. Tang. 2007. Hierarchical organization of prostate cancer cells in xenograft tumors: the CD44+α2β1+ cell population is enriched in tumor-initiating cells. *Cancer Res* **67**:6796–6805.

Peng, X., W. Guo, T. Liu, X. Wang, X. Tu, D. Xiong, S. Chen, Y. Lai, H. Du, G. Chen, G. Liu, Y. Tang, S. Huang, and X. Zou. 2011. Identification of miRs-143 and -145 that is associated with bone metastasis of prostate cancer and involved in the regulation of EMT. *PLoS ONE* **6**:e20341.

Reya, T., S.J. Morrison, M.F. Clarke, and I.L. Weissman. 2001. Stem cells, cancer, and cancer stem cells. *Nature* **414**:105–111.

REFERENCES

Shimono, Y., M. Zabala, R.W. Cho, N. Lobo, P. Dalerba, D. Qian, M. Diehn, H. Liu, S.P. Panula, E. Chiao, F.M. Dirbas, G. Somlo, R.A. Pera, K. Lao, and M.F. Clarke. 2009. Downregulation of miRNA-200c links breast cancer stem cells with normal stem cells. *Cell* **138**:592–603.

Silber, J., D.A. Lim, C. Petritsch, A.I. Persson, A.K. Maunakea, M. Yu, S.R. Vandenberg, D.G. Ginzinger, C.D. James, J.F. Costello, G. Bergers, W.A. Weiss, A. Alvarez-Buylla, and J.G. Hodgson. 2008. *miR-124* and *miR-137* inhibit proliferation of glioblastoma multiforme cells and induce differentiation of brain tumor stem cells. *BMC Med* **6**:14.

Trang, P., P.P. Medina, J.F. Wiggins, L. Ruffino, K. Kelnar, M. Omotola, R. Homer, D. Brown, A.G. Bader, J.B. Weidhaas, and F.J. Slack. 2010. Regression of murine lung tumors by the *let-7* microRNA. *Oncogene* **29**:1580–1587.

Trang, P., J.F. Wiggins, C.L. Daige, C. Cho, M. Omotola, D. Brown, J.B. Weidhaas, A.G. Bader, and F.J. Slack. 2011. Systemic delivery of tumor suppressor microRNA mimics using a neutral lipid emulsion inhibits lung tumors in mice. *Mol Ther* **19**:1116–1122.

Vallejo, D.M., E. Caparros, and M. Dominguez. 2011. Targeting Notch signalling by the conserved *miR-8/200* microRNA family in development and cancer cells. *EMBO J* **30**:756–769.

Visvader, J.E. and G.J. Linderman. 2008. Cancer stem cells in solid tumors: accumulating evidence and unresolved question. *Nat Rev Cancer* **8**:755–768.

Wang, Y., Y. Yu, A. Tsuyada, X. Ren, X. Wu, K. Stubblefield, E.K. Rankin-Gee, and S.E. Wang. 2011. Transforming growth factor-β regulates the sphere-initiating stem cell-like feature in breast cancer through miRNA-181 and ATM. *Oncogene* **30**:1470–1480.

Wiggins, J.F., L. Ruffino, K. Kelnar, M. Omotola, L. Patrawala, D. Brown, and A.G. Bader. 2010. Development of a lung cancer therapeutic based on the tumor suppressor microRNA-34. *Cancer Res* **70**:5923–5930.

Xia, H., W.K. Cheung, S.S. Ng, X. Jiang, S. Jiang, J. Sze, G.K. Leung, G. Lu, D.T. Chan, X.W. Bian, H.F. Kung, W.S. Poon, and M.C. Lin. 2012. Loss of brain-enriched *miR-124* microRNA enhances stem-like traits and invasiveness of glioma cells. *J Biol Chem* **287**:9962–9971.

Yang, Y.P., Y. Chien, G.Y. Chiou, J.Y. Cherng, M.L. Wang, W.L. Lo, Y.L. Chang, P.I. Huang, Y.W. Chen, Y.H. Shih, M.T. Chen, and S.H. Chiou. 2012. Inhibition of cancer stem cell-like properties and reduced chemoradioresistance of glioblastoma using microRNA145 with cationic polyurethane-short branch PEI. *Biomaterials* **33**:1462–1476.

Yin, G., R. Chen, A.B. Alvero, H.H. Fu, J. Holmberg, C. Glackin, T. Rutherford, and G. Mor. 2010. TWISTing stemness, inflammation and proliferation of epithelial ovarian cancer cells through MIR199A2/214. *Oncogene* **29**:3545–3553.

Yu, F., H. Deng, H. Yao, Q. Liu, F. Su, and E. Song. 2010. *MiR-30* reduction maintains self-renewal and inhibits apoptosis in breast tumor-initiating cells. *Oncogene* **9**:4194–4204.

Yu, F., H. Yao, P. Zhu, X. Zhang, Q. Pan, C. Gong, Y. Huang, X. Hu, F. Su, J. Lieberman, and E. Song. 2007. *Let-7* regulates self renewal and tumorigenicity of breast cancer cells. *Cell* **131**:1109–1123.

Yu, X.F., J. Zou, Z.J. Bao, and J. Dong. 2011. *miR-93* suppresses proliferation and colony formation of human colon cancer stem cells. *World J Gastroenterol* **17** 42:4711–4717.

Zhang, Y., T. Chao, R. Li, W. Liu, Y. Chen, X. Yan, Y. Gong, B. Yin, W. Liu, B. Qiang, J. Zhao, J. Yuan, and X. Peng. 2009. MicroRNA-128 inhibits glioma cells proliferation by targeting transcription factor E2F3a. *J Mol Med (Berl)* **87**:43–51.

39

THERAPEUTIC MODULATION OF MicroRNAs

Achim Aigner and Hannelore Dassow

Rudolf-Boehm-Institute for Pharmacology and Toxicology Clinical Pharmacology
University of Leipzig, Leipzig, Germany

I. Introduction: Physiological and Pathophysiological Roles of MicroRNAs 640
II. General Approaches for Therapeutic MicroRNA-Based Intervention 642
    A. Viral Delivery 643
    B. Non-Viral Strategies 643
    C. miRNA Inhibition 644
III. Therapeutic MicroRNA Replacement 644
    A. Delivery of *let-7* and *miR-34a* 644
    B. Non-Viral Delivery of *miR-143*, *miR-145*, and *miR-33a* 648
    C. Viral Delivery of miRNAs 651
    D. Other Examples of Viral and Non-Viral Delivery 652
IV. Therapeutic Studies in MicroRNA Inhibition 653
V. Perspectives, Issues, and Future Directions 655
Acknowledgments 657
References 657

ABBREVIATIONS

AAV   adeno-associated virus
AML   acute myeloid leukemia
AMO   anti-miRNA oligonucleotide
ASO   antisense oligonucleotide

MicroRNAs in Medicine, First Edition. Edited by Charles H. Lawrie.
© 2014 John Wiley & Sons, Inc. Published 2014 by John Wiley & Sons, Inc.

HCV	hepatitis C virus
HDL	high-density lipoprotein
miR	microRNA
LNA	locked nucleic acid
NSCLC	non-small cell lung cancer
PEI	polyethylenimine
RISC	RNA-induced silencing complex
RNAi	RNA interference
VLDL	very low density lipoprotein

I. INTRODUCTION: PHYSIOLOGICAL AND PATHOPHYSIOLOGICAL ROLES OF MicroRNAs

In the last years, microRNAs (miRNAs) have been recognized as an abundant class of endogenous RNAs that are small in size (~22 nt) and posttranscriptionally regulate gene expression. More specifically, their interaction with a target mRNA causes either translational repression or mRNA cleavage with subsequent degradation (Bartel 2004; Mathonnet et al. 2007; Pillai et al. 2007). Instrumental are nucleotides 2–8 of the miRNA, the so-called seed region, and their partial or complete complementarity to their target mRNA. Importantly, however, even in the case of a perfect seed pairing identified *in silico*, experiments are required to confirm actual interactions between the miRNA and the target mRNA and thus biological activity (Didiano and Hobert 2006).

While the expression of most miRNAs is driven by their own promoter, many miRNA genes are located in intron regions, thus leading to the transcription of the miRNA with the corresponding host gene. For example, in tumors, aberrant miRNA expression may be due to the fact that in most cases, fragile genomic regions appear to host miRNA genes, which may well result in amplification, deletion, or translocation (Calin et al. 2004). Beyond that, other processes leading to miRNA silencing have been described, including promoter DNA methylation or loss of histone acetylation/methylation (Saito et al. 2006), or oncogenes that affect miRNA processing or miRNA expression.

In various pathologies, miRNAs have been identified that are aberrantly expressed. For example, in cancer, miRNA profiles have been extensively analyzed in the last years, leading to in-depth insight into miRNA signatures of various tumor entities (for review, see e.g., Calin and Croce 2006; Shenouda and Alahari 2009; Cho 2010a, 2010b). This provides the basis for the possible use of miRNAs as tumor markers, tumor subclassifiers, prognostic markers, predictors of therapeutic outcome, and, notably, therapeutic targets or therapeutic compounds. Likewise, the aberrant expression and possible therapeutic potential of miRNAs in many other pathologies have been explored, for example, in the nervous system (Hutchison et al. 2009). Regarding miRNA functions as markers and predictors, the reader is referred to other chapters of this book or other excellent reviews. Instead, this chapter will focus on the next step, the therapeutic exploration of miRNAs.

While at the first glance, the miRNA mechanism seems to resemble RNA interference based on small interfering RNAs (siRNAs), major differences should be noted. Some of those refer to miRNA biogenesis, which relies on the transcription of miRNA genes into so-called pri-miRNAs, that is, primary transcripts of variable length (usually 1–3 kb) (Lee et al. 2004; Rodriguez et al. 2004) (Figure 39.1). Further processing in the nucleus by ribonucleases Drosha and DGCR8 leads to 70- to 100-nt hairpin intermediates called pre-miRNAs (Lee et al. 2003; Landthaler et al. 2004). Upon export into the cytoplasm

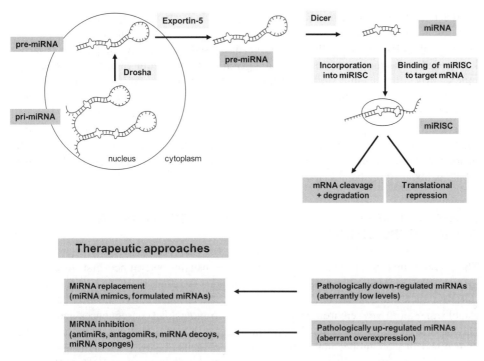

Figure 39.1. General approaches for therapeutic miRNA-based intervention.

mediated by exportin 5 (Bohnsack et al. 2004), they are finally processed by the nuclease Dicer into the mature 18–25 bp miRNA (Hammond et al. 2000). Similar to RNAi, the guide strand is then incorporated into an RNA-induced silencing complex (RISC) that also contains Argonaute and other proteins. The miRNA guide strand is instrumental of RISC target recognition (Hutvagner and Zamore 2002; Bartel 2009). Other differences between miRNA action and siRNA-based RNAi refer to the fact that partial complementarity is sufficient for miRNA action. Consequently, any given miRNA will be able to simultaneously target several target mRNAs.

Indeed, it can be assumed that miRNAs may regulate >100 target genes at the same time. On the other hand, the regulation of any given gene can be mediated by several miRNAs. Thus, any "one drug—one target" approach will not apply to miRNAs, with important implications for therapy. Rather, miRNAs repress thousands of target genes, thus leading to the regulation of up to 60% of the human genes (Friedman et al. 2009). Advantages or disadvantages of this broader action and limited specificity will be discussed later in the text. However, as might be expected, miRNAs are involved in the coordination of a wide variety of cellular, physiological, and pathophysiological processes, and thus represent an interesting, rather novel class of drugs or targets.

Despite the diversity of miRNA action, many miRNAs have been identified as pathologically relevant. Tumor suppressor miRNAs are frequently down-regulated in tumors and inhibit oncogenes. This allows for therapeutic approaches aiming at miRNA replacement (Figure 39.1). In contrast, oncogenic miRNAs are often overexpressed in tumors and are able to down-regulate tumor suppressor genes (Medina and Slack 2008; Cho 2010a; Garofalo and Croce 2011). In this case, therapeutic interventions may be based on any approach to inhibit the given miRNA. These approaches can be extrapolated to many other diseases. It should be mentioned, however, beyond the development of strategies for the delivery of miRNAs or miRNA inhibitors discussed later in the chapter, an in-depth knowledge of the functional relevance of a given miRNA is required to develop a rational miRNA-based therapeutic strategy. Addressing the latter issue, direct target genes have been described for many miRNAs *in vitro*, and tissue culture or *in vivo* studies have identified cellular processes linked to the miRNA-associated regulation of these targets. However, since computational predictive algorithms work rather poorly, and miRNAs work in a tissue-specific manner, the identification of a target gene always requires experimental validation. Notably, since miRNAs are able to regulate several target genes simultaneously, whole pathways may be affected.

II. GENERAL APPROACHES FOR THERAPEUTIC MicroRNA-BASED INTERVENTION

The possible therapeutic potential of a miRNA is usually determined first in cell-based systems. Beyond cell culture, the next step includes the analysis of *in vivo* functions of miRNAs upon implantation of *ex vivo* virally infected or non-virally transfected cells, thus establishing an *in vivo* environment with regard to molecular consequences and affected phenotypes. However, the actual therapeutic use of miRNAs or miRNA inhibitors requires their direct *in vivo* application, with different approaches being available (Figure 39.2). While delivery issues still have to be addressed, the strategy of miRNA replacement therapy essentially relies on the restoration of physiological miRNA levels by treating with molecules (miRNAs) that are already present under normal physiological conditions. Surprisingly, in general, only mild effects of miRNAs on their target genes are observed, even when the forced increase of miRNA levels clearly goes beyond that of physiological conditions. Additionally, side effects generally do not seem to be a problem either (Esquela-Kerscher et al. 2008; Kota et al. 2009; Takeshita et al. 2010; Wiggins et al. 2010; Ibrahim et al. 2011), Side effects, either non-specific ("off-target effects") or specific based on the simultaneous interference with unwanted target genes, although apparently phenotypically neutral, still need to be analyzed in great detail. Other issues that have come to light from the context of therapeutic siRNA strategies, such as sequence-dependent stimulation of the innate immune system (Robbins et al. 2009), should also be considered.

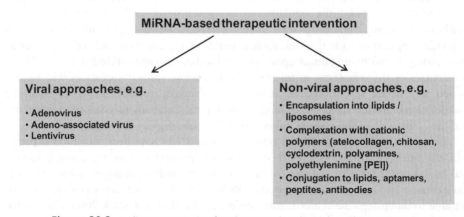

Figure 39.2. Delivery strategies for therapeutic miRNA-based intervention.

A. Viral Delivery

Viral approaches have been explored for miRNA replacement and proved successful for therapeutic intervention (see later in the text). Despite the use of viruses that do not insert into the genome, thus possibly avoiding insertional mutagenesis, other aspects related to viral delivery, such as often poor pharmacokinetics, safety issues will have to be considered, including the induction of toxic immune responses, as well as problems with reproducible large-scale production and loading capacities (see e.g., Aigner 2008; Itaka and Kataoka 2009 and references therein). Due to their inherent ability to transport genetic material into cells, several different viruses have been developed as gene therapy vectors (see e.g., Thomas et al. 2003 for review). The five main classes fall into two groups, according to whether the viral genome is integrated into the host cellular chromatin (lentiviruses and oncoretroviruses) or persists in the cell nucleus as extrachromosomal episomes (adenoviruses, adeno-associated viruses, and herpes viruses). With regard to gene therapy, oncoretroviruses have been the most widely used in clinical trials, although they can only transduce dividing cells since their entry into the nucleus relies on the breakdown of the nuclear membrane. In contrast, lentiviruses can also transduce non-dividing cells due to their ability to penetrate the intact nuclear membrane, and have emerged as important vector systems. Adenoviruses provide an extremely efficient transduction of most tissues, but show potent immunogenicity. More recent improvements have aimed at reducing the immunogenicity. Adeno-associated viruses are non-inflammatory and non-pathogenic.

B. Non-Viral Strategies

In contrast, non-viral systems largely avoid these issues, and in consequence, have been extensively studied for their use in the delivery of DNA or RNA molecules (including miRNAs). Such strategies include the encapsulation of nucleic acids in lipids or liposomes, complexation with cationic polymers or lipids, or covalent conjugation of the nucleic acid to lipids, aptamers, peptides, antibodies, or other binding partners for improved pharmacokinetics and/or cellular uptake (see e.g., Aigner 2008 for review). When the formulation does not provide protection against degradation, RNA molecules are instead often chemically modified (see later in the chapter). Major issues for this class of delivery agents include low efficacies of delivery or non-specific (toxic and immunogenic) effects, again indicating that a careful monitoring of unwanted effects is important. That said, many of these approaches have been tested for miRNA delivery (see later in the chapter) and have shown promising results.

Nanoparticulate systems are usually in the range of ~ 20 to several 100 nm. They protect their nucleic acid "payload," which is particularly important in the case of RNA molecules, mediate the delivery to the target organ, and allow cellular entry by endocytosis. Cationic polymers like atelocollagen, chitosan, cyclodextrin, polyamines, or polyethylenimines, as well as various liposomes or other lipid-based nanoparticles, have been explored (Figure 39.2). Of note, while many systems have been initially developed for DNA delivery and later on adapted for small RNAs, efficacies do not necessarily correlate (Hobel et al. 2011). This is probably due to differences in physicochemical properties, such as the stability of polymeric nanoplexes (Malek et al. 2008) and different requirements regarding the intracellular localization of the nucleic acid (nucleus vs. cytoplasm). Likewise, *in vivo* properties are often poorly reflected by *in vitro* analyses, thus making it difficult to assess the *in vivo* applicability only based on cell culture experiments

(Whitehead et al. 2012). Still, a large set of nanocarrier systems of different chemical compositions, sizes, and surface charge (zeta potential) have been explored, or especially developed, for miRNA therapies and will be discussed in the next section.

C. miRNA Inhibition

For miRNA inhibition, the delivery of miRNA inhibitors, including anti-miRs, antagomirs, miRNA decoys, or miRNA sponges, have been employed. Anti-miRNA oligonucleotides (AMOs) for miRNA inhibition are single-stranded oligonucleotides directly complementary to the miRNA to be inhibited (Garzon et al. 2010). They prevent miRNA activity by competing with the target 3′-UTR mRNA site for miRNA binding. Notably, one oligonucleotide can also contain multiple AMO units, which are thus able to simultaneously silence multiple-target miRNAs or miRNA seed families (Lu et al. 2009; Wang 2011). Chemical modifications have been introduced in order to reduce AMO degradation, increase stability, and avoid cleavage by the RISC nuclease, to improve pharmacokinetics, and sometimes also to mediate delivery. Thus, anti-miRs may contain 2′-O-methyl or 2′-O-methoxyethyl phosphorothioate modifications of the backbone. Furthermore, cholesterol-conjugated, 2′O-methyl-modified single-stranded RNA analogues with phosphorothioate linkages have been described and termed "antagomirs" (Krutzfeldt et al. 2005). Another modification leading to enhanced binding affinity is the inclusion of so-called locked nucleic acids (LNAs) (Vester and Wengel 2004), whereby the ribose moiety is modified with an extra bridge connecting the 4′ carbon and the 2′ oxygen. This locks the ribose in the 3′-endo conformation, which leads to increased hybridization properties.

In contrast to anti-miRs/antagomirs, miRNA sponges use multiple complementary 3′-UTR mRNA sites for a given miRNA to competitively bind the miRNA to be inhibited (Ebert et al. 2007; Ebert and Sharp 2010). Demonstrating the functionality of miRNA sponges, otherwise non-metastatic MCF-7 cells stably expressing a *miR-31* sponge were orthotopically implanted or intravenously injected into mice and showed a significant induction of lung metastases (Valastyan et al. 2009). In contrast, highly metastatic 4T1 mouse mammary tumor cells expressing *miR-9* levels showed markedly decreased lung metastasis in a mouse breast cancer model when *ex vivo* infected with *miR-9* sponges (Ma et al. 2010a). A similar approach is the use of a lentiviral vector that expresses multiple complementary binding sites for the targeted miRNA, thus acting as an "anti-miRNA decoy." The possibility of an efficient inhibition of miRNAs was demonstrated by the functional knockdown of *miR-223*, comparable with the corresponding knockout mice (Johnnidis et al. 2008; Gentner et al. 2009) (Table 39.1).

The following sections provide an overview of studies on miRNA replacement and on miRNA inhibition, thus highlighting the therapeutic potential of different strategies *in vivo*. This covers a wide range of pathologies, approaches for intervention, and strategies for miRNA drug delivery.

III. THERAPEUTIC MicroRNA REPLACEMENT

A. Delivery of *let-7* and *miR-34a*

In the tumor context, the miRNA *let-7* targets several important oncogenes including HMGA2, MYC, NOTCH, p16, and RAS, and is underexpressed in many solid tumors

TABLE 39.1. Overview of Therapeutic Studies Aiming at the *In Vivo* Inhibition of miRNAs

Specific Compound	Targeted miRNA	*In Vivo* Model	Formulation/Mode of Application	Therapeutic Effect	Reference
2′-O-methyl RNA oligonucleotides	*let-7*	*C. elegans* larvae	Direct microinjection	Altered phenotypes consistent with loss of *let-7* activity	Hutvagner et al. (2004)
2′-O-methoxyethyl phosphorothioate ASO	*miR-122*	Mouse liver		Reduced plasma cholesterol levels, increased hepatic fatty acid oxidation, decrease in hepatic fatty acid and cholesterol synthesis rates, increased activation of the central metabolic sensor AMPK	Esau et al. (2006)
2′-O-methoxyethyl phosphorothioate ASO	*miR-122*	Diet-induced obesity mouse model		Reduced plasma cholesterol levels, improvement in liver steatosis, reduction in lipogenic genes	Esau et al. (2006)
Antagomirs	*miR-122*; *miR-16*, *miR-192*,*miR -194*	Mice	Intravenous	Reduction of the corresponding miRNA in various organs	Krutzfeldt et al. (2005)
Antagomir	*miR-122*	Mice	Intravenous	Altered expression, effect on cholesterol biosynthesis	Krutzfeldt et al. (2005)
Antagomir	*miR-10b*	Mice	Intravenous	Suppression of lung metastasis (no reduction of primary tumor)	Ma et al. (2010b)
Antagomir	*miR-21*	Mouse pressureoverload-induced myocardial disease model	Injections by means of an implanted jugular vein catheter	Reduction of cardiac ERK-MAP kinase activity, inhibition of interstitial fibrosis attenuation of cardiac dysfunction	Thum et al. (2008)
Antagomir	*miR-134*	Mice	Intracerebroventricular	Neuroprotective and prolonged seizure-suppressant effects	Jimenez-Mateos et al. (2012)
LNA-anti-miR	*miR-192*	Mouse model of diabetic nephropathy		Decrease of renal fibrosis, attenuated proteinuria in diabetic mice	Putta et al. (2012)

(*Continued*)

TABLE 39.1. (*Continued*)

Specific Compound	Targeted miRNA	*In Vivo* Model	Formulation/Mode of Application	Therapeutic Effect	Reference
LNA-anti-miR	*miR-15*	Mice during ischemia-reperfusion injury	Intravenous	Protection against ischemia-induced cardiac injury	Hullinger et al. (2012)
LNA-anti-miR	*miR-122*	Non-human primates (African green monkeys)	Intravenous	*miR-122* depletion, lowering of plasma cholesterol	Elmen et al. (2008a)
LNA-anti-miR	*miR-122*	Mice	Intravenous	Functional *miR-122* antagonism (low cholesterol phenotype)	Elmen et al. (2008b)
LNA-anti-miR	*miR-122*	Chronically hepatitis C virus-infected chimpanzees	Intravenous	Long-lasting suppression of HCV viremia, improvement of HCV-induced liver pathology	Lanford et al. (2010)
(2′-F/MOE)-modified, phosphorothioate-backbone-modified antisense *miR-33*	*miR-33*	Non-human primates (African green monkeys)	Intraperitoneal	Increased hepatic expression of ABCA1 and other target genes involved in fatty acid oxidation, reduced the expression of genes involved in fatty acid synthesis, sustained increase in plasma HDL levels, suppression of VLDL-associated triglycerides	Rayner et al. (2011)
Chemically modified antimiR	*miR-182*	Mouse model of melanoma liver metastasis	intraperitoneal	Lower burden of liver metastases (Huynh et al. 2011)	Huynh et al. (2011)
let-7 inhibitor cocktail (*let-7a,d,f*) + anti-Fas monoclonal antibody	*let-7*	Subcutaneous colon carcinoma xenografts	Intratumoral	Regression of tumors due to sensitization of cells to Fas-related apoptosis	Geng et al. (2011)

(Henry et al. 2011). Consequently, its therapeutic application has been explored in several studies. In an early report (Esquela-Kerscher et al. 2008) employed an established orthotopic mouse lung cancer model based on the conditional Lox-Stop-Lox K-ras mouse strain, that harbors a conditionally activatable allele of oncogenic K-ras and develops lung adenocarcinomas when treated with a recombinant adenovirus expressing Cre recombinase (Jackson et al. 2001). Intranasal coadministration of Ad-cre (for the induction of tumor formation) and an adenovirus expressing a *let-7a* RNA hairpin (Ad-*let-7*) that yields mature *let-7* miRNA versus Ad-scr (adenovirus expressing a scrambled miRNA as negative control) revealed after 7 weeks a 66% reduction in tumor formation in the lungs of mice treated with *let-7*. The authors' conclusions regarding *let-7* repressing lung tumorgenesis was in line with xenograft studies in H460 or A540 NSCLC cells. Upon transient *ex vivo* transfection and subcutaneous injection, a reduced/delayed growth of tumor xenografts was observed (Esquela-Kerscher et al. 2008). Concomitantly, in the reverse experiment in the same tumor model, Trang et al. (2010) described increased tumor burden with more prominent bronchiolar papillary hyperplasia upon intranasal administration of an anti-*let-7g* anti-miR designed with 2'-O-methyl and phosphorothioate modifications according to Krutzfeldt et al. (2005). Perhaps therapeutically more relevant, in an advanced K-ras-activated non-small-cell lung carcinoma (NSCLC) mouse model, lentiviral *let-7* administration was shown to lead to remission of lung tumors (Trang et al. 2010). The same study also explored the intratumoral injection of the synthetic miRNA *let-7* complexed with the lipid-based transfection reagent siPORTamine into established subcutaneous H460 NSCLC tumor xenografts. Upon repeated injection of 6.25 µg miRNA every 3 days, a robust decrease in tumor growth was observed over 10 days in the *let-7* group as compared with negative controls (Trang et al. 2010). Inhibition of the growth upon intratumor injection of *let-7a* mimics was also shown in subcutaneous colon tumor xenografts. To this end, miRNA mimics were mixed with the transfection reagent Entranster and, upon establishment of tumor xenografts, 5 µg miRNA mimics were daily injected (Wang et al. 2012).

The systemic application of *let-7* in the K-ras autochthonous NSCLC mouse model was explored in another study. When mice were intravenously injected with synthetic *let-7b* miRNA conjugated with a neutral lipid emulsion every second day for a total of eight injections at a concentration of 1 mg/kg, a significantly lower tumor burden was observed as compared to the same treatment with a negative control miRNA. This effect was based on reduced proliferation but not on increased apoptosis (Trang et al. 2011). Similarly, the neutral lipid emulsion (NLE)-mediated delivery of *miR-34a* led to a significant 60% reduction in tumor burden in the same tumor model, based on reduced proliferation as well as on increased apoptosis (Trang et al. 2011). This was consistent with previous findings demonstrating that *miR-34a* is suppressed in various tumor types, including NSCLC (Gallardo et al. 2009; Wiggins et al. 2010). Wiggins et al. also demonstrated that local intratumoral injection of 100 µg *miR-34a* oligonucleotide, formulated in a lipid-based delivery vehicle, into established subcutaneous H460 tumor xenografts at days 12, 15, and 18, prevented the outgrowth of viable tumors. Immunohistochemical analysis revealed that this effect was due to inhibition of proliferation and the induction of apoptosis, leading to large tumor areas filled with cell debris (Wiggins et al. 2010). Furthermore, repeated injection of the same amount of *miR-34a* into the tail vein led to similar antitumor effects in H460 xenografts, while in A549 NSCLC xenografts fivefold lower amounts of formulated *miR-34a* were sufficient to elicit tumor inhibition. Upon systemic injection of formulated *miR-34a*, no elevation of cytokines or liver and kidney enzymes in serum was observed, indicating that the formulation was well tolerated and did not induce an immune response

(Wiggins et al. 2010). Finally, *miR-34a* was explored for systemic delivery into experimental lung metastases of murine B16F10 melanoma. To this end, a LPH (liposome-polycation-hyaluronic acid) nanoparticle formulation, modified with tumor-targeting single-chain antibody fragment (scFv), was generated. *miR-34a* delivery by the GC4-targeted nanoparticles, injected on days 10 and 11 with two consecutive intravenous administrations (0.3 mg miRNA/kg each), led to significant down-regulation of the target gene survivin in the metastatic tumor and reduced tumor load in the lung as compared with negative controls. Notably, the targeting moiety was critical in this experiment since the delivery of *miR-34a* in negative control targeted nanoparticles did not show any effects. Furthermore, while combined siRNAs against c-Myc, MDM2, and VEGF formulated in the same system showed tumor growth/metastasis inhibition as well, the combination of siRNAs and *miR-34a* co-delivered by GC4-targeted nanoparticles could additively inhibit tumor growth and enhance the therapeutic effects in B16F10 lung metastasis model (Chen et al. 2010) (Table 39.2).

B. Non-Viral Delivery of *miR-143*, *miR-145*, and *miR-33a*

For therapeutic intervention in colon carcinoma, the *miR-143/145* family has been explored because both miRNAs frequently show lower expression in cancer and adenoma specimens. Since this down-regulation, however, does not seem to show a correlation to clinical features (Wang et al. 2009; Akao et al. 2010) and low *miR-143/145* levels occur already at the early phase of adenoma formation, it was suggested that *miRs 143* and *145* are not involved in tumor progression but rather in the initiation of tumorigenesis (Akao et al. 2010). However, the therapeutic potential of both miRNAs has been demonstrated in mouse-based therapy studies. When mixed with cationic liposomes and injected intratumorally or intravenously, a synthetic *miR-143* analog containing aromatic benzene–pyridine (BP) moieties, *miR-143BP*, exerted significant tumor-suppressive effect on tumor xenografts (Akao et al. 2010). In this study, the sequence of the passenger strand in the duplex was changed, and the 3′-overhang was chemically modified according to a previous study with siRNAs (Ueno et al. 2008) in order to increase activity and nuclease stability. The *ex vivo* transfection of these *miR-143BPs* into macrophages, which were subsequently intravenously injected into mice, was also explored (Akao et al. 2011). Injection of these macrophages led to secretion of the *miR-143BPs* as microvesicles and effective delivery of the miRNA to the tumor xenografts. This paper thus introduced monocytes/macrophages and their secreted microvesicles as new RNA delivery method (Akao et al. 2011).

Another study explored the *miR-143/145* family as a miRNA replacement therapy without relying on chemical modifications (Ibrahim et al. 2011). For protection and cellular uptake, the *miR-145* duplexes were formulated in polymeric nanoparticles based on a low molecular weight polyethylenimine (PEI F25- LMW) (Werth et al. 2006). Polyethylenimines (PEIs) are linear or branched polymers that are available at various molecular weights. Based on their partial protonation under physiological conditions, electrostatic interactions with nucleic acids will lead to the formation of nanoscale complexes (Boussif et al. 1995). PEI-based complexes ("polyplexes") are internalized via caveolae- or clathrin-dependent routes prior to their release from endosomes due to the so-called proton sponge effect (Behr 1997). Thus, PEIs confer the protection, cellular uptake and intracellular release of nucleic acids. This also applies to small RNA molecules, such as siRNAs or miRNAs, which allows the systemic application of siRNAs *in vivo* (Aigner et al. 2002; Urban-Klein et al. 2005; Grzelinski et al. 2006; Hobel et al. 2010). Dependent on the route

TABLE 39.2. MiRNA Replacement Therapy

miRNA	Chemical Composition/ Formulation	Mode of Application	In Vivo Model	Therapeutic Effect	Reference
miR-34a	Lipid-based delivery vehicle	Local, systemic	NSCLC mouse models	Blockage of tumor growth	Wiggins et al. (2010)
let-7	siPORTamine transfection reagent	Intratumoral	Established NSCLC tumors in mice	Reduction of tumor burden	Trang et al. (2010)
let-7	miRNA precursor expression vector/lentivirus	Intranasal	K-ras-induced tumors	Reduced tumor size	Trang et al. (2010)
let-7	In vivo transfection reagent Entranster	Intratumoral	Subcutaneous CRC DLD1 xenografts	Reduced tumor growth	Wang et al. (2012)
miR-26a	Adeno-associated virus	Systemic	Hepatocellular carcinoma mouse model	Inhibition of cancer cell proliferation and tumor-specific apoptosis, protection from disease	Kota et al. (2009)
let-7, miR-34a	miR mimics/neutral lipid emulsion	Intravenous	Mouse models of lung cancer (Kras-activated autochthonous mouse model of NSCLC)	Decrease in tumor burden	Trang et al. (2011)
miR-29b	miRNA mimics/lipofectamine	Intratumoral	K562 xenografts	Suppression of tumor growth	Garzon et al. (2009)
miR-143	miR-143BPx	Intravenous	Subcutaneous colorectal cancer xenografts	Tumor suppression	Kitade and Akao (2010)
let-7	Adenovirus expressing a let-7a RNA hairpin (Ad-let-7)	Intranasal	Established orthotopic mouse lung cancer model	Reduction of tumor formation in the lungs of animals expressing a G12D activating mutation for the K-ras oncogene	Esquela-Kerscher et al. (2008)
miR-101	Adenovirus	Intratumoral	Subcutaneous gastric tumor xenografts	Tumor growth inhibition	Wang et al. (2010)

(Continued)

TABLE 39.2. (*Continued*)

miRNA	Chemical Composition/ Formulation	Mode of Application	In Vivo Model	Therapeutic Effect	Reference
miR-33a	Polyethylenimine complexation	Intraperitoneal	Subcutaneous colon carcinoma xenografts	Tumor growth inhibition	Ibrahim et al. (2011)
miR-145	Polyethylenimine complexation	Intraperitoneal	Subcutaneous colon carcinoma xenografts	Tumor growth inhibition	Ibrahim et al. (2011)
miR-15a/16	Virus particles	Intratumoral	Subcutaneous prostate carcinoma xenografts	Tumor growth inhibition	Bonci et al. (2008)
miR-15a/16-1	miR-encoding plasmid / DOTAP/cholesterol liposomes	Intravenous	Subcutaneous colon carcinoma xenografts	Inhibition of tumor growth and angiogenesis	Dai et al. (2012)
miR-34a	Liposome-polycation-hyaluronic acid nanoparticles, scFv for tumor targeting	Systemic	Mouse melanoma lung metastasis model	Reduced tumor load in the lung	Chen et al. (2010)
miR-196a	Adeno-associated virus	Skeletal muscle of the left quadriceps femoris	Spinal and bulbar muscular atrophy mouse model	Amelioration of SBMA phenotypes, increased survival	Miyazaki et al. (2012)
miR-16	Atelocollagen	Intravenous	Bone metastasis model	Inhibition of prostate tumor growth in bone	Takeshita et al. (2010)
miR-34a	LPH (liposome-polycation-hyaluronic acid) nanoparticle formulation, modified with tumor-targeting single-chain antibody fragment (scFv)	Intravenous	Lung metastasis mouse model	Reduced metastatic tumor load	Chen et al. (2010)
miR-502	miR-502 precursor / siPORTamine transfection reagent	Intratumoral	Established CRC tumors in mice	Inhibition of tumor growth	Zhai et al. (2012)
Artificial miRs against p21	Adenovirus (cocistronic expression with p53)	Intratumoral	Subcutaneous colon carcinoma xenografts	Tumor growth inhibition	Idogawa et al. (2009)

of administration, different biodistribution profiles are obtained. In addition, chemical modifications of PEI and the coupling of targeting ligands have been explored for a more target-specific uptake of PEI-based complexes into selected organs or tissues.

In subcutaneous LS174T colon carcinoma xenografts in mice, the intratumoral injection of PEI-complexed *miR-145* three times per week resulted in significant reduction of tumor growth (Ibrahim et al. 2011). Likewise, systemic (i.p.) application of low molecular weight PEI/miRNA complexes led to the delivery of intact miRNA molecules into the xenograft tumors where they caused profound antitumor effects. The delivery of *miR-145* increased apoptosis and decreased tumor proliferation, with concomitant repression of c-Myc and ERK5 as novel regulatory targets of *miR-145*. Notably, in contrast to Akao et al. (2011), prior *in vitro* analyses revealed no effect of *miR-143* replacement in colon carcinoma cells.

Using the same PEI delivery system, *miR-33a* was explored as well (Ibrahim et al. 2011). Two independent studies had reported that *miR-33a/b* posttranscriptionally repress the ATP-binding cassette transporter A1 (ABCA1), an important positive regulator of high-density lipoprotein (HDL) synthesis and reverse cholesterol transport, and *miR-33* thus contributes to the regulation of cholesterol homeostasis in liver cells (Najafi-Shoushtari et al. 2010; Rayner et al. 2010). In this context, antagonists of endogenous *miR-33* were suggested as a potential therapeutic strategy to mitigate cardiometabolic diseases. However, more recently, *miR-33a* was shown to directly regulate the Pim-1 kinase (Thomas et al. 2012). Pim-1 belongs to a family of constitutively active serine/threonine kinases, is overexpressed in various tumors, and has been linked to poor prognosis. Its role as a proto-oncogene is based on several Pim-1 target proteins involved in apoptosis, cell cycle regulation, signal transduction, and transcriptional regulation, as well as on Pim-1 acting synergistically with the oncogenic transcription factor c-myc on different levels. Most recent studies show the functional relevance of Pim-1 in colon carcinoma (Weirauch et al. 2013). Taken together, this implicated that *miR-33a/b* may play a role in tumor progression. Indeed, the systemic injection of PEI-complexed *miR-33a* led to antitumor effects comparable with the direct RNAi-mediated knockdown of Pim-1 by PEI/siRNA complexes (Ibrahim et al. 2011), thus introducing *miR-33a* replacement as a novel therapeutic targeting method for Pim-1. It should also be noted that this study directly compared siRNA-mediated Pim-1 knockdown through RNAi with therapeutic intervention based on a Pim-1-specific miRNA, and led to similar results despite miRNA effects being generally considered as milder. Also, *miR-33a* levels in human colon samples (matching pairs of normal and tumor tissue) were found to be very similar between normal and tumorous tissue. This also suggests that miRNA expression levels may not always be tightly correlated with their (patho-) physiological functions.

C. Viral Delivery of miRNAs

Virus-mediated miRNA delivery has been explored in multiple subcutaneous xenograft studies. For example, Kota et al. demonstrated that in hepatocellular carcinoma (HCC), using adeno-associated virus (AAV), the systemic administration of *miR-26a* led to profound antitumor effects (Kota et al. 2009). More specifically, a self-complementary scAAV vector system was constructed to evaluate the therapeutic potential of *miR-26a* in an inducible liver cancer mouse model. The AAV were intravenously injected into the tail veins of the mice (1×10^{12} vector genomes [vg] per animal) with a single injection, at 11 weeks of age, a time point at which animals typically have multiple small- to medium-sized tumors. The assessment of the tumor burden 3 weeks after treatment revealed protection from fulminant disease in 8 of 10 mice of the treatment group, with only small tumors or a complete absence of tumors upon gross inspection. Treatment efficacy correlated with viral transduction rates as determined by the analysis of the expression of the vector-encoded reporter gene EGFP. Further analyses also revealed that antitumor effects were based on an inhibition of cancer cell proliferation and induction of tumor-specific apoptosis (Kota et al. 2009).

The intratumoral injection of miRNA expressing virus particles was performed in two studies. In subcutaneous MKN45 tumor xenografts, the effects of *miR-101* on gastric tumor progression was analysed (Wang et al. 2010). Adenovirus particles (1×10^9 PFU/100μL) were injected intratumorally five times within 2 weeks, and tumor growth inhibition was observed in the treatment group as compared with the negative control virus group. Upon harvesting of the tumors, a moderate increase in *miR-101* levels was confirmed (Wang et al. 2010). Bonci et al. studied miRNAs *miR-15a* and *miR-16-1*, which have been shown to target bcl2, CCND1, and WNT3A, and to be significantly decreased in cancer cells of advanced prostate tumors (Bonci et al. 2008). The reconstitution of the expression of both miRNAs in LNCaP-derived prostate tumor xenografts was performed by a single injection of lentivirus particles containing a lentiviral *miR-15-16* vector. Within 1 week of treatment, *miR-15-16*-treated tumors underwent growth arrest and considerable volume regression thereafter, while no changes were observed in the empty-vector virus negative controls. In the treatment group, the histological analysis of the residual masses indicated the presence of diffuse necrosis with rare areas containing surviving cells. In combination with *in vitro* data, this established *miR-15a* and *miR-16* to act as tumor suppressor genes in prostate cancer through the control of cell survival, proliferation, and invasion (Bonci et al. 2008).

D. Other Examples of Viral and Non-Viral Delivery

A non-viral *miR-15a/16-1* replacement based on liposomes was described by Dai et al. (Dai et al. 2012). Upon intravenous injection of a *miR-15a/16-1*-encoding plasmid formulated in 1,2-dioleoyl-3-trimethylammoniumpropane (DOTAP)/cholesterol liposomes at amounts of 10μg DNA every 2 days for 3 weeks, a significant inhibition of subcutaneous tumor growth and angiogenesis in subcutaneous colon carcinoma tumor xenografts was observed (Dai et al. 2012). Another non-viral approach for *miR-16* replacement was published by Takeshita et al. using atelocollagen (Takeshita et al. 2010). Atelocollagen is generated through pepsin treatment of type I collagen of calf dermis (Ochiya et al. 1999, 2001; Sano et al. 2003), which removes the so-called telopeptides, that is, immunogenic N- and C-terminal ends, thus reducing immunogenicity. The complexation with atelocollagen protects RNA molecules and has been shown to allow the *in vivo* delivery, for example, of siRNAs. Upon three intravenous (tail vein) injections of 50μg *miR-16* complexed with atelocollagen into tumor-bearing mice every third day, an inhibition of bone-metastatic human prostate tumor growth in the mouse bone site was observed (Takeshita et al. 2010).

So far, even when chemically modified, we have discussed the use of miRNAs that are based on the naturally occurring sequences. In contrast, Idogawa et al. chose an approach using artificial miRNAs, that is, designed miRNAs with non-natural sequences. They constructed a recombinant adenovirus that enabled the cocistronic expression of p53 and of artificial miRNAs that target p21, and analyzed antitumor effects upon intratumoral injection into established subcutaneous tumor xenografts from colorectal carcinoma cells. Whereas the expression of p21-specific miRNAs alone resulted even in an increase in the tumor volume of SW480-derived tumors, tumor inhibition was observed upon adenovirus-mediated expression of p21-specific miRNAs together with p53 (Idogawa et al. 2009).

Non-viral, intratumoral delivery of miRNA mimics using the transfection reagent lipofectamine has been described for *miR-29b* in a xenograft leukemia model (Garzon et al. 2009). *miR-29b* expression has been shown to be deregulated in many cancers, including in primary acute myeloid leukemic (AML) blasts. For the assessment of effects on tumorigenicity, the AML K562 cell line was inoculated subcutaneously in both flanks of immunocompromised nude mice, prior to treatment of the mice with 5 μg *miR-29b* or scrambled negative control oligonucleotides, diluted in lipofectamine, four times over 10 days. Profound antitumor effects were observed, with two tumors going into complete remission (Garzon et al. 2009). In another study, a *miR-502* precursor similar to Trang et al. (2010) was formulated with siPORTAmine. When 6.25 μg were injected every 3 days into established HCT-116 tumor xenografts for a total of three times, inhibition of tumor growth was detected (Zhai et al. 2012).

Beyond a direct contribution to the process of tumorigenesis, aberrant miRNA expression can also be functionally relevant in tumors by influencing the sensitivity of tumor cells toward chemo- or radiotherapy. For example, in colon carcinoma *miR-21*, overexpression significantly reduced 5-FU-induced G2/M damage arrest and apoptosis (Valeri et al. 2010), and *let-7* inhibitors sensitized cells toward Fas-related apoptosis (Geng et al. 2011). Thus, miRNA inhibition or miRNA replacement therapy may be a putatively promising strategy when combined with established chemotherapy.

Finally, miRNA replacement has also been studied in the context of neurodegenerative diseases. Miyazaki et al. employed the adeno-associated virus (AAV) vector-mediated delivery of *miR-196a* for early intervention in a spinal and bulbar muscular atrophy (SBMA) mouse model. Upon injection of 10^{11} vg of the construct into the skeletal muscle of the left quadriceps femoris of AR-97Q mice, an amelioration of SBMA phenotypes and increased survival was observed (Miyazaki et al. 2012).

IV. THERAPEUTIC STUDIES IN MicroRNA INHIBITION

In addition to the down-regulation of miRNAs described earlier, many miRNAs are overexpressed under pathological conditions, and consequently, miRNA inhibition has also been extensively studied. As outlined earlier, several strategies are available, which will be described here in more detail (Table 39.1).

In an early proof-of-principle study in *Caenorhabditis elegans*, Hutvagner et al. demonstrated that the injection of of a 2′-O-methyl oligonucleotide complementary to *let-7* was able to induce a *let-7* loss-of-function phenocopy (Hutvagner et al. 2004). More specifically, using low flow and low pressure, the oligonucleotide solution was injected into the body cavity of the larvae and, after the injected animals reached adulthood, altered phenotypes consistent with loss of *let-7* activity were observed.

More extensively modified anti-miRs, 2′-O-methoxyethyl phosphorothioate antisense oligonucleotides (ASO) directed against *miR-122* have been employed as well (Esau et al. 2006). In normal mice, i.p. injection of 12.5–75 mg/kg *miR-122* ASO, dissolved in saline, twice weekly for 4 weeks resulted in reduced total cholesterol and triglyceride levels in the plasma of *miR-122* ASO-treated mice at all doses tested, while glucose levels remained unaltered. These effects were based on reduced plasma cholesterol levels and increased hepatic fatty-acid oxidation, while hepatic fatty-acid and cholesterol synthesis rates were decreased. In a diet-induced obesity mouse model (high-fat diet fed for 19 weeks), the lowest dose (12.5 mg/kg) already decreased plasma cholesterol levels due to *miR-122* inhibition and led to a significant improvement in liver steatosis, accompanied by reductions in several lipogenic genes. This introduced *miR-122* as an attractive therapeutic target for metabolic disease, based on its role as a key regulator of cholesterol and fatty-acid metabolism in the adult liver (Esau et al. 2006). Similar findings had already been obtained before by Krutzfeldt et al., who instead utilized antagomirs (chemically modified, cholesterol-conjugated single-stranded RNA analogues complementary to miRNAs; see earlier in the chapter). Intravenous injection led to silencing of *miR-122* and to reduced plasma cholesterol levels by affecting cholesterol biosynthesis. Antagomirs against other miRNAs, *miR-16, miR-192*, and *miR-194,* confirmed the specific reduction of the targeted miRNA in various organs, thus demonstrating antagomir potency in miRNA silencing for future therapeutic use (Krutzfeldt et al. 2005).

Indeed, in a later study, antagomirs were explored for *miR-21* inhibition in a mouse pressure overload-induced myocardial disease model (Thum et al. 2008). When mice subjected to pressure overload of the left ventricle by transverse aortic constriction (TAC) were treated with antagomir-21 for three consecutive days, cardiac *miR-21* expression was repressed for up to three weeks, indicating a longer-lasting effect. Injections of antagomir-21 (80 mg per kg body weight) by means of an implanted jugular vein catheter led to reduced cardiac ERK-MAP kinase activity, and inhibition of interstitial fibrosis attenuation of cardiac dysfunction. The significant attenuation of the impairment of cardiac function, as well as regression of cardiac hypertrophy and fibrosis indicated the therapeutic efficacy for silencing of *miR-21* by antagomirs in a cardiovascular disease setting. Notably, this also validated an miRNA, *miR-21,* which as a tumor-promoting miRNA has been associated with tumor progression and metastasis (see e.g., Aigner 2011 and references therein) and is often upregulated in tumors, as a disease target in heart failure.

Antagomirs have also been explored in the field of neurodegeneration. For example, upon intracerebroventricular injection of antogomirs against *miR-134,* neuroprotective and prolonged seizure-suppressant effects were observed (Jimenez-Mateos et al. 2012).

Using LNA-modified anti-miRs, the earlier-mentioned *miR-122* inhibition has been further explored. Elmen et al. described a systemically administered 16-nt, unconjugated LNA-anti-miR oligonucleotide complementary to the 5′ end of *miR-122* for specific, dose-dependent silencing of *miR-122*. In line with previous studies, a low cholesterol phenotype was observed (Elmen et al. 2008b). In a parallel study, the same group demonstrated the simple systemic delivery of an unconjugated, PBS-formulated LNA-modified anti-miR as sufficient to antagonize the liver-expressed *miR-122* in non-human primates. More specifically, intravenous injections of 3 or 10 mg/kg LNA-anti-miR to African green monkeys led to the depletion of mature *miR-122* and dose-dependent lowering of plasma cholesterol (Elmen et al. 2008a).

Finally, a LNA-modified anti-miR was used for *miR-122* inhibition in chronically hepatitis C virus-infected chimpanzees (Lanford et al. 2010). Treatment by a weekly intravenous injection of 5 mg/kg locked nucleic acid (LNA)-modified oligonucleotide (SPC3649) complementary to *miR-122* for 12 weeks led a to long-lasting suppression of HCV viremia, detected 3 weeks after onset of the therapy and improvement of HCV-induced liver pathology. Not surprisingly, taking into account previous studies (see earlier in the chapter), antiviral effects were accompanied by markedly lowered serum cholesterol levels. No rebound in viremia during the 12-week treatment and no adaptive mutations in the two *miR-122* seed sites of HCV were observed.

In an oncological setting, chemically modified anti-miRs targeting *miR-182* were employed in a mouse model of melanoma liver metastasis (Huynh et al. 2011). *MiR-182* has been described previously as prometastatic and frequently overexpressed in melanoma (Segura et al. 2009). Treatment of mice that had received intrasplenic injections of A375 melanoma cells with anti-*miR-182* oligonucleotides synthesized with 2′ sugar modifications and a phosphorothioate backbone (25 mg/kg, administered by i.p. injection twice weekly) led to a lower burden of liver metastases compared with negative control treatment mice.

In the context of heart disease, anti-miR-mediated inhibition of *miR-15b* was explored in mice during ischemia-reperfusion injury. Upon intravenous injection of anti-miRs at the onset of reperfusion after 75 minutes of ischemia, a decrease in infarct size was observed. This indicated that LNA-anti-miRs modulating cardiac miRNAs may interfere in cardiac remodeling and protect against cardiac ischemic injury (Hullinger et al. 2012).

MiR-33 has been demonstrated to be involved in cholesterol homeostasis. 2'-fluoro/methoxyethyl (2'-F/MOE)-modified, phosphorothioate-backbone-modified antisense *miR33* inhibited *miR-33a* and *miR-33b* with equal efficacy (Rayner et al. 2011). Subcutaneous injection of 5 mg/kg twice weekly (first 2 weeks) and then weekly (remainder of the study) increased hepatic expression of ABCA1 and other target genes involved in fatty acid oxidation, reduced the expression of genes involved in fatty acid synthesis and induced a sustained increase in plasma high-density lipoprotein (HDL) levels. Furthermore, the suppression of the plasma levels of very-low-density lipoprotein (VLDL)-associated triglycerides was observed. Thus, it is concluded that the pharmacological inhibition of *miR-33a* and *miR-33b* may represent a promising therapeutic strategy to lower VLDL triglyceride and to raise HDL levels for the treatment of dyslipidaemias that increase cardiovascular disease risk (Rayner et al. 2011).

Therapeutic effects of antagomirs were also observed with regard to the inhibition of metastasis in a mouse metastasis model (Ma et al. 2010b). When treating mice twice-weekly with intravenous doses of 50 mg/kg antagomir for 3 weeks, starting 2 days after tumor cell implantation in order to block the early steps of the metastatic process, no reduction in primary mammary tumor growth was observed. In contrast, however, the formation of lung metastases was markedly suppressed (Ma et al. 2010b). This approach may well be tested for other miRNAs that have been identified as relevant in metastasis (see Aigner 2011 for review), apart from effects on primary tumors.

As in miRNA replacement, miRNA inhibition may be combined with other therapeutic interventions. Interestingly, this was demonstrated for *let-7*, which has also been used a candidate for miRNA replacement: when a *let-7* inhibitor cocktail comprising *let-7a*, *let-7d*, and *let-7f* was combined with an anti-Fas monoclonal antibody, a regression of subcutaneous colon carcinoma xenografts upon intratumoral injection was observed. This effect was explained by the sensitization of the tumor cells to Fas-related apoptosis (Geng et al. 2011).

V. PERSPECTIVES, ISSUES, AND FUTURE DIRECTIONS

In the past years, miRNAs have emerged as a novel class of molecules involved in a surprisingly large number of physiological and pathophysiological processes. Strikingly, any given miRNA may well regulate >100 target genes at the same time. While limited specificities are generally considered as disadvantage in drug development, certain pathologies, such as cancer, may well be perceived as a "pathway disease" rather than the result of a single molecular event, and miRNA-based therapies may lead to the reprogramming of whole molecular pathways (Check Hayden 2008; Bader et al. 2010). Still, caution should be used, considering that miRNA replacement therapy may result in artificially high levels of a given miRNA, which could then also bind to seed regions that are not a target under physiological conditions. Thus, it needs to be defined for any miRNA if correct dosing is an issue or if the therapeutic window is rather broad. In contrast, miRNA inhibitors will require rather high concentrations at their site of action for complete miRNA inhibition. Here, it is more important to define which degree of miRNA inhibition is really necessary for the desired therapeutic effect. So far, however, studies indicate that miRNA overdosing in miRNA replacement therapy appears to be not so critical with regard to unwanted side effects, and, on the other side, that submaximal inhibition of overexpressed miRNAs is sufficient for a therapeutic benefit.

In this context, it should be noted that a given target gene may well be regulated by many miRNAs simultaneously, and miRNA effects on a given target can in general be expected to be rather mild. Thus, one would expect that more than one miRNA is needed for any successful therapeutic intervention, which, however, is not supported by the literature. Thus, while neither the "one drug—one target" nor the "one target—one drug" approach applies to miRNAs, several studies demonstrate therapeutic efficacies of miRNA-based therapies. Dependent on the pathology and the miRNA selected, however, this may well involve, or even rely on, the combination with other therapeutic strategies, as outlined earlier in some examples. Thus, the detailed functional analysis of any given miRNA in its pathological setting will always have to be the first step in its therapeutic exploration. In this context, it should also be noted that certain miRNAs appear to be relevant in several pathologies. For example, *miR-21* has been shown to be involved in myocardial disease (Thum et al. 2008), to be tumorigenic (Aigner 2011), and to promote fibrosis of the kidney (Chau et al. 2012). Thus, the development of strategies for therapeutic intervention (in this case: inhibition) may be putatively beneficial in various respects. In other cases, however, the functional roles of a given miRNA in different diseases may suggest opposite approaches: while *miR-33a* has been shown to be involved in cholesterol homeostasis and regulation of VLDL/HDL levels, with its pharmacological inhibition being a putative therapeutic strategy in the treatment of dyslipidemias that increase cardiovascular disease risk (Rayner et al. 2011), its forced up-regulation by miRNA replacement therapy exerts antitumor effects (Ibrahim et al. 2011).

Despite promising data on the effects of single miRNAs, combination therapy approaches may also include the combined use of several miRNAs or miRNA + siRNA formulations for enhanced therapeutic effects. This is particularly straightforward when using nanoparticles, which can easily harbor more than one class of RNA molecule as a payload. Their targeted delivery by means of ligands may further improve pharmacokinetics with regard to accumulation in the target organ. Just one example for both of the strategies is the use of targeted nanoparticles containing *miR-34a* and siRNAs in a melanoma lung metastasis model (Chen et al. 2010).

Still, miRNA-based therapies based on their reconstitution or repression meet the same issues as seen before in antisense, ribozyme, and RNAi technologies. Each time, an initial hype was followed by disillusion. On the other hand, miRNA therapies may well benefit from previous experiences and developments, especially with regard to the delivery strategies and the analysis of unwanted side effects. It should also be noted that considerable progress has been made regarding chemical modifications of RNA molecules (see, e.g., Lennox and Behlke 2011 for review). While siRNAs, in the ideal situation, mediate the efficient knockdown of a single target gene, miRNAs show milder effects on several targets, thus rather leading to the reprogramming of molecular pathways (Check Hayden 2008; Bader et al. 2010). Thus, despite similarities in the molecular structure of the effector molecule, both mechanisms are different. Probably due to the shorter timeline, since the first studies on the therapeutic exploration of miRNAs have been published, miRNA therapies have not been explored extensively. In fact, only in 2010, Santaris Pharma were the first to announce the advancement of Miravirsen (SPC3649), a miRNA targeted drug for treatment of hepatitis C virus (HCV) infection, into Phase 2 studies.

Notably, miRNA therapies will require no less effort than any other drug development with regard to pharmacodynamics (identifying of optimal targets and testing efficacies) and the analysis of unwanted side effects (off-target effects, unwanted specific effects, immune stimulation). From a pharmacokinetic viewpoint, they are even more challenging compared with most low molecular weight drugs. On the other hand, the therapeutic exploration of a class of regulators clearly and causally involved in so many physiological and pathological processes brings us a huge step closer to individualized, tailor-made therapeutic strategies.

ACKNOWLEDGMENTS

Original work from the Aigner Lab on miRNA replacement therapy was supported by grants from the Deutsche Forschungsgemeinschaft (German Research Foundation; Forschergruppe "Nanohale" AI 24/6-1 and DFG single grant AI 24/9-1), the Deutsche Krebshilfe (German Cancer Aid, Grants 106992 and 109260), and the Youssef Jameel Foundation. The authors apologize to colleagues whose excellent work has not been cited due to space or time restrictions.

REFERENCES

Aigner, A. 2008. Cellular delivery in vivo of siRNA-based therapeutics. *Curr Pharm Des* **14**:3603–3619.

Aigner, A. 2011. MicroRNAs (miRNAs) in cancer invasion and metastasis: therapeutic approaches based on metastasis-related miRNAs. *J Mol Med* **89**:445–457.

Aigner, A., D. Fischer, T. Merdan, C. Brus, T. Kissel, and F. Czubayko. 2002. Delivery of unmodified bioactive ribozymes by an RNA-stabilizing polyethylenimine (LMW-PEI) efficiently down-regulates gene expression. *Gene Ther* **9**:1700–1707.

Akao, Y., A. Iio, T. Itoh, S. Noguchi, Y. Itoh, Y. Ohtsuki, and T. Naoe. 2011. Microvesicle-mediated RNA molecule delivery system using monocytes/macrophages. *Mol Ther* **19**:395–399.

Akao, Y., Y. Nakagawa, I. Hirata, A. Iio, T. Itoh, K. Kojima, R. Nakashima, Y. Kitade, and T. Naoe. 2010. Role of anti-oncomirs *miR-143* and -145 in human colorectal tumors. *Cancer Gene Ther* **17**:398–408.

Bader, A.G., D. Brown, and M. Winkler. 2010. The promise of microRNA replacement therapy. *Cancer Res* **70**:7027–7030.

Bartel, D.P. 2004. MicroRNAs: genomics, biogenesis, mechanism, and function. *Cell* **116**:281–297.

Bartel, D.P. 2009. MicroRNAs: target recognition and regulatory functions. *Cell* **136**:215–233.

Behr, J.P. 1997. The proton sponge: a trick to enter cells the viruses did not exploit. *Chimia (Aarau)* **51**:34–36.

Bohnsack, M.T., K. Czaplinski, and D. Gorlich. 2004. Exportin 5 is a RanGTP-dependent dsRNA-binding protein that mediates nuclear export of pre-miRNAs. *RNA* **10**:185–191.

Bonci, D., V. Coppola, M. Musumeci, A. Addario, R. Giuffrida, L. Memeo, L. D'Urso, A. Pagliuca, M. Biffoni, C. Labbaye, M. Bartucci, G. Muto, et al. 2008. The *miR-15a-miR-16-1* cluster controls prostate cancer by targeting multiple oncogenic activities. *Nat Med* **14**:1271–1277.

Boussif, O., F. Lezoualc'h, M.A. Zanta, M.D. Mergny, D. Scherman, B. Demeneix, and J.P. Behr. 1995. A versatile vector for gene and oligonucleotide transfer into cells in culture and in vivo: polyethylenimine. *Proc Natl Acad Sci U S A* **92**:7297–7301.

Calin, G.A. and C.M. Croce. 2006. MicroRNA signatures in human cancers. *Nat Rev Cancer* **6**:857–866.

Calin, G.A., C. Sevignani, C.D. Dumitru, T. Hyslop, E. Noch, S. Yendamuri, M. Shimizu, S. Rattan, F. Bullrich, M. Negrini, and C.M. Croce. 2004. Human microRNA genes are frequently located at fragile sites and genomic regions involved in cancers. *Proc Natl Acad Sci U S A* **101**:2999–3004.

Chau, B.N., C. Xin, J. Hartner, S. Ren, A.P. Castano, G. Linn, J. Li, P.T. Tran, V. Kaimal, X. Huang, A.N. Chang, S. Li, et al. 2012. MicroRNA-21 promotes fibrosis of the kidney by silencing metabolic pathways. *Sci Transl Med* **4**:121ra118.

Check Hayden, E. 2008. Cancer complexity slows quest for cure. *Nature* **455**:148.

Chen, Y., X. Zhu, X. Zhang, B. Liu, and L. Huang. 2010. Nanoparticles modified with tumor-targeting scFv deliver siRNA and miRNA for cancer therapy. *Mol Ther* **18**:1650–1656.

Cho, W.C. 2010a. MicroRNAs in cancer—from research to therapy. *Biochim Biophys Acta* **1805**:209–217.

Cho, W.C. 2010b. MicroRNAs: potential biomarkers for cancer diagnosis, prognosis and targets for therapy. *Int J Biochem Cell Biol* **42**:1273–1281.

Dai, L., W. Wang, S. Zhang, Q. Jiang, R. Wang, L. Dai, L. Cheng, Y. Yang, Y.Q. Wei, and H.X. Deng. 2012. Vector-based *miR-15a/16-1* plasmid inhibits colon cancer growth in vivo. *Cell Biol Int* **36**:765–770.

Didiano, D. and O. Hobert. 2006. Perfect seed pairing is not a generally reliable predictor for miRNA-target interactions. *Nat Struct Mol Biol* **13**:849–851.

Ebert, M.S., J.R. Neilson, and P.A. Sharp. 2007. MicroRNA sponges: competitive inhibitors of small RNAs in mammalian cells. *Nat Methods* **4**:721–726.

Ebert, M.S. and P.A. Sharp. 2010. MicroRNA sponges: progress and possibilities. *RNA* **16**:2043–2050.

Elmen, J., M. Lindow, S. Schutz, M. Lawrence, A. Petri, S. Obad, M. Lindholm, M. Hedtjarn, H.F. Hansen, U. Berger, S. Gullans, P. Kearney, et al. 2008a. LNA-mediated microRNA silencing in non-human primates. *Nature* **452**:896–899.

Elmen, J., M. Lindow, A. Silahtaroglu, M. Bak, M. Christensen, A. Lind-Thomsen, M. Hedtjarn, J.B. Hansen, H.F. Hansen, E.M. Straarup, K. McCullagh, P. Kearney, et al. 2008b. Antagonism of microRNA-122 in mice by systemically administered LNA-antimiR leads to up-regulation of a large set of predicted target mRNAs in the liver. *Nucleic Acids Res* **36**:1153–1162.

Esau, C., S. Davis, S.F. Murray, X.X. Yu, S.K. Pandey, M. Pear, L. Watts, S.L. Booten, M. Graham, R. McKay, A. Subramaniam, S. Propp, et al. 2006. *miR-122* regulation of lipid metabolism revealed by in vivo antisense targeting. *Cell Metab* **3**:87–98.

Esquela-Kerscher, A., P. Trang, J.F. Wiggins, L. Patrawala, A. Cheng, L. Ford, J.B. Weidhaas, D. Brown, A.G. Bader, and F.J. Slack. 2008. The *let-7* microRNA reduces tumor growth in mouse models of lung cancer. *Cell Cycle* **7**:759–764.

Friedman, R.C., K.K. Farh, C.B. Burge, and D.P. Bartel. 2009. Most mammalian mRNAs are conserved targets of microRNAs. *Genome Res* **19**:92–105.

Gallardo, E., A. Navarro, N. Vinolas, R.M. Marrades, T. Diaz, B. Gel, A. Quera, E. Bandres, J. Garcia-Foncillas, J. Ramirez, and M. Monzo. 2009. *miR-34a* as a prognostic marker of relapse in surgically resected non-small-cell lung cancer. *Carcinogenesis* **30**:1903–1909.

Garofalo, M. and C.M. Croce. 2011. microRNAs: master regulators as potential therapeutics in cancer. *Annu Rev Pharmacol Toxicol* **51**:25–43.

Garzon, R., C.E. Heaphy, V. Havelange, M. Fabbri, S. Volinia, T. Tsao, N. Zanesi, S.M. Kornblau, G. Marcucci, G.A. Calin, M. Andreeff, and C.M. Croce. 2009. MicroRNA 29b functions in acute myeloid leukemia. *Blood* **114**:5331–5341.

Garzon, R., G. Marcucci, and C.M. Croce. 2010. Targeting microRNAs in cancer: rationale, strategies and challenges. *Nat Rev Drug Discov* **9**:775–789.

Geng, L., B. Zhu, B.H. Dai, C.J. Sui, F. Xu, T. Kan, W.F. Shen, and J.M. Yang. 2011. A *let-7/Fas* double-negative feedback loop regulates human colon carcinoma cells sensitivity to Fas-related apoptosis. *Biochem Biophys Res Commun* **408**:494–499.

Gentner, B., G. Schira, A. Giustacchini, M. Amendola, B.D. Brown, M. Ponzoni, and L. Naldini. 2009. Stable knockdown of microRNA in vivo by lentiviral vectors. *Nat Methods* **6**:63–66.

Grzelinski, M., B. Urban-Klein, T. Martens, K. Lamszus, U. Bakowsky, S. Hobel, F. Czubayko, and A. Aigner. 2006. RNA interference-mediated gene silencing of pleiotrophin through polyethylenimine-complexed small interfering RNAs in vivo exerts antitumoral effects in glioblastoma xenografts. *Hum Gene Ther* **17**:751–766.

Hammond, S.M., E. Bernstein, D. Beach, and G.J. Hannon. 2000. An RNA-directed nuclease mediates post-transcriptional gene silencing in Drosophila cells. *Nature* **404**:293–296.

Henry, J.C., A.C. Azevedo-Pouly, and T.D. Schmittgen. 2011. MicroRNA replacement therapy for cancer. *Pharm Res* **28**:3030–3042.

Hobel, S., I. Koburger, M. John, F. Czubayko, P. Hadwiger, H.P. Vornlocher, and A. Aigner. 2010. Polyethylenimine/small interfering RNA-mediated knockdown of vascular endothelial growth factor in vivo exerts anti-tumor effects synergistically with Bevacizumab. *J Gene Med* **12**:287–300.

Hobel, S., A. Loos, D. Appelhans, S. Schwarz, J. Seidel, B. Voit, and A. Aigner. 2011. Maltose- and maltotriose-modified, hyperbranched poly(ethylene imine)s (OM-PEIs): Physicochemical and biological properties of DNA and siRNA complexes. *J Control Release* **149**:146–158.

Hullinger, T.G., R.L. Montgomery, A.G. Seto, B.A. Dickinson, H.M. Semus, J.M. Lynch, C.M. Dalby, K. Robinson, C. Stack, P.A. Latimer, J.M. Hare, E.N. Olson, et al. 2012. Inhibition of *miR-15* protects against cardiac ischemic injury. *Circ Res* **110**:71–81.

Hutchison, E.R., E. Okun, and M.P. Mattson. 2009. The therapeutic potential of microRNAs in nervous system damage, degeneration, and repair. *Neuromolecular Med* **11**:153–161.

Hutvagner, G., M.J. Simard, C.C. Mello, and P.D. Zamore. 2004. Sequence-specific inhibition of small RNA function. *PLoS Biol* **2**:E98.

Hutvagner, G. and P.D. Zamore. 2002. A microRNA in a multiple-turnover RNAi enzyme complex. *Science* **297**:2056–2060.

Huynh, C., M.F. Segura, A. Gaziel-Sovran, S. Menendez, F. Darvishian, L. Chiriboga, B. Levin, D. Meruelo, I. Osman, J. Zavadil, E.G. Marcusson, and E. Hernando. 2011. Efficient in vivo microRNA targeting of liver metastasis. *Oncogene* **30**:1481–1488.

Ibrahim, A.F., U. Weirauch, M. Thomas, A. Grunweller, R.K. Hartmann, and A. Aigner. 2011. MicroRNA replacement therapy for *miR-145* and *miR-33a* is efficacious in a model of colon carcinoma. *Cancer Res* **71**:5214–5224.

Idogawa, M., Y. Sasaki, H. Suzuki, H. Mita, K. Imai, Y. Shinomura, and T. Tokino. 2009. A single recombinant adenovirus expressing p53 and p21-targeting artificial microRNAs efficiently induces apoptosis in human cancer cells. *Clin Cancer Res* **15**:3725–3732.

Itaka, K. and K. Kataoka. 2009. Recent development of nonviral gene delivery systems with virus-like structures and mechanisms. *Eur J Pharm Biopharm* **71**:475–483.

Jackson, E.L., N. Willis, K. Mercer, R.T. Bronson, D. Crowley, R. Montoya, T. Jacks, and D.A. Tuveson. 2001. Analysis of lung tumor initiation and progression using conditional expression of oncogenic K-ras. *Genes Dev* **15**:3243–3248.

Jimenez-Mateos, E.M., T. Engel, P. Merino-Serrais, R.C. McKiernan, K. Tanaka, G. Mouri, T. Sano, C. O'Tuathaigh, J.L. Waddington, S. Prenter, N. Delanty, M.A. Farrell, et al. 2012. Silencing microRNA-134 produces neuroprotective and prolonged seizure-suppressive effects. *Nat Med* **18**:1087–1094.

Johnnidis, J.B., M.H. Harris, R.T. Wheeler, S. Stehling-Sun, M.H. Lam, O. Kirak, T.R. Brummelkamp, M.D. Fleming, and F.D. Camargo. 2008. Regulation of progenitor cell proliferation and granulocyte function by microRNA-223. *Nature* **451**:1125–1129.

Kitade, Y. and Y. Akao. 2010. MicroRNAs and their therapeutic potential for human diseases: microRNAs, miR-143 and -145, function as anti-oncomirs and the application of chemically modified miR-143 as an anti-cancer drug. *J Pharmacol Sci* **114**(3):276–280.

Kota, J., R.R. Chivukula, K.A. O'Donnell, E.A. Wentzel, C.L. Montgomery, H.W. Hwang, T.C. Chang, P. Vivekanandan, M. Torbenson, K.R. Clark, J.R. Mendell, and J.T. Mendell. 2009. Therapeutic microRNA delivery suppresses tumorigenesis in a murine liver cancer model. *Cell* **137**:1005–1017.

Krutzfeldt, J., N. Rajewsky, R. Braich, K.G. Rajeev, T. Tuschl, M. Manoharan, and M. Stoffel. 2005. Silencing of microRNAs in vivo with "antagomirs". *Nature* **438**:685–689.

Landthaler, M., A. Yalcin, and T. Tuschl. 2004. The human DiGeorge syndrome critical region gene 8 and Its D. melanogaster homolog are required for miRNA biogenesis. *Curr Biol* **14**:2162–2167.

Lanford, R.E., E.S. Hildebrandt-Eriksen, A. Petri, R. Persson, M. Lindow, M.E. Munk, S. Kauppinen, and H. Orum. 2010. Therapeutic silencing of microRNA-122 in primates with chronic hepatitis C virus infection. *Science* **327**:198–201.

Lee, Y., C. Ahn, J. Han, H. Choi, J. Kim, J. Yim, J. Lee, P. Provost, O. Radmark, S. Kim, and V.N. Kim. 2003. The nuclear RNase III Drosha initiates microRNA processing. *Nature* **425**:415–419.

Lee, Y., M. Kim, J. Han, K.H. Yeom, S. Lee, S.H. Baek, and V.N. Kim. 2004. MicroRNA genes are transcribed by RNA polymerase II. *EMBO J* **23**:4051–4060.

Lennox, K.A. and M.A. Behlke. 2011. Chemical modification and design of anti-miRNA oligonucleotides. *Gene Ther* **18**:1111–1120.

Lu, Y., J. Xiao, H. Lin, Y. Bai, X. Luo, Z. Wang, and B. Yang. 2009. A single anti-microRNA antisense oligodeoxyribonucleotide (AMO) targeting multiple microRNAs offers an improved approach for microRNA interference. *Nucleic Acids Res* **37**:e24.

Ma, L., J. Young, H. Prabhala, E. Pan, P. Mestdagh, D. Muth, J. Teruya-Feldstein, F. Reinhardt, T.T. Onder, S. Valastyan, F. Westermann, F. Speleman, et al. 2010a. *miR-9*, a MYC/MYCN-activated microRNA, regulates E-cadherin and cancer metastasis. *Nat Cell Biol* **12**:247–256.

Ma, L., F. Reinhardt, E. Pan, J. Soutschek, B. Bhat, E.G. Marcusson, J. Teruya-Feldstein, G.W. Bell, and R.A. Weinberg. 2010b. Therapeutic silencing of *miR-10b* inhibits metastasis in a mouse mammary tumor model. *Nat Biotechnol* **28**:341–347.

Malek, A., F. Czubayko, and A. Aigner. 2008. PEG grafting of polyethylenimine (PEI) exerts different effects on DNA transfection and siRNA-induced gene targeting efficacy. *J Drug Target* **16**:124–139.

Mathonnet, G., M.R. Fabian, Y.V. Svitkin, A. Parsyan, L. Huck, T. Murata, S. Biffo, W.C. Merrick, E. Darzynkiewicz, R.S. Pillai, W. Filipowicz, T.F. Duchaine, et al. 2007. MicroRNA inhibition of translation initiation in vitro by targeting the cap-binding complex eIF4F. *Science* **317**:1764–1767.

Medina, P.P. and F.J. Slack. 2008. microRNAs and cancer: an overview. *Cell Cycle* **7**:2485–2492.

REFERENCES

Miyazaki, Y., H. Adachi, M. Katsuno, M. Minamiyama, Y.M. Jiang, Z. Huang, H. Doi, S. Matsumoto, N. Kondo, M. Iida, G. Tohnai, F. Tanaka, et al. 2012. Viral delivery of *miR-196a* ameliorates the SBMA phenotype via the silencing of CELF2. *Nat Med* **18**:1136–1141.

Najafi-Shoushtari, S.H., F. Kristo, Y. Li, T. Shioda, D.E. Cohen, R.E. Gerszten, and A.M. Naar. 2010. MicroRNA-33 and the SREBP host genes cooperate to control cholesterol homeostasis. *Science* **328**:1566–1569.

Ochiya, T., S. Nagahara, A. Sano, H. Itoh, and M. Terada. 2001. Biomaterials for gene delivery: atelocollagen-mediated controlled release of molecular medicines. *Curr Gene Ther* **1**:31–52.

Ochiya, T., Y. Takahama, S. Nagahara, Y. Sumita, A. Hisada, H. Itoh, Y. Nagai, and M. Terada. 1999. New delivery system for plasmid DNA in vivo using atelocollagen as a carrier material: the Minipellet. *Nat Med* **5**:707–710.

Pillai, R.S., S.N. Bhattacharyya, and W. Filipowicz. 2007. Repression of protein synthesis by miRNAs: how many mechanisms? *Trends Cell Biol* **17**:118–126.

Putta, S., L. Lanting, G. Sun, G. Lawson, M. Kato, and R. Natarajan. 2012. Inhibiting microRNA-192 ameliorates renal fibrosis in diabetic nephropathy. *J Am Soc Nephrol* **23**(3):458–469.

Rayner, K.J., C.C. Esau, F.N. Hussain, A.L. McDaniel, S.M. Marshall, J.M. van Gils, T.D. Ray, F.J. Sheedy, L. Goedeke, X. Liu, O.G. Khatsenko, V. Kaimal, et al. 2011. Inhibition of *miR-33a/b* in non-human primates raises plasma HDL and lowers VLDL triglycerides. *Nature* **478**:404–407.

Rayner, K.J., Y. Suarez, A. Davalos, S. Parathath, M.L. Fitzgerald, N. Tamehiro, E.A. Fisher, K.J. Moore, and C. Fernandez-Hernando. 2010. *MiR-33* contributes to the regulation of cholesterol homeostasis. *Science* **328**:1570–1573.

Robbins, M., A. Judge, and I. MacLachlan. 2009. siRNA and innate immunity. *Oligonucleotides* **19**:89–102.

Rodriguez, A., S. Griffiths-Jones, J.L. Ashurst, and A. Bradley. 2004. Identification of mammalian microRNA host genes and transcription units. *Genome Res* **14**:1902–1910.

Saito, Y., G. Liang, G. Egger, J.M. Friedman, J.C. Chuang, G.A. Coetzee, and P.A. Jones. 2006. Specific activation of microRNA-127 with downregulation of the proto-oncogene BCL6 by chromatin-modifying drugs in human cancer cells. *Cancer Cell* **9**:435–443.

Sano, A., M. Maeda, S. Nagahara, T. Ochiya, K. Honma, H. Itoh, T. Miyata, and K. Fujioka. 2003. Atelocollagen for protein and gene delivery. *Adv Drug Deliv Rev* **55**:1651–1677.

Segura, M.F., D. Hanniford, S. Menendez, L. Reavie, X. Zou, S. Alvarez-Diaz, J. Zakrzewski, E. Blochin, A. Rose, D. Bogunovic, D. Polsky, J. Wei, et al. 2009. Aberrant *miR-182* expression promotes melanoma metastasis by repressing FOXO3 and microphthalmia-associated transcription factor. *Proc Natl Acad Sci U S A* **106**:1814–1819.

Shenouda, S.K. and S.K. Alahari. 2009. MicroRNA function in cancer: oncogene or a tumor suppressor? *Cancer Metastasis Rev* **28**:369–378.

Takeshita, F., L. Patrawala, M. Osaki, R.U. Takahashi, Y. Yamamoto, N. Kosaka, M. Kawamata, K. Kelnar, A.G. Bader, D. Brown, and T. Ochiya. 2010. Systemic delivery of synthetic microRNA-16 inhibits the growth of metastatic prostate tumors via downregulation of multiple cell-cycle genes. *Mol Ther* **18**:181–187.

Thomas, C.E., A. Ehrhardt, and M.A. Kay. 2003. Progress and problems with the use of viral vectors for gene therapy. *Nat Rev Genet* **4**:346–358.

Thomas, M., K. Lange-Grunweller, U. Weirauch, D. Gutsch, A. Aigner, A. Grunweller, and R.K. Hartmann. 2012. The proto-oncogene Pim-1 is a target of *miR-33a*. *Oncogene* **31**:918–928.

Thum, T., C. Gross, J. Fiedler, T. Fischer, S. Kissler, M. Bussen, P. Galuppo, S. Just, W. Rottbauer, S. Frantz, M. Castoldi, J. Soutschek, et al. 2008. MicroRNA-21 contributes to myocardial disease by stimulating MAP kinase signalling in fibroblasts. *Nature* **456**:980–984.

Trang, P., P.P. Medina, J.F. Wiggins, L. Ruffino, K. Kelnar, M. Omotola, R. Homer, D. Brown, A.G. Bader, J.B. Weidhaas, and F.J. Slack. 2010. Regression of murine lung tumors by the *let-7* microRNA. *Oncogene* **29**:1580–1587.

Trang, P., J.F. Wiggins, C.L. Daige, C. Cho, M. Omotola, D. Brown, J.B. Weidhaas, A.G. Bader, and F.J. Slack. 2011. Systemic delivery of tumor suppressor microRNA mimics using a neutral lipid emulsion inhibits lung tumors in mice. *Mol Ther* **19**:1116–1122.

Ueno, Y., T. Inoue, M. Yoshida, K. Yoshikawa, A. Shibata, Y. Kitamura, and Y. Kitade. 2008. Synthesis of nuclease-resistant siRNAs possessing benzene-phosphate backbones in their 3′-overhang regions. *Bioorg Med Chem Lett* **18**:5194–5196.

Urban-Klein, B., S. Werth, S. Abuharbeid, F. Czubayko, and A. Aigner. 2005. RNAi-mediated gene-targeting through systemic application of polyethylenimine (PEI)-complexed siRNA in vivo. *Gene Ther* **12**:461–466.

Valastyan, S., F. Reinhardt, N. Benaich, D. Calogrias, A.M. Szasz, Z.C. Wang, J.E. Brock, A.L. Richardson, and R.A. Weinberg. 2009. A pleiotropically acting microRNA, *miR-31*, inhibits breast cancer metastasis. *Cell* **137**:1032–1046.

Valeri, N., P. Gasparini, C. Braconi, A. Paone, F. Lovat, M. Fabbri, K.M. Sumani, H. Alder, D. Amadori, T. Patel, G.J. Nuovo, R. Fishel, et al. 2010. MicroRNA-21 induces resistance to 5-fluorouracil by down-regulating human DNA MutS homolog 2 (hMSH2). *Proc Natl Acad Sci U S A* **107**:21098–21103.

Vester, B. and J. Wengel. 2004. LNA (locked nucleic acid): high-affinity targeting of complementary RNA and DNA. *Biochemistry* **43**:13233–13241.

Wang, C.J., Z.G. Zhou, L. Wang, L. Yang, B. Zhou, J. Gu, H.Y. Chen, and X.F. Sun. 2009. Clinicopathological significance of microRNA-31, -143 and -145 expression in colorectal cancer. *Dis Markers* **26**:27–34.

Wang, F., P. Zhang, Y. Ma, J. Yang, M.P. Moyer, C. Shi, J. Peng, and H. Qin. 2012. NIRF is frequently upregulated in colorectal cancer and its oncogenicity can be suppressed by *let-7a* microRNA. *Cancer Lett* **314**:223–231.

Wang, H.J., H.J. Ruan, X.J. He, Y.Y. Ma, X.T. Jiang, Y.J. Xia, Z.Y. Ye, and H.Q. Tao. 2010. MicroRNA-101 is down-regulated in gastric cancer and involved in cell migration and invasion. *Eur J Cancer* **46**:2295–2303.

Wang, Z. 2011. The concept of multiple-target anti-miRNA antisense oligonucleotide technology. *Methods Mol Biol* **676**:51–57.

Weirauch, U., N. Beckmann, M. Thomas, A. Grunweller, K. Huber, F. Bracher, R.K. Hartmann, and A. Aigner. 2013. Functional role and therapeutic potential of the pim-1 kinase in colon carcinoma. *Neoplasia* **15**(7):783–794.

Werth, S., B. Urban-Klein, L. Dai, S. Hobel, M. Grzelinski, U. Bakowsky, F. Czubayko, and A. Aigner. 2006. A low molecular weight fraction of polyethylenimine (PEI) displays increased transfection efficiency of DNA and siRNA in fresh or lyophilized complexes. *J Control Release* **112**:257–270.

Whitehead, K.A., J. Matthews, P.H. Chang, F. Niroui, J.R. Dorkin, M. Severgnini, and D.G. Anderson. 2012. In vitro-in vivo translation of lipid nanoparticles for hepatocellular siRNA delivery. *ACS Nano* **28**:6922–6929.

Wiggins, J.F., L. Ruffino, K. Kelnar, M. Omotola, L. Patrawala, D. Brown, and A.G. Bader. 2010. Development of a lung cancer therapeutic based on the tumor suppressor microRNA-34. *Cancer Res* **70**:5923–5930.

Zhai, H., B. Song, X. Xu, W. Zhu, and J. Ju. 2012. Inhibition of autophagy and tumor growth in colon cancer by *miR-502*. *Oncogene* **32**:1570–1579.

40

LOCKED NUCLEIC ACIDS AS MicroRNA THERAPEUTICS

Henrik Ørum

Santaris Pharma, Hørsholm, Denmark

I.	Introduction	664
II.	Design of LNA-Based Inhibitors of miRNAs	665
III.	Activity in Experimental Animals	666
IV.	LNA-antimiRs/Tiny LNAs in Pre-Clinical Development	667
V.	LNA-antimiR in Clinical Development	668
VI.	Discussion	669
	Conflict of Interest	670
	Acknowledgment	670
	References	670

ABBREVIATIONS

2′-OMe	2′-O-methyl
ALT	alanine aminotransferase
AST	aspartate aminotransferase
DAAs	direct-acting antivirals
HCV	hepatitis C virus
HDL	high-density lipoprotein
IFN-α	pegylated interferon-alpha
LNA	locked nucleic acid
LPS	lipopolysaccharide
MOE	2′-O-methoxyethyl
mRNA	messengerRNA

MicroRNAs in Medicine, First Edition. Edited by Charles H. Lawrie.
© 2014 John Wiley & Sons, Inc. Published 2014 by John Wiley & Sons, Inc.

I. INTRODUCTION

MicroRNAs (miRNAs) are gaining momentum as novel targets for therapeutic intervention. Interest is being driven by mounting evidence that associates a diversity of human diseases to faulty expression of different miRNAs. Interest is equally driven by the broad regulatory functions of miRNAs, which in turn enable miRNA targeted drugs to affect entire pathways and networks in a biologically balanced way. This approach to human medicine is distinctly different from the traditional approach of inhibiting single pathway constituents, raising hopes that miRNA targeted drugs may prove effective against diseases where current treatment options are limited or nonexisting.

As described in detail in other chapters of this book, miRNAs exert their regulatory role by base-pairing to partially complementary sequences in the 3′-UTR of target mRNAs, causing translational repression or mRNA degradation. Accordingly, single-stranded oligonucleotides, which can be designed to specifically bind to individual miRNAs, thereby inhibiting their interaction with the cognate mRNA targets, offer a potentially attractive means for developing novel drugs against a variety of human diseases.

Locked nucleic acid (LNA) is the best-known representative of a novel class of bicyclic RNA analogues, in which the 2′- and 4′-position of the ribose has been linked by different molecular moieties. In LNA, a methylene group connects the 4′-C to the 2′-O (Obika et al. 1997; Singh et al. 1998b), but other connecting groups can also be employed, such as those used in amino-LNA (4′–CH_2–NH–2′) (Singh et al. 1998a), thio-LNA (4′–CH_2–S–2′) (Singh et al. 1998a), ethylene nucleic acid (ENA) (4–CH_2CH_2–O–2′) (Morita et al. 2001), and cET-LNA (4′–$CH(CH_3)$–O–2′) (Seth et al. 2010), to mention but a few (Figure 40.1). When incorporated into oligonucleotides, LNA substantially increase the resistance against nuclease digestion and the affinity toward complementary RNA sequences, compared with all previously used chemistries (Obika et al. 1997; Singh and Wengel 1998; Singh et al. 1998; Frieden et al. 2003). In fact, the increase in affinity offered by LNA has enabled the development of a new, shorter-than-usual, class of antisense drugs that are substantially more potent than previous generations of antisense drugs based on lower affinity chemistries, such as 2′-O-methyl (2′-OMe) and 2′-O-methoxyethyl (MOE) (reviewed in Koch and Ørum 2007; Koch 2013). Typically, LNA-based antisense drugs are between 13 and 16 nucleotides in length, yet are substantially more potent in experimental animals than the 20+ mer oligonucleotides typical of other classes of antisense drugs, thus underpinning that both the affinity and the size of the oligo are key drivers of potency (Koch and Ørum 2007; Straarup et al. 2010; Koch 2013).

The benefit of reducing the size of antisense drugs, perhaps somewhat counterintuitively, also applies to specificity. Assuming that a sequence can be identified that is unique to the target mRNA, using the entire human transcriptome as reference (which is typically

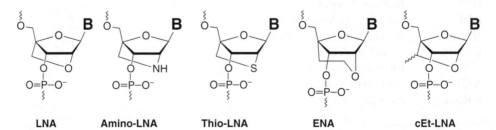

Figure 40.1. Molecular structure of LNA and some of its analogues. Locked nucleic acid (LNA), amino-LNA, thio-LNA, ethylene nucleic acid (ENA), and constrained ethyl (cET).

possible with sequences >13 nucleotides), the ability of the complementary oligonucleotide to effectively inhibit the target mRNA, while leaving incorrect targets unaffected, rests predominantly on the difference in affinity (T_m) for the correct and incorrect targets (i.e., ΔT_m). This parameter (ΔT_m) increases as the size of the oligo decreases, which is due to the increased destabilization of mismatches on shorter duplexes, and can be manipulated by oligonucleotide design to provide optimal discrimination at the human physiological temperature (Koch and Ørum 2007; Straarup et al. 2010; Koch 2013).

II. DESIGN OF LNA-BASED INHIBITORS OF miRNAs

Lacking a detailed understanding of the structure-activity relationship between miRNAs and their cognate mRNA targets, early attempts to inhibit miRNAs made use of oligonucleotides that were designed to be complementary to the entire mature miRNA sequence, thus shutting down any opportunity for hybridization-based interactions with the target mRNAs (Hutvágner et al. 2004; Krützfeldt et al. 2005; Davis et al. 2006). Indeed, in some of the earliest reports, the miRNA inhibitor was even larger than the miRNA, presumably in an attempt to mimic the longer, natural mRNA target, thereby seeking to exploit any unknown, but potentially stabilizing, interactions that non-hybridizing parts of the mRNA might have with protein components of the miRNA particle. Using a 15-mer oligonucleotide comprising a mixture of LNA and DNA nucleotides (Figure 40.2), Elmén et al. in 2007 demonstrated that such approaches were in fact not necessary, and that only a part of the miRNA sequence needed to be blocked in order to effectively inhibit its interaction with target mRNAs, thus opening the field of miRNA therapeutics to the potency and specificity advantages of smaller-than-usual, LNA drugs (Elmén et al. 2007). Since then, short LNA/DNA oligonucleotides, termed antimiRs, have been used successfully to selectively and potently inhibit a variety of different miRNAs in both cell cultures and experimental animals, *vide infra*.

Exploring the smaller-than-usual concept to its fullest, Obad et al. demonstrated in 2011 that as little as the seed sequence of the miRNA needed to be blocked in order to facilitate effective inhibition of the intended target (Obad et al. 2011). These fully LNA modified oligonucleotides (7–8 nucleotides long) were termed tiny LNAs (Figure 40.2) and provided the possibility to use a single LNA oligonucleotide to inhibit an entire family of miRNAs, which outside the shared seed sequence often has little sequence conservation. According to the miRBase 17, more than 25% of all human miRNAs (511 of 1733) belongs to families that share seed sequences and targets (Wang et al. 2012), and for a few of these

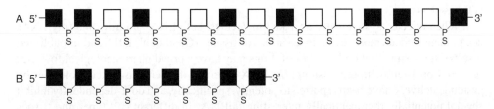

Figure 40.2. Design of (A) LNA-antimiRs and (B) tiny LNAs. Black and white boxes designate an LNA and a DNA monomer, respectively. PS designate a phophorothioate internucleoside linkage. LNA-antimiRs are typically used in the size range of 12–15-mers and typically contain about 50% LNA nucleotides. Tiny LNAs are exclusively composed of LNA nucleosides and are typically used as 8 mers.

families, the expected functional redundancy have been demonstrated (le Sage et al. 2007; Mu et al. 2009). Going forward, the tiny-LNA format could thus prove therapeutically relevant for a sizeable number of human diseases with miRNA involvement.

Because of their very small size, tiny LNAs encounter many fully complementary sites in non-target RNAs where they could cause undesired pharmacological effects. Microarray analysis of mRNAs from both cultured cells and the liver of mice treated with tiny LNAs against *let-7* or *miR-122*, however, showed the expected up-regulation of mRNAs harboring the miRNA seed site sequences, but no detectable effect on mRNAs containing target sequences complementary to the tiny LNAs (Obad et al. 2011). The finding that off-target effects of tiny LNAs may in fact be limited was further supported at the proteomic level, where the expected effects were observed on proteins encoded by mRNAs containing *miR-122* target sites but no significant effects were found on proteins encoded by mRNA containing perfect matched target sites for the tiny LNA. Assuming that at least some of these sites in non-target mRNAs are located within the coding region and are accessible to hybridization, these data suggest that mere binding of the tiny LNA is not enough to prevent translational "read through."

III. ACTIVITY IN EXPERIMENTAL ANIMALS

The first example of *in vivo* inhibition of a miRNA by an LNA-antimiR was reported by Elmén et al. in 2007, who used a 15-mer LNA antimiR, delivered intravenously, to inhibit the liver-specific *miR-122* in mice (Elmén et al. 2007). Inhibition of *miR-122*, which is involved in cholesterol and lipid metabolism (Esau et al. 2006), was observed both directly through detection of the *miR-122:LNA-antimiR* complex and indirectly through derepression of several direct mRNA targets and reductions in plasma cholesterol. This finding was subsequently extended to non-human primates, where a further optimized 15-mer LNA-antimiR injected intravenously at either 1, 3, or 10 mg/kg on day 1, 3, and 5 was shown to effectively and dose-dependently bind to endogenous *miR-122*, leading to derepression of several direct mRNA targets and reductions in plasma cholesterol of a magnitude similar to that observed in mice (Elmén et al. 2008). Consistent with the long half-life of LNA-antimiRs in the liver of monkeys, the effect was long lasting with significant reductions in plasma cholesterol being evident several months after the last administration of both the mid- and top dose. In the same paper, the authors also evaluated the effect of different LNA-antimiR designs. As expected from the mode of action of LNA-antimiRs, (i.e., sequestration of the miRNA in a stable and inactive LNA-antimiR : miRNA complex), the authors reported that the activity of a series of 15-mer LNA-antimiRs, containing different numbers of LNA nucleotides, was proportional to their affinity for the miRNA in cell based assays.

When injected systemically as naked molecules, both LNA-antimiRs and tiny LNAs accumulate to detectable levels in a variety of tissues (Obad et al. 2011 and unpublished results), suggesting that both classes of drugs may have a broad pharmacological footprint *in vivo*. Consistent with this assumption, LNA-antimiRs delivered as naked molecules by systemic delivery have been reported to potently, specifically, and dose-dependently inhibit several potentially therapeutically interesting miRNAs in different cell types and tissues *in vivo*. Najafi-Shoushtari et al. successfully exploited an LNA-antimiR to inhibit *miR-33a* in the liver of mice, in a study designed to assess its role in high-density lipoprotein (HDL) synthesis and reverse cholesterol transport (Najafi-Shoushtari et al. 2010). Likewise, Worm et al. used an LNA-antimiR to effectively inhibit *miR-155* expression in splenocytes in

LPS-treated mice, in a study designed to assess the role of the miRNA in inflammatory responses (Worm et al. 2009). Also, in a study designed to assess a possible role of the *miR-219* in cognitive and behavioral disturbances, an antimiR was used to mediate effective inhibition of the miRNA in the central nervous system upon ICV injection (Kocerha et al. 2009).

Tiny LNAs delivered intravenously to mice have also been reported to be broadly active *in vivo*. Obad et al., in the initial paper introducing the tiny LNA format, demonstrated pharmacological activity of tiny LNAs against *miR-21* in the kidney, liver, lung, and in 4T1 cells in an orthotopic breast tumor model (Obad et al. 2011). Patrick et al. subsequently used a tiny LNA to inhibit *miR-21* in the heart of mice in a study seeking to determine its role in preventing cardiac hypertrophy and fibrosis in response to pressure overload induced by transaortic constriction (Patrick et al. 2010). In a study designed to assess the role of *miR-155* in graft versus host disease, Ranganathan et al. reported that a tiny LNA effectively inhibited *miR-155* expression in allogenic donor T cells from C57BL/6 mice intraveneously transferred into lethally irradiated B6D2F1 recipient mice (Ranganathan et al. 2012). Also, Leucci et al. used a tiny LNA to inhibit *miR-9* in Hodgkin's lymphoma cells injected intravenously into immune-compromised mice, in a study designed to clarify its potential role in assisting tumor growth (Leucci et al. 2012). Finally, He et al. used a tiny LNA targeting all nine distinct *let-7* miRNA family members to study their role in μ-opioid receptor signaling in the CNS of mice (He et al. 2010), thus extending the initial demonstration of *in vitro* inhibition of the *let-7* family by a single tiny LNA by Obad et al. in 2011 to the *in vivo* situation.

IV. LNA-ANTIMIRS/TINY LNAS IN PRE-CLINICAL DEVELOPMENT

The myocyte specific *miR-208a*, plays an essential role in pathological cardiac remodeling (van Rooij et al. 2007), and an LNA-antimiR against this miRNA is currently being developed for the treatment of chronic heart failure by Miragen Therapeutics (http://www.miragentherapeutics.com). In mice, subcutaneous delivery of the LNA-antimiR produced a specific and sustained reduction in *miR-208a* in the heart. Similarly, in Dahl-sensitive rats with chronic hypertension, induced by a high salt diet, treatment with the LNA-antimiR significantly reduced cardiomyocyte hypertrophy and periarteriolar fibrosis, thus blunting functional deterioration and significantly delaying lethality compared with control groups (Montgomery et al. 2011). In both species, the LNA-antimiR was found to be well tolerated, with animals behaving normally throughout the study as determined by activity level and grooming behavior. Also, no signs of overt liver toxicity was observed in rats as determined by serum levels of aspartate aminotransferase (AST) and alanine aminotransferase (ALT) in treated versus control animals. Notably, no effect of *miR-208a* inhibition was observed on cardiac conductance, thus reducing a concern raised by a previous publication that had indicated a role for *miR-208a* in maintaining proper heart electrophysiology (Callis et al. 2009). Interestingly, a recent paper by Grueter et al. reported that *miR-208a*, via its regulation of the MED13 mRNA, is involved in regulating systemic energy homeostasis, suggesting that *miR-208* might also be an appropriate target for treating metabolic disorders (Grueter et al. 2012).

The *miR-15* family, comprise six miRNAs (*miR-15a* and *b*, *miR-16-1* and *2*, *miR-195*, and *miR-497*), all of which comprise the same seed sequence and are strongly up-regulated in cardiac ischemia and heart failure (Small et al. 2010). A tiny LNA directed against the conserved seed sequence is currently in preclinical development for the treatment of postmyocardial infarction remodeling (http://www.miragentherapeutics.com). In mice, the tiny LNA, delivered intravenously as a single injection, resulted in a dose-dependent accumulation in heart tissue that was paralleled by a dose-dependent inhibition of those *miR-15* family members that were measured, *miR-15b/16/195* (Hullinger et al. 2012). Alternative routes of administration, such as subcutaneous and intraperitoneal dosing, gave similar results. Interestingly, activity—albeit much less—was also noted upon delivery by oral gavage, confirming our own observations that small LNA oligonucleotides may have the potential for oral delivery (manuscript in preparation). The therapeutic relevance of inhibiting the *miR-15* family during ischemia-reperfusion injury was investigated in mice. Injection of the tiny LNA at the onset of reperfusion led to a significant decrease in infarct size compared with either saline or control oligo-treated mice as determined 24 hours after reperfusion. Moreover, hemodynamic analysis showed a significant increase in left ventricle end-diastolic pressure in the control versus treated animals, indicating an enhanced cardiac function in the tiny LNA treatment group. The authors also investigated the response of the *miR-15* family to ischemic damage of the heart in female Yorkshire pigs and the subsequent response to treatment with the tiny LNA. Similar to mice, the *miR-15* family was shown to be up-regulated in the infarcted region of the heart, and similar to mice, treatment with the tiny LNA caused an efficient and dose-dependent silencing of *miR-15* family members ,with the effect being comparable in different portions of the heart. In both mice and pigs, the tiny LNA was well tolerated and did not produce overt signs of toxicity. Also, no histological abnormalities or changes in serum levels of liver enzymes (ALT and AST) were observed in pigs.

V. LNA-ANTIMIR IN CLINICAL DEVELOPMENT

Hepatitis C virus (HCV) infection is a leading cause of liver failure and hepatocellular carcinoma and affects an estimated 170 million people worldwide. Until recently, HCV therapy has relied on extended treatments (typically 48 weeks), with a combination of pegylated interferon-α (IFN-α) and ribavirin—a combination that is only partly effective and often associated with serious side effects (Chisari 2005). Several new drugs that directly targets virally encoded proteins (direct acting antivirals, DAAs) are in development, and two such drugs, Telaprevir and Boceprevir, both of which target the viral protease, has recently been approved by the FDA. Added to standard of care, these drugs have improved cure rates and shortened typical treatment times, but have also added new toxicities that complicates treatment (Casey and Lee 2012). As with the other DDAs in development, both Telaprevir and Boceprevir are sensitive to the very high mutational rate of the virus that rapidly generates drug-resistant variants. Accordingly, the long-held ambition for an interferon-free HCV treatment based on DAAs will likely require careful combination of several different DAAs with mutually agreeable and additive/synergistic properties.

In 2005, Jopling et al. demonstrated that the liver-specific *miR-122*, which is involved in cholesterol and lipid metabolism (Esau et al. 2006), was required for HCV replicon accumulation in cultured liver cells, and that the mechanism involved direct binding of the miRNA to two closely spaced target sites in the 5'non-coding region of the HCV genome (Jopling et al. 2005). Further work has since refined our understanding of the mechanism and confirmed that *miR-122* is indeed an essential host factor for HCV replication and is exploited by all HCV subtypes (Jopling et al. 2008; Machlin et al. 2011). Therapies that target essential host factors for HCV may provide a higher barrier to viral resistance than that achievable by DAAs, thus providing an attractive solution to one of the key issues facing the development of effective new treatments.

Miravirsen, an LNA-antimiR that targets *miR-122*, is currently being developed by Santaris Pharma (http://www.santarispharma.com) as a novel, subtype independent, treatment for chronic hepatitis C viral infection. The compound has been extensively profiled in preclinical studies in mice, cynomolgus monkeys, green African monkeys, and chimpanzees (Elmén et al. 2008; Lanford et al. 2010; Hildebrandt-Eriksen et al. 2012). In all species, treatment with Miravirsen caused a dose-dependent reduction in free *miR-122* as measured by detection of the Miravirsen:*miR-122* complex, a reduction in plasma cholesterol levels of up to 35–40%, and an up-regulation of direct *miR-122* targets. Moreover, consistent with the long half-life of Miravirsen in the liver of monkeys and chimpanzees (~5-weeks), the observed pharmacological effects were still significant several months after last dose (Elmén et al. 2008; Lanford et al. 2010; Hildebrandt-Eriksen et al. 2012). In chimpanzees, chronically infected with HCV subtype 1a or 1b, once weekly intravenous dosing with Miravirsen for 12 weeks, led to a substantial, dose-dependent and sustained decrease in viral load that reached 2.6 orders of magnitude in the high-dose group (Lanford et al. 2010). In support of the hypothesis that targeting an essential host factor creates a very high barrier to viral breakthrough, no evidence of viral resistance was found in any of the treated animals either during or after cessation of dosing. In all animal species, Miravirsen was reported to be safe, with unremarkable toxicity profiles in both 4- and 13-weeks toxicity studies in rats and monkeys (Hildebrandt-Eriksen et al. 2012 and unpublished results) and excellent tolerability in the 6-months pharmacology study in chimpanzees (Lanford et al. 2010). In May 2008, Miravirsen became the first miRNA targeted drug to enter human clinical trials. Since then, it has completed two safety studies in healthy volunteers, a drug–drug interaction study with standard of care, and a 4-week proof-of-concept, monotherapy study in treatment naive patients chronically infected with HCV genotype 1a or 1b. In all studies Miravirsen has been found to be safe, with no serious adverse effects or dose-limiting toxicities. In fact, adverse effects have been infrequent and mostly mild and have not resulted in any discontinuations (Janssen et al. 2011 and unpublished results). In HCV patients, treatment with Miravirsen, by subcutaneous injection, resulted in a highly significant, dose-dependent reduction in viral load that averaged 2.9 logs in the top dose group of 7 mg/kg and continued to drop after completion of therapy. Notably, several patients (one out of nine in the 5 mg/kg and four out of nine in the 7 mg/kg dose group) became HCV undetectable during the study (Janssen et al. 2011). Consistent with Miravirsen's very high barrier to viral resistance in chimpanzees, there was no evidence of viral breakthrough at any point during the phase II study. Further trials of Miravirsen as a single agent and in combination with DAAs are planned.

VI. DISCUSSION

LNA-based miRNA therapeutics has come a long way in a relatively short time span, with several drugs in preclinical development and one drug in phase II clinical trials. That said, it is still early days, and there will undoubtedly be many and major challenges ahead before this fascinating new branch of human medicine becomes a clinical reality. In particular, much work needs to be done on the vast majority of miRNAs before they can be confidently selected for serious therapeutic efforts. This include a much deeper understanding of their complex biological function, their disease relevance, and the safety profile associated with their inhibition, both short term and long term in particular. Miravirsen has so far proven both safe and effective in several clinical trials, so a cautious optimism is warranted in regard to its prospects for becoming an important new drug in the fight to eradicate HCV. Whether the clinically successful targeting of *miR-122* by Miravirsen can be reproduced with LNA-antimiRs or tiny LNAs against miRNAs involved in other diseases remains to be seen, but there is little doubt that the coming years in LNA-based miRNA therapeutics are going to be exciting for all those involved.

CONFLICT OF INTEREST

The author is an employee of Santaris Pharma A/S, which owns the intellectual property rights to LNA-antimiRs and tiny LNAs, and which develops drugs on the basis of these formats.

ACKNOWLEDGMENT

Work on Miravirsen at Santaris has been supported by the Danish National Advanced Technology Foundation.

REFERENCES

Callis, T.E., K. Pandya, H.Y. Seok, R.H. Tang, M. Tatsuguchi, Z.P. Huang, J.F. Chen, Z. Deng, B. Gunn, J. Shumate, M.S. Willis, C.H. Selzman, and D.Z. Wang. 2009. MicroRNA-208a is a regulator of cardiac hypertrophy and conduction in mice. *J Clin Invest* **119**:2772–2786.

Casey, L.C. and W.M. Lee. 2012. Hepatitis C therapy update. *Curr Opin Gastroenterol* **28**: 188–192.

Chisari, F.V. 2005. Unscrambling hepatitis C virus-host interactions. *Nature* **536**:930–932.

Davis, S., B. Lollo, S. Freier, and C. Esau. 2006. Improved targeting of miRNA with antisense oligonucleotides. *Nucleic Acids Res* **34**:2294–2304.

Elmén, J., M. Lindow, A. Silahtaroglu, M. Bak, M. Christensen, A. Lind-Thomsen, M. Hedtjärn, J.B. Hansen, H.F. Hansen, E.M. Straarup, K. McCullagh, P. Kearney, and S. Kauppinen. 2007. Antagonism of microRNA-122 in mice by systemically administered LNA-antimiR leads to up-regulation of a large set of predicted target mRNAs in the liver. *Nucleic Acids Res* **36**: 1153–1162.

Elmén, J., M. Lindow, S. Schütz, M. Lawrence, A. Petri, S. Obad, M. Linholm, M. Hedtjärn, H.F. Hansen, U. Berger, S. Gullans, P. Kearney, P. Sarnow, E.M. Straarup, and S. Kauppinen. 2008. LNA-mediated microRNA silencing in non-human primates. *Nature* **452**:896–900.

Esau, C., S. Davis, S.F. Murray, X.X. Yu, S.K. Pandey, M. Pear, L. Watts, S.L. Booten, M. Graham, R. McKay, A. Subramaniam, S. Propp, B.A. Lollo, S. Freier, C.F. Bennett, S. Bhanot, and B.P. Monia. 2006. *miR-122* regulation of lipid metabolism revealed by in vivo antisense targeting. *Cell Metab* **3**:87–98.

Frieden, M., H.F. Hansen, and T. Koch. 2003. Nuclease stability of LNA oligonucleotides and LNA-DNA chimeras. *Nucleosides Nucleotides Nucleic Acids* **22**:1041.

Grueter, C.E., E. van Rooij, B.A. Johnson, S.M. Deleon, L.B. Sutherland, X. Qi, L. Gautron, J.K. Elmquist, R. Bassel-Duby, and E.N. Olson. 2012. A cardiac microRNA governs systemic energy homeostasis by regulation of MED13. *Cell* **149**:671–683.

He, Y., C. Yang, C.M. Kirkmire, and Z.J. Wang. 2010. Regulation of opioid tolerance by *let-7* family microRNA targeting the μ opioid receptor. *J Neurosci* **30**:10251–10258.

REFERENCES

Hildebrandt-Eriksen, E.S., V. Aarup, R. Persson, H.F. Hansen, M.E. Munk, and H. Ørum. 2012. A locked nucleic acid oligonucleotide targeting microRNA-122 is well tolerated in cynomolgus monkeys. *Nucleic Acid Ther* **22**(3):152–161.

Hullinger, T.G., R.L. Montgomery, A.G. Seto, B.A. Dickinson, H.M. Semus, J.M. Lynch, C.M. Dalby, K. Robinson, C. Stack, P.A. Latimer, J.M. Hare, N.E. Olson, and E. van Rooij. 2012. Inhibition of *miR-15* protects against cardiac ischemic injury. *Circ Res* **110**:71–81.

Hutvágner, G., M.J. Simard, C.C. Mello, and P.D. Zamore. 2004. Sequence-specific inhibition of small RNA function. *PLoS Biol* **2**:e98.

Janssen, H.L., H.W. Leerink, S. Zeuzem, E. Lawitz, M. Rodriguez-Torres, A. Chen, C. Davis, B. King, A.A. Levin, and M.R. Hodges MR. 2011. A randomized, double-blind, placebo (plb) controlled safety and ant-viral proof of concept study of miravirsen (MRI), an oligonucleotide targeting miR-122, in treatment naïve patients with genotype 1 (gt1) chronic infection. *Hepatologi* **54**:430A.

Jopling, C.L., S. Schutz, and P. Sarnow. 2008. Position-dependent function for a tandem microRNA *miR-122* binding site located in the hepatitis C virus RNA genome. *Cell Host Microbe* **4**: 77–85.

Jopling, C.L., M. Yi, A.M. Lancaster, S.M. Lemon, and P. Sarnow. 2005. Modulation of hepatitis C virus RNA abundance by a liver-specific microRNA. *Science* **309**:1577–1581.

Kocerha, J., M.A. Faghihi, M.A. Lopez-Toledano, J. Huang, A.J. Ramsey, M.G. Caron, N. Sales, D. Willoughby, J. Elmen, H.F. Hansen, H. Orum, S. Kauppinen, P.J. Kenny, and C. Wahlestedt. 2009. MicroRNA-219 modulates NMDA receptor-mediated neurobehavioral dysfunction. *Proc Natl Acad Sci U S A* **106**:3507–3512.

Koch, T. 2013. LNA antisense: a review. *Curr Phys Chem* **3**(1). Special issue: quantum nanobiololy and biophysical chemistry. Jalkanen KJ and Jensen GM, Eds.

Koch, T. and H. Ørum. 2007. Locked nucleic acid. In *Antisense drug discovery*, Second Edition. S.T. Crooke, Ed. CRC Press, Boca Raton, FL, pp. 520–564.

Krützfeldt, J., N. Rajewsky, R. Braich, K.G. Rajeev, T. Tuschl, M. Manoharan, and M. Stoffel. 2005. Silencing of microRNAs in vivo with antagomirs. *Nature* **438**:685–689.

Lanford, R.E., E.S. Hildebrandt-Eriksen, A. Petri, R. Persson, M. Lindow, M.E. Munk, S. Kauppinen, and H. Ørum. 2010. Therapeutic siliencing of microRNA-122 in primates with chronic hepatitis C virus infection. *Science* **327**:198–201.

le Sage, C., R. Nagel, D.A. Egan, M. Schrier, E. Mesman, A. Mangiola, C. Anile, G. Maira, N. Mercatelli, S.A. Ciafrè, M.G. Farace, and R. Agami. 2007. Regulation of the p27kip1 tumor suppressor by *miR-221* and *miR-222* promotes cancer cell proliferation. *EMBO J* **26**: 3699–3708.

Leucci, E., A. Zriwil, L.H. Gregersen, K.T. Jensen, S. Obad, C. Bellan, L. Leoncini, S. Kauppinen, and A.H. Lund. 2012. Inhibition of *miR-9* de-represses HuR and DICER1 and impairs Hodgkin lymphoma tumour outgrow *in vivo*. *Oncogene* **31**:5081–5089.

Machlin, E.S., P. Sarnow, and S.M. Sagan. 2011. Masking the 5′ terminal nucleotides of the hepatitis C virus genome by an unconventional microRNA-target RNA complex. *Proc Natl Acad Sci U S A* **108**:3193–3198.

Montgomery, R.L., T.G. Hullinger, H.M. Semus, B.A. Dickinson, A.G. Seto, J.M. Lynch, C. Stack, P.A. Latimer, E.N. Olson, and E. van Rooij. 2011. Therapeutic inhibition of *miR-208a* improves cardiac function and survival during heart failure. *Circulation* **124**:1537–1547.

Morita, K., C. Hasegawa, M. Kaneko, S. Tsutsumi, J. Sone, T. Ishikawa, T. Imanishi, and M. Koizumi. 2001. 2′-O,4′-C-ethylene-bridged nucleic acids (ENA) with nuclease-resistance and high affinity for RNA. *Nucleic Acids Res Suppl* **1**:241–242.

Mu, P., Y.-C. Han, D. Betel, E. Yao, M. Squatrito, P. Ogrodowski, E. de Stanchina, A. D'Andrea, C. Sander, and A. Ventura. 2009. Genetic dissection of the mir-17-93 cluster of microRNAs in Myc-induced B-cell lymphomas. *Genes Dev* **23**:2806–2811.

Najafi-Shoushtari, S.H., F. Kristo, Y. Li, T. Shioda, D. Cohen, R.E. Gerszten, and A. Näär. 2010. MicroRNA-33 and the SREBP host genes cooperate to control cholesterol homeostasis. *Science* **358**:1566–1569.

Obad, S., C.O. dos Santos, A. Petri, M. Heidenblad, O. Broom, C. Ruse, C. Fu, M. Lindow, J. Stenvang, E.M. Straarup, H.F. Hansen, T. Kock, D. Pappin, G.J. Hannon, and S. Kauppinen. 2011. Silencing of microRNA families by seed-targeting tiny LNAs. *Nat Genet* **43**:371–378.

Obika, S., J.A.K. Morio, D. Nanbu, and T. Imanishi. 1997. Synthesis and conformation of 3′-0,4′-C-methyleneribonucleosides, novel bicyclic nucleoside analogues for 2′,5′-linked oligonucleotide modification. *Chem Commun* 1643–1644.

Patrick, D.M., R.L. Montgomery, X. Qi, S. Obad, S. Kauppinen, J.A. Hill, E. van Rooij, and E.N. Olson. 2010. Stress-dependent cardiac remodeling occurs in the absence of microRNA-21 in mice. *J Clin Invest* **120**:3912–3916. Doi:10.1172/JCI43604

Ranganathan, P., C.E. Heaphy, S. Costinean, N. Stauffer, C. Na, M. Hamadani, R. Santhanam, C. Mao, P.A. Taylor, S. Sandhu, G. He, A. Shana'ah, G.J. Nuovo, A. Lagana, L. Cascione, S. Obad, O. Broom, S. Kauppinen, J.C. Byrd, M. Caligiuri, D. Perrotti, G.A. Hadley, G. Marcucci, S.M. Devine, B.R. Blazar, C.M. Croce, and R. Garzon. 2012. Regulation of acute graft-versus-host disease by microRNA-155. *Blood* **119**:4786–4797.

Seth, P.P., G. Vasquez, C.A. Allerson, A. Berdeja, H. Gaus, G.A. Kinberger, T.P. Prakash, M.T. Migawa, B. Bhat, and E.E. Swayze. 2010. Synthesis and biophysical evaluation of 2′,4′-constrained 2′O-methoxyethyl and 2′,4′-constrained 2′O-ethyl nucleic acid analogues. *J Org Chem* **75**:1569–1581.

Singh, S.K., R. Kumar, and J. Wengel. 1998a. Synthesis of novel bicycle [2.2.1] ribonucleosides: 2'-amino and 2'thio-LNA monomeric nucleosides. *J Org Chem* **63**:6078.

Singh, S.K., P. Nielsen, A. Koshkin, and J. Wengel. 1998b. LNA (locked nucleic acids): synthesis and high-affinity nucleic acid recognition. *Chem Commun* 455–456.

Singh, S.K. and J. Wengel. 1998. Universality of LNA-mediated high-affinity nucleic acid recognition. *Chem Commun* 1247–1248.

Small, E.M., R.J. Frost, and E.N. Olson. 2010. MicroRNAs add a new dimension to cardiovascular disease. *Circulation* **121**:1022–1032.

Straarup, E.M., N. Fisker, M. Hedtjarn, M.W. Lindholm, C. Rosenbohm, V. Aarup, H.F. Hansen, H. Orum, J.B. Hansen, and T. Koch. 2010. Short locked nucleic acid antisense oligonucleotides potently reduce apolipoprotein B mRNA and serum cholesterol in mice and non-human primates. *Nucleic Acids Res* **38**:7100–7111.

van Rooij, E., L.B. Sutherland, X. Qi, J.A. Richardson, J. Hill, and E.N. Olson. 2007. Control of stress-dependent cardiac growth and gene expression by a microRNA. *Science* **316**:575–579.

Wang, X.W., N.H.H. Heegaard, and H. Ørum. 2012. MicroRNAs in liver disease. *Gastroenterology* **142**:1431–1443.

Worm, J., J. Stenvang, A. Petri, K.S. Frederiksen, S. Obad, J. Elmén, M. Hedtjärn, E.M. Straarup, J.B. Hansen, and S. Kauppinen. 2009. Silencing of microRNA-155 in mice during acute inflammatory response leads to derepression of c/ebp Beta and down-regulation of G-CSF. *Nucleic Acids Res* **37**:5784–5792.

INDEX

AAV. *See* Adeno-associated virus
ABL1, 377
Acetaminophen (APAP), 513
Acetylcholinesterase (AChE), 545, 547
Activator protein 1 (AP1), 481
Acute lymphoblastic leukemia (ALL), 8–9, 361
  aberrant DNA methylation in, 371–373
    chromosomal translocations and, 373, 377–379
    miRNA biomarkers arising from, 374–376
    miRNA deregulation by, 374–379
    miRNA functional involvement after, 376–377
Acute myeloid leukemia (AML), 43, 274, 355
  deregulated miRNA and transcription factor coregulation in, 358–360
  diagnostic, prognostic, and predictive significance of miRNA expression in, 360–363
  therapeutic miRNAs in, 363–364
AD. *See* Alzheimer's disease
Adaptive immunity
  B lymphocytes in, 48–49
  T lymphocytes in, 49–52
ADARs. *See* Adenosine deaminases that act on RNA
Adeno-associated virus (AAV), 651
Adenosine deaminases that act on RNA (ADARs), 4
Adenovirus, 121–122, 643
Adult stem cells, 28–29
Aging, 77–78. *See also* Senescence
  future research perspectives on, 85
  invertebrate, 78–80
  mammalian
    aging-associated pathway management during, 81–85
    DNA modulation during, 81–83

    miRNA expression changes during, 80–81
Ago proteins. *See* Argonaut proteins
Air, 252
AIs. *See* Aromatase inhibitors
Akt signaling pathway, 473
Albuminuria, 502–503
Alcoholic liver disease (ALD), 511–513
ALL. *See* Acute lymphoblastic leukemia
ALS. *See* Amyotrophic lateral sclerosis
Alzheimer's disease (AD), 539–540, 549
  inflammatory processes and
    at neuronal–immune interface, 546–547
    overreactive immune system, 545–546
  miRNAs in
    prevalence of, 540–541
    quest for more effective treatment, 541–542
    regulation in neurodegenerative diseases, 542–544
  neuronal miRNAs in
    behavior and, 545–546
    as controllers, 544–545
    function at synapse, 543–544
    memory and, 542–543
  therapy for
    diagnostic biomarkers, 547
    different approaches to miRNA manipulations, 547–548
    limitations of, 548–549
Amino-LNA, 664
AML. *See* Acute myeloid leukemia
AMOs. *See* Anti-miRNA oligonucleotides
Amplification, in genetic dysregulation, 7–8
Amyotrophic lateral sclerosis (ALS), 542
Androgen receptors (ARs), 313
Angiogenesis, 480, 484–485
  miR-210 and, 273
  in tumors, 612–613

ANRIL, 249, 253, 255, 258–259
Antagomirs, 363–364, 561–562, 644–645, 654–655
Antiestrogens, resistance to, 295–297
Anti-miRNA oligonucleotides (AMOs), 644
Anti-miRs, 644
Antipsychotics
 for bipolar affective disorder, 559–560
 limitations of, 555–556
Antisense non-coding RNAs (AS ncRNAs), 248
Aortic aneurysms, 485
AP1. See Activator protein 1
APAP. See Acetaminophen
Apoptosis
 circulating miRNAs in, 570–572
 in malaria, 191–192
 miR-210 and, 274
AREs. See AU-rich elements
Argonaut (Ago) proteins, 5, 6–7, 10–11, 543, 595–596
Armitage, 544
Aromatase inhibitors (AIs), resistance to, 295
Arrhythmia, 484
ARs. See Androgen receptors
AS ncRNAs. See Antisense non-coding RNAs
Asymmetric cell division, 29
Atelocollagen, 652
AU-rich elements (AREs), miRNA binding to, 204
Avian herpesviruses, 137–138
 diseases associated with, 138–139
 identification of miRNAs encoded by
  DEV miRNAs, 140–141, 143
  HVT miRNAs, 140–143
  ILTV miRNAs, 140–141, 143
  MDV-1 miRNAs, 139–142
  MDV-2 miRNAs, 140–142
 miRNA targets of, 145–146
  cellular targets, 147–148
  viral targets, 146–147
 viral orthologues of host miRNAs, 143–145
5-Aza-2′-deoxycytidine, 9
5′-Azacytidine, 377

B lymphocytes
 in adaptive immunity, 48–49
 development of, 94–96, 405–406, 420–423
 in innate immunity, 45–46
 miR-29 expression in, 391
BACE1-AS ncRNA, 253
Basal-like breast cancer, 288–289
B-cell leukemias, 208

B-cell non-Hodgkin's lymphomas, 403–405
 B-cell differentiation and, 405–406
 Burkitt's lymphoma, 406–407, 409
 CLL, 383–384, 407, 409–410
  characteristics and clinical outcomes of, 384–385
  DNA methylation in, 394–395
  miR-15a/16-11 in, 7–8, 208, 231, 386–391, 393–395, 410
  miR-29 in, 386–387, 389, 391–393, 409–410
  miR-34(b/c) in, 386–387, 389, 393–394, 409–410
  MiR-181 in, 386–387, 389, 391–392, 409–410
  miRNA association with, 385
  miRNA dysregulation in, 12
  miRNA functional role in, 386–387
  miRNA signatures in, 385–386
  SNPs in, 394–395
 DLBCL, 407, 409–410, 419–420
  B-cell differentiation stage-specific miRNA expression and, 420–423
  miR-17-92 cluster in, 425–427
  miR-155 in, 422–423, 425–426
  miRNA biomarkers in, 427–429
  miRNA expression in, 422–425
  miRNA roles in, 425–427
 follicular lymphoma, 407, 410–411
 MALT lymphoma, 408, 411–412
 MCL, 408–409, 411
 NMZL, 408, 412
 SMZL, 412
 therapeutic miRNAs in, 412–413
BCL2-TRAF2 transgenic mice, 390–391
BCR–ABL1 fusion oncogene, 377
BCSCs. See Breast cancer stem cells
BDNF. See Brain-derived neurotrophic factor
Behavior, 545–546
Biliary diseases, 515–516
BIM, 48–51
Biogenesis, 3–5, 102–103
 in carcinogenesis, 225–226
 in CTCLs, 452–453
 HIV-1 effects on, 170–171
 regulation of, 10–11, 66–67, 206
 of viral miRNAs, 124–125
Biomarkers. See Markers/biomarkers
Bipolar affective disorder, 554, 559–560
 pharmacology for, 561
BL. See Burkitt's lymphoma
Blood–brain barrier, antagomirs and, 562–563

BLV. *See* Bovine leukemia retrovirus
Boceprevir, 668
Bovine leukemia retrovirus (BLV), 124–125, 127–128
Bowel disease, 576, 580
Brain, immune system and, 545–546
Brain cancer stem cells, 630–631
Brain-derived neurotrophic factor (BDNF), 557–558, 560
BRCA1 expression, 83
Breast cancer, 287–289
  biomarkers of
    aberrant miRNA expression and, 289–291
    circulating miRNAs, 575, 577–578
    tumor subtype prediction with, 291–292
  ErbB receptor signaling in, 297–298
    miRNA influence on HER2-targeting therapies, 301–302
    miRNA regulation of EGFR signaling, 298–299
    miRNA regulation of ErbB2 and ErbB3 signaling, 299–301
  ERs in, 288–289, 292–293
    estrogenic regulation of miRNA expression, 294–296
    miRNA markers of ER expression, 291–292
    miRNA regulation of, 293–294
    miRNAs in endocrine resistance, 295–297
  HER2 in, 288–289
    endocrine therapy resistance and, 296–297
    miRNA biomarkers of, 291–292
    miRNA regulation of, 299–301
    therapies targeting, 298, 301–302
  miR-125(a/b) in, 7–8, 290, 297, 299, 300
  miR-210 in, 270–271, 275, 290, 578
  miRNA polymorphisms in, 208–209
  therapeutic miRNAs in, 213
Breast cancer stem cells (BCSCs), 627–630
Bromocriptine, 559
Burkitt's lymphoma (BL), 406–407, 409
Bx332409, 255

CAD. *See* Coronary artery disease
CAGRs. *See* Cancer-associated genomic regions
cALCL. *See* Primary cutaneous anaplastic large-cell lymphoma
Cancer. *See also specific cancers*
  angiogenesis in, 612–613
  biomarkers of, 14–15, 212
    circulating miRNAs, 575–580
  biosynthesis defects of miRNAs in, 11
  cell communication between tumors and surrounding cells, 609
  DDR pathway in, 81
  diagnosis of, 233–234, 633–634
  DNA methylation in, 372–373
  epigenetic dysregulation of miRNA expression in, 8–9, 210
  genetic dysregulation of miRNA expression in, 7–8
  genetic variation in
    lncRNA alterations, 258–259
    miRNA alterations, 208–209
  hypoxia in
    HIFs in, 269, 275–280
    miR-210 in, 270–276, 279–280
    miRNA regulation under, 269–270
    in solid tumors, 268–269
    transcriptional regulation in, 269
  metastasis in, 319–320, 613–615
  miRNA dysregulation in, 12
  miRNA expression profiling in, 14, 211, 468
    breast cancer, 289–292
    prostate cancer, 313–317
  oncogenic miRNAs, 207–208, 223–224, 641–642
    clinical diagnosis, prognosis, and therapy based on, 233–234
    in prostate cancer, 314–316
    role in cancer, 225–230
    targets of, 224–225, 228
  principles of miRNA involvement in, 201–214
  prognosis of, 233–234, 633–634
  senescence and, 62
  therapeutic miRNAs in, 212–213, 233–234, 633–634
  tumor suppressor miRNAs, 207–208, 223–224, 641–642
    clinical diagnosis, prognosis, and therapy based on, 233–234
    in prostate cancer, 316–317
    role in cancer, 225–228, 230–233
    targets of, 224–225, 228
Cancer stem cells (CSCs), 615–617, 625–626
  in cancer diagnosis, prognosis, and therapy, 633–634
  miRNA regulation of, 626–630
    breast, 627–630
    glioma and brain, 630–631
    other, 632–633
    prostate, 631–632

Cancer-associated genomic regions (CAGRs), 207
Cardiac fibrosis, 480, 483–484
Cardiac hypertrophy, 480–482
Cardiomyocyte death, 480–483
Cardiovascular disease, 477–480, 487
   angiogenesis and vascular diseases, 484–485
   arrhythmia, 484
   cardiac fibrosis, 483–484
   cardiac hypertrophy, 480–482
   circulating miRNA biomarkers in, 576, 581
   heart failure, 485–486
   lipid metabolism, 486–487
   myocardial ischemia and cell death, 481–483
CCA. *See* Cholangiocarcinoma
CCND1, 249, 251, 257
CD8+ T cells, 50–51
CD47, 532
CDK6-RB pathway, 376
Cell communication
   circulating miRNAs in, 592–594
   EVs in, 608–609
   between tumors and surrounding cells, 609
Cell cycle
   miR-210 and, 273–274
   pluripotent state maintenance through control of, 30–31
   reprogramming and, 34–35
Cell fate, 31–33
Cellular messengers, circulating miRNAs as, 589–591, 599–600
   cellular export of miRNA, 594
   delivery of miRNA to recipient cells, 596
   extracellular miRNAs and, 590–594, 596–599
   intercellular communication, 592–594
   lipid-based miRNA carriers, 594–595
   protein miRNA carriers, 595–596
Cellular morphology, of senescent cells, 63
Cerebral malaria, 191–193
CeRNA hypothesis. *See* Competing endogenous RNA hypothesis
Cervical cancer, 575, 578
cET-LNA. *See* Constrained ethyl locked nucleic acids
Chlorpromazine, 561
Cholangiocarcinoma (CCA), 511, 515
Cholangiocytes, 510, 515
Choroideremia, 469
Chromosomal abnormalities, in CLL, 393–394
Chromosomal translocations, in ALL, 373, 377–379
Chronic lymphocytic leukemia (CLL), 383–384, 407, 409–410
   characteristics and clinical outcomes of, 384–385
   DNA methylation in, 394–395
   miR-15a/16-11 in, 7–8, 207–208, 231, 386–391, 393–395, 410
   miR-29 in, 386–387, 389, 391–393, 409–410
   miR-34(b/c) in, 386–387, 389, 393–394, 409–410
   MiR-181 in, 386–387, 389, 391–392, 409–410
   miRNA association with, 385
   miRNA dysregulation in, 12
   miRNA functional role in, 386–387
   miRNA signatures in, 385–386
   SNPs in, 394–395
Chronic myelogenous leukemia (CML)
   deregulated miRNA and transcription factor coregulation in, 360
   diagnostic, prognostic, and predictive significance of miRNA expression in, 360–363
Circulating miRNAs
   as biomarkers, 569–570, 581–582
   biological role and, 570–572
   in bowel disease and diabetes, 504, 576, 580
   in breast and ovarian cancer, 575, 577–578
   in cardiovascular disease, 576, 581
   in colorectal cancer, 330–332, 575
   in gastric cancer, 575, 578
   in liver and kidney disease, 576, 580–581
   in lung cancer, 575, 578–579
   in pregnancy, 576, 580
   in prostate cancer, 575, 579
   quantification for, 572–576
   as cellular messengers, 589–591, 599–600
   cellular export of miRNA, 594
   delivery of miRNA to recipient cells, 596
   extracellular miRNAs and, 590–594, 596–599
   intercellular communication, 592–594
   lipid-based miRNA carriers, 594–595
   protein miRNA carriers, 595–596
   in renal cell carcinoma, 611–612
Claudin-low breast cancer, 289
CLL. *See* Chronic lymphocytic leukemia

CLOCK gene, 560
CML. *See* Chronic myelogenous leukemia
CNS injury, in malaria, 191–193
CNVs. *See* Copy number variants
Colon cancer stem cells, 629
Colonosphere, 629
Colorectal cancer (CRC), 329–330
 biomarkers in
  circulating miRNAs in plasma/serum, 330–332, 575
  miRNAs in tumor tissue, 332–336
 prognosis of, 234
Competing endogenous RNA (ceRNA) hypothesis, 203–204, 214
Connective tissue growth factor (CTGF), 483
Connexin43 (Cx43), 484
Constrained ethyl locked nucleic acids (cET-LNA), 664
Copy number variants (CNVs), 208
Coronary artery disease (CAD), 482–483
CpG islands, 373
CRC. *See* Colorectal cancer
CREB activation, 544
CSCs. *See* Cancer stem cells
CTCLs. *See* Cutaneous T-cell lymphomas
CTGF. *See* Connective tissue growth factor
CUDR, 255
Cutaneous T-cell lymphomas (CTCLs), 449–450, 457–458
 aberrant miRNA biogenesis in
  DICER and RISC complex expression and, 453
  microprocessor complex expression and, 452
  processing modulators and, 452–453
 cALCL, 451–452, 454–457
 classification of, 450–451
 functional consequences of aberrant miRNA expression in, 457
 miRNA expression profiling in, 453–457
 mycosis fungoides, 450–452, 455–457
 Sézary syndrome, 207, 450–451, 453–458
Cx43. *See* Connexin43
Cytoadherence, in malaria, 186
Cytokines, in malaria, 186–187
Cytomegalovirus. *See* Human cytomegalovirus

DAAs. *See* Direct acting antivirals
Dasatinib, 361
DCs. *See* Dendritic cells
DDR pathway. *See* DNA damage response pathway
Deadenylation, miRNA control of, 6–7

Deafness, 469–472
Decoy activity, of miRNAs, 6
Degradation, miRNA control of, 6–7
Deletion, in genetic dysregulation, 7–8
Dendritic cells (DCs), 45–47
Destabilization, miRNA control of, 6–7
DEV. *See* Duck enteritis virus
DFNA50, 469
DFNA50 hearing loss, 469–472
DGCR8, 4–5, 11, 30
Diabetes, 495–496
 circulating miRNA biomarkers in, 504, 576, 580
 gene expression changes and, 496–497
 miRNAs in β-cell functions, 497–499
 miRNAs in complications of, 502–503
 miRNAs in insulin target tissues, 499–502
Diabetic nephropathy, 502–503
Diabetic retinopathy, 503
DICER, 4–5, 10–11
 in cancer, 11
 in CTCLs, 453
 deletion of
  embryonic stem cells effects of, 30
  immune system effects of, 42, 45, 47–48, 50–51
  senescence effects of, 67–68
 in hematopoiesis, 104
 in HIV-1 infection, 170
 in memory, 543
Differentiation. *See also* Hematopoiesis
 miR-210 and, 274
 miRNA regulation of, 31–33
Diffuse large B-cell lymphoma (DLBCL), 407, 409–410, 419–420
 B-cell differentiation stage-specific miRNA expression and, 420–423
 miR-17-92 cluster in, 425–427
 miR-155 in, 422–423, 425–426
 miRNA biomarkers in, 427–429
 miRNA expression in, 422–425
DiGeorge syndrome, 468
DILI. *See* Drug-induced liver injury
Direct acting antivirals (DAAs), 668
Disrupted-in-schizophrenia-1 (DISC1), 556
DLBCL. *See* Diffuse large B-cell lymphoma
DLEU7, 389–391
DNA damage response (DDR) pathway, 78–79, 81–83
DNA methylation
 in ALL, 371–373
  chromosomal translocations and, 373, 377–379

DNA methylation (cont'd)
  miRNA biomarkers arising from, 374–376
  miRNA deregulation by, 374–379
  miRNA functional involvement after, 376–377
  in cancer, 372–373
  in CLL, 394–395
  in genetic dysregulation of miRNA expression, 8
DNA modulation, during aging, 81–83
DNA repair, *miR-210* and, 273
DNA viruses, 122–124. *See also specific DNA viruses*
DNA-demethylating agents, 9
DND1, 11–12
Dopamine
  in bipolar affective disorder, 559–560
  in schizophrenia, 555
Dopaminergic neurons, 542
Dox treatment, *miR-146(a/b)* after, 486
Drosha, 4–5, 10–11
  in cancer, 11
  in HIV-1 infection, 170
Drug-induced liver injury (DILI), 511, 513
Duchenne muscular dystrophy, 469
Duck enteritis virus (DEV), 138–141, 143
Dystrophic neurites, 544–545

Early-onset anterior polar cataracts, 472–473
EBV. *See* Epstein-Barr herpesvirus
ECM. *See* Extracellular matrix
EDICT syndrome, 472–473
EGFR signaling, regulation of, 298–299
Embryonic stem cells (EScs), 28–29
  pluripotency of, 30
    cell cycle control and, 30–31
    differentiation fate choice and, 31–33
ENA. *See* Ethylene nucleic acid
Endocrine therapy, for breast cancer, 292
  resistance to, 295–297
Endothelial activation, in malaria, 186, 191–192
Enhancer RNAs (eRNAs), 248, 250
Epidermolysis bullosa, 469
Epigenetic regulation. *See also* DNA methylation; Histone modification
  lncRNAs role in, 252–253
  by miRNAs, 210
  of miRNAs, 8–9, 210
  in prostate cancer, 320–321

Epstein-Barr herpesvirus (EBV), 121–128, 597
  in HL, 436–438
  latency of, 161
ErbB receptor signaling
  breast cancer and, 297–298
  miRNA influence on HER2-targeting therapies, 301–302
  miRNA regulation of EGFR signaling, 298–299
  miRNA regulation of ErbB2 and ErbB3 signaling, 299–301
  miRNA regulation of ErbB2 and ErbB4 signaling, 486
ERK-MAPK. *See* Extracellular signal-regulated kinase/mitogen-activated protein kinase
eRNAs. *See* Enhancer RNAs
ERs. *See* Estrogen receptors
Erythroblastic leukemias, 208
Erythrocytes
  development of, 97
  malaria infection of, 185–186
ES cell-specific cell cycle-regulating miRNAs (ESCC), pluripotency regulation by, 31–33
ESes. *See* Embryonic stem cells
Esophageal cancer, 575, 580
Essential thrombocythemia (ET), 109–110
Estradiol, in breast cancer, 293–296
Estrogen, miRNA regulation by, 294–296
Estrogen receptors (ERs), 288–289, 292–293
  estrogenic regulation of miRNA expression, 294–296
  miRNA markers of, 291–292
  miRNA regulation of, 293–294
  miRNAs in endocrine resistance, 295–297
ET. *See* Essential thrombocythemia
Ethylene nucleic acid (ENA), 664
EVF2, 251
EVs. *See* Extracellular vesicles
Exosomes, 594–595
Exportin-5, 4
Extracellular matrix (ECM) proteins, 483
Extracellular miRNAs, 590–593
  cellular export of, 594
  functional roles of, 596–599
Extracellular signal-regulated kinase/mitogen-activated protein kinase (ERK-MAPK) signaling pathway, 483
Extracellular vesicles (EVs), 607–608
  cancer stem cells and, 615–617
  in cell communication, 608–609

in renal carcinomas, 609–610
  circulating miRNAs and, 611–612
  miRNAs and, 610–611
in tumor angiogenesis, 612–613
in tumor metastasis, 613–615

Fibroblast growth factor 20 (FGF20), 547
Fibrosis. See Cardiac fibrosis; Liver fibrosis
FL. See Follicular lymphoma
Fludarabine, 386
5-Fluorouracil (5-FU), 335
Follicular lymphoma (FL), 407, 410–411
14q32 cluster, in ALL, 8
Fragile X syndrome, 468
5-FU. See 5-Fluorouracil
Fuchs corneal dystrophy, 473
Fulvestrant, 295, 297

G1/S transition, 31
GABAergic system
  dysregulation of, 559
  in schizophrenia, 555
Gallid herpesvirus 1. See Infectious laryngotracheitis virus
Gallid herpesvirus 2. See Marek's disease virus-1
Gallid herpesvirus 3. See Marek's disease virus-2
Gap junction protein α1 (GJA1), 484
GAS5, 257
Gastric cancer, 575, 578
Gastric cancer stem cells, 628
GBM. See Glioblastoma multiforme
Gefitinib, 300
Gene expression arrays, miRNA targetome research with, 13
Gene expression regulation, by miRNAs, 3
  epigenetic mechanisms of, 210
  mechanisms of, 6–7
Gene silencing, 3
Genetic regulation, of miRNA expression, 7–8
Genetic variations. See also Mutations; Single nucleotide polymorphisms
  in cancer
    lncRNA alterations, 258–259
    miRNA alterations, 208–209
  in lung cancer susceptibility and survival, 343–345
    NBS1 SNPs affecting *miR-629* binding, 349
    REV3L 3'-UTR SNP affecting *miR-25* and *miR-32* regulation, 348–349

SNP in *KRAS* 3'-UTR affecting *let-7* binding, 347
SNPs in *MYCL1 miR-1827* binding site, 347–348
SNPs in pre-miRNA flanking region, 346
SNPs in pre-miRNA sequence, 345–346
GeromiRs, 77–78
  future perspectives on, 85
  invertebrate, 78–80
  mammalian
    aging-associated pathway management, 81–85
    DNA modulation, 81–83
    miRNA expression changes, 80–81
Gga-*mir-155*, 141, 143–144, 147
Gga-*mir-221*, 144–145
GJA1. See Gap junction protein α1
Glioblastoma, 628–629
Glioblastoma multiforme (GBM), 630
Glioma, 7–8
Glioma cancer stem cells, 628, 630–631
Glucose transporter 4 (GLUT4), 486
Glutamate, in schizophrenia, 555
Granulocytes
  development of, 96–97
  in innate immunity, 43, 46
Granuphilin, 499

H19, 254–255, 258
Haloperidol, 561
HCC. See Hepatocellular carcinoma
HCMV. See Human cytomegalovirus
HDL. See High-density lipoprotein
Head and neck cancers, 270–271, 275
Heart disease. See Cardiovascular disease
Heart failure, 485–486
Helper CD4+ T cells, 50–51
Hematologic disease. See also specific diseases
  miRNA dysregulation in, 102–103
Hematopoiesis, 91–93, 97–98, 103–104, 355. See also Myelopoiesis
  erythrocyte development, 97
  hematopoietic lineages, 92–93
  lymphocyte development, 94–96, 405–406, 420–423
  megakaryocyte development, 97, 103–108
  monocyte and granulocyte development, 96–97
Hematopoietic stem cells (HSCs), 28, 92–93, 103–104
Hepatic stellate cells (HSCs), 516

Hepatitis B. *See* Viral hepatitis
Hepatitis C. *See* Viral hepatitis
Hepatocellular carcinoma (HCC), 13, 274, 511, 514
  circulating miRNAs in, 575
  viral delivery of miRNAs for, 651–652
Hepatocytes, 510
HER2, 288–289. *See also* ErbB receptor signaling
  endocrine therapy resistance and, 296–297
  miRNA biomarkers of, 291–292
  miRNA regulation of, 299–301
  therapies targeting, 298
    miRNA influence on, 301–302
HER2 enriched breast cancer, 288–289
Hereditary disorders, 465–467, 473
  genomic rearrangements affecting miRNA sequence, 468–469
  mutations affecting 3′-UTR of mRNAs, 467–468
  mutations in genes involved in miRNA processing and function, 468
  point mutations resulting in monogenic-based disorders, 469–473
Hereditary hearing impairment, 469
Herpes simplex virus (HSV), 121–124, 126, 128, 161
Herpesvirus of turkey (HVT), 138–145
Herpesviruses, 121–129. *See also* Avian herpesviruses; *specific herpesviruses*
Heterozygosity loss, in genetic dysregulation, 8
HHV. *See* Human herpesvirus
HIFs. *See* Hypoxia-inducible factors
High-density lipoprotein (HDL), 486–487, 501–502
Histone deacetylase inhibitors, 9
  in CLL, 395
Histone modification, in genetic dysregulation, 8–9
History, of miRNAs, 2–3
HITS-CLIP, miRNA targetome research with, 13
HIV-1 infection. *See* Human immunodeficiency virus type 1 infection
HL. *See* Hodgkin's lymphoma
Hodgkin Reed Sternberg (HRS) cells, 436–437, 441
Hodgkin's lymphoma (HL), 435–437, 444
  clinical value of miRNAs in, 443
  functional miRNA studies in, 441–443
  future research perspectives in, 443

miRNA profiling studies in
  HL cell lines, 438–441
  microdissected HRS cells, 441
  total tissue samples, 441
Host immune response
  HCMV miRNA roles in evasion of, 158–159
  in malaria, 186–193
Host-virus interactions, 119–121
  future prospects in, 129–130
  host miRNAs in, 121
  viral miRNAs in, 121–122
    biological roles of, 126–129
    of DNA viruses, 122–124
    expression of, 124–126
    functional convergence of, 127–128
    future prospects in, 129–130
    of RNA viruses, 122–124
    in vivo, 129
HOTAIR, 249, 252–253, 255, 258
HPV. *See* Human papillomaviruses
HRS cells. *See* Hodgkin Reed Sternberg cells
HSCs. *See* Hematopoietic stem cells; Hepatic stellate cells
HSV. *See* Herpes simplex virus
HULC, 256, 258
Human cytomegalovirus (HCMV), 121–122, 125, 127–128, 153–154
  creation of latency-conducive environments by miRNAs, 155
  cellular miRNA regulation, 159–160
  HCMV miRNA target identification, 160
  host immune evasion, 158–159
  repression of IE gene expression, 157
  identification of miRNAs, 155–156
  summary of current knowledge, 156, 160–162
Human herpesvirus (HHV), 121–124
Human immunodeficiency virus type 1 (HIV-1) infection, 123, 165–168
  HIV-1 regulation of host miRNAs, 170
    biogenesis pathway perturbation, 170–171
    expression profile perturbation, 171
  host miRNA regulation of HIV-1, 172–173
  miRNAs and siRNAs in, 168–169
  small ncRNAs in virally infected cells, 173–174
    artificially encoded sncRNAs, 175–176
    naturally encoded sncRNAs, 174–175
  tdsmRNAs in, 168–169, 176–177
Human papillomaviruses (HPV), 123

HuR, 11
HVT. *See* Herpesvirus of turkey
Hvt-*mir-H14-3p*, 144–145
Hypertriglycemia, 486
Hypoxia, 267
  drugs targeting, 268–269
  HIFs in, 269, 275–280
  in malaria, 189, 191
  miR-210 in, 270–276, 279–280
  miRNA regulation under, 269–270
  in solid tumors, 268–269
  transcriptional regulation in, 269
Hypoxia-inducible factors (HIFs), 269
  regulation of, 275–280, 485

IE genes. *See* Immediate early genes
IGF-1. *See* Insulin-like growth factor
ILTV. *See* Infectious laryngotracheitis virus
Iltv-mir-I5, 146
Iltv-mir-I6, 146
Imatinib, 361
Immediate early (IE) genes, HCMV miRNA effects on, 157
Immune system
  brain and, 545–546
  viral evasion of, HCMV miRNA roles in, 158–159
Immunity, 41–43
  adaptive
    B lymphocytes in, 48–49
    T lymphocytes in, 49–52
  future research directions in, 52
  innate
    DCs in, 45–47
    granulocytes in, 43, 46
    mast cells in, 48
    monocytes and macrophages in, 44–46
    NK and NKT cells in, 45–47
Immunoprecipitation (IP) techniques, 13
Immunosuppression, 275
Induced pluripotency, 33–36
Infectious disease. *See* Viral infections
Infectious laryngotracheitis virus (ILTV), 138–141, 143, 146
Inflammation, 546–547
Inhibitors
  of miRNAs, 15, 212–213, 234, 644–645, 653–655
  miRNAs as, 212–213, 644–645
Innate immunity
  DCs in, 45–47
  granulocytes in, 43, 46
  mast cells in, 48

monocytes and macrophages in, 44–46
NK and NKT cells in, 45–47
Insulin, in aging, 84–85
Insulin target tissues, miRNAs in, 499–502
Insulin-like growth factor (IGF-1)
  in aging, 84–85
  MiR-1 and, 480
Intercellular communication, circulating miRNAs in, 592–594
Intergenic miRNAs, 205–206
Intragenic miRNAs, 205–207
Intronic miRNAs, 205–207, 481–482
IP techniques. *See* Immunoprecipitation techniques

Kaposi's sarcoma-associated herpesvirus (KSHV), 121–122, 124–129, 143, 161
KCNJ2. *See* Potassium inwardly rectifying channel, subfamily J, member 2
Kcnq1ot1, 252
Keratoconus, 472–473
Kidney disease, 576, 580–581
KLF15. *See* Kruppel-like factor-15
KRAS, lung cancer and, 347
Kruppel-like factor-15 (KLF15), 486
KSHV. *See* Kaposi's sarcoma-associated herpesvirus

Lapatinib, 298, 300
Large intergenic non-coding RNAs (lincRNAs), 247–248
Latency, HCMV miRNA roles in, 153–154
  creation of latency-conducive environments by miRNAs, 155–160
  identification of miRNAs, 155–156
  summary of current knowledge, 156, 160–162
LDL. *See* Low-density lipoprotein
Lentiviruses, 643
Let-7 family
  in aging, 84
  in ALL, 374, 378
  in AML, 361
  in angiogenesis, 484
  with antipsychotics, 560
  in B cell differentiation, 406
  in breast cancer, 293–294, 296, 301
  in Burkitt's lymphoma, 406
  in cancer, 208
  cell cycle control by, 31
  in colorectal cancer, 333, 334
  in CSCs, 627–628, 631–634
  delivery of, 644, 647–650

Let-7 family (*cont'd*)
  in diabetes, 497, 501
  differentiation regulation by, 32–33
  in follicular lymphoma, 411
  in HIV-1 infection, 175
  in HL, 439, 442, 444
  inhibition of, 645–646, 655
  in insulin target cells, 498
  in liver diseases, 514–516
  LNA inhibition of, 667
  in lung cancer, 210, 211, 213, 345, 347, 579
  in malaria, 188, 192
  in megakaryocytopoiesis, 105, 108
  in myeloid disorders, 361
  in NMZL, 412
  in prostate cancer, 314, 316–321, 579
  regulation of, 9–11
  replacement therapy with, 649
  in reprogramming, 34–35
  in senescence regulation, 67
  in T cell function, 51
  target genes of, 13
  in tumor metastasis, 615
  as tumor suppressor, 226, 228, 230–231
Leukemias, 8, 13, 103. *See also specific types*
  circulating miRNAs in, 575
  deregulated miRNA and transcription factor coregulation in, 358–360
  diagnostic, prognostic, and predictive significance of miRNA expression in, 360–363
  therapeutic miRNAs in, 363–364
Lin-4, 356
  in aging, 78–79
Lin-14, 356
  in aging, 79
Lin-28(a/b), 501
LincRNA-p21, 249, 251, 253, 257–258
LincRNAs. *See* Large intergenic non-coding RNAs
Lipid metabolism, 480, 486–487
Lipid-based miRNA carriers, 594–595
Lipofectamine, 653
Lipoproteins, 595
Lithium, 559–561
Liver cancer, 208. *See also* Hepatocellular carcinoma
  markers of, 258
  therapeutic miRNAs in, 213
Liver diseases, 509–511, 517
  alcoholic liver disease, 512–513
  biliary diseases, 515–516
  circulating miRNA biomarkers in, 576, 580–581
  drug-induced liver injury, 513
  fibrosis, 516
  HCC, 13, 274, 511, 514, 575, 651–652
  miRNA-based therapeutic approaches to, 516–517
  non-alcoholic fatty liver disease, 513
  viral hepatitis, 511–512
Liver fibrosis, 516
Liver steatosis, 486, 654
LNAs. *See* Locked nucleic acids
LncRNAs. *See* Long non-coding RNAs
Locked nucleic acids (LNAs), 644–646, 654, 669–670
  activity of, 666–667
  in clinical development, 668–669
  design of, 665–666
  molecular structure of, 664
  in pre-clinical development, 667–668
  tiny, 665–667
Long noncoding RNAs (lncRNAs), 245–246
  in cancer, 254
    genetic alterations in lncRNAs, 258–259
    tumor suppressor and oncogenic lncRNAs, 254–258
  classifications of, 247–250
  functions and mechanisms of, 250
    epigenetic regulation, 252–253
    posttranscriptional regulation, 253–254
    transcriptional regulation, 251–252
  future research perspectives in, 259
  general features of, 246–247
Low-density lipoprotein (LDL), 486–487, 501–502
Luminal breast cancer, 288–289
Lung cancer, 8, 210, 211
  circulating miRNAs in, 575, 578–579
  genetic variations affecting susceptibility and survival in, 343–345
    NBS1 SNPs affecting *miR-629* binding, 349
    REV3L 3′-UTR SNP affecting *miR-25* and *miR-32* regulation, 348–349
    SNP in *KRAS* 3′-UTR affecting *let-7* binding, 347
    SNPs in *MYCL1 miR-1827* binding site, 347–348
    SNPs in pre-miRNA flanking region, 346
    SNPs in pre-miRNA sequence, 345–346
  prognosis of, 234
  therapeutic miRNAs in, 213

Lymphocytes
  in adaptive immunity, 48–49, 49–52
  development of, 94–96, 405–406, 420–423
  in innate immunity, 45–46, 45–47
Lymphomas, 103, 129. *See also specific lymphomas*
  circulating miRNAs in, 575, 577

Macro ncRNAs, 250
Macrophages, 44–46
Major depression, 554
Major immediate early (MIE) genes, HCMV miRNA effects on, 155, 157
Malaria, 183–185
  future research directions in, 193
  miRNAs in
    host immune response and, 188–190
    mosquito miRNAs, 188
    parasite miRNAs, 187–188
  pathogenesis of, 185
    cytokines and host immune response, 186–187
    endothelial activation and dysfunction, 186
    microvascular flow reduction, 186
    sequestration and cytoadherence, 186
  pathophysiological mechanisms in host response to
    apoptosis and programmed cell death, 191–192
    CNS injury, 191–193
    endothelial activation, 191–192
    hypoxia, 189, 191
MALAT-1, 253–255, 258
MALT lymphoma, 408, 411–412
Mantle cell lymphoma (MCL), 408–409, 411
Marek's disease virus (MDV), 122, 126–129
Marek's disease virus-1 (MDV-1), 138–148
Marek's disease virus-2 (MDV-2), 138–142, 144–145
Marginal zone lymphoma, 408, 411–412
Markers/biomarkers
  in AD, 547
  in ALL, 374–376
  in breast cancer, 289–292, 575, 577–578
  in cancer, 14–15, 212
    circulating miRNAs, 575–580
    circulating miRNAs as, 569–570, 581–582
    biological role and, 570–572
    in bowel disease and diabetes, 504, 576, 580
    in breast and ovarian cancer, 575, 577–578
    in cancer, 575–580
    in cardiovascular disease, 576, 581
    in colorectal cancer, 330–332, 575
    in gastric cancer, 575, 578
    in liver and kidney disease, 576, 580–581
    in lung cancer, 575, 578–579
    in pregnancy, 576, 580
    in prostate cancer, 575, 579
    quantification for, 572–576
  clinical usefulness of miRNAs as, 14–15
  in CLL, 385–386
  in colorectal cancer
    circulating miRNAs in plasma/serum, 330–332, 575
    miRNAs in tumor tissue, 332–336
  in DLBCL, 427–429
  lncRNAs as, 258
  miR-210 as, 275
  in myeloid disorders and AML, 360–363
  platelet miRNAs as, 109–110
  in prostate cancer, 258, 318–319, 575, 579
  in RCC, 610–612
  of senescent cells, 63
Mast cells, 48
Matrix metalloprotease-2 (MMP-2), 483
MCL. *See* Mantle cell lymphoma
MDSs. *See* Myelodysplastic syndromes
MDV. *See* Marek's disease virus
MDV-1. *See* Marek's disease virus-1
Mdv1-mir-M3, 148
Mdv1-mir-M4-3p, 146–147
Mdv1-mir-M4-5p, 141, 143–144, 146–148
Mdv1-mir-M7-5p, 147–148
Mdv1-mir-M9, 141
MDV-2. *See* Marek's disease virus-2
Mdv2-mir-M21, 144–145
MECP2. *See* Methyl CpG-binding protein 2
Medicine, present and future of miRNAs in
  clinical usefulness of miRNAs, 14–15
  functional consequences of miRNA dysregulation in disease, 12–13
  non-coding RNA role in physiological function and disease, 13–14
Medulloblastoma, 628–629
MEG3, 257–258
Megakaryocytes, 97, 103–108
Mei-26, 11
Melanoma, 575
Meleagrid herpesvirus 1. *See* Herpesvirus of turkey
Membrane-derived microparticles (MPs), 595
Memory, 542–543

Messenger RNA (mRNA)
  deadenylation of, 6–7
  degradation of, 6–7
  destabilization of, 6–7
  translational inhibition and, 6–7
Metastasis, 613–615
  in prostate cancer, 319–320
Metformin, 279
Methyl CpG-binding protein 2 (MECP2), 557
MF. *See* Mycosis fungoides
MI. *See* Myocardial ischemia
Microparticles, platelet-derived, 110
Microprocessor complex, in CTCLs, 452
MicroRNAs (miRNAs). *See also* Neuronal miRNAs; Therapeutic miRNAs
  artificial, 652–653
  biogenesis of (*See* Biogenesis)
  characteristics of, 570
  expression of, 206, 640
  function of, 6–7, 102–103, 205
    miRNA dysregulation consequences for, 12–13
    regulation of, 11–12
  gene expression regulation by, 3
    epigenetic mechanisms, 210
    mechanisms of, 6–7
  genomic location of, 205–207
  history of, 2–3
  in medicine
    clinical usefulness of miRNAs, 14–15
    functional consequences of miRNA dysregulation in disease, 12–13
    non-coding RNA role in physiological function and disease, 13–14
  nomenclature of, 5–6
  physiological and pathophysiological roles of, 640–642
  regulation of, 7
    epigenetic, 8–9, 210
    function regulation, 11–12
    genetic, 7–8
    under hypoxia, 269–270
    synthesis and processing regulation, 10–11
    transcription factors and miRNA regulatory networks, 9–10
  targets/targetomes of (*See* Targets/targetomes)
  transcription of, 3–5, 205–206
Microvascular flow, in malaria, 186
MIE genes. *See* Major immediate early genes
Mimics, miRNAs as, 15

MiR-1
  in AD, 546
  in aging, 81, 84
  in arrhythmia, 484
  in cardiac hypertrophy, 480–481
  in cardiovascular disease, 581
  differentiation regulation by, 33
  in lung cancer, 579
  in prostate cancer, 314, 317–318
  in synapses, 545
  transdifferentiation and, 35–36
  in tumor metastasis, 614
MiR-7(a/b)
  in breast cancer, 298
  in CTCLs, 454
  in MS, 529
  in schizophrenia, 558
MiR-9
  in AD, 545–546
  in aging, 84
  alcohol and, 561
  in ALL, 374–376
  in B cells, 406, 422
  in brain function, 557
  as cellular messenger, 597
  differentiation regulation by, 32–33
  in granulocyte response, 43, 46
  in HL, 438–439, 441–444
  in immune system, 546
  LNA inhibition of, 667
  in monocyte and macrophage response, 44–46
  in prostate cancer, 319
  in RCC, 610–611
  in T cell function, 51
  transdifferentiation and, 35–36
MiR-10(a/b)
  in ALL, 374–375, 378
  in breast cancer, 290–291, 297, 577
  in cancer, 208
  in colorectal cancer, 333
  in hematopoiesis, 97
  inhibition of, 213, 645
  in myeloid disorders and leukemias, 361–362, 364
  in NAFLD, 513
  in tumor metastasis, 614
MiR-15(a)/16-11 cluster
  in angiogenesis, 480, 484
  in B-cell non-Hodgkin's lymphomas, 413
  in breast cancer, 300
  in CLL, 7–8, 207–208, 231, 386–391, 393–395, 410

genetic variations in, 208
in hematopoiesis, 95
in HL, 437
LNA-antimiR for, 667–668
in malaria, 192
in MCL, 411
MYC regulation of, 9
replacement therapy with, 650
as tumor suppressor, 226, 228, 231, 386–389, 652

MiR-15(a/b)
in B cell function, 49
in biliary diseases, 515
in breast cancer, 297
in diabetes, 580
in HL, 439
inhibition of, 646, 655
LNA-antimiR for, 667–668
regulation of, 9
in senescence, 65, 68
in tumor suppression, 652

MiR-16(a)(1/2)
in aging, 80, 82
in AML, 362
antagomirs against, 562
in bowel disease, 580
in breast cancer, 297
in CSCs, 627
in follicular lymphoma, 410
in hematopoiesis, 97
in HL, 438–439, 441
inhibition of, 645
LNA-antimiR for, 667–668
in MS, 529
in myeloid disorders, 362
in myelopoiesis, 357–358
in prostate cancer, 319–320
replacement therapy with, 650
in tumor suppression, 652

MiR-17-92 cluster
in aging, 82
in angiogenesis, 480, 484
in B cells, 48–49, 405–406
in breast cancer, 294
in Burkitt's lymphoma, 406
in cancer, 208
in CSCs, 630, 633
in CTCLs, 454, 457
in DLBCL, 425–427
in HCC, 514
in hematopoiesis, 93–95
HIF regulation by, 277–278
in HIV-1 infection, 171

in HL, 437, 442–444
in malaria, 189, 192
in MCL, 409, 411
MYC regulation of, 9
in myeloid disorders and leukemias, 362, 364
in myelopoiesis, 359–360
as oncogene, 227–228
in T cell function, 50–51
in tumor angiogenesis, 612
in viral infection, 121

MiR-17(3p/5p), 94
in breast cancer, 290, 297
in colorectal cancer, 331, 333
in coronary artery disease, 482
in CTCLs, 457
differentiation regulation by, 33
in DLBCL, 424
in gastrointestinal cancer, 578
in HCC, 514
in hematopoiesis, 96
in HL, 439
in malaria, 192
in MCL, 409, 411
in myeloid disorders and leukemias, 364
in myelopoiesis, 357–358

MiR-18(a/b), 94. See also MiR-17-92 cluster
in aging, 82
in breast cancer, 293–294, 296, 301
in colorectal cancer, 333
in DLBCL, 428
in HL, 439, 441

MiR-19(a/b), 94. See also MiR-17-92 cluster
in breast cancer, 294
in cancer EVs, 617
in colorectal cancer, 333
in DLBCL, 424, 428
in HL, 439
in platelet microparticles, 110
in senescence regulation, 67

MiR-20(a/b), 94. See also MiR-17-92 cluster
in aging, 83
in angiogenesis, 480, 484
in breast cancer, 294, 297
in colorectal cancer, 333, 335
in CSCs, 627
in DLBCL, 424
in follicular lymphoma, 410
in gastrointestinal cancer, 578
in hematopoiesis, 96
HIF regulation by, 275–276
in HL, 439, 441
in MCL, 409, 411

MiR-20(a/b) (cont'd)
   in prostate cancer, 319, 579
   in tumor angiogenesis, 612–613
MiR-21
   antagomirs for, 654
   in aortic aneurysms, 485
   in B-cell leukemias, 208
   in B-cell non-Hodgkin's lymphomas, 412
   in biliary diseases, 515
   in breast cancer, 290, 294, 296–297, 300–302, 578
   cancer screening with, 580
   in cardiac fibrosis, 480, 483
   in cardiomyocyte death, 480–481
   in CLL, 385–386
   in colorectal cancer, 333, 335
   in CTCLs, 454, 457
   in DC differentiation and function, 45–47
   in diabetes, 497–498
   in DLBCL, 409, 422–423, 428
   in heart failure, 486
   in HIV-1 infection, 171
   in HL, 437–439, 441
   inhibition of, 645, 656
   in liver diseases, 581
   LNA inhibition of, 667
   in lung cancer, 234, 578
   in lymphoma, 577
   modulation of, 485
   in MS, 531
   in myeloid disorders and leukemias, 361, 364
   in myelopoiesis, 359
   in NK and NKT cell function, 46–47
   as oncogene, 227–229
   in prostate cancer, 212, 314–315, 318–319, 579
   in RCC, 610–611
   in senescence regulation, 66
   in SMZL, 412
   in T cell function, 51
   in tumor angiogenesis, 612–613
   in tumor metastasis, 613–614
MiR-22
   in aging, 80, 83
   in breast cancer, 293, 296, 297, 301
   in colorectal cancer, 334
   in CSCs, 627, 630
   HIF regulation by, 277–278
   in MS, 525
   in senescence regulation, 66

MiR-23(a/b)
   in B cell differentiation, 405
   in bowel disease, 580
   in breast cancer, 296
   in DLBCL, 428
   in HCC, 514
   in MS, 529
   in senescence regulation, 66
   in tumor angiogenesis, 612
MiR-24(a)
   in aging, 82, 83
   with antipsychotics, 560
   in breast cancer, 296
   in DLBCL, 428
   DND1 regulation by, 12
   in hereditary diseases, 467
   in HL, 438–439
   in myelopoiesis, 357–358
   in schizophrenia, 558
   in senescence, 65, 68
   target genes of, 13
MiR-25
   in biliary diseases, 515
   in HL, 439
   in lung cancer, 345, 348–349
   in MS, 531
   in senescence, 65, 68
MiR-26(a/b)
   in breast cancer, 302
   in Burkitt's lymphoma, 406
   in cardiomyocytes, 481
   in follicular lymphoma, 410
   in glioma, 7–8
   in HIV-1 infection, 171
   in liver cancer, 213
   in liver diseases, 514–515
   in NMZL, 412
   in PV, 110
   replacement therapy with, 649
   in schizophrenia, 558
   in senescence regulation, 66
   viral delivery of, 651–652
MiR-27(a/b)
   in aging, 80–81
   in ALD, 512–513
   in angiogenesis, 480, 484
   in breast cancer, 294, 296
   in cardiac hypertrophy, 480–481
   in CTCLs, 454
   in DLBCL, 428
   in HL, 439
   in malaria, 188

in megakaryocytopoiesis, 105, 107–108
in prostate cancer, 314–316, 318
regulation of, 9–10
in tumor angiogenesis, 612
in viral infection, 121
MiR-28(3p)
  in diabetes, 580
  in DLBCL, 424, 428
  in HIV-1 infection, 172, 175
  in megakaryocytopoiesis, 105, 107
  in senescence regulation, 66
miR-29 transgenic mice, 392–393
MiR-29(a/b/c)
  in AD, 542
  in aging, 81–82
  in aortic aneurysms, 485
  in B cells, 422
  in bowel disease, 580
  in breast cancer, 297
  in cancer EVs, 617
  in cardiac fibrosis, 480, 483
  in CLL, 386–387, 389, 391–393, 409–410
  in colorectal cancer, 331, 333
  in CTCLs, 454
  in diabetes, 497–500, 502–503, 580
  in DLBCL, 424
  epigenetic mechanisms of, 210
  in follicular lymphoma, 410
  in gastrointestinal cancer, 578
  in heart failure, 486
  in HIV-1 infection, 169, 171, 173, 175
  in HL, 438–439
  in liver diseases, 515–516
  in malaria, 192
  in MCL, 409, 411
  in MDV-2, 144–145
  in myeloid disorders and leukemias, 361, 364
  in myelopoiesis, 359
  in NMZL, 412
  non-viral delivery of, 653
  in nucleus, 203
  in prostate cancer, 320–321
  replacement therapy with, 649
  in schizophrenia, 558
  in senescence, 66, 68
  in T cell function, 51–52
  as tumor suppressor, 228, 231–232
MiR-30(a(3p/5p)/b/c(1)/d/e(3p))
  in aging, 80
  with antipsychotics, 560
  in B cells, 406, 422

  BDNF and, 558
  in brain function, 557
  in breast cancer, 297, 302
  in cardiac fibrosis, 480, 483
  in CSCs, 627–628
  in CTCLs, 454
  in DLBCL, 428
  in follicular lymphoma, 411
  in heart failure, 486
  in HL, 439, 441
  in lung cancer, 346, 579
  in NK and NKT cell function, 47
  in prostate cancer, 319, 579
  in schizophrenia, 558–559
  in senescence, 66, 68
MiR-31
  in aging, 82
  in breast cancer, 290
  in colorectal cancer, 333, 335–336
  HIF regulation by, 277, 279–280
  in malaria, 192
  in T cell function, 51
  as tumor suppressor, 228, 232
MiR-32
  in breast cancer, 296
  in lung cancer, 345, 348–349
  in MS, 529
  in prostate cancer, 314, 317–318
MiR-33(a/b)
  in aging, 84
  in CSCs, 632
  encoding of, 486
  inhibition of, 646, 655
  in lipid metabolism, 480, 486
  metformin effects on, 279
  non-viral delivery of, 648, 651
  in plasma cholesterol levels, 501–502
  replacement therapy with, 650
MiR-34(a/b/c)
  in aging, 79–82, 84
  in ALL, 374–377
  in B cell differentiation, 405–406
  in breast cancer, 291, 297, 577
  in Burkitt's lymphoma, 406
  in cancer, 208
  in carcinogenesis, 226
  in CLL, 386–387, 389, 393–394, 409–410
  in colorectal cancer, 334
  in CSCs, 628, 631–633
  in DC differentiation and function, 45
  delivery of, 644, 647–650

MiR-34(a/b/c) (cont'd)
   in diabetes, 497–501
   in DLBCL, 428
   in liver diseases, 513, 515–516, 580
   in lung cancer, 213
   in megakaryocytopoiesis, 105, 107
   in MS, 530, 532
   in prostate cancer, 213, 314, 316–320
   replacement therapy with, 649–650
   in senescence, 64–65, 67–68
   as tumor suppressor, 228, 232
   in viral infection, 121
MiR-71, in aging, 79
MiR-92(a/b), 94. *See also MiR-17-92* cluster
   in angiogenesis, 484
   in breast cancer, 294
   in cancer EVs, 617
   as cellular messenger, 597
   in colorectal cancer, 331, 333
   in coronary artery disease, 482
   in CTCLs, 454
   in DLBCL, 424
   in gastrointestinal cancer, 578
   in HCMV latency, 159–161
   in HL, 439
MiR-93
   in aging, 80
   in breast cancer, 297
   in CSCs, 629, 633
   in CTCLs, 454–456
   in DLBCL, 428
   in HL, 439
   reprogramming and, 35
MiR-96
   in hereditary diseases, 469–472
   in HL, 441, 443, 444
   in platelet reactivity, 110
MiR-96-182-183 family, in hereditary diseases, 469–470
MiR-98
   in Burkitt's lymphoma, 409
   in malaria, 192
MiR-99 family, in aging, 83
MiR-100
   in CSCs, 632
   in DLBCL, 428
   in prostate cancer, 320
MiR-101(a/c)
   in aging, 80
   in B cells, 422
   in breast cancer, 297
   in follicular lymphoma, 410
   in HCC, 514
   in prostate cancer, 314, 318, 320–321
   replacement therapy with, 649, 652
MiR-103
   in diabetes, 497–498, 500
   in HL, 439
MiR-106(a)-92 cluster, in myelopoiesis, 359
MiR-106(a)-363 cluster, in DLBCL, 426–427
MiR-106(a/b)
   in bowel disease, 580
   in breast cancer, 294, 297
   in colorectal cancer, 333
   in CSCs, 631
   in DLBCL, 424
   in gastrointestinal cancer, 578
   in hematopoiesis, 96
   in HL, 439
   in MS, 531
   in prostate cancer, 314, 316, 318
   in RCC, 610–611
   reprogramming and, 35
   in schizophrenia, 558
   in senescence regulation, 65
MiR-106(b)-25 cluster
   in aging, 84
   in CSCs, 630
   in DLBCL, 426–427
   expression of, 207
   in MS, 531
MiR-107
   in bowel disease, 580
   in CSCs, 627
   in diabetes, 497–498, 500
   HIF regulation by, 276, 278
   in malaria, 193
   in myelopoiesis, 357–358
   in prostate cancer, 319
MiR-122
   antagomirs against, 562
   in breast cancer, 291, 577
   in CTCLs, 457
   hepatitis and, 668
   inhibition of, 645–646, 653–654
   in lipid metabolism, 480, 486
   in liver diseases, 511–514, 517, 580–581
   LNA-antimiR for, 666, 669–670
   in myeloid disorders and leukemias, 363
   in plasma cholesterol levels, 501–502
   regulation of, 11
   in senescence regulation, 67
MiR-124(a)
   in AD, 546
   in ALL, 9, 374–376
   in brain function, 557

INDEX

in breast cancer, 299
in cancer, 211
in CSCs, 629–630
in diabetes, 497–499
in immune system, 546
for liver disease, 516
in MALT lymphoma, 411
in MS, 532
transdifferentiation and, 35–36
MiR-125(a/b)
in B cells, 49, 405–406, 422
in breast cancer, 7–8, 290, 297, 299, 300
in colorectal cancer, 333
in hematopoiesis, 93, 96–97
in hereditary diseases, 469
in HIV-1 infection, 172
in leukemia, 8
in megakaryocytopoiesis, 105, 107–108
in monocyte and macrophage response, 44–46
in MS, 531
in prostate cancer, 314–315, 318–319
in SMZL, 412
SNPs in, 469
MiR-126
in AML, 361–362
in angiogenesis, 480, 484–485
in bowel disease, 580
in breast cancer, 578
as cellular messenger, 597, 598
in coronary artery disease, 482
in diabetes, 580
function of, 207
in malaria, 192
in mast cell function, 48
in megakaryocytopoiesis, 106
in myeloid disorders, 361–362
in platelet microparticles, 110
in prostate cancer, 320
in RCC, 611
in SMZL, 412
in tumor angiogenesis, 612, 613
in tumor metastasis, 614
as tumor suppressor, 228, 233
MiR-127(3p)
in cancer, 210
in DLBCL, 428
in myeloid disorders and AML, 362
in RCC, 611
MiR-128(a/b)
in AML, 361
antipsychotics and, 560–561

in CSCs, 627, 629–630, 632
in HL, 441
in myeloid disorders, 361
in myelopoiesis, 357–358
in senescence regulation, 66
MiR-129, in heart failure, 486
MiR-130(a/b)
in angiogenesis, 480, 484
in hematopoiesis, 97
HIF regulation by, 276, 278
in megakaryocytopoiesis, 105, 107
in prostate cancer, 314, 316
reprogramming and, 35–36
in schizophrenia, 557–558
in senescence regulation, 65
in substance P synthesis, 557
in tumor angiogenesis, 612
MiR-132
in AD, 544–547
in ALL, 374–375
in HCMV latency, 159, 161
in immune system, 547
in insulin-secreting cells, 499
in monocyte and macrophage response, 44–46
in tumor angiogenesis, 612
in viral infection, 121
MiR-133(a1)/miR-1(2), 480
MiR-133(a2)/miR-1(1), 480
MiR-133(a/b)
in arrhythmia, 484
in cardiac fibrosis, 480, 483
in cardiac hypertrophy, 480–481
in cardiovascular disease, 581
in colorectal cancer, 334
in diabetes, 497
differentiation regulation by, 33
in heart failure, 486
in platelet microparticles, 110
in prostate cancer, 314, 317
transdifferentiation and, 35–36
MiR-134
in AD, 546
antagomirs for, 654
in brain function, 557
differentiation regulation by, 32–33
in hippocampal neurons, 543
in HL, 441
inhibition of, 645
in synapses, 545
MiR-135(a)
in aging, 84
in HL, 442–444

MiR-138
    in AD, 546
    in aging, 82
    in follicular lymphoma, 410
    in HIV-1 infection, 175
    in HL, 441
    in senescence regulation, 65
    in synapses, 545
MiR-139, in SMZL, 412
MiR-140(3p)
    in DLBCL, 428
    in hereditary diseases, 467
    in HL, 439
    nicotine and, 561
MiR-141
    in breast cancer, 578
    in cancer EVs, 617
    in colorectal cancer, 331
    in CSCs, 627, 631–632
    in gastrointestinal cancer, 578
    in prostate cancer, 319–320, 579
    in RCC, 610–611
    in senescence, 65, 68
MiR-142(3p)(5p)(s)
    in colorectal cancer, 336
    in DC differentiation and function, 45, 47
    in DLBCL, 428
    in hematopoiesis, 92–94, 96
    in HL, 438–439
    in MS, 531
    in myeloid disorders and leukemias, 364
    in myelopoiesis, 355, 356, 358
    in T cell function, 51
MiR-143
    in aging, 82
    in ALL, 374, 378–379
    in colorectal cancer, 334–335
    in CSCs, 629, 632
    in CTCLs, 454
    in diabetes, 497–498, 500
    in HCC, 514
    in myelopoiesis, 357–358
    non-viral delivery of, 648, 651
    in prostate cancer, 314, 318
    replacement therapy with, 649
MiR-144
    in aging, 80
    with antipsychotics, 560
    in colorectal cancer, 336
    in DLBCL, 424
    in hematopoiesis, 93, 97
    in MS, 529
    in myelopoiesis, 355

MiR-145
    in aging, 82, 85
    in AML, 361
    in angiogenesis, 480, 485
    in breast cancer, 213, 290, 293, 296, 301
    in Burkitt's lymphoma, 409
    in carcinogenesis, 226
    in colorectal cancer, 334–336
    in coronary artery disease, 482
    in CSCs, 629–630, 632
    in CTCLs, 454
    in diabetes, 497
    differentiation regulation by, 32–33
    HIF regulation by, 277, 279
    in megakaryocytopoiesis, 105
    in myeloid disorders, 361
    non-viral delivery of, 648, 651
    in prostate cancer, 314, 316, 318–321, 579
    in RCC, 611
    replacement therapy with, 650
    as tumor suppressor, 228, 233
MiR-146(a/b(5p))
    in aging, 83
    as cellular messenger, 598
    in DC differentiation and function, 45–47
    in diabetes, 497–499
    in DLBCL, 422–423, 428
    after Dox treatment, 486
    in HCC, 514
    in heart failure, 486
    in lung cancer, 345–346
    in malaria, 192
    in megakaryocytopoiesis, 105–107
    in monocyte and macrophage response, 44–46
    in MS, 525, 531
    in myelopoiesis, 358
    in platelet microparticles, 110
    in senescence regulation, 65
    in SMZL, 412
    in T cell function, 50–52
    in tumor metastasis, 614
MiR-147
    in breast cancer, 299
    in monocyte and macrophage response, 44–46
MiR-148(a)
    in biliary diseases, 515
    in CLL, 386
    in DLBCL, 428
MiR-149, in lung cancer, 346

MiR-150
  in AML, 362
  in B cells, 49, 405–406, 422
  as cellular messenger, 597
  in colorectal cancer, 334
  in CTCLs, 454
  in hematopoiesis, 93–94, 97
  in HIV-1 infection, 172
  in HL, 437–438
  in malaria, 188
  in MALT lymphoma, 411–412
  in MDV-1, 141
  in megakaryocytopoiesis, 104–106
  in MS, 531
  in myeloid disorders, 362
  in myelopoiesis, 355, 358, 360
  in NK and NKT cell function, 46–47
  in T cell function, 51
  in tumor angiogenesis, 613
MiR-151
  in cancer EVs, 617
  in colorectal cancer, 336
  in DLBCL, 428
MiR-152
  in ALL, 374, 378
  in liver diseases, 512, 515
MiR-155
  in AD, 546–547
  in ALD, 513
  in AML, 43
  in B cells, 49, 405–406
  in B-cell leukemias, 208
  in B-cell non-Hodgkin's lymphomas, 412
  in breast cancer, 290–291, 577–578
  in Burkitt's lymphoma, 409
  in CLL, 386, 410
  in coronary artery disease, 482
  in CTCLs, 454–457
  in DC differentiation and function, 45–46
  in DLBCL, 422–423, 425–426, 428
  in granulocyte response, 43, 46
  in hematopoiesis, 93–95, 97
  in HL, 437–439, 441–444
  in immune system, 546–547
  LNA inhibition of, 666–667
  in lung cancer, 234
  in lymphoma, 129
  in MALT lymphoma, 412
  in MDV-1, 141, 143, 147–148
  in megakaryocytopoiesis, 104–106
  in monocyte and macrophage response, 44–46
  in MS, 525, 530–532
  in myeloid disorders and leukemias, 361–362, 364
  in myelopoiesis, 355, 357–358, 360
  in NK and NKT cell function, 46–47
  as oncogene, 144, 228–229
  in RCC, 610–611
  in SMZL, 412
  in T cell function, 50–52
  in viral infection, 121, 126, 129, 141, 143–144
MiR-171, in breast cancer, 296
MiR-181(a/b/c/d)(1)
  in aging, 84
  in AML, 361–362
  in B cells, 49, 405–406, 422–423
  in breast cancer, 296, 297
  in CLL, 386–387, 389, 391–392, 409–410
  in colorectal cancer, 333
  in CSCs, 628, 630, 633
  in CTCLs, 454–456
  in DLBCL, 428
  in HCC, 514
  in hematopoiesis, 92–94
  in malaria, 192
  in megakaryocytopoiesis, 105, 108
  in MS, 529
  in myeloid disorders, 361–362
  in myelopoiesis, 356–358
  in NK and NKT cell function, 46–47
  in schizophrenia, 558
  in senescence regulation, 65
  in T cell function, 50, 52
MiR-182
  in aging, 83
  in ALD, 512–513
  in breast cancer, 297
  in CSCs, 630
  in heart failure, 486
  in hereditary diseases, 470
  in HL, 443–444
  inhibition of, 646, 655
  in senescence regulation, 66
  in T cell function, 51
  in tumor metastasis, 614
MiR-183
  in ALD, 512–513
  in colorectal cancer, 333
  in CSCs, 630
  differentiation regulation by, 32–33
  in hereditary diseases, 470
  in HL, 443–444
  in senescence regulation, 65
  in tumor metastasis, 614

MiR-184
  in hereditary diseases, 472–473
  in MALT lymphoma, 412
MiR-185
  in B cell function, 48
  in colorectal cancer, 333
  HIF regulation by, 279–280
MiR-186, in HL, 439
MiR-189
  in hereditary diseases, 467
  in Tourette syndrome, 554
MiR-191
  in bowel disease, 580
  in CTCLs, 454
  in HL, 439
MiR-192
  in aging, 82
  in diabetes complications, 502
  in DILI, 513
  inhibition of, 645
  in liver diseases, 581
MiR-193(a(3p)/b), in breast cancer, 293, 296, 299, 301
MiR-194
  in follicular lymphoma, 410
  inhibition of, 645
  in tumor angiogenesis, 612
MiR-195
  in breast cancer, 291
  in cardiac hypertrophy, 480–481
  in colorectal cancer, 334
  in DLBCL, 428
  in HL, 439
  in liver fibrosis, 516
  LNA-antimiR for, 667–668
  in prostate cancer, 319
  in schizophrenia, 558–559
MiR-196(a/b)
  in ALL, 374–375, 378
  in gastrointestinal cancer, 578
  in lung cancer, 345–346
  in myeloid disorders and leukemias, 361–362, 364
  in myelopoiesis, 359
  replacement therapy with, 650
  in viral hepatitis, 512
MiR-198
  in HIV-1 infection, 173
  in schizophrenia, 559
MiR-199(a(1)(3p/5p)/b(5p))
  in aging, 84
  in ALD, 512–513
  antipsychotics and, 561

  in bowel disease, 580
  in breast cancer, 578
  in CSCs, 629, 631, 633
  in CTCLs, 454–456
  in DLBCL, 428
  function of, 207
  in HCMV latency, 159
  HIF regulation by, 276, 278, 280
  in liver diseases, 512–513, 516
MiR-200(a/b/c)
  in ALL, 374, 378
  in bowel disease, 580
  in breast cancer, 290, 294, 296, 297
  in cancer EVs, 617
  in CSCs, 627, 629, 632
  in CTCLs, 454
  in diabetes complications, 502–503
  differentiation regulation by, 32–33
  in liver diseases, 512–513, 515–516
  in MALT lymphoma, 412
  in prostate cancer, 319
  in RCC, 610–611
  regulation of, 9–10
  reprogramming and, 34–36
  in tumor metastasis, 614
MiR-203
  in ALL, 374–375, 377
  in breast cancer, 296
  in colorectal cancer, 333, 335
  in CTCLs, 455–456
  differentiation regulation by, 32–33
  in MALT lymphoma, 411
  in prostate cancer, 314, 316, 318, 320
  as tumor suppressor, 228, 233
MiR-204
  in aging, 84
  in biliary diseases, 515
MiR-205
  in breast cancer, 290, 297, 299–300
  in CSCs, 627, 630
  in CTCLs, 455–456
  in MALT lymphoma, 412
  in prostate cancer, 314, 316, 318, 320
  reprogramming and, 35–36
  in tumor metastasis, 614
MiR-206
  in AD, 546
  in aging, 85
  in ALS, 542
  in breast cancer, 290, 293, 295–296, 301
  in diabetes, 497

in schizophrenia, 559
in substance P synthesis, 557
in tumor metastasis, 614
MiR-208(a/b)
  in cardiac hypertrophy, 480–481
  in cardiovascular disease, 581
  LNA-antimiR for, 667
  transdifferentiation and, 35–36
MiR-210
  in aging, 82
  angiogenesis and, 273, 480, 484
  apoptosis and cell differentiation and, 274
  in biliary diseases, 515
  as biomarker, 275
  in breast cancer, 270–271, 275, 290, 578
  in cancer, 208
  cell cycle and, 273–274
  DNA repair and, 273
  in heart failure, 486
  HIF regulation by, 275–276
  in hypoxia, 270–276, 279–280
  immunosuppression and, 275
  in kidney diseases, 581
  in malaria, 189, 192
  mitochondrial functions and, 271
  in RCC, 610–611
  targets of, 271–272
  in tumor angiogenesis, 612
MiR-211
  in heart failure, 486
  in HIV-1 infection, 175
MiR-212
  in ALD, 513
  in ALL, 374–375
  in colorectal cancer, 336
  in heart failure, 486
MiR-213, in breast cancer, 296
MiR-214
  in aging, 80
  in ALD, 512–513
  in breast cancer, 578
  in cardiomyocyte death, 480, 482
  in CSCs, 633
  in CTCLs, 454–457
  in MS, 531
  in T cell function, 51
MiR-215
  in aging, 82
  in colorectal cancer, 334–335
  in MS, 531
MiR-216(a), in diabetes complications, 502

MiR-217
  in aging, 84
  in diabetes complications, 502
  in senescence regulation, 65
MiR-218
  in cervical cancer, 578
  expression of, 207
  in prostate cancer, 320
MiR-219
  in AD, 546
  antipsychotics and, 561
  in circadian rhythm, 545, 560
  LNA inhibition of, 667
  in MS, 529
MiR-220(c), in colorectal cancer, 333
MiR-221
  in angiogenesis, 480, 484–485
  after angioplasty, 485
  with antipsychotics, 560
  in breast cancer, 290, 293, 295–297, 301
  in CLL, 410
  in colorectal cancer, 331
  in DC differentiation and function, 45
  in DLBCL, 424, 428
  in HCC, 514
  in hematopoiesis, 93, 97
  for liver disease, 517
  in lymphoma, 579
  in mast cell function, 48
  in MDV-1, 145
  in myelopoiesis, 357–358
  in NMZL, 412
  as oncogene, 228–230
  in prostate cancer, 314–315, 318–321, 579
  in tumor angiogenesis, 612
MiR-222
  in AML, 361–362
  in angiogenesis, 480, 484–485
  after angioplasty, 485
  in breast cancer, 290, 293, 295–297, 301
  in CLL, 386, 410
  in DLBCL, 409, 422–424, 428–429
  in erythroblastic leukemias, 208
  in HCC, 13
  in hematopoiesis, 93, 97
  in leukemia, 13
  in liver cancer, 208
  in mast cell function, 48
  in myeloid disorders, 361–362
  in myelopoiesis, 357–358
  as oncogene, 228–230

MiR-222 (cont'd)
   in prostate cancer, 314–315, 318, 321
   target genes of, 13
   in tumor angiogenesis, 612
MiR-223
   in AML, 361–362
   in B cells, 49, 406, 422
   in CLL, 409
   in CTCLs, 454
   in diabetes, 580
   in granulocyte response, 43, 46
   in hematopoiesis, 92–93, 96
   in HIV-1 infection, 169, 172–173
   in liver diseases, 581
   in malaria, 188
   in MALT lymphoma, 412
   in MDV-1, 141
   in MS, 531
   in myeloid disorders, 361–362
   in myelopoiesis, 355–360
   in NMZL, 412
   in platelet microparticles, 110
MiR-224
   in AML, 362
   in colorectal cancer, 334
   expression of, 207
   in myeloid disorders, 362
MiR-238, in aging, 79
MiR-239, in aging, 79
MiR-246, in aging, 79
MiR-273, regulation of, 9
MiR-284, in synapses, 545
MiR-290 cluster
   cell cycle control by, 31
   differentiation regulation by, 32–33
   reprogramming and, 34–36
MiR-294, in reprogramming, 34
MiR-296
   differentiation regulation by, 32–33
   in tumor angiogenesis, 612, 613
MiR-300, in senescence regulation, 66
MiR-301(a/b)
   in CSCs, 631
   reprogramming and, 35–36
   in T cell function, 50
MiR-302(b/c)
   in aging, 84
   in breast cancer, 293, 296, 301
   cell cycle control by, 31
   differentiation regulation by, 32–33
   in DLBCL, 428
   reprogramming and, 34–36
   in senescence regulation, 65

MiR-320
   in aging, 85
   in biliary diseases, 515
   in cardiomyocyte death, 480–481
   in colorectal cancer, 336
MiR-322, in ALD, 512–513
MiR-323(3p), in pregnancy, 580
MiR-324, in DLBCL, 423
MiR-326
   in CTCLs, 455–456
   in HIV-1 infection, 175
   in MS, 525, 531–532
   in T cell function, 50
MiR-328
   in arrhythmia, 484
   binding of, 203
   decoy activity of, 6
MiR-329, in HIV-1 infection, 175
MiR-330, in DLBCL, 428
MiR-331(3p)
   in breast cancer, 299, 300
   in Burkitt's lymphoma, 409
   in DLBCL, 424, 428
MiR-335
   in breast cancer, 578
   in tumor metastasis, 614
MiR-338, in MS, 529–530, 532
MiR-339, in hematopoiesis, 97
MiR-342
   in breast cancer, 296, 297
   in CTCLs, 454, 457
   in Sézary syndrome, 207
MiR-345, in SMZL, 412
MiR-346
   in breast cancer, 297
   in schizophrenia, 559
MiR-362(3p), in bowel disease, 580
MiR-363
   in Burkitt's lymphoma, 409
   in DLBCL, 422–423, 428
MiR-365, in colorectal cancer, 334
MiR-367, in reprogramming, 34
MiR-371-302 cluster, in CSCs, 630
MiR-373(3)
   binding of, 204
   in reprogramming, 34
MiR-370, in biliary diseases, 515
MiR-375-377 cluster, in HCC, 514
MiR-375-377 cluster, in CSCs, 630
MiR-372
   cell cycle control by, 31
   as oncogene, 228, 230
   reprogramming and, 34–36

MiR-373
  in aging, 82
  in biliary diseases, 515
  binding of, 203
  in breast cancer, 290
  in hypoxia, 270, 273, 280
  as oncogene, 228, 230
  in prostate cancer, 314, 318–319

MiR-375
  in breast cancer, 296
  in diabetes, 497–498
  in prostate cancer, 314, 317–319, 579
  in senescence regulation, 66

MiR-377, in diabetes complications, 502

MiR-378
  in AML, 361
  in angiogenesis, 480, 484
  in hematopoiesis, 97
  in myeloid disorders, 361
  in myelopoiesis, 358
  in NK and NKT cell function, 47
  in RCC, 611

MiR-382, in HIV-1 infection, 172

MiR-421
  in aging, 82
  in biliary diseases, 515
  in DLBCL, 423

MiR-423, in heart failure, 486

MiR-424
  in hematopoiesis, 93, 96
  HIF regulation by, 276, 278
  in myelopoiesis, 358
  in tumor angiogenesis, 612

MiR-425
  in DLBCL, 428
  in HL, 439

MiR-429
  in ALL, 374, 378
  in CSCs, 627, 632

MiR-432, in ALL, 374, 378

MiR-449(a)
  epigenetic mechanisms of, 210
  in prostate cancer, 314, 318, 320–321

MiR-451
  in AML, 361–362
  in colorectal cancer, 334
  in CSCs, 629–630, 633
  in DLBCL, 424, 428
  in hematopoiesis, 93, 97
  in malaria, 188
  in myeloid disorders, 361–362
  in myelopoiesis, 358, 360
  in platelet microparticles, 110
  in RCC, 611

MiR-452, in CSCs, 631

MiR-454(3p), in DLBCL, 424

MiR-468, in aging, 80

MiR-470
  in aging, 84
  differentiation regulation by, 32–33

MiR-486(5p)
  in aging, 84
  in CTCLs, 454–457
  in lung cancer, 578, 579

MiR-489
  in aging, 84
  in breast cancer, 296

MiR-491
  in DLBCL, 428
  in MS, 532

MiR-494
  in biliary diseases, 515
  in senescence regulation, 67

MiR-495, in hypoxia, 270, 280

MiR-497, LNA-antimiR for, 667–668

MiR-498, in colorectal cancer, 334, 336

MiR-499
  in cardiac hypertrophy, 481
  in cardiovascular disease, 581
  in lung cancer, 345–346, 579

MiR-500
  in DLBCL, 422, 428
  in myelopoiesis, 357

MiR-502
  replacement therapy with, 650
  for tumor inhibition, 653

MiR-503
  in ALL, 374, 378
  in angiogenesis, 480, 484

MiR-505, in senescence regulation, 66

MiR-506, in biliary diseases, 515–516

MiR-508(5p), in CSCs, 632

MiR-512(3p), in senescence regulation, 65

MiR-515(3p), in senescence regulation, 65

MiR-516(a(3p)), in prostate cancer, 319

MiR-518(a/b)
  in DLBCL, 423
  in MALT lymphoma, 411

MiR-519(c)
  in aging, 84
  in angiogenesis, 480, 485
  in DLBCL, 428
  HIF regulation by, 277–278, 485
  in senescence regulation, 66

MiR-520(c)
  in breast cancer, 290
  in prostate cancer, 314, 318–319
MiR-526, in heart failure, 486
MiR-532(3p), in bowel disease, 580
MiR-539, in MALT lymphoma, 411
MiR-550, in MALT lymphoma, 411
MiR-551, in DLBCL, 424
MiR-565, in HL, 439
MiR-574(3p)(5p)
  in CTCLs, 454
  in DLBCL, 422, 428
  in prostate cancer, 319
MiR-590, in DLBCL, 423
MiR-592, in MS, 529
MiR-598, in aging, 84
MiR-605, in aging, 82
MiR-608, in DLBCL, 428
MiR-622, in prostate cancer, 319, 579
MiR-629, in lung cancer, 345, 349
MiR-637, in DLBCL, 428
MiR-650, in cancer EVs, 617
MiR-652, in myelopoiesis, 357–358
MiR-660, in megakaryocytopoiesis, 107–108
MiR-663(b), in CTCLs, 455–457
MiR-669(b/c), in aging, 80, 84
MiR-681, in aging, 84
MiR-705, in ALD, 512–513
MiR-708
  in ALD, 512–513
  in prostate cancer, 314, 318–319
MiR-709, in aging, 80
MiR-711, in CTCLs, 455–456
MiR-720(a)
  in aging, 80
  in DLBCL, 424
MiR-721
  in aging, 80
  reprogramming and, 35–36
MiR-885(p), in senescence regulation, 65
MiR-935, in myelopoiesis, 358
MiR-1224, in ALD, 512–513
MiR-1233, in RCC, 611
MiR-1246, in platelet microparticles, 110
MiR-1260, in DLBCL, 424
MiR-1280, in DLBCL, 424
MiR-1285, in prostate cancer, 319, 579
MiR-1827, in lung cancer, 345, 347–348
Miravirsen, 657, 669–670
MiR-H1-5p, in HIV-1 infection, 174–175
MiR-H2-3p, in HSV-1 latency, 161
MiR-H6, in HSV-1 latency, 161

MiR-K12-11
  in KSHV infection, 161
  in viral infection, 129
MiR-M21-1, in viral infection, 129
MiR-M23-2, in viral infection, 129
MiR-N367-3p, in HIV-1 infection, 174–175
MiRNA decoys, 644
MiRNA sponges, 644
MiRNAs. See MicroRNAs
MiR-TAR-5p/3p, in HIV-1 infection, 174–175
Mirtrons, 5, 206–207
MiR-UL112-1, in HCMV latency, 157–159, 161–162
MiR-UL148D, in HCMV latency, 158–159
MiR-US4-1, in HCMV latency, 158
MiR-US25-1, in HCMV latency, 157, 161
Mitochondrial functions, 271
MLL gene, 377–379
MMP-2. See Matrix metalloprotease-2
Molecular apocrine breast cancer, 289
Monoclonal antibodies, in breast cancer, 298, 300–302, 578
Monocytes
  development of, 96–97
  in innate immunity, 44–46
Monogenic-based disorders, point mutations resulting in, 469–473
MPDs. See Myeloproliferative neoplasms/disorders
MPNs. See Myeloproliferative neoplasms/disorders
MPs. See Membrane-derived microparticles
mRNA. See Messenger RNA
Multiple sclerosis (MS), 523–525, 533
  miRNAs in, 525–529
  pathophysiology of
    degenerative component, 532
    in EAE, 532–533
    immune system, 529–532
    myelination, 529–530
Mutations. See also Single nucleotide polymorphisms
  affecting 3′-UTR of mRNAs, 467–468
  in genes involved in miRNA processing and function, 468
  in genetic dysregulation of miRNA expression, 7–8
  resulting in monogenic-based disorders, 469–473
MYCL1, lung cancer response to SNPs in, 347–348
Mycosis fungoides (MF), 450–452, 455–457

Myelination, 529–530
Myelodysplasia, 103
  circulating miRNAs in, 575, 579–580
Myelodysplastic syndromes (MDSs), 274, 355
Myeloid disorders, 353–356
  deregulated miRNA and transcription factor coregulation in, 358–360
  diagnostic, prognostic, and predictive significance of miRNA expression in, 360–363
  therapeutic miRNAs in, 363–364
Myelopoiesis, 353–356
  miRNA signaling in
    expression patterns in normal stem/progenitor and differentiated cells, 357–358
    miRNA and transcription factor coregulation, 358–360
    significance of myeloid biology in mammalian miRNA discoveries, 356
Myeloproliferative neoplasms/disorders (MPNs/MPDs), 103, 355. *See also* Leukemias
  deregulated miRNA and transcription factor coregulation in, 358–360
Myocardial ischemia (MI), 480–483
MyomiRs, 481

NAFLD. *See* Non-alcoholic fatty liver disease
Nanoparticulate systems, 643–644
Natural antisense transcripts (NATs), 248
Natural killer (NK) cells, 45–47
  development of, 94–96
Natural killer T (NKT) cells, 46–47
NBS1, lung cancer response to polymorphism in, 349
NcRNA. *See* Non-coding RNA
Neural progenitor cells (NPCs), 32
Neuritic structure, 544–545
Neuronal miRNAs
  behavior and, 545–546
  as controllers, 544–545
  function at synapse, 543–544
  memory and, 542–543
Neuronal–immune interface, 546–547
Neurotransmitters, in psychosis, 554–556
Neutropenia, 358–359
Neutrophils, 43
NK cells. *See* Natural killer cells
NKT cells. *See* Natural killer T cells

NLPHL. *See* Nodular lymphocyte predominant HL
NMDA receptors, in schizophrenia, 555
Nodal marginal zone lymphoma (NMZL), 408, 412
Nodular lymphocyte predominant HL (NLPHL), 436
Nomenclature, of microRNAs, 5–6
Non-alcoholic fatty liver disease (NAFLD), 511, 513
Non-coding RNA (ncRNA), 102
  role in physiological function and disease, 13–14
Non-small-cell lung cancer (NSCLC), 344
Non-viral strategies, for delivery of miRNAs, 643–644
Normal breast-like breast cancer, 288–289
NPCs. *See* Neural progenitor cells
NRON, 254
NSCLC. *See* Non-small-cell lung cancer
Nucleus, miRNA in, 203

Olanzapine, 561
Oncogenes
  in ALL, 377–379
  lncRNAs as, 254–258
  miR-155 as, 144, 228–229
  miRNAs as, 207–208, 223–224, 641–642
    clinical diagnosis, prognosis, and therapy based on, 233–234
    in prostate cancer, 314–316
    role in cancer, 225–230
    targets of, 224–225, 228
Oncoretroviruses, 643
Ovarian cancer, 575, 577–578

p16/RB senescence pathway, 60–61, 63, 65
p21 NAT, 255
p53, in DDR pathway, 81–82
p53/p21/Arf senescence pathway, 60–61, 63, 64–65
PACT, 4–5
Palmitate, 499
Pancreatic β cells
  in diabetes, 496
  regulation of, 497–499
Pancreatic cancer stem cells, 628
PANDA, 256
PAR-CLIP, miRNA targetome research with, 13
Parkinson's disease (PD), 542
PBC. *See* Primary biliary cirrhosis

P-bodies, in HIV-1 infection, 169
PCa. *See* Prostate cancer
PCAT-1, 256, 258
PCGEM1, 256
PCLDs. *See* Polycystic liver diseases
PCSCs. *See* Prostate cancer stem cells
PD. *See* Parkinson's disease
PDCD4. *See* Programmed cell death 4
PEI. *See* Polyethylenimine
Peroxisome proliferator-activated receptor-α (PPAR-α), 513
Pertuzumab, 298
4-Phenylbutyrate, in ALL, 377
Phosphatase and tensin homologue (PTEN), 48–51, 483, 485
Plasmodium parasites. *See* Malaria
Plasticity, 29
Platelet reactivity, miRNA biomarkers for, 110
Platelets, 101–102
 future research directions in, 110–111
 megakaryocytopoiesis, 103–108
 microparticles and miRNAs, 110
 miRNAs of, 108–109
 as biomarkers, 109–110
Pluripotency, 29
 cell cycle control and, 30–31
 differentiation fate choice and, 31–33
 induced, 33–36
Point mutations. *See also* Single nucleotide polymorphisms (SNPs)
 monogenic-based disorders resulting from, 469–473
Polycystic liver diseases (PCLDs), 511, 515
Polycythemia vera (PV)
 deregulated miRNA and transcription factor coregulation in, 360
 miRNA biomarkers for, 109–110, 362
Polyethylenimine (PEI) delivery system, 648, 651
Polymorphisms. *See* Single nucleotide polymorphisms
Polyomaviruses, 122–123, 126–128
Posterior polymorphous dystrophy, 473
Posttranscriptional regulation
 lncRNAs role in, 253–254
 of senescence, 64
  miRNA biogenesis regulation, 66–67
  p16/RB pathway regulation, 65
  p53/p21/Arf pathway regulation, 64–65
  SASP regulation, 65
  transcription factor and posttranscriptional regulator modulation, 65–66

Posttranscriptional regulators, in senescence regulation, 65–66
Potassium inwardly rectifying channel, subfamily J, member 2 (KCNJ2), 484
Potency, 29
PPAR-α. *See* Peroxisome proliferator-activated receptor-α
PPMS. *See* Primary progressive multiple sclerosis
PPT, for breast cancer, 293
PR expression. *See* Progesterone receptor expression
PR610, 268–269
Pregnancy, circulating miRNA biomarkers in, 576, 580
Premature senescence, 60–61
Pre-miRNAs, 4, 6
Primary biliary cirrhosis (PBC), 511, 515–516
Primary cutaneous anaplastic large-cell lymphoma (cALCL), 451–452, 454–457
Primary progressive multiple sclerosis (PPMS), 524
Pri-miRNAs, 4
Processing, of miRNAs, 3–5
 regulation of, 10–11
Profiling of miRNA expression
 in breast cancer, 289–292
 in cancer, 14, 211
 in CLL, 385–386
 in CTCLs, 453–457
 in Hodgkin's lymphoma, 438–441
 in prostate cancer, 313–317
 in schizophrenia, 557
Progenitor cells, 28, 92–93, 355
Progeria, 81–82
Progesterone receptor (PR) expression
 in breast cancer, 288–289
 miRNA markers of, 291–292
Programmed cell death. *See* Apoptosis
Programmed cell death 4 (PDCD4), 481
Promoter-associated lncRNAs, 248–249
Promoters, miRNA binding to, 203–204
Prostate cancer (PCa), 212, 311–313
 circulating miRNAs in, 575, 579
 epigenetic machinery of pathogenesis in, 320–321
 markers of, 258, 318–319, 575, 579
 miRNA profiling in, 313
  oncogenic miRNAs, 314–316
  tumor suppressor miRNAs, 316–317
 miRNAs and miRNA targets in, 317–318

pathogenesis of, 313
therapeutic miRNAs in, 213, 321
tumor progression and metastasis in, 319–320
Prostate cancer stem cells (PCSCs), 628–629, 631–632
Protein kinase C, 560
Protein miRNA carriers, 595–596
Pseudogenes, 248, 250
Psychosis, 553–554
bipolar affective disorder, 559–560
etiology of, 556–557
future miRNA therapy for, 561–563
neurotransmitters in, 554–556
schizophrenia, 557–559
treatment of, 560–562
Psychostimulants, 559
PTEN. *See* Phosphatase and tensin homologue
PTENP1, 250, 257
PV. *See* Polycythemia vera

Quetiapine, 561

RBPs. *See* RNA-binding proteins
RCC. *See* Renal cell carcinoma
Red blood cells. *See* Erythrocytes
Regulation. *See also* Gene expression regulation; Posttranscriptional regulation; Transcriptional regulation
of miRNAs, 7
epigenetic, 8–9, 210
function regulation, 11–12
genetic, 7–8
under hypoxia, 269–270
synthesis and processing regulation, 10–11
transcription factors and miRNA regulatory networks, 9–10
Regulatory T cells, 51
Relapsing-remitting multiple sclerosis (RRMS), 524
Renal cell carcinoma (RCC), 607–610
angiogenesis in, 612–613
cancer stem cells in, 615–617
circulating miRNAs and, 611–612
metastasis in, 613–615
miRNAs and, 610–611
Replicative senescence, 60–61
Reprogramming, 33–36
Retroviruses, 121–125. *See also specific retroviruses*
Rett syndrome, 557
REV3L, lung cancer and, 348–349

RISC. *See* RNA-induced silencing complex
Risperidone, 561
Rituximab, 420
RNA inhibition, 212–213
RNA silencing, 3
RNA viruses, 122–124. *See also specific RNA viruses*
RNA-binding proteins (RBPs)
miRNA function and, 11–12
in senescence regulation, 65–67
RNA-induced silencing complex (RISC), 4–5, 543–544
in CTCLs, 453
regulation of, 11–12
RRMS. *See* Relapsing-remitting multiple sclerosis

SASP. *See* Senescence-associated secretory phenotype
Schizophrenia, 554, 557–559
diagnosis of, 554–555
etiology of, 556–557
miRNAs and, 557–559
pharmacology for, 561
SCLC. *See* Small-cell lung cancer
Secondary progressive multiple sclerosis (SPMS), 524
Selective estrogen receptor modulators (SERMs), resistance to, 295–297, 300
Self-renewal, 29
miRNA regulation of, 31–33
Senescence, 59
aging and, 83–84
of cultured cells
premature, 60–61
replicative, 60–61
detection of
cellular morphology, organization, and enzymatic activity changes, 63
markers of cell damage and tumor suppressor networks, 63
SASP, 64
future perspectives on, 68–69
posttranscriptional regulation of, 64
miRNA biogenesis regulation, 66–67
p16/RB pathway regulation, 65
p53/p21/Arf pathway regulation, 64–65
SASP regulation, 65
transcription factor and posttranscriptional regulator modulation, 65–66
in vivo, 61–63
cancer and, 62
miRNA impact on, 67–68

Senescence-associated secretory phenotype (SASP), 61–62, 64
  regulation of, 65
Sequestration, in malaria, 186
SERMs. *See* Selective estrogen receptor modulators
Serotonin, in schizophrenia, 555
17p–13q–11q deletions, in CLL, 393–394
Sézary syndrome, 207, 450–451, 453–458
SILAC. *See* Stable isotope labeling by amino acids in culture
Single nucleotide polymorphisms (SNPs), 209, 469
  in CLL, 394–395
  in genetic dysregulation of miRNA expression, 8
  in lung cancer susceptibility and survival
    KRAS 3′-UTR SNP affecting *let-7* binding, 347
    MYCL1 miR-1827 binding site SNPs, 347–348
    NBS1 SNPs affecting *miR-629* binding, 349
    pre-miRNA flanking region SNPs, 346
    pre-miRNA sequence SNPs, 345–346
    REV3L 3′-UTR SNP affecting *miR-25* and *miR-32* regulation, 348–349
Sip1 NAT, 255
SiRNAs. *See* Small interfering RNAs
SIRT1, 500–501
Sirtuin pathway, 84
Slit and Trk-like family member 1 (SLITRK1), 554
Small interfering RNAs (siRNAs), 640
  in HIV-1 infection, 168–169
  with miRNAs, 656–657
Small-cell lung cancer (SCLC), 344
SmallRNAs, history of, 2–3
SMZL. *See* Splenic marginal zone lymphoma
SNPs. *See* Single nucleotide polymorphisms
Solid tumors, hypoxia in, 268–269
Somatic stem cells. *See* Adult stem cells
Spastic paraplegia (SPG31), 467
Splenic marginal zone lymphoma (SMZL), 412
SPMS. *See* Secondary progressive multiple sclerosis
Sprouty homolog1 (SPRY1), 483, 502
SRA-1, 256
SREBPs. *See* Sterol regulatory element binding proteins
SRSs. *See* Suppressors of RNA silencing

Stable isotope labeling by amino acids in culture (SILAC), miRNA targetome research with, 13
Stem cells, 28–29. *See also* Cancer stem cells
  aging and, 84
  HSCs, 28, 92–93, 103–104
  pluripotency of
    cell cycle control in, 30–31
    differentiation fate choice and, 31–33
  reprogramming and, 33–36
  transdifferentiation and, 35–37
Sterol regulatory element binding proteins (SREBPs), 486–487, 501–502
Streptozotocin, 503
Stress-induced senescence. *See* Premature senescence
Substance P, 557
Suppressors of RNA silencing (SRSs), in HIV-1 infection, 170
Synapse, 542
  formation of, 544–545
  miRNA function at, 543–544
  pruning of, 557
Synthesis. *See* Biogenesis

T lymphocytes
  in adaptive immunity, 49–52
  development of, 94–96
  in innate immunity, 45–47
  in MS, 524–525
T1D. *See* Type 1 diabetes
T2D. *See* Type 2 diabetes
Tamoxifen, 295–297, 300
TAR RNA-binding protein 2 (TARBP2), 468
Targets/targetomes, 12–14, 205
  of avian herpesvirus miRNAs, 145–146
  in breast cancer, 290
  in cancer, 224–225, 228
  of HCMV miRNAs, 160
  in lung cancer, 347–349
  of *miR-210*, 271–272
  in prostate cancer, 317–318
  of viral miRNAs, 126–129
T-cell lymphomas. *See* Cutaneous T-cell lymphomas
TCL1, in CLL, 391–392
TCL1 transgenic mice, 392
TdsmRNAs. *See* Transfer RNA-derived small RNAs
Telaprevir, 668
Texel sheep model, 467

TFs. *See* Transcription factors
TGFβ 1. *See* Transforming growth factor beta 1
Therapeutic miRNAs, 547–548, 642–643
   in B-cell non-Hodgkin's lymphomas, 412–413
   in cancer, 212–213, 233–234, 633–634
   inhibition by miRNAs, 212–213, 644–645
   inhibitors of miRNAs, 15, 212–213, 234, 644–645, 653–655
   limitations of, 548–549
   mimics of miRNAs, 15
   in myeloid disorders and leukemias, 363–364
   non-viral strategies, 643–644
   perspectives, issues, and future directions of, 655–657
   in prostate cancer, 213, 321
   replacement of miRNAs
     let-7 and *miR-34a* delivery, 644, 647–650
     miR-143, *miR-145*, and *miR-33a* non-viral delivery, 648, 651
     other delivery examples, 652–653
     viral delivery, 642–643, 651–652
Thio-LNA, 664
T-ICs. *See* Tumor-initiating cells
Tiny locked nucleic acids (tiny-LNAs)
   activity of, 666–667
   design of, 665–666
TKIs. *See* Tyrosine kinase inhibitors
TMV. *See* Tumor EVs
TNF-α. *See* Tumor necrosis factor-α
Tourette syndrome, 554
TP53, in CLL, 393–394
TP53 pathway, in ALL, 376–377
TRAF2. *See* BCL2-TRAF2 transgenic mice
Transcribed ultraconserved regions (T-UCRs), lncRNAs produced from, 248, 250
Transcription factors (TFs)
   miRNA regulation by, 9–10
   in senescence regulation, 65–66
Transcriptional regulation
   in hypoxia, 269
   lncRNAs role in, 251–252
Transdifferentiation, 35–37
Transfer RNA-derived small RNAs (tdsmRNAs), in HIV-1 infection, 168–169, 176–177
Transforming growth factor beta 1 (TGFβ 1), 502
Translational activation, 6–7
Translational repression, 6–7
Translocation, in genetic dysregulation, 7–8

Trastuzumab, 298, 300–302, 486, 578
TRBP, 4–5, 11
T-reg cells, 529
Trichostatin A, 9
Trifluoperazine, 561
TRIM-NHL proteins, 11
T-UCRs. *See* Transcribed ultraconserved regions
Tumor EVs (TMV), 613–615
Tumor necrosis factor-α (TNF-α), 485, 547
Tumor suppressor networks, in senescent cells, 63
Tumor suppressors
   lncRNAs as, 254–258
   miR-15a/16-11 as, 226, 228, 231, 386–389, 652
   miRNAs as, 207–208, 223–224, 641–642
     clinical diagnosis, prognosis, and therapy based on, 233–234
     in prostate cancer, 316–317
     role in cancer, 225–228, 230–233
     targets of, 224–225, 228
Tumor-initiating cells (T-ICs), 626
Tumors. *See* Cancer
Twf1 expression, 480
Type 1 diabetes (T1D), 496
Type 2 diabetes (T2D), 496
Tyrosine kinase inhibitors (TKIs)
   in ALL, 377
   in breast cancer, 298, 300
   in myeloid disorders and AML, 361

UCA1, 256
Ultraconserved genes (UCGs), 204
3′-Untranslated region (3′-UTR), mutations affecting, 467–468

Vaccines
   for viral hepatitis, 511–512
   viral miRNAs in, 129
Valproate, 560–561
Vascular adhesion molecule 1 (VCAM-1), 485
Vascular diseases, 484–485
Vascular endothelial growth factor (VEGF), 503
Vascular smooth muscle cell (VSMC), *miR-221* and *miR-222* and, 485
VCAM-1. *See* Vascular adhesion molecule 1
VEGF. *See* Vascular endothelial growth factor
Very low-density lipoprotein (VLDL), 486–487
Vesicles. *See* Extracellular vesicles
Viral delivery, of miRNAs, 642–643, 651–652

Viral hepatitis, 511–512
  MiR-122 and, 668
Viral infections, 119–121. *See also specific viral infections*
  future prospects in, 129–130
  host miRNAs in, 121
  viral miRNAs in, 121–122
    biological roles of, 126–129
    DNA viruses, 122–124
    expression of, 124–126
    functional convergence of, 127–128
    future prospects in, 129–130
    RNA viruses, 122–124
    in vivo, 129
VL30-1, 255
VLDL. *See* Very low-density lipoprotein
VSMC. *See* Vascular smooth muscle cell

X inactive specific transcript (Xist), 252

ZAP-70, in CLL, 393–394
Zebularine, 9